I0888131

Industrial and Environmental Microbiology

This book provides a comprehensive overview of the role of microorganisms in industrial processes, environmental management, and the production of valuable industrial products. It reviews the role of microorganisms in industries such as food and beverage, pharmaceuticals, detergents, textiles, leather, paper and pulp, cosmetics, and bioenergy. It also examines the role of microbial indicators in assessing water quality and wastewater treatment. Additionally, the book explores the role of microbes in biogeochemical cycles and bioremediation. Towards the end, it discusses the potential of microbes in waste management, bioconversion of waste into bioenergy, microbial consortia for waste valorisation, bioprocess engineering, synthetic biology, circular economy approaches, and environmental monitoring and assessment. This book will be useful for researchers and professionals in the fields of microbiology, environmental sciences, biotechnology, and industrial processes.

Key Features

- Explores the application of microbes in different industries, including food and beverage, pharmaceuticals, cosmetics, and bioenergy
- Examines the potential of microbes in waste management and environmental monitoring
- Highlights the microbial regulation of different biogeochemical cycles and the anthropogenic influences that can disrupt their balance
- Presents innovative integration of microbial bioremediation, bioprocess engineering, and metabolic engineering for industrial waste management
- Discusses microbial enzymes used in industries and different enzyme immobilisation techniques
- Reviews the role of microorganisms as indicators of water quality and their effectiveness in assessing the presence of contaminants

Industrial and Environmental Microbiology

Zeeshan ur Rahman and Lebin Thomas

CRC Press
Taylor & Francis Group
Boca Raton London New York

CRC Press is an imprint of the
Taylor & Francis Group, an **informa** business

Designed cover image: Zeeshan ur Rahman and Lebin Thomas

First edition published 2026
by CRC Press
2385 NW Executive Center Drive, Suite 320, Boca Raton, FL 33431

and by CRC Press
4 Park Square, Milton Park, Abingdon, Oxon, OX14 4RN

CRC Press is an imprint of Taylor & Francis Group, LLC

ISBN: 9781032644769 (hbk)
ISBN: 9781032651088 (pbk)
ISBN: 9781032651095 (ebk)

DOI: 10.1201/9781032651095

Typeset in Times
by Newgen Publishing UK

Dedication

The late Professor Ved Pal Singh

(Former Professor, Department of Botany, University of Delhi)

We thank you for your unwavering support and showing the path of wisdom.

Contents

UNIT IV Microbial Enzymes and Their Immobilisation

UNIT V Microorganisms in Environments and Their Analysis

Preface

All industrially important microorganisms are remnants of the natural environment, and industrially grown microbes significantly contribute to environmental well-being. This cooperation for human ingenuity and ecological benefits has increased the demand for knowledge in Industrial Microbiology and Environmental Microbiology. These two fields often interplay and address each other for microbiological challenges and innovation. However, in the current educational landscape, few resources integrate their concepts and applications in one layout. This highlights a clear need for a textbook on Industrial and Environmental Microbiology, which provides comprehensive knowledge on various functional aspects of microbiology.

Building upon the requirement, this book attempts to consolidate information on the key concepts, technologies, and advancements in the fields of Industrial and Environmental Microbiology. The syllabi of different courses in BSc (Hons) Botany and Life Sciences, and BSc and MSc Microbiology (as per the guidelines of the University Grants Commission/ Higher Education Commission, India) have inspired the content of the book, targeting undergraduate and postgraduate students. However, the detailed coverage of the subject matter can extend its appeal to researchers, academicians, industrialists, and environmentalists for a deeper understanding of interlinked descriptions.

The book adopts a student-centered approach, providing theoretical foundations with real-world examples to address global challenges. We conceived a lucid writing style and tried to produce an explicable understanding while maintaining the standards of academic rigour. The book approaches a logical progression to cover the basic concepts and then extends to advanced knowledge. Each chapter vividly elaborates the concept and principle of subject matter, considering a mixed type of readers with diverse interests. The content is organised into small, easy-to-read sections and subsections, which allow readers to navigate the content based on their specific needs. Numerous visual aids, such as Boxes, Tables, Figures, and chemical reactions, help in the understanding of various topics. The book has additional features such as summaries, suggested readings, review questions, and glossaries that involve active engagement and enhance learning.

The book is organised into seven units that cover various aspects of microbial function and applications, as follows.

- Unit I explores the diverse applications of microorganisms for industries and the environment (Chapters 1 and 2). Subsequently, Chapter 3 delves into microbial integrity in academia, research activities, and industrial biotechnology.
- Unit II is built upon the intricacies of fermentation technology. This begins with a general description of the fermentation process and its principles (including microbial metabolism) (Chapter 4). The topics on upstream processing, focusing on various preparations required before the main fermentation process, are exclusive to Chapter 5. Chapter 6 discusses different types of fermentation processes based on the nature of substrates, operational modes, and aerated conditions. Chapter 7 delineates a description of the sizes and shapes of fermenters and their different components. Thereafter, Chapter 8 gathers information on various types of fermenters employed for multiple applications.
- Unit III encompasses the topics related to industrially relevant microorganisms, optimisation conditions of fermentation processes, downstream processing, and production of industrial products. Chapter 9 conveys information on natural and engineered microorganisms, emphasising containment for safe biotechnology. Chapter 10 sheds light on microbial growth phases and various parameters that optimise fermentation processes. Chapter 11 outlines a general workflow of downstream processing operations. This further constitutes a foundation for the different techniques involved in downstream processing (Chapter 12). Towards the end of this Unit, Chapter 13 provides an overview of the production of industrially relevant products, including an enzyme (α-amylase), an organic acid (citric acid), an alcohol (ethanol), an antibiotic (penicillin), and a vitamin (vitamin B12).
- Unit IV highlights the roles of industrially important enzymes, including their applications and immobilisation techniques. Chapter 14 provides an overview of the general characteristics of enzymes and their kinetic parameters, enzyme biotechnology, and applications of key industrial enzymes. Chapter 15 gives insights into enzyme immobilisation and its various methods. Chapter 16 addresses notes on the properties of immobilised enzyme systems, dealing with kinetic studies, physicochemical factors, protein engineering strategies, and bioreactor designs. Additionally, Chapter 17 discusses large-scale applications of immobilised enzymes, citing examples of glucose isomerase and penicillin acylase.
- Unit V contains detailed information on microbial life and their diverse habitats in the environment, exploring various methods for their investigation. Chapter 18 floats over the distribution of airborne microorganisms and their implications for human health and the environment. Chapter 19 digs into microbial communities of the soil, which contribute to agriculture and soil health. Chapter 20 dives into microbial life in aquatic environments and its relevance to ecosystem balance. Later on, Chapter 21 reveals different techniques employed for the study of microorganisms from environmental samples, including culture-dependent and -independent methods. Cultivation in microbiology

is crucial for identifying microbial potential, benefiting various sectors, as described in Chapter 22.

- Unit VI gives an account of water pollution and its parameters, and the implications of microorganisms in wastewater treatment (WWT). Chapter 23 unfolds the reports of water pollution sources and wastewater management strategies. Chapter 24 extends the description of various physicochemical and biological parameters used for the analysis of water samples, giving special emphasis on microbial monitoring. Further, Chapter 25 furnishes an understanding of the workflow of wastewater treatment, involving physical, chemical, biological (mainly microbial-related processes), and integrated methods.
- Unit VII throws light upon the significance of microorganisms in biogeochemical cycles, bioremediation of soil, industrial waste management, and environmental monitoring. Chapter 26 opens with remarks on some technical terms of biogeochemical cycles and mentions different compartments of the environment. Chapter 27 provides in-depth information on the carbon cycle and the critical roles of microorganisms for nourishment and maintenance of life. Chapter 28 outlines the details of the intricate relationship between the nitrogen cycle and microorganisms that have roles in the chemodiversity, mobility, and bioavailability of nitrogen. Chapter 29 builds a synopsis of the roles of microorganisms in sulphur cycles, showing the importance of the balance of different sulfur species in ecosystems. Chapter 30 unveils the principles and applications of various bioremediation strategies, with a focus on *ex situ* and *in situ* biotreatment options. The last chapter (Chapter 31) has perspectives on waste management schemes, focusing on the microbial workforce to modify waste into profitable products, and environmental monitoring.

We have taken our utmost care to ensure accuracy and to avoid any errors in the book. However, if there is any oversight, we request our educators to convey the potential concerns. We also welcome expert feedback and constructive criticism on our writing and encourage suggestions from students to improve understanding and engagement.

Acknowledgements

We are highly grateful to our knowledgeable reviewers for their constructive feedback on our chapters. We greatly appreciate their interest and the time they put into reviewing the chapters. We extend our deepest gratitude to Dr Bhavik Sawhney, Senior Editor (Life Sciences) at CRC Press, Taylor & Francis Group, for believing in our project and making it possible. We also sincerely thank Dr Madhurima Kahali, Editor II (Life Sciences), and Ms Ashavari Bhattacharya, Editorial Assistant at CRC Press, for their invaluable feedback on our book proposal.

LIST OF REVIEWERS

Chapter title	Reviewers	
1.1. Scope of Microorganisms in Industries	Dr Piyush Mathur Assistant Professor Department of Botany Banaras Hindu University, Varanasi, India	Dr Rajesh Kumar Assistant Professor Mahatma Gandhi Government Arts College Pondicherry University, Mahe, India
1.2. Scope of Microorganisms in Environment	Dr Heikham Evelin Associate Professor Department of Life Sciences Manipur University, Imphal, India	Dr Manoj Kumar Singh Assistant Professor Department of Botany University of Allahabad, Prayagraj, India
1.3. Microbial Research and Industrial Biotechnology	Prof Rup Lal Professor (Retired) Department of Zoology University of Delhi, New Delhi, India	Dr Ruchi Vir Assistant Professor Zakir Husain Delhi College University of Delhi, New Delhi, India
2.1. The Basics of Fermentation	Prof Ishwar Singh Professor Hansraj College University of Delhi, New Delhi, India	Dr SAI Choudhury Associate Professor Demoria College Guwahati University, Kamrup, India
2.2. Upstream Processing	Prof RC Kuhad Former Vice Chancellor Central University of Haryana, Mahendragarh, India Professor (Retired) Department of Microbiology University of Delhi, New Delhi, India	Dr Ramanjaneyulu Golla Assistant Professor Department of Biochemistry REVA University, Bengaluru, India
	Dr Vineet K Singh Assistant Professor Acharya Narendra Dev College University of Delhi, New Delhi, India	
2.3. Different Types of Fermentation Processes	Prof A Archana Professor Swami Shraddhanand College University of Delhi, New Delhi, India	Dr Poonam Dehariya Assistant Professor Department of Botany Dr Harisingh Gour Vishwavidyalaya, Sagar, India
	Dr Sarita Kumari Assistant Professor Maitreyi College University of Delhi, New Delhi, India	
2.4. Fermenters and Their Basic Components	Prof Yogender Pal Khasa Professor Department of Microbiology University of Delhi, New Delhi, India	Dr Priyanka Srivastava Assistant Professor Sri Guru Tegh Bahadur Khalsa College University of Delhi, New Delhi, India
2.5. Different Types of Fermenters	Prof S N Tripathi Professor Deen Dayal Upadhyaya College University of Delhi, New Delhi, India	Dr Savindra Kumar Assistant Professor Zakir Husain Delhi College University of Delhi, New Delhi, India

Chapter title	Reviewers	
3.1. Microorganisms in Fermentation	Dr Shikha Srivastava Assistant Professor Bhaskaracharya College of Applied Sciences University of Delhi, New Delhi, India	Dr Probin Phanjom Assistant Professor Department of Biosciences Assam Don Bosco University, Guwahati, India
3.2. Fermentation Conditions	Prof Awadhesh Kumar Professor Department of Botany Mizoram University, Aizawl, India	Dr Rajesh Kumar Associate Professor Hindu College University of Delhi, New Delhi, India
3.3. Downstream Processing and Uses	Dr Satya Prakash Churasia Assistant Professor Acharya Narendra Dev College University of Delhi, New Delhi, India	
3.4. Important Techniques in Downstream Processing	Dr Rishi Gupta Application Scientists Anton Paar India Pvt Ltd., Gurugram, India	Dr Pratibha Pant Assistant Professor Hansraj College University of Delhi, New Delhi, India
3.5. Production of Industrially Important Products	Dr Neetu Chaudhary Assistant Professor Gargi College University of Delhi, New Delhi, India	Dr Sweta Yadav Assistant Professor Swami Shraddhanand College University of Delhi, New Delhi, India
4.1. Microbial Enzymes of Industrial Interest	Prof Ratnum Kaul Wattal Professor (Retired) Zakir Husain Delhi College University of Delhi, New Delhi, India	Prof Deepak Vyas Professor Department of Botany Dr Harisingh Gour Vishwavidyalaya, Sagar, India
4.2. Enzyme Immobilisation	Dr Mahaswetta Saikia Associate Professor Hansraj College University of Delhi, New Delhi, India	Dr Archana Singh Assistant Professor Kirori Mal College University of Delhi, New Delhi, India
4.3. Properties of Immobilised Enzyme Systems	Dr Alok Kumar Assistant Professor Ramjas College University of Delhi, New Delhi, India	Dr Anil K Mavi Assistant Professor Sri Aurobindo College University of Delhi, New Delhi, India
4.4. Large-Scale Applications of Immobilised Enzymes	Dr Yamal Gupta Assistant Professor Kirori Mal College University of Delhi, New Delhi, India	Dr Rubina Chongtham Assistant Professor Deshbandhu College University of Delhi, New Delhi, India
5.1. Distribution of Airborne Microorganisms	Prof Chirashree Ghosh Professor Department of Environmental Studies University of Delhi, New Delhi, India	Dr Krishna Kumar Choudhary Assistant Professor Department of Botany, MMV Banaras Hindu University, Varanasi, India
	Dr Saurabh Sonwani Assistant Professor Zakir Husain Delhi College University of Delhi, New Delhi, India	
5.2. Study of Microorganisms in Soil	Prof HK Kheri Professor (Retired) Department of Botany University of Allahabad, Prayagraj, India	Dr Sarthak Malhotra Assistant Professor Dayal Singh Evening College University of Delhi, New Delhi, India
5.3. Study of Microorganisms in Water	Prof Eva S Lindström Professor Department of Ecology and Genetics/Limnology Uppsala Univeristy, Uppsala, Sweden	Dr Pooja Gokhale Sinha Assistant Professor Sri Venkateswara College University of Delhi, New Delhi, India
5.4. Methods for the Characterisation of Microorganisms	Dr Meenakshi Sharma Assistant Professor Daulat Ram College University of Delhi, New Delhi, India	Dr Vivek Kr Choudhary Scientist In-charge Tughlaqabad Biodiversity Park Biodiversity Parks Programme, CEMDE University of Delhi, New Delhi, India

Chapter title	Reviewers	
5.5. Microbial Cultivation	Prof Anupam Dikshit Professor (Retired) Department of Botany University of Allahabad, Prayagraj, India	Dr Devayani Muley Associate Professor Zakir Husain Delhi College University of Delhi, New Delhi, India
6.1. Water Pollution	Dr Anand Pandey Assistant Professor Avvaiyar Govt. College for Women Pondicherry University, Karaikal, India	Dr Sharad K Tripathi Head of the Department and Assistant Professor Department of Life Sciences Arni University, Kangra, India
6.2. Water Quality Parameters and Indicators	Dr Sandeep Kaushik Assistant Professor Department of Environmental Sciences Indira Gandhi National Tribal University, Amarkantak, India	Dr Ankush Assistant Professor Zakir Husain Delhi College University of Delhi, New Delhi, India
6.3. Wastewater Treatment System	Dr Nellie Laisram Assistant Professor Dayal Singh College University of Delhi, New Delhi, India	Dr Anurag Maurya Assistant Professor Shivaji College University of Delhi, New Delhi, India
7.1. Biogeochemical Cycles and Microorganisms	Dr Tabassum Jehan Associate Professor Zakir Husain Delhi College University of Delhi, New Delhi, India	Dr Devender Mudgil Assistant Professor Zakir Husain Delhi College University of Delhi, New Delhi, India
7.2. Nitrogen Cycle and Microorganisms	Dr Shantanu Mandal Assistant Professor Sri Venkateswara College University of Delhi, New Delhi, India	Dr Shrey Bodhankar Assistant Professor Department of Agricultural Microbiology Anurag University, Hyderabad, India
7.3. Nitrogen Cycle and Microorganisms	Prof Rajni Gupta Professor Kirori Mal College University of Delhi, New Delhi, India	Prof Madhulika Singh Professor Swami Shraddhanand College University of Delhi, New Delhi, India
	Dr M Wahid Ansari Associate Professor Zakir Husain Delhi College University of Delhi, New Delhi, India	
7.4. Sulfur Cycle and Microorganisms	Dr Pradeep Kumar Associate Professor Department of Botany University of Lucknow, Lucknow, India	Dr Preeti Kaur Assistant Professor Sri Guru Tegh Bahadur Khalsa College University of Delhi, New Delhi, India
7.5. Bioremediation of Contaminated Soil	Laurate Prof Ravi Naidu Professor and Managing Director Cooperative Research Centre for Contamination, Assessment and Remediation of the Environment (CRC CARE) University of Newcastle, Newcastle, Australia	Dr Shashi Kant Shukla Assistant Professor Centre of Science and Society, IIDS University of Allahabad, Prayagraj, India
	Dr Yamini Agrawal Assistant Professor Department of Botany University of Delhi, New Delhi, India	
7.6. Industrial Waste Management and Environmental Monitoring	Prof Rajkumari Sanayaima Devi Professor Deen Dayal Upadhyaya College University of Delhi, New Delhi, India	Dr Rohit Kumar Mishra Associate Professor & Former Coordinator Centre of Science & Society, IIDS University of Allahabad, Prayagraj, India
	Dr Vaibhav Srivastava Assistant Professor Department of Botany University of Allahabad, Prayagraj, India	

About the Authors

Zeeshan ur Rahman is working as an Assistant Professor at Zakir Husain Delhi College, University of Delhi. He received his BSc and MSc from the University of Allahabad, India, and MPhil and PhD from the University of Delhi, India. He was awarded Junior Research Fellowship (JRF) and Senior Research Fellowship (SRF) from the Council of Scientific and Industrial Research (CSIR), India, for his PhD research work, and is a recipient of the prestigious Erasmus Mundus Scholarship funded by the European Union. He is also appointed as a Research Fellow at INTI International University, Malaysia. He has been committed to microbial research and training for the last fifteen years, and his areas of specialisation include environmental microbiology, biogeochemistry, bioremediation, and environmental monitoring, resulting in several articles published in international and national journals and chapters in several edited books. He has served key roles in organising various national and international levels of events such as conferences, seminars, and workshops.

Lebin Thomas is currently serving as an Assistant Professor in the Department of Botany, Dr Hari Singh Gour Vishwavidyala, Sagar. He completed his higher education at the University of Delhi. He was awarded Junior Research Fellowship (JRF) and Senior Research Fellowship (SRF) from the Council of Scientific and Industrial Research (CSIR), India, for his PhD research work. His past teaching experiences include service in college at the University of Delhi, India. He is deeply interested in research in industrial microbiology on various topics related to industrial enzyme production, optimisation studies, solid-state fermentation, and biodegradation of agricultural residues. He has published numerous research and review articles in national and international journals, as well as contributing chapters to various edited books with collaborative research. Academic counselling, journal and book reviewing and organising various extracurricular activities (such as faculty development programs, workshops, webinars, seminars, and conferences) are among his notable administrative responsibilities. He is also a Life Member of the International Society of Plant Morphologists (ISPM), Department of Botany, University of Delhi.

Unit I

Scope of Microorganisms and Microbial Research

1 Scope of Microorganisms in Industries

1.1 INDUSTRIAL MICROBIOLOGY

Industrial microbiology is a discipline of applied microbiology that aims to manufacture valuable and fine-quality products using beneficial microorganisms in industrial setups. Microorganisms have tremendous scope for commercial exploitation, as almost every aspect of human health benefits from microbially-derived products. The industrial production of many products by microorganisms is unique and cannot be obtained from other sources. Also, many engineered microorganisms are used as suitable platforms to derive plant and animal products. Mass cultivation of microorganisms is a strict requirement in industrial microbiology. This can be a critical prerequisite for metabolite production. Fermentation and bioprocessing are the two major players in the successful production of a desired product. Fermentation helps in meeting the demand to deliver sufficient quantities of fine-quality products. The controlled operation in bioprocessing ensures the quality assurance of a product. In the fermentation industry, the broad categories of microbially derived products can be

- Living or dead cellular biomass
- Partial fraction of cells
- Different extracellular, intracellular, and membrane-bound enzymes
- Primary and secondary metabolites
- Microbially-derived transformed chemicals
- Products derived from mutation and recombination

The modern world sees microbial processes as sustainable solutions to many conventional techniques, such as in the food and beverage, textile, pharmaceutical, leather, and detergent industries. Many new industrial setups have increased reliance on cutting-edge methods of microbial processes, considering their eco-friendly approach and economic value. This area of biomanufacturing is also distinguished as **"white" biotechnology**. The products developed by biologically mediated processes are less energy-intensive, readily biodegradable, and less waste-generating.

The first step for an industrial setup is the scrutiny of suitable microorganisms. An industrial trial begins with basic research that leads to the isolation and screening of appropriate microorganisms capable of producing desirable products. The selected strain is optimised for both its mass cultivation and secondary metabolite production. These operations are distinct elements of upstream processing. A suitable strain is transferred into the fermentation process to manufacture the products. The principle of fermentation in industrial microbiology revolves around the controlled growth of microorganisms to convert substrates into desired products through their metabolic pathways. Most products are manufactured in a well-designed fermenter or bioreactor. The fermenters are configured to maximise the yield and efficiency of the desired products. This is followed by downstream processing, which aims to purify and recover the products. Units II and III provide in-depth information on upstream processing, fermentation process, and downstream processing.

1.2 INDUSTRIAL MICROORGANISMS AND THEIR CHARACTERISTICS

Virtually all microorganisms possess the ability to produce diverse secondary metabolites and enzymes. However, not all microorganisms qualify as producer organisms. Screening high-yielding strains from various microbial isolates is a crucial need for industrial setups. Such microorganisms are termed industrial microorganisms. But it is far more important for an industrial microorganism to possess many other features besides just being able to synthesise bulk quantities of products of interest. The full knowledge of other characteristics of an industrial microorganism is also a desired parameter because a less well-known strain can result in low economic value and less stability for commercial exploitation.

Industrially relevant microorganisms are diverse and are isolated from various sources. Certain yeasts and moulds derive more products than any other fungi. Similarly, a soil-dwelling bacterial genus, *Streptomyces,* can produce more than 7,600 different secondary metabolites. Even an *in-silico* study suggests this organism can be a source of as many as 150,000 metabolites. *Escherichia coli* is an ideal bacterium for deriving a variety of recombinant products.

Both natural and genetically engineered microorganisms can be used as industrial microorganisms. However, most industrially relevant microbial strains are genetically modified. They outperform wild-type strains in many beneficial aspects.

DOI: 10.1201/9781032651095-2

Microorganisms selected as metabolic specialists for industrial applications encompass some important characteristics-

- Microbial cultivars should be easily propagated and rejuvenated from their inactive forms. They should be stable for long-term use.
- The large-scale yield of a product is an obvious desire for microbial implications.
- Inoculants should observe simple reproductive methods for fast multiplication. For example, given the right conditions, binary fission in many bacteria produces billions of cells in about 10 h. Likewise, many fungi produce a huge number of asexual spores in a brief period of incubation for mass culturing.
- Industrial microorganisms should grow on various substances as the industrial production of many products depends on the waste streams of other industries for the raw materials.
- Incurred costs during bioprocessing due to the use of microorganisms should be minimal to maintain the economic values.
- The microbial strains should be devoid of any pathogenic concern for plants, humans, and animals. They should also not possess any other raised environmental concerns because mass culturing during industrial production may not ensure effective safety measures.
- Genetic manipulation in industrially important species offers a clear advantage for desired facilities. A stable engineered strain increases the yield and provides other ease during fermentation and downstream processes.

Industrial microorganisms have an impact across various industrial sectors (Figure 1.1). Their integrity and applications in different industries are discussed as follows.

1.3 FOOD AND BEVERAGE INDUSTRY

Microorganisms have had a profound influence on food and beverages since ancient times. The whole or partial microbial biomass can be used as a food source. Microbial masses that are used in the culinary industries have food-grade status and approval, such as 'Generally Recognised as Safe' (GRAS). The metabolic activities of many bacteria, fungi, and microalgae are central to the development of a variety of foods. Food-grade microorganisms contribute in some way to the changes in the food materials. They can modify the textures, colours, scents, and tastes in foodstuffs. The microbial components can alter the structure and chemicals of food to make them highly nutritive and delectable. Many microbial cells transfer nutritional ingredients and complex sets of enzymes in foodstuffs to improve their value. Many nutritive bacterial and yeast cells are the sources of food and feed. Some specific microorganisms produce health-promoting compounds for humans, having nutraceutical properties. Microbial biomasses are highly promising resources for the development of future food.

Based on the bioactive materials, the microbial augmentation in the food and beverages can be of three different types. These include whole microbial cells, microbial enzymes, and microbial deposits, discussed as follows.

1.3.1 Whole Microbial Cells

The whole microbial cells can be used as biomass, a source of nutrients, or to alter the nature of foodstuffs. The additional side reactions of selected microorganisms add distinct aroma, flavour, and digestibility to the food. In bacteria, lactic acid bacteria (LAB) have prominent roles in food fermentation and typically include heterogeneous groups of bacteria such as *Lactobacillus*, *Leuconostoc*, *Pediococcus*, and *Streptococcus*. The dairy industry greatly depends on the activities of LAB. These bacteria turn milk sour by converting its sugars into lactic acid. LAB utilises at least 11–12 different enzymes in a series of biochemical reactions (including the glycolytic pathway) for this transformation. A commercial dairy product yoghurt, is a result of milk fermentation using *Streptococcus thermophilus* and *Lactobacillus delbrueckii* (Figure 1.2a). Kefir is fermented milk, known for its exceptional nutrient density, developed using symbiotic cultures of bacteria and yeast (Figure 1.2b). LAB are also extensively used in fermented meat products (e.g., dry sausages), fermented fish products, fermented vegetables (e.g., Sauerkraut and Kimchi made from cabbage) and cereal products (e.g., Bushera made from sorghum flour) (Figure 1.2c). Kovászos uborka is a Hungarian summer pickle made up of fermented cucumber. In this pickle, the use of bread in the pickling promotes the LAB growth for fermentation reactions (Figure 1.2d). Modern livestock production uses LAB (such as *Lactobacillus*, *Streptococcus*, and *Pediococcus*) as feed to balance their intestinal flora and develop healthy stocks.

The use of fungal cultures for the fermentation of a variety of foodstuffs is also evident. A single-cell fungus (yeast), *Saccharomyces cerevisiae,* has a notable role in preparing bread and liquor (Figure 1.2e) (detailed in Chapter 13). The yeast extract left over from beer brewing can be fortified and used as a savoury food spread. Tempeh is a well-known Indonesian food, which is a product of the fermentation of soybeans using *Rhizopus oligosporus* or *Rhizopus oryzae* (Figure 1.2f). Fungi like *Mucor miehei* and *Aspergillus niger* are notable sources of microbial rennet for cheese making, which is used as a substitute for animal rennet. A filamentous fungus, *Penicillium roqueforti,* is well known for its ability to ripen blue-veined cheese (Roquefort cheese) (Figure 1.2g).

Some microbial cells have 30% protein in their biomass and offer dietary benefits. They can be dried and consumed as single-cell proteins (SCPs). SCPs are a key source of protein extract, which can be derived from bacteria, cyanobacteria, yeasts, moulds, and microalgae. The dried biomass of *Arthrospira platensis* (often known as spirulina) (a cyanobacterium) is a popular source of SCP. A green alga, *Chlorella,* is also very nutritious for health.

"Probiotics are live microorganisms that, when administered in adequate amounts, confer a health benefit on the host"— a modified definition of FAO/WHO. The fundamental essence of the word "probiotics" is microbial,

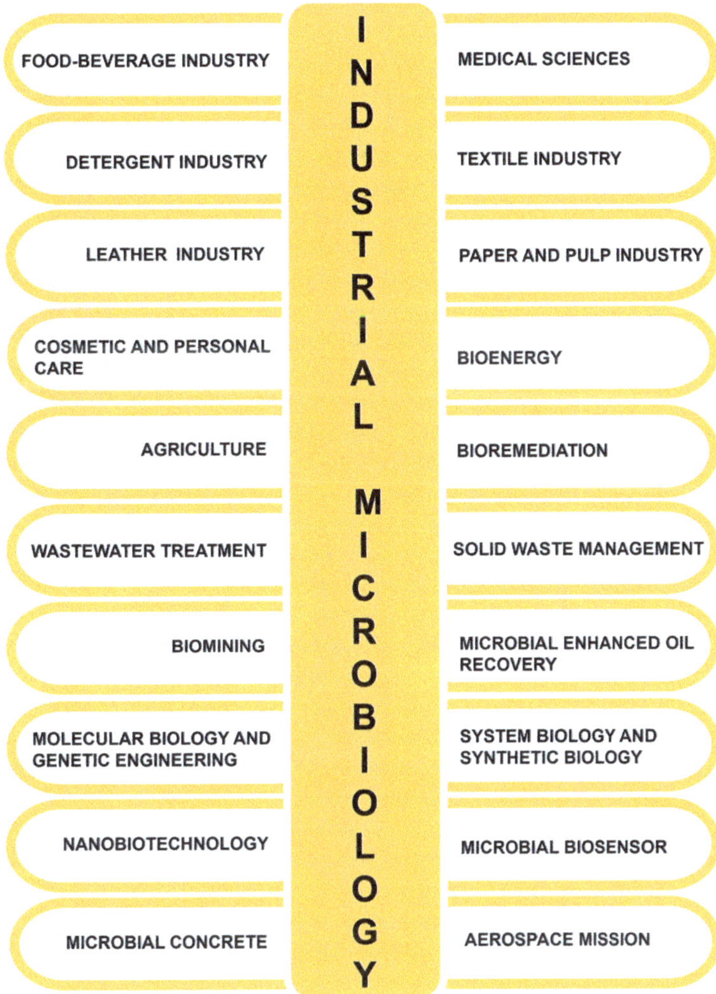

FIGURE 1.1 Scope of industrially relevant microorganisms in different sectors.

viable, and beneficial to health (Figure 1.2h). The ideal microorganisms in probiotic preparations have GRAS status and are resistant to bile, acid, and pancreatic juice. Also, they exhibit low intestinal permeability, induce lactic acid production, and are tolerant to acidic and basic environments of the stomach and duodenum, respectively. Such reputations can be found in some bacterial species such as *Lactobacillus*, *Bifidobacterium*, *Escherichia*, *Enterococcus*, *Bacillus* and *Streptococcus* and some yeasts such as *Saccharomyces boulardii* and *S. cerevisiae*. Probiotics confer the reestablishment of microbial equilibrium in the gut for health improvement and disease management. They have also been known to prevent invasion by pathogens, improve immune host response, and treat clinical conditions. An alternative to probiotics, prebiotics are the supply of non-digestible food ingredients that favour the growth of beneficial bacteria with probiotic properties in the gut. The mixture of probiotics and prebiotics is known as **synbiotics**, which act synergistically to improve the survival and implantation of beneficial microorganisms in the hostile environment of the gastrointestinal tract.

Fungal mycelia are important sources of future food and feed, being rich in **mycoprotein** and fibre. They are composed of multiple fungal hyphae. The fungal mycelia of *Fusarium venenatum,* under the trade name **Quorn™,** is already an established substitute for meat (Figure 1.2i). Fungal mycelia are composed of highly digestible proteins, fibre, and low fat, which mark their value as exceptionally healthy food. The production of plant-based meat products is already an established technology for sustainable food production. In this line, the use of microbial forms such as rhizobia and yeasts to replace animal-based food that mimics and tastes like meat is also an early entrant of the food industry.

Some fungi even bear visible fruiting bodies and grow aboveground, called mushrooms. They are macrofungi, but their basic structure is the same microscopic filament, hyphae. The fruiting bodies are specialised structures produced in specific conditions and are an important source of diet for humans around the world. More than 2,000 mushroom species have confirmed nutritional values. They are rich in vitamin B-complex (particularly riboflavin) and contain high, low-calorie sources of fibre, protein, and antioxidants. Some other

FIGURE 1.2 Examples of microbially derived global and regional foodstuffs. a. Yoghurt. b. Kefir. c. Sauerkraut. d. Kovászos uborka. e. Bread. f. Tempeh. g. Blue-veined cheese. h. Probiotics. i. Quorn™. (Image b obtained from Flickr/Aaron Gustafson[1], c from Flickr/ccbarr[2], f from Wikimedia/Christian Franke[3], g from Flickr/dewet[4] and with permission of Quorn®.)

prominent cultivated species are *Agaricus bisporus* (white or button mushroom), *Morchella*, *Pleurotus* (oyster), *Lentinula edodes* (shiitake), and *Volvariella volvacea* (paddy straw). Particularly, *Morchella* and *Pleurotus* are highly valued by gourmet chefs (Figure 1.3a, b). Two different fungi, *Laetiporus sulphureus* (commonly known as chicken of the woods) and *Fistulina hepatica* (beefsteak fungus), taste like meat and can be used as meat replacements (Figure 1.3c, d). Some fungi produced a different type of fruiting body in subterranean regions (i.e., underground spaces) called truffles. Some truffles are highly valuable and are very expensive. For example, white truffle (*Tuber magnatum*) and black truffle (*Tuber melanosporum* and *Tuber aestivum*) are sold at an average price of $3386.24 and $721.25 per kg, respectively (Figure 1.3e, f).

FIGURE 1.3 Fruiting bodies of edible fungi. a. *Morchella* (true morel). b. *Pleurotus* sp. (oyster). c. *Laetiporus sulphureus* (chicken of the woods). d. *Fistulina hepatica* (beefsteak fungus). e. *Tuber magnatum* (white truffle). f. *Tuber aestivum* (black truffle). (Image a from US Army Corps of Engineers, c reprinted from Luangharn et al. (2014)[5], © 2014 Mae Fah Luang University, d. Panoramio/Henk Monster[6], and e obtained from Wikimedia/Mortazavifar.[7])

1.3.2 Microbial Enzymes

Enzyme treatments have filled many needs in the food and beverage industry and produce a wide variety of products such as bakery products, dairy products, baby foods, canned fruits, soft drinks, and ice creams. Microorganisms are diverse sources of enzymes (Table 1.1). Many plant-derived and animal-based substances are suitable raw materials for microbial enzyme action. For example, enzymes, α-amylase and glucoamylase act upon nonfood matter, starch, and convert it into glucose that serves as an important resource for the preparation of different foodstuffs and beverages, including alcohol, sweeteners, ascorbic acid, lysine, etc. Cell wall degrading enzymes can be used to extract intracellular products from plants. The perfect consistency, stability, and quality of bread in the baking industry relies on the synergistic effects of enzymes such as α-amylase, xylanase, lipase, and glucose oxidase. The applications of the enzyme asparaginase in the baking industry reduce the formation of a potentially carcinogenic substance, acrylamide. α-amylase reduces the viscosity of gelatinous starchy material into an easy flowing liquid. A different type of enzyme, xylanase, also helps reduce the viscosity, which allows for the separation between starch and gluten in wheat flour slurry. The processes of liquefaction and saccharification, which break down complex carbohydrates—are essential to the alcohol industry and are enzyme-dependent. Liquefaction relies on α-amylase that acts on α-1,4-glycosidic bonds of starch and produces maltodextrins. Industrial saccharification is a real turning point for the transformation of carbohydrates into simple sugar molecules for alcohol fermentation and employs glucoamylase enzymes. Proteases can be used to hydrolyse milk proteins to prepare cheese, increase the shelf life of fruits and vegetables (by denaturing ripening enzymes), and tenderise meat.

In the food industry, interesterification is a process of rearrangement of the fatty acids in the fat products to control their physical characteristics. Enzyme interesterification is a suitable alternative to chemical methods and requires lipase. The enzymatic treatment is not only cost-effective but also maintains the product quality without producing trans fatty acids. The enzymatic degumming can be performed using microbial phospholipases to reduce the gums (i.e., phospholipids) from the crude vegetable oils. A different preparation is Durazyme syrup, which contains two enzymes, diastase and pepsin, from *A. oryzae* and is used for the inability to digest carbohydrates and proteins and as an appetite stimulant.

1.3.3 Microbial Extracts

A wide variety of microbial extracts are new-generation food additives in the food and beverage industry. They can belong to different categories like carotene, flavonoids, pyrroles,

TABLE 1.1
A list of different microbial enzymes and their applications in the food and beverage industry, and their microbial sources

Microbial enzymes	Applications in the food and beverage industry	Examples of microbial sources
α-amylases	Brewing, baking, starch liquefaction, corn syrup production, and juice clarification	Bacteria: *Bacillus subtilis*, *Bacillus stearothermophilus*, *Bacillus licheniformis*, and *Bacillus amyloliquefaciens* Fungi: *Aspergillus niger*, *Aspergillus awamori*, and *Aspergilus oryzae*
Proteases	Meat tenderisation, hydrolysis of whey protein, and baking	Bacteria: *B. subtilis, B. licheniformis, Streptomyces* Fungi: *A. oryzae* and *A. niger*
Lipases	Cheese ripening, baking, production of lipolysed milk products, and transesterification in fat and oil	Bacteria: *B. subtilis, B. licheniformis,* and *Bacillus thermocatenulatus* Fungi: *A. niger, A. oryzae, Rhizopus oryzae, Rhizopus delemar, Penicillium roqueforti, Penicillium camemberti, Candida rugosa,* and *Candida antarctica*
Lactase (β-galactosidase)	Lactose-free product development and prebiotic food ingredient	Bacteria: *Bifidobacterium* Fungi: *Kluyveromyces lactis*, *Kluyveromyces fragilis*, and *Saccharomyces fragilis*
Glucose oxidase	Baking, desulphuring of eggs, and food preservation	Bacteria: *Streptomyces* spp. Fungi: *A. niger*
Xylanases	Bear quality improvement, baking, and juice clarification	Bacteria: *B. subtilis, B. licheniformis, Bacillus pumilus, Streptomyces lividans, Streptomyces thermoviolaceus,* and *Clostridium thermocellum* Fungi: *A. niger*, *A. oryzae*, *Trichoderma reesei*, *Humicola insolens*, *Penicillium janthinellum* and *Penicillium funiculosum*
Asparaginase	Baking (to reduce acrylamide level) and for leukaemia treatment	Bacteria: *Erwinia chrysanthemi* Fungi: *Penicillium brevicompactum, Streptomyces albidoflavus,* and *Streptomyces griseus*
Pectinases	Juice clarification	Bacteria: *B. subtilis, B. licheniformis, Bacillus pumilus, Erwinia carotovora, Streptomyces lydicus*, and *Streptomyces albidoflavus* Fungi: *Candida tropicalis*
Catalase	Food preservation and cheese production	Bacteria: *Micrococcus luteus*, *B. subtilis* and *B. licheniformis* Fungi: *Candida boidinii*
Cellulases	Brewing, juice clarification, and animal feed	Bacteria: *B. subtilis, B. licheniformis, B. amyloliquefaciens, Clostridium thermocellum, Thermomonospora fusca, Streptomyces lividans, and Streptomyces thermoviolaceus* Fungi: *Trichoderma reesei, Humicola insolens, Penicillium funiculosum* and *Penicillium janthinellum,*
Glucoamylase (or amyloglucosidase)	Baking, brewing, and dextrose production	Bacteria: *B. subtilis, B. licheniformis,* and *Clostridium thermosulfurogenes* Fungi: *Penicillium expansum, R. oryzae, A. niger,* and *A. oryzae*
Laccases	Removal of polyphenols in wine and baking	Bacteria: *B. subtilis* and *B. licheniformis* Fungi: *Trametes versicolor, Pycnoporus cinnabarinus, Pleurotus ostreatus,* and *Fomitopsis pinicola*

polysaccharides, organic acids, amino acids, vitamins, etc. Microbial enzymes in the food and beverage industry offer a major augmentation and are already discussed separately. Many microorganisms are sources of vitamins B12, B2, B9, and B7, which can be widely used in food fortification, animal feed, and as a supplement. For example, selected bacteria, *Propionibacterium* sp. and *Pseudomonas denitrificans*, are cultured in a nutrient-rich medium that supports the biosynthesis of vitamin B12. Microbial extracts can be potential sources of nutritional supplements, thickening agents, colourants, flavour enhancers, flavour agents, and acidulants. The commercial production of food-grade additives from microorganisms requires fermentation and extraction facilities. In the current global market of food-grade colourants, riboflavin from *Bacillus subtilis* and *β*-carotene from *Blakeslea trispora* and *Dunaliella salina* are the sources of yellow pigments, lycopene from *B. trispora* and astaxanthin from *Haematococcus pluvialis* of red colour, and canthaxanthin from *Haematococcus lacustris* of orange colour. Xanthan gum, curdlan, and gellan gum can be obtained from *Xanthomonas campestris*, *Agrobacterium* spp., and *Sphingomonas elodea*, respectively and are extensively used as thickening or gelling agents. They serve as an alternative to agar. Yeast hydrolysate from yeast cells and monosodium glutamate (MSG) from *Corynebacterium glutamicum* and *Brevibacterium* spp. are natural flavour enhancers, commercially produced using microbial fermentation. Microbially-derived γ-decalactone, diacetyl, and methyl ketones are potential sources of flavour compounds. The fermentation products like citric acid, lactic acid, and acetic acid are acidulants that give tart, sour, or acidic flavour to foods and drinks.

1.4 PHARMACEUTICAL INDUSTRY AND MEDICAL SCIENCE

Microorganisms in the pharmaceutical industry and medical science can be used for antimicrobial compound production (e.g., antibiotics), vaccine manufacturing, biotherapies, and biocontrol activities (Figure 1.4). About half of the drugs and medicines available in the pharmaceutical market are microbial products and their derivatives. Many antibiotics are secondary metabolites of bacteria and fungi, which act as broad-spectrum agents against many infectious bacteria. Examples of those include penicillin from *Penicillium chrysogenum*, chloramphenicol from *Streptomyces venezuelae*, rifamycin from *Amycolatopsis*, and teixobactin from *Eleftheria terrae*. However, the rapid emergence of antibiotic resistance in bacterial pathogens has reduced the efficacy of natural antibiotics. Some natural antibiotics and microbial enzymes can be modified to derive semisynthetic antibiotics, which combat antibiotic resistance. For example, a microbial enzyme penicillin acylase converts natural penicillin into a precursor, 6-aminopenicillanic acid (APA), for the semisynthesis of a variety of β-lactam antibiotics (e.g., ampicillin, amoxicillin, and piperacillin). Bacteriocins are an antibacterial peptide class of bacterial metabolites that are present in up to 90% of prokaryotic taxa. These compounds control the growth of closely related strains and exhibit narrow inhibitory spectra. The commercial potential of bacteriocin production is found in LAB.

The genetic engineering approaches have shown tremendous applications in vaccine development. They can be live or attenuated recombinant viral particles and induce immune reactions. Dead or fractionated viruses also play similar roles in preventive measures of diseases. A different approach enables the production of antibodies in genetically modified microbial systems. The production of monoclonal antibodies (mAbs) through biotechnological tools represents a promising way of treatment of infectious diseases. Many hormones, anticoagulants, and high-value proteins can also be produced through genetic manipulation using *E. coli* and *S. cerevisiae* as suitable biofactories. In contrast, many pathogenic microorganisms have been used in warfare for their potential to incapacitate or harm populations (detailed in Box 1.1).

Many biotherapies are based on live biotherapeutic products (LBPs) and biologics. These include live-administered microorganisms in the human body through oral, topical, or injectable methods. LBPs strictly intend to treat, cure, or prevent a disease, which shows the marked differences from probiotics. Probiotics are used as dietary supplements and do not claim to treat disease. Genetic engineering in LBPs offers a diverse array of medical benefits. The genetically modified *Escherichia coli* Nissle 1917 (EcN) has attracted significant attention in the treatment of phenylketonuria and solid tumours. The modified *Lactococcus lactis* stimulates mucosal immunity against respiratory tract pathogens. Genetically attenuated *Listeria monocytogenes* is a popular vector for cancer immunotherapy. *Bifidobacterium longum* has protective effects against inflammatory bowel disease. In recent years, faecal microbiota transplantation (FMT) has received great attention for transplanting faecal matter from a healthy donor to a recipient to restore a healthy microbial balance in the gut. This has conferred a successful treatment of recurrent *Clostridium difficile* infections and promised for other health conditions.

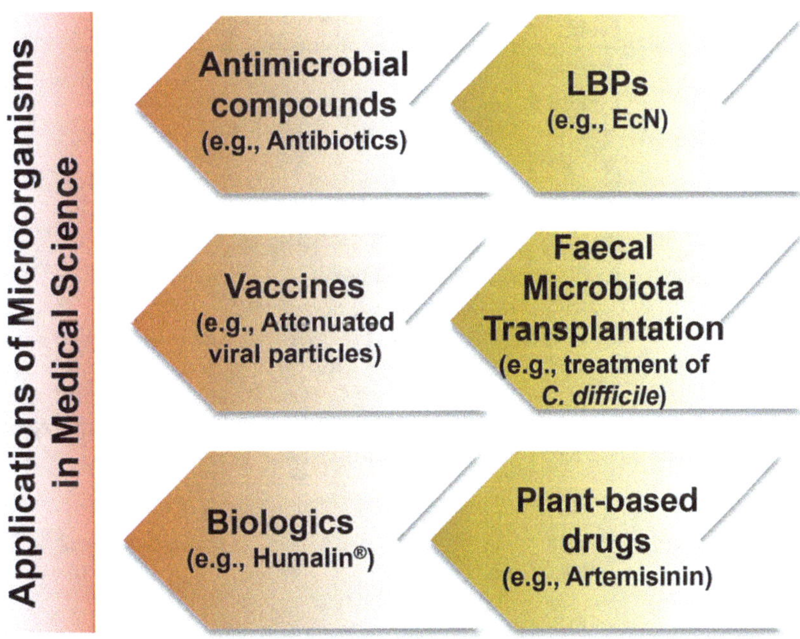

FIGURE 1.4 Various applications of microorganisms in medical science (LBPs stands for live biotherapeutic products).

The improved strategies in biotechnology have promised to manufacture a variety of plant-based drugs using microbial platforms at industrial scales. For example, some fungal endophytes (e.g., *Aspergillus fumigatus* and *Taxomyces andreanae*) are natural sources of the anticancer drug paclitaxel (or taxol), which has emerged as a promising alternative to plant sources. The genetically engineered yeast has been used to produce a precursor of a plant product, artemisinin, used in malaria treatment. Many products such as calcitonin, proinsulin, glutamate, and lysine, have been obtained from their native strains after some metabolic engineering. For example, the metabolic pathway of *Corynebacterium glutamicum* can be genetically modified for glutamate production. Chapter 13 gives a detailed account of the use of different microorganisms in medical sciences.

BOX 1.1 BIOLOGICAL WARFARE

Biological warfare (or bioterrorism) is an intentionally caused event of mass destruction using pathogens. While this is a highly unethical and illegal practice, some dreadful events in the past have created indiscriminate suffering and long-term ecological and health impacts. Many pathogens belonging to bacteria, viruses, fungi, and protozoa are potential bioweapons. Even many secondary metabolites produced by microorganisms are potent toxins, posing a high risk for bioterrorism. The advent of new tools such as DNA recombinant technology and gene editing has paved the way to modify and engineer infectious agents to enhance their vigour. The notable diseases that can cause major losses are anthrax, botulism, smallpox, aflatoxins, and brucellosis. They can be easily disseminated through aerosol, food, water, vector, and injection. Based on the power of dissemination and their impact on health, pathogens can be categorised at three different priority levels: Category A (highest priority), Category B (second highest priority), and Category C (third highest priority).

The natural history of biological terrorism dates back to the 14th century BC when the Hittites sent infected rams with tularaemia disease (caused by the bacterium *Francisella tularensis*) to their enemies. Thereafter, there have been many incidents of attempts to spread diseases by different means in different periods. A prominent case occurred during a medieval blockade of Kaffa in Ukraine (1346)— the Mongols tossed plague (caused by a bacterium, *Yersinia pestis*) on Kaffa. In the 20th century, people witnessed biological weapon programmes to build up battle-ready bioweapons in different parts of the world. For example, the United States of America (USA) military stations had developed facilities for the production of saxitoxin and botulinum toxins, which had microbial origins. However, such programs were terminated in 1969 after the ban on biological warfare. A dreadful example of bioterrorism in the past is an incident of "anthrax letters" (the deadly spores of *Bacillus anthracis*) delivered to the media offices and senators in the USA (2001). Box Table 1.1.1 provides a list of different bioagents of bioterrorism, including information related to associated diseases, route of infection, category, and any related historical account.

BOX TABLE 1.1.1

A list of some notable events of biological warfare using different microbial agents

Time	Locations	Events	Microbial agents
14th century BC	Modern-day Turkey	Epidemic of tularemia caused by Hittites	*Francisella tularensis*
1346	Modern-day Ukraine	Medieval blockade of Kaffa tossed by Mangol	*Yersinia pestis*
1495	Naples, Italy	Spanish mix wine with the blood of leprosy patients to sell to their French foes	*Mycobacterium leprae*
1763	Modern-day Pittsburgh, USA	Distribution of smallpox-infested blankets by British army officers	Smallpox
1920–1991	Modern-day Russia	Soviet biological weapons program	*Bacillus anthracis, Yersinia pestis, Francisella tularensis, Staphylococcus aureus, Burkholderia mallei, Clostridium botulinum, Brucella sp., Coxiella burnetii*, Venezuelan equine encephalitis virus (VEE), Smallpox, Marburg virus, and Orthopoxvirus
1984	The Dalles, Oregon, USA	Food poisoning through Salad bars	*Salmonella enterica* Typhimurium
1985–1991	Iraq	Biological weapon program	*Bacillus anthracis*, *Clostridium botulinum*, and Aflatoxin
2001	USA	Anthrax letters delivered to the media offices and senators	*Bacillus anthracis*

Mass destruction using bioweapons can cost as little as 0.5% of conventional weapons. However, a biological attack can lead to high mortality rates and cause anxiety and social disruption. Bioterrorism is often difficult to recognise because it may mimic a natural outburst and provide an easy escape. The high potential for secondary transmission and the outset of an epidemic or a pandemic outbreak of pathogenic microbial agents are serious concerns for public health. Such intentions challenge national security. Although international laws have outlawed this kind of research facility, the preparedness for the detection, protection, and decontamination of bioweapon outbreaks is still necessary under careful observation. The disastrous effects of SARS-associated coronavirus-2 (SARS-CoV-2) in the current transition have already shown our vulnerability to bioterrorism.

1.5 DETERGENT INDUSTRY

The multitude of hydrolytic enzymes like proteases, α-amylases, lipases, cellulases, pectinases, and mannanases are notable subcomponents in the detergent. The hydrolytic activity of enzymes offers solutions to remove dirt and stains in the laundry and automatic dishwashing. The cocktails of enzymes produce greater benefits through synergism. Most of these enzymes used in household and professional cleaning are alkaline and active at low temperatures. Microbial proteases can remove protein stains such as blood, egg, and human sweat through proteolysis. The detergent-compatible alkaline proteases are obtained from alkalophilic strains of *Bacillus*, and they are important components in commercial laundry detergents like Tide®, Ariel® and Biz®. In this category, "subtilisin" (first identified in *B. subtilis*), a distinct class of extracellular serine endopeptidase, is a prominent ingredient in detergents. The keratinolytic proteases promise to remove the blood stains, which digest peptide bonds in globin protein and remove protein stains (Figure 1.5). The cleaning of digest starchy stains like residues from chocolate, gravies, and pasta from fabrics depends on α-amylases.

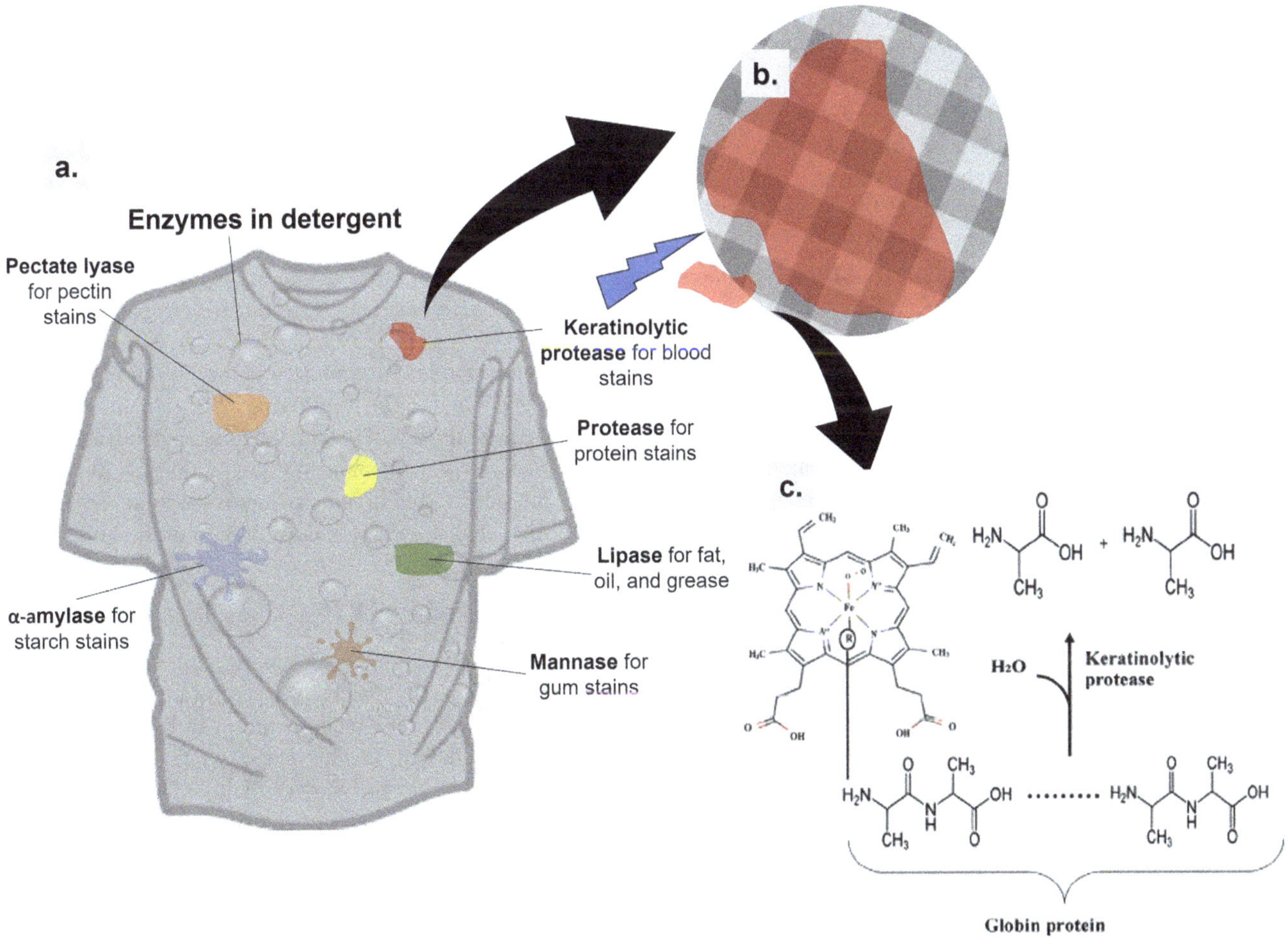

FIGURE 1.5 An illustration of the removal of different strains from cloth using an enzyme cocktail in detergents. a. Removal of different stains using enzymes. b. An enlarged view of the removal of blood stains using keratinolytic protease. c. A reaction showing the effect of keratinolytic protease on the globin protein of blood. (Image c adapted from Paul et al. (2014)[8] with permission of © 2013 Elsevier Ltd.)

The chief producers of commercial α-amylases for the detergent industry are different species of *Bacillus* (*e.g., Bacillus licheniformis, B. subtilis, Bacillus cereus*). Tough oil or grease stains are the main substrates for the action of enzyme lipases, which perform the hydrolysis of mono-, di- or triglycerides into glycerol and fatty acids. Lipases in a detergent can also be used to replace phosphate-based chemicals. Detergent composition containing cellulases improves the smoothness of fabric and removes soil dirt. Commercial cellulases can be derived from diverse fungal and bacterial sources. Other enzymes, such as mannases and pectinases, are useful to remove hard-to-remove stains from various foodstuffs. The commercial formulations of a detergent may even include biosurfactants that claim better wetting and spreading, solubilisation, enzyme activity, foaming, and dispersion during detergency.

1.6 TEXTILE INDUSTRY

The modern fashion world sees sustainability in the textile industry through microbial enzymes. Various enzymes have substituted chemicals for the pretreatment of raw fabric and claim to save lots of water and energy, reducing overall environmental impact. Most of the enzymes utilised in textile industries are hydrolases that improve the hydrophilicity of the fabrics and upgrade the wettability, dyeability, finishing, and grafting of textiles. A process of desizing, where the protective layer of starch is removed from the surface of yarn, can be performed using starch-degrading enzymes like α-amylases, diastases, and glucoamylase. Scouring in textiles is a process to get rid of natural wax and non-fibrous impurities from the fabrics involving the treatment of heavy chemicals. Pectinase is a suitable replacement for the chemical scouring of textile materials like cotton knits, woven yarn, and towels. Bleaching is a step to decolourise the natural colour of fabrics and use hydrogen peroxide (H_2O_2). Removal of H_2O_2 after its treatment from the textile is essential; otherwise, dyes do not colour the fabrics. In this situation, the sustainable solution lies in the formulation of catalase, which cleans up even the minute traces of bleach. Alternatively, different enzymes like glucose-oxidase, laccases, and peroxidases can be used in place of chemical bleach. The countless microorganisms, primarily fungi, provide the pigments to colour wool, silk, cotton, and synthetic materials. Examples of fungal-derived colourants are substances like anthraquinones, azaphilones, quinones, and carotenoids. Fungi, *Monascus purpureus* and *Dermocybe sanguinea,* are the source of red pigments, commercially used to colour synthetic fibres. Biopolishing intends to use cellulases to remove loose fibre ends and reduce fuzz and pills from cellulose-based fabrics. Denim abrasion and finishing are processes to achieve a faded and aged look, which can also be accomplished using cellulases (Figure 1.6a). The surface of wool fibres can be modified using proteases to prevent the shrinking, pilling and gumming of fibres. The synergisms of diverse enzymes offer multiple activities for the treatment of textiles. For example, a patented product, CombiPolish® (Novozymes), claims bleach clean-up and biopolishing in one step. Besides, enzymes like esterases, lipases, and cutinases in various combinations can be suitable formulations for modifying synthetic fabrics like polyester.

1.7 LEATHER INDUSTRY

The conversion of raw skins and hides into leather is considered one of the most polluting sectors due to the heavy use of chemicals. This generates sizable solid wastes, liquid wastes, and toxic gases. The use of microbial enzymes as an alternative to conventional chemical treatment in leather processing is a viable solution to overcome the hazard of tannery effluents. The reliance on microbial enzymes in the leather industry (or tanneries) is witnessed in a series of operations such as soaking, unhairing, bating, dyeing, degreasing, and effluent treatment. The proteolytic and lipolytic activities are most desirable at different stages of treatment. Soaking is the first step of leather manufacturing, where the use of amylases, proteases, and lipases has earned special attention in the removal of dirt, flesh, and blood from hides and skins. These enzymes can be outsourced from bacteria like *B. subtilis*, *B. licheniformis*, and *Pseudomonas fluorescens*. The next step is unhairing, where the processing of hide using proteases is a paradigm shift from chemical processing for removing hairs and furs from the hides. The unhairing is followed by bating. This involves a process of skin softening and removing scud and unwanted proteins. Modern commercial bating preparations consist of formulations of microbial proteases. The next step is degreasing, which removes the excess natural fat from the hide using organic solvents, which can be replaced with the potential usage of lipases. A further insight into microbial preparation in the leather industry is the pollution abatement of tannery effluents. The salt-loving microorganisms can be used in the treatment of chemical oxygen demand (COD), biochemical oxygen demand (BOD), total dissolved solids (TDS), inorganics, and organics. Figure 1.6b illustrates all the steps of the tanneries that are moved by microbial treatment.

1.8 PAPER AND PULP INDUSTRY

The raw materials of paper making are cellulosic fibres obtained from plant materials, which are susceptible to microbial treatment. The manufacturing of paper can employ microbial enzymes at various stages. The first step is the removal of bark from the woody plants. The process relies on mechanical processes along with some enzymes. Enzymes like pectinases and xylanases assist in removing pectins and hemicellulose from the outer part of the wood log. The next biologically derived step can be pulping, which is achieved by chemical and mechanical processes and is expandable to microbial activities. Fungi like *Coriolus versicolor*, *Pleurotus* sp., *Phanerochaete chrysosporium*, and *Schizophyllum commune* can be used for pulping. Solid-state fermentation (SSF) conditions promote the synthesis of lignin-degrading enzymes to act upon the raw matrix. Inbuilt enzymes such as lignin peroxidase, laccase, and manganese peroxidase facilitate the removal or loosening of the lignin content in the raw material and transform it into fibrous stuff pulp. Steam sterilisation of the raw materials is an essential step to remove cellulose-degrading microorganisms to protect the cellulose strength of plant fibres. Also, many woody plants contain large amounts of resinous ducts, which are composed of fats, resin

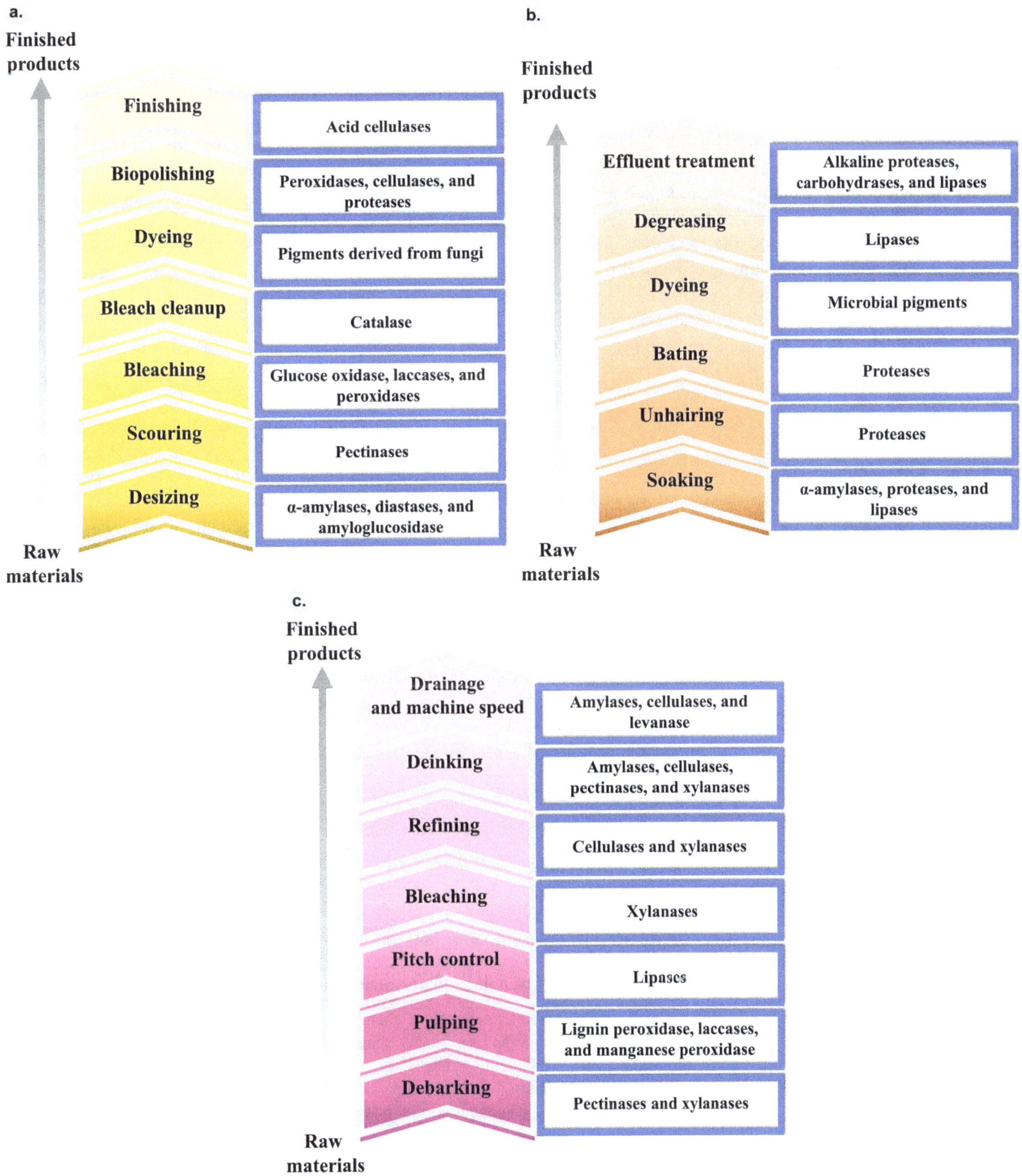

FIGURE 1.6 Applications of microbial enzymes at different steps of industrial processing. a. Textile manufacturing. b. Tanneries. c. Paper and pulp industry.

acids, sterols, and waxes and are essentially removed in pitch control methods. Microbial lipases can be sustainable choices for their removal. The next step is biobleaching, which acts on hemicellulose to release the trapped lignin. The thermostable, alkaline, stable xylanases are appropriate to break the network of xylan that acts as a physical barrier. This process allows the influx of bleaching agents for further detachment of lignin content. Cellulases can be added to improve the smoothness and tensile strength of paper.

In the recycling of waste paper, a process called deinking that enables the removal of ink from the fibre surface can also add sustainability to its operation through enzyme applications. Cellulases, often in combination with other enzymes like amylase, hemicellulases, or pectinases, are suitable choices for deinking. An enzyme called levanase, which acts on the extracellular polysaccharide levan in the slime layer, can be used to control slime formation during paper production. Levanase, in combination with amylases and cellulases, has improved activity to remove slime formation. In a different application, these enzyme formulations can be used to clean paper machines for smooth operation and energy efficiency (Figure 1.6c).

1.9 COSMETIC AND PERSONAL CARE INDUSTRY

Various microbial compounds like pigments, antioxidants, surfactants, vitamins, and enzymes derived from bacteria, fungi, and microalgae are prominent sources of cosmetic products (Table 1.2). These biological alternatives have many advantages over chemically synthesised products owing to their biodegradable, renewable, and nontoxic nature. The fermented extract/lysates of various probiotics such as *Lactococcus, Lactobacillus, Bifidobacterium longum, Bacillus coagulans*, and *Saccharomyces* are notable ingredients of cream, cleanser, shampoo, balm, serum, exfoliate, and soap bar. Protease-based enzyme blends are available in toothpaste, creams, shampoos, and skin cleansers. Lipase- and antioxidant-based formulations can be found for hair care products. Exopolysaccharides (EPS) are approved materials in skin formulations to maintain a hydrated environment. They can be obtained as dextran from *Leuconostoc*, xanthan from *Xanthomonas*, and alginates from *Pseudomonas*. The cyclic oligosaccharides cyclodextrins are used to enhance the aroma of perfumes and deodorants. These compounds are commercially produced using different strains of *Brevibacterium* spp. and *Bacillus* spp. Many microbial proteins and enzymes have properties that can be used to exfoliate and treat skin disorders. Microbially-derived biosurfactants are used in cleansing products, body massage oils, and anti-wrinkle cosmetics. An enzyme keratinase, has hair removal properties. *B. licheniformis* is a potent producer of keratinase. A microalga, *Nannochloropsis oculata,* produces a carotenoid, zeaxanthin, which claims skin whitening treatment. Some microalgae and cyanobacteria produce ultraviolet (UV)-photoprotective agents like mycosporine-like amino acids (MAAs) for sunscreen lotions and creams. A metabolite of the fungus *Schizophyllum*, schizophyllan excels at treatment against UV-induced immunosuppression. Vaginal administration of *Lactobacillus* spp. neutralises damage from urogenital pathogens and promises female intimate care. In addition, many fungal secondary metabolites, such as L-ergothioneine, chitin-glucans, and ceramides, have distinct uses in skin care, antioxidants, and hair products.

1.10 BIOENERGY INDUSTRY

In recent years, the energy sector has shown reliance on microbial processes due to the fast receding of fossil fuels (particularly crude oils and gases). Plant- and algal-based biomasses and other carbonic materials are reliable, eco-friendly, and cost-effective feedstocks for bioenergy production. Several microorganisms can act on these biomasses and convert them into renewable energy resources. The significant interests in bioenergy lie in biofuel, bioelectricity, bioproducts, and heat production (Figure 1.7).

Biofuels exist in liquid and gaseous forms. The use of biofuel in the transport sector is seen as one of the most viable options for reducing carbon dioxide (CO_2) emissions. The syntheses of different biofuels obey different microbial routes. One category of biofuel is alcohols such as methanol, ethanol, isopropanol, butanol, hexanol, and long-chain alcohols. They

TABLE 1.2
A list of various microbial-based ingredients used in cosmetic and personal care products

Microbial-based ingredients	Cosmetic and personal care products
Ferment extract/lysates	Creams, cleansers, shampoos, balms, serums, exfoliators, and soap bars
Proteases	Toothpaste, creams, shampoos, skin cleansers, and exfoliators
Lipases and antioxidants	Hair care products
Dextran, xanthan, alginates, and cyclodextrins	Skin formulations, perfumes, and deodorants
Biosurfactants	Cleansing products, body massage oils, and anti-wrinkle cosmetics
Keratinase	Hair removal cream
Carotenoid and zeaxanthin	Skin whitening cream
Mycosporine-like amino acids (MAAs)	Sunscreen lotions and creams
L-ergothioneine, chitin-glucans, and ceramides	Skincare, antioxidants, and hair products
Schizophyllum	UV-induced immunosuppression formulations
Lactobacillus spp.	Female intimate care products

are the metabolic products of different microorganisms. Bioethanol is the most popular among all. It is mainly produced by fermentation of simple sugars using a yeast, *Saccharomyces cerevisiae*. Butanol is commonly produced by bacteria like *Clostridium* spp., *Ralstonia eutropha*, *E. coli*, and *Bacillus* spp. Methanol is obtained from methanotrophic bacteria. A different category of biofuel is biodiesel (e.g., isoprenoid- and fatty acid-based esters). Oleaginous bacteria, yeast, and microalgae are potential sources of fatty acid methyl esters (FAMEs) (i.e., biodiesel). Oleaginous bacteria and yeast accumulate as high as 80% FAME of cellular dry weight (CDW).

Biogases are renewable sources of energy and can be used as an alternative to natural gases. Biogases include gases like methane (CH_4), CO_2, H_2, and other trace gases produced by microbial activities. They can be used for cooking, heating, and vehicle transportation. The industrial-scale CH_4 production involves the anaerobic digestion of organic materials in the digester. Organic wastes such as manure and municipal wastes are the common renewable sources for CH_4 production. Many plant biomass sources can also serve as feedstocks for the same. The complex bacterial and archaeal communities that live within organic wastes break down the complex organic materials under anaerobic conditions and produce CH_4 and other gases (detailed in Chapter 31).

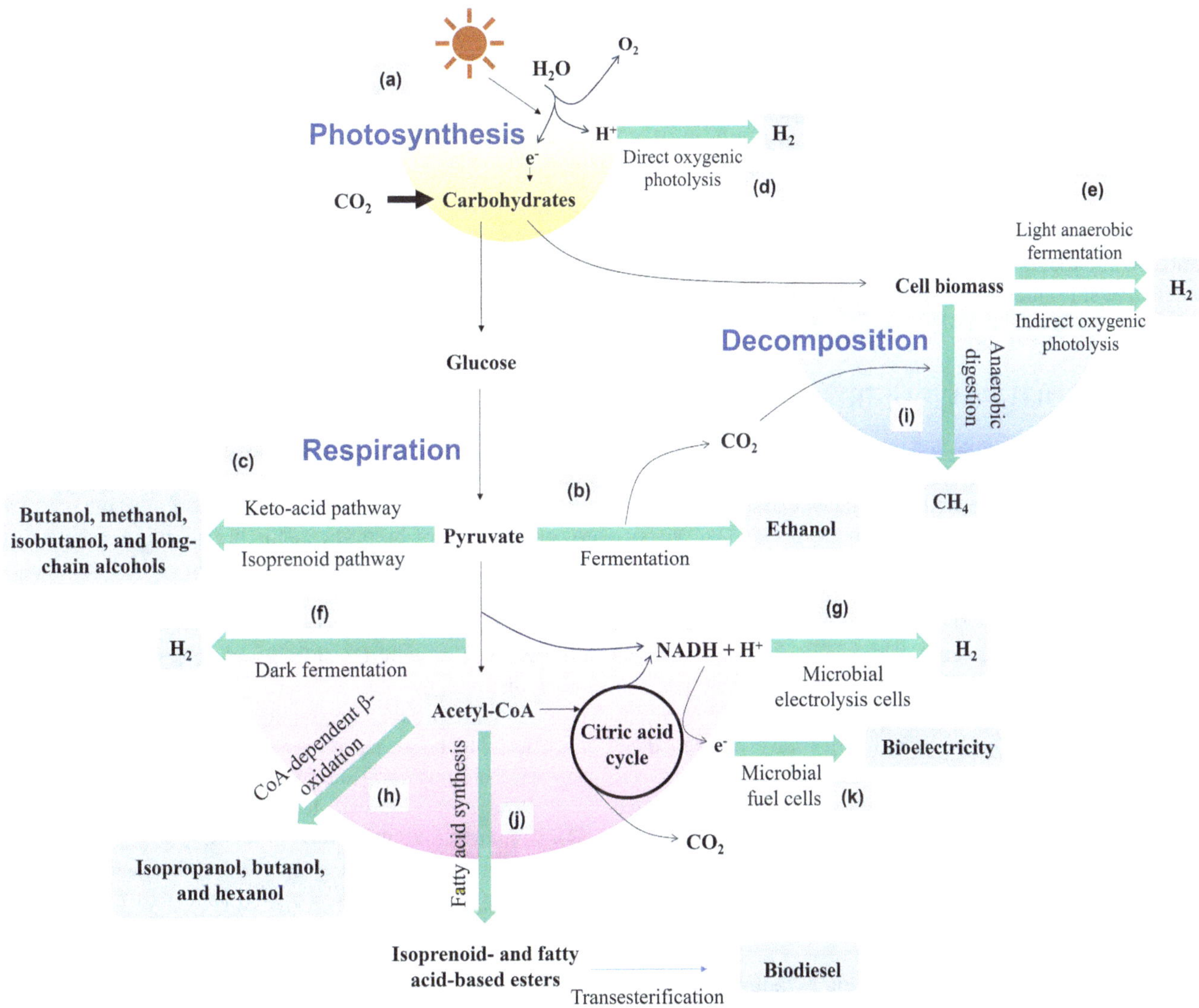

FIGURE 1.7 An illustration of microbial pathways relevant to the production of various bioenergy forms. The organic carbon fixed by photosynthetic activity using carbon dioxide (CO_2) serves as the feedstock for bioenergy production (a). Anaerobic transformation of pyruvate through fermentation results in the production of ethanol (b). The keto-acid and isoprenoid pathways in microorganisms develop butanol, methanol, isobutanol, and long-chain alcohols (c). Some distinct pathways (such as direct and indirect oxygenic photolysis, light anaerobic fermentation, dark fermentation, and microbial electrolysis cells) enable the production of hydrogen (H_2) gas (d, e, f, and g). The CoA (coenzyme A)-dependent β-oxidation utilises acetyl CoA and produces isopropanol, butanol, and hexanol (h). The anaerobic digestion of biomass generates methane (CH_4) (i). The fatty acid synthesis produces isoprenoid- and fatty acid-based esters using acetyl-CoA, which can transform into biodiesel (j). Bioelectricity production relies on electrons released during glucose oxidation during aerobic respiration (k).

Biohydrogen is a distinct energy source and can be produced by different microbial routes, such as photofermentation, dark fermentation, and microbial electrolysis. Photofermentation comprises three different pathways: direct oxygenic photolysis, indirect oxygenic photolysis, and light anaerobic fermentation. Phototrophs like green-sulphur bacteria, purple sulphur bacteria, and purple non-sulphur bacteria are the owners of this type of fermentation reaction. In contrast, dark fermentation is an activity of heterotrophs, which produces H_2 in the absence of light and oxygen (O_2). A different method of H_2 production involves microbial electrolysis cells (MECs). A MEC utilises a small external energy input and electrochemically active microorganisms and transforms organic waste into H_2. The device contains two independent chambers, the anode and cathode, separated by an ion-selective membrane. The anode chamber is responsible for microbial metabolism, while the cathode chamber features electrolysis that generates H_2 gas.

A different but related device is microbial fuel cells (MFCs) that convert chemical energy into electrical energy. This system allows bacterial metabolism to generate electricity (electric current). It is a rapidly evolving technology that offers dual benefits of waste degradation through oxidation and produces an electric current. When electrons liberated from biochemical reactions at the anode flow through a conductive substance to the cathode, the circuit is completed, and electricity is generated.

1.11 AGRICULTURE INDUSTRY

Many bacteria and fungi are beneficial for plants and are called plant-growth-promoting microorganisms (PGPMs). These microorganisms establish mutualistic associations with plant parts and produce plant-growth-promoting (PGP) traits in the rhizosphere. They are potential alternatives to chemical fertilisers, pesticides, and herbicides (Figure 1.8). The 21st century showcases these microorganisms and their products as eco-friendly and cost-effective solutions in crop management. Growing interest in organic farming has contributed to the expansion of PGPM production at industrial scales. In recent years, many agriculture-based industries have devoted their interest to the commercial production of microbial products. PGPMs come in various formulations, comprising carrier materials, nutrients, and other chemicals for their exposure to plant seeds, plant surfaces, and soil. Even some beneficial microorganisms display antagonistic or synergistic interactions with pathogenic microorganisms and protect plants from their infection.

Microorganisms in the agriculture industry can be applied as microorganism-derived substances and living microorganism agents. They are referred to as **biostimulants**. Microbiologically derived substances are inanimate biological extracts that may contain growth-promoting compounds, trace elements, and hormones for plants. This can even include organic acids (e.g., indole acetic acid, humic acid, and fulvic acids). Their augment improves soil structure, nutrient availability, and water retention. Live microbial agents include diazotrophs (nitrogen-fixing organisms), plant growth-promoting rhizobacteria (PGPR), and mycorrhizal fungi. Many of them are plant-associated microbiomes that enhance nutrient acquisition in plants. The commercial formulation of beneficial microorganisms in agriculture is known as **biofertilisers/bioinoculants**.

Biofertilisers can have broad to narrow spectra for their effects on crops and soil characteristics. *Azotobacter*, *Azospirillum*, rhizobia, actinomycetes, and blue-green algae are popular examples of biofertilisers. Numerous PGP bacteria and fungi possess capabilities for mineral solubilisation, phytohormone production, anti-metabolite production, antibiotic production, and organic matter decomposition (detailed in Chapters 2 and 19). Their careful implication can even reinstate the fertility of degraded land. A mosquito fern, *Azolla,* hosts a nitrogen-fixing cyanobacterium, *Anabaena azollae*, which is a well-accepted biofertiliser in agricultural land. An agriculture-oriented company, Pivot Bio, has developed a biofertiliser called PROVEN® 40, which can be applied on-seed or in-furrow for crops like corn for nitrogen fixation. The active ingredients of this formulation consist of diazotrophs, *Kosakonia sacchari* and *Klebsiella variicola*. Arbuscular mycorrhizal fungal (AMF) inoculation is another environmentally friendly biological method that offers numerous advantages to plants and soil. MycoApply® DR is an example of a granular formulation of AMF for organic farming.

A different active microbially derived component in crop management is biopesticides. They prevent yield loss and act against those organisms that damage the plant (pests). These compounds can be obtained from bacteria, fungi, and protozoa. A subcategory of biopesticides is bioinsecticides that kill or inactivate insects. A prominent example of bioinsecticide is the dry preparation of cry protein from *Bacillus thuringiensis* (Bt), which is toxic to insect larvae. A marketed preparation of crystal (Cry) protein, Dipel®, claims to control lepidopteran pests on over two hundred crops. Spinosad is a different bioinsecticide produced by Actinomyces, *Saccharopolyspora spinosa*, which is highly potent against numerous insect species. Entomopathogenic fungi like *Metarhizium anisopliae* and *Beauveria bassiana* feed on the cuticles of insects and are approved microbial pesticides. Some commercial preparations of microbial pesticides are DeVine®, Collego®, and Dr. BioSedge®.

Biofungicides can be used to combat fungal diseases in plants. A soil fungus, *Trichoderma,* is a well-known biofungicide that colonises plant roots and soil and antagonises pathogenic fungi. Some chitinase-producing bacteria, like *Bacillus*, *Aeromonas*, and *Serratia,* are potential decomposers of chitin (the important constituent of fungal cell walls) and retard the growth of fungal pathogens. Certain microbial metabolites are potent bioherbicides that possess growth-suppressive traits toward weeds. Their formulations can be obtained from some phytopathogenic fungi (e.g., *Colletotrichum gloeosporioides*)

FIGURE 1.8 Applications of PGPMs (plant-growth-promoting microorganisms) in agriculture and crop management.

and bacteria (e.g., *Xanthomonas campestris* pv *poannua*). Table 1.3 provides a partial list of microorganisms that can be employed in agricultural industries for various purposes.

1.12 BIOREMEDIATION

Bioremediation involves the treatment of environmental contaminants, mainly using microorganisms. Environmental contaminants can be categorised into classes, including inorganic or organic compounds. They contaminate the environment from various industrial and natural events. Among various microorganisms, bacteria and fungi are the major players in the treatment of contaminants, while archaea, microalgae, and protozoa have also contributed to the cleanup program to some extent. Biodegradation, immobilisation, and mobilisation are the main strategies to restore contaminated environments. Biodegradation is a common process for the breakdown of organic wastes. The effective degradation requires aeration, which stimulates indigenous microbial activities to digest and degrade various compounds. Treatment of non-perishable contaminants such as heavy metals and radionuclides using microbially induced processes relies on mobilisation (e.g., bioleaching) and immobilisation schemes (e.g., biosorption and bioaccumulation). The efficiency and credibility of bioremediation options lie in their cost-effective, eco-friendly, and sustainable approaches.

Two main strategies for treating contaminated soil are *in situ* and *ex situ* bioremediation. In *in situ* treatment, contaminants are treated at the contaminated sites. This can be mainly achieved by biostimulation and bioaugmentation. Biostimulation is a process of stimulating natural microbiota at contaminated sites by adding some chemicals for detoxification of chemicals. Whereas bioaugmentation involves the augmentation of microbial cultures (indigenous or foreign) at the site for contamination removal. A different approach, microbe-assisted phytoremediation, is based on synergistic actions of plant bodies and microorganisms, which have promising results for restoring contaminated environments.

Ex situ bioremediation involves the excavation of contaminated materials from their natural sites for biotreatment. This can rely on biopile, windrows, land

TABLE 1.3
A list of some microorganisms used in agriculture and crop management

Microorganisms	Applications
Rhizobium spp., *Mesorhizobium* spp., and *Bradyrhizobium* spp.	Nitrogen fixation in legumes
Azospirillum spp.	Nitrogen fixation in non-leguminous plants
Bacillus thuringiensis	Biological pest control
Trichoderma spp.	Biocontrol of soil-borne pathogens.
Mycorrhizal fungi	Nutrient and water uptake
Bacillus subtilis	Biocontrol of plant pathogens, plant growth promotion.
Azotobacter spp.	Free-living nitrogen fixation
Enterobacter spp.	Growth promotion and biocontrol.
Bacillus megaterium	Phosphate solubilisation and plant growth promotion
Fusarium oxysporum	Biocontrol agent
Streptomyces spp.	Antimicrobial activity

farming, and bioreactors. Oily sludge or contaminated land and water can be effectively treated using specialised vessels such as stirred tank bioreactors and membrane bioreactors.

Microorganisms that can restore the environment exist in so much diversity. However, those that have the potential for commercial exploitation are limited. Examples of some commercial formulations are mixed biostimulants such as BiNutrix® and VB591™ (BioNutraTech, Inc.) for the treatment of various contaminants. Oil Spill Eater II (OSEII) is a US Environmental Protection Agency (EPA)-approved aqueous-based bioremediation agent to clean up environmental contaminants. This is composed of biological enzyme additives such as biosurfactants and nutrients. Biosurfactants capture oil and create small micelles (oil droplets). This process improves the bioavailability of oil to indigenous bacteria for its degradation. Nutrients present in OSEII stimulate indigenous bacteria for enhanced removal of oil spills. A proteobacterium native to the marine environment, *Alcanivorax borkumensis*, has demonstrated the potential to degrade oil in seawater environments. In mid-2010, during the *Deepwater Horizon* oil spill disaster in the Gulf of Mexico, this bacterium was used to accelerate the digestion of hydrocarbons. Chapter 30 gives a detailed account of bioremediation of soil for the removal of various contaminants under various setups.

1.13 WASTEWATER TREATMENT

The ever-increasing volumes of wastewater generated from industries, households, agriculture and other sources are a major concern worldwide. Microbe-based technologies are promising and viable methods for their treatment. Removing impurities from wastewater before it is released into the natural environment is a complex and multi-stage process. These stages include pre-treatment, primary treatment, secondary treatment, tertiary or advanced treatment, and sludge treatment/stabilisation (detailed in Chapter 25). Microorganisms in wastewater treatment mainly participate during the secondary treatment. Microbial processes remove dissolved and suspended organic matter in wastewater through aerobic and anaerobic pathways. The roles of microorganisms in secondary treatment vary depending on the types and forms involved. Activated sludge and biofilters are the two most commonly used processes for aerobic treatment (Figure 1.9). The anaerobic biological treatment methods include a septic tank, up-flow anaerobic sludge blanket (UASB) process, and expanded granular sludge bed (EGSB) process.

In aerobic treatment, aeration in wastewater is provided through stirring and bubbling, which encourages the growth of bacteria and activates them for biotreatment. The augmentation of some nutrients in the wastewater can also stimulate microorganisms for the waste-removal process. The plethora of hydrolases and lyases (enzymes) produced from heterotrophic bacteria and fungi catalyse the biodegradation of organic matter. In this process, degradation can be continued until the maximum amount of degradable fraction of organics is mineralised (i.e., conversion into CO_2). Bacteria such as *Zoogloea* spp., *Nitrospira* spp., and *Docdonella kunshanensis* are dominantly present in wastewater for aerobic treatment. The proteobacterial ammonia oxidisers, such as *Nitrosomonas* and *Nitrosospira,* achieve effective ammonia (NH_3) oxidation and transform it into nitrite (NO_2^-) in wastewater. Nitrite oxidising bacteria, such as *Nitrobacter* and *Nitrococcus,* convert NO_2^- into nitrate (NO_3^-). Further, bacteria like *Alcaligenes*, *Bacillus*, *Hyphomicrobium*, and *Pseudomonas* take part in denitrification. A different process, anaerobic ammonium oxidation (anammox), transforms the NO_2^- and NH_3 directly into nitrogen gas using anammox bacteria (such as Candidatus *Brocadiae* and Candidatus *Kuenenia*). In addition, polyphosphate-accumulating bacteria such as Candidatus *Accumulibacter phosphatis* and *Tetrasphaera* ensure the removal of organic phosphorus by accumulation and degradation.

Microbial anaerobic processes reduce the sludge volume by degrading organic waste into CH_4 (a value-added by-product that can be used as a combustible fuel) and CO_2 under an anoxic environment. Bacteria like *Actinomyces*, *Bifidobacterium*, *Clostridium*, and *Propionibacterium* frequently participate in anaerobic digestion. Methanogenic archaea (such as *Methanosarcina, Methanothrix,* and *Methanobacterium*) survive on substrates like H_2, CO_2, and acetate and generate CH_4. In a distinct wastewater treatment setup, microbial integrity in wastewater treatment renders

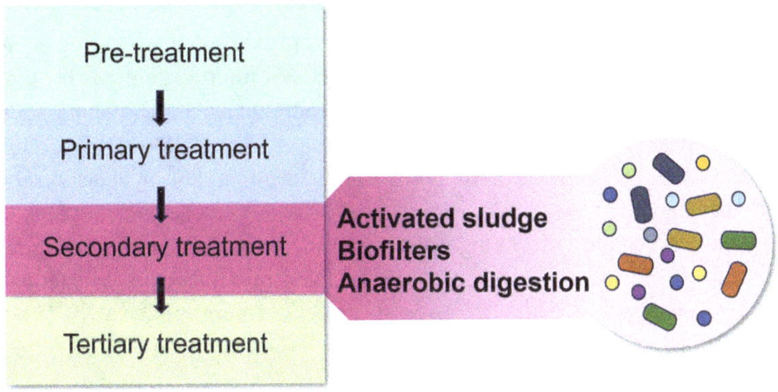

FIGURE 1.9 A flowchart of wastewater treatment showing different options for secondary treatment through microbial-assisted processes.

electricity production. This can be primarily achieved using microbial fuel cells (MFCs), which take advantage of native microflora mainly composed of *Geobacter*, *Shewanella*, *Rhodopseudomonas*, and other bacteria. This creates a distinct microbial and constitutes a different subdiscipline of microbiology called electromicrobiology (Box 1.2). A successful biotreatment of wastewater removes about 85–95% of BOD and suspended particles from the wastewater.

BOX 1.2 ELECTROMICROBIOLOGY

Electromicrobiology is the study of the electrical interplay between microorganisms and extracellular electron acceptors (EEAs). This involves a biochemical transformation dependent or independent of any catalytic potential and facilitates a delivery scheme of the electrons based on the faradic exchange of electrons or mediated electron transfer. This bioelectrochemical pathway acts as the communication option for microorganisms through direct contact between cells or through appendages such as pili and abiotic conducting substances. These microorganisms are known as electroactive microorganisms (EAMs). They can be of two different types: electrogens and electrotrophs. Electrogens are those microorganisms that donate electrons to the EEAs. In a different process, microorganisms accept electrons from EEAs and are identified as electrotrophs. There are no specific markers that characterise all EAMs. Cytochrome-based strategies for electron exchange are common, but non-cytochrome-based activity also exists. These microorganisms in an artificial setting can develop a bioelectrochemical system (BES), which can team up with either an electrogen at the anode or electrotrophs at the cathode. This facility enables the degradation of organic compounds or redox reactions of inorganic contaminants using a biotic reaction on either side or both sides. This setup is also pronounced as a microbial electrochemical system (MXC) and illustrated in Box Figure 1.2.1.

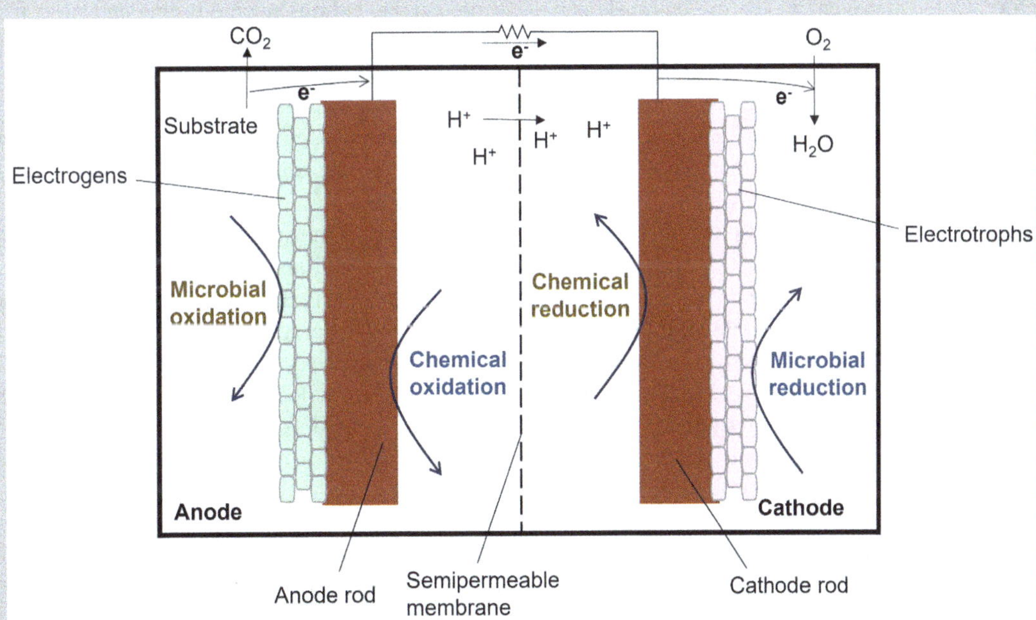

BOX FIGURE 1.2.1 A microbial electrochemical system (MXS) showing microbial involvement on both anodic and cathodic sides and electricity production.

EAMs are phylogenetically diverse and can be reported from many prokaryotes, including bacteria and archaea, but some fungi also exhibit the properties of EAMs. So far, more than hundreds of microorganisms have received the designation of EAMs. The most devoted EAMs are members of the families Geobacteraceae and Shwanellaceae. EAMs can be found in a variety of environments, including freshwater, seawater, wastewater, sludge, paddy fields, compost, and even mammalian gut and oral plaque. Microorganisms in the BES mainly exist as consortia having intricate symbiotic relationships.

Electromicrobiology has ecological perspectives, contributing to biogeochemical cycles, metal corrosion, and gut microbiomes. The multi-species microbial assemblage of environmental samples can be transformed into a technology for human-centric benefits. The bioelectrical process can be translated into a **microbial fuel cell (MFC)**, which converts chemical energy into electrical energy using electrodes as the physical supports. An MFC consists of an anode and a cathode, which are separated by a cation exchange membrane and connected through an external electrical circuit. The anode consists of EAMs (electrogens) and transfers electrons to the cathode. The cathode observes abiotic reactions using electrons. This setup completes a circuit and produces an electric current. Wastewater and sludge containing anaerobic microbial consortia are the most common sources of EAM in the anode chamber. Oxygen gas can be used as an electron acceptor in the cathode chamber. In some electrochemical systems, the anode involves a biofilm-mediated reaction, which has a strong and dynamic structure and demonstrates an improved electrocatalytic rate. The combined process of biological and electrochemical systems of electromicrobiology has emerged as an option for wastewater treatment, resource recovery, and producing electricity.

1.14 SOLID WASTE MANAGEMENT

The unwanted by-products or residual materials in solid and semi-solid forms pose serious challenges to human health and the environment. Their excessive volume on land and water seeks solutions for their management through recycling, treatment, and disposal. The biological outsets of waste transformation into useful products are an effective scheme of solid waste management. Some microbial schemes employ waste materials as feedstocks under the process of waste valorisation. The vast reservoirs of microorganisms and their metabolic activities in microbial consortia increase the robustness of the valorisation. The microbial solution of waste management contributes toward the adoption of circular economy approaches. Microbial implications for waste management focus on sustainable development and eco-friendly solutions. Bioenergy, fertilisers, animal feed, pharmaceuticals, biosurfactants, and biodegradable polymers are the key value-added products from microbial conversion of waste streams. Microbial processes can be suitably used to treat solid waste through anaerobic digestion and composting. Compost can be used as a soil conditioner in organic farming. Agricultural and industrial wastes are suitable feedstock for mushroom cultivation (Figure 1.10). Microorganisms like *Cupriavidus necator* can use wastes from the sugar and fruit industry and produce a biopolymer, **poly-3-hydroxybutyrate (PHB)** (a polyester), which can be used as a suitable alternative to plastic. In recent years, some industries have recruited filamentous fungi (e.g., *Ganoderma lucidum*) and yeasts (e.g., *Metschnikowia pulcherrima*) on industrial wastes to produce stuff that resembles leather without using animal hides. A detailed overview of the unconventional applications of microorganisms in waste management can be found in Chapter 31.

1.15 BIOMINING

Biomining is a technique to recover valuable metals or minerals from ores or mines using microbial activities. It is a reasonably eco-friendly alternative solution to some steps of traditional mining. Metals like copper (Cu), nickel (Ni), cobalt (Co), zinc (Zn), and gold (Au) are the primary targets for biomining. These metals mainly exist as metal sulphides in the mines, and biomining typically involves the oxidation of metal sulphides present in solid minerals. The primary goal of biomining is to provide ideal conditions for the growth of naturally occurring microorganisms. Native chemolithoautotrophs like *Acidithiobacillus ferrooxidans* and *Acidithiobacillus thiooxidans* are the main participants in the bioreactions, although some fungi can also contribute to their extraction. The use of foreign microorganisms is extremely difficult or not possible because biomining operates under harsh and non-sterile situations.

Biomining occurs mainly through two different processes: bioleaching and biooxidation. Bioleaching facilitates the solubilisation of metals, whereas biooxidation activity disrupts the surrounding minerals, facilitating an easy recovery of a desirable metal. Bioleaching is well-suited for the recovery of Cu and Ni, while the biooxidation process is employed for the extraction of Au.

Biomining exists in three different operational setups, including dump leaching, heap leaching, and agitated leaching. Dump leaching involves stacking low-value ore or waste rock on a leach pad, which is then bioleached by stimulating the growth of native bacteria. In heap leaching, the large particles in ores are crushed into small particles, which are then stacked into a heap, followed by bioleaching. The small particles allow better gaseous exchange and penetration

FIGURE 1.10 A sustainable approach to mushroom cultivation through waste valorisation.

FIGURE 1.11 Commercial tank bioleaching operations. a. Sao Bento BIOXw reactor (Brazil). b. Tamboraque BIOXw plant (Peru). c. Laizhou Mintek–BacTech Plant (China). d. Sansu BIOXw plant (Ghana). (Image adapted from Rawlings et al. (2003)[9] with permission of © 2002 Elsevier Science Ltd.)

by microorganisms for the bioreaction. Both dump leaching and heap leaching are old methods of biorecovery of metals. Agitated leaching is a relatively modern, controlled operation and requires a bioreactor. This technology relies on the proper shaking of small particles for the proper mixing of all the contents, which ensures sufficient aeration in the system to speed up the recovery of metals.

Biomining makes up a very minor contribution to the mining sector. This process is slower compared to conventional practices and can take days to months to complete. Bioming is best suited to extract metals from low-grade ores. This technique is used after conventional mining operations to recover whatever is left over. Chile has emerged as a leading industrial bioleaching practising country. In recent years, bioleaching has also been employed to recycle waste materials (like e-waste) for the recovery of precious metals. Figure 1.11 shows commercial tanks for bioleaching operations from different parts of the world.

1.16 MICROBIAL ENHANCED OIL RECOVERY (MEOR)

Microbial Enhanced Oil Recovery (MEOR) is an alternative strategy for the recovery of mineral oil from the field, relying on the stimulation of native microbiota (or injected microorganisms) in the reservoir. This is considered tertiary and an advanced biological-based technology for oil recovery. In the primary phase, an artificial lift device, like a jack pump, pulls the oil from the oilfield. Whereas the secondary method allows the injection of water or gas into the reservoir to increase pressure and displace the oil, driving it onto the surface. The primary and secondary procedures favour only partial recoveries, with efficiency of up to 10% and 20–40% oil existing in the field, respectively. However, the tertiary phase can pull as much as 50% of the remnant oil trapped in the reservoir after the primary and secondary phases. In the third category, the addition of nutrients in the field stimulates the growth of indigenous microorganisms. Biodegradable materials like molasses, glucose, acetate, crude oil, NO_3^-, urea, peptone, phosphorus salt, etc., are inexpensive and commonly used nutrients. Differently, the potential microorganisms can be augmented in the field for oil recovery.

Microbial metabolites like biopolymers, biosurfactants, enzymes, biogases, solvents, and biogenic acids are possible MEOR bioagents. These metabolites are principally released by bacteria like *Bacillus*, *Pseudomonas*, *Acinetobacter*, *Xanthomonas*, *Enterobacter*, *Leuconostoc*, *Clostridium*, etc. The microbial stimulation in the reservoir

TABLE 1.4
A list of microbial agents and their applications in MEOR (microbial enhanced oil recovery)

Microbial agents	Applications in MEOR
Biopolymers and biomass	Plugging high-permeability zones, achieving permeability adjustment, and oil degradation
Biosurfactants	Reducing oil viscosity and emulsifying oil
Enzymes	Highly selective catalytic activity of oil degradation, rock wettability, and reducing interfacial tension
Biogases	Increasing pressure, oil swelling, and reducing interfacial tension
Solvents	Reducing the oil viscosity and oil dissolution
Biogenic acids	Emulsifying oil and improving permeability

induces the production of stable oil-water emulsions and reduces interfacial tension and bio-plugging (Table 1.4). MEOR technology has received wide attention in recent years due to its eco-friendly approach. Also, this is an inexpensive and energy-efficient technique for sustainable development.

1.17 MOLECULAR BIOLOGY RESEARCH AND GENETIC ENGINEERING

Both microbial cells and their molecular deposits are empirical tools of molecular biology and genetic engineering. The simple structural organisation, less complicated biochemistry, and easy propagation of prokaryotic cells make them the favourite experimental subjects in biological research. The finding of the genetic code, the basics of DNA replication, transcription and translation, and the identification of genetic operons were initially made using bacterial cells. A Gram-negative bacterium, *E. coli,* has served as a model organism in the discoveries of molecular genetics. This single-celled organism has served as the "workhorse" of molecular biology and has paved the way to fetch at least twelve Nobel prizes in the past. Several key findings for genetics studies in the eukaryotic cells were realised using a filamentous fungus, *Neurospora crassa*. A unicellular yeast, *S. cerevisiae*, and a slime mould, *Dictyostelium discoideum,* are also amenable in molecular biology research. Table 1.5 provides a list of different model microorganisms for various research studies in genetics, molecular biology, and other biological fields.

The development of the polymerised chain reaction (PCR) was a revolutionary step in molecular studies. One of the key ingredients of the PCR reaction is an amplification enzyme, *Taq* polymerase, which is obtained from a thermophilic bacterium, *Thermus aquaticus*. This enzyme is known for its stability at high temperatures during the denaturation step for amplification. A different enzyme, *Pfu* DNA polymerase, obtained from a hyperthermophilic archaeon, *Pyrococcus furiosus,* is a suitable alternative to *Taq* polymerase. A different variant of PCR is reverse transcription PCR, which can be used to make copies of complementary (c) DNA from RNA using the enzyme reverse transcriptase. This is produced from a cloned sequence of Moloney Murine Leukaemia Virus (M-MuLV) in *E. coli*.

In DNA recombinant technology, many microbiological tools assist an organism in achieving desirable features through genetic modification. The genetic elements of viruses and bacterial plasmids are useful vectors to carry nucleic acid elements for constructing genetic libraries and transgenic organisms. Bacterial artificial chromosome (BAC) and yeast artificial chromosome (YAC) are human-engineered vectors inspired by the F-plasmid of bacteria and from a conjugant of yeast, *S. cerevisiae* and bacterial plasmids, respectively. They are typically used for the cloning of long fragments of DNA. A cosmid is a hybrid plasmid that carries a Lambda phage cos sequence.

Classical genetic methods of mutation have developed many new strains of bacteria and fungi for industrial applications. Mutations can be induced using different chemicals (e.g., ethyl methane sulphonate (EMS)) and physical mutagens (e.g., UV, X, and gamma rays). A different method to induce mutation is error-prone PCR amplification. Among bacteria, members of actinomycetes have been favourite subjects for mutational study in industrial microbiology. The development of the first superior penicillin-producing mutant, *Penicillium chrysogenum* X-1612, through X-ray mutagenesis, was developed using mutational genetics.

Restriction enzymes are a boon for molecular cloning obtained from various bacteria. They are used as markers in DNA fingerprinting and methylation analysis. More than 800 commercial enzymes are available to create nicks at different sites of DNA fragments. A different enzyme, DNA ligase, allows the binding of two DNA fragments. It is commercially obtained from bacteriophage T4 and different bacteria. Table 1.6 provides a list of some molecular tools derived from various microorganisms for genetic engineering.

CRISPR-Cas9 technology is a precise and facile method for site-specific genome editing in cells, tissues, and organisms. CRISPR-Cas9 is an adaptation of bacterial/archaeal defence mechanisms during viral invasion, where CRISPR is Clustered Regularly Interspaced Short Palindromic Repeats. A CRISPR array derives from viral DNA during infection and becomes a part of the bacterial genome to recognise similar viruses. CRISPR technology can be used to alleviate genetic disorders, treat human diseases, and improve livestock breeding. In plant biotechnology, this editing program has shown tremendous potential to increase plant yield and quality and combat plant pathogens in crop species.

TABLE 1.5
Examples of model microorganisms used in biological research

Type of microorganisms	Name of microorganisms	Fields in biological research
Bacteria	*E. coli*	Molecular biology and genetics
	Bacillus subtilis	Bacterial physiology
	Staphylococcus aureus	Bacterial pathogenesis and drug resistance
	Synechocystis PCC 6803	Photosynthesis
	Azotobacter vinelandii	Nitrogen fixation
Archaea	*Methanococcus maripaludis*	Methanogenesis
	Halobacterium salinarum	DNA repair
Fungi	*Neurospora crassa*	Genetics
	Saccharomyces cerevisiae	Genetic engineering
	Aspergillus nidulans	Genetics and production of secondary metabolites
	Schizosaccharomyces pombe	Cell cycle
	Rhizophagus irregularis	Arbuscular mycorrhizae
Microalgae	*Chlamydomonas reinhardtii*	Photosynthesis
	Nannochloropsis sp.	Biodiesel production

TABLE 1.6
List of various molecular tools derived from microorganisms used in genetic engineering

Types of biomolecules	Activities	Name of biomolecules	Microbial sources
Restriction endonucleases	Cleave DNA in specific regions	EcoRI	*Escherichia coli*
		BamHI	*Bacillus amyloliquefaciens*
		HaeIII	*Haemophilus aegyptius*
		HindIII	*Haemophilus influenzae*
		AluI	*Arthrobacter luteus*
Ligases	Join two strands of DNA	HiFi Taq DNA ligase	*E. coli*
		T4 DNA ligase	Bacteriophage T4
		Ampligase	A thermophilic bacterium
Vectors	Used in the cloning of foreign genes in host cells	Plasmids like pBR 322, pUC19	*E. coli*
		Ti and Ri plasmids	*Agrobacterium tumefaciens* and *Agrobacterium rhizogenes*
		Phage λ vectors, M13	Bacteriophages
		Phagemid and cosmid	Derived from plasmids and bacteriophages
		Yeast episomal plasmids (YEp), Yeast artificial chromosome (YAC)	*Saccharomyces cerevisiae*
DNA polymerases	DNA replication in PCR	Taq DNA polymerase	*Thermus aquaticus*
		Pfu DNA polymerase	*Pyrococcus furiosus*
		Tth DNA polymerase	*Thermus thermophilus*
RNA polymerases	DNA-directed synthesis of RNA	T7 RNA polymerase and Qβ replicase	Bacteriophages
Nucleases	Digest nucleic acids	DNases	Haemolytic streptococci and haemolytic staphylococci
		RNase H	Recombinant *E. coli*
		S1 nuclease	*Aspergillus oryzae*
Reverse transcriptases	Form a complementary DNA strand from a single-stranded RNA template	AMV- RT	Avian Myeloblastosis Virus
Phosphatases	Removal of terminal mono-esterified phosphates from deoxyribose-oligonucleosides	Alkaline phosphatase	*E. coli*

1.18 SYSTEM BIOLOGY AND SYNTHETIC BIOLOGY

System biology is the successor of molecular biology, which focuses on the interactions among different components of biological systems. This gives a holistic approach to studying organisms using engineering, informatics, and biology. This field of science expands its focus on modelling, prediction, and experimentation of biological systems. The best systems to investigate system biology are microbial cells. *E. coli* and *S. cerevisiae* are the two most well-characterised microbial systems for studying gene regulation, metabolic pathways, and cellular processes. The discoveries of broad biological principles in these model organisms uncovered many missing links to many biological properties. Moreover, the field of systems biology lays the scientific groundwork for synthetic biology.

Synthetic biology (SynBio) refers to the development of artificial life in which biological processes and cells can be disassembled and reassembled to create unique systems that perform useful activities. Modern researchers have initiated the design and construction of new biological systems using engineering principles. Microorganisms are ideal systems for the study of synthetic biology. In 2010, a group of researchers built the first synthetic cellular life, *Mycoplasma mycoides*. The procedure involved destroying and substituting the original DNA with computer-designed, laboratory-synthesised DNA. In 2019, a live creature that is entirely synthetic was also designed. This new form of life displayed the artificial synthesis of four million base pairs in *E. coli*. In the immediate years, scientists created a yeast strain (*Saccharomyces cerevisiae*), which had almost 50% synthetic DNA in its genome. It was also the first attempt to synthesise any eukaryotic genome. This initiative now aims to synthesise the whole yeast genome, consisting of 16 linear chromosomes and comprising roughly 6,000 genes. This ambitious project is called "The Synthetic Yeast Genome Project" (Sc2.0), which attempts to build the world's first eukaryotic organism from scratch. This research field has the prospects of improving diagnosis and producing therapeutic biological medicines. Engineered living cells have demonstrated potential to deal with the constraints of agriculture, animal welfare, bioenergy, and food security.

1.19 NANOBIOTECHNOLOGY

The new world has recently opened up its feathers in the field of nanotechnology in the production of near-atomic scale materials (in the range of 1 to 100 nanometres). The implication of microbiology in nanotechnology adds new avenues to the "green synthesis" of nanoparticles (Figure 1.12). The biogenic synthesis route of their production involves the absorption/adsorption of metal ions onto microbial cells, followed by

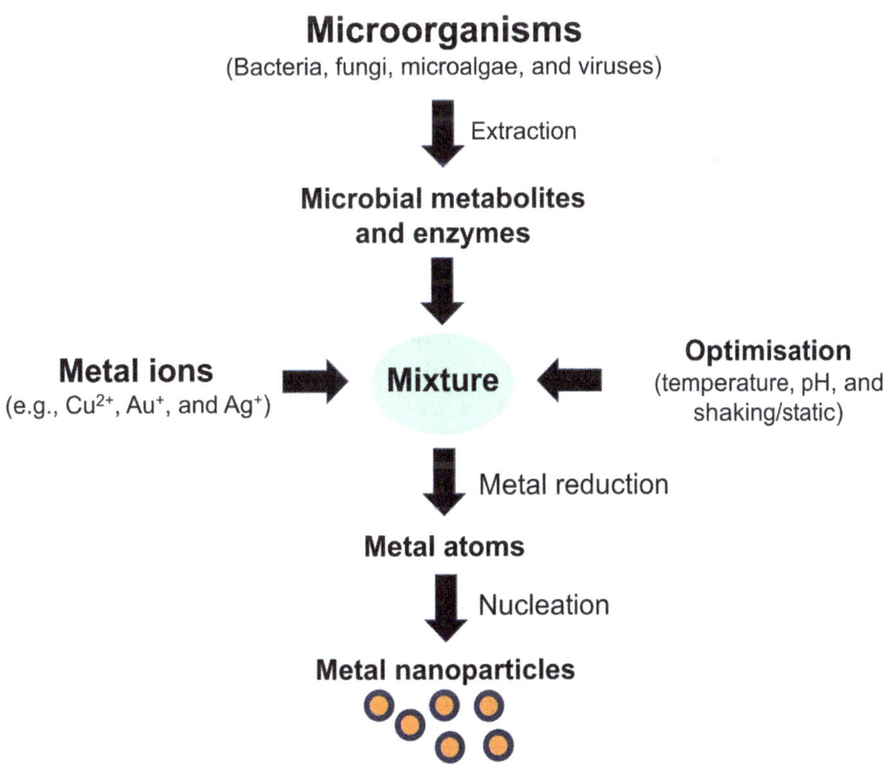

FIGURE 1.12 Biosynthesis of nanoparticles using microbial components.

enzyme-mediated reduction of the trapped metals. This is an eco-friendly, efficient, and easily scaled-up technology for the precise development, manipulation, and production of microbial nano-factories. Microbial nanoparticles contain novel features and suitability owing to their large surface/volume ratio, minute size, biocompatibility, and stability.

The intersection of nanotechnology and microbiology foresees tremendous scope in the food industry, agriculture, genetics, biotechnology, and pharmaceuticals. This potential confluence shows an application for high-throughput detection of the potential pathogens in contaminated food and water. Nanomaterials can help improve the shelf life, storage, and packaging of food. Nanotechnology in the agriculture sector offers a great deal to address the current challenges of crop production and soil management. The production of nanofertilisers and nanopesticides is a well-sound attempt to enhance the capacity of plants to absorb nutrients and maintain the health of plants in fields, respectively. Nanobiotechnology has delivered promised benefits in medical sciences. Nanoparticles covalently linked to proteins or nucleic acids can detect viruses. Nanomaterials can be employed as vehicles for enzyme immobilisation and in drug and gene delivery. Quantum dots (QD) and nanobiosensors have shown usefulness in biomedical imaging. A new frontier is the design of biochips for medical diagnosis at the molecular level.

1.20 MICROBIAL BIOSENSORS

A microbial biosensor is an analytical device which couples a microbial system with a physical transducer to generate measurable signals of biochemical processes. They are simple, portable, sensitive, and cost-effective devices, having good detection limits and rapid response. Whole microbial cells, purified enzymes, or other cell deposits can be employed in the fabrication of biosensors. Bioluminescence-based microbial biosensors rely on optical signals and are constructed through genetic engineering in microorganisms. The signalling proteins in biosensors can be green fluorescent protein (GFP), aequorin, firefly luciferase, and bacterial luciferase (Lux) (detailed in Chapter 31). Biosensors have tremendous scope in environmental monitoring and the identification of target molecules. They have also been employed in drug discovery and disease detection. Recent developments in microbial biosensor designs delineate their potential applications across diverse fields, including water quality, food processing, quality control, and fermentation.

1.21 MICROBIAL CONCRETE

Microbial concrete is a type of mineral produced by bacteria that can be used in building materials. The principle behind microbial concrete lies in a biological process called microbially induced calcite precipitation (MICP). This property is reported from some spore-producing bacteria (e.g., *Bacillus, Sporosarcina pasteurii*), which can be employed for crack-repair jobs. Bacterial spores are mixed with some nutrients (like calcium lactate) and transformed into micropellets/microcapsules. They are combined with concrete material before being applied in construction and building. These spores remain in a dormant stage for a very long time. However, when there is any crack and water seepage, these spores germinate and multiply using nutrients. A bacterial enzyme, urease, elevates the pH of the surrounding area by digesting the urea into NH_3 and CO_2. Carbon dioxide at high pH binds with calcium ions and forms a stable structure of calcium carbonate ($CaCO_3$). This microbial self-healing concrete process can be used as an eco-friendly and cost-effective solution to repair cracks and fissures in tunnels, water channels, retaining wall panels, and highways. Microbial concrete offers dual benefits through its application in infrastructure maintenance along with CO_2 sequestration.

1.22 AEROSPACE MISSIONS

The applications of microorganisms in aerospace are a scheme of bioregenerative life support systems (BLSS) for long-duration space missions. Some specific microorganisms provide life support systems and generate resources such as oxygen, food, and water for spacecraft. They can also be feasibly implemented in wastewater treatment and atmosphere revitalisation. Microorganisms to be used for BLSS can be engineered using synthetic biology tools, enabling them to acquire new or improved functions for interplanetary travel missions. Photoautotrophs like *Synechococcus* spp. and *Synechocystis* spp. can be employed to produce O_2 and remove CO_2 in the spacecraft. They produce a lot of biomass that can be an important source of proteins, carbohydrates, fatty acids, and other nutrients. Several photoautotrophs (e.g., *Nostoc*, *Lyngbya*, and *Leptolyngbya*) can be employed as **biophotovoltaics** to generate electricity. In this process, photoautotrophs utilising solar energy split water into electrons and protons, which can be transferred into an MFC for electricity production. The repurposing of wastes using microorganisms is a scheme for energy production and nutrient recovery. Wastewater treatment can be employed using anaerobic digestion. Decomposable solid wastes, such as inedible plant materials, can be degraded through SSF. Microorganisms can also contribute to the in-space manufacturing of engineered biomaterials (such as bioplastic, gel-based materials, and rubber).

SUMMARY

- Industrial microbiology is a sub-discipline of applied microbiology that helps produce microbially-derived products for commercial exploitation. Industrially relevant microorganisms are easily propagated, nonpathogenic, metabolic specialists, potent producers, amenable to genetic modification, responsive to flexible nutrition solutions, and pliant to harvest products at low prices.

- Lactic acid bacteria (LAB) and yeasts have profound roles in the food and beverage industry. Many food-grade microorganisms are used to change the nature of foodstuffs. Probiotics are oral supplements of microorganisms. Chemical extracts and enzymes also have wide applications in the food industry.
- In medical applications, microorganisms and their metabolites are key sources of therapeutic compounds (e.g., antibiotics), vaccines (e.g., monoclonal antibodies), and biologics (e.g., anticancer drug paclitaxel).
- The detergent, textile, leather, and paper and pulp industries have increased reliance on microbial enzymes, moving towards sustainable options and eco-friendly approaches. Also, the fermented extracts of diverse microorganisms are suitable choices in cosmetic and personal care products.
- Many microbial processes meet the demands for the production of biofuel, heat, bioelectricity, and bioproducts through microbial action on different wastes. Of different sources, bioethanol is the most common bioenergy form due to its large-scale production.
- In agriculture, the biological substances derived from microorganisms or living biotic agents can be used as biofertilisers, biopesticides, biofungicides, and bioherbicides. These agriculture-important microorganisms are suitably used in organic farming.
- Bioremediation is a process of detoxifying environmental contaminants using microorganisms. Activated sludge and biofilters are suitable methods for the removal of organic contaminants from wastewater. In waste management, microorganisms can transform waste streams into value-added products through a process called waste valorisation.
- Biomining is an industrial operation of metal extraction from ore or mines. This can be achieved through microbial processes such as bioleaching and biooxidation. Various setups in biomining are dump leaching, heap leaching, and agitated leaching.
- Microorganisms are critical tools of molecular biology, genetic engineering, system biology, synthetic biology, and nanotechnology. Microbial biosensors measure signals for environmental monitoring and assessment.
- Microbial concrete is a biodegradable concrete produced by a microbially induced calcite production process (MICP), which can be used to heal cracks in infrastructure projects. Microorganisms can also be used as bio-regenerative, sustainable life support systems for space settlements in aerospace.

DIFFERENT TYPES OF PRACTICE QUESTIONS

Fill in the blanks

I. ………………….. are live microorganisms that, when administered in adequate amounts, confer a health benefit on the host.

II. …………………… are secondary metabolites of bacteria and fungi, which act as broad-spectrum agents against many infectious bacteria.

III. Bioethanol is mainly produced by fermentation of simple sugars using ……………………………………

IV. The commercial formulation of useful microorganisms in agriculture is known as …..…………

V. ……………………… is a process of stimulating natural microbiota at contaminated sites by adding some chemicals for detoxification of chemicals.

VI. Microorganisms in wastewater treatment mainly participate during the …………………..

VII. An archaeon……………………….. produces a thermostable polymerase enzyme for PCR.

SOLUTION: I. Probiotics, II. Antibiotics, III. yeast/ Saccharomyces cerevisiae, IV. biofertilisers/ bioinoculants, V. Biostimulation, VI. secondary treatment, VII. Pyrococcus furiosus

Match the following

i. *Saccharomyces cerevisiae*	a. Single-cell protein
ii. *Arthrospira*	b. Lactic acid
iii. *Streptomyces*	c. Ethanol
iv. *Leuconostoc*	d. Nitrogen
v. *Rhizobium*	e. Antibiotics

SOLUTION: i-c, ii-a, iii-e, iv-b, v-d

Very short answer type of questions

I. Define industrial microbiology.
II. What is biofuel?
III. What is microbial concrete?
IV. Write the names of two commercial preparations of microorganisms or their derivatives used in the agriculture industry.
V. Expand PGPM and MEOR.
VI. Name a chemical compound derived from microorganisms that can be used as a bioplastic.
VII. Write an application of microorganisms in nanotechnology.

Short answer type of questions

I. What are the different microbial components used in industrial microbiology?
II. Write important characteristics of industrially useful microorganisms.
III. What are the main interests of microbial agents in medical science?

IV. Write a short note on the scopes of microbial amylases and proteases in industrial applications.
V. Write about the roles of microorganisms and their extracts in cosmetic and personal healthcare.
VI. What are the desirable parameters of microbial enzymes used in the detergent industry?
VII. What is biomining? How does it contribute to sustainable development?
VIII. How can system biology and synthetic biology enhance microbial efficiency in industrial applications?

Long answer type of questions

I. Give a detailed account of the roles of microorganisms in the food and beverage industry.
II. Microorganisms offer sustainable solutions to many industries. Give your statement with suitable examples.
III. How do microorganisms play roles in advancing biological sciences?
IV. Bacteria and fungi are the main recruiters for industrial applications. Justify this statement with suitable examples.

NOTES

1 www.flickr.com/photos/aarongustafson/197059871
2 www.flickr.com/photos/ccbarr/1670419128
3 www.commons.wikimedia.org/wiki/File:Tempeh_raw.jpg
4 www.flickr.com/photos/dewet/2320117002
5 Luangharn T, Karunarathna SC, Hyde KD, Chukeatirote E. Optimal conditions of mycelia growth of *Laetiporus sulphureus* sensu lato. *Mycology*. 2014;5(4):221–7.
6 https://web.archive.org/web/20161102190449/http://www.panoramio.com/photo/124332482
7 https://commons.wikimedia.org/wiki/File:White-truffle-pic.jpg
8 Paul T, Das A, Mandal A, et al. An efficient cloth cleaning properties of a crude keratinase combined with detergent: towards industrial viewpoint. *Journal of Cleaner Production*. 2014;66:672–84.
9 Rawlings DE, Dew D, du Plessis C. Biomineralization of metal-containing ores and concentrates. *Trends in Biotechnology*. 2003;21(1):38–44.

SUGGESTED READINGS

Abdel-Azeem AM, Yadav AN, Yadav N, Usmani Z. *Industrially important fungi for sustainable development*. Springer International Publishing; 2021.

Adams GO, Fufeyin PT, Okoro SE, Ehinomen I. Bioremediation, biostimulation and bioaugmentation: a review. *International Journal of Environmental Bioremediation & Biodegradation*. 2015;3(1):28–39.

Adrio JL, Demain AL. Microbial enzymes: tools for biotechnological processes. *Biomolecules*. 2014;4(1):117–39.

Antoni D, Zverlov VV, Schwarz WH. Biofuels from microbes. *Applied Microbiology and Biotechnology*. 2007;77(1):23–35.

Bandyopadhyay A, Saha A, Ghosh D, Dam B, Samanta AK, Dutta S. Microbial repairing of concrete & its role in CO_2 sequestration: a critical review. *Beni-Suef University Journal of Basic and Applied Sciences*. 2023;12(1):7.

Barrangou R, Doudna JA. Applications of CRISPR technologies in research and beyond. *Nature Biotechnology*. 2016;34(9):933–41.

Barras V, Greub G. History of biological warfare and bioterrorism. *Clinical Microbiology and Infection*. 2014;20(6):497–502.

Benner SA, Sismour AM. Synthetic biology. *Nature Reviews Genetics*. 2005;6(7):533–43.

Creative Enzymes. Application of Enzymes in Detergent Industry. www.creative-enzymes.com/resource/application-of-enzymes-in-detergent-industry_61.html

Gupta PL, Rajput M, Oza T, Trivedi U, Sanghvi G. Eminence of microbial products in the cosmetic industry. *Natural Products and Bioprospecting*. 2019;9(4):267–78.

Gurung N, Ray S, Bose S, Rai V. A broader view: microbial enzymes and their relevance in industries, medicine, and beyond. *BioMed Research*. 2013;2013.

Haki GD, Rakshit SK. Developments in industrially important thermostable enzymes: a review. *Bioresource Technology*. 2003;89(1):17–34.

Koehle AP, Brumwell SL, Seto EP, Lynch AM, Urbaniak C. Microbial applications for sustainable space exploration beyond low Earth orbit. *NPJ Microgravity*. 2023;9(1):47.

Lovley DR, Holmes DE. Electromicrobiology: the ecophysiology of phylogenetically diverse electroactive microorganisms. *Nature Reviews Microbiology*. 2022;20(1):5–19.

Madhu A, Chakraborty JN. Developments in application of enzymes for textile processing. *Journal of Cleaner Production*. 2017;145:114–33.

Mohammadi K, Sohrabi Y. Bacterial biofertilizers for sustainable crop production: a review. *ARPN J Agriculture of Biological Science*. 2012;7(5):307–16.

Neelakantan S, Mohanty AK, Kaushik JK. Production and use of microbial enzymes for dairy processing. *Current Science*. 1999:143–8.

Nikolova C, Gutierrez T. Use of microorganisms in the recovery of oil from recalcitrant oil reservoirs: current state of knowledge, technological advances and future perspectives. *Frontiers in Microbiology*. 2020;10:481081.

Prasad R, Bhattacharyya A, Nguyen QD. Nanotechnology in sustainable agriculture: recent developments, challenges, and perspectives. *Frontiers in Microbiology*. 2017;8:1014.

Raveendran S, Parameswaran B, Ummalyma SB, Abraham A, Mathew AK, Madhavan A, Rebello S, Pandey A. Applications of microbial enzymes in food industry. *Food Technology and Biotechnology*. 2018;56(1):16.

Sauer U. Evolutionary engineering of industrially important microbial phenotypes. *Metabolic Engineering*. 2001:129–69.

Shrestha UB, Bawa KS. Trade, harvest, and conservation of caterpillar fungus (*Ophiocordyceps sinensis*) in the Himalayas. *Biological Conservation*. 2013;159:514–20.

Singh R, Kumar M, Mittal A, Mehta PK. Microbial enzymes: industrial progress in 21st century. *3 Biotech*. 2016;6(2):1–5.

2 Scope of Microorganisms in the Environment

2.1 ENVIRONMENTAL MICROBIOLOGY

Environmental microbiology is a branch of applied microbiology that deals with the functional aspects of microorganisms in the environment. It is still a growing field and requires an interdisciplinary approach to reveal the direct and indirect microbial impact on the environment and human health. Both beneficial and pathogenic microorganisms of the environment are included. Beneficial microorganisms contribute to ecosystem services and environmental sustainability. On the other hand, the examination of pathogenic microorganisms is relevant for public health, agriculture, and wildlife conservation.

A different related discipline is **microbial ecology**, which is concerned with interactions and adaptation in diverse ecosystems. Contemporary research in microbial ecology pays attention to how microbial communities experience shifts in response to various environmental factors. A different ecological perspective of microbial studies is the microbial acclimatisation to altered conditions in the environment. In a broader sense, this discipline fills the knowledge gap between microorganisms and their environment. In the environment, microorganisms cooperate with various physical, chemical, and biological factors and establish communication within their communities. This paves the way for microorganisms to performs diverse activities for environmental well-being and human welfare.

Environmental microbiology offers many benefits through microbial ecological studies that focus on the diversity (number of species), distribution (living space), and abundance (how many) of microorganisms in the environment. Environmental bioprocesses and ecosystem services are the two most important concepts in environmental biotechnology. Environmental bioprocesses are the bio-assisted reactions of microorganisms in the environment for maintaining ecosystem health, whereas ecosystem services are the environmental bioprocesses that derive human benefits. These areas expand the ecological explanations of natural microbiota to their applications in various domains. Although there are no clear distinctions between microbial ecology and environmental microbiology, microbial ecology mainly perceives information on microbial processes of the environment, whereas environmental microbiology deals with beneficial/harmful aspects of microorganisms.

2.2 MICROORGANISMS AND THE ENVIRONMENT

Microbial life on Earth is extraordinarily diverse and ubiquitous. Microorganisms exhibit diverse abilities to establish interactions with their ecosystems. These include adaptation and modification in their surroundings. Many microorganisms exhibit metabolic flexibility, which allows them to survive across various landscapes. Differently, microorganisms with narrow metabolic capacities are susceptible to environmental change, and they have limited distribution.

Earth is home to about 10^{30} microorganisms. This figure even outnumbers the total number of stars present in the Universe. The total number of microbial species reported on Earth is estimated to be around 10^{10}. These tiny dwellers belong to bacteria, archaea, fungi, microscopic algae, and protozoa, which reside in various compartments of the environment. They have a massive impact on the environment and health. However, the identified number of microbial species is only close to 400,000. So, these calculations suggest that the identified microorganisms represent only a very small fraction of the microorganisms present on the entire planet. This low proportion of known microorganisms is due to the limitations of culturing techniques. The information on a large fraction (about 99%) of the microbial diversity is derived from cultivation-independent methods (detailed in Chapter 21). Metagenomic studies have given much more insight into microbial community compositions (MCCs) in environmental samples than any other method. Also, some new high-throughput culture techniques (involving culturomics) have proven useful in cultivating previously uncultured bacteria in artificial media.

2.3 MICROBIOLOGY OF DIFFERENT ECOSYSTEMS

Microorganisms are crucial to every ecosystem on Earth because they are the engine for ecosystem functions. They

DOI: 10.1201/9781032651095-3

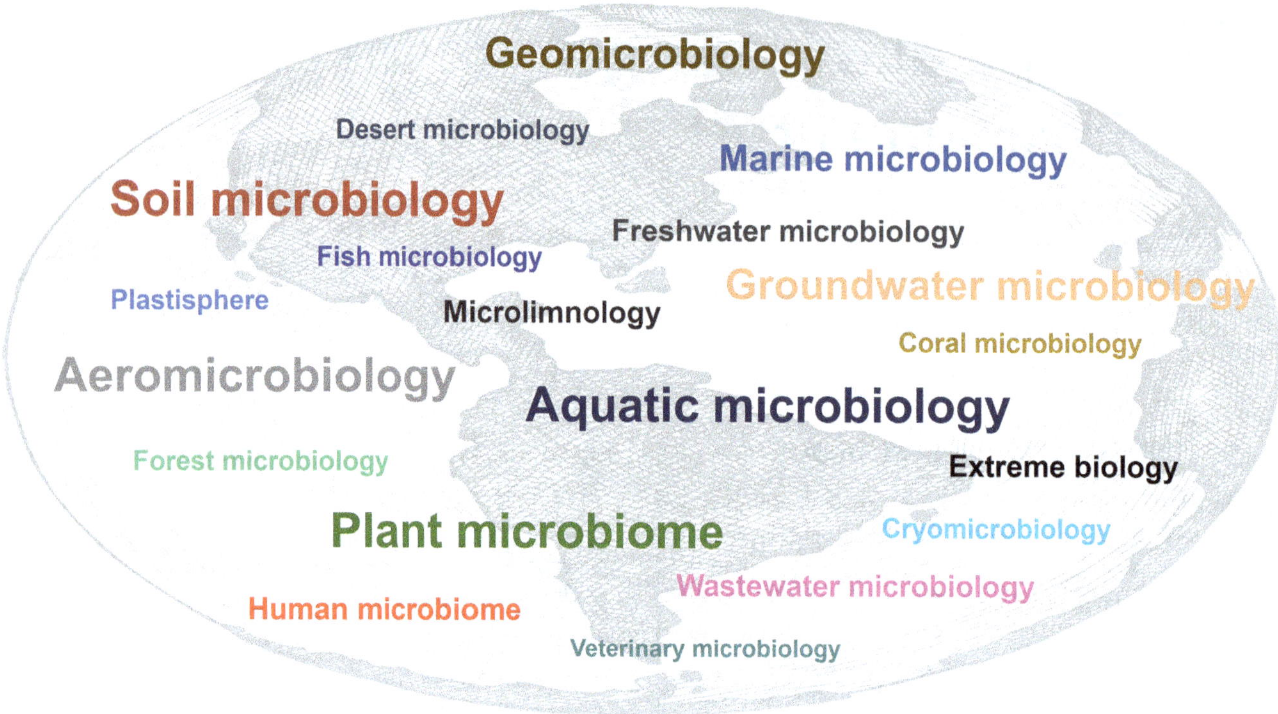

FIGURE 2.1 Various branches of environmental microbiology.

exhibit special attributes and survive using resources from their natural settings. Some in-built microbial tools, like horizontal gene transfer (HGT), chemotaxis, and biofilm formation, enhance cooperation and communication for their persistence in heterogeneous environments. HGT is a process of transferring genetic information between organisms that do not have a parent-offspring relationship. This process has raised concern over antibiotic resistance in bacteria. Chemotaxis is the ability of microorganisms to move in response to chemical stimuli. Biofilm formation is a process of irreversible attachment of microorganisms onto a solid surface. Quorum signalling, a process of cell-cell communication, plays an important role in regulating these attributes by coordinating cell activities, leading to the expression of specific genes. These phenomena demonstrate high levels of complexity in social microbial interactions.

Microbiology of different environments offers different perspectives on ecological processes and biotechnological applications. This opens a field to examine microorganisms in different habitats for their unique adaptation under distinct environmental conditions. This type of study also gives insight into their various implications. Based on the diverse habitats for microorganisms and their interaction with their native elements, environmental microbiology has several branches (Figure 2.1), which are briefly discussed as follows.

Aeromicrobiology is a recognised field of microbiology that deals with airborne microorganisms. Several microorganisms are suspended in the atmosphere and travel through the wind. They are collectively referred to as aeromicroflora. They have both positive and negative effects under different aspects. For example, the wild yeast floating in the air is responsible for brewing and traditionally making bread. However, when the milk is left open for a long duration, it results in curdling due to contamination by certain bacteria. Many diseases in plants and humans are caused by exposure to airborne pathogens (e.g., anthrax and chickenpox in humans and powdery mildew and late blight in plants). Occupational settings such as agriculture, road traffic, wastewater treatment plants (WWTPs), and healthcare facilities release a huge number of airborne microorganisms, which pose significant health risks, leading to a variety of respiratory and skin diseases. Chapter 18 provides detailed information on their implications for environmental and human health.

Soil is recognised as a nonrenewable resource, which is essential for many plants and animals, as well as all humans. The soil environment also provides a special niche for microbial growth and activities, and the concerned subdiscipline is **soil microbiology**. The soil microorganisms are particularly important for plant productivity and soil formation (i.e., pedogenesis). The soil stores around 75% of the carbon, which is mainly processed by soil microorganisms. In due course of evolution, the plants were able to stand on land due to **mycorrhizal association**. Our agriculture system is highly dependent on soil health, influenced by their native microbiota. Heterotrophic soil microorganisms (mainly bacteria and fungi) are particularly important for the decomposition of litter and the production of soil organic matter, and humus. A fungus, *Armillaria ostoyae,* spread over thousands of km below ground, has received wide attention because of its long stretch (detailed in Box 2.1). Chapter 19 sheds light on the characteristics of soil microbiota and their association with crop production.

Water is present in diverse locations such as open seas, above and below ground, and glaciers and ice caps. It is also present in low concentrations as vapour in the atmosphere. Water is essential for all living beings and so as for any microorganisms. The aquatic ecosystems team up with diverse microorganisms and influence their environments. This creates a distinct branch of microbiology called **aquatic microbiology** (detailed in Chapter 20). Aquatic habitats exhibit tremendous spatial heterogeneity in both horizontal and vertical spaces and harbour distinct microbial communities. Also, the temporal dynamics have an influence on MCCs and their function in aquatic environments. In the last two decades,

BOX 2.1 *Armillaria ostoyae*: THE LARGEST LIVING ORGANISM ON EARTH

Armillaria ostoyae is a humongous fungus belonging to the order Agaricales and is genuinely composed of tiny hyphal forms. Commonly known as the "shoe-string" or "honey fungus", this organism develops a contiguous fungal mycelium. It grows beneath the soil, colonises and kills the roots of conifers, and survives on its decay. Even being invisible to the eyes, they are far more massive than a blue whale in terms of weight, volume, and size. The mycelial network of *A. ostoyae* is spread over an area of about 2,400 acres (9.7 km^2) located in the Malheur National Forest, Oregon, USA (Box Figure 2.1.1). The whole network of this fungus belongs to a single genet and is estimated to weigh as high as 35,000 tons. This fungus is credited as the largest living organism on Earth. It has a predicted spread rate of up to 3.3 feet per year. The fungus also produces a small fruiting body called "honey mushrooms", which emerges above the soil, is short-lived, and grows during fall. It is also considered one of the most ancient living fungi, estimated to be 1,900–8,600 years old.

Since the actual portion of the fungus is hidden beneath the soil surface and due to the indefinite shape, size, and branching of mycelial life form, this organism was far out of sight from the discovery to be the largest living organism on Earth. Molecular evidence (e.g., sequence analysis) revealed that hyphae samples from different locations belonged to the same genetic stock. The current state of knowledge confirms that all the hyphae of *A. ostoyae* from the forest are genetically identical and place direct communication with one another, establishing the same network.

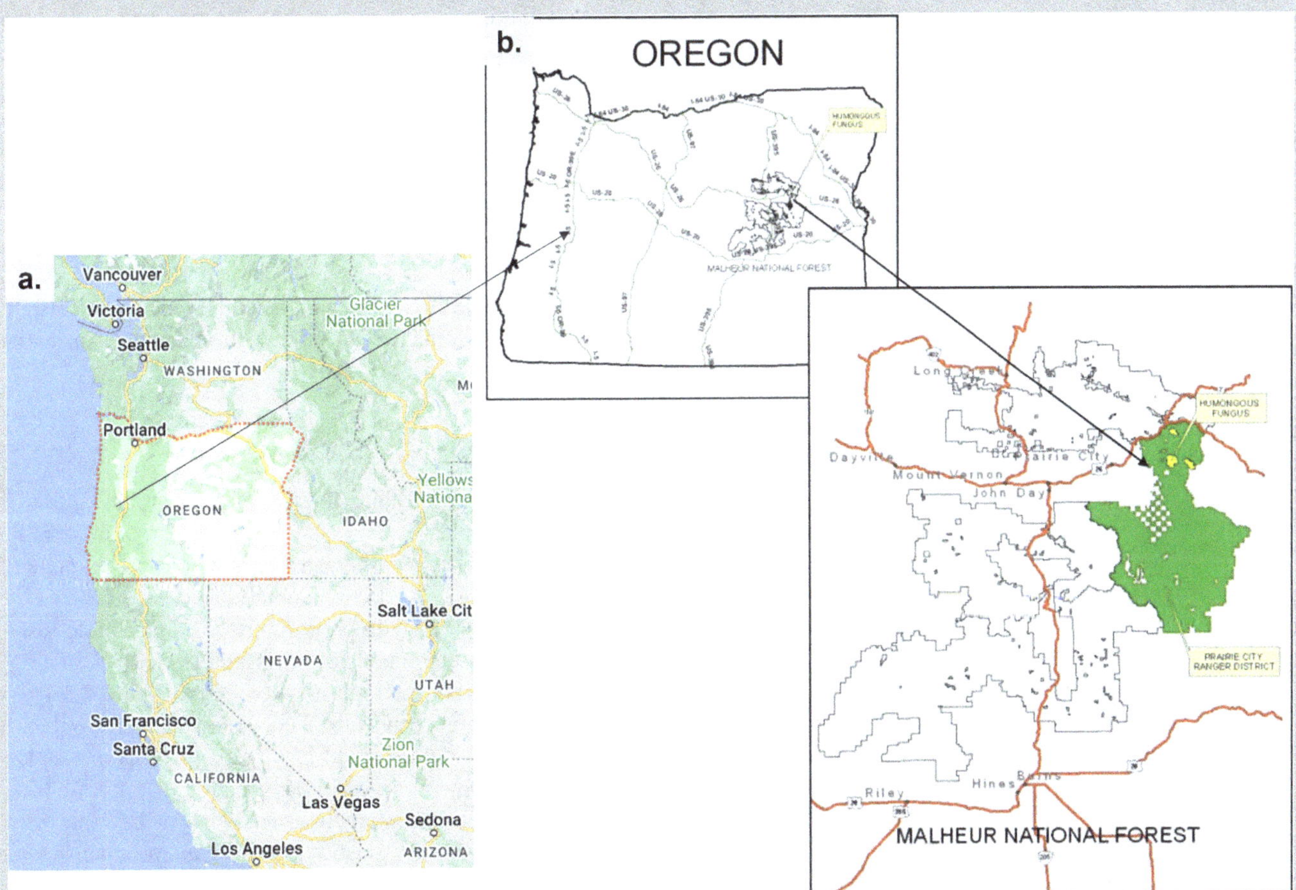

BOX FIGURE 2.1.1 Geographical distribution of *Armillaria ostoyae* in Malheur National Forest, Oregon, USA. a. A Google map showing the region of the western USA b. Vicinity maps of Oregon and Malheur National Forest (Image b reproduced from USDA[1], © 2008 USDA.)

a massive amount of plastic waste has accumulated in marine ecosystems due to anthropogenic activities, which has created a unique site underwater. This has created a different pressure on aquatic environments and attracted a distinct microbial assembly, constituting a **plastisphere** (detailed in Box 2.2).

Based on the different physicochemical characteristics within aquatic environments, there are different sub-branches of aquatic microbiology, such as marine microbiology, freshwater microbiology, etc. About 97% of the water on Earth is present in the marine ecosystem, which is saline in nature. It covers almost 70% of the Earth's surface and contains abundant biodiversity. Marine ecosystems consist of an unseen majority of microorganisms, which have essential roles in ecosystem functioning and environmental change. The study of these micro-sized inhabitants in marine environments is called **marine microbiology**. Marine microorganisms are the key players in climate change, primary production, and the flow of nutrients and energy. They also play a major role in maintaining the vitality of the marine environment. As life has evolved in the ocean, the marine environments are important research sites for studying the evolution of life on Earth. Marine microorganisms are chief participants in transforming carbon, nitrogen, sulphur, and other elements in the ocean. About 70–90% of the biomass in marine systems derives from microbial cells. Oceanic autotrophic microorganisms produce enormous amounts of fixed carbon and oxygen gas (O_2) that support other life forms. According to one estimate, one in every five human breaths (in terms of O_2 consumption) is produced by marine phytoplankton. Planktonic microorganisms are also the source of food for higher organisms. Corals in marine ecosystems inhabit many coral-associated microorganisms and establish symbiotic relations, which creates a distinct branch, **coral microbiology**.

Microorganisms of marine ecosystems possess unique metabolic capabilities and produce valuable compounds, obtaining support from biotechnological innovations. Several marine bioresources, such as energy, medicines, and enzymes, can be derived from microbial assistance, which are the essentials of **blue biotechnology**. This approach supports the concept of "Blue Economy", proposed by the World Bank, which refers to a policy of sustainable use of ocean resources for economic growth, improved livelihoods, and jobs.

The inland aquatic ecosystems are also diverse and are composed of springs, rivers, streams, lakes, groundwater, wetlands, and reservoirs. The study of these habitats is called **limnology**. The regions cover freshwater as well as saline water from both natural and manmade aquatic systems. The study of microorganisms inhabiting such an environment is addressed in **microlimnology**. A distinct field of aquatic microbiology is **freshwater microbiology**, where the study is restricted to microorganisms residing in freshwater ecosystems. Only ~3% of water in the world is freshwater. Of which ~69% is bound in glaciers and ice caps, and the remaining ~30% and ~1% exist in groundwater and surface water, respectively. Surface water has a relatively small coverage, and it is also highly accessible. This is one main reason why the information on microbial activities in surface water is far better than in any other aquatic environment. Habitats below the groundwater feature distinct environments, providing a niche for specialised microorganisms. This subfield of microbiology is called **groundwater microbiology** (detailed in Chapter 20). Wetlands can be natural or artificial ecosystems, which have very high productivity due to microbial dominance. Microbial communities in the wetlands are the major drivers of ecosystem functions and play roles in the degradation of contaminants and water purification.

Water pollution is a global problem. All types of aquatic ecosystems are under serious threat due to contamination. The main concern has arisen due to the input of diverse chemicals such as industrial waste, sewage, antibiotics, fertilisers, pesticides, and many more. The presence of pathogens, microbial toxins, and spores in aquatic bodies raises concerns for wildlife and human health. These chemicals and biological contaminants can have a strong impact on native

BOX 2.2 PLASTISPHERE

The excessive use of plastic in our day-to-day lives has led to the accumulation of high amounts of plastic waste in the environment, including aquatic and terrestrial ecosystems. A high amount of plastic debris is transported through rivers, which is dumped into oceans. The breakdown of plastic debris into small particles produces microplastics (size range 1 µm-5 mm). Many zooplankton ingest plastic debris and digest it into smaller particles. Some plastic particles float on the water's surface and act as a vehicle for pathogen and drug resistance. The hydrophobic surface of plastic debris attracts microbial communities and develops biofilms. Microorganisms colonised on plastic debris are distinct from their surrounding environments. Such microbial communities on plastic debris constitute a plastisphere (Box Figure 2.2.1). Microbial colonisation of plastic debris is a quick process, but biofilm development can be time-consuming. Heterotrophic bacteria are primary colonisers of plastic debris. Some inhabitants of plastic debris could also belong to diatoms and cyanobacteria if natural light exists. The mature plastisphere consists of diverse microbial forms, including autotrophs, heterotrophs, symbionts, and predators. These microorganisms establish various interactions and play a key role in ecological cycles and the food web of an ecosystem. Some inhabitants are potential plastic-degrading microorganisms. They produce enzymes like peroxidases, lipases, esterases, amidases, oxidases, and laccases in the biofilms, which convert plastic macromolecules into simple metabolites.

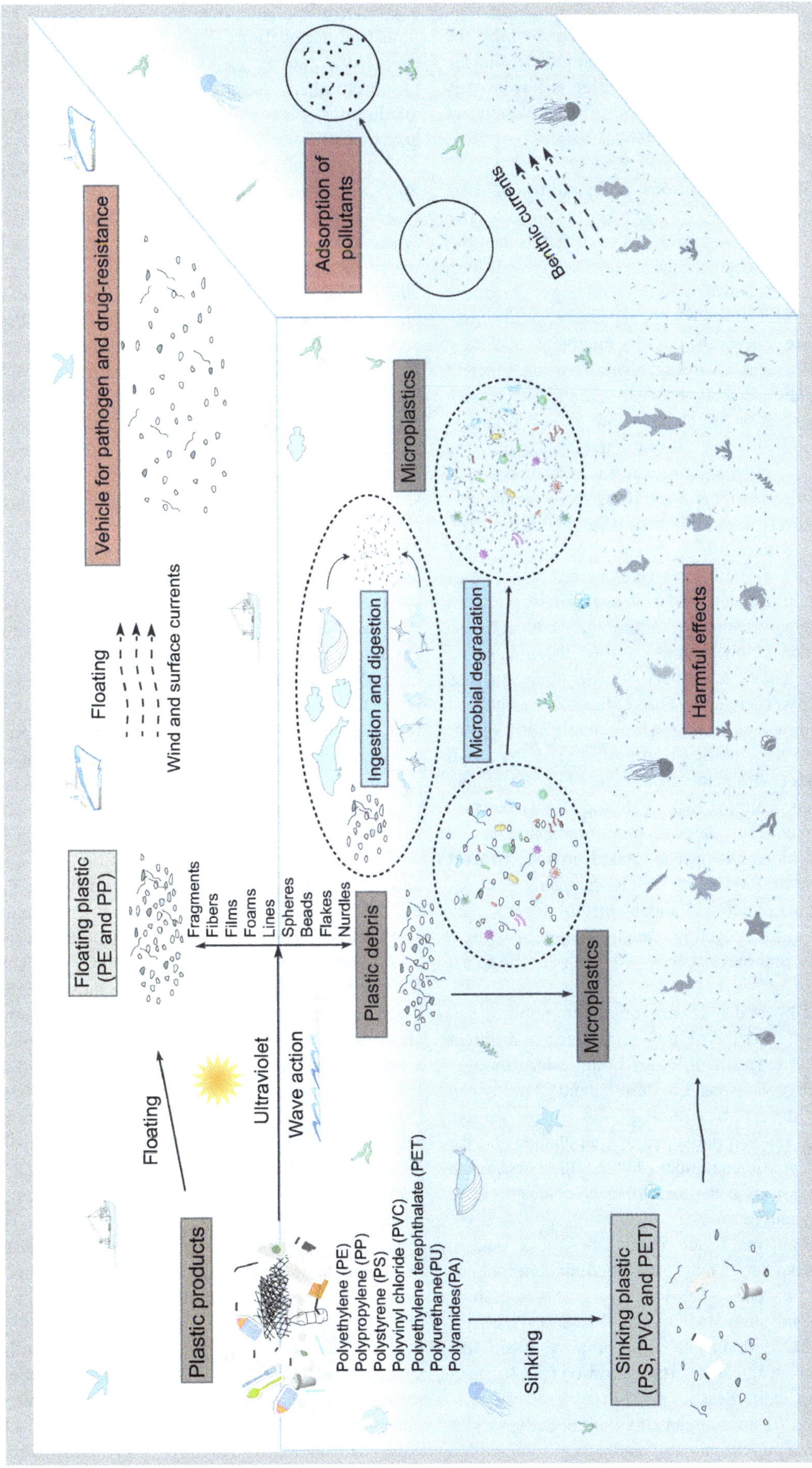

BOX FIGURE 2.2.1 Plastisphere in the ocean ecosystem and the possible fate of microplastics. (Images reproduced from Du et al. (2022)[2] with permission from Elsevier.)

microorganisms and may also alter ecosystem function. However, some native aquatic microbial species exhibit the potential for the treatment of diverse contaminants and are used in wastewater treatment. **Wastewater microbiology** covers all aspects of microorganisms that thrive in wastewater for their diverse roles (detailed in Unit VI).

Geomicrobiology lays the groundwork at the interface of earth science and microbiology, which focuses on the geoactive roles of microorganisms with minerals and metals beneath the surface of the Earth (detailed in Chapter 26). Another term for this field of study is **subsurface microbiology**. Terrestrial systems, sediments, groundwater, and deep oceans are the most pertinent subsurface environments for microbial activities. In biogeochemical cycles, microorganisms have an inherent role in mobilising the elements that help maintain ecosystem balance. The mobilisation of elements in the environment is also necessary for life. In the natural environment, the metabolic activities of microorganisms result in the dissolution of minerals and cause leaching. Several microbial inhabitants of the geosphere have roles in bioremediation and biomining. A distinct field of geomicrobiology revolves around the study of the long timescale in the Earth's ecosystems. This type of investigation yields information on the evolution of life on Earth.

An unprecedented number of microorganisms colonise both the above- and below-ground parts of the plants and live in symbiosis. They are present outside plant surfaces and inside the plant tissues. The collective insight of microbial communities interacting symbiotically with a plant body is referred to as a **plant microbiome**. The microbial communities interact with plants in diverse ways, which can be a driver for plant health, growth, and productivity. Section 2.12 provides an insight into the plant microbiome and its role in agriculture and biotechnology. A related branch is **forest microbiology**, which focuses on microorganisms residing in forest ecosystems. Microbial inhabitants of forests play dominant roles in nutrient cycling, decomposition of organic matter, soil fertility, and overall ecosystem health of the forest ecosystems.

The cryosphere hosts distinct microbial communities and microbial activities, and their study constitutes a different subdiscipline called **cryomicrobiology/polar microbiology**. Microbial life is also widespread in deserts, and its study can be referred to as **desert microbiology**. Environments like desert soils, rocks, and salt flats have unique challenges like low precipitation and arid conditions, and the microorganisms residing in such environments possess mechanisms to cope with water scarcity and heat/cold.

Some microorganisms reside inside the rocks, shells, minerals or corals and derive nutrition from their components. Representatives are called **endoliths** or "rock-eating" microorganisms and can belong to bacteria, archaea, fungi, and protozoa. Endoliths can also be phototrophs like cyanobacteria, lichens, and algae, which grow on rock surfaces and derive nutrients from the inner part of rocks.

A distinct group of microorganisms permanently resides in highly extreme environments, and they are called **extremophiles**. They possess unique features and can have industrial applications. Extremophiles have many exclusive categories, such as thermophiles, hyperthermophiles, psychrotrophs, psychrophiles, halophiles, acidophiles, alkaliphiles, piezophiles, radioresistant extremophiles, and anaerobes (Table 2.1). Thermophiles are high temperature-loving organisms and reside in the temperature range of 50–80 °C. Hyperthermophiles thrive in temperatures above 80 °C. A hyperthermophilic archaeon, strain 121 (*Geogemma barossii*), isolated from an inlet of the Pacific Ocean, marks the upper-temperature limit for life. This living form can multiply even at 121 °C (a temperature suitable for autoclaving) and remain viable even at 130 °C. Psychrotrophs and psychrophiles grow in cold environments, of which the former can survive even at temperatures above 20 °C, and the latter are permanent residents of cold environments. Some green algal members of Chlamydomonaceae are cold-loving organisms and impart red colour in the snow (Figure 2.2a). Halophiles are salt-loving microorganisms. The sea environment is a very vast habitat for the halophiles. They also reside on land in saline lakes. A unicellular green alga, *Dunaliella* sp., is a common resident of hypersaline lakes and ponds. This genus is responsible for the pink colour of lakes and ponds due to the mass production of β-carotenoid pigment (Figure 2.2b). Microorganisms can also be reported from environments having a pH range from 1 to 14. The low and high pH environments harbour acidophiles and alkaliphiles, respectively. Piezophiles are microorganisms that withstand high pressure and are found in the deep sea. Radioresistant extremophiles can survive in high radiation doses (e.g., *Dienococcus radiodurans*). Anaerobes stay in anaerobic environments (e.g., *Clostridium*). Microorganisms that survive in strictly anaerobic conditions are obligate anaerobes, while those that prefer aerobic conditions but can switch to anoxic conditions are facultative anaerobes (e.g., *Bacillus*). A different category of anaerobes is microaerophiles, which require low levels of O_2 (e.g., *Campylobacter*).

A biofilm is an assemblage of microbial cells embedded in a self-produced extracellular polymeric substance (EPS) matrix. They can attach to the surface of biological and non-biological surfaces and develop a slime layer. This architecture provides a unique environment for microbial growth and metabolism. Biofilm-associated cells are more tolerant to desiccation than their freely suspended counterparts due to social cooperation, resource capture, and quorum sensing (cell-to-cell signalling) (Box 2.3). This complex way of living together provides benefits in terms of resistance to toxic chemicals and offers protection from grazing. Microorganisms that develop biofilms can belong to bacteria, fungi, microalgae, and protists. Among all, bacteria are the most common in biofilms. The aquatic boundaries are their favourite habitats. They can have positive and negative consequences depending on the site of development. For example, in an engineered biotreatment system of wastewater, a fixed-film bioreactor relies on biofilm formation for an efficient biodegradation process. On the other hand, biofilm growing on the body parts of humans is a serious pathological condition, and its treatment is a serious

TABLE 2.1
A list of various types of extremophiles growing in different environmental conditions with suitable examples of microorganisms

Types of extremophiles	Environmental conditions	Examples of microbial taxa*
Thermophiles	Microorganisms thriving at high temperatures (between 50–80 °C)	• *Thermus aquaticus* • *Sulfolobus* • *Thermomyces*
Hyperthermophiles	Microorganisms thriving at very high temperatures (above 80 °C)	• *Thermotoga* • *Geogemma barossii* (Strain 121) • *Methanopyrus kandleri*
Psychrotrophs	Cold-tolerant microorganisms growing at temperatures around 15–20 °C	• *Pseudomonas* • *Aspergillus* • *Alternaria*
Psychrophiles	Cold-loving microorganisms growing at temperatures between -20 and 20 °C	• *Chlamydomonas priscuii* • *Rusavskia elegans*
Halophiles	Microorganisms growing in moderate salt concentrations	• *Salinibacter ruber* • *Halorubrum* • *Dunaliella salina* • *Wallemia ichthyophaga*
Acidophiles	Microorganisms thriving in acidic environments (pH ≤ 4.0)	• *Acidithiobacillus ferrooxidans* • *Ferroplasma acidarmanus* • *Acidothrix acidophila*
Alkaliphiles	Microorganisms thriving in basic environments (pH ≥ 8.0)	• *Thiohalospira alkaliphila* • *Sodiomyces alkalinus*
Piezophiles	Microorganisms surviving at pressures of 40–110 MPa	• *Shewanella benthica* • *Pyrococcus yayanosii* • *Thermococcus barophilus*
Radioresistant microorganisms	Microorganisms growing in high levels of the radiation environment	• *Deinococcus radiodurans* • *Chroococcidiopsis* • *Thermococcus gammatolerans*
Anaerobes	Microorganisms growing in the absence of oxygen	• *Clostridium botulinum* • Methanogens
Xerophiles	Microorganisms surviving in low water availability	• *Trichosporonoides nigrescens* • *Xeromyces bisporus*

* Microbial taxa represented by different colours (red- bacteria, blue- archaea, black- fungi, green- microalgae, and yellow- lichens).

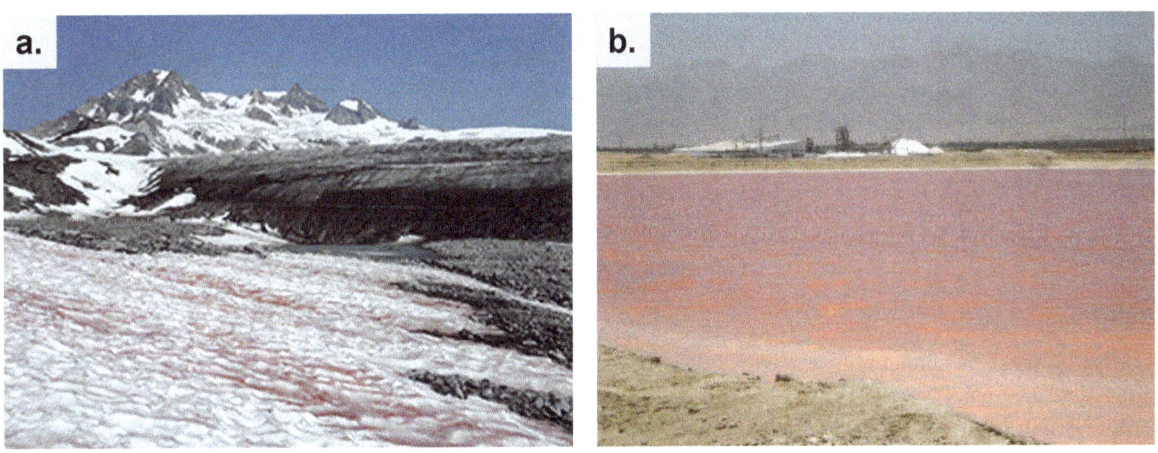

FIGURE 2.2 Examples of microalgae extremophiles. a. Members of Chalamydomonaceae, giving a red colour to snow in the Coast Range of British Columbia, Canada. b. The presence of archaeal communities and *Dunaliella* cells, imparting pink (or red) colour in a saltern crystalliser pond (Eilat, Israel). (Images a and b reproduced from Engstrom et al. (2020)[3] and Oren (2016)[4], respectively.)

BOX 2.3 QUORUM SENSING

Quorum sensing is a process of cell-cell communication in bacteria to detect and respond to cell population density through gene regulation. Such a motion aids in the coordination of bacterial activity, such as symbiosis, virulence, HGT, motility, conjugation, sporulation, antibiotic production, and biofilm development. For the same reason, bacteria produce a special type of extracellular chemical signal compounds called autoinducers. However, the signal for cell-cell communication requires a minimal threshold stimulatory concentration of autoinducers. When cell density is low, autoinducer molecules diffuse away, and the minimum threshold limit is not met, leading to no stimulation (Box Figure 2.3.1). When the number of cells grows, so does the concentration of autoinducers. This leads to a change in gene expression in response to changes in environmental conditions. The autoinducers in Gram-positive and Gram-negative bacteria have different chemical natures. The former group produces small peptides, while the latter group mainly relies on acyl-homoserine lactones (AHLs) for the same. The variation in the chemical natures of the autoinducers in different taxa belonging to the same groups also exists. Some chemical signals facilitate stimulation by a wide range of bacteria, enabling communication among different species.

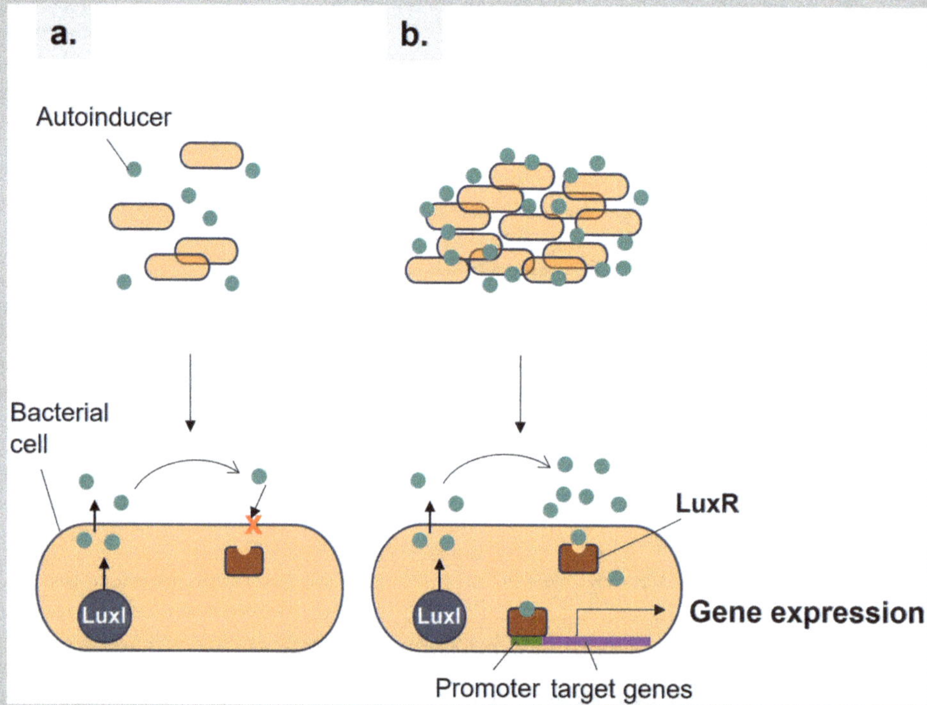

BOX FIGURE 2.3.1 Mechanisms of quorum sensing. a. When the microbial cells are suspended, the autoinducer molecules produced by cells diffuse away and do not induce any gene expression. b. When cells are aggregated, the autoinducers released by the group organisms exceed the threshold limits, allowing gene expression.

concern because it exhibits a high tolerance to antimicrobial compounds.

Human microbiome: The human body is a wonderful site for the dwelling of microorganisms. They often establish symbiotic relationships with different parts of a body and can remain as lifelong microbiome signatures. Microorganisms have a huge impact on humans. Human activities influence MCCs and their function in the environment. The cutaneous skin is always in direct contact with a huge number of microorganisms in the air. Microorganisms from the surroundings consistently attempt to invade the hosts through inhalation of air and ingestion of food items. The initial microbial colonisation in humans occurs during birth while passing through the vaginal fluid. Also, humans release immense amounts of inside microorganisms into the outer environment through breathing and excretion.

Though various microorganisms, including bacteria, archaea, fungi, protozoa, and viruses, find residence in the human body, the proportion of bacteria is much more than any other microbial form. Therefore, the operational definition of microbial cells in the context of the human microbiome generally refers to bacterial cells. A healthy human body hosts around 3.8 x 10^{13} microorganisms, which is even more than the total number of human cells in the body (i.e., around 3.0 x 10^{13}). Also, a human microbiome consists of about 500 more genes than what is present in a human genome (i.e., 20–25,000). A rough estimate suggests that there could be as many as 232 million microbial genes in the collective human microbiome. However, the total weight of all the microbial cells does not exceed 0.2 kg. The gastrointestinal (GI) tract, followed by the skin, is the most common site for microbial colonisation in the human body. In addition, microbes are also abundantly present in saliva, oral mucosa, conjunctiva, and vagina. The vaginal part is dominated by *Lactobacillus* spp., which maintains a low pH to avoid infection by other microorganisms. A specific bacterium, *Staphylococcus epidermidis*, is responsible for the cause of smelly feet. This bacterium breaks down the amino acid leucine present in sweat into isovaleric acid, which has a specific odour.

Human-microorganism interaction has a strong impact on wellness and diseases in humans. Gut bacteria are potential controllers of energy harvesting, digestion, and immune defence. Bacterial groups like Firmicutes, Bacteroidetes, and Actinobacteria are the dominant inhabitants of the gut. Recent research studies claim that the gut microbiota and the brain are linked through a bidirectional communication pathway called the Microbiota-Gut-Brain Axis (MGBA). An interdisciplinary effort of the National Institutes of Health (NIH)-funded Human Microbiome Project (HMP) has also greatly improved our understanding of the human microbiome and its essential roles in human health.

2.4 MICROBIAL INTEGRITY FOR ENVIRONMENTAL BENEFITS

Microbial integrity plays a critical role in coping with environmental matters. The plethora of biochemical processes allows microorganisms to survive in different types of ecosystems. In one aspect, microbial-driven reactions provide nutrition for cellular growth and their multiplication. A different aspect declares their relevance in ecosystem functioning. Microbial activity is closely related to everything that humans require to survive (Figure 2.3). However, our understanding of many microbiological processes remains elusive due to their tremendous complexity. For example, many macrofungi and mycorrhizae grow in the wild, and their cultivation in the laboratory is highly challenging. An entomopathogenic fungi, *Ophiocordyceps sinensis,* growing in the wild, is the most precious fungi in the world (Box 2.4), but its cultivation in an artificial environment has not achieved any commercial success.

The manifold benefits of microbial forms can be derived through cultivation practices. This can rely on monocultures or consortia. Monoculture is the production of a single species of microorganisms. This has gained popularity for the unidirectional approaches. The empire of industrial microbiology stands alone due to the advent of monoculture technology. A consortium is a mixed population of two or more microbial groups that live symbiotically and act synergistically. The utilisation of consortia has increased interest in holistic systems-level methods for wastewater treatment, waste management, and bioenergy generation. The use of highly integrated multispecies microbial communities in place of monocultures is a major shift in environmental microbiology that is more efficient, sustainable and profitable.

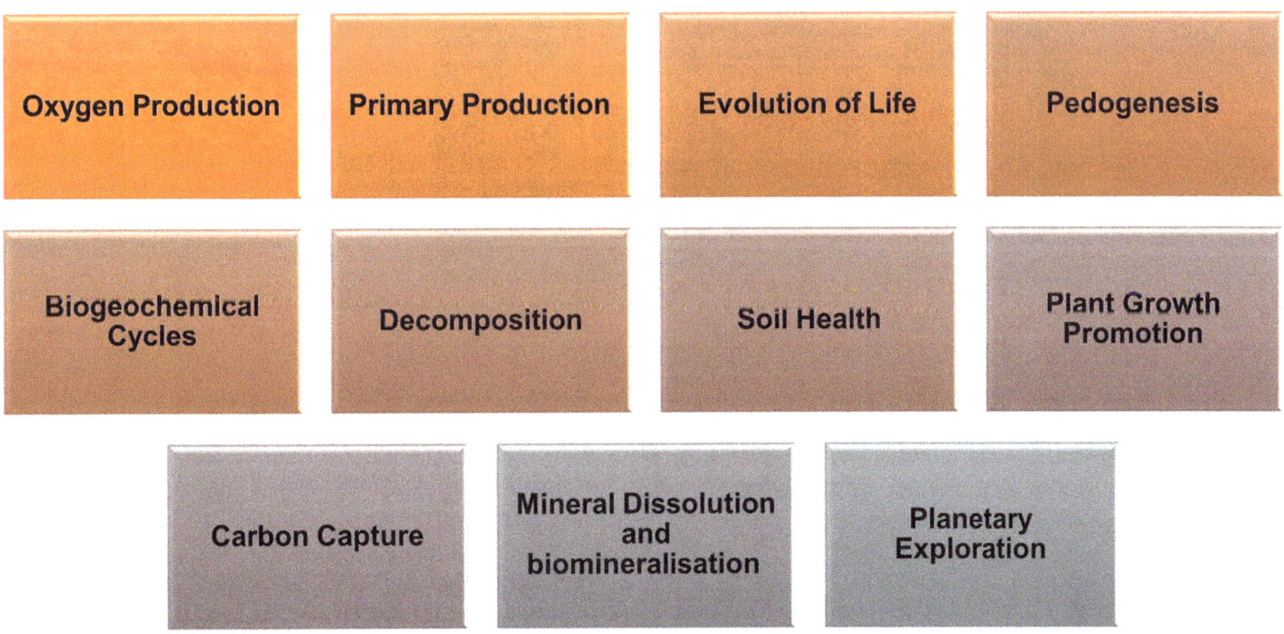

FIGURE 2.3 Scope of microbial activities for environmental well-being and human welfare.

BOX 2.4 *OPHIOCORDYCEPS SINENSIS*: THE MOST PRECIOUS FUNGUS IN THE WORLD

Ophiocordyceps sinensis (or *Cordyceps sinensis*), belonging to the phylum Ascomycota, is the most precious fungus in the world. It is a pathogenic fungus which infects insects (entomopathogenic). The common English name is caterpillar fungus. In the Tibetan region, it is identified as yartsa gunbu— the literal meaning is "summer grass, winter worm". This rare Asian fungus fetches around £80,500 per kg (or 50,000 USD a pound), which makes it more expensive than gold. This fungus has high-priced medicinal value and status symbolism in China. *O. sinensis* produces a very rare adenosine derivative, cordycepin, which has therapeutic and nutraceutical potential and is used in traditional Chinese medicine. This fungus is also claimed as an aphrodisiac and touted as "Himalayan Viagra". It is also known for its anticancer and antitumor activities.

O. sinensis is a psychrophilic fungus and grows best around temperatures of 18 °C. This fungus infects the larvae (caterpillars) of ghost moths through mycelium and kills the insects in the soil subsurface. Mycelium still grows and invades the host body. Lately, the fruiting body emerges above ground from the dead moth and stands upright. The size of the fruiting body is approximately 2–6 cm (Box Figure 2.4.1). This type of symbiosis is endemic to alpine grass and shrublands of the Tibetan Plateau and the Himalayas at an altitude between 3,000 and 5,000 m. Their fruiting bodies are rare and hard to find in the wild, and cannot be cultivated under controlled conditions. Their production is mainly dependent on natural production. However, climate change, overexploitation, and overharvesting of the fungi have resulted in the demise of its major wild population. Over the past ten years, the price of caterpillar fungus has increased tremendously (up to 230%) owing to its growing demand. The total annual production of caterpillar fungus is about 83.2 tonnes. Countries like China, followed by Nepal, are the largest producers and exporters of caterpillar fungus. The remote territories of the Tibetan Autonomous Region (TAR), India, and Bhutan also account for notable productions. Its harvest is an important source of cash income in the growing areas. This fungus has received a conservation status as an endangered species due to its mass hunting and habitat loss.

BOX FIGURE 2.4.1 A fruiting body of *Ophiocordyceps sinensis* attached to insect larvae. a. In the wild. b. A single filament. (Images a and b adapted from Wang et al. (2018)[5] and Zhang et al. (2021)[6], respectively.)

Natural microorganisms are important for the well-being of the Earth and human society in many different ways. These are discussed in many sections as follows.

2.5 OXYGEN PRODUCTION

Humans and animals cannot survive without O_2. Also, oxygenic respiration is equally important for plants and many microorganisms. The contribution of green plants to produce O_2 is well recognised. However, one would possibly be amazed to learn that microorganisms like cyanobacteria and microalgae in marine ecosystems are the major producers of O_2 in the environment. They supply more than half of the O_2 that is produced on Earth. An estimate even suggests that the ocean could be responsible for producing as much as 80% of O_2 on the planet—the contributors include phytoplankton,

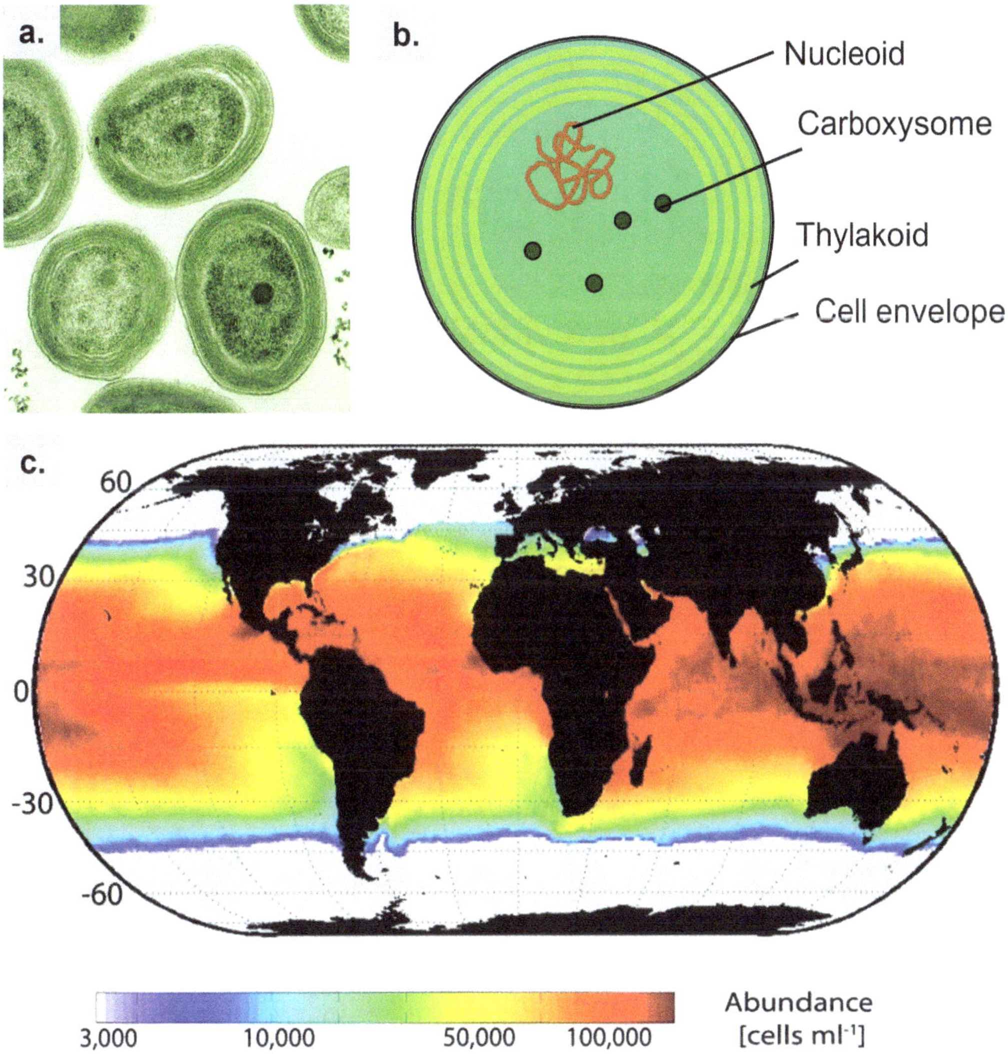

FIGURE 2.4 Structure and distribution of *Prochlorococcus*. a. Transmission electron micrograph. b. Diagrammatic view. c. Global distribution of *Prochlorococcus*. (Images a and c adapted from Thompson L of Chisholm Lab and Nikki Watson from MIT, 2007 and Flombaum et al. (2013)[7] from © 2013 National Academy of Sciences, respectively.)

algae, and drifting plants. Phytoplankton on the ocean surface have a much more prominent role in O_2 production. The microbial role in oxygen evolution has no human intervention and is a wonderful example of a self-sustained process. Photosynthesis using chlorophyll molecules in the presence of sunlight is an obvious pathway for O_2 production in these tiny life forms. One creature of the oceans, *Prochlorococcus* (Phylum: Cyanobacteria), produces about the same amount of O_2 produced by a combined effort of all the tropical rainforests. This amount is nearly 20% of O_2 production in the Earth. Figure 2.4 provides an insight into the structure of *Prochlorococcus* and its global distribution.

Few bacteria and archaea can even evolve O_2 independent of photosynthesis, which means without using light. However, their input for O_2 production is much less compared to photosynthetic activity. This includes an ammonia-oxidising archaeon, *Nitrosopumilus maritimus*, which releases O_2 in oxygen-depleted deep marine environments through a novel pathway. Some methane eating microorganisms (like *Methylomirabilis oxyfera*) inhabiting the bottom of rivers and lakes release gases like nitrogen (N_2) and O_2 through denitrification. Bacterial reduction of chlorates (ClO_3^- and ClO_4^-) yields chlorite (ClO_2^-), which can be broken down into chloride and O_2 using a chlorite dismutase enzyme. This pathway exists in some bacteria and archaea from the deep subsurface of aquatic environments. In addition, enzymes like catalase and superoxide dismutase generate a low amount of O_2 while preventing oxidative damage (from $OH^{\cdot}$, O_2^-, and H_2O_2) inside the cells.

2.6 PRIMARY PRODUCTION

Carbon is one of the main building block elements in the biological system, mainly available through primary production. This process is restricted in autotrophic organisms, including plants and some microorganisms. Plants contribute to primary

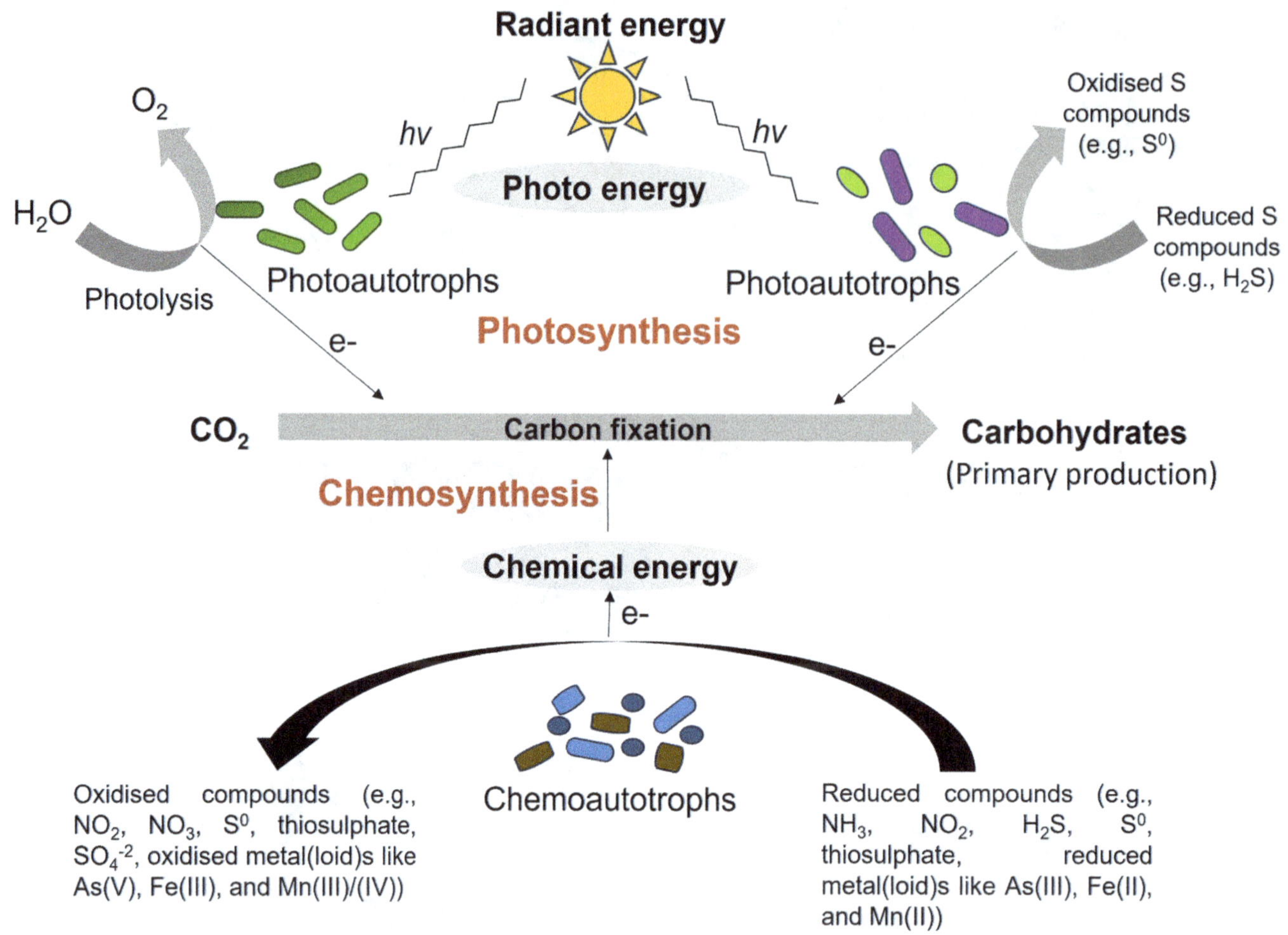

FIGURE 2.5 Diagram showing two major pathways (i.e., photosynthesis and chemosynthesis) for primary production in microorganisms. In photosynthesis, the energy comes from sunlight and electrons are derived either from the photolysis of water or the oxidation of reduced sulphur (S) compounds. However, in chemosynthesis, both energy and electrons are obtained by chemical reactions. (*hv* represents a photon.)

production through photosynthesis (in carbon fixation), but autotrophic microorganisms possess diverse pathways to fix carbon. These microbes use a variety of electron and energy sources (Figure 2.5). Based on the energy sources, they can be photoautotrophs and chemoautotrophs. Photoautotrophic microorganisms fix carbon using energy from sunlight and electrons mainly from inorganic chemicals. They can be oxygen-evolving microorganisms such as cyanobacteria and algae. These microorganisms derive electrons from water hydrolysis and release O_2. Differently, photosynthesis can also be a non-oxygen-evolving process, using electrons from the oxidation of sulphur compounds. Examples of nonoxygenic photosynthetic organisms are green sulphur bacteria, purple sulphur bacteria, purple nonsulphur bacteria, Chloroflexi, acidobacteria, and heliobacteria. Chemoautotrophs meet the energy requirements for carbon fixation from chemical reactions rather than sunlight. Methanogens, sulphur oxidisers and reducers, nitrifiers, anammox, and heavy metal-oxidising bacteria are notable examples of chemoautotrophs.

Photosynthesis by phytoplankton in oceans rules for the production of organic matter and generates more net primary productivity than any other ecosystem. As the plant system is the main source of primary production in terrestrial ecosystems, phytoplankton play dominant roles in the ocean for the same. Autotrophic microorganisms are responsible for about half of primary production worldwide. Cyanobacteria, diatoms, coccolithophores, dinoflagellates, and green algae in aquatic systems have key roles in primary production. Cyanobacteria alone contribute to about 25% of marine primary production. Diatoms contribute as high as 40% of carbon fixation in the ocean. Coccolithophorids are single-celled eukaryotic organisms and are important primary producers in the ocean. They display a remarkable ability to calcify, which is responsible for the formation of calcite oozes or chalk deposits on the seafloor. This process also tunes with the structure of the White Cliffs of Dover, UK, which emerged from the ocean some 70 mya (during the late Cretaceous period) due to the vast buildup of coccolith (Figure 2.6). Many cyanobacteria and green algae are also residents of freshwater ecosystems and soil and contribute to primary production. These include filamentous cyanobacteria like *Anabaena*, *Nodularia*, *Microcystis*, and *Oscillatoria*, green algae like *Chlamydomonas*, *Chlorella*, and diatoms. From an ecological point of view, the primary production

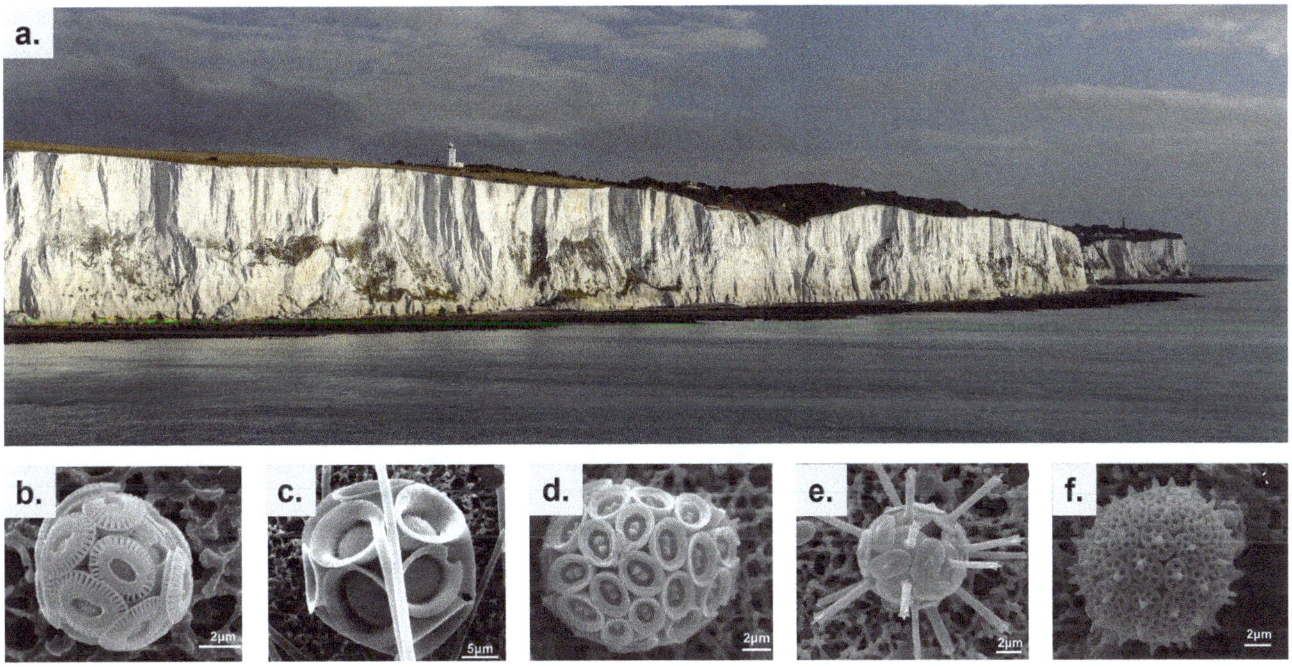

FIGURE 2.6 a. The White Cliffs of Dover, UK, formed due to the deposition of coccolith produced by a type of marine phytoplankton coccolithophores. b-d. Ultrastructure of coccolithophores. b. *Emiliania huxleyi*. c. *Pontosphaera syracusana*. d. *Syracosphaera mediterranea*. e. *Rhabdosphaera clavigera*. f. *Syracosphaera mediterranea* HOL wettsteinii type. (Images b-f adapted from Triantaphyllou et al. (2018)[8] © 2018 PLoS One.)

of phytoplankton fuels a heterotrophic lifestyle. This begins with the grazing of phytoplankton by protozoans, followed by their consumption of higher tropical organisms. Also, primary production by microorganisms contributes immensely to the global carbon cycle.

2.7 ORIGIN AND EVOLUTION OF LIFE

The microbial forms have ruled the Earth since their inception and still have a significant impact on the environment. They have pioneered life on Earth, and all other living forms descend from ancient life. However, the Origin of life on early Earth was highly underwhelming and confined. Several theories have been proposed for the origin of life. The two most popular theories are panspermia and abiogenesis. A theory of panspermia postulates that life on Earth began with "seeds of life" (i.e., microbial forms) distributed to this planet by comets, meteorites, asteroids, and dust from space. In contrast, the theory of abiogenesis advocates the origin of life from non-living matter.

Earth is 4.5 billion years old, but the conditions suitable to support life on Earth appeared around 4.3 billion years ago (bya). Chemical signals reported on the rock from 4.0 bya are the earliest evidence of prehistoric life. The single-celled primordial organisms growing in high temperatures under anaerobic conditions in deep-sea vents preceded all current life descends. They are portrayed as the last universal common ancestor (LUCA) (evolved at a point around 3.8 bya) and are considered the most common recent ancestors of present life. LUCA exhibits some remarkable features, such as primordial heredity, self-reproduction, self-assembly, and autocatalysis, which allowed early-life ancestors to diversify, persist, and proliferate through the course of evolution. However, the oldest fossil records of life and LUCA do not belong to the same genetic stock. Cyanobacteria fossils from Western Australian Archaean strata, which are over 3.4 billion years old, are direct evidence of the earliest life forms.

The passage of abiogenesis to biogenesis leads a path from evolutionary microbiology to microbial ecology. The genesis of photosynthesis in cyanobacteria was a momentous event in the course of evolution and was responsible for the Great Oxygenation Event (GOE) on Earth around 2.4 bya. In this event, gases like CH_4 were displaced, and O_2 became a major component of the atmosphere. The increase in the amount of O_2 transformed more than half of the minerals present on Earth and evolved oxygenic life. Nitrogen in life was present even before the GOE. The biologically active forms of nitrogen constituted nucleic acids, proteins, and energy currency. The evolution of biological nitrogen fixation (BNF) can be traced back to as early as 3.2 bya. Nitrogen fixation influenced the expansion of GOE. Both events are the key innovations of ancient lifeforms, which persisted, expanded, and established the modern Earth's biosphere. Some deep sites on Earth still reflect the early open ocean conditions where significant reducing conditions prevail and harbour microorganisms similar to those that seeded life on planets.

Aerobic respiration and eukaryotic organisms successively evolved around 2.3-2.2 bya. Membrane-bound organelles in eukaryotic cells originated around 1.4-2 bya due to endosymbiosis between prokaryotic microorganisms and eukaryotic cells. Mitochondria evolved before chloroplasts. The former structure was possibly developed from the consumption of aerobic prokaryotes by an amoeba-like cell, while the latter evolved through the ingestion of photosynthetic prokaryotic cells in eukaryotic cells. The setups of early synthesis of cell-bound organelles constituted an **endosymbiosis theory**, which is also known as **symbiogenesis**.

The genesis of the first vascular land plants occurred about 460 mya due to the symbiotic association of arbuscular mycorrhizae (AM) in roots. During the colonisation of land plants, the root penetrated the soil, and AM attached to the root body, supplying essential nutrients for plant growth. Ectomycorrhizae evolved much later (i.e., about 150 mya). Other plant microbe symbioses, such as *Frankia*-actinorhizal symbiosis and rhizobia-legume symbiosis, appeared 80–60 mya. Table 2.2 lists various microbiologically-associated events on Earth in the geological time scale.

Microbial activities have continued to evolve since their inception and are still evolving. The modern world can be seen as a result of the Anthropocene, where human activities have become a dominant force in shaping the environment of Earth. In this phase, the evolution of microorganisms has been seen to dominate for antimicrobial resistance, pollutant degradation, host-pathogen dynamics, and adaptation to a changing climate. The artificial evolution of microbial cells through research knowledge has accelerated the rate of microbial evolution and shifted its natural direction. The release of modified cells into the environment carries the risk of unintended ecological variations. The stability of modified microorganisms in natural environments and the interactions between the engineered and nonengineered organisms have also potentially altered the ecological dynamics.

2.8 PEDOGENESIS

Pedogenesis is a natural process of soil formation from the combined effects of physical, chemical, and biological activities on parent materials. It is a dynamic process and induces several variations in the physicochemical conditions of the soil. Such changes also bring variations in the composition and function of microbial communities. Microorganisms, including bacteria, archaea, fungi, and lichens, have a central role in pedogenesis. A pedological perspective terms them as **soil engineers**. They turn the underlying hard rock into productive soil through microbial weathering. Lichens are pronounced pioneer communities for the ecological succession on bare rock. In the first line of their interaction, the rock surface is disintegrated and fragmented below the lichen body. This process is termed physical biodeterioration. Some lichens also exhibit chemical weathering; they release organic acids, like oxalic acid and release minerals from rock surfaces. In ecological succession, the decomposition of lichen bodies contributes to the accumulation of organic matter in soil formation. The roles of bacterial communities

TABLE 2.2
Geological time scale for various microbiologically-associated events

Geological time periods		Time scale (mya)	Microbiological-associated events
Phanerozoic eon	Cenozoic era	60	Evolution of legume-rhizobia symbiosis
	Mesozoic era	80	Evolution of *Frankia*-actinorhiza symbiosis
		150	Appearance of ectomycorrhiza
	Palaeozoic era	400	First yeasts
			First lichens
		460	Appearance of arbuscular mycorrhiza (AM)
Proterozoic eon	-	1200	Evolution of chloroplast
		1450	Evolution of mitochondria
		1700	First eukaryotic algae
			Evolution of multicellular organisms
		2200	First eukaryotes
		2300	Evolution of aerobic respiration
		2400	Great Oxygenation Event (GOE)
			First fungal-like mycelia
Archaeon eon	-	2700	First oxygen-evolving organisms
		3200	Evolution of biological nitrogen fixation (BNF)
			First sulphate-reducing bacteria (SRB)
		3400	First cyanobacteria
		3500	First bacteria and archaea (methanogens)
			Initiation of anoxygenic photosynthesis and chemosynthesis
		4000	Origin of life (Last Universal Common Ancestor (LUCA))
Hadean eon	-	4300	First chemical signals to prehistoric life
		4500	Formation of Earth

Notes: mya stands for million years ago.

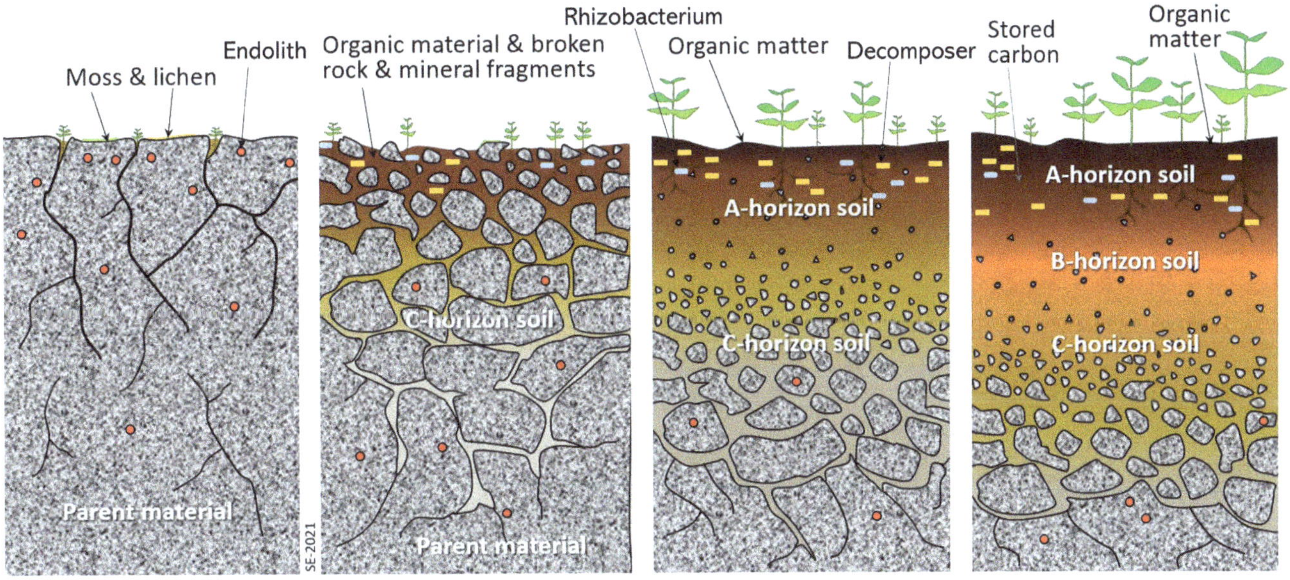

FIGURE 2.7 Different steps of pedogenesis showing the roles of lichen and moss. (Image reproduced from Environmental Geology © Steve Earle.)

in soil formation are also evident. The endolithic (rock-eating) bacteria and fungi release natural metabolites and solubilise minerals from rocks. They can also oxidise or reduce the minerals in the rock and facilitate their solubilisation and mobilisation. The acids and alkali released by microbial activities alter the pH of the substratum, which influences the growth and metabolic activity of microorganisms and other organisms. Figure 2.7 illustrates different steps of the biological process in pedogenesis.

A mature soil is composed of many layers with varying physicochemical properties. A gram of topsoil contains millions of microorganisms that support soil health. In soil, microorganisms and the humus make topsoil rich and fertile. Many of them are decomposers and degrade a variety of organic matter present on land. A significant amount of native soil microbiota belongs to nitrogen-fixing organisms and fungal networks that help maintain the vitality of the soil.

2.9 BIOGEOCHEMICAL CYCLES

A biogeochemical cycle is an illustration of the circulation of an element or compound through biologically influenced chemical and geological activities. The study of biogeochemistry provides an analysis of global chemical change. The atmosphere, the lithosphere, the hydrosphere, and the biosphere are critical compartments for biogeochemical reactions (Figure 2.8).

In the biogeochemical cycle, microbial integrity is essential to maintain a life support system. Microbial metabolism, coupled with elemental reactions, shapes the dynamics of ecosystems. The myriads of microbially-mediated reactions shift the chemical composition of the environmental compartments. The microbial-mediated reactions can produce two types of coupling biochemical pathways: assimilatory and dissimilatory pathways. Both pathways play roles in building biological systems. The assimilatory pathway captures elements from the environment and transfers them into cellular components such as nucleic acids, proteins, lipids, carbohydrates, etc. The dissimilatory pathways use nutrients in redox reactions and produce energy for their metabolism.

The foundation of biological macromolecules is based upon six key elements, including carbon, nitrogen, phosphorus, oxygen, hydrogen, and sulphur. These elements in biological systems are circulated through distinct biogeochemical cycles; however, the integrated cycles in parts also exist. Some common biogeochemical cycles that have ecological significance are the carbon cycle, nitrogen cycle, sulphur cycle, and phosphorus cycle. The roles of microorganisms in these cycles are central. For example, microbe-catalysing reactions in the nitrogen cycle are crucial for nitrogen fixation, nitrification, ammonification, denitrification, and mineralisation. Microorganisms crucial in the carbon cycle stretch biochemical pathways of respiration, photosynthesis, and decomposition. The multitude of microbial activities of biogeochemical cycles derive functional benefits for agriculture, bioremediation, bio-mining, wastewater treatment, etc. Also, the biogeochemical cycles are critical to estimate the impact of human activities. Unit 7 provides a detailed account of a general view of biogeochemical cycles and the roles of microorganisms on different cycles (including carbon, nitrogen, and sulphur).

2.10 DECOMPOSITION

Many natural microorganisms upcycle carbon wastes into simple and useful products through decomposition. The organisms involved in this process are called decomposers. This is a slow natural process, but essential to recycle carbon

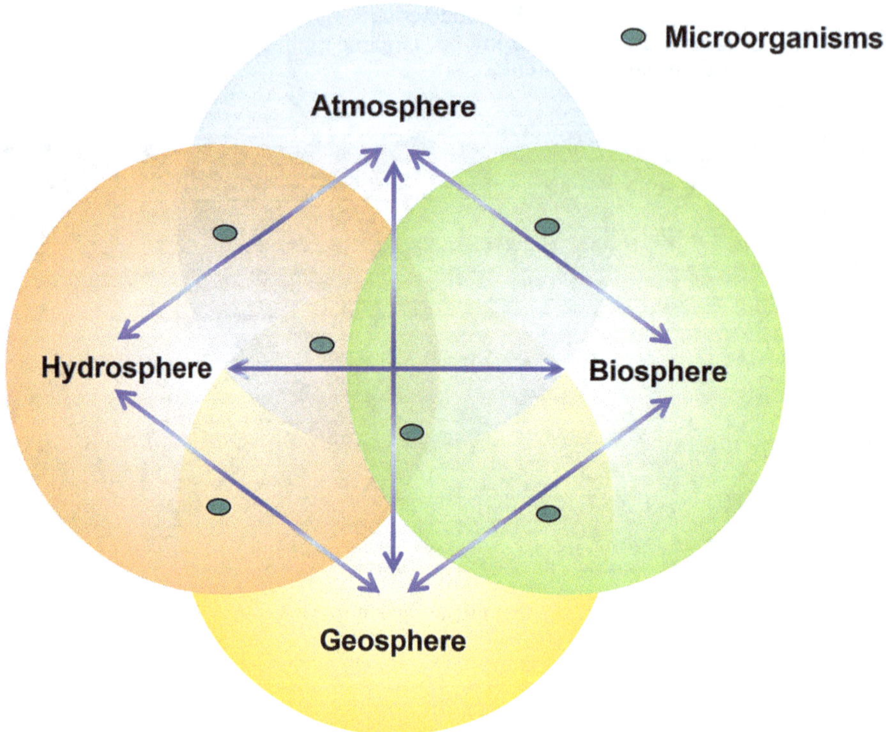

FIGURE 2.8 An illustration of the flow of chemicals in different environmental compartments involving diverse microbial activities.

and other nutrients. The organic matter decomposition can also be influenced by several environmental factors. The process proceeds through a breakdown of complex organic matter into simple organic matter by a certain group of microorganisms. The products of this reaction can further be degraded into simpler compounds by other groups of microorganisms. This process can be repeated until they reach a final step called mineralisation, where a simple organic compound finally releases its carbon as carbon dioxide (CO_2) in the atmosphere.

Biological decomposition can be an aerobic or anaerobic process, depending on the O_2 requirements. Aerobic decomposition is common in topsoil and shallow aquatic bodies. Many soil fungi are saprotrophs and are the primary consumers of organic matter. Heterotrophic aerobic bacteria possess a variety of enzymes to break down complex compounds into simple organic substances. Protozoans feed on organic matter, including bacteria and other microorganisms. They digest the organic components and release them into simple forms, contributing to soil health. Decomposition in anaerobic conditions (e.g., flood plains, aquifers, wetlands, deep seas, and compact soil) results in CH_4 production. The representative microorganisms are called methanogens, belonging to a different domain of life, Archaea. A different type of microorganism, methanotrophs, facilitates a reaction that transforms CH_4 into CO_2 (Figure 2.9).

The decomposition process is one of the initial activities in **detritus food chains**, where detritivores and decomposers act upon the detritus (i.e., dead and decaying organic matter) and provide support for the growth of organisms at trophic levels. The microbial colonisation of detritus and their feeding on dead organisms helps clean the environment, reintroduces vital elements, and establishes a new community.

Microbiologically influenced decomposition plays a critical role in composting, anaerobic digestion, and producing organic fertilisers for agriculture. Composting typically speeds up the natural aerobic decomposition process under a given set of environments. Decomposers convert agricultural wastes into bio-manure and make nutrients available to the soil. The resulting decomposed matter is called **black gold** for gardening and horticulture practices. The decomposed matter not only improves soil health and lessens erosion but also reduces CH_4 emissions.

2.11 SOIL HEALTH

Microorganisms are essential to maintain the fertility and quality of the soil. They are more abundant near plant roots (i.e., rhizosphere) than in the bulk soil. The bulk soil is composed of a highly heterogeneous microbiome with diverse functional traits, but has relatively low microbial density. In contrast, the microbial community assembly in the rhizosphere is less heterogeneous but highly dense. Many microbes residing in the rhizosphere can be representatives of bulk soil microbial reservoirs selected by plant recruitment. However, the chemo diversity in the soil due to the chemicals released from plant roots and litter attracts a wide array of microorganisms from diverse environments.

AM fungi make the soil porous and enable the nutrients available for plant acquisition. Many microorganisms live in the vicinity of plant roots and solubilise the nutrients

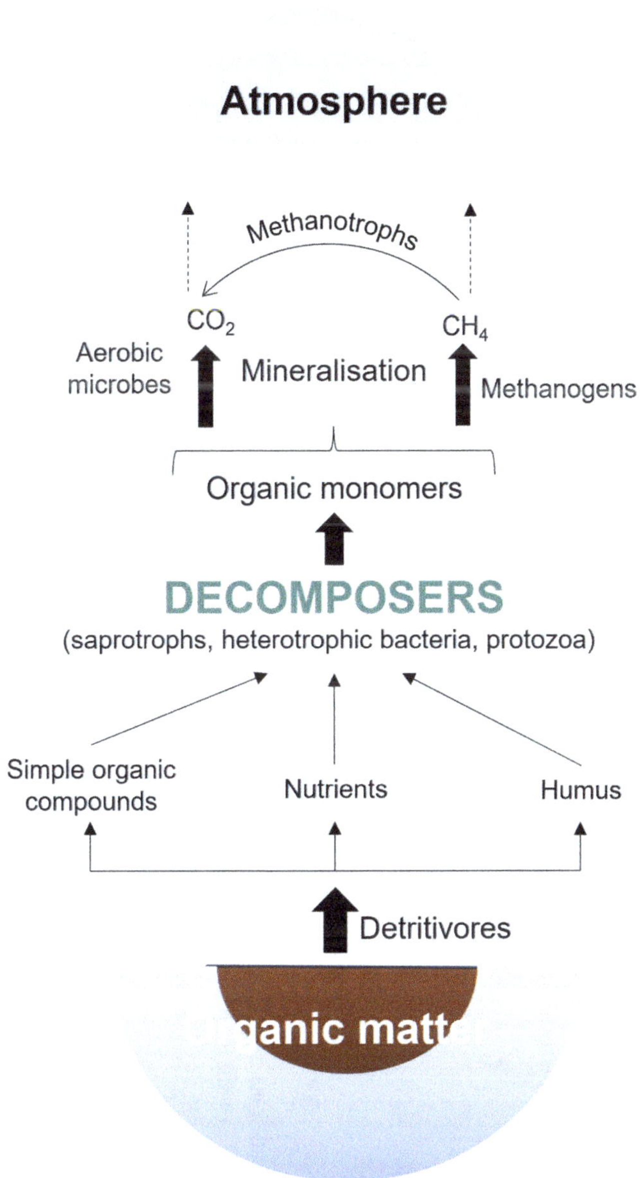

FIGURE 2.9 Mechanisms of organic matter decomposition. The biodegradation involves different types of microorganisms (called decomposers) at different steps of decomposition.

and transport them for plant growth. The high density of microorganisms and their association with plant roots in soil improves biological activities in soil, which improves moisture content, aeration, CO_2 concentration, nutrient content, porosity, macroaggregation, and soil texture (Figure 2.10). Most colonisers of healthy soil possess traits related to the nitrogen cycle, carbon cycle, and other biogeochemical cycles. These properties make soil highly fertile and resourceful. The bioavailable nutrients attract microorganisms and improve soil health. Some soil microorganisms secrete exopolysaccharides that serve as glue. This metabolite facilitates binding between soil particles and creates large aggregates. A well-aggregated soil has many benefits, such as improved water infiltration, enhanced nutrient availability, and increased soil aeration. A different useful property of soil microorganisms is the detoxification of environmental contaminants such as heavy metals, radionuclides, and organic contaminants.

2.12 PLANT GROWTH PROMOTION

The plant body is home to numerous microorganisms, and this constitutes a plant microbiome. Microorganisms colonise the above-ground and below-ground plant parts. In the above-ground region, plant parts such as stems, leaves, flowers, and fruits are distinct residences of microorganisms and are termed the **phyllosphere**. A different group of microorganisms abode near the roots below ground in the **rhizosphere**. The microbial residents of the rhizosphere are called **rhizobacteria**. Microorganisms that colonise inside the plant body are termed **endophytes** (Figure 2.11).

Microorganisms living around plants offer both beneficial and harmful activities. Most beneficial activities of microorganisms for plants are carried out by bacteria, whereas most plant pathogens belong to fungi. A distinct group of bacteria found in the rhizosphere owns plant-growth-promoting traits, known as **plant-growth-promoting rhizobacteria (PGPR)**. They produce various metabolites such as indole acetic acid (IAA), siderophore, ACC deaminase, and hydrogen cyanide for plant growth and development. They also enable phosphate solubilisation and many other beneficial activities. Some beneficial bacteria and fungi are even natural enemies of plant pathogens and are called **biocontrol agents**. Biological nitrogen fixation fills the void of nitrogen in microorganisms and plants. Nitrogen fixation is reported in only certain bacteria and archaea, which are collectively termed **diazotrophs**. Diazotrophs suck N_2 gas and convert it into ammonia (NH_3) using the nitrogenase enzyme under oxygen-limited conditions. These nitrogen-fixing microorganisms exist in diverse assemblages in natural habitats. For example, *Azotobacter* is a free-living nitrogen-fixing bacterium. *Azospirillum* is also free-living but can be found closely associated with grasses. A cyanobacterium, *Nostoc*, fixes N_2 in both free-living and symbiotic associations with different plant parts. *Rhizobium* and *Frankia* perform nitrogen fixation in nodulated roots of legumes and actinorhizal plants, respectively (detailed in Chapter 28). A recent report of nitrogen fixation in a red alga from the sea is an exception to the fact that nitrogen-fixing ability is restricted to prokaryotic organisms.

Mycorrhizae are a fungus-plant alliance in soil. Almost 90% of the land plants establish this association for plant growth and root development. In this association, fungal partners extract and solubilise the nutrients (e.g., nitrogen, phosphorus, and potassium) and water from the soil and transport them to plant roots through their hyphae. The returned gift of fungi from plants is the supply of organic matter for their metabolism (detailed in Chapter 19). This below-ground symbiosis develops a complex mycelial network and facilitates communication between different plants. This fungal network also protects plants against various biotic and abiotic stresses.

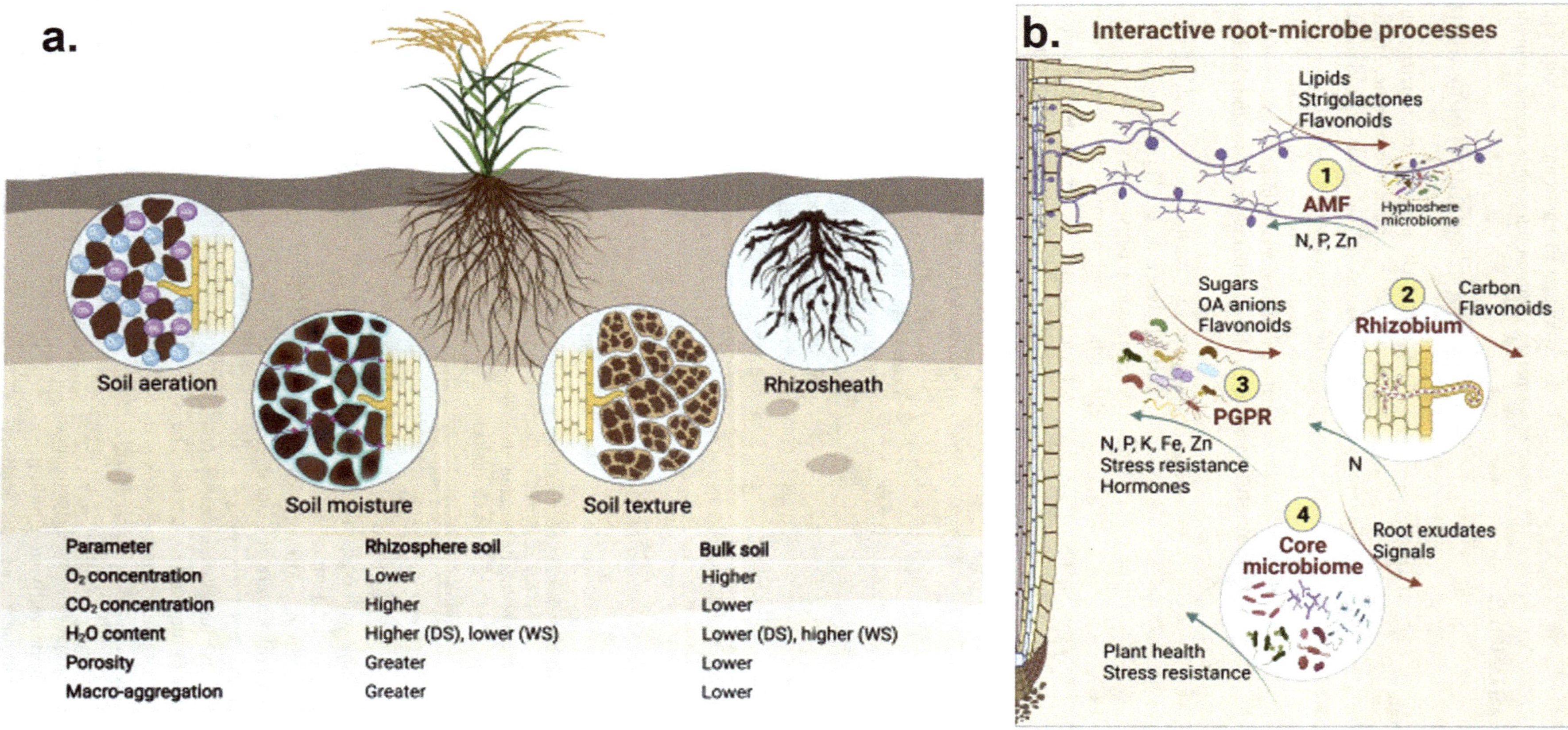

FIGURE 2.10 a. Various benefits of soil microorganisms in the rhizosphere for soil health and the key differences between rhizosphere and bulk soil. (DS is drying soil, and WS is wet soil) b. An illustration of root-microbe interaction in the rhizosphere. (AMF is arbuscular mycorrhizal fungi, and PGPR is plant-growth-promoting rhizobacteria) (Images adapted from Wang et al. (2022)[9] with permission.)

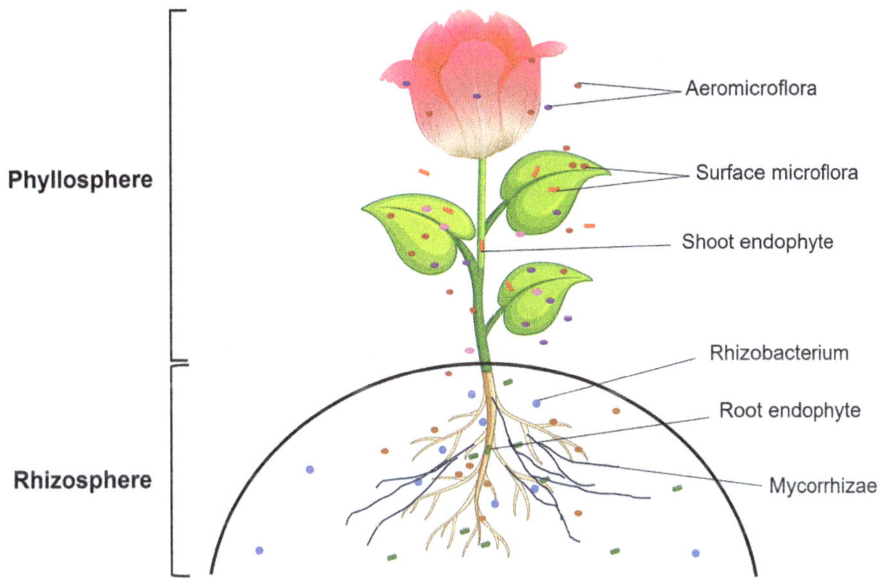

FIGURE 2.11 Illustration of plant-associated microorganisms. (Image of the plant body reproduced from Freepik/brgfx.)

PGP microorganisms are commercially available as bioinoculants/biofertilisers for agricultural applications. These preparations are promising alternatives to chemical fertilisers, which are sustainable and eco-friendly solutions to ameliorate soil health and stimulate plant growth.

2.13 CARBON CAPTURE

Carbon capture is a process of CO_2 sequestration into valuable goods, which helps in mitigating the effects of global warming. This process stores carbon in solid or dissolved forms for its long-term stabilisation. Carbon capture is a scheme to achieve net zero emissions by balancing the amount of CO_2 released from anthropogenic activities with CO_2 assimilation. It involves two basic approaches, such as biological and geological processes.

Biological processes mainly rely on photosynthesis and methanogenesis. Carbon sequestration through photosynthesis is an overdominant process which transforms CO_2 into triose and hexose carbohydrates and stores them in organic matter. The harvesting of cyanobacteria and microalgae has gained enormous attention for the same, having no requirement for arable land. These microorganisms exhibit high growth rates and a high capability to sequester carbon, which can be used as a valuable feedstock for renewable energy.

In anaerobic environments, methanogens are potential CO_2 sequesters and can be employed in biotechnology applications. For example, microbial methanogenesis can transform as much as 19% CO_2 into CH_4 in an enhanced oil recovery using CO_2. This is an *in situ* reaction that involves hydrogen gas (H_2) and CO_2— a process known as hydrogenotrophic methanogenesis. Methanogens are the sole contributors to this CO_2 loss mechanism.

Differently, an anaerobic bacterium, *Clostridium thermocellum*, possesses a unique pathway for CO_2 fixation, which allows CO_2 sequestration along with the fermentation of cellulose and produces formate. In recent years, a group of scientists engineered a carbon-negative fermentation route in an acetogen, *Clostridium autoethanogenum*, to convert CO_2 into industrially important chemicals. Its genome has been engineered to produce acetone and isopropanol.

Microbes that capture CO_2 are potential climate mitigation tools and also generate high-value products such as biofuel, food, and fertilisers and play a role in the circular economy. Engineering technology has improved microbial coping strategies to achieve added benefits. In the industrial revolution, many technological setups have shown interest in utilising CO_2 as a resource in the manufacturing of diverse goods.

2.14 MINERAL DISSOLUTION AND MINERALISATION

Natural ecosystems often tune their environments through biologically mediated redox reactions between microorganisms and minerals. These reactions are basic requirements for energy in acidophilic chemolithotrophic bacteria and archaea, which in turn aid mineral dissolution and metal recovery. Most biologically mediated reactions of mineral dissolution rely on the principle of oxidising metal oxides and metal sulphides, resulting in the mineral leaching. Bioleaching is another term for this bioprocess. The ability for mineral dissolution depends on the catalytic potential of microorganisms. One of the best-studied microorganisms for mineral dissolution is an iron-, sulphur-oxidising bacterium, *Acidithiobacillus ferrooxidans*. This bacterium survives on oxidising ferrous ions (e.g., pyrite (FeS_2)) and reduced sulphur compounds (e.g., S^{2-} and S^0). The oxidation of FeS_2 involves a complex chemical reaction and releases sulphuric acid, which helps in metal dissolution. A sulphur-oxidising bacterium, *Acidithiobacillus thiooxidans*, also has a similar role in mineral dissolution. Differently, the

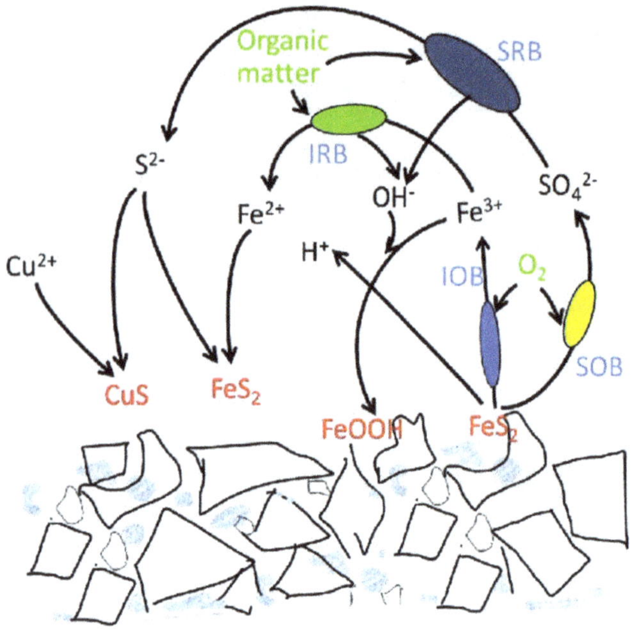

FIGURE 2.12 Dissolution of Cu (copper) involving SRB (sulphate-reducing bacteria), SOB (sulphur-oxidising bacteria), and IRB (iron-reducing bacteria). (Fe^{2+}: ferrous, S^{2-}: sulphide, SO_4^{2-}: sulphate, FeS_2: pyrite, and FeOOH: Iron(III) oxide-hydroxide) (Image reproduced from Phyo et al. (2020).[10])

consortium of mineral-degrading microorganisms comprising iron-oxidising *Leptospirillum ferrooxidans* and *L. ferriphilum*, *Acidiphilium*, and *Sulfobacillus* has collective efforts for metal bioleaching. Some archaea, including *Ferroplasma acidiphilum*, *Acidiplasma cupricumulans*, and *Sulfolobus metallicus*, also actively participate in mineral dissolution. Figure 2.12 illustrates the process of dissolution of copper (Cu) involving microbial consortia.

A contrary approach to bioleaching is biomineralisation. Biomineralisation is the process of converting soluble metal sulphates into insoluble mineral concentrates. This type of process is intentionally accelerated for wastewater treatment and solid waste management. The insoluble minerals can be precipitated, followed by their recovery through filtration or any other techniques. A different scheme of biomineralisation is microbially induced calcite precipitation (MICP), which is also known as bioprecipitation. This process creates a barrier of calcium carbonate ($CaCO_3$) around the minerals and prevents their leaching.

Bioleaching and biomineralisation methods are suitable alternatives to conventional energy-intensive methods for the recovery of elements from mineral pools, such as ore and wastewater. Bioleaching in metal recovery is a well-established commercial solution to extract metals from low-grade ores at low capital and operating costs. Gold and copper display a pronounced tendency for biomining. In recent years, bioleaching has even widened the possibility of recovering metals like cadmium (Cd), lead (Pb), zinc (Zn), nickel (Ni), rare earth elements, and radionuclides from low-grade deposits.

2.15 PLANETARY EXPLORATION

Planetary exploration is the study of planets, moons, and other celestial bodies in our solar system and beyond. The search for extraterrestrial life is one of the primary goals of planetary exploration. The key challenges for living beyond Earth include radiation, microgravity, extreme temperatures (both hot and cold), drought, and no air. These hostile situations can be found to a certain extent at different corners of Earth, which allow the inhabitation of specific microorganisms. The earth-bound extremophiles in permanently inhospitable physicochemical conditions serve as the key analogues to life on other planets (Figure 2.13). The adaptation mechanisms of extremophiles in extreme situations make them experimental subjects for the study of extraterrestrial life. This constitutes a unique field called **astromicrobiology** (**or exo microbiology**), which amalgamates astrobiology with environmental microbiology. Underlying research efforts include interest in the scrutiny of life beyond the Earth, the origin and evolution of life, and the scope for living on other planets.

A polyextremophilic bacterium, *Deinococcus radiodurans*, is distinguished by its extraordinary capacity to withstand high levels of radiation, cold, dehydration, and even vacuum. This bacterium is a favourite material for exobiology research. The iron-loving bacterium, *Tepidibacillus decaturensis*, found almost 2 km beneath the Earth's surface, is a proposed model organism for Martian life. A Gram-positive endospore-forming bacterium, Candidatus *Desulforudis audaxviator* (CDA), has incredible survival abilities and is possibly the most extreme microorganism known so far. CDA is a solo resident of 2.8 km beneath the Earth's surface and has survived millions of years of isolation. This bacterium is the most solitary organism on Earth and is known to comprise a complete ecosystem by itself. It acquires energy from the natural radioactive decay in minerals and retains the ability to manufacture all its organic matter using inorganic substances. This independently living creature survives under conditions previously considered impossible to exist, and now gives signs of life on other planets.

Mars is the prime target for the search for life beyond Earth because it shares similarities with prebiotic Earth. The extreme environments of polar deserts, alpine regions, volcanoes, submarine canyons and gullies, hydrothermal vents, salt marshes, dunes, and impact craters often give a glimpse of Mars and are considered its planetary analogue environments. Analysing these environments and their microbial inhabitants is of paramount significance in describing how life beyond Earth may look like. Sending microorganisms in spaceflight and characterising them in response to different extreme stimuli is another research interest in planetary exploration. In an investigation, the aggregated cells of *D. radiodurans* showed tolerance to very high ultraviolet (UV) radiation in space for three years. Several investigations (such as Microbial Observatory-1, Microbial Tracking-2, and BioRisk-MSV) have carried out research on microbial monitoring in space stations and found that many microorganisms exhibit active metabolism and can reproduce and produce offspring.

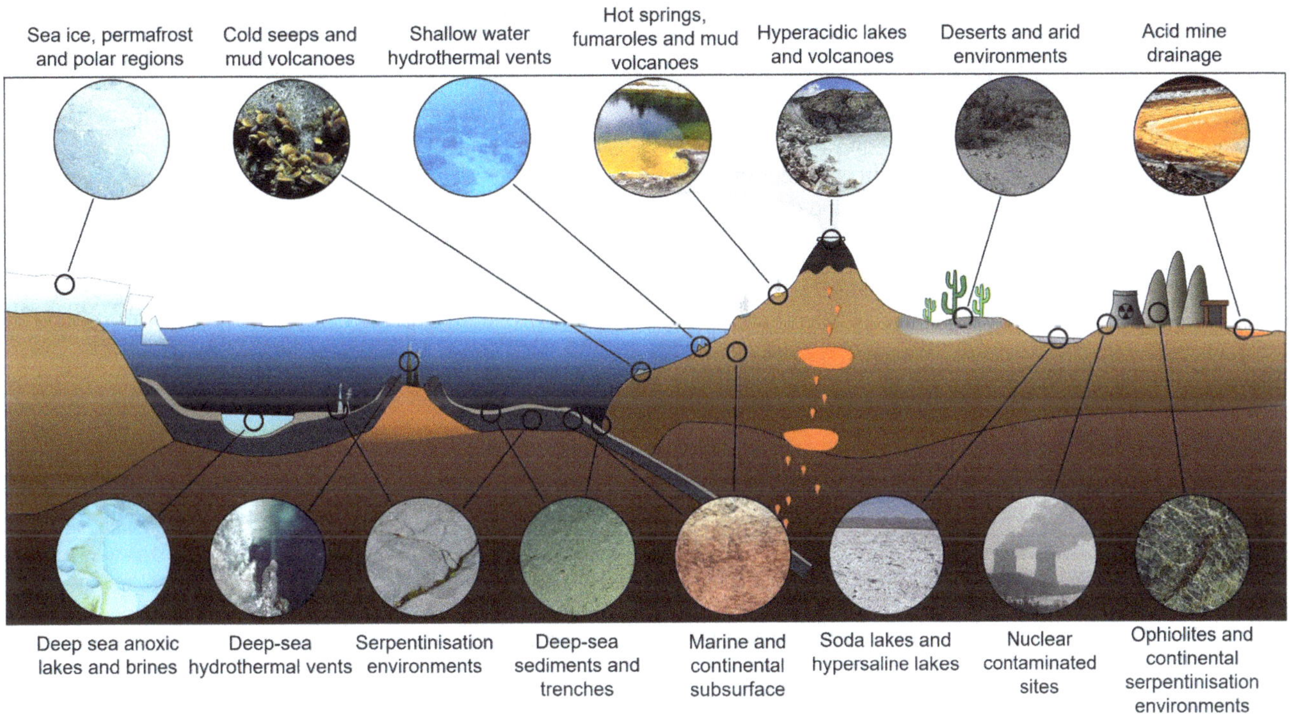

FIGURE 2.13 Various planetary analogues on Earth for the study of extraterrestrial life. (Image adapted from Merino et al. (2019).[11])

SUMMARY

- Environmental microbiology is the study of microbial communities in the environment for environmental well-being and human welfare. It has many branches, such as aeromicrobiology, soil microbiology, aquatic microbiology, wastewater microbiology, geomicrobiology, human microbiome, etc.
- In the environment, photosynthetic microorganisms yield a high quantity of O_2 gas and influence primary productivity. Phytoplankton in marine ecosystems contribute to more than half of the global O_2 supply.
- Biogenesis addresses the context related to the origin of life and evolutionary processes. Pedogenesis is the process of soil formation, where microorganisms play key roles in the development and health of soil.
- The global biogeochemical cycles explore the options for the mobilisation of diverse elements using biochemical activities. Microorganisms play a key role in mineral dissolution, mineralisation, and stabilisation.
- A variety of microorganisms, termed decomposers, feed on dead and decaying matter and are responsible for the decomposition of organic matter in the environment. A robust soil biota is the key component for a vital living ecosystem.
- Microorganisms inhabiting the aboveground/ belowground and outside/inside of a plant stimulate plant growth promotion activities for plants. They can also be pathogenic to plants and humans.
- Many indigenous microorganisms are responsible for the bioleaching of minerals in rock and ores. A distinct microbial group can sequester carbon (i.e., CO_2) into industrially relevant goods, mitigating the greenhouse effect. In planetary exploration, Earth-bound extremophiles can be the key analogues for extraterrestrial life.

DIFFERENT TYPES OF PRACTICE QUESTIONS

Fill in the blanks

I. .... is the recognised field that deals with microorganisms suspended in the atmosphere.

II. A distinct group of microorganisms permanently resides in highly extreme environments, and they are called ....

III. A .... is an assemblage of microbial cells embedded in a self-produced extracellular polymeric substance (EPS) matrix.

IV. The most common site of microbial residence in the human body is ....

V. Decomposition in .... emits methane gas.

VI. .... is the study of the scrutiny for life beyond the Earth, the origin and evolution of life, and the scope for living on other planets.

SOLUTION: I. Aeromicrobiology, II. extremophiles, III. biofilm, IV. gastrointestinal (GI) tract, V. anaerobic condition, VI. astromicrobiology/ exo microbiology

Match the following

i. Endoliths	a. Plant microorganisms
ii. LUCA	b. Detritus food chains
iii. Endophytes	c. Rock eating microorganisms
iv. Symbiogenesis	d. Common ancestral cell
v. Decomposers	e. Membrane-bound organelles

SOLUTION: i-c, ii-d, iii-a, iv-e, v-b

Match the following

i. *Dienococcus radiodurans*	a. Iron-oxidising bacterium
ii. *Chlamydomonas priscuii*	b. Halophilic green alga
iii. *Prochlorococcus* sp.	c. Psychrophilic green alga
iv. *Dunaliella salina*	d. Radioresistant bacterium
v. *Acidithiobacillus ferrooxidans*	e. Entomopathogenic fungus
vi. *Ophiocordyceps sinensis*	f. Marine cyanobacterium

SOLUTION: i-d, ii-c, iii-f, iv-b, v-a, vi-e

Very short answer type of questions

I. How does environmental microbiology cope with microbial ecology?
II. Define plastisphere.
III. How does microbiology contribute to the blue economy?
IV. What is a plant microbiome?
V. Write the names of any two industrial sectors which employ microbial consortia.
VI. What are soil engineers?
VII. What is the main role of dissimilatory reactions in biological systems?

Short answer type of questions

I. How have microorganisms pioneered life on Earth?
II. What are the roles of decomposers in the environment?
III. What are the major differences in the microbial composition of bulk soil and the rhizosphere?
IV. What are the differences between bioleaching and biomineralisation?
V. Why are oxygen production and primary production in microorganisms closely related but not entirely?
VI. What activities do bacteria play in carbon capturing to mitigate the effects of greenhouse gas effects?

Long answer type of questions

I. Write about how the microbial-driven reactions in the environment have tremendous potential for both environmental benefits and human welfare.
II. What are the different extreme situations on Earth for the colonisation of extremophiles? What are their implications for the environment? How do they play a role in the study of planetary exploration?
III. Give a detailed account of how soil microorganisms contribute towards pedogenesis and soil and plant health.

NOTES

1 Schmitt CL, Tatum ML. The Malheur National Forest Location of the World's Largest Living Organism [The Humongous Fungus]. 2008. U.S. Department of Agriculture (USDA) www.fs.usda.gov/Internet/FSE_DOCUMENTS/fsbdev3_033146.pdf

2 Du Y, Liu X, Dong X, Yin Z. A review on marine plastisphere: biodiversity, formation, and role in degradation. *Computational and Structural Biotechnology Journal*. 2022;20:975–88.

3 Engstrom CB, Yakimovich KM, Quarmby LM. Variation in snow algae blooms in the coast range of British Columbia. *Frontiers in Microbiology*. 2020;11:569.

4 Oren A. Probing saltern brines with an oxygen electrode: what can we learn about the community metabolism in hypersaline systems? *Life*. 2016;6(2):23.

5 Wang W, Wang K, Wang X, et al. Investigation on natural resources and species conservation of *Ophiocordyceps sinensis*, the famous medicinal fungus endemic to the Tibetan Plateau. Protein and Cell. 2018;9(8):671–3.

6 Zhang H, Yue P, Tong X, et al. mRNA-seq and miRNA-seq profiling analyses reveal molecular mechanisms regulating induction of fruiting body in *Ophiocordyceps sinensis*. *Scientific Reports*. 2021;11(1):12944.

7 Flombaum P, Gallegos JL, Gordillo RA, et al. Present and future global distributions of the marine Cyanobacteria *Prochlorococcus* and *Synechococcus*. *Proceedings of the National Academy of Sciences*. 2013;110(24):9824–9.

8 Triantaphyllou MV, Baumann KH, Karatsolis BT, et al. Coccolithophore community response along a natural CO_2 gradient off Methana (SW Saronikos Gulf, Greece, NE Mediterranean). *PLoS One*, 2018;13(7):e0200012.

9 Wang L, Rengel Z, Zhang K, et al. Ensuring future food security and resource sustainability: insights into the rhizosphere. *Iscience*. 2022;25(4):104168.

10 Phyo AK, Jia Y, Tan Q, et al. Competitive growth of sulfate-reducing bacteria with bioleaching acidophiles for bioremediation of heap bioleaching residue. *International Journal of Environmental Research and Public Health*. 2020;17(8):2715.

11 Merino N, Aronson HS, Bojanova DP, et al. Living at the extremes: extremophiles and the limits of life in a planetary context. *Frontiers in Microbiology*. 2019;10:780.

SUGGESTED READINGS

Ahn YJ, Lee JA, Choi KR, Bang J, Lee SY. Can microbes be harnessed to reduce atmospheric loads of greenhouse gases? *Environmental Microbiology*. 2023;25(1):17–25.

Azam F, Malfatti F. Microbial structuring of marine ecosystems. *Nature Reviews Microbiology*. 2007;5(10):782–91.

Backer R, Rokem JS, Ilangumaran G, Lamont J, Praslickova D, Ricci E, Subramanian S, Smith DL. Plant growth-promoting rhizobacteria: context, mechanisms of action, and roadmap to commercialization of biostimulants for sustainable agriculture. *Frontiers in Plant Science*. 2018;9:1473.

Brune KD, Bayer TS. Engineering microbial consortia to enhance biomining and bioremediation. *Frontiers in Microbiology*. 2012;3:203.

Cockell CS. Geomicrobiology beyond Earth: microbe–mineral interactions in space exploration and settlement. *Trends in Microbiology*. 2010;18(7):308–14.

Editorial. Microbiology by numbers. *Nature Reviews Microbiology*. 2011;9:628.

Flemming HC, Wingender J, Szewzyk U, Steinberg P, Rice SA, Kjelleberg S. Biofilms: an emergent form of bacterial life. *Nature Reviews Microbiology*. 2016;14(9):563–75.

Kaviya N, Upadhayay VK, Singh J, Khan A, Panwar M, Singh AV. Role of microorganisms in soil genesis and functions. In Varma, A., Choudhary, D. (eds). *Mycorrhizosphere and Pedogenesis*. Springer, 2019:25–52.

Lugtenberg B, Kamilova F. Plant-growth-promoting rhizobacteria. *Annual Review of Microbiology*. 2009;63:541–56.

Macalady JL, Hamilton TL, Grettenberger CL, Jones DS, Tsao LE, Burgos WD. Energy, ecology and the distribution of microbial life. *Philosophical Transactions of the Royal Society B: Biological Sciences*. 2013;368(1622):20120383.

Merino N, Aronson HS, Bojanova DP, Feyhl-Buska J, Wong ML, Zhang S, Giovannelli D. Living at the extremes: extremophiles and the limits of life in a planetary context. *Frontiers in Microbiology*. 2019;10:780.

Munn CB. *Marine microbiology: ecology & applications*. CRC Press; 2019.

NOAA 2023. *How much oxygen comes from the ocean?* www.oceanservice.noaa.gov/facts/ocean-oxygen.html

O'Malley MA, Parke EC. Philosophy of microbiology. In Zalta, EN (ed). *The stanford encyclopedia of philosophy* (Fall 2020 Edition), Department of Philosophy, Stanford University; 2020. https://plato.stanford.edu/archives/fall2020/entries/microbiology/.

Paul E, Frey S, editors. *Soil microbiology, ecology and biochemistry*. Elsevier; 2023.

Schulz S, Brankatschk R, Dümig A, Kögel-Knabner I, Schloter M, Zeyer J. The role of microorganisms at different stages of ecosystem development for soil formation. *Biogeosciences*. 2013;10(6):3983–96.

Sender R, Fuchs S, Milo R. Revised estimates for the number of human and bacteria cells in the body. *PLoS Biology*. 2016;14(8):e1002533.

Tyne RL, Barry PH, Lawson M, Byrne DJ, Warr O, Xie H, Hillegonds DJ, Formolo M, Summers ZM, Skinner B, Eiler JM. Rapid microbial methanogenesis during CO_2 storage in hydrocarbon reservoirs. *Nature*. 2021;600(7890):670–4.

West SA, Diggle SP, Buckling A, Gardner A, Griffin AS. The social lives of microbes. *Annual Review of Ecology, Evolution, and Systematics*. 2007;38:53–77.

Zahed MA, Movahed E, Khodayari A, Zanganeh S, Badamaki M. Biotechnology for carbon capture and fixation: *Critical Review and Future Directions. Journal of Environmental Management*. 2021;293:112830.

3 Microbial Research and Industrial Biotechnology

3.1 ACADEMIC DISCIPLINES

Microbiology is a dynamic and ever-growing discipline of natural sciences. Academia has played a pivotal role in providing a wealth of information on microbial aspects and fostering professionals in microbiology. Microbial studies are integral parts of education and research in almost every aspect of biological sciences. Microorganisms are highly diverse and include organisms from all three domains of life. Microbiology can be classified into several sub-disciplines, such as bacteriology, mycology, phycology, virology, protistology, and lichenology. Based on functional aspects, microbiology can be categorised as industrial microbiology, environmental microbiology, medical microbiology, agricultural microbiology, food microbiology, wastewater microbiology, and veterinary microbiology. These categories of microbiology have many subcategories. For example, in environmental microbiology, the microbiology of different ecosystems has many subcategories, such as aeromicrobiology, terrestrial microbiology, aquatic microbiology, desert microbiology, etc. (detailed in Chapter 2).

Regardless of the fields in biology, such as botany, zoology, parasitology, and agricultural sciences, microbiological concepts are crucial components in their course curricula. Even many advanced fields of biology, such as molecular biology, genetics, biochemistry, biotechnology, bioinformatics, etc., cannot be understood without integrating microbiological elements (Table 3.1).

The academic discipline of microbiology offers three primary functions: educating and training students, conducting fundamental and applied research, and offering expert support to a broader community. An effective education in microbiology necessitates a balance between educational resources and hands-on laboratory experiences. The hands-on training is vital and heavily depends on instrumentation and techniques. Cultivation practices are the key to the classic microbiological study. The proper verification of taxonomic and functional features of microorganisms requires upskilling in biochemical testing and molecular analysis. Microscopic studies provide valuable insights into their structural and functional details. Furthermore, microbiological training equips individuals with the knowledge and skills to address the protocols in food safety, water quality, and human health testing related to microbial parameters.

A simple microbiology laboratory typically consists of at least an autoclave, a laminar flow, and an incubator. Other basic requirements are culture media, glassware (e.g., Petri plates, flasks, spreaders, and measuring cylinders), spirit, cotton plugs, etc. Molecular analysis requires a thermal cycler, electrophoresis gel, sequencing facility, and bioinformatics analysis. These facilities are also essential for healthcare systems, food industries, and environmental sectors. The study of uncultured microorganisms in the environment has received wide attention. This has involved a lot of effort in metagenomics analysis through molecular studies. The introduction of new methods and cutting-edge insights in microbiology has paved the way for many groundbreaking discoveries. This has also contributed to widening industrial setups and environmental engineering. Some important instruments used in the fermentation industry and microbial research are mentioned in Chapters 6, 7, and 21.

Microorganisms inhabit all types of ecosystems and biological organisms. They establish close associations for various biological activities. Also, an organism's fitness is greatly influenced by its microbiota. Modern biological studies have adopted a holistic approach to research investigation because no organism can be considered an autonomous entity. A host organism, along with its microorganisms, collectively makes up a holobiont, and its study involves integrated research. In academia, microbiology has set a landmark in biological sciences, providing a wealth of information related to ecological processes and human health.

3.2 RESEARCH ACTIVITIES

Microbial technology is seen as one of the biggest opportunities of the 21st century. It offers a green solution to environmental issues. This new world perceives sustainability in agriculture, energy products, and environmental management through genetic manipulation in microorganisms. Humans have advanced their living standards based on microbially-driven reactions. The domain of pure and applied science research has immensely contributed to producing the fundamental corpus of knowledge on various microbial processes. Conventional microbiological research has risen above traditional culturing practices in the modern world. The integrated perspective has become a new requirement of microbial research. Technological advancements have generated new methods to

DOI: 10.1201/9781032651095-4

TABLE 3.1
Some relevant fields that integrate microbiology elements in academia

Based on functional aspects	Based on microbial groups	In basic biological sciences	In advanced courses
Industrial Microbiology	Bacteriology	Botany	Biotechnology and Bioengineering
Environmental Microbiology	Virology	Zoology	Molecular biology
Medical Microbiology	Mycology	Ecology	Genetics
Agriculture Microbiology	Phycology	Parasitology	Cell Biology
Food Microbiology	Protistology/protozoology	Food Science	Biochemistry
Wastewater Microbiology	Lichenology	Environmental Science	Bioinformatics
Veterinary Microbiology		Forensic Science	Biophysics
			Nanotechnology

tap the microbial potential. Molecular methods are the latest tools for the scrutiny of microbial regimes, which have not been obvious through classical methods. The previous two chapters (Chapters 1 and 2) have provided a comprehensive overview of the scopes of microbiology under various regimes.

Humans have tried to control the impressions of microorganisms in many aspects. However, the reverse seems more convincing for larger responsibilities. An estimate suggests that about 10^{12} microbial species are living on this planet. Only roughly 1–5% of total microbial species can be artificially examined for biotechnology. Only a limited number have been slightly manipulated by research tools. Many technological breakthroughs have now shed a deeper understanding of microbial diversity and its potential. This new information has made us realise where we stand.

In recent years, omics technologies have stepped stones in biology and entered mainstream research and technology to explore the roles, relationships and actions of biological systems at the molecular level. Based on molecular types, omics technologies have many variations, including genomics, transcriptomics, proteomics, and metagenomics. They involve high-dimensional biology, providing a comprehensive view of the molecules to understand the biological structure and function. The revolution in "omics" technology has provided crucial insight into microbial diversity and function. The omics-based bioengineering is, as ever, in cooperation with system biology for microbial manipulation.

In genetic engineering, a microorganism is a raw material which can be transformed into a robust system using restriction enzymes as biological scissors and ligases as glues. Combinatorial biosynthesis is an approach to reengineering biosynthetic pathways to control or alter natural products. Synthetic biology grafts new abilities in microorganisms using engineering principles. Metabolic engineering has been employed to improve the production of metabolites. Protein engineering is associated with modifications in protein structure and functions for valuable applications. New genetic traits that support a broad range of distinctive capabilities have turned out to be more relevant than previously anticipated. Computational biology is set for mathematical modelling and simulations to identify new biological phenomena.

The modern microbiology framework offers easy and fast access to research knowledge for product development and commercialisation. The multitude of research institutes, laboratories, and industries are charged with microbial impute. Microbiologists have received grants to address queries in microbiology, spanning from ecological to economic points of view. Some well-known research institutes that are focused on microbiological research are:

- Alimentary Pharmabiotic Centre (APC) Microbiome Ireland, Ireland
- Australian Wine Research Institute (AWRI), Australia
- Centre for Electromicrobiology, Aarhus University, Denmark
- Centre for Microbial Ecology and Technology (CMET), Belgium
- Centre for Marine Microbiology and Genetics Research (CMMG), Australia
- Centre for Microbial Oceanography: Research and Education (C-MORE), Hawaii
- Centre for Microbiology and Environmental Systems Science (CMESS), Austria
- Centre of Excellence for Plant and Microbial Sciences (CEPAMS), (a collaboration between John Innes Centre, United Kingdom (UK) and the Chinese Academy of Sciences (CAS), China)
- Collaborative Research Institute for Innovative Microbiology, Japan
- Institute of Microbial Technology (IMTech), India
- Institute of Microbiology, ETH Zurich, Switzerland
- Institute of Microbiology of the Chinese Academy of Sciences (IMCAS), China
- Leibniz Institute, Deutsche Sammlung von Mikroorganismen und Zellkulturen GmbH (DSMZ) (German Collection of Microorganisms and Cell Cultures), Germany
- Max Planck Institute for Marine Microbiology (MPIMM), Germany
- Max Planck Institute for Terrestrial Microbiology (MPI-TM), Germany
- National Institute of Molecular Biology and Biotechnology (NIMBB), Philippines

- National Microbiology Laboratory (NML), India
- National Oceanic and Atmospheric Administration (NOAA), United States of America (USA)
- Pasteur Institute, France
- Swiss National Centre of Competence in Research (NCCR) Microbiomes, Switzerland
- United States Department of Energy (US DoE), USA
- Waksman Institute of Microbiology, USA
- Waterloo Centre for Microbial Research (WCMR), Canada
- Westerdijk Fungal Biodiversity Institute (WFBI), Netherlands

3.3 PROFESSIONAL SOCIETIES

Professional societies in microbiology are often recognised as non-profit associations of professionals to support and promote microbiological research and knowledge. The primary duties of societies include advancing microbiology and directing its development. These organisations advocate for high integrity in scientific research. They also aim to enhance communication in microbial studies. Many societies inherit journals, textbooks, edited books, and magazines to expand and strengthen knowledge. Other noble works include promoting public awareness through science fairs, exhibitions, public outreach campaigns, and social movements. Besides, they play prominent roles in organising conferences, workshops, seminars, and annual meetings.

The members of the societies are professionals from relevant fields, including researchers, educators, policymakers, and health practitioners. The elected representatives of societies form a governing body and build several committees with different aims and objectives. The strategic framework of the societies is directed through office bearers who guide and empower their members through various activities. The organisational structure offers many professional and networking opportunities for its members. Many societies recognise scientific achievements and honour their contributors through awards and prizes. Some renowned societies also call for research proposals and offer funding. Other assets of their cooperation may include providing grants for research training, meeting attendance, etc. A major portion of these funding needs is met by requesting yearly membership charges from their members.

Some of the world's leading professional societies in microbiology are:

- American Society for Microbiology (ASM)- This society was set up in 1899 as the Society of American Bacteriologists. It was renamed the American Society for Microbiology (ASM) in 1960 to cover broad aspects of microbial studies. ASM is recognised as one of the world's oldest microbiological societies. It has been instrumental in developing other biological membership organisations.
- International Union of Microbiological Societies (IUMS)- The foundation of the International Union of Microbiological Societies (IUMS) was laid as the International Society of Microbiology in 1927. IUMS is a member of the International Science Council (ISC). The association has three main divisions, namely the Bacteriology & Applied Microbiology Division (BAM), the Mycology & Eukaryotic Microbiology Division (MEM), and the Virology Division. IUMS claims to maintain contacts and exchange knowledge with different national and regional scientific organisations across the globe. IUMS makes a significant contribution to microbial systematics.
- Microbiological Society- The Microbiological Society was established in 1945 in the UK. Sir Alexander Fleming was the first elected president. It is one of the largest societies in Europe and offers memberships across the globe.
- Federation of European Microbiological Society (FEMS)- It is another European platform of microbiological experts, founded in 1974. FEMS works in collaboration with many international and national societies of the related fields, with over 50 different groups. Most of the member societies come from various parts of Europe.
- Society for Applied Microbiology (SfAM)- Set up in 1931, the Society for Applied Microbiology (SfAM) is known as the oldest microbiology society in the UK. The core vision of society is to promote the improvement of microbial use for biotechnological applications.
- International Society for Microbial Ecology (ISME)- Formerly, it was established as a committee in 1970 called the International Commission for Microbial Ecology (ICME). In 1998, the committee was registered as a society and named the International Society for Microbial Ecology (ISME), with its head office in Geneva, Switzerland. ISME is a world-leading organisation in the field of microbial ecology and organises symposia on a wide range of topics every two years.

The list of microbiology societies is very long. They also exist at national, regional, and Institute levels. Some other notable mentions are:

- Society for Industrial Microbiology and Biotechnology (SIMB), USA
- Federation of Microbiological Societies of Japan (FMSJ), Japan
- Japanese Society of Food Microbiology (JSFM), Japan
- Australian Society for Microbiology (AuSM), Australia
- Canadian Society of Microbiologists (CSM), Canada
- Microbiological Society of Korea (MSK), South Korea
- Swiss Society for Microbiology (SSM), Switzerland
- Association of Microbiologists of India (AMI), India
- New Zealand Microbiology Network (NZMN), New Zealand

- International Scientific Association for Probiotics and Prebiotics (ISAPP), USA
- Portuguese Society of Microbiology/Sociedade Portuguesa de Microbiologia (SPM), Portugal- This society has received wide recognition in the last few years for its initiative to celebrate International Microorganism Day on 16th September (Box 3.1)

Some societies are theme-oriented and reserve their interest in selected topics such as mycology, probiotics, virology, medical microbiology, etc.
Examples of some learned societies of mycologists are:

- British Mycological Society (BMS)- It was established in 1896 and is possibly the most ancient society in microbial science.

BOX 3.1 INTERNATIONAL MICROORGANISMS DAY (IMD)

International Microorganisms Day (IMD) is a recent initiative to celebrate the positive aspects of microorganisms on the 17th of September. This date coincides with the day in 1683 when Antonie van Leeuwenhoek (the Father of Microbiology) sent his first letter to the Royal Society with the first description of the unseen world of microorganisms. Although this information hardly made sense to the authorities, the evidence, anyway, laid the foundation stone for a new science, microbiology.

IMD is an enterprise of the Portuguese Society of Microbiology and includes many fascinating features (Box Figure 3.1.1). It began in 2017 and has received support from many other parts of the world. The main goals of celebrating this day are to raise awareness and understanding of the critical functions that microorganisms play in the environment and health. The observation of this day supports the notion of combined microbiology-related activities worldwide. Microbiology-literate professionals at different locales have taken the initiative to mark the celebration differently. Events like lectures, seminars, workshops, exhibitions, and talent hunts on microbial themes have laid the groundwork for showcasing the diversity and importance of microorganisms for the general public and people from the scientific community. Such events have also struck the attention of educators, decision-makers in government and non-government organisations, business executives, and other influential groups.

A related but distinct event, World Microbiome Day, is celebrated on the 26th of June every year and marks the representation of microbiome research. This venture was started by APC Microbiome Ireland in 2018.

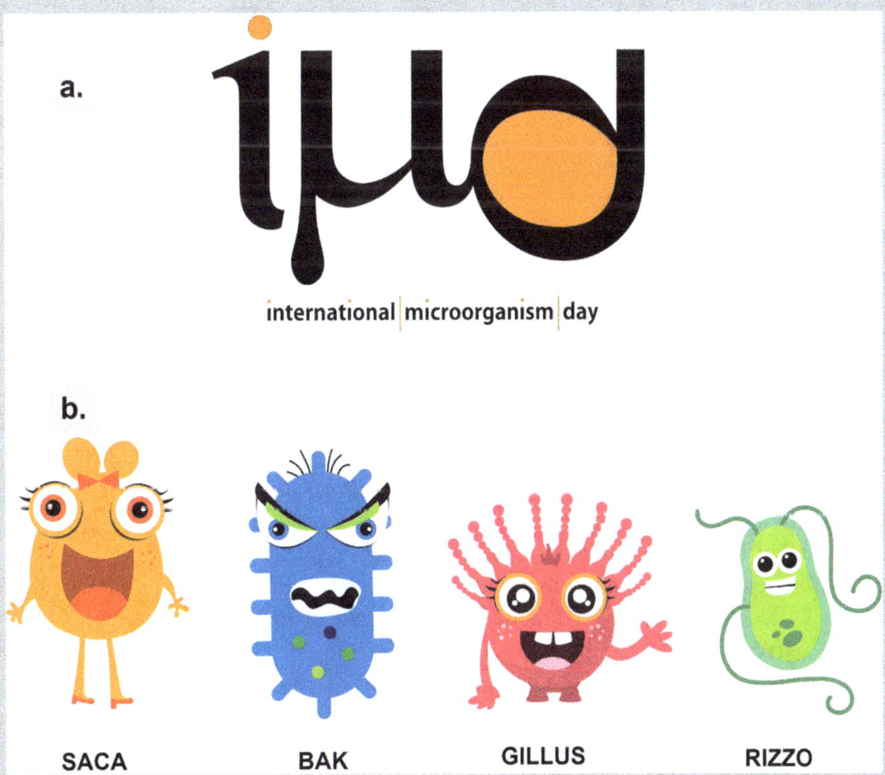

BOX FIGURE 3.1.1 a. Logo of #InternationalMicroorganismDay. b. Different mascots. (SACA, BAK, GILLUS, and RIZZO represent *Saccharomyces* yeasts, *Bacillus* genus, *Aspergillus* fungi, and rhizobia bacteria, respectively.) (Images reproduced from www.internationalmicroorganismday.org/; SACA, BAK, and GILLUS © Musse Ecodesign and RIZZO © @eliza_coli.)

- Mycological Society of America (MSA)
- International Society for Human and Animal Mycology (ISHAM)
- Mycological Society of Japan (MSJ)
- Mycological Society of India (MJI)

Some notable associations of virologists are:

- World Society for Virology (WSV)
- European Society for Clinical Virology (ESCV)
- American Society for Virology (ASV)
- International Society for Antiviral Research (ISAR)

3.4 MICROBIOME PROJECTS

Microbiome projects are large-scale scientific efforts aiming to provide accurate sequence information of entire microorganisms in distinct ecosystems. They have offered multifaceted research on microbial community analysis involving interdisciplinary efforts. They have built the most comprehensive baseline datasets for multiscale microbial diversity. The independent surveys of sites at an unprecedented scale and common standardised protocols are the key facets to meet the requirements of microbiome projects. Metagenomics and whole genome sequencing (WGS) are the two most widely used techniques to determine microbial compositions. Bacterial, archaeal, and eukaryotic microbial groups have all been subjects to close examination in microbiome projects.

These projects have broken many barriers in microbial research to provide a thorough understanding of ecological trends. The Earth Microbiome Project (EMP), the Human Microbiome Project (HMP), the International Census of Marine Microbes (ICoMM), the Metagenomics of the Human Intestinal Tract (MetaHIT), and the Chinese Academy of Sciences Initiative of Microbiome (CAS-CMI) are some well-known microbiome projects. They have taken a 5–10-year timeframe to complete their research targets. They received funding from various sources. Their research outputs are available through open-access publications and databases for further comparisons and cross-study meta-analyses. The fundamental research outputs are also available as baseline datasets for expanding scientific investigation across various disciplines of communal ecology, nutritional sciences, biogeochemistry, etc.

EMP is a collaborative research effort to build a microbial map for Earth (www.earthmicrobiome.org/). Launched in 2010, the project was fueled by crowd-sourced samples representing diverse environmental types and geographic locations. In the first phase of EMP, a total of 27,751 samples from 97 different studies arrived from 43 countries of all seven continents. The standard protocols of DNA sequencing and mass spectrometry involved uniform processing of the samples and produced a global Gene Atlas. The main repertoires of EMP are 16S rRNA and 18S rRNA amplicon data. The second phase of the EMP aimed to expand the research on coupled metagenomic and metabolomic profiling of fresh 500 individual samples.

A distinct project is HMP, launched by the US National Institutes of Health (NIH), which built a global catalogue of microorganisms for the human body (www.hmpdacc.org/hmp/). HMP has completed its two phases: HMP1 (2007–2014) and Integrative Human Microbiome Project (iHMP) (2014–2016). HMP1 included about 5,000 samples of different body parts from 300 healthy adults, including men and women. The primary vision of the HMP1 was the scrutiny of microbial taxonomic and functional diversity that lived inside the body and on the surface. iHMP characterised the human microbiota and improved our knowledge of how the microbiome affects human health and disease. Several omics technologies, such as metatranscriptomics, metabolomics, and IgA-sequencing, were employed to study microbiome-associated conditions.

ICoMM is a global initiative to evaluate the diversity, distribution, and abundance of marine microbial life (www.coml.org/international-census-marine-microbes-icomm/). This field project was started by the Census of Marine Life in 2010 and took ten years to complete. About 1,700 researchers from 70 different nations participated in the ICoMM programme. Over 1,200 different samples were included in the sequence analysis. The most intriguing finding of the project was that the microbial diversity of marine ecosystems surpassed previous estimates by one or two orders of magnitude. The investigators found that some microbial species accounted for widespread distribution, while some represented the low abundance at distinct locations, called "**the rare biosphere**". The rare biospheres have been anticipated for their key roles in maintaining microbial diversity and regulating ecological processes in the marine environment.

3.5 OUTSOURCED FACILITIES

Modern microbiological investigation stands on the premise of sophisticated instruments. The setup of these facilities craves high capital costs, infrastructure, handling support, and enough space. Also, the latest instrument facility requires upgrades and maintenance to generate high-standard data. Therefore, the maintenance of such facilities by research laboratories is out of the scope. These requirements in research fields are mainly outsourced by the centralised facility. Different government and nongovernment agencies have developed centres for various technical services, where research laboratories can establish collaboration with them or use their services on a payment basis.

The most common demand for outsourced facilities in microbiological research is molecular biology tools and techniques, and pilot-scale fermentation facilities. Many molecular studies require expensive and sophisticated equipment for the analysis and manipulation of biological molecules. The central facilities provide a broad range of services, from the extraction of nucleic acids to sequencing and data delivery, including genome analysis. Sequencing technology has gone through many routes of innovation and is still a part of the outsourced facility. Sanger sequencing set foot in the late 20th century and shifted the focus of microbial research toward molecular sequence analysis. The sequencing technology was replaced by the high-throughput technology (also known as next-generation sequencing (NGS)). It includes different variants: pyrosequencing and Illumina sequencing. Especially the latter techniques brought the "genomic revolution" and unlocked the genetic potential of microorganisms. A new arrival in the sequencing facility is a portable benchtop sequencer called nanopore technology, which has partially withdrawn the job services of third parties. However, this technique is still in the developmental stage and is not ready to replace the existing techniques. Other important outsourced services in molecular biology and genetic engineering are probe synthesis, labelling for nucleic acids, and detection by hybridisation. Protein sequencing is also an important part of molecular analysis, which mainly relies on mass spectrometry-based methods.

In fermentation projects, scaling up fermentation processes from laboratory to pilot scale frequently necessitates the use of outsourced facilities. These facilities offer various customisation and flexibility on specific requirements such as batch, fed-batch, or continuous processes. They can also assist in studying other parameters. This type of partnership bridges the gap between laboratory research and commercial production. These establishments also allow researchers and industries to reduce the need for capital investment in equipment and infrastructure for product development.

In addition, techniques like electron microscopy, bioimaging, cell sorting, mass spectrometry, nuclear magnetic resonance (NMR), chromatography, and X-ray-based analysis are also often important requirements in microbiological research and involve highly sophisticated instruments. These techniques are also procured from outsourced facilities and require skills for their operations.

3.6 RESOURCES

3.6.1 Culture Collection

Culture collection is a stock of microbial cells and cell lines in vital and pure forms, which can be revived for various purposes. These storehouses are also known as biological resource centres (BRC). The concept of culture collection emerged after Koch introduced the pure culture technique in 1881. Frantisek Král, a technician who rose to the rank of professor in 1890 at the German University of Prague (Czech Republic), was the first to preserve microorganisms and make them available on a payment basis. After he died in 1911, this practice continued and was maintained at the University of Vienna until it was destroyed during World War II. After some time, other countries like Belgium, the Netherlands, etc., also embraced the culture collection systems.

Microbial stocks in culture collections are mainly microbial cultivars recovered from various biological and environmental samples. The microbial stocks in culture collections reside along with a wealth of information regarding their sources, characteristics, and growth requirements. The major objectives of the culture collection are the long-term storage, conservation, and supply of cultures. Culture stocks are crucial bioresources for biomedical research and biotechnological applications. With new requirements, the purpose of the culture collection has expanded beyond its original need. Many culture stocks serve as model organisms and reference materials to recognise biological phenomena.

The ideal condition for long-term storage of microbial forms in culture collections is the lyophilised cells coated on porous beads stored in cryogenic vials at -70 °C. Such preparation keeps cultures pure and viable for decades. The culture can be revived using individual cell-coated beads inoculated on synthetic media. The freeze-dried aliquots can also be stored for a very long time. Certain cultures demand periodic reculturing. The upkeep of the genetically altered strains is identical to that of natural strains. However, the preservation of pathogenic strains necessitates careful observation and specialised facilities. Chapters 6 and 22 provide a thorough understanding of preservation techniques for microorganisms.

The World Federation for Culture Collection (WFCC) is a central platform for information related to microbial culture collection. It provides a framework for international cooperation among cultural collections and offers comprehensive details on services and the holding of cultures worldwide. World Data Centre for Microorganisms (WDCM) is a crucial database of available culture collections and keeps a record of each submission. It assigns a unique identification code for the submissions. These codes enable tracking of the specific strains in a database. As per the statistics of WDCM, about 3.4 million cultures of microorganisms and cell lines are registered in 861 culture collections spread across 80 nations (Figure 3.1). In culture facilities, bacterial stocks have made up over 52% of the storage and fungi, roughly 29% (Figure 3.2). Other types of microorganisms in the culture collection belong to viruses, microalgae, protozoans, and lichens.

The number of isolated microorganisms has increased at an exponential rate over the coming years, and several collection centres have been built in various locations across the globe. Universities, government/semi-government institutions, and industries are the main microbiological repositories of culture collection. These centres come in a wide range of sizes and storage capacities depending on the amenities available at the centres. Examples of some well-established centres for culture collections in the world are:

- American Type Culture Collection (ATCC), USA
- Northern Regional Research Laboratory (NRRL), USA
- Leibniz-Institut DSMZ, Germany
- Japan Collection of Microorganisms (JCM), Japan
- Belgian Coordinated Collections of Microorganisms (BCCM), Belgium
- National Collections of Industrial Food and Marine Bacteria (NCIMB), UK
- Oceanographic Institute of the University of São Paulo, Brazil

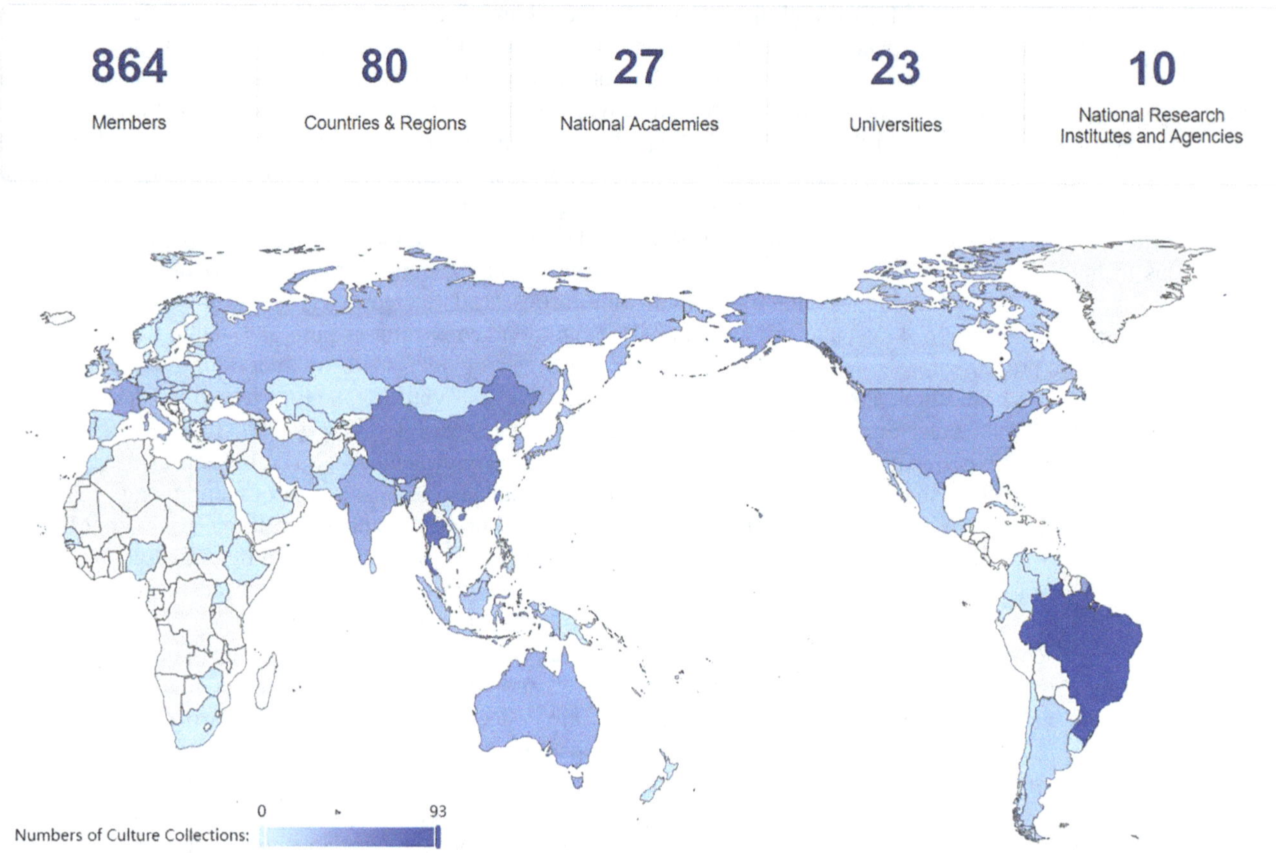

FIGURE 3.1 Number of Culture Collections in different parts of the world. (World Data Centre for Microorganisms (www.wdcm.org/) in Jan 2025.)

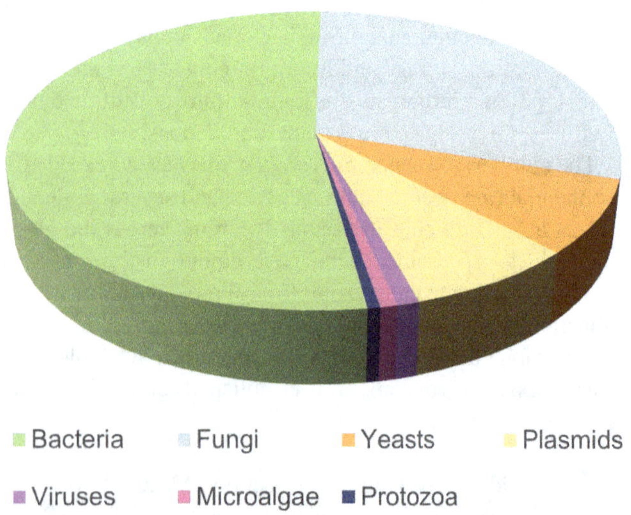

FIGURE 3.2 Proportion of different microbial groups in culture collections worldwide. As per an estimate, there are more than 4.1 million different microorganisms, including plasmids, in the storage facilities. (Data retrieved from World Data Centre for Microorganisms- Culture Collections Information Worldwide (WDCM- CCINFO) in Dec 2024.)

- Centre of General Microorganisms Culture Collection (CGMCC), China
- Collection de L'Institut Pasteur (CIP), France
- National Collection of Industrial Microorganisms (NCIM), India
- Microbial Type Culture Collection and Gene Bank (MTCC), India
- Collection de Champignons et Actinomycètes Pathogènes (UMIP), France
- National Culture Collection for Pathogens (NCCP), Korea
- SERI Microalgae Culture Collection, USA
- CNPSo, Culture Collection of Diazotrophic and Plant Growth Promoting Bacteria, Brazil
- Thailand Institute of Scientific and Technological Research (TISTR), Thailand
- Israel National Culture Collection of Algae (INCCA), Israel
- Korean Culture Centre of Microorganisms (KCCM), South Korea
- Korean Culture Collection of Aquatic Microorganisms (KoCAM), South Korea
- National Collection of Yeast Cultures (NCYC), UK

3.6.2 Databases

A wide array of research studies has generated an unprecedented amount of biological data. In the era of computer science and its application, digital libraries have offered great support to save biological information. These databases are called biological databases. They have many superior features such as high storage capacity, no physical boundary and multiple access across any part of the world. Also, they are highly structured, curated, and regularly updated. Biological databases maintain information at low maintenance costs and offer easy data retrieval. Biological databases are numerous and diverse. There are databases of scientific literature, taxonomy, culture stocks, disease, chemicals, biochemical pathways, molecular structure, etc.

The various stored information, data from microbiology-related studies, is extremely pervasive. A thorough investigation of microorganisms and their sequences has generated a massive amount of microbiological data. Some databases are specific to microbial information. For example, the National Microbiology Data Centre (NMDC), China (www.nmdc.cn), stores data on microbial sequences and structure and has obtained information on 120 culture collections in 46 countries. BacDive (www.bacdive.dsmz.de/) is a metadatabase for standardised and systematic access to bacterial and archaeal phenotypic information. Microbial Database for Activated Sludge (MiDAS) Field Guide (www.midasfieldguide.org/guide/search) disseminates information on the physiology and ecology of the native microorganisms in wastewater treatment systems through collaborative research. MycoBank (www.mycobank.org/) maintains documents for nomenclatures, descriptions, and illustrations of numerous fungi.

Sequence analysis in microbiology is growing as a basic requirement for the identification, function, and taxonomic verification of microorganisms. The basic research output of sequence analysis is sequence data. The submission of sequence data to a database is a strict requirement for research studies. This has generated enormous volumes of nucleotide and amino acid sequences in databases. The current estimate of storage is around 2 billion sequences. The main repositories of nucleotide sequence data are:

- GenBank, National Centre for Biotechnology Information (NCBI)- National Library of Medicine (NLM), National Institute of Health (NIH)
- European Nucleotide Archive (ENA), European Molecular Biology Laboratory (EMBL)- European Bioinformatics Institute (EBI)
- DNA Data Bank of Japan (DDBJ), National Institute of Genetics (NIG)

These three databases are linked to each other through the International Nucleotide Sequence Database Collaboration (INSDC).

Integrated Microbial Genomes & Microbiomes (IMG/M) (link: www.img.jgi.doe.gov/) is a very crucial web user interface for the genome browsing and annotation of microorganisms. It was developed by the US Department of Energy (DOE)- Joint Genome Institute (JGI) and has established collaboration with the NMDC and The Department of Energy Systems Biology Knowledgebase (KBase) for read-based analysis workflow and Uniprot search and ID mapping, respectively. Many software packages provide essential tools and a graphical interface for visualising and exploring genomic data. IMG/M houses a vast collection of microbial sequences and supports the analysis of genomes, metagenomes, proteins, and pathways in microorganisms. IMG/M also includes some key features, such as the Genomes OnLine Database (GOLD), which is a repository of microbial genome projects. MycoCosm (link: www.mycocosm.jgi.doe.gov/mycocosm/home) is a web fungal portal of IMG/M, which provides data access, visualisation, and analysis tools for comparative genomics of fungi. In addition, IMG/M has two specialist data marts for specific assignments: IMG/ABC (for biosynthetic gene clusters and new secondary metabolites) and IMG/VR (for viral data). Microbial genome data can also be retrieved using browsers like

- NCBI Genome- This browser reveals genome sequences of more than 700,000 prokaryotes, 37,000 eukaryotes, and 75,000 viruses.
- Ensembl Genome, EBI
- University of California, Santa Cruz (UCSC) Genome Browser

Amino acid sequences are retrieved from Protein Information Resources (PIR), EMBL-EBI, the Swiss Institute of Bioinformatics (SIB), Protein Research Foundation (PRF), and Protein Data Bank (PDB). Of various protein sequence databases, UniProtKB (Universal Protein Resource Knowledgebase or UniProt) is a user-friendly database which retrieves information on amino acid sequences, structures, and functions. GenPept is a different platform for the retrieval of amino acid sequences, which is basically a translated version of a subset of GenBank. PDB is primarily used to study the 3D structure of proteins, but it also stores amino acid sequence data side by side. A different user interface provides information on amino acid sequences by translating annotated coding regions of nucleotides from nucleotide databases.

3.6.3 Tools for Microbiological Data Analysis

The unprecedented volume of data on microbial sequences necessitates bioinformatics tools for its analysis and interpretation. Several free, open-source platforms are available for data integration and analysis. Many databases include built-in tools that enable data analysis. A classic example is gene sequence analysis through the Basic Local Alignment Search Tool (BLAST). BLAST is an integral tool of NCBI and many other databases, which aligns and compares a query nucleotide/amino acid sequence with sequences in a database to find homology. BLAST is used for the identification of unknown taxa, gene prediction, phylogenetic analysis, and sequence mapping.

Modern bioinformatics analysis integrates datasets from various sources and involves complex analysis, which requires knowledge of programming languages such as Python, R, Perl, MATLAB, and others. These languages allow researchers to increase functionality, create unique programmes, simulate complicated biological systems, and automate repetitive operations of bioinformatics tools. These languages are frequently employed in analytic pipelines to implement complex algorithms and automate repetitive tasks. Examples of software that are important for sequence analysis in microbiology include Quantitative Insights into Microbial Ecology (QIIME), QIIME 2.0, and the Microbial Genomes Atlas (MiGA). The de novo assembly is a novel technique of constructing a genome/transcriptome from short nucleotide sequences, which requires very powerful tools like MEGAHIT, MetaVelvet, IDBAUD, and metaSPAdes. In addition, some distinct bioinformatics tools procure information from metagenomic analysis, community profiling, comparative analysis, and genome annotation.

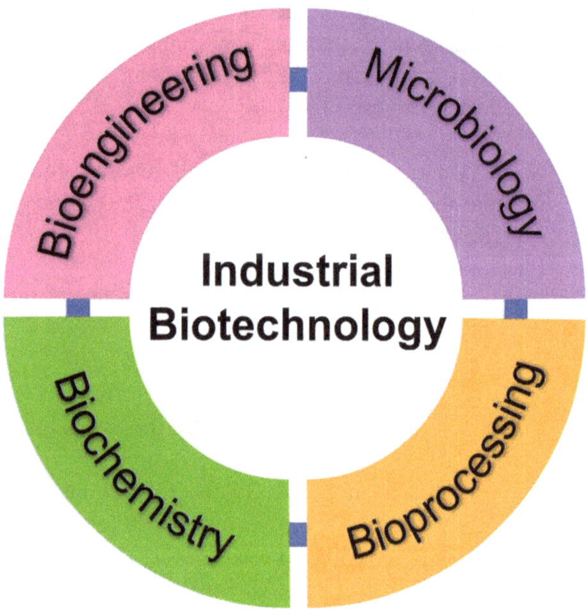

FIGURE 3.3 Different components of industrial biotechnology.

3.7 INDUSTRIAL BIOTECHNOLOGY

Industrial biotechnology is a branch of applied science that employs microorganisms in the development of various bioproducts for commercial deliveries. This field is also known as "white biotechnology". It includes four main components, *viz.* microbiology, biochemistry, bioengineering, and bioprocessing (Figure 3.3). Industrial biotechnology offers many exciting opportunities for developing sustainable and environmentally friendly production methods for a wide range of products. Human welfare and economic prosperity are the key goals of industrial biotechnology.

A biotechnology-oriented industry has a typical organisational setup consisting of facilities of research and development (R&D), production house, supply chain, quality control, regulatory compliance, sales, marketing, and finance (Figure 3.4). Many divisions of industrial setup in microbiology are charged with high microbial impute. R&D devotes time to microbial research, product development, and analytical solutions. This may or may not be an integrated facility of industrial management. For example, some companies like Novozymes and DSM-Firmenich have a high reputation for microbial enzyme research and manufacturing enzyme-based products. The production facilities in microbiology-related industries require fermentation plants and techniques for upstream and downstream facilities. The supply chain meets the demands of feedstock, microbial culture, and other resources. The quality control department ensures product quality and safety for the final product. Regulatory compliance focuses on controlling environmental pollution

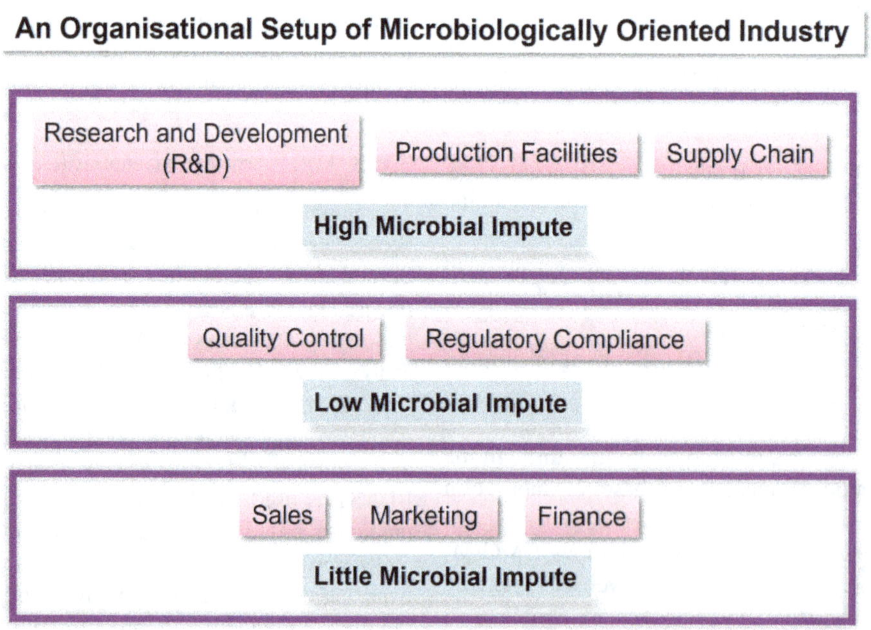

FIGURE 3.4 A typical organisational setup of a microbiologically oriented industry, showing different levels of microbial impute at different steps.

and food safety regulations. Sales and marketing efforts run parallel to the product development and contribute to its successful commercial exploitation. Finance is an important department because business in the biotechnology sector requires substantial financial resources to translate basic scientific findings into practical goods.

Modern manufacturing of industrial biotechnology was set up in the early 20th century. The idea of using tiny microbial cells as biofactories has revolutionised the techniques for manufacturing biological products. All the microbial-derived products are the result of state-of-the-art strategies. Early efforts in biosynthesis through microbial routes involved the production of organic compounds like alcohols and ketones. Other progresses in this field were the production of antibiotics and protein drugs.

In the modern world, the common bids on microbially related products are enzymes and secondary metabolites. Many biotech companies promise deliveries in the environmental and agricultural sectors. Bioengineering tools have added fuel to systems biology approaches, allowing us to gain several benefits from many microorganisms that are not their natural traits. Immobilisation techniques have also guaranteed stability, efficiency, and maximum utilisation of enzymes for industrial applications. Additionally, computational biology has revealed new prospects in microbial technology and helped overcome the challenges in real-world production.

One of the recent advances in microbial technology is the production of biofuels using agricultural wastes involving engineered microorganisms, such as *Acetivibrio thermocellus* and *Saccharomyces cerevisiae*. Microorganisms like *Cupriavidus necator* have shown the potential to transform food waste into biodegradable plastic, called polyhydroxyalkanoates (PHAs).

3.7.1 Technology Transfer

Technology transfer is the practice of sharing scientific innovation between different organisations to advance research and commercialisation. This can happen in multifarious establishments. The most realistic approach to technology transfer is between two facilities in the same industry. In this framework, the research partner devotes time and intellect to scientific innovation, and a business partner deals with large-scale production, storage, distribution, and sales. These grounds have served as the basis for establishing a myriad of microbiology-based industries. Because everyone in the business generally abides by the same standards, procedures, and regulations, the likelihood of technology transfer in such a setup is very high. Even some biotech industries have a reputation for their research facilities, which create new horizons and foster science and technology (Figure 3.5).

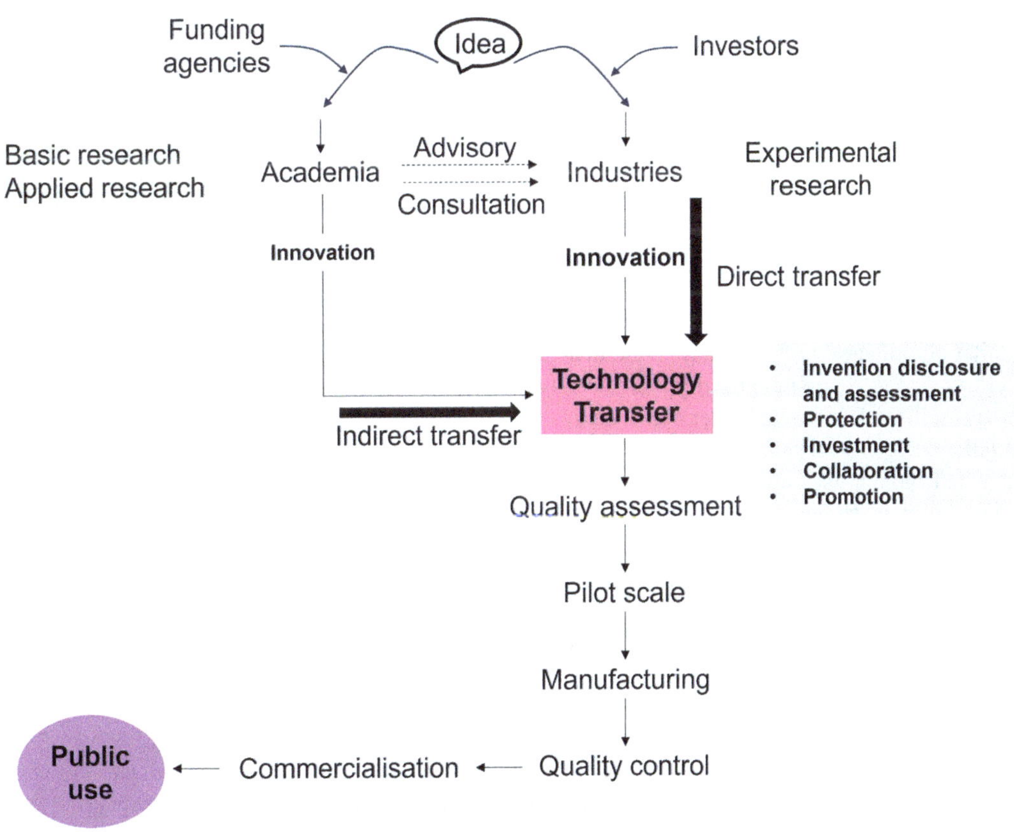

FIGURE 3.5 Technology transfer between academia and industries and different steps of the commercialisation process of a product after technology transfer.

A tie-up between a university and a company bridges the gap between academia and industry through technology transfer. Their collaboration supports scaling up the laboratory-based experiment to a pilot scale and then to a large-scale manufacturing unit. Academic entrepreneurship and start-up-driven innovation significantly contribute to advancing innovative bio-industrial technologies (Figure 3.5). However, technology transfers between a university and a biotechnology business entail a complicated structure due to their diverse approaches. The technological, administrative, legal, and personnel perspectives are the potential barriers to the deployment of technology between two different organisations. Most academic-based researchers are unaware of the processes involved in technology transfer and lack a basic understanding of public relations. Also, academicians are more concerned with publishing their studies than they are with creating a business plan to market their ideas.

The commercialisation of biotechnology products is likewise beset with several other constraints. A fundamental challenge to incorporating an industrial setup is obtaining regulatory approval. This involves not just product approval but also methods of production. The average cost for Food and Drug Administration (FDA) approval for foods/food additives, drugs, medical devices and biologics often exceeds \$250 million. In the USA, the Animal and Plant Health Inspection Service (APHIS) and the Environmental Protection Agency (US EPA) have the authority to provide licences for non-consumable biotechnology items. Due to safety concerns and the procedure involved, only a few of the numerous "new" biotechnology enterprises have achieved significant success. Despite several challenges, biotechnology industries have continued to translate basic scientific discoveries into real-world solutions for the food, health, energy, and environmental sectors.

3.7.2 Patenting in Microbial Technology

Patenting is a legal mechanism to protect an invention, allowing its inventor to exclude others from its unauthorised exploitation. Under compliance with legal terms, the inventor receives an exclusive right to pursue a third-party interest in making, using, and selling the patented invention. The patent laws were initially designed for mechanical and chemical inventions, but when biopharmaceuticals and biotechnology emerged as major fields of technology, the original laws of intellectual property rights (IPR) were amended.

When a microorganism has potential industrial value, patenting can be a strategy to guarantee a return on biotechnological inventions. However, the topic of patenting a microbial strain has been debatable for a long time because the discoveries of novel microorganisms and their utilities are not patentable subject matters. A mere identification and characterisation of microorganisms is not an invention and therefore, they do not qualify for a patent. This means a natural microorganism and a 'product of nature' cannot be patented. However, a microbiological process-based invention has no dispute about its patenting. The first patent for any microbial process was the process of fermenting beer, filed by Louis Pasteur on January 28, 1873. However, patenting microbial strains in 1980 was a prominent landmark for industrial biotechnology after the decision of the famous Diamond-Chakrabarty case in the US Supreme Court. The verdict of the court granted permission for patenting genetically modified microbial strains. The summary of the case is discussed in the next paragraph.

An Indian-origin scientist, Ananda Mohan Chakrabarty (Father of Patent Microbiology) in the USA developed a genetically engineered bacterium, *Pseudomonas putida* (superbug), which carried at least two plasmids for the degradation of hydrocarbons. This modified bacterium was a robust system for oil degradation. When he applied for a patent, his application was rejected. The objection was that the patent was sought for a living component. This case was argued with Sidney A. Diamond, Commissioner of Patents and Trademarks, at the US Court of Customs and Patent Appeals. The court overruled the Chakrabarty-favouring judgement. A further appeal in the Supreme Court reconsidered the entire case and passed a historic judgement in favour of Chakrabarty. The judgement perceived that an engineered microorganism represented a "manufacture" or "composition of matter" within that statute (Figure 3.6). Hereinafter, numerous patents on microorganisms that have undergone genetic modification have been approved.

3.7.3 Ethical Issues in Industrial Biotechnology

Bioethics is a multidisciplinary approach to moral sensitivities, which comprises directions of social sciences, law, and policy, as well as biological sciences and technology. Ethical integrity in scientific or healthcare missions navigates the safe delivery of useful products and the dissemination of accurate and authentic information to enable their proper usage. Ethical issues also address the upstanding of anonymity, privacy, property, public health, and scientific misconduct. The safety of the investigators and other group members, environmental protection, and experiment validation are of utmost concern under ethical matters.

The broad discipline of life sciences research and its diverse biotechnology enterprises present distinct ethical challenges. The discovery of the first vaccine by Edward Jenner in 1798 is one of the most pressing examples of a direct violation of ethical norms in the healthcare sector, regardless of his objective or excuse, where the investigator used an eight-year-old child as a test subject and injected him with coxpox to induce immunity against smallpox.

Current ethical issues of industrial biotechnological applications revolve around five different subject matters, including sustainability, naturalness of bioproducts, innovation trajectories, economic justice, and risk management (Figure 3.7). The present-day industrial biotechnology empire stands on the ground of sustainability and economic aspects. Many microbial-based products claim to provide sustainable food, reduce greenhouse gas emissions, and produce liquid fuels that are independent of petroleum. However, these

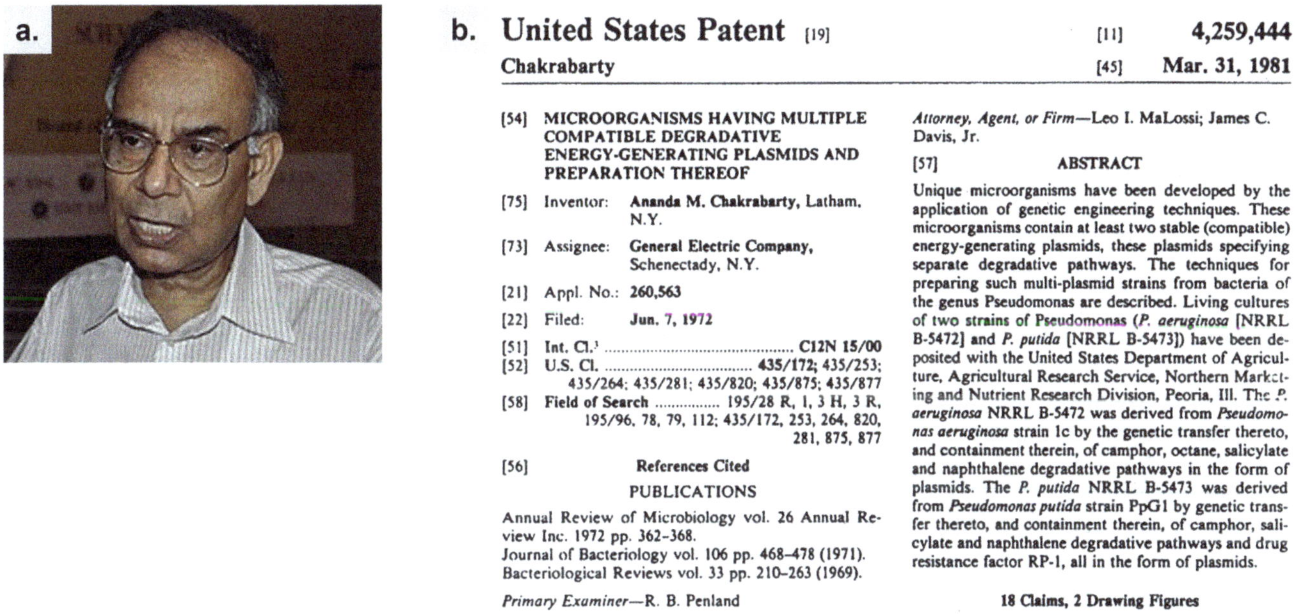

United States Patent [19]
Chakrabarty

[11] **4,259,444**
[45] **Mar. 31, 1981**

[54] **MICROORGANISMS HAVING MULTIPLE COMPATIBLE DEGRADATIVE ENERGY-GENERATING PLASMIDS AND PREPARATION THEREOF**

[75] Inventor: **Ananda M. Chakrabarty**, Latham, N.Y.

[73] Assignee: **General Electric Company**, Schenectady, N.Y.

[21] Appl. No.: **260,563**

[22] Filed: **Jun. 7, 1972**

[51] **Int. Cl.**[3] **C12N 15/00**
[52] **U.S. Cl.** **435/172**; 435/253; 435/264; 435/281; 435/820; 435/875; 435/877
[58] **Field of Search** .... 195/28 R, 1, 3 H, 3 R, 195/96, 78, 79, 112; 435/172, 253, 264, 820, 281, 875, 877

[56] **References Cited**

PUBLICATIONS

Annual Review of Microbiology vol. 26 Annual Review Inc. 1972 pp. 362–368.
Journal of Bacteriology vol. 106 pp. 468–478 (1971).
Bacteriological Reviews vol. 33 pp. 210–263 (1969).

Primary Examiner—R. B. Penland

Attorney, Agent, or Firm—Leo I. MaLossi; James C. Davis, Jr.

[57] **ABSTRACT**

Unique microorganisms have been developed by the application of genetic engineering techniques. These microorganisms contain at least two stable (compatible) energy-generating plasmids, these plasmids specifying separate degradative pathways. The techniques for preparing such multi-plasmid strains from bacteria of the genus Pseudomonas are described. Living cultures of two strains of Pseudomonas (*P. aeruginosa* [NRRL B-5472] and *P. putida* [NRRL B-5473]) have been deposited with the United States Department of Agriculture, Agricultural Research Service, Northern Marketing and Nutrient Research Division, Peoria, Ill. The *P. aeruginosa* NRRL B-5472 was derived from *Pseudomonas aeruginosa* strain 1c by the genetic transfer thereto, and containment therein, of camphor, octane, salicylate and naphthalene degradative pathways in the form of plasmids. The *P. putida* NRRL B-5473 was derived from *Pseudomonas putida* strain PpG1 by genetic transfer thereto, and containment therein, of camphor, salicylate and naphthalene degradative pathways and drug resistance factor RP-1, all in the form of plasmids.

18 Claims, 2 Drawing Figures

FIGURE 3.6 a. A portrait of Ananda Mohan Chakrabarty. b. The first United States patent for a genetically modified bacterium to Prof Chakrabarty. (Images a and b reproduced from Wikimedia/Biswarup Ganguly[1] and Google patent, respectively.)

FIGURE 3.7 Main ethical concerns of industrial biotechnology.

benefits are not without drawbacks. In this regard, the topic of biofuels is highly debatable. The production of biofuels depends on certain crops, which are required as the raw materials for microbial conversion. The cultivation of energy-intensive crops covers vast agricultural land, which competes with feeding zones and jeopardises nutritional stability. This has sparked ethical concerns about food security.

A bio-based economy, a stepping stone towards a circular economy, has also sparked ethical challenges. Many microbial-based products have come into existence due to their sustainability benefits. However, the naturalness of the microbially-derived products is subject to scrutiny due to engineering practices. Until recently, genetic alteration in industrial biotechnology has drawn less public attention than plant and animal biotechnology. However, the growth of industrial biotechnology products has grasped wide public concerns over technological aspects and the use of genetically modified organisms (GMOs). Consumer acceptance of GMOs is one of the major barriers to modern biotechnological setups. Also, the spread of altered genes within wild populations is a potential risk to natural ecosystems and human health.

Under the innovation trajectory (i.e., the development path through the innovation process), new techniques and innovation-related instruments have accelerated the ease of availability of biobased products in biotechnology. Ethical problems with microbial innovation trajectories occur when microbial processes are altered. A natural environment exposed to genetically engineered bacteria can have unanticipated effects due to novel microbial activities. Furthermore, innovations involving GMOs may have unexpected health consequences. The options of DNA recombinant technology and gene editing for an infectious agent also raise concerns about biothreat, but risk management agencies prevent any scientific inquiry at the cost of maleficence. International regulations have already made it illegal to research pathogenic microorganisms. Ethical issues in clinical microbiology are more stringent than in other related fields. A distinct ethical matter in biotechnology is economic justice, which is related to IPR (discussed in the previous section).

3.7.4 Biosafety

Biosafety refers to the practices of safe handling of infectious microorganisms for human health and the environment. A microbiology laboratory/industry is a one-of-a-kind space that demands special attention to appropriately protect those who work with microorganisms. In the workspace, sophisticated laboratory equipment, standard procedures, and modern infrastructure are the basic requirements for the safe containment of microorganisms. The phrase, "One safety rule for all microbiology laboratory practices", is a very effective biosafety approach that assumes that all microorganisms can be potential infectious agents. This underscores the importance of microbial risk assessment, considering all microorganisms with great caution, regardless of their known pathogenicity. In addition to the risk of infection from microorganisms, chemicals, and radioactive elements can also pose threats.

A regular examination of biosafety in the workstations can help prevent dangerous events. Microbiology-related practices should distinguish themselves from domestic life. A lab coat, closed-toe shoes, safety glasses, and disposable gloves are a minimal requirement for microbiology research. The disinfection of workspaces and body parts before and after work is a common practice in microbiology work ethics. Safe handling of tools and biomaterials strictly requires sterilisation techniques and a microbiological safety cabinet. In addition, aseptic techniques for inoculating specimens and accurate labelling of all components are fundamental and essential requirements in any microbiology-related experiments. The proper handling of glassware, open flames, and sharp objects is yet another concern.

The Centres for Disease Control and Prevention (CDC), in partnership with the National Institutes of Health (NIH), has proposed a guideline for biosafety practice and policy for Microbiological and Biomedical Laboratories (BMBL). The initial edition of this advisory document was released in 1984. It has continued to set best practices in its sixth edition in 2020. (www.cdc.gov/labs/pdf/SF__19_308133-A_BMBL6_00-BOOK-WEB-final-3.pdf). Based on the microbial agents used for research and industries, biological laboratories can be categorised into four biological safety levels (BSL1 to BSL4). Chapter 9 provides details of biosafety levels.

SUMMARY

- Pure and applied scientific research in academia has produced essential information for biotechnology innovation. The technological revolution and microbiome initiatives have immensely contributed to many novel findings in applied microbiology and biotechnology.
- Microbial societies are a forum for professionals from related fields to communicate, share, and promote microbiology-related information. They also take the initiative to organise various events (such as conferences, seminars, etc.) and provide research grants and fellowships.
- Culture collections, databases, and toolkits are essential resources for microbial research and innovation. Culture collections preserve live cultures in storehouses for reuse. Databases store information on sequences and organisms and provide support for data integration and validation using various toolkits.
- Microbiome projects are extensive scientific initiatives of community-level studies in microbiology, which have furnished sequence data of microorganisms from different habitats, such as the human body and the marine ecosystem.
- Industrial biotechnology focuses on the commercial deliveries of microbially-related products. Many divisions of industrial setups, such as research and development (R&D), manufacturing, supply chain, quality control, and regulatory compliance, involve microbial input for product development.
- Technology transfer in industrial biotechnology is a process of sharing information related to fermentation and bioprocessing from one organisational setup to others, leading to the commercialisation of bioproducts obtained from microorganisms.
- In industrial biotechnology, ethical matters cover matters related to scientific conduct and validation, property rights, and environmental protection. Biosafety in industrial biotechnology involves safe methods of production and risk assessments of bioproducts on the environment and health.

DIFFERENT TYPES OF PRACTICE QUESTIONS

Fill in the blanks

I. .... was set up in 1899 as the Society of American Bacteriologists.
II. The digital library that offers support to save biological information is known as a ....
III. .... is a stock of microbial cells and cell lines in vital and pure forms, which can be revived for various purposes.
IV. The first US patent for a genetically modified (GM) organism was ....
V. .... refers to the practices of safe handling of infectious microorganisms with an emphasis on human health and the environment.

SOLUTION: I. American Society for Microbiology (ASM), II. biological databases, III. Culture collection, IV. *Pseudomonas putida*, V. Biosafety

Short answer type of questions

I. Write the names of any two microbiological organisations/institutes from the Indian subcontinent.
II. Write the names of two theme-oriented societies in microbiology.
III. What is "the rare biosphere"?
IV. Expand ATCC, NRRL, and MTCC.
V. Write the names of any computational tools for microbiological data analysis.

Short answer type of questions

I. What are the key roles of academia in microbiology?
II. What are the important instruments and devices required for a basic hands-on laboratory exercise in microbial cultivation?
III. How do professional societies in microbiology contribute to fostering microbial information?
IV. What are the roles of outsourced facilities in microbial research?
V. What is technology transfer in industrial microbiology?
VI. Write a short note on patenting in microbial technology.
VII. What are the different ethical issues in industrial biotechnology?

Long answer type of questions

I. What are Microbiome Projects? Give an account of three important microbiome projects.
II. What is industrial biotechnology? What is the organisational setup of the microbiological industry?

NOTE

1 https://commons.wikimedia.org/wiki/File:Ananda_Mohan_Chakrabarty_-_Kolkata_2009-11-08_3121.JPG

SUGGESTED READINGS

Balachandra Nair R, Ramachandranna PC. Patenting of microorganisms: systems and concerns. *Journal of Commercial Biotechnology*. 2010;16:337–47.

Bennett JW, National Research Council (US) Board on Research Data and Information. Microbiology in the 21st Century. In *Designing the Microbial Research Commons: Proceedings of an International Symposium* 2011 (Vol. 2).

Caktu K, Turkoglu E. Microbial culture collections: the essential resources for life. *Gazi University Journal of Science*. 2011;24(2):175–80.

Chen IM, Chu K, Palaniappan K, Ratner A, Huang J, Huntemann M, Hajek P, Ritter S, Varghese N, Seshadri R, Roux S. The IMG/M data management and analysis system v. 6.0: new tools and advanced capabilities. *Nucleic Acids Research*. 2021;49(D1):D751–63.

Díaz-Rodríguez AM, Salcedo Gastelum LA, Félix Pablos CM, Parra-Cota FI, Santoyo G, Puente ML, Bhattacharya D, Mukherjee J, de los Santos-Villalobos S. The current and future role of microbial culture collections in food security worldwide. *Frontiers in Sustainable Food Systems*. 2021;4:614739.

Gavrilescu M, Chisti Y. Biotechnology—a sustainable alternative for chemical industry. *Biotechnology Advances*. 2005;23(7–8):471–99.

Jong S. Academic organizations and new industrial fields: Berkeley and Stanford after the rise of biotechnology. *Research Policy*. 2008;37(8):1267–82.

Robertson LA. Historical microbiology, is it relevant in the 21st century? *FEMS Microbiology Letters*. 2015;362(9):fnv057.

Schaechter M, Kolter R, Buckley M. *Microbiology in the 21st century: where are we and where are we going?* American Society for Microbiology; 2004.

Thompson LR, Sanders JG, McDonald D, Amir A, Ladau J, Locey KJ, Prill RJ, Tripathi A, Gibbons SM, Ackermann G, Navas-Molina JA. A communal catalogue reveals Earth's multiscale microbial diversity. *Nature*. 2017;551(7681):457–63.

Van Eldere J. Changing needs, opportunities and constraints for the 21st-century microbiology laboratory. *Clinical Microbiology and Infection*. 2005;11:15–8.

World Data Centre for Microorganisms. www.wdcm.org/

World Federation for Culture Collections. www.wfcc.info/

Unit II

Fermentation Technology

4 The Basics of Fermentation

4.1 FERMENTATION: A MULTIFACETED TERM

The term fermentation was derived from the Latin word *fervere*, which means boiling appearance. In the ancient past, this referred to the production of carbon dioxide (CO_2) bubbles when extracts of fruits or malted grains were used to produce alcohol. Initially, fermentation was believed to be a chemical process instead of a biological process. The fact that fermentation is a biological process was not proven until the French chemist Louis Pasteur identified and verified the process in 1857.

Today, fermentation is a common word in our daily lives. However, this term has different meanings for different disciplines. In biochemistry, fermentation is a powerful and flexible process of acquiring energy in some microorganisms under oxygen-limiting conditions. The physiology field describes fermentation as a mode of respiration that serves as an alternative metabolic pathway for survival in anaerobic situations. Fermentation in biochemistry and physiology suggests the same thing, but with different interpretations. In industrial biotechnology, the term fermentation combines basic microbial principles with bioprocessing (the process of utilising living organisms, cells, or their components to generate valuable and diverse products). Also, fermentation in industrial biotechnology does not discriminate between aerobic and anaerobic conditions. Of all the contexts, the last one has achieved major attention in recent years due to its enormous potential in industrial applications. Industrial fermentation necessitates an instrument called a fermenter, which develops controlled conditions for producing bulk microbial mass and its metabolites. The science of fermentation is also known as zymology, which is a distinct branch of applied science that studies the biochemical process of fermentation. It deals with an understanding of fermentation and its broad applications.

Metabolites are the key indicators for determining the intensity of the fermentation. However, those metabolites can vary widely based on sources of feedstock, microorganisms, and types of fermentation. The metabolic products developed in fermentation have many nutritional and health benefits for humankind. The physiology, biochemistry, and microbiology of fermentation are vast and have tremendous scope in food and dietary sciences, industrial enzyme production, the bioenergy sector, the pharmaceutical industry, agriculture applications, and environmental biotechnology. Even the gases released from fermentation guarantee the processing of diverse products.

The natural process of fermentation can be controlled by knowledge of microbiology and engineering, transforming it into a technique called **fermentation technology**. Fermentation technology regulates the synthesis of marked products in precise concentrations with distinct flavours and aromas. The intimate interdependence of humans and the microbial world has led us to believe that life would be very challenging without fermentation technology. Even modern theories claim that the acquisition of the fermentation process by the early hominins played an important role in brain expansion and gut reduction (Box 4.1).

4.2 RESPIRATION *VS.* FERMENTATION

Respiration and fermentation both involve manifold biochemical reactions that release energy from organic molecules. Respiration is fundamental to life and found in all domains of life. The respiration process mainly involves the oxidation of organic substances using exogenous oxidants (primarily oxygen gas (O_2)) to yield a significant amount of energy for the growth and development of organisms. Based on the requirement of O_2, respiration can be aerobic and anaerobic. All multicellular organisms utilise aerobic respiratory pathways to obtain energy. Aerobic respiration has high energy efficiency, involving the complete oxidation of glucose into CO_2 and water. This involves three biochemical pathways: glycolysis, the citric acid cycle, and oxidative phosphorylation. The energy output of this metabolic route is adenosine triphosphate (ATP) production (i.e., energy currency) through substrate-level phosphorylation and chemiosmosis. However, anaerobic respiration is unique to certain microorganisms and may or may not include all phases of aerobic pathways. Those anaerobic microorganisms that follow many aerobic respiration routes mark differences mainly in terms of terminal electron acceptors. This calls for molecules like nitrate (NO_3^-), sulphate (SO_4^{2-}), and arsenate (As(V)) in place of O_2. Examples of microorganisms that follow such anaerobic respiration are *Desulfovibrio vulgaris*, *Paracoccus denitrificans*, *Shewanella oneidensis*, *Chrysiogenes arsenatis*, and *Sulfolobus acidocaldarius*.

In a different situation, the oxidation of organic substrates is not associated with the electron transport chain (ETC)

DOI: 10.1201/9781032651095-6

and does not create an electrochemical gradient. Rather, the organic substrate itself serves as an electron acceptor without altering the total oxidation state of the substrate. This process is termed fermentation. Figure 4.1 gives a schematic representation of the metabolic reactions of respiration and microbial fermentation. Fermentation is considered to be less efficient in extracting energy from the substrate. The action of fermenting substrate as both electron donor and acceptor results in lower free energies than that of respiration utilising ETC. A major amount of the substrate in fermentation is used for NAD^+ regeneration (i.e., a carrier of electrons and hydrogen), while only a small amount contributes to cellular carbon frameworks. However, this process has been pivotal in human history for developing diverse products, shaping both cultures and health. Recent research studies have claimed that fermentation could have relatively higher energy efficiency than previously estimated (Box 4.2). Table 4.1 provides a comparative account of aerobic respiration, anaerobic respiration, and fermentation.

4.3 HISTORICAL BACKGROUND

The knowledge of fermentation dates back to the dawn of human civilisation. It is one of the oldest forms of food processing technology that is used even today. The earliest record of fermentation was found in approximately 13,000 BC at Neolithic Natufian burial sites in the Eastern Mediterranean, Hilazon Tachtit (Israel), belonging to a semi-sedentary foraging community. Analysis of the stone mortars from this site gave archaeological support for the use of cereals to produce alcoholic beverages. The following pieces of evidence in the line of fermentation by ancient human civilisations appeared from different times of the Neolithic age (10,000–4,500 BC), showing archaeological sites with artefacts, clay tablets,

BOX 4.1 ROLE OF FERMENTATION IN THE HUMAN BRAIN DEVELOPMENT

A group of early hominins, Australopiths, resided in Africa between 4.1 and 1.4 million years ago (mya) and are most likely early ancestors of humans. Australopiths differed from *Homo* in many aspects, including that they lacked a huge brain. The human lineage's brain expansion, which is called encephalization, began with the emergence of *Homo* and continued until the advent of *Homo neanderthalensis* and *Homo sapiens*. Several theories have been put forth to explain encephalisation, which is mainly associated with food resource utilisation in the body.

In several explanations, fermentation has played a major role in human brain development. Proposed mechanisms include internal and external fermentation. The human gut is a machine for internal fermentation, where digestion is performed by the native microbiomes. This allows the breakdown of organic food matter by microbial enzymes. They produce numerous acids and alcohols and release essential nutrients for human growth and the human brain. Internal fermentation can increase food digestibility by lowering non-nutritive components. The colon is the primary site for internal fermentation, which harbours about one trillion (10^{12}) bacteria per ml. The small intestine also keeps a very high density of bacteria (i.e., about 1 million (10^6) per ml) for the digestion of food.

An External Fermentation Hypothesis is a recent concept which postulates that the consumption of fermented food by early hominins might also have triggered a major role in initial encephalisation. External fermentation provides numerous adaptive benefits to hominins. This can be attributed to the fact that fermented foods can reduce digestive energy expenditure. This would have also helped with energy expenditure in metabolically expensive brain tissues. The relative reduction of the human colon over the evolutionary period could be associated with increased reliance on fermented foods.

External fermentation benefits from improved macronutrient absorption and enhanced nutrient digestibility. Fermented food supplies micronutrients essential for brain development and function. Fermentation of carbohydrates outsources B-complex vitamins (e.g., thiamin, riboflavin, and niacin). Anti-nutritional factors (ANFs) present in cereals, grains, seeds, legumes, and tubers form complexes with metal cations (like iron (Fe), zinc (Zn), magnesium (Mg), and calcium (Ca)) and prevent their absorption. External LAB-driven fermentation supports the degradation of ANF to enhance the absorption of minerals. Fermentation can make toxic foods edible. For example, when the toxin cyanogenic glycoside found in the bitter cassava tuber is treated with *Lactobacillus*, some enzymes of the bacterium hydrolyse the toxin and make it palatable. Fermentation also creates a favourable condition for the body of foodstuffs through the production of various acids, which allows other microorganisms to aid in hydrolysing toxins. External fermentation assists internal fermentation as consumed microflora from fermented foods can colonise the gut and increase its capacity to ferment ingested food into nutrients and energy. Ingested probiotic bacteria contribute to the health of the body through the production of bacteriocins (toxins that competitively inhibit pathogens) or anti-pathogenic metabolites. Probiotics provide immune benefits by lessening the ability of pathogenic bacteria to enter the bloodstream and preventing them from colonising the colonic intestinal wall. Certain microorganisms have even been found to produce neurotransmitters like serotonin and gamma-aminobutyric acid (GABA) directly in the gut (e.g., *Lactobacillus plantarum* and *Lactococcus lactis* subsp. *cremoris*), which can influence the gut-brain axis. Even studies have shown that gut microorganisms can reduce anxiety-like behaviours, boost mental state, and improve cognitive abilities by influencing the gut-brain axis.

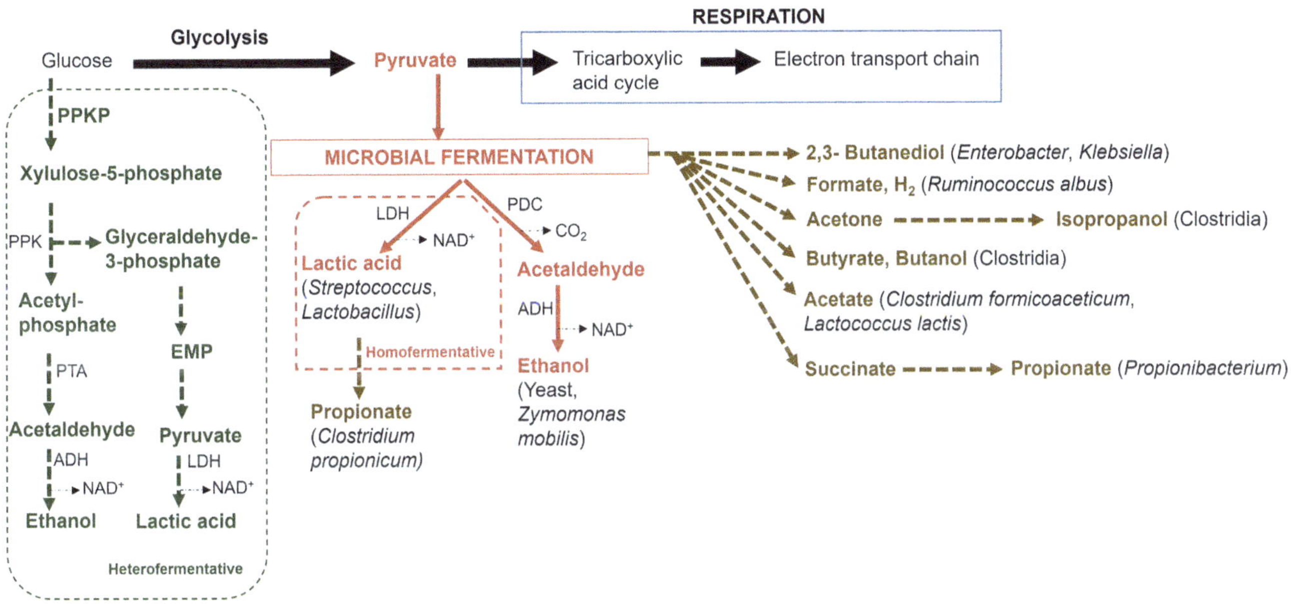

FIGURE 4.1 The schematic representation of the metabolic reactions of microbial respiration and fermentation. The respiration pathway takes advantage of tricarboxylic acid and the electron transport chain for energy conservation. Pyruvate is the key intermediate for the fermentation pathway, produced from glycolysis. Pyruvate can be transformed into lactic acid (using a homofermentative pathway) and ethanol using lactic acid bacteria (LAB) and yeasts, respectively (red colour). The heterofermentative LAB uses a distinct pentose phosphoketolase pathway (PPKP) for glucose catabolism and produces mixed products, including lactic acid, ethanol, and other products (green colour). A range of microorganisms utilise other metabolic pathways for fermentation reactions and form diverse products (brown colour). (LDH- lactate dehydrogenase; PDC- pyruvate decarboxylase; ADH- alcohol dehydrogenase; PPK- pentose phosphoketolase; PTA- phosphotransacetylase)

TABLE 4.1
Respiration *vs.* fermentation

Parameters	Aerobic respiration	Anaerobic respiration	Fermentation
Organisms	Prokaryotes and eukaryotes	Prokaryotes	Prokaryotes and eukaryotes
Cell component involved	Cytoplasm and mitochondria	Cytoplasm	Cytoplasm and hydrogenosomes
Tricarboxylic acid cycle	Present	May be present	Not present
Cofactors involved	NADH + H^+, $FADH_2$	NADH + H^+, $FADH_2$	NADH + H^+
Electron transport chain	Present	Usually present	Not present
Terminal electron acceptor	O_2	NO_3^-, SO_4^{2-}, Mn(IV), AsO_4^{3-}, fumarate, dimethyl sulfoxide	Mainly pyruvate
ATP source	Substrate-level phosphorylation and oxidative phosphorylation	Substrate-level phosphorylation and oxidative phosphorylation	Only substrate-level phosphorylation
ATP yield from glucose	32	Variable, but less than aerobic	2 (new research estimates much more than this.)
Products	CO_2 and H_2O	NO_2^-, H_2S, Mn(II), AsO_3^{3-}, succinate, dimethyl sulphide	Lactate, ethanol, CO_2, and many others

utensils, etc. In this period, fermentation was predominantly used to produce alcoholic beverages and food products. Records from China from around 7,000 BC showed that humans produced drinks of fermented honey, rice, and dates.

Possibly, the first leavened bread was made in southern Mesopotamia circa 6,000 BC using airborne wild yeast. Fermented dairy products also have a long history, with evidence indicating that inhabitants in the Sahara region of North Africa raised cattle around 6,000 BC and were consuming milk after its processing, likely through fermentation. Since ancient times, milk fermentation has been performed primarily for preservation and to reduce lactose levels, making it more palatable. Indian ancient religious scriptures, the Rig Veda and Upanishad (c. 6,000–4,000 BC) also included one of the earliest mentions of a fermented milk product (i.e., dahi). The fermented milk Yoghurt originated around 5,000 BC in Mesopotamia and was spread throughout Europe before becoming a staple of the global human diet.

The presence of tartaric acid, a natural product of grapefruit fermentation in a ~5,400 BC old jar from European Mesopotamia and grape pressings in Greece are strong indicators of winemaking in ancient times (Figure 4.2a). A Sumerian tablet, dated back to approximately 4,000 BC, includes the depiction of beer drinking, which shows people sipping beer from the same vessel through reed straws (Figure 4.2b). Egyptians in 3,500 BC used fermentation for brewing, bread leavening, and making dairy products of butter, cheese, and yoghurt. Ancient clay tablets of 3,100–3,000 BC Mesopotamian (Babylonian) civilisation also depicted beer as the staple drink (Figure 4.2c).

With growing knowledge of fermentation, the fermentation process became popular in many different cultures and for various purposes. This included cucumber pickling in the Middle East (2,000 BC), the production of fermented beverages (mead or honey wine) in Asia during the Vedic period (1,700 BC), preservation of meat sausages in ancient Babylonia (1,500 BC), cereal-legume food fermentation in several parts of the world (1,000–500 AD), and fermentation of vegetables and tea in ancient China (300–200 BC). Fermentation also played a crucial role in developing other products, such as East Asian soy sauce production and Greek fruit fermentation. Even many ancient societies relied on the fermentation process to preserve food during difficult seasons for ceremonial feasts and to enhance the food's sensory qualities. Many fermented food products developed in prehistory, made from rice, soybeans, vegetables, and fish, are still manufactured and consumed in various parts of the world.

In prehistory, the empirical implementation of fermentation by humankind presumably occurred with random assessments and careful observations, though without any scientific comprehension. Yeasts are considered to be the earliest industrial microorganisms utilised by humans. The early fermentation systems for producing alcohol and bread are assumed to have evolved as a result of natural microbial contamination in flour, other milled grains, milk, and sugar-containing fruit. The crucial milestone that fermentation was a biological process came much later in the 18th and 19th centuries, during research on cellular glycolysis and the fermentation of glucose/sugar into ethanol. In 1780, the

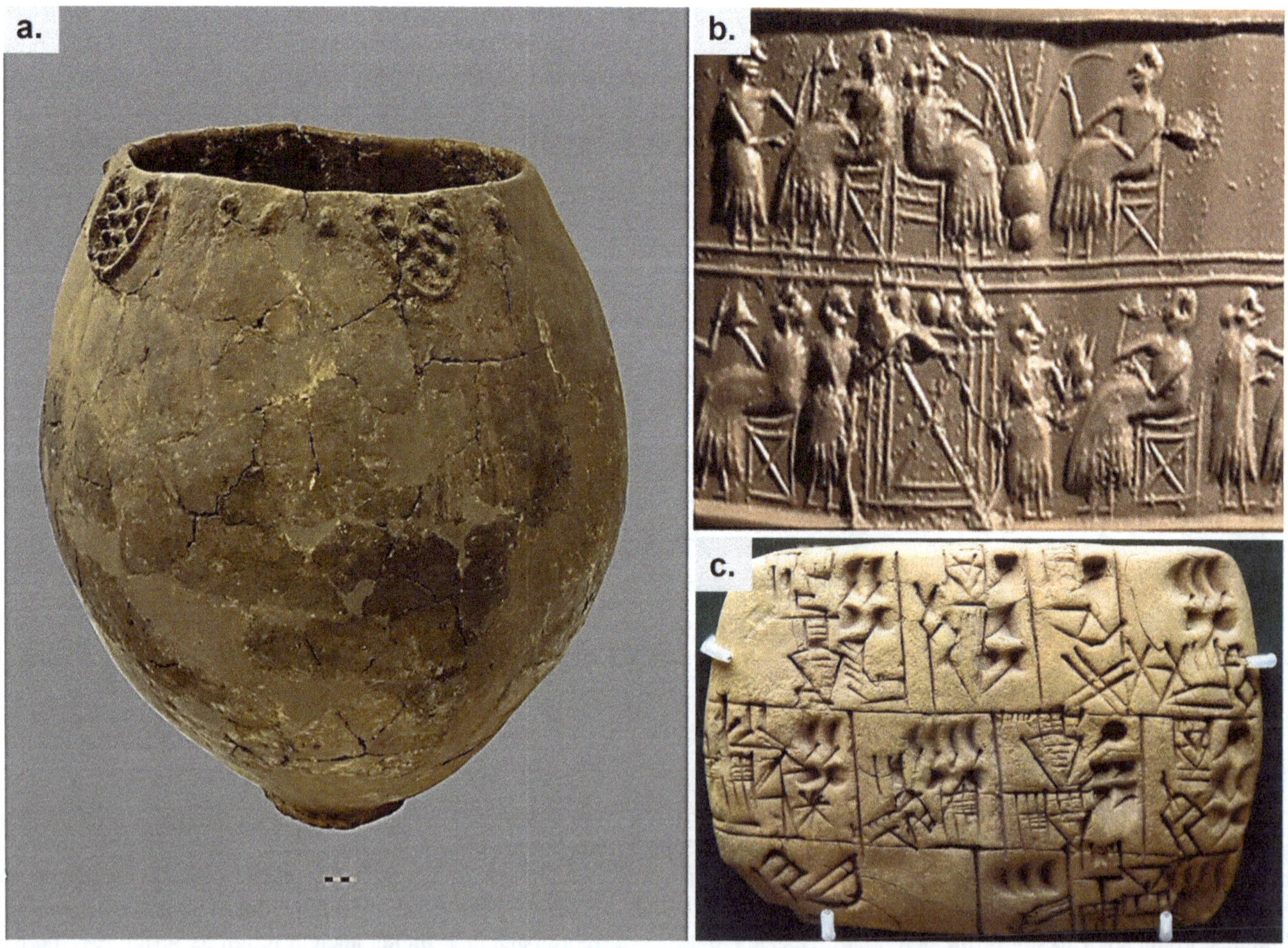

FIGURE 4.2 a. A large wine jar with grape cluster motifs around the mouth from the early Neolithic period. b. A Sumerian tablet of approximately 4,000 BC, showing people sipping beer from the same vessel through reed straws. c. A proto-cuneiform clay tablet, depicting beer (triangular clay vessel) as the staple drink of Mesopotamia from the late Uruk period around 3,100–3,000 BC. In this image, an upright jug with a pointed base is the symbol for beer. (Image a from McGovern et al. (2017)[1], reproduced with with permission, Photograph by Mindia Jalabadze, courtesy of the National Museum of Georgia, b adapted from Damerow (2012)[2], and c from Wikimedia/Osama SMA.[3])

Swedish chemist KW Scheele discovered and documented the presence of lactic acid in sour milk. The chemists LJ Thenard (1803) and JJ Colin (1825) also noted yeast growth during fermentation. In 1810, Joseph L Gay-Lussac experimentally concluded that insoluble yeast cells from air were responsible for the fermentation of grape juice. However, leading organic chemists Justus von Liebig, Friedrich Wöhler, and Jöns J Berzelius of that period contradicted the involvement of yeasts in fermentation. Lately, the independent efforts of Charles Friedrich T Kützing (1837), Theodor Schwann (1837), and Cagniard-Latour (1838) confirmed that yeasts were living and reproducing organisms that were essential for the fermentation of sugar.

Further, **Louis Pasteur** (1858, 1860) provided indisputable proof that fermentation was essentially connected to the vital process of the living yeasts. Pasteur observed that yeasts in a defined growth medium produced ethanol and CO_2 as the major fermentation products, along with some secondary products such as succinic acid and glycerol. Also, he distinguished the yeast for alcoholic fermentation from lactic acid fermentation. He is called the **Father of Modern Microbiology** for his notable contributions to the field of microbiology (Figure 4.3a and b). In the same period, German chemist **Eduard Buchner** (1897) successfully carried out glucose fermentation using a cell-free yeast extract (proteinaceous mass) to produce ethanol (Figure 4.3c). This was a milestone in the discovery of intracellular enzymes and enzymatic reactions. The thorough examination of yeast extract fermentation by British biochemists Arthur Harden and William Young (1906) showed that several low molecular weight components were additionally required for enzymatic activity. These components were organic (e.g., NADH + H^+/NAD^+, AMP, ADP, and ATP) and inorganic cofactors (inorganic phosphate).

FIGURE 4.3 a. A portrait of Louis Pasteur. b. Louis Pasteur's famous double flask for demonstrating anaerobic fermentation. c. A photo of Eduard Buchner (left) and his assistant (right). d. A portrait of Chaim Weizmann. e. A portrait of Charles Pfizer. (Images a and b courtesy of Science History Institute, reproduced with permission[4,5] and c, d, and e from Wikimedia Commons/Rolf Ukrow[6], Wikimedia Commons/Brown-Suarez Photo[7], and Wikimedia Commons/United States Public Domain[8], respectively.)

In the coming years, the use of microorganisms for aerobic and anaerobic fermentation became common for broader applications. The appliances of fermentation have become vital tools in industrial biotechnology for the manufacturing of diverse compounds such as antibiotics, secondary metabolites, and other products.

4.4 MICROORGANISMS FOR FERMENTATION

Fermentation generally implies the anaerobic respiration of microorganisms. Lactic acid bacteria (LAB) and yeasts are two prominent fermentative microorganisms concerning anaerobic reactions. Lactic acid and ethanol are their respective principal products. Some derivative pathways also exist, and their derived products could be acetic acid, butyric acid, formic acid, acetone, isopropanol, etc. These processes also release gases like CO_2 and hydrogen gas (H_2) as byproducts. Regardless of the type of fermentative microorganisms involved, the primary substrates for the fermentation reaction are simple carbohydrate compounds like glucose, fructose, and sucrose. Table 4.2 provides a comparative account of these two types of fermentation.

LAB are diverse Gram-positive, non-spore-forming rod/cocci bacteria belonging to the order Lactobacillales. Bacteria like *Lactobacillus*, *Leuconostoc*, *Streptococcus*, and *Pediococcus* are some common examples of LAB (Figure 4.4a and b). They are aerotolerant anaerobes, which means they prefer anaerobic conditions but can adapt to oxygenic environments. Based on the end products, LAB can be categorised as homofermentative and heterofermentative. Homofermentative bacteria produce only a single product, lactic acid, whereas heterofermentative bacteria produce mixed products, including the primary product, lactic acid. The side products from heterofermentative reactions include acetic acid, formate, ethanol, diacetyl, and acetoin. These two variants of fermentation in LAB also exhibit some differences in biochemical pathways, as mentioned in Table 4.3.

A different microorganism, *Saccharomyces cerevisiae* (commonly known as yeast), is a unicellular fungus belonging to the division Ascomycota, which observes a facultative anaerobic lifestyle. They are spherical to oval-shaped cells and reproduce asexually by budding (Figure 4.4c). *S. cerevisiae* is a notable producer of ethanol through fermentation reactions. The ability to switch between

TABLE 4.2
Differences between lactic acid and alcoholic fermentation

Parameters	Lactic acid fermentation	Alcohol fermentation
Microorganism involved	Lactic acid bacteria (LAB)	Yeasts and a few bacteria
Cell types	Prokaryotes	Prokaryotes and eukaryotes
Number of reactions	Single step reaction	Two-step reactions
Types	Homo- and heterofermentative	Only one type
Electron donor	Internal (homofermentative) or external (heterofermentative)	Internal
Main enzymes involved	Lactate dehydrogenase	Pyruvate decarboxylase (PDC) and alcohol dehydrogenase (ADH)
Main products	Lactic acid	Ethanol
Organisms	*Lactobacillus*, *Lactococcus*, and *Leuconostoc*	*Saccharomyces cerevisiae* and *Zymomonas mobilis*
Applications	Production of yoghurt, curd, cheese, sauerkraut, kimchi, kefir, etc.	In the bakery, liquor production, biofuel industry, and disinfectants

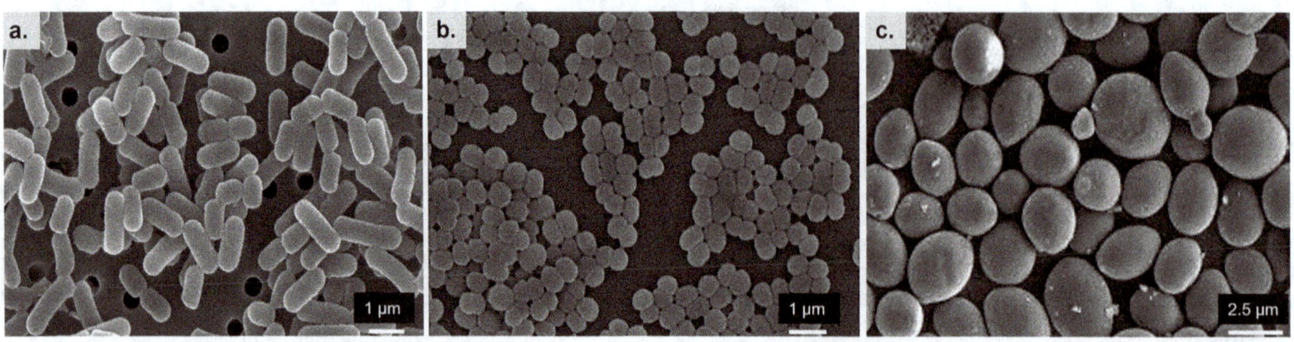

FIGURE 4.4 Scanning electron micrograph (SEM) of different microorganisms used in fermentation. a. *Lactobacillus plantarum*. b. *Pediococcus* sp. c. *Saccharomyces cerevisiae*, showing budding cells. (Images a, b, and c adapted from Muthusamy et al. (2020)[9], Katepogu et al. (2022)[10], and Karimy et al. (2020)[11], respectively.)

TABLE 4.3
Differences between homofermentative and heterofermentative metabolisms of the lactic acid bacteria (LAB)

Parameters	Homofermentative	Heterofermentative
Initial pathway	Glycolysis	Pentose phosphoketolase pathway (PPP)
Electron donor source(s)	Internal	Internal and external
Electron donor	NADH + H^+	NAD(P)H + H^+
Product(s)	Lactic acid	Lactic acid, acetic acid, ethanol, and CO_2
ATP production	2	1
Number of lactic acid/glucose	2	1
Dairy industry	As a starter culture	Uncommon for industrial applications
Examples	*Lactococcus* and *Streptococcus*	*Leuconostoc*, *Lactobacillus fermentum*, and *Lactobacillus brevis*

aerobic and anaerobic metabolism allows this microorganism to thrive in diverse environments. They are also the sources of different enzymes, making them highly valuable for numerous industrial applications.

In the context of industrial microbiology, the list of microorganisms used in fermentation is extensive and even includes many aerobic microorganisms. They are mainly characterised based on the values they produce and their mass cultivation in fermenters. Examples of some aerobic bacteria are *Bacillus* spp., *Pseudomonas* spp., *E. coli*, and *Streptomyces*. They are the common workhorses for the large-scale fermentation of different products. Moulds like *Aspergillus niger*, *Penicillium* spp., *Rhizopus stolonifer*, *Trichoderma* spp., etc. are important sources of diverse biological compounds. Many microorganisms are used to produce a variety of fermented foods that form major constituents of the human diet worldwide. A different part of the book gives a detailed account of industrially relevant microorganisms (Chapter 9).

4.5 FERMENTATION ENZYMES

Enzymes found in microorganisms are the key players that bring chemical changes to substrates during fermentation. Fermentation processes typically involve a series of enzymatic reactions with many common primary reactions. As many as ten different enzymes of the glycolytic reactions turn glucose into pyruvate, which serves as the primary substrate for fermentation pathways, resulting in the production of diverse fermentation products. The fate of pyruvate can be determined by the microorganisms involved and the environmental conditions. For example, LAB secrete lactate dehydrogenase (LDH) and produce lactic acid (Figure 4.5a). On the contrary, yeast cells use two different enzymes, including pyruvate decarboxylase (PDC) and alcohol dehydrogenase (ADH). They help metabolise pyruvate and release ethanol (Figure 4.5b). A bacterium, *Propionibacterium,* performs propionic acid fermentation, which requires three specific enzymes, including transcarboxylase, propionyl-CoA carboxylase, and methyl malonyl-CoA mutase. Similarly, industrial microorganisms are the sources of many other metabolites, producing different enzymes.

In the late 1980s, Edward Buchner made the first claim that the cell-free yeast extract was predominantly responsible for alcoholic fermentation. This cell-free yeast extract consisted of several enzymes responsible for glycolysis and fermentation. This enzyme complex was called zymase. This research work fetched Buchner the Nobel Prize in 1907. Yeast was also found to be a source of another enzyme, invertase (i.e., β-fructofuranosidase), which is indispensable when the fermented substrate is sucrose. Sucrose is a disaccharide and cannot be directly used for fermentation. Yeast cells secrete invertase outside the cells, allowing its hydrolysis into glucose and fructose, which can be readily absorbed by cells for glycolysis followed by fermentation reactions.

Based on the locations, enzymes can be secreted outside as extracellular enzymes, stored inside as intracellular enzymes, bound to the cell wall, or located in the periplasmic space. Microbial enzymes like α-amylase, lipase, and cellulase can be obtained from both extracellular and intracellular sources. Microorganisms that produce extracellular enzymes have a high demand due to low-cost production. However, some industrially important enzymes like ADH and penicillin acylase are specifically located inside the cell. An enzyme, alkaline phosphatase, that is employed in molecular biology is recovered from the periplasmic space of bacteria.

Fermentative enzymes, being proteinaceous, exhibit high sensitivity toward temperature, pH, moisture content, substrate concentration, O_2 level, cofactors, redox potential, harvesting times, etc. Therefore, the biochemical reaction of fermentation requires a strict range of physicochemical conditions. For example, ethanol production by yeasts works best at temperatures of 30–35 °C and pH 4-6. The lactic acid production also hangs in similar ranges of pH and temperatures. The slight variation in the physicochemical parameters can witness adjustments in the physiological states in microbial culture, resulting in the shift of metabolic pathways and a change in the profile of fermentation products.

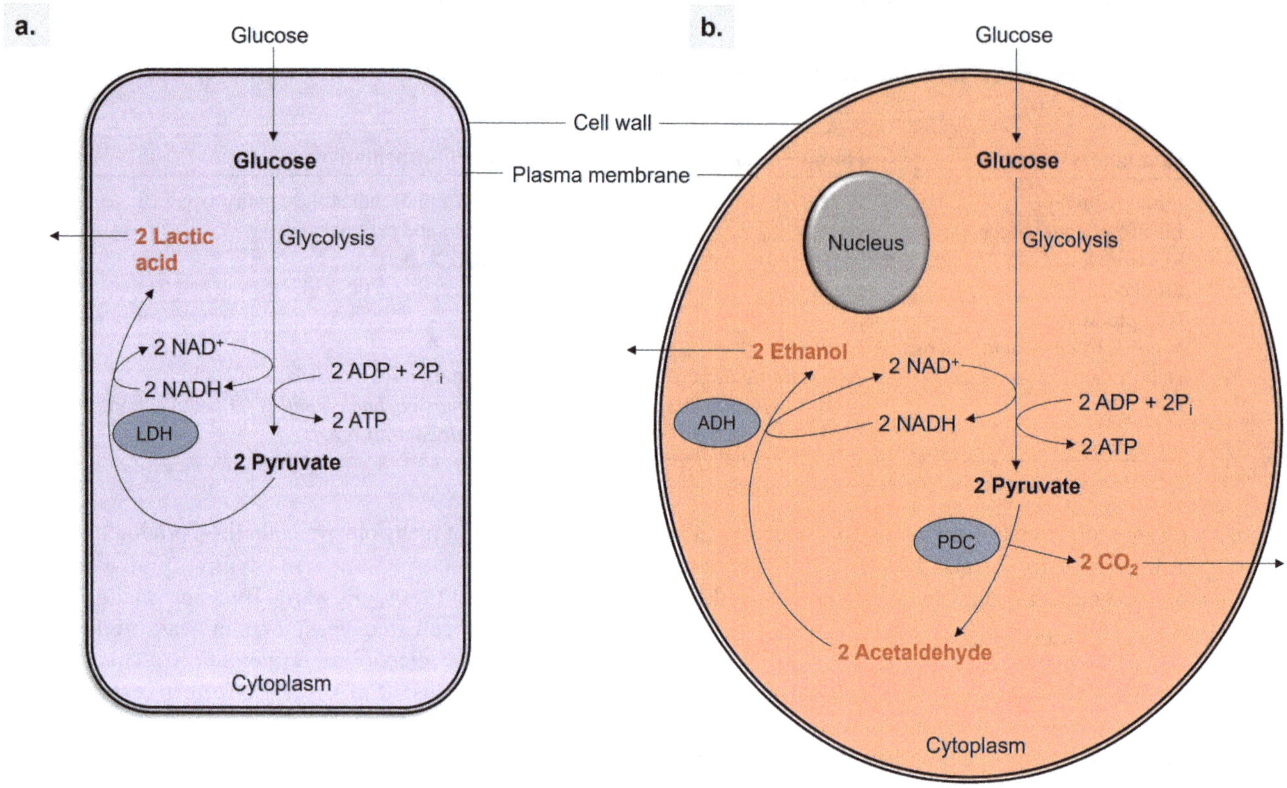

FIGURE 4.5 Fermentation pathways in microorganisms showing substrate-level phosphorylation. a. Homofermentative pathway in lactic acid bacteria (LAB) producing lactic acid. b. Ethanol fermentation pathway in a yeast cell. (LDH- lactate dehydrogenase, PDC- pyruvate decarboxylase, and ADH- alcohol dehydrogenase)

4.6 BIOCHEMISTRY OF FERMENTATION

Fermentation is considered to be among the oldest biochemical pathways. In the primordial Earth, it was prevalent before the emergence of O_2 in the atmosphere. The fermentation process runs under strict biochemical parameters. The key biochemical activities in the fermentation process are:

- Energy production using substrate-level phosphorylation
- An unaltered ratio of NADH + H^+/NAD^+
- Reaction under oxygen-limiting conditions

The common initial pathway for most fermentation reactions is glycolysis. In a series of reactions, glucose is initially broken down into two molecules of pyruvate along with the production of NADH + H^+ and ATP (2 molecules of each), passing through various oxidation, reduction, isomerisation, and phosphorylation reactions. Following that, depending on the specific microorganisms involved, the complex reactions of pyruvate under an oxygen-free environment become a gateway to the formation of fermentative products (Figure 4.1). The NADH + H^+ generated from glycolysis is consumed during the enzymatic reduction of pyruvate, and the oxidised NAD^+ can be re-employed in glycolysis. The substrate-level phosphorylation of glycolysis (i.e., ATP) becomes the exclusive source of energy because fermentation does not involve an ETC, except in certain cases (Box 4.2).

In alcohol production, pyruvate is first converted into acetaldehyde by an irreversible, non-oxidative decarboxylation reaction using an enzyme, PDC. The acetaldehyde undergoes a reduction reaction to produce ethanol using NADH + H^+ in the presence of ADH. On the other hand, the formation of lactic acid from pyruvate is a single-step reaction, which directly involves a reduction reaction using lactate dehydrogenase and NADH + H^+.

In some fermentation processes, the pentose phosphate pathway (PPP) runs in parallel to the glycolytic pathway, which can contribute to the overall metabolic flexibility and efficiency of the microbial cells. Particularly, the heterofermentative LAB exhibit variations in the use of traditional glycolytic pathways. They prefer to employ a PPP and generate a variety of side products such as ethanol, acetic acid, CO_2, etc. This fermentative pathway has a pivotal role in the synthesis of amino acids and nucleotides.

4.7 INDUSTRIAL FERMENTATION

Industrial fermentation is a process of desirable chemical change in substances using microbial cells or microbial enzymes. An industrial fermentation process aims to produce the desired bioproduct in the largest possible quantities and in the simplest and cheapest possible way. The ancient and empirical knowledge of fermentation is the foundation for the development of large-scale production industries. The first

BOX 4.2 ENERGY CONSERVATION DURING FERMENTATION IN AN ANAEROBIC SITUATION

In biochemical terms, ATP is the primary energy currency of the cell and is mainly derived through substrate-level phosphorylation and oxidative phosphorylation in distinct steps of the respiratory pathways. Substrate-level phosphorylation occurs when a nucleoside diphosphate (ADP or GDP) is phosphorylated by a substrate to produce a nucleoside triphosphate (ATP or GTP). Oxidative phosphorylation is the phosphorylation of ADP to ATP using the free energy produced from redox reactions in the ETC. The oxidative phosphorylation typically produces 28–30 molecules of ATP per glucose molecule. On the contrary, substrate-level phosphorylation yields only two molecules of ATP during glycolysis and the Krebs cycle. In fermentation reactions, the anaerobic situation does not allow for oxidative phosphorylation. A general view about energy production in fermentative bacteria has been that they solely rely on substrate-level phosphorylation. This calculation makes them highly unproductive and inefficient energy conservers. However, recent investigations reported that fermentative bacteria can have additional routes to enhance ATP production in anaerobic situations. They possess an ion-motive force (IMF) to conserve energy in the cells. IMF consists of a chemical gradient of H^+ or Na^+ across the membrane and generates ATP during fermentation.

One fermentative anaerobic bacterium, *Streptococcus cremoris,* possesses the ability to transport two lactic acid molecules (from glucose fermentation), accompanied by two H^+. This establishes an electrochemical H^+ gradient and produces 0.5 ATP from an ATP synthase. The Na^+ gradients for energy conservation in fermentative bacteria can occur through biotin-containing decarboxylase or reduction of NAD^+: ferredoxin oxidoreductase (Rnf). In the first category, the bacterial Na^+ pump couples the decarboxylation of a substrate (like oxaloacetate and succinate) with the translocation of Na^+ ions from the cytoplasm to the external medium for ATP production. Examples of biotin-containing decarboxylases are *Clostridium symbiosum* and *Acidaminococcus fermentans*. In the second category, anaerobic bacteria generate Na^+ motive force using Na^+ pump ferredoxin: NAD^+ oxidoreductase (Rnf complex). The Rnf complex is reported in plasma membranes of bacteria like *Rhodobacter capsulatus*, *Clostridium kluyveri*, *Clostridium ljungdahlii*, and *Acetobacterium woodii*.

decade of the 20th century was the period of the industrialisation of fermentation technology. The industrial synthesis of acetone, butanol, and ethanol (ABE) using the bacterium *Clostridium acetobutylicum* was the first pay-off deal in industrial microbiology. This process was a critical need for acetone production in the manufacture of cordite, which was an explosive propellant during World War I. This microbially-mediated process was developed by a Russian-British-Israeli biochemist, **Chaim Weizmann** (1913) (Figure 4.3d). His work on the fermentation process laid the groundwork for modern biotechnology and industrial microbiology. For his contributions to industrial microbiology, he was recognised as the **Father of Industrial Fermentation**. The German-American entrepreneur and chemist Charles Pfizer invented citric acid fermentation (Figure 4.3e) in 1923. The large-scale production of citric acid using *Aspergillus niger* was a significant milestone when its demand was very high. The industrial biosynthesis of penicillin during World War II set a revolution in the pharmaceutical industry.

Fermentation technology uses several biochemical pathways that confront conventional chemical synthesis. In recent years, fermentation technology has seen significant expansion beyond its traditional applications in the pharmaceutical and food and beverage industries. The growing need for efficient and sustainable production methods is a major driver of this trend across a wide range of industries. In industrial biotechnology, fermentation is a highly profitable process for industries and meets the requirements of Sustainable Development Goals (SDGs). Table 4.4 sheds light on the differences between metabolic and industrial fermentation.

Fermentation technology primarily pays attention to desirable fermentation, which ensures that only a particular microorganism grows on the substrates. Depending on the product of interest, the same organisms can be desirable or undesirable in the fermentation reaction. For example, when yeast grows on milk, it causes spoilage of milk, while the propagation of LAB on it produces lactic acid. Also, fermentation from an industrial perspective can take the benefits of diverse bacteria and moulds rather than just LAB and a few yeasts. Many physicochemical factors (like moisture, redox potential, pH, temperature, O_2 concentration, nutritional requirements, and inhibitors) play important roles in the production of desirable metabolites.

4.8 PROCESS OF INDUSTRIAL FERMENTATION

An industrial fermentation involves three sequential events: upstream processing, fermentation process, and downstream processing (Figure 4.6). They all play indispensable roles in achieving a successful outcome. Each of these events has distinct steps or stages. Table 4.5 identifies the main differences between the upstream and downstream processes. The upstream process typically involves microbial cultivation, genetic engineering, optimisation studies, pretreatment of raw materials, and media preparation. Appropriate sterilisation and cleaning are also necessary steps of upstream processing. The next chapter gives a detailed account of distinct activities associated with upstream processing (Chapter 5).

This is followed by the fermentation process, which is a time-bound reaction between microorganisms and nutrient

TABLE 4.4
Differences between metabolic and industrial fermentation

Parameters	Metabolic fermentation	Industrial fermentation
Objectives	Produce energy for microbial metabolism	Large-scale manufacturing of distinct products using microbial activities
Reactions involved	Enzymatic reduction of endogenous organic metabolites	Biosynthesis of primary and secondary metabolites using exogenous and endogenous substrates
Occurrence	Varying natural environments	Controlled physicochemical conditions of the fermenter
Performed by	Facultative anaerobic bacteria and yeasts	Industrially important microorganisms
O_2 requirement	Oxygen-limiting conditions	Aerobic or anaerobic conditions
Microbial metabolism	Chiefly primary	Primary and secondary
Fermentation product usefulness	Typically, byproducts for the microorganisms	Highly valuable in commodities of food, animal feed, pharmaceuticals, biofuels, etc.

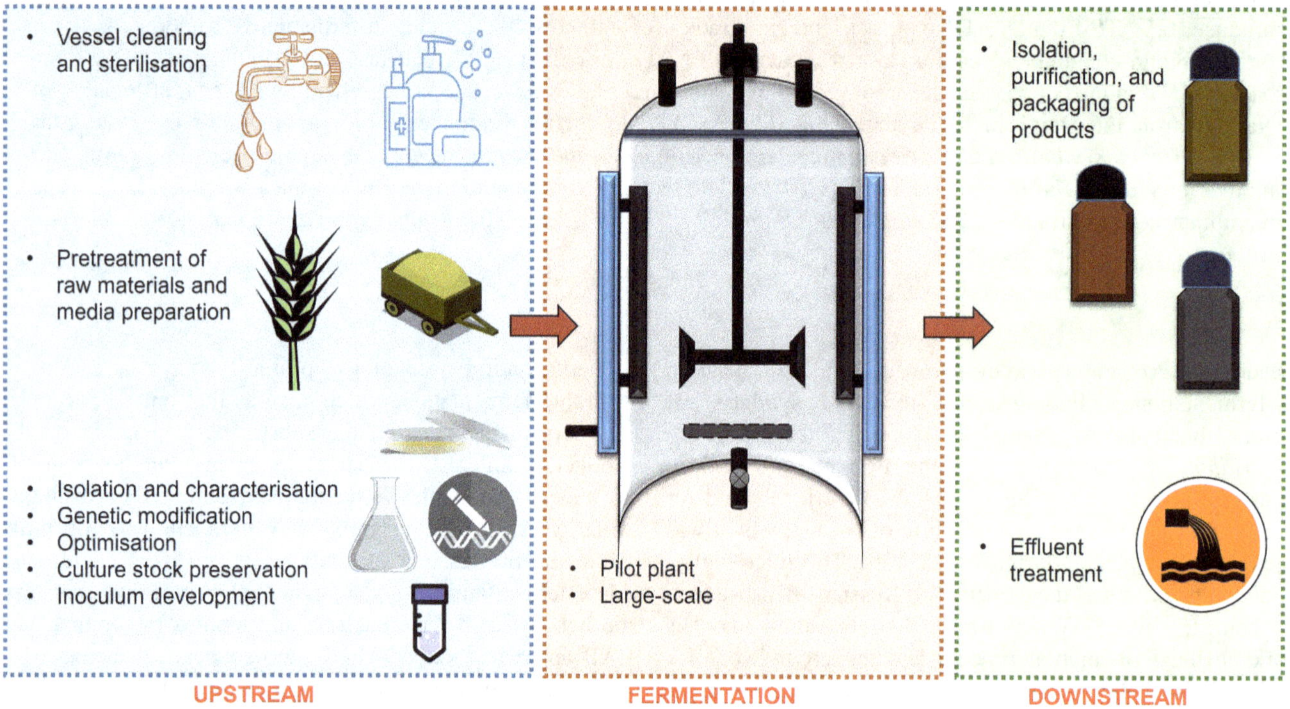

FIGURE 4.6 Three main events of industrial fermentation: upstream processing, fermentation process, and downstream processing.

TABLE 4.5
Differences between upstream and downstream processes in industrial fermentation

Parameters	Upstream processing	Downstream processing
Occurrence	Before the fermentation process	Everything after the fermentation process
Primary objective	Preparation for the fermentation reaction	Purification and recovery of the product
Essential to	Start fermentation	Complete a successful bioprocessing
Processes involved	Sterilisation, isolation and characterisation of pure microbial strains, microbial stock preservation, media formulation, and fermentation optimisation	Microbial biomass separation, intracellular or extracellular product purification, and packaging, storage of the product, and safe disposal of waste effluents
Optimisation	To establish fermentation parameters (nutritional and physicochemical)	To achieve maximum yield of product and obtain high-level pure products
Inoculum stage	Inoculum was built up successively from small to large volumes for fermentation	Separation of microbial biomass from the fermentation broth
Incurring cost	The fermentation can be made cost-effective with appropriate statistical optimisation	Processing is typically expensive, particularly when the purity requirements of a product are high

mixtures under well-optimised physicochemical conditions. Industrial-scale fermentation exists in three different modes of operation, including batch, fed-batch, and continuous processes. These schemes mainly differ concerning how nutrients are supplied and how the fermented products are obtained. Depending on the substrate types and volumes of water used in the nutrient media, the process can be distinguished as submerged fermentations (SmF) and solid-state fermentations (SSF). Aerobic fermentation typically demands a continuous supply of sterile oxygenated air for microbial growth. On the other hand, anaerobic fermentation takes place in the presence of nitrogen gas (N_2), CO_2, or by-products of the fermentation reaction. The anaerobic process is relatively slower than that of aerobic processes. Various technical aspects of the fermentation process are discussed in Chapter 6.

The stringent fermentation criteria call for a special instrument called a fermenter, which not only meets the physicochemical parameters of the bioreaction but also allows microbial proliferation, ensuring a contamination-free environment. A fermenter is indispensable for fermentation technology and large-scale production. Industrial fermenters can vary in shape, size, and type, and can effectively develop microbial biomass, metabolites (primary and secondary), modified products, and recombinant proteins. The fermenters are designed to tightly monitor and control the optimal ranges of several parameters, such as pH, temperature, dissolved oxygen (DO), nutrient concentrations, quality of product, microbial biomass, and presence of contamination. During fermentation, the inoculated microorganism metabolises the nutrient substrates, increases its biomass, and produces different products according to the growth phase. Chapters 7 and 8 provide a comprehensive overview of various types of fermenters and their configurations.

After the fermentation process is completed, the product is recovered and purified from the complex mixture of fermented broth. The concentration of desired products in the fermented broth is relatively low because the fermentation products, the cellular mass, residual nutrients, and byproducts remain in the same solution. Recovery of microbial products from the harvested fermentation broth is a multi-step process and involves a complex setup. This process is called downstream processing. The main steps of downstream processing are separation, recovery, purification, formulation, and packaging. Depending on the product and its applications, downstream processing may also need an aseptic environment. An efficient and economical bioprocessing of products can contribute to the success of their commercial exploitation. However, the overall production costs of the desired products can increase several-fold due to several factors, such as the number of steps involved in downstream processing, the use of sophisticated techniques, and the high purity requirements. The internal factors of fermentation productivity can also raise product costs. Each step during processing releases effluents or waste, which are also subjected to treatment or valorisation. These processes contribute to the overall expenditure involved in the production cost. Chapters 11 and 12 give full details of downstream processing and different techniques involved in the recovery of products.

SUMMARY

- Fermentation is a type of respiration in some microorganisms, which recovers energy (i.e., ATP) from glycolytic substrate-level phosphorylation. However, the same term in industrial microbiology means large-scale production of useful products from diverse microorganisms in a fermenter.
- Since the Neolithic era, human civilisations have used fermentation to produce alcoholic beverages and food items. The modern world also prepares fermented food and beverage using the same principle, but with much more scientific upgrades and on a large scale.
- LAB and yeasts are two potential examples of microorganisms used in fermentation through anaerobic reactions. Their respective products are lactic acid and ethanol. Fermentation reactions are enzyme-mediated and require strict physicochemical conditions.
- Microbial fermentation in industrial biotechnology involves the bulk production of microbial biomass and metabolites using diverse pathways, irrespective of aerobic and anaerobic conditions. This calls for a specific vessel, termed a fermenter, which maintains optimum conditions for fermentation reactions.
- The industrial fermentation can be categorised into three broad steps: upstream processing, fermentation process, and downstream processing. Upstream processing is the preparatory stage, the fermentation process is the central stage for microbial applications and product formation, and downstream processing is everything that occurs after fermentation.

DIFFERENT TYPES OF PRACTICE QUESTIONS

Fill in the blanks

I. In an anaerobic fermentation, bacteria obtain energy (i.e., ATP) from .... during glycolysis.
II. The earliest record of fermentation dates back to .... period.
III. A scientist .... provided indisputable proof that fermentation was essentially connected to the vital process of the living yeasts.
IV. .... identified that fermentation reactions do not require living cells but desire cellular enzymes.
V. Ethanol is a fermentation product manufactured by a microorganism ....
VI. .... is regarded as the father of industrial fermentation.
VII. The science of fermentation is also known as ....

SOLUTION: I. Substrate-level phosphorylation, II. Neolithic, III. Louis Pasteur, IV. Eduard Buchner, V. *Saccharomyces cerevisiae/ yeast*, VI. Chaim Weizmann, VII. zymology

Very short answer type of questions

I. Expand LAB.
II. Write the names of two microbial metabolic products and their sources manufactured at large scales during World Wars.
III. Write the names of enzymes that participate in ethanol fermentation.
IV. Define fermentation technology.
V. What are the three key events of industrial fermentation?

Short answer type of questions

I. Write a short note on respiration *vs*. fermentation.
II. Give an account of prehistoric records of microbial applications for human benefits.
III. Industrial fermentation extends beyond the use of anaerobic microorganisms. Comment on this statement.
IV. Differentiate between homo- and heterofermentative processes with suitable examples.
V. Write the differences between upstream and downstream processes of industrial fermentation.
VI. Why are zymase and invertase essential for ethanol fermentation? Where are they located?
VII. The final product of glycolysis is the primary substrate of different fermentation reactions. Elaborate.

Long answer type of question

I. How does the word "fermentation" have different meanings in different streams? How has industrial fermentation evolved from traditional fermentation knowledge to modern fermentation technology?

NOTES

1 McGovern P, Jalabadze M, Batiuk S, Callahan MP, Smith KE, Hall GR, Kvavadze E, Maghradze D, Rusishvili N, Bouby L, Failla O. Early neolithic wine of Georgia in the South Caucasus. Proceedings of the National Academy of Sciences. 2017;114(48):E10309–18.
2 Damerow, P. 2012. Sumerian Beer: The Origins of Brewing Technology in Ancient Mesopotamia. Cuneiform Digital Library Journal. 2012 (2). www.cdli.mpiwg-berlin.mpg.de/articles/cdlj/2012-2.
3 www.commons.wikimedia.org/wiki/File:Clay_tablet,_beer_for_the_workers._Temples_issued_workers_with_daily_rations_of_barely_beer,_the_staple_drink_of_Mesopotamia._Late_Uruk_period,_3000-3100_BCE._Purchased_via_Christie%27s_in_1989.jpg
4 Vallery-Radot René. "Portrait of Louis Pasteur." The Life of Pasteur. London, England: Constable, 1886–1911. www.digital.sciencehistory.org/works/av441ed
5 Fisher Scientific (Firm), and Louis Pasteur. "Double Flask for Demonstrating Anaerobic Fermentation," 1939–1967. Fisher-Pasteur Memorial Collection, Box 1. Science History Institute. Philadelphia. www.digital.sciencehistory.org/works/d791sg323
6 www.commons.wikimedia.org/wiki/File:E._Buchner_WS_1911-12.jpg
7 www.commons.wikimedia.org/wiki/File:Weizmann_1948.jpg
8 www.commons.wikimedia.org/wiki/Category:Charles_Pfizer
9 Muthusamy K, Soundharrajan I, Srisesharam S, et al. Probiotic characteristics and antifungal activity of *Lactobacillus plantarum* and its impact on fermentation of Italian ryegrass at low moisture. *Applied Sciences*. 2020;10(1):417.
10 Katepogu H, Wee Y-J, Almaary KS, Elbadawi YB, Gobinath R, Chinni SV, Lebaka VR. Isolation and Characterization of *Pediococcus* sp. HLV1 from fermented idly batter. *Fermentation*. 2022; 8(2):61. www.doi.org/10.3390/fermentation8020061
11 Karimy MF, Damayanti E, Suryani AE, et al. A simple method for analysis of *Saccharomyces cerevisiae* morphology by applying a high vacuum mode of the scanning electron microscopy and without chemical fixatives. *InIOP Conference Series: Earth and Environmental Science*. 2020;462(1):012048. IOP Publishing.

SUGGESTED READINGS

Alba-Lois L, Segal-Kischinevzky C. Beer & wine makers. *Nature*. 2010;3(9):17.

Bryant KL, Hansen C, Hecht EE. Fermentation technology as a driver of human brain expansion. *Communications Biology*. 2023;6(1):1190.

Buckel W. Energy conservation in fermentations of anaerobic bacteria. *Frontiers in Microbiology*. 2021;12:703525.

Damerow P. Sumerian beer: the origins of brewing technology in ancient Mesopotamia. *Cuneiform Digital Library Journal*. 2012;2:1–20.

De Vlieger JJ. *Green plastics for food packaging*. Woodhead Publishing Ltd; 2003.

Grüning NM, Ralser M. Glycolysis: how a 300yr long research journey that started with the desire to improve alcoholic beverages kept revolutionizing biochemistry. *Current Opinion in Systems Biology*. 2021:100380.

Hu J, Guo B, Li Z, Wu Z, Tao W. Freezing pretreatment assists potassium ferrate to promote hydrogen production from anaerobic fermentation of waste activated sludge. *Science of the Total Environment*. 2021;781:146685.

Lecomte SM, Achouak W, Abrouk D, Heulin T, Nesme X, Haichar FE. Diversifying anaerobic respiration strategies to compete in the rhizosphere. *Frontiers in Environmental Science*. 2018;6:139.

Mannaa M, Han G, Seo YS, Park I. Evolution of food fermentation processes and the use of multi-omics in deciphering the roles of the microbiota. *Foods*. 2021;10(11):2861.

Mentel M, Martin W. Energy metabolism among eukaryotic anaerobes in light of Proterozoic ocean chemistry. *Philosophical Transactions of the Royal Society B: Biological Sciences*. 2008;363(1504):2717–29.

Moo-Young M, Chisti Y. Bioreactor applications in waste treatment. *Resources, Conservation and Recycling*. 1994;11(1–4):13–24.

Nout MJ. Fermented foods and food safety. *Food Research International*. 1994;27(3):291–8.

Saz HJ. Facultative anaerobiosis in the invertebrates: pathways and control systems. *American Zoologist*. 1971;11(1):125–33.

Spector MP. Metabolism, central (intermediary). *Encyclopedia of Microbiology* (Third Edition) 2009:242–264.

Steinkraus KH. Nutritional significance of fermented foods. *Food Research International*. 1994;27(3):259–67.

5 Upstream Processing

5.1 UPSTREAM PROCESS: A PREPARATORY STAGE

Upstream processing refers to all the procedures that occur prior to fermentation. The establishment of successful fermentation is conditional upon this preparatory stage. This calls for a delicate balance of control and adaptation of microorganisms in a sterilised environment. Based on the frequency of the activities required during production, upstream processes can be categorised into two broad groups, including non-routine and routine processes (Figure 5.1). Non-routine processes refer to those activities that are performed only once or a few times during an industrial setup. Activities like isolation and characterisation of microorganisms, genetic manipulation, and fermentation optimisation are unrepeated tasks in the fermentation industry that are likely to be held only once. These approaches have scientific vigour and require expertise in the subject matter. On the other hand, routine processes are those activities that require effort in every cycle of production. These include sterilisation, raw material processing, nutrient media preparation, and inoculum development. These processes are handled by technicians and require careful observation. Stock preservation is considered a non-routine process, which is performed at regular intervals but less frequently than other routine processes.

Upstream processing in fermentation involves a series of chronological steps. However, some steps run parallel to each other. At some point, these parallel tasks converge with each other and assemble into a single route. A careful observation of different steps that are built upon each other is the key to the success of the fermentation process. For example, modifications to the strain may necessitate adjustments to the media formulation or the fermentation conditions.

In various preparations, sterilisation is the first and foremost step and the most basic requirement for any cultivation practices, which prevent contamination and maintain the yield and quality of products. The isolation and characterisation of microorganisms are the next but consequential steps in the selection of high-performing strains for industrial applications. Microorganisms isolated from natural habitats (like soil, water, and plant bodies) are important sources of novel strains and metabolites for the fermentation industry. They can be isolated on synthetic media and rigorously screened to determine their capacity to produce the desired products. They can further be genetically modified to improve their potential applications. The useful microorganisms are stored as stocks to protect their long-term viability and maintain their morphological, physiological, and genetic traits. Upstream processing extends studies on the optimisation of fermentation production media and process conditions. These studies are further expanded on laboratory and pilot scales to improve microbial productivity before being implemented in large-scale production. A wide range of optimisation strategies that employ classical to modern statistical and mathematical techniques provide solutions to complex problems across various fields and improve productivity. They can be utilised singly or in combination to determine the optimal parameters needed for fermentation. Ensuring that conditions are optimal for microbial growth and productivity, the large-scale fermentation can be administered.

An adequate amount of nutrients and essential substrates is required to prepare fermentation media. Some pretreatment of raw materials may also be needed to utilise the full potential of resources.

The initiation of fermentation requires a metabolically active microbial culture, which is called a fermentation inoculum or seed. The inoculum can be developed in sufficient amounts by rejuvenating the stock culture in the fresh sterile medium.

Upstream processing carefully navigates through various complex processes to achieve precise outcomes in the fermentation process and downstream processing. This chapter provides an overview of various preparations required during upstream processing.

5.2 STERILISATION

Sterilisation is one such process that ensures the complete removal of microorganisms from the environment. This is one of the common requirements essential for all types of culturing practices because a solution, bioreactor, or glassware contaminated with unknown microorganisms could be detrimental to microbial cultivation. When a contaminant becomes an admixture to the production microorganism, it can reduce the quality of the final biomass product. An antagonising

DOI: 10.1201/9781032651095-7

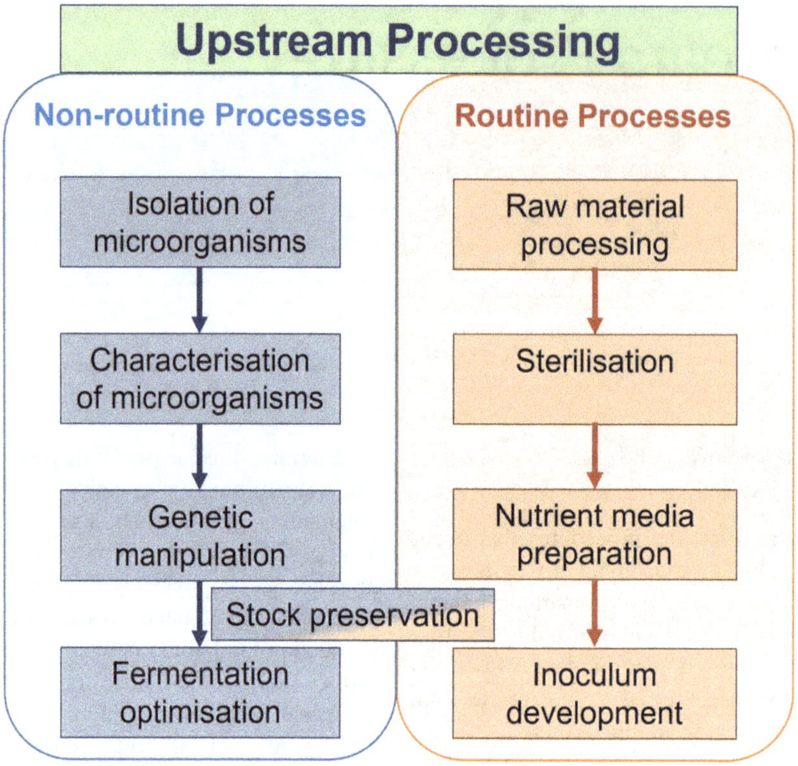

FIGURE 5.1 Different non-routine and routine processes of upstream processing.

effect of the contaminant on production microorganisms may cause the degradation of fermentation products, resulting in a loss of productivity. Furthermore, contaminants can make downstream processing difficult. Therefore, a process that can destroy or eliminate all microbial life forms, including resistant bacterial spores, becomes necessary for microbial culture preparation.

A different related process, disinfection, can be used in place of sterilisation. Disinfection of an object is mostly achieved using chemicals. However, disinfection does not guarantee a completely aseptic environment because it does not effectively act on resistant microorganisms and bacterial spores. Also, the use of chemicals can have an impact on the reaction mixtures and impede bioprocessing. Therefore, sterilisation is a stand-alone technique that achieves complete aseptic conditions for controlled biological operations. The terms sterilisation and disinfection are used interchangeably, owing to their interests for the same goal.

In the laboratory, sterilisation is strictly needed for microbial isolation, purification, and characterisation. This is equally important for small- and large-scale fermentation and even sometimes for downstream processing. Every glassware (such as Petri dishes, flasks, beakers, test tubes, and bottles) to be used for culturing must be aseptic. Inoculating loops, inoculating needles, and work surfaces are sterilised by flaming. In-hand exercise calls for hand sanitation to minimise the risk of contamination. Different body parts of the fermenter, such as the vessel, media and inocula feedlines, sampling port, and harvest lines, need cleaning and sterilisation after every use. Culture media and other solutions should be sterilised. Even the inlet air should also be germ-free when it passes inside the vessel.

The aseptic conditions for culturing practice depend on various sterilisation techniques. Based on the techniques and agents involved, it can be divided into physical and chemical methods (Figure 5.2). They are further discussed below.

5.2.1 Physical Methods

Physical sterilisation methods allow disruption of microbial structure and function using physical forces. The widely applied methods are wet heating, dry heating, radiation, and filtration (Figure 5.2), which are mentioned as follows.

5.2.1.1 Wet Heating (or Autoclaving)

Wet or moist-heating (steam under pressure) is the most widely used and reliable method of sterilisation. When saturated steam comes into contact with a cooler surface, it undergoes a phase transition from gas to liquid (condensation). This process releases the latent heat of vaporisation, resulting in a rapid and substantial transfer of heat energy to the surface. The released latent heat causes a rapid increase in the temperature of the surface, effectively eliminating microorganisms. The high temperature of pressurised steam has multiple effects on the viability of microorganisms, including denaturation of proteins, nucleic acids, and lipids and disruption of cellular structures. Steam sterilisation effectively kills bacteria, fungi, viruses, protozoa, prions, and spores. In particular, the

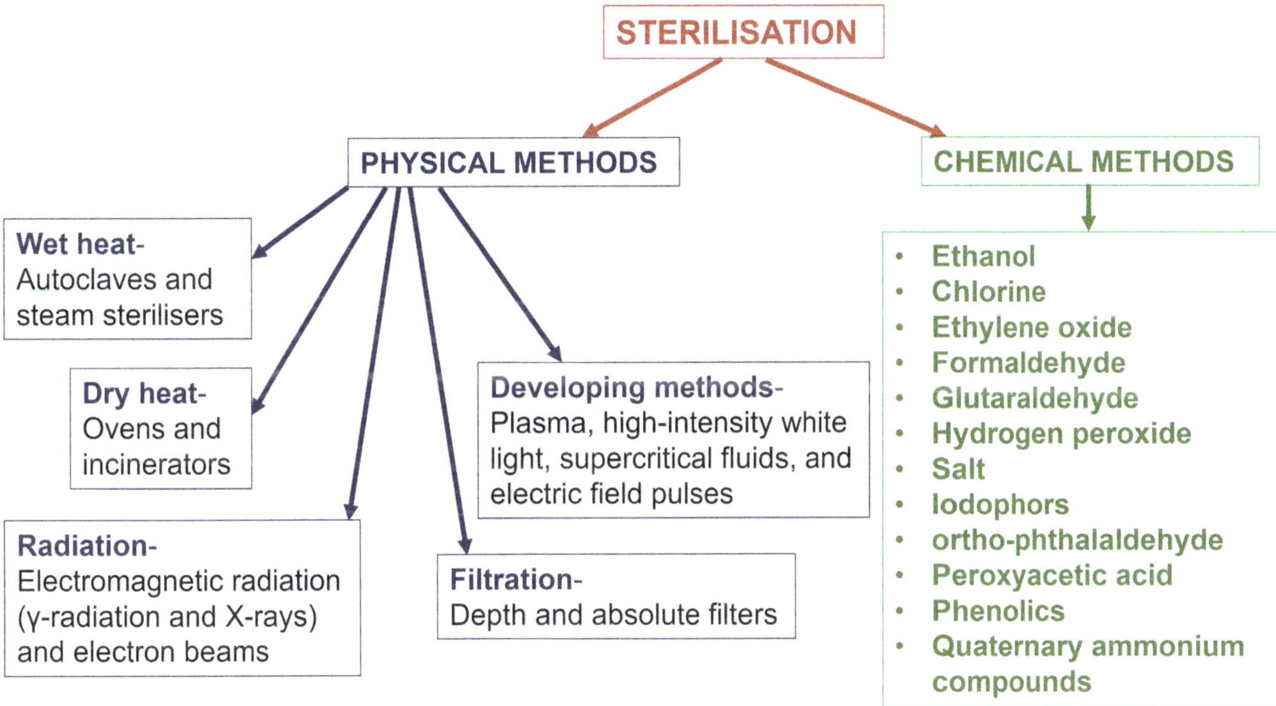

FIGURE 5.2 Various physical and chemical methods of sterilisation.

inactivation of bacterial spores requires very high temperatures that penetrate multiple protective layers and protein-protecting agents of dipicolinic acid to disrupt their vital content. Steam sterilisation is performed using sophisticated instruments like an autoclave, steam steriliser, or pressure vessel that can safely withstand high pressures and temperatures.

An autoclave is a very commonly used instrument for sterilisation in both research and industry. The sterilisation of an object using an autoclave is called autoclaving. The autoclave was first designed by a French microbiologist, Charles Chamberland (1879). The main body of the autoclave is a jacketed pressure chamber and consists of an airtight lid and an electric heating system. The safety valves on the lid act as a fail-safe device to prevent any catastrophic system failure. A pressure gauge indicates the amount of pressure applied during the operation. Autoclaves typically operate at 15 psi pressure and a temperature of about 121 °C for 15–30 min.

At sea level, the atmospheric pressure is 14.696 psi (pounds per square inch), water boils at 100 °C and produces steam. However, the pressure inside an autoclave is typically higher than atmospheric pressure. According to the principles of gas laws, an increase in the chamber pressure elevates the boiling point of the water. The operation at high pressure ensures that there is no evaporation from the liquid medium. The quality of steam is also a critical factor for effective sterilisation, which is determined by the water quality. When water is contaminated, organic and inorganic compounds present in water can be carried out along with the steam during autoclaving, which can be deposited inside the autoclave and onto the items being sterilised.

Steam has properties of efficient uniform distribution, penetrating and preventing unsterilised cold spots inside the autoclave. At the initial stage of autoclaving, freshly developed steam (i.e., water vapour) replaces the impure air and non-condensable gases (oxygen dissolved in water, carbon dioxide (CO_2), and nitrogen (N_2) in the chamber. A steam release valve facilitates the removal of steam from the chamber before and after a sterilisation procedure. Depending on the working principles for steam penetration in the chamber, autoclaves can be classified into-

- Upward displacement autoclave: It is a simple design unit where water is heated (with an electric heater or on a burner) at the base of the vessel to produce steam that rises upwards (Figure 5.3a). The preexisting air and steam are expelled under pressure for a given duration from a safety valve, followed by its closure. The chamber is heated for a specific duration to produce enough steam, which eventually raises the pressure and boiling point of water. After the operation is completed, the machine is allowed to cool down before being opened. This type of setup is typical of small laboratory autoclaves and pressure cookers. However, an autoclave can have low efficiency, particularly with complex heavy loads, because overloading prevents an efficient steam distribution, compromising the sterilisation process.
- Downward (or gravity) displacement autoclave: This equipment consists of a jacketed pressure vessel, which is preheated (electrically or with steam) before the main

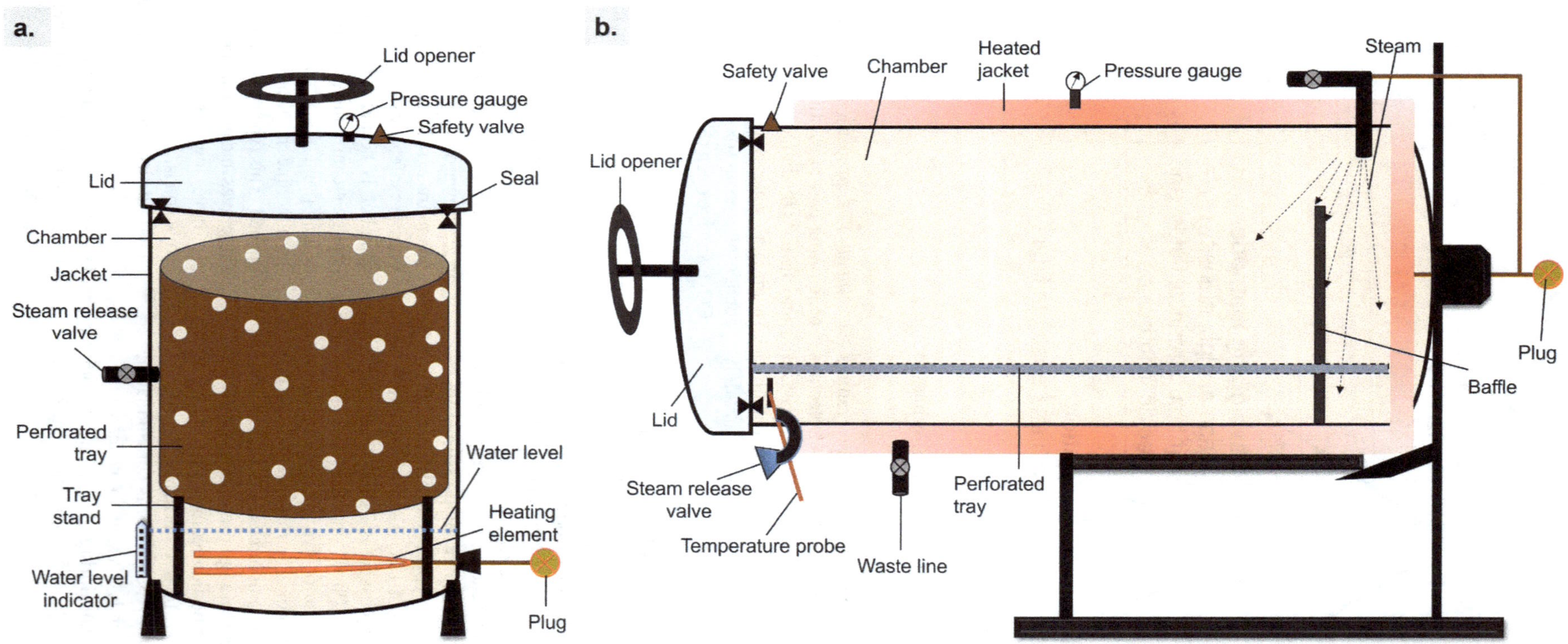

FIGURE 5.3 The schematic designs of different autoclaves. a. An upward displacement autoclave. b. A downward displacement autoclave.

operation. After preheating the load, steam is gently introduced into the top of the chamber (Figure 5.3b). The incoming steam passes through a tortuous path of a water separator to remove any accompanying water. The baffle attached to the walls of the chamber reduces the velocity of the entering steam to prevent mixing with the preexisting air. The less-dense steam fills the top of the vessel, which pushes preexisting air downward and forces it to go out from a valve at the chamber's base. After the chamber has attained the required temperature and pressure, the steam supply is shut off. The load rests on the platform for a specific duration. When the process is completed, the load is cooled by injecting air into the chamber. This kind of setup varies in size from small bench-top models to large laboratory and industrial scales.

- Vacuum and pressure-pulsing autoclave: In this type of unit, a vacuum pump is attached to the chamber, which develops a low pressure and forces the air to evacuate the vessel. Subsequently, vacuum pulses are introduced to ensure an even distribution of steam and eliminate any air pockets. The chamber maintains a set temperature and pressure for a specified duration to achieve sterilisation. When an operation is completed, the vessel chamber is allowed to cool down by releasing pressure or pulling the vacuum. This type of autoclave is very useful in conditions that require high degrees of sterility and where ordinary autoclaving is inadequate.

Autoclaving is much more effective and energy-efficient than dry heating sterilisation. This is because steam penetrates the items more deeply, transfers heat more efficiently and sterilises objects more quickly than dry heat. Wet heating also operates at a much lower temperature compared to dry heating. On an equimolar basis, steam provides at least seven times more heat energy than dry heat at 121 °C.

Autoclaving can be used to sterilise water, nutrient media, instruments, reusable devices, production units, and labware. The successful sterilisation in an autoclave involves the use of chemical and biological indicators. Chemical indicators include autoclave tape, chemical indicator strips, and integrators. They provide a visual indication that the item has been sterilised. Biological indication involves spore tests using vials or strips containing spores of highly heat-resistant bacteria (such as *Geobacillus stearothermophilus*). After the autoclave cycle, the spore strip is inoculated on a sterile medium and incubated to see if the spores have survived. A negative result (no growth) indicates successful sterilisation. Autoclaves can also be used to decontaminate laboratory-scale fermenter vessels and the different chemical and biological wastes. However, autoclaving is not suitable for those substances that are flammable, heat-sensitive, reactive, corrosive, or radioactive. Autoclaving of seawater alters its chemical composition due to precipitation of native salts. It also causes changes in pH, loss of dissolved gases, and changes in the composition of trace elements. A sealed container is also an autoclave-incompatible item because the high pressure inside the container causes it to explode. Table 5.1 provides a list of various advantages and disadvantages of steam sterilisation.

Industrial-scale fermentation employs a large volume of media, which calls for special facilities for sterilisation. Sterilisation for large-scale industries can be classified into *in situ* and *ex situ* methods. In an *in situ* sterilisation, the fermentation medium is sterilised inside the fermentation tank. The heat can be supplied into the surrounding water jacket and inside the vessel. The steam is injected inside the vessel, and the exit valve is closed to allow pressurisation and heating of the vessel above 100 °C (Figure 5.4a). In an *ex situ* option, a batch of the medium is sterilised in a separate chamber and then transferred to a steam-sterilised fermenter through sterile pipelines (Figure 5.4b). The use of a separate vessel for medium sterilisation offers several benefits, such as reduced strain on the fermenter, increased productivity, and energy efficiency. A different *ex situ* method is continuous sterilisation, which involves the sterilisation of feed using heat exchangers, followed by its supply into the fermenter. This mode of operation has advantages like scale-up automation, medium-quality maintenance, and reduced fermenter corrosion. Certain commercial autoclaves also incorporate a stirring mechanism to ensure the uniform distribution of heat and prevent the formation of medium clumps.

5.2.1.2 Dry Heating

Dry heating is a method of using hot air or heat (without humidity) for sterilisation. The hot air is promptly effective against bacteria, moulds, yeasts, protozoa, enveloped viral particles, and endotoxins. High temperatures of hot air oxidise cellular and viral particle components, which results in the loss of structure and function of biomolecules, including proteins, lipids, and nucleic acids. Viral particles rapidly disintegrate and lose their outer envelopes. Further, the boiling of intracellular water from high heat has effects similar to those of moist-heat sterilisation, including protein degradation, precipitation of cellular components, and loss of biomolecular function. However, bacterial spores and infectious prions are resistant to dry heat.

A hot air oven is a dry-heat steriliser that consists of a load-holding chamber and feeds electrically heated air. This operates in a temperature range of 50–300 °C, although some industrial and laboratory ovens can raise temperatures beyond this limit. The heating element has high resistance, thermal conductivity, and electrical insulation, which produces heat when an electric current is passed through it. A thermostat connected directly to the heater prevents fluctuation from the desired temperature in the hot air oven. A shock-resistant external shield (of aluminium or stainless steel), thick fibreglass or glass wool (in the space between the outer and inner chamber), and an asbestos door gasket on the side of the single door insulates the internal environment and prevents heat loss from the oven.

In a natural convection-type oven, the air is heated at the bottom, which rises towards the top. When it starts to cool

TABLE 5.1
Advantages and disadvantages of different sterilisation techniques

Techniques	Advantages	Disadvantages
Steam sterilisation	• Quick and highly effective • Inherent safety features • Exhibit broad-spectrum antimicrobial activity • Flexible and predictable • Depend on pressure, temperature, and contact duration • Materials sterilised in a proper container for subsequent storage or sterile handling	• Dependence on steam penetration • Inadequate for the release of trapped air or non-condensable gases • Not recommended for items susceptible to high pressure and temperature • Require a cooling period before handling • Organic and inorganic dirt prevent the penetration of steam • Not effective for bacterial endotoxins
Dry heat sterilisation	• Efficacy towards microorganisms and viruses • Useful to depyrogenate endotoxins or pyrogens • Simple loading and unloading • Convenient for sterilising heat-resistant, moisture-insensitive items • No issue of corrosion of the internal surface of the chamber • Suitable for the incineration of waste materials	• Not suitable for heat-labile materials and solutions • Energy inefficient • Not suitable for highly resistant bacterial spores • Thermal stratification • Inadequate airflow due to overloading • Release of hazardous byproducts (such as dioxins, furans, and chlorine monoxide) during operation
Radiation sterilisation	• Rapid and highly reliable • No need for heat or steam • Chemical and residue-free process • Cooling not required • Suitable for a range of materials	• Costly • Mandate safety measures to prevent accidental radiation exposure • Damage sensitive materials • Require sufficient radiation dosage and contact • Radioactive waste management and disposal concerns • Not effective against radiation-resistant microorganisms • Affect the colour, odour, and taste of foodstuffs and pharmaceuticals
Filtration	• Suitable for heat-sensitive materials • Effectively filter a wide range of microbes • Convenient usage	• Clogging of filters • Costly and frequent maintenance of filters • Ineffective for sterilising viscous or large volumes of liquids
Chemical treatment	• Sterilise heat-sensitive materials • Faster sterilisation • Vapours can effectively penetrate tiny spaces of materials	• Potentially toxic to humans and the environment • Can corrode metal instruments • Require special protocols, ventilation and protective equipment • Need careful disposal

down, it descends toward the bottom. The continuous natural heating and cooling within the device cause a non-uniform distribution of temperature. A relatively modern design is a forced convection oven with an installed motorised fan/blower, which uniformly distributes the hot air in the entire interior of the chamber (Figure 5.5). The forced convection allows constant sample contact with the hot air for instant heat absorption and sterilisation. The duration of sterilisation is inversely proportional to the temperature. For cooling the load, air is passed through a high-efficiency particulate air (HEPA) filter.

Dry heat sterilisers are suitable for sterilising laboratory glassware, heat-resistant chemicals, and the disposal of industrial wastes. However, it is strictly inappropriate for any solutions due to evaporation loss. This is also not suitable for flammable, radioactive, or volatile substances. Dry heating is highly energy inefficient and generally not recommended if autoclaving is possible. The various advantages and disadvantages of dry heating are given in Table 5.1.

5.2.1.3 Radiation

Electromagnetic radiation (e.g., gamma (γ)-rays, X-rays, and ultraviolet (UV) rays) and electron beams are the widely used ionising-radiation for sterilisation. These rays have high penetrating efficiency and sufficient energy to release electrons (ionisation) from the atoms of a compound. This disrupts the structural and functional attributes of biomolecules like lipids, proteins, and nucleic acids. Ionising radiations exert lethal effects on cells, including bacteria, by inducing DNA damage through both direct interaction with DNA molecules and indirect effects of water-derived free radicals. The ionisation of water and oxygen in the medium generates free radicals, ozone, and other reactive species, which have antimicrobial effects. These species may also react with the surface and

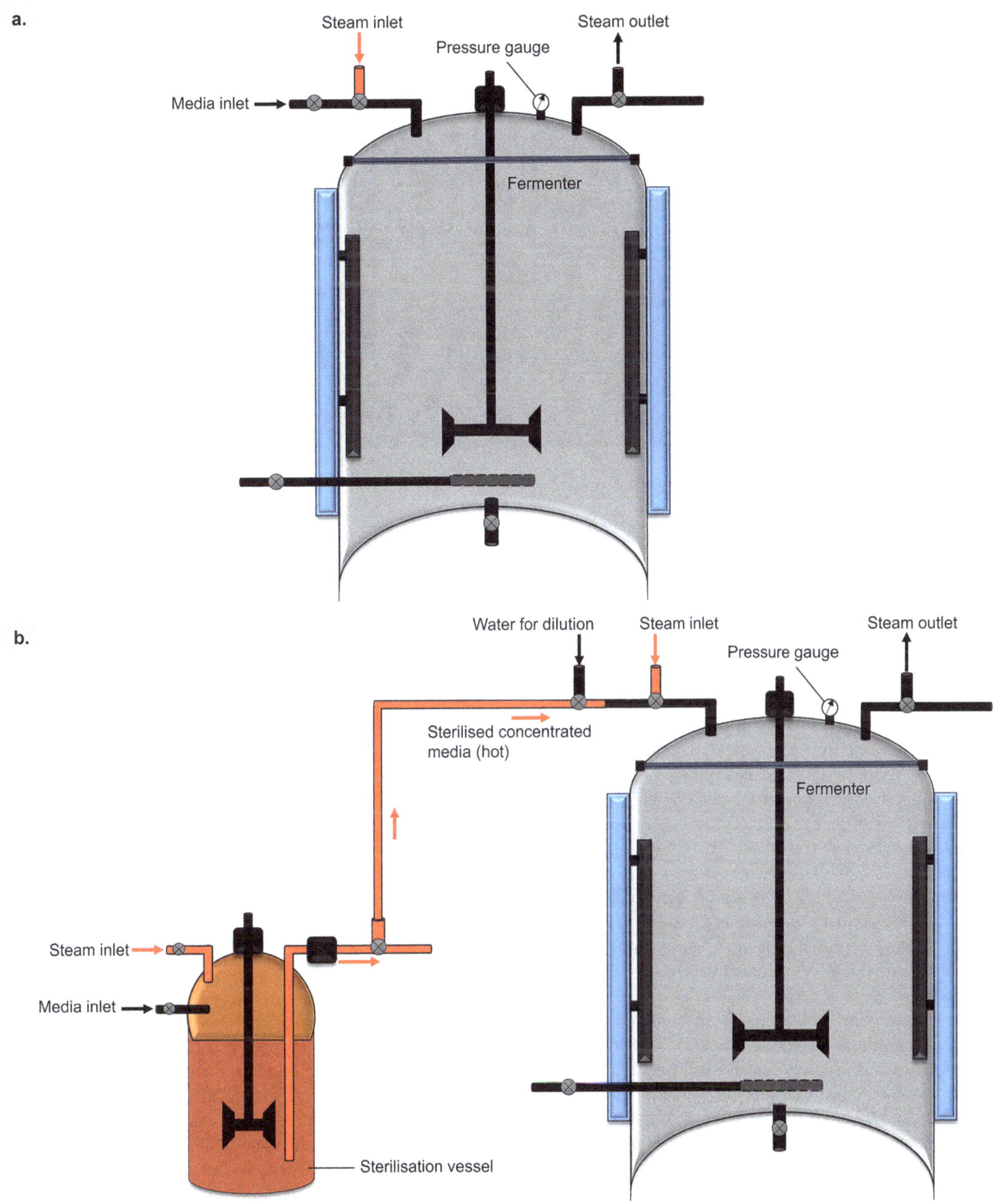

FIGURE 5.4 The schematic representations of different approaches for sterilising the media. a. An *in situ* approach for sterilising the media inside the fermenter. b. A separate sterilisation vessel connected to the main fermenter for an *ex situ* batch sterilisation of media.

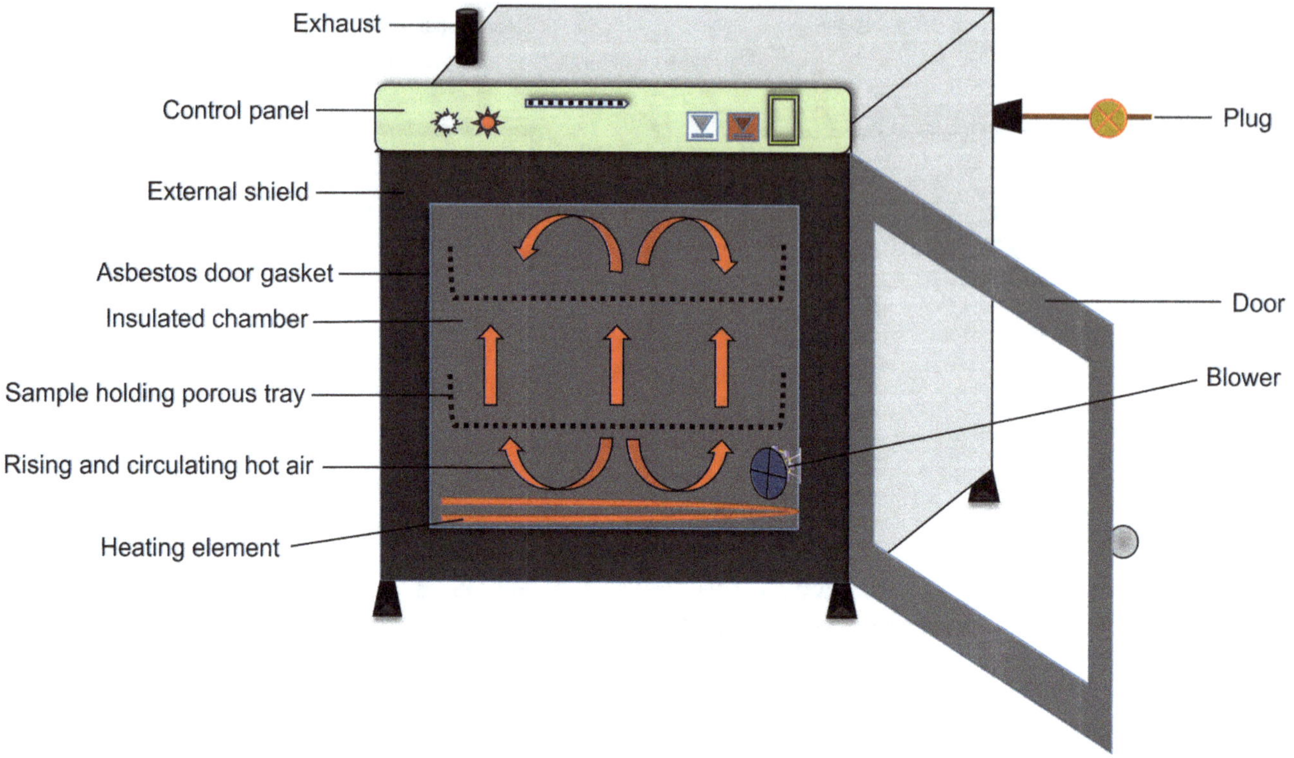

FIGURE 5.5 A diagram of a forced convection hot air oven used for dry heat sterilisation.

internal structures of the cells and have negative effects on cellular organisations.

The sterilisation via radiation utilises high-energy photon sources or high-energy electron accelerators (like direct current accelerators and radio frequency powered accelerators) to achieve effective microbial inactivation. High-energy photons of γ-radiation can be outsourced from the decay of isotopes (or radionuclides) like Cobalt-60 (^{60}Co), Caesium-137 (^{137}Cs), and Iridium-192 (^{192}Ir). Biocidal X-rays are produced by colliding high-speed electrons with a particular solid metal (tungsten or lead) in a glass vacuum tube. The UV rays can be produced using a germicidal lamp. These radiation sources are housed in a shielded compartment called a concrete shielded irradiator.

Treatment with radiation can be done in a batch or continuous mode. A range of 1–100 kGy (kiloGray) ionising-radiation doses is usually optimal for disinfecting cosmetics, water, wastewater, air/gas, pharmaceutical products, and prepackaged devices. This is also a suitable choice for pasteurisation and insect killing. The UV germicidal lamp fitted under laminar flow provides a germ-free environment inside the cabinet for culturing. The advantages and disadvantages of radiation for sterilisation can be found in Table 5.1.

5.2.1.4 Filtration

Filtration or filter sterilisation is a preferred choice for sterilisation when the components of culture media are heat-labile and denatured or modified during moist and dry heat sterilisation. A filtration unit mainly consists of filter membranes having several minute pores. A 0.22 μm pore size of the membrane filter is optimum for sterilising the liquid content. This method effectively separates all bacterial cells from liquid, as the smallest bacteria range in size from 0.20–0.30 μm. A good filter is made up of materials that have a low adsorption of liquid components. When the solution is passed through the filter membrane, microorganisms and other large particles settle on the filter surface. The solution that passes through the pores becomes aseptic. The mechanisms of filtration of suspended particles in a fluid occur according to four different principles:

- Inertial impaction- In this type, particles move in straight lines along with fluid to impact on filter fibres.
- Diffusion- In this type, particles move randomly and deviate from the fluid flow pattern to impact filter fibres.
- Electrostatic attraction- In this type, charged particles get attracted to the opposite charges on the filter surface.
- Interception- In this type, a single particle following the fluid flow comes within one particle radius of a filter fibre and is captured.

Based on pore distribution, filters can be of two types: depth and absolute filters (Table 5.2). A depth filter retains particles within its porous filtration media rather than just on their surface. In the depth filter, the liquid moves through an asymmetrical, three-dimensional fibre system. They are typically composed of thick, porous materials, such as bonded

TABLE 5.2
Differences between depth and absolute filters

Characteristics	Depth filters	Absolute filters
Type	Non-fixed pore filters	Fixed pore filters
Filter system	Loosely packed and cemented together	Fitted onto stainless steel and steam-sterilisable modules
Filter pore size	Not defined but ranging from 0.01 to >15.0 µm	Defined and customised from 0.01 to 5.0 µm
Filtration mechanism	Inertial impaction, diffusion, and electrostatic attraction	Mainly interception
Flow pattern	Tortuous routes	Constant flow patterns
Filtration efficiency	Restricted	Absolute
Economic consideration	Highly cost-effective	Expensive
Applications	Filtration of large masses of particles, such as cell debris, suspended solids, and fermented broth	Filtration of low-mass particles, like harmful particles, toxins, incoming air, and exhaust gases

borosilicate glass microfibre, cellulose, or polypropylene. On the other hand, an absolute filtration system consists of a series of different hydrophilic membrane filter cartridges fitted onto stainless steel and steam-sterilisable modules. Absolute filters are typically made from materials like polyethersulfone (PES), polytetrafluoroethylene (PTFE), and nylon. These filter cartridges provide a sequential filtration of particles in a decreasing pore size pattern. This is a surface filtration process in which the particles larger than the pore size remain on the filter surface, and the small-sized particles pass through. It is a widely used method for broth filtration. The absolute filters and HEPA filters work on a similar principle. The filtration process is also vital for the downstream processing of many important industrial products. A section in Chapter 12 provides a detailed account of the filtration technique for downstream processing.

5.2.1.5 Other Methods

Plasma, intense white light, supercritical fluids (SCFs), and electric field pulses are distinct tools of physical sterilisation. In plasma sterilisation, an adequate amount of energy in the form of heat or an electromagnetic field ionises gas to generate plasma, which is highly reactive and consists of unstable free radicals. These species tend to react with atoms and molecules to add or remove electrons from diverse materials. Plasma exposure damages the cell components of microorganisms. The excited electrons additionally release photons in the UV wavelength range that further contribute to antimicrobial effects. However, the reactive properties of plasma can also damage metal and plastic surfaces.

High-intensity white light is a high energy-containing light (ranging from 0.01 to 50 J/cm^2), which has rapid bactericidal, fungicidal, virucidal, sporicidal, and cytocidal effects. This can be used in short-duration pulses during sterilisation. It causes effects on nucleic acids, which develop dimers and photoproducts and facilitate strand breakage. Intense white light pulses can be involved in disinfecting and sterilising foods, food-packaging materials, liquids, vaccines, and biopharmaceuticals. This method is compatible with a wide range of materials, including those that are heat-sensitive and susceptible to chemical damage. However, pulsed light does not penetrate the cell deeply, making it unsuitable for bulk quantities of soil and highly viscous liquids. Further, it consumes high power and could produce toxic products from various photochemical/photothermal reactions.

SCFs can be used for the extraction and purification of various chemicals, precision cleaning of intricate or complex hydrophobic materials, and sterilisation of thermolabile liquids, including foods and pharmaceutical products. These fluids exist above their critical temperature and pressure and demonstrate properties of both liquid and gas. Supercritical CO_2 is the most commonly used SCF for sterilisation due to its favourable properties. It can diffuse into the cells and extract or disrupt the lipids by altering the intracellular pH. However, their applications require costly equipment and high pressures that have associated safety risks.

Electric field pulses are high-voltage electric fields (ranging from 1 to 20 kV/cm) that demonstrate antimicrobial activities on vegetative bacteria and yeasts. These pulses create compression and pore formation in the electrified membrane, which causes leakage of cytoplasm. This technique can be more effective when used in combination with SCFs, ozonation, hydrogen peroxide (H_2O_2), or other biocides.

Tyndallization is a relatively old technique of physical sterilisation, which is also known as fractional sterilisation or intermittent sterilisation. It relies on steam or wet heat to sterilise media for thermolabile substances (such as sugars). This procedure requires heating the material several times, with an interval in between. In classic Tyndallization, the materials are exposed to steam at 100 °C for 20–45 min at each session for three successive days. The initial steam exposure at 100 °C kills vegetative bacteria. Following that involves 37 °C moist incubation during the interval period that allows any dormant endospores to germinate into vegetative bacterial cell forms. These vegetative cells are subsequently destroyed within two days of exposure to steam. Vegetative cells tend to deactivate more easily compared to endospores. However, this process is unreliable and time-consuming, making it unsuitable for most large-scale applications.

5.2.2 Chemical Methods

Some chemical compounds are potent antimicrobial agents. They can be used for disinfecting or sterilising materials sensitive to high heat (in wet and dry heat sterilisation) and irradiation. The widely utilised chemical agents are alcohols, chlorine, ethylene oxide, formaldehyde, glutaraldehyde, H_2O_2, food acids, ozone, quaternary ammonium salts, etc. (Figure 5.2). These chemicals have different effects on microorganisms, which are discussed as follows.

- Alcohols like ethanol and isopropyl alcohol (60–90%, v/v) have bactericidal, fungicidal, and virucidal effects but do not destroy bacterial spores. Ethanol causes dissociation or dissolution of the lipid membrane and denaturation of proteins. It is occasionally used to disinfect external surfaces of equipment and as a sanitiser.
- Chlorine disinfectants or hypochlorites have a broad antimicrobial activity. Their microbicidal mechanisms include a combination of factors. One condition generates oxidative stress and disrupts respiratory components, sulphydryl enzymes, and amino acids. They additionally inhibit protein synthesis, oxygen uptake, and ATP production and induce DNA breakage. These chemicals are widely used for disinfecting healthcare facilities and water purification.
- Ethylene oxide alkylates nucleic acids (on purine bases) and proteins (on sulfhydryl, hydroxyl, carboxyl, and primary amino groups). Alkylation disrupts DNA replication, cellular growth, and reproduction in microorganisms.
- Formaldehyde is used as a high-level disinfectant and sterilant. Formalin (a 37% aqueous solution of formaldehyde) alkylate amino and sulfhydryl groups of proteins and the nitrogen atoms of purine bases in nucleic acids. Vapours of heated paraformaldehyde (a solid polymer of formaldehyde) are employed for the gaseous decontamination of laminar flow biological safety cabinets.
- Glutaraldehyde is also a notable chemical sterilant. An alkaline (pH 7.5–8.5) aqueous solution of glutaraldehyde has biocidal properties and is sporicidal. Glutaraldehyde alters the synthesis of DNA, RNA, and protein by the alkylating sulfhydryl, hydroxyl, carboxyl, and amino groups. It is used for surface disinfection of equipment in healthcare settings and instruments in laboratories.
- Both liquid and gas forms of H_2O_2 produce destructive hydroxyl free radicals that significantly disrupt membrane lipids, DNA, and other essential cell components.
- Salts like sodium chloride (NaCl) are a traditional preservative used in household food items like pickles, cheeses, meat items, savoury snacks, and other food items. Excess amounts of salt in the medium cause osmosis, inhibiting microbial growth and reproduction. Such a process requires salt concentrations above 10%. Food items preserved by this method last for several months.
- Iodophor is a combination of iodine and a solubilising agent or carrier. The free iodine of diluted iodophor is microbicidal but requires prolonged contact to disrupt fungi and bacterial spores. Iodine has antiseptic and disinfectant properties due to its capacity to interfere with microbial cell walls. Additionally, it can disrupt the structure and synthesis of nucleic acids and proteins.
- ortho-phthalaldehyde is a high-level disinfectant with excellent microbicidal activity, material compatibility, and stability at a wide pH range (pH 3-9). It is a lipophilic compound that interacts with amino acids and proteins of cells and disrupts spore germination.
- Peracetic (or peroxyacetic acid) has a rapid action against bacteria and fungi, including yeasts. It removes organic materials, disrupts the cell wall permeability, and oxidises sulfhydryl and sulphur bonds in proteins.
- Phenol and its derivatives disrupt the cell wall to cause leakage of essential metabolites and penetrate the cells to precipitate proteins and inactivate enzyme systems.
- Quaternary ammonium compounds like alkyl dimethyl benzyl ammonium chloride, alkyl didecyl dimethyl ammonium chloride, and dialkyl dimethyl ammonium chloride are widely used disinfectants and cleaning agents. These compounds cause cell membrane disruption, protein denaturation, and inactivation of energy-producing enzymes.

5.3 ISOLATION AND CHARACTERISATION OF MICROORGANISMS

Isolation of microorganisms refers to the process of cultivation and purification of a microbial form from an environmental sample onto a synthetic medium. The methods of microbial cultivation in an artificial environment are manifold. These comprise spread plating, streak plating, pour plating, stab culturing, broth culturing, etc. Microorganisms are isolated using standard media. Traditional methods of characterising microorganisms involve analysis of the morphological, metabolic, immunological, genetic, ecological, and pathogenic parameters. Modern molecular techniques like genomics, transcriptomics, proteomics, and metabolomics shed light on the metabolic capabilities of microorganisms. Chapters 21 and 22 provide in-depth information on the isolation and characterisation methods of microorganisms.

5.4 MICROBIAL GENETIC MANIPULATION

Microbial genetic manipulation is an approach to graft new abilities into producer microorganisms. This can be achieved through cutting, splicing, and integrating specific DNA segments in the microbial cells. A genetically engineered microorganism (GEM) is defined as a microorganism endowed with new phenotypes and the ability to efficiently express desirable proteins or molecular compounds for a specific reason.

The construction of GEM in industrial biotechnology is considered an upstream stage. Genetic engineering techniques have made significant contributions to the development of diverse GEMs, which are capable of producing a wide range of products from prokaryotic and eukaryotic cells. The approach of microbial engineering can also be used to make downstream processing less complicated (Box 5.1). A few of the notable techniques that assist microbial genetic manipulation include direct DNA synthesis, polymerase chain reaction (PCR) methods, restriction enzyme digestion, recombination methods such as CRISPR (clustered regularly interspaced short palindromic repeats), associated protein 9 (Cas9) genome editing and the Red/ET recombination technology, etc. The techniques of conjugation, cloning, electroporation transfection, heat shock, lysogeny, protoplast fusion, and transduction are methods for gene incorporation and genetic recombination.

The high-throughput omics technologies, including genomics, transcriptomics, proteomics, epigenomics, metabolomics, metagenomics, meta-transcriptomics, meta-proteomics, and meta-metabolomics, have made some notable contributions to microbial genetic modification. Omics technologies produce plenty of information about important genes, regulatory pathways and elements, proteins, RNAs, and adaptive mechanisms that enhance the functional attributes of microorganisms at a relatively low cost. Advanced sequencing technologies and bioinformatics analysis have excavated many functional genes and gene clusters from nonculturable microorganisms, allowing precise manipulation of bacteria and fungi for diverse applications.

The key advantages of genetic modification include the production of multiple foreign products in the same strain, easy manufacturing scale-up, long-term effects in colonisation sites, low adverse effects, and the production of non-native products. In addition, the GEM in medical science can actively supply molecules, express foreign genes, restrain pathogenic bacteria, enhance endogenous gene expression, or destroy tumour cells. Chapter 9 provides a detailed account of genetic manipulation in industrially useful microorganisms and their applications.

5.5 MICROBIAL STOCK PRESERVATION

The number of isolated microorganisms is steadily increasing as contemporary cultivation techniques are being advanced. However, their culturing would be inadequate without suitable preservation techniques. When microorganisms are isolated and screened, they are purified and maintained as a stock culture for future use. This calls for reproducible and sustainable preservation techniques that do not interfere with microbial integrity for long-term storage. Storage at a below-freezing point is one of the most common facilities for preserving microorganisms. The active microbial forms pass through different growth phases. Cultures at the stationary phase are the best stage for long-term preservation because cells are well-nourished and usually the most resistant. The stored cells often have a tendency to get lethal or sublethal injury. Therefore, the high cell density in the stock solution also enhances the survival percentage of cells during preservation.

When the preserved microorganisms are revived, they are subjected to evaluation based on qualitative and quantitative parameters. This can be evaluated by their ability to reproduce, particularly with quantitative measurements such as colony-forming units (CFU) or optical measurements, which detect microbial density in the solution. They can be further assessed for their phenotypic and genotypic properties. This type of assessment is important because microorganisms stored for long periods are prone to genetic alteration due to selective pressure and mutation.

Some well-known preservation methods of microbial cultures are discussed below.

- Periodic subculturing- A common and convenient method of maintaining the microorganisms in regular use is growing them periodically on the surface of an agar-containing medium. The storage facility requires a normal refrigerator with a temperature below 10 °C. This method is suitable for microorganisms, which are less resistant to extreme freezing or drying. Anaerobic microorganisms are typically stored in stab or broth cultures. However, periodic subculturing on agar

BOX 5.1 MICROBIAL ENGINEERING FOR THE EASE OF DOWNSTREAM PROCESSING

Modern biomanufacturing of microbial products demands smart techniques, which not only overcome the burden of cost, enhance productivity, process simplification, and improve product quality but also reduce their impact on the environment, meeting the sustainable parameters. Chemical engineering has already established a landmark for such requirements and is still progressive in various parameters. The next level of integrity involves modification in the producer organisms. The advancement of microbial engineering makes it possible to do less complicated downstream processing. For the same reason, genetic engineering, protein engineering, synthetic biology, and systems biology have offered new tools to alter the morphology and physiology of producer microorganisms. Morphology engineering aims to improve cell aggregation during downstream processing. This can allow the development of large spheres of cells for their easy recovery. The genes involved in the synthesis and regulation of bacterial cell wall components, such as transpeptidase, transglycosylase, transmembrane protein, and cytoskeleton protein, can be manipulated to alter cell morphology. The insertion of the lysis gene in the cell offers better and more controlled lysis of the cell during cell disruption. The deletion, addition, or frameshift mutations of some genetic elements can affect cell division. Physiological engineering allows the insertion of new genes for the coproduction of different metabolites. In recent years, immobilisation techniques have garnered interest to speed up downstream processing, which is discussed in other parts of the book (Chapters 15, 16, and 17).

medium has a short lifespan. With each subculture, the likelihood of genetic change and morphological variation also increases.

- Oil immersion- Certain microorganisms, such as *Bacillus*, *Pseudomonas*, and *Escherichia,* growing on agar surfaces can be covered with mineral oil for short-term storage. While it does not affect the morphological or physiological characteristics, long-term storage may alter the genetic makeup. Subculturing from storage is a tedious method, and therefore, this is not a preferred storage method.
- Distilled water suspension- A suspension of microbial spores and cells in distilled water can be used for short-term preservation. This is considered to be suitable for a few microorganisms like *Pseudomonas*, sporulating fungi, and *Saccharomyces cerevisiae*.
- Spray drying- Spray drying is an efficient dehydration technique for preserving microbial cultures at an industrial scale. This technique has a lot of commercial applications for drying various products in downstream processing. In this process, a homogeneous solution of microbial culture is atomised into a drying chamber at high temperatures. This allows separation of water from the solid particles. Spray drying produces a dry culture, containing about 4–7% moisture and maintaining good storage stability. It is a cost-effective technology for manufacturing probiotics. Protective carrier matrices like skim milk or polydextrose-based prebiotic substances can be added to protect the cells from high drying temperatures. However, the elevated temperatures during spray drying induce heat shock and osmotic cellular stresses on sensitive microorganisms. The details about spray drying are given in Chapter 12.
- Lyophilisation- Lyophilisation or freeze drying is the most preferred method for archiving and long-term preservation of microorganisms. Most microorganisms survive drying, and the lyophilised cultures can be easily stored. They can be shipped at room temperature without significant loss of cell viability. The lyophilisation process mainly involves freezing the sample (microbial culture suspension) in glass vials or ampoules, followed by sublimation and drying. The microbial cultures from the stress period of excess cooling can be protected by using a variety of suspending medium combinations, such as lactose, nutritional broth, serum and glucose broth, skim milk, sucrose, etc. Detailed information on lyophilisation can be obtained from Chapter 12.
- Cryopreservation- Cryopreservation is the process of preserving living cells at very low temperatures for long-term storage. This requires temperatures below −80 °C, or typically near the liquid nitrogen (-196 °C). Liquid nitrogen is used extensively for this purpose as it provides the lowest practical temperature without causing any genetic alteration to cells. However, microorganisms can variably survive during ultra-rapid freezing and thawing. During cryopreservation, cells experience two primary injuries: osmotic shock and mechanical injury. Excessively slow cooling can destroy the cell membrane lipoprotein complex by increasing the electrolyte concentration of the cell solution, giving an osmotic shock. On the other hand, excessively rapid cooling produces large intracellular ice. Depending on cooling rates, cryopreservation can be employed through rapid cooling (i.e., vitrification) and slow freezing. The vitrification process employs high concentrations of cryoprotectants, which prevent osmotic shock and mechanical damage due to ice formation. Cryoprotective agents or cryoprotectants are often mixed in microbial cultures to prevent freezing-induced osmotic pressure injuries to the cells. Cryoprotectants are water-soluble, non-toxic, and non-reactive. They can be categorised into osmotic and nonosmotic types. Examples of osmotic cryoprotective agents are glycerol, ethylene glycol, propylene glycol, dimethyl sulfoxide (DMSO), and acetamide. These agents penetrate the cells to maintain osmotic pressure. Examples of non-osmotic cryoprotectants are albumin, polyethylene glycol (PEG), polyvinylpyrrolidone (PVP), and sucrose. However, these agents do not penetrate the cell walls.

In addition, some new preservation techniques, such as immobilisation and high-voltage electrohydrodynamic processes, have emerged to complement the conventional preservation methodologies (detailed in Box 5.2). The microbial stock cultures provide many benefits to microbial technology, including biodiversity conservation, identification, research and training programs, genetic manipulation, and industrial production. Several private and governmental microbial culture collections and resource centres have maintained microbial stocks, as described in Chapter 3.

5.6 RAW MATERIALS AND THEIR PROCESSING

Raw materials are unprocessed materials that can be used to produce diverse products. The choices of raw materials for industrial fermentation are manifold. The selection of the best raw material for an industrial setting is based on several important criteria. These include high availability, reliable quality, easily fermentable sources of carbon and nitrogen, the correct ratio of nutrients for the target microorganism, cost-effectiveness, the absence of contaminants that could interfere with fermentation, the capacity to generate the desired product in large quantities, and few by-products. Agro-industrial wastes and industrial by-products are modern resources to be used as raw materials to produce diverse products in industrial-scale fermentation (Figure 5.6). This can also include specially formulated feedstock to enhance the economic feasibility of manufacturing systems while reducing their environmental impact. These substances are plentiful and are high in proteins, carbohydrates, and minerals that microorganisms can ferment and assimilate. They have unknown chemical makeups but consist of essential ingredients for conventional microbial fermentation media.

BOX 5.2 MICROENCAPSULATION: A NOVEL PRESERVATION TECHNIQUE

Microencapsulation is a relatively novel approach to microbial preservation. In this process, cells are entrapped in water-soluble polymers like acacia gum or pullulan, which spontaneously polymerise and trap cells in a protective, stable film. This can be kept for a long time without any cold storage. These polymers confine water in their trap and keep the bound cells hydrated. Cell entrapment in inorganic matrices, like a silicon dioxide-derived matrix, prevents microbial genetic instability and loss of metabolite production. Cell immobilisation into porous hydrophilic polymers provides mechanical strength and thermal stability for unstable microorganisms. Microencapsulation into a gel matrix (like calcium alginate) is prolific for the controlled release of probiotic cells in the human gut.

High-voltage electrohydrodynamic processes, such as electrospinning and electrospraying, are novel microencapsulation techniques for preserving the viability of sensitive microorganisms. Electrohydrodynamic processes utilise a wide range of polysaccharides (chitosan, cellulose, or alginate) or protein-based (whey protein concentrate) support matrices for cell entrapment. The atomisation process develops nanoscale and microscale fibres, which maintain high protein stability and functionality. The high surface area-to-volume ratios and permeability of these fibrous materials allow the entrapped cells to obtain nutrients while retaining their metabolic activity. The processes have potential applications for encapsulating probiotic strains and in the development of functional foods.

Corn steep liquor is a by-product of the corn wet-milling industry, which serves as an organic nitrogen source for fermentation. This material also provides complex polysaccharides, reducing sugars, and lactic acid. This is a key supplement in several fermentation media for producing enzymes, antibiotics, ethanol, lactic acid, and polysaccharide pullulan. Molasses is a viscous residue of sugarcane or sugar beets from the sugar-making process. This material is rich in carbohydrates (sucrose), nitrogenous matter, vitamins (B and biotin), and minerals. Bacteria and fungi can be grown in molasses-containing media to produce alcohol, baking yeast, citric acid, and food additives (such as lysine and monosodium glutamate). Malted barley aqueous extract contains about 90% carbohydrate and is suitable for the cultivation of actinomycetes, yeasts, and filamentous fungi. Sulphite waste liquor is a by-product of the wood pulp industry that can be used to grow yeast. A watery by-product of the dairy industry is whey, which contains lactose and helps lactic acid bacterial (LAB) growth. Glycerol waste released from the biodiesel industry can be modified into several value-added products such as ethanol, citric acid, succinic acid, 1,3-propanediol, poly(3-hydroxybutyrate) (PHB), etc., employing diverse microorganisms.

Solid-state fermentation (SSF) substrates are mostly agro-industrial residues. These include substances like straw and bran (of wheat, rice, and lentil), husk (of rice, green gram, and black gram), corn cob, millet grain, and groundnut shell. The plant oil-seed cakes of palm, maize, coconut, groundnut, sunflower, castor, and olive are good sources of proteins, carbohydrates, and other nutrients. The pineapple leaf, raspberry seed powder, tea waste, coffee pulp, fruit waste, fruit peels (like banana peels and pomegranate peel), grass powder, and sawdust are some popular low-cost, highly available raw materials for SSF.

Raw materials often require some pretreatment to extract the essential substrates and nutrients before the fermentation process. Pretreatment methods can be physical, chemical, or biological processes, depending on the types of raw materials (Figure 5.6). Physical treatment, like milling, steam explosion, or heating, improves microbial access to the cellular content in the solution. Milling reduces the size of particles of lignocellulosic biomass. Chemical methods can be acid hydrolysis, alkaline treatment, and solvent extraction. For example, protein hydrolysates produced by the acidic digestion of agro-industrial materials release peptones, which serve as the nitrogen source in the fermentation media. Enzymatic treatments (using enzymes such as amylases, cellulases, and xylanases) have received wide attention for the pretreatment of raw materials, owing to their sustainability and eco-friendly approaches. For example, enzymes convert complex polysaccharides into fermentable sugars. Some raw materials are recalcitrant and require treatment by more than one method. A different purpose for the pretreatment is the removal of inhibitory agents (like lignin in lignocellulosic mass) or toxins from raw materials.

5.7 OPTIMISATION

The optimisation of variables (factors), including nutrient media (sources and concentrations of carbon, nitrogen, minerals, inducers, etc.) and fermentation conditions (such as temperature, pH, agitation rate, and aeration), is a critical procedure of upstream processing for large-scale fermentation. Optimisation studies significantly improve productivity and product activity, which renders low production costs. There are various techniques of optimisation, each having its different advantages and disadvantages (Table 5.3). These range from the classical, lengthy technique that checks one parameter at a time to the more modern, rapid, and economical mathematical/statistical techniques that can validate several interacting parameters in one step.

The statistical techniques mostly involve different designs of experiments (DOE), which systematically apply statistics to experimentation. The experiments are designed in a random order that maximises accuracy and reduces the chance of incorrect inference. These methods allow

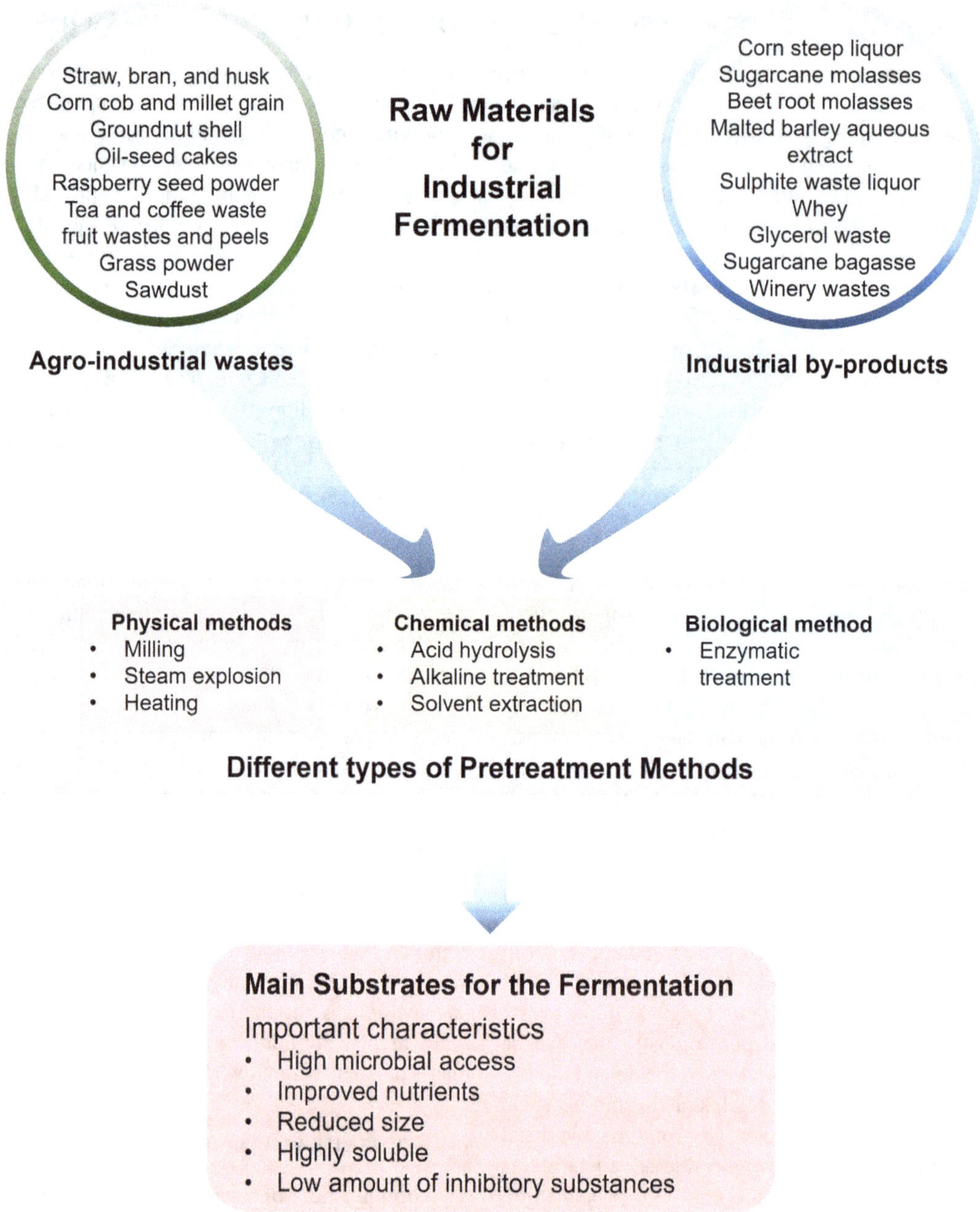

FIGURE 5.6 Various raw materials used in industrial fermentation and their pretreatment.

for the identification of those factors that are important for fermentation. The screened relevant factors can then be further optimised in a different set of DOE to analyse for their optimum concentrations or interaction effects. Software programs like Design-Expert, Minitab, R Studio, JMP, and Statistica are suitable platforms that can design, calculate, and analyse experimental runs for statistical optimisation. The advantage of statistical optimisation applies to a wide range of experiments. All factors can be varied simultaneously, which helps determine their interactions to maximise experimental efficiency. An interaction effect occurs when a third variable affects the relationship between an independent and dependent variable. Mathematical models can also be built to predict the involvement of interactions between variables.

The brief characteristics of various types of classical and statistical DOE optimisation techniques are as follows-

a. One factor at a time (OFAT) technique- It is the classical method for the optimisation of fermentation parameters. This involves manual interventions where only one factor is varied at a time while keeping all the other factors constant. For example, if the effect of different carbon sources (e.g., glucose, fructose, sucrose, lactose, maltose, and cellulose) on enzyme production is studied independently, all the other

TABLE 5.3
Advantages and disadvantages of different optimisation techniques

Optimisation Technique	Advantages	Disadvantages
One factor at a time (OFAT)	• Ease • Convenience • Appropriate for beginners	• Check the effect of one parameter on an occasion • Laborious, tedious, and expensive
Plackett Burman Design (PBD)	• Identify the main effects of essential factors with fewer experiments • Used in the initial screening • Economical	• Lack of accountability for an interaction effect between different factors • Lack of precision for the smallest design
Taguchi Design	• Analysis of many factors with a reduced number of experiments • Identify factors with the most effect • Used at initial development stages	• Difficult to resolve the interaction effects of the factors • All variable combinations of the factors cannot be checked.
Response Surface Methodology (RSM)	• Economical screening designs • Flexible sequential designs • Maximum information in a minimum experimental trial • High statistical efficiency • Analyses interaction effect between different factors • Estimate the nonlinearity of responses in the given data set	• Valid only for the selected ranges of the parameters • Invalid for regions other than the studied ranges of factors and requires many confirmation trials • The distribution of response variables is assumed to be normal
Artificial Neural Network (ANN)	• Used for optimisation, control and evaluating changes in response variables • Used to develop predictive models • Useful for good decision-making related to design and functioning	• Hard to interpret the reasons for optimised values • Computationally intensive and requires strong processors • Require a large amount of data for optimisation • Difficult to map the output signals
Bayesian Optimisation (BO)	• Detailed prior knowledge of the processes is not needed • The model gets constantly refined with more data collection • Multiple parameters can be adjusted	• Require significant expertise • Relies heavily on the chosen surrogate model • High computational cost

parameters (nitrogen source, pH, and temperature) are kept constant. In this setting, the concentrations of the medium can be adjusted within a predetermined range. Several standardised physical parameters, including pH, temperature, agitation, and aeration requirements of the fermentation process using OFAT evaluation, can also be determined. The OFAT is further divided into removal experiments and supplementation studies based on the methodology used. The removal type of experiment involves removing medium components from the production medium and observing their effects on metabolite or product production. Supplementation experiments are typically conducted to assess the effects of different carbon and nitrogen supplements on metabolite production.

The ease and convenience of the OFAT technique make it suitable for initial screening and optimisation, particularly when information about fermentation media and parameters is not known. The OFAT has been the most popular option for creating the medium composition or production of new metabolites. It is employed in the early phases of industrial settings because of its simplicity. However, this requires an extensive number of experiments to determine the optimum values. Also, estimating any interaction effect between different factors is difficult.

b. Factorial Design- It is a classical type of DOE based on the analysis of variance (ANOVA), including various models. It can analyse the effect of two or more independent variables (factors) on one dependent variable. The factorial design experiments can vary with the number of factors and their levels involved. For instance, if the effects of two levels (presence and absence) of three factors (carbon, nitrogen, and trace element) are studied on one dependent variable (the production of an enzyme), this experiment is a 2^3-full factorial design containing eight experimental runs. This can analyse the optimal values of the main factors and also allows estimation of their interaction effects. However, the number of factors and their levels increases the experimental runs, which often makes full factorial designs redundant. This can be improved by the use of economical fractional factorial designs, which involve a subset of the complete experimental runs of a full factorial design to provide necessary information on the important requirements of optimisation.

c. Plackett Burman Design (PBD)- It is a type of DOE used for screening several variables concurrently. The PBD determines those variables that are important for fermentation while screening out the least important factors (noise). The design has real and dummy variables. The real variable can vary at two levels (high and low levels), whereas the dummy variable remains constant (to estimate error). In a PBD experiment, N number of experimental runs are used to test up to N-1 variables. The N is any experiment in multiples of 4 (or 4N), but it starts with a minimum of 8 runs. When an experiment has 4-7, 8-11, 12-15, 16-19, and 20-23 variables, it generates 8, 12, 16, 20, and 24 experimental runs, respectively. For instance, if there are six variables, the PBD will have eight experimental runs (six real and two dummy variables). However, PBD analysis can not estimate the potential interaction between the different variables. Therefore, this tool can estimate only the main effects of process-related variables (with >90–95% confidence levels).

d. Taguchi Design- It is a factorial-based technique that assigns the experimental variables using a systematic orthogonal array. An orthogonal array conducts a sequence of experiments under various conditions that promote a balanced design with equal weightage to all variables. Taguchi's Design experiment optimises the effective variables to minimise variation and obtain target values of an output. The study uses a minimum number of robust designs of experiments to examine the effects of various variables on the mean and variance of fermentation efficiency. For instance, if there are six variables with three levels each, a Taguchi orthogonal array will have eighteen experimental runs, denoted with L_{18} (3^6). This is a much-reduced number of experiments compared to a fractional factorial design, where there will be 729 experiments to perform. The Taguchi method offers advantages like cost-effectiveness, robustness, and ease of explaining the results, but is limited to an optimal combination of variables and levels.

e. Response Surface Methodology (RSM)- This is a robust statistical optimisation technique that uses different factorial DOE and multiple regression analysis to optimise fermentation productivity. RSM obtains the best productivity, which is formulated under a set of mathematical equations. The optimisation occurs in several phases, which start with an initial screening of factors with a factorial design or PBD. It then proceeds to a second follow-up design (fitted with quadratic regression models) to explore the possible high-order univariate effects and interactions of the screened variables. The central composite design (CCD) or the Box-Wilson design is an experimental RSM design that is extensively used for the optimisation of the variables and responses with a limited number of experiments.

   RSM has become the core of industrial microbiology experimentation with the ability to estimate the experimental variability, provide the adequacy of the proposed model and experimental data, predict the observed response precisely, easily identify outlier data, and maintain high efficiency with economical cost and duration. For example, *Bacillus drentensis* accumulated more polyhydroxybutyrate bioplastic (5.55 g/l) after optimisation using CCD than under the non-optimised fermentation condition (0.17 g/l). In a different study, an RSM with BBD-optimised L-methionine production from *Alcaligenes* spp. in submerged fermentation achieved a 1.4-fold increase in production compared to the initial condition. The outputs of RSM can also be visualised through 3D surface plots and contour plots (Figure 5.7).

f. Artificial Neural Network (ANN)- This is a computational model consisting of numerous artificial neurons that can handle complex tasks with applications in various domains, including fermentation. These designs are useful for modelling complex procedures, predicting the responses to changes in environmental variables, and detecting the relationships among variables. The ANN model has several processing elements (nodes, neurodes, or perceptrons) that are arranged in a layer or vector for receiving the inputs and delivering outputs according to their determined activation functions. In this arrangement, the output of one layer becomes an input to the other layers after passing through one or more layers of hidden nodes, thereby simulating the electrical excitation of a nerve cell. ANN can be employed for the modelling, prediction, and optimisation of fermentation processes. Neural networks predict fermentation process indexes using historical and real-time monitoring data, enabling automatic control and optimisation of the process. For instance, the production of biohydrogen in dark microbial fermentation was enhanced by utilising a novel hybrid strategy incorporating a merged RSM (with BBD) and ANN model. This methodology is highly robust, cost-effective, and reliable for extensively analysing the critical fermentation parameters affecting microbial hydrogen and other metabolite production.

g. Bayesian Optimisation (BO)- It effectively optimises the parameters of the fermentation process using a statistical approach. In contrast to traditional optimisation techniques, BO does not analyse every potential parameter combination. It makes use of a probabilistic model to determine the most promising conditions for optimising product yield. The BO reduces the number of necessary experiments, especially when handling intricate and non-linear interactions between

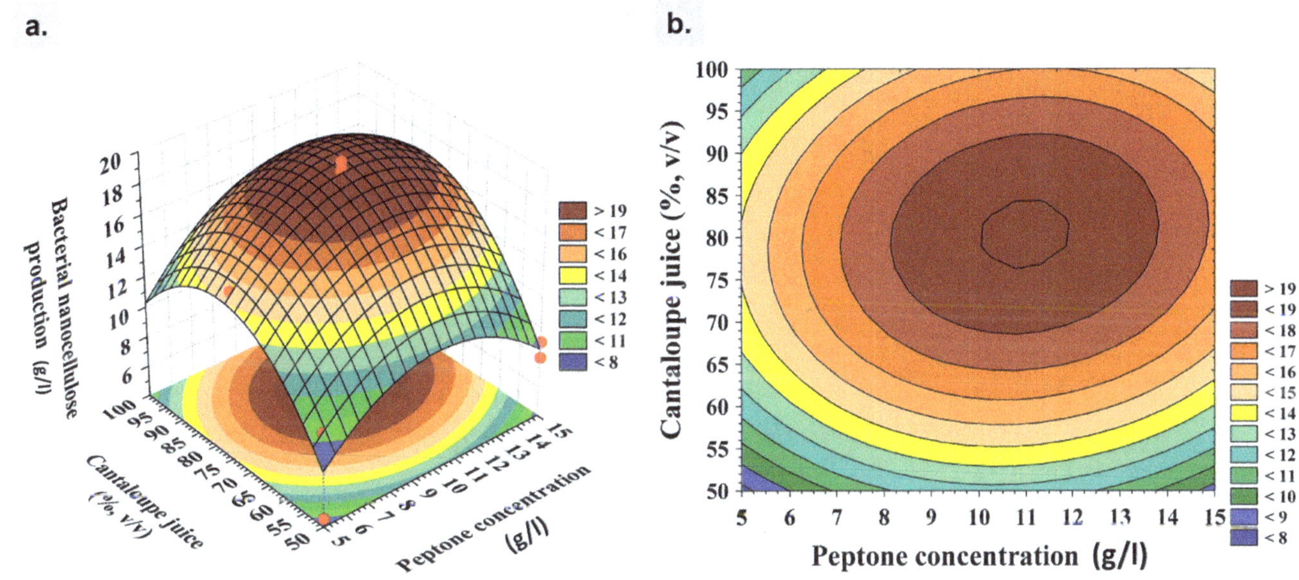

FIGURE 5.7 3D plot (a) and contour graph (b) of the mutual effects of peptone concentration and Cantaloupe juice on bacterial nanocellulose production through response surface methodology (RSM) analysis. The study found that the highest amount of nanocellulose was produced at moderate concentrations of peptone and cantaloupe juice. (Images adapted from El-Naggar et al. (2023).[1])

fermentation process variables. In fermentation processes, BO is ideally suited for optimising when the exact correlation between input parameters (such as temperature, pH, and nutrient concentrations) and the expected outcome (product yield) is not entirely known. The BO method has applications in optimising fermentation media composition, determining optimal conditions, and optimising genetic modifications in microbial strains to enhance product production. This method offers several benefits, including reduced experimentation time and resources, increased product yield, and enhanced predictions with more data collection. However, laboratory-scale optimisations with the BO method may not necessarily apply to large-scale industrial fermenters for scale-up considerations.

5.8 PREPARATION OF INDUSTRIAL FERMENTATION MEDIA

A nutrient medium in the fermentation industry has great importance for the economic success of various bioprocesses. The preparation of industrial fermentation media requires different ingredients for the growth of microorganisms. The nutritional requirements of different microorganisms are different, and therefore, media compositions for the cultivation of different microorganisms are distinct. Nutrients that are stimulatory for one microorganism might be inhibitory to other microorganisms. Formulation of media that can be used for industrial fermentation also greatly varies owing to different requirements. Most fermentation media have basic requirements of carbon, nitrogen, trace elements, and micronutrients. Agricultural and industrial wastes are economical sources of these nutrients and can be used in media after some processing (Section 5.6). The production of specific metabolites requires the addition of specific precursors. A standard fermentation medium additionally contains buffers, salts, and chelating agents. Some fermentation processes require nutrients at specific intervals during fermentation to produce specific metabolites. The composition of culture media is discussed in Chapter 22.

5.9 INOCULUM DEVELOPMENT

Inoculum development is a method of rejuvenating the inactive microbial form into a healthy, active, and suitable reproducible form. This preparation is an essential prerequisite to initiating large-scale production. An inoculum is essentially of a pure form and free of any external contamination. This culture has an active metabolism that possesses the ability to increase its cell count. Based on their growth stages, the culture development produces inoculum, starter culture (preculture), and seed. These terms distinguish from each other but are often used interchangeably. The inoculum is built up successively from small to large volume, with several purity checks at all stages prior to its inoculation inside the fermenter (Figure 5.8). This process ensures a minimum lag period and it raises the culture for the main fermentation reactions.

The first step is the revival of a stock culture that has been maintained for long periods. The stock is initially grown in a solid medium or in a liquid medium to develop a preculture. After sufficient growth, this preculture is inoculated into a large seed fermenter that produces seed. The scale-up from laboratory shake flasks to a seed fermenter allows the inoculum to become well-adapted to fermentation conditions. Seed culturing is used for the proliferation of cells rather than metabolite production. Sometimes, the inoculum is built up in a series of

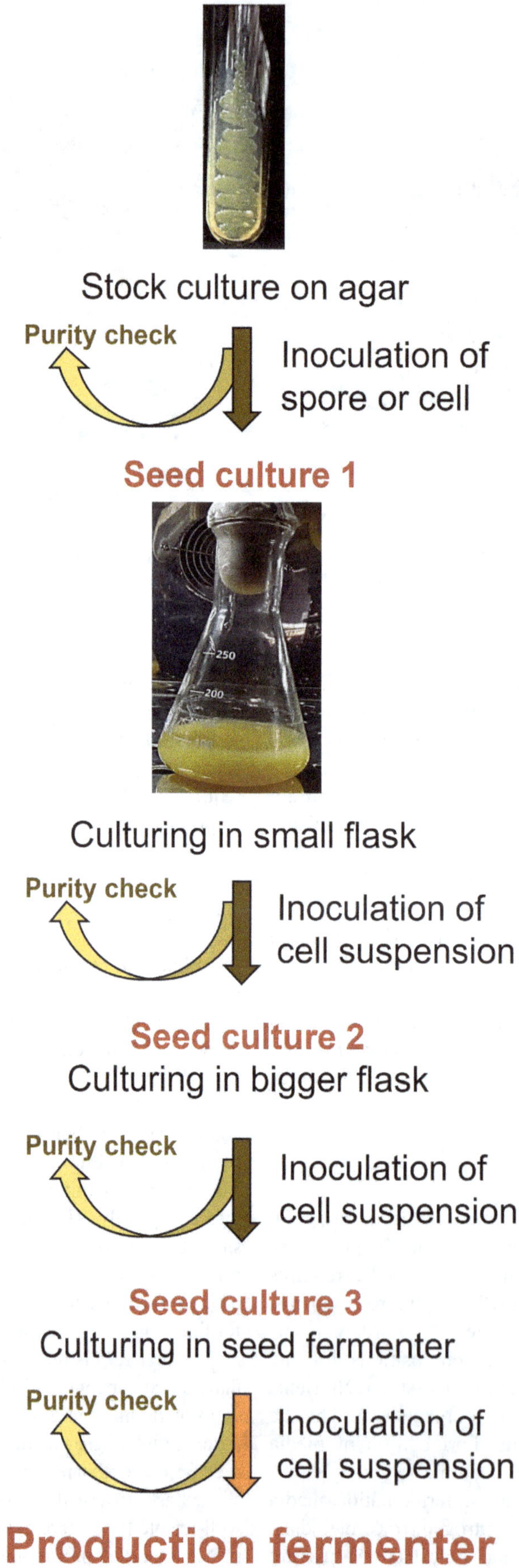

FIGURE 5.8 A schematic representation of successive inoculum development from stock culture to large-scale fermentation with purity check at all stages.

seed fermenters, each larger than the previous one, reducing microorganism stress and ensuring ideal growth circumstances. Scale-up ratios typically range from 1:10 to 1:20, implying that each subsequent fermenter is around ten to twenty times larger than the previous one. The seed is transferred to a larger production-scale fermenter for metabolite secretion. When the seed is inoculated in a large-volume production fermenter, the culture is in a highly active form.

Certain important factors that matter in the inoculum development are-

- Media composition- The nutrient medium should support the maximum growth of the culture. The medium composition used for seed formation could be different from that of the fermentation medium. However, a sufficiently similar medium of seed culture and fermentation would facilitate a better adaptation and considerably minimise the lag phase for microbial growth during large-scale fermentation.
- Morphological type- Some microorganisms exist in more than one morphological form, which gives an option to choose the suitable type as inoculum. They play a great role in growth characteristics, inoculum quality, and the efficiency of the scale-up process. For example, asexual spore suspensions of filamentous bacteria and fungi have gained much attention as inocula because filamentous and mycelial forms pose several challenges due to their complex growth patterns. However, when the fermentation process requires cellular forms of bacteria and yeasts, the inoculum is made up of pure cell suspensions.
- Physiological conditions- The seed culture must be in a precise physiological state before inoculation in the main fermenter. An inoculation of culture for the scale-up mainly uses exponential growth. This can be determined by analysing morphology, biomass, dry weight, turbidity, respiration, residual nutrient concentration, and CO_2 production rate.
- Inoculum size- The appropriate inoculum size ensures that the fermentation process starts robustly and proceeds efficiently. This parameter not only optimises growth and product development but also minimises the lag phase of the fermentation. The quantity of inoculum required for fermentation can vary based on several factors, such as the type of microorganisms involved, the fermentation medium, and the main objectives of the reaction. The inoculum concentration for large-scale fermentation typically varies from 3 to 10% of the total fermentation medium volume. Slow-growing microorganisms often require a large inoculum size to reduce the total duration of the fermentation. Chapter 10 provides further details about inoculum size.

SUMMARY

- Upstream processing involves all the procedures that occur before the fermentation process, including sterilisation, isolation and screening of useful microorganisms, preservation of pure microbial culture, genetic modification, fermentation optimisation, nutrient media preparation, and inoculum development.
- Sterilisation is a method of eliminating all types of microbial forms through various physical and chemical means. Examples of physical methods are wet heating (autoclaving), dry heating, radiation, and filtration. Chemical treatments are disinfection methods and rely on various chemical agents (e.g., ethanol and chlorine).
- Isolation of microorganisms from environmental samples is the main task of microbial cultivation. The potential microorganisms are preserved in stock for long-term use. The most common method for preserving microorganisms involves their storage at below-freezing temperatures.
- The useful microorganisms are subject to genetic manipulation to enhance their production ability of desired products, and due to other economic considerations. This can be mainly achieved through mutation and genetic engineering.
- Optimisation (classical or statistical techniques) of media and physicochemical conditions is a critical upstream processing step to boost productivity before large-scale manufacturing of microbial fermentation products.
- The preparation of liquid feed is a crucial need for the fermentation process. This consists of a main substrate to produce desirable metabolites and various nutrients to support microbial growth. The main components of commercial media for fermentation include preprocessed waste materials/byproducts from industrial waste, owing to economic considerations.
- One last step of upstream processing is inoculum development. This mainly involves rejuvenating inactive microbial forms into active growth forms using suitable culture conditions. In industrial fermentation, an inoculum is built up gradually from small to large volumes before inoculating inside the fermenters.

DIFFERENT TYPES OF PRACTICE QUESTIONS

Fill in the blanks

I. .... refers to all the procedures that occur prior to fermentation.

II. A preferred method for sterilising the heat-labile components of fermentation media is ....

III. .... is a technique that involves freezing of microbial suspension, followed by its sublimation and drying.

IV. Food items that contain excess amounts of salt, like NaCl as preservatives, disrupt microorganisms due to ....

V. In .... method, microbial cultures are stored at or near the temperature of liquid nitrogen.

VI. .... is a factorial-based statistical optimisation technique that assigns the experimental variables using a systematic orthogonal array.

SOLUTION: I. Upstream processing, II. filtration/ filter sterilisation, III. Lyophilisation/ freeze drying, IV. osmosis, V. cryopreservation, VI. Taguchi Design

Very short answer type of questions

I. Name any three procedures related to upstream processing.
II. Write the names of any two chemical disinfectants.
III. A fermenter contaminated with unknown microorganisms could be detrimental to microbial cultivation. Why?
IV. Name two software programs used for statistical optimisation of fermentation.
V. What is the response surface methodology (RSM)?
VI. What is an inoculum in fermentation technology?

Short answer type of questions

I. What is the scope of the filtration technique in sterilisation?
II. Distinguish between dry and moist heating treatment for sterilisation.
III. Can the compositions of growth and fermentation media be different? Support your answer with suitable examples.
IV. What are the scopes of microbial genetic manipulation in industrial biotechnology?
V. How is the inoculum built up successively from small to large volume?

Long answer type of questions

I. What is sterilisation and its significance in fermentation technology? Describe any two physical methods of sterilisation, including their advantages and disadvantages.
II. Why would the repeated microbial culturing for fermentation be inadequate without suitable preservation techniques? Give a detailed account of various preservation techniques.

NOTE

1 El-Naggar NE, Mohammed AA, El-Malkey SE. Bacterial nanocellulose production using Cantaloupe juice, statistical optimization and characterization. Scientific Reports. 2023;13(1):51.

SUGGESTED READINGS

Alonso S. Novel preservation techniques for microbial cultures. Novel food fermentation technologies. In Ojha K, Tiwari B (eds). *Novel Food fermentation technologies. Food engineering series* 2016 (pp. 7–33). Springer.

Berenjian A, editor. *Essentials in fermentation technology*. Springer; 2019.

Berovic M. Sterilisation in biotechnology. *Biotechnology Annual Review*. 2005;11:257–79.

Calcott PH. Cryopreservation of microorganisms. *Critical Reviews in Biotechnology*. 1986;4(3):279–97.

Chaudhary SC, Dodd PW. Economic importance of sterility in industrial fermentation processes. *Advances in Biochemical Engineering/Biotechnology*. 1994;51:35–8.

Guo N, Wei Q, Xu Y. Optimization of cryopreservation of pathogenic microbial strains. *Journal of Biosafety and Biosecurity*. 2020;2(2):66–70.

Heckly RJ. Preservation of microorganisms. *Advances in Applied Microbiology*. 1978;24:1–53.

Keratimanoch S, Takahashi K, Kuda T, Okazaki E, Geng JT, Osako K. Effects of Tyndallization temperature on the sterility and quality of kamaboko. *Food Chemistry*. 2022;366:130692.

Lincoln RE. Control of stock culture preservation and inoculum build-up in bacterial fermentation. *Journal of Biochemical and Microbiological Technology and Engineering*. 1960;2(4):481–500.

Lindskog EK. The upstream process: principal modes of operation. In Jagschies G, Lindskog E, Lacki K, Galliher PM (eds.) *Biopharmaceutical processing* 2018 (pp. 625–635). Elsevier.

Liu Y, Feng J, Pan H, Zhang X, Zhang Y. Genetically engineered bacterium: principles, practices, and prospects. *Frontiers in Microbiology*. 2022;13:4046.

McDonnell GE. *Antisepsis, disinfection, and sterilization: types, action, and resistance*. Wiley; 2020.

Mohkam M, Taleban Y, Golkar N, Berenjian A, Dehshahri A, Mobasher MA, Ghasemi Y. Isolation and identification of novel L-methioninase producing bacteria and optimization of its production by experimental design method. *Biocatalysis and Agricultural Biotechnology*. 2020;26:101566.

Murgatroyd K. *The freeze drying process. Good pharmaceutical freeze-drying practice*; 1997: (pp. 1–58), CRC press.

Penkhrue W, Jendrossek D, Khanongnuch C, Pathom-Aree W, Aizawa T, Behrens RL, Lumyong S. Response surface method for polyhydroxybutyrate (PHB) bioplastic accumulation in *Bacillus drentensis* BP17 using pineapple peel. *PloS One*. 2020;15(3):e0230443.

Prakash O, Nimonkar Y, Shouche YS. Practice and prospects of microbial preservation. *FEMS Microbiology Letters*. 2013;339(1):1–9.

Russell C, Hansen T, Paulsson M. Sterilisation methods. In Bouwman-Boer Y, Fenton-May V, Brun P (eds.) *Practical pharmaceutics: An international guideline for the preparation, care and use of medicinal products* 2023 (pp. 731–748). Springer International Publishing.

Singh V, Haque S, Niwas R, Srivastava A, Pasupuleti M, Tripathi C. Strategies for fermentation medium optimization: an in-depth review. *Frontiers in Microbiology*. 2017;7:2087.

Soccol CR, Pandey A, Larroche C, editors. *Fermentation processes engineering in the food industry*. CRC Press; 2013.

Stanbury PF, Whitaker A, Hall SJ. *Principles of fermentation technology*. Elsevier; 2013.

Wang Y, Yang G, Sage V, Xu J, Sun G, He J, Sun Y. Optimization of dark fermentation for biohydrogen production using a hybrid artificial neural network (ANN) and response surface methodology (RSM) approach. *Environmental Progress & Sustainable Energy*. 2021;40(1):e13485.

6 Different Types of Fermentation Processes

Industrial fermentation types are typically determined by the nature of substrates, operational modes, and aerated conditions (Figure 6.1). Fermentation processes utilise various substrates such as liquids, solids, and gases to produce a diverse range of products. Submerged fermentation (SmF) is a microbial cultivation technique in which the medium is a liquid nutrient with an excess of free-flowing water. Solid-state fermentation (SSF) is a method of microbial cultivation in moist solid substances without utilising excess free liquid volume. Surface fermentation involves growing microorganisms on nutrient substrates, while gas-phase fermentation is a novel method for converting gases into useful products. Fermentation processes can be operated in three modes: batch, fed-batch, and continuous, depending on the nutrient flow during the process. Oxygen (O_2) presence or absence drives different types of fermentation reactions. The presence of O_2 encourages growth and metabolite production of aerobes, while anaerobes are deterred from growing in oxic conditions.

6.1 FERMENTATION BASED ON THE NATURE OF SUBSTRATES

Industrial fermentation can use different substrates, including solids, liquids, and gases, to produce various products. The fermentation reactions of these substrates require distinct setups for their effective utilisation and product formation. Based on the substrate types, fermentation can be categorised as SmF, SSF, and gas-phase fermentation (Figure 6.1), which are discussed in the following sections.

6.1.1 Submerged Fermentation (SmF)

SmF is the process of growing microorganisms in a medium containing plenty of free fluid in a tank fermenter. The fermentation medium consists of an excess volume of free-flowing water and approximately 5% nutrients of the total volume (Figure 6.2a). This provides a sufficient space for microbial growth and metabolite production. The ingredients of the liquid medium include both soluble and insoluble constituents. The soluble constituents of the medium can be simple carbohydrates, inorganic salts, antibiotics, and some specific metabolites. The insoluble constituents can be complex substances, such as wastes and leftovers from forestry, agriculture, and food production, which require some pretreatment before being used in SmF. The pretreatment of raw materials is an upstream process and is elaborated in Chapter 5. SmF is widely used for the production of microbial biomass and metabolites. Alcohols, amino acids, antibiotics, biogas, enzymes, organic acids, polymers, and vitamins are commonly developed products using SmF. In particular, bacteria are considered more suitable for SmF than any other microbial form because they have high moisture requirements and can actively secrete metabolites into the external liquid medium. Filamentous fungi like *Aspergillus niger*, *Penicillium chrysogenum*, and *Trichoderma reesei* can also be suitably used in SmF because of their excellent production and secretion capacities.

The substantial volume of free-flowing liquid in industrial SmF necessitates a large fermenter vessel equipped with a control panel to regulate various parameters. SmF offers accurate monitoring and online automation during the entire operation. This provides a precise amount of aeration, agitation, biomass formation, dissolved oxygen (DO), foam control, pH and temperature optima, product formation, substrate utilisation, and viscosity inside the vessel. Agitators and spargers in SmF provide a thorough homogeneity of culture and medium. When the fermentation is performed without agitation, productivity is often lower due to insufficient mixing of air and nutrients. Excessive agitation, on the other hand, can damage microbial cells, impairing fermentation output. Furthermore, SmF can be operated in batch, fed-batch, or continuous modes. After fermentation, the product is separated from the fermented broth using a variety of downstream processes such as centrifugation, filtration, precipitation, and sedimentation.

The SmF procedures are most common for industrial applications. The key advantages of SmF technology are its high yield in a short duration, reduced labour cost, low production costs, and easy product purification. The use of well-defined culture media for the fermentation process ensures good reproducibility. However, the volumetric yield of the products in SmF is relatively low, considering the total volume of the broth used. The entire process requires complex instruments for fermentation and bioprocessing. Due to feedback inhibition, the end product of the reaction can hamper the overproduction of the desired product in SmF.

DOI: 10.1201/9781032651095-8

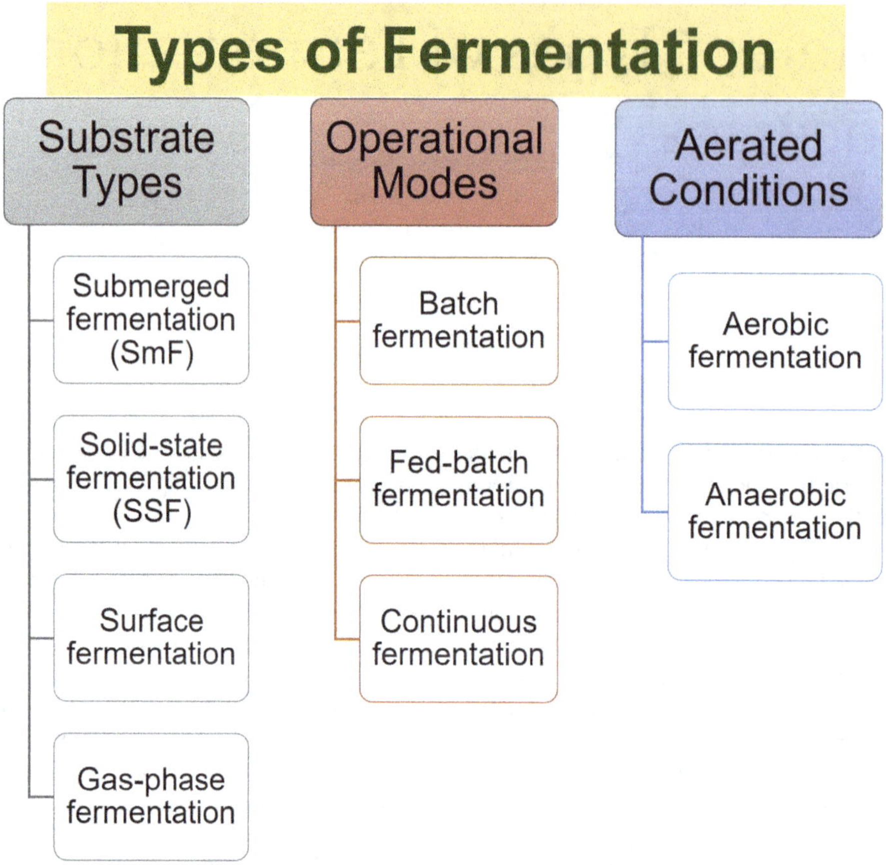

FIGURE 6.1 Different types of fermentation based on substrate types, operational modes, and aerated conditions.

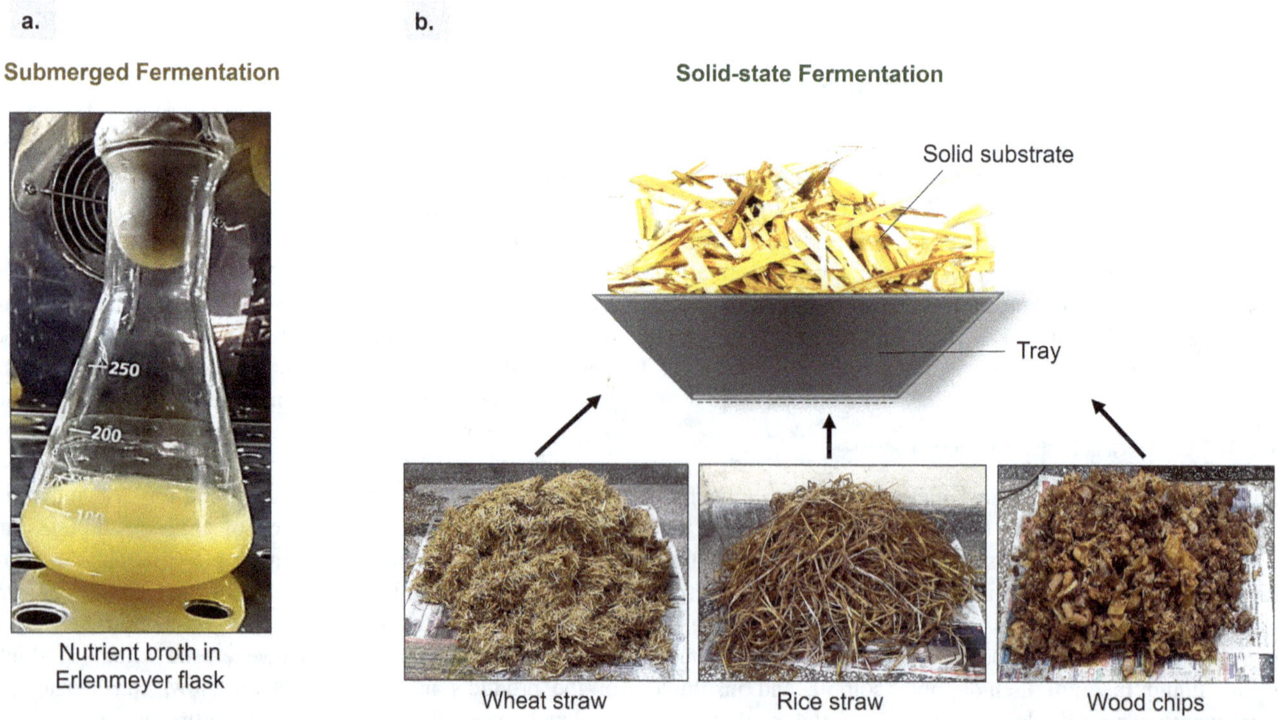

FIGURE 6.2 Fermentation based on liquid and solid nutrient media. a. Submerged fermentation (SmF). b. Solid-state fermentation (SSF) using agroindustrial waste.

Also, an SmF process produces a large volume of effluents, which require suitable disposal or treatment.

6.1.2 Solid-State Fermentation (SSF)

SSF is a method of microbial fermentation in moist solid substances without using excess volume of free liquid (Figure 6.2b). SSF has been traditionally used in the preparation of many fermented foods. Examples include the production of soy sauce using *Aspergillus oryzae* on soybean (Koji fermentation) and tempeh using *Rhizopus oligosporus* on soybean. Over the last three decades, SSF has gained interest in developing more industrial bioprocesses because of its low energy requirement and eco-friendly character. Applications of SSF are expanded to modern foodstuffs (such as the production of cheese, sauerkraut, and kimchi), enzyme production, bioleaching of metals, organic acid production, pharmaceuticals (such as antibiotics and biosurfactants), composting, and sewage treatment (Table 6.1).

Although SSF has no demand for free liquid, the process requires about 30–85% moisture or water to moisten the solid substrate bed, mobilise and provide nutrients, maintain an active growth of microorganisms, and scavenge metabolic waste. Water rests in substrate gaps or exists as thin layers, but does not exceed the saturation capacity of the solid substrate. The solid materials are usually insoluble and play the roles of both

TABLE 6.1
A list of different products manufactured using different solid substrates and producer microorganisms through solid-state fermentation (SSF)

Industrial sectors	Products	Examples of solid substrates	Examples of producer microorganisms
Food	Cheese	Coagulated milk curds	Lactic acid bacteria (LAB), *Penicillium roqueforti*, and *Penicillium glaucum*
	Polysaccharides	Rice and corn bran	*Cordyceps militaris*
	Sauerkraut	Shredded cabbage	*Lactobacillus brevis* and *Lactobacillus plantarum*
	Soy sauce	Defatted soybeans and roasted wheat	*Pediococcus halophilus* and *Aspergillus oryzae*
	Flavours with pleasant aromatic smells (ethyl acetate-pineapple, ethyl lactate-fruity)	Steamed sorghum grains	*Lactobacillus*, *Saccharomyces*, *Pichia*, *Geotrichum*, and *Candida*
	Tempeh	Dehulled soybeans	*Rhizopus oligosporus* and *Lactobacillus* spp.
	King oyster mushroom	Barley straw and sugar beet pulp	*Pleurotus eryngii*
	Oriental reishi mushroom	Cotton stalks, rice straw, and sugarcane bagasse	*Ganoderma lucidum*
	Natural pigment (e.g., yellow-monascin)	Saba banana peel	*Monascus purpureus*
Enzymes	Cellulase, xylanase, and ligninase	Corn cob	*Saccharomyces cerevisiae*
	Amylase	Wheat bran	*Bacillus subtilis*
	Laccase	Tea residues	*Trametes versicolor*
	Tannase	Pomegranate peel	*Bacillus velezensis*
	Chitinases	Wheat bran with colloidal chitin	*Trichoderma koningiopsis*
	Inulinase	Apple pomace	*Mucor circinelloides*
	Thermostable proteases	Wheat bran	*Mycothermus thermophiles*
Agriculture	Gibberellic acid (growth hormone)	Cow manure with glycine	*Paecilomyces* sp. ZB
	Biopesticide conidia	Rice husk or potato peel	*Beauveria bassiana*
	Biopesticide spores	Digestate and biowaste	*Bacillus thuringiensis*
Antibiotics	Cephalosporin C	Wheat rawa	*Acremonium chrysogenum*
	Rifamycin SV	Ragi bran with cotton cake	*Amycolatopsis mediterranei*
	Neomycin	Raspberry seed powder with dextrin	*Streptomyces marinensis*
Organic acids	Citric acid	Cassava peels and corn cobs	*Aspergillus niger*
	Itaconic and fumaric acids	Wheat bran and corn cobs	*Aspergillus oryzae*
Biofuel	Ethanol	Steam-exploded corn stover	*Zymomonas mobilis*
Pharmaceutical	Ergot alkaloids	Sugar-cane pith bagasse	*Claviceps purpurea*
Textile	Bioremediation of dyestuffs	Distillery industry waste-yeast biomass	*Bacillus cereus*
Bioremediation	Biosurfactant as a bio-dispersant of burned motor oil	Wheat bran	*Serratia marcescens*

a solid support and a nutrient source for the microorganisms. Typically, low-cost and abundantly available substrates, such as agricultural residues (e.g., bran, corn stover, leaf litters, and straw) and industrial wastes (e.g., sugarcane bagasse, fruit pomace, husk, oil cakes, wood chips, sawdust, and tea waste) are commercial raw materials for SSF (Figure 6.2b). The heavy, nutritionally complex, heterogeneous, and recalcitrant solid materials often necessitate some physical, chemical, or enzymatic pre-treatment. The pre-treatment reduces the size of the solid substrates and makes their nutrients more amenable to microorganisms for fermentation.

Conventional SSF is carried out in simple-to-construct fermenters. They can be with mechanical agitation (as in a rotating drum), non-mechanical agitation (as in a forced airflow system and fluidised bed fermenter), or without agitation (as in tray fermenter, packed bed, and column reactors) (detailed in Chapter 8). The product is recovered by flowing the water into the fermented medium, followed by pressing, leaching, and filtration. Moreover, with a small amount of water added to the substrate, the product can be simply extracted in a relatively more concentrated form. The high concentrated productivity in SSF also reduces the waste effluent volume and treatment requirements sufficiently.

SSF offers several economic advantages over the conventional SmF. These include the usage of abundant and inexpensive agro-industrial substances, direct spore seeding, and low volume of medium per unit weight of the substrate. These activities yield high enzyme activity while minimising energy requirement, catabolite repression, and bacterial contamination risk. The sustainable use of agro-industrial wastes as solid substrates in SSF promotes their biodegradation while transforming them into biofuels and other high-value-added products. This process reduces the load of environmental pollution on the environment, owing to its cost-effective treatment and promotes a circular economy approach. SSF processes mainly rely on filamentous fungi, although some processes involve yeasts or bacteria.

SSF is a lucrative and high-value agricultural venture to cultivate edible mushrooms, which can turn industrial and agricultural wastes into food for humans (Figure 6.3). Mushroom cultivation demands relatively less effort, takes little time, and requires small acreage. It is something that less affluent businessmen can also embrace. In recent years, some new schemes have shown directions to employ the wastes generated from mushroom cultivation in the manufacturing of many commodities such as textiles, leather, cosmetics, and biofuel.

Despite many advantages, industrial applications of SSF are very limited. The complexity and heterogeneity of the solid substrates hamper microbial biomass development, substrate utilisation, gas (CO_2 and O_2) transfer, and product formation. Further, without adequate mixing of mass and heat, the temperature within the substrate bed varies, which deleteriously affects microbial growth and product formation.

FIGURE 6.3 Cultivation of oyster mushrooms (*Pleurotus ostreatus*) through solid-state fermentation (SSF) using agricultural lignocellulosic waste (wheat straw.)

Poor knowledge about the design and operation of large-scale SSF fermenters has also reduced their value for scale-up. Table 6.2 provides the differences between SmF and SSF.

6.1.3 Surface Fermentation

Surface fermentation is a process of growing microorganisms on the surface of a nutrient substrate. The substrate used for surface fermentation may be liquid or solid, spread on aluminium or stainless steel trays. The setup is relatively simple and does not require complex equipment. This calls for non-agitated or static trays stacked in fermentation chambers with controlled temperature and humidity, which receive sterile air for aeration. The lack of agitation decreases shear stress, which can be advantageous for microbial forms that are sensitive to mechanical force. During fermentation, microorganisms form a layer on top of the nutrient medium and develop products. The microorganisms involved in surface fermentation typically have limited mobility due to biofilm formation. This allows an easy extraction of products. If the nutrition medium is a liquid feed, the solvent may be evaporated to increase the product's concentration.

The advantages of surface fermentation include cost-effectiveness, the use of simple fermenter chambers, the non-requirement of medium agitation, easy control of aeration, low losses at product recovery, and less energy consumption. Despite these advantages, surface fermentation is rarely used in industries, as microorganisms are exposed to unequal distribution of aeration and nutrients. The major drawback is finite resource utilisation. Surface fermentation requires more manual effort for setup, monitoring, and harvesting than automated agitated systems. Maintaining aseptic conditions in surface fermentation is also challenging due to the high surface area exposed to the environment. The high costs of building investment and personnel wages, long duration of fermentation, and low productivity also impede its large-scale development for industrial applications.

6.1.4 Gas-Phase Fermentation

Gas-phase fermentation is a relatively novel approach for transforming gases like carbon monoxide (CO), carbon dioxide (CO_2), and hydrogen (H_2) into useful products through microbial applications. This process is popularly termed as **synthesis gas (syngas) fermentation**. Acetogenic bacteria like *Clostridium ljungdahlii* and *C. autoethanogenum* are the most commonly involved organisms in gas fermentation processes. They mainly rely on the Wood-Ljungdahl pathway or the acetyl-CoA pathway. These processes are the sources of many biofuels like ethanol, butanol, and 2,3-butanediol. Non-acetogenic organisms like *Escherichia coli* or yeasts have shown potential for gas-phase fermentation after some genetic modification. Differently, aerobic hydrogen-oxidising bacteria (also called Knall gas bacteria) (e.g., *Cupriavidus necator*) facilitate the conversion of H_2 O_2 and CO_2 into polyhydroxyalkanoates (PHA), which can be used as a substitute for plastics and are generally called **bioplastics**.

The productions of CO_2-based fuels and chemicals using microbial routes offer new ways to valorise waste gases, contributing to sustainability and waste reduction. This approach has also been used to produce feed for animals. In recent years, industrial setups such as Bio Base Europe Pilot Plant (BBEPP) and Avecom, Belgium, have gained confidence in the applications of gas fermentation to produce many

TABLE 6.2
Differences between submerged fermentation (SmF) and solid-state fermentation (SSF)

Parameters	SmF	SSF
Substrates	Dissolved nutrients	Undissolved solids like bran, bagasse, straw, and pulp
Amount of water	Large volumes of an aqueous medium	Moistened with a low amount of water
Medium components	Mostly defined	Complex and heterogeneous, such as agricultural, forestry and industrial residues
Operated in	Batch, fed-batch, and continuous modes	Mostly batch mode
Product quality	Constant	Variable
Main phase	Liquid	Solid
Microbial metabolism	Fast	Slow
Requirements of space, agitation, and energy	Large	Small
Inoculum-to-substrate ratio	Small	Large
Fermenters involved	Large, such as stirred tanks, airlifts, or tower fermenters	Small, such as surface, drum, or tray fermenters
Agitation in fermenters	Mechanical (with impellers) or non-mechanical (with forced air)	Typically non-agitated (static), but can be agitated with forced air or by rotating the fermenter drum
Automation	Exist for accurate monitoring and controlling of various fermentation parameters	Often difficult
Movement of mass and heat	Adequate for homogenous microbial growth and product formation	Not adequate
Waste effluents generated	Large	Small

commodities (such as 2,3 butanediol, single-cell proteins, bioplastics, organic fertilisers, etc.) that help meet the balance between human and environmental needs.

6.2 OPERATIONAL MODE OF INDUSTRIAL FERMENTATION

The operation of industrial fermentation can broadly vary depending on the flow of nutrients during fermentation (Figure 6.1). This can operate in three modes: batch, fed-batch, and continuous fermentation (Table 6.3). Batch fermentation maintains a static mode, keeping the same constituents during the entire process. Fed-batch mode observes a continuous inflow of medium, leading to an increase in total volume in the vessel. Continuous mode perceives continuous inflow and outflow of the broth, maintaining a constant volume in the vessel. Despite these differences, fermentation in all three modes requires inoculation only once, at the beginning of the process, because adding multiple inoculants causes variable growth phases that could lead to deviations in the production of desired products or cause inhibitory activities. One instrument can have all three operational modes, mainly by handling the inlet and outlet valves, though some specific components are also required for each type. Detailed information on all three types of fermentation processes is given below.

6.2.1 Batch Fermentation

Batch cultivation is a traditional and simple process of fermentation. This involves a very straightforward mode of operation, where microorganisms are inoculated in a specific volume of fermentation medium for a predetermined time

TABLE 6.3
Differences among batch, fed-batch, and continuous fermentation

Parameters	Batch fermentation	Fed-batch fermentation	Continuous fermentation
Run for	Definite period	Definite but relatively long period	Very long period
Addition of fresh medium during fermentation	No	Periodically	Continuously, followed by the removal of the fermented broth
System	Closed and discontinuous	Partly open and semi-continuous	Open and continuous
Chance of contamination	Low	High	Very high
Growth phases of microorganisms during fermentation	All, including lag, exponential, stationary, and deceleration	All but the exponential or stationary phase is prolonged	Continuously maintained in the exponential or stationary phase
State of fermentation	Steady, as biomass and broth viscosity increase in a constant volume of the medium	Quasi-steady, as biomass and volume increase, although substrate addition equals consumption	Steady, as the biomass concentration, product formation, and medium volume remain constant
Demands for cooling, oxygen, and pH control	High	More than batch	Constant
The exponential phase is described using	$dx/dt = \mu x$, where x- concentration of biomass (g), t- duration (h), μ- specific growth rate (h^{-1})	$dx/dt = GY$, where G- substrate feed rate ($g\ L^{-1}\ h^{-1}$), Y- yield factor (g biomass produced g^{-1} substrate consumed)	$dx/dt = \mu x - Dx$, where μx- growth, Dx- output, D- dilution rate (h^{-1})
Specific growth rate (μ)	Primarily increases and then decreases	Constant for a certain duration and then decrease	Constant and controlled by the dilution rate
Product formation	For a brief duration	Prolonged for a particular period	Extended indefinitely
Productivity	Less, as there is more system downtime, including periods of exhaustive labour in every batch for medium making, sterilisation, harvesting, and cleaning	More than batch, as there is less system downtime and intensive labour	More than batch and fed-batch, as there is less system downtime with constant labour demand
Automation for substrate addition	Frequently absent	Present	Present, with chemostat, turbidostat, or auxostat
Can be conducted	In the single-stage	As a cyclic culture (by taking the residual culture to start another fed-batch fermentation), in addition to a single-stage	In a single-stage, multiple stages, and recycle system

before the broth is collected for downstream processing. Microbial growth in batch fermentation follows sequential stages, including lag, logarithmic (or exponential), deceleration, stationary, and death phases (detailed in Chapter 10). Metabolites are formed during the exponential and stationary phases. Primary metabolites are generated during the exponential growth phase, whereas secondary metabolites are essentially synthesised in the stationary phase. After the fermentation process is completed, the product is harvested. Thereafter, the fermenter is cleaned and ready for the next round of fermentation (Figure 6.4a).

Batch fermentation is a closed system in which fermentation broth is not added to the fermenter once the fermentation has begun and is not removed until the fermentation is finished. However, the fermenter continuously receives sterile air/oxygen, acid or base buffers, and an antifoam agent during the entire operation. The closed, non-continuous nature of batch fermentation considerably lowers the risk of external microbial contamination. However, batch fermentation is an unsteady system because the concentration and composition of nutrients, biomass, and metabolites vary throughout the fermentation process.

Batch fermentation is widely used in the food, chemical, and pharmaceutical processing industries. The products manufactured by batch cultivation are flavouring amino acids (e.g., aspartate and monosodium glutamate), organic acids (e.g., lactic acid, citric acid, and acetic acid), alcoholic beverages (e.g., beer, wine, distilled spirits of brandy, whisky, and rum), biomass (e.g., single cell protein (SCP), yeasts, and lactic acid bacteria (LAB)), etc. A batch process is believed to be more effective than any other mode of production due to ease of control and operations, low investment and maintenance costs, minimal chances of contamination, and high yield. A batch fermentation exhibits operational flexibility, which provides a facility to run batches under diverse conditions and with varying compositions, allowing experimentation and optimisation. However, batch fermentation is highly labour-intensive because it requires frequent emptying, cleaning, sterilisation, and refilling of fermenters. The addition of fresh seed culture in each cycle prolongs the lag phase for overall functioning. All these factors lengthen the overall manufacturing process, lowering productivity. Maintaining consistent productivity across different batches in batch fermentation necessitates a larger initial capital investment than the continuous mode of operation. Furthermore, the

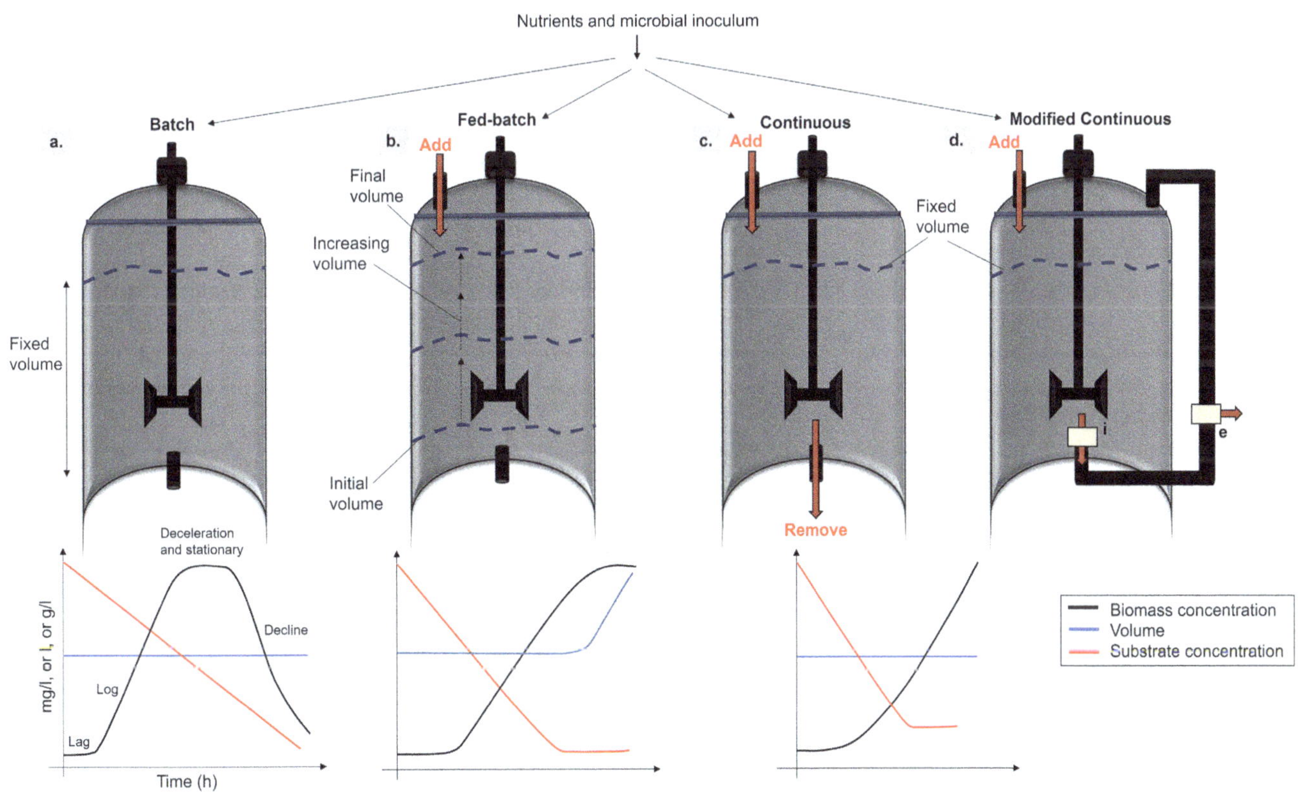

FIGURE 6.4 The different operational modes of fermentation. a. Batch fermentation. The vessel maintains a constant fixed volume, and the fermented broth is harvested after the fermentation is completed. b. Fed-batch fermentation. After the initiation of fermentation, there is a subsequent addition of fresh sterile nutrient (or growth limiting substrate) into the fermenter without any removal of the fermented broth, resulting in a volume increase (upward facing arrows) and an extended active growth phase. c. Continuous fermentation. After the initiation of fermentation, the fresh nutrient medium is regularly added (at a suitable dilution rate) with equal removal of fermented broth for product recovery, thereby maintaining a steady culture volume. d. Modified continuous fermentation. In this mode, there can be internal (i) or external (e) feedback systems to recover microbial cells from the fermenter.

entire contents of the fermenter may be contaminated if any harmful metabolites are produced within.

6.2.2 Fed-Batch Fermentation

The fed-batch fermentation is a variation of batch fermentation where substrate and nutrients are added while the fermentation is taking place. It is an advanced process in which the feeding of media does not accompany any recovery of fermented broth from the fermenter. Feeding can be intermittent or continuous, making the fed-batch fermentation semi-open or semi-continuous (Figure 6.4b). The microbial growth phases in a fed-batch fermentation are substantially the same as those in a batch culture and usually consist of a lag phase, an exponential phase, a stationary phase, and a death phase. However, a primary difference with a fed-batch fermentation is that to prolong the exponential phase and optimise product production, nutrients are supplied gradually throughout the fermentation process. Fed-batch fermentation establishes a quasi-steady state, having an equal consumption of the added substrate by a virtually constant cell concentration[1]. Microbial culture is maintained in one growth phase for a particular duration. The fermentation process is terminated when the volume of the fermentation broth reaches about 75% of the fermenter volume. Fed-batch fermentation can be carried out in two modes: variable and fixed systems. The variable-fed batch system involves the regular addition of the liquid feed or any limiting substrate in the feed, leading to an increase in the overall volume of the medium. However, the fixed-fed batch system features the augmentation of a highly concentrated limiting substrate at a lower rate, resulting in a negligible volume increase.

Fed-batch production has been a popular option for many years because of its robustness and high product yields both per run and per time. It has now established itself as the industry standard for production scales. Many biotherapeutics that are now on the market or in the clinical stages are developed in fed-batch fermentation. A fed-batch culture of recombinant *Escherichia coli* notably produces human insulin, which was traditionally the first "golden molecule" in the biotech realm. A wide variety of products, including amino acids, antibiotics, biomass, cell proteins, enzymes, growth hormones, nucleotides, organic acids, organic solvents, recombinant proteins, and vitamins, are also produced using fed-batch fermentation.

The fed-batch fermentation is preferentially employed when a high concentration of a nutrient becomes toxic to microbial culture. For example, microbial ethanol production requires high sugar concentrations, but high sugar concentrations create excessive osmotic pressure. This could inhibit the growth of the fermenting microorganisms, reduce the fermentation yield, and prolong the fermentation period. In this situation, the fermentation process is initiated with a low sugar concentration that favours robust microbial growth at the trophophase. After the sugar is consumed, more sugar is added periodically or at a defined rate to produce ethanol. This entire setup aims to prevent the Crabtree effect (i.e., preference for fermentation over respiration that affects biomass production) and achieve high cell density at the initial stage. In a different situation, fed-batch fermentation allows for controlled feeding of nutrients in the production of penicillin. A controlled addition of the nutrients enables the process's yield and productivity to be increased. The periodic augmentation of carbon (e.g., glucose) and nitrogen sources in the fermenter maintains their optimum levels for fermentation and prevents any catabolite repression from their high concentrations.

This operational mode has several advantages, such as controlled feed flow of nutrients, optimal microbial growth, extensive biomass accumulation, enhanced production of secondary metabolites, prevention of substrate inhibition, decreased broth viscosity, and effective management of catabolite repression. Some disadvantages associated with fed-batch fermentation include the requirement of considerable operator skills, analysis of the nutritional demands of microorganisms, and control of the feed flow.

6.2.3 Continuous Fermentation

Continuous fermentation is an open system process with a constant influx of fresh sterile feed and efflux of the fermented broth. As the fresh medium is continuously introduced, microbial growth in continuous fermentation usually only shows two different phases: a lag phase and a sustained exponential phase. The continuous operation allows a steady-state growth rate while maintaining a continuous production rate. Under an equilibrium condition, the amount of new biomass produced equals the amount of biomass leaving the fermenter. When fermentation starts, it works in batch mode. However, once the culture enters the exponential growth phase or meets substrate limitation, the fresh nutrient medium is added to the fermenter while an equivalent amount of fermented culture broth is removed. Consequently, the total volume of the nutrient medium in the vessel remains the same (Figure 6.4c). A low dilution rate of the incoming fresh medium during continuous fermentation can prolong exponential growth and increase the yield of microbial biomass (Box 6.1). In continuous fermentation, the loss of microbial biomass during product harvesting can be controlled by internal or external feedback. The former approach employs a filter which prevents the loss of microbial biomass, while the latter facilitates recovery of the biomass (with sedimentation or centrifugation) from the effluent and returns a portion of concentrated biomass into the fermenter (Figure 6.4d). Continuous fermenters come in many configurations, such as airlift reactor, continuous stirred tank reactor (CSTR), fluidised-bed reactor (FBR), packed-bed reactor (PBR), and up-flow anaerobic sludge blanket (UASB) reactor.

Continuous fermentation can be operated in single-stage or multistage fermenter systems (Figure 6.5). A single-stage fermentation consists of a single fermenter, which operates in continuous mode by balancing the input of nutrient solution with the output of harvested culture. Multistage continuous culture fermentation (MCCF) is a process in which several

BOX 6.1 PRODUCTIVITY OF BATCH, FED-BATCH, AND CONTINUOUS FERMENTATION

Productivity (R) of fermentation refers to the output of biomass or metabolic products per unit duration of fermentation. The productivity of fermentation processes can be estimated using specific formulas, considering factors like the production rate, volume, and time. Depending on different modes, the method of calculating R can vary.

For **batch fermentation**, the productivity (R_{batch}) can be estimated as

$$R_{batch} = (x_{max} - x_o)/(t_i - t_{ii})$$

where x_{max} is the maximum cell concentration at the stationary phase
x_o is the initial cell concentration during inoculation
t_i is the duration at which an organism grows at μ_{max} (maximum specific growth rate)
t_{ii} is the duration at which the organism is not growing at μ_{max}, including lag, deceleration, periods of batching, sterilisation, and harvesting.

In batch fermentation, the negative contribution of t_{ii} contributes to the overall cycle time and can significantly impact the overall productivity of the process.

For **fed-batch fermentation**, the formula of productivity ($R_{fed\text{-}batch}$) can be adjusted to the total time spent for fermentation and volumetric efficiency as

$$R_{fed\text{-}batch} = (x_{max} - x_o)/(t_{batch} + t_{fed\text{-}batch}) \times V_f/V_i$$

V_f is the final volume after feeding
V_i is the initial volume before feeding
t_{batch} is the total batch time, including all the non-productive time intervals
$t_{fed\text{-}batch}$ is the time period during which fresh feed is intermittently added to the fermenter

In this situation, the overall productivity is obtained by multiplying fermentation productivity by fermenter volumetric efficiency.

For **continuous fermentation**, productivity (R_{cont}) can be estimated as

$$R_{cont} = D\bar{x}[1 - (t_{iii}/T)]$$

where D is the dilution rate
$\bar{x}$ is the steady-state cell concentration
t_{iii}/T represents the fraction of the total time spent in a non-productive phase of the fermentation process

In continuous fermentation, D represents the reciprocal relationship between the culture volume (V) and the medium flow (F) into the fermenter as

$$D = F/V$$

As per the equation, the productivity of continuous fermentation (R_{cont}) is directly proportional to the D and the average biomass concentration ($\bar{x}$), which can be adjusted downwards based on the fraction of time spent during the lag phase (t_{iii}/T).

tanks or stages are connected in series, and each stage has a distinct situation. The initial stage is the growth phase, producing maximum cell growth using nutrient-rich media. This is followed by the transition stage, where succeeding fermenters gradually shift the culture conditions to favour metabolite synthesis. The final setup is the production stage, which is optimal for high-yield metabolite production. This type of operation is suitable where the secondary metabolite is partially or non-growth associated.

The continuous fermentation systems come in several variations, such as chemostat, turbidostat, and auxostat. A chemostat system is a tool to maintain a continuous supply of growth-limiting nutrients in the fermenting medium. When the concentration of a chemical constituent decreases below the required limit, a device sends a signal to allow the flow of nutrients into the fermenter. This recycling provides more productivity of cells, particularly when the available substrate (immiscible hydrocarbon, sulfite waste liquor, etc.) in the

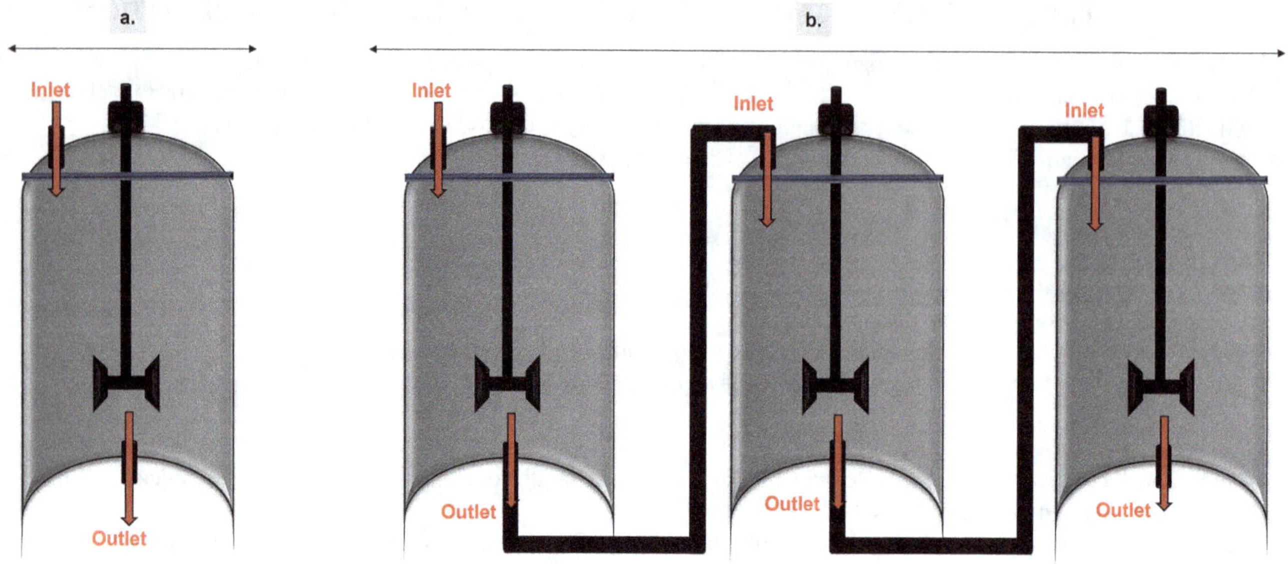

FIGURE 6.5 Different types of continuous fermentation systems. a. Single fermenter system. b. Multistage fermenter system.

nutrient solution is low. Turbidostat maintains the turbidity of the medium by measuring cell density. Turbidostat tries to maintain the cell concentration in the reactor using a photoelectric cell. When the biomass concentration increases, the turbidity of the fermenting broth also increases, which generates a signal that switches on the pump to supply fresh sterile medium into the fermenter. Besides, a substrate and pH auxostat manage the substrate and pH in the medium, respectively, for the controlled growth rate of a microbial population. An auxostat is typically far more stable at high dilution rates than a chemostat.

Continuous fermentation has several advantages and offers many benefits for commercial-scale production. It can be suitably used for the industrial production of SCP, fodder yeast, baking yeast, and various metabolic products such as antibiotics, beer, organic solvents, and vinegar. This is also a suitable choice for the wastewater treatment plants. In recent years, this mode of operation has been used to raise the productivity of acetone-butanol-ethanol (ABE) fermentation. Continuous fermentation maintains the optimum conditions for long-term stable product synthesis. The procedure is cost-effective and streamlines culture scale-up based on a consistent power-to-volume ratio. Continuous culturing enhances the efficiency, productivity, and viability of microbial processes by minimising downtime and reducing the need for frequent harvesting, cleaning, and sterilisation. It can also prevent product inhibition due to excess toxin buildup in the vessel. For example, the accumulation of ethanol during the fermentation process hampers the activity of the source microorganisms (i.e., yeast cells) that produce ethanol. One way to mitigate this failure is by extending the fermentation process to continuous mode. Continuous culture of microorganisms yields a relatively low product concentration but higher productivity than batch culture (Box 6.1). Furthermore, there is a negligible batch-to-batch variation in the product quality, which eventually reduces production costs.

Despite continuous fermentation being a useful method, it is still less commonly employed than batch fermentation for industrial applications. A major concern is that prolonged incubation with continuous fermentation tends to alter the behaviour and lower the activity of the microbial culture. This can also lead to long-term sterility issues due to the growth of undesirable microorganisms from the outer environment. The filamentous microorganisms can overgrow along the walls, blocking the piping and valves, which can impede homogeneous exposure to nutrients. Any sporulation by immobilised cells could provide an internal inoculum source, resulting in variable growth phases in the vessel. Furthermore, the withdrawal of fermentation broth can result in excessive nutrient washout and increase the risk of genetic drift in the culture. However, the washout issues can be resolved to a certain extent by immobilising the microorganisms or enzymes onto a solid support, preventing them from leaving the fermenter.

6.3 FERMENTATION IN DIFFERENT AERATED CONDITIONS

The presence or absence of DO in the medium is a major driving force during fermentation. This is an essential prerequisite for the cultivation of fermentation microorganisms and the production of desired metabolites. Aerobic microorganisms require high DO in the medium for optimal growth and product formation. On the other hand, anaerobes deter from the O_2 presence. Even some metabolic products are sensitive to the presence of DO and are prone to their oxidation (e.g., H_2, butanol, and CH_4). Some microorganisms, such as LAB and yeast, are facultative anaerobes, which can live in both aerobic and anaerobic conditions using different metabolic pathways. These microorganisms, however, prefer aerobic pathways for optimal energy production but can switch to anaerobic conditions in an anoxic environment and produce specific metabolites. The fermentation process based on the requirement of DO can be categorised as aerobic and anaerobic fermentation (Figure 6.1), which are further discussed as follows.

6.3.1 Aerobic Fermentation

Aerobic fermentation is the mainstream of industrial biotechnology, owing to its versatility, efficiency, and ability to produce a wide range of valuable products. In this process, an adequate supply of O_2 is essential to achieve high cell growth for the production of metabolites and consumption of organic matter. Many industrial products, such as enzymes, antibiotics, vitamins, organic acids, biochemicals, and biosurfactants, are the outcomes of aerobic fermentation. This process works equally well for wastewater treatment and bioremediation of soil, where aerobic microorganisms degrade organic pollutants.

The choices of fermenter designs for the aerobic process depend on the specific applications, economic aspects, and microorganisms involved. In the large-scale fermenter, O_2 transfer becomes a limiting factor due to its excess demand. This calls for a special device, called a sparger, in the fermenter with fine pores, fitted at the base, which is meant to provide efficient O_2 concentrations. Fermenter also uses a high rate of agitation, which provides sufficient O_2 movement into the liquid, a large gas-liquid interfacial area, and a longer residence duration of gas bubbles in the liquid. Mechanical agitation effectively disperses and breaks the gas bubbles into smaller ones and provides homogeneity and mixing. Further, the high pressure in the headspace of the fermenter can also increase O_2 solubility and movement in the liquid phase. A different type, an airlift fermenter, circulates air in the medium without mechanical agitation. In this setup, the movement of bubbles creates a loop for circulation, ensuring the uniform distribution of cells and nutrients. This eliminates the requirement for an impeller and is most useful for cultivating delicate cells, which could get injured by impeller blades during the fermentation process. A bubble column fermenter also relies on air bubbles coming from an air sparger located at the bottom. This has a less complex design compared to an airlift fermenter. Both airlift fermenters and bubble column fermenters produce low shear stress, making them suitable for sensitive cultures. A different chapter provides in-depth information on the design and applications of various types of fermenters (Chapter 8).

Because aerobic bacteria grow more quickly than anaerobic bacteria, fermentation periods are relatively short, and productivity is high. Also, aerobic conditions during fermentation favour high energy efficiency. However, the major limitations of aerobic fermentation are the poor distribution and low solubility of O_2 in the broth medium due to poor mixing. Some operations may produce foam, which could spill and cause contamination or loss of culture. Also, aerobic fermentation is highly susceptible to contamination from airborne microorganisms.

6.3.2 Anaerobic Fermentation

An industrial fermentation process that occurs in oxygen-absentia is called anaerobic fermentation. Industrial anaerobic fermentation is a versatile process that can produce high-value products, such as solvents, gases, and organic acids, by decomposing various biomasses, including organic waste. It has been applied to many important applications such as the production of ethanol (using *Saccharomyces cerevisiae* or other yeasts), essential amino acid L-Tryptophan (using *Corynebacterium glutamicum* and *Escherichia coli*), lactic acid (using *Lactobacillus delbrueckii*), acetic acid using clostridia (*Clostridium aceticum*, *C. thermoaceticum*, and *C. formicoaceticum*), volatile fatty acids, H_2, as well as nutraceuticals and antimicrobial molecules. Anaerobic fermentation is also employed for anaerobic digestion of waste sludge. This process is considered a viable technology for the environmentally friendly treatment of agricultural or organic waste materials into biogas or bioenergy.

Anaerobic fermentation calls for unique fermenter designs. The most typical setups for such fermentation are a UASB reactor and an expanded granular sludge bed (EGSB) reactor (detailed in Chapter 8). Many typical fermenters, such as CSTR, FBR, fixed bed reactors, and membrane bioreactors (MBR), have versatile reactor architectures that can support both aerobic and anaerobic processes by adjusting the O_2 concentrations and configuring the system to match the specific needs for fermentation.

When the anaerobic fermentation process starts, the fermenter requires an exhaust pump to remove O_2 from the headspace. This is followed by the pumping of inert gases like argon (Ar) or nitrogen gas (N_2). The emergence of certain gaseous by-products like CO_2 or H_2 during fermentation can also fill the space. This applies to those microorganisms which are obligate anaerobes. However, certain microorganisms like yeast require initial aeration to build up high cell biomass before anaerobic product formation. To further maintain the anaerobic environment, the sterilised medium is kept viscous and stationary and inoculated at the fermenter's bottom. Anaerobes grow slowly in the fermentation vessel, involving their unique enzymes and catabolic pathways to utilise different substrates. The low amount of biomass production in anaerobic fermentation ensures the effective utilisation of the carbon content and gives a high product yield (i.e., alcohol).

The advantages of anaerobic fermentation are low energy requirement to keep cells in suspension, minimum contamination under unfavourable conditions, low nutrient requirements, high efficiency in reducing organic waste, development of economically valuable byproducts (like H_2 and CH_4), and low sludge production. Anaerobic fermentation allows for the conversion of many agricultural residues and industrial wastes into useful products while reducing and recycling their residues from the environment. Many organic resources, including lignocellulosic biomass, animal dung, municipal solid waste, and waste from food processing, are suitable feedstocks for anaerobic fermentation. Anaerobic digestion produces digestate, which can be utilised as a fertiliser in agricultural fields. Anaerobic fermentation strictly requires viscous media to create anaerobic conditions, especially for cultivating obligate anaerobes. However, microbial communities can become unstable and are difficult to model with varying environmental conditions (like exposure to O_2) and the availability of nutrients. Furthermore, the metabolic pathways and genetic manipulation in anaerobic microorganisms are less well understood than in aerobic microorganisms.

SUMMARY

- SmF is a method of product formation in a liquid nutrient medium using microorganisms. This mode of operation requires sufficient mixing through mechanical or non-mechanical agitation.
- SSF is the process of growing microorganisms in a sufficiently moist solid substrate, but no free-flowing fluid. Agriculture and food waste are suitable feedstocks for the SSF. Surface fermentation involves the cultivation of microorganisms on the surface of a liquid or solid substrate.
- Gas fermentation utilises microorganisms to convert gaseous substrates, such as CO, CO_2, H_2, and CH_4, into valuable chemicals and biofuels. These gases can be obtained from industrial processes.
- Batch fermentation is a closed system in which the substrate and microorganism are added to the fermenter and are not removed until the fermentation is complete.
- Fed-batch fermentation is a semi-open system or modified form of batch fermentation, wherein growth-limiting substrates are periodically or continuously fed into the fermenter at a controlled rate without any recovery of fermented broth before termination of the fermentation process.
- Continuous fermentation is an open system in which a sterile substrate is added to the fermenter while removing an equal volume of fermented broth. It can vary depending on control parameters, such as chemostat, turbidostat, biostat, etc.
- Aerobic fermentation is the primary industrial biotechnology method because of its versatility, efficiency, and production of various valuable products like enzymes, antibiotics, vitamins, organic acids, biochemicals, and biosurfactants.
- Anaerobic fermentation requires unique fermenters and involves removing O_2 and pumping inert gases into a fermenter while maintaining a viscous, stationary medium inoculated at the bottom. It is utilised for producing ethanol, L-tryptophan, lactic acid, acetic acid, and volatile fatty acids.

DIFFERENT TYPES OF PRACTICE QUESTIONS

Fill in the blanks

I. .... is a method of microbial cultivation in a liquid nutrient medium in a stirred tank reactor.
II. Solid-state fermentation requires about 30–85% .... to maintain an active growth of microorganisms.
III. .... fermentation establishes a quasi-steady state, having an equal consumption of the added substrate by a virtually constant cell concentration.
IV. In .... fermentation, the fresh medium is periodically added with an equal removal of fermented broth.
V. From an industrial perspective, the stationary growth phase in fermentation is referred to as ....
VI. .... calls for a unique fermenter design having an exhaust pump to remove oxygen from the headspace.

SOLUTION: I. Submerged fermentation (SmF), II. moisture, III. Fed-batch, IV. Continuous, V. Idiophase VI. the anaerobic fermentation

Match the following

i. Koji fermentation	a. *Clostridium ljungdahlii*
ii. Fed-batch fermentation	b. *Agaricus bisporus*
iii. Mushroom cultivation	c. *Saccharomyces cerevisiae*
iv. Syngas fermentation	d. *Aspergillus oryzae*
v. Ethanol fermentation	e. *Penicillium chrysogenum*

SOLUTION: i-d, ii-e, iii-b, iv-a, v-c

Very short answer type of questions

I. How does solid-state fermentation offer an economic advantage over conventional submerged fermentation?
II. What is syngas fermentation?
III. Define chemostat.
IV. Expand MCCF and CSTR.
V. Bacteria are considered more suitable for SmF than any other type of microorganisms. Give a suitable reason.
VI. Why is batch fermentation considered a closed, unsteady system?
VII. Why is surface fermentation rarely used in industries?

Short answer type of questions

I. Differentiate submerged fermentation from solid-state fermentation.
II. Despite continuous fermentation being a useful operation, it has limited industrial applications. Why?
III. Outline the differences between trophophase and idiophase in industrial fermentation.
IV. What are the applications of batch fermentation in an industrial-scale operation?
V. What are the various raw materials and their products in solid-state fermentation?
VI. When an enzyme production is sensitive to substrate concentration and requires specific control operation, which type of fermentation would be the most suitable and why?

Long answer type of questions

I. What are the different modes of operation of industrial fermentation based on the flow of nutrients? Provide a comparative account of them.
II. How do the choices of different modes of operations in fermentation impact the overall cost-effectiveness and sustainability of industrial-scale production?

NOTE

1 The cell concentration in microbiology represents the number of cells in a given volume of liquid, which can be generally measured as cells/ml using flow cytometry or in optical density (OD) using spectrophotometer.

SUGGESTED READINGS

Bengelsdorf FR, Straub M, Dürre P. Bacterial synthesis gas (syngas) fermentation. *Environmental Technology*. 2013;34(13–14):1639–51.

Cinar A, Parulekar SJ, Undey C, Birol G. *Batch fermentation: modeling: monitoring, and control*. CRC Press; 2003.

Kadam AA, Kamatkar JD, Khandare RV, Jadhav JP, Govindwar SP. Solid-state fermentation: tool for bioremediation of adsorbed textile dyestuff on distillery industry waste-yeast biomass using isolated *Bacillus cereus* strain EBT1. *Environmental Science and Pollution Research*. 2013;20:1009–20.

Kaewlaoyoong A, Cheng CY, Lin C, Chen JR, Huang WY, Sriprom P. White rot fungus *Pleurotus pulmonarius* enhanced bioremediation of highly PCDD/F-contaminated field soil via solid state fermentation. *Science of the Total Environment*. 2020;738:139670.

Krishna C. Solid-state fermentation systems—an overview. *Critical Reviews in Biotechnology*. 2005;25(1–2):1–30.

Li SY, Srivastava R, Suib SL, Li Y, Parnas RS. Performance of batch, fed-batch, and continuous A–B–E fermentation with pH-control. *Bioresource Technology*. 2011;102(5):4241–50.

Li T, Chen XB, Chen JC, Wu Q, Chen GQ. Open and continuous fermentation: products, conditions and bioprocess economy. *Biotechnology Journal*. 2014;9(12):1503–11.

Lindskog EK. The upstream process: principal modes of operation. In Jagschies G, Lindskog E, Łącki K, Galliher P (eds.) *Biopharmaceutical processing* 2018 (pp. 625–635). Elsevier.

Louhasakul Y, Wado H, Lateh R, Cheirsilp B. Solid-state fermentation of Saba banana peel for pigment production by *Monascus purpureus*. *Brazilian Journal of Microbiology*. 2023;54(1):93–102.

Maxon WD. Continuous fermentation: A discussion of its principles and applications. *Applied Microbiology*. 1955;3(2):110–22.

Minihane BJ, Brown DE. Fed-batch culture technology. *Biotechnology Advances*. 1986;4(2):207–18.

Owoade A, Alshami AS, Levin D, Onaizi S, Malaibari ZO. Progress and development of syngas fermentation processes toward commercial bioethanol production. *Biofuels, Bioproducts and Biorefining*. 2023;17(5):1328–42.

Pandey A. Solid-state fermentation. *Biochemical Engineering Journal*. 2003;13(2–3):81–4.

Patakova P, Linhova M, Rychtera M, Paulova L, Melzoch K. Novel and neglected issues of acetone–butanol–ethanol (ABE) fermentation by clostridia: Clostridium metabolic diversity, tools for process mapping and continuous fermentation systems. *Biotechnology Advances*. 2013;31(1):58–67.

Righelato RC. Anaerobic fermentation: alcohol production. *Philosophical Transactions of the Royal Society of London. B, Biological Sciences*. 1980;290(1040):303–12.

Roukas T, Kotzekidou P. Production of citric acid from brewery wastes by surface fermentation using *Aspergillus niger*. *Journal of Food Science*. 1986;51(1):225–8.

Wang S, Wu Q, Nie Y, Wu J, Xu Y. Construction of synthetic microbiota for reproducible flavor compound metabolism in Chinese light-aroma-type liquor produced by solid-state fermentation. *Applied and Environmental Microbiology*. 2019;85(10):e03090–18.

Watson TG. The present status and future prospects of the turbidostat. *Journal of Applied Chemistry and Biotechnology*. 1972;22:229–243.

Yang Y, Sha M. *A beginner's guide to bioprocess modes–batch, fed-batch, and continuous fermentation*. Eppendorf Inc. 2019;408:1–6.

Yuan Y, Hu X, Chen H, Zhou Y, Zhou Y, Wang D. Advances in enhanced volatile fatty acid production from anaerobic fermentation of waste activated sludge. *Science of the Total Environment*. 2019;694:133741.

7 Fermenters and Their Basic Components

7.1 FERMENTERS AND THEIR CHARACTERISTICS

A fermenter is an engineered instrument for fermentation using suitable microorganisms and substrates. It is an enclosed vessel that provides a precisely regulated, optimal environment for small- to large-scale microbial growth and biosynthesis of their metabolic products. Another common interchangeable term for the fermenter is the bioreactor. Although both terms have the same working principle, the latter form is generally used for the mass cultivation of diverse biological organisms, including microorganisms, insects, plants, animals, and mammalian cells.

Fermenters come in various sizes and forms, along with a variety of accompanying components for different functions. Some components of fermenters can be customised as per the requirements. Of various popular fermenters, a mechanically agitated fermenter is the most basic type and is commonly used in microbial fermentation. Figure 7.1 illustrates a schematic representation of a typical mechanically agitated fermenter (or mechanically stirred tank bioreactor) design showing various components. Chapter 8 provides a detailed account of different variants of fermenters with their specifications. Regardless of its various types, fermenters have some essential elements that provide a means to control and promote biological activities. These characteristics are:

- The fermenter should be able to perform different fermentation reactions for distinct requirements.
- The operation should run in a controlled environment.
- The vessel should reliably function aseptically for a long duration.
- There should be inlet facilities to transfer inoculum, nutrients, air, defoamers, and pH buffers.
- The outlet valves are needed for gas exhaustion and product harvesting.
- An adequate supply of agitation and aeration inside the vessel is required for microbial proliferation and substrate utilisation.
- Evaporation loss should be minimal.
- There should be sensors to monitor the fermentation events. A control panel should be installed to regulate the temperature, pH, and dissolved gases to optimise biological activity.
- A sampling port is required to access samples from the fermentation broth under typical operating conditions for bioprocess evaluation.
- A fermenter should meet the standard containment regulations.
- The construction materials of the fermenters should be inexpensive and productive.
- Energy efficiency is highly desirable.
- The process should employ the minimum amount of labour in functioning, harvesting, cleaning, and maintenance activities.

7.2 FERMENTER SCALE-UP

Fermenters come in a wide range of volume capacities for distinct purposes. Fermentation essentially starts at a small laboratory scale and then proceeds to an intermediate-sized pilot scale, followed by large-scale production (Figure 7.2). This step-wise process is called fermenter scale-up (Table 7.1). When a new product has a commercial application, the main challenge is the technology transfer from a laboratory scale to a commercial scale. Taking economic considerations into account, the development of industrial-scale fermentation necessitates the fermentation of multiple production scales. On the laboratory scale, the studies on microbial growth by controlling fermentation conditions provide crucial information on optimal microbial activity and product formation. However, microbial productivity predetermined and optimised at a laboratory scale shows substantial differences for pilot fermenters or large production fermenters due to critical variations in fermentation parameters. Therefore, a series of laboratory-scale trials provides the baseline data for statistical optimisation to determine the best conditions for microbiological growth and fermentation. These optimal fermentation conditions are then realised with semi-technical and pilot-scale fermenters. The final step is correlation studies of different scale-ups to evaluate the merits and performance risks of fermentation for large-scale production.

A laboratory fermenter can have a typical size range of 500 ml to 100 l. Non-toxic, corrosion-resistant borosilicate glass is a common construction material for laboratory fermenters. The transparent surface of the body material provides close observation of the growth behaviour, metabolism, and genetic regulation of fermentation microorganisms. Such apparatus is widely used for academic research and industrial bioprocess development. The experiment in the laboratory fermenter

DOI: 10.1201/9781032651095-9

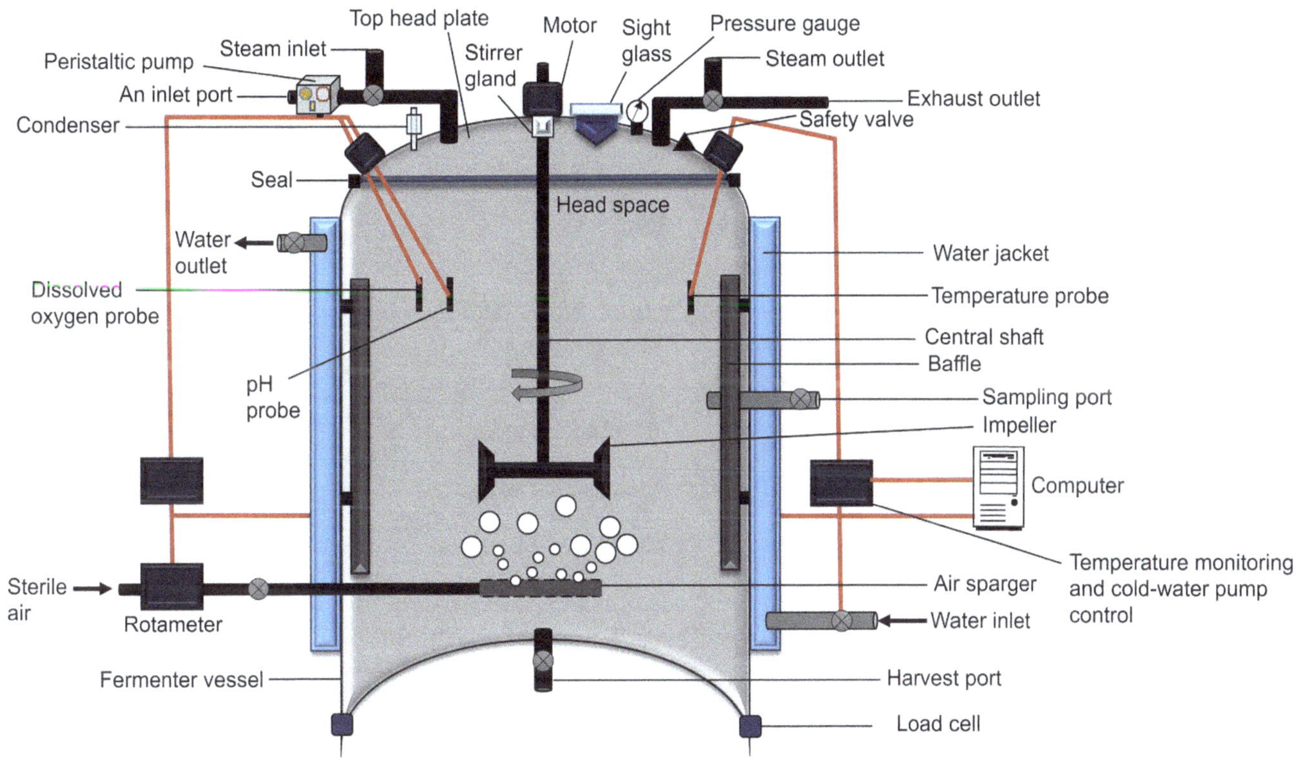

FIGURE 7.1 A schematic diagram of a typical mechanically agitated fermenter showing various components.

FIGURE 7.2 Fermenters of different sizes. a. Laboratory scale fermenter. b. Pilot scale fermenter. c. Production scale stainless steel fermenter. (Image c from iStock/DedMityay.[1])

supports several feed strategies for customisation. In addition, laboratory fermenters can precisely control and monitor fermentation conditions. They come in modular designs with the facility to easily add or remove components. The development of new bioprocesses and methodologies calls for a laboratory fermenter.

A pilot plant is an intermediate pre-commercial production system which analyses a new fermentation setup before large-scale manufacturing. Pilot-scale fermenters come in a wide range of sizes for laboratories to industries and exist in size ranges of 10-1,000 l. However, the volume capacity of the pilot scale can increase beyond this size range and up

TABLE 7.1
A comparative account of fermenters of different scales

	Laboratory scale fermenters	Pilot scale fermenters	Production scale fermenters
Type	Small scale	Small to large scale	Large scale
Total volume capacity	500 ml-100 l	10-10,000 l	>10,000 l
Construction material	Borosilicate glass (rarely stainless steel)	Stainless steel	Stainless steel
Fermenter sterilisation	Autoclaving	Mostly *in situ*	*In situ*
Temperature control	Thermostatically controlled bath, internal heating coils, heating jacket, silicone heating jacket	Internal coils in which hot or cold water is circulated	Internal coils
Pressure	0.5 bar	2 bar	2 bar
Limits for the agitator	80% of critical speed	70% of critical speed	65% of critical speed
Baffles	Four	Four to six	Six to eight
Aeration spargers	Porous spargers, combined sparger-agitator	Orifice and nozzle spargers	Nozzle spargers
Dissolved oxygen (DO) concentration monitoring	Galvanic electrodes	Polarographic electrodes	Polarographic electrodes
Containment	For <10 l vessels, there are containment shelves and good manufacturing practices. Regulation protocols are used for>10 l vessels	The GILSP* is used for non-pathogenic and non-toxic strains. For other hazard groups, higher containment levels are considered	The GILSP and higher containment levels are considered according to the microbial hazard group.
Accessories	Few	These have similarities in size, shape, and accessories of large production fermenters, but can be varied according to the development of new procedures	Many
Function	For the research and development of fermentation methods	For the evaluation of new fermentation technology before large-scale setups and for training and education purposes	For the industrial production of fermentation products
Media sterilisation	Media is placed in the fermenter and autoclaved	*Ex situ*	*Ex situ* and *in situ* in different setups
Surface area to liquid (volume) ratio	Relatively large	Relatively small	Relatively small

* GILSP stands for Good Industrial Large-Scale Practice.

to 10,000 l to bridge the gap between pilot and industrial scales. The construction materials and design of a pilot-scale fermenter are essentially engineered similarly to large-scale fermenters (i.e., stainless steel) so that the related challenges for mass production can be studied. The main benefits of the pilot studies are cost-effective cataloguing, modification of various limitations, and mitigation of the associated risks before constructing large-scale fermenters. In a pilot study, the fermentation parameters can be analysed with semi-empirical computer modelling and simulation methods that provide valuable scientific and engineering data about material stability properties and the regulation of optimum conditions. The data generated from pilot-scale studies helps identify further research objectives and develop new ideas for a more constructive design of high-scale production systems. Also, this kind of setup can be a useful display of the functioning of large-scale plants for education and training purposes.

The large-scale industrial fermenters can have a capacity of more than 10,000 l. However, many industrial facilities even have production at a working volume of 1,000 l and above. The large-scale fermenters are made of stainless steel, which provides resistance to corrosion and rusting. Increasing fermenter volume during the scale-up affects a variety of fermentation parameters. These parameters include equipment usage, nutrient concentration, inoculum, homogeneity, surface area to volume ratio, non-linear reaction kinetics, fluid dynamics, etc. As the process is scaled up, some additional pieces are installed in the fermenter vessel to monitor the physicochemical parameters and achieve other facilities. The same also necessitates more power to provide aeration, hydrodynamic pressure, and agitation to maintain a homogeneous fermentation broth. The large-sized fermenter contains a large space, resulting in a lower surface area to volume ratio, which requires high volumes of inoculum, nutrients, and buffer. The properties of some materials in large-scale fermenters are also likely to change with increasing volume, which alters the fluid dynamics of liquids and gases inside the vessel. Fermenter scale-up can further lead to non-linear reaction kinetics. This often displays more complex and chaotic systems than the pilot scale. The behaviour of a chaotic system is highly unpredictable and exhibits fluctuating concentrations of intermediate species. Moreover, in large-scale fermenters, the natural microbial strains demonstrate tolerance to variations of fermentation parameters due to their evolutionary adaptations, but may possess low productivity. On the other hand, the recombinant strains achieve desirable levels of robustness under specific conditions, but they bear low tolerance to altered parameters.

7.3 ASPECT RATIO

Fermenters can be taller or broader, depending on their geometric dimensions, particularly the height and diameter. The ratio of the fermenter height (H) to the fermenter diameter (D) is called the aspect ratio (Figure 7.3). An optimum aspect ratio of a fermenter depends on the design requirements of the specific mass movement, dissolved gas uniformity, cost of equipment, electric power consumption, heat movement, and site restrictions. Many modern fermenters have a high aspect ratio that makes them tall and wide. They are usually cylindrical and are considered a type of tank fermenter. The aspect ratio of these fermenters can be 3:1 or 3:2 (Figure 7.3a). A different type of vessel with an aspect ratio of more than 3:1 is an elongated fermenter. A non-mechanically stirred tower fermenter can have an aspect ratio of 6:1 for the tubular section or 10:1 overall (Figure 7.3b). For some applications, the aspect ratio of a tower fermenter can reach up to 16:1. On the other hand, a rotary drum fermenter has a cylindrical shape with an aspect ratio of 1:5 or up to 1:10.

A suitable aspect ratio varies with the types of fermentation reactions. An aspect ratio of 3:1 is appropriate for hydrogen (H_2) production and many other metabolites in a stirred tank reactor. A fluidised-bed fermenter with an aspect ratio of about 10:1 can be used for ethanol production using the flocculent culture of *Saccharomyces cerevisiae*. Low aspect ratio fermenters provide adequate mixing and lower shear rates that cause less product degradation. A stirred tank fermenter of a 3:2 aspect ratio is typically used for biomass culturing. This ratio facilitates adequate mixing and aeration for maximum cellular growth. However, the optimum ratio of a fermenter for anaerobic fermentation is usually 1:1. Fermenters with a low aspect ratio can also be used for polysaccharide production (e.g., dextran, xanthan gum, and pullulan).

As the ratio of the fermenter is changed from cylindrical to elongated, the vessel layout also undergoes several modifications. Tall and thin fermenters have large surface areas for heat movement, and therefore, they do not require internal heating coils. The internal composition of nutrients and dissolved oxygen (DO) might become less uniform in elongated fermenters. Elongated fermenter models are more expensive due to high installation costs. The design of longer agitator shafts is relatively difficult. When the agitator shaft operates at a critical speed or natural frequency, it develops high vibration from the unbalanced hydraulic forces, which leads to structural and functional issues, including bending.

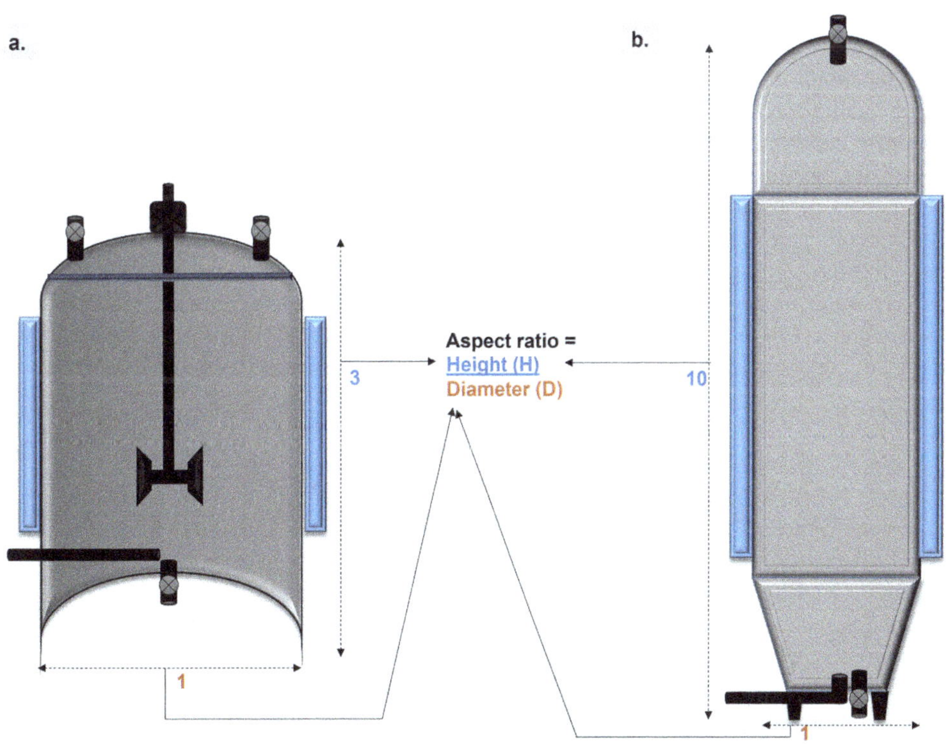

FIGURE 7.3 The aspect ratios of two different fermenters. a. Tank type, showing a 3:1 aspect ratio. b. Column type showing a 10:1 aspect ratio.

This can be overcome by using more than one impeller attached to the shaft on the vertical axis. A steady bearing on the top of the tall fermenter tank supports the agitator shaft. Also, when the shaft rotates above the critical speed, there is no noticeable vibration. High gas velocity in the lower part of the elongated fermenter increases the driving force of mass movement, resulting in high total hydrostatic pressure. Although higher hydrostatic pressure makes the agitator powerless and increases gas solubility, it can also affect pH and metabolic activity in the broth, compromising cell viability.

7.4 MECHANICAL CONSTITUENTS

The mechanical constituents of a fermenter are those components that provide an optimum temperature, aeration (for aerobic fermentation), and agitation. A typical fermenter can have additional assets such as baffles, seals, foam prevention (foam breaker), and accessory (piping and valves) systems. These components are mostly regulated by a controlling system, which acts on the probe's signals. Fermenters are equipped with polarised electrodes in an innovative approach to direct product distribution during fermentations (detailed in Box 7.1). The following subsections provide details on different mechanical constituents and their functions.

7.4.1 Temperature Control

Temperature exerts ultimate impacts on the fermentation process. A small change in the temperature inside the vessel alters microbial growth and productivity. The heating and cooling systems attached to the fermenters maintain the requisite temperature conditions in response to signals sent by a probe. The small-scale fermenters do not generate enough heat for the reactions, and the necessary heat can be provided using a heating jacket or a thermostatically controlled water bath. In contrast, large-scale fermentation generates excess heat due to microbial metabolic activity, mechanical agitation, and aeration power. These vessels have a covering of a cooling device, such as a converted refrigerator, chest freezer, or cooling jacket. Most commercial tanks have water or cooling jackets composed of coolant channels, which run at the outer surface of the fermenter to cycle cold water. In some large-scale fermenters, the cooling coils reside inside the fermenter in addition to the cooling jackets (Figure 7.4).

7.4.2 Sparger

The supply of sufficient sterile air to the microorganisms is an essential requirement for their active metabolism, particularly in aerobic submerged fermentation. A sparger is a device typically installed at the bottom of the fermenter below the top agitator to meet the demand for air supply inside the fermenter. It is connected to a blower, which blows the air from outside. A special type of filter (high-efficiency particulate air (HEPA) filter) is attached to the blower to remove the microscopic particles and purify the air. The sterile air produced from the blower systems passing through the sparger is released inside the vessel. A fermenter that consists of a sparger has a highly efficient and controllable mode of aeration. The agitation produced by aeration also facilitates the proper mixing of air bubbles throughout the vessel. In the airlift fermenter, aeration through the sparger is enough to provide agitation when the medium has low viscosity and contains low concentrations of total solids. A rotameter can be attached to the sparger to measure the rate of airflow inside the fermenter.

Spargers can have upward-facing or downward-facing pores. The upward-facing pores of the sparger tend to get clogged with the settling of substrate/product and microbial biomass during fermentation, which can inhibit proper aeration in the vessel. In contrast, when the pores on the sparger are oriented downwards, fewer air bubbles would impinge onto the agitator, and their pores would not get clogged. The size of the air bubble released from the sparger has a positive correlation with the pore size and pipe cavity volume. The air bubbles of small sizes have a comparatively larger surface area than the air bubbles of large sizes. The large surface area of small bubbles facilitates greater oxygen (O_2) diffusion from the bubble to the medium. According to the arrangement of pipes and pores, spargers can be of different types, which are discussed below.

- Porous sparger: It consists of a single pipe made of fritted glass, ceramics, or metal having multiple fine pores through which air passes (Figure 7.5a). The pores can have a diameter of 1 to 500 μm. The small pores of

> **BOX 7.1 ELECTRO-FERMENTATION**
>
> Electro-fermentation (EF) technology is emerging as an innovative, sustainable approach to trigger variations in yield, titer, or product distribution of traditional fermentation processes. The EF technology design is based on the application of a polarised electrode interface during fermentations. These electrodes serve as electron donors, acceptors, or a means to control the oxidation-reduction potential of the fermentation broth. Anodic electro-fermentation (AEF) utilises the electrode as an electron acceptor, while cathodic electro-fermentation (CEF) acts as an electron donor. A widely adopted laboratory scale design of the EF system consists of two compartments separated by an ion transducing membrane, each forming an electrochemical half cell with an anode and a cathode. The usage of EF has enhanced the production of various products, including ethanol, butyrate, acetoin, propionate, acetone, lactate, L-lysine, propanediol, etc. Critical analysis of the underlying biochemical, microbiological, and engineering aspects is yet required for the EF technology to proceed from the laboratory to the market.

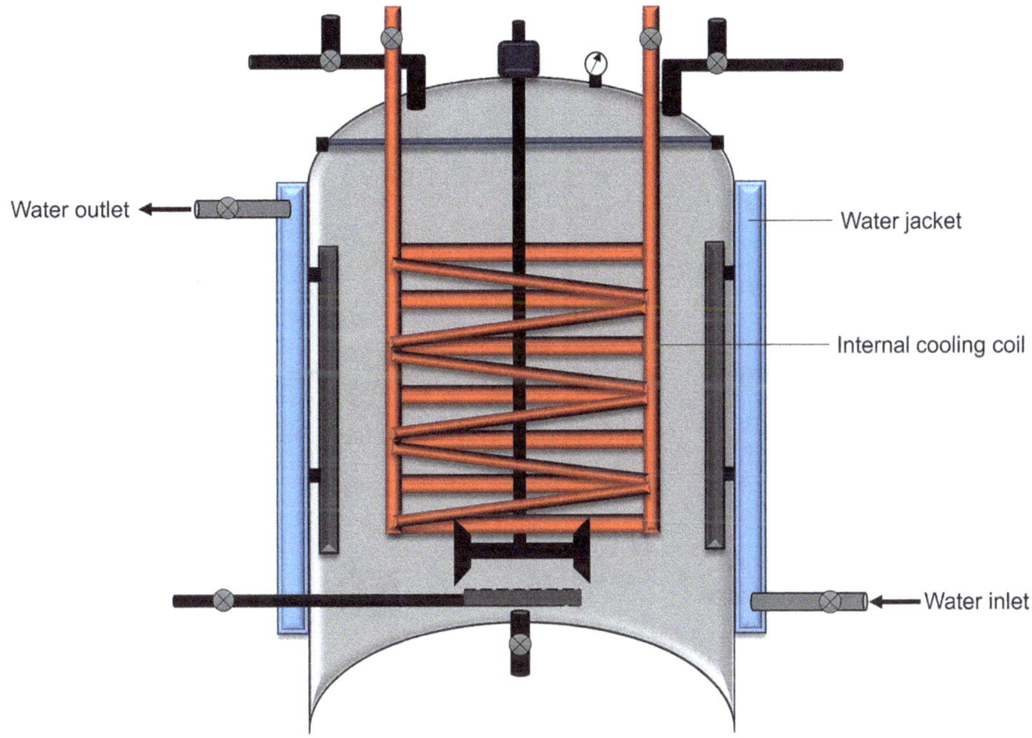

FIGURE 7.4 A schematic diagram of an industrial fermenter showing temperature control components, which include a water jacket and an internal coil that passes hot or cold water from inside.

the porous sparger tend to get plugged with microbial biomass, causing a pressure drop and reducing product productivity. This type of sparger is frequently used in laboratory-scale fermenters.

- Orifice or ring sparger: It consists of a metal pipe containing a few but larger pores or orifices (about 6 mm) than a porous sparger. This type of sparger comes in different arrangements. In one arrangement, several perforated pipes can be attached to a central disc that gives the appearance of crosses (Figure 7.5b-d). In another arrangement, perforated circular pipes or discs give the appearance of rings. The larger pores minimise the plugging from microbial growth and the accompanying pressure drop. This type of sparger is common in many types of fermenters.
- Nozzle sparger: It consists of a metal, bronze, stainless, or carbon steel pipe having one big opening (of about 150 mm). The large pores do not get clogged with microbial biomass and provide a bulky flow of air bubbles for efficient gas dispersion and aeration (Figure 7.5e). Also, they facilitate easy mixing of fermentation broth to provide a homogenous composition. This type of sparger can be installed in both small and large-scale mechanically agitated fermenters. The large-sized bubble can be converted into small bubbles using an agitator.

Bottom sparging maintains the fermentation broth in a highly turbulent state. The sparger's performance has an impact on the hydrodynamics of the reactor, including liquid circulation patterns. A good design of the sparger ensures efficient mixing and reduces dead zones inside the fermenter. On the contrary, a malfunction of sparging can cause flooding due to air bubbles, which ceases the radial flow of bubbles, leading to inefficient gas dispersion in the lower portion of the vessel. The flooding causes gas to flow upwards in the centre of the vessel while the broth is pushed toward the walls.

7.4.3 Agitator

Agitation is the process of bulk mixing of fermentation broth, which ensures a homogenous and uniform environment in the entire fermenter vessel. This activity develops a proper suspension of nutrient particles in the broth, prevents microbial aggregation, and disseminates heat. Agitation improves O_2 dispersion in small bubbles and prevents air bubble coalescence. It also allows air bubbles to flow in a more tortuous path to delay their escape from the broth and reduces the thickness of the liquid layer at the gas/liquid interface. The agitation in mechanically agitated fermenters can be achieved using a mechanical device called an agitator, consisting of impellers (or propellers) and a central shaft. An impeller is a rotating device or rotor that consists of several blades (or vanes). The rotation of the shaft and blades alters the flow and pressure of fermentation broths or gases by drawing them in with an intrinsic sucking force to impelling them out of the other side.

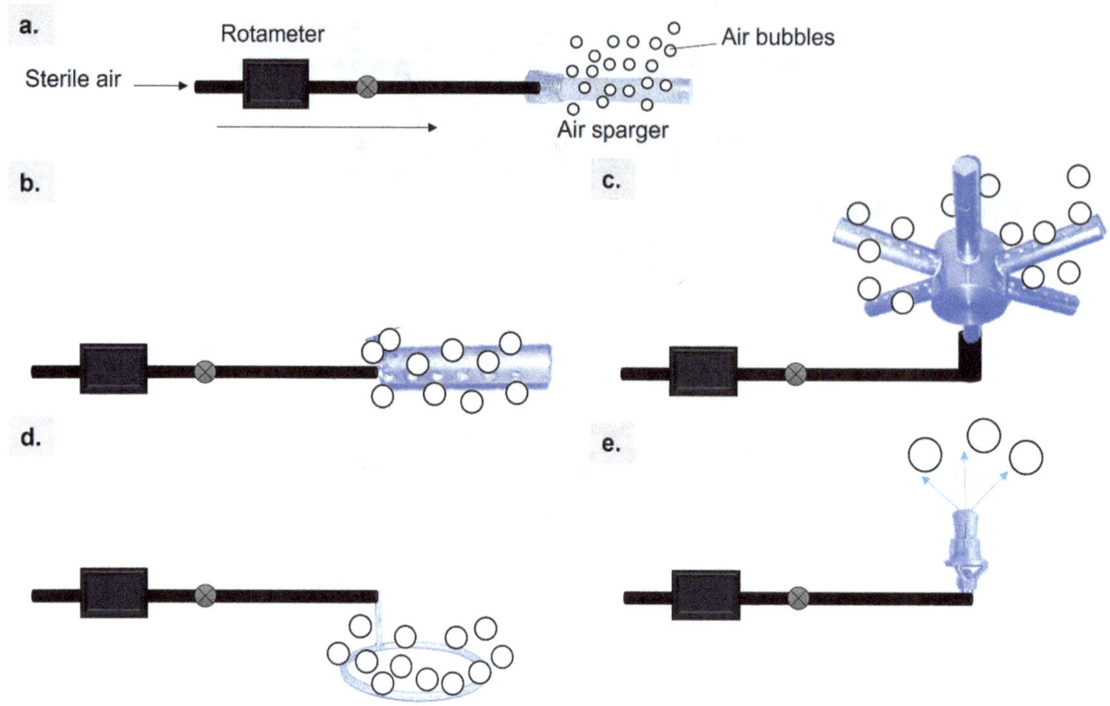

FIGURE 7.5 Schematic diagrams of different types of air spargers. a. Porous sparger. b. Orifice sparger with a single pipe. c. Orifice sparger with multiple pipes (cross pattern). d. Orifice sparger with a single pipe (ring pattern). e. Nozzle sparger.

A different type of impeller is a propeller, which consists of three or four blades that are attached at an angle to a hub on the central shaft. The rotation of propeller blades creates an extrinsic circular motion that pushes the broth away from the blades to develop a forward movement. Both impellers and propellers are typically one-third the diameter of a fermenter and are powered by an external motor attached to the top plate of the fermenter. In different settings, agitators can be attached to the lateral or bottom walls of the fermenters. Central shafts with a short length are considered more suitable for high agitation rates. They have comparatively low whipping effects or vibration during rotation. Industrial agitators have a variety of designs that usually vary according to their blade arrangement, ability to direct airflow patterns, and mixing efficiency. Some popular variants of agitators are discussed below.

- Open turbine agitator: In this type, the blades are directly attached to a hub, which is mounted on the central shaft without any surrounding wall. This type of agitator mainly develops an axial bulk flow pattern where the radial dispersion of air bubbles is less efficient. These agitators are usually attached to small pumps. Based on the pitch (the angle of the blade), an open turbine agitator can be of two different types. When the blades of the agitator are at a fixed pitch, it has a rotating fan-type structure. One of its forms is a three or more-bladed structure with high efficiency in producing a large flow with less energy, which is called a hydrofoil agitator (Figure 7.6a). In the second type, the pitch of the blades is rotated around their long axis, and it is called a variable pitch propeller (Figure 7.6b). A modern design is the top-mounted vertical scaba propeller (Figure 7.6c). It has three blades that provide greater radial flow than axial flow for bulk blending of viscous broths and accommodate high superficial velocity (Vs). It also prevents the flooding of air bubbles at the bottom. This type of agitator is more efficient and easier to maintain than others.
- Semi-closed agitator: In this type, the blades are attached to a disc, which is connected to a central shaft. The presence of a disc adds strength to the agitator. A semi-closed agitator creates a radial bulk flow pattern as the air bubbles from the sparger get displaced toward the vanes after hitting the underside of the disc. Flooding of the agitator occurs at a high Vs (around 120 m/h), thereby making them suitable for high agitation of low-viscosity fermentation broths. The agitator disc can have different arrangements of blades. In the disc turbine type, a series of rectangular blades is arranged perpendicularly along the disc's circumference (Figure 7.6d). In a different arrangement, the vaned disc type has a series of rectangular blades arranged in a perpendicular plane along the underside of the disc (Figure 7.6e).
- Closed agitator: In this type, the blades have a back and front wall around the blades that provide an increased strength (Figure 7.6f). This type of agitator is primarily used with large pumps for radial bulk mixing of broth components. However, high concentrations of suspended solids in the broth can result in clogging inside the

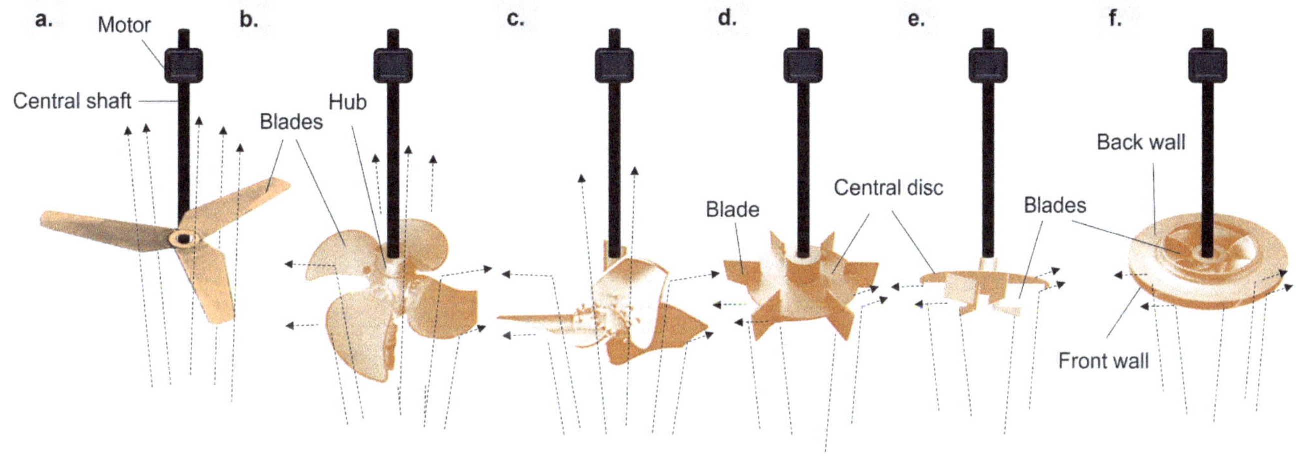

FIGURE 7.6 Schematic diagrams of the different types of agitators. a. Fixed pitch hydrofoil. b. Variable pitch propeller. c. Scaba propeller. d. Disc turbine. e. Vaned disc turbine. f. A closed agitator. Dashed arrows represent the bulk flow pattern caused by the blades.

agitator during mixing, thereby causing difficulty in its cleaning.

In a mechanically agitated fermenter, high agitation rates cause discernible vortexing of the fermenting broth, which eventually hurts fermentation productivity. During vortexing, the fermentation broth is directed with its circular motion towards the fermenter walls, and the top of the vessel creates a partially free area. In this situation, nutrient particles and microbial cells also do not get mixed properly and tend to remain in one position relative to the whirling broth. Some air bubbles coming from the sparger also escape without getting dissolved in the broth. This can result in low aeration in the vessel due to broth splashing at the top of the fermenter. The proper design of the blade tip shape enhances the axial flow efficiency, which can minimise the vortex effect. However, most fermenters install baffles to avoid the limitation of vortexing.

7.4.4 Baffle

A baffle is typically a flat vertical metal strip/plate attached radially to the inside wall of the fermenter to increase the mixing rate. When low-viscosity fluids are agitated in a stirred fermenter vessel without baffles, the rotating impeller facilitates a tangential movement of liquid. The fluid travels in circles with a high circumferential velocity, resulting in inadequate mixing. The free surface of the circular motion of the fluid creates a vortex. The vortex draws the liquid downward in the centre and hurls it against the fermenter's walls, eventually dropping at the bottom (Figure 7.7a). This causes an uneven distribution of nutrients, O_2, and temperature and reduces the overall efficiency of fermentation processes. When a fermenter contains baffles, the circular flows or tangential velocities of fluid are significantly reduced, preventing the creation of vortices and making the liquid surface nearly flat. Additionally, the axial flow becomes significantly stronger, which enhances the rate of mixing. The baffles are positioned at regular intervals around the circumference of the vessel, which further prevents the formation of a vortex in the fermentation broth (Figure 7.7b). This process also assists in the aeration and agitation during fermentation.

Baffles are almost one-tenth the size of a fermenter. Wide baffles can slightly increase the agitation effect compared to the narrow types. Furthermore, baffles are designed to have a little spacing with the fermenter wall in order to support cleaning and prevent any microbial growth around them. The number of baffles varies according to the size of the fermenters. Small fermenters can have four, while large fermenters possess six or eight baffles. However, excessive baffling would prolong or disrupt the mixing of the liquid. When the feed is highly viscous, the fermenter does not need baffles because the strong friction of the fluid on the cylindrical wall and the low agitation rate prevent vortex formation. The utilisation of a baffle-free fermenter for viscous fluid also improves mixing efficiency and reduces shear stress.

Apart from the vertical baffles (Figure 7.8a), some fermenters also use horizontal (Figure 7.8b) and inclined (Figure 7.8c) baffles. Horizontal baffles are typically found in long fermenters and can redistribute the flow of liquid. They have perforations that are mainly used to provide a barrier and disperse the liquid uniformly. The inclined baffles are positioned at an angle that is either fully horizontal or vertical. The baffle inclination helps reduce power consumption and the swirling of fluid at the liquid's free surface.

7.4.5 Seals

The different openings and joints of the fermenter are occupied with reliable components of the seals to provide aseptic conditions. Sensors, entry ports (of inoculum, nutrient medium, pH buffers, and anti-foam), and sampling ports are the main sites for sealing. Seals are mostly made of nitryl, silicone, or butyl rubber. The seals are inserted into the grooves of the top plate (lid) and fermenter vessel using an appropriate compressible gasket, lip seal, or O-ring. Fermenters are usually

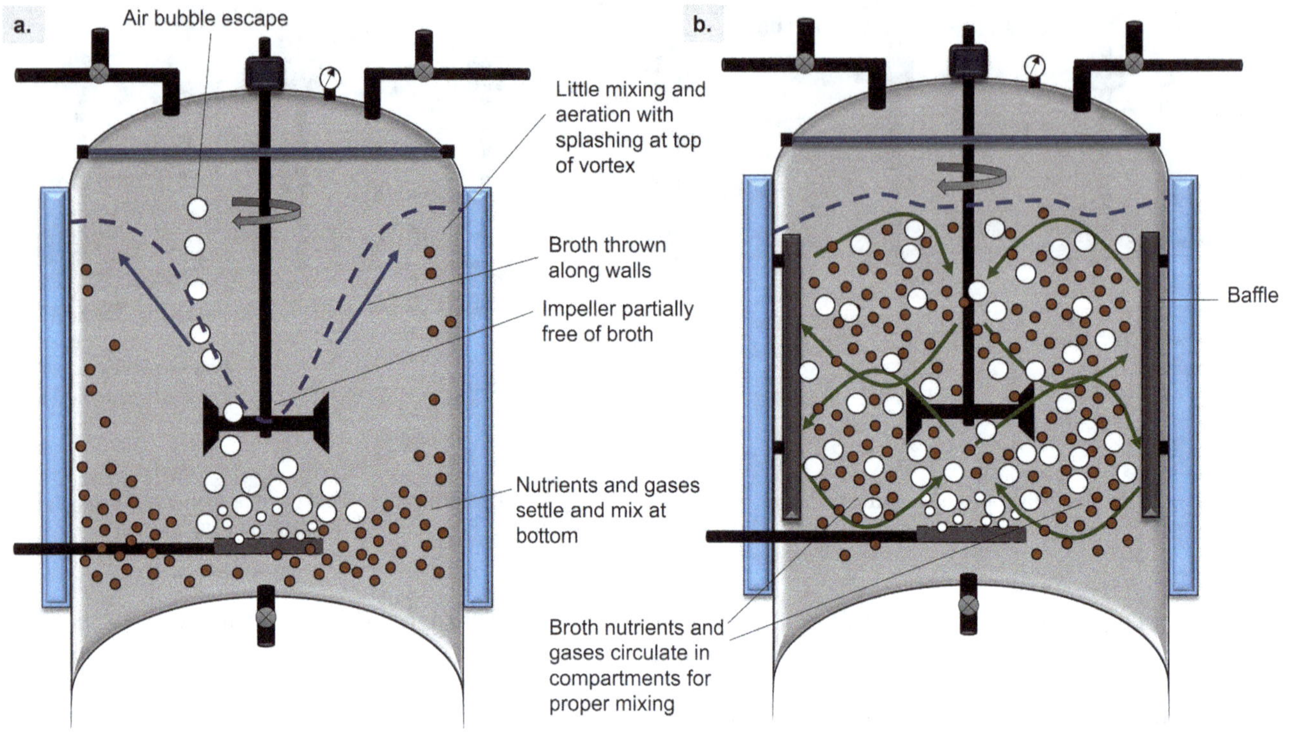

FIGURE 7.7 Schematic diagrams of the fermenter without and with baffles, showing flow patterns of air bubbles and nutrients. a. Vortex formation in the absence of a baffle. b. Compartmentalised mixing in the presence of baffles. (White circles represent air bubbles, and small red circles depict nutrients)

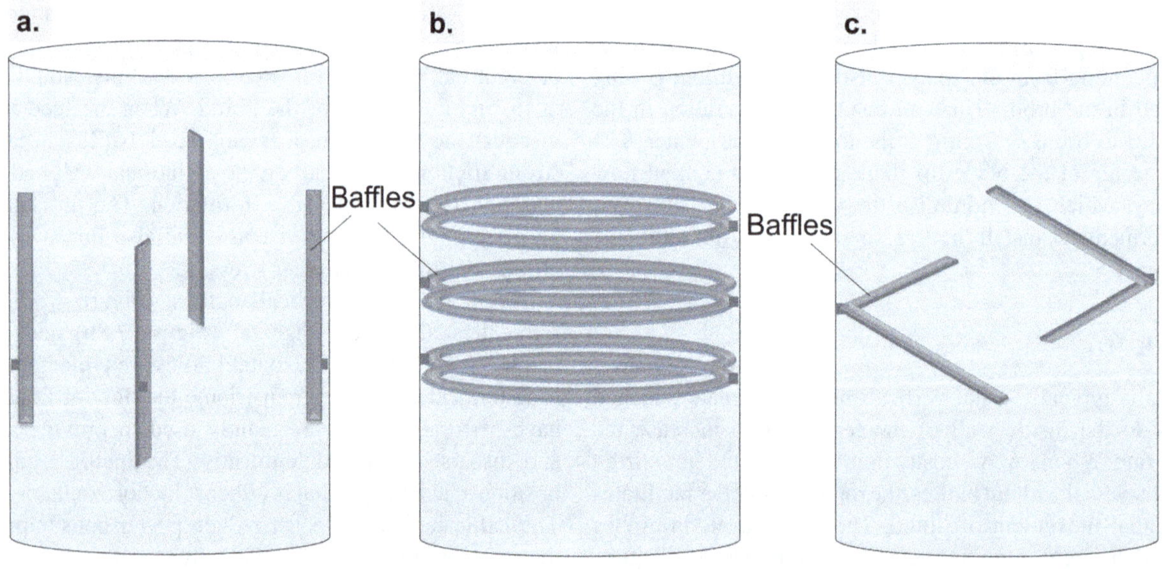

FIGURE 7.8 Schematic diagrams of the different types of fermenter baffles. a. Vertical. b. Horizontal. c. Inclined.

loosely sealed to allow the escape of fermentation gases from agitation systems. An aseptic seal consists of stirrer glands and bearings required at the entrance of the agitator shaft. The stirrer glands are designed to introduce the agitator shaft into the fermenter vessel with a vapour-tight seal. At the same time, bearings restrain the free rotation of the shaft around a fixed axis and reduce friction with their rolling motion.

7.5 FOAM CONTROL

During high loads of fermentation, the denatured proteins generate thousands of bubbles in the liquid. These bubbles deposit themselves at the liquid-air interface and cause foaming. Different protein sources, such as yeast extract, beef extract, corn steep liquor, and soybean meal in the fermentation media, are responsible for the foaming. Excessive foaming affects fermentation productivity. It interferes with internal sensors that control fermentation parameter monitoring. The working volume of the fermenter is also reduced, lessening the dispersion of solids and heat. Intense foaming even moistens the air filters and can affect the sterility of the vessel due to microbial infection. Moreover, microbial cells are prone to attach to these bubbles, separating them from the fermentation broth. Also, those microbial cells are prone to autolysis.

The use of a defined medium can avoid foaming to a certain extent. However, it is mainly controlled by injecting antifoams (or defoamers) into the medium. They can proactively prevent foam formation or reactively eliminate the existing foam. A distinct probe present in the fermenter can detect foaming. This probe is a stainless steel rod that is insulated except at the tip. It is positioned towards the fermentation broth surface. When foaming occurs, it touches the tip of the probe, which generates a current in the circuit. This triggers the pump, which releases an antifoam into the fermenter from a reservoir and prevents the formation of foam or ruptures it. Polypropylene glycol and castor oil are the most commonly used antifoams in fermentation processes. The former compound is a synthetic polymer, while the latter is a natural plant metabolite. Both compounds are hydrophobic and have limited water solubility. They are safe for their use in food, pharmaceuticals, and cosmetics. Some fermenters are equipped with a mechanical foam breaker (like a horizontally rotating disc, propeller, or brush) attached to the top part of the agitator central shaft at the broth surface. When the probe detects foam formation, the foam breaker gets activated and promotes its coalescence. However, the energy consumption of the mechanical system is high, and its installation can increase the product cost.

7.6 MONITORING COMPONENTS

A well-defined environmental condition for microbial biomass and product formation requires careful monitoring of the various parameters. The frequently monitored components in a fermenter are the various physical and chemical parameters. Physical parameters include temperature, pressure, agitator shaft power, rate of agitator rotation, foaming, weight, and air/liquid flow rate. Chemical parameters encompass pH, O_2, exhaust gas, solute, and inoculum concentrations. These employ monitoring devices like sensors or sensing probes in the fermenters to correct any deviation from optimum conditions during fermentation. The sensing probes that detect these parameters could be of three different types based on their position and use. One group of sensors penetrates the interior of the fermenter vessel to monitor temperature, pH, DO, pressure, and foaming. The second type of probe acts on withdrawn samples, such as exhaust gas and fermented broth. The third type of sensor does not come in contact with fermentation broth, such as tachometers and load cells. All monitoring sensors have characteristics like accuracy, ease of calibration, stability, sterilisation resistance, productive design material, reliable output signals, and cost-efficiency.

A thermistor is a highly accurate temperature sensor used in a fermenter. It is a resistance thermometer composed of sintered semiconductor material, where resistance is proportional to slight variations in temperature. A pH meter in the fermenter involves a calomel electrode. It uses a silver metal/silver chloride (AgCl) electrode to measure potential, while a saturated KCl solution stabilises the electrode response. Dissolved oxygen probes measure O_2 diffusion across a membrane, producing an electrical signal upon entry into a sensor. Galvanic DO probes generate voltage when O_2 diffuses across their thin semi-permeable membrane. A diaphragm pressure gauge measures pressure using a flexible material, such as thin metal, leather, rubber, or cloth. The resilience characteristics and the diaphragm's repeated pressure curvature are used to measure pressure differences. A piezoelectric transducer generates a measurable voltage as a function of the force or pressure applied to it. A torsion dynamometer detects large powers transmitted along the agitator shaft of a turbine or motor vessel. A tachometer measures the rotation speed of an agitator shaft in units of revolutions per minute (RPM) or revolutions per second (RPS). The rotameter (also called a flowmeter or variable area flow meter) is an instrument to detect the flow rate of a liquid or a gas. A rotameter consists of a glass or plastic tube, a metal float, a scale, inline fittings, and optional outputs. It has a mechanical means of flow measurement based on the free upward and downward movement of the float within the flow tube. A load cell is used to monitor the weight of the fermenter's contents over time. Table 7.2 enlists some important probes used in the fermenter to detect various physical and chemical parameters.

The monitoring of microbial strain and product quality in the fermentation broth is also obligatory. The accurate monitoring of fermentative microorganisms provides information on the requirement of maintaining constant fermentation conditions, the progress of the fermentation, contamination, microorganism degradation, and the final stage of harvesting the broth. The growth of microbial biomass inside the fermenter is primarily monitored using samples taken at regular intervals. The monitoring typically involves estimation of dry weight, cell density/turbidity (using light scattering at 600 nm or near wavelength in a spectrophotometer), cell number (using a Coulter counter), cellular components like NAD^+ (using fluorometric measurement), etc. The metabolic activities of the viable cells and their concentrations can be traced by cell viability assays. One common method includes measuring the level of ATP in the broth, which is an indicator of cellular energy and viability. A different method employs some online recurrent neural networks that can predict biomass concentrations with input variables of nutrient feed rate, fermentation broth volume, and DO. Cell growth can also be monitored using different sophisticated instruments to detect specific components. For example, inductively coupled plasma mass spectrometry (ICP-MS) and atomic absorption

TABLE 7.2
A list of different probes for the monitoring of fermentation parameters

Fermentation parameters	Probes
Temperature	Thermometers: mercury-in-glass, bimetallic, thermocouples, metal resistance, and thermistors
pH	Combined glass reference electrodes of silver or silver chloride with potassium chloride electrolyte and calomel or mercury electrodes connected to a pH meter
Gas flow	Variable area meter of a rotameter
Liquid flow	Electrical flow transducer, Metering pump: motorised syringe, peristaltic pump, piston pump, and diaphragm pump
Pressure	Bourdon tube gauge, diaphragm gauge, and piezoelectric transducer containing quartz
Agitator shaft power consumption	Watt meter attached to a motor, torsion dynamometers, and strain gauges
Agitator rotation or stirring	Tachometer
Foaming	Stainless steel rod, which is insulated except at the tip
Weight of fermenter	Load cell
Dissolved oxygen	Galvanic electrodes and polarographic electrodes
Inlet and exit gas	Oxygen: Paramagnetic gas analysers and CO_2: Infrared analyser
Microbial biomass	UV-vis spectrophotometer, microscopic analysis, and culturing
Microbial metabolites	Mass spectrometer, dielectric spectroscopy, and enzyme electrodes
Elements, ions, and functional groups	Inductively coupled plasma mass spectrometry (ICP-MS), atomic absorption spectrometry (AAS), near-infrared spectroscopy (FT-IR), and microelectrodes

spectrometry (AAS) measure different biological elements (e.g., Cu, Mn, Ni, and Zn) in the sample. Microelectrodes recognise ions such as ammonium ions (NH_4^+), calcium ions (Ca^{2+}), magnesium ions (Mg^{2+}), phosphate ions (PO_4^{3-}), potassium (K^+), and sulphite (SO_3^{2-}). Enzyme electrodes monitor organic acids, organic solvents, macromolecules, and antibiotics. Near-infrared spectroscopy identifies medium fat, microbial teichoic acid, antibiotics, etc. A mass spectrometer detects partial gas pressure, dissolved gases, and volatiles. Dielectric spectroscopy can be used to measure bacterial concentration with their dielectric permittivity and conductivity.

Modern designs of the fermenter call for automated process monitoring, data collection, data storage, and error detection integrated with a computer system. The probes send signals to the controlling components, which instruct the mechanical components of the fermenter for the process optimisation. A useful electronic device in the fermenter is the proportional-integral-derivative (PID) controller. A PID controller regulates process parameters within specified limits, like adjusting the stirrer rate in the fermenter to maintain DO levels. PID controllers are widely used in the biopharmaceutical industry to ensure effective process control in both upstream and downstream bioprocessing. A relatively new scheme of monitoring the fermentation process is an online system using sensors, which continuously assesses the fermentation parameters in real-time. Online monitoring offers more accurate results than manual measurements because of more real-time data collection, reducing lengthy procedures and inaccuracies. Intelligent monitoring control systems, combining statistical software and artificial intelligence (AI), are revolutionising vaccine production through online monitoring and control of variables in fermenters.

7.7 ACCESSORY COMPONENTS

Along with the essential mechanical and monitoring components, fermenters have many accessory components that contribute towards productive fermentation. These include various piping systems, steam sterilisation facilities, sterile filters for exhaust gases, numerous pumps or valves for controlling the flow of liquids and gases, a feed inlet port, a sampling port, an outlet/drainage port, and a steam trap. The peristaltic pump is an accessory component that gently pumps nutrients and feeds inside the fermenter. It maintains the flow of liquid inside the fermenter for a stabilised fermentation process. A feed port is a silicone tube that supplies the fermenter with nutrients and acid/alkali. The steam trap is an automated valve that lets steam flow at a controlled pace while draining the condensate. An internal float in many fermenters monitors the liquid level inside the fermenter. Pressure relief valves regulate fermenter pressure to prevent equipment failure by gradually opening in response to pressure increases. A safety valve acts as a fail-safe device that releases pressure immediately when the pressure reaches a predetermined threshold during system failure. A sampling port with seals is connected to the fermenter to collect the sterile samples of fermentation broth. The final products are extracted from the outflow port of the fermenter. This is an aseptic outlet that is often found at the bottom of the fermenter to collect the products and fermented medium. A fermenter condenser, made of glass or stainless steel, removes moisture from exhaust gases and reduces liquid evaporation. The condenser is connected to the head plate using an adjustable height probe adapter. A condenser tends to cool the air leaving the fermenter, causing it to condense on its surface and return as a liquid to the vessel. Some important accessory components and their functions are mentioned in Table 7.3.

TABLE 7.3
A list of various accessory components of fermenters and their functions

Accessory fermenter components	Functions
Piping system	To provide steam, inoculum, nutrients, supplements, pH buffers, samples, and sterile air
Feed ports of silicone tubes and peristaltic pump	To add nutrients and pH buffers
Sampling ports	To obtain fermentation broth samples
Gate valves, globe valves, needle valves, and diaphragm valves	For regulating the flow of steam or liquid
Piston valves, plug valves, and ball valves	For regulating the flow of steam
Butterfly valves	To regulate the flow of liquid
Ball valves and pinch valves	For regulating the flow of mycelial broths
Check valves: swing check, lift check, combined stop and check with variants	To prevent accidental backflow of liquid or gas in a broken pipe
Pressure relief valves incorporated into pipelines	To reduce high pressure to lower values
Pressure retaining valves	To maintain pressure in the pipeline
Safety valves incorporated into air or steam line	To ensure that pressure would never increase the prescribed safe upper limit
Vacuum release valves	To ensure that fermenters do not collapse when a vacuum occurs during cold rinsing of hot fermenter or carbon dioxide absorption while cleaning with caustic solutions
Condenser	To cool and condense the exhaust air
Steam traps	To remove accumulating steam condensate in pipes

7.8 CONTAINMENT COMPONENTS

All fermenters are essentially equipped with containment components as a safeguard. These components offer protection to the inside and outside environments from contamination. This is very important because if any foreign microorganisms find a chance to contaminate liquid feed inside the fermenter, it can cause major loss to the entire bioprocess. On the other hand, the release of any dangerous microorganisms from fermenters into the atmosphere raises concerns for public health and the environment. Containment components include biosafety manuals, seals, medical surveillance, exhaust treatment, a HEPA filter, washing facilities, disinfecting facilities, protective clothing, etc. Also, the containment component of a fermenter design includes a comprehensive risk assessment of the microorganisms involved. The central role of containment involves preventing the escape of viable microbial cells from fermenters during and after fermentation. Some fermenters install sight glasses for visual inspection of the fermentation process. Modern fermenters are also often integrated with Steam-in-Place (SIP) and Clean-in-Place (CIP) systems for sterilisation and cleaning without disassembling the instrument. In SIP, a superheated steam (between 120–135 °C) is introduced to the interior of the vessel and pipes for a certain duration to eliminate any microorganisms. However, CIP involves heated water or cleaning agents with some turbulence to dislodge the waste residues.

SUMMARY

- A fermenter is an engineered device to regulate the fermentation process, providing an optimal environment for microbial cell growth and biosynthesis. A fermenter calls for many specific characteristics, such as a large vessel, piping for inlets and outlets, monitoring and controlling parameters, and various containment protocols.
- Industrial-scale fermentation requires multiple production scales, starting from small lab to intermediate pilot and large-scale production, known as fermenter scale-up. The scale-up process ensures streamlined operations, implementing process optimisation, cost efficiency, and commercial viability.
- The aspect ratio refers to the ratio of the fermenter's height to diameter, which varies depending on fermentation reactions. Modern fermenters have high aspect ratios, typically cylindrical, with a 3:1 or 3:2 ratio. Elongated fermenters have ratios over 3:1, while non-mechanically stirred tower fermenters can reach 16:1.
- A fermenter has several mechanical components, such as a sparger, agitator, baffle, seal, and foam controller. A sparger supplies sterile air inside the fermenter, an agitator is responsible for mixing the content, a baffle prevents vortex effects during agitation, sealing helps maintain aseptic conditions, and a foam controller manages foam control.
- Microbial growth and product formation in a fermenter require careful monitoring of physicochemical parameters using sensors or probes. These devices send signals to controlling components and correct deviations from optimum conditions.
- A fermenter also consists of accessory components such as piping systems, steam sterilisation facilities, sterile filters, valves, feed ports, steam traps, and pressure relief facilities for efficient fermentation.

- Fermenters essentially contain some containment components to prevent contamination of the fermentation process by foreign microorganisms and also protect the environment from the exposure of fermentation microorganisms to the outer environment.

DIFFERENT TYPES OF PRACTICE QUESTIONS

Fill in the blanks

I. .... is an aseptic contained vessel for microbial industrial fermentation.
II. .... device introduces sterile air into the fermenter.
III. The load cell is used to monitor .... of the fermenter.
IV. A bulk mixing of fermenting broth and aerating gases is obtained with......................
V. Vortexing of fermentation broth during high agitation rate is prevented with ....
VI. The gas flow into a fermenter is monitored with .... probe.
VII. A .... surrounds the outer wall of the fermenter vessel for temperature regulation.

SOLUTION: I. Fermenter, II. Sparger, III. weight, IV. agitator, V. baffles, VI. rotameter, VII. water or cooling jackets

Match the following

i. Fermenter	a. Allows bulk mixing of fermentation broth
ii. Seal	b. Introduces sterile air into the fermenter
iii. Sparger	c. Disrupts vortexing of fermentation broth
iv. Agitator	d. Involves biosynthesis of microbial products
v. Baffle	e. Prevents the leakage of feed and microorganisms

SOLUTION: i-d, ii-e, iii-b, iv-a, v-c

Match the following

i. Rotameter	a. Agitator shaft power monitoring
ii. Diaphragm gauge	b. Monitoring the rate of agitator rotation
iii. Torsion dynamometers	c. Monitoring of gas flow in a fermenter
iv. Tachometer	d. Microbial biomass detection
v. Dielectric spectroscopy	e. Monitoring of pressure in a fermenter

SOLUTION: i-c, ii-e, iii-a, iv-b, v-d

Very short answer type of questions

I. What kind of material is mostly used for the construction of large-scale fermenters?
II. Which monitoring component is used to measure the power consumption of fermenters?
III. What is the use of a load cell in the fermenter?
IV. Name any two antifoam agents used in fermenters.
V. Write the names of three different ports in a fermenter.

Short answer type of questions

I. What is fermenter scale-up and its significance in industrial production?
II. What is the use of aeration in the fermenter? How is aeration done inside the fermenter vessel?
III. What are the roles of an agitator in a fermenter?
IV. Vortexing of the fermenting broth during agitation can be a disadvantage in a fermenter. Give two reasons. How can that be prevented?
V. Write a short note on how fermenters are tightly sealed to maintain an aseptic environment.
VI. Why does foaming affect fermentation productivity, and how can that be controlled?

Long answer type of questions

I. What is a fermenter? How is it different from a bioreactor? What are the important characteristics of a fermenter?
II. Give a schematic representation of a typical mechanically agitated fermenter. Describe any three essential mechanical components of a fermenter.
III. What are the different probes for fermentation? How are they regulated?

NOTE

1 www.istockphoto.com/photo/stainless-steel-brewing-equipment-iron-reservoirs-or-tanks-and-pipes-in-modern-beer-gm1160867255-317894776

SUGGESTED READINGS

Ager B, Moses V. The regulation of biotechnology in Europe. In Springham DG, Moses V, Cape RE (eds.) *Biotechnology* 2020 (pp. 123–132). CRC Press.

Alves SS, Vasconcelos JM. Optimisation of agitation and aeration in fermenters. *Bioprocess Engineering*. 1996;14(3):119–23.

Benz GT. Bioreactor design for chemical engineers. *Chemical Engineering Progress*. 2011;107(2126):13.

Chen LZ, Nguang SK, Li XM, Chen XD. Soft sensors for on-line biomass measurements. *Bioprocess and Biosystems Engineering*. 2004;26(3):191–5.

Flores- Cosío G, Herrera- López EJ, Arellano- Plaza M, Gschaedler- Mathis A, Kirchmayr M, Amaya- Delgado L. Application of dielectric spectroscopy to unravel the physiological state of microorganisms: current state, prospects and limits. *Applied Microbiology and Biotechnology*. 2020;104:6101–13.

Ji Y, Song A, Li C, Cao Y. Role of sparger configuration in determining flotation performance under oscillatory air supply. *Processes*. 2021;9(4):638.

Kamla Y, Bouzit M, Ameur H, Arab MI, Hadjeb A. Effect of the inclination of baffles on the power consumption and fluid flow in a vessel stirred by a Rushton turbine. *Chinese Journal of Mechanical Engineering*. 2017;30(4):1008–16.

Karanth NG, Manchanda AC. Useful sparger arrangement for fermentations involving heavy mycelial suspensions. *Biotechnology and Bioengineering*. 1980;22(9):1985–6.

Karcz J, Major M. An effect of a baffle length on the power consumption in an agitated vessel. *Chemical Engineering and Processing: Process Intensification*. 1998;37(3):249–56.

Kumaresan T, Joshi JB. Effect of impeller design on the flow pattern and mixing in stirred tanks. *Chemical Engineering Journal*. 2006;115(3):173–93.

Montague G. *Monitoring and control of fermenters*. IChemE; 1997.

Sewalt V, Shanahan D, Gregg L, La Marta J, Carrillo R. The generally recognized as safe (GRAS) process for industrial microbial enzymes. *Industrial Biotechnology*. 2016;12(5):295–302.

Singhal G, Verma V, Bhagyawant SS, Srivastava N. Fermentation technology prospecting on bioreactors/fermenters: design and types. In Kuila A, Sharma V (eds.) *Principles and applications of fermentation technology* 2018 (pp. 65). Wiley.

Stanbury PF, Whitaker A, Hall SJ. *Principles of fermentation technology*. Elsevier; 2013 Oct 22.

Tomita F. Regulation and safety of recombinant microorganisms. In Murooka Y, Imanaka T (eds.) *Recombinant microbes for industrial and agricultural applications* 2020 (pp. 15–26). CRC press.

Vaidyanathan S, Macaloney G, Vaughan J, McNeil B, Harvey LM. Monitoring of submerged bioprocesses. *Critical Reviews in Biotechnology*. 1999;19(4):277–316.

Virdis B, Hoelzle R, Marchetti A, Boto ST, Rosenbaum MA, Blasco-Gómez R, Puig S, Freguia S, Villano M. Electro-fermentation: sustainable bioproductions steered by electricity. *Biotechnology Advances*. 2022;59:107950.

8 Different Types of Fermenters

8.1 FERMENTER CONFIGURATIONS

The specific types of fermentation processes require specific configurations in fermenters to optimise the fermentation process and obtain desired outcomes. Based on the states of substrate and biocatalysts (liquid and solid) used in fermentation, fermenters can be designed for three different categories: solid substrate-liquid biocatalyst, liquid substrate-solid biocatalyst, and both substrate and biocatalyst in liquid phase. The first category is best known as solid-state fermentation (SSF). In the second category, the substrate is present in the solution, and microbial cells or enzymes are immobilised in a bed. The third category is the example of submerged fermentation (SmF), where both substrate and biocatalyst are in solution. In the SmF, it can be further divided into mechanically and non-mechanically agitated types. The fermentation process in anaerobic conditions calls for fermenters to have specific configurations (Figure 8.1). Furthermore, some special types of fermenters use a membrane system and a photobioreactor. These different types of fermenters are discussed in detail as follows.

8.2 FERMENTERS FOR SOLID-STATE FERMENTATION (SSF)

The fermenters used for SSF are relatively much simpler than those utilised for SmF. Industrial-scale fermenters for SSF can be non-agitated and agitated. The non-agitated type is the tray fermenter. The agitated types include PlaFractor™ fermenter, rotating horizontal drum fermenter, and stirred drum fermenter. They can also be installed at pilot and large scales. The following sections provide in-depth information about their geometric configurations and applications.

8.2.1 Tray Fermenter

A tray fermenter is a simple apparatus that operates with minimal physical parameters and without forced aeration (Figure 8.2a). It consists of a chamber with many stacked wooden, metal, or plastic trays that are perforated or non-perforated. The perforation on the tiers improves the air circulation and prevents overheating. This setup is suitable for small or medium-scale operations. The substrate on the tray is spread as a thin layer to receive maximum air and optimal temperature. In distinct situations, solid substrates are moistened with the nutrient medium and can be packed with support materials, such as seed shells and silica. The temperature and humidity of the chambers are monitored and controlled for optimal growth and production.

Tray fermenter is used for the production of fermented foods (e.g., cheese, tempeh, tofu, miso, koji, and soy sauce), enzymes (e.g., cellulase, laccase, pectinase, polygalacturonase, and protease), organic compounds (e.g., citric acid, ethanol, fumaric acid, lactic acid, and lovastatin), antibiotics (e.g., iturin A, meroparamycin, paromomycin, surfactin, and tetracycline), phenolic compounds (e.g., terpenoids, flavonoids, and alkaloids), and aromatic compounds (e.g., pyrazines). SSF is also used to grow fungal spores, and those grown on tray fermenters have longer shelf lives and high quality.

However, fermentation in a tray fermenter is a highly labour-intensive process due to the preparation, loading, monitoring, emptying, and cleaning of individual trays. Further, a build-up and inefficient dissipation of metabolic heat within the trays during fermentation exceeds the temperatures more than the optimum. The unregulated temperature increase can impede microbial growth optima and decrease product yield. Additionally, the solid substrates have a low thermal conductivity that further affects and reduces the moisture concentration. The microbial growth as biofilms on the substrate could result in an unavoidable low oxygen (O_2) concentration. The sterilisation of a large amount of substrates away from the fermenter unit makes containment and maintenance of sterile procedures difficult. Also, the production of many commercial products, such as antibiotics and cytotoxic pharmaceuticals, could cause allergic reactions while handling the products.

8.2.2 PlaFractor™ Fermenter

It is a next-generation automated tray fermenter developed at Biocon, India (1998) on a pilot scale, which was later developed for a large-scale industrial application. The production of the immunosuppressant drug Mycophenolate Mofetil was the first commercial call to use this technology. The PlaFractor™ fermenter overcomes several limitations of the conventional system and has delivered consistent, predictable, and repeatable outcomes. This is similar to the tray fermenter but uses an automated system logged by the computer. The PlaFractor™ fermenter contains tray modules. The trays are

DOI: 10.1201/9781032651095-10

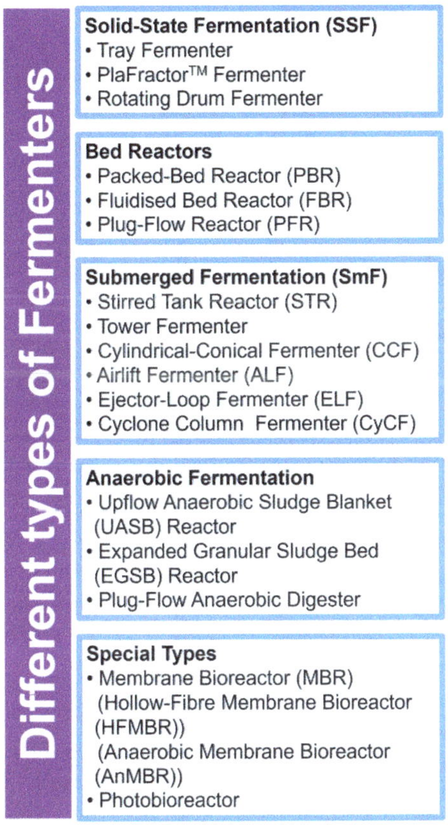

FIGURE 8.1 The different types of fermenters used in various industries.

stacked and interconnected, along with communicating and non-communicating channels at their baseplates (Figure 8.2b). Communicating channels supply steam, moisture, aeration, and chemicals, while noncommunicating channels carry heating and cooling fluids to control the temperature.

This fermenter has configurations for both upstream and downstream solid matrix fermentation processes. The operation immediately starts with the transfer of a solid matrix into the PlaFractor™. An *in situ* sterilisation setup, consisting of a series of valves, allows steam to sterilise the contents. This is followed by the cooling of the chamber for the inoculation. A different pre-sterilised valve injects the inoculum into the modules. The PlaFractor™ fermenter is also equipped with a mixing arm for agitation. They are engineered to move slowly and smoothly, creating a gentle stirring motion. The mixing arms of each module revolve around the central axis to properly mix the solid matrix and inoculum. Temperature control and airflow are maintained using noncommunicating and communicating channels, respectively. The piping arrangement supplies the nutrients or feed required during fermentation. When the fermentation process is over, an integrated system supports product recovery. An appropriate valve injects the extraction solvent into the module. The mixing arrangement transfers the product into the solvent, which can be recovered from a different value. Further, residual solvent from the matrix can be recovered through evaporation.

8.2.3 Rotating Drum Fermenter

The rotating drum fermenter is a mechanically agitated type of fermenter. It consists of a horizontal or near-horizontal cylindrical drum that rotates (with an electric motor) around the central axis for slow agitation (Figure 8.3). A different variant, a stirred drum fermenter, is equipped with a moving paddle mixer or scrapers for substrate mixing. Rotating drums are typically longer, with an aspect ratio that can reach as wide as 1:10 at an industrial scale. About 10–40% of the fermenter volume is filled with a substrate bed, which is mixed continuously or intermittently at a particular rotational speed. The air is supplied from the top of the fermenter. The efficiency of mixing can be enhanced by incorporating baffles inside the fermenter. A controlled rotation offers a uniform mixing of the contents and heat movement in the vessel.

The rotating drum fermenter is low power-consuming compared to other types of fermenters and has excellent gas-liquid mass movement rates. It is used for producing microbial enzymes (such as ligninolytic, fucoidanase, amylase, pectinase, phytase, and L-asparaginase), antibiotics (like penicillin), and organic acids (like citric acid). In a different application, this equipment can be used for the manufacturing of an anisotropic bacterial nanocellulose hydrogel membrane for tissue engineering and filtration technologies.

Drum fermenters have a few disadvantages. When the rotational speed is slow, the mixing of the substrate bed and microbial mass is poor, leading to a slumping flow. On the other hand, the high rotational speed causes deleterious shearing effects, impacting the microbial cells. Manual tasks of loading and unloading the solid substrate into the fermenter demand high manpower. The gas-flow pattern in the headspace causes significant axial temperature gradients in the substrate bed. The scale-up of the fermenter is limited by the vessel size due to its complex design.

8.3 BED REACTORS

In this category, enzymes or cells are bound with an immobilising matrix to provide a rigid structure for the biocatalyst in the fermenter. The feed solution (containing substrates and nutrients) flows through the bed, allowing the biocatalyst to interact with substrates and carry out the desired fermentation. Immobilisation of a biocatalyst in bed maintains its high density and stability for extended periods, improving the efficiency of the fermentation reaction. The immobilisation process reduces the loss of biocatalysts during the operation. This type of setup works best for continuous fermentation. Some well-known fermenter configurations that employ bed formation are discussed as follows.

8.3.1 Packed-Bed Reactor (PBR)

A packed-bed reactor (PBR) is a non-agitated fixed-bed reactor/fermenter that is a cylindrical tube made of glass or metal. The centre of the column consists of porous solid support material as a bed packed onto a perforated base

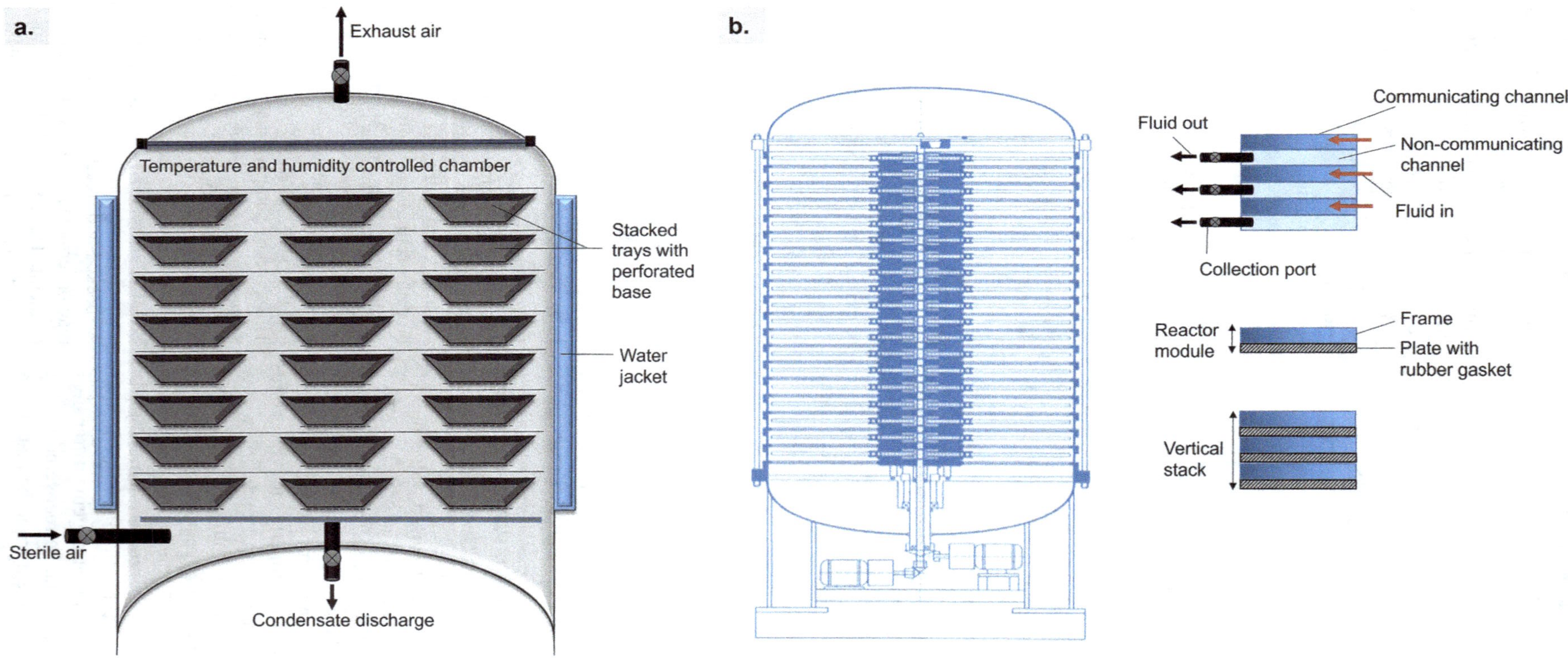

FIGURE 8.2 Fermenters for Solid-State Fermentation (SSF). a. Schematic representation of the tray fermenter. b. Schematic representation of the PlaFractor™ fermenter (Image b adapted from Arora et al. (2018)[1] with permission of © 2018 Elsevier B.V.)

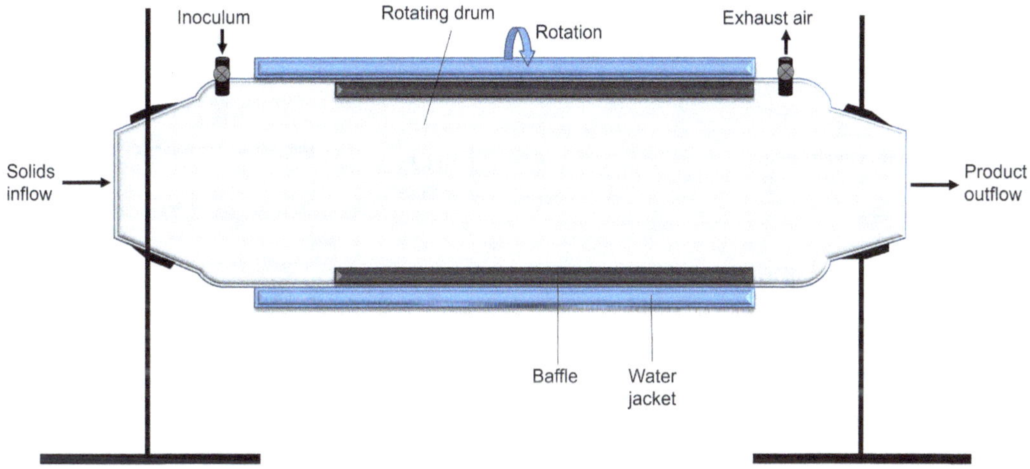

FIGURE 8.3 A schematic representation of a rotating drum fermenter for solid-state fermentation (SSF).

(Figure 8.4a). The microorganism is inoculated into the packed bed, where it grows and covers the bed in the form of a mat or biofilm. In a different setup, the solid support can be preloaded with immobilised enzymes before packing the bed inside the column. When the biocatalyst is attached to a solid support, a liquid feed is fed into the column through an inlet. As the feed passes through the bed, it is transduced into the product. Based on the direction of the flow of feed and the air, PBR can be classified into three different types-

a. Trickle bed co-current down flow: In this widely used reactor, the liquid feed and air are supplied from the top of the reactor onto the packed microbial bed (Figure 8.4a). The products and exhaust gases pass from the microbial bed and settle at the reactor base, where they are collected from their respective outlets.
b. Trickle bed countercurrent flow: In this reactor, the liquid substrate is supplied from the top of the reactor, while the air is injected from the reactor base. This allows the substrate and air to flow in opposite directions to get mixed within the packed bed. The product is collected at the base, and the exhaust gases are released from the top of the reactor.
c. Packed bubble flow co-current up flow: In this system, both liquid substrate and air are supplied from the reactor base in the same direction. The product and exhaust gases pass from the bed and are collected at the top of the reactor from their respective outlets. This mode of operation ensures prolonged contact of the substrate with microorganisms and complete wetting of the catalyst bed.

PBR can be configured for both laboratory and large-scale fermentation processes. At the laboratory scale, the column is connected to air bubblers, with a rotameter adjusting the airflow. This column is placed in a water bath for a temperature control system. The laboratory scale PBR is used to study the effect of forced aeration on growth, metabolism, biomolecule production, O_2 consumption, and carbon dioxide (CO_2) release of the microorganisms. At a large-scale production, the reactor can be jacketed to pass cooling water for heat movement control.

This type of fermenter can be used for vinegar production using ethanol as feed and *Acetobacter* spp. growing on beechwood chips as a solid biocatalyst. This kind of setup is also used for the production of glucoamylase and protease using *Aspergillus awamori* from waste bread pieces, cellulases using *Aspergillus uvarum* from olive oil and wine processing waste, and volatile aroma compounds using *Kluyveromyces marxianus* from cassava bagasse. This process has immense potential in the dairy industry to produce nutraceuticals using immobilised ß-galactosidase. This fermenter has also gained interest in the treatment of gas liquor and sewage.

The main advantages of this setup are an easy mode of operation, high mass transfer and reaction rates, and quality control. However, the high rate of airflow can reduce bed moisture and cause shear stress on the microorganisms, resulting in low-quality products. Also, a unidirectional airflow could end up overheating the column, rendering microbial biomass inactive. The pores inside the solid support tend to get clogged after a long run, resulting in bed compaction and liquid logging. This shrinkage causes a pressure drop along the bed, increasing temperature and decreasing the O_2 uptake rate. Some spots in the bed can tear apart and create big tunnels, causing the liquid substrate to travel faster. In this situation, the feed may not have spent enough time in contact with the biocatalyst to form a product.

8.3.2 Fluidised Bed Reactor (FBR)

The fluidised bed reactor (FBR)/fluidised bed fermenter is a non-mechanically agitated, pneumatic (that uses compressed air) type of fermenter. It was developed by the German industrial chemist Fritz Winkler (1920). The fermenter vessel consists of solid granular material of the immobilised enzymes or cells as a bed, which is supported by a porous distributor plate from the lower side (Figure 8.4b). A fluid (gas or liquid) substrate is forcefully passed from the distributor by an upward flow

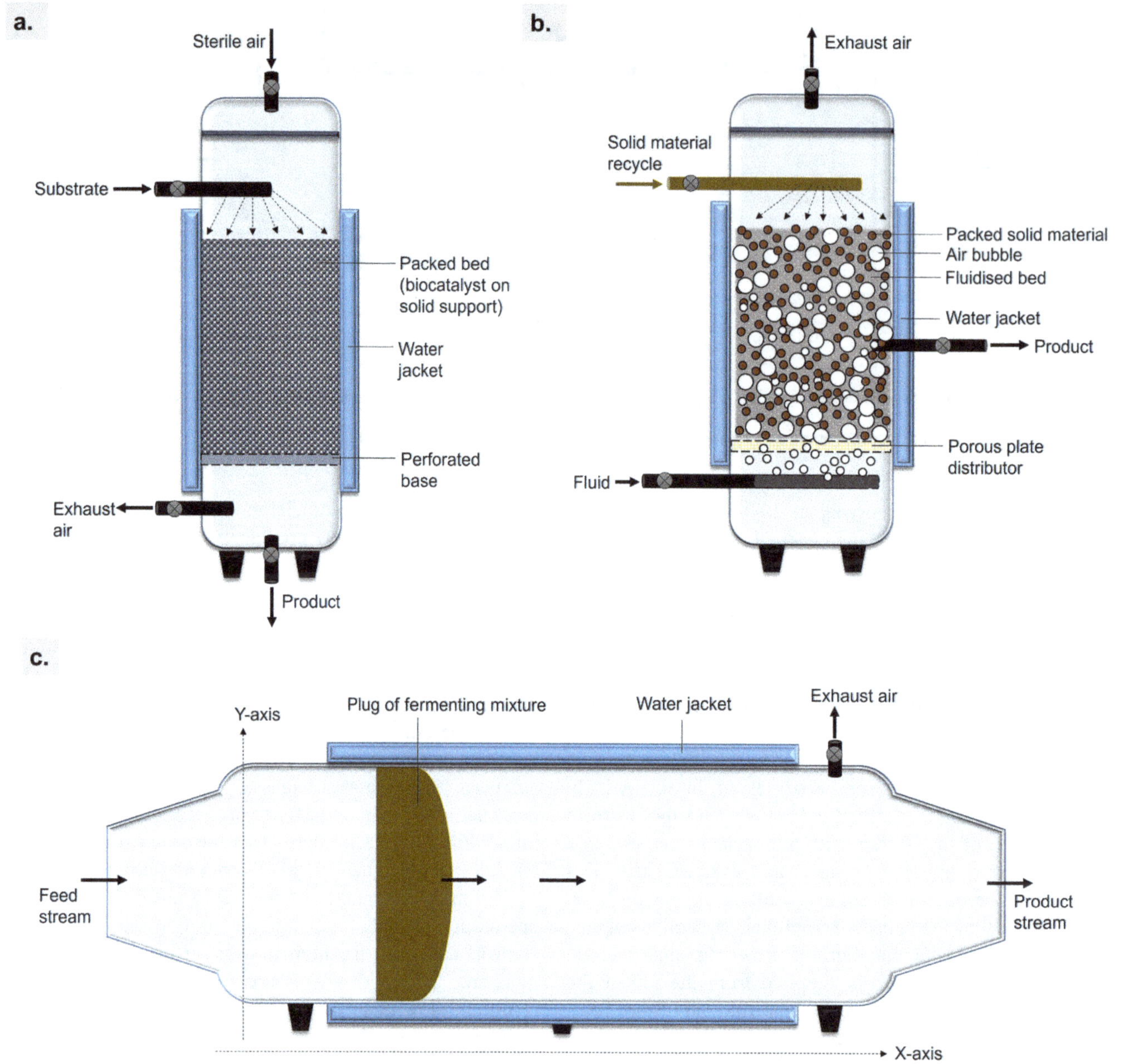

FIGURE 8.4 The schematic representation of bed reactors. a. Trickle bed co-current downflow type of packed-bed reactor (PBR). b. Fluidised bed reactor (FBR). c. Plug-flow reactor (PFR).

toward the solid biocatalyst material. Subsequently, the bed gets suspended and behaves like a fluid. This process is called fluidisation, which ensures an even distribution of the fluid in the packed material. The pressurised high mass transfer rate encourages a constant movement of all the particles in both upward and downward directions. The voids between the particles in the bed provide sufficient spaces for fluid movement. This also improves O_2 and nutrient distribution and heat transfer during fermentation. Fluidisation, therefore, makes FBR different from a PBR (Table 8.1). Depending on the fluid velocity, three conditions are possible-

- At low fluid velocities, the fluid passes slowly from the voids of the unmoving solid bed. In this situation, the fermenter functions similarly to a PBR.
- At intermediate fluid velocities, the force of the passing fluid balances the weights of the solid bed material. It observes a minimum fluidisation velocity, which represents the incipient fluidisation of the bed.
- At high fluid velocities, the force of the passing fluid surpasses a minimum fluidisation velocity. Solid bed material expands and swirls around, operating much like a mechanically agitated tank. Fluid properties (such as density and viscosity) and solid phase properties (including density, size, shape, cohesion/adhesiveness, and coefficient of restitution) contribute significantly to fluidisation behaviour.

This type of bioreactor is used in wastewater treatment using microbial cells or enzymes. The FBR has several

TABLE 8.1
The differences between a packed-bed reactor (PBR) and a fluidised bed reactor (FBR)

Parameters	PBR	FBR
Velocity of the liquid medium passed from the bed of solid catalyst	Low	High enough
Fluidisation of a packed bed	Not possible	Possible
Mixing and dispersion achieved within the bed	Improper	Good
System	Heterogeneous	Homogeneous
Catalyst availability for the reaction	Little	High, as they are well mixed in the fluid and provide a larger surface area
Monitoring and control of fermentation parameters	Often difficult	Easy
Occurrence of gradients in nutrients, products, and temperature.	Frequent	Negligible
Mass and heat movement among bulk particles	Less	High
Particle sampling and removal of active fractions during fermentation	Difficult	Easy
Scale-up	Difficult	Achievable
Bed compaction and channelling	Possible	Not possible
Costly and power-consuming	Less	More
Reactor is ideally suited for	Batch productivity	Fast pyrolysis reactions and continuous productivity

advantages. An intrinsic fluid-type behaviour of the solid bed material facilitates uniform particle mixing. There is a high fluid-solid interface area and contact, which eliminates radial and axial concentration gradients and yields an efficient reaction. Fluidisation prevents the local hot or cold spots in the reaction bed and develops a uniform or isothermal temperature distribution within the fermenter. There is a high bed-to-surface heat movement coefficient of the fermenter. It offers easy control of pH and temperature. This process has certain disadvantages. For example, solid bed fluidisation demands excessive pumping power, which increases production costs. Expansion of the solid bed at fluidisation requires a large-sized fermenter vessel. Fluidisation of the solid bed can result in wearing and erosion of the fermenter interior, demanding frequent maintenance. Fine solid particles might become entrained in the fermenter due to high fluid velocity, making separation difficult and expensive. The process is prone to significant catalyst attrition. A fluctuation or sudden loss of fluid pressure and fluidisation can reduce the solid bed surface area, which could have serious implications for heat movement, particularly for exothermic reactions. Product back-mixing and fluidisation patterns are difficult to predict. Fermenter scale-up is often difficult because the hydrodynamics of FBR are quite complex. Also, the knowledge of the flow behaviour of material and heat in the fluidised bed system is limited.

8.3.3 Plug-Flow Reactor (PFR)

The plug-flow reactor (PFR) is a narrow, rectangular, non-mechanically agitated, one-dimensional flow reactor/fermenter. It is a type of insulated, heated anaerobic digester, which is often made up of reinforced concrete, steel, or fibreglass. Both horizontal and vertical configurations are possible. The typical length-to-width ratio of the reactor lies in the range of 3.5:1 to 5:1. Depending on the scale, the size can vary from a few centimetres to metres. The reactor with increased length provides a long hydraulic retention time (HRT), giving reactants sufficient time to form a stable product. This kind of setup does not allow the mixing of the reaction mixtures in the axial direction (forward or backwards), but radial mixing is evident. The solid or semisolid fermentation raw material enters at a high concentration from the feed inlet located at one end of the reactor (Figure 8.4c). After fermentation, the product is discharged from the other end of the reactor. The material is gradually pushed toward the outlet when fresh material is added to the reactor. The reaction mixture in the PFR flows like plugs from inlet to outlet without backflow of the medium along the long axis (the X-axis). The flow rate of the nutrients plays a significant role in the fermentation reaction. As the medium flows through the reactor, the long axis of the vessel experiences a spatial variation that maintains concentration gradients of nutrients, O_2, products, and byproducts. Water circulating through a surrounding jacket maintains the temperature of the fermenter.

The advantages of this type of reactor are a simple structure, good stability, and low energy consumption. The equipment is free from the requirement of any mechanical mixing. The substrate concentration does not drop immediately to the level of the exit point but rather decreases gradually towards the axial direction. This type of bioprocessing can make use of immobilised cells or enzymes. The PFR can suitably use semi-solid raw materials, including sludge, vegetable waste, cow dung, etc. This mode of operation plays a role in steady-state bioenergy production, combustion of different chemicals, and sulphuric acid decomposition. There are also certain disadvantages to this fermenter design. One of the major concerns is the precipitation of the reactants, which imbalances the volume of the medium in the reactor, resulting in thick crusts of precipitate on the wall. Maintaining a consistent temperature, pH, and O_2 supply in a large reactor is also difficult. Viscous reactants cause a high-pressure drop inside the vessel. The requirement of a continuous flow of inoculum in the vessel is challenging.

8.4 FERMENTERS FOR SUBMERGED FERMENTATION (SMF)

The fermenters used in SmF are much more common than those used in SSF. Also, they come in various configurations based on the specific requirements. The fermenters for submerged aerobic fermentation employ air sparging into the fermentation broth, which can be agitated either mechanically or non-mechanically. Mechanically agitated fermenters have agitators for agitation, whereas non-mechanically agitated fermenters are pneumatic, which introduces compressed or pressurised air bubbles in the fluid. The pneumatic fermenters can be bubble column or loop types. In bubble column fermenters, the fermentation broth is sparged with compressed air co-currently or counter-currently to the liquid feed flow. In the loop fermenters, the fermentation broth is circulated between the low-density gassed and high-density un-gassed regions. Besides these types, various other combinations and modifications in fermenter designs for SmF also exist. Their various configurations are described in the following sections.

8.4.1 Stirred Tank Reactor (STR)

The stirred tank reactor (STR) is among the most conventional and widely used fermenters or bioreactors. The typical aspect ratio of the STR vessel for microbial fermentations is in the range of 3:1 to 5:1. A cylindrical vessel of STR contains one or more agitators placed on a motor-driven central shaft for mechanical agitation. These agitators provide efficient homogenisation of fermenting broth with proper mixing of air, heat, liquid, and mass of cells and nutrients. Around the periphery of the STR vessel, baffles are equidistantly placed to prevent the swirling and vortexing of the broth. For animal cell culture applications, the vessels are often unbaffled with about a 2:1 aspect ratio. A sparger is located below the impeller to supply sterile air in the vessel. The fermenter design of STR is suitable for all three modes of operation, including batch, semi-continuous, and continuous fermentation. The key components of an STR are elaborated in Chapter 7.

The most basic kind of reactor is a batch STR. The main components of this system are a vessel and a mechanical agitator. In this setup, the sterile nutrients and microorganisms or enzymes are mixed, which allows for converting substrate into products. However, the products rest in the reactor until the end of the fermentation process. The temperature optima for the reaction are controlled using a cooling jacket on the vessel. The batch reactor type is ideal for high-viscosity substrate solutions and is typically used for small amounts of production.

A continuous stirred tank reactor (CSTR) is a mechanically agitated vessel that consists of the same parts as a batch STR but with a slightly more complex design. The CSTR is more efficient than the batch STR due to its continuous harvesting system with a substrate/nutrient inlet and product outlet. The substrate/nutrient is continuously pumped into the reactor while the same amount of fermented product stream is simultaneously removed to maintain the volume constant. A number of ancillaries, such as pumping, material handling equipment, and vessel ventilation systems, provide support for the continuous supply of feedstocks and removal of products. The CSTR operates at a steady state, maintaining a uniform composition of the feed. The reactor uses constant stirring to evenly distribute reactants and maintain a uniform reaction rate, allowing precise control over temperature, pressure, and the flow rates of the solution. A perfusion culture bioreactor is another modified STR used for continuous fermentation processes. In this type, the fermented broth is regularly removed from the reactor while retaining the microbial cells in the fermenter (Box 8.1).

BOX 8.1 PERFUSION CULTURE BIOREACTOR

A perfusion culture bioreactor is a modified, gently stirred medium-sized tank used for continuous fermentation processes. In this culture technique, the fermented broth is regularly removed from the reactor while retaining the microbial cells in a cell retention device. The cell retention device is a vital part of the perfusion culture technique, ensuring efficient, high retention rates without damaging cells or affecting protein productivity and selectively retaining viable cells. Tangential flow filtration (TFF) and alternating tangential flow filtration (ATF) are the most commonly used processes for filtration from cell retention devices that use ceramic or stainless steel filters. The biological cells retained in the cell retention device can be fed back to the fermenter. The vessel is replenished with fresh media for continuous fermentation. During filtration, the filter is spun to avoid cell blockage; therefore, this type of culture is also known as spin culture. Cell densities in such a system are 10–30 times higher than those obtained in a conventional fermenter. The recovered broth from filtration is continually pumped to a product reservoir. The fermenter maintains high cell concentrations while maintaining constant working volume and lowering the risk of cell washout due to high dilution rates. Furthermore, this can achieve high product yields as the retrieved broth is free of cells.

This culture technique combines the benefits of fed-batch and continuous culture systems with simplified logistics, ensuring a steady and stable culture environment, high cell density and productivity without substrate inhibition. Also, they are advantageous because of their smaller scale compared to batch cultures. The perfusion culture bioreactor is notably a good choice for mammalian cell processes and plant secondary metabolite production. Due to low residence time in the bioreactor, quick harvesting and purification, and avoidance of toxic metabolite accumulation, the perfusion culture bioreactor can be used for unstable protein production.

Conventional STR has several advantages, such as easy scale-up, well broth mixing, amenability to customisation with numerous new impeller designs, and adherence to good manufacturing practice (GMP) regulations. The STR is extensively used in microbial fermentation, including the brewing industry, culturing of biological cells, and the production of enzymes or antibodies. They are used in many industries (chemical, pharmaceutical, food, and metallurgical processes) and for the treatment of municipal and industrial wastewater. Also, the STR has certain disadvantages, such as high-power consumption, unsuitability for shear-sensitive cells, reduced stability of agitator shafts (for tall vessels), and safety concerns about the sealing of vessels.

8.4.2 Tower Fermenter

A tower fermenter is an elongated type of fermenter with an aspect ratio of at least 6:1 or a total aspect ratio of about 10:1. This fermenter has a simple design. It allows for good rates of O_2 supply. The substrate can be fed either from the top or the bottom of the fermenter. In a downflow tower fermenter, the substrate is fed at the top of the fermenter and flows towards the base. In an up-flow floc tower fermenter, the substrate is fed at the bottom of the fermenter (Figure 8.5). This fermenter is a type of bubble column fermenter. It is a non-mechanically agitated fermenter because there is no agitator. The sterile air is pumped with high force at the base of the fermenter. A unidirectional flow maintains the gradient of nutrients and gases for efficient fermentation. In tower fermenters, the sparged air facilitates both aeration and agitation. In a distinct setup, the tower fermenter can also be operated as a multistage system, where the column body vessel is divided into many individual compartments with several horizontal plates having small perforations. In this system, feed can also be injected into different compartments. A production scale vertical up-flow tower fermenter for brewing can also be equipped with an air sparger, yeast separator (settling zone or clarifying tube), and horizontal perforated baffle (or distribution) plates. Two or more attemperator panels (water-jacketed tubes) are attached to the tower fermenter for temperature control.

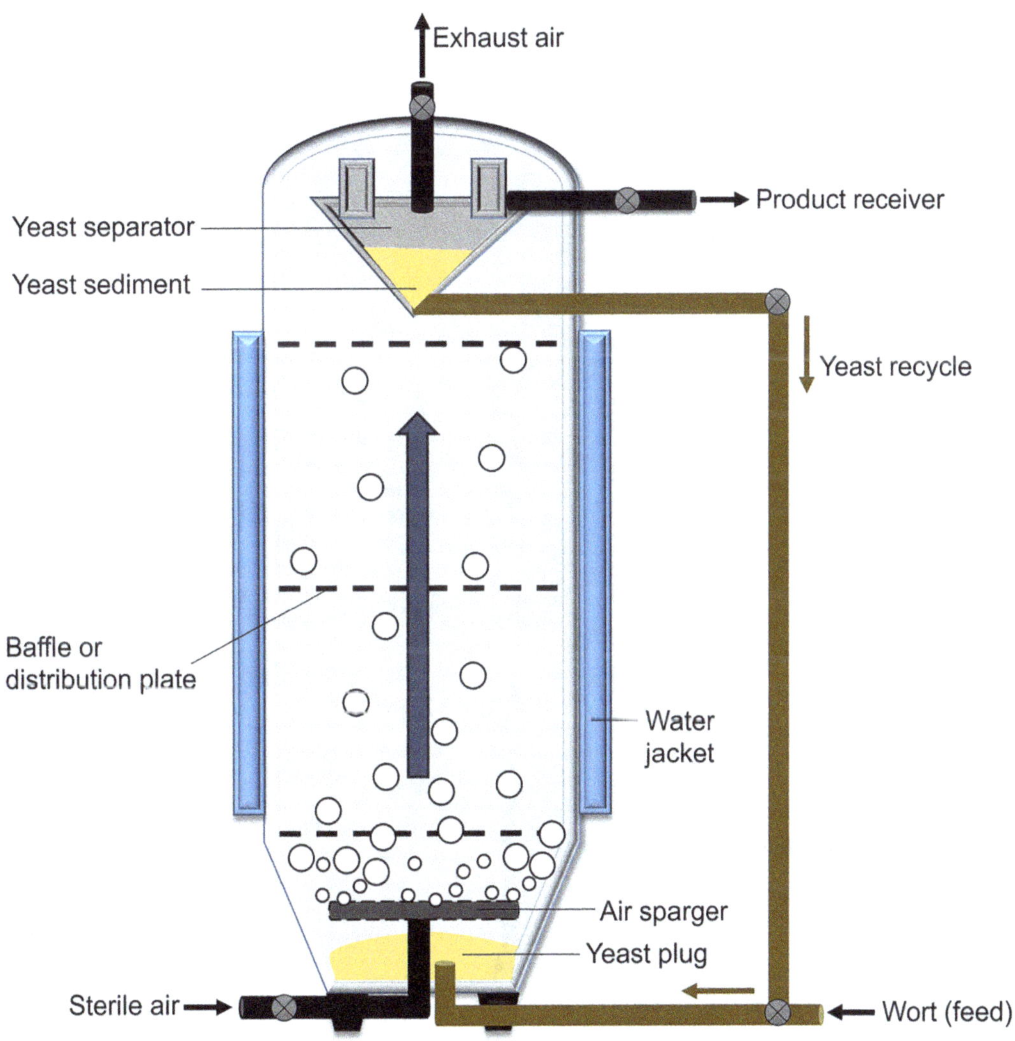

FIGURE 8.5 A schematic representation of an up-flow floc tower fermenter.

The tower fermenter has applications for microbial production of citric acid, tetracycline, vinegar, acetic acid, bacterial biomass, and fungal biomass. This fermenter has the potential for continuous culturing of *E. coli* and yeast, as well as the treatment of activated sludge. The tower fermenter has been used to produce fermented beer in continuous mode on a commercial scale, similar to that produced using a batch procedure. This fermenter also provides a low-cost technology for single-cell protein (SCP) production. Tower fermenters have high O_2 movement rates (about 280 mM O_2/l/hr) that can yield high protein (about 39%).

Tower fermenters offer advantages, particularly in continuous fermentation processes. They often require a relatively short duration for fermentation and cleaning. Certain disadvantages of this fermenter are long start-up duration, technical complexity, and the need for highly skilled personnel. Furthermore, an extended delay between fermentation processes can produce variability in the product quality.

8.4.3 Cylindrical-Conical Fermenter (CCF)

The cylindrical-conical fermenter (CCF) was developed by a Swiss inventor, Leopold Nathan (1927). This design is commonly used for beer production (including fermentation, maturation, and storage) in modern breweries. The CCF consists of an upper cylindrical tank with a conical bottom (with an internal cone angle of 60–70 °) and is commonly fabricated from stainless steel (Figure 8.6). The typical aspect ratio of CCF is around 4:1. Fermenter is equipped with cooling/heating coils or attemperator jackets to maintain the temperature regime required during inoculation, fermentation, and pasteurisation. Natural mixing primarily occurs through CO_2 emission and thermal currents during fermentation, without incorporating mechanical agitators.

There are several advantages of CCF. The convection currents produced by the rising gas bubbles (of CO_2) offer sufficient mixing. Recovery of yeast after fermentation is easy, as it settles down at the cone of the fermenter. Tanks can have several cooling zones or a complete jacket lining to dissipate the heat and control the temperature. The same vessel can also be used for primary and secondary ageing of the beer. Cleaning of the vessel is also relatively easy.

8.4.4 Airlift Fermenter (ALF)

The airlift fermenter (ALF) is a non-mechanical and pneumatically agitated tower-type fermenter developed by Scholler and Seidel (1940). A widely used ALF design consists of an internal hollow draft tube (or draught or discharge tube) immersed in the fermentation broth (Figure 8.7a). The central draft tube divides the fluid volume into two interconnected zones: the riser and the downcomer. This inside region of this

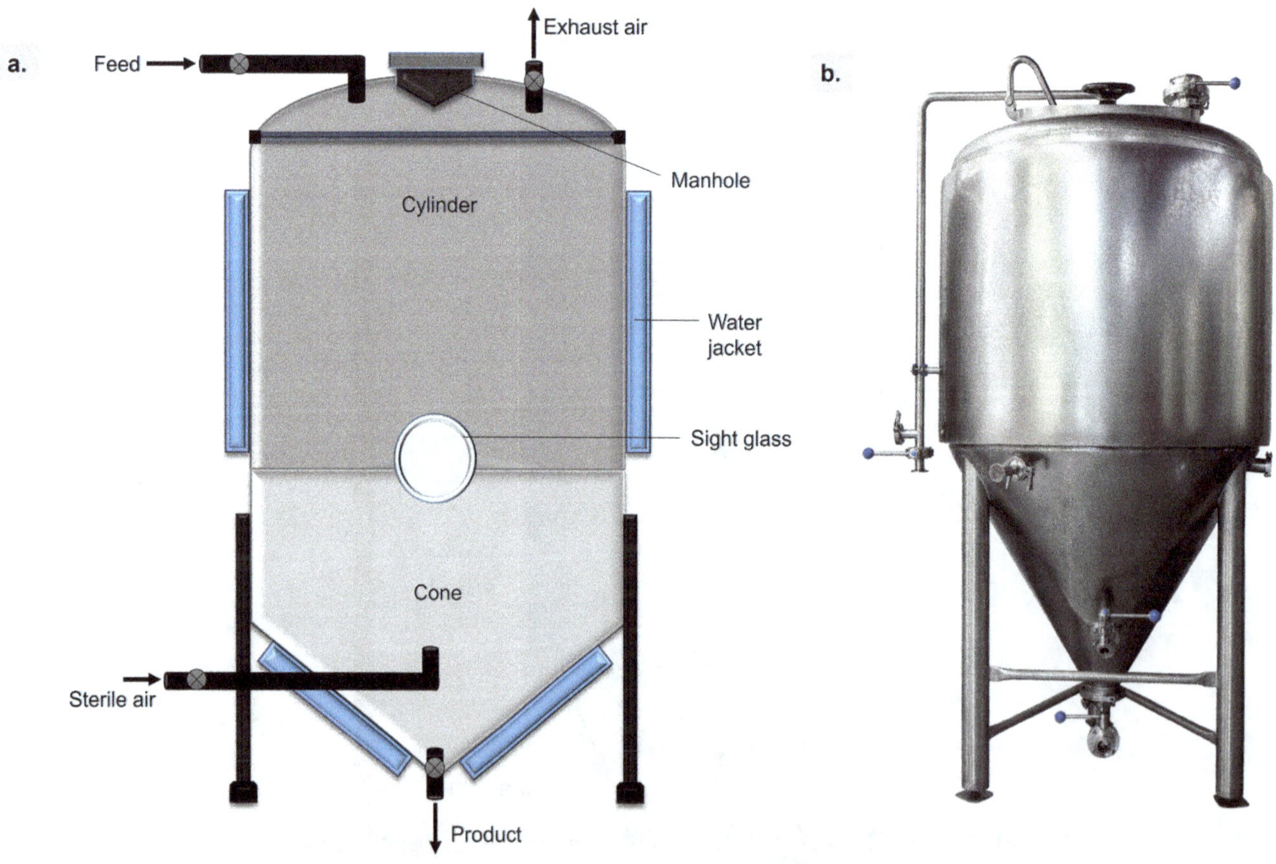

FIGURE 8.6 Cylindrical-conical fermenter. a. A schematic representation. b. A 500 l isobaric beer fermenter. (Source of image b: iStock/panchof.[2])

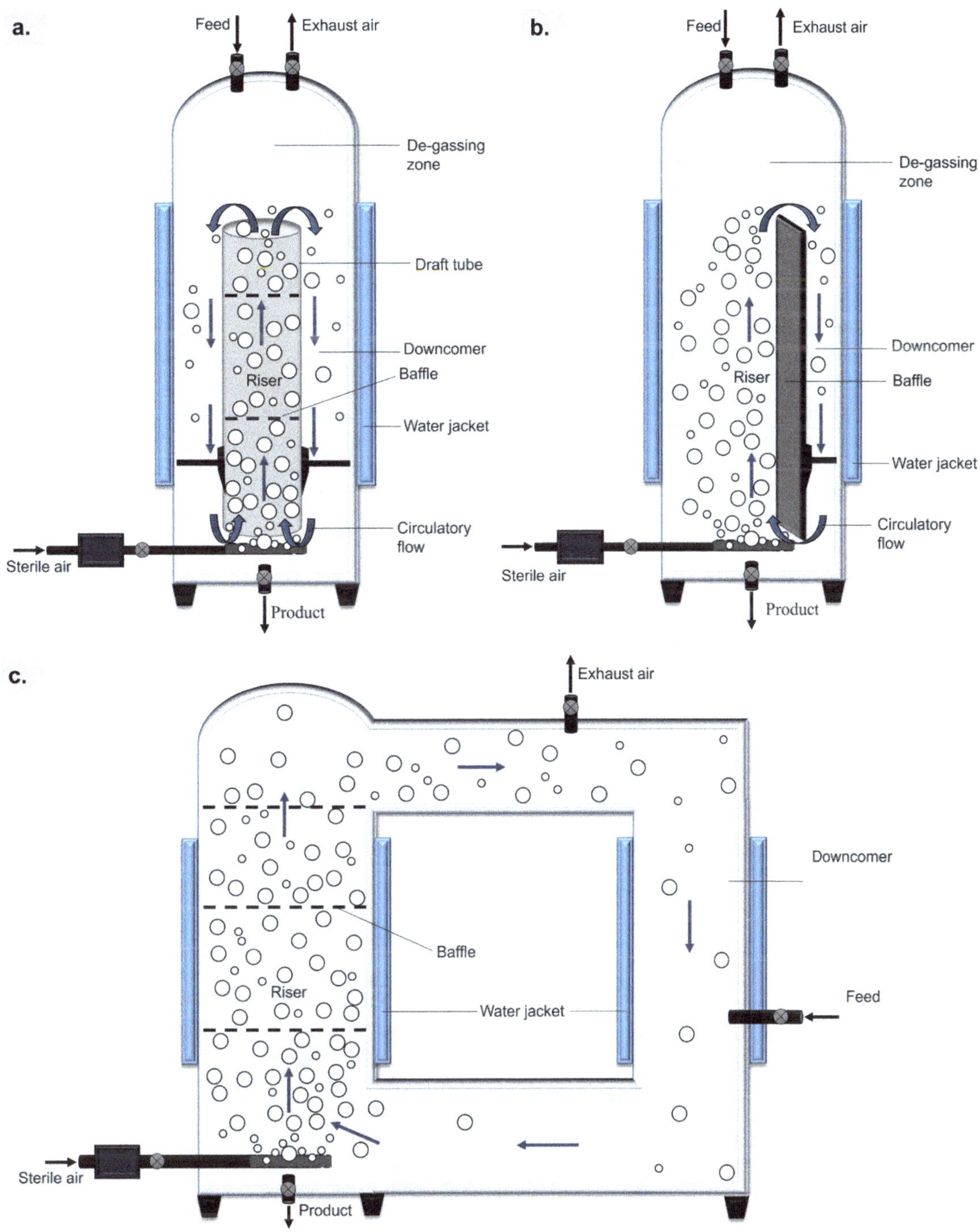

FIGURE 8.7 The schematic representation of the airlift fermenter (ALF). a. Internal loop ALF with a central draft tube. b. Internal loop ALF with a baffle. c. External loop ALF.

draft tube is the riser zone, which is sparged with air. The other surrounding region of ALF is the downcomer zone, where the fermentation broth descends after emerging from the riser zone at the top.

During operation, compressed sterile air is pumped in with a large force at the bottom of the draft tube. The gas and fermentation broth mixture gets displaced upwards inside the draft tube and exits the tube from the top. This gas-fermentation broth mixture (gassed liquid) in the riser zone has less density than the surrounding liquid. As the mixture moves upwards inside the draft tube, gas escapes at the top of the fermenter, and the broth comes in contact with the downcomer (degassed liquid) zone, which has a higher density than the riser zone. The difference in the density of the liquid column in the riser (liquid ascending) and downcomer (liquid descending) zones provides a driving force for the circulation of the medium. An internal loop or circulatory pattern of broth movement in the vessel ensures adequate aeration and mixing of the fermenting broth along with temperature regulation.

A modified internal loop ALF has a vertical baffle plate in place of a draft tube (Figure 8.7b). In this configuration, one side of the vessel gets air sparged (the riser zone), and the other side becomes the downcomer zone. Another design is an external loop ALF, where the liquid circulates between two conduits that are connected from the top and bottom. These conduits are the riser and downcomer zones, respectively (Figure 8.7c).

In the ALF, turbulence is induced by the fluid flow, which ensures adequate mixing, particularly for denser microbial broths. Gas flow-derived mixing reduces the shear stress or attrition effects and increases the liquid-solid and gas-liquid mass movement. The external loop ALF improves fluidisation characteristics for immobilised cells and avoids cell washout during bed expansion. The O_2 mass movement coefficient of an ALF is higher than that of a mechanically agitated STR. The energy cost of an ALF is typically one-sixth of that of a STR. The internal draft tube or the external loop develops a circulatory pathway of air bubbles.It prevents any bubble coalescence, which increases mass and heat movement rates and provides a more favourable environment for cell growth.

The ALF has successful applications in cell culturing (e.g., bacteria, yeasts, microalgae, plant cells, and animals) and biocatalysis (with free or immobilised enzymes). This fermenter can be used for SCP production, microbial enzyme production, bacterial cellulose production, exothermic fermentations (methanol production), synthesis of fungal mycoprotein and itaconic acid, synthesis of plant secondary metabolites, production of monoclonal antibodies from hybridoma cells, and in wastewater treatment. The major disadvantages of ALF are the cost and energy associated with the supply of a large volume of compressed sterile air.

8.4.5 Ejector-Loop Fermenter (ELF)

The ejector-loop fermenter (ELF) is a novel, non-mechanically agitated design of ALF that overcomes the requirement of an air compressor. ELF consists of an external-loop tower fermenter (made of plexiglass), a gas-liquid separator, a plate-heat remover, and a gas-liquid ejector (Figure 8.8). The last two items are mainly equipped to maintain a constant fermentation temperature and provide aeration in the feed, respectively. When the process starts, the feed is forced into the ejector, which is mixed with sterile air. The gas-liquid mixture from the ejector is supplied at the bottom of the fermenter. This mixture moves along the fermenter column and reaches the gas-liquid separator, where the liquid is degassed, and air bubbles escape at the top. Degassed liquid gathers at the bottom of a gas-liquid separator and flows from the plate-heat remover. Degassing of the recycled liquid stream is an essential step to enhance

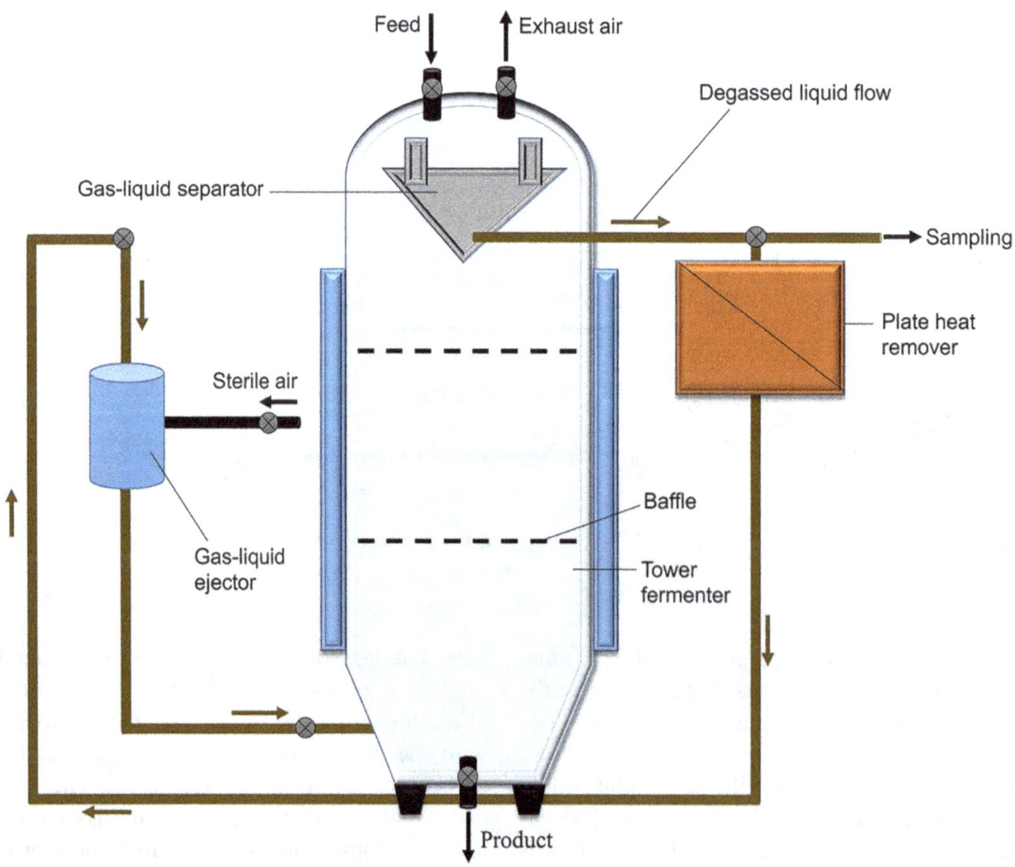

FIGURE 8.8 A schematic representation of an ejector-loop fermenter (ELF).

the pumping efficiency. The centrifugal pump forwards the recycled feed towards the ejector in a recirculating loop.

The ELF is used in microbial fermentations as well as industrial chemical processes for gas/liquid reactions. It has beneficial effects on the production of biomass, SCP, and recombinant proteins. The use of an ejector is more cost-effective than the use of an air compressor. ELF can achieve higher mass movement rates than STR at a low power input. An external plate-heat removal is more suitable for heat movement than those in STR and ALF. The fermentation and maintenance in ELF incur minimal costs. Fermenters generate less noise compared to other types of fermenters and are also amenable to continuous fermentation mode.

8.4.6 Cyclone Column Fermenter (CyCF)

The cyclone column fermenter (CyCF) was developed by Dawson (1974). The fermenter vessel is non-mechanically agitated and consists of a vertical glass column, which has a tangential feed port at the cyclone head. This specific arrangement of the feed port injects feed with high kinetic energy. The culture moves into the vessel in a whirling pattern. Sometimes, the helical baffles arranged on both walls also guide the flow pattern (Figure 8.9). Nutrients and air or any required gas mixture are sparged at the bottom of the column, causing them to move upwards in a countercurrent direction to the flow of the culture. At the base of the fermenter, a pump continuously recycles the culture from the bottom to the top of the column, thereby forming a closed loop. The exhaust gases are released from the top of the vessel.

The CyCF has potential application in mixed microbial fermentation for the production of natural gas from sewage sludge. The CyCF provides many advantages for the fermentation process. Circulation of the culture around a closed loop improves aeration and mixing of cells with feed. Both batch and continuous operations are possible. This type of fermenter generally does not produce high foam during fermentation. Also, the cyclone effect prevents microbial growth on the fermenter walls. However, the process is sensitive to variations in feed composition or flow rates. The design of the recirculation pump and flow rate significantly affect biomass production and cell morphology. For example, *Aspergillus niger* can have highly branched hyphae when there is no recirculation. On the other hand, they can form shorter cells with a distinct main hypha and several short branches at the increased speed of the centrifugal pump.

8.5 FERMENTERS FOR ANAEROBIC FERMENTATION

Anaerobic fermenters have non-complex mechanical equipment and are widely used for the microbial treatment of industrial and municipal wastewater. In the fermenter, microorganisms multiply and aggregate to form immobilised

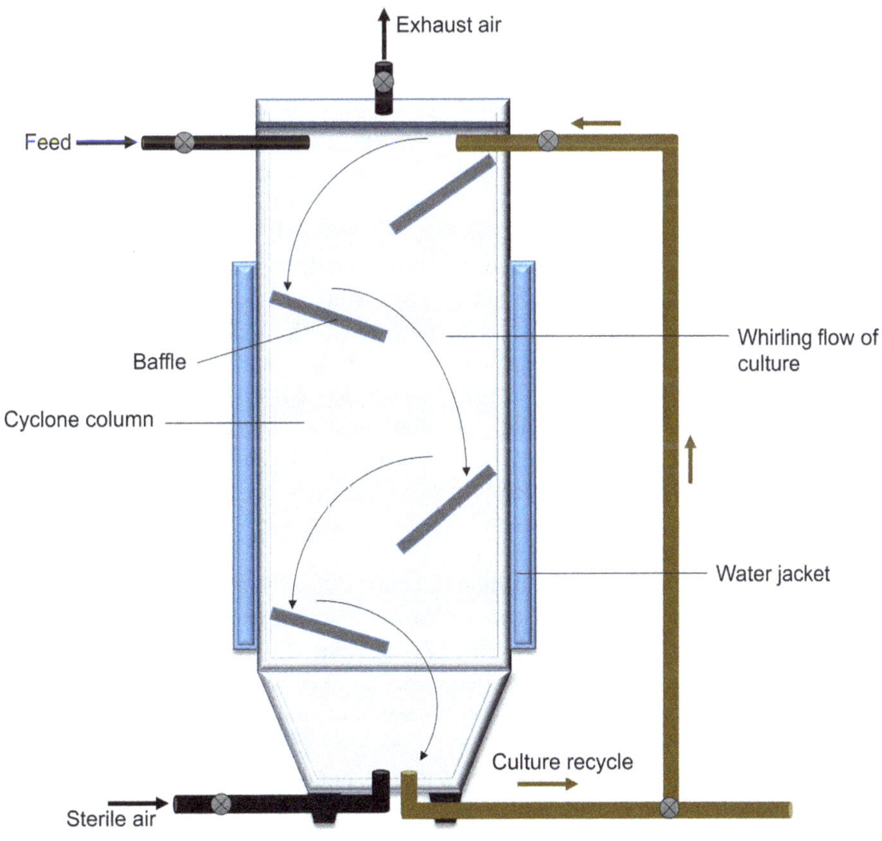

FIGURE 8.9 A schematic representation of a cyclone column fermenter (CyCF).

granules and settle at the bottom of the reactor in the form of sludge. These reactors have simple construction, anaerobic environment maintenance, high loading rate intake, and the ability to withstand organic and toxic substances. The fermenters may be non-agitated or mechanically agitated. Some commonly used fermenters for anaerobic fermentation are discussed as follows.

8.5.1 Upflow Anaerobic Sludge Blanket (UASB) Reactor

The up-flow anaerobic sludge blanket (UASB) reactor is the most widely used anaerobic system for domestic and industrial wastewater treatment. It was developed by Lettinga and associates in 1980. The reactor consists of three internal zones: the bottom settling zone, the middle fermentation or digestion zone, and the top separation zone (Figure 8.10a). The bottom settling zone of the reactor has a bed of granular active sludge consisting of an anaerobic microbial consortium. Influent wastewater is fed from the base of the reactor and flown towards the bed at a sufficiently high up-flow velocity. Upflowing wastewater causes the bed to expand into a sludge blanket at the top. The sludge blanket develops a fermentation zone where anaerobic bacteria hydrolyse or solubilise the complex organics, followed by the synthesis of organic acid. Methanogens use acid products and produce methane (CH_4). This gas floats with effluent (treated water) to the top of the reactor. The top separation zone contains a characteristic separation system that segregates three-phase materials of solid-sludge particles, treated water, and CH_4. Horizontal baffles alter the direction of the uprising granules and promote their settling into the sludge bed. Biogas bubbles are disengaged from the rising sludge particles and collected at the disengagement zone, thereby causing the sludge to settle towards the bed at the bottom. The good settling properties of anaerobic sludge essentially provide an internal recycling of the sludge. Even flow distribution of influent wastewater with a sufficiently high up-flow velocity, and the agitation due to microbial gas production, ensures adequate mixing between sludge and wastewater. The process relies on the slow growth rate of the anaerobic microorganisms for effective utilisation of organic matter, and the use of granular sludge in UASB appears beneficial to achieve high biomass concentration in the system.

UASB technology can be used to treat domestic wastewater and industrial effluents of food and beverages, breweries, pulp and paper, etc. In the wastewater treatment plant, the UASB reactor can have the capacity to treat more than 1.75 million m^3 of wastewater per day. In this system, about 55–70% of the total organic mass in the wastewater produces biogas, leading to the accumulation of low amounts of sludge (about 5%). The UASB technology is acclaimed for energy recovery and low sludge management costs. However, UASB systems are less sensitive to nutrient removal (such as nitrogen and phosphorus). An anaerobic ammonium oxidation (anammox) in a UASB reactor can show some nitrogen removal ability at some temperature variations. Also, the low up-flow velocity of the reactor prevents effluent recirculation, causing toxic substance accumulation and process failure.

8.5.2 Expanded Granular Sludge Bed (EGSB) Reactor

The EGSB reactor is a type of UASB reactor with modified hydrodynamics. It is used for anaerobic treatment of low-strength soluble and complex wastewater. After the traditional UASB, it is the second most popular high-rate anaerobic digester. The EGSB also relies on the use of granular sludge, which maintains stability and seamlessly agglomerates biofilms. However, the design features a faster upward-flow velocity for wastewater, allowing partial expansion of the granular sludge bed and segregation of small suspended particles. The EGSB is superior to UASB due to its ability to treat low- to high-strength wastewaters with high organic and hydraulic loadings. It is capable of treating acidified wastewaters under psychrophilic conditions (4–10 °C) and also wastewaters containing lipids and toxic compounds like formaldehyde and lauric acid. Further, the EGSB reactor involves recirculation of effluent to dilute reactor contents, enabling the treatment of wastewater containing tannin, phenol, and phthalate, which UASB cannot handle.

The EGSB reactor is typically a cylindrical tower with a reactor height of about 15–20 meters (m). It is distinguished by a high aspect ratio, typically up to 5:1. The reactor contains an inlet water distribution system (water distributor), reaction zone, three-phase separation zone (three-phase separator), gas collecting chamber, and outlet channel system (Figure 8.10b). It is also equipped with a unique effluent return mechanism. When the wastewater enters the reactor through a sewage pump, the sludge settled at the bottom of the tank traps the organics, facilitating their treatment and absorption. High hydraulic and gas loads combine sludge and organic matter, resulting in full expansion of the sludge, high mass transfer rate, and enhanced anaerobic reaction rate. The biogas produced during fermentation rises to the top and is separated from sludge, sewage, and other materials by a three-phase separator. The EGSB reactor has a high superficial velocity (4–10 m/h) for effluent recirculation, hydraulic mixing and substrate-sludge contact. However, the high velocity reduces the contact time for large aggregates and coarse suspended particles in the sludge for their full treatment.

8.5.3 Plug-Flow Anaerobic Digester

The plug-flow anaerobic digester works on a similar principle to PFR, which ensures plug flow with minimal back-mixing. The major difference is that this operation takes place in anaerobic conditions and is specifically used for wastewater treatment and biogas production. This type of setup was developed in South Africa (1957). The equipment contains a horizontal thin tank with an aspect ratio of about 1:5. The inlet and outlet are located at opposite ends. The treatment involves a continuous operation at a temperature range between 35–55 °C with around 15–20 days of HRT. The water

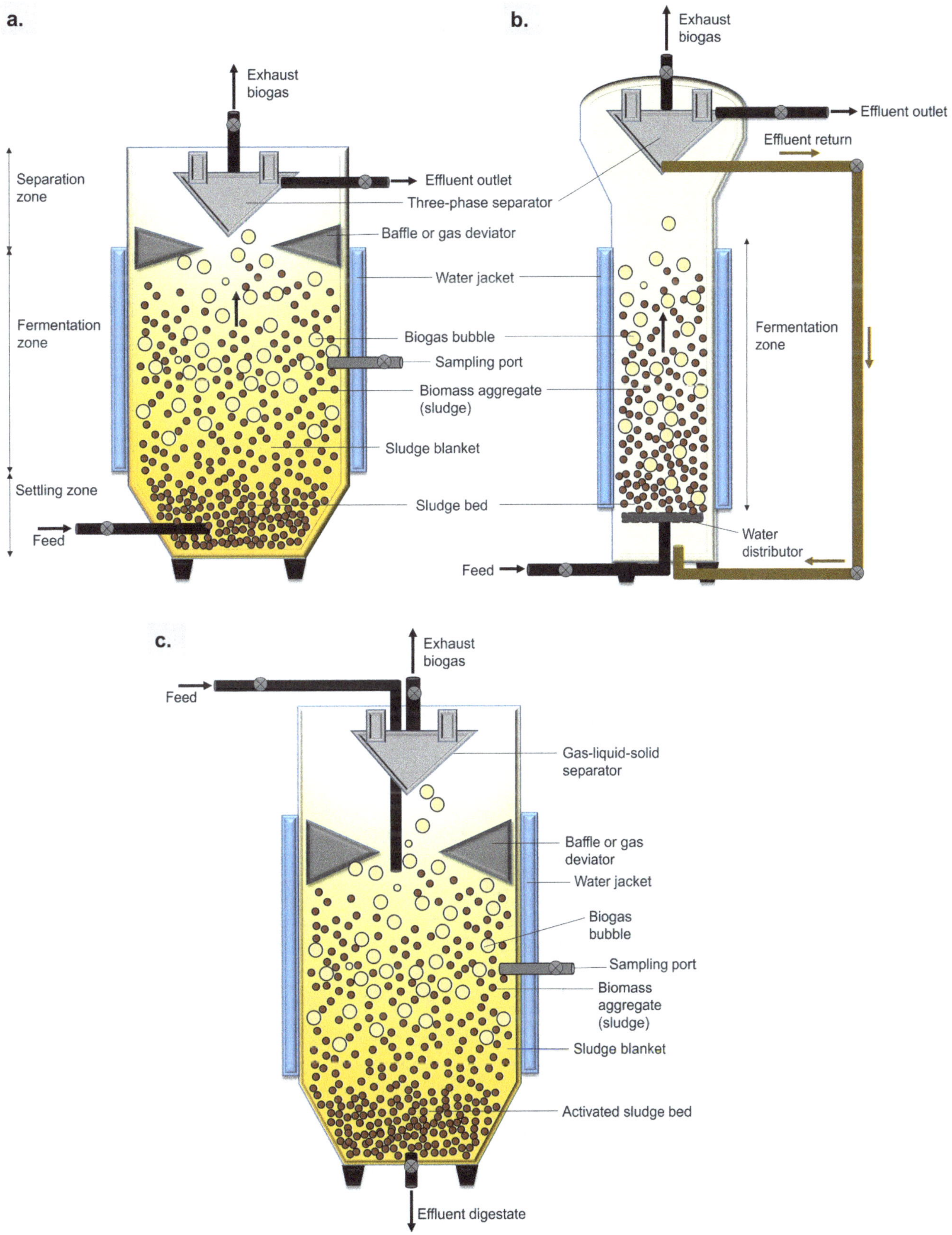

FIGURE 8.10 A schematic representation of fermenters for anaerobic reactions. a. An up-flow anaerobic sludge blanket (UASB) reactor. b. An expanded granular sludge bed (EGSB) reactor. c. A down plug-flow anaerobic reactor (DPAR).

bath jacket with circulating hot water on the surface of the digester maintains mesophilic temperatures for the growth of mesophilic microorganisms. Treatment does not require any stirring device. A typical concentration of total solids (TS) in the reaction mixture can range from 10–15% to maintain fluidity in the tank.

A modern design of plug-flow anaerobic digester is the down plug-flow anaerobic reactor (DPAR), designed by Chen et al. (2015) for the treatment of swine manure as a continuous dry biogas fermentation process. The tank is elongated, the manure-feeding inlet lies on the top, and an outlet is located at the bottom (Figure 8.10c). The TS concentration in an inlet feedstock can be 20–35%. During the operation, sludge is activated at a temperature of around 25 °C for 5 days, followed by the transfer of fresh swine manure at a fixed loading rate. The treatment can run for 190 days, during which fresh swine manure is continuously fed into the reactor. The same amount of fermented digestate is simultaneously discharged. The plug-flow reactor possesses several advantages for treatment applications. The reactor is energy-saving, as it successfully performs the treatment at ambient temperature without requiring much heating or stirring, which assures low operating costs.

8.6 SPECIAL TYPES OF FERMENTERS

The special types of fermenters include membrane bioreactors and photobioreactors. The membrane bioreactors rely on membrane systems attached to the biocatalysts. These reactors come in different configurations that are used for a variety of purposes, like wastewater treatment and microbial cultivation. A photobioreactor is utilised for the growth of phototrophs like microalgae, cyanobacteria, and certain plants.

8.6.1 Membrane Bioreactor (MBR)

A membrane bioreactor (MBR) is a tank that confines the physical space for biocatalysts, combining a semipermeable membrane system with the biological reaction. Membrane bioreactors usually combine microfiltration and ultrafiltration with activated sludge for municipal and industrial wastewater treatment applications. MBR can utilise various membrane materials, such as organic-based polymeric membranes and ceramic membranes. The type of membrane used in wastewater treatment can enhance the retention of different types of pollutants. Membrane bioreactors come in two configurations: submerged and side stream. Submerged bioreactors features submerge membranes in vessel, while side stream bioreactors add membranes outside the reactor. In these systems, the bacteria that break down the organic molecules are supplied with O_2 by an aeration system. In addition, the aeration also keeps particles in suspension, scours and cleans the membrane surface, and enhances cell production and biodegradability. Submerged MBR configurations also consist of filtration elements above the aeration system in the bioreactor vessel or separate tank, while side-stream MBR technology positions them outside the aerobic tank. Approximately 99% of the organic load in wastewater gets eliminated using by this procedure. Advanced treatment of wastewater could be possible due to membrane-based technologies. However, they come with a considerable energy expense. Another configuration of membrane bioreactor is the Hollow-Fibre Membrane Bioreactor (HFMBR). The HFMBR is suitable for a range of industrial processes, including *in situ* cultivation of microbial consortia from natural environments (Box 8.2).

When a low-cost treatment that requires energy conservation, an MBR with anaerobic processes is utilised.

BOX 8.2 HOLLOW-FIBRE MEMBRANE BIOREACTOR (HFMBR)

Hollow-Fibre Membrane Bioreactor (HFMBR) is a chamber consisting of a bundle of hollow fibres (Box Figure 8.2.1). These fibres have microlitre-to-millilitre-sized lumens. The fibre length can be modified from a long to a short to increase or decrease the volume. These hollow fibres allow the selective passage of certain substances while retaining others. During cultivation, the fibres of HFMBR are submerged in a liquid phase and incubated for a certain period. The fibres are charged with feed using syringes. The fibres are porous structures and selectively restrict the mobility of the metabolites, nutrients, and signal molecules based on the particle size and chemical properties. This provides a suitable environment for the growth of autochthonous microorganisms outside the hollow fibres.

This bioreactor can be suitably used for the cultivation of microorganisms that can not be cultivated using classical culturing techniques. HFMBR fills the gap for untapped consortia in natural environments and allows for their cultivation in a controlled environment. This can be used for enumerating microorganisms from samples of sediments from tidal flats and activated sludge from wastewater treatment plants. HFMBRs are suitably used for enzyme reactions and yeast and bacterial culturing. Mammalian cells in HFMBR can make costly biochemicals such as interferons, hormones, growth factors, viruses, and monoclonal antibodies.

HFMBR provides an array of advantages. HFMBR facilitates high cell densities owing to its high surface-to-volume ratio. Microorganisms needing syntrophic partners or any intraspecific or interspecific interactions can survive together and multiply in HFMBR. The HFMBR system allows for precise control over the supply of substrates and other growth factors for microbial growth. A controlled supply of liquid and gaseous nutrients in the reactor provides a stable microenvironment for microbial growth. Also, the selective permeability of fibres can facilitate both cell retention and product recovery. However, the scaling up of HFMBR systems has a notable challenge, including the risk of membrane fouling, uniform distribution of nutrients and oxygen and temperature control.

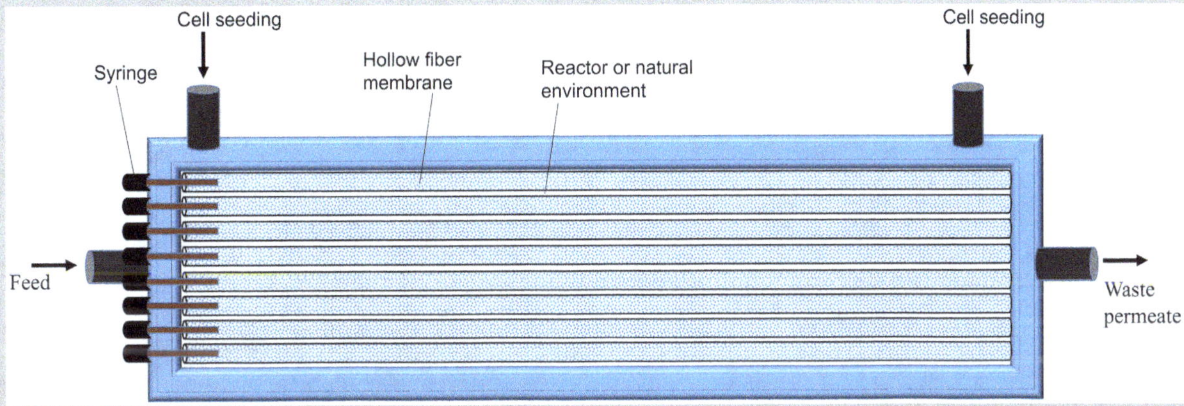

BOX FIGURE 8.2.1 A schematic representation of the hollow-fibre membrane bioreactor (HFMBR) used as an *in situ* cultivation device for environmental microorganisms.

One configuration is the anaerobic membrane bioreactor (AnMBR), which was developed by Dorr-Oliver (1980). This is a membrane filtration technology for biomass retention in anaerobic environments, where bacteria and archaea degrade organic materials into CO_2 and CH_4. It consists of an external membrane microfiltration unit, which is placed before or after an anaerobic reactor. The AnMBR technology involves two stages: wastewater enters an anaerobic bioreactor unit for degradation, and the remaining liquid is separated into smaller solid particles in a membrane unit. The solids from wastewater are retained at the membrane and returned to the reactor sludge. The AnMBR is an efficient technology for urban wastewater management and comes in three variations-

I. Submerged AnMBR: The membrane unit is integrated directly into the bioreactor unit. The raw influent enters the membrane unit before proceeding through the anaerobic process. The submerged membrane unit is placed at low negative pressures (Figure 8.11a). Filtration during the treatment is driven by the pressure differences between the wastewater in the reactor (at high pressure) and inside the membrane unit (at low pressure).
II. External/pressurised AnMBR: The membrane unit is placed externally to the main reactor unit (Figure 8.11b). Wastewater, after anaerobic treatment, flows under pressure from the rector to the membrane unit in a cross-flow or tangential flow. The treated effluent (permeate) is obtained on one side of the membrane. Retained matter (retentate) on the membrane's other side gets collected and recirculated to the reactor.
III. Externally submerged AnMBR: The reactor contains a submerged membrane unit placed externally. The influent wastewater is pumped into this externally submerged chamber, filtered into permeate and retentate. The retentate containing the organic load is recycled into the reactor, where it is anaerobically treated. The external membrane unit is kept at low negative pressure and immersed in a biomass-filled tank (Figure 8.11c).

Anaerobic microorganisms degrade the organic substances anaerobically with enzymatic hydrolysis, acidogenesis, acetogenesis, and methanogenesis. Full-scale AnMBR plants are applied for treating various types of wastes, including petrochemical wastewater, alcohol production stillage, municipal sludge, residues from food processing (dairy, potato, confectionery, etc.), and biogas production from anaerobic digestion of biomass. Advantages of AnMBR include selective retention of microorganisms and solids at the membrane (to enable high loading space), higher quality of the effluent, reduced production of sludge, and decoupled retention durations of solids and hydraulics. The limitations of AnMBR are the high cost of the filter material, blockages within the reactor, organic fouling in the membrane from bacterial accumulation that increases maintenance costs, increased energy consumption, and loss of anaerobic microorganisms in the recirculated effluent with a high pump flow rate.

8.6.2 Photobioreactor

A photobioreactor is usually a closed system which supports and manages the physiological requirements for the growth of phototrophs. This is used for the cultivation of microalgae, cyanobacteria, and some plants. The vessel is made up of glass or transparent plastic to meet the demands of light. The light in the photobioreactor can be maintained throughout the day and night by integrating artificial and solar light systems (Figure 8.12). The artificial light source is activated when the solar light intensity falls below a predetermined threshold. The sparger supplies aeration and CO_2 inside the chamber. Culture can be stirred using an agitator. A temperature range of 25–40 °C is usually optimal for the cultivation of phototrophs

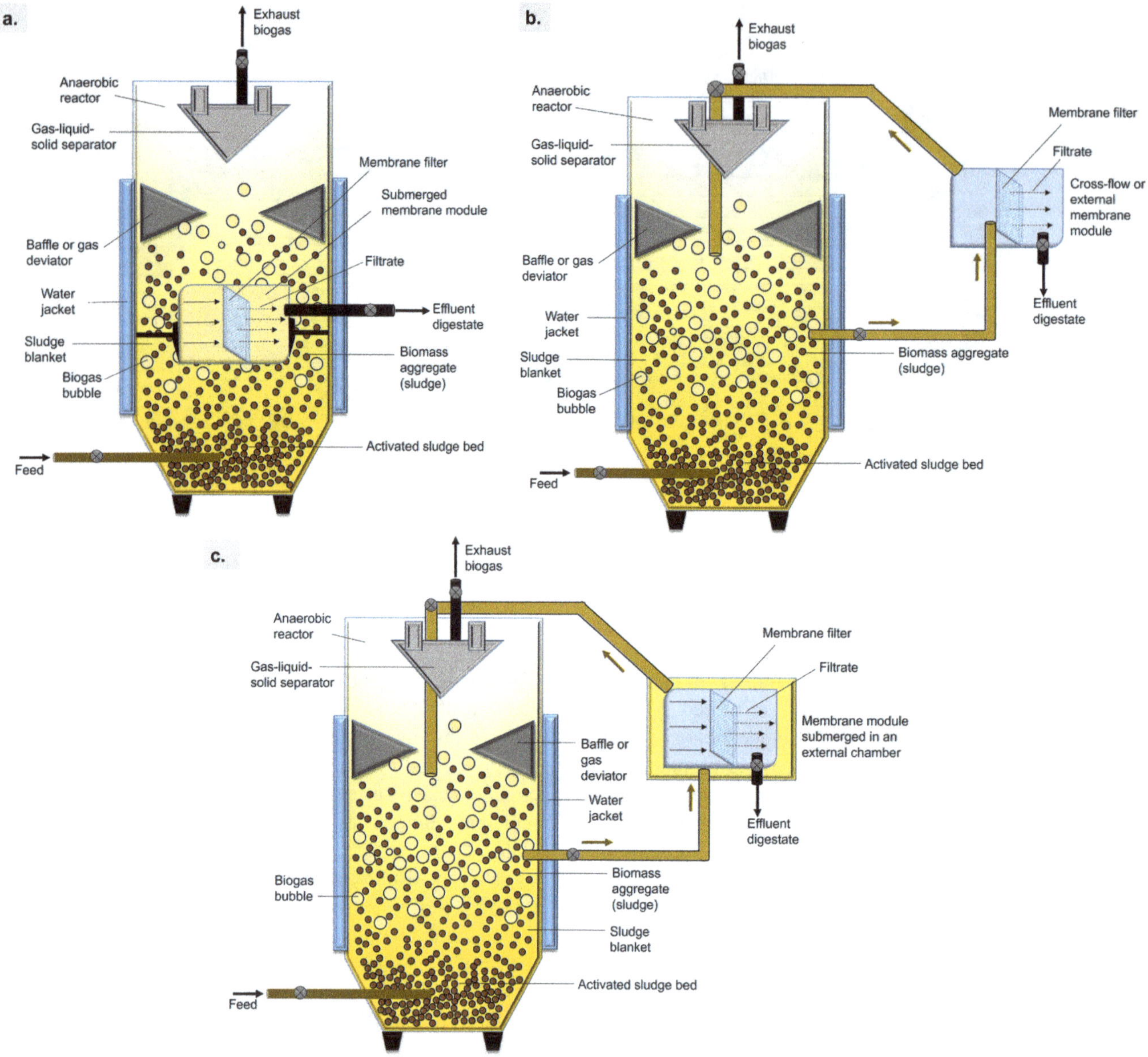

FIGURE 8.11 A schematic representation of anaerobic membrane bioreactors (AnMBR). a. Submerged AnMBR. b. External/pressurised AnMBR. c. Externally submerged AnMBR.

inside photobioreactors. Algal cultures absorb CO_2 and produce O_2 in the presence of sunlight, but high O_2 concentrations inside the vessel can be detrimental to algal growth. The excess O_2 is exhausted through a degasser port. The culture can also be cycled in solar receivers using centrifugal or airlift pumps, ensuring continuous cell circulation. Based on configurations, photobioreactors can be of different types, such as tubular photobioreactors, flat plate photobioreactors, Christmas tree photobioreactors, and column photobioreactors. A different type of setup for the large-scale production of microalgae is an open system, such as a natural open pond or artificial raceway pond.

Photobioreactors have compact designs for controlled growth conditions and reduced contamination risks. The setup has great value for producing high-value algal metabolites such as polysaccharides, carotenes, pigments, polyunsaturated fatty acids, etc. Sustainable production of renewable biofuel, wastewater treatment, and $hydrogen_2$ also demands photobioreactor setups. The regulated growth environment in a photobioreactor supports the production of high biomass per unit of substrate. Photobioreactor designs are built at the pilot stages and industrial scales. An algae-based industry in Israel has developed large-scale tubular reactors to produce different microalgae such as *Haematococcus*, *Phaeodactylum*, *Porphyridium*, *Nannochloropsis*, and *Euglena* to produce various commercial products. However, photobioreactors face design challenges due to limited practical usage and high capital and operational costs, making them uneconomical from an industrial perspective.

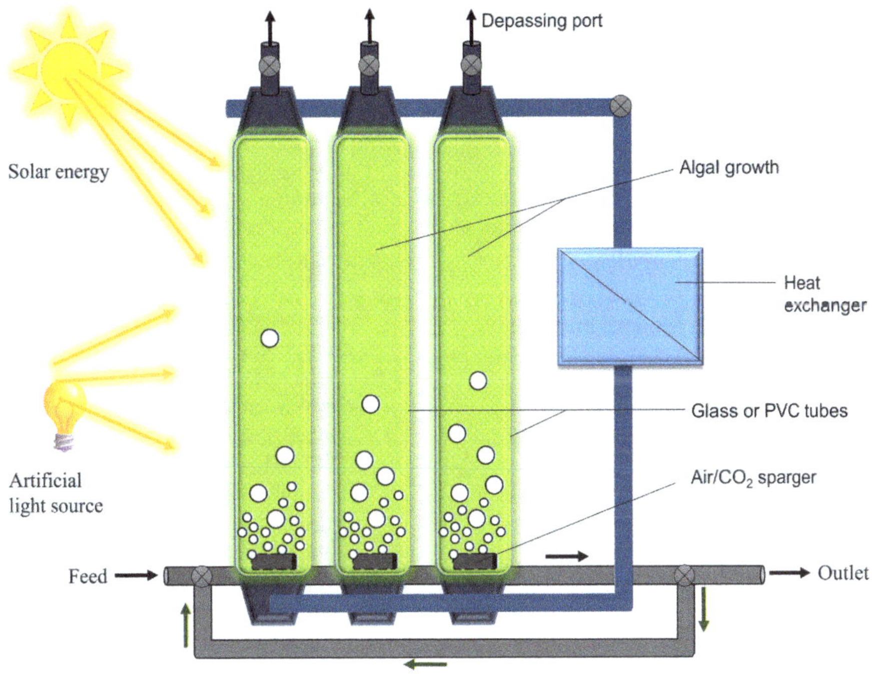

FIGURE 8.12 A schematic representation of a multi-tubular airlift photobioreactor.

SUMMARY

- Based on the state of the substrate and biocatalysts used in the fermentation process, fermenters can be of different types. These include fermenters for solid-state fermentation (SSF), bed reactors, and fermenters for submerged fermentation (SmF).
- Fermenters used for SSF are substantially simpler in comparison to those used for liquid-submerged fermentation. Industrial scale fermenters for SSF can be non-agitated (including tray fermenter and PlaFractor™ fermenter) and agitated (rotating horizontal drum, stirred drum) types.
- Bed reactors contain a rigid structure of immobilised enzymes called beds. The feed solution moves through the bed and allows for the reaction. Examples of bed reactors include packed-bed reactor (PBR), fixed-bed reactor (FBR), and plug-flow reactor (PFR).
- Fermenters used for SmF include both feed and biocatalysts in the solution. They can be mechanically agitated or non mechanically agitated. Various types of fermenters are stirred tank reactor (STR), tower fermenter, cylindrical-conical fermenter (CCF), cyclone column fermenter (CyCF), airlift fermenter (ALF), and ejector loop fermenter (ELF).
- Fermenters used for anaerobic processes are non-complex mechanical equipment and are widely utilised for the microbial treatment of industrial and municipal wastewater. These include the upflow anaerobic sludge blanket (UASB), expanded granular sludge bed (EGSB) reactor, and plug-flow anaerobic digester.
- Some fermenters have special configurations and are used for distinct processes. These include membrane bioreactors (MBR), hollow-fibre membrane reactors (HFMR), and photobioreactors.

DIFFERENT TYPES OF PRACTICE QUESTIONS

Fill in the blanks

I. In a packed-bed reactor, ... is attached to a solid support.
II. In the fluidised bed fermenter, ... is passed at high velocity from the distributor by the upward flow.
III. The non-mechanically agitated pneumatic fermenters rely on .... for agitation.
IV. The mechanically stirred tank bioreactor uses .... for efficient homogenisation of fermentation broth.
V. The tower fermenter has a total aspect ratio of about ....
VI. In an airlift fermenter, ... divides the internal fluid volume into two interconnected zones.
VII. A .... is a type of reactor which supports the growth of photoautotrophs.

SOLUTION: I. microbial cell/ enzymes, II. fluid (gas/ liquid), III. air bubbles, IV. agitators, V. 10:1, VI. draft tube, VII. photobioreactor

Match the following

i. Expanded granular sludge bed reactor	a. Unidirectional flow of broth
ii. Cylindrical-conical fermenter	b. Microbial biofilm
iii. Tower fermenter	c. Solid-state fermentation
iv. Rotating drum fermenter	d. Circulatory flow of broth
v. Airlift fermenter	e. Modern breweries
vi. Packed-bed reactor	f. Anaerobic Fermentation

SOLUTION: i-f, ii-e, iii-a, iv-c, v-d, vi-b

Very short answer type of questions

I. Name any two fermenters used for solid-state fermentation.
II. What are pneumatic fermenters?
III. Name the fermenter that develops riser and downcomer zones.
IV. What is fluidisation in the fluidised bed reactor?
V. Expand the UASB and EGSB.

Short answer type of questions

I. Explain the components and applications of the fluidised bed fermenter.
II. Give the differences between fixed bed and fluidised bed reactors.
III. The expanded granular sludge bed reactor is a type of up-flow anaerobic sludge blanket reactor with modified hydrodynamics. Explain.
IV. Photobioreactors have a compact design and are important for producing high-value algal metabolites. Elaborate.
V. Which bioreactor allows for *in situ* cultivation of diverse environmental microorganisms? Elaborate.
VI. Describe anaerobic membrane bioreactors.

Long answer type of questions

I. Elaborate on any mechanically stirred and pneumatic types of fermenters.
II. Explain any two types of fermenters used for anaerobic fermentation.

NOTES

1 Arora S, Rani R, Ghosh S. Bioreactors in solid state fermentation technology: Design, applications and engineering aspects. Journal of Biotechnology. 2018;269:16–34.
2 www.istockphoto.com/photo/beer-fermenter-gm1443090717-482180967?searchscope=image%2Cfilm

SUGGESTED READINGS

Ault RG, Hampton AN, Newton R, Roberts RH. Biological and biochemical aspects of tower fermentation. *Journal of the Institute of Brewing*. 1969;75(3):260–77.

Aoi Y, Kinoshita T, Hata T, Ohta H, Obokata H, Tsuneda S. Hollow-fiber membrane chamber as a device for *in situ* environmental cultivation. *Applied and Environmental Microbiology*. 2009;75(11):3826–33.

Banerjee DK, Fedorak PM, Hashimoto A, Masliyah JH, Pickard MA, Gray MR. Monitoring the biological treatment of anthracene-contaminated soil in a rotating-drum bioreactor. *Applied Microbiology and Biotechnology*. 1995;43(3):521–8.

Chen C, Zheng D, Liu GJ, Deng LW, Long Y, Fan ZH. Continuous dry fermentation of swine manure for biogas production. *Waste Management*. 2015;38:436–42.

Chen CS, Chan E, Wang SL, Gong CS, Chen LF. Ethanol fermentation in a tower fermenter using self-aggregating *Saccharomyces uvarum*. *Applied Biochemistry and Biotechnology*. 1994;45(1):531–44.

Dawson PS. Cyclone column fermentor. In *Biotechnology and Bioengineering Symposium;(United States)* 1974 (Vol. 4). National Research Council.

Dhillon GS, Brar SK, Kaur S, Verma M. Bioproduction and extraction optimization of citric acid from *Aspergillus niger* by rotating drum type solid-state bioreactor. *Industrial Crops and Products*. 2013;41:78–84.

Dourado F, Fontão AI, Leal M, Rodrigues AC, Gama M. Process modelling and techno-economic evaluation of an industrial airlift bacterial cellulose fermentation process. In Lee KY (ed.) *Nanocellulose and sustainability* 2018 (pp. 1–16). CRC Press.

He X, Li H, Zhu J. A value-added step towards promoting the serviceability of fluidized bed bioreactor in treating wastewater with low carbon to nitrogen ratio. *Science of the Total Environment*. 2021;750:141665.

Jatoi AS, Aziz S, Soomro SA. Biological assisted organic sulfur removal from low rank indigenous coal using airlift bioreactor. *Bioprocess and Biosystems Engineering*. 2021;44(2):417–27.

Jiraprasertwong A, Maitriwong K, Chavadej S. Production of biogas from cassava wastewater using a three-stage upflow anaerobic sludge blanket (UASB) reactor. *Renewable Energy*. 2019;130:191–205.

Kamilakis EG, Allen DG. Cultivating filamentous microorganisms in a cyclone bioreactor: the influence of pumping on cell morphology. *Process Biochemistry*. 1995;30(4):353–60.

Kim J, Choi H, Lee C. Formation and characterization of conductive magnetite-embedded granules in upflow anaerobic sludge blanket reactor treating dairy wastewater. *Bioresource Technology*. 2022;345:126492.

Kumar M, Kuroda K, Patel AK, Patel N, Bhattacharya P, Joshi M, Joshi CG. Decay of SARS-CoV-2 RNA along the wastewater treatment outfitted with Upflow Anaerobic Sludge Blanket (UASB) system evaluated through two sample concentration techniques. *Science of the Total Environment*. 2021;754:142329.

Lechowska J, Kordas M, Konopacki M, Fijałkowski K, Drozd R, Rakoczy R. Hydrodynamic studies in magnetically assisted external-loop airlift reactor. *Chemical Engineering Journal*. 2019;362:298–309.

Li J, He J, Si B, Liu Z, Zhang C, Wang Y, Xing XH. A pilot study of biohythane production from cornstalk via two-stage anaerobic fermentation. *International Journal of Hydrogen Energy*. 2020;45(56):31719–31.

Li MC, Song Y, Shen W, Wang C, Qi WK, Peng Y, Li YY. The performance of an anaerobic ammonium oxidation upflow anaerobic sludge blanket reactor during natural periodic temperature variations. *Bioresource Technology*. 2019;293:122039.

Li X, Chen Y, Zheng Z, Gao M, Wang Z, Zhang K, Liu H, Zhan X. Power-saving airlift bioreactor with helical sieve plates: developmental and performance studies. *Chemical Engineering Research and Design*. 2020;158:1–11.

Liu E, Wilkins MR. Process optimization and scale-up production of fungal aryl alcohol oxidase from genetically modified *Aspergillus nidulans* in stirred-tank bioreactor. *Bioresource Technology*. 2020;315:123792.

Meneses DP, Paixao LM, Fonteles TV, Gudina EJ, Rodrigues LR, Fernandes FA, Rodrigues S. Esterase production by *Aureobasidium pullulans* URM 7059 in stirred tank and airlift bioreactors using residual biodiesel glycerol as substrate. *Biochemical Engineering Journal*. 2021;168:107954.

Mohammadi M, Younesi H, Najafpour G, Mohamed AR. Sustainable ethanol fermentation from synthesis gas by *Clostridium ljungdahlii* in a continuous stirred tank bioreactor. *Journal of Chemical Technology & Biotechnology*. 2012;87(6):837–43.

Moresi M, Bartolo Gianturco G, Sebastiani E. The ejector-loop fermenter: description and performance of the apparatus. *Biotechnology and Bioengineering*. 1983;25(12):2889–904.

Patinvoh RJ, Mehrjerdi AK, Horváth IS, Taherzadeh MJ. Dry fermentation of manure with straw in continuous plug flow reactor: reactor development and process stability at different loading rates. *Bioresource Technology*. 2017;224: 197–205.

Prince IG, Barford JP. Tower fermentation using *Zymomonas mobilis* for ethanol production. *Biotechnology Letters*. 1982;4(8):525–30.

Rodríguez-Jasso RM, Mussatto SI, Sepúlveda L, Agrasar AT, Pastrana L, Aguilar CN, Teixeira JA. Fungal fucoidanase production by solid-state fermentation in a rotating drum bioreactor using algal biomass as substrate. *Food and Bioproducts Processing*. 2013;91(4):587–94.

Ross BS, Pott RW. Hydrogen production by immobilized *Rhodopseudomonas palustris* in packed or fluidized bed photobioreactor systems. *International Journal of Hydrogen Energy*. 2021;46(2):1715–27.

Sukačová K, Lošák P, Brummer V, Máša V, Vícha D, Zavřel T. Perspective design of algae photobioreactor for greenhouses—A comparative study. *Energies*. 2021;14(5):1338.

Tran KQ. Fast hydrothermal liquefaction for production of chemicals and biofuels from wet biomass– The need to develop a plug-flow reactor. *Bioresource Technology*. 2016;213:327–32.

Tregidgo M, Dorn M, Lucas C, Micheletti M. Design and characterization of a novel perfusion reactor for biopharmaceuticals production. *Chemical Engineering Research and Design*. 2023;194:344–57.

Unit III

Microbial Production of Industrial Products

9 Microorganisms in the Fermentation Industry

9.1 NATURAL MICROORGANISMS

Microorganisms are ubiquitous in the natural environment. They are incredibly diverse and crucial for ecosystem balance. All compartments of the Earth, including terrestrial, aquatic, and aerial environments, are residences of useful bacteria, archaea, fungi, and microalgae. Natural microorganisms are important sources of enzymes and metabolites for human welfare. However, not all natural microorganisms qualify as producer organisms for industrial applications. The beneficial properties of microorganisms are determined by their stability, production capacity, and unique metabolites. For example, the primary metabolites are principally common in all natural microorganisms, but only a few selected strains have the potential to be commercially employed. In contrast, the traits of secondary metabolites are present in limited microbial species, and there are even fewer choices of suitable strains for industrial applications.

The most useful products are mainly obtained from bacteria and fungi. Filamentous bacteria, moulds and yeasts are rich sources of antibiotics and enzymes. In particular, *Streptomyces* can produce as many as 200 different metabolites. Lactic acid bacteria (LAB) such as *Lactococcus*, *Streptococcus*, *Pediococcus*, *Oenococcus*, *Leuconostoc*, and *Lactobacillus* are indispensable for dairy products such as yoghurt (curd), cheese, buttermilk, and kefir. They also contribute to the fermentation of many food products, such as vegetables, meat, fish, and cereals. Bacteria like *Lactobacillus* and *Bifidobacterium* are used as probiotics to promote gut health and aid digestion. Yeast cells (*Saccharomyces cerevisiae*) are commonly employed in producing alcoholic beverages (such as beer, vodka, wine, etc.). Yeast cells are equally useful in the manufacturing of bakery products. The principle of alcohol fermentation by yeast is also applicable to biofuel production. The production of vinegar from fruits and grains is a fungal-bacterial-mediated process and has been in use since ancient times. Filamentous fungi such as *Rhizopus stolonifer* and *Aspergillus* spp. are producers of many organic acids such as fumaric acid, citric acid, glutamic acid, and oxalic acid. *Penicillium chrysogenum* is the natural source of the antibiotic penicillin. Many soil bacteria (particularly belonging to actinomycetes) are sources of many other antibiotics. Some fungi (e.g. *Agaricus bisporus*, *Pleurotus ostreatus*, *Morchella* spp., *Tuber*, etc.) develop fruiting bodies such as mushrooms and truffles, which have culinary values and are known for their taste, texture, and nutritional content. Some natural bacteria and fungi serve as the platforms for genetic engineering.

In terrestrial environments, healthy microorganisms team up with soil particles and plant body parts for diverse benefits. Plant parts like roots, shoots, and leaves are suitable residents of microbial communities. Many native microorganisms possess plant growth-promoting traits and can be employed as biofertilisers. Some well-known examples of biofertilisers are *Azotobacter*, *Azospirillum*, and *Rhizobium*. Also, some wild-type microorganisms are suitable biological control agents that help suppress the growth of harmful insects and pathogens in agriculture. In a different role, many microbial inhabitants at contaminated sites facilitate the detoxification of environmental contaminants. These different types of microorganisms can be grown on synthetic media in fermenters and applied to various field applications. This can be a scheme of natural attenuation, which involves intrinsic remediation and relies on the natural capabilities of microorganisms to break down contaminants without human intervention. Differently, some external factors can be applied at the contaminated sites to accelerate their removal efficiencies.

Aquatic environments also host many useful microorganisms. Microorganisms isolated from water sediments and near shores are important sources of enzymes, secondary metabolites, antibiotics, and food items. In aquaculture, they help maintain water quality, suppress pathogens, and serve as feed supplements. Many natural microorganisms found in contaminated water can be used to treat heavy metals and organics. They can also be used as bioengineers for wastewater treatment and oil spill cleanup.

Some specific microorganisms reside in extreme situations such as high/low temperature, high/low pH, saline environments, and anoxic conditions. They are characteristics of unique lifestyles and produce industrially relevant enzymes and metabolites. Particularly, heat-tolerant microorganisms are in high demand in industries due to the high-temperature requirements for many applications. They can be isolated from hot springs and hydrothermal vents. Many useful species contain thermophilic enzymes such as protease, lipase, cellulase, and amylase. Heat-tolerant lactic acid bacteria (LAB) are commercially employed in lactic acid fermentation. Thermophilic yeasts are used in bakeries and breweries. A heat lover, *Thermus aquaticus* produces heat-stable Taq polymerase essential for polymerase chain reaction

DOI: 10.1201/9781032651095-12

(PCR) reactions. A hyperthermophile, *Pyrococcus furiosus,* produces similar thermophilic enzymes. Many extremophiles are expression systems for stable recombinant proteins. In addition, enzymes used in the textile sector and the paper and pulp industry also require efficiencies under harsh conditions. Enzymes derived from extremophiles, such as acidophiles, basophiles, thermophiles, and halophiles, are particularly valuable in these sectors. For example, *Bacillus alcalophilus* is useful for expressing proteins under alkaline conditions.

Many industrial-scale operations require the cooperation of numerous natural microorganisms, which are often referred to as consortia. They offer multiple roles in reaction systems for an improved output. When the reaction system has a complex environment, the indigenous microorganisms of the reaction mixture offer better treatment than the single microbial system. This approach is amenable in wastewater treatment, biomining, bioelectricity production, anaerobic digestion, and composting.

9.2 GENETICALLY MODIFIED ORGANISMS (GMOS)

Genetically modified organisms (GMOs) are those organisms that have experienced an intentional variation in the genome due to the addition, elimination, or modification of the DNA sequences. They are also pronounced as non-natural microorganisms. Genetic manipulation is a common tool to modify natural producers to stimulate product formation for industrial applications. This is mainly attributed to hyperproduction, easy purification, and other economic considerations. Differently, microbial cells can be used as cell factories to produce recombinant proteins and foreign biologically active compounds and their analogues.

Many natural producer organisms in industrial microbiology have been modified using induced mutation. This is a classical method of genetic modification and is still an indispensable tool in food and beverage, owing to the low ethical concerns about genetic modification. In the past, many industrial biotechnology empires, such as penicillin and citric acid production, were raised using mutant strains. Detailed information on induced mutation is covered in a different part of the chapter (section 9.5).

In recent years, microorganisms have been directly engineered for broad applications in the pharmaceutical and enzyme industries. The first genetically recombinant pharmaceutical product was human insulin, developed in 1982, which involved inserting the human insulin gene into *Escherichia coli.* Since then, multitudes of recombinant proteins have been derived using various microbial expression systems. Their products are commercially available for therapeutic applications. The development of GMOs has also greatly benefited the industrial enzyme market. More than half of the industrial enzymes in the market are developed using GMOs. They are suitable alternatives to many animal- and plant-derived enzymes. For example, in cheese manufacturing, the coagulation of milk protein requires chymosin or rennin, which is traditionally separated from the calf rennet. The heterologous expression of these proteins in *E. coli* or fungi has transformed the entire process of their manufacturing.

Bacteria and fungi have sparked the production of many novel products using microbial technology. The selection of suitable hosts depends on several parameters, including specific traits, genetic compatibility, genome architecture, plasmid construction, genetic stability, growth conditions, and safety concerns. A Gram-negative bacterium, *E. coli,* represents a robust platform for developing diverse bioproducts. *E. coli* has played a pioneering role in recombinant DNA technology, laying the foundation of modern biotechnology. This continues to be the first choice of microbial host for genetic engineering and is employed for manufacturing more than one out of four available bioengineered drugs in biopharmaceuticals. The most commonly involved strains for biotechnological applications are *E. coli* K12 and *E. coli* B (BL21 and BL21(DE3)). The former strain has an excellent capacity to express foreign traits under stress conditions. In contrast, the latter strain has tremendous potential for expressing periplasmic proteins and promoting elevated heterologous protein expression.

Gram-positive bacteria lack an inclusion body and endotoxin and are relatively safe for biotechnology projects. Many industries identify them as their first choice for expression systems. The monolayer cell wall of the Gram-positive bacteria allows an easy secretion of proteins and metabolites outside the cell and enables relatively simple purification. Examples of Gram-positive bacteria that have an appeal to be genetically modified for industrial applications include *Lactococcus lactis*, *Streptomyces* spp., and *Bacillus* spp. Genetically modified *L. lactis* is fairly common in the food and dairy industry. It has probiotic properties and is generally recognised as safe (GRAS).

Bacillus is best known for enzyme production. They dominate the market for enzyme production, comprising about half of the entire industrial sector. Their ability to adapt to broad pH and temperature allows them to thrive in varying conditions. They also have a high secretory capacity profile that renders them attractive in the fermentation industry. *Bacillus subtilis* is highly amenable to genetic modifications and is used as a cell factory in food, agriculture, biomaterials, and medicines. This bacterium, however, lacks plasmid stability and has natural proteases, which prohibit its application in genetic engineering for the expression of recombinant proteins. *Streptomyces* spp. are dominant sources of natural metabolites, and they can be genetically manipulated to obtain their abundant supply. Some species, like *S. lactis* and *S. pyogenes,* are typically used as heterologous protein expression platforms for the production of biologics. Many bacteria offer a suitable platform to produce specific proteins, including antigens for vaccines. The genetically modified bacteria can also be directly administered to the human body. They act as live vaccine vectors and stimulate an immune response in the host organism without causing disease.

The main advantages of using bacteria as microbial hosts come from their rapid growth, easy manipulation, deep knowledge of their omics, availability of genetic tools, and cost efficiency. However, the synthesis of many important plant- and human-related metabolites using bacterial systems is not feasible due to limited post-translational modification (e.g., non-native glycosylation patterns), differences in protein folding, translation error, and ethical concerns. Particularly, post-translational modification in eukaryotic proteins is an important requirement for their proper functioning. However, prokaryotic cells are often devoid of such modification, and therefore, they do not achieve much success in expressing eukaryotic proteins. This limitation has been effectively addressed using eukaryotic fungal models.

A yeast, *S. cerevisiae*, is a suitable eukaryotic platform for genetic engineering for the wide production of neutral metabolites and biologics. Compared to other fungi, *S. cerevisiae* has advantages for protein expression due to its GRAS status, compatibility for large-scale fermentation, and ease of upstream and downstream processing. This fungus promises heterologous expression of plant and human genes to meet the current demand in the food and drug industries. For example, *S. cerevisiae* was genetically modified to retain and express genes of the biosynthetic pathway of paclitaxel (an anticancer agent) (brand name, Taxol®) from a plant, *Taxus brevifolia*. In a different study, at least twelve genes related to artemisinin biosynthesis from the medicinal plant *Artemisia annua* L. were cloned in *S. cerevisiae* for the large-scale microbial production of the artemisinin precursors. A company, Impossible Foods, is commercially producing a vegan alternative to beef hamburgers using an engineered yeast, *Pichia pastoris*, which is modified to produce leghemoglobin (a heme-containing plant protein of legume plants). A different yeast, *Yarrowia lipolytica*, is a suitable alternative host for glycoprotein production and performs glycosylation similar to human cells. The multicellular fungi *Aspergillus niger*, *A. oryzae*, and *Hansenula polymorpha* are other common cell factories in fermentation industries due to their GRAS status and adjustable cultivation conditions. These microorganisms can be successfully used for the production of industrial enzymes, antibodies, vaccines, and human serum albumin.

Although genetic engineering techniques have achieved remarkable success, they have certain limitations. Off-target effects often lead to unpredictable consequences. The genetic stability of the genetically modified species is a matter of great concern. Their amendment in consumable goods requires intensive regulatory scrutiny to assure their safety. GMOs render low interest as food additives due to restrictive food legislation, stringent approval, consumer perception, and labelling regulations. However, the applications of GMOs are much more common in pharmaceuticals, biofuel industries, and other business sectors. Also, recent years have witnessed better public acceptance of many food products prepared using GMOs. Despite several challenges, ongoing research continues to shed more light on genetic engineering technologies for human welfare and a better future.

9.3 STRAIN IMPROVEMENT STRATEGIES

Genetic modification techniques have a long history of strain improvement for fermentation technology. A natural microbe produces metabolites only in the quantity required for its metabolic activities. Their production in synthetic environments can have challenging outcomes. Genetic modification techniques can generate new characteristics in microorganisms that favour their overproduction, meeting the demand for industrial fermentation. Though the strain development program focuses on human-oriented benefits, it usually upsets the normal growth regulation of microorganisms. A biological trait can be derived from unigene or multigene systems. The genetic manipulation of a trait derived from multigene systems is much more complicated than the unigene system. A producer microorganism contains at least five distinct genetic loci in the genome from which metabolite synthesis can be controlled (Figure 9.1). These include

- Structural genes that code for product synthesis
- Regulatory genes that control the expression of structural genes
- Resistance genes that code for resistance of the host to the antibiotics
- Permeability genes that regulate the accumulation, discharge, and expulsion of the product
- Regulatory and structural genes of the pathway that stimulate the synthesis of precursors and cofactors

These genetic loci are the primary targets which influence the metabolic potential of an organism. Furthermore, the gene modification linked with competing routes can also significantly impact production levels.

The advances in genetic information on a producer microorganism through genomic studies have allowed scientists to identify the potential DNA sequences responsible for any useful traits. The in-depth knowledge of sequence analysis has led to a great expansion in targeting DNA sequences for their manipulation. Different methods for genetic modification are covered in the next section.

9.4 GENETIC MODIFICATION IN INDUSTRIAL MICROBIOLOGY

Genetic modification in industrially relevant microorganisms can be achieved through natural recombination, sexual reproduction, artificial mutation, and genetic engineering. Protoplast fusion, mutation, and genetic engineering can be applied to a wide range of microorganisms.

Natural recombination, including genetic recombination or sexual reproduction, is the natural ability to create genetic variation. These processes are the main driving forces behind evolution and novel metabolic pathways. Genetic

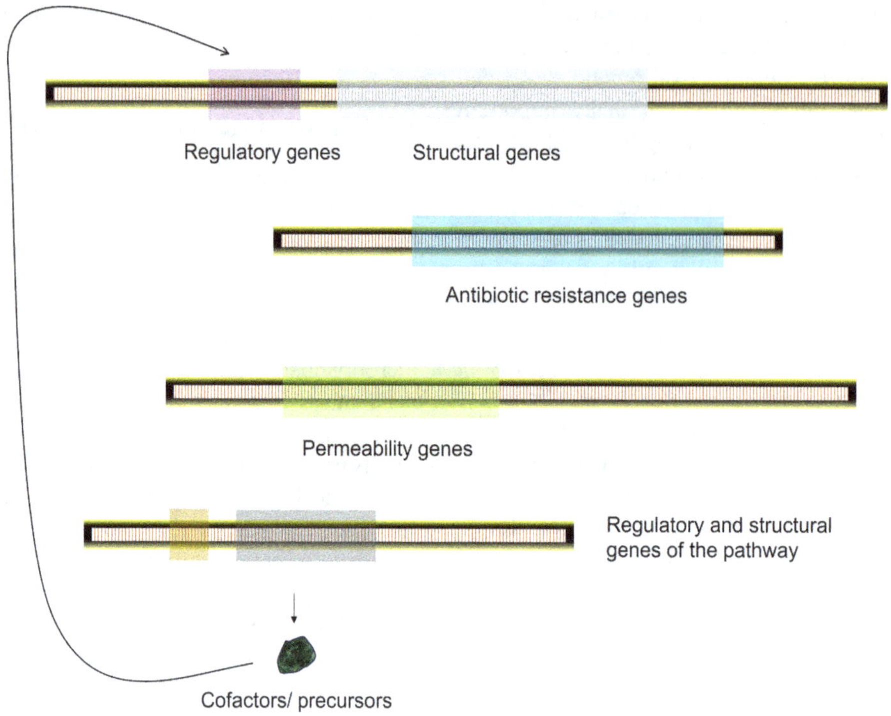

FIGURE 9.1 A schematic representation of important genetic loci in the genome of a microbe for metabolic synthesis of industrially relevant products.

recombination is mainly reported in bacterial systems, while sexual reproduction can be involved in organisms like fungi and microalgae. Genetic recombination in prokaryotes can be achieved through transformation, conjugation, and transduction. Transformation facilitates the uptake of exogenous double-stranded DNA from the environment, which is inherent to many bacterial species. Conjugation is a process of bacterial mating through close physical contact and the transfer of a DNA sequence from donor to recipient. Transduction allows the transfer of new DNA in bacteria through phages.

Genetic variability in eukaryotes arises through processes like genome shuffling and crossing over. Some fungi observe a different type of genetic recombination called parasexuality when sexual reproduction is not reported. Examples of industrially relevant fungal species in which parasexuality is reported include *Aspergillus niger*, *Penicillium roqueforti*, and *Penicillium chrysogenum*. Protoplast fusion is a hybridisation technique where two protoplast cells are fused, and their entire genome combines into a single cell. This process can be applied to some eukaryotic organisms. In industrial microbiology, this technique can be used to improve microalgal strains. Protoplast fusion exhibits potential for strain modification, but with limited efficiency and restricted range.

Genetic manipulation in industrial microbiology has a long history, even before the discovery of the DNA structure. The early studies on genetic modification relied on classical genetics, including induced mutation and genetic recombination. Induced mutation in producer microorganisms raised benefits for commercial-scale production. Following that, modern tools of DNA recombinant technology came into existence. This process completely revolutionised the industrial schemes and significantly contributed to a cost-competitive bio-based economy. DNA recombinant technology allows direct manipulation of DNA, facilitating new combinations of genes. Contrarily, mutation involves an alteration in the position of bases within the natural order of the DNA sequence. Clustered regularly interspaced palindromic repeats (CRISPR)/Cas9 gene editing method is the latest edition of the molecular tool that allows genetic modification *in vivo*. This technique has offered much ease in the genetic modification of eukaryotic systems. Figure 9.2 provides information on many important genetic modification techniques for industrial biotechnology. All these methods of genetic modification are further discussed in various sections below.

9.5 MUTAGENESIS

A mutation is a sudden change in the genotype of a microorganism under some natural or artificial treatment. These modifications could occur as a result of a single nucleotide substitution, multiple nucleotide substitutions, addition, or deletion. The change in genotype also bestows modifications in its phenotypic traits. After mutation, survivors can be detected by selection, followed by screening. Beneficial strains are recovered for long-term storage, and non-useful strains are eliminated. This process can be used to produce mutant proteins for commercial interest. In the

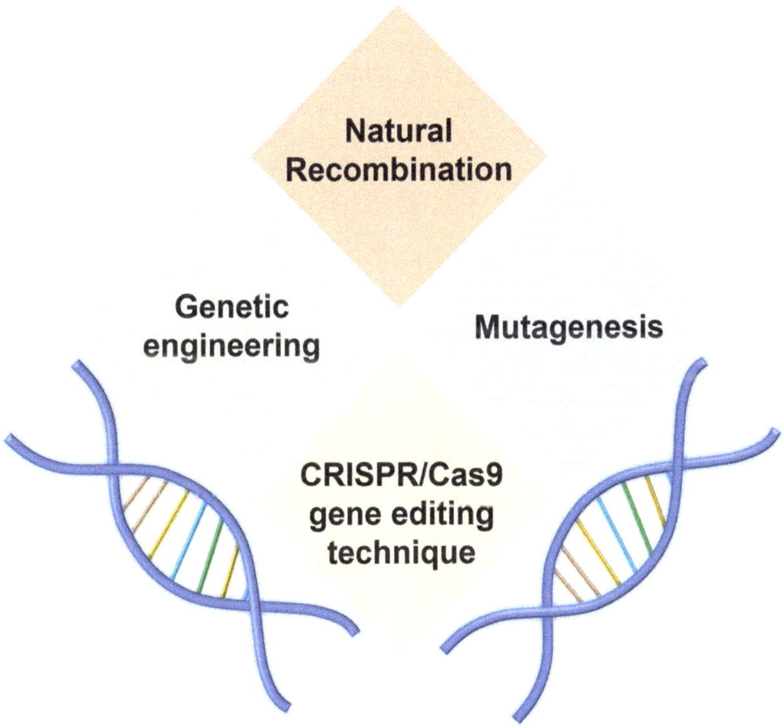

FIGURE 9.2 Different techniques for strain improvement.

1940s, when the world demanded high titers of penicillin from microbial sources, the mutation played a key role in strain improvement in industrial microbiology. The penicillin-producing fungus, after X-ray mutagenesis, generated mutant strains with significantly higher penicillin production. The strain development programs using mutation continued to expand and are still vital for improved productivity of industrial cultures.

Mutations can be induced using physical and chemical mutagens. Physical mutagens can be radiation like UV rays, X-rays, and gamma rays. Examples of some mutagens in the chemical category are 5-bromouracil, 2-aminopurine, acridine orange, nitrosoguanidine (NTG), 4-nitroquinoline 1-oxide, and methyl methanesulphonate (MMS). The genetic variation in an organism can also be created by mutagenic polymerase chain reaction (using error-prone PCR) or repeated oligonucleotide-directed mutagenesis.

Based on the specificity of the site of action, mutations can be of two types: random and site-directed. Random mutation targets unspecific regions in the sequence. In random mutation, a sizable number of cells are exposed to a mutagen. However, the survival rate of mutated strains is meagre because most of the genetic variations produced after mutation are detrimental due to the impairment of control mechanisms. A successful mutant may carry one to ten mutation sites in its genetic constituents. Differently, the site-directed mutagenesis produces variation in specific amino acids in the protein. This method allows scientists to precisely alter one or more nucleotides at a particular site for a controlled variation in the protein structures and function.

When an organism loses the ability to synthesise certain metabolites due to genetic modification, the mutant is called an **auxotroph**. For example, an auxotroph of *Corynebacterium glutamicum* cannot build homoserine and produces lysine in large quantities. Homoserine, acting as a precursor, has a role in controlling the feedback inhibition on the lysine biosynthetic pathway. In the absence of homoserine, the bacterium can continue to manufacture lysine using a precursor from the synthetic medium.

Mutation is also an integral part of directed evolution, which has received wide recognition in protein engineering. The procedure can be iterated with multiple rounds of genetic modification, followed by screening for the desired metabolic phenotype. Mutational studies play a role in identifying molecular sites for the function of genetic loci, bioprocesses, and bioproducts.

9.6 GENETIC ENGINEERING

Genetic engineering tools involve deliberate modifications of the genetic elements for artificial variations in microbial systems. These organisms are specifically pronounced as genetically engineered microorganisms (GEMs). In the fermentation industry, genetic engineering provides a foundation to improve the productivity and strength of producer strains for industrial setups. The genes related to enzymes or metabolic processes can be the primary sites of engineering in cells. The genetic makeup of an enzyme can be altered to enhance its properties, such as overexpression, thermal stability, enzymatic efficiency, and resistance to inhibitors. However, an organism tailored by genetic engineering encounters more regulatory obstacles than a mutation-based strain because the former technique involves a deliberate insertion of foreign genetic material into a host cell. Genetic engineering in microorganisms is mainly performed by recombinant DNA technology and CRISPR-Cas9, which are discussed in the upcoming sections.

9.6.1 Recombinant DNA Technology

Recombinant DNA technology has completely revolutionised microbial technology and other fields by enabling the insertion of the DNA of interest for various purposes. Recombining a host cell into a GMO is a long process and involves DNA manipulation outside the host cells. The constituents of genetic recombination allow the transfer of genes between similar, related taxonomic species or cross-kingdom species. Based on the types of genes involved, recombination can be of two different types: homologous and heterologous expression. When the recombination of genes occurs between the same species or within the species, the process is called a homologous recombination. This can be used for the hyperproduction of a trait by introducing multiple copies of similar genes. However, when the transfer of genes occurs between completely different species, the recombination is referred to as heterologous expression. With such an adjustment, plant and human-based products can be expressed in microbial cells. Heterologous expression strictly involves the addition of new genes in the host cell. Heterologous expression is a much more complicated process than homologous expression and requires many additional establishments at the molecular level for host compatibility with foreign genes (Table 9.1). A detailed understanding of heterologous expression in microorganisms is discussed in section 9.6.1.1.

The recombination process involves several steps. The first step is the gene identification for a desired trait from an organism, followed by its excision using restriction enzymes. The next step is amplification of the gene of interest to produce many copies. These gene copies are combined with other genetic elements such as promoters, poly A sequence, ribosomal binding site, affinity tag, and markers in a vector (like plasmid, cosmid, yeast artificial chromosome (YAC), and bacterial artificial chromosome (BAC)). The vector is then transferred into a host cell for the multiplication of cloned constructs and their expression (Figure 9.3). In heterologous expression, the expression of foreign genes may require two different types of vectors: the cloning vectors and expression vectors. This requires an additional step of gene transfer from the cloning vector to the expression vector. The cloning vectors are responsible for replicating gene elements, whereas the expression vectors facilitate their transcription and translation. The expression vectors are equipped with host-compatible expression systems for heterologous protein production.

9.6.1.1 Microbial Expression System for Heterologous Proteins

The process of heterologous protein expression is the main ground for expanding the empire of biopharmaceutical industries. The recombinant proteins licensed for therapeutic use include more than 400 different products. Several bacteria and yeasts are the host organisms for synthesising human insulin, interferons, a recombinant vaccine for hepatitis B, hirudin (an anticoagulant), etc. Before the discovery of heterologous protein production using microbial expression systems, the production and purification of a small amount of protein were highly cumbersome, and it required large volumes of biological fluids or high amounts of animal and plant materials. The scale-up of bioproduction using recombinant technology gives a high protein yield through low-cost culture methods in a flexible system.

In a recombination process, the expression of foreign proteins in a microbial system is highly technical. The main principle behind this expression is the introduction of DNA encoding the target protein and other necessary regulatory elements into a plasmid, which is then transferred into a bacterial cell. Transformed cells proliferate and generate the protein of interest. The bacterial system also allows the expression of more than one protein in the same host using the same or different plasmids. The transfer of eukaryotic genes into

TABLE 9.1
A comparative account of homologous and heterologous expressions

	Homologous expression	Heterologous expression
Objective	Overexpression or regulation of a protein	Expression of a protein in a new host
Expression system	Same as the source organism	Different from the source organism
Genetic compatibility	High	Low
Complexity	Low	High
Challenges	Low because the gene is native to the organism	High because of differences in post-translational requirement, codon requirement, and removal of intron regions
Expression level	Often higher and regulated by native elements	Vary and require optimisation for the new host
Regulatory elements	Typically compatible with host regulatory systems	Require modification or addition of regulatory elements
Desirability	Overproduction	Ease of production

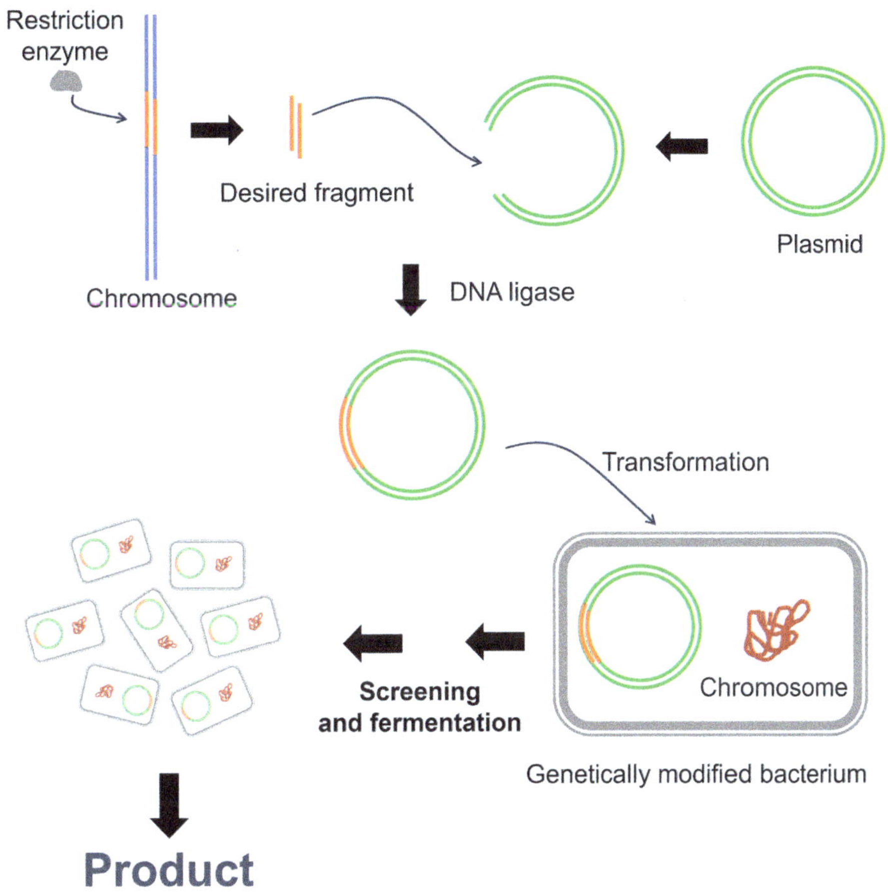

FIGURE 9.3 Illustration of genetic modification through recombinant DNA technology.

the prokaryotic system involves the removal of intron regions (non-coding regions). These noncoding regions in eukaryotic cells are removed by RNA splicing after transcription. However, the prokaryotic system is incapable of RNA splicing. Therefore, the eukaryotic genes require modification in the prokaryotic format for their heterologous expression in a prokaryotic host. An enzyme reverse transcriptase can act upon eukaryotic mRNA and transform it into complementary DNA (cDNA) (that does not contain non-coding regions) before cloning. However, the bacterial expression system encounters some major drawbacks in expressing eukaryotic proteins. For example, they do not have post-translational modifications such as glycosylation. Therefore, many proteins with such requirements can not be expressed in bacteria. The large DNA fragment can also not be stably maintained in bacteria. Moreover, the microenvironment of bacteria can lead to protein instability and build an aggregate of proteins, resulting in the formation of an inclusion body.

Yeast-like S. cerevisiae and P. pastoris are suitable alternatives to eukaryotic platforms and bacterial expression systems. They are superior to bacterial systems in terms of heterologous production of proteins with post-translational modification. Synthesis of cDNA is still a requirement to remove intron sequence, as yeast cells have a limited ability to splice RNA sequences. Also, these platforms may not produce suitable glycosylation patterns.

9.6.2 CRISPR/Cas9 Gene Editing

CRISPRs are clustered regularly interspaced short palindromic repeats that are fairly common among prokaryotes. They are components of the microbial immune system that have a role in defence mechanisms against invading viruses and other foreign genetic materials. This defence mechanism inspired the development of a new genome editing tool called CRISPR/Cas9 gene editing. This method allows for an *in vivo* method of genetic engineering in eukaryotic organisms and has made genetic modification relatively simple and precise.

The CRISPR/Cas9 gene editing system has two fundamental components: CRISPR-associated (Cas9) proteins and synthetic guide RNA (sgRNA). Cas9 is an endonuclease and can be found in many bacteria. Cas9 derived from *Streptococcus pyogenes* is the most well-characterised and widely employed in genome editing. A sgRNA contains two subunits: a CRISPR-RNA (crRNA) and a trans-activating CRISPR RNA (tracrRNA). crRNA recognises and complements the target DNA, whereas a tracrRNA binds with the Cas9 protein.

Genome editing through a CRISPR system involves three basic steps: recognition, cleavage, and repair. In the first step, Cas9 associates with a sgRNA and forms a complex, which recognises and causes double-strand breaks (DSB) at a DNA target (also called protospacer). This DSB in eukaryotic cells initiates DNA repair via intrinsic cellular mechanisms called

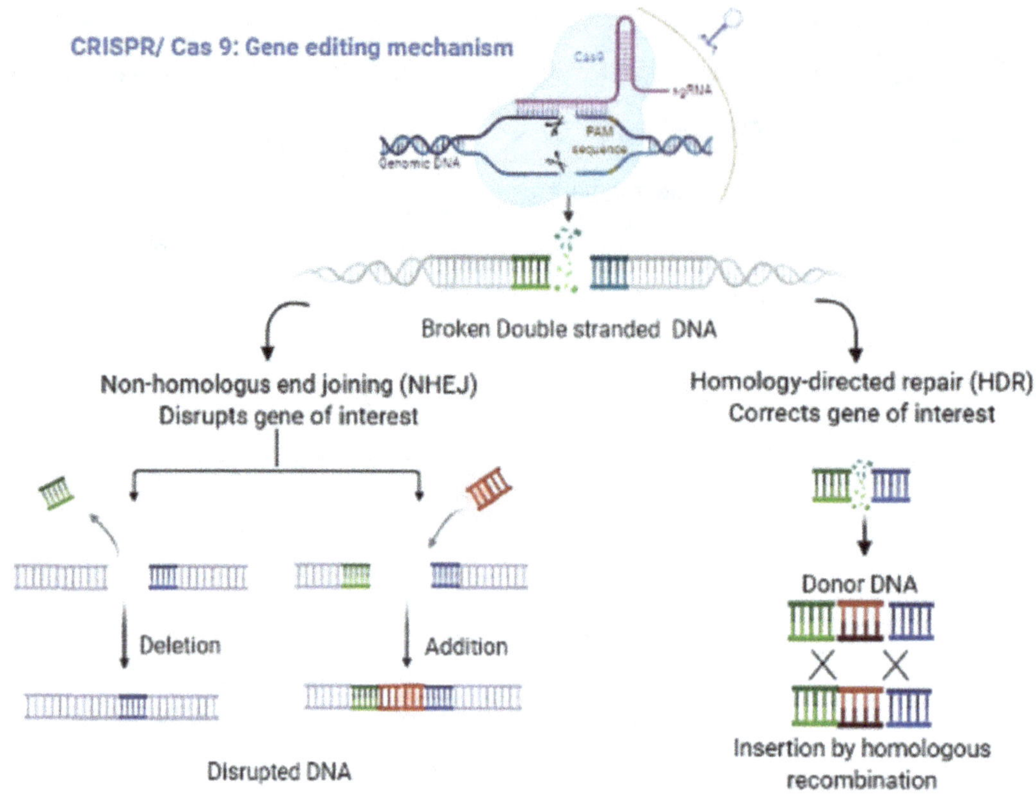

FIGURE 9.4 Mechanism of gene editing using the CRISPR/Cas9 technique. (Image adapted from Bharathkumar et al. (2022)[2] with permission of © 2022 Springer Nature.)

non-homologous end joining (NHEJ). NHEJ allows the ligation of two broken ends with tiny insertions or deletions (indels), with some possibilities to add new or remove the existing traits in the cell. However, this repair system in prokaryotes requires homologous templates, and the process is called a homology-directed repair (HDR), which generates limited possibilities to engineer the cells (Figure 9.4). This route in prokaryotes can be used to create mutations in an organism. In fermentation technology, the CRISPR/Cas9 gene editing system has shown potential for strain improvement, particularly for producer fungi and microalgae, by modifying traits, such as growth rates, product yields, metabolic pathways, reducing byproduct formation, and tolerance to environmental stressors.

9.7 CONTAINMENTS FOR SAFE BIOTECHNOLOGY

The use of microorganisms is essential to many large-scale industrial processes. However, their safety for human beings and the environment is of utmost concern. The Organisation for Economic Co-operation and Development (OECD), United States of America (USA), has developed a framework for its containment in industrial biotechnology. This mainly focuses on two different types of containment: physical and biological. Physical containment looks after the physical boundary for microorganisms, whereas biological containment ensures the biological safety of various microorganisms (including GMOs) and the handling of potentially dangerous organisms.

As proposed by the Centers for Disease Control and Prevention (CDC), USA, containment can be categorised into four main biosafety levels (BSLs) (BSL-1 to BSL-4) with their specific safety measures for safe laboratory practices. The containment regulations become stringent only at the levels above the first category, allowing minimal or complete prevention of microbial release.

In a different approach, the development of products using GMOs requires a similar set of guidelines and principles under a good industrial large-scale practice (GILSP). Based on the potential hazards associated with the use of GMOs, there are four levels of contaminants, which comprise GILSP, containment categories 1, 2, and 3 (Table 9.2). These categories align with the four BSLs. The four levels of GILSP-based guidelines for unit operations of microorganisms are discussed below-

- GILSP- This category contains harmless microorganisms that can be used on a large scale without integrating any containment methods except for aseptic functioning. This category is equivalent to BSL-1. Examples of microorganisms in this category are non-pathogenic strains of *E. coli* and *Saccharomyces cerevisiae*.
- Containment Category 1- This class addresses low-risk microorganisms and requires the implementation of Good Microbiological Techniques (GMT), including restricted access to the facility. This class of microorganisms are equivalent to BSL-2. They are

TABLE 9.2
Descriptions of biosafety level (BSLs) and good industrial large-scale practice (GILSP)-based categories for safe handling of microorganisms in laboratory and industrial applications, respectively

	Biosafety levels			
	BSL-1	BSL-2	BSL-3	BSL-4
	GILSP-based categories			
	GISLP	Containment Category 1	Containment Category 2	Containment Category 3
Description	• No contamination • Defined organisms • Unlikely to cause disease	• Containment • Moderate risk • Disease of varying severity	• High containment • Aerosol transmission • Serious/ potentially lethal disease	• Max containment • "Exotic", high-risk agents • Life-threatening disease
Examples of organisms	*E. coli* and *Saccharomyces cerevisiae*	*Staphylococcus aureus* and *Salmonella* spp.	*Mycobacterium* spp. and COVID-19*	Ebola virus and Marburg virus
Pathogen type	Agents that present minimal potential hazard to personnel and the environment	Agents associated with human diseases that pose moderate hazards to personnel and the environment	Indigenous or exotic agents that present a potential for aerosol transmission and agents causing serious or potentially lethal disease	Dangerous and exotic agents that pose a high risk of aerosol-transmitted laboratory infections and life-threatening disease
Autoclave requirement	None	None	Pass-through autoclave with bioseal required in the laboratory room	Pass-through autoclave with bioseal required in the laboratory room
Other facilities	• Open bench top • Standard microbiological practices	• Biosafety cabinet	• Biosafety cabinet • Physical containment • Respiratory protection	• Biosafety cabinet • Physical containment • Respiratory protection • Positive pressure personnel suit • Rigorous decontamination procedures

* COVID-19 (Coronavirus disease).
Source: Modified from Bathula and Rakhimol (2017).[1]

managed by employing autoclaving, disinfection, and aerosol control. Examples include *Staphylococcus aureus* and *Salmonella* spp.

- Containment Category 2- This class deals with medium-risk microorganisms that may cause severe threats to humans from their exposure (equivalent to BSL-3). This type of containment is based on preventing the release of microorganisms rather than minimising their release. In addition to the facilities required in Category 1, this requires protective suits for workers, restricted entry at the workplace, sealed equipment, biosafety cabinets, and high-efficiency particulate air (HEPA) filters for air exhaust. Examples of those microorganisms are *Mycobacterium tuberculosis* and severe acute respiratory syndrome coronavirus-2 (SARS-CoV-2).
- Containment Category 3- This is the category of high-risk microorganisms that have lethal effects (equivalent to BSL-4). They are rarely used in industrial applications due to their high risk (e.g., vaccine development). The workplace is strictly isolated, airlocked, and sealable in addition to the containment protocols of the above categories. Examples are Ebola and Marburg viruses.

9.8 GENERALLY RECOGNISED AS SAFE (GRAS) MICROORGANISMS

GRAS is the status of a substance to be used as a food additive in foods and beverages, set forth by the US Food and Drug Administration (FDA), USA. A food additive is a component of food or a substance that can be used to affect the characteristics of food. The nature of food additives and their conditions of use in foods or beverages are matters of great concern for their safety. A US legislation grants the FDA the right to control food safety and set requirements for food additives. The FDA oversees and monitors the safety of food ingredients and substances and establishes food safety guidelines and frameworks. The FDA certification ensures the products' safety based on a common consensus among experts from a rigorous evaluation of the food additives for intended use in the product for consumption by the general public. Manufacturers have the responsibility to present scientific evidence to back their claims of the safety of their food additives. The products that comply with the regulatory standard of GRAS ensure that the possibility that a product and/or its constituents pose any health concerns as a result of food additives is quite low, and the food additives have been used within specified safety limits.

The history of the term GRAS dates back to the early 20th century. In 1958, the FDA passed the Food Additive Amendment to the Federal Food, Drug, and Cosmetic Act (FDCA) for the safe usage of food additives. Sections 201(s) and 409 of the FDCA issued a directive that any substance with implications for intentional use in food mandates FDA premarket evaluation and approval unless it is GRAS. Common food additives with "general recognition" and a long history of use (before 1958) do not require technical evidence of safety and are covered as GRAS under 21 Code of Federal Regulations (CFR) 170.30(c) and 170.3(f). These substances with a long history of safe use are exempted from the rigorous premarket approval process.

Under CFR Title 21, sections 170.3 and 170.30, the scrutiny of the substance requires a team of experts for its investigation through scientific procedures for adequate safety and standards under the condition of intended use. The evaluation of a substance for GRAS status can be affirmed by experts from an independent organisation or the FDA. A substance already having a GRAS designation does not require any pre-approval to become a component of food. However, the GRAS status of any substance can be revoked in a situation when it does not fit within safety parameters. There have been time-to-time amendments in the FDA regulations to ensure product safety. The FDA frequently updates its affirmations and notices with new information. Food and beverage industries keep up-to-date with the most recent FDA regulations and guidelines regarding GRAS substances to maintain the safety of the products they produce and market.

GRAS regulatory status is equally applied to microorganisms or chemicals produced from microorganisms to be used as food additives. The FDA's codified Title 21 CFR parts 131, 133, 136, 137, 172, 173, and 184 include partial lists of food additives generated from microorganisms. The list of GRAS microorganisms is extensive, and numerous independent agencies have classified them. The full listing of notices and response letters from 1998 to date can be found in the inventory of GRAS notices (GRN) under the US Food and Drug Administration. Table 9.3 enlists some useful GRAS microorganisms. In this list, *Saccharomyces cerevisiae, Candida utilis*, *Aspergillus niger*, and *Rhizopus oryzae* are some of the well-known GRAS fungi. Examples of GRAS bacteria include *Streptomyces natalensis*, *Xanthomonas campestris*, and *Micrococcus lysodeikticus*.

Microbial enzymes are also subject to the GRAS designation. A variety of enzymes used in baking, beverage production, and dairy have a long history of safe usage and are exempt from premarket evaluation. However, the discovery of new enzymes that open up new opportunities for the food and beverage industries requires critical analysis. This involves the assessment of four parameters, including the enzyme characteristics, its source, production method, and safety concerns. The methods involved in downstream processing and formulation for the manufacturing of a product are also critically examined. The exposure assessment and margin of safety of the food substances are other notable parameters of their safety evaluation.

For the GRAS status, microorganisms should essentially be non-toxigenic and non-pathogenic to become a part of

TABLE 9.3
Examples of generally recognised as safe (GRAS) microorganisms

Bacteria	Yeasts	Filamentous fungi	Mushrooms	Microalgae
Xanthomonas campestris	*Saccharomyces cerevisiae*	*Aspergillus niger*	*Agaricus bisporus*	*Schizochytrium sp.* FJRK-SCH3
Streptomyces natalensis	*Saccharomyces fragilis*	*Aspergillus oryzae*	*Hericium erinaceus*	*Chlamydomonas reinhardtii* strain TAI114
Propionibacterium freudenreichii	*Kluyveromyces lactis*	*Penicillium roqueforti*	*Inonotus obliquus*	*Chlorella sorokiniana*
Bacillus subtilis	*Candida utilitis*	*Mucor javanicus*		*Haematococcus pluvialis*
Micrococcus lysodeikticus	*Candida lipolytica*	*Rhizopus niveus*		*Crypthecodinium cohnii*
Leuconostoc lactis	*Komagataella phaffii*	*Trichoderma reesei*		*Arthrospira platensis*
Lactobacillus acidophilus		*Neurospora crassa*		

TABLE 9.4
A list of some genetically modified organisms (GMOs) showing generally recognised as safe (GRAS) status

Pullulanase from *Bacillus deramificans* in *Bacillus subtilis*
Myoglobin from cattle in *Pichia pastoris*
α-amylase from *Thermoactinomyces vulgaris* in *Bacillus subtilis* strain AR-651
Subtilisin from *Bacillus clausii* in *Bacillus subtilis*
Phospholipase A2 enzyme from wild pig in *Yarrowia lipolytica*
Lipase from *Aspergillus tubingensis* in *Trichoderma reesei*

food additives. As per the current Good Manufacturing Practices (cGMP), the production of metabolites using pure culture is highly recommended. Submerged fermentation and solid-state fermentation (SSF) are suitable choices for large-scale production of microbial metabolites. The evaluation of whether the producer microorganism is wild-type or recombinant is also a primary concern to achieve GRAS status. The applications of GMOs require strict assessment based on the host cells, the inserted gene, and the transformation process (Table 9.4).

A similar assessment of safe microbial use for large applications in the food and feed industry in the European Union (EU) countries falls under the Qualified Presumption of Safety (QPS) under the authority of the European Food Safety Authority (EFSA). Other nations, such as Canada, Australia, Japan, etc., also have similar guidelines for the safe usage of microorganisms and their metabolites in the food and beverage industry.

SUMMARY

- Natural microorganisms are permanent inhabitants of diverse environments. They are key sources of important metabolites and enzymes with applications in human welfare and environmental benefits.
- GMOs are unnatural organisms that contain modified DNA sequences. Genetic modification in industrially relevant microorganisms can be achieved mainly through artificial mutagenesis and genetic engineering. GMOs are amenable to the fermentation conditions in industrial biotechnology.
- Artificial mutation is a classical method of genetic modification in microbial cells that can be random or site-specific. Mutation can be employed to achieve high titers of metabolites from microorganisms. This is also an integral part of protein engineering.
- Genetic engineering is the deliberate modification of genes using recombinant DNA technology and CRISPR/Cas9 gene editing. Genetic engineering can be achieved through homologous and heterologous expression. *E. coli* and *S. cerevisiae* are the most commonly used host systems for genetic engineering.
- In industrial microbiology, heterologous expression is a process of expressing genes from various microorganisms, plants, and animals into suitable microbial expression systems. This process is the key to producing many plant-based and animal-based products in microorganisms.
- In industrial biotechnology, the safe usage of microorganisms in large-scale industrial processes calls for stringent microbial containment and safety assessment. A GILSP-based guideline provides a comprehensive framework for microbial containment, ensuring safeguards to human health and the environment.
- In industrial microbiology, GRAS is the status for certain biological entities, including microorganisms, for their safe and intended use as food additives in the food and beverages industry, as proposed by the US FDA. The list of GRAS substances includes whole microbial cells or any metabolites derived from microorganisms.

DIFFERENT TYPES OF PRACTICE QUESTIONS

Fill in the blanks

I. .... is a growing *in vivo* method of genetic engineering in eukaryotic organisms.
II. .... expression is a genetic engineering-based technique for expressing foreign genes in host organisms.
III. The first genetically engineered pharmaceutical product launched was ...., produced in ....
IV. *Azotobacter*, *Azospirillum*, and *Rhizobium* are some notable examples of ....
V. Common food additives with a long history of use before......... . do not require technical evidence of safety to be considered GRAS.
VI. In four biosafety levels, ... deals with high-risk microorganisms such as Ebola virus and Marburg virus.

SOLUTION: I. CRISPR/Cas9 gene editing, II. Heterologous, III. human insulin, 1982, IV. biofertilisers, V. 1958, VI. biosafety level 4

Very short answer type of questions

I. Expand GRAS and GILSP.
II. What is the key difference between cloning and expression vectors?
III. Define auxotroph with a suitable example.
IV. Write the names of two chemical mutagens.
V. Write the names of two genetically engineered microorganisms used in the food industry.

Short answer type of questions

I. What are the roles of natural microorganisms in industrial microbiology?
II. Why do genetic modifications in microorganisms have stringent guidelines in industrial biotechnology?
III. Write a short note on GRAS.
IV. Differentiate between homologous and heterologous expressions in genetic engineering with suitable examples.
V. Mutation is a classical method of genetic modification in industrial microbiology. Comment on it.
VI. What are the primary targets in the genetic loci of a microbial genome to control the metabolic pathways in microbial technology?

Long answer type of questions

I. Elaborate on different methods of creating GMOs in industrial biotechnology.

NOTES

1 Bathula SR, Rakhimol A. Global trends in biorisk management. BioRisk. 2017;12:1–23.

2 Bharathkumar N, Sunil A, Meera P, et al. CRISPR/Cas-based modifications for therapeutic applications: a review. Molecular Biotechnology. 2022;64(4):355–72.

SUGGESTED READINGS

Adrio JL, Demain AL. Genetic improvement of processes yielding microbial products. *FEMS Microbiology Reviews*. 2006;30(2):187–214.

Ding W, Zhang Y, Shi S. Development and application of CRISPR/Cas in microbial biotechnology. *Frontiers in Bioengineering and Biotechnology*. 2020;8:711.

Frommer W, Ager B, Archer L, Brunius G, Collins CH, Donikian R, Frontali C, Hamp S, Houwink EH, Küenzi MT, Krämer P. Safe biotechnology: III. Safety precautions for handling microorganisms of different risk classes. *Applied Microbiology and Biotechnology*. 1989;30:541–52.

Kour D, Rana KL, Thakur S, Sharma S, Yadav N, Rastegari AA, Yadav AN, Saxena AK. Disruption of protease genes in microorganisms for the production of heterologous proteins. In Gupta VK (ed.) *New and future developments in microbial biotechnology and bioengineering* 2019 (pp. 35–75). Elsevier.

Pham JV, Yilma MA, Feliz A, Majid MT, Maffetone N, Walker JR, Kim E, Cho HJ, Reynolds JM, Song MC, Park SR. A review of the microbial production of bioactive natural products and biologics. *Frontiers in Microbiology*. 2019;10:1404.

Rosano GL, Ceccarelli EA. Recombinant protein expression in Escherichia coli: advances and challenges. *Frontiers in Microbiology*. 2014;5:172.

Sewalt V, Shanahan D, Gregg L, La Marta J, Carrillo R. The generally recognized as safe (GRAS) process for industrial microbial enzymes. *Industrial Biotechnology.* 2016;12(5):295–302.

Terpe K. Overview of bacterial expression systems for heterologous protein production: from molecular and biochemical fundamentals to commercial systems. *Applied Microbiology and Biotechnology*. 2006;72(2):211–22.

U.S. Food & Drug Administration. Generally Recognized as Safe (GRAS). www.fda.gov/food/food-ingredients-packaging/generally-recognized-safe-gras#:~:text=Under%2021%20CFR%20170.30(c,a%20significant%20number%20of%20consumers

10 Fermentation Conditions

Fermentation is a complex process of product formation involving a specific microorganism and fermentation medium in a controlled environment. The particular outcomes of fermentation reactions are highly dependent on the types of microorganisms involved and fermentation conditions. Most industrial fermentation reactions require an active growth phase of microorganisms to meet the criteria for producing a desired product. Many physical and chemical parameters are strict requirements for the fermentation process. Inoculum size is a critical parameter in microbial fermentation. The fermenter design also plays a very important role in the production of specific metabolites. This chapter provides information on various parameters that optimise fermentation processes.

10.1 MICROBIAL GROWTH PHASES

During the fermentation process, a microorganism passes through several growth phases of its life cycle, known as the microbial growth phase. Microbial growth refers to the process of increasing cell density in the medium. The asexual modes of reproduction, such as binary fission and budding, are the common means for microbial population growth in bacteria and fungi. The concept of microbial growth curve is mainly applicable to bacterial forms. Many fungi also exhibit some growth patterns and can be exploited through growth phases. However, fungal growth can be more complex due to factors like hyphal growth and spore formation. The typical sequential growth stages of diverse microorganisms include lag, logarithmic (or exponential), deceleration, stationary, and death phases (Figure 10.1).

The lag or delay phase occurs immediately after inoculation. It is an initial period of adaptation during which inoculated microbial cells (seed) are acclimated to a new environment without any significant increase in their growth. Cells increase their size and mass and steadily allocate their energy to cell division and metabolic activities. In this phase, metabolic activities are responsible for the synthesis of certain proteins required for the next phase. The phenotypic characteristics of the inoculum, medium composition, and physical conditions significantly influence the length of the lag phase. A productive fermentation desires a short lag phase, which calls for active and young cells in the inoculum. The inoculum size should be big (0.5–5% by volume) for a short lag period in fermentation. If the medium composition and the environmental conditions for seed culture are identical to the production fermentation, a microbial form generally requires less time to adapt to the new environment. However, some complex metabolic pathways require a prolonged lag phase, which can be achieved by augmenting the medium with growth-limiting factors. Low temperatures tend to reduce the metabolic rate of cells, lengthening the lag phase. Also, if the inoculum size is small or has a low fraction of viable cells, the lag phase is generally long. The lag period often increases with inoculum age. When the medium consists of a large amount of inoculum, or the inoculated cells are in the late exponential phase of growth (with uniform population composition), the lag phase of the microorganism is short. The cells have adapted to their new environment and multiply at a constant specific growth rate. When microorganisms consume substrate at a rapid rate, the exponential development of microorganisms can be formulated as

$$dN/dt = \mu N$$

where N is the number of cells (biomass),
t is the time, and
μ is the specific growth rate.

The lag phase is followed by the logarithmic (or log) or exponential phase. It is the second phase of the typical bacterial growth curve, which has three sub-stages, including early, mid and late phases. In the early exponential phase, the inoculum undergoes a period of exponential growth, characterised by rapid cell division and maximal metabolic activity. The cells have adapted to their new environment and multiply at a constant specific growth rate. Towards the mid-exponential phase, the cells divert much of the energy resource into mRNA translation and global protein synthesis.

The exponential phase can be used to calculate the generation time (g), which is the length of time that a bacterial population takes to double in size, because microbial growth is predictable. Healthy and actively growing microbial cells achieve the exponential phase rapidly. There is a predominance of primary metabolism, which produces primary metabolites

DOI: 10.1201/9781032651095-13

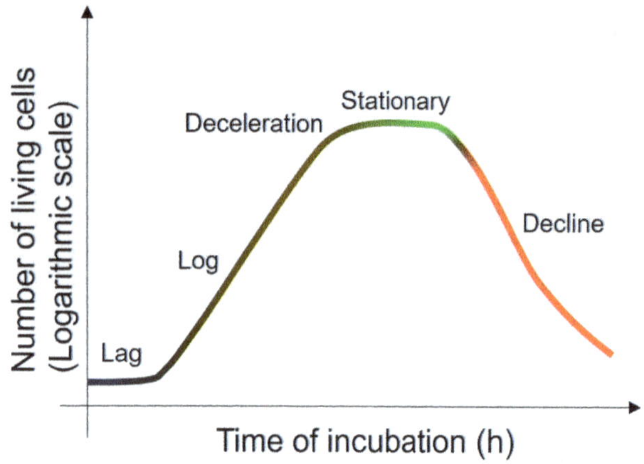

FIGURE 10.1 Microbial growth phases in fermentation. Normal microbial growth curve, showing a sequential pattern of lag, log (logarithmic or exponential), deceleration, stationary, and death/decline phases.

or growth-related products. All cell components grow simultaneously during this phase of balanced growth (a pseudo-steady state). The number of cells at the beginning of the exponential phase and the number of cells after a certain amount of time have a certain relationship that can be represented as

$$N = N_0 \times 2^n$$

where N is the final cell concentration
N_0 is the initial cell concentration, and
n represents the number of generations that took place during the given time frame.

$$\text{Then}, g = \frac{60 \min \times h}{n}$$

For instance, when a medium contains about 100 bacterial cells, and they divide every 30 min, the generation time of microorganisms can be calculated. In 3 h, this bacterium will divide 6 times or n = 6. According to the formula, there will be 6,400 cells ($2^6 \times 100$) after 3 h. Similarly, g for the population is 60 min × 3 h/ 6 = 30 min per generation.

The deceleration phase, which is the late part of the exponential phase, is distinguished by a slow decline in the rate of cell division. Numerous environmental challenges, including food scarcity and endoplasmic reticulum (ER) stress, arrest the cells and encounter their growth as the growth phase is close to achieving maximal cell concentration. In the deceleration phase, microbial growth tends to slow down due to the buildup of harmful growth byproducts, a decrease in space, or the depletion of one or more critical nutrients. The cell's internal organisation and molecular mechanisms undergo modifications (with repression and induction) to boost its chances of surviving in the oncoming harsh environment. The deceleration phase lasts briefly and causes unbalanced growth, characterised by altered cell composition and size.

The prolonged deceleration phase gives rise to the stationary phase. It begins with the exhaustion of essential nutrients or an oligotrophic environment in the external medium. In this phase, there is metabolite deregulation, and the total growth rate is almost zero. This means there is no cell division. Cells might, however, still be metabolically active and produce endospores (in some Gram-positive bacteria) and non-growth-related secondary metabolites (such as antibiotics and hormones). Cellular reserves are now broken down to create new building blocks and energy-producing monomers. This process is called endogenous metabolism. The cells constantly expend their energy (or maintenance energy) to sustain vital metabolic processes, including repairing cellular damage. Production and accumulation of an inhibitory product can also occur in this phase.

A death phase (or decline phase) starts during or even before the stationary phase. Cell lysis occurs, resulting in a reduction of the number of viable cells and biomass concentration. Some cells may still grow on the byproducts of cell lysis, resulting in cryptic growth. When the dead cells are lysed, intracellular nutrients are released into the external medium for use by the still-living organisms. This subfraction of the population survives for an extended period and creates variation in the growth of the original strain, having altered metabolic capabilities.

10.1.1 Diauxic Growth

If a growth medium consists of multiple carbon sources, a microorganism can induce more than one lag phase in the system. For example, when the medium has two carbon sources and one of them is easier to metabolise by the inoculum, the cellular growth occurs in two phases. This type of condition is called diauxic, diphasic, or diauxie growth. A diauxic growth curve of an organism exhibits two growth peaks, which is illustrated in Figure 10.2. Microorganisms have evolved intricate genetic regulatory systems that enable them to respond to changes in their environment and optimise their growth. This might involve upregulating genes for nutrient utilisation while downregulating genes for less essential functions.

In the initial stage, cells preferentially metabolise the sugar (carbon 1), allowing for quicker growth. The cells do not use the second sugar (carbon 2) unless the first one is exhausted. There is frequently a second lag period after the utilisation of carbon 1, during which cells make the enzymes required to metabolise the second sugar. The second sugar is metabolised when the cellular machinery responsible for its metabolism is activated during the lag phase. For instance, *Lactococcus lactis* initially produce enzymes to metabolise glucose when cultivated in the presence of both glucose and maltose. It eventually switches to metabolising maltose when the glucose supply is depleted. Similarly, *Escherichia coli* growing in a medium containing glucose and lactose will initially consume glucose after a normal lag phase. When glucose is

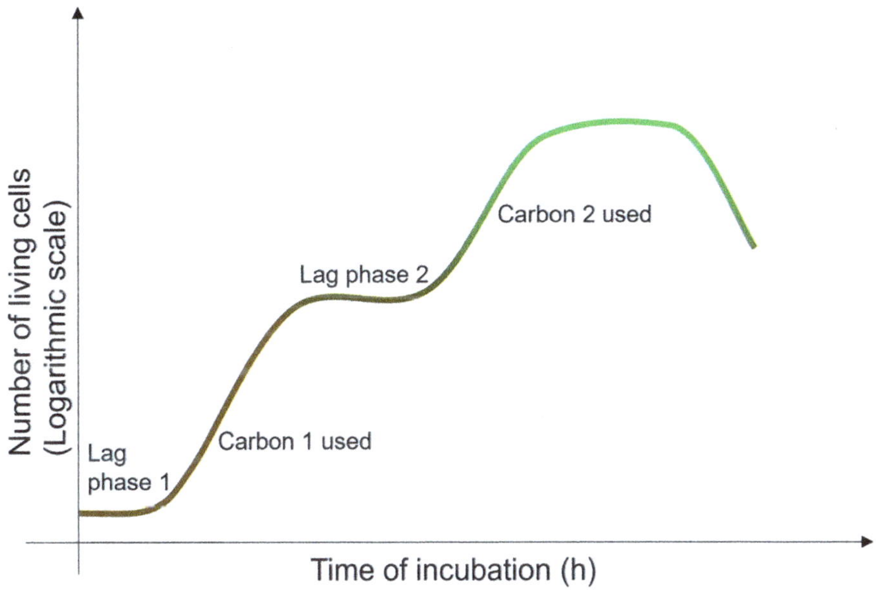

FIGURE 10.2 Diauxic growth curve of a bacterium cultured in two different sugar or carbon sources. When the readily metabolisable carbon 1 is utilised, there is a brief lag period during which the bacterium synthesises the enzymes required to consume carbon 2.

TABLE 10.1
Differences between trophophase and idiophase stages of microbial metabolite production

Parameters	Trophophase	Idiophase
Refers to	The logarithmic/exponential phase of growth	The stationary and deceleration phases of growth
Microbial growth	Rapid	Slow or non-growing
Growth-linked activities	Growth-linked primary metabolites, which are essential for the growth of microbial cells	Non-growth-linked secondary metabolites, which do not have an obvious function in cell metabolism and growth
Product specificity	Non-specific and reported in various microorganisms	Specific to certain microbial species
Examples of products	Proteins, nucleic acids, lipids, and carbohydrates	Antibiotics, glycosides, terpenes, steroids, polyketides, and aromatic metabolites

exhausted, *E. coli* synthesises the β-galactosidase enzyme to utilise lactose after the second large phase. Also, *Clostridium acetobutylicum* exhibits diauxic growth when cultivated on glucose and cellobiose combinations. Growth kinetics of the fermentation process show that there is a few-hour adaptation period for glucose consumption before cellobiose can be used.

10.2 STAGES OF MICROBIAL METABOLITE PRODUCTION

In a fermenter, the production of microbial metabolites usually follows the sequential stages: inoculation (lag phase), exponential growth, stationary phase, and decline phase. However, exponential growth and the stationary phase are more important for metabolite formation.

From an industrial perspective, exponential and stationary growth phases are referred to as trophophase and idiophase, respectively. During these stages of growth, there are significant differences in the behaviour and products of the culture (Table 10.1). While secondary metabolites (idiolites) are mostly synthesised in idiophase, when cell development slows down and stress conditions exist, primary metabolites are generated during the trophophase. For example, the unicellular yeast *Saccharomyces cerevisiae* secretes maximum ethanol during the trophophase. In contrast, the fungus *Penicillium chrysogenum* gives a high titer of penicillin (a secondary metabolite) when the cells enter idiophase.

Trophophase metabolites, including pyruvate, amino acids, fatty acids, nucleotides, and carbohydrates, are vital cellular components for basic cell function and reproduction. Idiolites are unique metabolites that typically have unusual chemical structures. While they are not necessary for the growth of the producing microorganism, they serve survival purposes. Only certain species are capable of producing secondary metabolites. Large macrolide rings, unsaturated bonds of polyacetylenes and polyenes, cyclic peptides composed of standard and modified amino acids, and β-lactam rings are some of the unique chemical linkages found in idiolites. Given the limited specificity of some enzymes involved in secondary metabolism, idiolites are usually generated as members of a distinct chemical family. Mycotoxins, antibiotics, pigments, and pheromones are a few of them.

In batch fermentations with nutritionally rich media, the production of significant amounts of idiolites occurs only after the majority of the cellular growth has taken place. Genes involved in idiolite production are typically not expressed at high growth rates. For this, the growth phase is trophophase, and the production phase is idiophase. Ideally, trophophase and idiophase happen at two different times in batch culture. However, they can overlap. Typical trophophase-idiophase dynamics in many idiolite fermentations take place in complex media that can support fast growth, but in defined media that favour slow growth, the two phases show overlap. Some nutritional factors in a defined medium may be growth-limiting from the onset of fermentation, favouring idiolite production during periods of slow development. The secondary metabolism enzymes are suppressed during growth. Penicillin acyltransferase and the side-chain activating enzyme of penicillin biosynthesis in *Penicillium chrysogenum* are examples of enzymes whose activity only becomes apparent during the idiophase. Similarly, ergot alkaloid synthesising enzymes dimethylallyl-tryptophan synthetase and chanoclavine-1-cyclase only emerge after growth in *Claviceps purpurea*. Also, the presence of certain carbon, nitrogen, and phosphate sources can control the start of idiolite synthesis. For instance, glucose is typically a good carbon source for growth, but it prevents the production of numerous secondary metabolites like penicillin and cephalosporin. Likewise, *Gibberella fujikuroi* produces the antibiotic bikaverin through nitrogen regulation, using glycine as a nitrogen source for fermentation, but the production commences upon exhaustion of glycine. Bikaverin production is also impeded by inorganic phosphate, which does not inhibit growth. Temperature, light, oxygen, and metal ions are additional variables that might affect the commencement of idiolite production.

10.3 PHYSICAL PARAMETERS

Physical parameters for the fermentation process mainly include temperature, agitation, aeration, pressure, and moisture (Figure 10.3). Temperature significantly affects cellular growth, product biosynthesis, and broth physical properties. Due to its relation with enzyme catalytic activities, temperature influences microbial metabolism and fermentation. The optimal temperatures for fermentation processes vary depending on the microorganisms involved and metabolic products. For example, 32–35 °C is ideal for yeast fermentation. In breweries, the fermentation temperature affects beer aroma, with low-temperature fermentation resulting in better taste, superior aroma, high ethanol production, and increased beer productivity. Most bacterial hydrogen (H_2) production has fermentation in the mesophilic range (20–45 °C). Also, the production of biogas during anaerobic fermentation often increases with increasing temperature. The range of temperatures for biogas generation might be either thermophilic (45–65 °C) or mesophilic (30–40 °C). A different insight for temperature requirements in the fermentation process is that the temperature optima for cellular growth and metabolite production can be different. For example, the optimal temperature for *Streptomyces kanasenisi* growth is 25 °C, while for glycoprotein production, it is 30 °C. High temperature is beneficial in many processes because of moderate cooling requirements and a low chance of microbial contamination from the outer environment. Production of organic acids, solvents, and other chemicals through microbial fermentation is a high-temperature operation. Thermotolerant microorganisms that can thrive at elevated temperatures are suitable for elevated-temperature bioreactions. *Bacillus subtilis*, *Geobacillus* species and *Thermomyces lanuginosus* are notable sources of thermostable enzymes, such as amylases and proteases. Low-temperature fermentation is mainly desirable in the food and beverage industry and the pharmaceutical industry. The best possible taste, flavour, and texture of bread develop from fermentation at 10 °C. The development of fat-free yoghurt with antioxidant activity requires a temperature of 22 °C, while smooth and clean lagers produced at 9–15 °C. The ideal fermentation temperature for sauerkraut is between 16–21°C. Wines with higher esters and lower terpenes are produced at 25 °C. Apart from enhancing flavour components and the accumulation of metabolic chemicals, low-temperature fermentation also inhibits the growth of other microorganisms in the fermentation medium. Fermentation at low temperatures relies on cryotolerant microorganisms, which are adapted to survive and grow at low temperatures. Examples include *Saccharomyces* species (e.g., *S. uvarum*, *S. kudriavzevii*, and *S. eubayanus*), *Lactobacillus* (e.g., *L. plantarum* and *L. sakei*), and *Leuconostoc*.

In fermentation processes, agitation plays a role in mixing the various components in the fermentation medium, shearing microbial cells, and product degradation. Through constant mixing, agitation not only enhances the flow of mass and oxygen across the various phases but also maintains a uniform condition of the medium's chemical and physical parameters. The very low agitation speed increases the viscosity of the fermentation broth, which decreases the mass distribution efficiency. However, extremely high agitation can cause shear stresses for microorganisms. This could cause morphological variations, alter growth and metabolite synthesis, and even damage cell structures. An agitation speed that is too high also results in heterogeneous mixing and increases power consumption. For example, the optimal agitation speed for *S. kanasenisi* for growth and the production of glycoprotein is 200 rpm. However, the speed of 300 rpm causes damage to the cell's structure and decreases glycoprotein production. Differently, 150–200 rpm is the optimal agitation rate for the growth of yeast cells, and 200 rpm gives the maximum ethanol yield. The agitation rate is also crucial for regulating the nutrients from the fermentation broth into cells and the removal of ethanol from cells into the fermentation broth. The rotational speed in ethanol fermentation also assists in consuming sugar and reducing the impact of ethanol on cells.

However, an excessive agitation rate inhibits cell metabolism, making it unsuitable for smooth ethanol production.

Aeration in fermentation provides the oxygen for cell growth. Also, aeration helps mix the fermentation broth, particularly in situations with low mechanical agitation rates. Oxygen supply in the fermentation medium is a strict requirement for aerobic microorganisms. However, an excessive supply of oxygen in the fermentation medium can be harmful to industrial fermentation. The high amount of aeration can reduce the volume of fermentation broth due to excessive evaporation and foaming. The specifications of the fermentation process determine the aeration rates. For example, aeration rates between 0.5 and 2.0 vessel volumes per minute (vvm) promote *S. kanasenisi* growth and produce glycoproteins in a linear pattern. The mushroom *Paecilomyces tenuipes* exhibits an enhanced density of mycelial biomass and production of exopolysaccharides at an aeration rate of 3.5 vvm. The production of α-amylase by *Bacillus subtilis* increases when the aeration rate is increased from 0.25 to 0.75 vvm. However, at 1.0 vvm, *Bacillus amyloliquefaciens* produces the most α-amylase. Aeration rates for yeast-based ethanol production can vary from 0.13 to 2.5 vvm. Additionally, *Saccharomyces cerevisiae*'s ergosterol synthesis is promoted by an ideal aeration rate of 0.31 vvm, which enhances ethanol tolerance.

Fermentative techniques can be carried out under different pressures to produce various commercial products with different traits. Fermentation under pressure necessitates a vessel for safe pressurisation. In ethanol production, pressure fermentation amplifies natural carbonation in the beer, increasing the CO_2 solubility in the liquid and reducing the need for additional carbonation. It is a method that uses

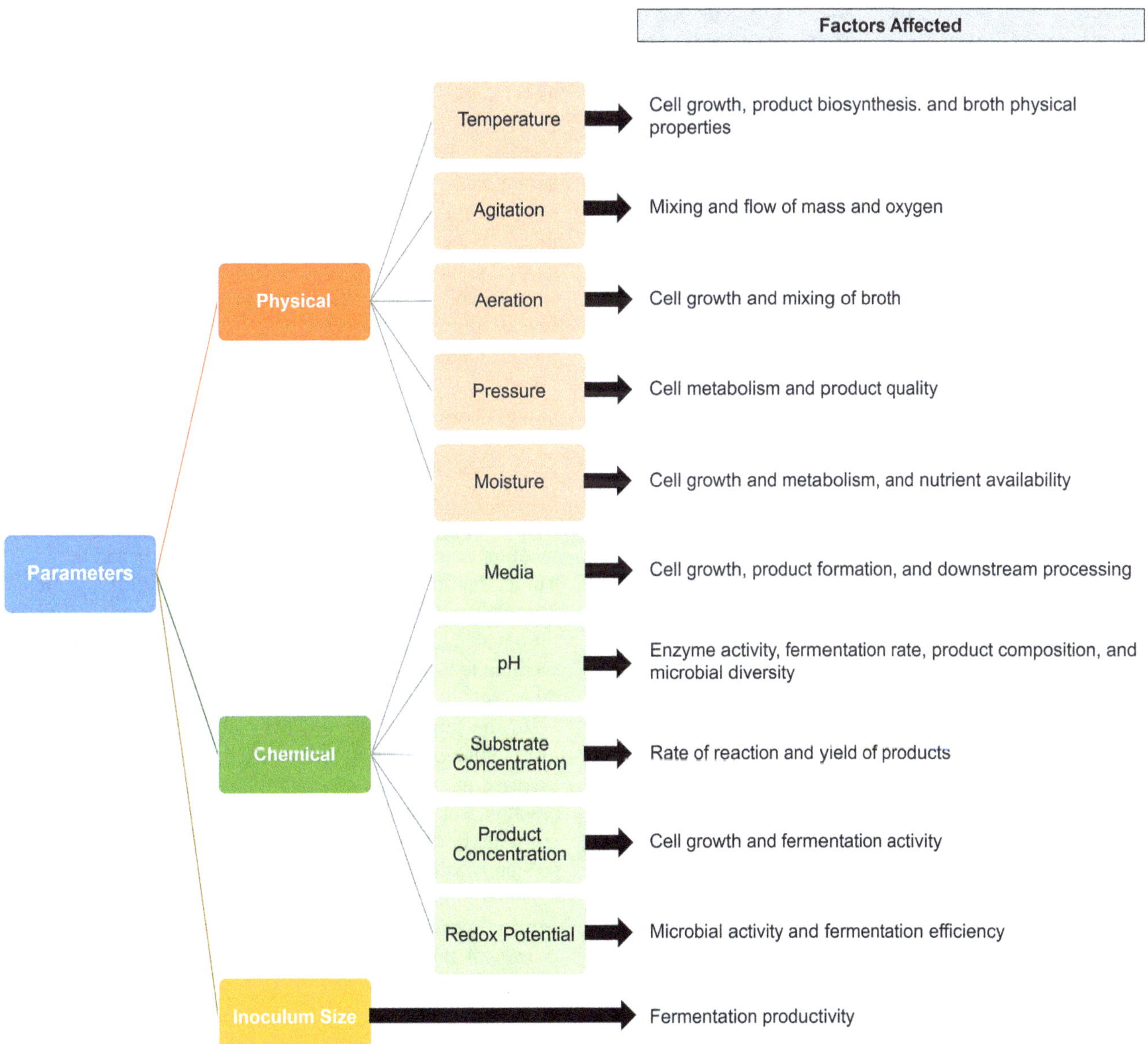

FIGURE 10.3 Roles of different parameters (including physical and chemical) in the fermentation process.

sealed vessels to ferment beer under elevated pressure, unlike traditional open fermentation methods that operate at atmospheric pressure. In acetone-butanol fermentation, pressure (1 bar) is critical for solvent production. When *Clostridium thermocellum* is grown at sub-lethal pressures (7–17 MPa), its fermentation products are diverted from acetate to ethanol. Some microorganisms can withstand high pressures, but excessive load can inhibit their growth. For example, the atmospheric pressure of around 7 atm inhibits yeast metabolism, and 30 atm kills all the yeast cells. Also, high pressure disrupts lactic acid fermentation, inhibiting microbial metabolism and growth.

Moisture plays an important role in maintaining the cellular volume and structural integrity of proteins and preserving the permeability of the plasma membrane. It also helps dissolve and disperse the nutrients. Up to 80% of the heat produced in fermentation could be eliminated by water evaporation. Solid-state fermentation (SSF) is highly dependent on moisture content because the process involves the absence or near absence of free water. In this process, the solid substrate is moistened only sufficiently to enable the growth and metabolism of microorganisms. SSF typically has a moisture content of 12–70%, with a typical amount of around 60%. Water loss due to evaporation and microbial consumption in SSF causes uneven distribution of the substrate and oxygen concentrations and creates water stress. For anaerobic fermentation, the moisture content is typically 85 w%. The stability of the anaerobic fermentation further depends on the organic loading rate (OLR) and hydraulic retention time (HRT), with high OLR leading to unbalanced fermentation and short HRT inhibiting the process. A low OLR (typically around 10.1 g chemical oxygen demand (COD)/l per day) and long HRT (about 15 days) are the optimal strategies for achieving constant and maximum biogas yields.

10.4 CHEMICAL PARAMETERS

Chemical parameters include pH, nutrient media, substrate and product concentrations, and redox potential (Figure 10.3). pH can influence enzyme activity, fermentation rate, final product composition, and microbial population diversity. This can also affect several microbiological characteristics. A pH range of 4–8 is suitable for yeast proliferation and fermentation, with 5.5 being the ideal. On the other hand, Lactic acid bacteria (LAB) typically prefer low pH environments, with an optimum pH range of 5–7 for fermentation. For example, *Lactobacillus helveticus* and *L. delbrueckii* subsp. *Bulgaricus* are popular dairy starters that grow best in a pH range of 5.5–5.8. However, LAB such as *Lactococcus lactis* subsp. *cremoris* and *Lactococcus lactis* subsp. *Lactis* requires an external pH range of 6.3–6.9. The variation in pH can also influence the type of product formed during fermentation. For instance, when pH is maintained at 5.5, *Clostridium pasteurianum* develops ethanol; however, a pH of 8.0 contributes to its three times the yield of biomass. Also, when *Clostridium tyrobutyricum* ferments xylose, the main product at pH 6 is butyric acid, but at pH 5, it promotes the production of lactate and acetate. The pH further has an important effect on the rate at which microorganisms grow in an anaerobic fermentative system. Anaerobic fermentation typically requires a pH between 6.8 and 7.2. At pH levels above 8.0 and below 6.0, for instance, biogas-producing methanogenic archaea develop at a significantly slower rate.

A different chemical parameter for the fermentation process is the fermentation medium, which is a basic requirement for the productivity and economic success of a bioprocess. It meets the nutritional needs of the microorganism and provides osmotic balance for the growth conditions. The medium formulation is a laborious and expensive process that focuses on the optimal chemical and nutritional environment to support the growth of microbial forms. Fermentation medium can significantly affect microbial growth, product formation, and downstream processing. The specific nutritional requirements of the microorganisms in industrial fermentation processes are just as diverse and complex as the microorganisms themselves. Fermentation media, whether chemically defined or undefined, can enumerate numerous microorganisms. The different types of fermentation media and their components are elaborated in Chapter 22. The fermentation media often consists of carbohydrates (such as sucrose, glucose, fructose, maltose, and maltotriose), nitrogen compounds (such as amino acids, peptides, and ammonium salts), vitamins (such as biotin, thiamine, pantothenic acid, nicotinic acid, riboflavin, inositol, pyridoxine, pyridoxal, and pyridoxamine) and minerals (such as phosphate, potassium, magnesium, calcium, sulphur, and zinc), buffer, and antifoam. Yeast can grow on a variety of substrates for fermentation and proliferation. These include cereal starches, sucrose-rich plants like sugarcane juice or molasses, fruits, etc. LAB have particular dietary requirements, typically a carbon supply like glucose, maltose, raffinose, or lactose, peptides and amino acids as a nitrogen source, tweens as a type of polysorbate emulsifiers, minerals (like iron (Fe), magnesium (Mg), calcium (Ca), potassium (K), or manganese (Mn)), vitamins (such as thiamine, biotin, nicotinic acid, riboflavin, or pantothenic acid), and fatty acids. The carbon-to-nitrogen mass ratio (C: N ratio) significantly impacts the performance of anaerobic fermentation. The C: N ratio with a range of 20:1-30:1 provides a balanced nutritional medium to anaerobic bacteria and is the optimal condition for their growth and stability.

In the fermentation process, substrate concentration can also influence the rate of reaction and product yield. The increase in substrate concentrations can increase the rate of reaction, but only to a certain limit. An excessive amount of substrate in the medium can inhibit the fermentation process (which is called substrate inhibition). For example, the initial concentration of substrates (like starch, sugar, cellulose, and hemicellulose) directly influences the substrate hydrolysis and growth of microbial cells. However, high concentrations of substrate make the slurry more viscous, making it difficult to mix and pump. Increased viscosity impedes the appropriate dispersion of enzymes and substrates, leading to incomplete hydrolysis. The high sugar concentrations in the medium create severe osmotic stress, affecting the rate of fermentation and microbial growth. It also lengthens the fermentation process and increases the product cost. For example, H_2 production through dark fermentation requires substrate concentration in a range of 6–100 g/l. However, an excess amount of substrate impedes bacterial activity in several ways, including elevated H_2 partial pressure, decreased intracellular pH, and the build-up of volatile fatty acids (VFAs).

In a different situation, the extracellular byproducts can have positive or negative effects on the fermentation process. Some byproducts can act as nutrients and improve microbial growth. However, certain byproducts in the medium cause product inhibition, interrupting microbial growth and fermentation activity. The high accumulation of toxic byproducts can exacerbate product inhibition. The loss of membrane integrity, proton motive force dissipation from partially dissociated product intake, inhibition of an enzyme, and the high energy demand are other reasons for the fermentation loss. In these situations, the productivity or yield of the fermentation process can drop or cease completely. A high concentration of product formation while maintaining high volume-specific productivity is a key challenge in the fermentation process. Yeasts that ferment glucose into ethanol represent a good example of noncompetitive product inhibition, where ethanol acts as an inhibitor at concentrations above 5%. High ethanol concentrations result in instability due to the destruction of the microorganism's lipid bilayer membrane and denaturation of the protein enzyme. In a different example, microbial cellulolytic hydrolysis of lignocellulosic biomass on an industrial scale yields high concentrations of cellobiose. The buildup of this byproduct inhibits microbial cellulase hydrolysis, which eventually reduces the glucose yields.

Redox potential (E_h), also referred to as oxidation-reduction or oxidoreduction potential (ORP), is a measure of the environmental capacity for oxidation or reduction. High E_h indicates high concentrations of electron-accepting species in the fermentation medium, which means the environment is more likely to oxidise substances. Microorganisms may thrive at low or high E_h and take responsibility for the oxidation or reduction of substances in the medium. A high concentration of oxygen is an indicator of high E_h, encouraging the growth of aerobic microorganisms by oxidising the organic matter, while a low oxygen concentration sets the stage for low E_h, which is ideal for the proliferation of anaerobes using different nutrients. In fermentation technology, the measurement of E_h provides insight into the metabolic activity of microorganisms, which has a direct role in product formation. E_h influences the choice between different metabolic routes, such as oxidative or reductive pathways, leading to the formation of distinct products. Variation in E_h can alter intracellular metabolic flow by altering enzyme activity. This can also affect the ratio of oxidised to reduced forms of various coenzymes inside the cell. Particularly, metal-containing enzymes are highly sensitive to changes in E_h. The presence of redox molecules, enzyme activity, industrial processes like heating or electroreduction, and treatment with gases (like nitrogen and hydrogen) can modify E_h. The E_h has key roles in manufacturing biofuels, the treatment of wastewater, the production of milk and cheese, and the assessment of food safety. Redox potential-controlled approaches are also commonly used in producing diverse microbial metabolites such as hydrogen, pyruvate, 1,3-propanediol, butanol, and 2,3-butanediol. The transformation of glycerol into 1,3-propanediol by *Klebsiella pneumoniae* requires E_h around -190 mV. In a different setup, a redox potential control at around -290 mV significantly increases butanol production in *Clostridium acetobutylicum*. The supplement of nicotinic acid increases the NAD(P)H levels, which enhances butanol production by *Clostridium sporogenes* during the fermentation reaction. The E_h, a crucial indicator of anaerobic fermentation, is essential for the growth of methanogenic archaea, which requires a low E_h ranging from -200 to -400 mV. Low E_h promotes high hydrogen generation capability in the anaerobic fermentation of *Enterobacter aerogenes* and accelerates NAD(P)H-dependent membrane hydrogenase activity. The E_h in the fermentation broth can be detected using an electrode. A new type of device, a nanosensor, has been developed that can enter individual cells and monitor intracellular E_h.

10.5 INOCULUM SIZE

The population of microbial cells added to the fermentation medium to start the fermentation process is called an inoculum. The concentration or size of the inoculum is a critical parameter in microbial fermentation. An optimum inoculum size is a strict requirement in industrial fermentation to accelerate the fermentation process and achieve high productivity. This also contributes to the quality of products and outcompetes potential contaminants. The size of the inoculum for any fermentation reaction can be varied. The inoculum size can influence the rate of fermentation, microbial growth, and nutritional value of the finished product. Also, the inoculum size can reduce the lag phase and accelerate fermentation rates at high cell densities. A low inoculum concentration results in prolonged fermentation cycles, reduced productivity, and a higher risk of contamination. On the contrary, the augmentation of an excessive amount of inoculum in the media requires high capital and operating expenditures, but without offering any additional benefits to the fermentation process. Also, when the inoculum size is increased, the growth rate is relatively constant, but the fermentation medium rapidly achieves nutritional limitation. For instance, a 5% inoculation induces yeasts to produce the maximum ethanol (Figure 10.4). The production of ethanol increases with increasing the concentrations of inoculum size until 5%, but further increases in the inoculum size doesn't enhance the fermentation process as it leads to substrate exhaustion.

The preparation of inoculum can be a multi-step process to obtain a healthy inoculum. The inoculum development process for different microorganisms can vary due to differences in their growth rates and metabolic requirements. Also, the media constituents for inoculum development can be different from the actual fermentation media because the main goal is to support a high growth rate rather than the production of metabolites. Many environmental factors, such

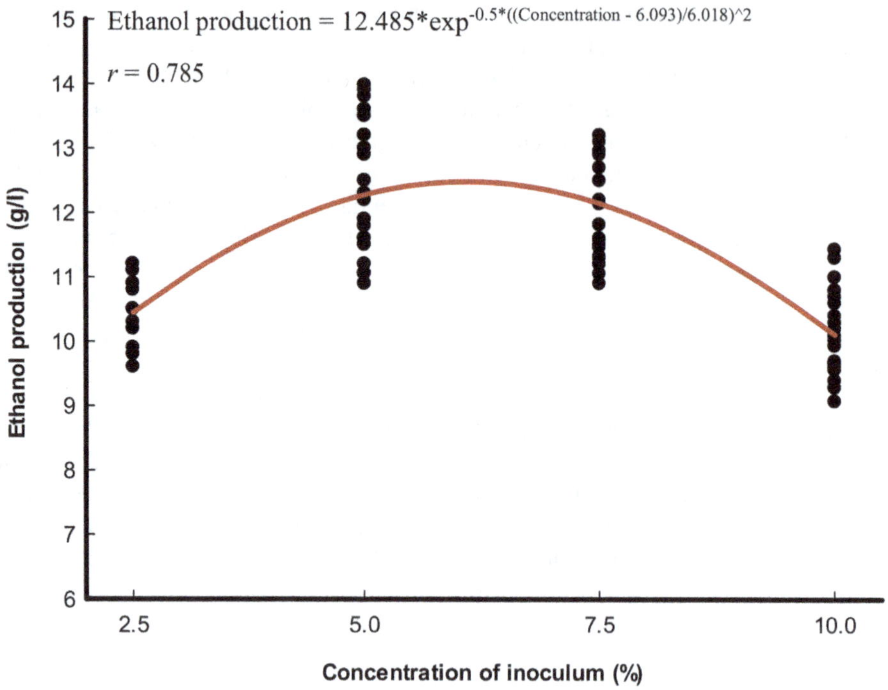

FIGURE 10.4 Effect of inoculum concentration on ethanol fermentation. The study examines the correlation between inoculum concentration and ethanol production by yeasts using pretreated starchy biowaste as feedstock in batch culture under specific conditions. The 5% concentration of yeast inoculum was found to be optimal for ethanol production. In the graph, black dots represent ethanol production, while a red line depicts the correlation trend among values of ethanol production at different concentrations of inoculum. (Image adapted from Hashem et al. (2021).[1])

as pH, temperature, and oxygen level, differ for inoculation according to the demand for increasing biomass and metabolic activities.

In the inoculum development process, a low aliquot from the stock or slant culture is transferred to a suitable liquid medium, followed by incubation for a certain period. The microorganisms rejuvenated from an inactive culture can be reinoculated into fresh media to achieve high biomass. In certain situations, inoculum development is carried out in a series of phases. The concentration used at each phase of inoculum development is usually between 0.5 and 5% of the medium volume. Using this process, a 20-200-fold increase in the inoculum volume can be achieved from its initial size. Ultimately, in the production stage, an active growth stage is transferred to the main fermenter for the production of the desired metabolites.

10.6 FERMENTER DESIGNS AND OPERATION

Fermenter designs have a great influence on the controlled operation of the fermentation process. A fermenter can be a highly sophisticated instrument that maintains an aseptic environment for product development. Fermenters meet the specific requirements of production organisms, the optimal fermentation conditions, product types, and production scales. The fermenters include several key features such as simplicity, sterilisation ability, flexibility, long-term stability, easy monitoring, control techniques, scale-up, upstream and downstream process compatibility, and antifoaming steps. They monitor several factors such as aeration, agitation, fluid velocity, cooling rate or heating intensity, feeding rate of nutrients and base or acid, substrate concentrations, dissolved oxygen and products. These parameters in the fermenter vessel provide a regulated environment and suitable conditions for the growth of different microorganisms and the development of various products. A fermenter is further designed to extract waste biomass and products from cultured microorganisms with drain or vent arrangements.

The fermentation processes can be conducted in fermenters of various sizes, from laboratory to industrial scale, for a range of products. Fermentation in fermenters can be operated in batch, fed-batch, and continuous configurations. The reactor design, operation, and the use of sophisticated modelling tools greatly improve the reaction rates in fermenters, enhancing productivity and lowering costs. Unit 2 provides a comprehensive account of the fermentation process and fermenter design.

10.7 MIXED CULTURE FERMENTATION (MCF)

Industrial fermentation mainly relies on pure microbial cultures to produce specific products, necessitating expensive, sterile production conditions and high-quality raw materials (Figure 10.5). A different scheme of beneficial process in

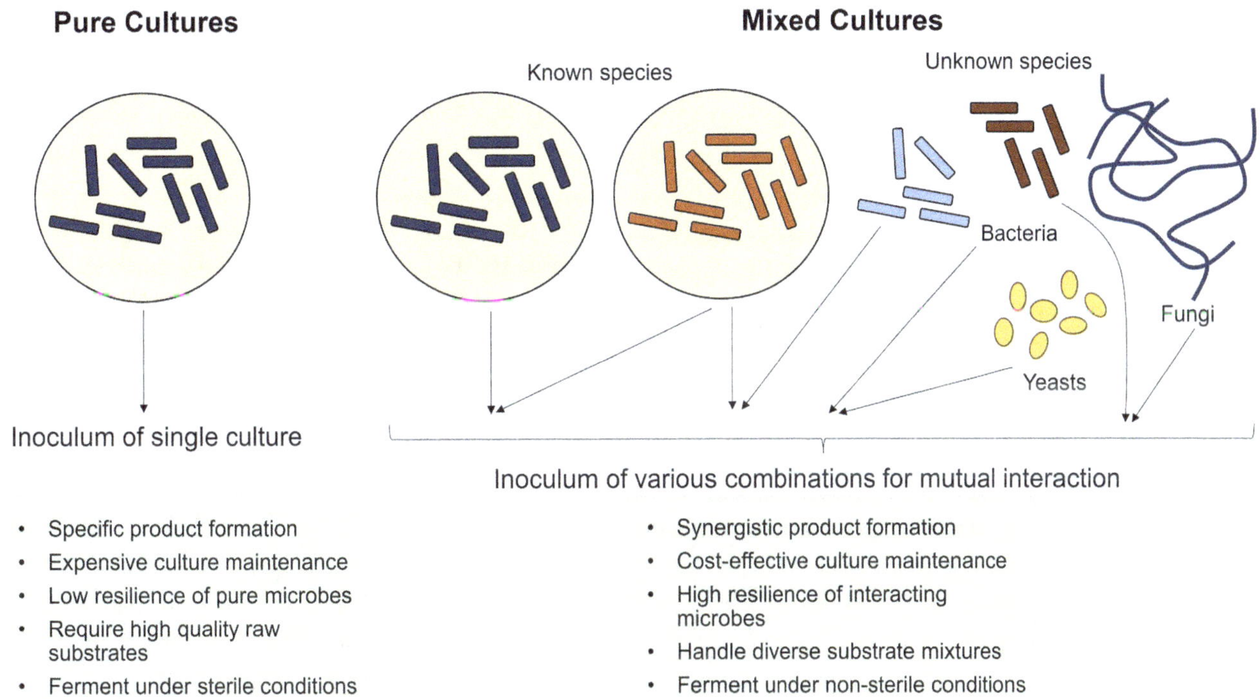

FIGURE 10.5 The differences between pure culture fermentation and mixed culture fermentation (MCF).

industrial microbiology is mixed culture fermentation (MCF), which utilises more than one type of organism that can act synergistically. The MCF technology has many advantages and applications. Mixed cultures do not need expensive culture maintenance because of their high resilience to disturbances and mutual interaction. The consortium cultures can handle diverse substrate mixtures and compositions. They have the potential to increase fermenter loading rates and facilitate continuous fermentation. High-quality raw materials can be replaced with even low-quality raw materials in order to reduce the cost. It can involve some pregrown cultures in pure forms, followed by their mixing for large-scale fermentation. The operation can also be run using open mixed cultures involving indigenous microorganisms under non-sterile conditions. This condition facilitates the fermentation process without risk of strain degeneration.

The microorganisms to be used in the inoculum for MCF can be composed of known or unknown species. A mixed culture of known species produces predictable outcomes and can be effectively employed in the food industry and pharmaceuticals. On the other hand, mixed cultures of unknown species develop varying outcomes. This type of approach is useful for environmental applications such as waste management and bioenergy sectors. Microorganisms in mixed culture can belong to a single microbial group (including different bacteria), a mixture of fungi and bacteria, or other types of combinations (Figure 10.5). For example, *Streptococcus thermophilus* and *Lactobacillus bulgaricus* act synergistically to produce yoghurt in milk fermentation. In this process, the former species lowers the pH of the milk, providing favourable conditions for the growth of another species for lactic acid production. Microbial consortia have roles in altering the flavour and texture of foodstuffs. MCF is widely used in the traditional processing of meat and dairy. This process also contributes to food preservation through natural fermentation. The consortia of several beneficial microorganisms can be used to produce probiotics for health benefits. MCF technology also contributes to the production and degradation of various chemicals in the biomanufacturing of various goods and bioremediation.

The bioenergy sectors foresee sustainability in mixed culture technology. Mixed-type cultures grow on a variety of non-sterile feedstocks with varying compositions for their effective utilisation and transformation into useful substances. For example, a microbial consortium that utilises starch and sugar in addition to cellulose is ideal when processing waste agricultural biomass. This process combines cellulolytic fungi and yeast to utilise starch and cellulose, resulting in a more productive system. In the agriculture industry, the augmentation of mixed culture on land enhances soil fertility and plant growth by promoting nutrient cycling, suppressing pathogens, and enhancing plant-microbe interactions. Commercial biofertilisers are composed of more than one type of microorganism, such as cultures of nitrogen-fixing and phosphate-solubilising bacteria. In a different setting, anaerobic consortia facilitate the effective conversion of organic waste into biogas, a renewable energy source, through the process of methanogenesis. MCF has tremendous scope in the bioprocessing of waste materials. The process of bioleaching depends on mixed cultures for the recovery of metals. Wastewater treatment through biological processes requires diverse native species that cooperate for the effective removal of contaminants. In a different application, the native microbial communities in the wastewater represent the mixed cultures, which convert organic waste into electricity in the microbial fuel cell (MFC) technology.

SUMMARY

- Industrial fermentation requires the active growth of microorganisms and industrial substrates to produce desired products. Industries prefer the asexual mode of reproduction to achieve fast multiplication of the cells. This employs a sequential growth phase including lag, logarithmic, deceleration, stationary, and death stages.
- Diauxic growth is a two-phase growth pattern of microorganisms observed in the presence of more than one carbon source. When the more easily usable carbon source is depleted, the cells modify their metabolic processes to utilise the second carbon source.
- Physical parameters significantly impact the effectiveness and outcome of fermentation. Temperature affects cell growth, product biosynthesis, and broth properties. Agitation and aeration provide mixing and oxygen, respectively. Even pressure can enhance the taste, aroma, and quality of certain products (e.g., beer) during fermentation.
- Chemical parameters, including nutrient media, substrate concentration, pH, product concentration, and redox potential, are crucial for effectively managing and enhancing the fermentation process. These parameters significantly affect growth rate, enzyme activity, fermentation rate, and final product composition.
- The inoculum is a population of cells added to a fermentation medium. It is prepared in a multi-step process before the actual fermentation process starts. The inoculum size significantly affects culture development and yield, typically ranging from 0.5% to 5% of the medium volume.
- Mixed culture fermentation involves more than one organism as an inoculum for the product formation. This inoculum may consist of known or unknown species. It is commonly used in traditional meat and dairy industries, food fermentations, bioenergy production, wastewater treatment, etc.

DIFFERENT TYPES OF PRACTICE QUESTIONS

Fill in the blanks

I. The lag phase of fermentation is followed by the …………………….. phase, during which the cell growth rate increases.
II. At a stationary phase of bacterial growth, the total growth rate is ……………………..
III. If a fermentation growth medium contains two carbon sources, the bacterium will show …………………….. growth curve.
IV. The population of cells or microorganisms added to the fermentation medium to start the fermentation process is known as ……………………………
V. In the mixed culture fermentation to make yoghurt, ……………………… and ………………………… bacteria are used to ferment milk.

SOLUTION: I. logarithmic or exponential, II. zero, III. diauxic, IV. inoculum, V. *Streptococcus thermophilus, Lactobacillus bulgaricus*.

Very short answer type of questions

I. What is the lag phase during fermentation?
II. Name any two physical parameters vital for the fermentation conditions.
III. Define pressure fermentation.
IV. What is product inhibition of fermentation?
V. What is the standard inoculum size range for fermentation?

Short answer type of questions

I. Temperature and pH significantly affect microbial fermentation. Explain.
II. How does the fermentation medium play a crucial role in the productivity of a bioprocess?
III. The mixed culture fermentation technology is an appealing alternative to traditional pure-culture-based technology. Elaborate.
IV. Why is a measurement of redox potential vital to fermentation efficiency and product quality?

Long answer type of questions

I. What is the microbial growth phase? What are the different phases in fermentation? Draw a microbial growth curve. Explain diauxic growth in microorganisms.
II. What is the fermentation condition? Explain various physical or chemical parameters that influence the effectiveness of fermentation.

NOTE

1 Hashem M, Alamri SA, Asseri TA, et al.. On the optimization of fermentation conditions for enhanced bioethanol yields from starchy biowaste via yeast co-cultures. *Sustainability*. 2021;13(4):1890.

SUGGESTED READINGS

Atasoy M, Eyice O, Schnürer A, Cetecioglu Z. Volatile fatty acids production via mixed culture fermentation: revealing the link between pH, inoculum type and bacterial composition. *Bioresource Technology*. 2019;292:121889.

Brown DE, Zainudeen MA. Effect of inoculum size on the aeration pattern of batch cultures of a fungal microorganism. *Biotechnology and Bioengineering*. 1978;20(7):1045–61.

Brückner R, Titgemeyer F. Carbon catabolite repression in bacteria: choice of the carbon source and autoregulatory limitation of sugar utilization. *FEMS Microbiology Letters*. 2002;209(2):141–8.

Deesuth O, Laopaiboon P, Laopaiboon L. High ethanol production under optimal aeration conditions and yeast composition in a very high gravity fermentation from sweet sorghum juice by *Saccharomyces cerevisiae*. *Industrial Crops and Products*. 2016;92:263–70.

Demain AL. Regulation of secondary metabolism in fungi. *Pure and Applied Chemistry*. 1986;58(2):219–26.

Doremus MG, Linden JC, Moreira AR. Agitation and pressure effects on acetone-butanol fermentation. *Biotechnology and Bioengineering*. 1985;27(6):852–60.

Elbeshbishy E, Dhar BR, Nakhla G, Lee HS. A critical review on inhibition of dark biohydrogen fermentation. *Renewable and Sustainable Energy Reviews*. 2017;79:656–68.

Gibbons WR, Westby CA. Effects of inoculum size on solid-phase fermentation of fodder beets for fuel ethanol production. *Applied and Environmental Microbiology*. 1986;52(4): 960–2.

Harrison DE. Mixed cultures in industrial fermentation processes. In Perlman D (ed.) *Advances in applied microbiology* 1978 (Vol. 24, pp. 129–164). Academic Press.

Hutkins RW, Nannen NL. pH homeostasis in lactic acid bacteria. *Journal of Dairy Science*. 1993;76:2354–65.

Kucharczyk K, Tuszyński T. The effect of temperature on fermentation and beer volatiles at an industrial scale. *Journal of the Institute of Brewing*. 2018;124(3):230–5.

Kunkee RE, Ough CS. Multiplication and fermentation of *Saccharomyces cerevisiae* under carbon dioxide pressure in wine. *Applied Microbiology*. 1966;14(4):643–8.

Link H, Weuster-Botz D. Upstream Processing/Medium Formulation and Development. In Young MM (ed.) *Comprehensive biotechnology* 2011. Elsevier.

Liu CG, Qin JC, Lin YH. Fermentation and redox potential. In Jozala AF (ed.) *Fermentation Processes*. IntechOpen, 2017;565.

Peters, D. Raw materials. In Ulber R, Sell D (eds). *White biotechnology. Advances in biochemical engineering/biotechnology* 2006 (vol. 105). Springer.

Roussel C, Ebel B, Munier E, Michelon D, Martin-Dejardin F, Beuvier E, De Coninck J, Gaudu P, Cachon R. Green strategies to control redox potential in the fermented food industry. *Food Research International*. 2022;156:111154.

Temudo MF, Muyzer G, Kleerebezem R, van Loosdrecht MC. Diversity of microbial communities in open mixed culture fermentations: impact of the pH and carbon source. *Applied Microbiology and Biotechnology*. 2008;80:1121–30.

Xu CP, Yun JW. Influence of aeration on the production and the quality of the exopolysaccharides from *Paecilomyces tenuipes* C240 in a stirred-tank fermenter. *Enzyme and Microbial Technology*. 2004;35(1):33–9.

Yeboah PJ, Ibrahim SA, Krastonov A. A review of fermentation and the nutritional requirements for effective growth media for lactic acid bacteria. *Food Science and Applied Biotechnology*. 2023;6(2):215–40.

Zhou Y, Han LR, He HW, Sang B, Yu DL, Feng JT, Zhang X. Effects of agitation, aeration and temperature on production of a novel glycoprotein GP-1 by *Streptomyces kanasenisi* ZX01 and scale-up based on volumetric oxygen transfer coefficient. *Molecules*. 2018;23(1):125.

11 Downstream Processing

11.1 AN INTRODUCTION TO DOWNSTREAM PROCESSING

Downstream processing is a multistep biochemical engineering practice for separating, purifying, concentrating, and recovering biological products in industrial biotechnology. This involves a series of unit operations that extract low-concentration products from the large volume of fermentation broth. The primary goal of the entire setup is to maximise the yield of products cost-effectively. The recovery of different products necessitates different combinations of unit actions. The properties of the products and media play a key role in choosing appropriate techniques. Different engagements related to media composition, types of fermentation (submerged/solid-state), involved biological units (enzymes/microbial cells), location (intracellular/extracellular), attachment on the solid support (immobilised/mobilised), and environmental conditions (pH and temperature) also influence the assignment of specific procedures for downstream processing. Fermentation products can be either the microbial cell itself, the intracellular content of microbial cells or the extracellular fluid. Downstream processing for any of these products can involve multiple steps using various techniques. Using different procedures, the same products may be obtained in either high quantity with low quality, or low quantity with high quality. The products required in high purity always involve additional steps and generally employ sophisticated techniques. The recovery of fragile products is even more complicated. Particularly, the pharmaceutical industries have strict regulatory compliance towards the purity and quality of its products. A large number of industrial activities call for specialised workforces to manage manual intervention. In a series of operations, the precision of each step is important for the subsequent level of treatment.

The expenses in downstream processing make a sizable contribution to the overall cost of a biotechnology product. Depending on the techniques and number of steps involved, downstream processing can raise the total cost of the products by 5–90%. The entire setup of downstream processing is energy-consuming, labour-intensive, and cumbersome. Its workflow depends on mechanical processes and requires heating and cooling during the operation. This expands the overall operational cost and also raises environmental concerns due to high energy demand. Renewable energy sources can help lessen their impact on the environment.

Monitoring warrants regular assessment of the processing at almost every step of operation. It contributes to the early detection of deviations or issues in the bioprocess. Any variation at one step can have a considerable impact on the subsequent steps and the quality of the final product. If impurities are not eliminated at the assigned step, the recovery of the target compound with high purity becomes very challenging. In such a situation, some additional steps may be involved to recover the desired parameters. Differently, the solution is repurposed or discarded. Failing to meet the quality assurance standards of the product leads to significant losses of resources, energy, and money. Therefore, the entire production procedure mandates critical analysis of various parameters to ensure the quality, quantity, and purity of the final products.

The process of fermentation and downstream processing in the food and beverage industry and the pharmaceutical industry has strict safety compliance. Many government and independent organisations have issued guidelines for the downstream regulations. For example, the Food and Drug Administration (FDA), United States of America (USA), is a regulatory authority for drugs, medical devices, and food additives. The European Medicines Agency (EMA) takes responsibility for the safety and efficacy of medicines in the European Union. The Genetic Engineering Appraisal Committee (GEAC) is a governmental body in India that looks for activities related to the large-scale use of hazardous microorganisms and the release of genetically engineered organisms and products into the environment.

Many important downstream procedures for biological products were initially developed on a lab scale and transformed into technologies that resulted in higher production capacities. Simultaneously, old recipes have been replaced by new methods, which sets the stage for more advanced separation techniques. Modern processing units benefit greatly from technological development, which depends on sophisticated equipment with varying degrees of complexity for varying scales of production. Automation and controlled operation have become essential requirements for ease of functioning, scalability, consistency, and repetitive tasks. In recent years, industries have increased reliance on sophisticated computer

DOI: 10.1201/9781032651095-14

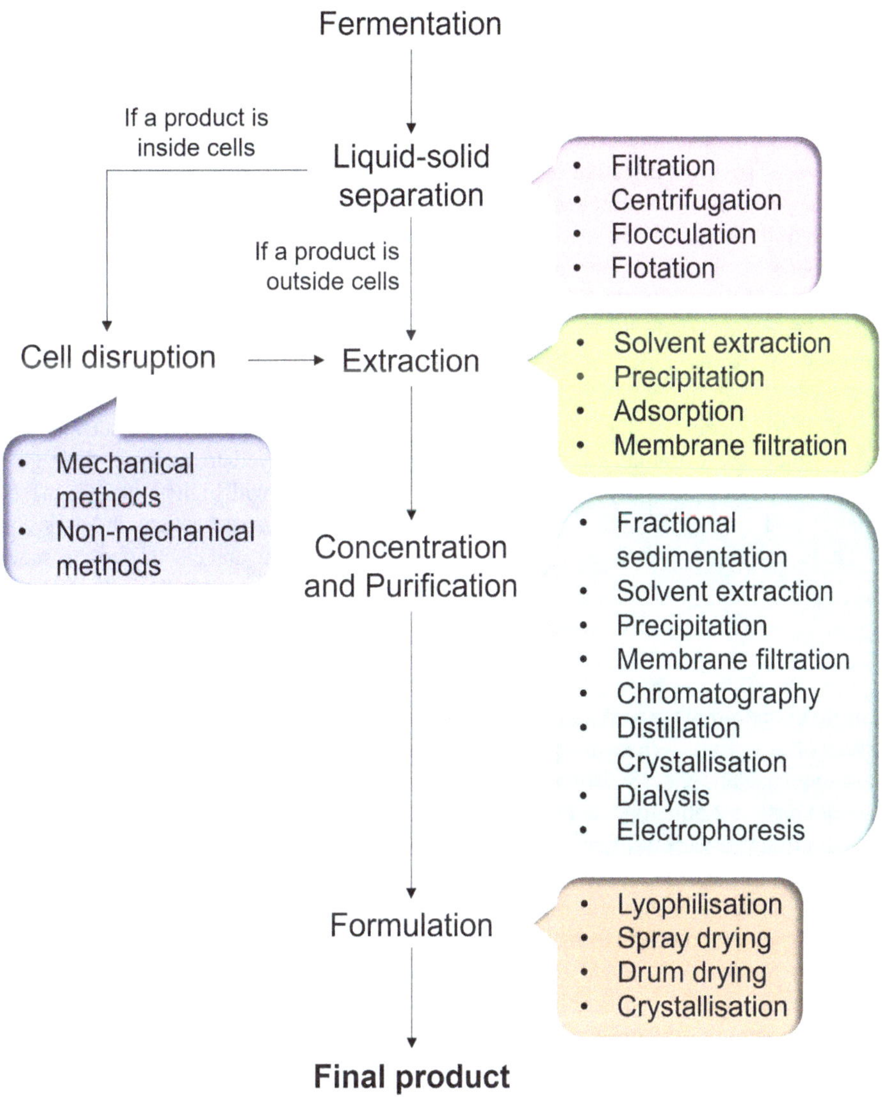

FIGURE 11.1 Different steps of downstream processing and the associated techniques.

programs and robotics to assist, control, and monitor recovery. This aspect also reduces the risk of human error and increases processing capacity. Integrated data analytics and machine learning algorithms have delivered information on process improvement, quality assurance, and predictive maintenance. Some modern industrial production designs integrate upstream and downstream processes inside the bioreactors, which boost product yield and omit some recovery steps. This approach also curtails the time and expenses involved in product development.

Downstream processing can be generalised for ease of understanding in microbial biotechnology. It can have many sequential key steps, including liquid-solid separation, extraction, concentration, purification, formulation, and packaging. Each step in downstream processing serves a specific purpose and relies on some distinct and common techniques. Figure 11.1 illustrates key steps involved in downstream processing and the associated techniques. These steps are also discussed in detail as follows.

11.2 LIQUID-SOLID SEPARATION

Liquid-solid separation is the first step of downstream processing. The microbial cells, cellular fragments, and insoluble ingredients can be the solid components in the medium. When the desired products are soluble components, the solid fractions of the fermentation broth are undesirable. On the other hand, the extracellular fluid is a bulk waste in the harvest of microbial biomass. Filtration and centrifugation are two commonly involved techniques in solid-liquid separation. The properties of solid (like particle size, density, and mechanical strength) and liquid (like viscosity and density) of the fermentation medium are the two main factors that aid in selecting the most appropriate techniques for their separation.

Filtration is a relatively simple operation and removes large suspended particles. This employs filters based on particle sizes. Microfiltration, having pore sizes in micrometres, is most appropriate for liquid-solid separation and can be employed to separate fungal mycelia, yeasts, and large bacterial cells. Centrifugation is a different technique, which relies on artificial gravity to recover particles from the solution. This is also appropriate for the solutions that possess density differences between solid and liquid. Both methods are accessible in batch and continuous modes for large-scale operations. These techniques warrant the effective removal of proteins, bacterial cells, and yeasts from the solution.

In addition, flotation and flocculation can be employed as the pretreatment methods to improve the liquid-solid separation. In flotation, gas bubbles are added to the medium. The low-density suspended particles are drawn to the bubbles and become stuck to them. These particles can then rise to the surface as foam and be removed. In flocculation, different chemical agents are added to encourage the formation of large aggregates (flocs), which are eventually separated by gravity sedimentation. This process is purposely used to remove particulate matter and microbial cells in domestic wastewater treatment. These techniques are discussed in detail in Chapter 12.

11.3 EXTRACTION

Extraction involves a series of operations to recover the desired products from a mixture of solutions. When the product is intracellular (e.g., L-asparaginase and glucose isomerase), the product recovery from biomass is obligatory and requires a cell disruption method. There are a variety of cell disruption methods suited to different microbial cells and their products. Two major categories are mechanical and non-mechanical cell disruption. The former type of method is more commonly employed in most industrial setups. However, if the desired product is extracellular and persists in the medium, the solution can be directly used for the extraction of valuable products. One of the most common methods used for extraction is the fractional sedimentation of the liquid phase using different solvents such as ethanol, acetone, inorganic salts, acid or base. This process prepares the crude extract of the product by transferring compounds of interest into a solvent phase. The crude extract is subjected to further treatment, such as solvent extraction, precipitation, adsorption, or membrane filtration, to enrich the concentrate. The solvent extraction consumes a high volume of organic solvents and separates the carbonic compounds based on their partition coefficients. Precipitation utilises chemicals (precipitants) to decrease the solubility of certain compounds in the solution, creating an insoluble solid (precipitate) and liquid. This process removes impurities from the extract. The adsorption technique employs adsorbates (e.g., resins, activated carbon, or silica gels), which selectively or non-selectively bind the impurities or target products, allowing separation of the pure product from the solution. Membrane filtration relies on semi-permeable membranes. This can be a pressure-driven physical process, which develops an absolute barrier for solid particles in the liquid stream.

11.4 CONCENTRATION AND PURIFICATION

These steps involve the processing of the extract to increase the product concentration and remove coarse particles. A part of the concentration and purification operations has already been accomplished during the early treatment steps of liquid-solid separation and extraction. Despite their efficiency, the product characteristics, the type of impurities, the desired final concentration of the product, and the economic feasibility may call for additional requirements to improve its concentration. For example, techniques like distillation and evaporation are correspondingly used to improve the ethanol content.

The purification step is specifically designed to raise the product's quality rather than its quantity. Even many concentration steps overlap with the purification steps, such as solvent extraction, precipitation, membrane filtration, chromatography, and crystallisation. Though these steps procure the high standard products, some additional techniques may be employed to improve their grades. For example, the biocompounds with high purity used in research and development studies and pharmaceutical industries are processed through chromatographic and electrophoretic methods. Chromatography allows for the selective separation and purification of biomolecules in highly pure forms. Chromatography is based on the differential affinities of molecules for a stationary phase (typically a solid or liquid) and a mobile phase (a liquid or gas). A chromatographic technique is useful for purifying recombinant human insulin. Electrophoresis is valuable for the purification of biomolecules based on size, charge, shape, isoelectric point (pI), and the binding abilities of target compounds using different types of gels. A distinct method of purification is dialysis. When the extract is immersed in a larger volume of buffer or solvent and is kept in the dialysis membrane, small particles pass through the pores of the dialysis membrane through passive diffusion and osmosis, but large particles are retained. This technique is particularly useful for proteins or other molecules that are sensitive to high salt concentrations.

11.5 FORMULATION

Formulation is the finishing step of downstream processing, which ensures high activity and stability of biotechnological products for long-term storage and distribution. The end products comply with precise quality standards and could be liquid suspension or solid (tablet, powder, or gel). Their development requires some modification in their nature to deal with environmental and biological stresses. This can be achieved using different additives and/or by drying the products. Chemical additives such as glycerol, sorbitol, ammonium sulphate salt, and polyethylene glycol (PEG) can act as stabilisers of purified compounds. Diluents, emulsifiers, lubricants, and suspending agents can be added to preserve and retain the effectiveness of derived constituents.

Drying ensures the removal of excess amounts of water and reduces the product volume. Since the biologically derived products are heat sensitive, heating methods are not suitable to remove the excess volume of solvents. Crystallisation, spray drying, and freeze drying are the most commonly employed drying methods in biotechnology. The biological-derived products, such as citric acid and antibiotics, are recovered using crystallisation. Spray drying stabilises probiotic bacteria and can be used for microencapsulation. This process is also useful for the drying of proteins and hormones. Freeze drying (or lyophilisation) is appropriate for the storage of living microorganisms. The fundamental principle of lyophilisation is the direct transformation of solid (ice form) into vapour through sublimation. It can be used in food and beverage industries, pharmaceutical applications, and culture collections. Drying vaccines through lyophilisation is a very common approach in the pharmaceutical industry. After drying through various techniques, the sensitive products may require a sterilisation treatment and are sealed in sterile packages.

Chapter 12 provides in-depth information on the principles and applications of different techniques involved in the downstream processing of various biotechnology products.

SUMMARY

- Downstream processing is a series of operations after the fermentation process to recover the target products from a large volume of broth. This involves many sequential steps to extract the biocompounds using different techniques.
- Liquid-solid separation is the first step of downstream processing. This mainly depends on filtration and centrifugation, and may involve some pretreatment steps like flotation or flocculation.
- Extraction and purification is a process of recovery of the target compound from the rest of the mixture through solvent extraction, precipitation, adsorption, and membrane filtration. Some additional techniques, such as chromatography and electrophoresis, can further concentrate and purify the product.
- The final step is the formulation, which involves techniques like evaporation, crystallisation, spray drying, and lyophilisation.

DIFFERENT TYPES OF PRACTICE QUESTIONS

Fill in the blanks

I. Downstream processing can raise the total cost of the products by ....%.
II. .... and .... are two commonly involved techniques in solid-liquid separation.
III. When the product is intracellular, the product recovery from biomass requires .... method.
IV. The fundamental principle of lyophilisation is the direct transformation of solid (ice form) into vapour through ....

SOLUTION: I. 5–90%, II. Filtration, centrifugation, III. cell disruption, IV. sublimation.

Very short answer type of questions

I. Define downstream processing.
II. Write the names of the main steps involved in downstream processing.
III. What is the main purpose of formulation in downstream processing?

Short answer type of questions

I. The choice of techniques for downstream processing depends on upstream processing. Give your comments on this statement.
II. Write a short note on concentration and purification in downstream processing.

Long answer type of questions

I. Elaborate on the sequential steps of downstream processing in industrial biotechnology, giving suitable examples of different techniques involved.

SUGGESTED READINGS

Anderson TM. *Industrial fermentation processes*. Encyclopedia of Microbiology (Third Edition); 2009.

Okafor N, Okeke BC. *Modern industrial microbiology and biotechnology*. CRC Press; 2017.

Puvanakrishnan R, Sivasubramanian S, Hemalatha T. *Upstream and downstream processing of bioproducts*. MJP Publishers; 2016.

Wang Y, Ling C, Chen Y, Jiang X, Chen GQ. Microbial engineering for easy downstream processing. *Biotechnology Advances*. 2019;37(6):107365.

12 Important Techniques in Downstream Processing

12.1 FILTRATION

Filtration is one of the most common techniques involved during an early stage of downstream processing. In microbial technology, this technique is used to separate solid components from broth. The unit operation of filtration equipment consists of a porous material called a filter medium. The porous medium can be derived from various natural and synthetic materials. Examples of natural substances include cellulose, activated carbon, gravel, sand, zeolite, and peat. The options for synthetic materials for filter media are polypropylene, polyethylene, nylon, polyester, cellulose acetate, and glass fibres. Synthetic filter media are comparatively more popular than natural materials because synthetic materials can be engineered to meet specific filtration requirements, such as capturing particles of defined sizes, temperature resistance, chemical resistance, low-density filtration, durability, and consistency. In the fermentation industry, the broth has a high viscosity and contains a variety of materials, which occasionally require a vacuum pump to generate positive pressure or vacuum for filtration. At a pressure drop, the solid particles settle down in the filter medium, forcing the fluid through the medium. The solid substance deposited at the surface of the filter medium is called filter cake.

Based on the pore sizes of the filter media, the process of filtration can be classified as microfiltration, ultrafiltration, and nanofiltration (Figure 12.1). Microfilters having a pore size range from 0.1 to 0.8 µm are appropriate for the recovery of microbial cells (such as bacterial cells and fungal mycelia) from the fermentation broth. The recovery of smaller particles requires ultrafiltration and nanofiltration. They are much more sophisticated techniques compared to microfiltration and have pore sizes around 2–100 nm and 0.5–2.0 nm, respectively (detailed in section 12.8). Many other factors, such as viscosity, temperature, and contaminants of the medium, also have a major influence on filtration efficiency.

In filtration, the recovered products can be either the culture filtrate or microbial biomass. Culture filtration is a liquid fraction, including dissolved and undissolved compounds, whereas microbial biomass represents the bulk solid mass of microorganisms in the fermentation media. If the desired product of microbial fermentation is intracellular, the first step is the recovery of biomass from the broth using prefiltration (i.e., a process of filtration to remove large debris before using fine filters). This is followed by cell lysis. The crude extract of the cell can be further filtered using fine filters to separate cell debris and liquid medium. The filtration process can also be employed for the selective precipitation of proteins and the removal of precipitated contaminants in successive purification steps.

In the filtration process, the rate of the initial deposition of solid particles on the filter medium determines its fate for permeability and the rate of flow in the long run. When the filtration begins, the rate of filtration is high at low-pressure drops because the filter medium is free from any deposition of solid particles. However, the filter flux declines over the period due to increased resistance to flow as the depth of the filter cake continues to increase during operation. An increase in the pressure drop can ensure some extra liquid flow, but this process causes more resistance in the filter medium. Many microbial cells are compressible and can make the filter cake less porous. Excessive pressure drop can mobilise the compressible microbial cells through the medium or tear the pores of the filter medium. The low filtration rates and product loss are major concerns when the fluid contains high concentrations of solid content. A low flow rate during the beginning of the filtration process is highly recommended because it can increase the life of the filter medium.

Based on the direction of the feed flow for the filter medium, the process can be categorised as dead-end filtration and cross-flow filtration.

- Dead-End Filtration- When the feed flows perpendicular to the surface of the medium, the internal pore of the medium is blocked due to the high accumulation of solid particles. This can result in dead-end filtration due to clogging (Figure 12.2a).
- Cross-Flow Filtration- In a different approach, the flow of broth solution can be kept tangential to the filter medium as cross-flow filtration (Figure 12.2b). This kind of arrangement produces less depth of cake on the filter medium and has less blockage on the filter pores as compared to where the flow is perpendicular.

In a different setting, the filter medium can be precoated with a filtration aid to prevent internal pore blockage, which improves its performance for a long duration (Figure 12.2c).

DOI: 10.1201/9781032651095-15

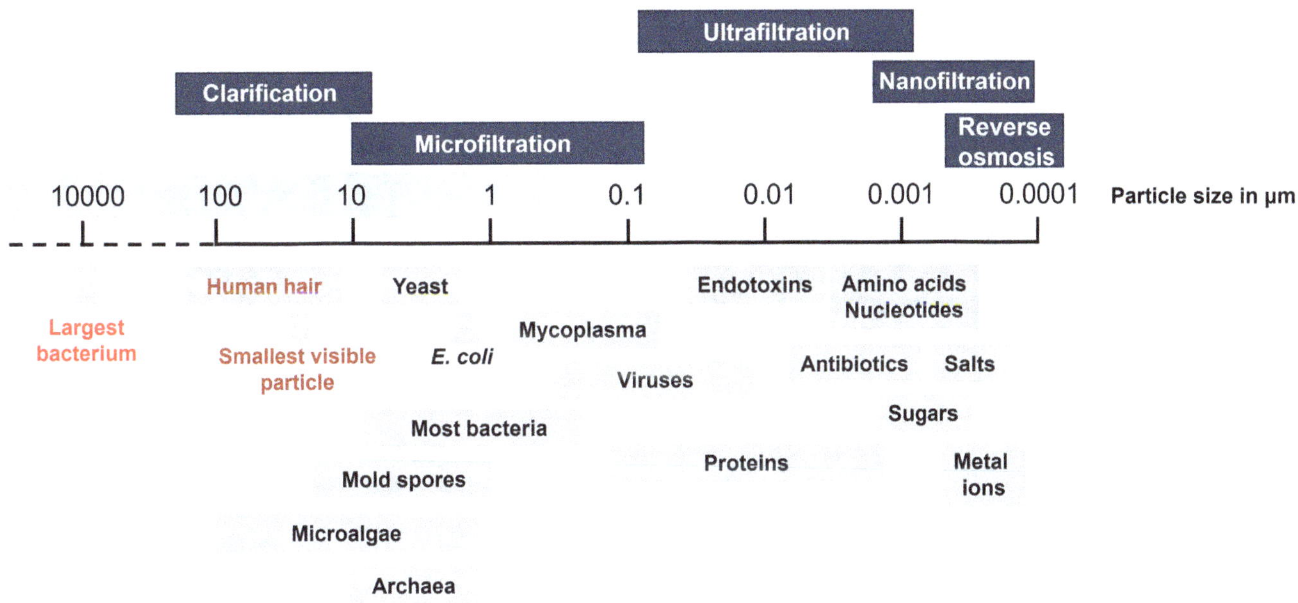

FIGURE 12.1 Different methods of filtration to separate different types of particles.

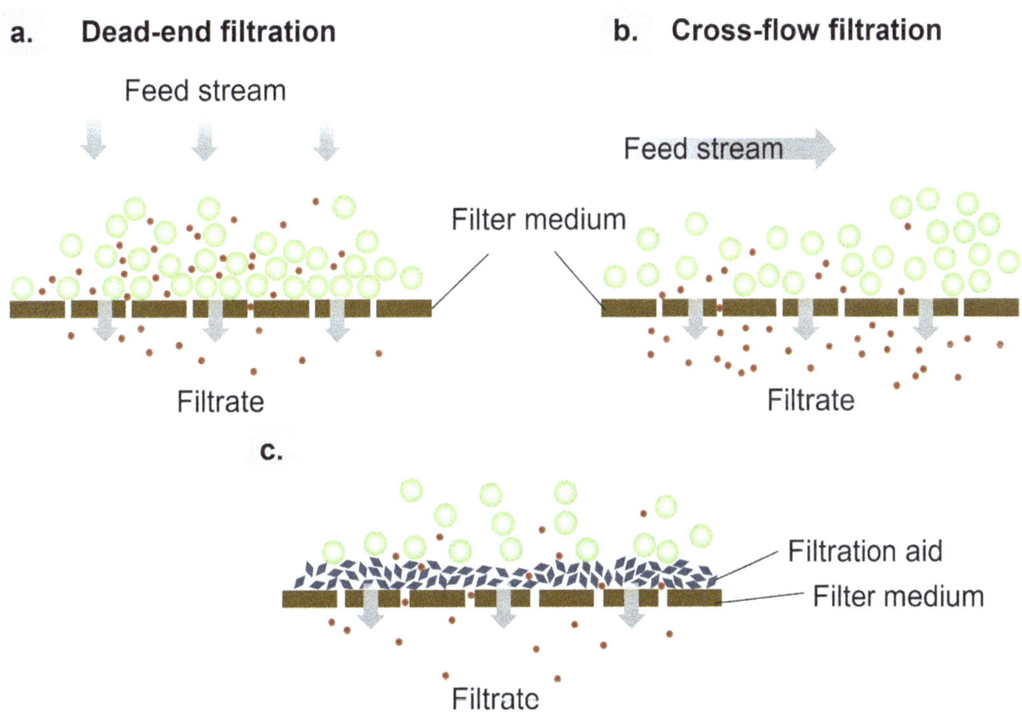

FIGURE 12.2 Different types of filtration. a. and b. Filtration based on feed flow. c. Use of filtration aid in the filtration unit.

Sometimes, the filter aid can be mixed with fermentation broth before the filtration process. The diatomaceous Earth (the siliceous shells of diatoms), also known as kieselguhr, is a suitable choice for a filtration aid. This material has high porosity, which ensures a high rate of flow. However, the use of filter aids in filtration units has some inherent limitations: i. porous materials can adsorb/absorb some amount of the liquid, ii. the filtrate is less clarified, iii. disposal of solid waste is challenging, and iv. the process is not suitable for the recovery of biomass.

Industrial-scale filtration units come in various configurations and are discussed as follows:

a. Filter Press (or Plate and Frame Filter)- It consists of a stack of plates and frames laid alternately in a horizontal framework. The filter medium is usually a

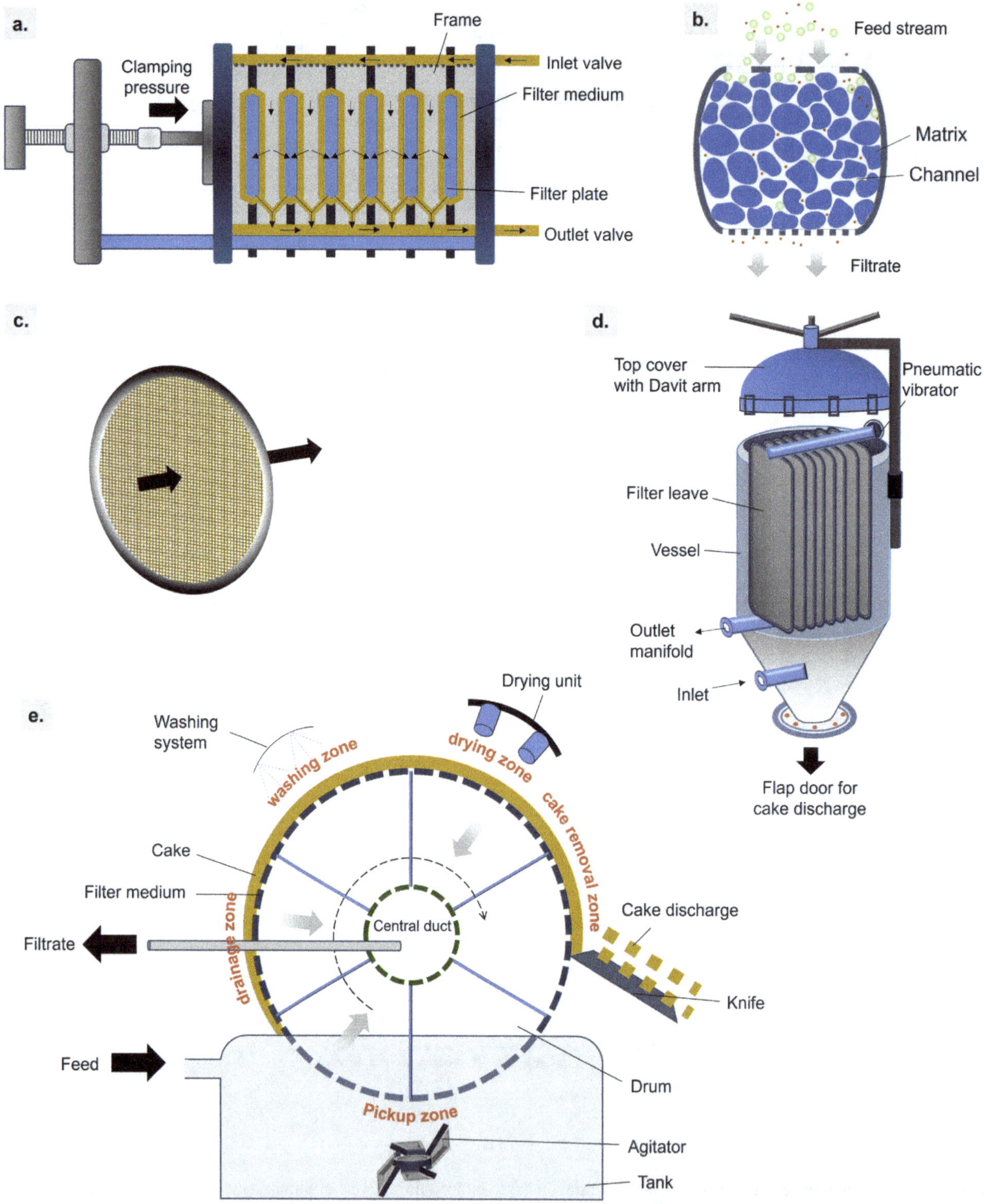

FIGURE 12.3 Different types of filtration devices. a. Filter press. b. Depth filter. c. Screen filter. d. Pressure leaf filter. e. Rotary drum vacuum filter.

filter cloth that is placed between the plate and frame (Figure 12.3a). The whole setup is tightly packed to avoid any leakage. The liquid feed enters through an inlet channel in all the filter frames and comes in contact with the adjacent filter media. The solid particles are retained in the filter frame, while the filtrate is collected in the filter plate, which is displaced using an outlet valve. This process is simple and economical. A filter press is useful for solutions having low solid content, but it runs in a batch operation. After its operation for a long time, the device is dismantled for cleaning, and the filter medium is replaced. However, the new design of the filter press is equipped with a self-washing process that cuts down manual intervention. A filter press in microbial technology is used in brewing to remove residual yeast cells from liquor before using the solution for centrifugation or rotary drum filtration.

b. Depth Filter- The filter medium is composed of a thickened substance or multiple layers of a filter sheet, which can retain the solid particles throughout the medium rather than just on the surface (Figure 12.3b). The constructive materials can be nonwoven fibres (such as cellulose, glass fibre, and polypropylene fibres) or granular materials (i.e., discrete large solid particles). These substances have high rigidity and provide a base structure to the medium. Filter aids and binders are used to improve the retention capacity of depth filters. Binders (e.g., polystyrene and acrylics) act as glue and help the filter medium bind together with filter aids. The filter does not have a fixed pore size, and the matrix provides an irregular and interlinked path across the medium. The pore-size rating is nominal, which means the filter is capable of preventing the flow of a minimum percentage of solid particles greater than the particles of a stated size. This reflects the degree of filtration. For example, the nominal rated filter of 85% of 15 μm indicates that the filter unit prevents 85% of all particles greater than 15 μm. The media can be engineered to retain the large particles toward an exterior end. They can also be modified to allow the small particles to travel some distance inside the matrix before they are captured. Depth filters are highly useful for the clarification of high solid content from high densities of cells, cell debris, and fungal mycelium from the broth solution.

c. Screen Filter- This consists of a uniform interwoven screen with a precise pore size. This acts like a sieve and is suitable for the removal of very fine particles in secondary filtration (i.e., in the second round of filtration using fine filters) (Figure 12.3c). Screen filters come in various sizes, and the pores in the screen filter are measured in mesh, indicating the number of pores found in an area of 1 inch^2. A screen of 150 mesh size has 150 pores in 1 inch^2. In downstream processing, this type of filter can be used to remove the remaining traces of microbial cells and colloidal particles after prefiltration.

d. Pressure Leaf Filter- This constitutes a closed filtration system in a vessel equipped with multiple filter leaves (Figure 12.3d). The leaves are surrounded by a wire or textile cloth cover. The vessel can have either a horizontal or vertical configuration. The system runs at high pressure under batch operation. Filter cakes are deposited on the leaves that can be removed through vibration or washing. This process is suitable for the removal of low solid content. Pressure leaf filters can be used for the polishing (i.e., the removal of odour and bitter taste) of liquor (or spirits).

e. Rotary Drum Vacuum Filter- This equipment mainly consists of a horizontal drum, a tank, an agitator, a washing system, a drying unit, and a discharge (Figure 12.3e). The drum is a hollow structure with a wide size range of 1 to 20 feet long. The inside body is segmented with internal piping, and all the filtrate pipes connect to a central duct around the rotating valve. The outer surface of the drum is covered with a filter medium or cloth. The tank is located at the base and filled with feed. An agitator inside the tank ensures uniformity of solid particles in the feed and avoids their settling at the bottom. The spray jet and dryer are the washing and drying devices, respectively, which are placed at the top of the drum.

During operation, the rotary drum is partially submerged in the bottom feed and rotates at a speed of less than 1 rotation per minute (rpm). This region is known as the pickup zone. The inside of the drum creates low pressure using a vacuum pump, which allows liquid to pass through the filter medium. The particle-free filtrate passes through the filtrate pipe and is transported to the central duct. The solid particles are deposited at the surface of the medium as a filter cake and move toward the drainage zone. In the washing zone, the water spray washes the filter cake and extracts the remaining filtrate. The cake now enters the drying zone, where the air current removes the moisture content of the cake. The dried cake is finally removed using a suitable discharge method. Air pressure from the body of the drum provides ease in the process of cake removal.

The rotary drum vacuum filter is popular for its continuous operation, production capacity, and robustness. This instrument has high efficiency, high process quality, and low energy consumption. It is suitable to filter the feed having 15–30% solid content. In fermentation technology, a rotatory drum vacuum filter is used to separate the fungal and bacterial mass from the fermentation broth in the pharmaceutical and beverage industries. One notable example is the recovery of penicillin extract from the fungal biomass of *Penicillium chrysogenum* after fermentation. This technique is also effectively used in algae dewatering and wastewater treatment.

12.2 CELL DISRUPTION

Cell disruption is a process of breaking the cell boundary to recover cell-bound products. The choices for the cell disruption methods are manifold. The selection of disruption methods can be based on the type of microorganisms involved, location and the types of products, generic application, and scale of production.

Microbial cell walls are made up of different constituents, and therefore, their breakage requires different techniques. The fungal body is chiefly covered with chitin, glucans, and glycoproteins. Bacterial cells (Gram-positive and Gram-negative forms) consist of different amounts of peptidoglycan

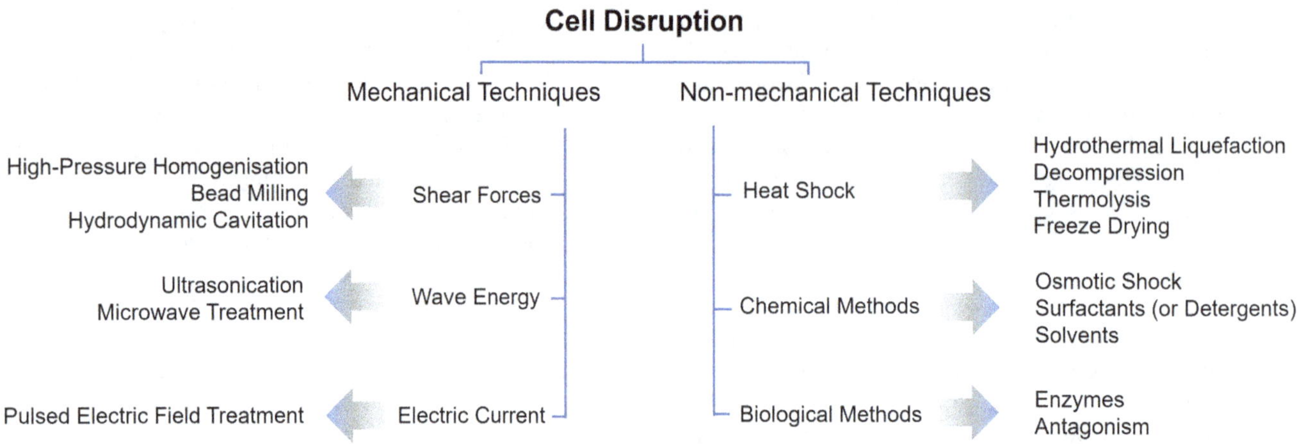

FIGURE 12.4 Various methods for cell disruption.

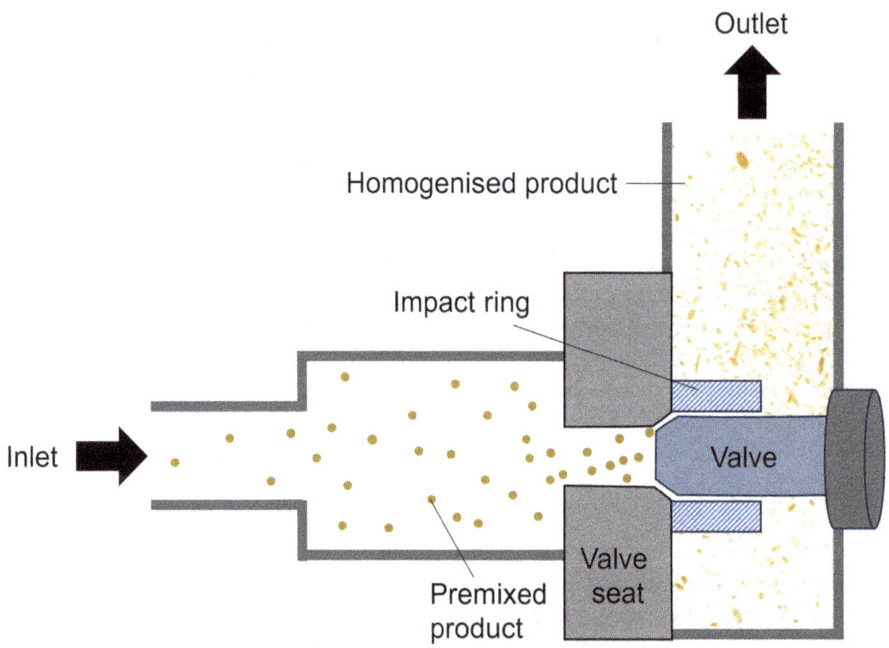

FIGURE 12.5 An illustration of a high-pressure homogeniser for cell disruption.

in their cell wall compositions. Microalgae display a thick wall and are mainly composed of a cellulose-pectin complex. Most protozoans do not possess a cell wall, and cell disruption is relatively easy. Some major and minor variations in the cell wall compositions among different organisms within the same group also exist and require different techniques for cell disruption. Moreover, cell disruption relies on distinct techniques to recover the extract from the periplasmic space, the membrane-bound, and the intracellular. Even some microbial cells require more than one disruption method to release maximum content. However, cell disruption methods may have some impacts on the yields and value of the derived products.

Based on the methods involved, cell disruption can be categorised as mechanical and non-mechanical techniques (Figure 12.4), which are discussed below.

12.2.1 Mechanical Techniques

Mechanical cell disruption applies physical forces on the cell to extract intracellular contents. The intense impact of mechanical forces disrupts the interaction between molecules inside the cells and breaks cell walls and cell membranes. Mechanical forces are most commonly employed for industrial applications due to the high product recovery yields. These forces are energy-intensive but less invasive. The disruption of cells takes place in nonspecific patterns. There is also a low probability that the homogenised solution will be contaminated during mechanical processing. Their operations are generally controllable and scalable. However, they are not compatible with shear-sensitive products such as nucleic acids and volatile intracellular compounds. Mechanical stress can be applied through shear forces, wave energy, electric current, and heat shock. These stresses on microbial cells can be applied through different techniques as follows.

12.2.1.1 Shear Forces

- ***High-Pressure Homogenisation***

This is a simple rupture technique through shear-force disruption, which applies a tangential force onto the cell wall. This is one of the most commonly used techniques for cell disruption at an industrial scale in the pharmaceutical industry. The instrument is called a high-pressure homogeniser and is mainly equipped with two devices: a high-pressure positive displacement pump and a homogenising valve. When the highly concentrated microbial fluid is injected into the homogeniser, it experiences a very high pressure under a displacement pump while passing through a narrow orifice in the homogenising valve. An intense pressure drop across the valve creates a massive shock for the cell and causes cavitation (the formation and subsequent collapse of vapour-filled cavities). The homogenised product hits the impact ring and exits through the valve (Figure 12.5). The whole operation runs at low temperatures (0–4 °C) to reduce the impact of heat generated due to pressure drop. The primary contributing factors to this shear disruption on cells are the magnitude of the pressure drop and valve designs. An industrial setup can achieve a maximum pressure of 2,000 bar, whereas an ultra-high-pressure homogeniser can raise the pressure as high as 4,000 bar. There are about 150 patents available for different valve designs for effective homogenisation. An optimal condition of high-pressure homogenisation can achieve more than 90% efficiency in releasing the intracellular content. The scheme is most useful for the disintegration of recalcitrant cells such as yeasts.

- ***Bead Milling (or Bead Beating)***

A bead mill is a cylindrical vessel consisting of a stirring unit (agitator). The vessel can be filled with a grinding medium called beads. Beads are made of quartz or metal with a 0.1–3 mm diameter. The vessel is surrounded by a cooling jacket to maintain its temperature. The liquid feed is mixed with beads in a bead mill and runs at a high speed. The solid shear due to stirring and grinding induces severe damage to cells and causes disruption. This procedure also facilitates efficient solvent access to the intracellular content of the cells. This method is useful for the destruction of recalcitrant cells and in lipid extraction. However, bead milling can damage some bioproducts because solid shear is not suitable for fragile substances.

- ***Hydrodynamic Cavitation***

This technology is based on bubble formation, expansion, and cavitation due to the motion of the fluid. When the fluid passes through the narrowest flow region in a jet, it creates microbubbles, which disintegrate as the pressure returns above the vapour pressure. The violent collapse of the cavities momentarily elevates pressure and temperature in the range of 100–5,000 atm and 250–14,500 °C, respectively and causes cell disruption. This process has very high energy efficiency and, therefore, can be employed for large-scale applications. This technique is comparable to high-pressure homogenisation, which also operates at high pressure and is an energy-intensive process.

12.2.1.2 Wave Energy

- ***Ultrasonication***

This process employs high-frequency sound waves (greater than about 20 kHz, i.e., 20,000 cycles per second) to shear bacterial and yeast cells. This can be performed using an ultrasonic bath or an ultrasonicator probe. The whole process runs under chilling conditions to maintain the local temperature. The energy from the sound waves causes cavitation, followed by shock-wave propagation. In this process, the sinusoidal movement of waves in the reaction system generates alternating high-pressure (compression) and low-pressure (rarefaction) cycles. This pressure-difference environment creates vacuum bubbles that collapse during compression (cavitation). The violent compression generates a shear force for the disruption of the cells. Although this method is incredibly effective, it costs a lot of money and energy to operate on a wide scale. This process requires approximately three times higher energy input compared to high-pressure homogenisers. Also, the tip of the probe is highly delicate and needs replacement after some rounds.

- ***Microwave Treatment***

This is also a wave energy-based cell disruption using microwave irradiance. This extraction procedure includes a shorter extraction time than ultrasonication. However, the energy requirement for microwave treatment is higher than that for ultrasonication, and the difference in disruption efficiencies is almost negligible.

12.2.1.3 Electric Current

- ***Pulsed Electric Field Treatment***

This is a relatively new approach to cell disruption and is mainly applied to the rigid cell wall of microalgae for lipid extraction. The process is based on electroporation. A bipolar pulse generator creates an electric field strength of ~30 kV at ~2,000 Hz in the reaction system, which develops a critical electrical potential across the cell boundary. The unequal electric charges on the boundary layer create permanent pores, allowing the escape of intracellular content. In this process, the electric field strength and pulse are directly proportional to the size and number of pores on the cell surface.

12.2.2 Non-Mechanical Techniques

Non-mechanical procedures depend on heat stock, chemical processes, and biological activities. The last two methods are particularly appropriate for delicate products, involving gentle procedures. Some non-mechanical methods are useful for the selective disruption of specific cellular structures or components. Various techniques of non-mechanical cell disruption are discussed as follows.

12.2.2.1 Heat Shock

- ***Hydrothermal Liquefaction***

This is a high-pressure (10–25 MPa) and high-temperature (200–370 °C) driven process to effectively depolymerise the biomolecules, which enables cell disruption. The excess internal pressure of the cell due to heating ruptures the cell wall, leading to cellular structure disintegration. The key steps involved in hydrothermal liquefaction are the drying and grinding of biomass feedstock, followed by liquefaction at high temperatures and pressure. This method is commonly involved in the separation of water-soluble molecules and lipids from wet microalgal biomass to produce liquid fuel.

- ***Decompression (or Steam Explosion)***

This thermo-mechanical process also operates at high temperatures (180 to 240 °C) under pressure (1 to 3.5 MPa) for 1–20 min. However, a pressure drop is the primary cause of cell wall disintegration due to explosive decompression of the biomass. This process is more suitable for the rigid cell wall of microalgae than for any other microorganisms. Under high pressure, steam penetrates the cellulosic wall of microalgae, and the sudden change in pressure cleaves the glycosidic bonds and hydrogen bonds. Both batch and continuous modes of operation are possible, as is the possibility of large-scale operations. This procedure can also be used in the pretreatment of lignocellulosic biomass and microalgal refinement.

- ***Thermolysis***

This involves the lysis of the cell wall due to excessive heating. Prolonged exposure of the microbial cells to elevated temperatures enables chemical decomposition of the cells. Though the process is relatively simple and inexpensive, treatment under high temperatures is generally not suitable for the recovery of microbial-based products.

- ***Freezing Thawing Process***

This process uses repeated cycling of freezing and thawing of the cell suspension. The freezing of the cell suspension in a dry ice/ethanol bath or freezer produces ice crystals, resulting in cell expansion. After that, instant thawing at room temperature breaks the ice crystals, leading to the contraction of the cell and subsequent cell disruption. This method works well for a variety of bioproducts and generates less heat than any other aforementioned heat-based treatments. A key advantage of freeze drying is the high efficiency in recovering recombinant proteins from the bacterial cells. However, the low yields of desired compounds, high energy consumption, and long drying time limit its applications as a commercial cell disruption method.

12.2.2.2 Chemical Methods

- ***Osmotic Shock***

Osmotic shock can be induced by varying the ionic strengths of the cell-containing solution. An instant change from high osmotic pressure to low osmotic pressure develops internal pressure on the cell wall and causes lysis. In this process, salt like NaCl (usually 1–2%) can be added to the solution, which generates a high osmotic pressure of the solution (hypertonic), and water flows outward from the cell through a semipermeable membrane. A sudden dilution of the solution causes the water flow to enter the interior of cells (hypotonic), resulting in cell swelling and bursting. Though the process is relatively simple, it has low efficiency and low scalability. This also requires a high amount of salt and water. The excess water consumption dilutes the desired products. The clean-up of salt is an additional burden in subsequent purification steps. This technique, to some extent, can be used to recover the periplasmic content of Gram-negative cells (having thin walls) and mammalian cells (which have no cell walls). Otherwise, this technique is not suitable for disrupting microbial cells with thick and rigid cell walls. Due to these reasons, osmotic shock is not appropriate for cell disruption at an industrial scale.

- ***Surfactants (or Detergents)***

Surfactants are detergents that have the ability to solubilise cell membranes. All three types of surfactants, including cationic, anionic, and nonionic surfactants, can be used to disrupt the membrane structure through lipid-liquid, lipid-protein, and protein-protein interactions. Examples of cationic and anionic surfactants are cetyltrimethylammonium bromide (CTAB) and sodium dodecyl sulphate (SDS), respectively. CTAB is most commonly employed for the extraction of plant proteins, where SDS is a strong lysis agent and is also active for numerous microbial cells. However, the high ionic strength may cause damage to proteins and enzymes. Nonionic detergents such as Triton X-100 and Tween (20, 80) are mild lysis agents and are commonly used for protein isolation. However, the use of surfactants is not a popular choice for the recovery of intracellular products at a large scale because the process is economically unfeasible and may cause protein denaturation.

- ***Solvents***

The treatment with various solvents such as hexane, chloroform, dichloromethane, ethanol, and acetone. These solvents solubilise the lipid contents of the cell wall and membrane, making cells permeable to release the contents inside. The formulations and concentrations of solvents strongly inspire recovery efficiencies. The pH-based cell lysis also exists. The hydroxide ion (OH^-) in a strongly alkaline condition (pH of 10.5-12-5) hydrolyses the cell wall and membrane of the cells. However, this requires a large volume of acid to neutralise the alkaline effects in the solution. This type of treatment is suitable for extracting alkali-stable compounds.

12.2.2.3 Biological Methods

- ***Enzymes***

The enzymatic treatment for cell disruption is highly site-specific. Lysozyme specifically acts on β-1-4-glycosidic bonds that create a link between N-acetylmuramic acid (NAM) and N-acetylglucosamine (NAG) in peptidoglycan. This enzyme is highly potent for the lysis of Gram-positive bacteria as they contain abundant peptidoglycan in the cell wall. However, Gram-negative bacteria are less prone to the effects of this enzyme because they possess only a few layers of peptidoglycan. Proteases (such as proteinase K or subtilisin) can be used to disturb several cell types if the desirable product is not a protein. Enzymes like cellulases, amylases, and pectinases show activities against cellulosic, starchy, and pectin materials, respectively, which are often present in microalgae and some specific microorganisms. The use of biosurfactants in the enzymatic formulation has shown improved product-recovery efficiencies during cell disruption. The use of multienzyme components (enzyme cocktails) has proven useful in effectively digesting the cell walls and membranes that are composed of multiple constituents like polysaccharides, lipids, and proteins. The combination of enzyme cocktails can vary based on the cell wall and membrane compositions of the target organisms. For example, an enzyme mixture, zymolyase (also known as lyticase) (source organism: *Arthrobacter lutes*), possesses β-1,3 glucanase and β-1,3-glucan laminaripentao-hydrolase activities, which are routinely used for the cell wall lysis of yeasts and fungi. The enzymatic method is an eco-friendly solution to cell lysis and has low energy consumption. However, the large-scale implications of the enzymes for cell disruption are limited due to their slow reaction rate and high cost.

- ***Antagonism***

In a different approach, some microbial cells, as control agents, fall prey and injure target organisms, enabling the release of their intracellular content. This is called an antagonistic treatment. The antagonistic effects can be produced by extracellular allelopathic compounds and enzymes released from the control agents. The control agents can be different bacteria, cyanobacteria, microalgae, and fungi. In a different method, the phages (viruses) can be used as predators of bacteria through their lytic cycles. However, the approaches of antagonism do not show any commercial value for cell disruption in industrial microbiology.

12.3 CENTRIFUGATION

Centrifugation is a mechanical process used to separate different components from the solution by applying centrifugal force (gravitational force or g-force). The spinning of a solution around an axis at high speeds generates a force that acts on the particles and increases the sedimentation rate. This is based on the principle of Stokes' law. According to this law, the rate of sedimentation is directly proportional to the acceleration of gravity. The difference in particle density and solvent density, the diameter of particles, and the viscosity of the medium are some other crucial factors that affect the sedimentation rate. It can be formulated as:

$$v = (p-L)\, d^2 g/18n$$

Where v is the sedimentation rate, p is the particle density, L is the solvent density, d is the diameter of the particle, g is the acceleration of gravity, and n is the viscosity of the medium. In this formula, g is calculated as

$$g = r\omega^2$$

Where r is the radius of the centrifuge rotor, and ω is the angular velocity.

The mechanical properties of the machine, such as angular velocity, the radius of the rotor, temperature, and residence time, also affect the sedimentation rate.

The operating device of centrifugation is called a centrifuge. It is suitable for both solid-liquid and liquid-liquid separation. Although the centrifuge is a crucial, sophisticated laboratory tool for research studies, it also has a broad range of industrial applications. In biotechnology, a centrifuge can be used to separate particles or macromolecules like cells, sub-cellular components, and proteins from fermentation broth. The suspended particles, which are difficult to collect through filtration, can be separated through centrifugation. This technique is also useful for separating different components based on the differences in their specific gravity (or relative density). This principle is used to remove bacterial spores from milk, which have very similar densities.

Large-scale applications have requirements such as large capacity, low cost, stable running performance, and high separation efficiency. The industrial scale centrifuges contain a solid bowl as the rotor. They are made up of high-quality stainless steel (corrosion and acid-resistant) and include many safety parameters. They can be fed in batch or continuous modes, but batch mode is preferable when the equipment runs at high speed. Depending on the requirements, industrial centrifuges come in different configurations. Some common industrial centrifuge designs are illustrated in Figure 12.6 and are discussed below.

- Perforated Bowl Basket Centrifuge- In this centrifuge, the rotor is a cylindrical basket with a perforated wall (Figure 12.6a). The working principle is filtration, where the liquid is separated through pores, and solid particles settle on the walls. The size of the pores in the wall is usually smaller than the solid particles. The basket is also lined with filter cloth (of nylon or cotton) to avoid clogging the pores. The solution is fed through a feed nozzle inside the basket. The machine runs at a speed of 1,000 rpm (about 150 × g) but can be increased as high as 4,000 rpm (about 2000 × g). However, the high speed of the particle can block the pores. The solid cake retained on the wall is discharged using a scraper. The long-term operation develops a hard cake on the wall that can be difficult to remove. This type of centrifuge is used to separate fungal mycelia from the broth. A commercial preparation of Chakka, Indian-style fermented milk that is made by draining the whey from the curd, needs a perforated bowl basket centrifuge.
- Tubular Bowl Centrifuge- This centrifuge applies enormously high centrifugal force and runs under a

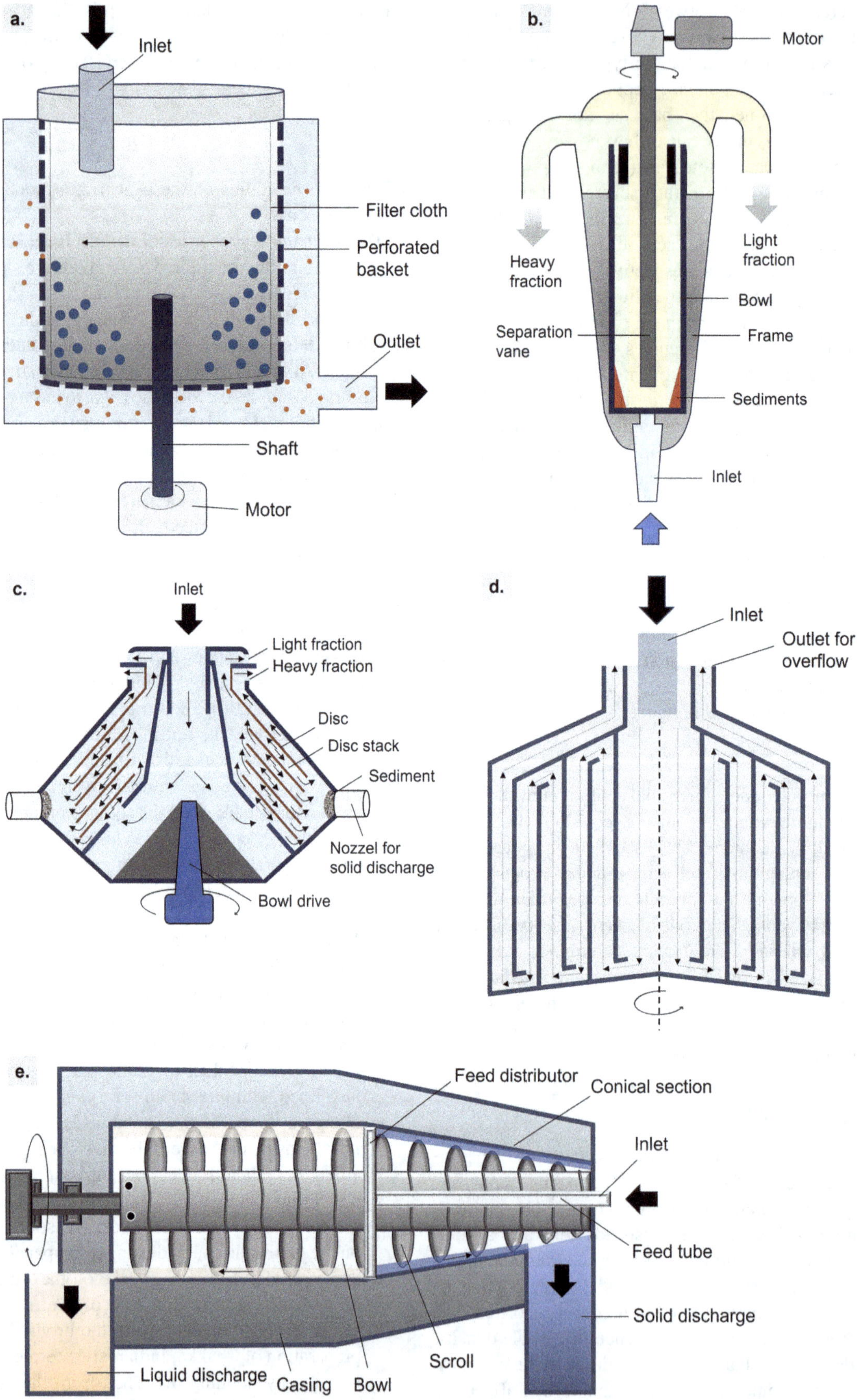

FIGURE 12.6 An illustration of industrial scale centrifuges. a. Perforated bowl basket centrifuge. b. Tubular bowl centrifuge. c. Disc bowl centrifuge. d. Multichamber centrifuge. e. Solid bowl centrifuge.

vacuum-sealed chamber. The bowl is a long hollow tube with an inlet at the bottom and two outlets at the top. The tube consists of a flexible shaft, which is driven by an electric motor. A specialised narrow tubular bowl centrifuge can achieve a much higher speed under low pressure, which is suitable for purifying vaccines and other minute particles. The machine can generate centrifugal fields in the range of 13,000–16,000 × g. The feed is pumped from the bottom, and sedimentation occurs by its flow from the lower region to the top. The heavy phase of the separation lies towards the periphery, and the light phase towards the centre. One outlet is used to withdraw the heavy phase, and the other collects the light phase (Figure 12.6b). This type of centrifuge can be used for cell harvesting.

- Disc Bowl Centrifuge- This is also called an inverted chamber bowl centrifuge, where the inlet is at the top, and two outlets are present at the periphery. The inside body contains a multitude of vertical stacks of thin discs in the shape of cones, which may be separated by spacers to build up the stack (Figure 12.6c). This centrifuge typically runs at speeds of 5,000–10,000 rpm (corresponding to 3,000–11,000 × g if the rotor radius is 10 cm). When the bowl spins, the disks also experience the rotation under which the fluid is separated into two phases: the heavy phase and the light phase. The heavy phase under the centrifugal force is displaced towards the outside wall, while the light phase shifts towards the centre. This device is useful for the removal of fluid having solid content up to 25% of the size range 0.05–120 µm. The discharge of the solid is either in continuous mode or as an intermittent jet. In microbial technology, this type of centrifuge can be used to remove yeast, bacterial cells, and other solid particles from the broth.
- Multichamber Centrifuge- This centrifuge consists of multiple chambers and is suitable for separating soluble solids from their low-content solution (Figure 12.6d). The solid capacity of the bowl is very high compared to other types of centrifuges. The operational speed is typically low, with a centrifugal force of around 10,000 × g. The solution inside the centrifuge is released from the top through a tube, which passes through different chambers, moving from the centre to the periphery in a circuitous fashion under high centrifugal force. Each chamber experiences a different amount of force due to its differences in the distance from the axis. The solid particles are deposited based on their density and size in different chambers. The chamber which lies farthest from the centre receives maximum force and can separate the smallest particle. The coarse particles are settled in the chambers near the axis. The solid particles are recovered after the centrifugation is completed.
- Solid Bowl Centrifuge (Scroll Centrifuge or Decanters)- This centrifuge has a conical, cylindrical design with a horizontal bowl. The liquid feed enters the bowl through a deep inlet. The strong centrifugal force inside the bowl separates the fluid into a heavy phase (solid) and a light phase (liquid). The conventional centrifuge can achieve a maximum centrifugal force of 6,000 × g. The bowl consists of a screw conveyor or scroll that also rotates but a little faster or slower than the bowl (Figure 12.6e). Their differential speed lies in the range of 0.5–100 rpm. The screw collects the solid particles and projects them towards the conical end for their discharge, whereas the liquid phase moves in the opposite direction due to the slope. A paring disc assists in releasing the liquid phase from the bowl using centripetal force. This machine is commonly employed in the recovery of desired products from fermentation broth and sludge dewatering during wastewater treatment.

12.4 FLOCCULATION

Flocculation involves a physical force to convert solid particles into bulky flocs that settle at the bottom. This process in industrial microbiology is used to remove microbial cells (or cell debris) from the fermentation broth. When the microbial cells are minute and their densities are slightly higher than the broth solution, the removal of biomass by centrifugation can be difficult, time-consuming, and expensive. Flocculation provides a solution for their removal from the solution involving physical factors or chemical flocculants. In this process, a gradual and controlled input of a factor produces an impact on the native particles that tends to destabilise them. The destabilised particles in the solution experience aggregation and create microflocs, which further agglomerate to form huge clusters or flocs. Flocs are generally bulky and settle at the bottom and can be easily separated.

Filtration becomes necessary after flocculation when some yeast cells float as scum on the fluid's surface or as suspended particles in the solution. Flocculation can be used as a facultative low-cost pretreatment during downstream processing. The most typical industrial application of flocculation is the recovery of yeast cells after alcoholic fermentation. Yeast flocculation depends on various parameters such as dissolved oxygen, pH, temperature, and nutrients. Flocculation using chemical flocculants is highly prominent for the bioprocessing of ethanol. A trace amount of calcium ions (Ca^{2+}) in the fermentation broth turns yeast cells into flocs. Some microalgae have natural abilities of auto-flocculation owing to extracellular polymeric substance production. Even in some situations, the prolonged mixing of the solution increases the collision of small particles, followed by their clumping into large particles and settling.

Flocculation is a reversible process, and it is termed deflocculation. The change in the physical and chemical conditions of flocs can interfere with the forces that hold particles together, altering the stability of the flocs. For example, the treatment of flocs with phosphate ions (PO_4^{3-}) encourages a reaction with Ca^{2+} and precipitates it as calcium phosphate ($Ca_3(PO_4)_2$) and the particles are once again dispersed in the liquid, releasing yeast cells. A different example of a deflocculating agent is sulphate ions (SO_4^{2-}), which induce the synthesis of calcium sulphate ($CaSO_4$). Ethylenediaminetetraacetic acid (EDTA) encourages chelation of Ca^{2+}, releasing yeast cells in the solution. If the cells are still active after recovery, they can be used for new reactions after deflocculation.

12.5 FLOATATION

Flotation is a distinct method to recover solids from liquids, allowing solid particles to float on the surface. This process runs on the principles of buoyancy and hydrophobicity. When the liquid feed is injected with gas bubbles (air or nitrogen gas), the solid particles (size range 10–200 μm) adhere to the gas bubbles by physicochemical interactions of adsorption and collision. This results in low-density components that rise and float at the liquid surface. The process may also use different reagents to improve the solid separation. For example, a substance having a hydrophilic surface can be modified into a hydrophobic surface by using collectors (or surfactants) such as long-chain hydrocarbon compounds, which makes the particle floatable. In addition, some chemicals can be added to the solution for froth stability. They are called a frother and a modifier (or conditioner).

The formation of gas bubbles in the solution is also possible through natural processes. For example, the production of ethanol and acetic acid triggers foam generation during the inbuilt natural metabolic pathways of microbial cells. The gas produced during the fermentation reaction helps develop froth in the medium. In a different scenario, an agitator with a pump can be installed to blow air from outside.

Compared to other techniques, flotation is one of the most cost-effective, energy-efficient, and resource-saving solid-liquid separation methods. This is a suitable method for separating yeasts, bacteria (e.g., *Bacillus subtilis* and *E. coli*), microalgae, and proteins after fermentation. A similar process can be employed to recover surface-active substances from the solution based on their surface activities.

12.6 SOLVENT EXTRACTION

Solvent extraction, liquid-liquid extraction, or partitioning is a technique which allows the selective transfer of a compound from one solvent to another. This is a relatively simple and sensitive process with high selectivity and fast mass transfer. When the original feed solution (fermented broth) consisting of enriched solute (F) is stirred with an extraction solvent (S), the solute of interest gets transferred from the feed stream in the extraction solvent as an extract (E). The feed solution after the removal of the solute component is called raffinate (R) (Figure 12.7).

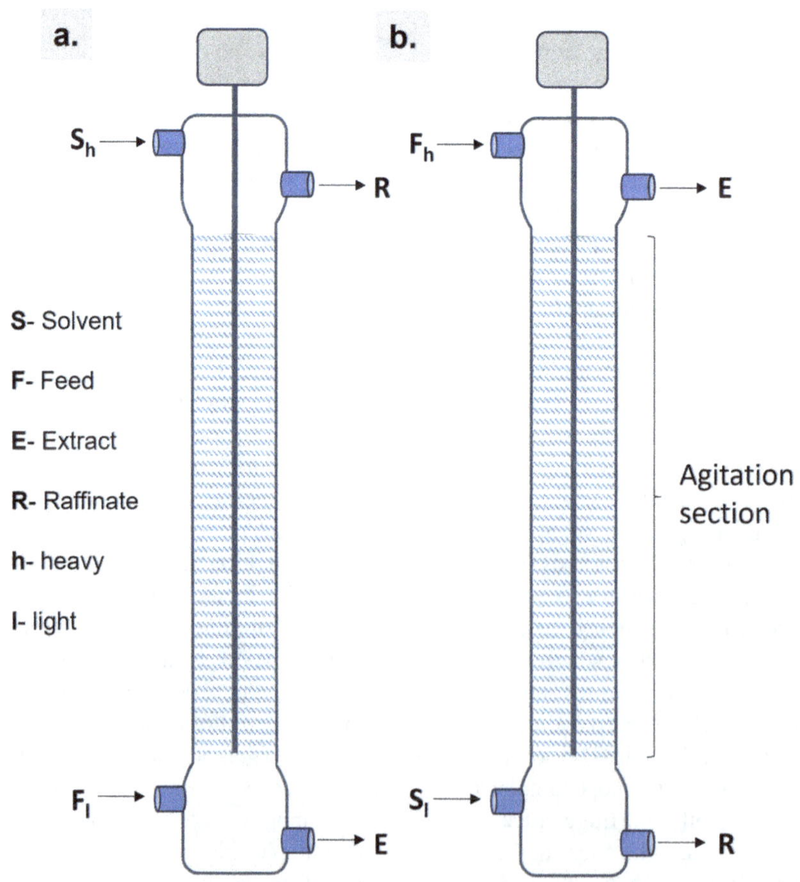

FIGURE 12.7 An illustration of solvent extraction. a. When the solvent phase is heavier than the feed, their respective inlets are positioned at the top and bottom, and therefore, outlets for raffinate and extract lie at the top and bottom, respectively. b. When feed is heavier than solvent, their inlets are situated at the top and bottom, respectively, and the outlets for effluent and raffinate also remain in the same order.

Figure 12.7. An illustration of solvent extraction. a. When the solvent phase is heavier than the feed, their respective inlets are positioned at the top and bottom, therefore, outlets for raffinate and extract lie at the top and bottom, respectively. b. When the feed is heavier than the solvent, their inlets are situated at the top and bottom, respectively, and the outlets for effluent and raffinate also remain in the same order.

$$F + S = E + R$$

Where F represents feed stream; S, extraction solvent; E, extract; and R, raffinate.

The extraction process chiefly involves two immiscible solvents: aqueous and organic solutions. The choice of organic solvent plays a key role in effective partitioning, allowing high selectivity and high solubility for the solute in an extraction solvent. Organic solvents used in liquid-liquid extractions are hydrophobic, are typically less dense than water, and form the upper phase. When these two liquids are vigorously shaken for perfect contact, the solute of interest transfers from one phase to another. At equilibrium, the solute distributed between two immiscible solutions is called the partition coefficient or distribution, denoted as KD. At equilibrium,

$$KD \rightarrow [X]l/[X]h \; or \rightarrow [X]org/[X]aq$$

[X]l is the concentration of component "X" in the lighter phase (or [X]org, in the organic phase), and [X]h is the concentration of "X" in the heavier phase (or [X]aq in the aqueous phase)

At equilibrium and a specific temperature, the KD of the solute in two known solvents is a constant value. According to the equation, a high KD means that the solute is more soluble in the organic phase than in the aqueous phase, whereas a low KD indicates the vice versa. For example, when the value of KD is above 3, it favours the effective separation of solute from the aqueous phase to the organic phase in a single stage of extraction. Likewise, the low KD value (e.g., KD < 1/3) ensures a higher solute concentration in the aqueous phase than in the organic phase. The equation also indicates that even with an elevated KD value on either phase, a low concentration of solute still resides in the feed. The extraction procedure can be repeated by adding more solvent, which is known as multistage extraction. Multistage extraction allows for the separation of an extended amount of solute from the solution.

Solvent extraction can be performed using a conical-shaped glass separatory funnel, which is a simple apparatus used for small laboratory-scale setups. An industrial scale separation through solvent extraction requires highly sophisticated instruments such as a continuous extraction system, a packed column extractor, or a liquid-liquid extraction tank.

Many antibiotics that have high partition coefficients in organic solvents can be purified and concentrated through solvent extraction. For example, the extraction of penicillin from the fermentation broth requires solvents like amyl acetate and butyl acetate. Erythromycin can be extracted using a different solvent, pentyl acetate. The extraction of lipophilic compounds from microorganisms and microalgae utilises organic solvents such as chloroform, hexane, acetone, petroleum ether, and methanol. This technique can also be used to remove impurities such as phenol and higher alcohol from ethanol fermentation when a feedstock, lignocellulose, is involved. This extraction technique can also be employed for the removal of different carboxylic acids, such as acetic acid, formic acid, succinic acid, and valeric acid, from alcohol fermentation using different organic solvents.

A similar principle is used to create an aqueous two-phase system (ATPS), which necessitates two immiscible aqueous phases for the purification of proteins. The two immiscible phases of ATPS are developed by mixing two different compounds (either two different polymers or a polymer and a salt) in the water itself. The extraction using ATPS takes a longer time than the organic-aqueous distribution. The extraction of proteins using organic solvents is not possible because it can disrupt their 3D configuration, which is essential for their activities. One example that includes ATPS in the recovery of an enzyme is the separation of glucose isomerase using immiscible aqueous solutions of polyethylene glycol and potassium phosphate.

Solvent extraction is a very popular technique in microbial bioprocessing, owing to its properties of low energy consumption, low capital investment, ease of automation, simplicity of the process, and continuous operation. The process ensures bulk purification using minimum extraction time. This method is also suitable for those compounds which are heat-sensitive and non-volatile. However, the high consumption of solvent during extraction increases operational costs and raises environmental concerns. Moreover, the dissolution of unwanted compounds during solvent extraction is a significant challenge during the subsequent purifying steps.

12.7 PRECIPITATION

In downstream processing, precipitation can be used to concentrate the desired products for fractionation or to remove impurities. The operation runs on the principle of solubility. This involves the conversion of a soluble substance into an insoluble precipitate, keeping the supernatant solution apart. The precipitates settle down at the bottom due to their high density. However, some fine solid particles remain dispersed as suspended particles in the solution. The morphology of the precipitate is undefined and disordered.

A typical precipitation reaction uses some specific chemical reagents called precipitants or precipitating agents. The choice of a suitable precipitant is determined by the nature of the substances to be separated and the nature of the solvent. For example, inorganic salts such as ammonium sulphate ($(NH_4)_2SO_4$) and sodium chloride (NaCl) are notable choices of precipitants for ionic precipitation. The addition of organic solvents like ethanol and acetone to the solution reduces the hydration of the proteins, resulting in protein aggregation and precipitation. Several metal ions (e.g., Cu^{2+}, Zn^{2+}, and Fe^{2+}), non-ionic polymers (e.g., polyethylene glycol (PEG)), ionic polymers (e.g., polyacrylic acid and polyethyleneimine) and ligands offer specific binding with different chemicals in the solution for their precipitation. Acid or base can also be used to adjust the precipitation or dissolution of pH-dependent chemicals. In a different method, an increase in the temperature of the solution can also induce precipitation of heat-sensitive proteins.

After precipitation, the precipitate is separated from the supernatant. The precipitate formed as a result can be either a product or an impurity. If the precipitate consists of the desired product, it can undergo further treatment, such as washing and drying, for final recovery. Otherwise, this can be discarded after separating from the supernatant. Later, precipitating agents can also be recovered by different techniques like filtration or centrifugation. The main benefits of precipitation are its low cost and high yield.

An example of precipitation in microbial biotechnology is the purification of succinic acid from fermentation broth. The addition of $Ca(SO_4)_2$ in a succinic acid-containing solution enables precipitation of calcium succinate. The precipitate can further be converted into a free acid using sulphuric acid (H_2SO_4). This preparation yields about 92% purity of succinic acid. This technique is also useful in the recovery of citric acid from solution. In one step, calcium hydroxide ($Ca(OH)_2$) is added to the solution to react with oxalic acid (a byproduct) to form a precipitate of calcium oxalate, which can be separated. In the same environment, citric acid is transformed into monocalcium citrate, which is retained in the solution. The pH of the mother liquor is adjusted to roughly 9.0 and a temperature of 90 °C, which promotes the precipitation of tricalcium citrate tetrahydrate. The recovered precipitate is highly purified and can be redissolved in 70% H_2SO_4, where Ca^{2+} combines with SO_4^{2-} to generate insoluble $CaSO_4$ (gypsum).

The precipitation technique is also highly reliable for the recovery of antibiotics like tetracycline, erythromycin, and rifamycin. However, the stability of precipitated products is a great concern for therapeutic applications in which product efficacy, safety, and consistency are paramount. A different technique, crystallisation, also involves a similar operation but works on a different principle and is used for the formulation (detailed in Section 12.12). Table 12.1 provides a comparative account between precipitation and crystallisation.

12.8 ULTRAFILTRATION AND REVERSE OSMOSIS

Ultrafiltration is a highly sophisticated process to filter out small molecules from a solution using a membrane filter having a very fine pore size (0.001–0.1 µm). When the solution passes through the membrane, suspended particles are removed from soluble compounds. Ultrafiltration can separate items like viruses, colloidal silica, milk proteins, gelatins, and endotoxin pyrogen. It is an indispensable process to separate cellular debris and bacteria from draft beer to produce a high-quality beer. Diafiltration is related to ultrafiltration, which is performed by adding solvent to a concentrated solution while removing permeate in a flow-through manner.

A related but more sophisticated filtration system is nanofiltration. This system has even smaller pore sizes and is used for purifying enzymes in the pharmaceutical industry. This process counts on a distinct filter medium called a nanofilter that consists of a charged membrane. This process fills the void between ultrafiltration and reverse osmosis. These filters can remove dissolved substances such as multivalent ions and divalent ions, but allow the passage of monovalent ions. This technique, along with ultrafiltration, is a reliable tool to recover antibiotics and hormones from the solutions. The treatment of feed solution through this filtration technique and even by reverse osmosis (RO) requires a pretreatment through microfiltration to remove large particles such as bacteria, viruses, and other substances. This approach ensures the long-term efficiency of nanofiltration membranes.

Reverse osmosis (or hyperfiltration) employs filters having a pore size of approximately 0.0001 µm, which is even smaller than ultrafilters and nanofilters. This operation requires a strong pressure (100–800 psi) against the natural flow of water from its high concentration to its low concentration. The artificial pressure induces the water movement in the reverse direction to osmosis and filters most of the minerals from the solution. Hyperfilters positioned in reverse osmosis have a filter efficiency of 95–99%. Such filters can even prevent the transport of small molecules like NaCl across the membrane and desalinate the water. But this process also removes healthy minerals like Ca^{2+} and magnesium ions (Mg^{2+}) from the water.

TABLE 12.1
Dissimilarities and similarities between precipitation and crystallisation

Dissimilarities	Precipitation	Crystallisation
Stage of treatment	As one of the initial steps	As one of the final steps
Product morphology	Amorphous and colloidal precipitate	Highly defined crystals
Process	Uncontrolled	Can be controlled
Similarities	Liquid-solid separation	
	A crucial step of the downstream process	
	Based on the principle of solubility	
	Use of chemicals and other parameters to alter the solubility	
	Economical and convenient	

12.9 CHROMATOGRAPHY

Chromatography is a powerful technique to separate a mixture of compounds in a solution based on their differential migration and selective retardation into two distinct phases: the stationary phase and the mobile phase. The stationary phase is a porous solid material through which the mobile phase can be either gas or liquid flows (Figure 12.8). Some components pass rapidly into the mobile phase, while others stay in the stationary phase for a long duration. The migration and retention of the feed components in the column arise due to differences in size, adsorption, affinity towards immobilised ligands, or molecular weights.

Based on the purpose, chromatographic separation can be analytical and preparative. The principle behind both techniques remains the same, but the latter setup is designed to handle larger sample volumes. Therefore, preparative chromatography involves large columns and high flow rates, allowing the separation of valuable products for industrial biotechnology. On the other hand, analytical chromatography is mainly used to identify and quantify components in a mixture. This process utilises a very low volume of the sample. The mass ratio of the sample to the stationary phase lies below 1:100,000 for chemical analysis.

Based on different requirements and molecular characteristics of components, a variety of chromatographic procedures are available. The most popular and commonly used chromatography techniques in microbial biotechnology include-

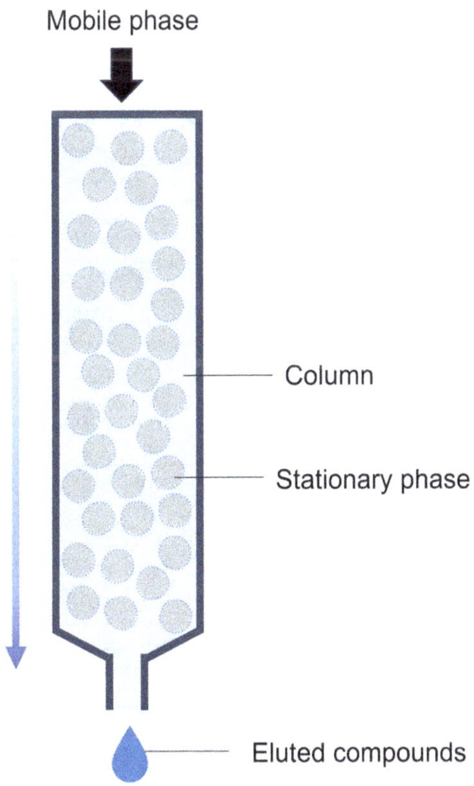

FIGURE 12.8 A general view of the chromatography technique.

- Adsorption Chromatography- In this process, the column is packed with adsorbents like silica gel, alumina, and charcoal for the selective adsorption of a solute. The mobile phase elutes from the column because of low interaction with the adsorbents.
- Partition Chromatography- This is a procedure of liquid-liquid extraction, where a compound is partitioned between two liquid phases. The stationary phase is developed as a liquid coated on a solid support. The porous silica particles are the most commonly used material for solid support.
- Ion-Exchange Chromatography- This separating technique is based on the affinity of ion exchangers to charged biological molecules at low ionic strength. Ion exchangers are solid materials or liquid solutions that generate a stationary phase. They display specific charges on the surface and can be of two types: cation and anion exchangers. Cation exchangers have negatively charged groups and attract cations, while anion exchangers include positively charged groups and attract anions. The commonly used cationic exchangers are carboxymethyl (CM)-cellulose, sulphopropyl (SP)-sepharose, and agarose. Examples of anion exchangers include diethyl aminoethyl (DEAE)-cellulose, quaternary ammonium (Q)-sepharose, and quaternary aminoethyl (QAE)-Sephadex. The transient variation in the pH of the mobile phase has a major influence on the separation through ion chromatography, as the net charge alters with pH change.
- Gel Chromatography- Gel chromatography or size exclusion chromatography is a separation method based on the difference in sizes or molecular weights of chemical substances using a bed of a porous, semisolid substance. The most commonly used beds are polyacrylamide and dextran-containing materials. In the gel, the smallest compounds tend to elute by the tail-end of the chromatograph, and the largest compounds remain by the front end. Medium-sized molecules in the column diffuse according to their sizes between the front and tail ends.
- Affinity Chromatography- This type of chromatography is the most powerful method of separation and can produce several hundred-fold purified compounds. The basic principle of the separation is a highly specific binding between an immobilised ligand (i.e., stationary phase) and a receptor in any mixture (i.e., mobile phase). This type of interaction is reversible. The variation in the pH, ionic strength or polarity of the elution buffer can elute the target compounds. This separation technique is mainly used in the purification of proteins.
- High-Performance Liquid Chromatography (HPLC)- It is widely used for quantitative analysis. The fundamental principle of high-performance liquid chromatography (HPLC) involves the distribution of sample compounds between a stationary phase (in a column) and a mobile phase (from the pump). The commercial scale operation is run using an advanced setup called a preparative

HPLC system. The preparative scale operation produces a large amount of pure compound quickly and at a very high flow rate (100 ml/min). A vacuum pump can be used to generate pressure for mobile phase delivery and controlled solvent flow. This technique can be used to obtain highly pure organic acids like citric acid, oxalic acid, and succinic acid from a sample mixture.

- Gas Chromatography- This is also an analytical tool which allows the detection and quantification of organic molecules or gases present in a sample mixture. The preparatory setup of gas chromatography exists for the separation of desirable compounds. The basic setup utilises a liquid as the stationary phase and a gas as the mobile phase.

Chromatographic techniques are used to produce high-quality products for pharmaceuticals, diagnostics, therapeutics, and research and development. Some typical applications of chromatography are the purification of antibiotics, amino acids, steroids, lipids, and mycotoxins. They also play important roles in the manufacturing of food-grade and medical-grade maltose, high-fructose corn syrup (HFCS), and highly pure forms of glucose and fructose. Continuous chromatography techniques are being developed to further improve throughput and reduce costs. However, chromatography methods are time-consuming and require technical expertise for their operations.

12.10 LYOPHILISATION

Lyophilisation or freeze drying is a technique of solvent removal (or dehydration) from the product by freezing the material. Many microbial-derived products are heat-labile and cannot be recovered through boiling. This drying technology bypasses the step of the liquid phase, allowing the frozen material to change directly from solid to vapour under the vacuum. This process ensures the removal of solvent from the medium. Lyophilisation can be divided into three sequential steps: freezing, sublimation (primary drying), and desorption (secondary drying) (Figure 12.9). A pretreatment step may additionally be required to remove excess water from the solution prior to lyophilisation.

In freezing, the liquid material is kept at a very low temperature, at -40 °C or -50 °C, for several hours for its complete transformation from liquid to solid. This step is the first stage to separate the solute from water by converting it into ice. Slow freezing of the material ensures the formation of large crystals, which warrant maximum dehydration from the product during sublimation. Cryoprotectant agents such as dextran, glycine, mannitol, and trehalose can be added to prevent cell damage from aggregation or shrinkage during freezing. A thermal treatment called annealing is followed by freezing. This can be accomplished by keeping the temperature between the ice melting temperature and the transition temperature of the frozen concentrate (Tg') for a defined period. Depending on the constituents present in the formulation, the temperature of the content is increased from -40 °C to -25 °C or even higher and then back to low temperature. The freezing-annealing cycle can influence the size distribution of ice crystals and the texture of the final product.

The next step is the aseptic loading of frozen material in the lyophiliser. The machine operates on the principle of sublimation, which generates a high vacuum at sub-ambient to elevated temperatures only below the triple point (a temperature at which solid, liquid, and gas phases coexist) of water. The triple point for water is 0.01 °C (32.02 °F) and 4.58 torr (0.00603 atm) (Figure 12.9a). The high vacuum in the equipment creates low pressure for the sublimation. The frozen material obtains heat for sublimation through conduction or radiation. This is a slow step and takes several hours. In this process, most materials can be dried off with roughly 95% water content.

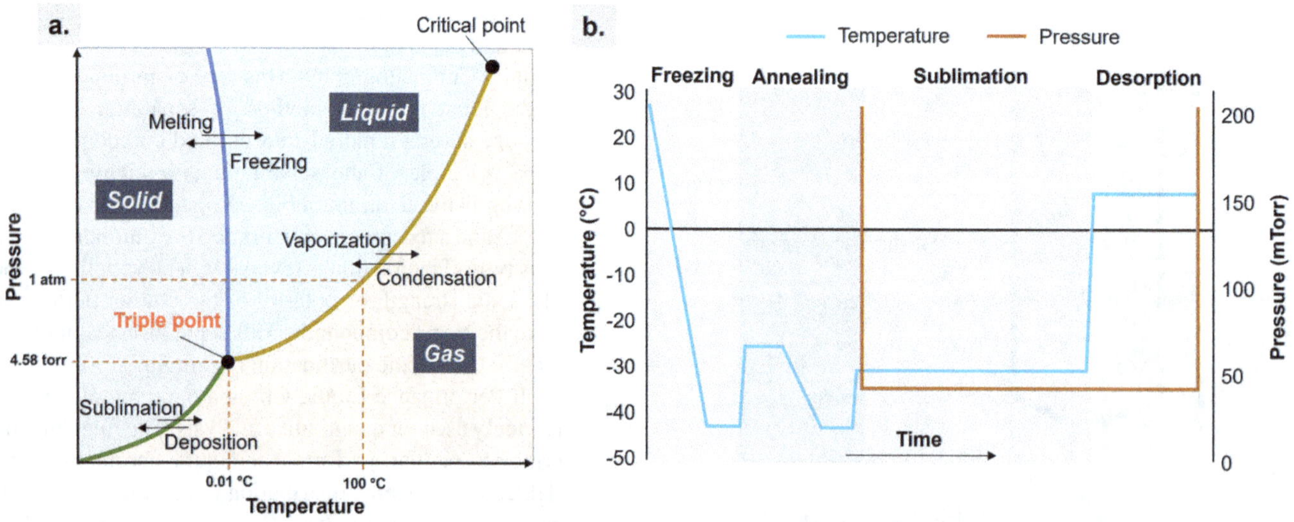

FIGURE 12.9 a. A phase diagram showing the triple point of water. b. The variation in temperature and pressure during different steps of lyophilisation.

After the primary drying, a low amount of unfrozen water (about 5%) remains associated with the product. A subsequent drying step, secondary drying, is necessary to remove the bound water, which is based on the principle of adsorption isotherms. In this process, the supply of additional heat increases the temperature of the content above 0 °C, which breaks the physicochemical interactions between water and product. This step can remove the remaining 1–5% residual moisture from the product. Figure 12.9b provides the details of different temperatures and pressures required during various stages of lyophilisation.

Although lyophilisation is relatively costly and slow compared to oven drying, it maintains the potency of the product. About 60–70% of microbial cells remain viable after lyophilisation. Also, this technique has simple aseptic handling, and the instrument can be equipped with a stoppering facility to seal the content after drying. The lyophilised product has a long shelf life and can be reconstituted after dissolution. In many commercial facilities, resource centres, and research institutes, microbial cells are stored as lyophilised cultures to provide extended viability. Lyophilisation can also be used to preserve ultrasensitive pharmaceutical products such as antibiotics, amino acids, proteins, sugars, and polymers.

12.11 SPRAY DRYING

Spray drying is a technique of converting fluid material into powdery products at low temperatures. This is the preferred method of drying over lyophilisation owing to its low-cost approach, high energy efficiency, and high process yield. Also, the drying process is very fast and continuous. Spray drying is based on the principle of convection. It initiates through a process called atomization, where a fluid material is fed into the chamber for its transformation into fine mist or droplets. When the hot air comes in contact with the atomised droplets, the water is released through evaporation and the solute is converted into a powder that settles at the bottom.

The equipment used in the process is a spray dryer, and it can be designed in any capacity. In commercial scales, the feed rate can be as high as 100 tonnes per h. The main part is a large cylindrical chamber with a conical bottom made up of stainless steel. The roof of the chamber is connected to the hot air supply system and the feed supply system. The hot air supply system consists of a fan, process air heater, and high-efficiency particulate air (HEPA) filter. Liquid feed is transported to the atomiser through a feed pump. The atomising device, an atomiser or nozzle, is fixed at the top of the chamber. The atomiser disperses the feed solution into a large number of nanosized droplets. Depending on the requirements, atomisers can be of three different types:

- Rotary atomiser- This consists of a high-speed rotating wheel, disk, or cup that converts the bulk liquid into droplets using centrifugal force. This atomiser produces homogeneous particles. The particle size can be controlled between 10–200 µm using the speed of the atomiser in the range of 10,000–30,000 rpm.
- Pressure atomiser- This type of atomiser relies on a pressure energy of 250–10,000 psi. The particle size produced from this type of nozzle is coarse and less homogeneous.
- Two-fluid atomiser- In this atomiser, liquid feed and hot gas are transported using the same nozzle, and the atomisation takes place using kinetic energy.

The lower part of the chamber is connected to a drum called a powder recovery system to collect the dry product. A cyclone in the dryer is situated between the chamber and

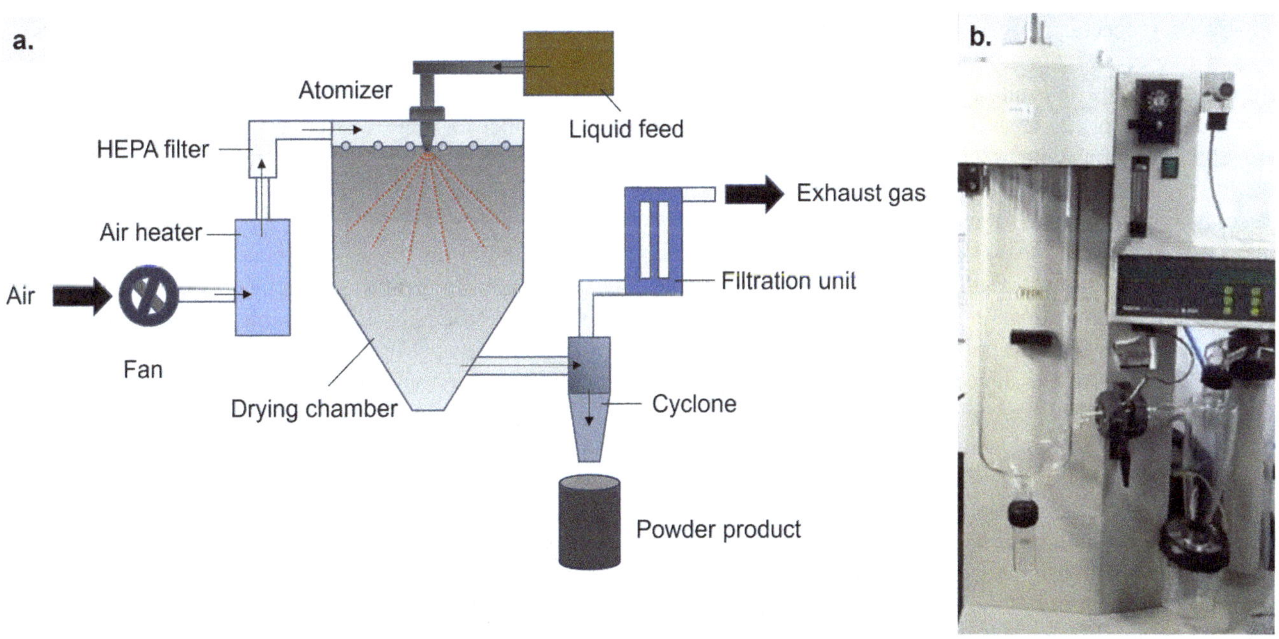

FIGURE 12.10 a. An illustration of spray drying. b. A setup of a spray dryer.

drum, which acts as a separator of solid product from the gas phase by applying a centrifugal force. An outlet from the cyclone shows the pathway for the exit gas that passes through the bag filter (Figure 12.10).

This technique is very useful for large-scale operations, but still has some limitations for the drying of microbial cells. The temperature of hot air in the chamber can reach up to 150–200 °C, which is usually not suitable for sensitive microorganisms and biological products. Many microbial cells experience injury during such treatment and can not be preserved through spray drying. However, lactic acid bacteria (LAB) such as *Lactobacillus lactis*, *Streptococcus thermophilus*, and *Bifidobacterium longum* are less sensitive to shear forces or heating while travelling from an atomiser. The spray-dried LAB are used as starter cultures. Bacteria harvested from the stationary phase of the growth ensure the maximum viability of cells during spray drying. Although the survival proportion of microbial cells after spray drying is relatively low compared to lyophilisation, the incurred cost of spray drying is about 80% less than lyophilisation. Also, the use of protectants or carriers such as skim milk, maltodextrins, and gum acacia can positively influence drying by preserving the sensitive properties of biomaterials. In addition to drying microbial cells, spray drying has tremendous applications in producing powdered milk, eggs, coffee, tea, enzymes, vitamins, and animal feeds.

12.12 CRYSTALLISATION

Crystallisation is one of the final steps of purification and recovery of desirable substances in the form of crystals. During crystallisation, the solution undergoes a physical transformation from its liquid phase to the solid phase. The recovered crystals are highly organised, rigid, and pure. The operation is based on the solubility principle, which states that decreasing the solute's solubility leads to crystallisation. Nucleation (or seeding) followed by crystal growth are the two major succeeding events that create the macroscopic sizes of fine crystals. In crystallisation, external factors like temperature, seeds, crystallisation agents, electric fields, and co-crystallising agents create a situation where the physical state of the solution contains more solute than the equilibrium solubility. This generates a supersaturation condition in which the solute transforms into crystals.

Known as one of the oldest methods of downstream processing, crystallisation is still very effective and economical for large-scale applications in industrial microbiology. This process is suitable for recovering many different microbially-derived products such as fumaric acid, succinic acid, malic acid, lactic acid, L-phenylalanine, β-lactam antibiotics, and enzymes. The combined efforts of temperature and pH can recover many organic acids in the form of crystals. An adjustment of the fermentation broth at pH 2 and 4 °C facilitates the crystallisation of the high-value succinic acid. Although the crystallisation facilitates only 70% recovery of succinic acid, the purity of the compound is very high (about 90%). Fumaric acid crystals can be developed by altering the pH of the fermentation broth between 0.5 and 3.5 (using 0.8 M H_2SO_4 solution) at 30 °C. The pH adjustment of the fermentation liquor to a range of 2.5 to 4.0 and heating to 60–70 °C can crystallise L-tryptophan. In a different situation, the addition of non-ionic detergents in the fermentation broth effectively reduces the solubility of L-tryptophan, resulting in its crystallisation. The cross-linked enzyme crystals (CLECs) technique is a distinct approach to the crystallisation of enzymes in pure forms. However, in the fermentation broth, the contamination of microbial cells, diverse byproducts, and nutrients can negatively affect the thermodynamic and kinetic characteristics of crystallisation.

SUMMARY

- Filtration is a method of solid-liquid separation, where a porous filter medium allows the passage of liquid and captures the solid particles. A vacuum pump can be used to create a pressure differential to speed up the filtration process.
- Centrifugation involves artificial gravity for solid-liquid or liquid-liquid separation using a centrifuge. Industrial centrifuges come in great variations, such as perforated bowl basket centrifuge, tubular bowl centrifuge, disk bowl centrifuge, multi-chamber centrifuge, and solid bowl centrifuge.
- Flocculation and flotation are preliminary steps of downstream processing. The former technique converts particles into bulky flocs that settle at the bottom, whilst the latter process injects gas bubbles in the fluid to make solid particles float on the surface.
- When the desired bioproduct is intracellular, cell disruption becomes necessary. Some prominent methods of cell disruption are high-pressure homogenisation, bead milling, ultrasonication, thermolysis, osmotic shock, and solvent and enzyme treatment.
- Solvent extraction is a method of mass transfer of compounds between two immiscible solvents: an organic phase and an aqueous phase. Vigorous mixing of these phases transfers the solute from one solvent to another for their recovery into a new phase.
- Precipitation runs on the principle of solubility by converting a soluble substance into an insoluble precipitate. Precipitate has an undefined and disordered structure and can be separated by sedimentation, filtration, or centrifugation.
- Chromatography is a valuable technique for achieving high purity of a biological compound by taking advantage of the different affinities or partitioning tendencies of the product in a sample between a stationary phase and a mobile phase.
- Lyophilisation and spray drying are the methods to transform a liquid product into a solid form. The former technique is driven through sublimation and desorption, whereas the latter works on the principle of convection. A different technique, crystallisation, induces insolubility in the liquid phase to convert it into a large, organised crystal.

DIFFERENT TYPES OF PRACTICE QUESTIONS

Fill in the blanks

I. .... is a process that applies Stokes' law to separate different components from the solution.
II. An enzyme .... can be used to hydrolyse peptidoglycan in Gram-positive bacteria to recover cell-bound products.
III. In solvent extraction, the feed solution from which the solute component has been withdrawn is called ....
IV. The value of .... remains constant for a solute distribution between two known solvents at equilibrium at a specific temperature.
V. In a precipitation reaction, a solid forms and separates from a solution using some specific chemical reagents called ....
VI. Spray drying works on the principle of ....

SOLUTION: 1 I. Centrifugation, II. Lysozyme, III. raffinate, IV. partition coefficient or distribution, V. precipitants or precipitating agents, VI. atomization

Match the following

i. Chromatography	a. Formulation
ii. Lyophilisation	b. Liquid-solid separation
iii. Bead Milling	c. Liquid-liquid separation
iv. Solvent extraction	d. Recovery of intracellular content
v. Filtration	e. Purification

SOLUTION: i-b, ii-a, iii-d, iv-c, v-e

Match the following

i. Sublimation	a. Solvent extraction
ii. Atomizer	b. Solid bowl centrifuge
iii. Screw conveyor	c. Spray dryer
iv. Aqueous two-phase system	d. Lyophilisation
v. Pickup zone	e. Rotary drum vacuum filtration

SOLUTION: i-d, ii-c, iii-b, iv-a, v-e

Very short answer type of questions

I. What is the use of filter aid in a filtration unit?
II. Define flocculation with a suitable example.
III. Mention the names of two different chemical methods for cell disruption.
IV. What are the main benefits of using enzymes in cell disruption?
V. Write the names of two microbiologically derived products that are purified using the precipitation technique.
VI. What is the difference between primary and secondary drying in lyophilisation?
VII. Expand HPLC.
VIII. What is bead milling and its applications in downstream processing?

Short answer type of questions

I. What is solvent extraction? How can this technique be used in the downstream processing of biological products?
II. What are the similarities and differences between crystallisation and precipitation in downstream processing?
III. Write a short note on rotary drum vacuum filtration, including a suitable diagram.
IV. How are ultrafiltration and reverse osmosis different from microfiltration? What are their different applications?
V. Discuss different methods of chromatography.
VI. What is a triple point? Which downstream technique relies on this term? And how?

Long answer type of questions

I. What is the role of cell disruption in downstream processing? Discuss different types of cell disruption methods.
II. Give a detailed account of three different methods that can be used in the formulation of a product in industrial biotechnology.
III. What is the role of centrifugation in downstream processing? Discuss different types of industrial-scale centrifuges.

SUGGESTED READINGS

Flickinger MC (ed.). *Downstream industrial biotechnology: recovery and purification*. Wiley; 2013.

Lee SY, Cho JM, Chang YK, Oh YK. Cell disruption and lipid extraction for microalgal biorefineries: a review. *Bioresource Technology*. 2017;244:1317–28.

Peighambardoust SH, Tafti AG, Hesari J. Application of spray drying for preservation of lactic acid starter cultures: a review. *Trends in Food Science and Technology*. 2011;22(5):215–24.

Santivarangkna C, Kulozik U, Foerst P. Alternative drying processes for the industrial preservation of lactic acid starter cultures. *Biotechnology Progress*. 2007;23(2):302–15.

Su W, Jiang Y, Zuo X, Li C, Wang H. Engineering nucleation/crystallization to intensify the enzymatic reactions and fermentation: a review. *Chemical Engineering Journal*. 2021:134186.

Sutherland K, Chase G. *Filters and filtration handbook*. Elsevier; 2011.

Verstrepen KJ, Derdelinckx G, Verachtert H, Delvaux FR. Yeast flocculation: what brewers should know. *Applied Microbiology and Biotechnology*. 2003;61(3):197–205.

13 Production of Industrially Important Products

13.1 PRODUCTION OF MICROBIAL ENZYMES

Modern industries ensure sustainability through the use of enzymes. Microbial enzymes have the properties to replace chemical treatments in various industries and also play roles in the treatment of domestic and industrial wastes. The depletion of nonrenewable resources, new requirements of industries, and in-depth knowledge of microbial enzymes have promoted their production on a mass scale. Various properties of enzymes, such as stability, catalytic potential, feasibility, consistency, and ease of production and optimisation, allow them to be used across several fields. They carry out reactions in a different range of environmental conditions of temperature, pH, salinity, and aeration. Other advantages of enzyme applications include improved quality of products, low manufacturing cost, less waste, and high energy efficiency. They are powerful tools of green chemistry and offer an eco-friendly solution. When enzymes are released into the environment, they can be degraded by natural microorganisms. The enzymes from microbial sources are immensely diverse. The rapid growth rate and exceptional performance of microorganisms in fermenters are the key to the commercial production of microbial enzymes.

The delights of microbial enzymes for industrial production came into existence in the 1980s and 1990s and have remained so ever since. The follow-up research made important breakthroughs in various requirements for industrial setups. These efforts are attributed to genetic engineering, metabolic engineering, and system biology. Directed evolution has been a critical toolbox that expedites the natural evolution process of enzymes to improve their characteristics. Metagenomics has led to the identification of many novel enzymes with unique properties (e.g., amylase, lipase, and cellulase) that are highly valuable for industrial applications. In metagenomics, the analysis of genetic sequences of environmental DNA allows the identification of diverse genes encoding enzymes and the discovery of enzymes that can have stability, activity, or specificity. Also, many industrial enzymes have been immobilised to increase their stability and reusability in harsh environmental conditions. Chapter 15 provides a detailed overview of immobilised enzymes and their industrial applications.

In recent years, the industrial enzyme market has been estimated at approximately $12 billion. With an increasing demand for enzymatic solutions, the global enzyme market has been projected for a compound annual growth rate of more than 6% in the coming years. The demand for enzymes is highest in the food and beverages industry, which aids a variety of applications in bakeries, dairy products, beverages, sugar syrups, processed meat, and nutraceuticals. The second largest consumer of microbial enzymes is the detergent industry in laundry and dishwashing detergents. Also, enzymes are the standard requirements in the feed industry, biofuel production, paper and pulp industry, textile industry, tanneries, and wastewater treatment. Several companies that are driven by enzyme production are Novozymes (Denmark), DuPont (USA), Genencor (Denmark), DSM (Netherlands), AB Enzymes (Germany), KAO (Japan), and BASF (Germany).

Microbial enzymes can catalyse various types of reactions, including redox reactions, hydrolysis, isomerisation, lysis, etc. Hydrolytic enzymes such as amylases, lipases, proteases, and cellulases are much more in demand than any other enzymes. Chapter 14 provides a detailed overview of microbial enzymes of industrial interest. Of all the different enzymes, amylases represent 25% of industrial enzymes on the market. They digest starch materials and can be categorised into three different types: α-, β-, and γ-amylases. Of all three types, α-amylases tend to be faster-acting and dominate over others. The following section gives an overview of the microbial production of α-amylases.

13.1.1 Production of α-Amylases

α-Amylases (endo-α-1,4-d-glucan hydrolases) (EC 3.2.1.1) are hydrolase enzymes that degrade starch into sugars. They act on α-1,4-glycosidic bonds, and the end products can be short oligosaccharides such as maltose, maltotriose, and branched oligosaccharides (Figure 13.1). All types of biological systems, including microorganisms, plants, animals, and humans, produce α-amylases. However, the commercial potential of α-amylases is found in bacteria and fungi, with thermostable amylases being in high demand because of their high stability. Their primary sources are various species of *Bacillus* and *Aspergillus*. Extremophilic amylases that operate

DOI: 10.1201/9781032651095-16

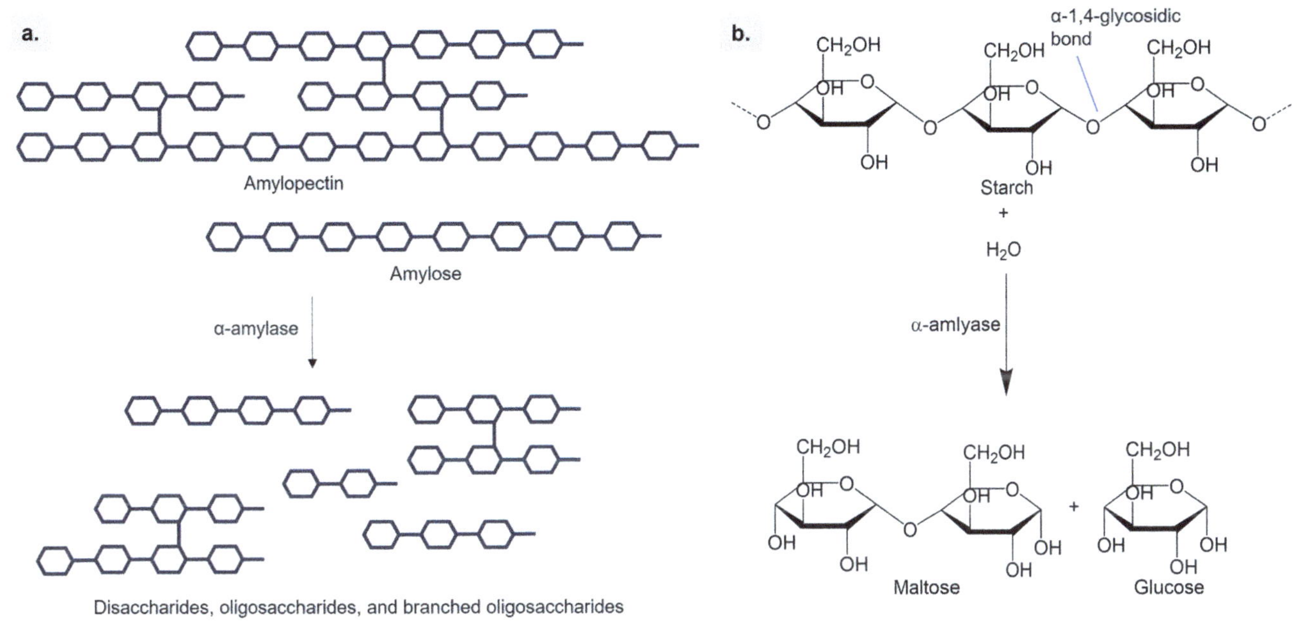

FIGURE 13.1 a. Degradation of amylopectin and amylose using α-amylase. b. Demonstration of starch hydrolysis at the molecular level.

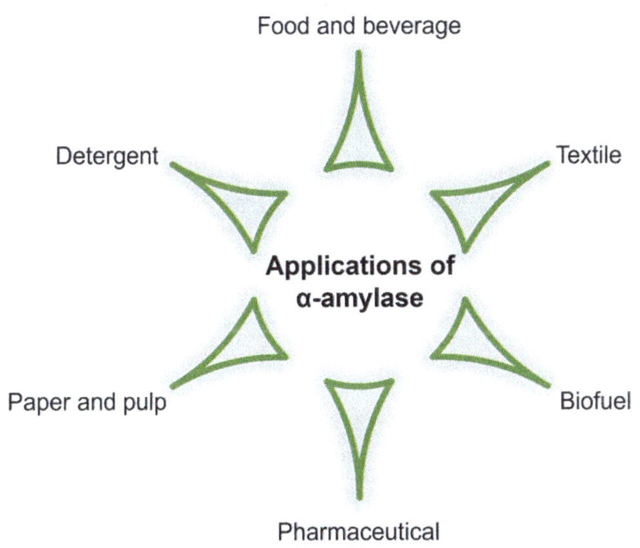

FIGURE 13.2 Applications of α-amylase in different industries.

under harsh conditions such as low temperature, high/low pH, and high salt concentrations have great value.

Thermostable amylase has particularly widespread applications in various industries (Figure 13.2). The food industry utilises thermostable amylases for baking, brewing, manufacturing fructose and glucose syrups, alcohol production, and cheese-making. The preparation of food-grade amylases mainly relies on fungal sources, of which *Aspergillus oryzae,* with its generally recognised as safe (GRAS) status, is the most common source. Other probable sources of food-grade amylases are *Bacillus amyloliquefaciens* and *Bacillus stearothermophilus*. In biofuel production, the bioconversion of raw material, corn starch, into dextrin (the main substrate) requires thermostable bacterial α-amylase and fungal glucoamylase. This step is known as liquefaction, a crucial pretreatment process in bioethanol production. The key sources of thermostable α-amylase are *B. licheniformis* and *B. subtilis*. In the paper and pulp industry, starch is used to coat paper, improving its strength. α-amylases cleave starch in the pulp, which reduces the viscosity of starch and facilitates the smoothening of paper. The textile industry uses starch as warp sizing to improve the strength of the thread, where α-amylase finds its application during the desizing process. In tanneries, soaking of skin or hides depends on a mixture of enzyme solutions, including an alkaline halophilic α-amylase, to accelerate the hydration and fibre splitting. *Halobacillus* sp MA-2 and *Halomonas meridiana* are natural sources of halophilic α-amylase. The cold active, high pH tolerant α-amylases are used in the detergent industry to digest starchy stains for laundry and automatic dishwashing. A genetically engineered α-amylase produced from *B. licheniformis* is an important ingredient in dishwashing detergent. A bacterium, *Alteromonas haloplanktis,* showed maximum production of α-amylase at 4 °C. Fungi such as *A. niger, A. awamori, A. kawachii, Thermomyces lanuginosus*, and *Penicillium* spp. can also produce significant amounts of α-amylase with diverse characteristics for their applications in different industries. Commercial α-amylases are available under different trade names for a wide range of applications. Some notable examples are Aquazym®, Bakezyme® P, Bioamylase® L, Duramyl, Fungamyl®, Natalase®, Purastar OxAm®, Stainzyme®, and Termamyl®.

The production of α-amylase at the industrial scale can be divided into stages of upstream processing, fermentation process, and downstream processing (Figure 13.3), as follows.

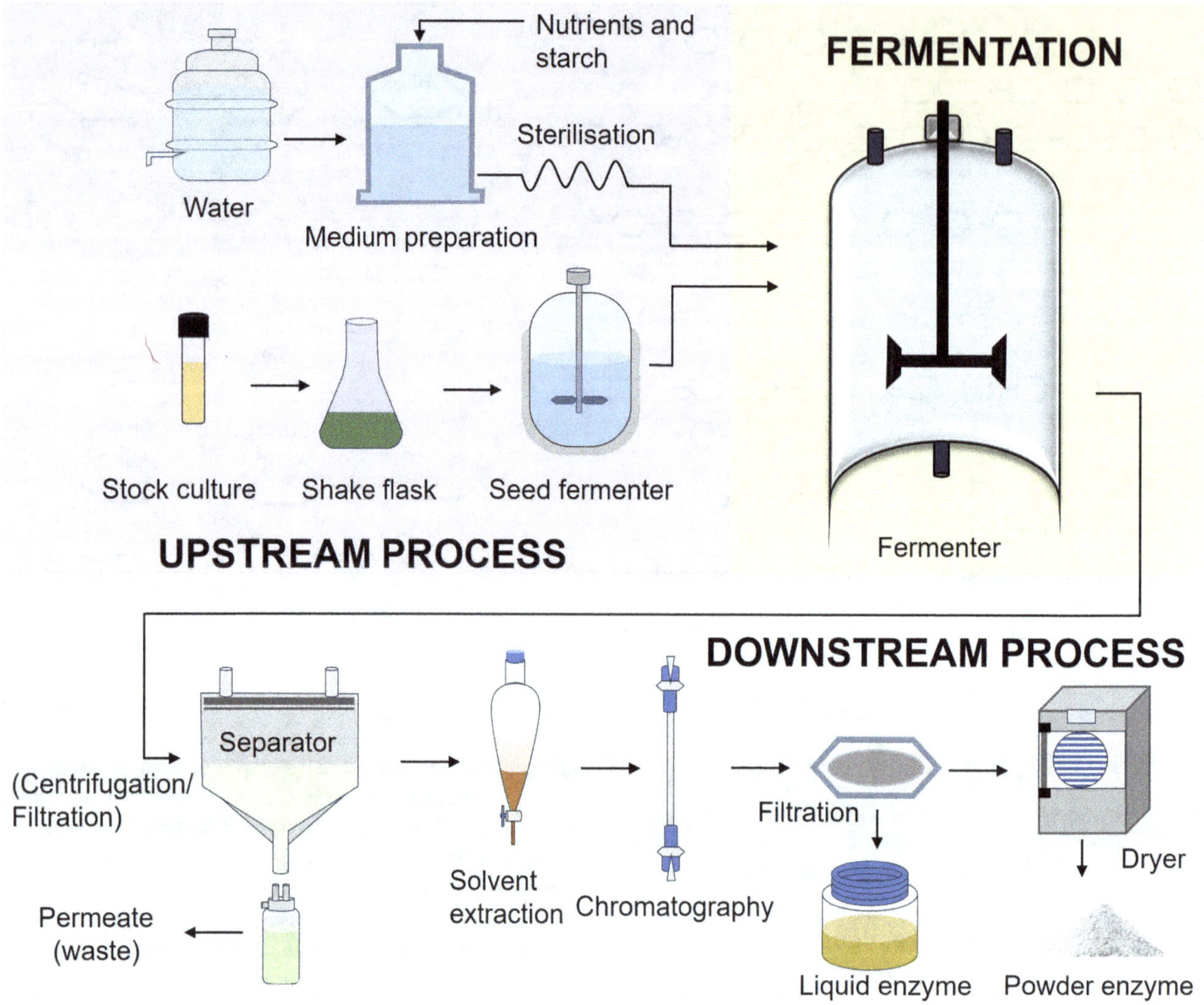

FIGURE 13.3 A general scheme of α-amylase production through submerged fermentation (SmF).

13.1.2 Upstream Processing

The upstream processing for the production of α-amylase requires many plans. This covers some non-routine and routine preparations. The non-routine preparations cover the isolation of suitable microorganisms and their genetic modification and optimisation of the situation for enzyme production. In contrast, routine preparations include seed development and culture medium preparation.

Isolation and characterisation of useful microorganisms are the primary steps toward the scrutiny of commercial sources of an enzyme. Also, a variety of commercially available enzymes are available in culture collections for their potential applications. The high-level expression in microorganisms can be achieved using directed evolution. Random mutagenesis and recombination techniques render producer organisms compatible for large-scale production. In industrial production, optimisation of the culture conditions is yet another crucial parameter for the high amount of enzyme secretion. Many factors influence enzyme production, including different carbon and nitrogen sources, pH, temperature, inducers, and fermenting microorganisms. Culture-based strategies for increased production include several statistical designs of one-factor-at-a-time (OFAT) experiments and factorial experiments (detailed in Chapter 5).

The production of α-amylase requires high substrate concentrations, which can be derived from various industrial and food wastes. Submerged fermentations (SmF) mainly rely on industrial processed by-products such as molasses and agricultural waste biomass. On the other hand, solid-state fermentation (SSF) utilises recycled waste materials such as paper pulp and agricultural waste biomass (e.g., bran and bagasse). These items contain high amounts of sugars, which serve as the substrates for enzyme production. The addition of a modifier or stabiliser in the fermentation medium improves the stability of the enzyme. A low concentration of calcium salt is important as it serves as a cofactor for the enzyme.

The inoculum for the fermentation is prepared using the stock, and its high growth is achieved through scale-up from flask culture to seed culture. Before full-scale production, the culture can be inoculated into a pilot-scale fermenter, where its growth conditions can be monitored and controlled.

13.1.3 Fermentation Process

The industrial-level production of α-amylase involves two distinct methods: SmF and SSF. SmF is a traditional method and requires a concentrated liquid nutrient medium. This type of fermentation involves a fast reaction and requires a regular supply of the substrate for a high biomass yield of α-amylase. The fermentation can be carried out using batch or fed-batch fermentation. Many studies claim that the synthesis of this enzyme in many microorganisms is inducible and, therefore, requires a high concentration of substrate. The medium requires a high level of dissolved oxygen to ensure aerobic metabolism for α-amylase production. The SmF also maintains several factors, such as pH, temperature, aeration, and moisture, to produce the optimum. The enzyme fermentation rate is maximum when the growth rate passes from the stationary phase. Both bacteria and fungi can be suitably used for α-amylase production under SmF. It is possible that the optimum conditions for growth and enzyme production are not associated. For example, a strain of *B. subtilis* showed maximum enzyme production at 37 °C, while it grew best at 42 °C. Most bacteria suitably grow and produce this enzyme in the temperature range of 37–48 °C, while for fungi, 30–55 °C. Some thermophilic strains claim to produce thermophilic enzymes. The high yield of α-amylase can be obtained after 48–72 h of incubation. Both upstream and downstream processing in SmF are relatively easy.

SSF is a modern technique of α-amylase production. SSF is suitable for α-amylase production using different fungi and some selected species of *Bacillus,* such as *B. subtilis*, *B. amyloliquefaciens*, and *B. licheniformis*. This process requires relatively simple equipment, consumes a small amount of water, and generates less waste. In fungi, α-amylase production by SSF takes a longer time (72–96 h) than its production by SmF. However, a similar procedure using some specific bacterial strains consumes less time. The main interest of SSF over SmF for α-amylase production comes from its low capital investment and high yield.

13.1.4 Downstream Processing

Many commercial applications of α-amylase rely on crude enzyme preparation due to economic considerations, as it requires a few steps in downstream processing. However, the pharmaceutical industry and scientific research require enzymes of high purity that can be obtained using additional steps. The costs of pure enzymes can increase due to the increased number of steps and involving sophisticated methods. Many producer microorganisms secrete α-amylase outside the cells. The first step of downstream processing is the extraction of cell-free culture supernatant from microbial cells using separators, rotary drum filters, or filtration. Removal of insoluble can involve a one-step filtration process when the fermentation microorganisms are fungi. However, the recovery from bacterial cells may involve an additional step of centrifugation after filtration. Cell-free culture supernatant contains active substances that can be concentrated, involving microfiltration and ultrafiltration. The alternative techniques for purification can be ammonium sulphate ($(NH_4)_2SO_4$) precipitation or solvent extraction using ethanol/acetone. The selection of techniques for purification is based on the targeted market and the quality and cost of the products. Enzyme purification from SSF is simpler than purification from SmF. The high-level purity of enzymes can be achieved using techniques such as gel filtration, ion exchange chromatography, affinity adsorption chromatography, reverse phase chromatography, etc. The purified enzyme can be dried to get a solid product. Further, the enzyme can be immobilised using different physical and chemical interactions for industrial applications (detailed in Chapter 15).

The activity of the final product is determined in terms of enzyme catalytic activity by measuring the amount of enzyme and its activity. The activity in the final product is measured by three different procedures under a standard assay condition, including the Iodine method, the dinitrosalicylic acid (DNS) method, and the Nelson-Somogyi method (Figure 13.4). The Iodine method utilises

α-amylase activity

	Iodine method	Dinitrosalicylic acid (DNS) method	The Nelson-Somogyi method
Reagents	Lugol iodine	Dinitrosalicylic acid	Copper reagents and arsenomolybdate
Absorbance at	580-600 nm	540 nm	620 nm
Measure	Starch	Reducing sugars	Reducing sugars
Detect	Amount of remaining substrate	Amount of product formed	Amount of product formed

FIGURE 13.4 Different methods of estimation of α-amylase activity.

Lugol iodine solution that forms a complex with starch and produces a blue-black colour. The absorbance measured at 580–600 nm detects the amount of starch degraded. The DNS method is based on the change in colour due to the attachment of DNS with reducing sugars after enzymatic reaction, which can be measured at 540 nm in a UV/vis spectrophotometer. The Nelson-Somogyi method is the most sensitive procedure and measures reducing sugar using copper reagents and arsenomolybdate at 620 nm.

13.2 PRODUCTION OF ORGANIC ACIDS

Organic acids are carboxyl group-containing compounds with terrific buffering or acidifying capacity. They are sparingly soluble in water and organic solvents. The high demand for organic acids on a global platform for various applications draws attention to their mass production, and their annual market values are in tonnage. Though chemical process engineering has devised synthetic routes for their production, fermentation is preferred for the large-scale production of some organic acids such as citric acid, lactic acid, gluconic acid, succinic acid, itaconic acid, kojic acid, and acetic acid (Table 13.1). Many organic acids reside as natural metabolites in the biological system and can be highly accumulative in some organisms. The natural sources of many organic acids are microbial strains. Glucose metabolism has a pivotal role in producing industrially relevant organic acids, and the formation of pyruvate is a central metabolic node to derive organic acids.

Organic acid production by fermentation requires organic substrates, such as carbohydrates, short-chain organic acids, and amino acids. These compounds can be obtained from various industrial byproducts like molasses, fruit and vegetable processing waste, food waste, whey from dairy production, and lignocellulosic biomass. In recent years, lignocellulosic mass from the agricultural industry has drawn major attention to its exploitation as a suitable feedstock for organic acid production because it is abundant, low-cost, and a promising renewable resource.

Some industrially relevant organic acids, such as citric acid, malic acid, and succinic acid, are primary metabolites. However, their high accumulations require rerouted pathways using a special set of enzymes found in some special microorganisms. Succinic acid is naturally

TABLE 13.1
A list of different organic acids, their key microbial source, important applications, and chemical structures

Organic acids	Key microbial sources	Important applications	Chemical structures
Citric acid	*Aspergillus niger*	Food and beverages, pharmaceutical products, cleaning agents, cosmetics, and water softeners	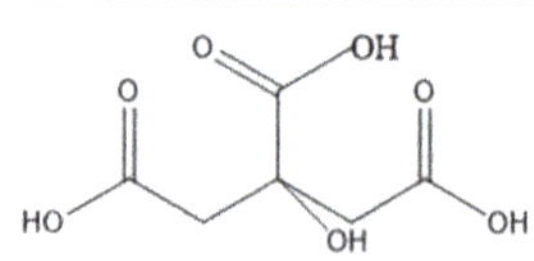
Lactic acid	*Lactobacillus* spp., *Streptococcus* spp., and *Bifidobacterium* spp.	Preservatives, pharmaceutical products, cleaning agents, and animal feed	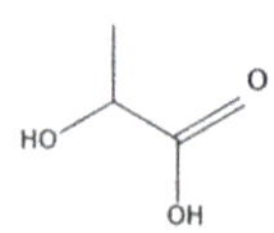
Succinic acid	*Actinobacillus succinogenes* and *Mannheimia succiniciproducens*	Flavouring agents, preservatives, pH modifiers, plant fertiliser, and cosmetic formulation	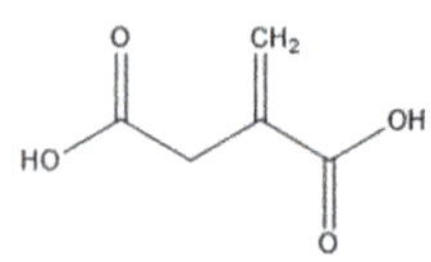
Itaconic acid	*Aspergillus terreus*	Drug development, hair and skin care products, and synthesis of biodegradable polymers	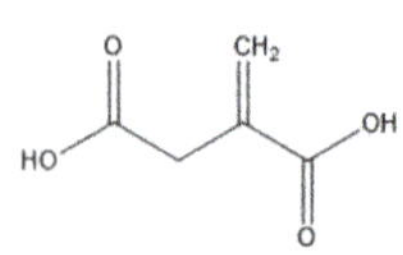
Fumaric acid	*Rhizopus arrhizus* and *Rhizopus oryzae*	Acidulants, flavour enhancement, pharmaceutical formulations, pesticide formulation, and cosmetics	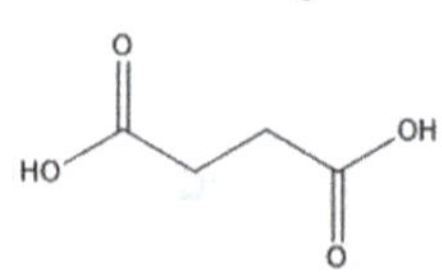
Acetic acid	*Acetobacter* spp.	Vinegar production, solvent, medical treatments, cleaning agents, cosmetics, and wastewater treatment	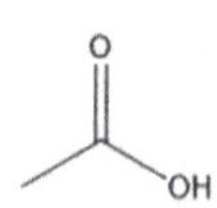
Gluconic acid	*Aspergillus niger*	Preservatives, acid regulators, mineral supplements, drug stabilisers, cleaning agents, and skin care products	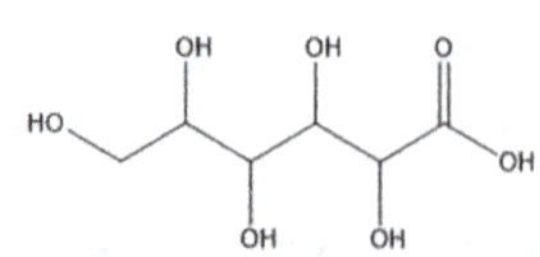

produced by *Actinobacillus succinogenes* and *Mannheimia succiniciproducens*, which have been induced in *E. coli* through genetic recombination. *Aspergillus niger* and *Aspergillus terreus* are the highly productive microbial strains of gluconic acid and itaconic acid, respectively. Lactic acid production occurs using lactic acid bacteria (LAB) that chiefly involve several species of *Lactobacillus* spp., *Streptococcus* spp., and *Bifidobacterium* spp. This type of fermentation is used to produce many other delicacies, such as yoghurt, kimchi (Korean cuisine), and sauerkraut (Chinese cuisine). Acetic acid is notably manufactured using *Acetobacter* from ethanol for the production of vinegar. Of all the different organic acids, citric acid has received wide interest in commercial production through the fermentation process and is discussed in detail in the next section.

13.2.1 Production of Citric Acid

Citric acid is a tricarboxylic acid found in small quantities in all aerobic organisms as a key intermediate in the Krebs cycle. This compound can be produced on commercial scales using some specific microorganisms. However, citric acid was primarily identified and extracted from citrus fruits. In the early nineteenth century and before, people used citrus fruits and pineapple for citric acid extraction. However, plant-based production was not sufficient to meet the ever-increasing demand for citric acid. The report of citric acid production through microbial fermentation by Wehmer in 1893 completely transformed its production in the next thirty years. The first pilot-scale microbial fermentation was set up by Pfizer in Belgium (1919), and large-scale production began in the USA in 1923. Though several methods of chemical synthesis for citric acid are also available, none of them can compete with the fermentation process. The recent global production of citric acid is estimated at more than 2 million tonnes/year. A recent forecast estimates its consumption to rise at a rate of 5% per year in the near future. The food and beverage industries have a major demand for citric acid as an acidulant, antioxidant, and flavouring agent. Other notable consumers of citric acid are pharmaceutical products, cleaning agents, cosmetics, and water softeners.

The natural producers of citric acid are certain bacteria and fungi. In bacteria, it is derived from *Arthrobacter paraffinens*, *Bacillus licheniformis*, and *Corynebacterium* spp. In fungi, it can be obtained from *Aspergillus* spp., *Penicillium janthinellum*, *Candida*, *Pichia*, *Kloeckera*, *Saccharomyces*, *Zygosaccharomyces*, and *Yarrowia*. However, the industrial workhorse of citric acid through fermentation is a filamentous fungus, *Aspergillus niger*. About 99% of citric acid fermentation worldwide employs *A. niger*. Other species of *Aspergillus,* such as *A. aculeatus*, *A. awamori*, *A. wentii*, *A. foetidus*, and *A. carbonarius,* are also notable producers of citric acid. A yeast, *Yarrowia lipolytica*, also has the potential for commercial applications due to its low sensitivity to trace metals and oxygen conditions for its production. However, the development of an undesirable by-product, isocitric acid, by the producer strains impedes downstream processing.

The citric acid accumulation in *A. niger* mainly takes place through the Krebs cycle (also known as the citric acid pathway) (Figure 13.5). This cycle involves a total number of eight steps and takes place inside the mitochondrial matrix. The key intermediate of citric acid production is pyruvate, which is an end product of the glycolytic pathway produced in the cytoplasm. The pyruvate enters the mitochondrial matrix and is decarboxylated to produce acetyl-CoA. The acetyl-CoA enters the Krebs cycle, which primarily reacts with oxaloacetic acid and produces citric acid. In the next step, citric acid is transformed into aconitate and isocitrate, followed by the removal of two carbon units as carbon dioxide (CO_2).

The accumulation of citric acid in *A. niger* can be achieved by deregulation of biosynthesis. Also, the production of citric acid can be enhanced by inhibiting its degradation and the deactivation of aconitase and/or isocitrate dehydrogenase in the Krebs cycle. A high glycolytic flux stimulates the synthesis of citric acid. The high oxygen concentration is also a strict requirement for the accumulation of citric acid in the medium.

A reroute pathway, anaplerosis, allows the replenishment of certain intermediates of the Krebs cycle. This metabolic pathway involves an anabolic reaction between phosphoenol pyruvate (PEP) and pyruvate to form oxaloacetate (OAA). Further, the cytosolic OAA is reduced to malate and transported to mitochondria using an antiporter. The antiporter is a membrane protein for the coupled transport of two different compounds across the membrane, but in the opposite direction. In this process, the transport of malate inside the mitochondria is coupled with the transport of citric acid in the cytoplasm. Inside the mitochondrial matrix, malate is oxidised in OAA. A different transport protein located on the cell membrane secretes the citric acid outside the cell (Figure 13.5).

Citric acid production can also be improved by metabolic engineering. This mainly relies on metabolic models and genetic engineering, which introduce new metabolic pathways or modify existing ones in the producer microorganisms. The main strategies for the strain improvements are overexpression of key enzymes like pyruvate carboxylase and citrate synthase, regulation of metabolic pathways such as feedback inhibition control and promoter engineering, and modification in genes for carbon source utilisation and byproduct reduction.

Industrial-scale citric production can be divided into three stages: upstream processing, fermentation process, and downstream processing (Figure 13.6), which are discussed as follows.

13.2.2 Upstream Processing

The careful strain selection and mutation (using chemical mutagens and radiation) of *A. niger* play a key role in improving citric acid production. The inoculation of microorganisms in the fermentation media is carried out using asexual spores called conidia. Conidia are produced in different types of nutrients, and the inoculant is prepared using spore suspension. The spore suspension is transferred to a preferment for seed development, where a low concentration of cyanide stimulates biomass growth in granular form. This granular form is used

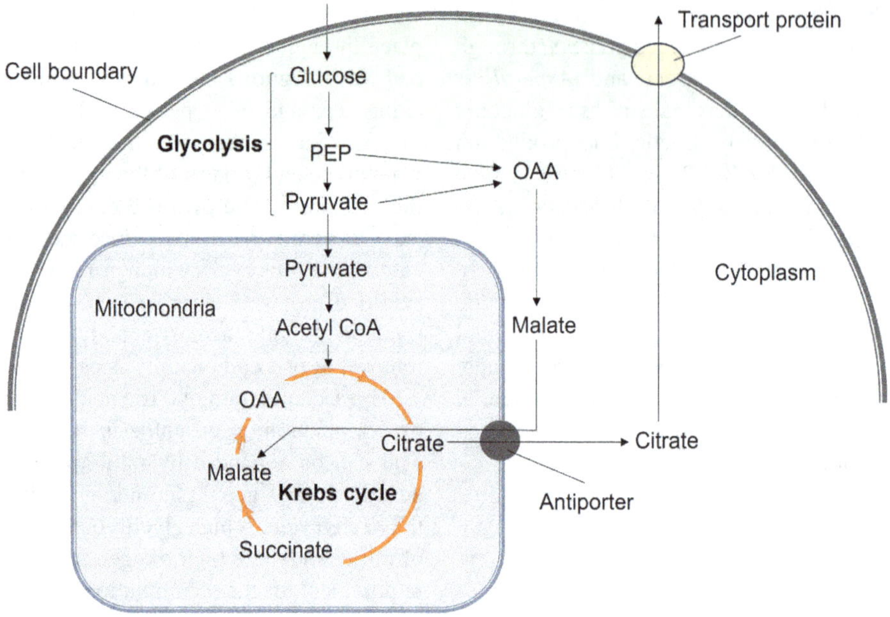

FIGURE 13.5 A general pathway of citric acid production in *Aspergillus niger*. (OAA: oxaloacetic acid, PEP: Phosphoenolpyruvate, CoA: Coenzyme A.)

as the main inoculum in the seed fermenter. The initial growth phase during the first 24–48 h requires a high pH above 5.5 for the inoculum development. Thereafter, the pH of the inoculum is reduced below 3.0 before inoculating into the main fermenter, as the optimal pH for the citric acid production is below 2.0. On the other hand, the high pH favours gluconic acid and oxalic acid production. Also, the chance of microbial contamination during citric acid production is low at low pH. Industrially relevant substrates for large-scale fermentation are starchy materials and molasses that have high sugar content (20–40%) and are inexpensive. The most suitable raw materials for submerged fermentation are hydrolysed starch, refined or raw sugar, beet or cane juice, palm oil, wood hemicellulose, etc. In addition, beet molasses, black strap molasses, and cane molasses can also be used after pretreatment, which is necessary to remove trace metals (Zn, Mn, Fe, Cu and other toxic heavy metals) from raw materials, as their presence in the medium can inhibit citric acid production. These impurities can be precipitated using potassium ferrocyanide and clarified using sulphuric acid (H_2SO_4). However, the augmentation of low concentrations of phosphate and nitrogen in the media promotes citric acid production.

13.2.3 Fermentation Process

The citric acid at the commercial scale can be manufactured by surface, solid-state, and submerged fermentation. Each of these methods has some conveniences and drawbacks (Figure 13.7). Surface fermentation is an ancient method of production that has a low energy requirement, but the process is extremely sensitive to contamination. A traditional Japanese method of citric acid production in SSF is called the "Koji process". In the Koji process, agro-industrial residues such as banana peel, cotton waste, apple pomace, and orange peel are used as solid substrates for the growth of *A. niger* under high moisture conditions (about 70%). This process produces high-quality citric acid and takes 4–5 days to complete the fermentation. In one aspect, the process can be inexpensive, considering it involves cost-free substrates, a pretreatment-independent process, and relatively simple downstream processing. However, citric acid production through SSF suffers from some limitations in production scale. The main problems are crude utilisation of the nutrients in the substrates, low penetration of oxygen, and the limited viability of microorganisms.

Submerged fermentation is considered the most suitable of all and accounts for 80% of the total global production. The fermentation reactions are performed as batch or fed-batch processes in airlift fermenters or stirred tank fermenters. The production scale can vary, typically ranging from 100 to 1,000 m^3 or even more. The production is maximum at temperatures around 28–30 °C and pH below 2. Spores germinate into filaments, which develop into compact aggregates or spherical pellets of the fungal filaments (intertwined mycelial form). Citric acid production takes place inside the fungal body and is released outside. Its production is maximum after 5–7 days of spore germination. In a stirred tank fermenter, an agitator can be installed for slow mixing to improve oxygen dispersion in the feed. The presence of antifoam restrains the formation of foam during fermentation. The final citric acid titer is obtained in the range of 150–200 g/l.

13.2.4 Downstream Processing

The first step of downstream processing involves the removal of the fungal mycelia and obtaining a clear fermented liquor. Heating of harvested broth at 70 °C for 15 min causes coagulation of protein molecules that facilitate easy removal of solid particles. The filtration of biomass is performed

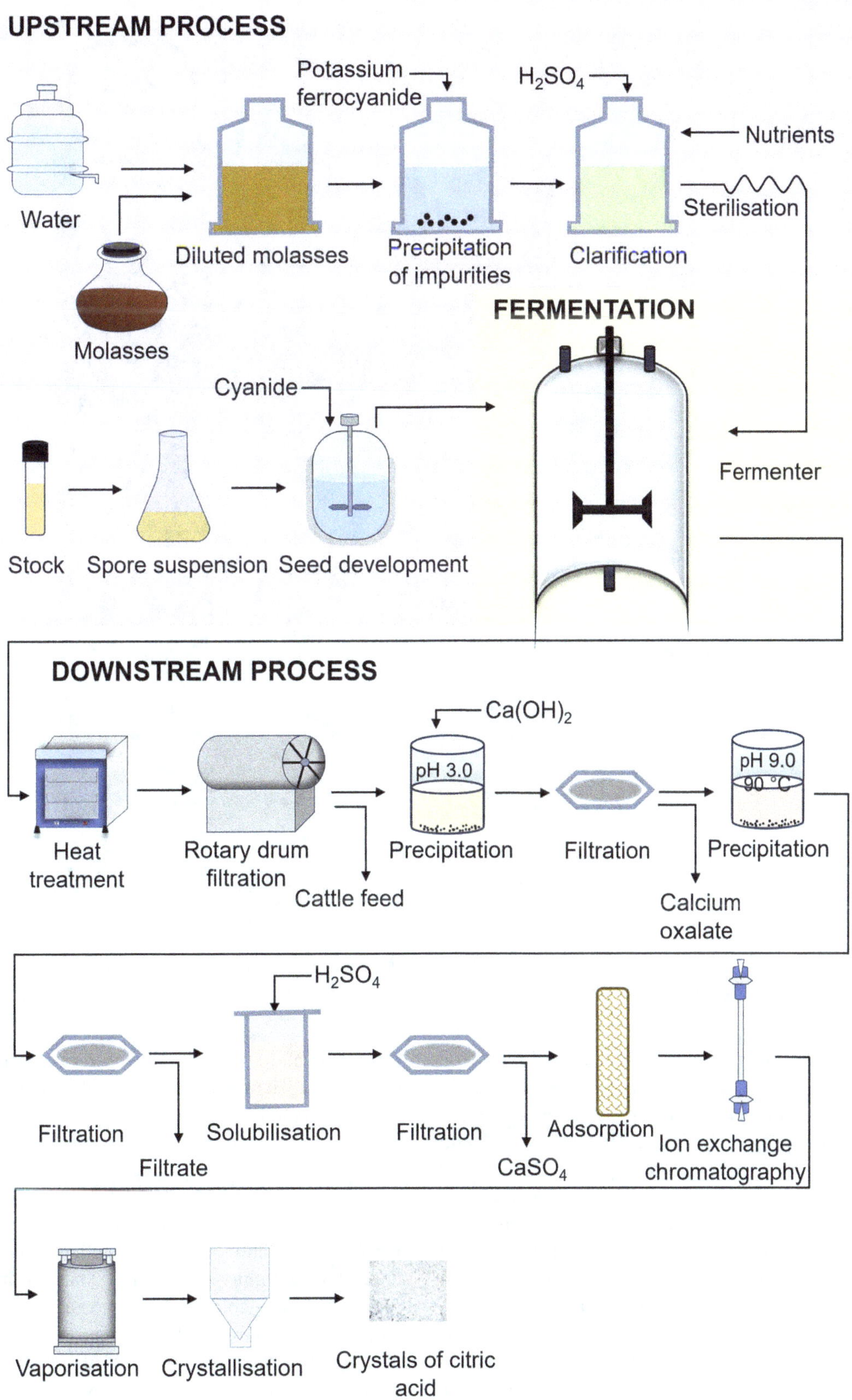

FIGURE 13.6 A general scheme of citric production through SmF.

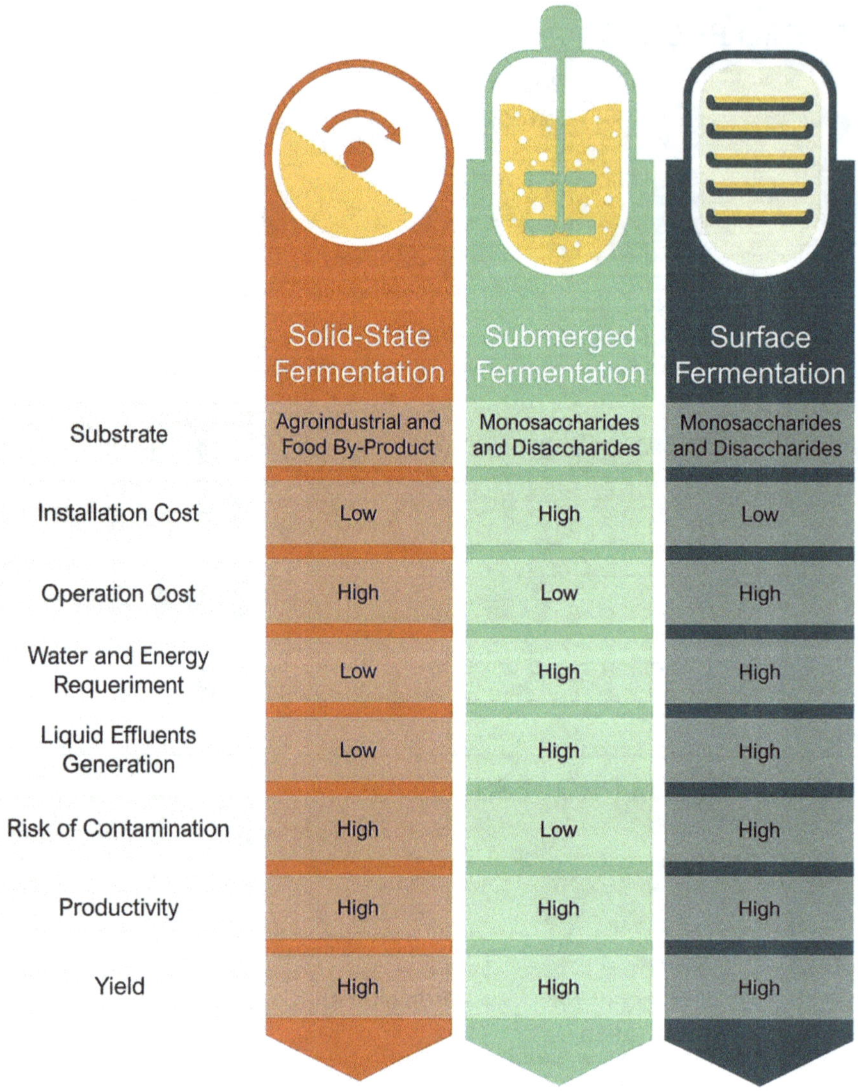

FIGURE 13.7 A comparative account of different modes of fermentation for citric acid production. (Image reproduced from Mores et al. (2021)[1] with permission of © 2020 Elsevier Ltd.)

using a rotary vacuum drum filter or belt discharge filter. The recovered mycelial mass can be used as a protein source in cattle feed after washing and drying. If the filtered solution consists of oxalic acid, it can be precipitated as calcium oxalate at pH around 3.0 using calcium hydroxide ($Ca(OH)_2$) and eliminated through filtration or centrifugation. In the same environment, citric acid is converted into mono-calcium citrate and retained in the solution. This mother liquor can be treated with $Ca(OH)_2$ at pH around 9.0, 90 °C for the precipitation of citric acid in the form of tricalcium citrate tetrahydrate. The recovered precipitate has high purity. This is solubilised back into the water using 70% H_2SO_4, where the calcium ions react with sulphate ions to form an insoluble gypsum ($CaSO_4$). The residual impurities and colouration can be removed in subsequent purification steps involving adsorption and ion exchange chromatography. Solvent extraction can be used as an alternative method for the purification and recovery of citric acid in place of calcium citrate precipitation and acid solubilisation. Further, the citric acid is concentrated by evaporation at 40 °C and crystallised in a vacuum crystalliser at 28–36 °C. The final product is anhydrous or monohydrate citric acid crystals.

13.3 PRODUCTION OF ETHANOL

Ethanol fermentation is the most ancient knowledge of microbial application. Evidence of ethanol fermentation dates back to the Neolithic period (11,000 BC) for the production of liquor. The total alcohol consumption across different regions continued to grow, and the current global annual average consumption is more than 5.0 litres of pure alcohol per person. A myriad of alcoholic beverages exist with distinct aromas, flavours, and alcohol contents. The choices of raw materials in different beverages are also distinct (Table 13.2). The process of alcohol fermentation is equally admissible in the preparation of bakery products. However, the fermentation products, ethanol and CO_2, are lost from the main finished products (bread) and leave the impression of a light and airy

TABLE 13.2
A list of different alcoholic beverages showing alcohol concentrations and the involved raw materials

Name of beverages	Alcohol concentrations (% in v/v)	Raw materials
Beer	4–8	Cereals such as barley
Toddy	5–10	Sap of palm trees
Wines	6–21	Grapes
Brandy	35–60	Grapes and other fruit juices
Rum	35–60	Sugarcane molasses
Whiskey	35–60	Malt of barley, corn, rye, and wheat
Vodka	40–95	Grains like wheat, corn, or potatoes
Gin	35–50	Juniper berries
Tequila	40–50	Blue agave plant

texture. A sufficient volume of ethanol is produced for use as a solvent and disinfectant.

Ethanol is also a promising substitute for fossil fuels and a renewable energy resource. It was first introduced in the 1920s. However, the real potential of ethanol was not realised until the oil crisis in the 1970s. Bioethanol as a biofuel gained popularity in the 1980s and is now regarded as an alternative fuel in many nations. The global annual production of ethanol in the biofuel industry in the last six years is above 26 billion gallons, of which the USA and Brazil alone surpassed 55% and 27% production, respectively (i.e., 82% in total) (Figure 13.8). The ethanol-powered car that entirely runs on ethanol is also designed. Ethanol-blended fuel is already in use in the existing vehicular models. About 10–15% ethanol in gasoline is widely used in different nations. For example, India is currently blending around 13% ethanol in petrol and has aimed to achieve a 20% blend of ethanol in petrol by the year 2025–26. Brazil blends about 27% ethanol with transport sector fuel, which is the highest of any country. Many countries are now planning to increase the mandate to 30%. Even some countries like the United States (US), Canada, Australia, France, and Sweden are using an 85% blend of ethanol in fuel for high-performance vehicles.

The feedstock for the production of bioethanol as biofuel can employ a variety of sugar-, starch-, and lignocellulosic-containing biomass. These may include waste byproducts of different industries, such as molasses from the sugar industry, sulphur waste liquor produced from the paper and pulp industry, and whey liquid waste from the dairy industry. Biofuel production in the US relies on corn (a temperate crop), while Brazil is dependent on sugarcane (a tropical crop). In many tropical and subtropical countries, the starchy root of cassava is a notable fermentable material. Based on the feedstocks, bioethanol production is categorised as the first, second, third, and fourth-generation biofuels (Figure 13.9). First-generation biofuel is made from edible energy crops such as sugarcane, sugar beet, corn, wheat, barley, rapeseed, and canola. Many of these substances are also important sources of food and are used in the preparation of alcoholic beverages. Therefore, first-generation biofuel competes with food resources. Second-generation biofuel is derived from lignocellulosic biomass, including woody crops and agricultural residues. These kinds of substances are potential sources for not being food-competitive, but depend on arable land. Third-generation biofuel relies on algae that feature additional benefits of fast growth, high energy content, and non-dependence on arable land. However, algae-based fuels could not achieve commercial success due to economic considerations. The most modern feedstocks, genetically modified algae, have only recently been introduced as raw materials for fourth-generation biofuels but have not entered the mainstream.

About 95% of global ethanol production is carried out through fermentation. A single-cell yeast, *Saccharomyces cerevisiae,* is the key microorganism for its large-scale production. This microorganism has a tolerance to high alcohol concentration and low pH. The markedly available *S. cerevisiae* comes under two different names: Baker's yeasts and Brewer's yeasts. They are two different strains and are suitable for distinct industries. Baker's yeast is used in baking and generates more CO_2 but less alcohol than brewer's yeasts that are used in breweries and distilleries. The strains of producer microorganisms in bioethanol production are different from those that are used in food and beverages. The key traits of bioethanol-producing microorganisms are that they have the ability to utilise a wide range of sugars (including pentoses from hemicellulose) and exhibit resistance to inhibitors. Also, they are extensively genetically modified to achieve the maximum yield.

Candida shehatae and *Pichia stipitis* are pentose fermenting yeasts that utilise xylose from hemicellulose (the second most abundant plant biomass after cellulose). The fermentation potential for industrial ethanol production is also reported in other yeasts such as *Schizosaccharomyces* spp., *Pichia* spp., and *Kluyveromyces fagilis*. A Gram-negative facultative anaerobic bacterium, *Zymomonas mobilis*, is also a potential source of ethanol production. This bacterium has received great interest in the biofuel industry, as the yield of ethanol by *Z. mobilis* can surpass that of *S. cerevisiae* under certain conditions. A new method of ethanol production involves the syngas (a mixture of carbon monoxide (CO) and hydrogen (H_2)) using an anaerobic autotrophic bacterium, *Clostridium ljungdahlii*.

FIGURE 13.8 An ethanol plant in South Dakota, USA. (Image adapted from iStock/Jacob Boomsma.[2])

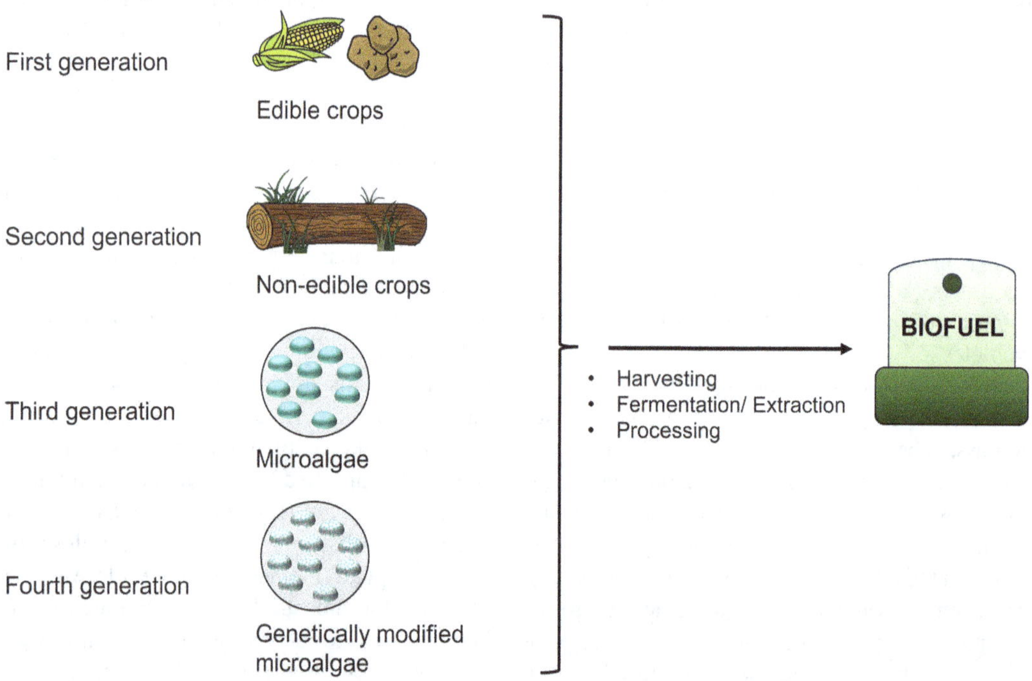

FIGURE 13.9 First, second, third, and fourth-generation biofuels based on different feedstocks.

The key substrates of alcohol fermentation are rich sources of monosaccharides (e.g., glucose) or disaccharides (sugar). These substrates enter the glycolytic pathway, where one glucose unit is broken down into two units of pyruvate. This is followed by a two-step reaction, including decarboxylation of pyruvate that produces acetaldehyde and then reduction of acetaldehyde to produce ethanol. These reactions utilise pyruvate decarboxylase and alcohol dehydrogenase, respectively (Figure 13.10). This metabolism predominates under suboxic or anoxic conditions. However, the high concentration of glucose (above 0.5%)

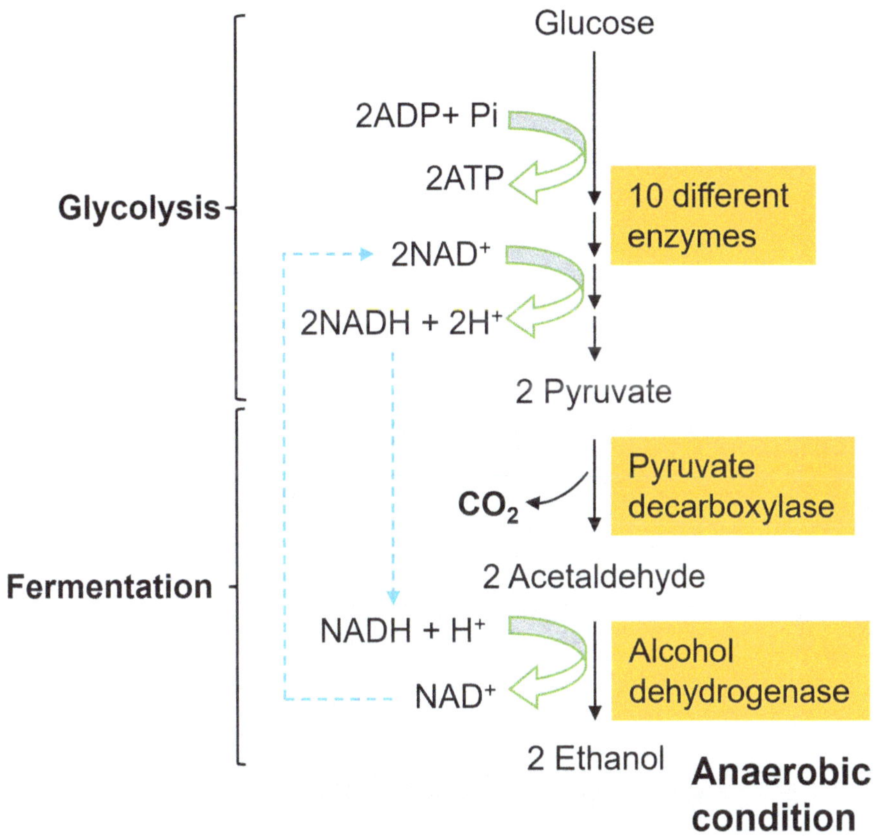

FIGURE 13.10 Reactions involved in ethanol fermentation.

promotes ethanol production regardless of oxygen concentration in the medium or a strict aerobic situation (i.e., Crabtree effect). In the Crabtree effect, glucose acts as a feedback inhibitor of aerobic respiratory metabolism. This suggests that the high substrate concentrations accelerate the glycolytic reactions and induce substrate-level phosphorylation. This fills the demand for ATP for microbial metabolism, independent of oxygen requirement. This effect is characterised by the suppression of aerobic metabolism, and therefore, the biomass yield is low during ethanol fermentation.

A simple scheme of ethanol production includes the pretreatment of biomass, followed by enzymatic hydrolysis, fermentation, distillation, and packaging (Figure 13.11). These processes involve several additional steps in between and assist in their production. The following subsections provide in-depth details on the different stages of ethanol production.

13.3.1 Upstream Processing

As the main substrates for the fermentation are simple sugars, all types of raw materials require some pretreatment to produce a high concentration of simple sugars. The pretreatment steps also remove inhibitors and reduce the complexity of downstream processing. The first step of pretreatment is milling, where the raw material is ground into fine, consistent-sized particles. A different process, overliming, facilitates the removal of the inhibitory substances from the lignocellulosic hydrolysates by their precipitation using $Ca(OH)_2$. The ground substances consist of starch-containing and cellulosic materials, which are subject to further treatment. The next step is liquefaction, which converts insoluble straw biomass into a liquified state at high temperatures in the presence of thermostable enzymes. It is followed by saccharification, which involves the hydrolysis of soluble carbohydrates into simple sugars using suitable enzymes. In enzymatic reactions, starch-containing substrates receive treatment from α-amylases and glucoamylases, while cellulose-dominating materials are broken down using cellulases. Lignocellulosic biomass is treated using cellulases and other enzymes, converting cellulose and hemicellulose into simple sugars. In a different approach, saccharification and fermentation can be performed simultaneously by adding glucoamylase along with yeast cells.

The fermentation microorganisms for ethanol are mainly *S. cerevisiae* and *Z. mobilis*. The selection of suitable microorganisms and their strains depends on the types of feedstocks used and applications. The genetic modifications in yeast and bacterial cells have enhanced ethanol production and improved resource utilisation. The lignocellulosic biomass consists of a high amount of hemicellulose, which is made of monomeric units of pentose and hexose sugars. The metabolic engineering of producing microorganisms allows the utilisation of pentose sugars, which is not an inbuilt property of any natural strains of *S. cerevisiae* and *Z. mobilis*.

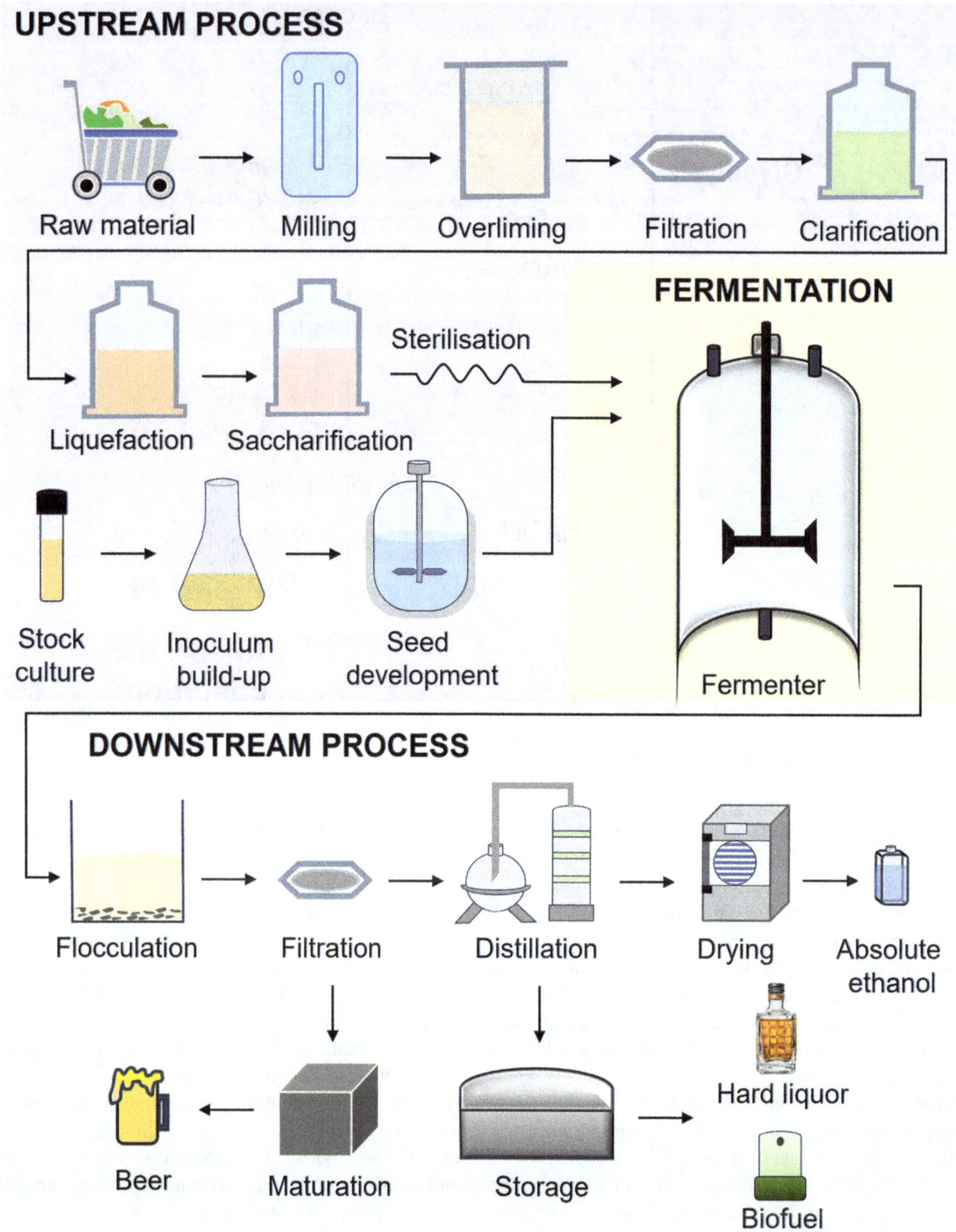

FIGURE 13.11 A general scheme of ethanol production.

The seed of yeast for large-scale fermentation is prepared using pure stock culture. The key constituents of the medium for seed development include sugars, as well as some chemicals, including ammonium salts, trace minerals, and iron. In the seed fermenter, the low glucose concentration and elevated oxygen levels promote biomass development for seed culture while producing a small quantity of alcohol. In the last step, the feedstock is sterilised and transferred to the main fermenter. Seed culture is also introduced in the fermenter to initiate the fermentation process.

13.3.2 Fermentation Process

Both submerged fermentation (SmF) and solid-state fermentation (SSF) can be used to produce alcohol. In SSF, the supporting environment is comparable to their native habitat of microorganisms, resulting in optimum metabolic activity. In comparison to SmF, SSF offers several benefits, such as reduced water consumption, a reduced need for sterilisation, direct use of solid substrates, and the use of a mixed culture of microorganisms. However, SSF hampers the scale-up of the process in an industrial setup because of low ethanol yields, slow growth rate, partial utilisation of raw materials, and inability to regulate environmental conditions. On the other hand, SmF is appropriate for commercial-level production. All types of operations, including batch, fed-batch, and continuous processes, are possible for ethanol production. The fermenter design is determined by the type of raw materials used as well as operating parameters such as agitation, cooling, titre volume, and loading. The modern setup of alcohol production relies on stirred tank fermenters, air-lift fermenters, fluidised bed bioreactors, and packed bed bioreactors, of which stirred tank fermenters are most widely used. An industrial plant can produce thousands of litres of ethanol. The optimum pH and temperature for ethanol fermentation can vary in different setups and depend on various factors. Mildly acidic conditions and temperatures around 28–35 °C are suitable conditions for large-scale production. Traditional brewing yeast requires 28–30 °C and completes the production in 30–50 h. In a typical liquid fermentation, the ethanol concentration can reach up to 9–15% (by volume). The high ethanol concentration in the medium shows inhibitory effects on the yeast cells.

13.3.3 Downstream Processing

Once the process of fermentation is over, yeast cells are separated through an off-cost process, flocculation. The flocs are the big clusters of yeast cells that are settled at the bottom or suspended in the solution, which can be recovered through sedimentation, filtration, or centrifugation for the next cycle. Yeast biomass can be used as a single-cell protein for animal' feed. The broth also consists of side products such as methanol, acetic acid, succinic acid, and glycols. In the maturation phase, the organic acids and alcohols (other than ethanol) are converted into ester compounds, which give a fruity aroma to the liquor.

In the brewery, this is followed by filtration and packaging and ethanol concentrations in beer are typically below 5%. On the other hand, hard drinks contain high concentrations of ethanol and require purification and concentration steps. This is performed by distillation, and such an industrial setup is called a distillery. This involves a physical separation between ethanol and water through condensation and evaporation. A typical distillation unit consists of a boiler, a column, and a condenser. The boiler is filled with the liquid and is heated to around 87 °C. This temperature is sufficient to convert ethanol into vapour due to its boiling point of 78.37 °C, but still less than the boiling point of water (100 °C). The evaporated ethanol rises through the column and reaches a condenser, where the cold surface condenses the vapour back into the liquid state. Distillation can be repeated more than once to produce ethanol in its highly pure form. However, a low volume of water also boils along with ethanol and evaporates at the same temperature. After the distillation, ethanol can achieve a maximum 95.4% purity. About 4.4% of the ethanol solution is water, which cannot be removed by simple distillation. It remains along with alcohol as an azeotropic mixture. This level of purity is sufficient for most of the applications of ethanol. The absolute ethanol (100%) can be obtained by "drying" it using lime or salt. Besides, a small quantity of benzene can also disrupt the azeotropic interaction between ethanol and water to achieve pure ethanol.

The production of hard liquor usually requires purity above 40–70%, and it is concerned with the distinct flavour and aroma. Ethanol produced for liquor must also meet food-grade standards. However, the ethanol produced for biofuel requires high purity and high concentrations, which can be achieved by performing repeated distillation.

13.4 PRODUCTION OF ANTIBIOTICS

Antibiotics are a special class of small organic compounds for treating infectious diseases. They are called modern medicines and have increased the average human lifespan by twenty-three years. In 1928, penicillin was the first discovery of a new class of drugs, antibiotics. It was recovered from a fungus, *Penicillium*, for the treatment of bacterial infection. Since then, many new antibiotics have been discovered from different microorganisms. With their mass production beginning in 1940, antibiotics saved millions of lives. The period from the 1940s to the 1960s bestowed many novel antibiotic classes on mankind and is known as the Golden Age of Antibiotic Discovery (Figure 13.12a). They have been mainly outsourced from selected bacteria and fungi. As per an estimate, about 64% of all natural antibiotics are produced by the members of actinomycetes (Figure 13.12b). Some well-known antibiotics produced by bacteria are streptomycin, streptothricin, tetracycline, vancomycin, erythromycin, and daptomycin. Moreover, microbial fermentation is important in the production of many therapeutic compounds, vaccines, and biologics (Box 13.1).

The production of antibiotics can be of three different types: natural, semisynthetic and synthetic. When the production of antibiotics entirely depends on the biological pathway, they are known as natural antibiotics. Natural antibiotics have become ineffective in the treatment of many infections due to the prevalence of antimicrobial resistance. Many common bacterial pathogens have evolved mechanisms to combat the effects of antibiotics. Also, in the last twenty-five years, there have been very few discoveries of natural antibiotics. Some recent reports of natural antibiotics are teixobactin (source: *Eleftheria terrae*), darobactin (source: *Photorhabdus khanii*), and evybactin (source: *Photorhabdus noenieputensis*). Semisynthetic antibiotics are chemically modified natural antibiotics and have the potential to combat the problem of antimicrobial resistance. Semisynthetic antibiotics also show a greater spectrum of activity against several resistant pathogens and have fewer side effects than natural antibiotics. Common examples of semisynthetic antibiotics include clarithromycin, azithromycin, and rifampin. Synthetic antibiotics are synthetic products manufactured purely from chemicals. The inspiration for many synthetic antibiotics comes from the carbon backbones of natural antibiotics with diverse chemical group assemblages. Some have altogether different structures and can fight, to a

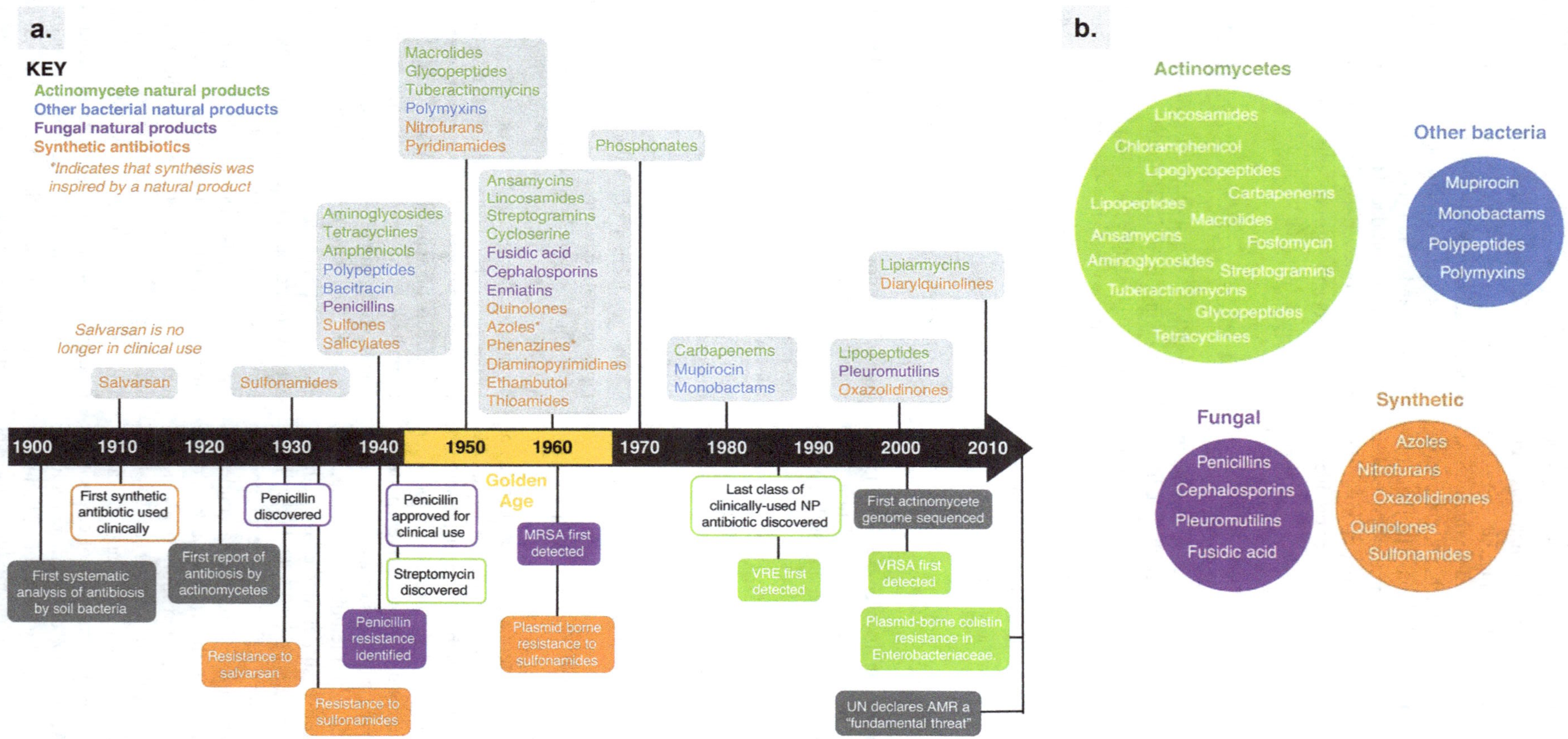

FIGURE 13.12 a. Timeline showing the development of antibiotics (MRSA: methicillin-resistant *Staphylococcus aureus*, VRE: vancomycin-resistant enterococci, VRSA: vancomycin-resistant *S. aureus*, AMR: antimicrobial resistance, NP: natural product, and UN: United Nations). b. Some clinically useful classes of antibiotics are derived from microorganisms and chemical synthesis. (Image adapted from Hutchings et al. (2019).[3])

BOX 13.1 PRODUCTION OF THERAPEUTIC COMPOUNDS, VACCINES, AND BIOLOGICS

Many microbial-derived, physiologically active chemical entities and their analogues are useful drugs for biopharmaceutical applications. They come in diverse structures and function in treating many ailments. Since the discovery of penicillin, a variety of natural metabolites have shown potential for their antibacterial, antifungal, immunosuppressive, and anticancer activities (Box Figure 13.1.1). Many bacteria and fungi are identified as natural sources of antibiotics. Microorganisms like LAB produce antimicrobial compounds and are useful in the food industry for food safety against foodborne pathogens. The antifungal compounds like Nystatin and Amphotericin B are derived from different species of *Streptomyces* for their effective use against *Aspergillus* spp. A naturally occurring antibiotic, Actinomycin D (Source: *Streptomyces* spp.), is a beneficial drug licensed by the US Food and Drug Administration (FDA) for the treatment of several forms of cancer. A different type of anticancer agent, epothilone, is generated by the mycobacterium *Sorangium cellulosum*. Rapamycin and Tacrolimus are powerful natural drugs which are also obtained from *Streptomyces*. They are used as immunosuppressive and anti-inflammatory agents.

In a different approach, microorganisms can be employed to derive laboratory-tailored compounds, which are usually large and complex biological products known as biologics. This area of biological manufacturing has substantial growth and utilises bacterial and fungal cells as cell factories to produce a wide array of recombinant products in biopharmaceuticals. This bioengineering method has broadened the therapeutic options for a wide range of medical ailments and disorders. Typical examples of biologics include vaccines, monoclonal antibodies, hormones, blood factor derivatives, allergens, recombinant therapeutic proteins, and enzymes (Box Figure 13.1.1). This started with the discovery of the first recombinant protein Humalin® (a genetically engineered insulin) in 1978 for the treatment of diabetes. This primarily involved inserting the human insulin gene into a plasmid of *E. coli*, followed by its transfer into cells. These recombinant cells were suitable for cultivation in a fermenter for the mass production of the protein. This was commercially available in the market in 1983 after approval by the FDA. Lantus® is a modern insulin-based drug produced using *E. coli*, which can be used to manage diabetes under specific needs. The pneumococcal vaccines, Prevnar7 and Prevnar13, are formed by the conjugation of purified capsular polysaccharides and nontoxic recombinant variants of diphtheria toxin of *Streptococcus pneumoniae* and *Corynebacterium diphtheriae*, respectively. Protropin® and Humatrope® are recombinant human growth hormones developed using a bacterial platform, *E. coli*. Recombivax is a recombinant vaccine synthesised using *S. cerevisiae* that prevents infection with the hepatitis B virus.

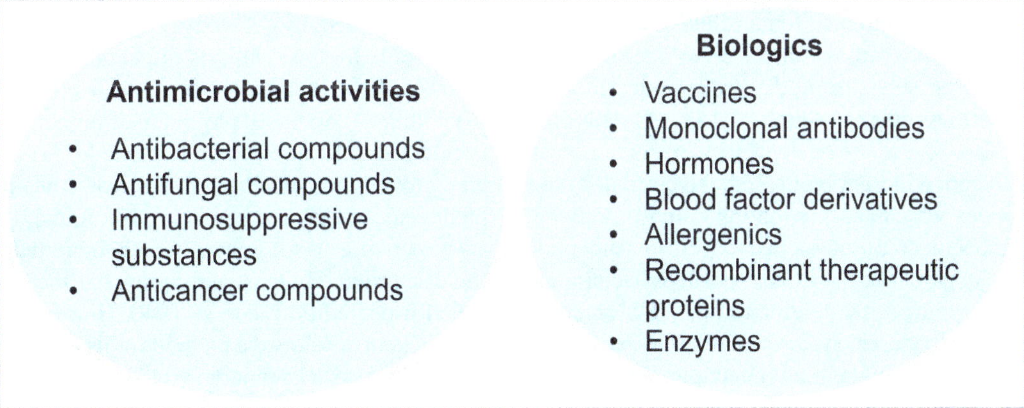

BOX FIGURE 13.1.1 Microorganisms in biopharmaceutical applications.

certain extent, those which have antimicrobial drug resistance. Fluoroquinolones and sulfonamides are examples of two distinct classes of synthetic antibiotics, which are entirely different from natural antibiotics. They are effective against a wide range of pathogens, including both Gram-positive and Gram-negative bacteria. Antibiotics exert two distinct effects on bacteria, viz., bacteriostatic and bacteriocidal. Bacteriostatic antibiotics cause the death of bacterial cells, whereas bacteriocidal antibiotics interfere with growth activities.

Based on the specific carbon core structures and specific groups, most natural antibiotics fall into some specific classes, such as β-lactam, aminoglycosides, tetracyclines, macrolides, lincomycins, and glycopeptides. Each class of antibiotics have a different site of action on infectious organisms and can have different sources (Table 13.3). Of various classes of antibiotics, β-lactam antibiotics act on cell wall synthesis and are mainly active against Gram-positive bacteria. Aminoglycosides are active against both Gram-positive

TABLE 13.3
Different classes of antibiotics, including their modes of synthesis with suitable examples and microbial sources

Class of Antibiotics	Synthesis	Name of Antibiotics	Sources
β-lactam (Penicillin)	Natural	Penicillin G Penicillin V potassium	*Penicillium chrysogenum*
	Semisynthetic	Oxacillin Ampicillin and Amoxicillin Carbenicillin and Piperacillin	The penicillin nucleus, 6-aminopenicillanic acid (6-APA)
β-lactam (cephalosporins)	Natural	Cephalosporin C	*Acremonium* spp.
	Semisynthetic	Cephalexin, Cephradine, and Cefadroxil Ceftaroline and Ceftobiprole	The cephalosporin nucleus, 7-aminocephalosporanic acid (7-ACA)
β-lactam (carbapenems)	Natural	Thienamycin (thienpenem)	*Streptomyces cattleya*
	Natural	Clavulanic acid	*Streptomyces clavuligerus*
Aminoglycosides	Natural	Streptomycin	*Streptomyces griseus*
		Kanamycin A	*Streptomyces kanamyceticus*
		Gentamicin	*Micromonospora purpurea*
Tetracyclines	Natural	Chlortetracycline	*Streptomyces aureofaciens*
		Oxytetracycline	*Streptomyces rimosus*
	Semisynthetic	Doxycycline	Using oxytetracycline nucleus
Macrolides	Natural	Erythromycin	*Saccharopolyspora erythraea*
		Rifamycin	*Amycolatopsis rifamycinica*
Lincomycins	Natural	Lincomycin	*Streptomyces lincolnensis*
	Semisynthetic	Clindamycin	Using lincomycin nucleus
Glycopeptides	Natural	Teicoplanin	*Actinoplanes teichomyceticus*
		Vancomycin	*Amycolatopsis orientalis*
Teixobactin	Natural	Teixobactin	*Eleftheria terrae*

and Gram-negative bacteria by inhibiting the activity of ribosomes for protein synthesis. Tetracyclines interfere with the initiation of protein synthesis in bacteria. Macrolides can reversibly bind large subunits of the prokaryotic ribosome and inhibit protein synthesis. Another class of antibiotics, lincomycins, also share the mechanism of action with macrolides. Glycopeptide antibiotics are glycosylated non-ribosomal peptides which act through inhibition of cell-wall synthesis. Teixobactin constitutes a macrocyclic depsipeptide. This compound is potent against Gram-positive bacteria and inhibits cell wall synthesis by binding to lipid II, a precursor to peptidoglycan. In recent years, antibiotics have been increasingly ineffective in treating infections as antibiotic resistance in microorganisms has spread globally. However, they have continued to be an important source of antimicrobial drugs for treating many diseases. The semisynthetic route is very popular in combating the new drug resistance. Among various natural antibiotics, penicillins have continued to be of interest in commercial-scale production.

13.4.1 Production of Penicillins

Penicillins are bactericidal antibiotics, which means these compounds kill bacteria. The core structure of penicillin is called a "penum", including a thiazolidine ring fused to the β-lactam core. The β-lactam core is linked to the variable acyl side chains, which have significant roles in the efficacy of antibiotics against various pathogens. Penicillins primarily inhibit the activity of transpeptidase, a bacterial enzyme responsible for the synthesis of peptidoglycan (the cell wall). Gram-positive bacteria are the main victims of the effects of penicillin. A Scottish physician-scientist, Fleming (1928), accidentally identified that a rare mould, *Penicillium notatum*, was growing on the Petri plates that contained colonies of pathogenic *Staphylococcus* sp. He found that this mould was secreting a substance, which inhibited the growth of this bacterium. He managed to prepare the crude extract and called it penicillin. Later, in 1940, Howard Florey and Ernst Boris Chain developed a method for its extraction in a purified form, which set a landmark in the development of a life-saving drug. The discovery of penicillin, along with its identification in purified form, was collectively recognised with a Nobel Prize (1945). The commercial-producing strain of penicillins was identified in *Penicillium chrysogenum* (*P. rubrum*) strain NRRL 1951 (discovered in 1943). This strain was bioengineered for penicillin titers. The industrial production of penicillin was not more than 1 g per cubic decimeter when initially produced. However, the identification of new strains, classical genetic improvements, and advanced extraction procedures have increased the antibiotics yield as high as 50 g per cubic decimeter, which is 50 times higher than its initial production.

When the culture media consists of phenylacetic acid, the product is benzylpenicillin (penicillin G), which is a prototype

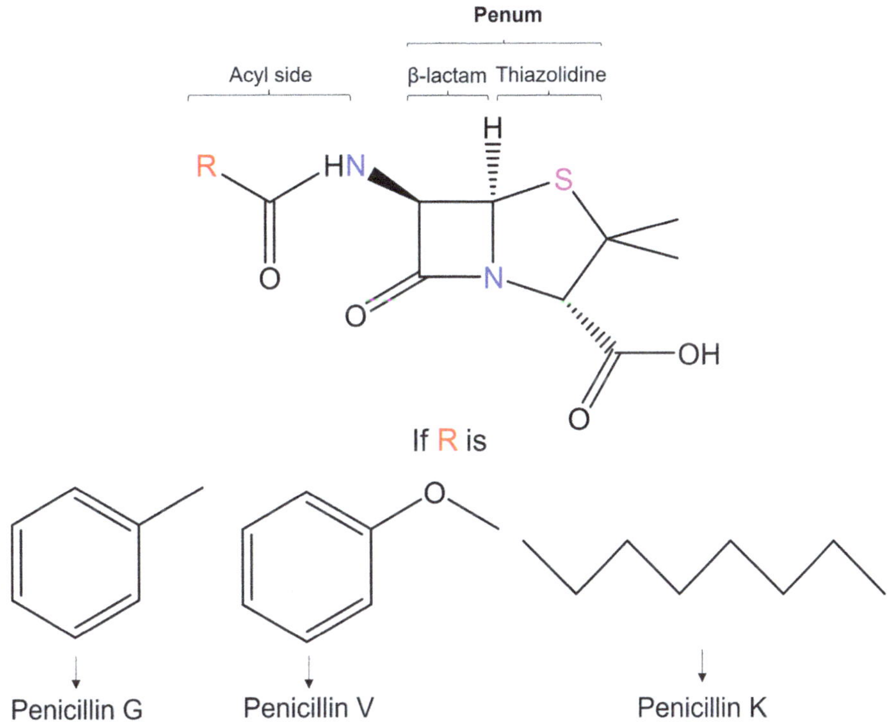

FIGURE 13.13 Chemical structure of Penicillin, where "R" can be a variable group.

of penicillin. If the precursor is phenoxy acetic acid, the fungus produces phenoxymethylpenicillin (penicillin V) (an oral derivative) (Figure 13.13). However, natural penicillins are rarely used in clinical applications. Many key pathogens have acquired resistance to penicillin by producing a special type of β-lactamase enzyme, penicillinase, which hydrolyses the β-lactam ring and deactivates the antibiotics.

Despite antimicrobial resistance, penicillin continues to be the main interest of industrial-scale production. Its main commercial interest exists in the production of 6-aminopenicillanic acid (6-APA), which is the central component for the synthesis of semisynthetic β-lactam antibiotics. An enzyme, penicillin acylase, allows the transformation of penicillin into 6-APA, which is subjected to chemical modification (detailed in Chapter 17). The chemical modification in 6-APA can produce four generations of β-lactam antibiotics, which have a wide spectrum of action. Of the forty-five newly discovered drugs which were under clinical trial for the US market in 2018, about thirteen products were β-lactam derivatives. Five derivatives are β-lactams, two represent hybrid variants, and seven compounds are β-lactamase inhibitors. Due to the prevalence of infectious diseases, the consumption of β-lactam and β-lactamase inhibitors is still increasing in middle and low-income countries. As per an estimate, the market size of the derivatives of β-lactam was valued at $27,126 million in terms of sales for 2018 and is projected to reach $34,170 million by 2028.

The industrial production of penicillins passes through three main stages: upstream processing, fermentation process, and downstream processing (Figure 13.14), which are discussed as follows.

13.4.2 Upstream Processing

The first step of penicillin production is the identification and screening of suitable strains. In the nonroutine preparation, the genetic and cellular modifications of the mould improve the yield several folds by amplification of the penicillin biosynthetic gene cluster, the increase in the number of intact, functional peroxisomes, and increasing the concentration of transporter proteins for the transport of penicillin. All the physicochemical conditions for penicillin production are optimised before the large-scale set-up. These can include parameters like feed optimisation, type of fermentation, and growth condition requirements. In the routine preparation, the preserved dry spores are germinated on a suitable medium. The grown vegetative body is transferred into a shaker flask (of 3-5 l) for the inoculum build-up. The developed mycelial content is transported into the seed tank (of about 500 l) for the pre-fermenter culture. The initial seed growth phases can be repeated two to three times before industrial-scale production.

The main feedstock for penicillin production is corn-steep liquor, a by-product of the corn starch industry. The sugarcane bagasse, having plentiful biomass with high contents of sugar, is also a suitable raw material for penicillin fermentation. The sugar content of the medium can be further enriched using various simple sugars such as sucrose, glucose, lactose, etc. The fermentation process additionally requires specific minerals, oil, and precursors. The addition of ammonium salts meets the requirements of a nitrogen source for fungal growth. When the broth consists of phenylacetic acid, the product is penicillin G. The high exogenous concentrations

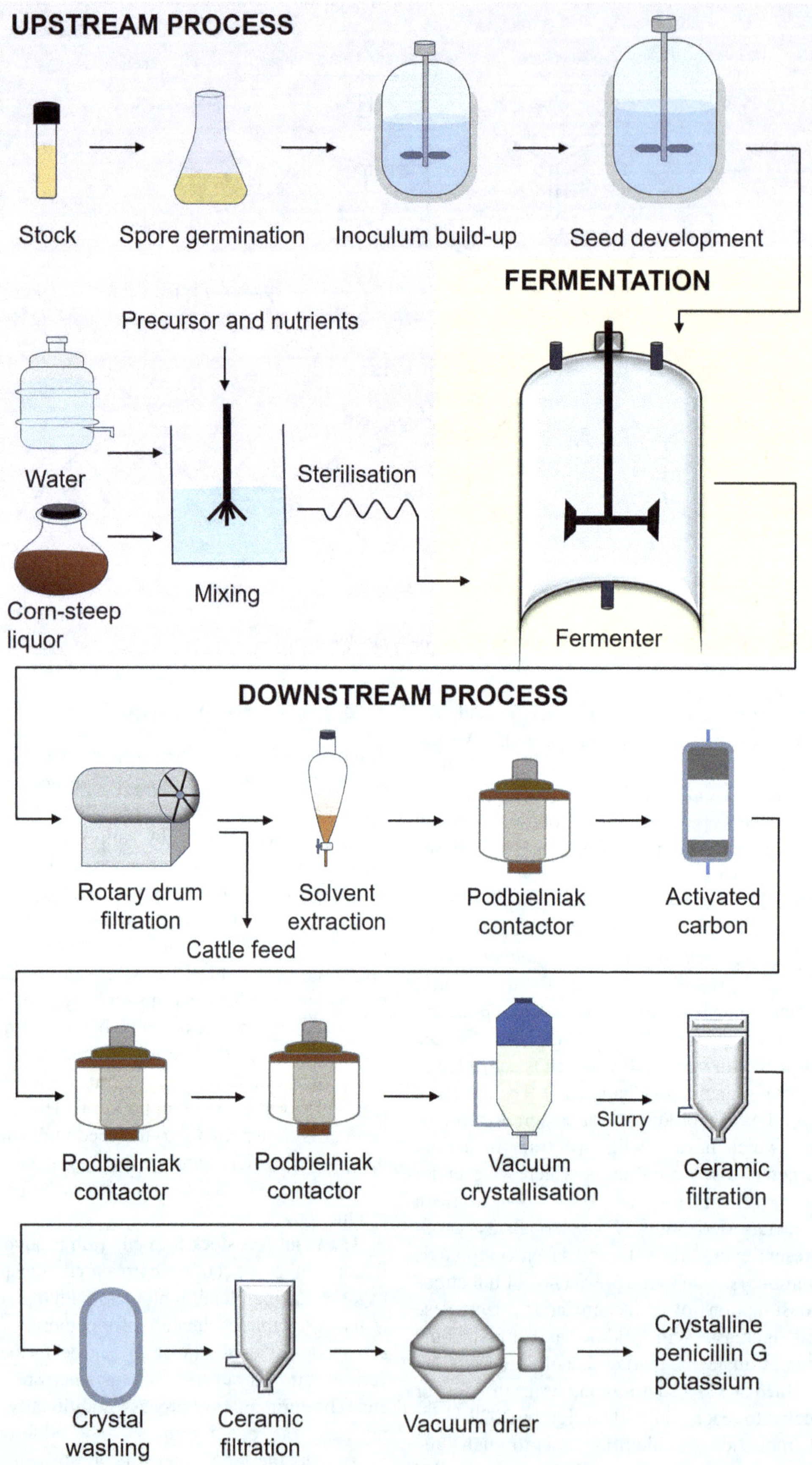

FIGURE 13.14 A general scheme of penicillin production through submerged fermentation (SmF).

of phenoxyacetic acid produce penicillin V. However, the presence of both precursors in the same medium blocks the pathway of penicillin V and produces only penicillin G.

13.4.3 Fermentation Process

The commercial production is carried out in deep-tank fermenters of 40,000 to 200,000 l of stainless steel tanks. The mode of fermentation is a fed-batch process. During the fermentation, a portion of 20–40% of the culture is recovered at intervals, and a new sterilised medium is added. This type of "batch-fill and withdraw" system is called "repeated fed-batch culture" and has a role in increasing the production of penicillin. The typical constituents of the medium for penicillin production are corn-steep liquor, glucose, lactose, minerals, oil, and precursor. Sugars and precursors are fed continuously. Only 10% of the carbon used in the medium is utilised for penicillin production, while the remaining 90% is exhausted in cellular maintenance and growth. The sugar in the medium also regulates the pH of the solution for the active production phase and maintains it around 6.4–6.8. The peak growth periods of the culture (i.e., around 40–50 h time-period of the cycle) require adequate oxygen levels inside the fermenter, which is added and churned using a sparger and agitator, respectively. Phenylacetic acid and phenoxyacetic acid are the respective precursors for penicillin G and penicillin V, which are attached to the side chains of the main backbones. A prolonged stationary phase of the culture is desirable for a high yield of many secondary metabolites, and so is applicable for penicillins. The production phase in the tank lasts 120–200 h. The monitoring of penicillin production during fermentation can be done by withdrawing a low volume of sample from the reaction mixture for high-performance liquid chromatography (HPLC) analysis.

13.4.4 Downstream Processing

Advanced technologies of downstream processing have achieved about 90% recovery yield of penicillins. The operational mode of recovery and purification is semicontinuous. Penicillin recovery can involve as many as ten different steps. The first step is broth filtration, which separates the solid and liquid from the harvested broth. The content is usually filtered using a rotary vacuum filter in continuous processing. This process removes large solid particles and microbial cells from the broth. This mother liquor can be placed for another round of rotary vacuum filtration to clarify the solution. The next step is solvent extraction (liquid-liquid extraction) using an extraction agent, butyl acetate or amyl acetate. At low pH (around pH 2), a very high amount of penicillin in the aqueous phase is transferred to the extract phase (solvent). The pH of the solution is lowered using H_2SO_4 or phosphoric acid. In the next step, penicillin is extracted from broth using Podbielniak Contactor (Pod). A pod is a cylindrical drum-mounted centrifuge that provides differential countercurrent extraction, which can be used when two solvents have very similar densities. This step is a perfect design for the recovery of penicillin, as it enriches the potency of penicillin ten times in the solution. The extraction of penicillin requires low temperatures (around 4 °C) to prevent its degradation at low pH. After the extraction, the extract phase is treated with activated carbon (0.25%) to remove impurities. Thereafter, penicillin is transferred back from the organic phase to the fresh aqueous phase using another cycle of extraction using Pod. This step is performed at high pH by adding potassium hydroxide (KOH) or sodium hydroxide (NaOH) to the solution, which converts the antibiotic into a penicillin salt. The organic solvent is recovered by distillation, which can be carried out in three different phases, including evaporation, vapour-liquid separation, and condensation. This is followed by ceramic filtration. The penicillin salt is obtained in crystal form through a process of crystallisation. This can be initiated by converting the solution into a supersaturation phase. A supersaturated solution contains an excess amount of solute in the dissolved form at normal temperature. The gentle cooling promotes the crystallisation of the solute. In a different method, the addition of a nonsolvent in the solution can decrease the solubility of the solute (penicillin), leading to its crystallisation. Differently, solvent evaporation can also induce crystallisation, but is an energy-intensive process. Further, the crystals are washed with an immiscible liquid to remove impurities. The wet penicillin is dried through lyophilisation, spray dryers, or vacuum band dryers in a final product. The solvent and precursor recovered from different steps of downstream processing improve the production process and reduce their environmental impact. In addition, the optimum recovery of penicillin can be achieved by modification in the said techniques or using alternative processes such as membrane alternatives and extraction decanters.

13.5 PRODUCTION OF VITAMINS

Vitamins and vitamin-like compounds are organic substances that are required in microquantities for the normal growth and development of living beings. There are only six different types of vitamins (A, B, C, D, E, and K). Vitamins A, D, E, and K are classified as fat-soluble, while vitamins B and C exhibit water solubility. Particularly, vitamin B exists in eight different forms, including biotin, folate, niacin, pantothenic acid, riboflavin, thiamine, vitamin B6, and vitamin B12. Each plays a unique role in the homeostasis of body function. They are mainly derived from the food we consume. The natural sources of most of the vitamins are fruit, vegetables, grains, meat, eggs, and dairy products. The human body can synthesise vitamins D and K to some extent.

Humans are becoming vitamin deficient due to changes in diet and lifestyle, necessitating the use of vitamin supplements and foods enriched with vitamins. Large-scale food fortification (deliberate modification of foods to improve nutritional content) is an affordable way to combat vitamin and mineral shortages. Moreover, vitamins have roles in numerous other applications, such as cosmetics, feed, medicine, and other fields. The rise in global demand for vitamins has sparked considerable interest in innovative production systems. The biotechnological routes of their production can involve different bacteria and fungi. Vitamins that are entirely developed using microbial fermentation include vitamin B2 (riboflavin), B12 (cobalamin), C, D2 (ergocaliferol), and K2 (menaquinone). The natural producers of vitamin B2 are fungi like *Ashbya gossypii* and *Candida flaeri,* and bacteria like *B. subtilis* and *Corynebacterium ammoniagenes*. The vitamin B12 biosynthetic pathway is found in only certain prokaryotes (both bacteria and archaea). Vitamin C can be produced using bacteria like *Acetobacter*, *Gluconobacter*, *Ketogulonicigenium*, *Erwinia*, and *Corynebacterium*. The GRAS yeasts, such as S. cerevisiae, Saccharomyces uvarum, and C. utilis, are the sources of vitamin D. Even some mushrooms are notable producers of vitamin D3. Vitamin K can be produced using *B. subtilis*, *Propionibacterium freudenreichii*, and *Flavobacterium* spp. Vitamin-like compounds such as ubiquinone, pyrroloquinoline quinone, gamma-aminobutyric acid (GABA), glutathione (GSH), and essential fatty acids (EFA) can be derived from diverse microbial sources. Some vitamins, like B3 (niacin) and B5, are manufactured using a combined approach of microbial fermentation and chemical synthesis. The commercial production of β-carotene compounds (i.e., precursor of vitamin A) through microalgae requires their cultivation in ponds or fermenters. Table 13.4 provides a list of different vitamins and their microbial sources for industrial applications.

In recent years, the demand for microbial-produced vitamins has significantly increased owing to consumer demand, environmental concerns, and their diverse applications. The continuous advancements in metabolic engineering and synthetic biotechnology have broadened the scope of microbial fermentation to develop new compounds of vitamins at industrial scales, promising improved efficiency, safety, sustainability, and regulatory compliance. Fermentation technology has also achieved many technological advancements, improving its production and ease of bioprocessing. Synthetic biology has advanced the biosynthesis of vitamins through metabolic engineering. A recent breakthrough is the development of novel microbial strains by assembling genes from various organisms and optimising their expression for B12 production. This process has constructed a new biosynthetic pathway in the microbe.

The following section provides a detailed understanding of vitamin B12 production using microbial fermentation.

13.5.1 Production of Vitamin B12

Vitamin B12 (or cobalamin (Cbl)) is the most complex type of vitamin. It is a metalorganic compound which has a central cobalt atom coordinated with four nitrogen atoms of a tetrapyrrolic corrin ring. Cobalt is attached to two other ligands: 5,6-dimethyl benzimidazole (DMB) in the α-axial position and a variable group in the β-axial conformation (Figure 13.15). The corrin ring has a resemblance to the porphyrin rings of chlorophyll and haemoglobin. Vitamin B12 supplements come in several forms, including cyanocobalamin (CNCble), adenosylcobalamin (AdCbl), methylcobalamin (MeCbl), and hydroxocobalamin (OHCbl). The trace amount of vitamin B12 is important for the synthesis of nucleic acids, red blood cells, and neurological function. The main sources of Cbl are meat, eggs, and dairy products. Particularly, vegetarians and vegans find alternative sources of this vitamin through fortified foods and Cbl supplements.

The chemical synthesis of Cbl was identified in 1973, but it is not feasible for industrial production due to the cumbersome steps (more than sixty steps) and very low yield. On the other hand, certain bacteria and archaea are the natural producers of Cbl, which can complete its biosynthesis in about

TABLE 13.4

A list of different vitamins, including information related to associated deficiency diseases, their microbial sources, and compounds produced

Vitamins	Major deficiency diseases	Microbial sources	Compounds from microbial sources
A	Blindness and nyctalopia	Microalgae (e.g., *Dunaliella salina*) and bacteria like *Blakeslea trispora* and *Yarrowia lipolytica*	β-carotene compounds (precursors)
B (including biotin, folate, niacin, pantothenic acid, riboflavin, thiamine, vitamin B6, and vitamin B12)	Beriberi, pellagra, and anaemia	Vitamins B2 from *Ashbya gossypii, Eremothecium ashbyii, Bacillus subtilis*, and *Corynebacterium ammoniagenes* B12 from *Propionibacterium freudenreichii* and *Pseudomonas denitrificans*	B2 (riboflavin), B12 (cobalamin), (partially B3 (niacin), and B5)
C	Scurvy	*Acetobacter*, *Gluconobacter*, *Ketogulonicigenium*, *Erwinia*, and *Corynebacterium*	Ascorbic acid
D	Rickets	*Saccharomyces cerevisiae*, *Saccharomyces uvarum*, *C. utilis*, and mushrooms	D2 (ergocaliferol) and D3
E	Ataxia	-	-
K	Osteoporosis, hemorrhagic disease, and cystic fibrosis	*B. subtilis*, *P. freudenreichii*, and *Flavobacterium* spp.	K2 (menaquinone)

Note: Bold text represents water-soluble vitamins, and the remaining forms are fat-soluble.

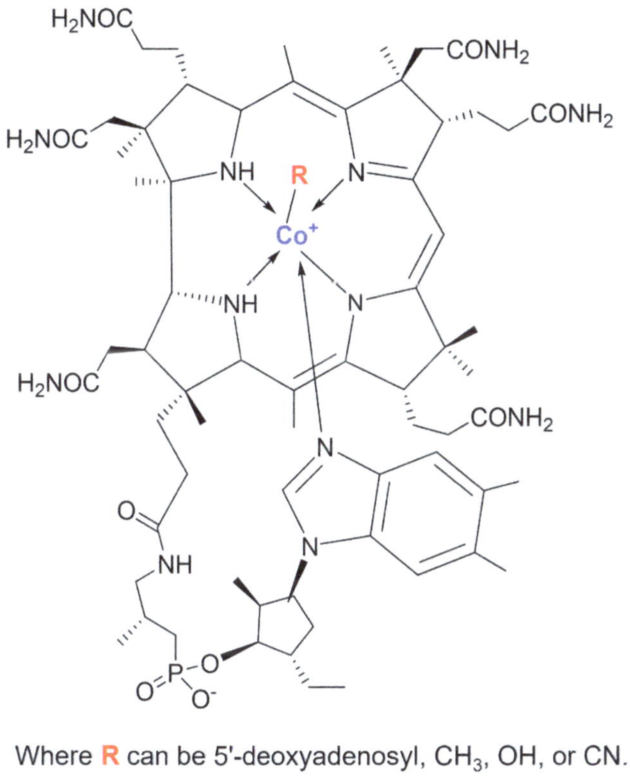

FIGURE 13.15 The chemical structure of vitamin B12.

thirty biotransformation steps. The industrial production of Cbl mainly relies on two different bacteria, *P. freudenreichii* (Gram-positive) and *Pseudomonas denitrificans* (Gram-negative). Both species inherit different pathways for their synthesis. The former species produces Cbl using an anaerobic or early cobalt insertion pathway. *P. freudenreichii* mainly uses different *CBI* genes. This type of pathway is also reported in other bacteria, like *Bacillus megaterium* and *Salmonella enterica*. However, *P. denitrificans* utilises aerobic or late cobalt insertion pathways, which are also found in microorganisms like *Ensifer casida* and *Sinorhizobium meliloti*. The genetic determinants in this pathway are *cob* genes.

There are two distinct pathways for vitamin B12 synthesis (Figure 13.16). Both pathways show key differences in their steps of ring contraction and cobalt chelation. In the anaerobic pathway, cobalt is inserted prior to ring contraction and is independent of the aerobic process. On the other hand, the aerobic pathway uses oxygen and a monooxygenase enzyme for ring contraction before cobalt insertion, and that is why the process is also known as the late cobalt insertion pathway. However, the synthesis of a tetrapyrrole ring implies a distinct biosynthetic pathway using precursor 5-aminolevulinic acid (ALA), independent of aerobic and anaerobic pathways. Likewise, the synthesis of DMB relies on different routes. The main stages of vitamin B12 production using different microorganisms are illustrated in Figure 13.17.

The non-genetically modified *P. freudenreichii* has the recognition of GRAS status and can be used as a food ingredient for the vitamin B12 supplement in fortified food. However, the volumetric production of cobalamin utilises genetically modified strains *P. denitrificans* and *P. freudenreichii* (non-GRAS strains). Pharmaceutical companies like Sanofi (France) and Huarong Pharmacy Corporation (China) are some of the leading producers of vitamin B12 using *P. denitrificans*.

13.5.2 Upstream Processing

The first step of upstream processing is the identification of a suitable strain. In the non-routine process, producer strains can be genetically modified for the overexpression of vitamin B12 or compatibility with the downstream processing. Techniques like random mutagenesis and genetic engineering have developed high-yielding strains of *P. denitrificans* for commercial applications. Synthetic biology has enabled the construction of heterologous biosynthetic pathways in *E. coli* to improve the production of vitamin B12. The non-genetically modified bacterium, *P. freudenreichii,* has achieved productivity improvement through optimisation of the production medium and changes in bioprocessing for its direct use as a food additive. The overexpression of some *cbl* genes in *P. freudenreichii* has triggered a high yield of cobalamin. The decrease in propionic acid production using inhibitors and through genetic modification has also shown positive effects on vitamin B12 production.

The industrial substrates for the fermentation include sugar beet molasses, corn steep liquor, and betaine, along with other supplements such as minerals, vitamins, and yeast extracts. The sugar beet molasses also contains a low concentration of betaine. Substances like ALA and Co^{2+} are critical precursors for its biosynthesis. The addition of DMB and Zn^{2+} shows positive effects on production. Betaine in the medium acts as a methyl donor for vitamin B12 and induces the biosynthesis of ALA, but its high concentration inhibits microbial growth. Therefore, the concentration of betaine in the fermenter is maintained by continuously feeding in the range of about 5–7 g/l. The low-cost medium usually has the property to change the pH of the solution, and so the feeding of betaine and glucose also controls the pH and maintains it around 6–7.

Before fermentation, the ingredients of the feed medium are sterilised, mixed, and transferred to the main vessel. A separate arrangement involves the preparation of the inoculum of the production strain from a fresh slant. The pre-cultured microorganism is gradually cultivated for two stages of seed growth in 150 l and 9 m^3 small fermenters before transferring the inoculum into the main vessel.

13.5.3 Fermentation Process

The full-scale operation of vitamin B12 production runs as submerged fermentation using *P. denitrificans* in a volume capacity of 120,000 l (i.e., 120 m^3 size fermenter) in a fed-batch or continuous manner. Only 5% of the working volume from the second seed culture is used for the inoculation of vitamin B12 production. The fermenter is equipped with three-bladed agitators and probes for pH, temperature, and dissolved oxygen (DO) to optimise the production. The optimal temperature for fermentation is 30 °C. The fermentation at the initial stage

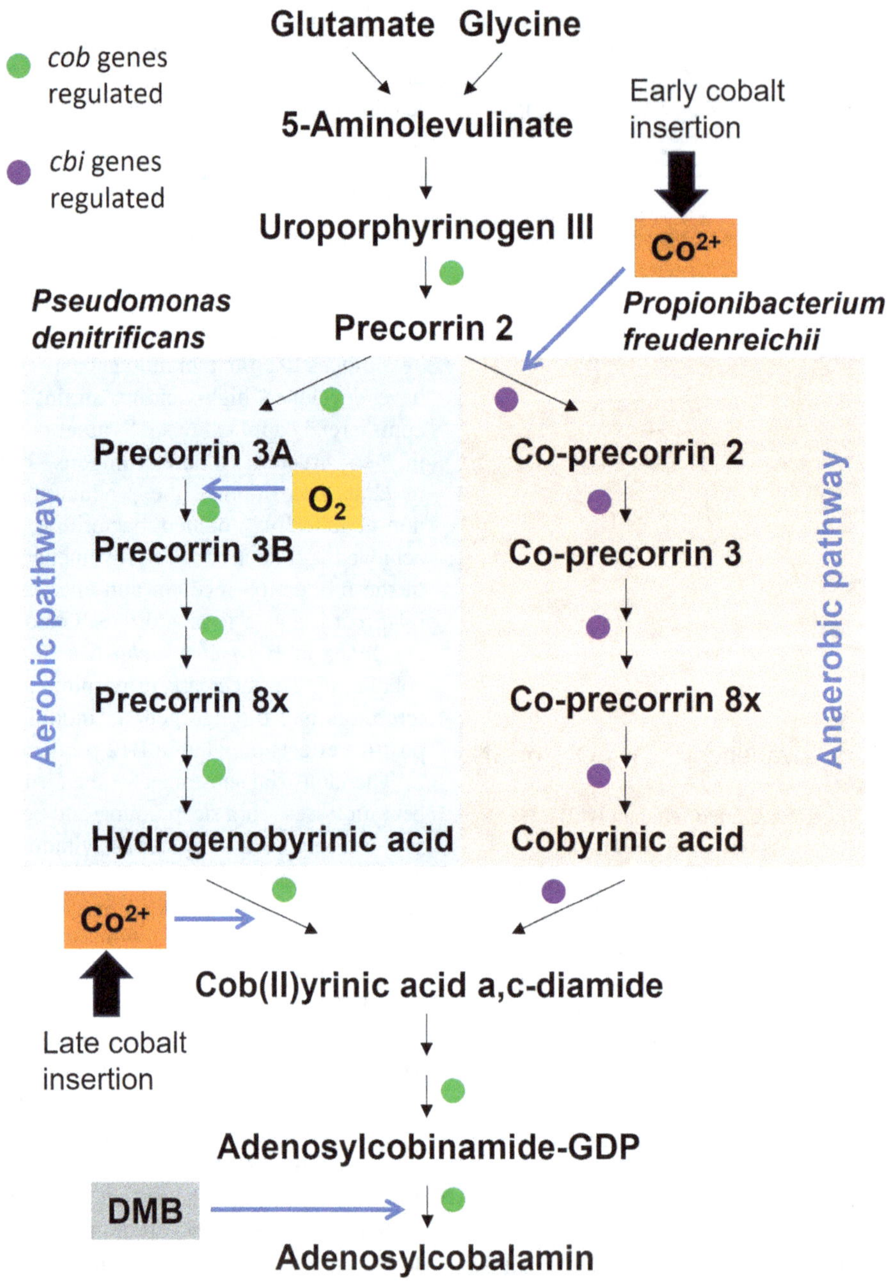

FIGURE 13.16 Two different biosynthetic routes of vitamin B12 (adenosylcobalamin) (Co^{2+}: Cobalt ion and DMB: 5,6-dimethylbenzimidazole).

requires a high oxygen transfer rate to develop sufficient bacterial growth. However, the maximum amount of vitamin B12 accumulates during the stationary phase of microbial growth. This requires low oxygen concentration and utilises multi-step oxygen control strategies to gradually reduce the oxygen concentration below 2%. This can be followed by the use of respiratory chain inhibitors, such as rotenone, which even causes more injurious effects on cell growth and triggers high production of vitamin B12. This process claims to achieve a high volumetric production of vitamin B12, with reported yields as high as 200–300 mg/l.

The fermentation cycle for the production of vitamin B12 in a volumetric production takes about 180 h with successive anaerobic and aerobic phases using *P. freudenreichii*. The initial phase requires anaerobic conditions for 72–96 h to run the early cobalt insertion pathway. After this period of cultivation, the medium requires gentle aeration for the synthesis of DMB and its attachment with the main ring for cobalamin synthesis. This period lasts for another 72–96 h. In contrast to *P. denitrificans*, which produces vitamin B12 throughout the fermentation cycle, the main biosynthesis of vitamin B12 using *P. freudenreichii* begins only after 72 h of incubation. The estimation of vitamin B12 in the medium is carried out using microbiological assay, spectrophotometric, or chromatographic techniques.

The food-grade propionibacteria can be directly inoculated into the foodstuff to produce vitamin B12. An *in situ* application of *P. freudenreichii* in cereal matrices improves

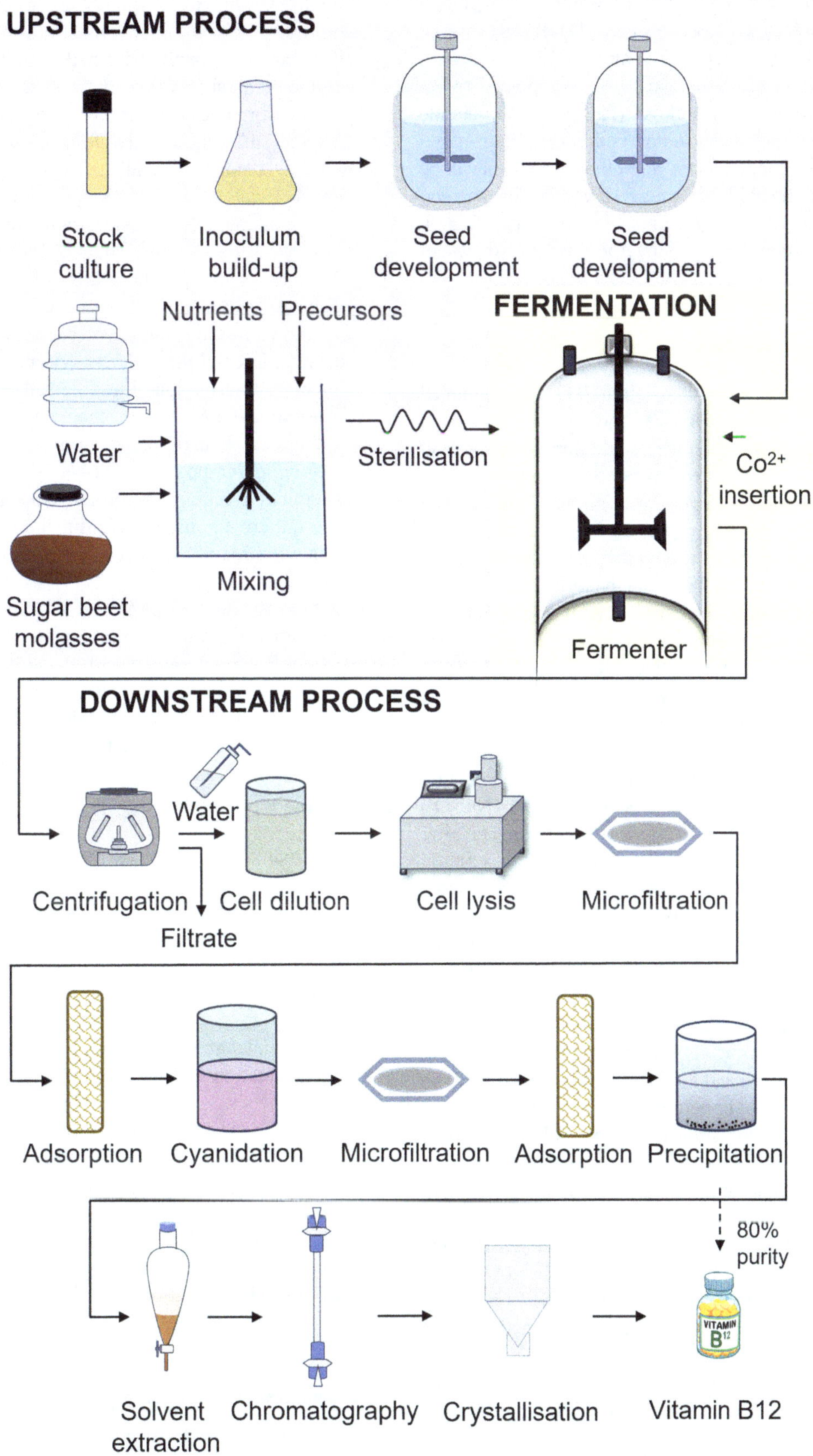

FIGURE 13.17 A general scheme of vitamin B12 production through submerged fermentation (SmF).

the bioavailability of vitamin B12 in vegan or vegetarian diets. However, in fortified food, the use of DMB (an essential compound for the synthesis of vitamin B12) as a food additive raises safety concerns. This can be replaced by its natural precursors, such as riboflavin (RF) and nicotinamide (NAM), for its synthesis under *in vivo* conditions. Plant-based foods like yoghurt, cheese, juices and other beverages can be fortified with vitamin B12. Vitamin B12 content in the food can range between 1.6–80 µg/kg fresh weight, depending on fermentation conditions, types of microorganisms, and plant matrices involved.

13.5.4 Downstream Processing

The downstream processing of vitamin B12 has remained unchanged for the last few decades. The recovery at a large scale has three main processes, *viz*., extraction, filtration, and purification. The first step is the recovery of microbial biomass as the main product is developed intracellularly. This can be achieved through centrifugation. Since a low concentration of vitamin B12 can be present in the medium, a separate technique can be used to extract the vitamin from the fermentation broth. The recovered cell is diluted in sterile water and is exposed to a pH of 6.5–8.5 and a temperature of 80–120 °C for 10–30 minutes. This ensures an effective cell lysis and release of their internal components. The lysate of the cells is passed through microfiltration, and the filtrate is subsequently purified through the adsorption process. The addition of potassium cyanide (KCN) or thiocyanate in the solution converts vitamin B12 into CNCbl in the presence of sodium nitrite ($NaNO_3$) and heat. The cyanidation can be performed before or after the microfiltration process. If performed later, the solution needs further treatment by microfiltration/nanofiltration, followed by adsorption to clarify the vitamin solution. The solution is now treated with zinc chloride ($ZnCl_2$) or zinc hydroxide ($Zn(OH)_2$), and vitamin B12 is precipitated. The next step is solvent extraction using organic solvents like acetone, tannic acid, or cresol. The preparation has 80% purity of vitamin B12, which is suitably used as animal feed. However, pharmaceutical and food applications require a high level of purity, which can be achieved by involving additional steps such as solvent extraction in organic solvents (carbon tetrachloride and butanol) and different adsorption techniques (e.g., ion exchangers, aluminium oxide, or activated carbon). Chromatographic techniques can be used for high purification. The final pure vitamin B12 compound is obtained through drum drying or crystallisation by adding phenol and water.

The downstream processing of vitamin B12-producing food-grade bacteria for the development of fortified foods is relatively simple and relies on techniques like centrifugation and spray drying. The collected dry microbial biomass can be directly inoculated into the foodstuff to produce vitamin B12.

SUMMARY

- Microbial enzymes are powerful tools and eco-friendly solutions for catalysis in modern industries. The industrial enzymes, α-amylases, represent great value for transforming carbohydrates into simple sugars. Their fermentation can be performed by SmF and SSF, and the crude preparation of the enzymes relies on filtration.
- Various organic acids offer benefits to industries for their terrific buffering and acidifying capacity. Citric acid is the most popular of all and can be chiefly produced by *A. niger* using surface, solid-state, and submerged fermentation. Its downstream processing mainly involves filtration, precipitation, and crystallisation.
- Ethanol is a valuable source of biofuel and liquor. Ethanol is typically produced using a variety of raw materials in the presence of yeast, *S. cerevisiae,* under anaerobic conditions. The recovery of ethanol mainly involves flocculation, filtration, and distillation.
- Antibiotics are modern medicines to cure bacterial infections. Penicillins are primarily produced using *Penicillium* spp. in fed-batch fermentation. The production requires a prolonged stationary phase of the culture for high penicillin titers. The recovery of penicillin mainly involves filtration, solvent extraction, Pod, and crystallisation.
- Vitamins are essential for numerous metabolic reactions. Vitamin B12 has a global demand, developed using *P. freudenreichii* and *P. denitrificans*. The full-scale operation for its production runs as SmF, and the recovery involves techniques like centrifugation, filtration, precipitation, adsorption, and drum drying.

DIFFERENT TYPES OF PRACTICE QUESTIONS

Fill in the blanks

I. The α-amylases can be produced by two distinct fermentation processes, including .... and ....

II. A metal .... serves as a cofactor for the α-amylase.

III. The industrial workhorse of citric acid through fermentation is a fungus ....

IV. The .... is a Japanese traditional way of citric acid production in solid-state fermentation.

V. A Gram-negative anaerobic bacterium, ..., is a potential source of ethanol production.

VI. β-lactam antibiotics act on .... and are mainly active against ....

VII. For food fortification, vitamin B12 is mainly produced using bacterium ....

SOLUTION: 1 I. Submerged fermentation (SmF), solid-state fermentation (SSF), II. Ca, III. *Aspergillus niger*, IV. Koji process, V. *Zymomonas mobilis*, VI. cell wall synthesis, Gram-positive bacteria, VII. *Propionibacterium freudenreichii*

Match the following

i.	Penicillin	a. Vacuum crystalliser
ii.	Ethanol	b. Immobilisation
iii.	Citric acid	c. Distillation
iv.	Vitamin B12	d. Podbielniak centrifugal contactor
v.	α-amylase	e. KCN treatment

SOLUTION: i-d, ii-c, iii-a, iv-e, v-b

Very short answer type of questions

I. Write the names of any two hydrolytic enzymes used in industrial biotechnology.
II. What are the applications of citric acid in the food industry?
III. What is liquefaction in ethanol fermentation?
IV. What is the Golden Age of antibiotic discovery?
V. What is food fortification? Give a suitable example.

Short answer type of questions

I. What are α-amylases? Write about various applications of microbial α-amylases in industries.
II. Write a short note on the fermentation process of citric acid.
III. What is the main application for the bulk production of penicillin? Discuss the techniques involved in its downstream processing.
IV. What are the notable microbial sources and essential conditions for ethanol fermentation?
V. Why is vitamin B12 production a complex process? What are the two different cobalt insertion pathways in vitamin B12 production?

Long answer type of questions

I. What are the applications of citric acid? Give a detailed account of its industrial production from microorganisms.
II. Bioethanol is a promising substitute for fossil fuels. Shed some light on the statement. How do microorganisms play a role in its synthesis? Prepare a flow chart for its industrial production.
III. What are the natural, semisynthetic, and synthetic antibiotics? What are their applications? Discuss the roles of microorganisms in antibiotic production with suitable examples.

NOTES

1 Mores S, de Souza Vandenberghe LP, et al. Citric acid bioproduction and downstream processing: Status, opportunities, and challenges. Bioresource Technology. 2021;320:124426.

2 www.istockphoto.com/photo/aerial-view-of-an-ethanol-plant-in-south-dakota-gm1139974082-304894318

3 Hutchings MI, Truman AW, Wilkinson B. Antibiotics: past, present and future. Current Opinion in Microbiology. 2019;51:72–80.

SUGGESTED READINGS

Balabanova L, Averianova L, Marchenok M, Son O, Tekutyeva L. Microbial and genetic resources for cobalamin (vitamin B12) biosynthesis: from ecosystems to industrial biotechnology. *International Journal of Molecular Sciences*. 2021;22(9):4522.

Behera BC. Citric acid from *Aspergillus niger*: A comprehensive overview. *Critical Reviews in Microbiology*. 2020;46(6):727–49.

Berovic M, Legisa M. Citric acid production. *Biotechnology Annual Review*. 2007;13:303–43.

Calvillo Á, Pellicer T, Carnicer M, Planas A. Bioprocess strategies for vitamin B12 production by microbial fermentation and its market applications. *Bioengineering*. 2022;9(8):365.

Elander RP. Industrial production of β-lactam antibiotics. *Applied Microbiology and Biotechnology*. 2003;61(5):385–92.

Fang H, Kang J, Zhang D. Microbial production of vitamin B12: a review and future perspectives. *Microbial Cell Factories*. 2017;16(1):1–4.

Hutchings MI, Truman AW, Wilkinson B. Antibiotics: past, present and future. *Current Opinion in Microbiology*. 2019;51:72–80.

Ledesma-Amaro R, Santos MA, Jiménez A, Revuelta JL. Microbial production of vitamins. In McNeil B, Archer D, Giavasis I, Harvey L (eds.) *Microbial production of food ingredients, enzymes and nutraceuticals* 2013 (pp. 571–594). Woodhead Publishing.

Li S, Yang X, Yang S, Zhu M, Wang X. Technology prospecting on enzymes: application, marketing and engineering. *Computational and Structural Biotechnology Journal*. 2012;2(3):e201209017.

Martens JH, Barg H, Warren MA, Jahn D. Microbial production of vitamin B12. *Applied Microbiology and Biotechnology*. 2002;58(3):275–85.

Mores S, de Souza Vandenberghe LP, Júnior AI, de Carvalho JC, de Mello AF, Pandey A, Soccol CR. Citric acid bioproduction and downstream processing: Status, opportunities, and challenges. *Bioresource Technology*. 2021;320:124426.

Sauer M, Porro D, Mattanovich D, Branduardi P. Microbial production of organic acids: expanding the markets. *Trends in Biotechnology*. 2008;26(2):100–8.

Show PL, Oladele KO, Siew QY, Aziz Zakry FA, Lan JC, Ling TC. Overview of citric acid production from *Aspergillus niger*. *Frontiers in Life Science*. 2015;8(3):271–83.

Soccol CR, Vandenberghe LP, Rodrigues C, Pandey A. New perspectives for citric acid production and application. *Food Technology and Biotechnology*. 2006;44(2):141–9.

Souza PM, Magalhães PD. Application of microbial α-amylase in industry-A review. *Brazilian Journal of Microbiology*. 2010;41:850–61.

Sundarram A, Murthy TP. α-amylase production and applications: a review. *Journal of Applied and Environmental Microbiology*. 2014;2(4):166–75.

Vandamme EJ, Revuelta JL. *Industrial biotechnology of vitamins, biopigments, and antioxidants*. Wiley; 2016.

Wang Y, Liu L, Jin Z, Zhang D. Microbial cell factories for green production of vitamins. *Frontiers in Bioengineering and Biotechnology*. 2021;9:473.

Unit IV

Microbial Enzymes and Their Immobilisation

14 Microbial Enzymes of Industrial Interest

14.1 ENZYMES AND THEIR CHARACTERISTICS

Enzymes are present throughout the biological system and are known to serve as biocatalysts. They play important roles in almost all physiologically mediated processes, from DNA replication to food digestion. They are primarily proteinaceous entities except for ribozymes. Twenty different amino acids are the building blocks of enzymes, which are connected through amide bonds. These bonds, in the context of proteins, are specifically called peptide bonds. A typical enzyme can be composed of 100 to 35,000 amino acids. They comprise a linear chain of amino acids that fold into a precise 3D structure to provide a highly ordered structure necessary to support a specific catalytic reaction.

An enzyme has a main role in accelerating the rate of reaction. An appropriate enzyme can increase the rate of reaction by million folds. However, the enzyme-catalysed reaction does not alter the equilibrium of the chemical reaction. When a substrate binds to the enzyme, it establishes a transient, non-covalent enzyme-substrate complex (reaction 14.1). This interaction reduces the activation energy needed for a chemical reaction, facilitates a transition state, and forms a product (Figure 14.1). After the reaction, the enzyme releases the products and is ready for the other round of reactions. A small amount of enzyme can catalyse the conversion of a large amount of substrate. Enzymes can be used repeatedly without being consumed or altered, which makes them highly remarkable for industrial applications.

$$E + S \rightarrow ES \rightarrow P + E \qquad \text{Reaction 14.1}$$

E, S, and P represent an enzyme, a substrate, and a product, respectively.

An enzyme is composed of both proteinaceous and non-proteinaceous entities. The inactive proteinaceous component of the enzymes is called apoenzyme. When an apoenzyme is attached to an obligatory non-proteinaceous component, this structure is called a holoenzyme (Figure 14.2). A holoenzyme is an active form of the enzyme and requires tightly or loosely bound components for its activity, which are called cofactors. They can be inorganic ions (e.g., Fe^{3+}, Mg^{2+}, Mn^{2+}, and Zn^{2+}) or small organic molecules. Some organic cofactors are tightly bound with enzymes and are permanently attached. They are called prosthetic groups (e.g., heme group in haemoglobin and flavin adenine dinucleotide (FAD) in redox enzymes). Some complex organic compounds, such as nicotinamide adenine dinucleotide (NAD^+), nicotinamide adenine dinucleotide phosphate ($NADP^+$), and FAD, are loosely bound to enzymes, which are called coenzymes. They donate or accept electrons, protons, or different chemical groups to the reactants. However, these are not a common requirement for all the enzymes. Many enzymes catalyse the reactions without using any prosthetic groups or coenzymes.

Enzymes catalyse a multitude of inorganic and organic substances. They operate under relatively mild conditions of pH, temperature, and pressure. This is important for maintaining their structural integrity and activity, which is crucial for their role in the cellular processes. Enzymes can be inducible or constitutive based on their production in the cells. An inducible enzyme requires the presence of an inducer for its production (e.g., β-galactosidase). On the contrary, a constitutive enzyme is produced at a constant rate in the cell, regardless of the availability of any compounds (e.g., citrate synthase).

14.2 CLASSIFICATION OF ENZYMES

The International Union of Biochemistry and Molecular Biology (IUBMB) has recommended a classification for all the enzymes based on their chemical reactions. In this classification, the enzymes have received a name and a number to identify them. More than 7,000 enzyme reactions are considered in the numerical classification scheme. As per the classification, enzymes are classified into seven main categories, *viz.*, oxidoreductase, transferase, hydrolase, lyase, ligase, isomerase, and translocase. The last top-level category, translocase, is a new addition to the enzyme classification and was included in 2018. All categories are further subdivided into subclasses by differences in their reaction mechanisms. A total of three subclasses exist for all the top levels of categories. Notions for the numerical representation of any enzyme include the letter EC (EC stands for Enzyme Commission) followed by four numbers. All the numbers are separated by a punctuation mark (.). A typical example is EC X.X.X.X, where X represents different numbers. A list of all the top-level categories,

DOI: 10.1201/9781032651095-18

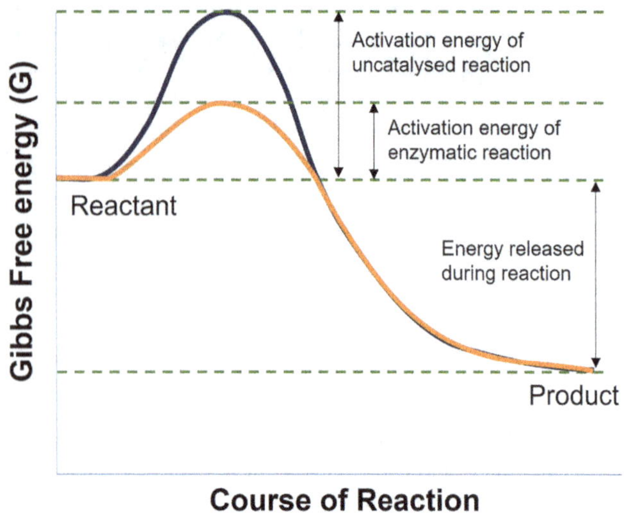

FIGURE 14.1 Effect of enzymes on the activation energy of a reaction.

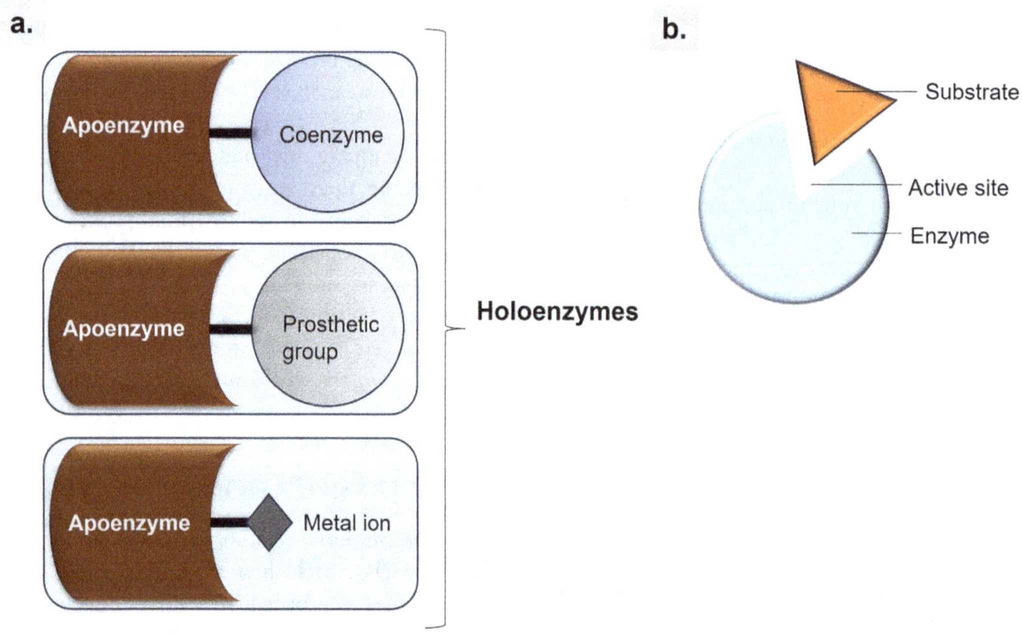

FIGURE 14.2 a. Important components of enzymes. b. An illustration of an active site on an enzyme for the binding of a substrate.

including their numbering system and the main catalytic reaction, is listed in Table 14.1.

14.3 ENZYME CATALYSIS

An enzyme has a specific location for the binding and catalysis of the substrate, called active sites. The other voluminous part of the enzyme supports the structure and position of the active sites. The active site accounts for only 10–20% of the overall enzyme volume and is the only catalytic region in the enzyme. The active site consists of only three or four amino acids critical for catalysis. Enzymes exhibit high substrate specificity due to the precise arrangement of amino acid residues. This specificity ensures that each enzyme interacts with a particular substrate or group of structurally related substrates. The specific combination and arrangement of amino acids within the active sites allow enzymes to recognise, bind, and catalyse chemical reactions for their target substrates. These amino acids interact with the substrate through various forces, such as hydrogen bonding, electrostatic interactions, and hydrophobic interactions.

Each amino acid present in the active site possesses specific functional groups for interaction with substrates. For example, amino groups (-NH_2) from lysine or arginine residues, carboxyl groups (-COOH) from aspartate or glutamate residues, and imidazole groups (-NH-) from histidine residues participate in acid-base catalysis. Hydroxyl groups (-OH) from serine, threonine, or tyrosine residues serve as nucleophiles or participate in hydrogen bonding with the substrate. Sulfhydryl groups (-SH) from cysteine residues take part in redox reactions or form disulfide bonds. Serine, threonine, or tyrosine residues act as nucleophiles and participate in phosphorylation reactions. Some enzymes require metal ions (such as zinc (Zn), magnesium (Mg), and iron (Fe)) to stabilise intermediates or participate in enzyme catalysis.

Many enzymes recognise specific substrates and exhibit high selectivity for substrates. This leads to their chemo-, stereo-, and regioselectivity. In some conditions, they detect functional groups of substrates such as phosphate, amino, or hydroxyl. In linkage-specific breakdown, they recognise special bonds in substrates, such as glycosidic bonds and amide linkages. The stereospecific enzymes identify steric or optical isomers such as L- or D-sugars/amino acids. These specificities ensure that they catalyse the right kind of reactions in a complex mixture of chemicals.

TABLE 14.1
A list of different types of enzymes in enzyme classification (EC)

EC	Type of enzymes	Reactions	Examples of industrial enzymes
EC 1.X.X.X	Oxidoreductase	The reaction utilises an electron or proton for the oxidation and reduction of a substrate	Catalase, laccases, and glucose oxidase
EC 2.X.X.X	Transferase	The reaction enables the transfer of a chemical group from one molecule to another	Glucosyltransferase and transglutaminases
EC 3.X.X.X	Hydrolase	The reaction undergoes hydrolysis of the substrate	Amylase, cellulase, lipases, proteases, xylanases, lactase, pectinase, penicillin acylase, mannanases, phytases, pullulanase, and cutinase
EC 4.X.X.X	Lyase	The reaction catalyses the breaking of the bond or the non-hydrolytic addition of groups	Pectate lyase and α-acetolactate decarboxylase
EC 5.X.X.X	Isomerase	The reaction involves the structural rearrangement of a molecule	Glucose isomerase
EC 6.X.X.X	Ligase	The reaction is required in the formation of chemical bonds between two macromolecules	DNA ligases
EC 7.X.X.X	Translocase	The reaction assists the movement of ions or molecules across membranes or their separation within membranes	SecA

The framework for understanding how enzymes function is a critical subject. Several models have been proposed to explain how an enzyme facilitates chemical reactions. Two prominent models of enzyme catalysis are the lock and key model and the induced fit model. According to the first model, a substrate has a complementary shape to the active site of the enzyme to bind and undergo a reaction. In the induced fit model, an enzyme experiences slight conformational changes when a substrate binds to it. The active site gently modifies the shape of the enzyme in order to fit the substrate more firmly. This conformational change helps to stabilise the transition state of the reaction and facilitates catalysis.

14.4 KINETIC PARAMETERS

Kinetic studies are fundamental to understanding the behaviour and mechanisms of enzymes. They unravel the information about the rate of reaction and catalytic mechanism of an enzyme through mathematical expressions. Knowledge of enzyme kinetics helps identify the quantitative and qualitative descriptions of enzymes and substrates for industrial applications. An important kinetic parameter is the catalytic constant, K_{cat}, also known as the turnover number. This parameter gives information about the maximal number of substrate molecules transformed into products per active site per unit of time under defined reaction conditions. However, this measurement does not consider the substrate concentration, which is very important for the rate of reaction and binding affinity. A different framework, Michaelis-Menten kinetics, is the principal model to evaluate enzyme dynamics and can be used for a variety of biochemical reactions. The maximal velocity (V_{max}) and Michaelis-Menten constant (K_m) are the two main parameters to predict Michaelis-Menten kinetics. V_{max} is defined as the maximum rate of reaction when all the active sites are saturated with substrate. K_m is a measure of affinity between the substrate and the enzyme and is not influenced by enzyme concentrations. Considering this, Michaelis-Menten kinetics can be formulated as

$$v = \frac{V_{max}[S]}{K_m + [S]}.$$

Where v is velocity, and S is substrate concentration.

However, many enzymes do not follow Michaelis-Menten kinetics due to coupled enzyme reactions or the involvement of multisubstrate and/or multi-product components. Such reactions can be more conveniently described by the Briggs-Haldane model and the Ping-Pong mechanism.

A different parameter to study the enzyme kinetics is the K_{cat}/K_m ratio, which is referred to as catalytic efficiency or specificity constant. This measurement allows easy comparison of the rate of reactions of different substrates on an enzyme. The high ratio of catalytic efficiency means a high rate of catalysis. The catalytic mechanism is also affected by the presence or absence of cofactors and inhibitors. They alter the reaction rate, which can be demonstrated by the Lineweaver–Burk plot (that is a derivative of the Michaelis-Menten equation). This plot is a derivative of the Michaelis-Menten equation, which can be formulated as

$$\frac{1}{v} = \frac{K_m + [S]}{V_{max}[S]} = \frac{K_m}{V_{max}}\frac{1}{[S]} + \frac{1}{V_{max}}.$$

14.5 HISTORICAL ACCOUNT OF ENZYME TECHNOLOGY

Since ancient times, humans, unknowingly, have taken advantage of enzymes for millennia, long before their structure and function were recognised. Archaeological studies have recorded evidence of enzymatic utility from the Neolithic era (nearly 13,000 years ago) for the production of alcohol. The curiosity about chemical changes in food material has always existed in human minds. The utility of enzymes has played a pivotal role in identifying many branches of biological sciences, including molecular biology, immunology, functional genomics, proteomics, and metabolomics, etc. The timeline of the study of enzymes or enzymology began in 1833 with the discovery of diastase. It was a group of enzymes recovered from an aqueous extract of malt for brewing. However, the word 'enzyme' was pioneered by German physiologist Wilhelm Kühne in 1878. It is derived from the combination of two Greek words, *en* and *zume*, which mean "within" and "yeast", respectively, for describing the alcohol production by yeast. The major outbreak in enzymology was the isolation and crystallisation of an enzyme by James B. Sumner (urease from jack bean) in 1927, which earned him the Nobel Prize in 1947.

14.6 ADVANTAGES OF ENZYME TECHNOLOGY

The synthetic routes of chemicals using catalysts are traditional pathways. However, they suffer from many drawbacks, such as high energy consumption, lower efficiency, less economical, and non-eco-friendly approaches. The chemical route can be replaced by enzyme technology. Enzymes are fragile but highly effective for chemical reactions. Enzyme-catalysed processes are versatile and considered sustainable green methodologies. Such importance is due to their non-toxicity and biodegradability. They can be derived using renewable sources such as plants and microbes, whereas the production of synthetic catalysts disrupts the nonrenewable reserves, like precious metals in the Earth's crust.

Enzymes offer safe and simple procedures that promote their applications in the mainstream repertoire. Many enzymatic operations require water as a solvent, which is capable of dissolving a wide range of substances. Also, the process requires low energy and has reduced water consumption. The production methods of enzymes are relatively shorter than the synthesis of organometallic catalysts. The recovery of products after enzymatic reactions involves relatively simple purification steps. The quality of the products from enzyme-based reactions is high standard. The enzyme-based processes generate less waste, and their treatment is relatively cost-effective.

The catalytic efficiency and specificity of the enzymes are the keys to their large-scale applications. Enzymes can even distinguish between stereoisomers. Industries using enzymatic reactions operate under batch-to-batch consistency. For a system, the enzyme is only 0.1–1.0% of the entire reaction mixture. Enzyme-mediated reactions usually produce fewer by-products. The ability of enzymes to act under mild conditions like ambient temperature, pH, and pressure of the cells contributes to their implication for low-cost operation. However, some enzymes (also called extremozymes) catalyse reactions at extreme temperatures. Enzymes that can act at very high or low pH also exist.

Though enzymes can be derived from plants or animals, microbial sources are of major interest for industrial applications. Microbial enzymes can be produced in bulk quantities, owing to the rapid growth of microorganisms on inexpensive media. Also, their production is more convenient and economical than that of other types of organisms. Enzymes derived from microorganisms are highly diverse and exist in different varieties for various applications. The stability of the microbial enzymes is relatively better than that of those derived from other types of organisms. Genetic engineering and protein engineering can be effectively employed to manipulate microbial enzymes with desired properties. The techniques of modifying enzymes at the genetic and protein levels provide tailor-made solutions to meet the specific needs of the reaction conditions and challenges. The production of microbial enzymes is uninfluenced by seasonal fluctuations due to operation under controlled parameters.

14.7 STATE-OF-THE-ART STRATEGIES OF ENZYME PRODUCTION

Enzymatic technology is currently one of the main requirements in the industrial process. The state-of-the-art strategy has enabled the production of enzymes using the correct type of microorganisms and their optimisation to meet the perfect requirements of the industries. Also, the recovery of enzymes for their reusability is an important criterion for economic value, which can be achieved by their immobilisation. The methods for the optimisation and modification of the enzymes for industrial applications are manifold (Figure 14.3), which are briefly mentioned as follows.

- Statistical approaches like Plackett-Burman design (PBD) and response surface methodology (RSM) can be used to optimise the physicochemical requirements of an enzyme.
- Chemical modifications of enzymes can tap them into powerful transformers. For example, the covalent conjugation of an enzyme with different functional groups can increase its stability and selectivity.
- The enzyme immobilisation approach achieves increased thermal tolerance and stability to pH and organic solvents for real-world technological processes. A suitable choice of solid matrix for the immobilisation of enzymes can reduce the cost and adoption of the recycling system (detailed in Chapter 15).

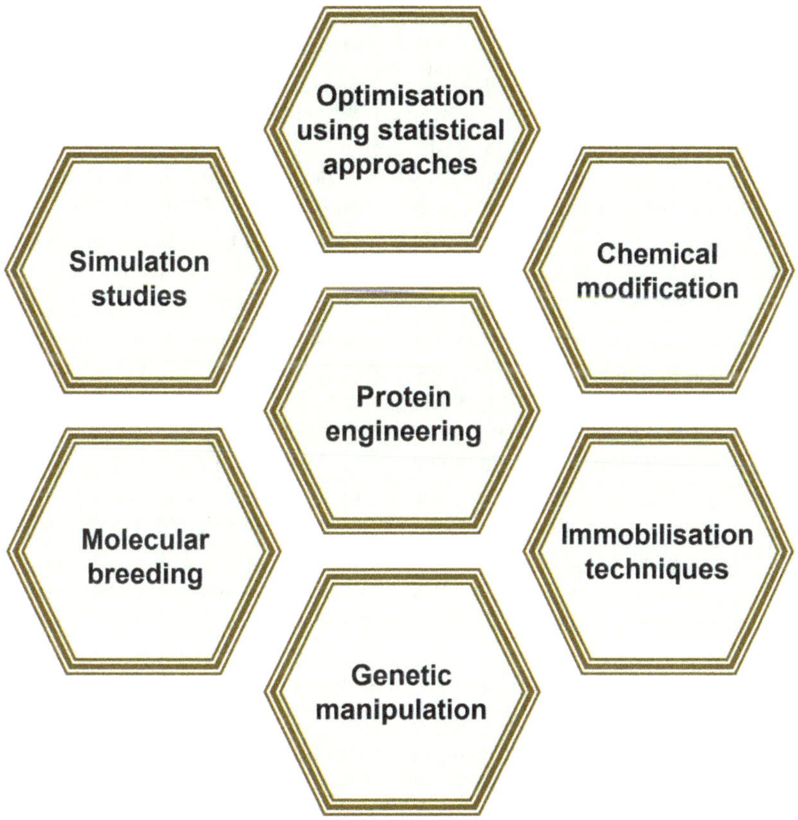

FIGURE 14.3 Methods of enzyme optimisation and modifications for industrial applications.

- Genetic manipulation can be employed to fine-tune the microorganisms. One of the major approaches is genetic engineering for the same. The insertion of genes encoding the enzymes into suitable hosts has widened the scope of enzyme technology and paved the way for innovation. Cloning and overexpression have met the demand for scale-up production. Many powerful genes of slow-growing organisms and higher organisms have been transferred into compatible microorganisms.
- Mutagenesis has tailored the enzymes to fulfil requirements like high activity, stability, and specificity under different physical and chemical conditions.
- Molecular breeding techniques like DNA shuffling tailor enzymes with novel functions or involve blending of desired properties. Modern gene synthesis is a technological breakthrough for the synthesis of customised solid-phase oligonucleotides, which involves a highly streamlined process for the synthesis of desired proteins.
- The contemporary rational techniques of protein engineering render insight into the design or functioning of enzymes using computational algorithms.
- The bioinformatics tools have enabled the identification of novel physiological traits in organisms. The computational tools help in elucidating the structure and functions of enzymes.

14.8 APPLICATIONS OF MICROBIAL ENZYMES IN INDUSTRIES

Enzymes used in industries are estimated to be worth USD 14.0 billion in 2024. They are expected to reach USD 20.4 billion by 2029 as a result of high demand and growing concern over traditional, non-sustainable procedures. The cost-effective approaches in enzyme technology are key to their large-scale production. In an enzymatic reaction, diffusion rate, reaction rates, solubility and hydrophobicity of reactants, and viscosity of the solution are the major criteria for scalable processes. Their ability to perform complex reactions under mild conditions makes them valuable tools.

Microbial cells are unique sources of diverse enzymes. In microorganisms, enzymes can be located inside, secreted outside the cells, or membrane-bound. In bacteria, they can even be found in the periplasmic regions. Humans have developed a deep understanding of microbial processes and can effectively harness the power of microbial enzymes for a wide range of applications. In recent years, the main growth in the use of industrial enzymes has been reported in the food industry, followed by the detergent industry. They are also extensively used in various other applications, such as the pharmaceutical industry, agroprocessing, textile industry, pulp and paper industry, tannery, cosmetic and personal care, agriculture, bioremediation, wastewater treatment, biofuel, waste management, biological research, nanotechnology, and biosensor technology (Figure 14.4).

Microorganisms are the key producers of industrially relevant enzymes. Microbial enzymes act on various substrates, including plant tissues, animal tissues, and microbial cells. The most common sources of microbial enzymes in industries are the fungus *Aspergillus* and the bacterium *Bacillus*. Both purified and crude enzymes are recovered for industrial applications. Purified enzymes have value for their specific product-oriented reactions. Crude enzymes are in high demand due to their nature as a multienzyme system, which meets the need for multiple reactions in a single system. Also, the production of crude enzymes incurs low costs. Of all types of enzymes, the hydrolase class is of major interest, accounting for more than 75% of enzymes employed in the industry. The list of microbial enzymes is very long, and their applications in industries are enormous. Many industries that are moved by the production of microbial enzymes are included in Table 14.2. Some popular industrially important enzymes of microbial origin and their important characteristics are discussed below.

Amylases (EC 3.2.1.X, where X is a variable number) are hydrolysing enzymes that act on α-1, 4 glycosidic bonds of starch. Amylases come in three different forms (α, β, and γ), each having different properties. α-Amylase is in high demand in industries compared to other forms. The hydrolytic products of amylases are oligosaccharides and glucose. This reaction is important in many sectors because glucose or simple carbohydrates are useful ingredients in several bioprocesses. Microbial amylases have wide commercial applications in brewing, the food industry, the detergent industry, the textile industry, biofuel production, and wastewater treatment. α-amylases can be derived from many microorganisms, but a bacterium, *Bacillus* and a fungus, *Aspergillus,* are the most commonly exploited for the same. Different species of *Bacillus* that produce α-amylases are *Bacillus subtilis, Bacillus stearothermophilus*, *Bacillus licheniformis*, *Bacillus amyloliquefaciens*, and *Bacillus dipsosauri.* Other bacterial forms, such as *Rhodothermus marinus*, *Corynebacterium gigantea*, *Chromohalobacter* sp., and *Halomonas meridiana,* also possess the ability to produce sufficient amounts of amylases. In fungi, this enzyme can also be produced from genera like *Penicillium* spp., *Streptomyces rimosus*, *Thermomyces lanuginosus*, *Pycnoporus sanguineus*, *Cryptococcus flavus*, *Thermomonospora curvata*, and *Mucor* spp. Detailed information about industrial production of α-amylase can be found elsewhere (Chapter 13).

Cellulases (EC 3.2.1.X, where X is a variable number) are a group of enzymes that hydrolyse β-1, 4 glycosidic linkages in cellulose, which is the major structural component in plants. Cellulases exist in several forms and synergistically act at a multitude of sites. Based on the reaction types, endocellulase (EC 3.2.1.4) acts randomly on the cellulose chain anywhere at the internal positions away from the end, exocellulase (EC 3.2.1.91 or 3.2.1.176) breaks the chains from either non-reducing or reducing end, and β-glucosidase (EC 3.2.1.21) contributes to the hydrolysis of cellobiose (product of exocellulase) to glucose. Cellulases are major industrial enzymes used in paper recycling, cotton processing, laundry detergents, food and feed industries, beer and wine production, agriculture, protoplast production, genetic engineering, soil nutrient recovery, and renewable fuel production. In

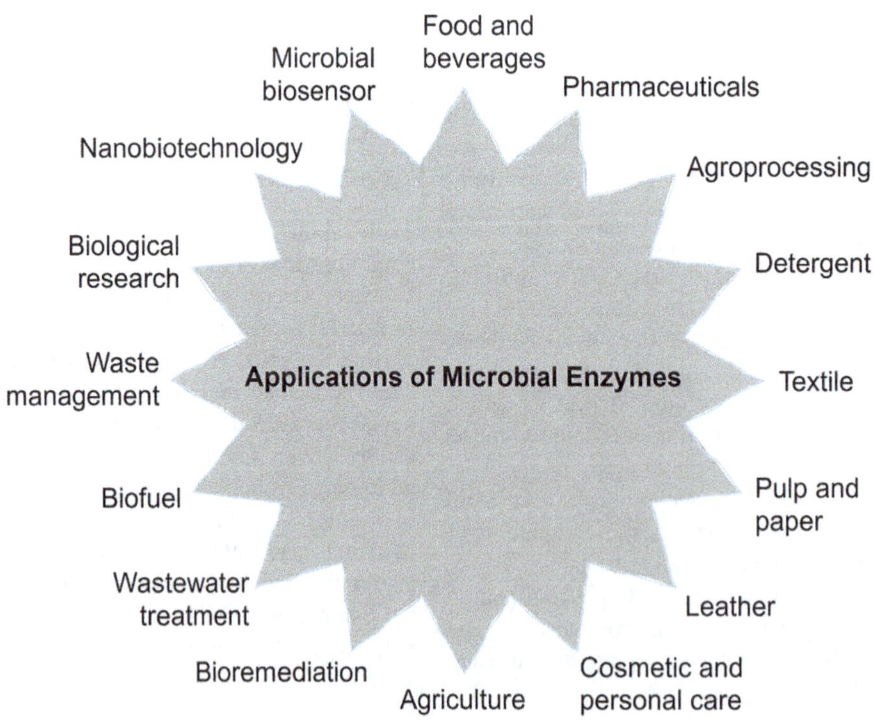

FIGURE 14.4 Applications of microbial enzymes in various industries.

TABLE 14.2
Some leading enzyme manufacturing companies from different parts of the world and their key applications

Company Name	Country of Origin	Key Applications
Novozymes	Denmark	Food processing, detergents, biofuels, and animal feed
DSM	Netherlands	Food and beverage, animal nutrition, and personal care
Chr. Hansen	Denmark	Food and dairy industries and human health
DuPont	USA*	Food processing, biofuels, and agriculture
AB Enzymes	USA	Food and feed industries
Aumgene Biosciences	India	Food, feed, and pharmaceutical industries
Enzytek Bio	China	Food processing and pharmaceuticals
Biocatalysts	UK	Speciality chemicals and food and beverage
Codexis	USA	Industrial biotechnology and pharmaceuticals

* USA and UK stand for the United States of America and the United Kingdom, respectively.

the textile industry, cellulases can be employed for bio-blasting of cellulosic cotton fabric and lyocell (rayon fabric), which is a process of removing loose fibres. Bio-blasting prevents pilling, which creates a ball of fuzz that gives a knotty appearance to the fabrics. The practical, ready-to-use formulations of cellulases are used for denim finishing to get a slightly worn look. The upgrading of fabrics using cellulase gives wash-proof and nongreasy smoothness and softness to fabrics. Cellulases are an ingredient of heavy-duty laundry detergents, which remove small dirt particles from fabrics by hydrolysing a few glycosidic bonds of cellulosic fibres. Many problems of stains on clothes are derived from modern food products that can also be removed using cellulases. A different cellulolytic commercial preparation involves the conversion of lignocellulosic into sugars for alcohol production (the second-generation fuel). A varied but well-known application of cellulases is in the pulp and paper industry as a surface cleaner during de-inking and tissue and fibre modifications. Fungi like *Trichoderma reesei*, *Trichoderma longibrachiatum*, *Penicillium*, and *Aspergillus* spp. are high-yielding strains of cellulases. Bacterial species that can hydrolyse cellulose are commonly found in the phylum Actinomycetes, including genera like *Asanoa*, *Dactylosporangium*, *Kitasatospora*, *Nonomuraea*, *Streptomyces*, and *Streptosporangium*. Several members belonging to Proteobacteria, Actinobacteria, and Firmicutes also produce cellulases. Examples of thermophilic cellulolytic bacteria are *Bacillus*, *Geobacillus*, *Acidothermus*, *Cellulomonas*, and *Caldocellum*. They provide compatible enzymes for harsh industrial processes.

Lipases (EC 3.1.1.X, where X is a variable number) are considered one of the most versatile and flexible enzymes for industrial applications. They perform reactions like hydrolysis of triacylglycerols, esterification, transesterification, and alcoholysis. Lipases show a very broad substrate specificity and high regio- and enantioselectivity. In the food industry, lipases can solubilise lipids and create various fatty acids, which aid in the production of aroma and flavour in the food. Lipases hydrolyse milk fat, resulting in unique tastes in cheese, butter, and cream. They can also be used to remove excess fat from meat and fish. A different type of lipase produces low-calorie dietary fats and baby formula. Many commercial lipases derived from microorganisms, such as Lipo Prime®, Lumafast, and Lipomax, are used in the detergent industry. Lipases are also essential in the paper and pulp industries for pitch control, stickies control, de-inking, and cleaning. In organic synthesis, they can be used to make biodegradable polyesters. Lipases can be employed in environmental cleaning, including wastewater treatment and oil spill cleanup. Bacterial and fungal cells are the most common sources of lipases. *Bacillus* and *Pseudomonas* species are the most notable bacterial sources for the production of profit-oriented lipases. *Aspergillus, Penicillium, Rhizopus,* and *Candida* are some notable fungal producers of lipases. *Candida antarctica* lipase B (CalB) is one of the most commercially important enzymes in industries. It has "Generally Recognised As Safe" (GRAS) status and is widely used in the food business. CalB can also be used as a pesticide in agriculture. Several patents on cutting-edge genetic modification and immobilisation tactics have altered the properties of lipases for their improved performance. For example, Novozym 435 (Novozymes) is a recombinant immobilised lipase, which is CalB produced in an engineered *Aspergillus niger* and immobilised onto an acrylic macroporous resin for broad commercial applicability.

Proteases (EC 3.4.X.X, where X is a variable number) are enzymes that cleave peptide bonds in the polypeptide chain of amino acids. They are highly heterogeneous and act on different residues of peptide bonds. When they react at the N-terminal or C-terminal end and release free amino acids or dipeptides, they are called exopeptidases. Examples of exopeptidases are aminopeptidases and dipeptidyl peptidase-4 (DPPIV). On the contrary, some proteases cleave internal peptide bonds, creating shorter peptides, they are termed endopeptidases. Trypsin, chymotrypsin, and pepsin are examples of endopeptidases. Based on the pH optimum of their activity, they can be acidic, neutral, or alkaline. Alkaline proteases are in high demand in many industries due to the elevated pH conditions of the reaction mixtures. Alkaline proteases in the leather industry help remove undesirable colours based on their elastolytic and keratinolytic activities. These activities can also dehair and soften hides in the leather industry. In the food industry, proteases have roles to enhance the nutritional and functional values of food proteins. They alter solubility, emulsification, and foaming capabilities in food items. The degradation of proteins by proteases also contributes to the development of tastes in foods. Papain-like proteases can be used to tenderise meat items. Fungal proteases aid in the clearance of calf rennet. Proteases can also be employed during mashing in fermentation process to boost the free nitrogen concentration for yeast growth and high ethanol output. Other uses for proteases include wine stability and clarification, and soy protein modification. In the textile industry, proteases are used in the finishing and cleaning of wool and silk fabrics. In a different setting, proteases are employed to digest the gelatin layer in photographic films. In the pharmaceutical industry, proteases can be used in the production of digestive enzyme supplements, thrombolytic factors, and wound healing agents. Proteases (e.g., subtilisin) in detergents help the cleaning of garments and kitchen utensils from protein-based stains (e.g., blood, sweat, and food). Around two-thirds of the commercial proteases available in the market are produced from microbial cells. *Bacillus* and *Pseudomonas* are the key producers of proteases. *B. subtilis*, *B. licheniformis*, *B. stearothermophilus*, *Bacillus nakamurai*, and *Bacillus tropicus* are noble producers of extracellular alkaline proteases. Fungal taxa like *Penicillium* and *Rhizopus* chiefly produce acidic proteases. Different species of *Aspergillus*, including *Aspergillus oryzae, A. niger, Aspergillus sojae, Aspergillus nidulans*, and *Aspergillus tamarii,* are important sources of alkaline and neutral proteases. Flavourzyme® from *A. oryzae* is a food-grade protease/peptidase complex composed of at least eight distinct enzymes, including endo- and exo-peptidases and α-amylase. Flavourzyme® is best suited for extraction, flavour production, and debittering in meat proteins.

Xylanases (EC 3.2.1.8) are a type of hydrolysing enzymes that catalyse depolymerisation reactions on xylan. This type of enzyme mainly acts on the heterogeneous β-1, 4 glycosidic linkages. Xylan is the third most abundant polysaccharide on Earth and can be a resourceful raw material to produce various goods using xylanases. The most common use of xylanases is in the paper and pulp industry for chlorine-free bleaching of wood pulp. Besides, balanced combinations of xylanase and other hydrolysing enzymes are important ingredients in the beverage industry (e.g., juice processing). It can also be used to form a strong network of gluten in dough for bread making. In a different application, xylanase contributes to biofuel production by converting the xylan fraction of hemicellulose into fermentable sugars. In bacteria, xylanases are mainly found in *Bacillus, Thermobifida, Cellulomonas, Staphylococcus, Arthrobacter, Microbacterium, Pseudoxanthomonas*, and *Rhodothermus*. This enzyme can also be produced by fungi like *Aspergillus* and *Trichoderma*. The industrial sources of xylanase are mainly produced from recombinant *Escherichia coli* strains using genes from *Bacillus* and *Thermobifida*. In fungi, *Trichoderma viride* is a commercial natural source of xylanase. Differently, xylanase can be obtained from a genetically modified *A. oryzae*, which expresses the enzyme of a thermophilic fungus, *Thermomyces lanuginosus*.

β-Galactosidase (Lactase) (EC 3.2.1.23) catalyses the hydrolysis of a disaccharide lactose, which breaks β-1, 4 glycosidic bonds and produces galactose and glucose. Another popular name for this enzyme is lactase. This enzyme is widely used in the dairy industry to increase the digestibility of milk for lactose-intolerant people. The byproduct, whey, released from the cheese industry can be treated with β-galactosidase for its applications in the production of sweet syrups and ethanol. β-galactosidase is also used in the manufacturing of baby food and powdered milk. The yeast *Kluyveromyces* spp. and a mould, *A. oryzae,* produce highly stable β-galactosidase and are suitable for large-scale applications. Some other fungal sources are *Penicillium simplicissum, Rhizobium meliloti, Rhizomucor* spp., and *Saccharopolyspora rectivirgula.* The bacterium *E. coli* is also a commercially important source of β-galactosidase. Some other bacteria that contribute to β-galactosidase production include *Erwinia aroideae, Klebsiella pneumoniae,* and *Lactobacillus acidophilus.* A probiotic bacterium, *Bifidobacterium,* also produces β-galactosidase, which can be directly supplemented for health-promoting effects.

Glucose Isomerase (GI) (or Xylose isomerase) (EC 5.3.1.5) is an enzyme that catalyses the isomerisation reaction between aldoses and ketoses. The key reactants involved in the reactions are D-xylose/D-xylulose and D-glucose/D-fructose. GI, in its immobilised form, has a very high demand in the food industry. The highly successful application of GI is the production of high fructose corn syrups (HFCS). The food and beverage industry utilises HFCS as an alternative sweetener to sucrose. The most common industrial sources of GI are *Streptomyces lividans* and *Streptomyces rubiginosus.* Some microbial species, such as *Actinoplanes* spp., *Arthrobacter* spp., *Bacillus coagulans, Flavobacterium* spp., *Lactobacillus* spp., and *Streptomyces* spp., also produce GI in significant quantities. A detailed description of this enzyme is mentioned in Chapter 17.

Penicillin Acylase (PA) (EC 3.5.1.11) is an enzyme which is the second most common enzyme used in immobilised form after GI. This enzyme utilises a biosynthetic product, penicillin, as the substrate to produce 6-aminopenicillanic acid (APA) or 7-amino deacetoxy cephalosporanic acid (ADCA) that acts as a substrate for the semisynthetic penicillins/cephalosporins. Some common semisynthetic penicillins are amoxicillin, ampicillin, cefadroxil, and cephalexin. Penicillin acylase can be derived from diverse bacterial or fungal taxa. Some bacteria that express penicillin acylase are *Streptomyces lavendulae, Alcaligenes faecalis*, and *Proteus rutgeri.* An example of a fungus that produces PA is *Fusarium oxysporum.* Detailed information on PA can be found in Chapter 17.

Glucose Oxidase (GOX) (EC 1.1.3.4) catalyses the oxidation of the β-anomer of D-glucose using oxygen (O_2) and releases hydrogen peroxide (H_2O_2). In the baking industry, glucose oxidase (GOX) can be used as an alternative to chemical oxidants like bromate for the strength of gluten. The food industry uses GOX as an antioxidant, preservative, and stabiliser. Even some commercial toothpastes and mouthwashes consist of multienzyme systems, including GOX, which serves as an antibacterial agent for oral/dental healthcare. In the pharmaceutical industry, GOX can be used to deliver glucose-sensitive drugs and for the quantitative determination of diabetes. In a different role, H_2O_2 produced from the GOX activity induces the polymerisation of aniline to produce polyamines, which are conductive polymers used in biomedical devices and biosensors. In the textile industry, GOX activity is associated with bleaching from H_2O_2. The microorganisms that produce GOX are fungi, such as *A. niger*, *Penicillium chrysogenum, Penicillium glaucum,* and *Penicillium adametzii.*

Laccases (EC 1.10.3.2) are broad-spectrum multi-copper oxidase enzymes that perform oxidation reactions for a wide variety of organic compounds. Oxygen is an essential co-substrate for the reaction and yields a water molecule. It is a promising green tool for industrial catalytic reactions. Laccase engineering is already an established technology in the textile industry, paper and pulp industry, and food industry. In the textile industry, it is used in the leaching of indigo-stained denim to create a fashion outlook. The paper and pulp industries can utilise this enzyme in pulp whitening, delignification, and pitch control. The enzymes also have a strong impact on the bioprocessing of different food industry by-products. Laccases have shown prominent roles in the degradation of phenolic compounds, dyes, and pesticides from wastewater. Many microbes, including bacteria and fungi, produce laccase. In bacteria, Gram-positive forms like *Bacillus, Streptomyces, Rhodococcus,* and *Lysinibacillus* are notable producers of laccases. Many fungi, such as *Phlebia radiata*, *Pleurotus ostreatus*, *Trametes versicolour*, *Phanerochaete chrysosporium*, *Theiophora terrestris*, and *Lenzites betulina,* also produce laccases. Bacterial laccases have higher stability at elevated temperatures than fungal laccases and remain active at elevated pH values.

Mannanases (EC 3.2.1.X, EC 3.1.1.X, where X is a variable number) are mannan degrading enzymes that break the main and side chains of mannan. This substance is widely available in nature and is mainly found in hemicellulose. Mannases have a highly heterogeneous nature and exist in many variants. endo-β-1,4-mannanase (EC 3.2.1.78) is the most commonly used variant in industrial applications. It is used as a gel breaker in the food industry. This enzyme is employed in laundry detergents to remove food stains. The important sources of mannanase are some Gram-positive bacteria like *Bacillus lentus* and *Paenibacillus lentus,* and fungi such as *A. niger* and *T. reesei*. Some popular trademarks of mannanases for industrial applications are Hemicell®, CTCzyme®, and Mannaway®.

Pectinases are the enzymes that break down the complex heteropolysaccharides, pectin/pectic acid. Pectinases have many variants, such as pectin esterase (EC 3.1.1.X), hydrolases (EC 3.2.1.X), and lyases (EC 4.2.2.X), where X can be a variable number. All these variants of pectinases have scope for different applications. A perfect example of pectinase usage is in the wine-making industry during maceration to obtain a high yield of juice, extract more aromatic constituents, release colours, and prevent a "pectin haze". Pectate lyase is used for bioscouring in the textile industry to increase the wettability of cellulosic fabrics. In this process, alkaline pectinases remove pectin, which is found between the fibre core and wax. In the detergent industry, pectinases have a role in removing hard-to-remove stains from clothes. These enzymes also hold promise in the extraction of vegetable oil, bleaching of pulp in the paper industry, and processing of food and animal feed. Most of the commercially produced pectinases are derived from fungi. *A. niger* is the leading producer of pectinases, though other fungi such as *Trichoderma*, *Penicillium* and some bacteria like *Bacillus* spp. can also be used for various applications. Pectinases are marketed under names such as Rapidase, Ultrazyme, Pectinex, and Panzym.

Phytases (EC 3.1.3.X, where X represents a variable number) are the enzymes that remove phosphorus from phytate/phytic acid. These enzymes have broad applications in the animal feed industry. Monogastric animals (like pigs and poultry) cannot utilise phytate or its derivatives, and this enzyme can be added to their diet for the digestion of phytate. This enzyme also provides micronutrients such as calcium (Ca), Mg, Fe, and Zn in animals due to their chelating properties. The large-scale production of phytase can be developed using different species of *Aspergillus*, *Peniophora*, and *Penicillin.* Phytases can also be produced by a variety of bacteria, including *B. subtilis*, *E. coli*, and *Enterobacter aerogenes. E. coli,* after genetic recombination, has demonstrated potential for the industrial-scale production of phytase. Their products are available in the market under different trademarks such as Allzyme®, Phyzyme®, and Avizyme®.

Pullulanase (EC 3.2.1.41) is a debranching enzyme that cleaves α-1,6 glucosidic linkage in glucan-polysaccharides such as amylopectin, pullulan, and related oligosaccharides. Industries manufacturing ethanol and sweeteners are the main consumers of pullulanase. This enzyme, along with glucoamylase, is commonly used during the saccharification process in the production of glucose/maltose syrups. Different species of *Bacillus* and *Klebsiella* are the key producers of this enzyme. This can also be produced from *Aerobacter aerogenes*, *Bacillus acidopullulyticus*, *Bacillus flavocaldarius*, *Fervidobacterium pennavorans*, and *Streptomyces* sp.

Catalase (EC 1.11.1.6) is an oxidoreductase enzyme that performs a disproportionation reaction of H_2O_2 and produces water and O_2. This reaction helps to eliminate excess H_2O_2 and prevent its potentially harmful effects. One of the most common industrial applications of this enzyme is in the removal of excess H_2O_2 during cheese production, particularly for varieties like Swiss cheese. The textile industry utilises this enzyme to eliminate peroxide after the bleaching of cotton. Catalase can also be used to clean contact lenses. This enzyme, along with glucose oxidase, has the properties of food preservatives and egg processing. Catalase can be used to extract animal proteins and prepare pet foods. Another aid of this enzyme is sustainable salmon farming by reducing H_2O_2 from water, which is used to decrease parasite and disease loads in marine water. Commercial production of catalases involves the cultivation of a fungus, *A. niger*, through solid-state fermentation (SSF). It can also be produced by some bacteria, like *B. subtilis* and *P. aeruginosa*.

Transglutaminases (EC 2.3.2.13) have the ability to crosslink and are involved in the formation of isopeptide bonds between the amino groups of glutamine and lysine residues. The cross-linking ability of transglutaminases gives them the nickname "biological glue". This type of enzyme is mainly obtained from a soil bacterium, *Streptomyces mobaraensis*. They can also be recovered from other sources such as *Streptoverticillium mobaraense*, *Streptoverticillium cinnamoneum*, and *Bacillus circulans*. In the food industry, transglutaminases are used to improve the texture and appearance of processed meat, poultry, and seafood products. These enzymes help bind different meat pieces and enhance the overall quality of food products. The enzymatic action of transglutaminases in the dough improves baking quality and bread texture. In the dairy industry, transglutaminase preparations have a positive impact on the quality and texture of yoghurt and cheese. Examples of commercially available microbially derived transglutaminases include Activa® from Ajinomoto in Japan and Moo Gloo from Modernist Pantry in the United States of America (USA).

Cutinase (3.1.1.74) is a type of esterase enzyme that especially splits cutin and other ester compounds. Cutin is a waxy polymeric lipid found in plants, composed of repeated units of ω-hydroxy acids and their derivatives. This enzyme acts on the carboxylic ester bonds and performs hydrolysis. Due to its ability to break down similar compounds, cutinase has several applications beyond its natural role. Cutinase can be used in laundry detergents and dishwashing detergents to help remove greasy or oily stains from fabrics and dishes, respectively. Cutinase can perform esterification (produce an ester compound from a carboxylic acid and an alcohol) and transesterification (exchange of ester groups between an ester and an alcohol, forming a new ester and alcohol). These additional capabilities of cutinase lead to its potential for biodiesel production, flavour and fragrance synthesis, and the modification of natural products. The yeast *Arxula adeninivorans* is a key source of cutinase for various industrial applications. This enzyme recovered from *Thermobifida fusca* is used for surface modification of synthetic fibres in the textile industry. A cutinase enzyme isolated from *Thermobifida cellulosilytica* can be genetically modified to improve the degradation of polyethylene terephthalate (PET) (a type of plastic). Cutinase from *Fusarium solani*, when expressed in *A. adeninivorans*, has been used in organic synthesis.

α-Acetolactate decarboxylase (ALDC) (EC 4.1.1.5) is a carboxy-lyase that cleaves the C-C bond in α-acetolactate, a compound that is released during beer production. A slow nonenzymatic oxidative decarboxylation converts α-acetolactate into diacetyl, which has an unpleasant flavour. Diacetyl is subjected to an enzymatic reaction (a yeast reductase) to produce acetoin, which has almost no taste. This whole process extends the maturation period of beer. The use of ALDC can reduce the maturation period by acting upon α-acetolactate for the direct conversion into acetoin, bypassing the diacetyl formation. ALDC can be outsourced from various microorganisms. It is mainly produced through a submerged culture of *B. subtilis*. ALDC can also be produced from *Bacillus brevis, Brevibacillus brevis, Enterobacter aerogenes, Enterobacter cloacae, K. pneumoniae, Streptococcus thermophilus*, and *Lactococcus lactis* subsp. *lactis*.

Several other enzymes have applications in many different fields. In this series, a microbial enzyme, naringinase, can be used to remove the bitterness of citrus juices, primarily from grapefruits. This is carried out by the hydrolysis of naringin, a bitter compound found in citrus fruits, improving the overall taste of the juice. A different enzyme, asparaginase, inhibits the proliferation of cancer cells by lowering the level of asparagine in the bloodstream. In the food industry, peroxidases can be used to enhance the flavour and nutritional value of various products. They can help eliminate undesirable compounds and improve the colour, texture, and overall quality of food items. Glucosyltransferases demonstrate potential in the synthesis of functional sweeteners in the food industry. This enzyme adds a glucose moiety to the disaccharide maltose and converts it into panose (a mildly sweet trisaccharide) that can be used in commercial juice processing. A different enzyme, nitrilase, is used in the bioremediation process to degrade nitrile compounds and cyanide and detoxify industrial wastes and environmental contaminants.

SUMMARY

- Enzymes are biocatalysts, which are essential for biologically mediated reactions. They have a prominent role in the green synthesis of various relevant products.
- Microorganisms are the main producers of industrially important enzymes, the fungus *Aspergillus* and the bacterium *Bacillus* being the most commonly utilised sources.
- The key attributes of microbial enzymes for industrial applications lie in their sustainable green methodologies, safe and simple procedures, and cost-effective approaches. The chemo-, stereo-, and regioselectivity add supremacy to produce specific products in industries.
- Microbial enzymes have shown immense potential for food and beverage processing, pharmaceutical production, biofuel generation, textile manufacturing, paper and pulp production, etc.
- The optimisation of enzyme production and activities involves state-of-the-art strategies (such as genetic manipulation, protein engineering, chemical modification, and immobilisation) and kinetic studies.
- Examples of some useful enzymes in industries include α-amylase, cellulases, lipases, proteases, xylanases, lactase, penicillin acylase, glucose oxidase, glucose isomerase, laccases, mannases, pectinases, phytases,

pullulanase, catalase, transglutaminases, cutinase, and α-acetolactate decarboxylase.

DIFFERENT TYPES OF PRACTICE QUESTIONS

Fill in the blanks

I. Some organic cofactors establish covalent bonds with enzymes and are permanently attached, which are called ....
II. .... is a measure of affinity between the substrate and the enzyme and is not influenced by enzyme concentrations.
III. The timeline of the study of enzymes or enzymology began with the discovery of ....
IV. The cross-linking ability of .... enzymes gives them the nickname "biological glue".
V. A microbial enzyme, ..., can be used to remove the bitterness of citrus juices, primarily from grapefruits.
VI. An enzyme that contributes to the digestion of phosphorus from phytate in monogastric animals is..................

SOLUTION: I. prosthetic groups, II. Michaelis- Menten constant (Km), III. diastase, IV. transglutaminases, V. naringinase, VI. phytase

Match the following

i. Glucose Isomerase	a. CalB
ii. Lipases	b. Subtilisin
iii. Proteases	c. Lactase
iv. β-Galactosidase	d. HFCS
v. Glucosyltransferases	e. Panose

SOLUTION: i-d, ii-a, iii-b, iv-c, v-e

Very short answer type of questions

I. What is an active site in enzymes?
II. What are the two most popular models that explain enzyme catalysis?
III. Write the formula for the catalytic efficiency of an enzyme.
IV. Write two major applications of the microbial lipase enzyme.
V. Why are crude enzymes in higher demand in industries than purified enzymes?
VI. Write the name of any one industrial enzyme belonging to the oxidoreductase class.

Short answer type of questions

I. Describe the applications of microbial enzymes in the food industry.
II. Write about the advantages of enzyme technology.
III. What are the reasons that enzymes from microbial sources are of major interest for industrial applications over any other sources?
IV. Write about state-of-the-art strategies of enzyme production.

Long answer type of questions

I. What are the advantages of using microbial enzymes in industrial processes compared to chemical catalysts? What are the different hydrolases used in industries?

SUGGESTED READINGS

Adrio JL, Demain AL. Microbial enzymes: tools for biotechnological processes. *Biomolecules*. 2014;4(1):117–39.

Choi JM, Han SS, Kim HS. Industrial applications of enzyme biocatalysis: current status and future aspects. *Biotechnology Advances*. 2015;33(7):1443–54.

Enzyme at work. Novozymes.

Hasan F, Shah AA, Hameed A. Industrial applications of microbial lipases. *Enzyme and Microbial Technology*. 2006;39(2):235–51.

Kirk O, Borchert TV, Fuglsang CC. Industrial enzyme applications. *Current Opinion in Biotechnology*. 2002;13(4):345–51.

Raveendran S, Parameswaran B, Ummalyma SB, Abraham A, Mathew AK, Madhavan A, Rebello S, Pandey A. Applications of microbial enzymes in food industry. *Food Technology and Biotechnology*. 2018;56(1):16.

Singh R, Kumar M, Mittal A, Mehta PK. Microbial enzymes: industrial progress in 21st century. *3 Biotech*. 2016;6(2):1–5.

Wei R, Zimmermann W. Microbial enzymes for the recycling of recalcitrant petroleum-based plastics: how far are we? *Microbial Biotechnology*. 2017;10(6):1308–22.

15 Enzyme Immobilisation

15.1 IMMOBILISATION

Immobilisation is the process of packaging an enzyme or a whole cell onto a support system or in a matrix. This process is mainly used to fix a free enzyme or cell into a solid heterogeneous or homogeneous compound. Immobilisation can be achieved using various techniques involving different types of bonds. The selection of support systems for immobilisation presents a multitude of options. Enzymes in an immobilised state are stable, robust, and restorable. They have higher turnover and prolonged availability over their soluble forms. Immobilised enzymes also reveal additional beneficial traits such as enhanced resistance and catalytic capabilities under stress conditions. The primary targets for immobilisation are microbial enzymes because of their diverse applications. The enzymes from plants and animals retained on diverse support systems are also practical biocatalysts.

"Enzyme immobilisation" technology involves an interdisciplinary approach. This calls for expertise in various disciplines, including bioprocess technology, protein chemistry, surface chemistry, material science, molecular biology, nanotechnology, biophysics, biotechnology, chemical engineering, and computational analysis. Immobilised enzymes have applications in the food and beverages industry, drug production, disease treatment, biosensor technology, bioremediation, wastewater treatment, biodiesel production, the agriculture sector, the detergent industry, the textile industry, the paper and pulp industry, tanning, etc.

15.2 A BRIEF HISTORICAL ACCOUNT

The records of first-hand use of immobilised enzymes date back to the early 19th century for acetic acid production and wastewater treatment. Subsequently, Nelson and Gryphon, in 1916, researched enzyme immobilisation to stabilise an invertase enzyme on coal or aluminium hydroxide. However, these studies proposed that immobilisation had no or little impact on enzyme activity. After some years, immunologists identified the role of immobilisation in the isolation of protein. The next two decades were a dormant stage for inventions on enzyme immobilisation. Some modern methods of enzyme immobilisation were developed in the post-1950s. The key progresses were the identification of diverse natural and synthetic polymers as support systems and the various methods of physical and chemical attachment. These age-old processes were further improved for their industrial applications. However, the commercial potential of immobilised enzymes was realised in 1969 when Tanabe Seiyaku Company in Japan scaled up the manufacturing of immobilised amino acid acylase recovered from a fungus, *Aspergillus oryzae*. Chibata et al. were the main investigators of this innovation, who used the ionic surface of the diethyl aminoethyl (DEAE)-Sephadex to bind the fungal amino acid acylase and resolve the optical properties of synthetic racemic D-L amino acids.

Following that came the commercial production of many immobilised enzymes. There was no turning back in the field of enzyme immobilisation with the discovery of multiple-enzyme immobilisation in 1985. The other advancements over the years included the identification of cofactor regeneration and cell immobilisation. In particular, the past four decades have witnessed a tremendous boom in immobilisation technology. The multidisciplinary approaches have offered improvement in various perspectives, such as multipoint attachment, efficiency, recycling times, development of novel support systems, and engineering by rational design. To date, thousands of patents are available on immobilisation techniques and numerous support systems. Also, immobilised enzymes are available under commercial rights.

15.3 IMMOBILISATION PRINCIPLES

The key principle of the immobilisation techniques is the anchoring of enzymes on a platform (Figure 15.1). The interaction between the enzyme and support systems can be random or position-specific. The physical and chemical characteristics of both enzyme units and support systems affect how they interact with each other. The degree to which enzymes and support systems are integrated depends largely on the charge density and the presence of hydrophobic and hydrophilic groups on both components. An enzyme has two important sites: a catalytic site and a non-catalytic site. The catalytic sites of the enzymes must be free for the reaction with substrates and are an absolute necessity for catalysis, while the non-catalytic sites can be used for binding with

DOI: 10.1201/9781032651095-19

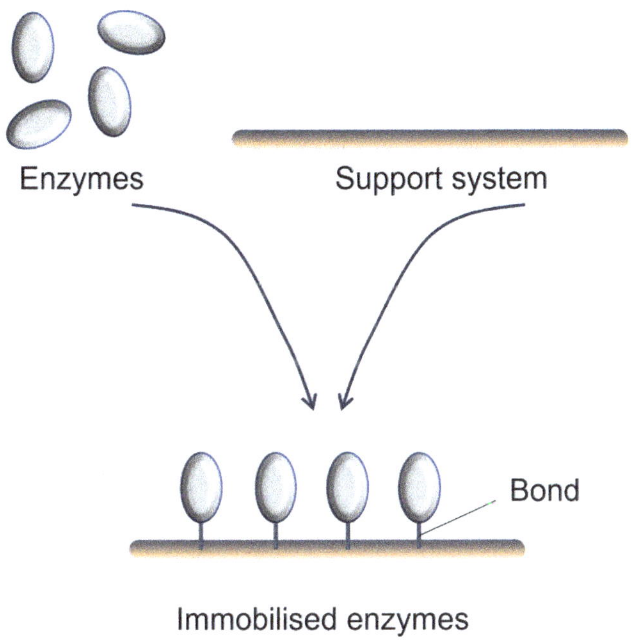

FIGURE 15.1 A simple process of enzyme immobilisation.

support systems. The functional groups of the side chain in the non-catalytic sites can be used to affix support systems.

The support systems can produce reversible (temporary bond) or irreversible (strong bond) types of binding with enzymes. The strong binding creates an effective and long-lasting immobilised system, whereas the temporary binding aims at the recycling of support systems after the enzyme loss. A distinct type of interaction involves the caging of enzymes inside the matrix, called entrapment. Some immobilised enzymes involve the binding of enzyme units with each other through linkers (in cross-linking) rather than attachment through any support systems.

The choice of support systems for immobilising enzymes is a critical parameter in determining the overall performance of immobilised enzymes. Also, the orientation and three-dimensional configuration of the enzyme in the immobilised system are paramount for enzymatic activity and stability. The protein folding and conformational alterations of proteins after immobilisation alter the kinetic characteristics of enzymes and reduce their functional ability. This is often a major drawback in constructing an effective immobilised biosystem. Parameters like immobilisation yield, efficiency, and activity recovery are other important factors for the stability and activity of enzymes and are discussed in Chapter 16.

15.4 CHARACTERISTICS OF IMMOBILISED ENZYMES

Immobilised enzymes act differently compared to free enzymes. They possess unique characteristics as mentioned below-

- An immobilised enzyme provides an excellent base to avoid enzyme leaching in the reaction mixture. When an enzyme is immobilised, it is simple to recover it from the solution. An immobilised form enhances the reusability of enzymes. It helps to reduce the cost of the process, as continuous production of enzymes is an expensive approach.
- The organised framework of carrier-bound or carrier-free enzymes allows easy diffusion of substrates and products. It may improve the conditions for enzyme activity in certain situations.
- Immobilisation facilitates high operational stability and intensification of the process.
- The large surface area of a carrier material contributes to the accumulation of many enzyme units in a confined space.
- The bound enzymes can exhibit substrate specificity and enantioselectivity.
- Immobilised enzymes can display high mechanical, physical, chemical, and biological strength.
- They help prevent contamination, allowing easy separation of products.
- The support systems shield enzymes from adverse environmental conditions and maintain their performance in harsh industrial operations. They can even exhibit the properties to demonstrate satisfactory performance even in unconventional settings.
- The support systems can be recovered/recycled for subsequent usage.
- An easy, amenable, and cost effective technology is preferred to scale up the immobilisation process.
- Immobilised enzymes should meet standard quality for product formation.

15.5 CARRIER MATERIALS

Carrier materials are mainly insoluble recyclable solids, and they can be any affordable substance. Their main characteristics are inertness, mechanical strength, high affinity for enzymes, porosity, regenerability, non-toxicity, and ability to enhance enzyme activity and specificity. Porous materials are preferred because they have a large surface area for high enzyme loading per unit mass and protect the enzymes from stress. There is no such thing as an ideal carrier material that can be used for all types of enzymes. Different materials may have different affinities for different enzymes. The best support system for immobilisation depends on the enzymes involved, specificity, hydrodynamic conditions, reaction media, desired purification, and the intrigued industries.

Since the support systems stay in direct contact with enzymes, their influence on enzyme catalytic characteristics is a notable obstacle. The safe use of support systems in the food and pharmaceutical industries is also a critical issue that necessitates comprehensive health and risk assessments. The Code of Federal Regulations (CFR) in the USA (generally recognised as safe (GRAS) status) and the International Organisation for Standardisation (ISO) (headquarter in Switzerland) are distinct agencies that have issued several recommendations and approved support systems in the food industry to be used as food additives.

A carrier material can be an inorganic compound, an organic compound, or a hybrid of both. Organic support is more commonly used than inorganic form. Based on their origin, they are classified into two types: natural and synthetic. Natural substances outperform synthetic materials in terms of sustainability, biodegradability, and cost-effectiveness. Furthermore, the use of natural compounds as support systems raises little or no ethical concerns. On the other hand, synthetic materials can be manufactured with desired qualities to achieve the best performance. Also, many natural substances can be tailored to improve their immobilising properties. Some notable examples of carrier materials employed for enzyme immobilisation are discussed as follows.

15.5.1 Organic Substances

Organic compounds derived from natural sources are important support systems for enzyme immobilisation because of their biocompatibility, low cost, and ease of availability. They possess some unique abilities for enzyme-carrier interactions, such as structural inhomogeneity, gelling strength, and stiffness. Polysaccharides, the long chains of carbohydrates (sugars), provide a stable environment for enzymes. These include water-insoluble, natural polysaccharides such as agarose, carrageenan, chitosan, alginate, cellulose, lignin, and starch (Table 15.1). They are amenable to chemical modification to enhance their performance or to introduce specific binding sites for the enzymes. Agarose is derived from seaweed, and the beaded structure of agarose is commercially available under the trade name Sepharose. Carrageenan is a natural polysaccharide extracted from red seaweed. Chitosan is derived from chitin, found in the exoskeleton of crustaceans. The deacetylation of chitin naturally produces chitosan, which creates a gel-forming cationic compound (commercially available under the brand names Chitopearl and Chitopharm). Calcium-alginate can be artificially developed by cross-linking calcium ions and alginate. Cellulose and its derivatives (e.g., cellulose acetate and carboxymethyl cellulose) provide a stable and hydrophilic surface for enzyme binding and have excellent biocompatibility. Gelatin is a biodegradable protein derived from collagen, which forms a stable matrix for enzyme immobilisation and can be chemically modified to enhance enzyme binding. A different protein, albumin, commonly found in blood plasma, can be a useful carrier material for biomedical applications.

Examples of synthetic polymer-based support systems for immobilisation include polyacrylamide (PAA), polyvinyl alcohol (PVA), polyurethane (PU), polyethylene glycol (PEG), polystyrene (PS), polyacrylonitrile (PAN), polyethyleneimine (PEI), and polyhydroxyalkanoates (PHA) (Table 15.1). Polymers like PAA or PS expose functional groups like amino groups, offering direct support to enzymes. Amberlite™ is an acidic cation exchange resin and can be used for immobilisation through ionic interaction. A different type of synthetic polymer, acrylic resins, supports both reversible and irreversible types of immobilisation. Eupergit C is a commercial surface-functionalised acrylic resin for covalent binding with enzymes. A different example of an acrylic resin-based carrier is Lewatit VP OC 1600, made from cross-linking poly(methyl methacrylate) and divinylbenzene, which is marketed by Lanxess (Germany).

15.5.2 Inorganic Substances

Both natural and synthetic inorganic minerals can be used as carrier materials. Chemical modification of natural materials is a practical strategy to enhance their performance. Examples include celite, amorphous silica, bentonite, and zirconia. Celite is the silicate skeleton of diatoms, a commonly used carrier material for enzymes. Compared to natural celite, calcined celite is more porous and, therefore, more effective for enhanced enzyme binding and offers a wider range of

TABLE 15.1
Different types of carrier materials with suitable examples

	Organic	Inorganic	Organic-inorganic framework
Natural	Agarose Carrageenan Chitosan Alginate Cellulose Lignin Starch	Celite Bentonite Zirconia Mica Titania kaolinite Alumina Clinoptilolite	-
Synthetic	Polyacrylamide (PAA), Polyvinyl alcohol (PVA), Polyurethane (PU), Polyethene glycol (PEG), Polystyrene (PS), Polyacrylonitrile (PAN), Polyethyleneimine (PEI), Polyhydroxyalkanoates (PHA)	Aluminosilicates	CuBDC (BDC = 1,4-benzene-dicarboxylate) CuPABA (PABA= *p*-aminobenzoic acid) HKUST-1 (Hong Kong University of Science and Technology-1) Zirconium-based MOFs (Zr-MOFs)
Semisynthetic	Cellulose acetate Carboxymethyl cellulose	-	Calcium-alginate

applications. The amorphous silica (composed of silicon and oxygen) possesses a large surface area and high porosity for adsorption and catalysis. But amorphous silica has low mechanical strength, and that can be improved by transforming it into hard silica granules. Silica granules have a well-defined crystal lattice structure and are more robust and stable to harsh conditions in specific applications because of their mechanical stability. Examples of manufactured silica granules that are commercially available include Aerosil (Degussa), Promaxon (Promat), Spherosil (Rhone-Poulenc), and Controlled Pore Glass (CPG). The amorphous materials with high silica oxide (SiO_2) content are another promising material, having the properties of high thermal, mechanical, and chemical stabilities for rigorous applications. Bentonite is a clay mineral with high adsorption and swelling capabilities. Bentonite-supported enzymes are used in various biotechnological and industrial applications. Zirconia is a ceramic material (crystalline oxide of zirconium) that can house enzymes on its surface and has unique properties, such as a wide surface area, chemical stability, and biocompatibility. Other examples of natural minerals include mica, titania, kaolinite, alumina, and clinoptilolite (Table 15.1). They consist of diverse physical properties and have a broad range of pore sizes for different immobilisation preparations.

Some metals can be used in enzyme immobilisation due to their versatility and ability to attach the enzyme. Metal ions like copper (Cu), nickel (Ni), or zinc (Zn) chelate the enzymes. The specific amino acid residues in the enzymes create potent coordination interactions with metals. However, metal-based materials do not have porosity, which is highly desirable in an enzyme immobilisation matrix. In distinct preparations, salts like potassium sulphate (K_2SO_4) and potassium chloride (KCl) can also be used as core materials to develop microcrystals for the deposition of enzymes.

15.5.3 Metal-Organic Frameworks

Metal-organic frameworks (MOFs) have gained significant attention in recent years due to their diverse applications in enzyme immobilisation. They are a class of porous materials consisting of metal ions or clusters coordinated with organic ligands, forming a three-dimensional network. The crystallographic structure, high tunability, and low density of MOFs offer uniform loading and minimum leaching of enzymes. Also, enzymes immobilised on MOFs have better tolerance to thermal and chemical conditions. The tetrahedral coordination of transition metals like Cu, Zn, iron (Fe), and cobalt (Co) with imidazolate linkers is the most popular MOF for enzyme immobilisation and is known as ZIFs (zeolitic imidazolate framework). Their structures exhibit topological similarity with aluminosilicate minerals and zeolites. Other notable MOFs used to immobilise microbial enzymes can be CuBDC (BDC = 1,4-benzene-dicarboxylate), CuPABA (PABA= *p*-aminobenzoic acid), HKUST-1 (Hong Kong University of Science and Technology-1), and Zirconium-based MOFs (Zr-MOFs) (Table 15.1).

15.5.4 Nanomaterials

A variety of nanostructures (sizes ranging between 1–100 nm at least in one dimension) can be used as support materials for enzyme immobilisation. They can exist in different forms, such as nanoparticles, fullerenes, nanotubes, nanohorns, nanosheets, nanolayers, nanotube arrays, and nanowire arrays (Figure 15.2). These nanomaterials can interact with enzymes through adsorption, covalent coupling, cross-linking, and entrapment. Nanomaterials contain many unique properties for enzyme immobilisation, including homogeneous nature, inertness, and large surface area. Other features include high pore volume, possible tailoring of particle size, oversimplified

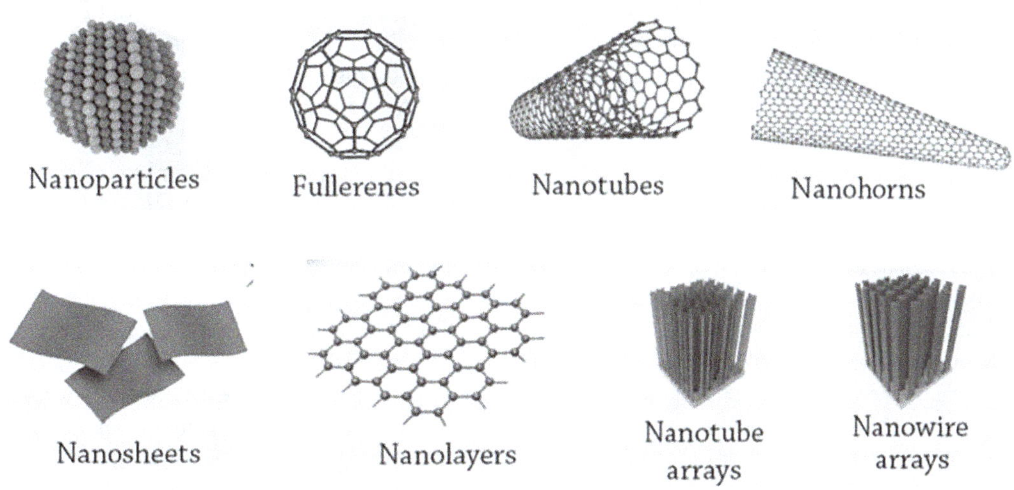

FIGURE 15.2 Different types of nanomaterials that can be used as support systems for enzyme immobilisation. (Images adapted from Joudeh and Linke (2022).[1])

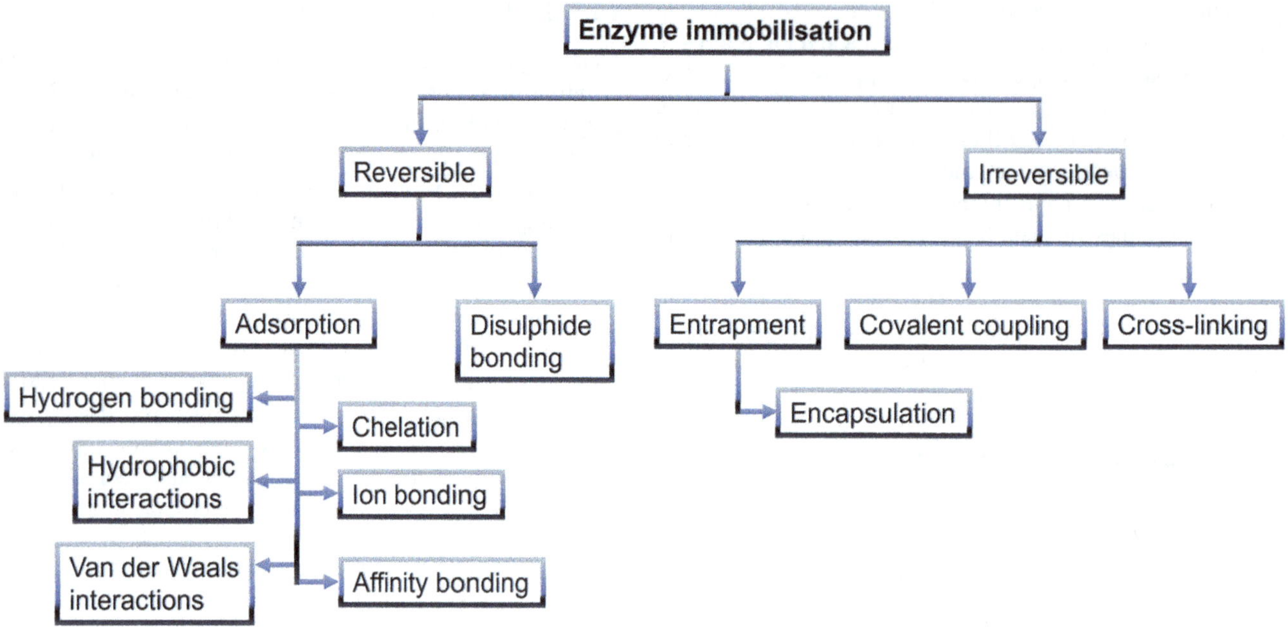

FIGURE 15.3 Different types of irreversible and reversible methods of enzyme immobilisation.

production of nano enzyme particles, well-defined core-shell, and minimum leaching of enzymes. Examples of nanomaterials-based support materials include silica nanoparticles, chitosan nanofibers, nano-sized magnetite, zirconia nanoparticles, and polystyrene nanoparticles. Carbon nanotubes are nonporous materials but exhibit strong bonding with enzymes, and the materials are highly thermal, chemical and mechanical resistant. In an experiment, a microbial enzyme, β-galactosidase, covalently attached to a carbon nanotube exhibited high stability and showed potential for reusability. The nanoparticles of gold or silver can have an affinity for adsorption or covalent bonding with different enzymes. The use of magnetic iron oxide nanoparticles (MNPs) as support materials can provide another benefit for their recovery using external magnetic fields. However, the development of nano-sized carrier particles is difficult and expensive due to their small dimensions. Also, the excessive use of nanomaterials has raised concerns about their significant accumulation in biological systems.

15.6 METHODS OF ENZYME IMMOBILISATION

Enzyme immobilisation can be broadly divided into reversible and irreversible methods based on the strength of the enzyme binding to the support systems. In reversible immobilisation, the support systems are loosely bound to the enzymes. This approach relies on weak intermolecular bonding, which can be disassembled by variations in the physical environment (e.g., pH, temperature, ionic strength, and polarity). The quaternary structure of the enzymes is least affected by reversible bonding. When the enzymatic activity is ceased, the carrier component can be recovered for reuse. Examples of reversible methods of enzyme immobilisation are adsorption and disulphide binding.

An irreversible method of enzyme immobilisation involves covalent bonding between an enzyme and a carrier material. Such preparation may require the activation of functional groups in one or both substances. The covalent type of bonding ensures a strong and stable bond between the enzyme and the support, preventing enzyme leaching and allowing for repeated use (Figure 15.3). This might involve modification of the functional groups to create reactive sites. The encapsulation method traps enzymes, which cannot be reversed. A different approach to irreversible immobilisation is cross-linking.

The choice of immobilisation method for an enzyme is determined by several criteria, such as enzyme properties, the intended use, stability requirements, and ease of separation. Each method has some advantages and disadvantages (Table 15.2). Further, the selection of the best method depends on assessments and quality control for industrial applications. The following subsections provide descriptions of both reversible and irreversible immobilisation methods.

15.6.1 Adsorption

Adsorption is a relatively simple and inexpensive approach to immobilisation. The carrier material in adsorption is called the adsorbent, and the enzyme unit is the adsorbate. A carrier modifier can be used to achieve tuning between an adsorbent and an enzyme. Adsorbents can be inorganic or organic substances. They can have different geometric configurations, such as membranes, beads, fibres, sponges, microcrystalline forms, or nanoparticles. The physical properties of adsorbents, such as surface area, pore volume, adjustable cavities, pore size distribution, and surface functional groups, are critical parameters that affect the mechanical stability and performance of immobilised enzymes.

TABLE 15.2
Advantages and disadvantages of the different enzyme immobilisation methods

Methods	Advantages	Disadvantages
Adsorption	• Simple • Inexpensive • Little conformational change in the enzyme • Requires mild conditions • Enzyme activity is often retained	• Weak binding • Leaching • Mass transfer limitations • Nonspecific binding
Disulphide Bonding	• Simple • Mild conditions • Reversible	• Weak binding • pH and ionic strength • Affect binding
Encapsulation	• Mild conditions • Protects the enzyme from denaturation • Used in continuous processes	• Mass transfer limitations • Reduced enzyme activity • Leakage of the enzyme
Covalent bonding	• Strong binding • Stable • Enzymes can be reused	• Complex • Expensive • Damage enzyme structure • Mass transfer limitations
Cross-linking	• Stable • Resistant to harsh conditions • Used in organic solvents	• Loss of activity • Mass transfer limitations • Difficult to control

The support systems provide substantial surface areas for high enzyme loading capacities and improved catalytic efficiency. The flexibility in the support system and different binding choices make the adsorption method a cost-effective strategy. Immobilisation of enzymes through this method also offers ease of operation. Even though adsorption is frequently a post-synthesis step, some procedures can involve its participation during downstream processing. Adsorption is suitable for a wide range of enzymes, including fragile and complex ones. It involves gentle procedures and does not impose harsh conditions that could denature the enzyme. The use of environmentally friendly support systems and the integration of sustainable practices enhance the attractiveness and viability of adsorption methods compared to other techniques. However, adsorption does have its defects. Nonspecific adsorption bonding may not be durable for large-scale applications, and the same may cause enzyme leakage. The irregular enzyme density reduces the overall strength of immobilised enzymes. Adsorption enzymes are soluble only in a few solvents. The aggregation of enzymes on adsorbate can reduce the enzyme activity.

Adsorption is a common technique for enzyme immobilisation due to its ease of use, affordability, and broad applicability despite several drawbacks. Physical adsorption involves different bonds between the enzyme and carrier substances, which influence the enzyme's activity and structure. These include nonspecific attachment, such as hydrogen bonding, hydrophobic interactions, van der Waals forces, affinity bonding, chelation, and ionic bonding (Figure 15.4), which are further discussed as follows.

- Hydrogen Bonding

Hydrogen bonding is a type of intermolecular force that occurs when a hydrogen atom is connected to an electronegative atom, such as oxygen or nitrogen. In enzyme immobilisation, this type of interaction is possible between the functional groups of an enzyme and a support material. The amino acid residues on the surface of enzymes, in particular, can develop hydrogen bonding with hydrophilic carrier molecules. Celite, microcrystalline cellulose (Avicel®), cellulose, lignin, and clay are some useful hydrophilic carrier materials for enzyme immobilisation through hydrogen bonding. This can even include the contribution of some polar compounds like sugars, polyethylene glycol (PEG) and albumin, which enhance enzyme stability. In addition, glycosylation of enzymes with sugar residues improves the enzyme's hydrophilicity, allowing for greater adsorption on carriers. The choice of support material and enzyme for adsorption through hydrogen bonding is critical for optimising the enzyme's activity, stability, and reusability. Other parameters that influence the stability and subsequent catalytic effectiveness of immobilised enzymes include pH, temperature, and ionic strength of the reaction media.

- Entropically Driven Hydrophobic Interactions

Enzymes and support systems with hydrophobic groups combine to provide a "hydrophobic effect" that promotes their aggregation by reducing exposure to water. In hydrophobic interaction, the water molecules around hydrophobic groups are organised in a less thermodynamically favourable

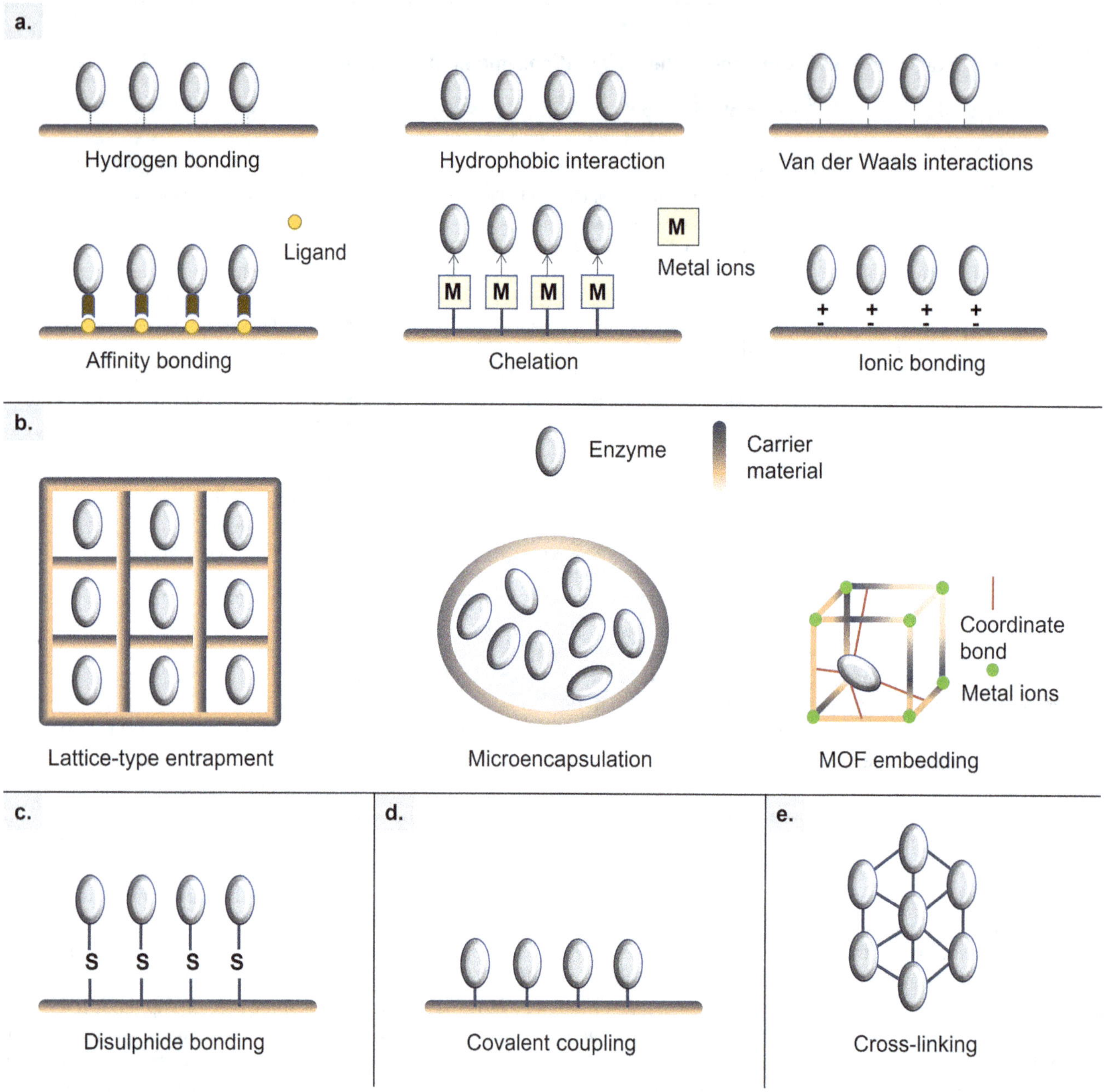

FIGURE 15.4 An illustration of different methods of enzyme immobilisation. a. Adsorption. b. Encapsulation. c. Disulphide bonding. d. Covalent coupling. e. Cross-linking.

way, allowing an increase in the entropy of the system. Many natural substances exhibit properties of hydrophobic interaction. In addition, carrier materials have been modified to acquire or modulate hydrophobic characteristics through covalent coupling or simply by coating. A routine step is the addition of a hydrophobic tail, which usually contains *n*-alkyl or aryl amino or carboxylic groups, Triton X-100, and Remazol blue. A lipase B from a yeast, *Candida antarctica* (CaLB), is a stable commercialised lipase immobilised by interfacial activation on a macroporous resin, Lewatit VP OC 1600. This composite is popularly known as Novozym 435 (N435).

- Van der Waals Interactions

Van der Waals interactions are weak intermolecular forces that arise due to temporary fluctuations in electron distribution within molecules. They can be combined with other immobilisation techniques, such as covalent bonding or hydrophobic interactions, to immobilise enzymes effectively. The aid of hydrophobic interactions is a popular choice, mimicking the conditions of permanent interfacial activation and preserving enzyme activity. This technique is useful for enzymes like lipases with a hydrophobic region on their surface that undergoes conformational changes upon activation.

Carrier materials like EP-100 polypropylene, Accurel MP1004 polypropylene, octyl-silica, and octyl-agarose are known for delivering van der Waals interactions and possess large surface areas. Notably, octyl-silica and octyl-agarose are natural beads modified with octyl groups to introduce hydrophobicity for Van der Waals interaction. However, the unassisted Van der Waals forces or in association with other weak forces have very little significance for enzyme immobilisation in the industrial sector. Enzyme leakage is a potential limitation. Even the slight variations in temperature or pH in the medium pose significant challenges to their effective utilisation.

- Affinity Bonding

Affinity-based immobilisation is a technique of specific and selective interactions among complementary molecules, enzymes, and affinity ligands. The choice of affinity ligands depends on the enzyme, the intended applications, and the desired level of enzyme activity, stability, and reusability. Ligands can be pre-coupled to support systems through covalent bonds, creating a platform for enzyme immobilisation (Figure 15.4a). Affinity binding allows for the controlled orientation of enzyme molecules on the carrier material's surface. This controlled orientation minimises conformational changes in the immobilised enzymes, which can be important for maintaining their activity. Tannin is an example of an affinity ligand that can interact with enzymes like xylanase, (R)-xylosidase, chitinase, and amino acid acylase. Tannins bind to these enzymes specifically due to their complementary interactions. The carrier material, chitosan, can serve as a stable platform for immobilising tannin and, subsequently, the enzymes.

One of the primary advantages of affinity-based immobilisation is its high retention of enzymes on the carrier material. This means that the enzymes remain bound and active for extended periods. Affinity-based immobilisation can be more resistant to heat-induced denaturation. This process can achieve high loading capacities, sometimes as much as five times greater than those achieved with ionic adsorbents. This allows for a higher density of enzymes on the carrier material. Metal-ligand interaction is a related immobilisation strategy that is based on chelation.

- Chelation

In chelation, a transition metal tied with a ligand through coordination bonds behaves as a support system (or chelator) for enzyme immobilisation (Figure 15.4a). Carboxymethylated amines, aspartic acids, and multidentate amino compounds are suitable examples of chelating ligands. These ligands have multiple binding sites for the coordination of metal ions like Co, Ni, Fe, Cu, Zn, and manganese (Mn). These metals bind to nucleophilic atoms (O, N, and S) of the enzymes and ligands. The charged and polar amino acids (histidine, lysine, phenylalanine, cysteine, and tyrosine) of the enzyme contribute to the immobilisation. In their interactions, steric factors allow some of the metal units in the support system to dangle, which can coordinate the binding of the enzyme residues. The binding behaviour of the metal with a nucleophilic group is greatly influenced by the degree of its exposure, the number and the separation distance. In some cases, the chelating particles can be stabilised on additional organic materials like cellulose, chitin, and silica-based carriers through covalent interaction. This provides additional support, which prevents the chelating ligands from leaching and increases stability.

The advantages of this technique include short and simple processing. This method reduces some steps of downstream processing, as this process can be involved during enzyme purification. The metal chelating carriers after the inactivation of enzymes can be recovered through a cost-efficient process. However, the reproducibility of the operational stability of immobilised enzymes is a major challenge due to the leakage of metal ions from the support system and inconsistent adsorption sites. Also, this method is not suitable for the food and pharmaceutical industries due to the risk of metal contamination.

- Ionic Bonding

Ionic bonding involves an electrostatic attraction between ions of opposite charges. In enzymes, amino acids can have positive and negative charges in the amino and carboxyl groups, respectively. These charged groups can interact with oppositely charged ionic ligands (or ion exchangers) (Figure 15.4a). Ion exchangers can either be positively charged (cation exchangers) or negatively charged (anion exchangers). The choice of an ion exchanger depends on the surface charge and charge distribution of the enzyme. Diethyl aminoethyl (DEAE)-Sephadex is a notable example of a commercial anion exchanger. It was first used in the 1960s to immobilise aminoacylase from *A. oryzae*. This marked an important advancement in enzyme immobilisation for commercial-scale applications. In a different example, glucose oxidase can be immobilised through ionic interaction between its negatively charged residues and positively charged groups (i.e., amino group) of chitosan.

Ionic bonding is relatively strong, making it the second strongest type of chemical bond after covalent bonding. This strength comes from the electrostatic forces between different ions. Ionic interactions are reversible, meaning they can be formed and broken under different conditions. The ionic strength (concentration of ions) and pH of the medium play significant roles in the interaction between enzymes and ion exchangers. The pore size of ion exchangers is also a critical parameter in the performance of immobilised enzymes. It can impact the stability and protection of enzymes by affecting their accessibility to the ion exchange sites. However, the ionic bond being reversible may not be suitable under certain environmental conditions.

15.6.2 Entrapment

Entrapment involves a physical restriction of an enzyme in a network/matrix and is an example of an irreversible type of immobilisation. The type of framework provides a confined space for enzymes where the interaction between an enzyme and the matrix is not overly strong. The enzyme units in entrapment are entrapped without being bound. The matrices are porous materials, which restrict the diffusion of enzymes but allow the passage of substrate and product through the polymeric network. The microenvironment inside the matrix provides optimum conditions (pH, temperature, and resistance to inhibitors and proteolysis) for enzymatic action. Matrices can be made of various polymeric materials such as alginate, gelatin, polyurethane, nylon, cellulose, collagen, polyacrylamide, carrageenan, and silicon rubber. Besides, nanostructured supports like nano-silica and carbon nanotubes create a robust enzyme network system. Some MOFs (like the zeolitic imidazolate framework (ZIF-8)) are alternative materials that are suitable support systems to entrap enzymes. Matrix materials used for entrapment contain properties like mild gelling behaviour, biocompatibility, non-toxicity, protein affinity, and chemical modifiability. A good carrier material also controls the pore size of the matrix, which is essential to accommodate and release the defined sizes of substrates and products, respectively, ensuring efficient diffusion. Enzymes entrapped with matrices develop three different types of structures:

- Lattice-type entrapment- In this form, enzymes are embedded within the interstitial spaces of a cross-linked polymer matrix, forming an interlaced crystal structure (Figure 15.4b).
- Microencapsulation- This involves enclosing enzymes within tiny hollow spherical structures (10–100 mm) that are covered by a porous membrane (Figure 15.4b).
- MOF embedding- MOF develops a coordination network of metal ions and organic linkers, which accommodates enzymes in their framework (Figure 15.4b).

The key advantages of entrapment are mechanical stability and low enzyme leaching. Entrapment helps in preventing the aggregation of enzymes. The process of entrapment is relatively fast. Also, the denaturation of enzymes is minimal because only a limited number of chemicals can directly interact with enzymes. However, the major drawback in the scale-up of entrapment is mass transfer limitations, which refer to the difficulties in the movement of substrate molecules to the active site of the enzyme and the subsequent diffusion of product molecules away from the active site. In contrast, the large pore size of the network increases the possibility of enzyme leakage. Also, the porous nature of the polymer matrix facilitates microbial growth, leading to contamination issues and reduced enzymatic activity. The poor loading capacity and wear of support systems during usage reduce its performance on the economic parameter. Despite many constraints, the entrapment method provides a balance between ease of implementation and maintaining enzyme activity, making it a valuable tool in various biotechnological processes. The entrapment method is also highly effective for whole-cell immobilisation.

A related but distinct technique of enzyme protection is enzyme encapsulation, which involves enclosing enzymes within a protective structure and promoting controlled environments for specific chemical processes. The physical barrier of the encapsulated structure shields enzymes from harsh environmental conditions, such as temperature extremes, pH changes, and chemical denaturants. This protection can extend the enzyme's stability and lifespan. Different biopolymers, such as alginate, maltodextrin, cellulose, chitosan, and gelatin, can be used for the encapsulation of enzymes. Encapsulated enzymes do not feature robust catalytic activity like traditional immobilised enzymes. Instead, this type of structure is used as a vehicle for targeted enzyme delivery.

15.6.3 Disulphide Bonding

Disulphide bonds are covalent bonds formed between cysteine residues of enzymes and support systems through the oxidation of their thiol (-SH) groups (Figure 15.4c). These bonds are relatively strong but reversible, making them unique among covalent couplings used in enzyme immobilisation. The formation of disulphide bridges can contribute to the stability and retention of enzymes on a surface. Support systems that are traditionally used for disulphide bonding include 2-pyridyl disulphide-agarose (PyS2-gel), thiosulphinate gel, and thiolsulphonate gel. These materials offer surfaces where disulphide bonds can be formed with the enzymes' cysteine residues. This method can be extended to even those enzymes and support systems also that do not contain thiol groups. Enzymes and support systems lacking thiol groups can be engineered to introduce cysteine residues for the reaction. One of the key advantages of the disulphide bond is its reversible nature. The presence of reducing reagents such as dithiothreitol (DTT) and β-mercaptoethanol in the medium facilitates the removal of enzymes from the support systems under mild conditions. These reagents break the disulphide bond, allowing the components to be separated. However, there are some drawbacks to this technique. pH alterations can lead to the leakage of enzyme residues due to the reversible nature of the disulphide bonds. Additionally, the process might require high enzyme usage, where the technique's efficacy could be influenced by the enzyme's susceptibility to degradation. Overall, disulphide bonding provides a relatively gentle attachment of enzymes over various irreversible couplings. It is particularly suitable for enzymes that can be degraded faster than the support systems, allowing for enzyme removal and reusing of the support systems.

15.6.4 Covalent Coupling with Carriers

This is a classical method of enzyme immobilisation. A covalent bond between the enzyme and carrier material is a powerful link and thermally stable. In this type of interaction, the nonessential groups of enzymes participate in the covalent coupling for their immobilisation (Figure 15.4d), which often includes:

- ε-amino acid of lysine
- Hydroxyl of serine and threonine
- Carboxyl of aspartic and glutamic acids

- Imidazole of histidine
- Phenolic of tyrosine
- Indole of tryptophan

These functional groups are strong nucleophilic groups, which react with electrophilic groups of support systems. A variety of support systems, such as cellulose, polyacrylamide, collagen, gelatin, amino benzyl cellulose, silica, cyanogen bromide (CnBr)-agarose, and CnBr-sepharose, bear an electrophilic group and provide a solid matrix for enzyme attachment through covalent coupling. The support systems can further be activated by introducing electrophilic groups onto carrier materials. Common activation methods include treatment of the surface with chemicals like carbodiimides, epoxides, or activated esters. They introduce functional groups like amino or carboxyl groups. This can create amide, ether, thioether, or carbamate bonds between the enzymes and support systems. The stability of the immobilised enzyme can be enhanced by appropriate post-coupling treatments. For example, any unreacted functional groups on the support material or the enzyme are blocked using blocking agents such as bovine serum albumin (BSA) or ethanolamine. This prevents nonspecific binding or undesired reactions. After the immobilisation process, immobilised enzymes are repeatedly washed to remove any loosely bound materials.

Immobilised enzymes through covalent coupling have known benefits in carrying out reactions in a variety of solvents. This can be particularly relevant for enzymes that are not stable in aqueous environments or that occur more effectively in non-aqueous solvents. Immobilisation via multiple covalent bonds can help maintain the conformation of the enzyme, preventing unfolding and denaturation. This is because the covalent attachments using similar bonds can restrict the enzyme's flexibility, stabilising its structure. Enzymes often rely on their precise spatial orientation and interactions with specific functional groups to carry out their catalytic functions. If the covalent attachment during immobilisation interferes with these sites, it can lead to reduced enzymatic activity or altered substrate specificity. Particularly, the covalent binding of the support systems to the amino groups at the active site of enzymes causes a major loss. This constraint can be alleviated to some extent by using affinity reagents for selective binding. Sometimes, specific epitopes or tags can be attached to enzymes to improve the selectivity of binding with carrier materials only at the desired sites.

15.6.5 Cross-linking

This process is also known as carrier-free immobilisation and does not require support systems to hold enzymes together. Cross-linking rather involves an intermolecular cross-linkage between adjacent enzyme molecules, which is accomplished via a cross-linking agent (or crosslinker). The crosslinkers possess multiple reactive sites, which facilitate binding between two or more functional units of similar or different types of materials (Figure 15.4e). In some conditions, crosslinkers can be used to bridge the gap between enzymes and support systems.

Glutaraldehyde is the most common, facile, inexpensive crosslinker. The functional group carbonyl of glutaraldehyde induces a condensed reaction with protein amines (Figure 15.5a). In some situations, the hydroxyl group in enzymes also participates in the cross-linking with glutaraldehyde (Figure 15.5b). The cross-linking has a minimum impact on the pre-organised superstructure of enzymes associated with catalytic activity. Cross-linking maintains the attachment between different molecules through covalent bonding. A different approach to cross-linking is physical binding, which calls for specific temperature, pH, or other environmental factors.

The integration of additional techniques in cross-linking can strengthen immobilised enzymes. The development of cross-linking enzyme crystals (CLEC) is one such process. It features a pre-cross-linking procedure that allows controlled precipitation of enzymes into microcrystals (Figure 15.6). This technique increases the protein's thermal and proteolytic

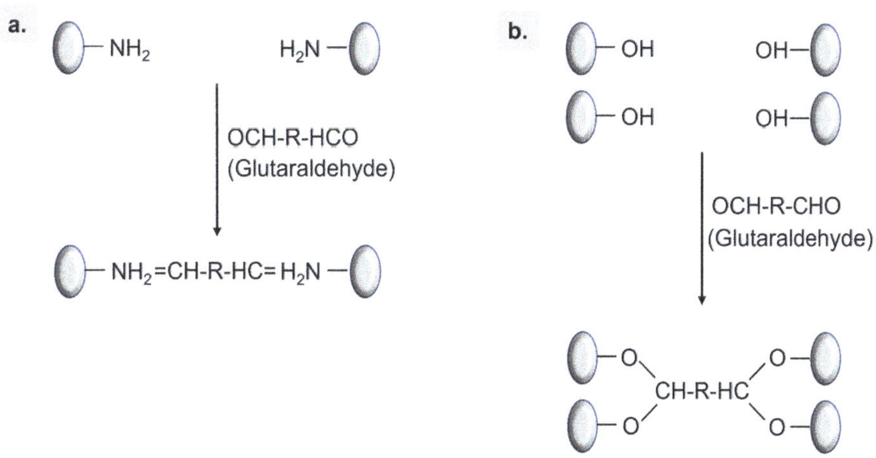

FIGURE 15.5 Mechanisms of cross-linking for enzyme immobilisation involving amino (a) and hydroxyl groups (b) of enzymes with the carbonyl group of glutaraldehyde (a crosslinker).

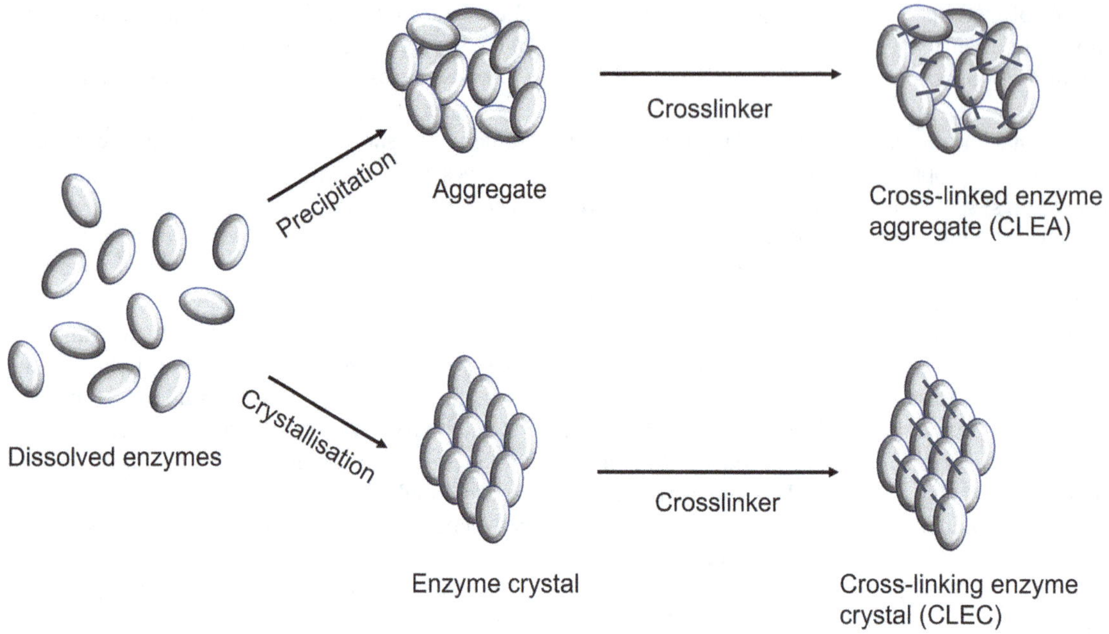

FIGURE 15.6 An illustration of the formation of cross-linked enzyme crystal (CLEC) and cross-linked enzyme aggregate (CLEA).

stability. The crystallised enzymes are insoluble in both aqueous and organic solvents. However, the high purity of crystallised enzymes is necessary for their effective applications, which is an inherent limitation for scale-up. In a different setting, cross-linked enzyme aggregate (CLEA) is a more contemporary form than CLEC that includes multipoint cross-linking as well as enzyme aggregation (Figure 15.6). This preparation does not necessitate a high purity of enzyme. In this preparation, the enzymes interact non-covalently to form aggregates, followed by cross-linking. Additional chemicals can be used to precipitate enzymes prior to the treatment of crosslinkers. Acetone, ammonium sulphate, and ethanol are common aggregation precipitants. This treatment increases the stability and stiffness of the enzymes. CLEAs are common industrial biocatalysts utilised in pharmaceuticals, food processing, wastewater treatment, and other applications. This approach is effectively utilised to immobilise penicillin acylase for the manufacturing of semisynthetic penicillins, which is an example of successful commercial preparation.

Cross-linking eliminates the limitation of the associated support systems. The cost of the expensive support systems is void. The choice of crosslinker and precipitants is based on their low cost. The carrier-free enzyme system has a low proportion of the noncatalytic unit, which expands the total amount of the catalytic unit in the solution and provides high space-time yields in the production process. Also, cross-linked enzymes are highly porous structures and do not have diffusional limitations. However, some crosslinkers can react with the active sites of enzymes, resulting in the loss of enzymatic activity. Large aggregates may have limited diffusional properties, which can hinder the availability of substrates to the enzyme's active sites. Additionally, excessive cross-linking can reduce the enzyme's flexibility, compromising catalytic efficiency.

15.7 MULTIENZYME IMMOBILISATION

Multienzyme immobilisation is an assembly of a multitude of enzymes in a support system. The inspiration comes from the schemes of the natural metabolic pathways of a biological cell. Such scaffolds can be designed and artificially synthesised to perform a set of chemical reactions *in vitro*. This facility utilises two or more enzymes from the same or different organisms that exhibit compatibility for different reactions under the same environment. Immobilised multi-enzymes can achieve several biocatalytic reactions in a one-pot system. The co-localisation of enzymes provides substrate channelling and efficient cascade reactions. In this process, the product of the previous reaction serves as the substrate of the next reaction.

The main advantage of a multienzyme system over a single-enzyme system is the unified processing of a complex reaction. Co-immobilisation allows minimum transport and fast conversion of the substrates for the multistep reactions, and reduces the mass transfer limitation and intermediate loss. The formation of by-products and toxins is minimal. Other benefits include the enhanced performance of enzymatic reactions by creating a controlled environment and potentially improved reaction rates and yields. However, multienzyme immobilisation is an imperfect technology due to many challenges in understanding the proposition of biocatalysis and kinetic studies of multiple enzymes. The production cost of multienzyme systems is also overwhelming. Downstream processing can become more intricate due to the need for specific recovery and high purity of enzymes. Despite several limitations, the multienzyme reaction has received great interest in producing industrially important products. The main interests of multienzyme immobilised systems lie in biofuel production, pharmaceutical synthesis, food and beverage processing, and environmental remediation. An immobilised multienzyme system can be constructed into three different types of designs:

- Random co-immobilisation- It is the simplest strategy of immobilisation, where enzymes are randomly attached to the carrier materials (Figure 15.7a). However, the crude arrangement of distinct enzymes might not guarantee their optimal interaction.
- Positional co-immobilisation- The assembly of enzymes has a specific orientation or sequence on the support systems, which can help in cascade enzymatic reaction rates (Figure 15.7b).
- Compartmentalisation- It is mainly achieved through entrapment. This process aims to mimic the natural organisation of enzymes within cells by creating distinct compartments of enzymes on the carrier materials (Figure 15.7c). This can improve the efficiency of multienzyme systems by recreating a microenvironment.

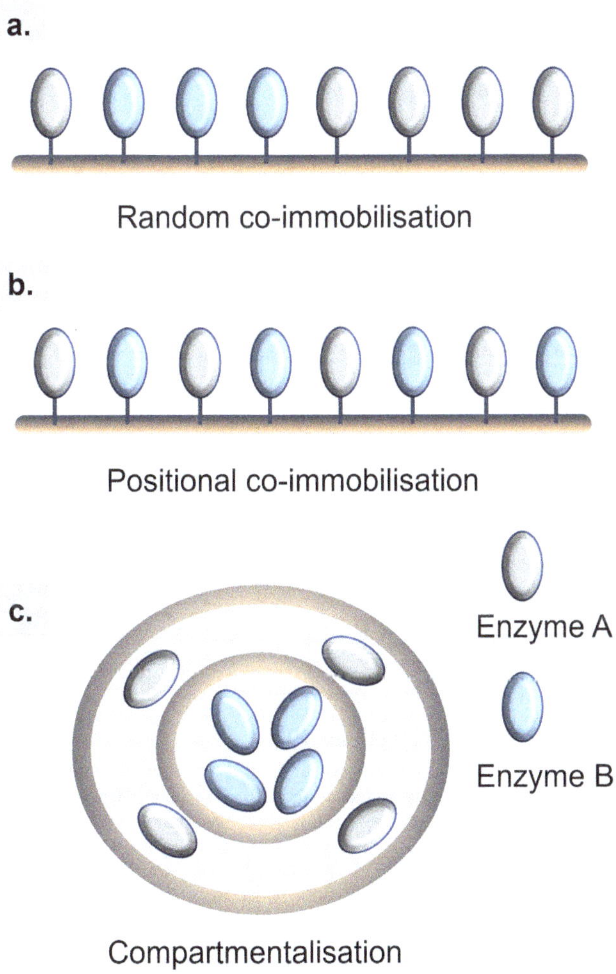

FIGURE 15.7 Different methods of enzyme co-immobilisation. a. Random co-immobilisation. b. Positional co-immobilisation. c. Compartmentalisation.

15.8 WHOLE-CELL IMMOBILISATION

An alternative strategy to exploit multi-enzyme facilities is the use of whole-cell immobilisation. This bridges the gap between microbial cells' natural functions and their practical applications. When the source of multi-enzymes is from the same cell, the immobilisation of whole cells is a better option. Both whole-cell immobilisation and enzyme immobilisation share the general concept of attaching biological components to a support matrix. The types of bonding/caging involved in the immobilisation of target cells and enzyme immobilisation are the same. Cell immobilisation involves adsorption, cross-linking, entrapment, encapsulation, and covalent binding using a variety of support materials. However, the process of cell immobilisation from enzyme immobilisation can be distinct, with differing goals and applications. Whole-cell immobilisation leverages the potential of the entire cells, while enzyme immobilisation primarily focuses on enhancing the performance of individual enzymes.

The use of whole-cell immobilisation reduces many steps of the downstream process and, therefore, the cost. Moreover, the whole cell provides a stable environment for enzymes, protecting them from denaturation and degradation. Enzymes within the cells remain shielded from harsh conditions and fluctuations in temperature, pH, and other factors. Immobilised whole cells retain the entire cellular machinery, including enzymes, co-factors, and regulatory mechanisms. The cells in an immobilised system can be living or nonliving. When the cells are alive, the inbuilt natural system is self-sufficient to regenerate coenzymes. However, whole-cell immobilisation suffers from major drawbacks, such as mass transfer limitations due to the physical barrier, cell inactivation, and limited substrate accessibility. Also, the undesirable side reactions of the whole cells can impede the desired reactions. Researchers continue to explore innovative techniques and materials to overcome these challenges and expand the applications of whole-cell immobilisation in different industries. For example, the physical/chemical treatment of the cells can improve the permeability of cells to a certain extent for the transfer of substrates and products. However, too much shearing on the covering of the cells may increase the chances of enzyme leakage.

Despite several challenges, immobilised cells can be employed for various enzymatic reactions (Table 15.3). For example, when the cells of *Rhizopus oryzae* are immobilised onto a suitable carrier material and exposed to Jatropha oil, the lipase enzymes they produce catalyse the transesterification reaction for biodiesel production. Since the cells are immobilised on a solid support, they can be reused for multiple reaction cycles. Other applications of immobilised cells are the production of enzymes nitrile hydratase and glucose isomerase for large-scale synthesis of acrylamide and HFCS, respectively. Wastewater treatment technologies have employed immobilised microbial cells to decompose pollutants and impurities. Immobilised cells can also be used to produce biofuels and various other commodities. They are particularly beneficial in the production of alcoholic drinks and

TABLE 15.3
Examples of immobilised cells for various enzymes in industries

Microorganisms	Enzymes	Industries	Processes
Rhodococcus rhodochrous J1	Nitrile hydratase	Chemical	Acrylonitrile hydrolysis to acrylamide
Streptomyces lividans and *Streptomyces rubiginosus*	Glucose isomerase	Food	High Fructose Corn Syrup (HFCS) production
Saccharomyces cerevisiae	Invertase	Food	Hydrolysis of sucrose to glucose and fructose
Aspergillus niger	Glucose oxidase	Food	Removal of glucose from egg white
E. coli	Penicillin acylase	Pharmaceutical	Semisynthesis of antibiotics

organic acids and can be utilised in both continuous and batch fermentation processes. The winemaking process heavily relies on immobilised yeast cells that offer better control in the fermentation process. The whole-cell immobilisation can minimise or eliminate yeast cryo-sensitivity, potentially increasing the shelf life of frozen dough products by preserving yeast viability and activity over extended storage periods. Differently, immobilised cells can be used to manufacture biosensors, which have various applications such as environmental monitoring, medical diagnosis, food safety, and pharmaceutical research.

SUMMARY

- Enzyme immobilisation is a technique to restrict the mobility of an enzyme by its attachment to the carrier material or matrix. This process reduces enzyme leaching in a reaction mixture and improves its activity and operational stability.
- Carrier materials are the support systems that capture free enzymes and immobilise them onto solid heterogeneous or homogeneous compounds. They can be inorganic materials, organic materials, or a hybrid of both. Nature-derived biodegradable materials beat synthetic materials in terms of sustainability and cost-effectiveness.
- Enzymes can be immobilised using reversible or irreversible methods. The reversible methods feature weak interactions and typically involve adsorption, entrapment, and disulphide bonding, whereas the irreversible method produces a powerful link through covalent coupling and cross-linking.
- The concept of multi-enzyme immobilisation deals with the idea of immobilising multiple enzymes onto a single support system. The assembly of different enzymes on the same platform reduces mass transfer limitation, minimises intermediate loss, and enhances reaction efficiency.
- Whole-cell immobilisation is a technique of confining whole cells onto a specific support system while allowing them to maintain their metabolic activity. The immobilised whole-cell can offer a stable environment and protection for the enzymes, but suffers from mass transfer limitations and undesirable side reactions.

DIFFERENT TYPES OF PRACTICE QUESTIONS

Fill in the blanks

I. ………………. are covalent bonds formed between two cysteine residues of enzymes.
II. ………………. is a method of immobilising enzymes by trapping them within a solid or gel matrix.
III. ………………. is a process of assembly of a multitude of enzymes confined to a single support system.
IV. In ………………., enzymes are physically adsorbed onto a solid support.
V. Cross-linking is a technique where enzymes are linked together using ……………….

SOLUTION: I. Disulphide bond, II. Entrapment, III. Multienzyme immobilisation, IV. adsorption, V. cross-linking agent/crosslinker

Match the following

i. Entrapment	a. Glutaraldehyde
ii. Disulphide bonding	b. Cysteine
iii. Chelation	c. Matrix/gel
iv. Cross-linking	d. Transition metal
v. Affinity bonding	e. Ligand

SOLUTION: i-c, ii-b, iii-d, iv-a, v-e

Very short answer type of questions

I. Write the name of a cross-linker used in immobilisation.
II. Expand CLEC and CLEA.
III. Which is the first commercial immobilised enzyme?
IV. What is the support system in enzyme immobilisation?
V. Write the names of any two natural organic carrier materials for enzyme immobilisation.
VI. What is the mass transfer limitation in immobilised enzymes?

Short answer type of questions

I. What is enzyme immobilisation? Write about the immobilisation principles.
II. How does covalent coupling contribute to enzyme immobilisation? What are the major advantages and disadvantages of this technique?
III. Provide a comparative account of enzyme immobilisation and whole cell immobilisation.
IV. What is multienzyme immobilisation? What are the applications of multienzyme systems?
V. How does whole-cell immobilisation differ from enzyme immobilisation? What are the methods to improve the stability and activity of immobilised whole cells?

Long answer type of questions

I. What are the different carrier materials for enzyme immobilisation?
II. Shed light on various methods of enzyme immobilisation. Also, discuss their advantages and disadvantages.

NOTE

1 Joudeh N, Linke D. Nanoparticle classification, physicochemical properties, characterization, and applications: a comprehensive review for biologists. Journal of Nanobiotechnology. 2022;20(1):262.

SUGGESTED READING

Brena B, González-Pombo P, Batista-Viera F. Immobilization of enzymes: a literature survey. In Guisan JM (ed.) *Immobilization of enzymes and cells: Third edition* 2013 (pp. 15–31). Springer.

Datta S, Christena LR, Rajaram YR. Enzyme immobilization: an overview on techniques and support materials. *3 Biotech.* 2013;3:1–9.

Gupta MN, Kaloti M, Kapoor M, Solanki K. Nanomaterials as matrices for enzyme immobilization. *Artificial Cells, Blood Substitutes, and Biotechnology.* 2011;39(2):98–109.

Hartmeier W. *Immobilized Biocatalysts.* Springer-Verlag; 1988.

Jesionowski T, Zdarta J, Krajewska B. Enzyme immobilization by adsorption: a review. *Adsorption.* 2014;20:801–21.

Liu DM, Chen J, Shi YP. Advances on methods and easy separated support materials for enzymes immobilization. *TrAC Trends in Analytical Chemistry.* 2018;102:332–42.

Mohamad NR, Marzuki NH, Buang NA, Huyop F, Wahab RA. An overview of technologies for immobilization of enzymes and surface analysis techniques for immobilized enzymes. *Biotechnology & Biotechnological Equipment.* 2015;29(2):205–20.

Sheldon RA, Basso A, Brady D. New frontiers in enzyme immobilisation: robust biocatalysts for a circular bio-based economy. *Chemical Society Reviews.* 2021;50(10):5850–62.

Sheldon RA, van Pelt S. Enzyme immobilisation in biocatalysis: why, what and how. *Chemical Society Reviews.* 2013;42(15):6223–35.

Xia H, Li N, Zhong X, Jiang Y. Metal-organic frameworks: a potential platform for enzyme immobilization and related applications. *Frontiers in Bioengineering and Biotechnology.* 2020;8:695.

Xu K, Chen X, Zheng R, Zheng Y. Immobilization of multi-enzymes on support materials for efficient biocatalysis. *Frontiers in Bioengineering and Biotechnology.* 2020;8:660.

Zhang S, Bilal M, Zdarta J, Cui J, Kumar A, Franco M, Ferreira LF, Iqbal HM. Biopolymers and nanostructured materials to develop pectinases-based immobilized nano-biocatalytic systems for biotechnological applications. *Food Research International.* 2021;140:109979.

16 Properties of Immobilised Enzyme Systems

16.1 KINETIC STUDIES OF IMMOBILISED ENZYMES

The kinetic studies of enzymes provide insights into their functional characteristics, such as the rate of reaction and catalytic mechanism. This type of investigation is also important for immobilised enzymes because immobilisation induces changes in the enzyme's microenvironment, leading to alterations in kinetic parameters when compared to the soluble form of the enzyme. Immobilisation affects enzyme stability and activity. A comparison of kinetic parameters obtained from immobilised enzymes with those of their soluble counterparts provides valuable information. Understanding the kinetics of immobilised enzymes and their behaviour under different conditions allows for the optimisation of reaction parameters to achieve desired outcomes in specific applications. The kinetic data of immobilised enzymes can be fit into several kinetic models to evaluate the effects of immobilisation and optimise the reaction conditions. Enzyme kinetics can be used to study the rate of substrate conversion or product formation using immobilised enzymes under various conditions such as temperature, pH, and storage. The long-term performance of an immobilised enzyme also necessitates monitoring its activity after several cycles of reaction and reactivation procedures. In this situation, enzyme kinetics can be used to compare the rate of reactions after different cycles.

One of the most recognised and straightforward theories of enzyme kinetics is the Michaelis-Menten equation. This equation is a fundamental model for enzyme kinetics, describing the relationship between substrate concentration (S) and initial reaction rate (V_0). This equation allows the determination of kinetic parameters such as Michaelis-Menten constants (K_m) and maximum velocity (V_{max}). K_m is referred to as the substrate concentration at which the velocity of the reaction is half V_{max}. A lower K_m indicates a higher affinity of the enzyme for its substrate (detailed in Chapter 14).

Figure 16.1 depicts differences in kinetic parameters, including K_m, rate of product formation, and specific activity of enzymes for a free and immobilised system. The study included a chromogenic substrate, *p*NPB (*p*-nitrophenyl butyrate), using a lipase enzyme. In the figure, a higher K_m value for the immobilised enzyme than the free enzyme indicates that the immobilised enzyme has a lower affinity for its substrate. This indicates that the immobilised enzyme requires a high amount of substrate to achieve the same rate of product formation as the free enzyme. The mass transfer limitation in immobilised systems is the primary reason for the difference in enzyme kinetics. When enzymes are immobilised, the rates of diffusion for substrate and product compounds to and from the enzyme's active site through the solid support are restricted. This can result in the increased K_m values of the immobilised enzyme compared to its soluble counterpart. However, in some rare situations, the K_m of an immobilised system might also decrease due to variations in the microenvironment (factors like altered pH, ionic strength, or hydrophobicity near the enzyme's active site).

The immobilisation process can also modify the values of catalytic constant (k_{cat}) and V_{max} of enzymatic reactions. k_{cat} measures catalytic efficiency and can be represented as the maximum number of substrates that an enzyme converts into products per unit of time when the enzyme is fully saturated with substrate. Immobilisation may reduce V_{max} due to mass transfer limitations. The specificity constant is an index of enzyme performance in real-world applications, which is used to compare the catalytic efficiency of different enzymes. This depends on two parameters, including k_{cat} and K_m. The specificity constant (k_{cat}/K_m) of an immobilised enzyme also experiences alteration due to variations in its parameters.

16.2 ESTIMATION OF THE PERFORMANCE OF IMMOBILISED SYSTEMS

Enzyme (total) activity and specific activity are the two most common probes for evaluating the performance of an enzyme. Enzyme activity refers to the number of moles of substrate transformed into products per unit of time and expressed in μmol/min (1 μmol/min = 1 enzyme unit (U)). Specific activity is defined as the enzyme activity per mg of total protein (unit is μmol/mg), which is an estimation of the enzyme purity.

These measurements are equally applicable to immobilised enzymes, similar to those of free enzymes. However, the immobilisation of enzymes onto a carrier material can take

DOI: 10.1201/9781032651095-20

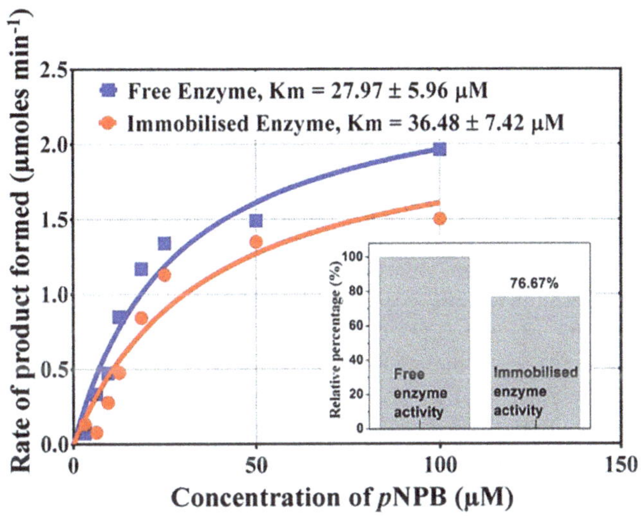

FIGURE 16.1 Plot of enzyme activity for lipase, determined by *p*NPB (*p*-nitrophenyl butyrate) assay for free and immobilised enzymes. The inset shows the percentage of specific activity of the immobilised enzyme relative to that of the free enzyme. A higher K_m value of immobilised enzyme than free enzyme indicates that the immobilised enzyme has a lower affinity for its substrate. This means that the immobilised enzyme requires a higher substrate concentration to achieve the same rate of product formation. (Image reproduced from Veeramani et al. (2014)[2] with permission of © Royal Society of Chemistry.)

some new deliberations into account due to some variations in the enzyme configurations. For example, the enzyme activity of immobilised enzymes can be low due to potential loss of activity during immobilisation. The specific activity also tends to be low because the total protein content used in the calculation corresponds to the amount of protein present in the immobilised form, including the weight of the support or matrix.

In addition, immobilisation yield, percentage efficiency, efficiency coefficient, and activity recovery are exclusive parameters used for immobilised enzymes. Immobilisation yield provides information on the success of immobilisation of an enzyme on a carrier material. It is generally expressed in percentage, as follows.

$$\text{Immobilisation yield (\%)} = ((A-B)/A) \times 100$$

Where A is the total enzyme activity in the solution before immobilisation and
B is the total residual enzyme activity in the solution after immobilisation.
In this formula, A-B is referred to as immobilised activity, which represents the "activity that is immobilised".
The percentage efficiency measures the percentage of enzyme activity in the immobilised states. This can be calculated as

$$\text{Percentage efficiency (\%)} = (C/(A-B)) \times 100$$

Where C is the total enzyme activity in an immobilised form.

Immobilisation yield provides an estimate of the proportion of the enzyme present in the solution that could have been immobilised. However, percentage efficiency gives information on how effectively enzyme activity is retained in its immobilised form. For example, when an enzyme is mixed with a carrier material, an immobilisation yield of 100%, but an efficiency of 0% could also be possible. This occurs when all the enzyme units present in the solution are immobilised, but they have lost their complete activity due to immobilisation. Practically, an immobilisation yield of 100% is also not achievable.

A related parameter is the efficiency coefficient or effectiveness factor. This is an index of the efficiency of a system, which measures the extent of mass transfer limitation of immobilised enzymes. The general expression symbol of the effectiveness factor is η, and the formula is as follows.

$$\eta = V_{max} \text{ of immobilised enzyme}/V_{max} \text{ of free enzyme}$$

Where V_{max} is the maximum velocity

Activity recovery is a numerical value of the percentage of the enzyme's original activity that is retained after immobilisation. This gives an idea about the success of the immobilisation process. It can be formulated as

$$\text{Activity recovery (\%)} = (C/A) \times 100$$

Protein loading provides information on the amount of enzyme loaded on the carrier. This is estimated as-

$$\text{Protein loading} = (X-Y)/Z$$

Where X is the total amount of enzyme in the solution before the addition of the carrier,

Y is the total amount of free enzyme left in the solution after the removal of the immobilised enzyme, and
Z is the total amount of carrier materials added to the solution for immobilisation.

The term protein loading is sometimes confusingly considered an equivalent parameter to immobilisation yield. The measurement of protein loading is based on the amount of protein and carrier material, while immobilisation yield is estimated using the specific activity of the enzyme. The estimation of protein concentration in protein loading does not give an actual representation of the concerned bound enzyme due to the presence of some protein impurities in the medium, which may also have an affinity for the carrier. The protein loading is usually an overestimation of the immobilisation yield, whereas the immobilisation yield is a precise estimation of the concerned bound enzyme.

In an immobilisation exercise of an enzyme, an investigator uses 50 mg of lipase and transfers it to a solution. The total enzyme activity in the solution is 1000 U. Further, a carrier material of 2 g is added to the solution for the immobilisation of enzymes. After 20 h of incubation, the immobilised enzymes are recovered. The total residual enzyme activity and residual protein concentration in the solution were 200 U and 10 mg, respectively. The enzyme activity in the recovered immobilised enzyme was 400 U. Under these observations, the key parameters of the immobilised system can be calculated as follows:

$$\text{Immobilisation yield} = ((1000-200)/1000) \times 100$$
$$= 80\%$$

$$\text{Percentage efficiency} = (400/(1000-200)) \times 100$$
$$= 50\%$$

$$\text{Activity recovery} = (400/1000) \times 100$$
$$= 40\%$$

$$\text{Protein loading} = (50\text{-}10)/2$$
$$= 20 \text{ mg/g carrier}$$

This can be interpreted as 80% of the enzyme activity present in the solution has left the solution and transferred into an immobilised system. However, an immobilised enzyme with a percentage efficiency of 50% indicates that half of the total enzyme activity from the solution is retained in the immobilised system, and the other half of the activity is lost due to enzyme inactivation and mass transfer limitation after enzyme immobilisation. An active recovery of 40% is a measure of the activity of the immobilised enzyme compared to the initial activity of the free enzyme. The protein loading of a 20 mg/g carrier can be explained as 20 mg of enzyme protein has been successfully immobilised on each gram of carrier.

In addition, specific productivity, turnover number, and operational time are some other relevant parameters which can be used to assess the potential of immobilised enzymes. A different useful measure is space-time yield, which is an estimate of product quantity per volume of immobilised system and time. This parameter is used to compare different carrier materials in enzyme immobilisation for cost-benefit analysis.

A different parameter of kinetics studies of immobilised enzymes is the quantitative estimation of the reusability of an immobilised system and its residual activity per cycle. For example, the commercially available immobilised enzyme Novozym 435 can be used as many as 200 times without losing its initial activity. Another enzyme, laccase in immobilised form, retains 60% activity after reusing the enzyme for eight cycles. Usually, an industrial process calls for a minimum of 50% residual activity (i.e., a threshold limit) in the immobilised enzyme for repeated usage.

16.3 FACTORS AFFECTING THE PERFORMANCE OF IMMOBILISED ENZYMES

Every enzyme is a different structural and functional unit, and its interactions with different carrier materials in an immobilised system vary. The complex interplay of various factors affects the performance of an immobilised system. Therefore, a one-size-fits-all approach to studying the immobilised enzymes is not feasible. This calls for a tailored approach to the study of the factors affecting the performance of immobilised enzyme systems. A systematic investigation through iterative optimisation is a suitable approach to enzyme immobilisation. Iterative optimisation is a process where insight into initial experiments provides continuous refinement for subsequent experiments to achieve the best possible outcome. Factors that can affect the performance of immobilised enzymes are carrier materials, enzymes, immobilisation process, and reaction systems. These factors have many subcategories (Table 16.1), which influence the performance of immobilised enzymes and are discussed in the following paragraphs.

Carriers have different chemical compositions and possess different functional groups, which give options for different types of immobilisation processes (discussed in Chapter 15). The physical and chemical parameters of carrier materials have a critical impact on enzyme immobilisation. The morphology, particle size, porosity, active surface area, and surface properties affect the orientation of the enzyme on the carrier and its accessibility to substrates and products. The large surface area of small-sized particles is the superior quality of an effective carrier. A high degree of hydrophobicity and swelling behaviour in the carrier materials are advantageous for binding with enzymes. The shedding and leakage of enzymes during recycling are considerable drawbacks for long-term reusability. This can be reduced by relying on interaction through covalent binding, which creates a permanent bond between enzymes and carriers. Although this type of interaction allows for multiple uses, the recovery of carrier materials after the deactivation of enzymes is not feasible.

TABLE 16.1
Factors affecting the immobilised enzymes in the reaction systems

	Parameters	Examples
Carrier	Chemical parameters	Composition, functional groups, swelling behaviour, and hydrophobicity
	Physical parameters	Morphology, particle size, porosity (pore structure, pore size distribution), active surface area, and surface charge
Enzyme	Biochemical properties	Availability, storage stability, toxicology, purity, conformational flexibility, molecular weight, size, active site, morphology, functional groups on the surface, and surface charge
	Kinetic parameters	Specific activity, pH and temperature profiles, stability towards reaction conditions (pH, solvents, temperature), and stabilising or activating additives
Immobilisation process	Immobilisation strategy	Adsorption, encapsulation, covalent coupling, and cross-linking
	Immobilisation conditions	pH, temperature, solvents, glycosylation, and stabilising agents
	Diffusion/mass transfer	Buffer effect, viscosity of reaction media, pore diffusion, and interfacial interaction
	Performance	Space-yield time, recycling, reusability, product removal, toxicity, and enzyme inhibition
	Process cost	Costs of the enzyme and support, waste disposal, and energy
Reaction System	Physical parameters	Types of media, viscosity of the mixture, reaction thermodynamics, microenvironment of the immobilised enzyme, abrasion (for stirred tanks), and flow resistance (for a fixed-bed reactor)
	Chemical parameters	pH, ionic strength, solvent effects, substrate type, its concentrations, presence of inhibitors, and solute carrier interaction

Note: Inspired by Hartmann and Kostrov (2013)[1] from the permission of © Royal Society of Chemistry.

Several factors of enzyme units, including biochemical properties and kinetic parameters, have a significant impact on the immobilised systems. Structure, size, weight, active sites, and functional groups of enzymes are crucial for the enzyme loading on carrier materials. The surface functional groups/charge density on both materials must be compatible for their effective binding. The presence of hydrophobic and hydrophilic regions on enzymes is particularly important for their specific orientation. The conformational flexibility of enzymes allows them to undergo dynamic changes in their structure to bind substrates, catalyse reactions, and release products. The conformational changes that arise due to immobilisation can result in an alteration in enzyme activity. The level of enzyme purity required for an immobilised system depends on the type of immobilisation process involved and the cost considerations. For example, cross-linking and encapsulation are less sensitive to enzyme impurities, whereas adsorption and covalent bonding require a sufficient level of purity. The size and chemical nature of an enzyme are substantial for the encapsulation and attachment to the matrix. However, interference from stabilisers, inhibitors, and other reactive chemical species can modify the stability and activity of anchored proteins. Glycosylation causes considerable variations in the surface properties of the enzyme, and therefore, the suitable choices of carrier materials for attachment to the glycosylated enzymes can also vary.

Immobilisation strategies and conditions play dominant roles in the immobilisation process, which affect their performance in reaction systems. There are two different types of physicochemical constraints for the immobilised enzymes: partition effects and diffusion barrier. The partition effect refers to the unequal solubilities around the immobilised enzymes. The diffusion barrier is a function of physical parameters, such as size and shape, which measure the hindrance of substrate molecules reaching the active site of the enzyme. The size of the pores (pore diameter and volume) of the carrier in the immobilised system should be in the size range of substrates and products so that they can attain an unhindered float. The mass transfer effects influence the kinetics of immobilised enzymes. The enzymes attached to carrier materials may result in a low reaction rate due to a low opportunity for contact between enzymes and solutes. The dilute substrate concentration is usually not effective for a uniform reaction using immobilised enzymes because the substrate may not penetrate deeply inside the carrier materials. The steady reasons for these limitations are preconceived external and internal diffusional restrictions. Particularly, the diffusion of substrates within porous materials is challenging. The rate of reaction is inversely proportional to the depth of the enzyme inside the particle. The diffusional restrictions can be minimised by reducing the particle size of the carriers. Preferential binding of enzymes on the outer shell of the porous materials gives reasonable reaction conditions. When the enzymes have high specificity, their low density on carrier material is a better choice for enhanced enzyme efficiency. In some situations, reaction and carrier-generated proton gradients bind to active sites of enzymes in the immobilised system, which can inhibit enzyme activity. This can be minimised by operating the reaction at a slightly higher pH than the optimum pH of enzyme activity. The use of a buffer can also minimise the effect of protons. An alternative solution to this problem is the co-immobilisation of a proton-consuming enzyme. High volumetric activity and operational stability of enzymes in an immobilised system can be achieved by their interfacial interaction at both molecular and system levels. For such an interaction, the properties of both enzymes and carrier materials are relevant.

The physical properties of the reaction system, such as the types of medium, the viscosity of the mixture, and reaction thermodynamics, also affect the rate of reactions of immobilised enzymes. The abrasion in stirred tank bioreactors and the flow resistance pattern in fixed bed bioreactors influence their stability and performance in the long term. The microenvironment of immobilised enzymes can be different from the bulk solution (i.e., the solvent in which the immobilised enzyme is suspended) due to many physical constraints. The variation in pH and ionic strength of the solution alters the performance of immobilised enzymes with respect to their activity and stability. In addition, the non-specific solute-carrier interactions result in enzyme inhibition or precipitation.

16.4 PROTEIN ENGINEERING STRATEGIES

Protein engineering is a strategy of protein modification, often through artificial amino acid variation at the genetic level. This also accompanies the evaluation of the functionality of the resulting variants. In enzyme immobilisation, protein engineering is mainly employed for the stable binding of the enzymes with carriers, maintaining protein performance, specificity, stability, activity, and purification yield under non-conventional environments. Based on the methods involved, protein engineering can be of three different types, including directed evolution, rational design, and semi-rational design (Figure 16.2). They are further discussed as follows:

16.4.1 Directed Evolution

Directed evolution aims for artificial modification in the genetic makeup of the cells, mimicking natural evolution. This type of modification depends on three sequential steps: gene diversity generation, screening/selection, and the iterations of beneficial mutants. The high-throughput screening techniques

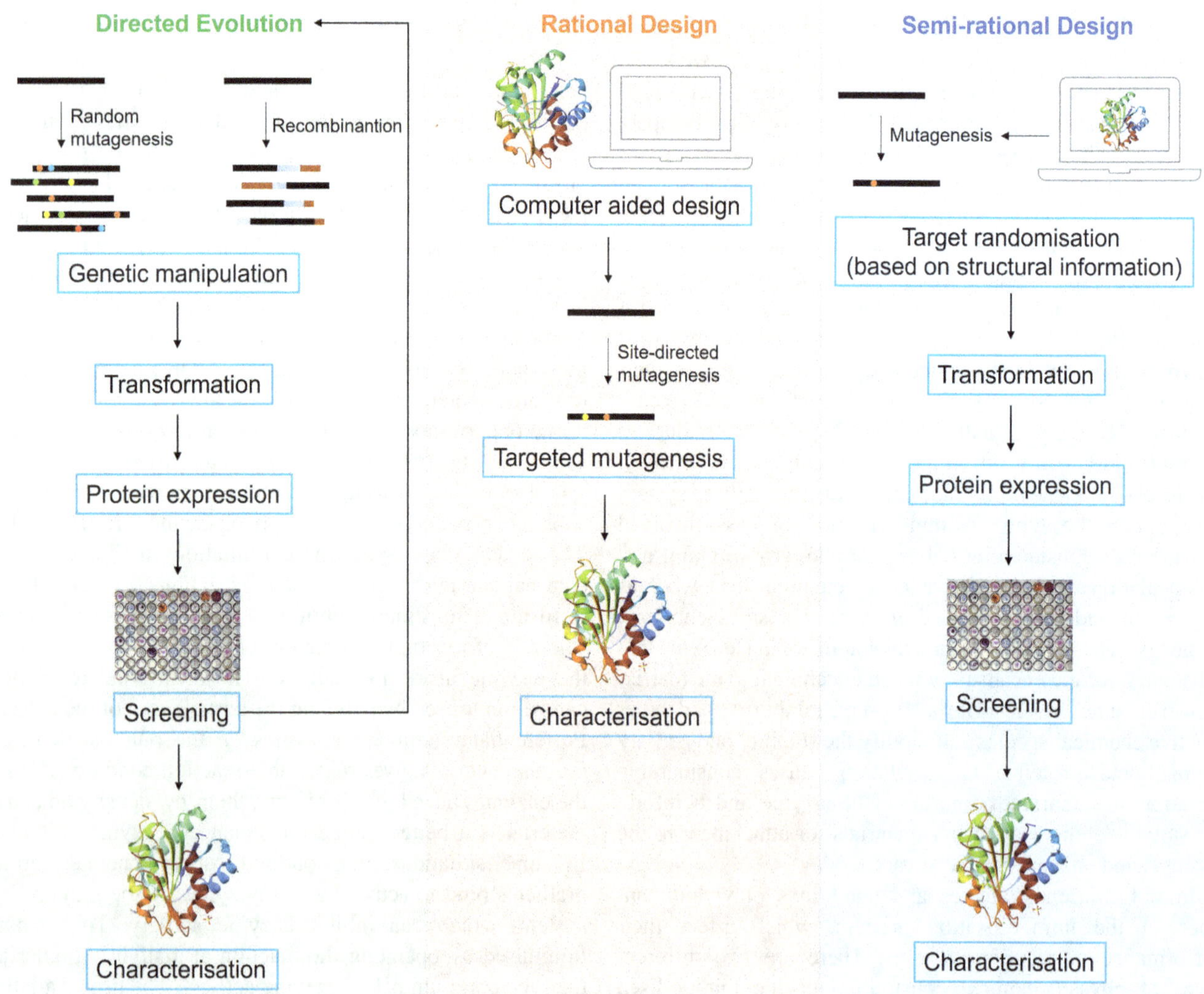

FIGURE 16.2 Process of protein engineering through directed evolution, rational design, and semi-rational design.

facilitate the identification of the target-specific traits from the gene libraries. Recombinant DNA technology and mutagenesis are two main strategies for generating gene diversity. They are highly sensitive methods and promise to create variations at the molecular level.

Recombinant DNA technology involves the integration of foreign DNA with the gene of interest to create new variants. The process involves enzymatic cleavage at the specific target of the host DNA and incorporates the new DNA segment(s), followed by ligation. The cloned cells are selected and harvested for the production of novel enzymes. The application of recombinant technology is less common for the enzyme immobilisation process compared to mutagenesis.

The mutagenesis can be achieved through whole gene random mutation or site-directed mutagenesis. Whole gene random mutation is a process of random mutation in the target genes. This can result in the generation of many variants and the development of mutagenesis libraries. However, site-directed mutagenesis (SDM) acts on specific sites and can replace one or a few amino acid(s) through codon exchange and produce a single or a few enzymes variant. The modification of a single amino acid of an enzyme can also change its catalytic property, stability, and binding affinity with the carrier. For example, the removal of lysine residues near the active sites of lipase in *Thermomyces lanuginosus* achieved high activity when the enzyme was immobilised. The removal of lysine from the surface of the enzyme exposed a high surface area for catalytic reactions. Otherwise, carrier materials, Eupergit or Accurel, are susceptible to covalent binding with lysine residues for immobilisation, which blocked the active site of enzymes for their catalytic reaction.

16.4.2 Rational Design

The rational design is a computational approach to predict and engineer protein performance. This relies on the knowledge of the 3D structure of the target proteins and their function. Different peer-reviewed public repositories such as Protein Data Bank (PDB), PubChem, ArrayExpress, MetaCys, BRENDA, and Gene Expression Omnibus (GEO) provide a strong foundation for research in rational design. It basically involves *in silico* hypotheses using mathematical algorithms and bioinformatic tools, and allows for precise predictions of protein behaviour under the influence of substituent amino acids. The manipulation of the enzyme using computer-aided support ensures targeted mutagenesis and allows the study of genetic modification on enzyme structure and function. This also reduces the effort of maintaining a large mutagenesis library and enables virtual screening. Modern machine learning programmes and artificial intelligence (AI)-driven tools have analysed enormous datasets of enzyme properties (including physicochemical characteristics, 3D structure, surface properties, dynamic behaviour, and reactivity of enzymes) to reveal more accurate predictions and design strategies. The main aim of the rational design is to reduce the shortcomings of the empiricism of enzyme technology.

Rational design of immobilised derivative (RDID) is a popular tool for designing and optimising the immobilised enzymes. Molecular dynamics studies of proteins under the influence of their surroundings can be studied using a software tool such as CAVER. However, a database that can provide comprehensive information on immobilised enzymes does not exist. This requires a collaborative effort to build a centralised database that contains standardised information on enzymes and support materials to optimise the immobilisation process under different evolutionary pressures and challenges for the development of modern biocatalysts.

16.4.3 Semi-Rational Design

Semi-rational design integrates the elements of rational design and techniques of directed evolution. This type of approach addresses the "diversity challenge" that is, the limitation of screening capabilities of a high number of generated libraries from random mutations in directed evolution. The integrated tool of rational design reduces the screening and selection efforts of mutagenesis libraries and sorts out only the meaningful, targeted residues. This is followed by experimental validation by introducing mutations at the targeted residues of enzymes. The combination of computation and experimental approaches accelerates the enzyme engineering process, leading to more predictable outcomes.

16.5 WORKFLOW FOR COMBINED APPROACHES

Protein engineering in enzyme immobilisation can be achieved using two different principles, including "sequential" and "integrating" approaches. A sequential approach of enzyme improvement for an immobilised system involves protein engineering of enzymes, followed by immobilisation of mutant enzymes. In this process, the target enzyme is genetically modified to create new variants. They are screened, and the selected variants are used as input for new iterative rounds to create more variations in enzymes. The improved enzyme is now immobilised using selected support materials. Based on the evaluation of experimental conditions, the immobilisation strategies can further be modified. The resultant immobilised enzyme is characterised (Figure 16.3a).

The integrating approach of protein engineering and enzyme immobilisation is based on a parallel, high-throughput format. In this scheme, the genetic variants of target enzymes are immobilised on suitable support materials using a predefined immobilisation strategy in a high-throughput format using a microtiter plate (MTP). The immobilised enzymes of each mutant are subject to evolutionary pressure or challenges to evaluate their performance. The improved variants are screened. The selected variants undergo new rounds of iteration for further genetic improvement. The best variant of the immobilised system is characterised (Figure 16.3b).

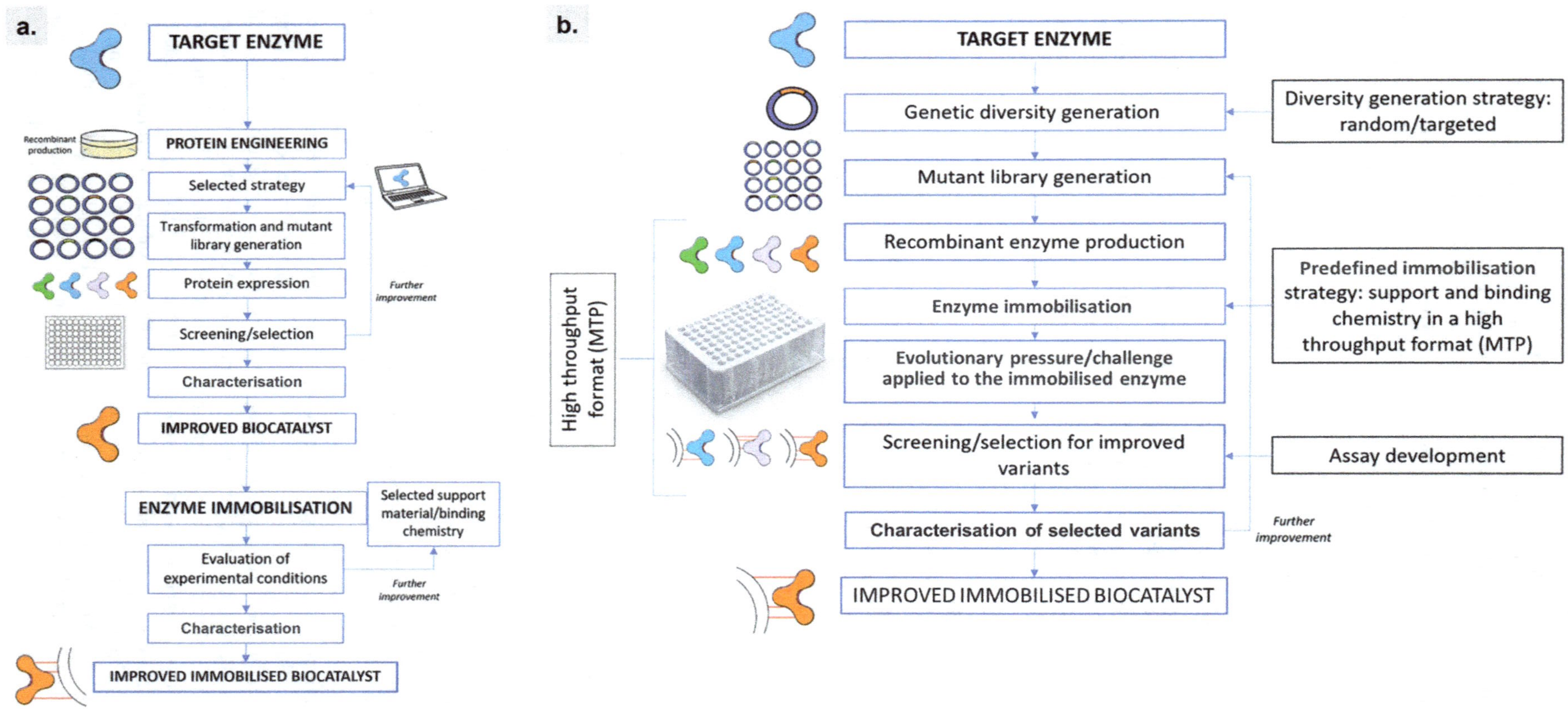

FIGURE 16.3 a. A process of enzyme improvement using a sequential approach of protein engineering and enzyme immobilisation. b. A process of enzyme improvement using an integrated approach of protein engineering and enzyme immobilisation. (Images adapted from Bernal et al. (2018)[3] with permission of © 2018 Elsevier Inc.)

TABLE 16.2
A list of techniques used for the conformational analysis of the immobilised enzymes

Conformational Analysis of the Immobilised Enzymes		
Microscopy	**Spectrometry**	**Other techniques**
• Scanning electron microscopy (SEM) • Field-emission scanning electron microscopy (FESEM) • Transmission electron microscopy (TEM) • Atomic force microscopy (AFM) • Scanning electrochemical microscopy (SECM)	• UV/Vis spectrophotometer • Fourier-transform infrared (FTIR) spectroscopy • Surface-enhanced infrared absorption spectroscopy (SEIRAS) • Synchrotron FTIR • X-ray photoelectron spectroscopy (XPS) • FT-Raman spectroscopy • Time-of-flight secondary ion mass spectroscopy (TOF-SIMS) • Förster resonance energy transfer (FRET)	• Thermal gravimetric analysis • Circular dichroism (CD) • Surface plasmon resonance (SPR) • Microcalorimetry • Neutron reflection

16.6 CONFORMATIONAL ANALYSIS

A comprehensive understanding of the structure, properties, and performance of immobilised enzymes calls for their surface and molecular analyses. A sophisticated biocatalyst requires high stability and activity, which is highly dependent on orientation and three 3D structures of the enzyme after immobilisation. The microenvironment, conformation, and protein refolding of the immobilised system are some notable parameters that matter for its proper function. These parameters are mainly studied using various techniques of microscopy and spectrometric methods (Table 16.2).

Microscopy provides a visual retreat of the immobilised materials, and electron microscopy is a suitable approach for their imaging at the nanoscale range. Scanning electron microscopy (SEM) and field-emission scanning electron microscopy (FESEM) give structural information on the surface topography of an immobilised system. Transmission electron microscopy (TEM) reveals the two-dimensional projections of the immobilised enzyme and also measures particle size. Atomic force microscopy (AFM) gathers data to identify the surface topography of immobilised enzymes at the molecular level and generate information on their properties, such as adherence, elasticity, association processes, and dynamics of the immobilised surface. Scanning electrochemical microscopy (SECM) tracks information on biocatalytic reactions and the apparent kinetics of the enzyme.

Spectrometry is an analytical method which tracks the changes in the absorbance of light for enzymatic reactions. The UV/Vis spectrophotometer can detect protein conformational variation in an immobilised system. Fourier-transform infrared (FTIR) spectroscopy can be employed to confirm the attachment of enzymes to carrier materials. In certain situations, attenuated total reflection (ATR) within infrared spectroscopy can be employed to study the chemical moieties at the surface of the immobilised enzymes. This process can enhance the sensitivity of detection, and the technique is pronounced as surface-enhanced infrared absorption spectroscopy (SEIRAS). A synchrotron FT-IR microscopy combines FT-IR with a bright light source of a synchrotron, which can be useful for high-resolution imaging of chemical species in an immobilised system. The fluorescence spectroscopic technique provides insight into the structural stability of enzymes on the solid support. An investigation through X-ray photoelectron spectroscopy (XPS) corresponds to the elemental analysis and its quantitative estimations in the immobilised system. FT-Raman spectroscopy contributes to the conformational studies of immobilised enzymes. A time-of-flight secondary ion mass spectroscopy (TOF-SIMS) performs highly sensitive surface characterisation, producing crucial information on the orientation and spatial distribution of enzymes within immobilised systems. Förster resonance energy transfer (FRET) is a powerful technique to gather information on quantitative estimation of conformational changes in the protein due to immobilisation.

A different technique, thermal gravimetric analysis, is based on thermal analysis, which sheds light on the thermal stability of the support system in an immobilised system. Circular dichroism (CD) operates on a similar principle to spectroscopy and can detect the conformation and stability of proteins (the integrity of the secondary and tertiary structures). Surface plasmon resonance (SPR) is an optical method to detect molecular interaction in real-time and offers a comprehensive understanding, producing kinetic and thermodynamic data. Microcalorimetry measures thermal activities in terms of heat, which can be used to detect the stability of the enzymes in immobilised systems. Neutron reflection is a surface-sensitive technique which uses neutrons to probe the surface and interfacial layers of immobilised systems. Different probes can be used to get insight into the microenvironment (e.g., pH and temperature) around the immobilised enzyme. Electrochemical and spectrometric methods provide data to monitor reaction rates and study enzyme kinetics.

16.7 BIOREACTOR CONFIGURATIONS

Immobilised enzymes for large-scale applications require specific configurations in the reactors (bioreactors or fermenters). The setups of these reactors can be designed in such a way that immobilised enzymes can be used repeatedly or continuously. The flow of solution containing substrate molecules in the bioreactor can be static or dynamic. The most common configurations of the bioreactors for immobilised enzymes are packed-bed bioreactor (PBR), fluidised bed bioreactor (FBR), continuous stirred tank bioreactor (CSTR), and membrane bioreactor (MBR) (Figure 16.4). These specific configurations have the quality of retaining immobilised enzymes and their activity for multiple uses.

- In the PBR, immobilised enzymes are packed within a solid support, and substrate flows through the packed bed. This is an ideal setup for glucose isomerase (GI) in high-fructose corn syrup (HFCS) production.
- The FBR utilises immobilised enzymes in the suspended form in the fluidised bed, and substrate flows through the bed. This setup is used for the pharmaceutical and food industries.
- The CSTR consists of the suspended solution having immobilised enzymes. Mechanical stirring allows uniform mixing of substrates and immobilised enzymes. This type of setup involves continuous substrate feeding and product removal. It has broad applications in enzyme hydrolysis of biomass (e.g., hydrolysis of lignocellulosic biomass using cellulases and lactose hydrolysis using β-galactosidase) and isomerisation reaction (e.g., GI).
- In MBR, immobilised enzymes rest on or behind a semi-permeable membrane, allowing diffusion of substrates and products through the membrane. It can be operational for enzymatic hydrolysis and wastewater treatment.

The structural, functional, and operational details of these bioreactors can be found in Chapter 8. In addition, the trickle-bed bioreactor (TBR) and monolithic bioreactor are other useful configurations for the applications of immobilised enzymes. A different type of setup is a microreactor, which is a miniaturised reaction system, sometimes keeping immobilised enzymes. Immobilised enzymes can be an important component of a lab-on-a-chip (LOC) device that is used for substance screening.

16.8 ADVANTAGES AND DISADVANTAGES OF ENZYME IMMOBILISATION

Enzyme immobilisation is a topic of immense interest due to its wide applications. An immobilised system offers numerous advantages. The cost contribution of an industrial operation is a key parameter for its success. The low enzyme consumption costs and high stability of enzymes in an immobilised system make the fermentation process highly cost-effective. The immobilised enzymes can be reused multiple times. They can be easily separated from the reaction mixture after the fermentation process using filtration or centrifugation. This also eliminates the complexity of downstream processing. An immobilised enzyme system has a well-balanced overall performance. Immobilisation improves the operational stability of unstable enzymes. Carrier materials, acting as physical barriers, provide a safeguard to enzymes from harsh physical and chemical stresses. Immobilisation protects the enzyme from thermal denaturation and maintains

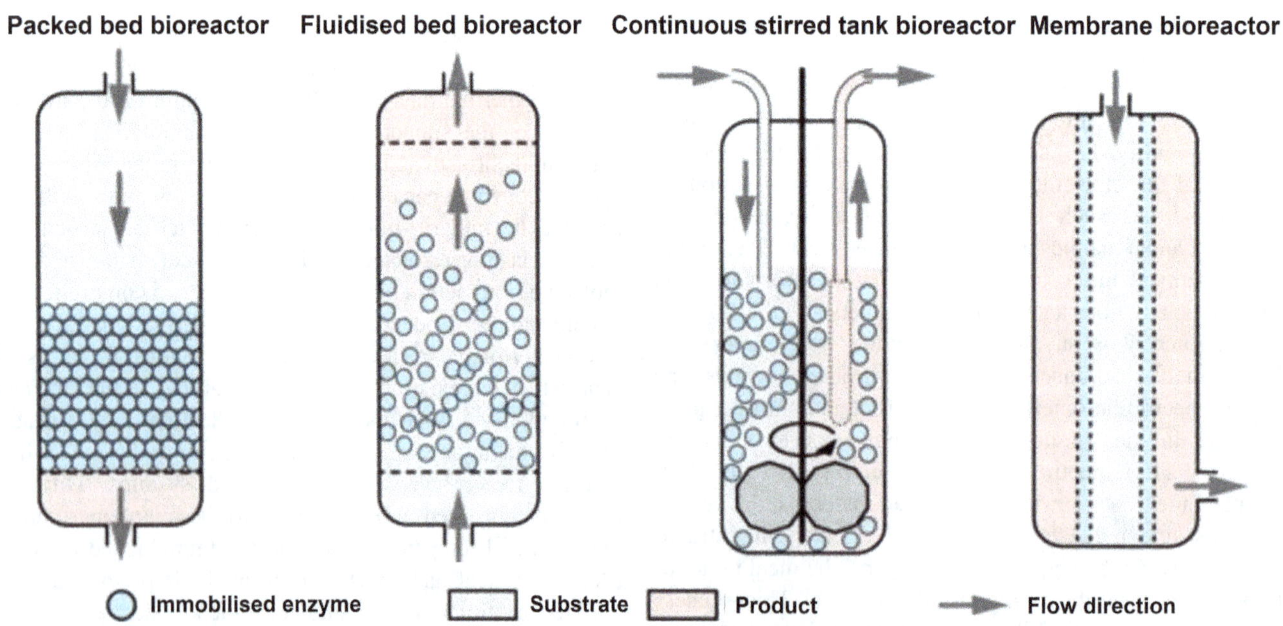

FIGURE 16.4 Different configurations of bioreactors for immobilised enzymes. (Images adapted from Dorau et al. (2021)[4] with permission of The author(s) © 2021, under exclusive licence to Springer-Verlag GmbH Germany.)

its performance at elevated temperatures. The use of biodegradable substances as carrier materials reduces the environmental impact of immobilised enzymes. Some carrier materials can be regenerated after enzyme inactivation, and their binding sites can be restored for the next round of applications, which promotes the sustainability of enzyme-based processes. The careful selection of carrier material in immobilised enzymes for industrial operation also reduces the exposure of enzymes to the skin and body of humans, exhibiting low or no allergenicity.

Though immobilisation is being undertaken for several technique improvements, eco-friendly approaches, and sustainable practices, it still suffers from some limitations. Some of the notable drawbacks are poor enzyme loading, high cost of carrier materials, shearing of enzymes in non-covalent binding, and long-term operational instability. Also, those attached through covalent binding exhibit poor recovery of carrier material after enzyme deactivation. Mass transfer limitation is a common challenge of immobilised enzyme systems, owing to the low rate of diffusion of substrate and product at the active site of enzymes. This process reduces the overall reaction rate and efficiency of enzymes. In recent years, protein engineering has addressed many limitations of immobilised systems and improved their performance for various applications.

SUMMARY

- The interactions between an enzyme and a support system influence the kinetic properties such as the Michaelis–Menten constant (K_m), catalytic constant (K_{cat}), maximum velocity (V_{max}), specificity constant (k_{cat}/K_m), and apparent affinity of the enzyme for the substrate (K_{mapp}).
- A multitude of measures, including immobilisation yield, percentage efficiency, efficiency coefficient, and activity recovery, can be used to study the performance of an immobilised enzyme.
- Some notable factors that affect the performance of an immobilised enzyme include interfacial and nonspecific enzyme-carrier interaction, physical properties of the reaction system (such as types of medium, viscosity of the mixture, and reaction thermodynamics), and physicochemical constraints (such as partition effects and diffusion barrier).
- The performance of an immobilised enzyme can be enhanced by protein engineering (incorporating directed evolution, rational design, or semi-rational design), conformational analysis, and bioreactor design.
- The main advantages of immobilised enzymes are operational stability, reusability, protection, easy downstream processing, etc. Notable disadvantages include loss of activity, poor enzyme loading, mass transfer limitations, costs, complexity, compatibility, etc.

DIFFERENT TYPES OF PRACTICE QUESTIONS

Fill in the blanks

I. The K_m of an immobilised enzyme is usually .... than its soluble form due to less accessibility of the substrate to the active site of the enzyme.
II. Specific activity is defined as enzyme activity per .... of total protein.
III. .... aims for artificial modification in the genetic makeup of the cells mimicking natural evolution.
IV. High .... of carrier materials promotes better binding with enzymes, leading to higher immobilisation efficiency.
V. The .... approach of protein engineering involves creating variations in enzyme properties by introducing mutations at targeted residues.
VI. A .... bioreactor allows continuous mixing of immobilised enzymes with substrates through mechanical stirring.

Fill in blanks: higher, II. milligram (mg), III. Directed evolution, IV. surface area/ hydrophobicity, V. site-directed mutagenesis, VI. stirred tank

Very short answer type of questions

I. Write the formula of the specificity constant of an immobilised enzyme.
II. What is rational design in protein engineering?
III. Define the efficiency coefficient for immobilised enzymes.
IV. Write the names of two spectroscopic techniques used for conformational analysis of immobilised enzymes.

Short answer type of questions

I. What are the different parameters used to estimate the performance of immobilised enzymes?
II. Write the differences between immobilisation yield and protein loading.
III. How does protein engineering contribute to enzyme technology for designing an effective immobilised system?
IV. What are the various advantages and disadvantages of enzyme immobilisation?
V. Write a short note on the different configurations of bioreactors used for immobilised enzymes.

Long answer type of questions

I. Give a detailed account of various factors that affect the performance of immobilised enzymes.
II. What is protein engineering? How does it play an important role in enzyme immobilisation? What are the sequential and integrating approaches of protein engineering in enzyme immobilisation?

NOTES

1 Hartmann M, Kostrov X. Immobilization of enzymes on porous silicas-benefits and challenges. *Chemical Society Reviews*. 2013;42(15):6277–89.
2 Veeramani MS, Shyam KP, Ratchagar NP, et al. Miniaturised silicon biosensors for the detection of triglyceride in blood serum. Analytical Methods. 2014;6(6):1728–35.
3 Bernal C, Rodriguez K, Martínez R. Integrating enzyme immobilization and protein engineering: An alternative path for the development of novel and improved industrial biocatalysts. Biotechnology Advances. 2018;36(5):1470–80.
4 Dorau R, Jensen PR, Solem C. Purified lactases versus whole-cell lactases—the winner takes it all. Applied Microbiology and Biotechnology. 2021;105:4943–55.

SUGGESTED READING

Bernal C, Rodriguez K, Martinez R. Integrating enzyme immobilization and protein engineering: An alternative path for the development of novel and improved industrial biocatalysts. *Biotechnology Advances*. 2018;36(5):1470–80.

Bolivar JM, López-Gallego F. Characterization and evaluation of immobilized enzymes for applications in flow reactors. *Current Opinion in Green and Sustainable Chemistry*. 2020;25:100349.

Chica RA, Doucet N, Pelletier JN. Semi-rational approaches to engineering enzyme activity: combining the benefits of directed evolution and rational design. *Current Opinion in Biotechnology*. 2005;16(4):378–84.

Hanefeld U, Gardossi L, Magner E. Understanding enzyme immobilisation. *Chemical Society Reviews*. 2009;38(2):453–68.

Liu Q, Xun G, Feng Y. The state-of-the-art strategies of protein engineering for enzyme stabilization. *Biotechnology Advances*. 2019;37(4):530–7.

Mohamad NR, Marzuki NH, Buang NA, Huyop F, Wahab RA. An overview of technologies for immobilization of enzymes and surface analysis techniques for immobilized enzymes. *Biotechnology & Biotechnological Equipment*. 2015;29(2):205–20.

Secundo F. Conformational changes of enzymes upon immobilisation. *Chemical Society Reviews*. 2013;42(15):6250–61.

Sheldon RA, van Pelt S. Enzyme immobilisation in biocatalysis: why, what and how. *Chemical Society Reviews*. 2013;42(15):6223–35.

Urban PL, Goodall DM, Bruce NC. Enzymatic microreactors in chemical analysis and kinetic studies. *Biotechnology Advances*. 2006;24(1):42–57.

17 Large-Scale Applications of Immobilised Enzymes

Enzymes that are immobilised have a sizable market. They are useful in a wide range of fields, including food, medicine, cosmetics, biosensors, wastewater treatment, etc. Hundreds of immobilised enzymes are commercially available for various applications (Table 17.1). Some enzymes require protein engineering treatment to become suitable for immobilisation on appropriate materials. Enzymes can be immobilised using diverse carrier materials. The same enzyme immobilised on different carrier materials can exhibit different properties and have different names. Also, the commercial names of immobilised enzymes are unique and may not correspond to their actual names.

17.1 EXAMPLES OF IMMOBILISED ENZYMES

Immobilised enzymes have received wide attention in industrial biotechnology for various commercial applications. Many enzymes in immobilised forms are currently being used in industrial operations. For example, immobilised glucose isomerase (GI), D-psicose epimerase, and β-galactosidase are standard requirements for the synthesis of sweeteners in the food and beverages industry. An immobilised naringinase is used to debitter citrus fruit juices. Domestic products like detergents utilise various immobilised enzymes to remove stains. Savinase® CLEA is an immobilised protease and is used as an ingredient of liquid and powder laundry formulations. This enzyme is recovered from the *Bacillus* spp. and immobilised through a cross-linking technique. In wastewater treatment, immobilised enzymes, particularly laccase, provide sustainable solutions for the removal of organic pollutants. The immobilised enzymes (such as amylase, protease, and lipase) in the textile industry have wide applications for desizing, washing, polishing, and finishing. The use of immobilised enzymes is also very apparent in medical sciences. The immobilised penicillin acylase (PA) is one of the very successful examples of a confined enzyme system in pharmaceutical industries for producing semisynthetic antibiotics. Immobilised transaminase enzymes are critical for manufacturing Sitagliptin, which serves as a medication used in antidiabetes therapy. Immobilised urease can be used in dialysis equipment to remove urea from the blood. Encapsulated enzymes have gained significant attention in the medical field owing to their roles in targeted drug delivery and controlled release. For instance, inborn metabolic deficiency is a genetic disorder due to impaired metabolism, which can be treated by supplementing the missing enzymes in immobilised forms. In different medical conditions, medical devices with immobilised lipase can be used for fat absorption in patients.

Lipase in immobilised forms has gained major attention due to its wide applications. This enzyme has been recovered from various microbial sources and has been stabilised on various support systems. The suitable choices of carrier materials for immobilisation and enzymes depend on the standard requirements of the industrial setup and desired applications. For example, Lipozyme® TL IM is a brand name for lipase from *Thermomyces lanuginosus* that is stabilised on a silicate and an organic binder. This preparation is used for interesterification reaction. Novozym® 40086 is a different immobilised lipase used for stereoselective transesterification and ester hydrolysis. *Rhizomucor miehei* is a key producer of this enzyme, which is confined to a resin carrier. A different comparable enzyme is *Candida antarctica* lipase B (CalB), which has wide applications in the production of biodiesel, medicinal chemicals, food additives, and personal care items. CalB has proved useful in the synthesis of various Active Pharmaceutical Ingredients (APIs), including Odanacatib and Sofosbuvir. Novozym 435 (N435), created by Novozymes, is the brand name for CalB immobilised on a resin called Lewatit VP OC 1600.

Two immobilised enzymes, GI and PA, have received commercial success and are manufactured in large-scale production systems. The following sections provide details of their sources, manufacturing, and immobilisation processes.

17.2 GLUCOSE ISOMERASE (GI)

GI (EC 5.3.1.5) is also known by other names, such as D-xylose ketol isomerase and D-xylose isomerase. This enzyme catalyses isomerisation reactions between D-glucose and D-fructose and between D-xylose and D-xylulose (Figure 17.1). It is also active against a wide range of other sugar substrates, including five and six-carbon ketose/aldose molecules, sugar alcohols and sugar phosphates. In traditional chemistry, this isomerisation reaction is economically not viable, as the synthetic routes require high pH and temperature. Also, the reactions produce colourful side products with objectionable

DOI: 10.1201/9781032651095-21

TABLE 17.1
A partial list of microbial enzymes used in immobilised forms for various applications

Enzymes	Key source organisms	Immobilisation methods	Major Products
Amylase	*Bacillus subtilis* and *Aspergillus niger*	Adsorption and entrapment	High-fructose corn syrup (HFCS) and glucose syrup
Protease	*B. subtilis* and *A. niger*	Covalent bonding, entrapment	Detergent additives, meat tenderisers, and amino acids
Lipase	*Candida antarctica* and *Rhizomucor miehei*	Covalent bonding and entrapment	Flavour compounds, biodiesel, and detergents
Glucose isomerase	*Streptomyces* spp.	Entrapment and cross-linking	HFCS
Lactase	*Kluyveromyces lactis* and *A. niger*	Adsorption and covalent bonding	Lactose-hydrolysed milk products
Penicillin acylase	*Escherichia coli*	Covalent bonding and entrapment	semisynthetic penicillins
Invertase	*Saccharomyces cerevisiae*	Adsorption and entrapment	Inverted sugar syrups
Glucoamylase	*A. niger*	Adsorption and covalent bonding	High-glucose syrups and dextrose
Cellulase	*Trichoderma reesei*	Adsorption and covalent bonding	Cellulosic ethanol and textile processing
Pectinase	*A. niger*	Adsorption and covalent bonding	Fruit juice clarification and wine production

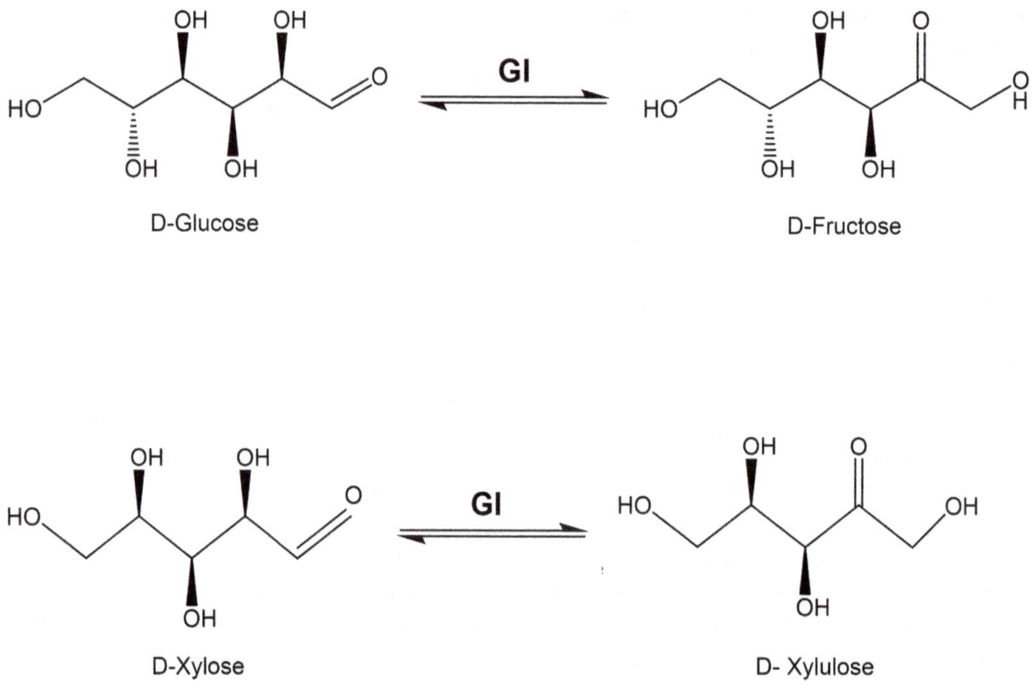

FIGURE 17.1 Isomerisation reactions between D-glucose and D-fructose and between D-xylose and D-xylulose using an enzyme GI (glucose isomerase).

tastes and odours. In contrast, enzyme-catalysed isomerisation is highly selective and economically feasible and does not form many byproducts.

GI can be obtained from a variety of organisms, including bacteria, fungi, plants, and animals. Commercially important sources of GI are bacteria like *Actinoplanes* spp., *Arthrobacter* spp., *Bacillus coagulans*, *Flavobacterium* spp., *Lactobacillus* spp., and *Streptomyces* spp. In fungi, GI is predominantly reported in *Aspergillus oryzae.* The primary site of enzyme production in bacteria is the intracellular space, while some strains secrete this enzyme extracellularly. GI isolated from wild-type strains prefer xylose as a substrate. However, the main commercial interest of GI lies in the conversion of glucose into fructose for the production of high-fructose corn syrup (HFCS). Therefore, protein engineering of enzymes has played a major role in increasing the selectivity of enzymes for the transformation of glucose.

HFCS is a sweetener, a mixture of glucose and fructose, which is an excellent substitute for granulated sugar. HFCS is widely used in the production of soft beverages, processed meals, cereals, and baked goods. The sweetening power of glucose is only 70-75% of the sweetening power of sucrose, whereas fructose, with double the sweetening power of sucrose, is considered the sweetest natural carbohydrate. Based on the sweetening capability, the overall production cost of HFCS through the fermentation process can be lower than that of sucrose. This is particularly relevant for temperate regions, which have an abundant supply of corn. A different key benefit of HFCS is that it does not produce any undesirable crystallisation in food and beverages.

Genetic manipulations in the GI-producing strains have played a key role in their effective application in HFCS production. The annual global production of HFCS using GI technology is estimated to have reached more than 10 megatonnes. This amount of HFCS is manufactured using more than 500 tonnes of immobilised GI. The main interest of GI for industrial HFCS production is due to its specificity, the ambient conditions for reactions, and effective immobilisation. The fructose concentrations in HFCS are usually maintained at 42% and 55%.

The hydrolytic products of a non-sweet substrate like cornstarch are the primary feedstocks for HFCS production. Other suitable raw materials can be cassava starch, raw cereals, bagasse, tapioca, inulin, and rice. The conversion of cornstarch into HFCS is a multi-enzymatic reaction. The first step is wet-milling, which involves separating maize kernels into their constituent parts, such as starch, protein, germ, and fibre. This is followed by liquefaction, a thinning reaction, which requires heat and enzyme treatments. In this step, thermostable bacterial α-amylase and fungal glucoamylase partially hydrolyse cornstarch and reduce its viscosity. This can be followed by saccharification, which involves the addition of amyloglucosidase and pullulanase to aid in further hydrolysis. Amyloglucosidase acts on the glycosidic linkages, while pullulanase targets α-1,6 glycosidic linkages in starch, effectively "debranching" into small oligosaccharides. These reactions also produce glucose or dextrose, which serves as a main substrate for an enzymatic reaction by GI (Figure 17.2). Industrial HFCS production using GI is typically carried out in a continuous fixed-bed bioreactor (FBR). This requires the preparation of a fixed bed of immobilised GI, and glucose solution is passed through a bioreactor for its conversion into fructose molecules.

GI can also be useful in the biofuel industry, as it can use xylose as the key substrate. The raw material for biofuel production is lignocellulosic biomass, which is abundant with xylose (5-35%). This high xylose composition inhibits its full utilisation for bioethanol production. Under these circumstances, the application of GI facilitates the transformation of xylose into a fermentative substrate, D-xylulose. This isomerised product can further be metabolised by yeast or bacteria to produce bioethanol. Additionally, the typical glucose-based reactions of the GI can also be utilised for bioethanol production. The isomerisation of glucose into a readily fermentable substance, fructose, improves the biofuel yields compared to the direct fermentation of glucose.

17.2.1 Production

The economically viable production of GI involves submerged aerated fermentation. The nutritional composition

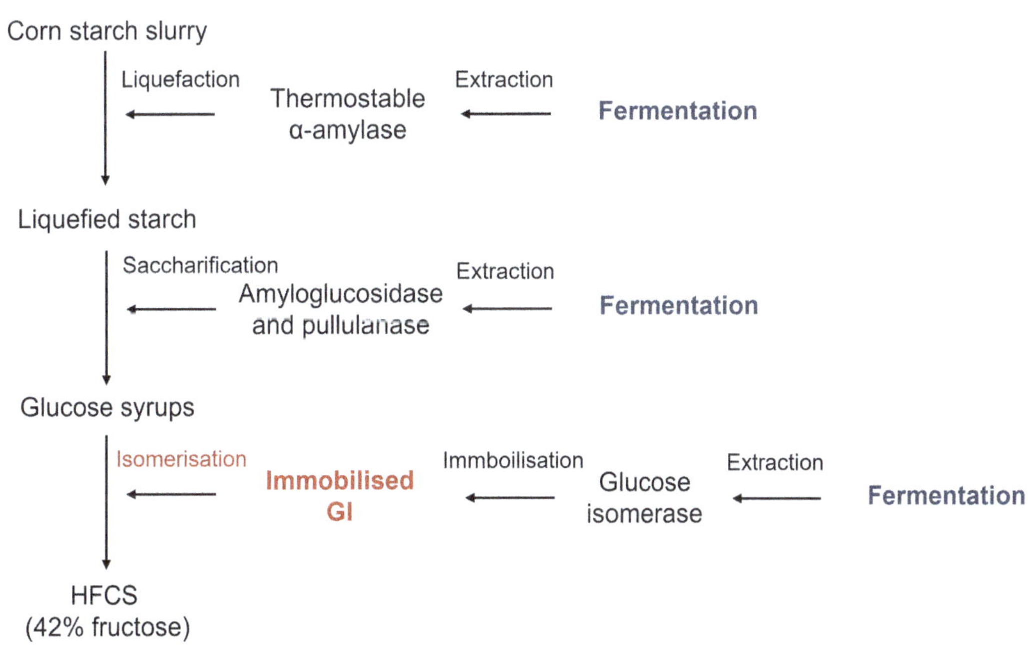

FIGURE 17.2 Pathway for the synthesis of HFCS (high-fructose corn syrup) using an industrial byproduct, cornstarch slurry, as raw material. This involves a multistep reaction involving various enzymes, including an immobilised GI (glucose isomerase), produced through different fermentation processes.

and physical requirements for enzyme production may vary depending on source organisms. GI in many bacteria is inducible and requires substrates like xylose/glucose for its production. The wild-type strains of *Actinoplanes missouriensis* have exceptional potential for constitutive production of GI. In a different approach, mutagenesis and genetic recombination in GI-producing microbes have been employed for the constitutive production of the enzyme. Some strains exhibit properties to use starch, glucose, glycerol, or sorbitol as inducers, replacing expensive xylose. Cornstarch is a suitable raw material for producing glucose molecules. Corn-steep liquor and soy flour are cheap nitrogen sources in the medium. Yeast extract, peptones, and ammonium salt can also be suitable alternative nitrogen sources. GI is cofactor-dependent and requires different divalent cations in the feed. For example, magnesium ion (Mg^{2+}) could be essential for the activation of the enzyme. Cobalt (Co^{2+}) is required to maintain the quaternary structure of the enzyme. The microbial source of the enzyme additionally determines which other metal ions could be required.

The reaction catalysed by GI is a reversible process, and the conversion of glucose into fructose is maintained under specific environmental conditions. Also, GI, which is biased toward the conversion of glucose into fructose, is in demand. The optimal condition for GI production in a fermenter lies in the range of pH 7.0-9.0 and 30-60 °C. Thermophilic bacteria are suitable choices for GI production due to the thermal stability of the enzyme for industrial applications.

After the fermentation process, the production of cell-free GI initially requires a cell disruption method to diffuse the intracellular materials into the enzyme extract. Cationic detergents or mechanical disruption (sonic oscillation or abrasive grinding) can be suitably used to release GI from the cells. Both methods are equally effective in releasing the enzymes; however, the enzyme activity is relatively higher when cationic detergents are used. A bacterium, *Escherichia coli*, has been engineered with thermotolerant GI, which could be used without cell disruption (due to induced extracellular secretion) and has a high transformation efficiency for fructose. Filtration or centrifugation can be suitably used to extract the enzyme from the cell extract medium. Further, solvent extraction, precipitation, or chromatographic separations can be employed to remove insoluble impurities. However, the intense purification of enzymes to a homogeneous state is not a strict requirement in industrial setups. The extracted enzyme can be immobilised on suitable carriers.

17.2.2 Immobilisation

Given that GI is water soluble and that the reaction takes place in an aqueous environment, the enzyme can be immobilised on a matrix that is insoluble in water for long-term use. The attachment of inert material on the surface of GI improves stability and thermal resistance (Figure 17.3). The first industrial application of GI was recognised in 1967 by the American company Clinton Com Processing Co., and the knowledge of its immobilisation was available on commercial rights in 1974. The increasing demand for HFCS production in the following years led to the development of new methods of immobilising the enzyme to meet the different requirements. GI technology, in its present state, has the largest market share in the food industry.

Both cell-free enzymes and complete cells that produce GI can be immobilised. Since GI is mainly an intracellular enzyme in many source organisms, immobilisation of the entire cell is also a popular choice to avoid the expensive procedure of enzyme purification. After fermentation, cells can be instantly immobilised using salts from the fermentation broth. Physical and chemical entrapment methods are other methods for the immobilisation of GI.

In the current industrial setup, cell-free GI is preferred over using entire cells. The immobilised cell-free enzymes help maintain effective flow characteristics in continuous operation. Immobilisation of GI can be achieved by adsorption, covalent binding, entrapment, and cross-linking using a variety of substances. Physical adsorption on ion exchange resins creates a stable immobilised enzyme system. A positively charged resin, diethylaminoethanol-cellulose (DEAE-C), is a suitable adsorbent for GI, where the negatively charged groups of the enzyme bind to the positively charged diethylaminoethyl group of the resin through electrostatic interactions. However, this type of interaction at a high temperature can result in permanent loss of enzyme activity. In a different method, natural polymers such as chitosan, bentonite, kaolin, or diatomite can be used as carrier materials to establish covalent bonding with enzymes. These materials can also be bound with glutaraldehyde for crosslinking. Natural polymers can be chemically modified to increase the number of functional groups at their surface, increasing the attachment points of enzymes. The synthetic, water-soluble polymer polyethylenimine (PEI) is another suitable choice for the assembly of enzymatic residues involving glutaraldehyde. Alumina, a metal oxide having properties like high surface area, mechanical stability, and biocompatibility, has shown great potential for GI immobilisation. In a distinct preparation, immobilised GI and immobilised glucoamylase can be mixed in the fermenter to manufacture fructose directly from liquefied starch. GI can also be co-immobilised with glucoamylase or cellulase using porous silica gel or polystyrene beads for one-step conversion of raw materials like starch or cellulose into glucose and fructose.

GI attached with various carrier materials at the industrial platform is available under different brand names. Two prominent examples are Sweetzyme IT and Gensweet™. Sweetzyme IT is a granulated immobilised GI of Novozymes produced by the bacterium *Streptomyces murinus*. One kg of Sweetzyme IT can manufacture more than 18 tonnes of dried syrup. Gensweet™ is a registered trademark of DuPont Industrial Biosciences (Genencor) and is a homogenate of glutaraldehyde crosslinked GI that contains natural polymers as inorganic carriers. The source organism for GI in Gensweet™ is *Streptomyces rubiginosus*. This immobilised form claims productivity of about 30 tonnes per kg.

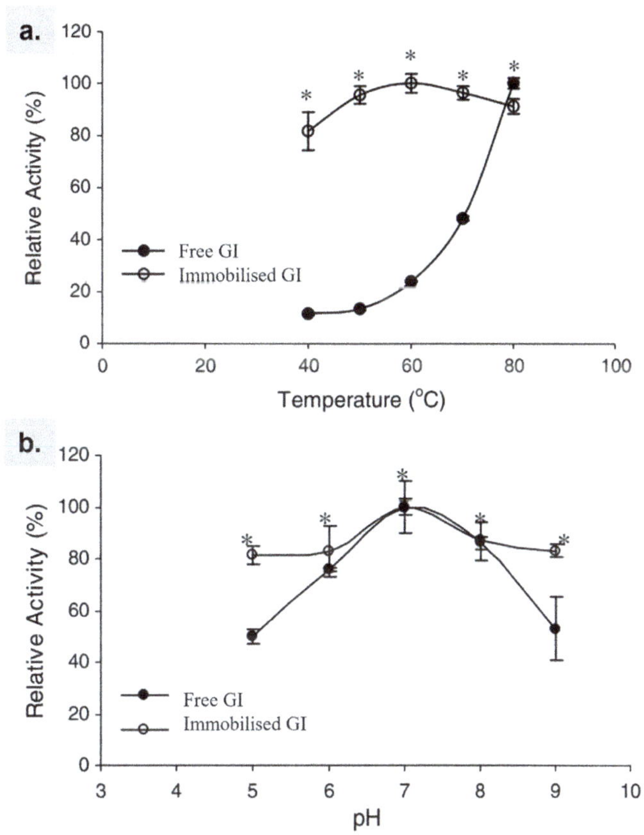

FIGURE 17.3 Graphs showing relative activity of free and immobilised GI (glucose isomerase) under different temperatures (a) and pH (b). (* indicates significant ($p < 0.001$) differences resulting from the Duncan tests between free and immobilised GI). (Images adapted from Tükel and Alagöz (2008)[1] with permission of © 2008 Elsevier Ltd.)

17.3 PENICILLIN ACYLASES (PAs)

β-lactam antibiotics are modern medicines for combating bacterial diseases. They are commonly used to treat bacterial infections by disrupting the synthesis of bacterial cell walls. The β-lactam ring in β-lactam antibiotics is the main structure for antimicrobial activity. Certain side chains on the β-lactam ring affect their pharmacological properties, including their spectrum of activity, stability, and effectiveness. However, due to the excessive and improper application of natural antibiotics, many bacteria have developed resistance to their effects. Researchers have found alternative solutions in synthetic and semisynthetic antibiotics that can combat antibiotic resistance. The synthesis of synthetic antibiotics involves a complete chemical pathway. In contrast, semisynthetic antibiotics rely on enzyme-based processes followed by chemical synthesis. The traditional chemical methods for the production of β-lactam antibiotics exist, but biological routes for their production are much more convenient and economical. In biological methods, natural penicillins are transformed into β-lactam antibiotic intermediates using microbial enzymes. These microbial products can act as precursors for the chemical synthesis of a new class of semisynthetic β-lactam antibiotics.

PAs (EC 3.5.1.11) are industrial enzymes that play a crucial role in the industrial production of semisynthetic β-lactam antibiotics. Figure 17.4 depicts key steps involved in the synthesis of semisynthetic β-lactam antibiotics. Both substrate (i.e., penicillin G) and enzyme (PA) are the products of different fermentation processes. In this reaction, a special class of PA, penicillin G acylase (PGA) is used in the immobilised form. This reaction produces a key intermediate, 6-APA, for the chemical synthesis of semisynthetic β-lactam antibiotics.

PAs belong to the subcategory of aminohydrolase of hydrolytic enzymes. PAs exist in many forms, such as PGA, penicillin V acylase (PVA), and penicillin K acylase (PKA) (Figure 17.5). They all exhibit a high level of substrate specificity. The tremendous and continuous interest in PAs for industrial applications is PGA. The fermentation products like penicillin G, penicillin V, and other natural derivatives derived from *Penicillium* spp. are the primary substrates for PAs (Figure 17.5). PAs help modify these antibiotics by selective N-deacylation of the side chains of a β-lactam ring, resulting in the formation of 6-aminopenicillanic acid (6-APA).

PAs are widely reported in bacteria, yeasts, and filamentous fungi, existing in more than forty different microorganisms. The first such enzyme was discovered as PGA in *Penicillium chrysogenum* and *Aspergillus oryzae* by Japanese investigators Sakaguchi and Murao (1950). The most robust PAs are derived from *E. coli* and *Bacillus megaterium*. Other important sources of PAs are *Streptomyces lavendulae*, *Alcaligenes*

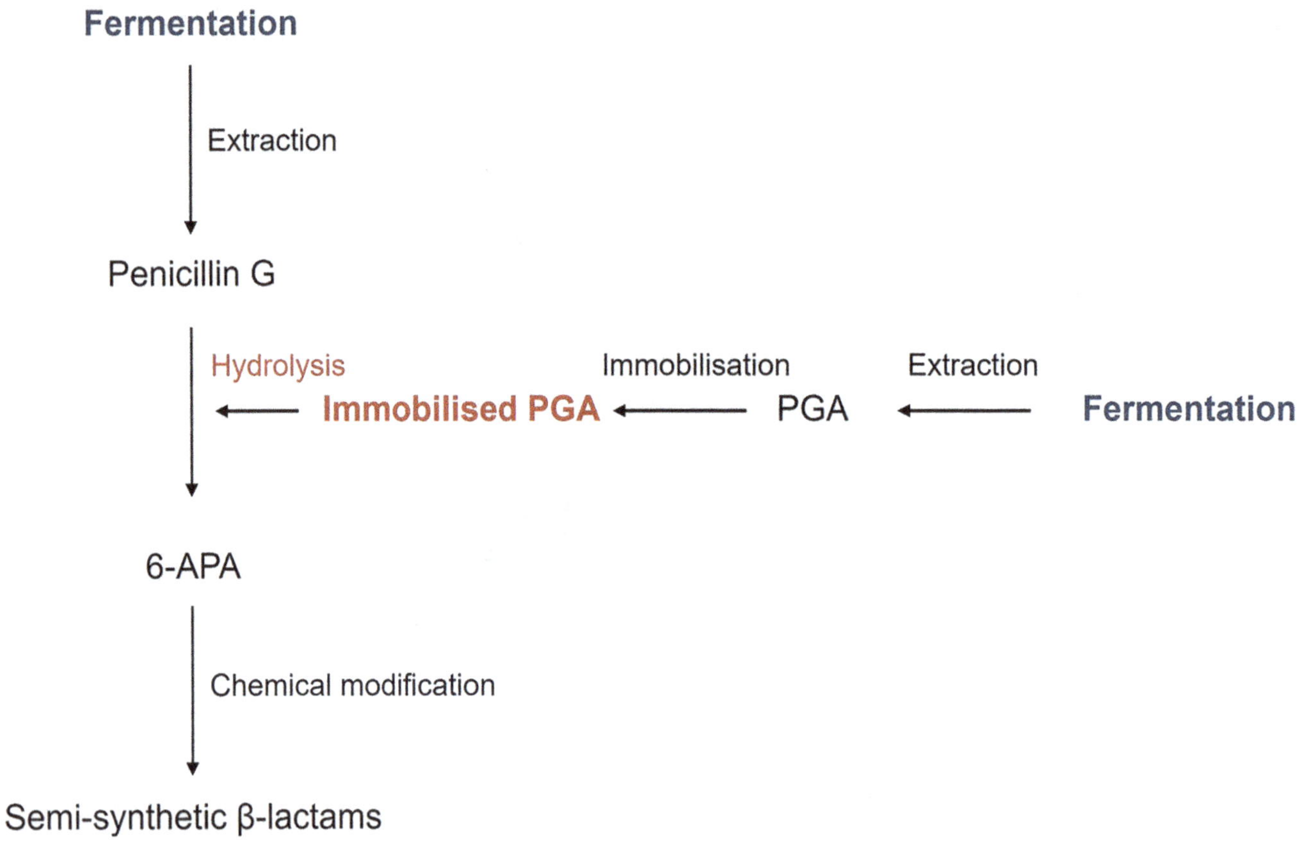

FIGURE 17.4 Pathway for the synthesis of semisynthetic β-lactams using fermentation products, penicillin G and an immobilised PGA (penicillin G acylase). (6-APA: 6-aminopenicillanic acid).

faecalis, and *Proteus rutgeri*. The enzymes can be found in extracellular, intracellular, or periplasmic regions, depending on the microbial species.

The final product of PAs, 6-APA, serves as the reactant for the chemical synthesis of the new generation of β-lactam antibiotics. One common method for the production of semisynthetic forms is the condensation of the side chains of the core structure with various D-(-)-amino acid derivatives. This condensation reaction introduces new functional groups or modifies existing ones, ultimately producing novel antibiotics with improved properties, such as an enhanced spectrum of activity or increased stability. Amoxicillin, ampicillin, and cloxacillin are representatives of semisynthetic penicillins produced from core β-lactam penicillins via a 6-APA intermediate. They are new-generation drugs and far more effective against bacterial infection than traditional, unmodified, natural antibiotics that are prone to drug resistance.

The new generation β-lactam antibiotics have broad-spectrum activity against both Gram-positive and Gram-negative types of bacteria. In contrast, natural penicillins exhibit activity mainly against Gram-positive bacteria. Around 65% of the global antibiotic market comprises β-lactam antibiotics. Their prevalence in the market reflects the widespread use of PAs. The annual production of PGA is as high as 30 million tonnes. The application of other related enzymes (e.g., PVA) for the production of APA is also in practice, but it has comparatively low stability, low activity, and high production cost. The expansion of the penicillin G ring can produce cephalosporin and its derivatives, which are less susceptible to β-lactamases (an enzyme that provides multi-resistance to β-lactam antibiotics). Cephalosporins are a different class of β-lactam antibiotics that are obtained from the fungus *Acremonium.* Natural cephalosporins can be converted into 7-aminocephalosporanic acid (7-ACA) and 7-aminodesacetoxycephalosporanic acid (7-ADCA) by cephalosporin acylase. These compounds can also act as precursors for the synthesis of various chemical derivatives.

17.3.1 Production

The production of PAs relies on wild or mutant-type strains of microorganisms. Manipulation of novel PA-encoding traits and site-specific mutagenesis are suitable strategies to enhance enzyme production. Genetically modified *E. coli* and *B. megaterium* are the most commonly used microbes for PA-encoding traits. Large-scale production of PA is performed in submerged fermentation (SmF) (in fed-batch, stirred fermenters). Production can be carried out at a scale of >5,000 l under administered conditions. Carbohydrates like

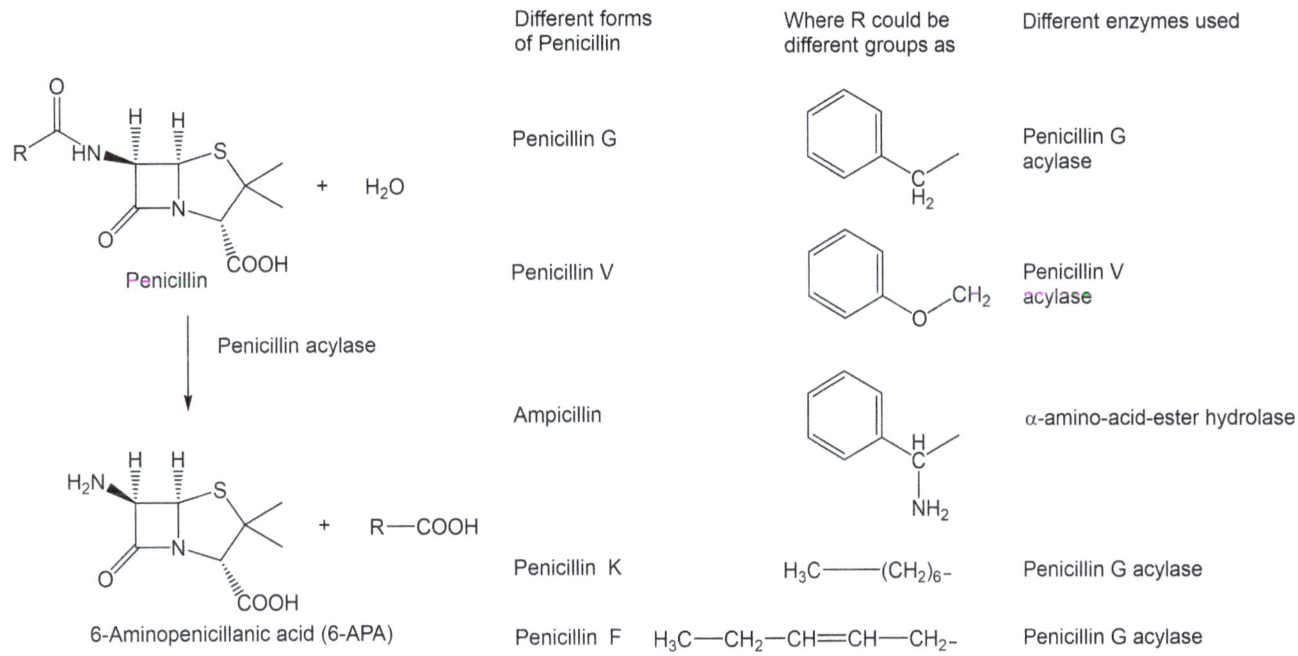

FIGURE 17.5 Synthesis of 6-APA (6-aminopenicillanic acid) using penicillin in the presence of enzyme penicillin acylase (PA). PAs exist in different forms and take part in the hydrolysis of different types of penicillins.

corn-steep liquor, cheese whey, and cellulose gum are cost-effective feedstocks in fermentation media for large-scale production. The nutrient-controlled metabolism and genotypic variation in the source organisms have a great impact on the production of PAs. A different fermentation process can be carried out for the production of natural penicillins (discussed in detail in Chapter 13), which serves as the main reactant for the enzymatic reaction.

The optimal incubation time for the production of PAs in many strains of *E. coli* is typically around 6-10 h under mild conditions. This duration varies based on the microbial species involved in the fermentation process. The recombinant *E. coli* expresses the *PGA* gene in the periplasmic region, which requires permeabilisation in order to release PGA into the surrounding fluid. The extraction and purification of enzymes involve physical and/or chemical methods. Osmotic shock, combined with the use of ethylenediaminetetraacetic acid (EDTA), is an effective method for extracting high concentrations of enzyme units from the cells. A different method for cell permeabilisation involves polar organic solvents along with aqueous solutions of ionic and non-ionic detergents. The cells can also be homogenised by breaking down the cell structures, which ensures thorough mixing of the contents. After cell permeabilisation, the resulting cellular debris and extraneous materials are removed using centrifugation or flocculation methods. The next step is recovery of the enzyme via precipitation with ammonium sulphate ($(NH_4)_2SO_4$). The next step is desalting, which enables the removal of excess salts and other small molecules from the sample. The partially purified enzymes can be directly immobilised onto carrier materials.

17.3.2 Immobilisation

PAs in immobilised states have a huge demand in the pharmaceutical sectors for the production of β-lactam antibiotic derivatives. Immobilising PAs could lead to more efficient and cost-effective processes for the production of 6-APA than free enzymes. They also have a role in the resolution of racemic mixtures and peptide synthesis. The consumption of immobilised PAs in the global commercial immobilised enzyme market is second only to GI. The binding of PAs to carriers can be achieved via adsorption, covalent coupling, entrapment, cross-linking, and encapsulation. Materials like Eupergit C, Sepabeads, glyoxal agarose, alumina beads, nylon fibres, silica support, and xerogel are the prominent carriers of PAs. However, the immobilisation of PAs can have an impact on their conformation, leading to damage to catalytic activities.

PAs immobilised through physical adsorption are prone to falling off during vigorous operation due to weak interaction between PAs and carriers. Multiple attachments through covalent coupling are a popular choice for their immobilisation, which achieves high stability even in hostile media. Eupergit C is a brand name for a type of epoxy-based porous acrylic polymer that offers strong covalent binding and high operational stability. Immobilisation of PA on this carrier retains approximately 60% of its activity even after 800 cycles. However, the diffusion limitation is a major drawback of using Eupergit C. Some studies have demonstrated that PAs immobilised on epoxy-activated Sepabeads can be more effective and stable than those on Eupergit C. Sepabeads are ideal carriers for PGA and can be attached primarily by adsorption; however, they can be chemically modified for efficient covalent bonding. Sepabeads-PGA derivatives exhibit resistance to inactivation by both high temperatures and high concentrations of organic solvents. In different situations, gelatin, along with polymers like alginate, amine, and chitosan, creates a biocompatible hydrogel matrix, which possesses free amino groups for binding with free enzymes.

A different immobilisation preparation of PAs is a carrier-free crosslinking method. This includes cross-linked enzyme aggregate-PA (CLEA-PA) and cross-linked enzyme crystal-PA (CLEC-PA). CLEA is more robust than CLEC for immobilising PAs. The preparation using CLEA has advantages in terms of its ability to perform kinetically controlled synthesis of 6-APA in both aqueous and organic solvents. Additionally, CLEA can utilise crude fermentation broth without the need for extensive purification steps. A different type of material, ethylene glycol dimethacrylate (EGDMA), can be used as a crosslinking agent to create a stable matrix for its immobilisation.

Under a different approach, immobilising whole cells, such as *E. coli*, that express these enzymes is also beneficial for certain processes. One such method involves using a polymatrix gel as a support material (e.g., alginate, agarose, and polyacrylamide). The immobilised cells can be permeabilised using appropriate agents to facilitate substrate and product transport because the enzyme is intracellular.

Immobilised enzymes require stability in different organic solvents because the synthesis of 6-APA using PAs involves a water-organic solvent mixture. The organic solvent can be butyl acetate or methyl isobutyl ketone. These solvents can have varying effects on the immobilisation efficiency and have an impact on enzymatic activity and, consequently, product yield. Some solvents destabilise the immobilised enzyme, while others can have protective effects. Therefore, the selection of carrier materials that are compatible with the solvent is a strict requirement for the stability and activity of immobilised enzymes.

Industrial production of 6-APA generally requires high temperatures for enhanced reaction rates, improved substrate solubility, and a lower viscosity ratio. Therefore, the thermostability of the system is an important parameter, as the optimum temperature for the catalysis of PAs is near 30-35 °C. Immobilisation of PAs increases the capacity for thermotolerance on carrier-supported enzymes (Figure 17.6). Immobilised enzyme systems with PA attached to a magnetic nanocomposite have been reported to exhibit great pH and thermal stability, retaining high activity after multiple usages.

Companies like Amicogen, GistBrocades, Recordati, Rhm, and Roche are commercial producers of immobilised PAs. In some cases, the immobilised enzyme can be co-immobilised with other enzymes on substances like polyethyleneimine, aldehyde-dextran, or sulphate-dextran. This can increase the stability of the multienzyme system in an organic solvent environment, enhancing the overall performance and longevity of the enzyme.

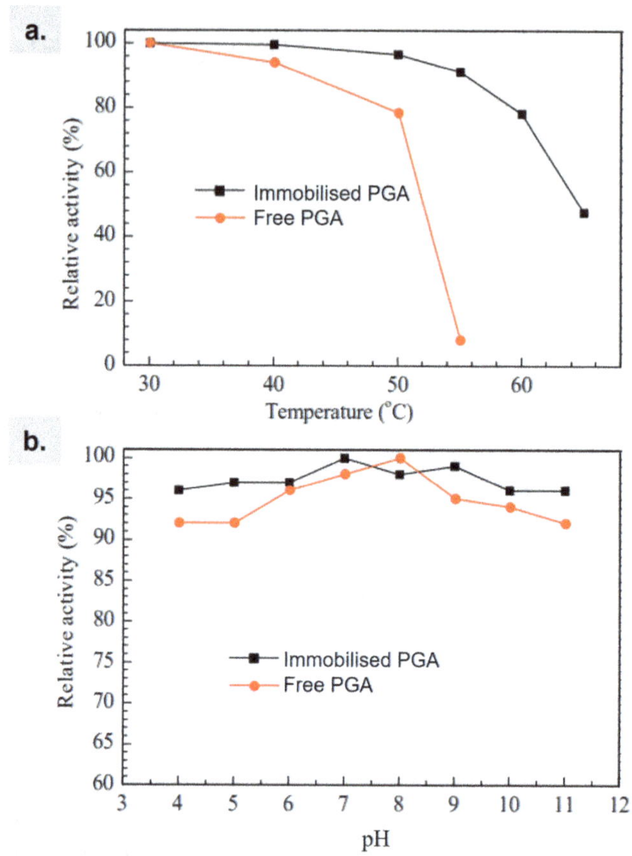

FIGURE 17.6 Graphs showing relative activity of free and immobilised PGA (penicillin G acylase) at different temperatures (a) and pH values (b). (Images adapted from Chen et al. (2018)[2] with permission of © 2018 Dalian Institute of Chemical Physics, the Chinese Academy of Sciences. Elsevier BV.)

SUMMARY

- Immobilised enzymes have a sizable market in food, medicine, cosmetics, biosensors, wastewater treatment, etc. Glucose isomerase (GI) and penicillin acylases (PAs) in their immobilised forms are the two most commonly utilised enzymes in industries.
- GI is an isomerisation enzyme that can be immobilised on diverse carrier materials for the commercial production of a sweetener, high-fructose corn syrup (HFCS) (a mixture of glucose and fructose), which is an excellent substitute for granulated sugar.
- GI is produced using submerged aerated fermentation. The harvest of microbial cells undergoes cell disruption to release intracellular GI. This content is partially purified, and crude extract can be immobilised through adsorption and crosslinking for various applications.
- Immobilised PAs are pharmaceutically important enzymes. This can be used to produce key intermediates of semisynthetic β-lactam antibiotics using natural penicillins.
- Both PAs and penicillins are the natural products of distinct fermentation processes. PAs are mainly

produced in fed-batch stirred tank fermenters. The enzyme is released from the cells using suitable cell disruption methods. After sufficient purification, it can be immobilised through adsorption, covalent bonding, and crosslinking.

DIFFERENT TYPES OF PRACTICE QUESTIONS

Fill in the blanks

I. Glucose isomerase performs isomerisation reactions between D-xylose and ....
II. Glucose isomerase is generally cofactor-dependent and requires divalent cations .... in the feed.
III. The main industrial application of glucose isomerase is in the production of ....
IV. Penicillin acylase is an enzyme used in the industrial production of .... antibiotics.
V. A different class of β-lactam antibiotics is ...., obtained from the fungus *Acremonium*.

SOLUTION: I. D-xylulose, II. magnesium (Mg2+) and cobalt (Co2+), III. high-fructose corn syrup (HFCS), IV. semisynthetic β- lactam, V. cephalosporins

Very short answer type of questions

I. Write the name of any immobilised lipase enzyme system.
II. Write the name of a bacterium which produces glucose isomerase.
III. What is a carrier-free crosslinking method for immobilising penicillin acylase?
IV. Give the names of any two carrier materials used for immobilising penicillin acylase.
V. Name any two semisynthetic penicillins.

Short answer type of questions

I. What are the different immobilised enzymes derived from microbial sources?
II. Explain the various applications of microbial glucose isomerase.
III. Elaborate on the different raw materials and microbial enzymes used for the production of HFCS. Provide a layout for the production of HFCS using a raw material.
IV. What is 6-APA? How is it produced using microbial routes? What are the main applications of it?

Long answer type of questions

I. What is glucose isomerase and its applications? Discuss the processes of glucose isomerase immobilisation.
II. Provide an overview of penicillin acylase, including its function and industrial applications. Discuss the techniques and advantages of penicillin acylase immobilisation.

NOTES

1 Tükel SS, Alagöz D. Catalytic efficiency of immobilized glucose isomerase in isomerization of glucose to fructose. *Food Chemistry*. 2008;111(3):658–62.
2 Chen X, Yang L, Zhan W, et al. Immobilization of penicillin G acylase on paramagnetic polymer microspheres with epoxy groups. Chinese Journal of Catalysis. 2018;39(1):47–53.

SUGGESTED READINGS

Arroyo M, De la Mata I, Acebal C, Pilar Castillon M. Biotechnological applications of penicillin acylases: state-of-the-art. *Applied Microbiology and Biotechnology*. 2003;60:507–14.

Basso A, Serban S. Industrial applications of immobilized enzymes—A review. *Molecular Catalysis*. 2019;479:110607.

Bhosale SH, Rao MB, Deshpande VV. Molecular and industrial aspects of glucose isomerase. *Microbiological Reviews*. 1996;60(2):280–300.

Chandel AK, Rao LV, Narasu ML, Singh OV. The realm of penicillin G acylase in β-lactam antibiotics. *Enzyme and Microbial Technology*. 2008;42(3):199–207.

Franssen MC, Steunenberg P, Scott EL, Zuilhof H, Sanders JP. Immobilised enzymes in biorenewables production. *Chemical Society Reviews*. 2013;42(15):6491–533.

Hartley BS, Hanlon N, Jackson RJ, Rangarajan M. Glucose isomerase: insights into protein engineering for increased thermostability. *Biochimica et Biophysica Acta (BBA)-Protein Structure and Molecular Enzymology*. 2000;1543(2):294–335.

Hassan ME. Production, immobilization and industrial uses of penicillin G acylase. *International Journal of Current Research and Review*. 2016;8(15):11.

Sambyal K, Singh RV. Exploitation of *E. coli* for the production of penicillin G amidase: a tool for the synthesis of semisynthetic β-lactam antibiotics. *Journal of Genetic Engineering and Biotechnology*. 2021;19:1.

Suekane M. Immobilization of glucose isomerase. *Zeitschrift für allgemeine Mikrobiologie*. 1982;22(8):565–76.

Unit V

Microorganisms in Environments and Their Analysis

18 Study of Airborne Microorganisms

18.1 AEROMICROBIOLOGY

The atmosphere is one of the harshest physical environments on Earth. The inadequate amount of nutrients and extreme situations, such as low temperatures and radiation, hamper the growth and reproduction of biological organisms in the atmosphere. Despite various limitations, tiny life forms like microorganisms, spores, pollen, and insects exist in the atmosphere. The atmosphere is an important route for their transport and dissemination over short and long distances. The study of life in the air is called **aerobiology**. F. C. Meier coined this term in the 1930s. A subdiscipline of aerobiology, which involves the study of suspended microorganisms in the air, constitutes **aeromicrobiology**. The presence of microbial life in the air was first suggested nearly 195 years ago, when Charles Darwin collected dust samples during his voyage on the *Beagle*, which were later found to contain microorganisms. Those detected life forms were mould spores at the Cape Verde Islands in the 1830s. However, the first experimental study on microorganisms in air was conducted by **Louis Pasteur**. His main objective was to disprove the theory of spontaneous generation, which led to the foundation of microbiology in the air. In a rudimentary experimental setup, he confirmed the occurrence and dispersion of microorganisms in the air.

A heterogeneous mixture of suspended airborne particles comprises biological and non-biological components, collectively known as **particulate matter (PM)**. As per the United States Environmental Protection Agency (US EPA), this term is used for a mixture of solid particles and liquid droplets found in the air. The two size categories of PM are PM_{10} and $PM_{2.5}$. They serve as prominent parameters to measure ambient air pollution. PM_{10} are particles with diameters less than or equal to 10 μm, having a coarse fraction, while $PM_{2.5}$ encompass particles with diameters less than or equal to 2.5 μm, having a fine fraction. PM is linked with health concerns for humans, where prolonged exposure to PM (having a small size) causes harm to the body.

A mixture of solid or liquid particles suspended in the air is called an **aerosol**. This term is used in a broader context, including various substances. Aerosols tend to collide with each other and form large particles. The process of transforming a large physical body into minuscule light particles and their release into the atmosphere is called **aerosolisation**. PM and aerosol are more or less similar terms. However, the former term mainly focuses on particles of a specific size range, while the latter term is used for the entire system of particles suspended in air.

18.2 BIOAEROSOLS

The suspended biological particles of the atmosphere are specifically called **bioaerosols**. They belong to a subcategory of aerosols, encompassing biological materials such as microorganisms, pollens, and animal and plant debris. Both dead and alive biological cells can be components of bioaerosols. The size of bioaerosols can range between 0.002-100 μm in diameter. The smallest size of bioaerosols is derived from viruses, while the largest size of airborne particles can be pollen grains. Based on size distribution, bioaerosols can be divided into three categories: ultrafine (nuclei mode) (<0.1 μm), fine (accumulation mode) (0.1 to 2 μm), and coarse (>2 μm) (Figure 18.1). The residence time of bioparticles in the air depends on their sizes. Light particles float in the air for an extended period compared to heavy particles. The density of bioaerosols is also a significant factor in their vertical dispersion in the atmosphere. Bioaerosols can influence the hydrological cycle and atmospheric chemistry owing to their support in cloud formation, ice crystal formation, and precipitation.

Bioaerosols can be made up of two different components, including biological and nonbiological components. The non-biological components include dust and droplets, which act as a 'raft' for the biological components (Figure 18.2). The proportion of biological components in the bioaerosol is generally around 50%. However, it could exceed 90%. During aerosolisation, the surface hydrophobicity of the microbial components plays a key role in their adhesion with non-biological particles. Bioaerosols themselves can be produced by the aggregation and adhesion of cells.

18.3 AIRBORNE MICROORGANISMS

Microorganisms occupy a fairly high proportion of the bioaerosol. These living forms in the air are called airborne microorganisms. The atmosphere is a temporary residence of microorganisms because the gaseous envelope of the atmosphere gathers limited nutrients and water to support the metabolic activities of microorganisms. Some bacteria can multiply using organic matter and water vapour from the

DOI: 10.1201/9781032651095-23

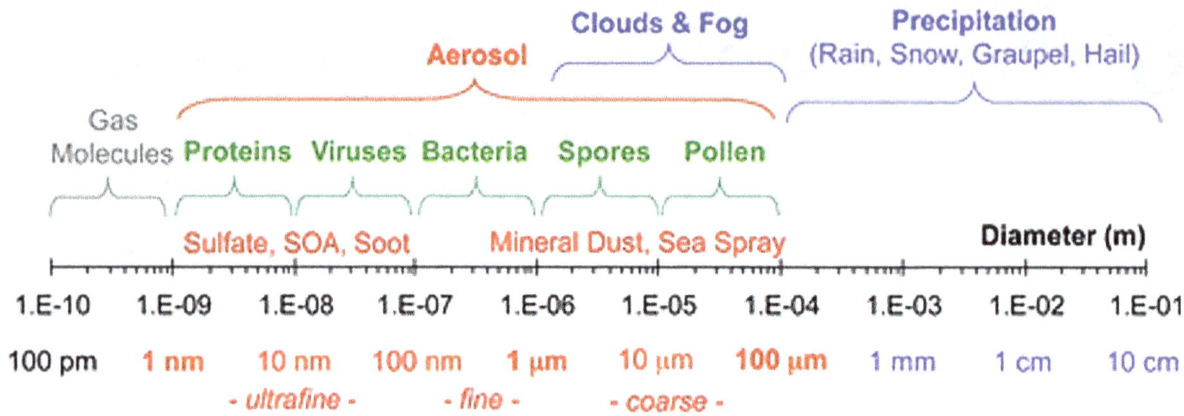

FIGURE 18.1 Size range of different particles in the atmosphere. (SOA: Secondary organic aerosol) (Image adapted from Pöschl and Shiraiwa, 2015, © 2015 American Chemical Society.[1])

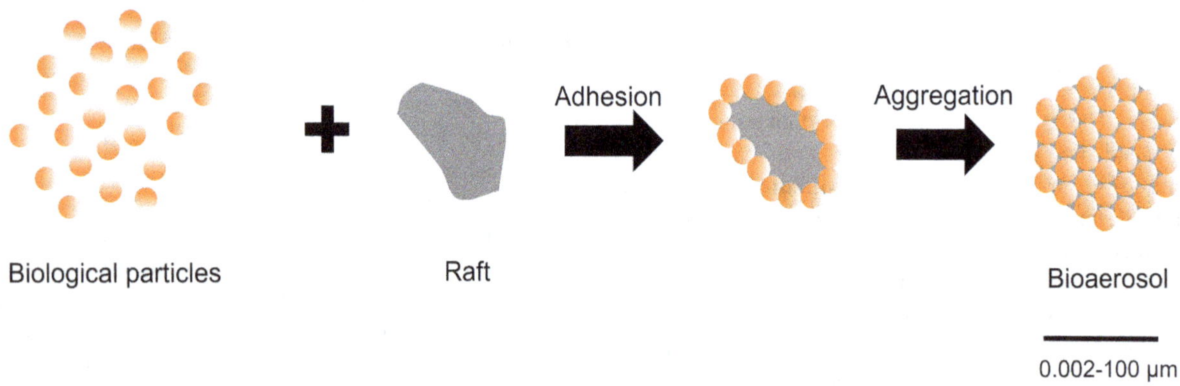

FIGURE 18.2 Schematic representation of bioaerosol formation in the atmosphere.

non-biological component of the aerosols, but the fraction of metabolically active bacterial cells in the air is very small. They can travel hundreds to thousands of kilometres (km) in the air, acquiring different protection mechanisms. The concentration of microorganisms in the air is calculated in units of numbers of cells/m^3 or mass of particulate matter in µg/m^3. Airborne microorganisms have a prominent role in microbial distribution on Earth. This mainly depends on three factors: (a) their fluxes between the atmosphere and surface environments, (b) the distance travelled by bioaerosols, and (c) their viability after deposition on the surface.

The atmosphere has many distinct vertical layers, including the troposphere, stratosphere, mesosphere, thermosphere, and exosphere (Figure 18.3). However, microorganisms mainly inhabit the lowest layer, the troposphere (especially below 5 km). The density of airborne microorganisms is maximum in the planetary boundary layer (PBL), which lies somewhere between 100 m and 2,000 m above the ground surface. This layer occupies the maximum amount of water vapour and is highly dense compared to any other layers. However, the survival of microbes is reported even from the stratosphere, a region intensely hit by harmful ultraviolet (UV) radiation. Examples of those thriving in extreme environments are bacteria like *Bacillus simplex* and *Staphylococcus pasteur,* and fungi like *Engyodontium album.*

18.3.1 Different Types of Airborne Microorganisms

Of all microbial agents, bacteria and fungi are very common in air samples. The unicellular forms and spores are more prevalent above ground because wind can easily lift them from land or oceanic sources. In the atmosphere, unicellular bacteria and their spores prefer to exist in aggregate forms, whereas the fungal spores can travel in single units.

Unlike the spores of any other organisms, bacterial spores are not reproductive units but a strictly durable structure to escape unfavourable situations such as high temperature, high UV irradiation, and desiccation. They are produced only in some Gram-positive cells and are popularly known as **endospores** (Figure 18.4). Spores of *Bacillus* and *Clostridium* have a high density in the air. The transformation of a normal cell into an endospore involves a "quorum-sensing" pathway, which is influenced by stress periods. In the sporulation process, a bacterial cell is highly differentiated into a dormant resting stage of the endospore. On the contrary, a favourable situation triggers the redifferentiation of the endospore into a normal cell. Endospores do not have any active growth

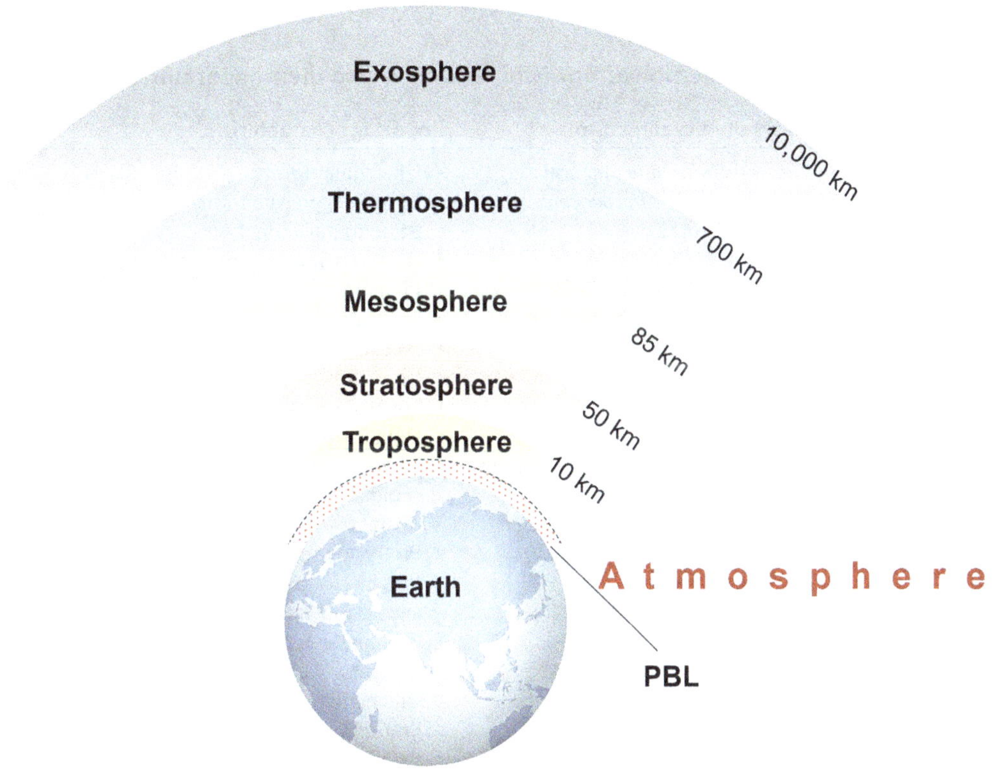

FIGURE 18.3 Different layers of the atmosphere of the Earth. PBL stands for the planetary boundary layer, which has the highest density of airborne biological particles. (Diagram is not to scale.)

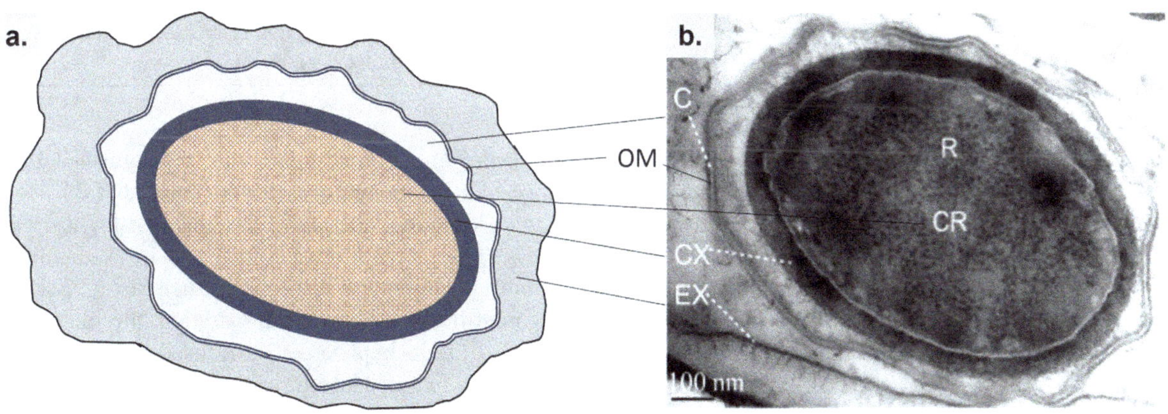

FIGURE 18.4 An endospore of *Bacillus subtilis*. a. Diagram. b. TEM image. EX: exosporium, C: coat, CX: cortex, CR: core, and R: ribosomes. (Image b adapted from Wilke et al. 2015.[2])

or metabolism, but can travel in the air for a long distance. Examples of non-sporulating Gram-positive bacteria that can float in air for extended periods are *Staphylococcus* and *Micrococcus* (Table 18.1).

Many Gram-negative bacteria can survive in the air for an extended period. They mainly belong to Alphaproteobacteria, Gammaproteobacteria, and Bacteroidia. Examples of common inhabitants of the atmosphere are *Escherichia coli*, *Pseudomonas*, *Klebsiella*, *Salmonella*, *Enterobacter*, and *Pantoea*. The non-spore-forming bacteria are common at low altitudes. However, they fail to survive in high-altitude regions due to the harmful effects of UV rays. Airborne bacteria acquire several mechanisms to thrive in the challenging environment of the atmosphere. For example, the secretion of exopolysaccharides (EPS) and other metabolites (as in membrane compositions) in some specific bacteria can help them encounter atmospheric stress for their prolonged survival. Several other adaptation mechanisms that allow them to endure stressful environments include DNA repair mechanisms, pigmentation, and scavenging for water and nutrients in the air. Molecular studies also confirmed genetic

TABLE 18.1
A list of different types of airborne microorganisms with examples and their important characteristics

Types of microorganisms	Examples of microorganisms	Important characteristics
Gram-positive bacteria	*Bacillus* and *Clostridium* (sporulating) *Staphylococcus* and *Micrococcus* (non-sporulating)	• Unicellular forms • Exist in aggregate forms • Secretion of exopolysaccharides (EPS) and other metabolites
Gram-negative bacteria	*Escherichia coli*, *Pseudomonas*, *Klebsiella*, *Salmonella*, *Enterobacter*, *Mycobacterium tuberculosis*, and *Pantoea*	• DNA repair mechanisms • Pigmentation • Scavenging for water and nutrients in the air • Mechanisms to alleviate cold shock and oxidative stress
Archaea	*Methanobrevibacter*, *Halobacterium*, and *Nitrososphaera*	• Unicellular forms • DNA repair mechanisms • High resistance to extreme situations
Fungi	*Cladosporium*, *Aspergillus*, *Penicillium*, *Alternaria*, *Aureobasidium*, Engyodontium, *Fusarium*, *Chaetomium*, *Acremonium*, *Drechslera*, and *Curvularia*	• Dissemination through asexual, non-motile bodies such as conidia, fungal fragment broken from their hyphae, and their sexual spores • Preferably travel in single units • Presence of protective pigments like melanins, mycosporines, and scytonemins • Potent allergens • Release volatile organic metabolites and mycotoxins
Protozoa	*Pneumocystis jirovecii*	• Unicellular • Protozoan cysts • Parasitic
Microalgae	*Chlorella*	• Aerosolised during dry conditions • Allergens
Viruses	Influenza virus, measles virus, SARS-CoV-2 (Severe Acute Respiratory Syndrome Coronavirus-2), and tomato spotted wilt virus	• Nanoscale size • Highly infectious and contagious • Inert particles in the air • Aerosolised through respiratory droplets/plant debris • Some have a protective covering

determinants of cold shock, oxidative stress, and UV repair enzymes in airborne microorganisms.

In fungi, the members of Dothideomycetes and Agaricomycetes are fairly common in the atmosphere. Protective pigments like melanins, mycosporines, and scytonemins protect airborne fungi against oxidative and desiccative stresses. These survival advantages are much more common in taxa thriving in the air than those living on the ground. However, most fungi reside as spores in the air. They are mainly asexual, non-motile bodies commonly known as **conidia**. Conidia are the primary means of dispersion and can be easily liberated in the air. They develop at the tip of specialised hyphae called conidiophores and detach after maturity. The conidia of *Cladosporium*, *Aspergillus*, *Penicillium*, and *Alternaria* are the common residents of the atmosphere (Figure 18.5). Besides, *Aureobasidium*, *Fusarium*, *Chaetomium*, *Acremonium*, *Drechslera*, and *Curvularia* are also globally distributed wind-dispersed fungal phylotypes. The size of most of the fungal aerosolised spores lies in the range of 2 to 50 μm in diameter, and they are potent allergens due to their respirable size range. The presence of fungal components in the air strongly depends on environmental factors such as humidity and temperature. The outdoor concentrations of airborne fungal spores lie between 200 and 10^6 spores/m^3. Fungal spores are distinguished by a latent stage of life similar to bacterial spores, but do not produce a thick covering; therefore, they are not highly resistant to stress. The production of fungal spores is mainly a means of multiplication and dispersion. Their sexual pores are also apparent in the air current. In addition, the fungal fragments broken from its hyphae get aerosolised and transported in the air. Many fungi, while travelling in the air, release volatile organic metabolites and mycotoxins. These metabolites can reside in the atmosphere for an extended period and induce allergenic and toxigenic effects.

A low number of archaea, protozoa, and microalgae also reside in the atmosphere. Archaea in the atmosphere are mainly reported from phyla Euryarchaeota, Thaumarchaeota, and Crenarchaeota. Examples of some notable taxa are *Methanobrevibacter*, *Halobacterium* and *Nitrososphaera*. Protozoa are rare and mainly exist as cysts. Microalgae like *Chlorella* can be aerosolised during the dry season and reside in the atmosphere for some periods.

The load of viruses in the atmosphere is less than 10%, but their existence greatly matters owing to pathogenicity concerns. In contrast to other life forms, viruses are usually not a stable component of the aeromicroflora because they are transmitted in the air from infected hosts or contaminated materials.

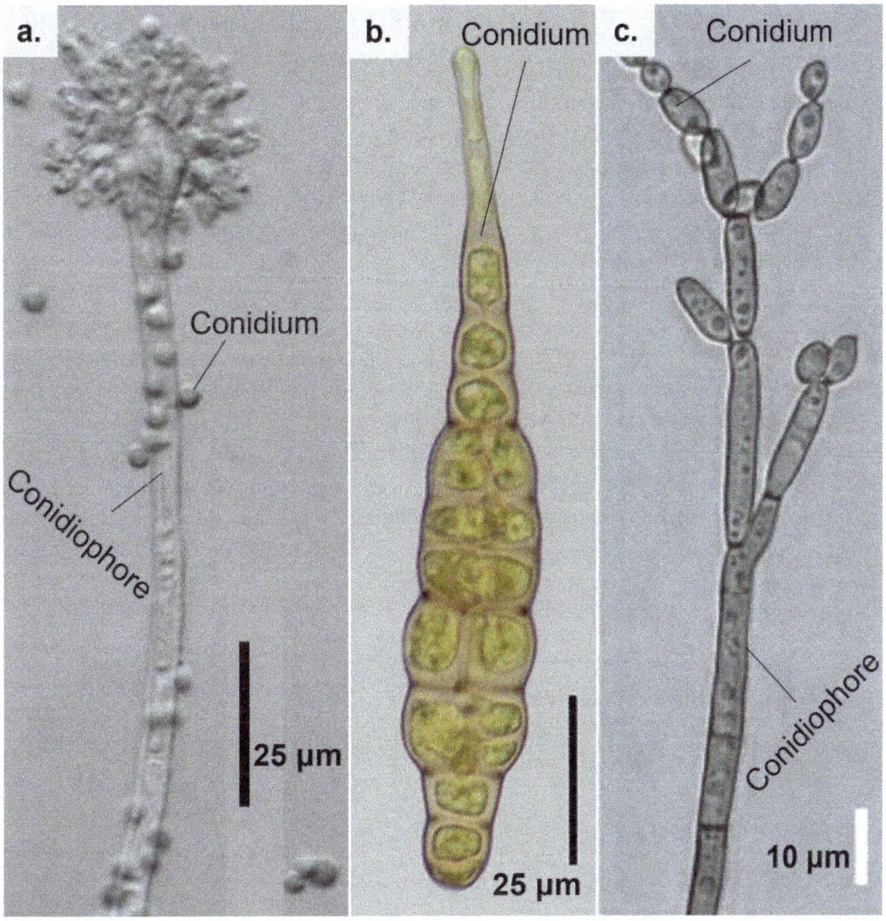

FIGURE 18.5 Morphological features of airborne fungal spores. a. *Aspergillus* sp. b. *Alternaria* sp. c. *Cladosporium* sp. (Images a, b, and c adapted from Mahata et al. 2022[3], Gou et al. 2022[4], and Yang et al. 2023[5], respectively.)

They remain as inert particles (i.e., virions) throughout their circulation in the air. Human viruses primarily disseminate into the environment through large respiratory droplets that do not travel long distances. Plant airborne viruses aerosolise plant debris. Among different types of viruses, ssDNA viruses like *Gemini-*, *Circo-*, *Nano-*, and *Microviridae* are widespread in the air. A low load of dsDNA, ssRNA and dsRNA viruses also exist in the air. Some notable examples of airborne viruses are mentioned in Table 18.1. In past years, even the coronavirus outbreak (having ssRNA) received wide attention due to its transmission via aerosols (Box 18.1).

18.4 INDOOR AND OUTDOOR MICROBIOMES

Indoor environments such as houses, schools, offices, and hospitals are closed compartments and maintain distinct microbial compositions. The study of microorganisms in indoor environments is called intramural aeromicrobiology. These local microorganisms in artificial environments are termed indoor microbiomes. Indoor air is highly dynamic and directly impacts the presence and concentrations of microorganisms. The study on indoor microbiomes is of increased public interest because their accumulation in indoor environments has an impact on indoor air quality (IAQ). Indoor microbiomes stay in close contact with the microorganisms of the outside environment and undergo an exchange in both directions. The outdoor environment is a major force for the dispersal of indoor air microorganisms. The microbial composition in the built environment varies so differently that even different rooms within the same compartment and the same room at different times can feature distinct microbial community compositions (MCCs). Several factors, such as the ventilation system, fan usage, weather, and resident activities, affect indoor air microbiota. Poor ventilation and excessive human activities in indoor facilities increase bioaerosol accumulation. The total concentration of airborne microorganisms in indoor environments can even be higher than in outdoor environments nearby. A high quantity of microorganisms in indoor environments is released from people, pets, and pests. Modern people spend, on average, more than 90% of their time in a built environment (inside buildings and enclosed vehicles). Therefore, many bacteria associated with indoor bioaerosols are representatives of human microflora. Indoor plants, plumbing systems, and air-conditioning systems are also distinct signatures of the indoor environment. Excess moisture on the wall promotes the growth of fungi. The black spots that landed on the damp wall could be due to

BOX 18.1 CORONAVIRUS DISEASE (COVID-19) AND ITS DISSEMINATION THROUGH THE AIR

The recent COVID-19 pandemic outbreak has led to the shutdown of the entire world. It has posed a global health crisis in demographics and emerged as the most disastrous situation since the era of the influenza pandemic of 1918. First reported in Wuhan, China, in December 2019, it reached the entire world in a few months. The infectious agent is severe acute respiratory syndrome coronavirus 2 (SARS-CoV-2). The aggressive nature of the virus triggered severe acute respiratory illness in humans, such as fever, cough, myalgia, and fatigue. This virus is highly contagious and transmitted by human-to-human contact. It can also be transmitted through both 'respiratory' droplets and 'airborne' infections in community settings. However, the World Health Organisation (WHO) took almost two years to admit that the transmission of SARS-CoV-2 could also be airborne.

When a person coughs, sneezes, or breathes heavily, it releases small liquid particles from the mouth, which can be of different sizes, including droplets and aerosols. In the air, the virus-laden droplets are larger than the usual aerosols (i.e., 5 μm). Therefore, droplets cannot travel more than 1 m and settle down on the earth, while aerosols can travel far and linger in the air. Likewise, the respiratory droplets of SARS-CoV-2 produced by coughing and sneezing consist of both small and large particle aerosols. Large particle aerosols (>0.1 mm diameter) settle on the surface of the body or get breathed in within a radius of 1 m. Small particle aerosols (<0.1 mm diameter) are respirable by short-range airborne transmission (up to several metres) (Box Figure 18.1.1). In crowded and inadequately ventilated indoor facilities, aerosols could carry the virus in long-range transmission.

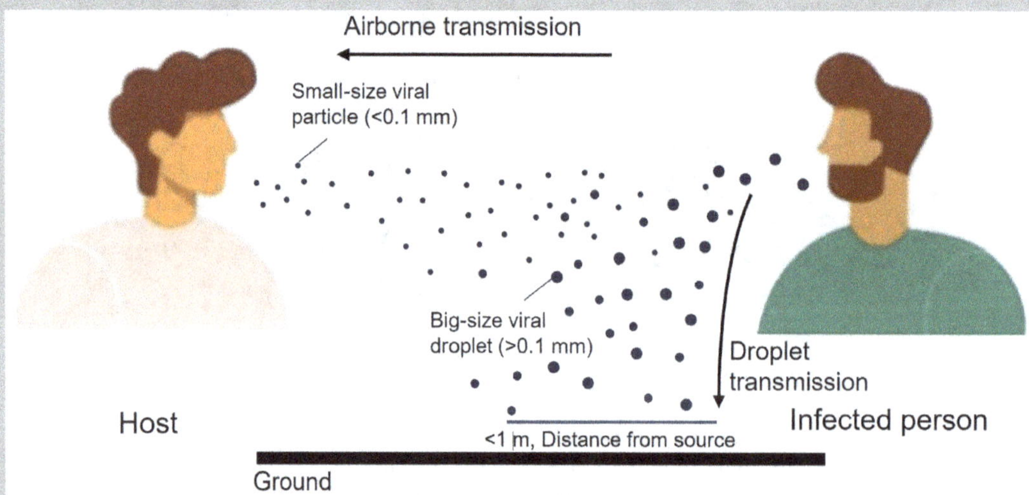

BOX FIGURE 18.1.1 Airborne transmission of respiratory microorganisms. (Human cartoons adapted from iStock/Anastasia Sydoruk.)

the growth of moulds (Figure 18.6). The porous surfaces of various materials, such as carpets and wood items, can trap microorganisms. Non-porous materials like glass, metal, and plastic have smooth surfaces and have fewer microorganisms. Figure 18.7a illustrates some notable sources of indoor microorganisms.

The study of microorganisms in outdoor environments is referred to as extramural aeromicrobiology. The collections of all the microorganisms in outer space are termed outdoor microbiomes. The outdoor microbiomes are spread in wide areas and harbour diverse microbial communities. The microbial compositions in outside environments vary geographically, such as urban areas, agricultural areas, and forest areas. The major sources of outdoor microorganisms are terrestrial and aquatic ecosystems. Plant and animal communities are notable drivers of outdoor microbiota. A major part of the plant body establishes direct contact with air, and its wide surface from the phyllosphere promotes the dispersal of plant-colonised microbes in the atmosphere. Animals transport microorganisms from their bodies, fur, and feathers through their various activities. Animal excreta, such as faeces, urine, and other bodily secretions, are also rich in microorganisms and nutrients, which can be distributed in a wide area through the air. A low concentration of inland or aquatic microbiota resides immediately above the ground/water surface. The pristine environments are also not free from microbial contamination due to their dissemination from local regions and over long distances. Certain outdoor microorganisms have raised concerns about diseases in plants, animals, and humans (detailed in section 18.9). The atmospheric features such as

FIGURE 18.6 Mould growing on the wall in an indoor environment. (Image credit: Flickr/Amanda Slater.[6])

wind, rain, humidity, and temperature have a strong influence on the composition and structure of the air microbiota. In addition, climate change, agricultural activities, urbanisation, and industrialisation could also induce significant impacts on the MCCs. Commercial and market areas, especially those which have high footfall, have a high density of microbes. Places in the proximity of vehicles, buses, and metros release various pollutants, including particulate matter, into the air, which support the growth of microbes. Figure 18.7b features some notable sources of outdoor microorganisms.

18.5 FACTORS INFLUENCING AIRBORNE MICROORGANISMS

Microorganisms form a diverse and dynamic pool of the atmosphere. Their temporal and spatial variabilities are quite apparent due to a wide array of influencing factors. The mean concentrations of bacteria, viral particles, or fungal spores in the atmosphere can be around 10^4–10^8 cells/m^3. Such concentrations over the sea can be lower by a factor of ~100–1000 compared to their concentrations over land (Figure 18.8). The microbial densities in fresh snowfall and cloud water can reach around 10^5 cells/m^3, while an estimate from desert dust is nearly 10^7 cells/m^3. A substantial influence on microbial distribution in the atmosphere is generated from surrounding landscapes (especially the perimeter of 50 km). Meteorological parameters, sources of microorganisms, physical forces of deposition, and anthropogenic activities are some of the major forces that can play roles in their distribution and survival in the atmosphere (Figure 18.9). These factors are discussed in detail as follows.

18.5.1 Meteorological Factors

- Wind Speed- Airborne microorganisms share a complex relationship with wind speed. A certain wind speed is required to lift the biological particles from the ground or water surface and for their movement in the air. The turbulent winds of dust storms can transport bacterial cells and fungal spores thousands of km. The aerosol emission from plant canopies can occur even at low wind speeds (0.4 m/s), whereas emission via sea spray requires at least 10 times higher wind speed. Wind speed also controls the size and density of microbial aerosols. Microbial aggregates are likely to break into small particles at high wind speeds, increasing the number of particles in the atmosphere.
- Humidity- High relative humidity positively affects the growth of bacteria and fungi in the atmosphere. Relative humidity of more than 70% in the atmosphere can increase the size of bioaerosols, enabling the clumping of particles. High humidity supports the germination of microbial spores and increases the chances of microbial contamination and allergic responses. However, low ambient humidity may cause acute dehydration in cells, causing the inactivation of cell death. The transmission of viruses is usually highest at low humidity.
- Rain- The splashing of rain droplets on the ground induces aerosol formation. Even elevated moisture content in the air can support nutrition requirements for microbial growth. However, the consequence of immediate rainfall promotes the settling of suspended particles. This event immediately restores air quality by reducing PM concentrations.

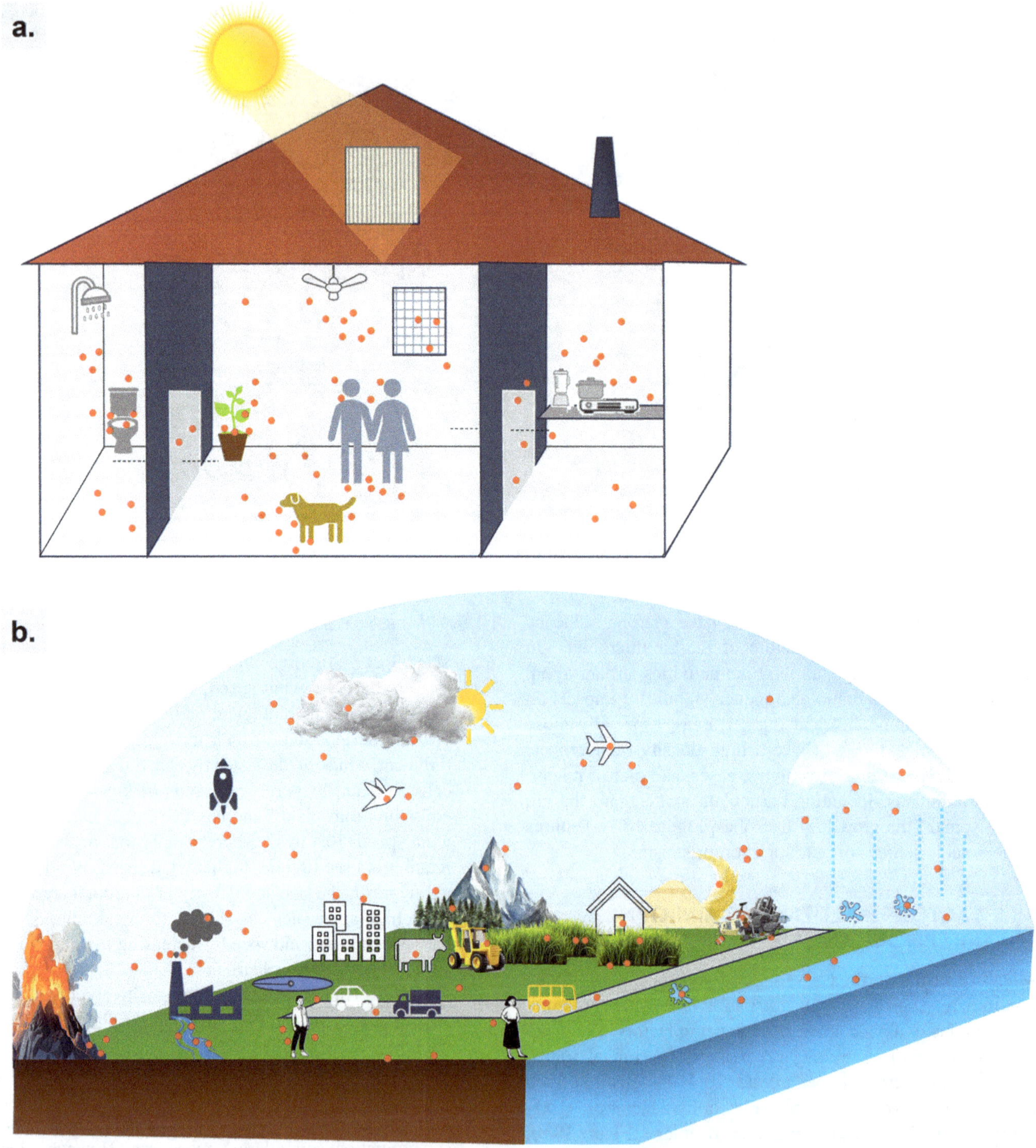

FIGURE 18.7 a. Illustration of indoor microbiota showing various sources such as humans, pets, plants, fans, ventilation, outdoor air, dust, shower, toilet, and kitchenware. b. Illustration of outdoor microbiota, showing various sources such as vegetation, vehicles, buildings, air traffic, animals, humans, industries, agriculture, sewage treatment plants, landfill sites, mountains, volcanoes, dunes, sea spray, and rain. The red dots in the image represent airborne microorganisms.

- Temperature- The atmosphere can inhabit a range of microorganisms, including psychrophiles, mesophiles, and thermophiles. Major variations in the temperature of an environment may result in the desiccation or inactivation of some of the microbial forms. Some bacteria differentiate into endospores to protect their cellular content. Due to protein denaturation, bacteria such as *Escherichia coli* and *Staphylococcus aureus* have reduced chances of survival at high temperatures.
- Radiation- Bioaerosols can absorb and scatter solar radiation, which can have an impact on the energy budget of the ground-atmosphere system. However,

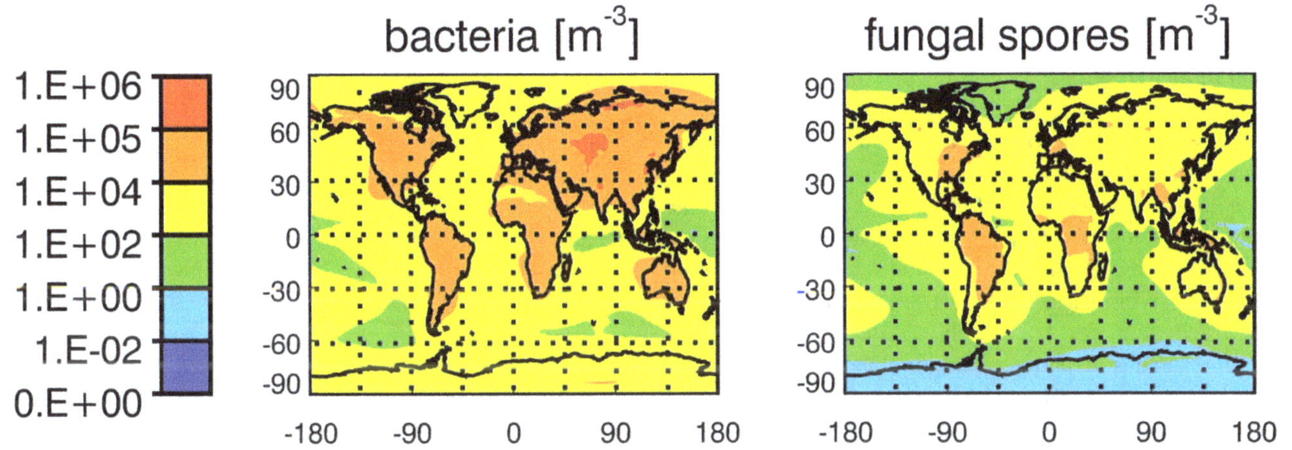

FIGURE 18.8 Geographical distribution of bacteria and fungal spores on Earth (simulated annual average near-surface). (Adapted from Hoose et al. 2010[7], © 2010 IOP Publishing Ltd.)

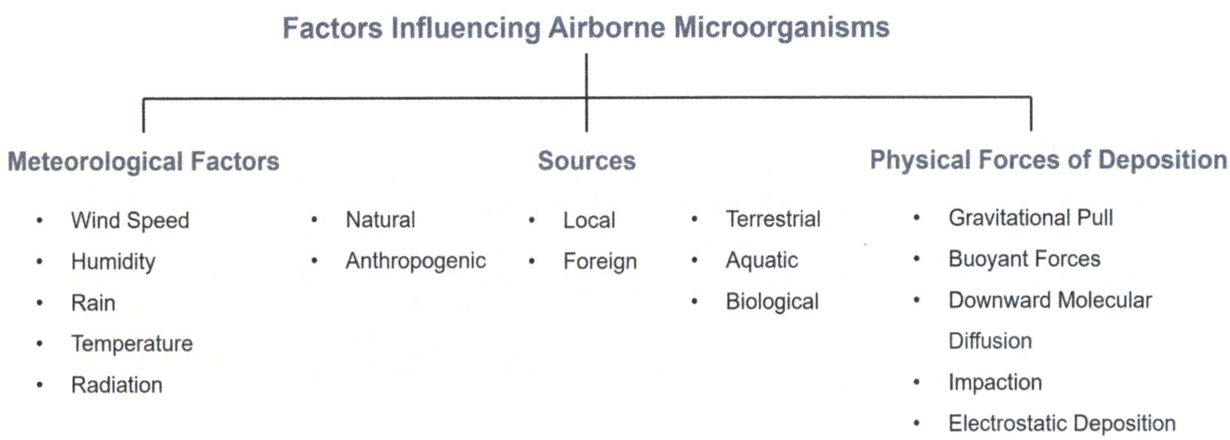

FIGURE 18.9 Factors influencing airborne microorganisms.

certain rays are detrimental to airborne microorganisms. UV and other ionising rays produce anomalies in the genetic constituents of microorganisms. This causes adverse effects on microorganisms and reduces their chances of survival.

The synergistic effects of several meteorological factors are more meaningful than their impacts by individual factors. However, the distinct meteorological variations in diurnal and seasonal cycles cause major shifts in the airborne microbial composition. Factors like relative humidity in summer, vapour pressure in autumn, air pressure in winter, and air temperature and relative humidity in spring are the major drivers for MCCs in the atmosphere. Some regions set a cyclic trend of microbial communities in the atmosphere. However, the violent environmental discrepancies can distract from the periodical variations of the atmosphere.

18.5.2 Sources

Airborne microorganisms come from diverse sources emitting from various natural and anthropogenic ecosystems. The dominant natural sources are vegetation, wind, volcanic eruption, and sea spray. The principal anthropogenic inputs of airborne microorganisms are hospitals, houses, schools, construction sites, and agricultural land. Even the treatment facilities, including composting and wastewater treatment plants, immensely contribute to bioaerosol accumulation in the local environment. In wastewater treatment facilities, activated sludge and trickling filters are potentially associated with turbulence and agitation in wastewater, leading to the aerosolisation of waterborne microorganisms. The high pedestrian volume, vehicular traffic, industrialisation, biomass burning, greenhouse gas emission, and waste generation and its management have induced more dissemination of microorganisms in the air than any other processes. The emission of particles close to the ground is greatly affected by turbulence. Air pollutants such as SO_2, O_3, NO_x, etc., share positive, negative, or neutral relations with airborne microorganisms. Chemical composition influences nutrient availability, toxicity, and hygroscopicity, impacting microbial survival and activity.

The human body is a good reservoir of airborne microorganisms, occupying more than 10^{12} microbial cells in the skin epidermis and 10^{14} cells in the oral cavity. Intense human density can increase the density of microorganisms in the air through respiration, shedding of skin cells, and excreta.

Both local and foreign microbial sources contribute to the overall microbial MCCs in the air. However, the assemblage of microorganisms in the atmosphere includes taxa mainly native to the surrounding environments. Local sources are rather more influential than the immediate surface interfaces for the microbial composition in the air. The productivity of the underlying surface habitats has a major influence on their distribution. For example, inland dust in urban regions, vegetation and soil above agricultural lands, and seawater above marine ecosystems are the primary sources of airborne microorganisms in their local environments. A substantial load of microorganisms can also be transported by long-distance air masses. Microorganisms found in open ocean environments, for instance, have the ability to travel thousands of km from their sources. Also, a substantial load of airborne microorganisms in marine environments could be traced to terrestrial origins.

Many microorganisms are vector-borne and use carriers for their transmission. For example, *Plasmodium* is transmitted through the mosquito, *Anopheles*. Some plant-associated bacteria find shelter in seeds and pollens and can travel in the air. The spread of microbial cells through vectors enables their dispersal over a hundred kilometres without requiring much adaptation to environmental stresses.

18.5.3 Physical Forces of Deposition

Forces like gravitational pull, downward molecular diffusion, impaction, and electrostatic deposition act upon all suspended particles in the air, affecting their settlement. The interplay of such forces influences the residence time of airborne microorganisms in the atmosphere. The gravitational force acts upon all materials heavier than gases. However, the buoyant forces on small particles dominate over gravity, allowing them to float in the air. The particle size has an important role in the dispersal, attachment, and community structure of airborne microorganisms. The natural air currents and eddies induce random downward diffusion of the particles, leading them to a permanent settlement on the surface. Impaction is the tendency of an airborne particle to escape the drag forces of the gas flow by which particles make contact with a surface. In impaction, particles can have two alternative pathways. The downward molecular diffusion of particles favours their gravitational settling. Differently, bouncing off the surface favours their re-suspension in the air. In a different process, the negatively charged surface of particles combines with positively charged particles, which would induce coagulation and condensation as electrostatic deposition. This process tends to increase the mass of particles, resulting in permanent deposition.

18.6 BIOAEROSOL CYCLE

The dispersal of biological particles is central to maintaining bioaerosol-ecosystem interactions. Different global and regional models have been proposed to study airborne transmission. This provides estimates of bioaerosol fluxes, transport processes, and atmospheric impact on particles. The biogenic particles can undergo various physical, chemical, and biological transformations in the air. Their transmission creates cycling of airborne bioparticles called bioaerosol cycling, which can be divided into three sequential steps, including emission, dispersal, and deposition (Figure 18.10). They are further discussed in detail as follows.

18.6.1 Emission

Emission is a process of releasing particles from the surface of an object and suspending them in the atmosphere. The anthropogenically disturbed regions contribute more fluxes of particles in the atmosphere than undisturbed ecosystems. Resuspension of settled particles can act as a secondary source of airborne particles. In certain conditions, wind beyond a threshold velocity induces bouncing off the surface and generates aerosols in the near-surface atmosphere. An estimated annual emission of bacterial cells in the atmosphere from terrestrial environments ranges between 10^{23} and 10^{24}, which can account for 470-1,100 × 10^9 kg. The total annual emission of fungal spores from terrestrial ecosystems is nearly 10^{24}, which is approximately equal to 50 × 10^9 kg fungal biomass. Airborne particles from seawater are released as sea sprays, produced by breaking waves in the underlying water.

Sources for emission can be of two different types: point and diffuse sources. When the discharge is traced back to a single origin, such as construction sites, food processing plants, wastewater treatment systems, and health care facilities, they represent point sources and display a general conical-type dispersion. However, diffuse sources are large and less well-defined areas and display particulate wave dispersion. For example, rural dwellings, agriculture, and forestry have no specific point of discharge of airborne microorganisms in the air and demonstrate a wide region for their dispersal.

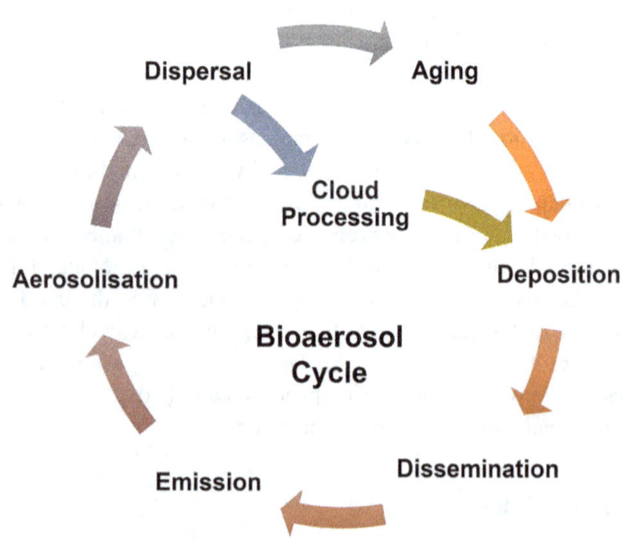

FIGURE 18.10 Bioaerosol cycle.

Based on occurrence, emission can be instantaneous or continuous. An instantaneous emission is a single event, whereas a continuous emission involves dispersion over an extended period. Sneezing and air expulsion are examples of instantaneous sources, and large fields and wastewater treatment plants represent continuous sources.

18.6.2 Dispersal

When ground and water sources launch microorganisms into the air, they reside in the atmosphere and travel through the wind. This is called the dispersal of airborne microorganisms. During their residence periods in the atmosphere, they experience turbulence. In the airway, they pass through different stages, including chemical transformation, stress, ageing, and desiccation. The circulation of bioaerosols across different temporal and spatial scales greatly depends on various meteorological, environmental, and anthropogenic factors. The air current, pattern, and strengths have significant impacts on their distribution in local and global regions. Forces like diffusion, inactivation, and gravity also act upon the airborne particles. The microbial diversity in the air is closely associated with its emission sources. However, this association progressively decouples over time and above PBL. The local and global dispersals of airborne particles depend on their particle sizes. Based on particle size and the range of transport, they can be categorised into four distinct forms:

- Submicroscale- They have a residence time of less than 10 min in the atmosphere and travel less than 100 m.
- Microscale- They have a residence time of 10 min to 1 h in the atmosphere, and the travel range is 100 m to 1 km.
- Mesoscale- They have a residence time of several days and travel up to 100 km.
- Macroscale- The residence time and the travel range are extended even further.

The process of dispersion and diffusion of airborne particles is extraordinarily complex. Scientists rely on physical and mathematical models to understand the physical and biological dynamics of airborne microorganisms.

18.6.3 Deposition

Deposition is an earthward movement of atmospheric constituents into terrestrial and aquatic ecosystems. Different physical forces contribute to the landing of particles from the air onto the Earth's surface (detailed in Section 18.5). The deposited particles are the prominent sources of contaminants in environmental compartments and could be pathogens for crops, animals, and humans. They can also be responsible for the spoilage of foodstuffs, wood, and other goods. Deposition can be of two different types: wet and dry deposition. Wet deposition involves mixing biogenic particles and suspended water in the air, resulting in sedimentation and deposition. Weather conditions like rain, snow, sleet, and fog scavenge the microorganisms from the atmosphere and transfer them to the surface. Dry deposition is mainly the free fall of the suspended particles on the floor. The dust-laden sedimentation and hailstorms are different forms of dry deposition on the ground. The deposition of suspended particles in the respiratory tract is also an example of dry deposition and is a major concern for respiratory diseases.

18.7 DETECTION OF AIRBORNE MICROORGANISMS

The detection of airborne microorganisms serves as valid evidence of microbial presence in the environment and provide insights into their density in the air. The two essential and consequential components in their detection are the sampling and sensing of airborne particles. Of these two steps, the first step, sampling, is more challenging. The main problem during sampling is caging the air. This involves capturing enough viable particles from the atmosphere. Based on the forces involved, the air sampling can be active and passive.

Active sampling employs a pumping machine that pumps the air through a collection device. The principle of active sampling is linked to mechanical forces. The collection efficiency depends on aerodynamics and inlet characteristics. The low abundance and loss of viability of microbial cells in the sampling device are the preliminary concerns. However, new sampling devices are designed to recover high volumes of air (several hundred cubic metres) in short timeframes. The methods of active sampling are manifold, and some important ones are briefly discussed as follows.

- The impactor is a widely used sampler for active sampling. The basic principle of the impactor is the impaction (or crowding) of the air over the collecting solid surface of the solid medium placed in the sampler (Figure 18.11a). The impactor can be used for size fractionation in a graduated manner using multiple plates with different aperture sizes. This creates a cascade system, and the technique is called a cascade impactor. The size ranges for different plates in a cascade impactor typically include coarse particles (> 2.5 μm), fine particles (1.0–2.5 μm), ultrafine particles (0.1–1.0 μm), and nanoparticles (< 0.1 μm). Figure 18.12 shows different body parts of a photograph of the six-stage Andersen cascade impactor. A different variant of the device relies on centrifugal force for the impaction of aerosol onto a solid surface.
- A different method is impingement, where a pump strikes the air in the impinger that hits the fluid (Figure 18.11b). The suspended particles are impinged into the liquid medium, which can be used for microbial cultivation. Both solid impactors and liquid impingers have received wide recognition for acquiring maximum air volume in a short time. They are also considered suitable for quantitative analysis of microbial load in the air.
- Thermal precipitation is a method where the air laden with aerosol is drawn over a thermal gradient on which particles tend to move away from hot surfaces and are

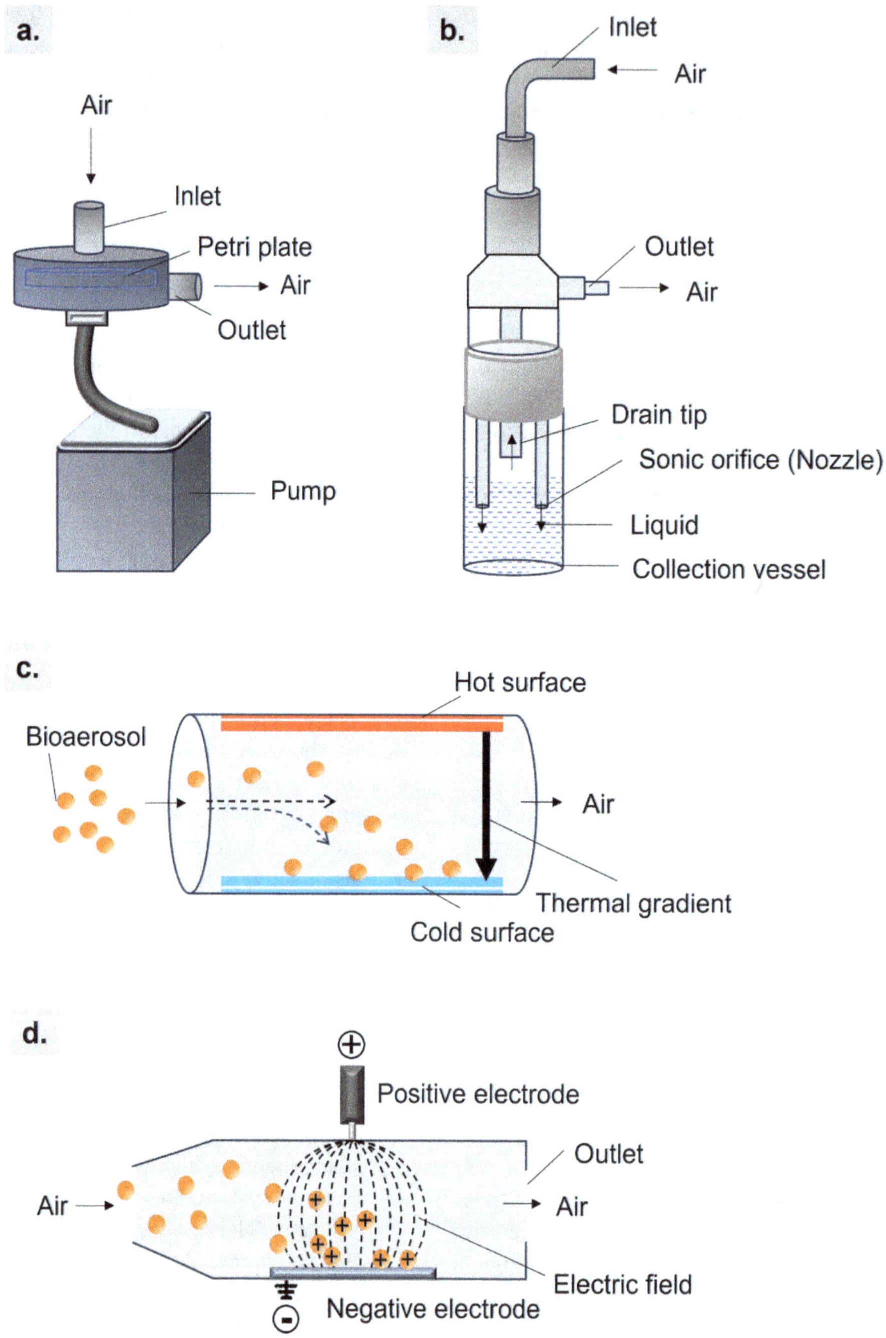

FIGURE 18.11 Different tools for the active sampling of airborne microorganisms. a. Impactor. b. Impinger. c. Thermostatic deposition. d. Electrostatic precipitation.

deposited on colder surfaces (Figure 18.11c). However, high temperatures tend to reduce the microbial viability in the samples.

- Electrostatic precipitation employs electrostatic forces to charge the aerosol at the inlet (Figure 18.11d). The charged particles inside the sampler are deposited on overcharged plates under the influence of electrical fields.

The deposited samples of airborne microorganisms in the sampler can be investigated using techniques such as microscopy, culturing, flow cytometry, quantitative polymerase chain reaction (qPCR), gene sequencing, and ATP-based bioluminescence. A relatively new technique, the Spectral Intensity Bioaerosol Sensor (SIBS), is an optical-based detection that provides size fractionated data in real-time. The culturing method can be suitable for isolating airborne microorganisms, but only a few are cultivable. The molecular ecological approaches are more reliable for estimating diversity and taxa identification.

In contrast, passive sampling depends on the kinetic energy and diffusion of gases. The common scheme for passive sampling is the settle plate method. This method is also known as the gravity method of air sampling. In this method, the sterilised culture media-filled Petri plates are exposed to the

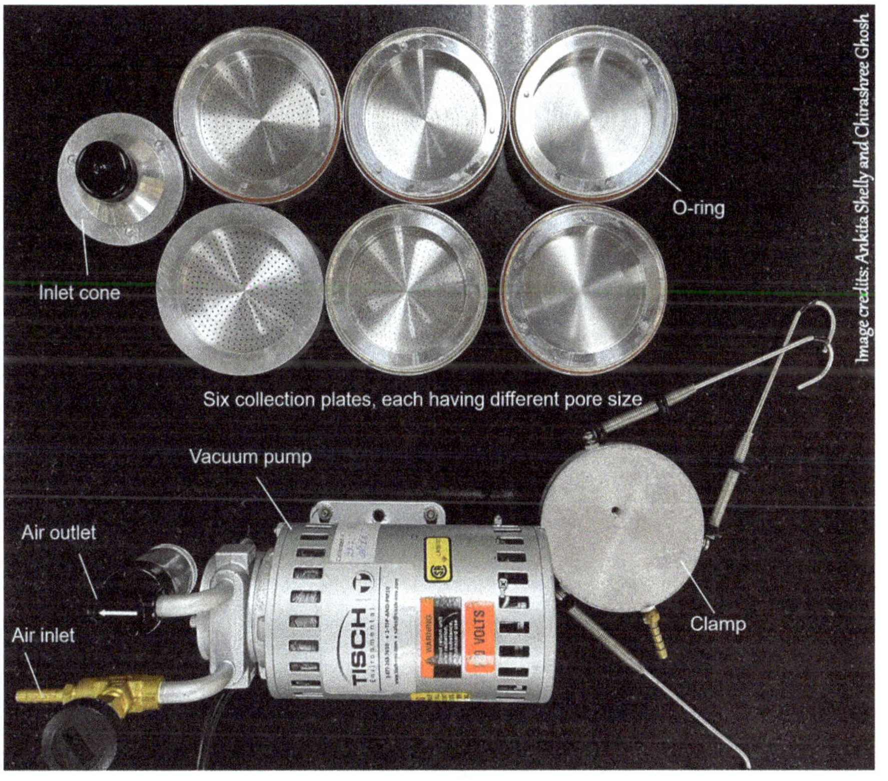

FIGURE 18.12 A photo showing different parts of a six-stage Andersen cascade impactor for air sampling. (Image credits: Ankita Shelly and Chirashree Ghosh/Department of Environmental Studies, University of Delhi.)

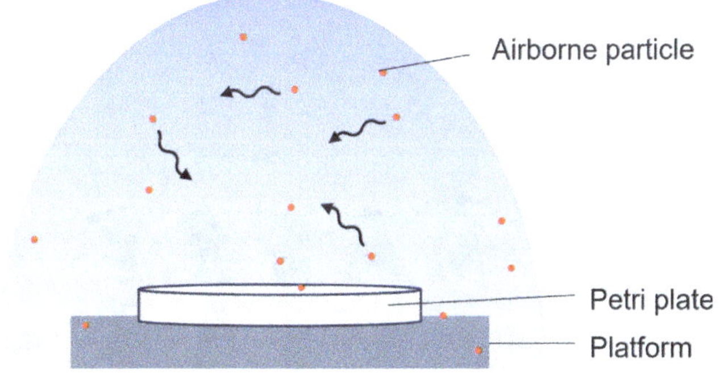

FIGURE 18.13 Illustration of passive air sampling on an agar plate for the cultivation of airborne microorganisms.

air for a definite period (Figure 18.13). The setup facilitates unassisted deposition of the microbial cells over open Petri plates. The incubation of plates for certain durations promotes the growth of deposited cells in the form of colonies. The number of colonies grown on a plate reflects the microbial load on the immediate environment. The ease of operation and inexpensive approach have led to this technique being widely used to detect airborne microorganisms. This is also useful to detect microbial contamination from the surface of various objects. However, it only provides a partial image of what is in an air environment. The low sampling flow rate and cultivation of very limited microorganisms on the culture media are the major limiting factors. Also, this method is considered qualitative because the sampler data does not specify the amount of air used.

18.8 ECOLOGICAL SIGNIFICANCE OF AIRBORNE MICROORGANISMS

The abundance and mass burden of microorganisms in the atmosphere are certainly much less than their presence in other environments. However, the poor distribution does not preclude their significance. Airborne microorganisms have dynamic effects on the Earth's system. They play prominent roles in atmospheric processes and ecological balance. Microorganisms residing in the atmosphere are potential agents of condensation nuclei for cloud formation. A variety of airborne microorganisms produce ice nucleating proteins (INPs) that induce ice nucleation at much warmer temperatures (at high sub-zero temperatures) than any other particles (at around -40 °C). Such a process affects cloud formation and precipitation and influences global precipitation patterns (Figure 18.14). However, recent estimates suggest that up to 15% of airborne species of bacteria responsible for cloud formation could go extinct due to climate change. A protein, InaZ, from *Pseudomonas syringae* is the model system of INP (Figure 18.15).

The metabolic activities of airborne microorganisms using atmospheric substrates have a limited but important contribution to overall atmospheric chemistry. The biological oxidation of organic compounds (e.g., phenol) by the active fraction of air microbiota (mainly composed of bacteria) at night is comparable with their photooxidation in the daytime. A limited number of autotrophic microorganisms (e.g., cyanobacteria and microalgae) reside in clouds and desert

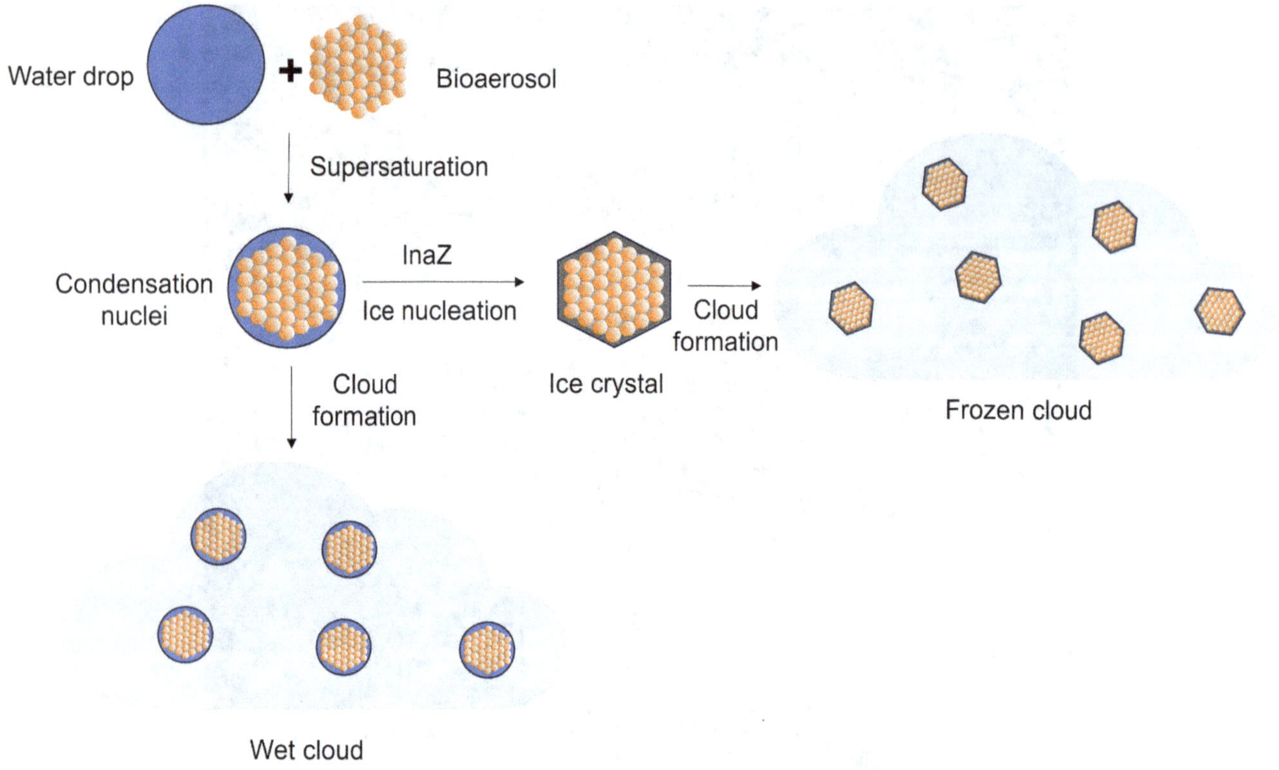

FIGURE 18.14 Role of bioaerosols in cloud formation. In the atmosphere, the interaction between a water drop and a bioaerosol in supersaturated conditions creates condensation nuclei. The presence of ice nucleating proteins (INPs) like InaZ induces ice nucleation of condensation nuclei and creates ice crystals, which further contribute to the formation of frozen clouds. A different situation favours the accumulation of condensation nuclei in the wet cloud.

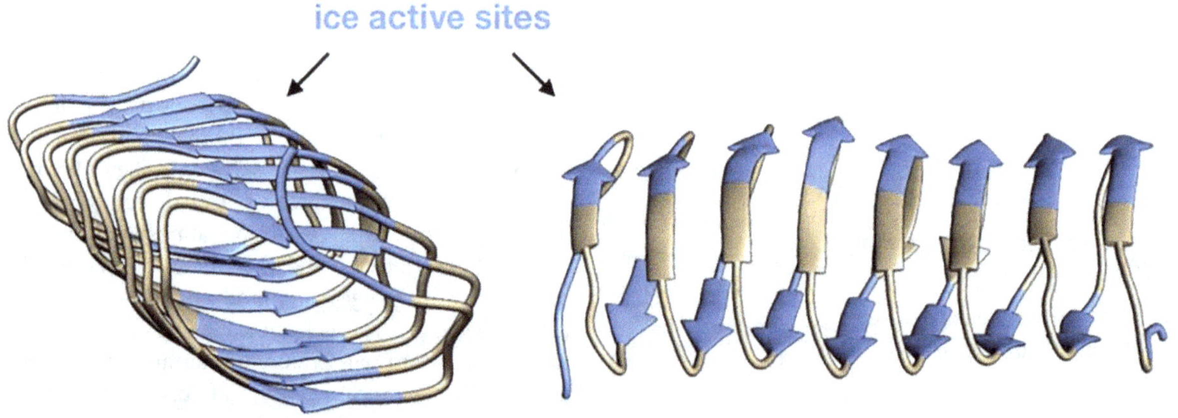

FIGURE 18.15 Ice-nucleation active protein (InaZ) without the N- and C-terminal domains. The blue sequences are ice-nucleation active sites. (Images adapted from Roeters et al. 2021.[8])

dust and can perform photosynthesis. These photoautotrophs receive protection against harmful rays in the atmosphere from their habitats. The microbial flux between the atmosphere and surface habitats has profound effects on terrestrial and marine ecosystems. Various airborne biological particles are reproductive units that have a role in their global distribution. They are involved in the biogeochemical cycling of nutrients, organic matter, and water. Nitrogen (N_2) and carbon dioxide (CO_2) gases are components of the atmosphere, and the microorganisms at the interface of the atmosphere and pedosphere perform nitrogen fixation and carbon fixation. In addition, they contribute to the dispersal of pathogens.

18.9 GLOBAL CONCERN

Exposure to airborne microorganisms can be a global concern for agriculture and public health. They can be responsible for poor air quality. Many airborne microorganisms can be invasive species in new regions when they are transported over long distances. Bioaerosols can affect the climate through cloud formation. Many airborne microorganisms are inherent pathogens and are associated with various diseases in plants, animals, and humans. Employees working in the agriculture, health, and waste management industries have health issues due to occupational exposure to bioaerosols. They exhibit adverse implications for agriculture and public health. In the distant past, airborne microorganisms had been intentionally used to disseminate toxins or pathogens for mass destruction.

Plants fall prey to many airborne bacteria, fungi, and viruses. Crop plants are highly susceptible to airborne diseases, and they cause severe loss every year. Fungi affect more crops than any other microorganisms. They mainly develop infections in food crops. For example, *Puccinia graminis tritici* causes stem rust in wheat and can disperse over long spatial and temporal scales. This fungus produces five different types of spores at distinct stages of its life cycle. These spores exhibit different abilities to suppress global wheat production using an alternative plant host, Berberry. Other examples of plant diseases due to airborne fungal pathogens are powdery mildew of cereals and grasses by *Erysiphe graminis*, rice blast disease by *Magnaporthe oryzae*, and late blight of potato/tomato by *Phytophthora infestans* (Figure 18.16). Examples of some airborne viruses that cause notable diseases in plants are the mottled browning of tobacco leaves by tobacco mosaic virus (TMV) and the "spotted wilt" disease of tomato by tomato spotted wilt virus (TSWV). In bacterial diseases, citrus canker by *Xanthomonas citri*, fire blight of apple and pear by *Erwinia amylovora*, angular leaf spot of cotton by *Pseudomonas syringae*, etc., can be easily propagated through wind. Airborne pathogens can also cause significant pre-harvest losses, which pose a significant challenge to food security and have a pressing impact on human health. In addition, many airborne microorganisms are responsible for food spoilage and can sour milk.

Humans and animals are continuously exposed to airborne particles through their respiratory systems, which are neutralised by activating the immune system. However, the high load of airborne particles in the air can impair the immune system and lead to adverse health effects. Their long-term exposure to hosts is linked to lung, cardiovascular, and neurological diseases and even newborn-related disorders. The particle size plays an important role in their accumulation inside the body. The small-sized particles attain a high penetration depth, whereas the large particles are stuck in the nose and mouth. The major impact of airborne diseases on human and animal health arises from endotoxins, β-glucans, mycotoxins, and allergens. Many airborne pathogens of humans spread from person to person by respiratory droplets. Moreover, the suspended particles can attach to different body parts or get inside through the oral and nasal cavities.

Examples of bacteria that generate concerns about airborne diseases in humans are tuberculosis caused by *Mycobacterium tuberculosis*, diphtheria caused by *Corynebacterium diphtheriae*, and anthrax caused by *Bacillus anthracis*. The illness of meningococcal meningitis in sub-Saharan Africa is correlated with airborne particles. Many airborne fungi, primarily from the order Mucorales such as *Mucor*, *Rhizopus*, and *Syncephalastrum*, cause a rare disease, mucormycosis, in immunosuppressed persons. Other examples of fungal diseases are Aspergillosis and Aspergilloma, produced by *Aspergillus* spp. Some airborne microbes are airborne allergens which cause a chronic inflammatory condition of the airways. This

FIGURE 18.16 a. Powdery mildew in *Poa pratensis* by *Erysiphe graminis*. b. Rice infection by the blast fungus *Magnaporthe oryzae*. c. Late blight on tomato by *Phytophthora infestans*. (Image a adapted from Wikimedia/Rasbak[9], Images b and c from Galhano and Talbot 2011[10] with permission of Elsevier Ltd, © 2011 British Mycological Society and Griffith et al. 1995[11] with permission of Elsevier Ltd, © 1995 British Mycological Society, respectively.)

kind of situation triggers asthma in humans. Airborne microbes have also shown a correlation with emphysema and chronic obstructive pulmonary disease (COPD). A cyanobacterium, *Nostoc muscorum* and a green alga, *Chlorella* sp., present in the air trigger allergic reactions in humans. Children contract measles from the mumps and rubella virus, which is disseminated through the air. Aphthovirus is a contagious airborne virus that causes cattle foot-and-mouth disease (FMD). Even the recent pandemic of the coronavirus illness (due to coronavirus disease 19 (COVID-19)), which is linked to severe acute respiratory syndrome, is also supported by a theory of transmission through the air (Box 18.1).

18.10 BIOAEROSOL CONTROL

Intense air pollution at public facilities and residential premises has aroused widespread public health concerns. This has increased attention to bioaerosol removal and control in the atmosphere. However, bioaerosol control is a challenging job. It primarily requires monitoring techniques to obtain information on bioaerosol levels and dynamics to identify their spread and potential health risks. Based on the data from the monitoring program, the intervention as a part of the prevention and control measures can be established. The strict regulation on industrial emissions and controlled discharge of waste materials have a positive impact on the air quality. In addition, waste management effectively contributes to controlling the dispersal of particles in the air. Improper waste disposal leads to an increase in the density of bioaerosols in the immediate airway. The buildup of organic matter from wastes can harbour and release microbes into the air. Careful disposal and treatment can prevent the dispersal of microbes in the air. An important precaution is minimising the sources of microbes from various objects. Regular cleaning of the surface reduces the microbial load on the objects that can become airborne through dust and disturbance.

The possible control measures of bioaerosols in the air include filtration, UV radiation, disinfection, plantation, steam condensation, and smart infrastructure, which are briefly discussed as follows.

- Filtration- This process involves the unidirectional airflow through the filtration unit, which prevents the passage of small particles across the filter. The pore size determines the efficiency of filters. The most commonly used filter is a high-efficiency particulate air (HEPA) filter, which has a pore size of ~0.3 μm and can filter many bacterial cells and fungal spores. A more sophisticated air filter, an ultra-low particulate air (ULPA) filter, can remove smaller particle sizes, including dust and allergens (of 0.12 μm diameter), with 99.99% efficiency. The best air filtration unit is generally composed of ULPA in combination with HEPA and prefilter to withdraw maximum particles from the air. However, this approach falls short of providing defence against airborne endotoxins and viruses in the nanosize range. For example, a coronavirus having a size range of 0.07-0.09 μm can pass through the HEPA and ULPA filters.

  In modern air purifiers, the filters are coated with titanium dioxide (TiO_2), which induces photocatalytic activity under UV and visible light. In this process, water molecules in the air react with TiO_2 and produce oxygen and hydroxyl radicals. These radicals oxidise organic compounds in germs and viruses to neutralise them. This photocatalytic oxidation technology has enormous potential for cleaning the air from germs and odours. The air filters can also be coated with an activated carbon layer, which assists in the adsorption of harmful gases. They are called carbon filters. The coating of chemicals like zeolite, potassium iodide, or potassium permanganate allows the removal of gases, especially those with small molecular weights (<46 g/mol).
- UV-C Radiation- The short wave of UV rays (i.e., UV-C rays) can be used to inactivate airborne microorganisms. UV-C irradiation mainly causes harm to the purine and pyrimidine bases of DNA and disrupts their organisation. A UV-C lamp, also called a "germicidal" lamp, is a common tool to treat microbial contamination in indoor facilities. In the research laboratory, a germicidal lamp and air filtration are preinstalled facilities in a laminar airflow unit to provide aseptic conditions for the treatment of biological materials. Hospitals and healthcare facilities also employ germicidal lamps in their operating rooms to reduce microbiological contamination. However, this facility alone is not strong enough to provide complete aseptic conditions. Also, direct exposure to high UV-C radiation poses significant potential health and safety risks for humans and plants. The prolonged exposure to the atmosphere to the UV-C radiation can induce the photolysis of O_2 and produce toxic ozone (O_3) gas.
- Disinfection- Chlorine dioxide (ClO_2) spray is a common disinfectant to control airborne microorganisms in institutions, hospitals, and offices. The use of a 70% alcohol-based spray is an alternative choice to get rid of actively suspended bioparticles. A different disinfectant is O_3. O_3 generators produce less than 0.6 ppm O_3, a safe concentration level. O_3 has superior disinfectant properties to chlorinated compounds for killing airborne microorganisms. Disinfectants can also aid in the removal of unpleasant odours from indoor facilities.
- Plantation- Many plants act as air purifiers for indoor and outdoor facilities. The beneficial plant-associated bacteria and secondary metabolites can interplay with the microbial fingerprints of humans and the outside environment. They can reduce the load of airborne pathogens in the built environment. The forestation and bushes in open areas create a physical barrier by reducing the dispersion of dust particles.
- Steam Condensation- In this process, air particles respond to steam and condense into droplets. The dust particles gravitationally settle on the Earth's surface, reducing their concentration in the atmosphere.

However, this is not a reliable method for air purification due to low efficiency. It can cause secondary issues such as mould growth, respiratory problems, and structural damage in buildings.

- Smart Infrastructure- Smart infrastructure, planning, and regulations help reduce air pollution. An effective ventilation system favours the dilution and displaces interior pollutants. An exhaust fan is a mechanical system that depressurises the building environment to decontaminate the air environment. Local exhaust is an economical and straightforward solution for air cleaning.

SUMMARY

- The atmosphere is a pool of diverse biological materials, including microorganisms. The study of microorganisms and toxins suspended in the air is called aeromicrobiology. The most dominant airborne microorganisms are bacteria and fungi.
- The living biological agents in the air find the support of non-living components, which act as the shaft. This composite is called a bioaerosol. Based on the particle sizes, they can also be termed particulate matter (PM).
- Airborne microorganisms are the dynamic pool in the atmosphere, influenced by dynamic meteorological factors, different sources, and physical forces of deposition. Based on their residence types, they can be categorised as indoor and outdoor microbiomes.
- The bioaerosol cycle is a process of emission, dispersal, and deposition of airborne microorganisms. The emission involves the uplifting of biological particles in the air. They interact with abiotic components to form bioaerosols and disperse in the atmosphere. After their transport, they are deposited on the Earth's surface.
- Airborne microorganisms can be detected through sampling and sensing. The sampling can be performed using impaction, impingement, filtration, and precipitation. The sensing involves quantification, identification, and characterisation of biological units using microscopic, spectroscopic, culture- and non-culture-based methods.
- Many airborne biological particles have cited implications for agriculture and human health, which demand immediate action and planning. The most common measures to control the bioaerosols in the atmosphere are elimination at the source, controlling the source, and controlling exposure.

DIFFERENT TYPES OF PRACTICE QUESTIONS

Fill in the blanks

I. The dormant, resting stage of specific Gram-positive cells in the air is ………………..

II. …………………… is an earthward movement of atmospheric constituents into terrestrial and aquatic ecosystems.

III. In bioaerosol, the …………….. subsets act as a 'raft' for the airborne biological components.

IV. The concentrations of microorganisms in the air are calculated in units of ……………………

V. The cultivation of airborne microorganisms in sterilised culture media-filled Petri plates exposed to the air for a definite period is called …………………

VI. A "germicidal" lamp emits ………………………, which is used to treat microbial contamination in indoor facilities.

SOLUTION: I. endospore, II. Deposition, III. non-biological components (dust and droplets), IV. cells/m^3 or mass of particulate matter in μg/m^3, V. settle plate method, VI. UV-C rays

Match the following

i. Mottled browning of tobacco leaf	a. *Corynebacterium diphtheriae*
ii. Anthrax	b. *Rhizopus*
iii. Mucormycosis	c. *Erysiphe graminis*
iv. Powdery mildew	d. *Bacillus anthracis*
v. Diphtheria	e. Tobacco Mosaic Virus

SOLUTION: i-e, ii-d, iii-b, iv-c, v-a

Match the following

i. Disinfectants	a. TiO_2
ii. Photocatalyst	b. Fungi
iii. Endotoxin	c. ClO_2
iv. Endospore	e. Gram-positive bacteria
v. Mycotoxin	f. Gram-negative bacteria

SOLUTION: i-c, ii-a, iii-f, iv-e, v-b

Very short answer type of questions

I. Define aeromicrobiology.

II. Expand on HEPA.

III. Differentiate between particulate matter and bioaerosol.

IV. Name any two extremophiles that reside in the atmosphere.

V. How do Gram-positive bacteria maintain their viability during stress periods?

VI. Name any two common bacterial and fungal inhabitants of the atmosphere.

Short answer type of questions

I. What is particulate matter? Discuss its two different size categories and their significance in the environment.
II. What are airborne pathogens? Discuss the roles of disease-causing airborne microorganisms in the environment.
III. Write a short note on the global concerns of airborne microorganisms.
IV. Give a comparative account of indoor and outdoor microbiomes.
V. What are the different factors that influence the concentrations and compositions of outdoor microorganisms?

Long answer type of questions

I. Define bioaerosol. What are the primary sources of airborne microorganisms?
II. What are the common methods to detect airborne microorganisms? Mention the challenges associated with the study of airborne microorganisms.
III. How does the bioaerosol cycle play a role in the dissemination of airborne microorganisms in diverse environments?

NOTES

1 Pöschl U, Shiraiwa M. Multiphase chemistry at the atmosphere–biosphere interface influencing climate and public health in the anthropocene. *Chemical Reviews*. 2015;115(10):4440–75.
2 Wilke RN, Hoppert M, Krenkel M, et al.. Quantitative X-ray phase contrast waveguide imaging of bacterial endospores. *Journal of Applied Crystallography*. 2015;48(2):464–76.
3 Mahata PK, Dass RS, Pan A, Muthusamy B. Substantive Morphological Descriptions, Phylogenetic Analysis and Single Nucleotide Polymorphisms of *Aspergillus* Species From *Foeniculum vulgare*. *Frontiers in Microbiology*. 2022;13:832320.
4 Gou YN, Aung SL, Htun AA, Huang CX, Deng JX. *Alternaria* species in section *Alternaria* associated with Iris plants in China. *Frontiers in Microbiology*. 2022;13:1036950.
5 Yang Y, Luo W, Zhang W, Mridha MA, Wijesinghe SN, McKenzie EH, Wang Y. *Cladosporium* Species Associated with Fruit Trees in Guizhou Province, China. *Journal of Fungi*. 2023;9(2):250.
6 www.flickr.com/photos/pikerslanefarm/51042653266/in/photostream/
7 Hoose C, Kristjánsson JE, Burrows SM. How important is biological ice nucleation in clouds on a global scale?. *Environmental Research Letters*. 2010;5(2):024009.
8 Roeters SJ, Golbek TW, Bregnhøj M, et al. Ice-nucleating proteins are activated by low temperatures to control the structure of interfacial water. *Nature Communications*. 2021;12(1):1183.
9 https://commons.wikimedia.org/wiki/File:Poa_pratensis_Erysiphe_graminis,_veldbeemdgras_met_echte_meeldauw_%282%29.jpg
10 Galhano R, Talbot NJ. The biology of blast: Understanding how *Magnaporthe oryzae* invades rice plants. Fungal Biology Reviews. 2011;25(1):61–7.
11 Griffith GW, Snell R, Shaw DS. Late blight (*Phytophthora infestans*) on tomato in the tropics. Mycologist. 1995;9(2):87–9.

SUGGESTED READINGS

Després V, Huffman JA, Burrows SM, Hoose C, Safatov A, Buryak G, Fröhlich-Nowoisky J, Elbert W, Andreae M, Pöschl U, Jaenicke R. Primary biological aerosol particles in the atmosphere: a review. *Tellus B: Chemical and Physical Meteorology*. 2012;64(1):15598.

Fröhlich-Nowoisky J, Kampf CJ, Weber B, Huffman JA, Pöhlker C, Andreae MO, Lang-Yona N, Burrows SM, Gunthe SS, Elbert W, Su H. Bioaerosols in the Earth system: climate, health, and ecosystem interactions. *Atmospheric Research*. 2016;182:346–76.

Ghosh B, Lal H, Srivastava A. Review of bioaerosols in indoor environment with special reference to sampling, analysis and control mechanisms. *Environment International*. 2015;85:254–72.

Ghosh C, Bisht K, Shelly A, Kumari R. Evaluation of size-segregated bioaerosols concentration, deposition fraction and inhalation dose in individuals of an educational institute: a case study for health risk assessment. *Air Quality, Atmosphere & Health*. 2024 Nov;17(11):2693–709.

Jones AM, Harrison RM. The effects of meteorological factors on atmospheric bioaerosol concentrations—a review. *Science of the Total Environment*. 2004;326(1–3):151–80.

Kasprzyk I. Aeromycology--main research fields of interest during the last 25 years. *Annals of Agricultural and Environmental Medicine*. 2008;15(1):1–7.

Kim KH, Kabir E, Jahan SA. Airborne bioaerosols and their impact on human health. *Journal of Environmental Sciences*. 2018;67:23–35.

Lewis D. Why the WHO took two years to say COVID is airborne. *Nature*. 2022;604(7904):26–31.

Mainelis G. Bioaerosol sampling: classical approaches, advances, and perspectives. *Aerosol Science and Technology*. 2020;54(5):496–519.

Mandal J, Brandl H. Bioaerosols in indoor environment-a review with special reference to residential and occupational locations. *The Open Environmental & Biological Monitoring Journal*. 2011;4(1):83–96.

Pepper IL, Gerba CP. Aeromicrobiology. In Pepper IL, Gerba CP, Gentry TJ (eds.) *Environmental microbiology* 2015 (pp. *89–110)*. Academic Press.

Smets W, Moretti S, Denys S, Lebeer S. Airborne bacteria in the atmosphere: presence, purpose, and potential. *Atmospheric Environment*. 2016;139:214–21.

Xu Z, Wu Y, Shen F, Chen Q, Tan M, Yao M. Bioaerosol science, technology, and engineering: past, present, and future. *Aerosol Science and Technology*. 2011;45(11):1337–49.

Zhai Y, Li X, Wang T, Wang B, Li C, Zeng G. A review on airborne microorganisms in particulate matters: composition, characteristics and influence factors. *Environment International*. 2018;113:74–90.

Zhao J, Jin L, Wu D, Xie JW, Li J, Fu XW, Cong ZY, Fu PQ, Zhang Y, Luo XS, Feng XB. Global airborne bacterial community—interactions with Earth's microbiomes and anthropogenic activities. *Proceedings of the National Academy of Sciences*. 2022;119(42):e2204465119.

19 Study of Microorganisms in Soil

19.1 SOIL ENVIRONMENT

Soil is composed of mineral particles, organic matter, air, water, and living beings. All these components in soil maintain a constant but slow interaction. Soil, with its highly enriched constituents, provides a wonderful habitat and nutrition for its residents. About 30% of the surface area of Earth is covered with land. Of these, 71% of the land has a thin layer of soil, which is habitable. The interaction between various non-biological and biological factors makes soil extremely complex and dynamic. This active site of the Earth's crust is called the "**pedosphere**". The study of soil in a natural environment is called **pedology**. Healthy soils teem with micro and macroscopic organisms. Soil is also a reservoir for the dispersal units and resting stages, such as gametes and spores. Approximately 86% of the total biomass of Earth (≈480 gigatons of carbon (Gt C)) is stored in land-based ecosystems.

Soil resides at an interface between Earth, water, and air. It is considered a non-renewable resource because its formation is extremely slow, often exceeding human lifespans. Soil anchors vegetation and provides physical support and nutrients to plants. In return, plants shed organic matter from their fall and decay for soil health. Plants are the key food sources for most soil residents. They are important components of food chains and the food web. They offer very crucial services to the soil ecosystem. Soil-dwelling invertebrates such as earthworms, collembolans, mites, nematodes, and protozoans contribute to nutrient cycling, aeration, and soil structure (Box 19.1). Soil provides a solid substrate and strong base for the movement and other activities of humans as well as wildlife. Humans use soil as a steady platform to build various artificial structures through engineering for recreational and landscaping purposes.

Based on the size of the particles, soil can have different textures, namely clay, sand, silt, peat, chalky, and loam. They possess distinct characteristics and develop special habitats for distinct life forms (Table 19.1). The vertical soil layers are also very heterogeneous due to geogenic activities (i.e., mainly breakdown of rocks and minerals), pedogenic processes (i.e., formation of soil), or a mixture of both. The depth of most of the soil ranges between a few centimetres and metres, which is minuscule compared to the radius of Earth, i.e., 6,000 km. The vertical section of soil is known as the **soil profile**, categorised into different layers called **horizons**. This makeup includes soil horizons O, A, B, C, and R, which have further subcategories (Figure 19.1). The O horizon is a thin organic surface layer rich in organic matter. These matters exist in various stages of decomposition. The A horizon is topsoil and provides maximum support for terrestrial life. The B horizon refers to the subsoil, a zone of accumulation that consists of iron oxides and clay minerals. The C horizon is a substratum or parent material of poorly weathered or unweathered rocks comprised of calcium carbonate ($CaCO_3$). The R is the deepest horizon among all layers and is a bedrock.

19.2 SOIL MICROBIOLOGY

Soil microbiology is the study of microorganisms in soil and their functions, interactions, and impacts on soil health and ecosystems. Microorganisms are the most dominant life forms in soil. Soil provides a wide variety of niche spaces for microorganisms. Soil microorganisms live in large groups, dwelling as colonies, biofilms, or aggregates. Micropores and macropores of the soil provide sufficient space for microbial residence (Figure 19.2). A single gram of healthy soil can inhabit approximately 10 billion (10^{10}) microorganisms (Table 19.2). This number surpasses the total number of humans present on the Earth, with an estimated current human population of around 8 billion. The microbial diversity in the soil can range from thousands of microbial species from all three domains of life. However, they comprise only about 1% of the total soil volume. Microbial diversity is one of the main drivers for the functioning of terrestrial ecosystems. Soil microorganisms can be treated as both a component and producer of soil organic matter (SOM) that rests in the soil for a very long period and enhances soil fertility, aeration, and water-holding capacity. Many soil microorganisms are capable of carbon mineralisation.

Based on physiological states, soil microorganisms can be categorised as active, potentially active, dormant, and dead cells. The active forms have a relatively low proportion (less than 5% in most cases). They are native species and use nutrients from their surroundings for their metabolic activities. The potentially active form of microorganisms is

DOI: 10.1201/9781032651095-24

BOX 19.1 THE DRILOSPHERE: A MICROBIAL HOTSPOT IN SOIL

The term drilosphere was pioneered by M.B. Bouché (1975). It is a unique microhabitat of soil impacted by earthworms. Microbial activities also have a fundamental role in the formation of the drilosphere. Particularly, earthworm-microbiota interactions in soil build a physical boundary of the drilosphere, keeping distinct properties of self-organising systems. Drilisophere is considered a microbial hotspot in soil. The soil particles in the drilosphere are associated with intense microbial exercises, responsible for different ecologically relevant events.

Some hot moments in the drilosphere are food preferences, digestion, excretion of organic matter, and the digging of earthworms. When earthworms live in soil, they have a high capacity for litter decomposition. This adds large quantities of organic matter to the soil. They can, however, immediately reduce a small proportion of microbial abundance in soil by ingesting native microorganisms along with litter. Meanwhile, worm poop plays an important role in fostering microbial activity by expanding soil nutrient concentration (Box Figure 19.1.1). The input of labile organic matter in soil triggers rapid microbial growth. The gut of earthworms also inhabits diverse microorganisms, which play a role in digestion. Gut microorganisms aid in the necessary enzymes required for the breakdown of recalcitrant substrates. The casting is an excellent soil microbial source, offering a full range of microorganisms through the gastrointestinal route and a high concentration of organic humus. The energy-rich mucus in casts sends repeated signals to dormant soil microorganisms. The digging by earthworms encourages the formation of burrows, tunnels, and middens, which provide unique habitats for distinct microorganisms. Such microbial hotspots exist at micro-scale levels yet have large cumulative effects on ecosystems and landscapes at macroscales.

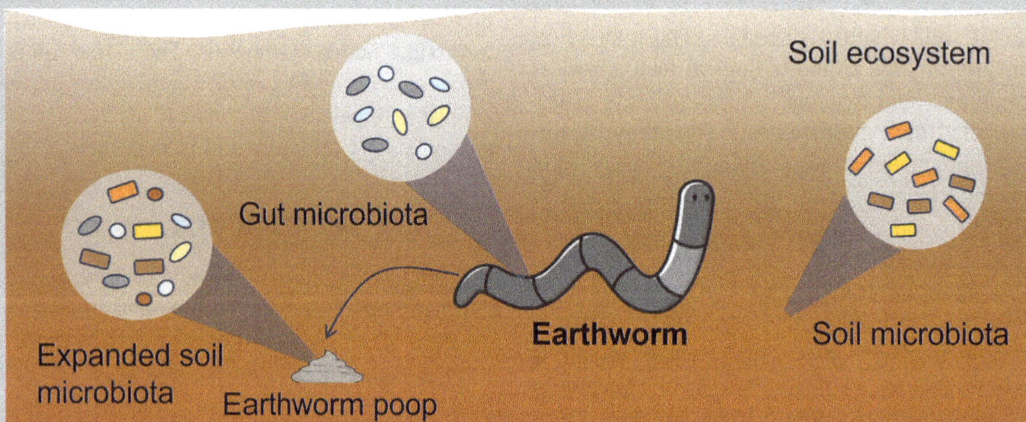

BOX FIGURE 19.1.1 Microbial composition in the drilosphere.

living but slow-growing because of unfavourable situations. Such microorganisms can become active within a few minutes or hours when situations improve and cover a sizable area of the soil. Microbial species having such a physiological state cover the maximal area in the soil and include as much as 40-60% of the total microorganisms. However, when a microbiota is under stress, microorganisms enter the dormant stage. It has much less physiological activity and takes a much longer time to set an active stage. On the other hand, dead cells have no metabolic activity, but they still contribute to the soil's overall carbon and nitrogen turnover.

The composition and function of microbial communities in soils also display spatial and temporal variabilities. In geographic space, microorganisms can be present in both vertical and horizontal regions. On a horizontal scale, the rhizosphere is the most notable site of microbial abundance and diversity. This region is dominated by plant growth and influenced by root exudates. The excess amounts of organic matter and nutrients favour the growth of beneficial microorganisms. Other regions with rich microbial diversity can be the detritusphere, biopores, and aggregate surfaces. Microorganisms can also be reported from different vertical layers of the soil profile. The abundance, activity, and diversity of microorganisms are all highest at the top level of the soil, which tends to decline with increasing soil depth. In vertical layers, soil microorganisms have a major influence on factors like texture, organic matter, porosity, oxygen (O_2), and carbon dioxide (CO_2).

A sufficient number of microorganisms also grow inside soil particles and rocks and live a distinct lifestyle. They are termed **endolithic microorganisms**. Endoliths can be heterotrophic and autotrophic organisms. Heterotrophic endoliths survive on organic compounds, while autotrophic endoliths harbour chemolithotrophic lifestyles using inorganic compounds. However, some microorganisms grow on the surface of the rock, penetrating the rock for nutrients. They

FIGURE 19.1 Soil horizons showing different layers.

can be photoautotrophs. For example, lichens growing on rock can perform photosynthesis. Endolithic habit represents an incredible example of resilience and adaptation, demonstrating survival strategies to live and thrive in hostile conditions. They also contribute to the mineral solubilisation and degradation.

19.3 MICROBIAL FORMS OF SOIL

Numerous types of soil microorganisms are bacteria, archaea, fungi, microalgae, and protozoans, with bacteria occupying the most prominent position, followed by fungi. Bacteria possess great adaptability in soil, showing diverse characteristics. They exist in 10^8-10^{10} cells per g of soil at 0-15 cm depth. Bacteria typically come in the size range of 0.5-10.0 µm, which overlaps with particle sizes of clay (< 0.2 µm) and silt (2-50 µm) (Table 19.1). Thin water films inside the soil provide sufficient space for bacterial growth and other activities. Many soil bacteria can easily be doubled within 15-30 min by binary fission. They can utilise diverse substrates found in soil. Many bacterial residents are important sources of enzymes, antibiotics, and other secondary metabolites. Many species exhibit tolerance to several physical and chemical stressors. Some possess the ability to acclimatise in altered environments. The members of the phyla Proteobacteria, Actinobacteria, and Acidobacteria are widespread in all types of soils. The members of Actinobacteria and Cyanobacteria in soil are responsible for the "earthy" smell of freshly ploughed soil. These bacteria release a volatile secondary metabolite, **geosmin**, which tends to accumulate in soil pores (Figure 19.3). When the first rain spouts, geosmin is released from the soil into the air. The special odour of geosmin is called "**petrichor**". Actinobacteria are a class of filamentous bacteria that build hyphal networks

TABLE 19.1
Approximate dimensions (µm) of different soil particles, biota, water-filled pores and water films

	Types	Approximate dimensions (µm)
Soil particles	Stones	>2,000
	Coarse sand	2,000–200
	Fine sand	200–50
	Silt	50–2
	Clay	2–0.2
Plant materials	Roots	1,000
	Fine roots	1,000–50
	Root hairs	15–7
Microorganisms	Fungal hyphae	10–3
	Actinomycetes	10–1.5
	Bacteria	10–0.5
	Viruses	0.2–0.05
Microfauna	Earthworms	5,000–2
	Mites	2,000–500
	Nematodes	2,000–500
	Protozoa	80–10
Water-filled pores	10 kPa	<30
	100 kPa	<3
	1,000 kPa	0.3
Water films	100 kPa	<0.0003
	1,000 kPa	Few molecules thick

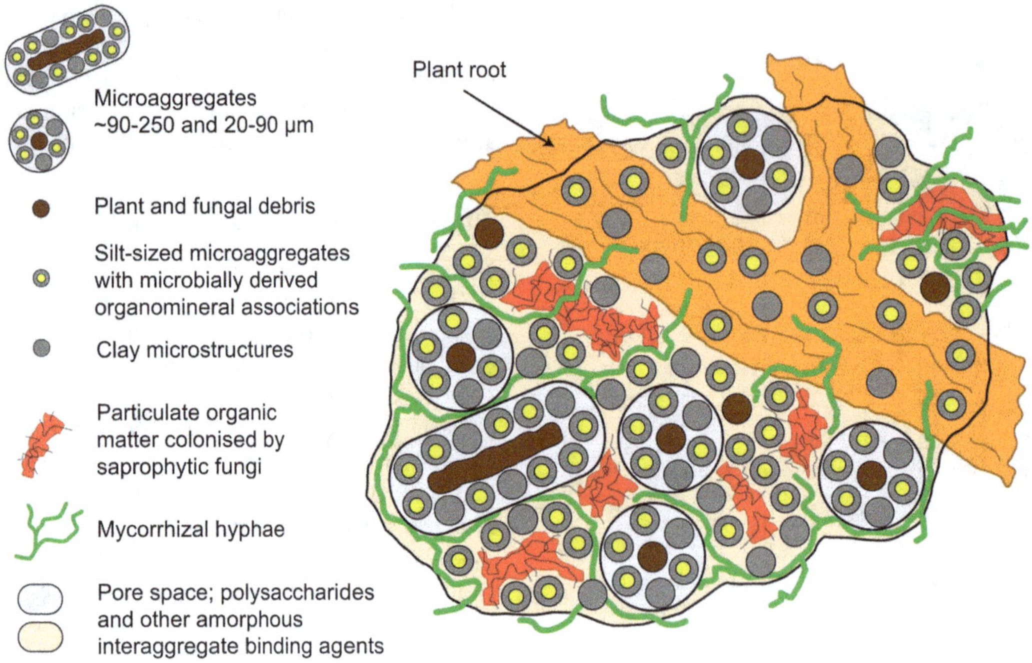

FIGURE 19.2 Schematic diagram of a soil macroaggregate. (Image adapted from Jastrow and Miller (1998)[1] with permission of CRC Press)

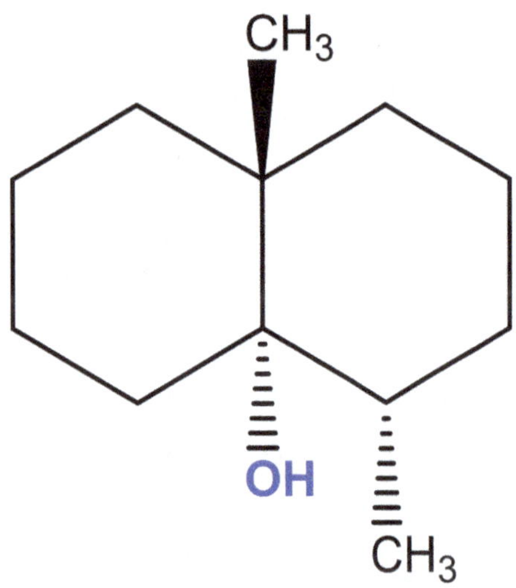

FIGURE 19.3 Chemical structure of geosmin.

in the soil. About 90% of the actinobacteria inhabiting the soil belong to *Streptomyces*. Cyanobacteria are found only on the soil surface because they are photoautotrophs. They require sunlight for photosynthesis and produce organic matter and O_2. The presence of cyanobacteria in soil plays a crucial role in controlling microbial abundance, diversity, and physiology. They are considered pioneers on the bare rock surfaces and also of the early Earth (details in Section 19.4).

In soil, archaea are less diverse than bacteria, but they exhibit high genetic and phenotypic flexibility. In temperate soil, the proportion of archaea can be as high as 10% of total microbial cells. The members of Euryarchaeota are common inhabitants of waterlogged and dry soils. They are methanogens and can produce methane (CH_4) in anoxic conditions. Saline soil harbours members of Haloarchaea that survive in halophilic environments. Many soil bacteria are ammonia-oxidisers and belong to the phylum Thaumarchaea. They convert ammonia (NH_3) into nitrite (NO_2^-). In soil ecosystems, archaea play roles in the ecological processes of carbon and nitrogen turnover. In addition, they contribute to the bioremediation of contaminated soil and produce unique enzymes and secondary metabolites.

Fungi develop a complex network of hyphae in soil. They are the second most abundant microorganisms after bacteria and constitute up to 30% of the microbial communities in healthy soil. Also, fungi account for a very high biomass due to their long mycelial network. They reside as decomposers, mutualists, and pathogens in the soil. They can produce different resting spores in the soil, which can germinate under favourable conditions. Moist conditions and low pH are often considered the best for most fungal taxa. Most fungi in the soil come from the phylum Ascomycota. The phylotypes of the phylum Basidiomycota are also fairly common in soil in terms of their abundance and richness. The most common soil fungi are moulds, yeasts, and mushrooms. Fungal taxa like *Cladosporium*, *Toxicocladosporium*, and *Alternaria* have cosmopolitan distributions. Some soil fungi establish mycorrhizal associations with plant roots. In forest areas, the mycelia of mycorrhizal fungi can extend up to several kilometres, whereas their filaments in agriculture spread over several meters. Due to their extensive growth, fungal biomass in soil can reach as high as 1,500 g/m^2. Many fungi exhibit rapid growth, high plasticity, and resistance to noxious chemicals and physical stresses, which enable them to survive even in contaminated regions.

TABLE 19.2
The abundance of different microorganisms in healthy soil

Microbial group	Abundance (cells/g of soil)
Bacteria	10^8–10^{10}
Fungi	10^5–10^7
Protozoa	10^3–10^6
Microalgae	10^3–10^4
Archaea	10^4–10^7

Protozoa are unicellular eukaryotic microorganisms and active forms of soil microfauna. More than 1,600 species of protozoa have been identified in the soil. One gram of soil can inhabit about 10^3-10^6 protozoan cells (Table 19.2). Flagellates, amoeba, sporozoans, and ciliates are the most common soil residents. Their size ranges between 5-500 μm in diameter, which is much bigger than the size of any other microorganisms. Many protozoans survive adverse conditions by forming a dormant stage called a "cyst". They mainly reproduce through the asexual method of binary fission. Soil protists exhibit functional versatility in their feeding habits. Many soil protozoans feed on bacterial cells and influence the bacterial community compositions (BCCs) and their function in soil. Soil containing numerous protozoa produces more nitrogen for plants than soil containing fewer or no protozoa. Approximately 30% of bioavailable nitrogen and high amounts of carbon in soil are produced by the protozoan consumption of bacterial cells. Most naked amebae and flagellates are regular consumers of bacteria. Protozoans can also feed on other microorganisms like fungi, microalgae, and smaller protozoans. This feeding habit is called **phagotrophy**. They also respond to various natural and anthropogenic activities, which make them suitable bioindicators of soil pollution.

Contrary to other microbial forms, microalgae grow only at the soil surface because they require sunlight. They often develop green or brown sticky slime patches in the soil and on walls. Excess moisture and humidity are basic requirements for their growth. Their abundance is comparatively much less than the abundance of heterotrophic microorganisms. A healthy soil supports around 10^2-10^4 microalgae per gram of soil. Soil microalgae inhabitants can belong mainly to four different phyla: Cyanophyta (blue-green algae), Chlorophyta (green algae), Bacillariophyta (diatoms), and Xanthophyta (yellow-green algae), the former two being dominant in soil. Cyanophyta is a prokaryotic group, and the other three phyla belong to eukaryotic microalgae. A Cyanophyta member, *Nostoc*, resides in soil and establishes a symbiotic relationship with plant roots (e.g., *Cycas*). The key representatives of eukaryotic soil algae are *Chlorella* and *Chlamydomonas* from Chlorophyta, *Navicula* and *Pinnularia* from Bacillariophyta, and *Botrydium* and *Vaucheria* in Xanthophyta.

19.4 MICROBIAL PROCESSES IN SOIL FORMATION

Soil formation is a prolonged process and involves a series of changes in bare rock. This process is called pedogenesis. Synthesis of a few centimetres of soil takes at least 50 years. The process is even slower in higher altitudes. Soil formation is a result of multiple abiotic and biotic transformations of geological parent materials. A pedologist, Hans Jenny (1941), proposed an equation for soil formation including the factors influencing soil formation as follows:

$$S = f(cl, o, r, p, t, \ldots)$$

Where S represents soil formation; f, function of; cl, climate; o, organisms (microorganisms, mesofauna, and soil biology); r, relief (or topography); p, parent material; t, time; and '…', other factors.

Soil formation is a gradual and continuous process that has evolved over geological time scales. The processes of rock disintegration appear relevant even for the early stages of ecosystem evolution when the early life forms on Earth's crust started to settle. The primary minerals and rocks on Earth were insoluble and non-bioavailable. The earliest record of the evolution of terrestrial ecosystems comes from the Ordovician period (around 450 million years ago (mya) when plants started to colonise the land. However, the development of contemporary soils occurred during the Devonian period (around 416 mya). This period also witnessed the first appearance of land plants, where microorganisms had already conquered space.

Weathering is the first step of soil formation, which involves the disintegration, decomposition, and chemical modification of parent rocks through physical, chemical, and biological processes. In physical processes, the parent materials experience transformation through temperature fluctuations, precipitation, grinding, and cycles of wetting and drying. Chemical decomposition brings notable changes in the structure and composition of the soil through hydrolysis, carbonation, redox reaction, and hydration. In biological methods, bacteria, fungi, and microalgae are the main architects of soil formation. The microbial action in the soil involves biological transformations of native components for nutrient development. They release inorganic/organic acids and chelating agents, which aid in mineral dissolution and rock degradation. They drive organic matter, fertility, and biogeochemical cycles in the soil. Biological methods also indirectly contribute to mechanical weathering, involving penetration, expansion, and contraction of biological objects. Decomposers are particularly important in the recycling of organic matter in soil. Humus is the dark, decomposed organic matter formed in the soil, which contains essential nutrients for soil health. Humus also improves the water-holding capacities of soil. Many farmers boost soil fertility using compost as a humus-rich fertiliser. Compost is a product of the action of native microflora and invertebrates on dead plant tissues.

FIGURE 19.4 Lichens growing on the surface of rocks. a. Crustose lichen. b. Foliose lichen. (Source of image b: Wikimedia/Daniel Schwen.[2])

The presence of sufficient humus in soil has a substantial impact on microbial growth and their activities, as well as the overall health and fertility of the soil. The dissolved minerals in the soil provide nutrients for the development of diverse organisms, accelerating soil formation. The roles of lichens, in particular, are very crucial for soil formation.

19.4.1 Role of Lichens in Soil Development

In ecological succession, lichens are the pioneer communities in the early stage of soil ecosystem development on bare, extremely dry terrains. Lichens are composite organisms of two symbiotic partners, cyanobacteria/microalgae (phycobiont) and fungi (mycobiont). The phycobiont performs photosynthesis and prepares food for the mycobiont. On the other hand, the mycobiont is responsible for producing diverse secondary metabolites for its growth and expansion. One of its forms, crustose lichens, is the pioneer community in the bare land. They adhere to the surface of barren rock and give a crust-like appearance (Figure 19.4a). Crustose lichens are so firmly attached to the substratum that they are practically inseparable from the surface. They release acids in their surroundings that disintegrate rock and aid in weathering. When these lichens die, they shed organic matter and nutrients into the ground. A different form, foliose lichens are flat, leafy-like, and attached to the substratum by rhizine (Figure 19.4b). The rhizines are distinct fungal hyphae which penetrate and physically break down the substratum. These lichen forms also produce organic matter through photosynthesis and release organic acids, adding layers of organic matter on the surface. In addition, lichens set the stage for the growth of other types of organisms.

19.5 FACTORS INFLUENCING MICROORGANISMS IN SOIL

Environmental factors have critical roles in shaping the soil microbial communities. This covers the effects on diversity, composition, and activities of native microbiota. Their impression of soil microbiota can have a substantial impact on nutrient cycling, carbon sequestration, and soil fertility. They can be categorised into physical, chemical, and biological factors (Figure 19.5). Many factors are interconnected and influence each other. They matter as cumulative effects of different factors. These factors are further discussed as follows.

19.5.1 Physical Factors

Different soil types (such as sandy, loamy, and clayed) have different properties, and they can be important drivers of microbial community structure and functions. For example, the pore space in soil is an important requirement for microbial abundance. This provides not only space but also water, air, and sufficient nutrients for their growth. Compact soils are often bad indicators of soil microbial health. This is because bulk density reduces the space and permeability in soil microenvironments. The increasing soil depth exhibits a negative correlation with the richness and diversity of microbial communities. Deep soils are less aerated and compact and have limited nutrients, which creates a less hospitable environment for microorganisms. Soil microorganisms are most dominant near plant roots and improve soil stability. However, soil erosion can pose adverse effects on the soil microflora. Major impacts of erosion-induced changes include poor microbial network complexity, reduced soil multifunctionality, relatively less abundance of microbial taxa, and impaired associations between different microbial taxa.

Soil microbiota experience variation in their assembly and abundance due to seasonal dynamics. The summer and autumn seasons induce comparatively higher microbial activities than other seasons. The autumn season in temperate regions adds a huge amount of organic matter to soil due to litter fall, which contributes to its decomposition through microbial activities. Microorganisms residing at the top layers are least active during winter seasons due to low temperatures. Typically, warm temperatures in soil have positive effects on microbial activities and accelerate biological degradation. The amount of moisture in the soil imparts notable effects on

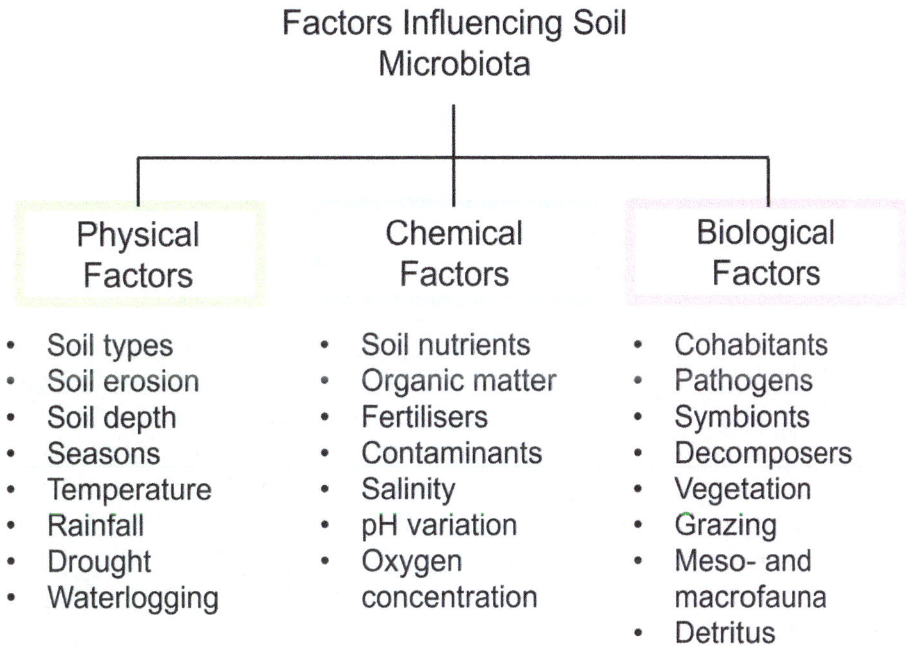

FIGURE 19.5 Different factors influencing soil microbiota.

microbial establishment. Natural rainfall controls the spatial coverage of soil moisture. The flooded rice soils harbour anaerobic microbial forms, particularly methanogens. The soil needs certain moisture levels for the dormant forms of microorganisms to transform into physiologically active forms. Drought has severe impacts on plant development, which is also an alarming constraint for rhizospheric microorganisms. The dry regions are occupied with small and distinct patches of habitats due to the physical isolation of water bodies. Loss of soil moisture can devastate microbial growth and activities by reducing intracellular water potential. The fragmented aqueous phases also reduce O_2 diffusion (for aerobic microbial activity) and provide protection against various stresses. Restricted substrate supply is another limiting factor for the growth of microorganisms in physically dry environments.

A mature soil microbial community is more resilient to disturbance than a young soil. Temporal variability has a relatively smaller contribution than spatial variability for the soil microbial communities. The temporal variability of soil microorganisms depends on climatic factors and seasonal changcs. Microbial scnsitivity to scasonal variation is also quitc apparent. Summer seasons favour the growth of fast-growing bacteria, while winter seasons retard metabolic activities, leading to poor growth. Spring seasons support more species richness than any other season. Rainfall creates conditions favourable for acid-tolerant microorganisms and decomposers.

19.5.2 Chemical Factors

The top level of soil tends to have high concentrations of soil nutrients and organic matter, resulting in extensive dominance of soil microorganisms. Different organic matter sources (plant residues, animal manure, compost, and synthetic matter) and their concentrations strongly influence microbial composition. Organic carbon is a source of energy and nutrients for soil microbial respiration. The high content of recalcitrant organic materials in soil impedes microbial growth. In agriculture, the application of organic wastes such as compost and manure enhances microbial activities for plant growth and crop yield. Volcanic eruptions also enrich the land surface with different nutrients and increase microbial diversity.

Nutrient cycling is a key factor that controls microbial diversity in soil and is linked to fulfilling the nutritional requirements of the plants. Prolonged use of chemical fertilisers (e.g., nitrogen, phosphorus, and potassium) in agricultural land has altered microbial community compositions (MCCs) and their function. Many micronutrients, such as iron (Fe), manganese (Mn), zinc (Zn), and copper (Cu) in soil, are essential requirements for the growth of microorganisms. However, their high concentrations are detrimental to many species. The presence of toxic chemicals and heavy metals (e.g., mercury (Hg), cadmium (Cd), and lead (Pb)) can lead to scvcrc loss of thc nativc microbial community. Salinity has also altered the microbial composition. An increase in soil salinity has induced the growth of halophiles. Soil pH fluctuation has demonstrated the influences of both qualitative and quantitative variations in the community composition. Soil oxygen levels have a vivid impact on their inhabitants. The well-oxygenated soils induce the growth of aerobic microorganisms and support organic decomposition. Oxygen diffusion is maximum at the topsoil and is greatly influenced by the physical properties of the soil, including texture, porosity, tortuosity, conditioning, etc. Low oxygen environments on land can arise due to O_2 displacement by water or other gases, high compaction and impermeable barriers. These contrasting habitats of an anoxic environment mainly harbour sulphate-reducing bacteria and methanogenic archaea.

19.5.3 Biological Factors

The overall microbial structure of the soil is a result of competitive interactions between microbial communities themselves and with other organisms. They establish relationships as pathogens, symbiotic mutualists, and decomposers with other organisms. Soil microorganisms can accomplish cell-cell communication with inter- and intra-kingdom species for their growth and reproduction. Many soil microorganisms obtain nutrition and metabolic support from other living species. An obligate pathogen can not survive without a host, while facultative pathogens can survive in the absence of suitable hosts (detailed in section 19.9). Symbiotic mutualism facilitates an exchange of physiological factors between different species and exhibits mutual dependency for their growth. An example of obligatory mutualism is mycorrhiza. The presence of dead organic components from plants, animals, and microorganisms encourages microbial proliferation through the decomposition process. The number of trophic levels in a food web also controls microbial community structure and diversity.

Soil microorganism-plant interactions play a key role in developing microbial communities in the rhizosphere and phyllosphere. Different plant species and their different growth stages can harbour distinct microbial compositions. The types of vegetation also mark notable variations in their structure and function. A small degree of change in the environment can have long-lasting effects on species interactions. The photosynthetically-derived matter is the main source of organic carbon in the vegetated land. Non-vegetated soils have low microbial species abundance and low microbial diversity. Organic carbon is usually a limiting factor for microbial diversity in non-rooted bulk soil.

The role of grazing on microbial communities is also evident. However, the effects of grazing on soil microbiota at different sites are inconsistent. Many soil-inhabiting faunas increase the fragmentation of coarse particles into fine particles and affect the spread of organic matter in the soil. They not only dig, ingest, digest, and disperse dead organic matter but also disseminate microorganisms. These organisms belong to microfauna (e.g., single-celled protozoans and small nematodes), mesofauna (e.g., mites, nematodes, and centipedes), and macrofauna (e.g., earthworms, termites, and ants). They have the ability to break down detritus (dead particulate organic material), which greatly influences fungal and bacterial activities for their turnover. They are responsible for the synthesis of labile and simple organic matter. The roles of earthworms in improving micro-habitats are also well established. They provide routes for the infiltration and transport of soil microorganisms and stimulate microbial growth and activities (detailed in Box 19.1).

19.6 SOIL ENZYME ACTIVITIES

Soil activities are essentially dependent on the diffusion of enzymes from source organisms and substrates and products of the reactions. The sources of soil enzymes can be various living and dead biological organisms, including microorganisms, plants, animals, and humans, of which the microbial counterpart is enormous. Other sources of soil enzymes can be the urine and faeces of human and animals. Many microbial enzymes are discharged into the soil, which become an integral component of the soil and remain associated with the soil for a long duration. Also, a variety of enzymes are permanently bound to cells and perform their activities in all three stages of life, including active, dormant, and dead cells. In general, enzyme activities are higher near the surface and lower at greater depths. They have a role in stabilising the soil matrix, organic decomposition, and nutrient cycling.

Soil enzyme activities act as indicators for soil quality and fertility. Soil enzymes are highly sensitive to ecosystem management and soil stress and display an early response to any changes. As many soil quality parameters depend on soil enzyme activities, their monitoring provides valuable information on biochemical reactions for soil health. The most notable indicator soil enzymes are dehydrogenase, phosphatase, arylsulphatase, and β-glucosidase. Many hydrolytic enzymes, such as amylase, chitinase, and urease, help in the degradation of organic wastes and are useful parameters for determining the soil quality index (SQI). SQI is a quantitative estimation of the overall health and function of soil by combining various soil properties. Microbial enzymes in soil also have an important influence on the biodegradation and biodeterioration of harmful environmental contaminants, as well as carbon storage. Also, soil enzyme activities in soil contribute to the balanced flow of nutrients in biogeochemical cycles (Table 19.3).

19.7 RHIZOSPHERIC MICROBIOME

A small region at the plant root-soil interface is the **rhizosphere**, which harbours great microbial diversity. This term was coined by Lorenz Hiltner in 1904. Rhizospheres exhibit tripartite interactions involving useful microorganisms, pathogens, and plants. Such concomitant activities make a rhizosphere playground and battlefield for the beneficial and deleterious effects of many microorganisms on plants. The microbial inhabitants of the rhizosphere are recognised as the **rhizosphere microbiome** or **rhizomicrobiome** (Figure 19.6). Compared to non-rooted bulk soil, the rhizosphere soil is the hotspot for microbial activities. The microbiomes of the rhizosphere soil offer diverse benefits for plant communities. The plant rhizosphere-microbe partnerships have direct impacts on plant growth and yield, as well as soil health. The effects of rhizomicrobiome are not limited to the roots; their effects are also apparent in other plant parts (such as shoots, leaves, and fruits). Bacteria are the most common microbial inhabitants of the rhizosphere and have various beneficial

TABLE 19.3
Different soil enzymes as indicators for soil quality, fertility, the enzyme reaction involved, and their roles in biogeochemical cycles

Soil enzymes	Enzyme reactions	Biogeochemical cycles
Dehydrogenase	Electron transport system (ETS)	Carbon
β-glucosidase	Cellobiose hydrolysis	Carbon
Phenol oxidase	Lignin hydrolysis	Carbon
Phosphatase	Release of the phosphate group	Phosphorous
Arylsulphatase	Release of the sulphate group	Sulphur
Cellulase	Cellulose hydrolysis	Carbon
Amylase	Starch hydrolysis	Carbon
Chitinase	Chitin hydrolysis	Carbon
Protease	Protein hydrolysis	Nitrogen
Urease	Urea hydrolysis	Nitrogen

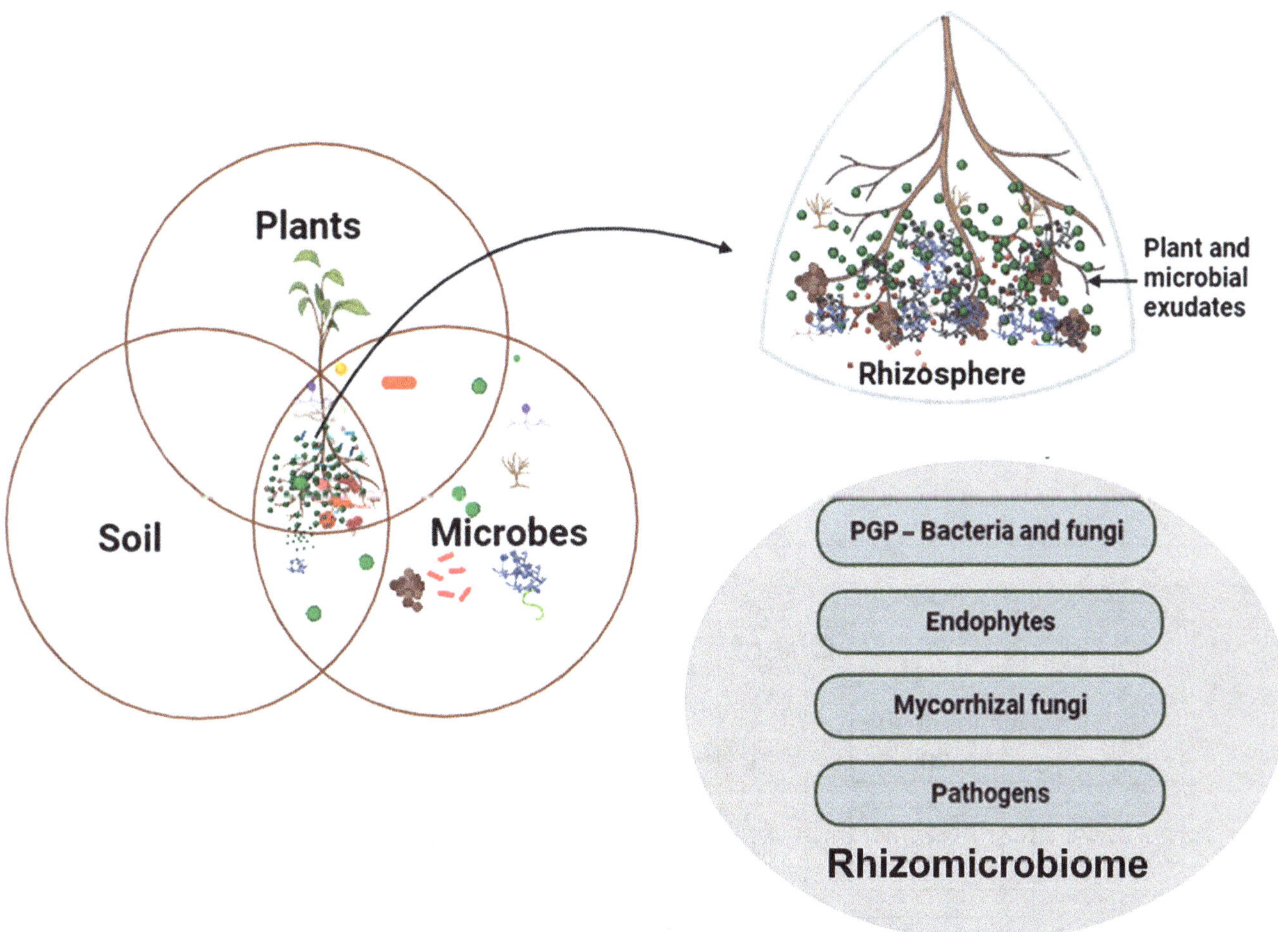

FIGURE 19.6 Plant-microbe interaction in soil and different types of microorganisms present in the rhizomicrobiome. (PGP stands for plant growth promoting) (Image adapted from Chauhan et al. (2023).[3])

effects. They are often pronounced as **plant growth-promoting rhizobacteria (PGPR)**.

The microecological region of the rhizosphere is a dynamic environment due to the diverse biotic and abiotic interactions and consists of three distinguished zones, including rhizoplane, ectorhizosphere, and endorhizosphere. Rhizoplane is the microenvironment on the surface of plant roots. It is a very thin layer where microorganisms are attached to the root surface through flagella, fimbriae, and mucilage. These microorganisms remain in direct contact with plant roots and soil particles simultaneously. Ectorhizosphere is rhizospheric soil away from the root surface, but it has exposure to root exudates. Soil pores of the ectorhizosphere inhabit many microorganisms and play a role in soil health. The endorhizosphere is a region within plant root tissues. A large number of microorganisms inhabit the root cells and are known as **endophytes**. The root tissues are the favourite sites for microbial residence, where they colonise the root cortex, epidermis, and root hairs. Root nodule formation and mycorrhizal association are two key examples of mutualistic relations between plant roots and microorganisms in the soil. During their interaction, plant roots secrete chemical exudates, which act as chemical signals for microbial colonisation. The microbial symbiont also releases some chemicals to establish communication with plants. In the endorhizosphere, plant roots secrete numerous secondary metabolites, which support microbial proliferation. These chemical exudates include carbohydrates, organic acids, amino acids, and other secondary metabolites, which are also collectively termed "**rhizodeposits**". In agriculture, rhizospheric microorganisms have tremendous scope as biofertilisers and can be used as an alternative to chemical fertilisers.

The following sections provide details on soil microorganisms for their interaction with plants and for soil health.

19.7.1 Plant Growth-Promoting Rhizobacteria (PGPR)

PGPR are bacteria that reside in the rhizosphere and can enhance plant growth. Bacteria with similar properties found in diverse locations can be recognised as plant growth-promoting bacteria (PGPB). For example, bacteria with plant growth-promoting properties isolated from leaves or stems are called PGPB, while those that colonise the rhizosphere are referred to as PGPR. They cover a wide range of taxa such as *Acinetobacter*, *Arthrobacter*, *Bacillus*, *Enterobacter*, *Herbaspirillum*, *Microbacterium*, *Paenibacillus*, *Pseudomonas*, *Ralstonia*, and *Stenotrophomonas*. The term PGPB is used in a broad scope, in which PGPR can be inclusive. Considering the diversity, richness, and activities of PGPR, they are of major importance for soil fertility and plant health. The interior of the main axis of the root, root hairs, and lateral roots are the preferred choices for their colonisation. The relationship between PGPR and plants is less specific and less obligate. However, certain host-specific symbionts, such as nodule-forming species, also exist. Those which reside in legume roots are called rhizobial endophytes.

PGPR perform multifarious activities to support plant growth under both normal and stressed environments. These activities can have direct and indirect benefits to the plants (Figure 19.7). The most pronounced direct activities of PGPR for plant growth are phytohormone production, nitrogen fixation, phosphate solubilisation, siderophore production, and hydrogen cyanide (HCN) production. PGPR produces phytohormones like auxins, cytokinins, and gibberellins, which act as molecular signals to manipulate plant growth. Bacterial indole-3-acetic acid (IAA) exerts strong effects on root development. IAA-producing bacteria include *Bacillus*, *Pseudomonas*, *Erwinia*, *Pantoea*, *Azospirillum*, and *Rhizobium*. Gibberellins help in the rapid growth of shoots, which is pronounced in bacteria like *Bacillus, Azotobacter*,

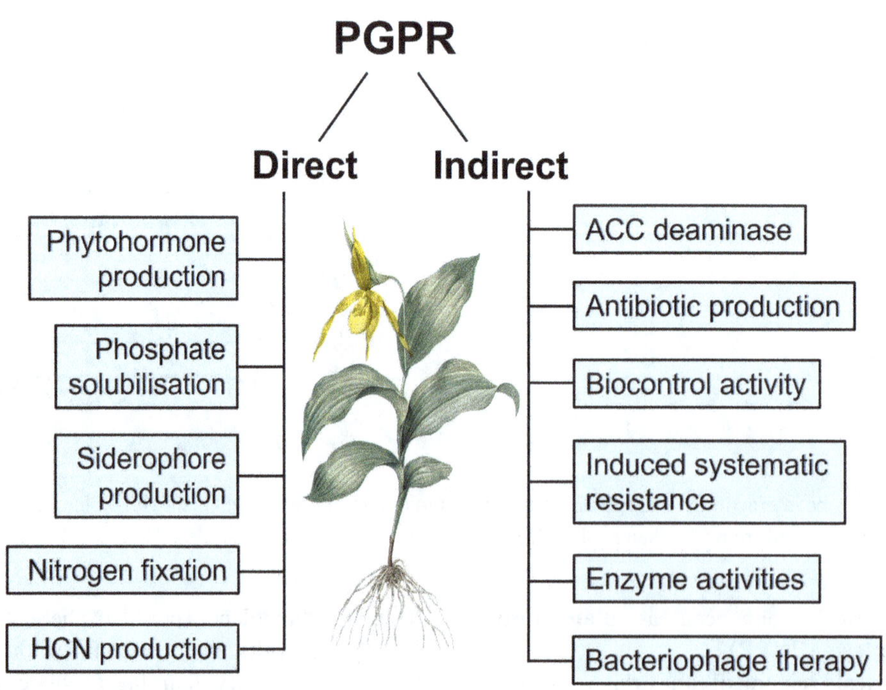

FIGURE 19.7 Direct and indirect benefits of PGPR. (PGPR stands for plant growth-promoting rhizobacteria, HCN for hydrogen cyanide, and ACC for 1-aminocyclopropane-1-carboxylate.)

Pseudomonas, *Burkholderia*, and *Acetobacter*. Soils typically contain phosphorus in non-soluble forms. Phosphorus-solubilising bacteria (PSB) produce special organic acids and phytases, which contribute to solubilising the phosphorus for plant intake from its nonreactive recalcitrant states. PSB can belong to a wide range of genera, such as *Arthrobacter*, *Bacillus*, *Enterobacter*, *Klebsiella*, *Pseudomonas*, *Serratia*, etc. Many other nutrients, such as Fe and Zn, also reside in nonmobile forms. Some specific PGPR facilitate their solubilisation by producing siderophores or organic acids. Siderophores are low-molecular-weight molecules that chelate Fe for plants. Examples of siderophore-producing bacteria are *Bacillus*, *Burkholderia*, and *Pseudomonas*. Some PGPR produce HCN that has antimicrobial properties, offering a role in suppressing plant pathogens.

Many other secondary metabolites and volatile organic compounds (VOCs) have prominent effects on stimulating growth and improving stress tolerance in plants. For example, *Bacillus megaterium* has roles in improving root architecture, increasing biomass, and elevating photosynthesis by secretion of polyamines. Cyanobacteria have proactive roles in improving the structure and stability of soil and its aggregation. Cyanobacteria liberate some adhesive materials, called extracellular polysaccharides (EPS), that warrant the binding of non-aggregated soil particles for water retention, aeration, and nutrient availability in the soil.

Nitrogen is one of the most important plant nutrient requirements. Nitrogen-fixing bacteria and archaea are the only sources of plant-available nitrogen. They are abundantly present in healthy soil and are collectively referred to as **diazotrophs**. Under strict anaerobic conditions, they use an enzyme complex called nitrogenase to convert air nitrogen gas (N_2) to ammonia (NH_3). Nitrogen fixation can be accomplished by both free-living and symbiotic organisms. Examples of free-living nitrogen-fixing bacteria are *Azospirillum*, *Azotobacter*, *Clostridium*, *Herbaspirillum*, and *Klebsiella*. Their favourite place of dwelling on land-based ecosystems is in the rhizospheric soil. Some nitrogen-fixing bacteria reside in plant roots, shoots, or leaves, and they establish a mutualistic relationship. The mutualistic symbiosis of the root-bacterial system benefits both partners. The bacterial partner fulfils nitrogen requirements in plants, whereas plant hosts provide them with fixed carbon, nutrients, and shelter. In terrestrial ecosystems, an example of a well-established symbiotic relationship is the root nodule, which has great ecological and agricultural importance. Root nodule hosts are certain legumes and actinorhizal plants. Legumes belong to the order Fables, while actinorhizal plants are reported from three different plant groups, including Fagales, Cucurbitales, and Rosales. Around 15-18 bacterial taxa (**rhizobia**) from the order Rhizobiales, such as *Rhizobium*, *Mesorhizobium*, *Bradyrhizobium*, etc., are native to various legume roots. However, a Gram-positive bacterium, *Frankia* (Phylum: Actinomycetes), is the only representative of root nodules within actinorhizal plants. In gymnosperms, cycads build specialised coralloid roots, which host cyanobacteria like *Nostoc* and *Anabaena*. These roots grow apogeotropically in the soil to acquire sunlight for photosynthesis in cyanobacteria, and these endogenous cyanobacteria fix nitrogen for plants. Some commercially available PGPRs are available under the trade names RhizoFix, RhizoMax, Nitrofix, and PhosphoMax for their applications in agricultural fields. Chapter 28 provides comprehensive information on biological nitrogen fixation (BNF).

In addition, PGPR display indirect ways to stimulate plant growth. For example, nitrite reductase in bacteria produces nitric oxide (NO), which acts as the signal molecule for the plant auxin pathway. Pseudomonads produce 2,4-diacetyl phloroglucinol (DAPG) in low concentrations to induce systemic resistance in plants. The secretion of an enzyme, 1-aminocyclopropane-1-carboxylate (ACC) deaminase, reduces the level of ethylene in plants (a stress hormone) by degrading ethylene precursor. This activity also demonstrates the potential to ameliorate salt stress. The bacterial role as biocontrol agents protects plants from various plant pathogens. Many bacterial exudates trigger the elicitation of biochemical and molecular defence responses to ensure their survival against pathogens and other stresses. Some microorganisms exhibit the properties of **quorum quenching**, which inhibits quorum sensing (QS) (i.e., cell-to-cell communication mechanisms) of pathogenic microbial cells and controls plant diseases. Plants can be exposed to nonpathogenic strains prior to infection to precondition them and induce resistance to a broad range of pathogens. This phenomenon is called **induced systemic resistance (ISR)**. This type of plant-microbe interaction improves plant immunity.

19.7.2 Plant Growth-Promoting Fungi (PGPF)

A group of nonpathogenic fungi in the rhizosphere constitutes plant growth-promoting fungi (PGPF). Some common PGPF reported in soil are *Aspergillus*, *Fusarium*, *Penicillium*, *Piriformospora*, *Phoma*, and *Trichoderma*. They provide a variety of benefits to plants by producing plant growth-promoting attributes such as plant hormones (like auxins, gibberellins, and cytokines), hydrolysing enzymes (like amylase, lipase, and protease), and secondary metabolites (like HCN and NH_3). They are the primary decomposers of soil, which helps in the degradation of plant debris and the recycling of nutrients in the soil. Some fungi display properties of transforming organic matter into more stable forms, which reduces the CO_2 emission into the atmosphere, reducing their impact on climate change. *Gibberella fujikuroi* is a plant pathogen, but one of its metabolites, gibberellin, exhibits plant growth-promoting traits. This fungus is commercially cultivated to produce gibberellins on a large scale. Some soil fungi can be agriculturally important because they suppress the growth of some phytopathogens. For example, *Trichoderma* and *Gliocladium* are notable biocontrol agents. They act against diseases like *Rhizoctonia solani* in radishes and *Botrytis cinerea* in apples, respectively. *Trichoderma*-based commercial biological soil fungicides are available in the market under tradenames such as Tusal®, Trichonativa®, Tri-cure®, etc. Even some non-pathogenic strains of phytopathogens like *Fusarium oxysporum* and

Pythium oligandrum can stimulate ISR in plants and improve plant health. Fungal endophytes in hypersaline environments and geothermal regions have roles in ameliorating stress responses in native plants. The presence of beneficial fungi in the soil is also attributed to soil improvements, soil nutrient content, root extension, and seed germination. Some fungi develop a mutual relationship with plants and establish a mycorrhizal association. Section 19.8 provides in-depth information on mycorrhizae and their roles in plant growth and soil health.

19.8 MYCORRHIZAE AND SOIL HEALTH

Mycorrhiza (pl. Mycorrhizae) is a mutualistic association between fungi and plant roots in soil ecosystems. The association was first recognised by T. Hartig (1840) and F. Kamienski (1882). However, the critical characteristics were first established by a German Plant Pathologist, Albert Bernhard Frank, in 1885. He also coined the term "Mycorrhiza", which is derived from a Greek word meaning fungus roots. This is a wonderful example of mutualism, where the fungal component enhances water and nutrient uptake in plants, and the host plants supply photosynthesis-derived carbon to fungi. The association is widespread in all kinds of land ecosystems, from tropical forests to agricultural lands to deserts. About 340,000 plant species (that encompass 85-90% of all vascular plants) can establish a mycorrhizal association. Only a few plant families appear to lack mycorrhizal association, and they are commonly perpetuated as a **'nonmycotrophic'** group. Examples of nonmycotrophy are families such as Brassicaceae, Chenopodiaceae, Amaranthaceae, Zygophyllaceae, and many hemiparasitic plants. The mycorrhizal association is also common in gymnosperms. In pteridophytes, this mutualistic association can be found in the members of lycopsids and ferns. Even some 'lower' land plants, such as hornworts and hepatics, produce structural analogies to mycorrhizal associations.

Mycorrhizal fungi are the soil-borne microorganisms and dominant components of the plant-soil microbiota. Approximately 50,000 different fungal species belonging to four different phyla, including Glomeromycota, Mucoromycota, Ascomycota, and Basidiomycota, can take part in this symbiotic relationship. They grow as very thin hyphae and produce spores. They are achlorophyllous and can be aseptate or septate, depending on the groups they belong to. A single fungus can colonise multiple plants and develop an underground hyphal network called **mycelia**. Based on the morphological characteristics of the fungal host contact zones, mycorrhizae can be categorised into two main groups, including ectomycorrhizae (ECM) and endomycorrhizae. Endomycorrhizae have many subcategories, such as arbuscular mycorrhizae (AM), orchid mycorrhizae (OCM), ericoid mycorrhizae (ERM), and monotropoid mycorrhizae. A different category is ectendomycorrhiza, which is specifically called arbutoid mycorrhiza. Figure 19.8 illustrates different types of mycorrhizal associations. These are further discussed as follows.

19.8.1 Ectomycorrhizae (ECM)

ECM is a type of association where the fungal hyphae can grow inside the plant roots but do not invaginate the cortical cells. A series of hyphal layers grows on the root surfaces, called the **mycelial mantle**. The mantle is a pseudo parenchymatous tissue formed by the proliferation and differentiation of attached fungi. The fungi-plant interface around the root epidermal and cortical cells is referred to as the **Hartig net**, which provides a large surface area for resource exchange (Figure 19.8b). The association has diverse assemblages of both plants and fungi. More than 20,000 species of fungi belonging to phyla of Basidiomycota, Ascomycota and Zygomycota can invade ~6,000 plant species for the ectomycorrhizal association. An estimate suggests that about 30% of the microbial biomass in forest soil is composed of ECM fungi. A cosmopolitan fungal species, *Laccaria bicolour* (Basidiomycota), is a model organism to study ECM relationships and was used for the first ECM fungal genome sequencing. Some fungi, such as *Pisolithus tinctorius* (Basidiomycota), can have a broad range of hosts in temperate forests. More than 60% of the ECM fungi develop fruiting bodies (e.g., mushrooms) above ground. Examples of mushroom-producing ECMs are *Amanita*, *Boletus*, *Russula*, and *Cortinarius*. Others can produce inconspicuous truffles (e.g., *Tuber*), crusts (e.g., *Stereum ostrea*), or sclerotia (e.g., *Ceriporia*) below ground. Boreal and temperate climates in both the Northern and Southern hemispheres are the primary colonising areas for ECM fungi. However, the association of ECM is reported in a narrow range of plant species (only 2%), and ECM fungi usually colonise only selected woody plant species. Trees like pines, beeches, oaks, eucalypts, dipterocarps, and poplars are common examples of ECM plants. Resource management is the primary mutual benefit for both partners. This form of symbiosis plays a critical role in the establishment of many tree species in many phytogeographic regions despite seasonal and landscape fluctuations.

19.8.2 Arbuscular Mycorrhizae (AM)

AM is the most dominant type of mycorrhiza, reported in about 72% of plant members. About 200,000 plant species are known to form AM. The chief characteristic of AM is that fungal hyphae penetrate the cortical cells and invaginate them. The hyphae inside the cells grow in a tree-shaped structure called **arbuscules** (Figure 19.8c). In certain situations, hyphae terminate in a specialised structure, **vesicles**, which act as storage organs. Another old name for this type of mycorrhiza is vesicular-arbuscular mycorrhiza (VAM). Since vesicles are inconsistent in many forms, the term AM is more acceptable. The taxonomic affiliation of arbuscular mycorrhizal fungi (AMF) belongs to Glomeromycetes. Glomeromycetes is a monophyletic group consisting of about 244 members, of which many genera contribute to the development of AM. Some common examples of AMF are *Glomus*, *Gigospora*, *Acaulospora*, *Funneliformis*, *Sclerocystis*, *Rhizophagus*, and *Scutellospora*. All AMFs are obligate mutualists and cannot

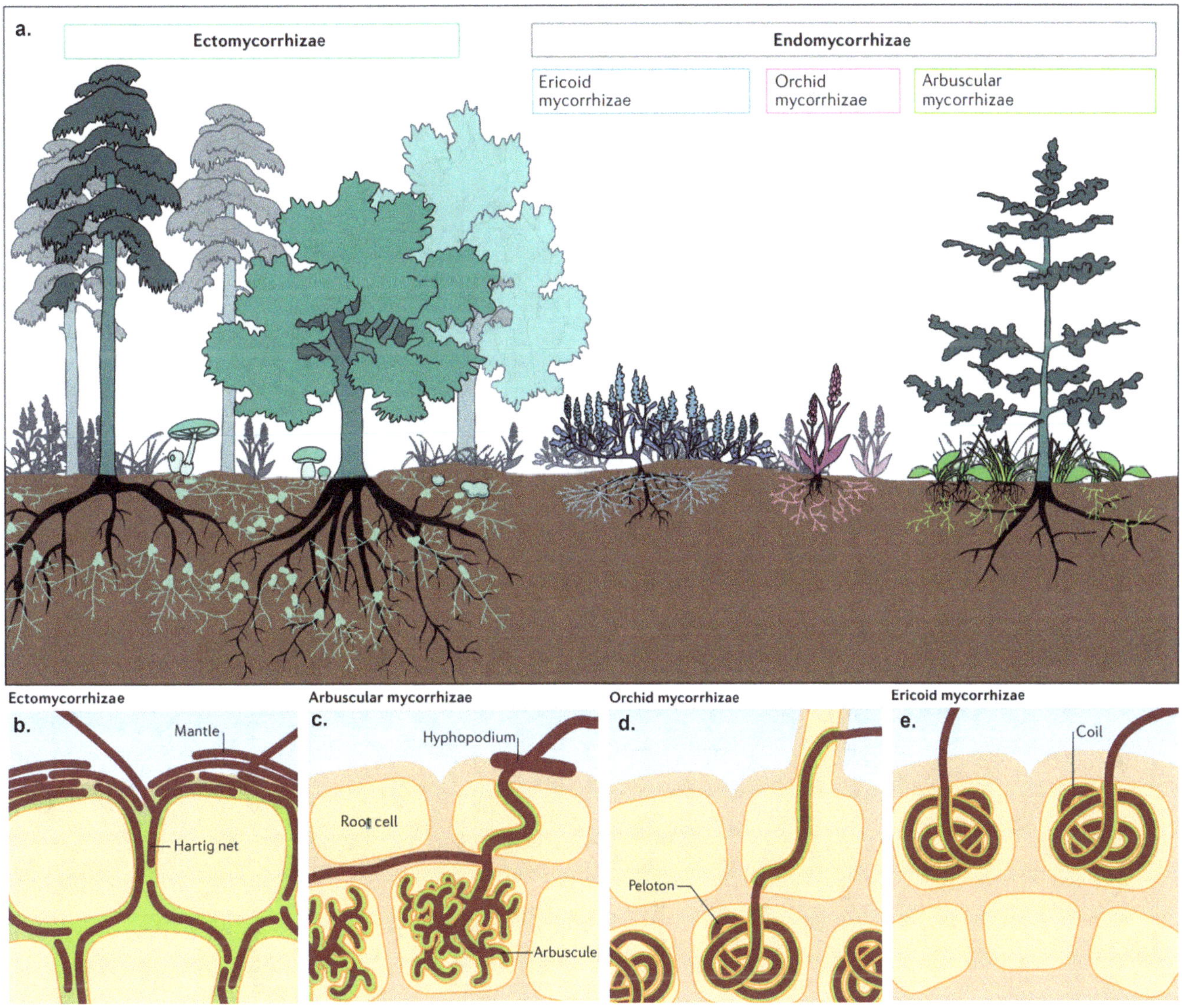

FIGURE 19.8 a. Mycorrhizal association with different plants. b-e. Cellular features of ectomycorrhizae, arbuscular, orchid, and ericoid mycorrhizae. (Images adapted from Genre et al. (2020)[4] by the permission of Nature Reviews Microbiology.)

complete their life cycle without a host. Therefore, the artificial cultivation of many such fungi has several challenges. However, advances in culturing techniques have allowed their cultivation through root organ culture.

This mutualistic association is considered very ancient and has been acclaimed for initiating plant life on the land ecosystem and the evolution of vascular plants. This type of association is also recognised as the "**mother of plant root endosymbioses**". Their evolution dates back to the early Devonian period (around 460 million years ago (mya)). Fossil records of *Aglaophyton major* (the most primitive vascular plant) from Rhynie Chert beds in Aberdeenshire, Scotland, provide strong evidence about the existence of AM in its root. Minute insights inside the root of *A. major* revealed arbuscule-like structures.

The development of AM symbiosis in plant roots is a complex and highly organised process (Figure 19.9). This is initiated by an exchange of signalling molecules between the host root and fungal component (i.e., fungal spores). Some specific root exudates, like **strigolactones**, stimulate the germination of the spores of AM fungi and trigger strong branching of the hyphae. AM fungi also release signalling molecules, mycorrhiza (Myc) factors, which recognise the hosts. AM fungi elicit calcium spiking in host plants and induce cascade signals, leading to the activation of cellular and transcriptional responses in the cells of the root. AM fungi develop extensive branching near the infection site and develop a specialised appressorium, **hyphopodium**. It is a flattened organ that exerts tremendous pressure on the root surface. The accommodation of fungal hyphae undergoes a dramatic reorganisation in the cytoplasm of root cells. This is achieved by a specialised aggregation of organelles and pre-penetration apparatus (PPA). This transient formation acts as an interface compartment that allows fungal hyphae to grow across it and approach the cortical cells for colonisation. AM fungal hyphae penetrate the cell wall of root cells and develop

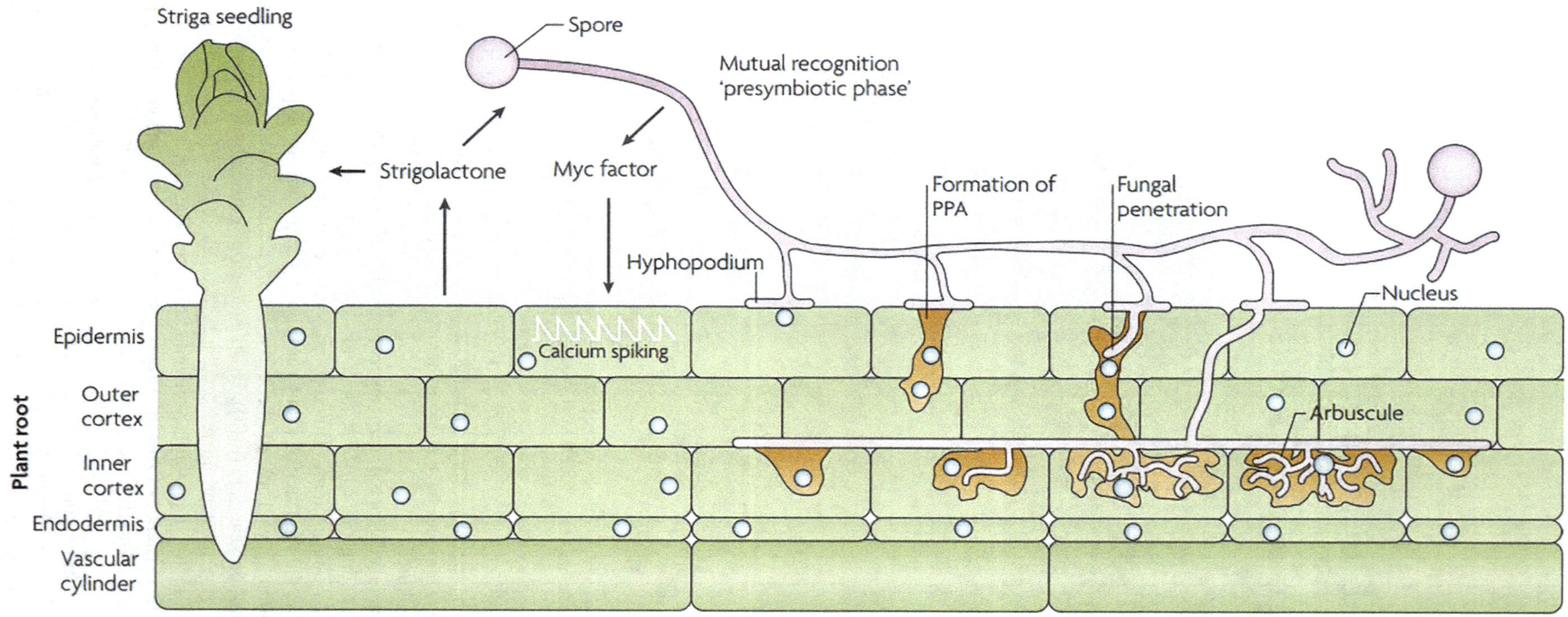

FIGURE 19.9 Steps of arbuscular mycorrhizal symbiosis in a plant root. (Images adapted from Parniske (2008)[5] with permission of Springer Nature.)

arbuscules. Arbuscules increase the surface area at the fungus-plant interface for enhanced resource exchange.

19.8.3 Orchid Mycorrhizae (OCM)

Orchids, the epiphytic plants, establish a crucial mycorrhizal symbiosis called OCM. They are the second most common type of mycorrhizae, which develops symbiosis with over 25,000 orchid species. Because early developmental stages of orchids follow a heterotrophic lifestyle due to being achlorophyllous, fungal colonisation is essential to provide carbon and other nutrients to the embryo. Orchids get their organic nutrients from soil organic matter channelled through fungal hyphae. Germinating seeds and seedling roots are the primary infection sites of OCM fungi. OCM fungi have prominent roles during the seedling and protocorm stages. In the infection, OCM fungi produce highly specialised coiled hyphal structures inside the cortical cells called **pelotons** (Figure 19.8d). The partner fungi mainly belong to the members of Basidiomycota. Examples of taxa include *Rhizoctonia*, *Tulasnella*, and *Russula*. A few species of Ascomycota can also infect orchids. The distribution of OCM fungi has a well-known impact on orchid growth and fitness.

19.8.4 Ericoid Mycorrhizae (ERM)

Ericoid mycorrhizal symbiosis is only found in plants of the family Ericaceae, which is restricted to 1% of angiosperm species. They are dominantly located in the Northern Hemisphere. Heathlands and the understory of boreal forests are notable habitats for colonisation by ERM. ERM fungi can grow in harsh edaphic conditions such as acidic soil, low soil moisture, poor mineral nutrients, and low temperature. During infection, the loose wefts of fungi cover the root hair surface, which penetrates the epidermal cells of the root and develops intercellular densely coiled hyphae (Figure 19.8e). Fungal invasion increases the volume of root hair by up to 80% and improves the nutrient intake from their surroundings. This type of association is temporary and facultative, as fungal hyphae are degraded after some time. Fungal partners mainly belong to the phylum Ascomycota. Some plants can also establish relationships with Basidiomycota members. Examples of some well-studied ERM fungal species are *Meliniomyces bicolour*, *Oidiodendron maius*, and *Rhizoscyphus ericae*.

19.8.5 Arbutoid Mycorrhizae

Arbutoid mycorrhizae share characteristics of both ecto- and endomycorrhizae, which develop a mantle-like structure at the root surface like ECM and can also penetrate the cortical cells like endomycorrhizae. The colonising fungi come from Basidiomycota and also include some species that form ectomycorrhizae. These fungi invade only selected plants such as *Arctostaphylos*, *Arbutus*, and *Pyrola* (belonging to the subfamilies of Ericaceae).

19.8.6 Monotropoid Mycorrhizae

Monotropoid mycorrhizal symbiosis is reported in the subfamily Monotropoideae (family Ericaceae), which includes about ten plants, all being non-chlorophyllous and mycoheterotrophic. Examples of those plants are *Monotropa*, *Sarcodes*, and *Pterospora*. In this association, plants not only receive nutrients from the soil but also receive a carbon supply from neighbouring trees. In this association, the underground fungal network of mycorrhizal channels photosynthetic materials from trees like beech, oak, and cedar to non-chlorophyllous plants. The well-known fungal partner of monotropoid mycorrhizae is *Boletus* (phylum: Basidiomycota). This fungus also forms ectomycorrhizal associations with many different trees. In monotropoid mycorrhiza, the fungal partner colonises the roots of *Monotropa*, similar to ectomycorrhiza, which develops a multilayer mantle and Hartig net. However, numerous small protuberance structures on the cortical cells, known as fungal pegs, distinguish monotropoid mycorrhizae from ectomycorrhizal associations. These are formed by short hyphae of the mantle and Hartig net, which grow as finger-like projections.

Significance of Mycorrhizae: Mycorrhizae form a mycelial network in the soil. This network not only exchanges signals between roots on the same plant but also establishes connections with neighbouring soil environments and other plants. Mycelial networks in soil microhabitats improve soil aggregation. The long filamentous branched hyphae of mycorrhizal association channelise the nutrition and water in the soil, which influences plant productivity and growth responses (Figure 19.10). These mycelial bridges also facilitate the transfer of chemical signals between different plants. The interplant signal exchange protects the plants from various stresses. Mycorrhizae are important for seedling establishment and root colonisation. They have roles in alleviating various biotic and abiotic stresses. Their association with soil ecosystems offers plant resistance to drought and heavy metals. Mycorrhizal fungi help as biocontrol agents against various pathogens and also modify competitive interactions between different plants. AM meets the demands of food security and sustainable agricultural practices. The absence of mycorrhizal networks in the soil produces adverse effects on the biomass and yield of the crop plants.

The associated fungi also have apparent roles in the biogeochemical cycles of terrestrial ecosystems. They are potential contributors to the carbon supply. AM fungi receive about 20% of the photosynthetic carbon from plants. Fungal hyphae provide nutrients like nitrogen and phosphorus to the host plants, contributing up to 80% of the total plant uptake. The mycorrhizal association assists in solubilising many nutrients, like Zn and Cu, in the soil for plant nutrition. Mycorrhizal fungi and rhizobacteria also share positive interactions and benefit plant growth. ECM play important roles in ecological restoration, landscaping, phytostabilisation, and arboriculture. OCM fill the need for conservation and traditional medicine. ERM show tremendous potential for natural attenuation, bioremediation, and fruit and flower production.

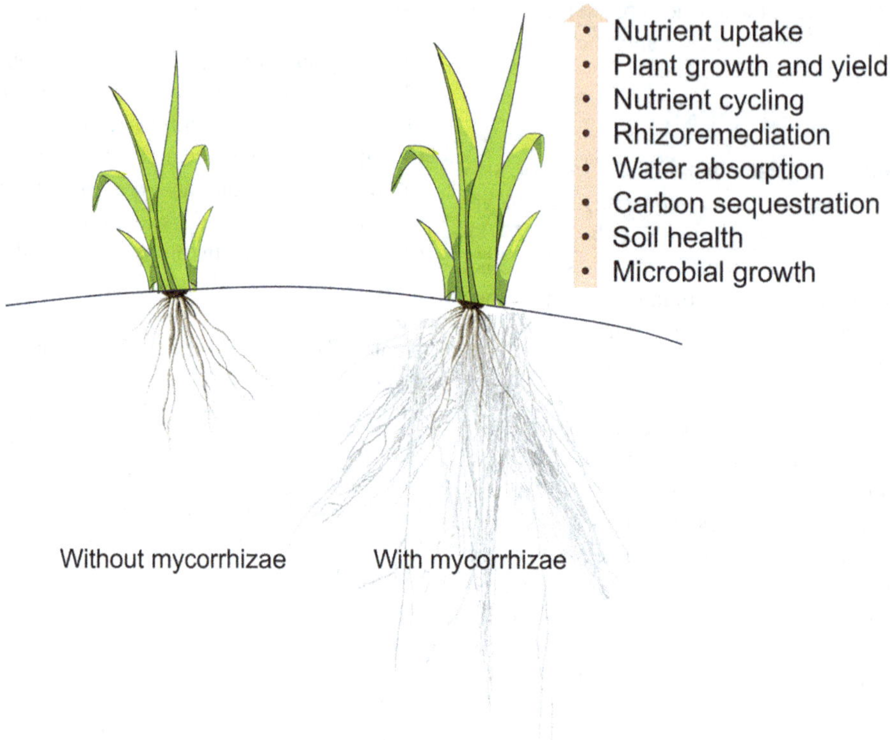

FIGURE 19.10 Plant root without and with mycorrhizal association. In a mycorrhizal association, the hyphal mycelial network extends far beyond the root area and offers several benefits.

19.9 SOIL-BORNE PATHOGENS

Soil is also a reservoir of different pathogens, which exert adverse effects on plants, animals and humans and are referred to as soil-borne pathogens. Plant diseases incur heavy economic losses each year. Some human pathogens also reside in the soil, which causes infection through contaminated food and water. They are further discussed in the following sections.

19.9.1 Plant Pathogens

Plants infected with soil-borne pathogens suffer from vascular wilt, rot, damping-off, stunting, etc. Plant pathogens cause potential economic losses by reducing the growth and yield of crops. Although many soil-borne diseases are climatically driven, agronomic practices have also contributed to the dispersal of pathogens. Roots are the primary site of infection for soil-borne pathogens. Plant diseases are often difficult to diagnose as the main infections are usually below ground. The above-ground symptoms are not distinct and may show similarities for abiotic and biotic stresses. Fungi are the most common types of causal organisms that spread the disease through spores (e.g., root rot of corn by *Rhizoctonia*). Examples of soil-borne root pathogenic fungi are *Fusarium*, *Verticillium*, *Pythium*, and *Phytophthora*.

Bacterial diseases for plants are less common because many bacteria are non-spore forming and do not survive in the soil for an extended period. The spread of soil-borne bacterial pathogens is somewhat limited because many of them are confined in the soil as aggregated or clustered forms. However, some soil-borne pathogens get attached through aerosols that can spread over a wide range of space through wind and storms. Some well-known soil-borne bacteria that can be pathogenic include *Xanthomonas* spp., *Ralstonia solanacearum*, and *Agrobacterium tumefaciens*. A list of soil-borne pathogens that raise concerns for plants is given in Table 19.4.

19.9.2 Human Pathogens

Many soil-borne microorganisms can infect humans and animals. The mode of infection can be direct inoculation, ingestion, or inhalation. They can be zoonotic diseases transmitted through vectors or carriers. Food-borne pathogens are the common mode of infection in humans. Severe infections can be life-threatening. Based on dependency on hosts to complete their life cycle, soil-borne pathogens can be categorised into two groups: edaphic pathogenic organisms (EPOs) and soil-transmitted pathogens/parasites (STPs). EPOs complete their life cycle without causing any disease in humans. However, immunosuppressed persons are prone to infection by EPOs. The causal organisms are called opportunistic pathogens. Examples of those include coccidioidomycosis by *Coccidiodes* spp. and blastomycosis by *Blastomyces dermatitidis*. Mucormycosis is a rare fungal disease caused by

TABLE 19.4
Examples of soil-borne plant diseases, their pathogens, and hosts

Diseases	Pathogens	Plant hosts
Fusarium wilt	*Fusarium oxysporum*	Tomato, eggplant, and pepper
Verticillium wilt	*Verticillium* spp.	Chrysanthemum, geranium, petunia, and sunflower
Phytophthora root rot	*Phytophthora* spp.	Tomato, pepper, eggplant, and avocado
Rhizoctonia root rot	*Rhizoctonia solani*	Soybean and wheat
Pythium root rot	*Pythium* spp.	Lathyrus, geranium, snapdragon, and viola
Bacterial wilt	*Ralstonia solanacearum*	Soybean, tobacco, tomato, potato, and banana
Crown gall	*Agrobacterium tumefaciens*	A wide range of plants
Hairy root disease	*Agrobacterium rhizogenes*	A wide range of plants

TABLE 19.5
Examples of soil-borne diseases in humans and the associated pathogens

Diseases	Pathogens
Coccidioidomycosis	*Coccidiodes* spp.
Blastomycosis	*Blastomyces dermatitidis*
Mucormycosis	*Rhizopus*, *Rhizomucor*, *Mucor*, *Apophysomyces*, and *Lichtheimia*
Tetanus	*Clostridium tetani*
Botulism	*Clostridium botulinum*
Anthrax	*Bacillus anthracis*
Pneumonic plague	*Yersinia pestis*

moulds. They belong to a class of Mucormycetes and include genera like *Rhizopus*, *Rhizomucor*, *Mucor*, *Apophysomyces*, and *Lichtheimia*. They usually inhabit soil and usually grow on decaying organic matter. However, human exposure to these moulds can affect individuals with weakened immune systems. During the recent pandemic of active severe acute respiratory syndrome coronavirus 2 (SARS-CoV-2) outbreak, the spread of mucormycosis in coronavirus patients emerged as a new health crisis.

In contrast, STPs essentially require the hosts to complete their life cycles. They are obligate parasites and remain in the soil in an inactive form for extended periods before infecting humans. Tetanus is a common soil-borne disease which spreads when a wound or an animal bite is exposed to dirt. The causal organism is an endospore-producing bacterium, *Clostridium tetani*. This bacterium has a worldwide distribution and is usually found in the dormant stage in soil. Other soil-induced infections in humans can be botulism, anthrax, and pneumonic plague caused by *Clostridium botulinum*, *Bacillus anthracis*, and *Yersinia pestis*, respectively (Table 19.5).

19.10 ECOLOGICAL AND ECONOMICAL IMPORTANCE OF SOIL MICROBIOTA

Microbial diversity in soil is a main driver for the functioning of terrestrial ecosystems. They have profound roles in ecosystem balance and economic growth. The biodegradation of dead biomass on the ground plays a key role in the global biogeochemical cycles of many elements. Microbial decomposition of soil organic matter helps maintain soil health. They have roles in eco-friendly, sustainable agriculture practices. Native microorganisms help regain the stability and resilience of stressed and disturbed soil. They also exert effects on climate change through metabolic activities. The natural processes of native microbiota in soil are responsible for natural attenuation. Their biocontrol activities save many plants from disease infection.

Under distinct roles, soil microorganisms provide a variety of resources to human society. Soil microorganisms are important sources of industrially relevant enzymes and secondary metabolites. Many plant-associated microorganisms are used as commercial biofertilisers in agriculture. The soil-borne pathogens *Agrobacterium tumefaciens* and *Rhizobium rhizogenes* (formerly known as *Agrobacterium rhizogenes*), which cause gall formation and hairy root disease in plants, respectively, are remarkable tools of biotechnology. *A. tumefaciens* consists of a unique plasmid (tumour-inducing or *Ti* plasmid) that facilitates gene transfer in plants and produces transgenic plants. *R. rhizogenes* can be used in hairy root culture to derive specialised metabolites or secondary metabolites in plants. In bioremediation approaches, soil microorganisms can stabilise or remove the contaminants in contaminated soil (detailed in Chapter 30). A variety of soil microorganisms can assist in waste decomposition and waste valorisation (detailed in Chapters 25 and 31). Besides, a large extent of soil and its microbiome remain unexplored, which could exhibit unique beneficial activities.

19.11 ANTHROPOGENIC IMPACT ON SOIL MICROORGANISMS

Intensive and extensive anthropogenic activities have created natural imbalances in the soil. Agricultural practices have had adverse impacts on natural habitats, as well as on native microbiota. Deforestation has weakened soil health through soil erosion and the removal of the rhizosphere. Overuse of agrochemicals has dramatically altered the microbial

BOX 19.2 MICROBIAL RESPONSE TO GLOBAL CLIMATE CHANGE

Global warming continues to have a profound impact on weather conditions, including soil health. The direct (like temperature and extreme events) and indirect (like elevated CO_2) feedback of this cause have induced long-term shifts for their inhabitants. The natural microbial communities have also received major shifts in their composition and function. New findings suggest that climate change has influenced not just microbial makeup but also microbial responses. Soil microorganisms are prominent consumers and producers of three greenhouse gases, including CO_2, CH_4, and nitrous oxide (N_2O).

In the natural environment, microbial decomposition of soil organic matter promotes the production of CO_2. Rising temperatures accelerate the rate of microbial respiration, leading to a high efflux of CO_2 into the atmosphere. Consequently, elevated CO_2 concentration in the surroundings causes a shift in MCC and function. Meanwhile, photosynthetic microorganisms are the key consumers of CO_2, sequestering the carbon and producing O_2.

Methanogens are responsible for the production of CH_4. They are strictly anaerobes and are only found in the archaeal phylum Euryarchaeota (e.g., *Methanococcus*, *Methanobacterium*, and *Methanosarcina*). These microorganisms can be reported from environments like wetlands, rice paddies, and the digestive tracts of animals and are responsible for CH_4 emissions. A different group of microorganisms consume CH_4 and are known as methanotrophs. This feature is reported in certain bacteria and archaea, thriving in anaerobic or aerobic situations. They utilise CH_4 as their primary carbon and energy source, reducing its impact on the environment.

A different greenhouse gas, N_2O, which has a much higher warming potential than CO_2 and CH_4, is mainly released from the terrestrial systems. The high concentration of N_2O in the environment is attributed to the excessive use of chemical fertilisers. A variety of microorganisms can utilise various nitrogenous substances and produce N_2O through different biochemical pathways such as ammonia oxidation, denitrification, nitrite oxidation, and nitrite reduction. Nitrous oxide also serves as a substrate for the last step of denitrification to release N_2.

Of the different regions of the world, the Arctic and Antarctic zones are the most vulnerable to climate change. The continuous increase in the average temperature of these regions has increased microbial activities. This, in turn, has resulted in the decomposition of the enormous reservoir of terrestrial carbon in thawing permafrost, leading to substantial CO_2 and CH_4 emissions. The fast soil carbon decomposition rates due to temperature rise are also apparent in boreal forest ecosystems that constitute approximately 1/3 of the terrestrial carbon stock. In temperate and tropical regions, the elevated temperature is attributed to microbial mineralisation of soil organic matter. Wetlands are natural sources of greenhouse gases (CH_4, N_2O, and CO_2). The increasing global sea level is responsible for expanding the wetlands in coastal areas. In the wetland, the primary cause of intense carbon breakdown and greenhouse gas emission is climate change, which is linked to varying microbial reactions.

Another extreme event related to microbial response to climate change is ocean acidification due to elevated concentrations of CO_2, which poses a significant threat to marine ecosystems and coral reefs. The altered microbial composition and function have also impaired natural nutrient cycling, resulting in deviations in mineral fluxes in the ecosystems that are out of equilibrium in terrestrial ecosystems. Climate change has also been found to have an impact on the distribution and transmission of pathogenic microorganisms, potentially increasing disease outbreaks in plants, humans, and wildlife.

composition in agricultural lands and their stretches. These chemicals are causing a great loss in the abundance of AM fungi. Also, soil disturbance and cultivation of non-mycotrophic crops have exhibited negative effects on mycorrhizal fungi. Frequent tillage practices pose serious destruction to habitats for microorganisms. The damaged soils are prone to compaction, which ultimately results in loss of air, water, and space in the soil. Also, tillage has induced the 'burning' of microbial supplements, such as organic matter.

The spread of contaminated water in the field is also a great concern for native microorganisms. Industrial effluents containing toxic chemicals have caused a significant loss of beneficial microorganisms and induced the growth of some tolerant species. The contamination of heavy metals like Hg, Cd, and As showed strong inhibitory actions on microbial metabolisms. The spread of sewage water in the soil has provoked the emergence of pathogens, spreading various diseases to plants, animals, and humans. The dual organisms, lichens, are sensitive to air pollutants and unable to colonise rocks under stress. Global warming has also notably induced a shift in MCCs and their functions (Box 19.2). Notably, microorganisms that withstand anthropogenic activities play roles in ecological restoration and ecosystem balance.

SUMMARY

- Soil is a heterogeneous mixture of living and nonliving components. It is one of the most active regions of Earth and supports diverse life systems. Soil microorganisms are dynamic components of the Earth's system and include bacteria, archaea, fungi, microalgae, and protozoans.

- Different soil characteristics such as temperature, pH, O_2, organic matter, nutrients, moisture content, porosity, redox reactions, and other biological activities influence soil microbiota. Soil enzyme activities are indicators of soil quality, and soil microorganisms are the main sources of these enzymes.
- Microbial responses to plant communities produce a variety of plant-beneficial and plant-antagonistic activities. Root nodule formation and mycorrhizae are two important mutualistic relations between plant roots and microorganisms in the soil.
- Plant growth-promoting rhizobacteria are rhizospheric bacteria that secrete specific metabolites for plant growth and soil health. Some contribute to nitrogen fixation in free-living and mutualistic conditions (e.g., rhizobia-legume partnership and *Frankia*-actinorhizal association).
- Mycorrhiza is a fungal association in plant roots in soil. They exist as ectomycorrhizae, arbuscular mycorrhizae (AM), orchid mycorrhizae, ericoid mycorrhizae, arbutoid mycorrhizae, and monotropoid mycorrhizae. About 90% of all land plants have a mycorrhizal association, with AM being the most dominant.
- Soil microorganisms provide a multitude of benefits for the environment and humans through their activities. They take part in pedogenesis, decompose organic matter, recycle nutrients, play roles in land restoration, improve agricultural production, degrade environmental contaminants, and care for soil health.
- Soil is also a reservoir of several soil-borne pathogens, which exert adverse effects on plants, wildlife, and humans. Intense human activities have altered the soil microbial communities and produced long-term negative effects on soil health.

DIFFERENT TYPES OF PRACTICE QUESTIONS

Fill in the blanks

I. A single gram of healthy soil can inhabit approximately .... microorganisms.
II. .... is also recognised as the "mother of plant root endosymbioses".
III. .... obtain their nutrients from the interior of the rocks.
IV. Several microorganisms reside inside the plant tissues, they are called ....
V. .... are the pioneer communities in bare land.
VI. The fungi develop underground hyphal networks called........................
VII. In certain conditions, the hyphae in arbuscular mycorrhiza terminate in a specialised nutrient storage structure called..............
VIII. .... is a Gram-positive bacterium that establishes mutualistic relations with plant roots and develops root nodules.

SOLUTION: I. 10 billion (1010), II. Arbuscular mycorrhiza, III. endolithic microorganisms, IV. endophytes, V. lichens, VI. mycelia, VII. vesicle, VIII. *Frankia*

Match the following

i. Arbuscules	a. Ectomycorrhiza
ii. Hartig net	b. Orchid mycorrhiza
iii. Pelotons	c. Monotropoid mycorrhiza
iv. Fungal pegs	d. Arbutoid mycorrhiza
v. Ecto- and endomycorrhiza	e. Arbuscular mycorrhiza

SOLUTION: i-e, ii-a, iii-b, iv-c, v-d

Very short answer type of questions

I. Expand PGPR and PGPB.
II. Write the names of two soil-borne diseases in plants.
III. Name any two microalgae that reside in a soil environment.
IV. Write the names of two enzymes that act as indicators for soil quality and fertility.
V. What is the first fossil record of mycorrhizal association?
VI. What are the key mutual benefits for plants and bacteria in root nodule formation?

Short answer type of questions

I. What is petrichor? How is it formed?
II. Write a short note on the soil as a habitat for diverse microorganisms.
III. How do soil microorganisms influence soil health?
IV. Discuss the multifarious activities of PGPR.
V. Write a short note on soil enzyme activities.
VI. Elaborate on different biological factors that influence soil microbiota.
VII. Give an account of the anthropogenic impact on soil microbiota.

Long answer type of questions

I. What are mycorrhizae and their different types? Give an account of the mycorrhizal association and soil health.
II. Discuss the ecological and economic importance of soil microbiota. What are the roles of microorganisms in pedogenesis?

NOTES

1 Jastrow JD, Miller RM. Soil aggregate stabilization and carbon sequestration: feedbacks through organomineral associations. In: *Soil processes and the carbon cycle 2018* (pp. 207–223). CRC press.

2 www.commons.wikimedia.org/wiki/File:Lichen_on_rock.jpg
3 Chauhan P, Sharma N, Tapwal A, et al. Soil microbiome: Diversity, benefits and interactions with plants. *Sustainability*. 2023;15(19):14643.
4 Genre A, Lanfranco L, Perotto S, et al. Unique and common traits in mycorrhizal symbioses. *Nature Reviews Microbiology*. 2020;18(11):649–60.
5 Parniske M. Arbuscular mycorrhiza: the mother of plant root endosymbioses. *Nature Reviews Microbiology*. 2008;6(10):763–75.

SUGGESTED READINGS

Ahemad M, Kibret M. Mechanisms and applications of plant growth promoting rhizobacteria: current perspective. *Journal of King Saud University Science*. 2014;26(1):1–20.

Aislabie J, Deslippe JR, Dymond J. Soil microbes and their contribution to soil services. Ecosystem services in New Zealand–conditions and trends. *Manaaki Whenua Press, Lincoln, New Zealand*. 2013;1(12):143–61.

Backer R, Rokem JS, Ilangumaran G, Lamont J, Praslickova D, Ricci E, Subramanian S, Smith DL. Plant growth-promoting rhizobacteria: context, mechanisms of action, and roadmap to commercialization of biostimulants for sustainable agriculture. *Frontiers in Plant Science*. 2018;9:1473.

Banerjee S, Schlaeppi K, van der Heijden MG. Keystone taxa as drivers of microbiome structure and functioning. *Nature Reviews Microbiology*. 2018;16(9):567–76.

Bar-On YM, Phillips R, Milo R. The biomass distribution on Earth. *Proceedings of the National Academy of Sciences*. 2018;115(25):6506–11.

Blagodatskaya E, Kuzyakov Y. Active microorganisms in soil: critical review of estimation criteria and approaches. *Soil Biology and Biochemistry*. 2013;67:192–211.

Bultman MW, Fisher FS, Pappagianis D. The ecology of soil-borne human pathogens. In Selinus S, Alloway BJ (eds.) *Essentials of medical geology* 2013 (pp. 477–504). Springer.

Cao H, Chen R, Wang L, Jiang L, Yang F, Zheng S, Wang G, Lin X. Soil pH, total phosphorus, climate and distance are the major factors influencing microbial activity at a regional spatial scale. *Scientific Reports*. 2016;6(1):1.

Cavicchioli R, Ripple WJ, Timmis KN, Azam F, Bakken LR, Baylis M, Behrenfeld MJ, Boetius A, Boyd PW, Classen AT, Crowther TW. Scientists' warning to humanity: microorganisms and climate change. *Nature Reviews Microbiology*. 2019;17(9):569–86.

Chamizo S, Mugnai G, Rossi F, Certini G, De Philippis R. Cyanobacteria inoculation improves soil stability and fertility on different textured soils: gaining insights for applicability in soil restoration. *Frontiers in Environmental Science*. 2018;6:49.

Delgado-Baquerizo M, Oliverio AM, Brewer TE, Benavent-González A, Eldridge DJ, Bardgett RD, Maestre FT, Singh BK, Fierer N. A global atlas of the dominant bacteria found in soil. *Science*. 2018;359(6373):320–5.

Fierer N. Embracing the unknown: disentangling the complexities of the soil microbiome. *Nature Reviews Microbiology*. 2017;15(10):579–90.

Flemming HC, Wuertz S. Bacteria and archaea on Earth and their abundance in biofilms. *Nature Reviews Microbiology*. 2019;17(4):247–60.

Foissner W. *Protozoa, reference module in earth systems and environmental sciences*. Elsevier; 2014. ISBN 9780124095489.

Geisen S, Mitchell EA, Adl S, Bonkowski M, Dunthorn M, Ekelund F, Fernández LD, Jousset A, Krashevska V, Singer D, Spiegel FW. Soil protists: a fertile frontier in soil biology research. *FEMS Microbiology Reviews*. 2018;42(3):293–323.

Genre A, Lanfranco L, Perotto S, Bonfante P. Unique and common traits in mycorrhizal symbioses. *Nature Reviews Microbiology*. 2020;18(11):649–60.

Hakim S, Naqqash T, Nawaz MS, Laraib I, Siddique MJ, Zia R, Mirza MS, Imran A. Rhizosphere engineering with plant growth-promoting microorganisms for agriculture and ecological sustainability. *Frontiers in Sustainable Food Systems*. 2021;5:16.

Hillel D, Hatfield JL, editors. *Encyclopedia of soils in the environment*. Elsevier; 2005.

Kapoor R, Evelin H, Mathur P, Giri B. Arbuscular mycorrhiza: approaches for abiotic stress tolerance in crop plants for sustainable agriculture. In Tuteja N, Singh S (eds.) *Plant acclimation to environmental stress* 2012 (pp. 359–401). Springer New York.

Khan A, Rao TS. Molecular evolution of xenobiotic degrading genes and mobile DNA elements in soil bacteria. In Dash HR, Das S (eds.) *Microbial diversity in the genomic era* 2019 (pp. 657–678). Academic Press.

Kuzyakov Y, Blagodatskaya E. Microbial hotspots and hot moments in soil: concept & review. *Soil Biology and Biochemistry*. 2015;83:184–99.

Leininger S, Urich T, Schloter M, Schwark L, Qi J, Nicol GW, Prosser JI, Schuster SC, Schleper C. Archaea predominate among ammonia-oxidizing prokaryotes in soils. *Nature*. 2006;442(7104):806–9.

Martin F, Kohler A, Murat C, Veneault-Fourrey C, Hibbett DS. Unearthing the roots of ectomycorrhizal symbioses. *Nature Reviews Microbiology*. 2016;14(12):760–73.

Metting B. The systematics and ecology of soil algae. *The Botanical Review*. 1981;47(2):195–312.

Parniske M. Arbuscular mycorrhiza: the mother of plant root endosymbioses. *Nature Reviews Microbiology*. 2008;6(10):763–75.

Qiu L, Zhang Q, Zhu H, Reich PB, Banerjee S, van der Heijden MG, Sadowsky MJ, Ishii S, Jia X, Shao M, Liu B. Erosion reduces soil microbial diversity, network complexity and multifunctionality. *The ISME Journal*. 2021;15(8):2474–89.

Raaijmakers JM, Paulitz TC, Steinberg C, Alabouvette C, Moënne-Loccoz Y. The rhizosphere: a playground and battlefield for soilborne pathogens and beneficial microorganisms. *Plant and Soil*. 2009;321(1):341–61.

Read DJ, Duckett JG, Francis R, Ligrone R, Russell A. Symbiotic fungal associations in lower land plants. *Philosophical Transactions of the Royal Society of London. Series B: Biological Sciences*. 2000;355(1398):815–31.

Steffan JJ, Derby JA, Brevik EC. Soil pathogens that may potentially cause pandemics, including severe acute respiratory syndrome (SARS) coronaviruses. *Current Opinion in Environmental Science & Health*. 2020;17:35–40.

Tate III RL. *Soil microbiology*. Wiley; 2020.

Torsvik V, Øvreås L. Microbial diversity and function in soil: from genes to ecosystems. *Current Opinion in Microbiology*. 2002;5(3):240–5.

van der Heijden MG, Martin FM, Selosse MA, Sanders IR. Mycorrhizal ecology and evolution: the past, the present, and the future. *New Phytologist*. 2015;205(4):1406–23.

20 Study of Microorganisms in Water

20.1 WATER IN THE ENVIRONMENT

The wonderful sight of Earth from space has a blue appearance. The reason is the abundance of liquid water on the Earth's surface and its blue colour— that is why the globe is also called a "blue planet". Approximately 71% of the Earth's surface is covered with water. Liquid water exists in the ocean, rivers, streams, lakes, underground waters, etc. The solid mass of water is found in snow and ice. A small quantity of water in the form of vapour is present in the atmosphere. All types of water on the planet constitute the **hydrosphere**. The frozen part of the hydrosphere is termed the **cryosphere**. Even after having a high volume of water but in solid form, Antarctica and the Arctic regions are the two biggest deserts on Earth (cold deserts), 1.5 times larger than the hot desert of the Sahara.

Water has been detected on other planets, namely Mars and Venus, as well as the moons of Jupiter and Saturn. Water in these extraterrestrial regions is present in either vapour or solid states, making it biologically unavailable. Liquid water is unique to the Earth's system and is essentially required for biological systems. It is a universal medium that supports biochemical reactions and, thus, life on Earth. It is also crucial for other processes such as ingestion, farming, mineral exploration, electricity generation, and many industrial activities.

Earth stores approximately 1.386 billion km^3 of water. About 96.5% of the water is present in marine ecosystems with high salt concentrations. A small volume of saline water (1.0%) is also present in "inland environments". Only about 2.5% of the total water on Earth is freshwater. Approximately 69% of the freshwater is frozen in ice caps, glaciers, and permanent snow. This accounts for the second biggest water reservoir on Earth. A small proportion of freshwater (30%) resides below ground as **groundwater**. Freshwater on the surface that comprises lakes, rivers, ponds, streams and wetlands occupies only about 1% of the total inland freshwater. Freshwater is the major support system for a large number of terrestrial life. Surprisingly, the total amount of freshwater found in the soil (3.5%) is greater than what is reported in the riverine system (0.46%). Also, a low volume of freshwater (0.22% of the inland freshwater) resides in the biosphere, stored in biomass. About 70-90% of the cell weight is composed of water. The distribution of water in various ecosystems is illustrated in Figure 20.1.

20.2 PROPERTIES OF WATER

The chemical name of water is dihydrogen monoxide (DHMO). It is made up of two atoms of hydrogen and one atom of oxygen, connected through covalent bonding. Water molecules are held together through an electrostatic dipole-dipole interaction that creates hydrogen bonding. This gives water molecules many distinctive properties, such as high surface tension, high boiling point (100 °C), low viscosity, high adhesion, and high specific heat capacity. The solid mass of water floats on the liquid water because the ice has less density than liquid water. The chemically pure water is tasteless, odourless, nearly colourless, and transparent. However, water bodies in the large reservoirs generally give a blue hue due to the reflection of light— a phenomenon called Rayleigh scattering.

Water is the only substance on Earth that naturally exists in solid, liquid, and gaseous states. The three states are regulated by temperature and pressure. Water can transform from one state to another under different physicochemical conditions. Water in the liquid and gaseous forms tends to flow from one reservoir to another. These processes regulate its movement into diverse ecosystems and create a water cycle (Figure 20.2). Some important processes of the water cycle are:

- Evaporation (vaporisation)- Transformation of liquid water into vapour (a similar process in plants is known as transpiration)
- Freezing- Conversion of water into ice
- Melting- Conversion of ice into water
- Sublimation- Direct conversion of ice and snow into water vapour
- Deposition- Direct conversion of water vapour into ice and snow
- Condensation- Formation of cloud
- Precipitation- Release of water from the cloud in the form of rain, sleet, snow, or hail
- Surface runoff- Water flowing over the ground surface
- Infiltration- Downwater entry of water in the subsurface of the ground

DOI: 10.1201/9781032651095-25

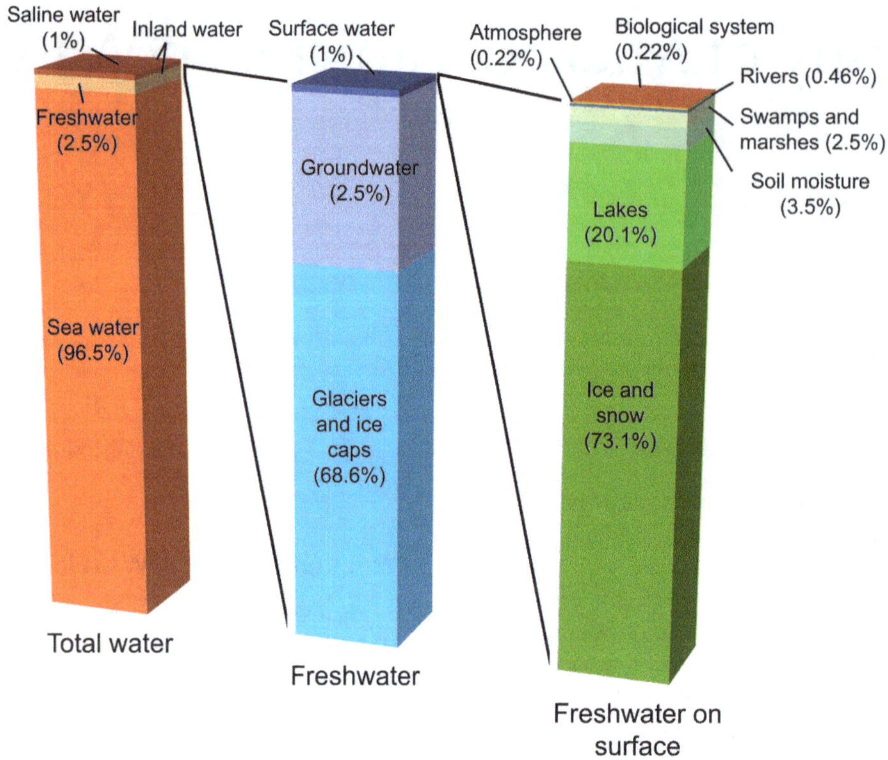

FIGURE 20.1 Distribution of water on Earth.

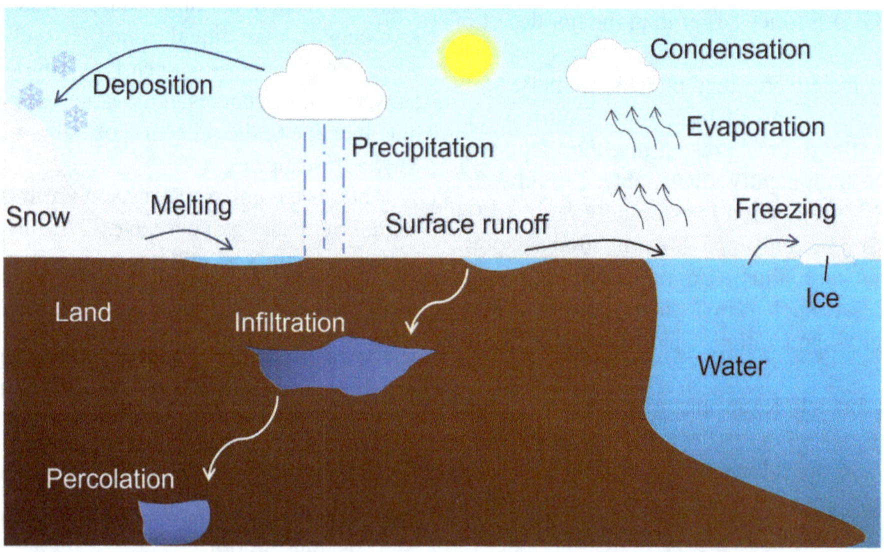

FIGURE 20.2 An illustration of the water cycle on Earth.

- Percolation- Transfer of water into deeper layers of the ground
- Capillary ascent- Upward entry of water in soil through capillary action

20.3 AQUATIC HABITATS

Two broad categories of aquatic habitats are inland water and seawater. The confluence of freshwater and marine water constitutes a third category, brackish water. The study of inland water is specifically called **limnology**. All three different types of aquatic environments support diverse life forms, including microorganisms, plants, and animals.

The aquatic environments have many distinct or indistinct boundaries, each with a defined set of physicochemical characteristics harbouring specific biological communities. Some regions are permanently covered with water, while some places have occasional or seasonal floods. Marine ecosystems are highly interconnected. They have a very high biodiversity from top to bottom and on all sides. Coral reefs in the ocean are vibrant ecosystems which are home to a variety of marine life forms. Deep-sea hydrothermal vents are located on the ocean floor and continually discharge geothermally heated water. The deepest point on the entire Earth is Challenger Deep, located in the Mariana Trench (11,000 m below sea level) and hosts many unique lifeforms.

Inland water also has great variations, including freshwater and saltwater. The Caspian Sea is an excellent example of an inland system and is, in fact, the largest lake on Earth, which is often mistakenly considered a marine ecosystem due to its doubtful name and spread over a wide geographical area (372,000 km^2). This is a brackish water lake and hosts many unique life forms, including diverse microorganisms. Even many inland water ecosystems are much saltier than marine ecosystems and are peculiar habitats for diverse organisms. The salt concentrations in the inland lakes can reach as high as 433 g/kg (more than twelve times the average of seawater) in Gaet'ale Pond, Ethiopia. The Dead Sea is also an example of an inland hypersaline aquatic system, which has very deep zones below sea level. These inland water systems, even under harsh environments, are also not free from microbial inhabitants. The halophilic bacteria and archaea are the dominant residents of these environments. The subglacial lakes in polar regions are unique sites for diverse cold-adapted living creatures. Soda Lakes (e.g., Mono Lake, California) are characterised by alkaline pH with a pH level of 9-12 and feature highly productive aquatic ecosystems due to the heavy dose of dissolved carbon dioxide (CO_2). The largest (by volume) and oldest freshwater Lake Baikal in Siberia, Russia, holds about 20% of Earth's unfrozen freshwater. This lake is one of the water bodies with low pollution levels in the world and has the deepest point in the inland water ecosystem (reaching a depth of 1,642 m from sea level). Even the deep regions of the lake have high oxygen (O_2) levels and can support the growth of aerobic life forms, including fish.

20.4 AQUATIC MICROBIOLOGY

Aquatic microbiology is the study of microscopic neuston, plankton, periphyton, and benthos in aquatic regions. They are the unseen heroes of aquatic ecosystems. These organisms belong to bacteria, archaea, microalgae, protozoa, fungi, and viruses and have very high densities in water (Table 20.1). Many microorganisms inhabit the surface microlayer at the air-water interface as **neuston**. In this habitat, microorganisms establish direct contact with the atmosphere from above and the hydrosphere from below. The surface microlayer is a physicochemically distinct zone and supports a much higher density of microorganisms than the underlying water column. These include some protozoans and bacterial communities. **Plankton** are a large number of microbial forms that exist in the drifting state of water bodies. Planktonic cells contain more than 80% water, which provides a nearly similar bulk density to aquatic systems. This makes many of them suspended at the surface and allows unconfined floating in the water. The plankton communities include a variety of microorganisms, such as phytoplankton, zooplankton, bacterioplankton, mycoplankton, and virioplankton. **Phytoplankton** are photoautotrophic, which include cyanobacteria, diatoms, dinoflagellates, coccolithophores, and photosynthesising protists. They are the key components for producing organic matter in aquatic environments. **Zooplankton** comprise microscopic heterotrophic lifeforms such as protozoans, some dinoflagellates, and marine microanimals. A mixture of autotrophic and heterotrophic microorganisms growing on submerged surfaces develops a complex community known as **periphyton**. Periphyton chiefly includes bacteria and microalgae and mainly resides on the surface of submerged aquatic plants. The primary productivity of periphyton supports a wide range of aquatic life through grazing and feeding. An array of microbial forms stays in benthic regions (at the lowest level of a body of water) and thrives under distinct physicochemical situations. Those organisms are called **benthos** (Figure 20.3a-b).

Both heterotrophic and autotrophic bacterial communities and archaeal communities constitute **bacterioplankton**. The fungal communities in the water make up **mycoplankton**. Viruses are also pervasive in aquatic systems by infecting planktonic communities. This organisation makes up **virioplankton**, and they are principally composed of bacteriophages and algal viruses.

Sometimes, microbial cells tend to aggregate over a substratum in the aquatic environment and evolve as self-sustaining systems (e.g., biofilm and microbial mat). They can be composed of more than one type of microorganism and display distinct phenotypic traits. A **biofilm** is a highly organised and complex assemblage of microbial cells that are firmly stuck to the surface of an inert substratum (Figure 20.4a). Microorganisms in this biological architecture secrete an extracellular polymer matrix that aids in colony formation and the development of a slimy coating on the solid body. **A microbial mat** is a multilayer sheet of diverse microbial communities attached to various substrata. It can be vertically stratified into distinct layers and is usually found at the interface between different materials in aquatic habitats. For example, **stromatolites** are layered deposits of calcareous materials of fossilised microbial mats (mainly

TABLE 20.1
Estimated number of various microbial groups in seawater

Types of microorganisms	Estimated number of microorganisms in seawater
Viruses	1E+31
Bacteria	1E+30
Archaea	1E+28
Microalgae	1E+27
Protozoa	1E+24
Fungi	1E+24

Note: E represents the exponent in scientific notation.

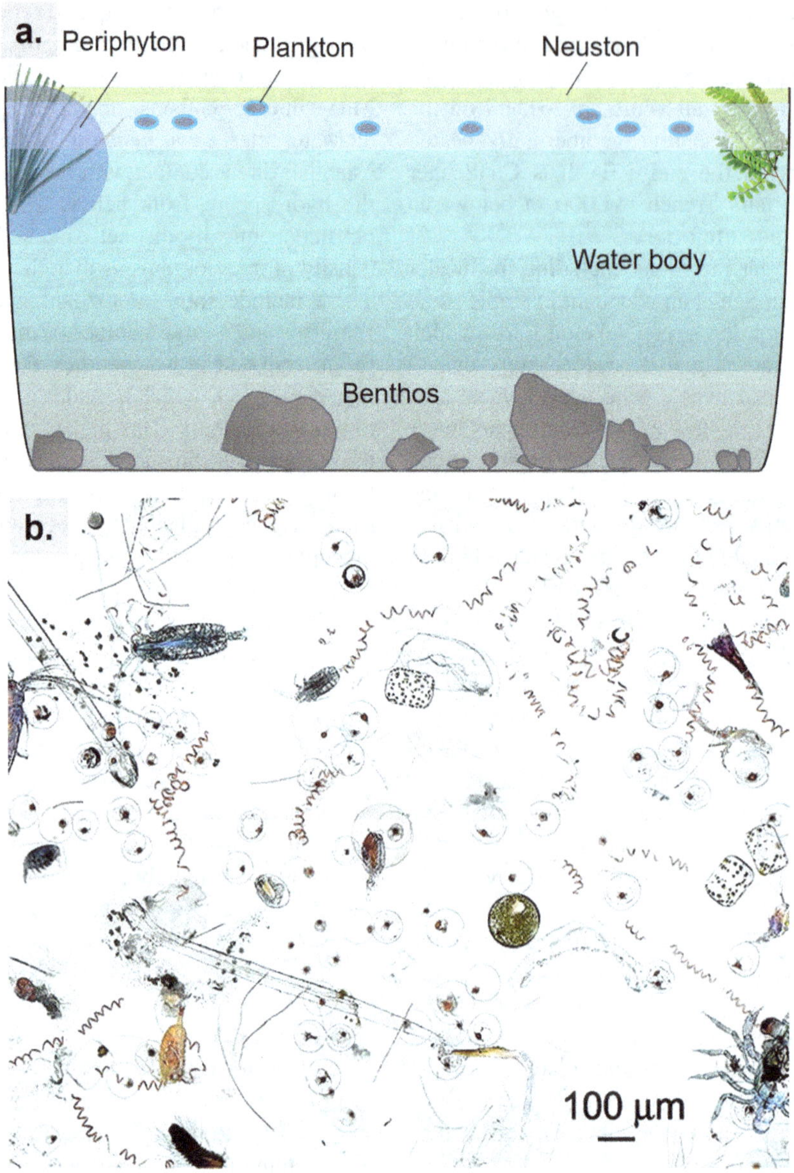

FIGURE 20.3 a. Different categories of microbial forms based on their habitats in aquatic systems. b. Marine planktonic community of surface water, showing diverse plankton, including cyanobacteria, diatoms, and zooplankton. (Image b adapted from Wikimedia/Nadaeau et al.[1])

FIGURE 20.4 a. Biofilm in San Francisco Bay, California, USA. b. Stromatolites, Hamelin Pool Marine Nature Reserve, Shark Bay in Western Australia. c. Algal bloom in Lake Erie, Canada, captured by satellite Landsat 8. (Image b obtained from Wikimedia/Paul Harrison.[2])

from photosynthetic microorganisms like cyanobacteria) that date back to 3.5 billion years. They are among the earliest evidence of life on Earth. Stromatolites have played a critical role as the architects of geological characteristics, especially shaping some sedimentary rocks (Figure 20.4b). In a distinct situation, some microalgae cover the surface of water body and create **algal bloom** (Figure 20.4c). The aquatic habitats are also home to some of the largest known bacteria. The vast area of aquatic regions represents the dark matter for microbial life. Much of this area has been unexplored, and also, the vast number of microorganisms in the ocean have failed to grow in synthetic environments.

20.5 FACTORS AFFECTING AQUATIC MICROBIOTA

A sum of different environmental factors, including physical, chemical, and biological parameters, play important roles in establishing microbial communities in aquatic environments. These factors can be distributed as vertical gradients and horizontal variations in the aquatic systems. The vertical strata form a different continuum of gradations or clines such as thermocline (temperature), pycnocline (density), halocline (salinity), and nutricline (nutrients) (Figure 20.5). These strata can be distinct into either uniform or layered stratification. The uniform stratum in aquatic ecosystems can result in homogeneous water columns, especially due to minimum variations in temperature and salinity. Such conditions prevail in small, shallow water bodies. Layered stratification is a condition of distinct layers in the water bodies. It is much more prominent in oceanic systems than in inland water ecosystems due to its high depth and width. The horizontal variation in aquatic regions is generally due to salinity, latitude (temperature and light intensity), nutrient inputs, and topography. Some important physical, chemical and biological factors that affect aquatic microbial life are discussed below.

20.5.1 Physical Factors

Variability due to water currents, depth, topographical features, transparency of the water column, and seasonality are critical drivers for microbial colonisation. The physical disruptions caused by tides and water currents are common features in almost all lotic aquatic systems (flowing aquatic environments), which makes it difficult for any species to exist for an extended period in the same location. A few periphyton attached to submerged surfaces might be less influenced by turbulence. However, large quantities of sand and gravel in upstream regions can still damage the periphyton layer due to strong water currents. In most aquatic habitats, periphyton constitutes an intricate combination of algae, cyanobacteria, heterotrophic microorganisms, and detritus that adhere to submerged solid substrata. Microorganisms living in aquatic environments with high air pressure and tides have strong mechanical strength. These microorganisms feature strong cell walls, specialised proteins, and adhesion mechanisms. On the other hand, low air pressure and tidal zones select microorganisms with adaptability to efficient nutrient uptake and stress tolerance mechanisms.

Penetration of sunlight in water is one of the most important factors that dictate the growth of phytoplankton and microalgae across the vertical layers. The intensity of the sun controls primary production by phototrophs. However, turbidity sets boundaries of the photic zone in the vertical layers. Temperature is a significant determinant of diversity. A change in the temperature pushes water to bend upon the

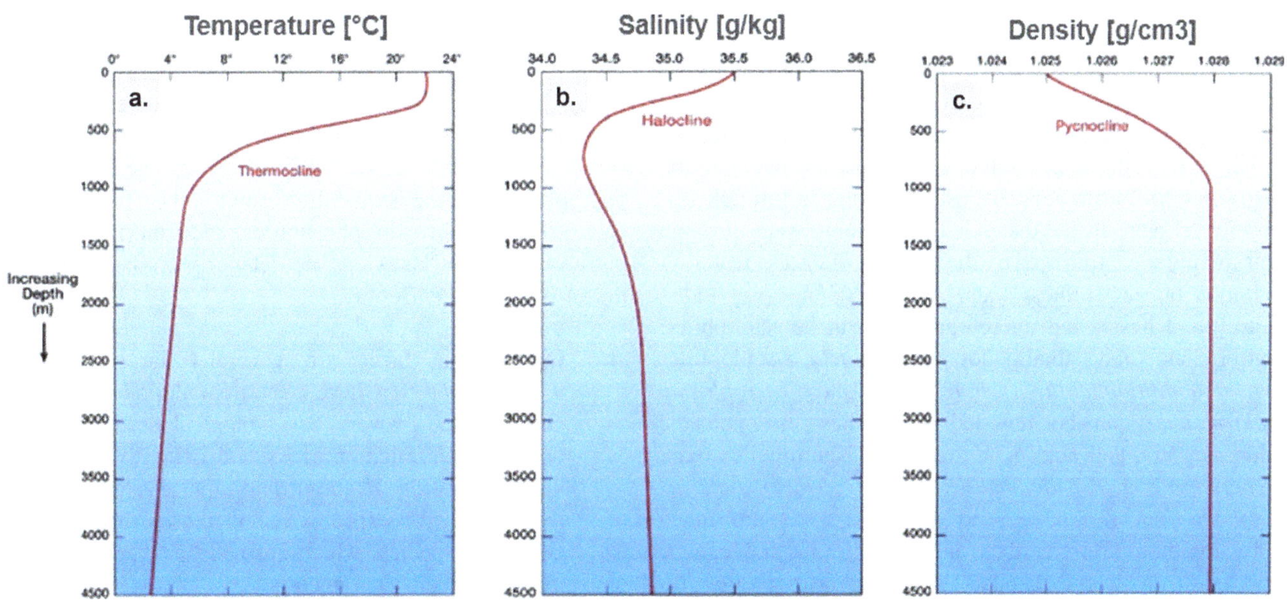

FIGURE 20.5 Vertical stratification in aquatic systems based on different environmental factors. a. Temperature. b. Salinity. c. Density. (Source: lecture2:buoyancy, Ocean Wiki.[3])

density. A thermocline is a stratum where the temperature fluctuates very rapidly with depth. It separates the warm surface water from cool deep water beneath (Figure 20.5a). The stratum at the lower regions of the water column just above the sediment also creates a distinct zone for high microbial counts and enriched microbial activities due to the resuspension of nutrients and cells from the sediments.

Seasonality can directly affect aquatic system dynamics due to several variations in the physicochemical parameters that induce microbial community shifts. Both wet- and dry seasons harbour distinct microbial communities and different functions. Heavy rainfall recharges wetlands, groundwater, and floodplains. The catchment areas have a notable impact on these low-lying regions for the quality and quantity of nutrients for water bodies and, therefore, on their native residents. A prolonged dry or hot season increases the evaporation of surface water, which tends to make the leftovers saltier and denser. This phenomenon may facilitate the mixing of microbial populations across different vertical layers as a result of the reduction of the water column from the top. However, water stagnation in the water column disrupts the nutrient distribution, which creates stratification. It fosters distinct microbial communities at different depths.

20.5.2 Chemical Factors

Chemical factors, such as pH, O_2 concentration, salinity, inorganic nutrients, organic matter, toxins, etc., in aquatic systems lead to variations in microbial community compositions (MCCs). Salinity is a very important factor that ranges somewhere between 0.5% to 37% in water. The extreme situation can even allow salt concentrations of about 290%, as in the Dead Sea. This parameter also bifurcates the living forms into two ecological groups: freshwater and marine organisms. Freshwater and marine habitats have quite different demands in terms of osmotic potential and specific conductance due to ionic concentration variations. Therefore, microbial inhabitants can adapt to only one set of conditions, either saline or freshwater. Increased salinity obliges halophilic microorganisms to perform osmotic movement of water out of their cytoplasm in high-salt environments. On the contrary, freshwater lifeforms maintain osmotic balance through low osmotic pressure inside the cells by absorbing water through their membranes, followed by the removal of excess water.

Based on pH values, water can be defined as acidic, neutral, and basic, and microorganisms can be acidophiles, neutrophiles, and alkaliphiles, respectively. Acidophiles (e.g., *Acidithiobacillus ferrooxidans* and *Ferroplasma acidarmanus*) can be reported from sites like volcanic fields, geysers, and acid mine drainages. Neutrophiles (e.g., *Escherichia coli, Pseudomonas aeruginosa, Vibrio cholerae, Halobacterium salinarum*, and *Chlorella spp.*) are much more common in water than any other forms and can be reported from diverse habitats. Alkaliphiles (e.g., *Bacillus alcalophilus*, *Thioalkalivibrio* spp. and *Halorubrum* spp.) are fairly common in soda lakes, alkaline springs, and deep-sea hydrothermal vents.

The presence and absence of O_2 in the aquatic environment support the growth of aerobes and anaerobes, respectively. The aerobic situation promotes the oxidation of organic and inorganic substances (e.g., ferrous (Fe(II)), Mn(II), ammonia (NH_3), and elemental sulphur (S^0) by native microbiota. Low O_2 concentrations encourage the growth of anaerobic microorganisms using alternative electron acceptors such as iron (Fe(III)), sulphate (SO_4^{2-}), nitrate (NO_3^-), etc. This type of reaction helps remove NO_3^- through the denitrification process. Other noteworthy electron acceptors for anaerobic respiration can be Cr(VI), As(V), and highly oxidised chlorinated chemicals like tetrachloroethene (PCE) and trichloroethene (TCE).

Microbial reductive dissolution of Fe(III) and manganese oxides is responsible for transforming insoluble metals into dissolved ionic forms. These chemicals add another layer of complexity to the biogeochemistry of groundwater and marine ecosystems. These substances negatively affect the environment and the makeup of microbial communities and are seen as environmental contaminants. Some microbial inhabitants can even use reduced forms of nitrogen, iron, and sulphur as electron donors to retrieve energy. For example, the high concentration of NH_3 at the benthic region promotes its oxidation using NO_3^- by anammox and denitrification.

The presence of various nutrients meets the dietary needs of microorganisms. The availability of dissolved organic matter (DOM) is vital for the existence of heterotrophic microorganisms. Both the quality and quantity of DOM are critical to the growth and conditioning of physiologically diverse aquatic microbial communities. The origin of DOM, whether autochthonous (native) or allochthonous (foreign), can also matter in the establishment of microbial communities. Anthropogenic input of diverse chemicals in water bodies presents an adverse impact on microbial compositions.

Autotrophs consume CO_2 and manufacture their food using distinct energy sources. They can be phototrophs (e.g., phytoplankton) and chemotrophs. A chemosynthetic lifestyle requires chemicals like hydrogen gas (H_2), hydrogen sulphide (H_2S), thiosulphate, tetrathionate, S^0, Fe(II), etc., in their metabolic processes to obtain energy.

One important aspect that restricts the growth of phytoplankton is the availability of nutrients. These nutrients chiefly include nitrogen, phosphorus, and iron. Based on the nutritional status, lakes can be categorised as oligotrophic, mesotrophic, and eutrophic.

- Oligotrophic lakes are generally large and deep water bodies that have a nutrient deficiency and are characterised by very low biological productivity and O_2 depletion. Such an ecosystem is highly sensitive to nutrient loading. High nutrient input causes significant loss of sensitive species but increases the growth of generalist species, which are resistant to variations in the water quality parameters.
- Mesotrophic lakes have moderate nutrient levels. Water has medium transparency due to an intermediate level of biological productivity. Possibly, water gets its

colour due to the presence of chromophoric dissolved organic matter (CDOM) and iron that could have an allochthonous origin.

- Eutrophic lakes are characterised by nutrient-rich water bodies having high levels of biological productivity. They are usually smaller and shallower than any other trophic state. The excess nutrients support high primary productivity and increase the demand for dissolved oxygen (DO) in water.

Excessive nutrients (like N and P) in the aquatic systems create the problem of **eutrophication**. Eutrophication is a natural process, but anthropogenic activities such as the discharge of industrial wastewater and fertiliser runoff have exacerbated it across larger areas. The overabundance of nutrients in the aquatic system encourages the growth of aquatic plants and algae. When the surplus algae and plant stuff die, a significant amount of organic matter is accumulated in the abyssal plain. The excess organic matter attracts the growth of decomposers, which use the remaining DO for their respiration from the surroundings, leading to hypoxia. Eventually, this becomes a reason for the death of aquatic wildlife and turns the area into a dead zone (Figure 20.6). When the primary contributors of this phenomenon are algae, this event is known as an **algal bloom** (Figure 20.4c). In some conditions, the massive growth of some algae on the coastline seawater produces a red hue called a red tide. This also raises concerns about the discharge of toxins into the water. The decomposition process releases many acids and excess CO_2 in water, which contribute to the acidification of the environment. However, the major cause of ocean acidification is the increased CO_2 concentrations in the atmosphere due to intense anthropogenic activities.

20.5.3 Biological Factors

In biological interaction, aquatic microorganisms establish communication with micro and macro cohabitants. It can be distinguished as mutualism, parasitism, commensalism, and ammensalism (Table 20.2). Mutualism is a type of interaction between two organisms where both partners receive benefits. Parasitism is a relationship between two organisms, where one organism takes advantage of another at the cost of its fitness. Commensalism involves a one-sided relationship between two species, enabling benefits for one without causing any harm to the other. Ammensalism is also a one-sided interaction but has negative effects, where one organism is harmed while the other is unaffected.

Microorganisms can derive biological affinities through various direct or indirect signals. The direct signals induce contact between cells or through appendages. The indirect signals convey information through chemical exudates, metabolic wastes, and the remains and decayed body parts of organisms. These signals set the stage for the growth and loss of biodiversity in aquatic regions. For example, the microbial symbiosis of coral reef ecosystems is crucial for their vitality and health. Microbes found in coral reef ecosystems have various roles to play in disease resistance, reef calcification, nutrient cycling, and symbiosis. Some fish coexist with bioluminescent microorganisms for communication, camouflage, or luring prey. Deepwater ecosystems exhibit a very strong relationship between microorganisms and other inhabitants to address the

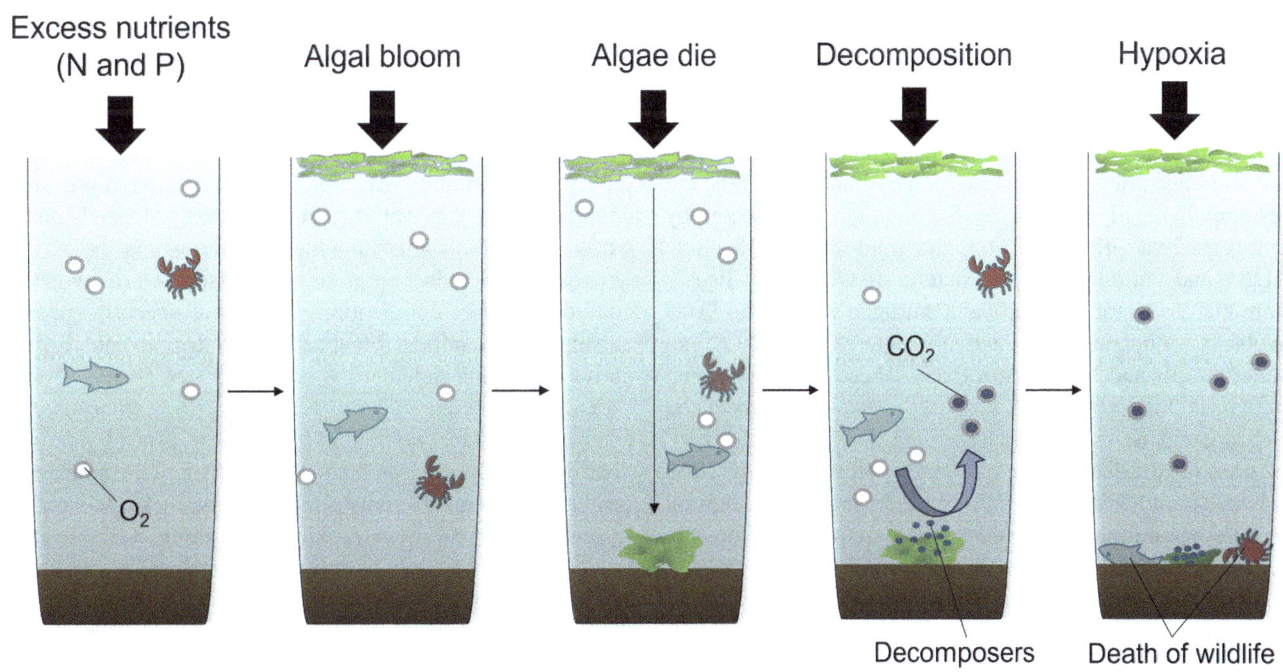

FIGURE 20.6 An illustration of the eutrophication process (where N and P are nitrogen and phosphorus, respectively, and O_2 and CO_2 are oxygen and carbon dioxide gases, respectively.)

TABLE 20.2
Different types of symbiosis and their examples in aquatic ecosystems

Types of interaction	Definition	Examples in aquatic ecosystems	Processes
Mutualism	Both partners receive benefits	Coral and zooxanthellae	Coral reefs provide protection, nutrients, and CO_2. Zooxanthellae supply fixed carbon
Parasitism	One organism takes advantage of another at the cost of its fitness	Chytrids and diatoms	Chytrids infect bloom, causing harm to diatoms in marine ecosystems
Commensalism	One partner gets benefits without causing any harm or benefits to another	Heterotrophic bacteria and cyanobacteria	Heterotrophic bacteria benefit from organic carbon released by cyanobacteria
Ammensalism	One partner is harmed, while the other is unaffected	Algal blooms and fish	Harmful algal blooms release toxins that kill fish, but algae remain unaffected

needs for primary production, nutrient cycling, and feeding. These cross-kingdom communications significantly contribute to the ecological functioning and biodiversity of their natural environments.

Water is an ideal medium for horizontal gene transfer, allowing the exchange of genetic material between different microbial species. This process is responsible for the spread of antibiotic and heavy metal resistance in microorganisms. For example, antibiotic-laden wastewater from livestock farms and agricultural fields, untreated or inadequately treated sewage, improper disposal of medical wastes, and wastewater from pharmaceutical manufacturing plants are the common sources of antibiotics in aquatic ecosystems.

Many microorganisms exhibit syntropic association, which involves cross-feeding habit between different species. This is a process of metabolic interdependence of one organism on the products of other organisms. This can be observed in methanogenic archaea and acetogens in oxygen-depleted sediments. In this cooperation, acetogens allow the degradation of organic matter into H_2 and CO_2. These can be consumed by methanogens to produce methane (CH_4). The low H_2 concentration in the environment makes the reaction thermodynamically favourable for organic degradation by acetogens. In a different setting, ammonia-oxidising bacteria (AOB) and nitrite-oxidising bacteria (NOB) also display syntropic association for efficient nitrogen processing. This involves a sequential oxidation of NH_3 into NO_2^- and NO_2^- into NO_3^- by AOB and NOB, respectively. These reactions are an energy-yielding process and support the growth of associated microorganisms. Many bacteria and fungi secrete exo-enzymes for the degradation of complex organic detritus. The early decay of leaf litter in running waters mainly depends on fungal communities. The availability of simple DOM favours the growth of heterotrophic bacteria.

Many microbial cells are subject to predation by higher trophic levels of organisms. In particular, zooplankton are the grazers of microalgae and bacteria. Protozoa capture and ingest bacterial cells by phagocytosis. This feeding habit has a dominant role in controlling the bacterioplankton population in aquatic systems. Many viruses (cyanophages, bacteriophages, and phycoviruses) are also responsible for controlling the growth of bacteria and algae.

20.6 DISTRIBUTION OF AQUATIC MICROORGANISMS

A plentiful number of microorganisms belonging to different groups are present in diverse aquatic environments. Microorganisms have inhabited aquatic ecosystems, possibly since their emergence. The deep ocean, in particular, bears the imprint of their ancient activities and evolution. The deep ocean sediments have a rich fossil record of ancient microbial life. In recent years, a group of scientists recovered living microbes from the seabed that was buried under the floorboard for more than 100 million years.

Based on the origin, microorganisms can be autochthonous or allochthonous. Autochthonous species are permanent, long-term residents of the environment, whereas allochthonous microorganisms are taxa imported from a foreign ecosystem. The microorganisms of foreign origin can become a part of the native community over time after adaptation to prevailing conditions. The survival of foreign organisms depends on two parameters. They either tune their metabolism in the existing environment or induce migratory responses toward favourable conditions. Also, competition of invasive species with pre-existing communities for space and nutrition can displace native organisms and alter the ecosystem's balance.

The dispersion and response to environmental heterogeneity are two major factors that control microbial distribution in an environment. An aphorism of the nineteenth-century microbiology roots given by Beijerinck and Becking, "*Everything is everywhere,* but *the environment selects*", is one concept that explains microbial distribution patterns in aquatic environments. The initial part of the sentence refers to the dispersal potential of microorganisms, and the latter part signifies the activity and growth in a particular environment. Microbial dispersal is a physical movement that is possible through both active and passive mechanisms. Passive dispersal through water currents, wind, and other organisms greatly contributes to community assembly. In contrast active dispersal processes involve microbial motility at the local scale and play roles in colonising favourable habitats. Planktonic forms can migrate vertically and have a wide distribution in vertical columns.

A different category of waterborne microorganisms is holoplankton and meroplankton, distinguished by their life cycle. Holoplankton spend their entire life in the water, whereas meroplankton are temporary residents of the aquatic ecosystems and are only planktonic for part of their life cycle. Holoplankton are common in all aquatic ecosystems, while meroplankton are less common in freshwater ecosystems than in marine ecosystems.

A water body is a medium to transport and disseminate microorganisms from one place to another. Bacteria present in aquatic systems can be fast swimmers with a speed of 1-1000 μm/s. Bacteria accomplish their directional movement responses using flagella on their cell surface, driven by molecular motors. The bacterial movement is greatly dependent on gradient length, cell size, turn angle, sensitivity to gradient, water current, etc. The abundance and fate of microorganisms in the aquatic system are determined by the physio-biochemical parameters of the system.

A continuous influx of inland water adds freshwater microorganisms to marine ecosystems. Air transport is also one of the major long-distance dispersal mechanisms. The colonisation of airborne microorganisms in the aquatic environment can be less significant on a short time scale due to distinct mechanisms for their survival. However, when aquatic microorganisms are released from the water environment, they reside as bioaerosols in the atmosphere. These airborne particles can be suspended in the air for up to four days and travel up to 11,000 km in the ocean. Differently, commercial shipping in the aquatic environment has also had a substantial impact on the global distribution of microbial species.

Based on key distinctions of the site characteristics and hydrological regime, aquatic microbial ecosystems can be distinguished into three different landscapes. These include microorganisms in inland ecosystems, marine ecosystems, and estuaries, which have further subcategories. They are discussed in detail in the following sections.

20.7 MICROORGANISMS IN INLAND ECOSYSTEMS

Well-known natural inland water-influenced environments are groundwater and surface water. A large number of microorganisms also thrive on glaciers, ice caps, snow, and ice. This discipline mainly includes studies related to freshwater ecosystems. However, the saline waters within the boundaries of the land are also not precluded. The study also extends to manmade stretches such as agricultural irrigation, artificial pools, canals, and dams. Most freshwater ecosystems are spatially discrete on the land and have distinct boundaries (Figure 20.7a). The freshwater environment can be categorised into four different ecological zones, as follows:

- Littoral zone- This region is the nearshore area. This is the shallow photic zone and receives a lot of nutrients from runoff. This zone supports high density and diversity of microbial forms.
- Limnetic zone- This region is the open water area of the ponds and lakes and has low concentrations of nutrients. It covers the depth where sunlight penetrates. A related term is the pelagic zone, which covers a broader scope and encompasses all depths.
- Profundal zone- This region is the aphotic zone of the free-standing water in the inland aquatic systems such as lakes and ponds.
- Benthic zone- This is the deepest region of the aquatic ecosystem, consisting of the sediment surface and some subsurface layers. This zone gathers nutrients and organic matter from the water surface and the water bed. The local inhabitants of the benthic zone are termed **benthos**. The same term is also used for the residents of the ocean floor. In the benthic region, microbial density and activity are noticeably high because nutrient and carbon levels are fairly elevated compared to the overlying water layers. Some microbial seeds of benthos can emanate from overhead and disperse in overlying water layers.

A different ecological classification of freshwater habitats is based on the flow regime of the water. It can be lotic and lentic. A lotic ecosystem is characterised by flowing water that is rope-shaped, and examples include rivers and streams. The constant water flow in this habitat constitutes life in the dynamic environment. A lentic ecosystem is standing water surrounded by boundaries forming circular, semicircular, rectangular or square-shaped structures, and examples include groundwater, ponds, and lakes. Although all types of aquatic systems exhibit some movement, the distinctions between lotic and limnetic ecosystems are based on the water current and unidirectional movement.

Based on their sources, inland aquatic systems can be bifurcated into surface water and groundwater. These ecological sites have distinct microbiological characteristics, which are discussed in the following sections.

20.7.1 Microorganisms in Surface Water

Examples of surface water are streams, rivers, floodplains, lakes, ponds, reservoirs, and wetlands. These sites possess different sets of physicochemical characteristics and, therefore, harbour different microbial communities. The local characteristics of water bodies, such as thermal influence and nutritional supplements, induce a notable impact on the microbial community at both temporal and spatial scales. The sediments and benthic regions are the hotspots of microbial abundance and diversity. The soil-water and rock-water interfaces are the ecotones and establish distinct MCCs.

Rivers and streams are examples of turbulent water flow. They comprise a large-scale variation in their physicochemical characteristics from source to mouth, and the variation in

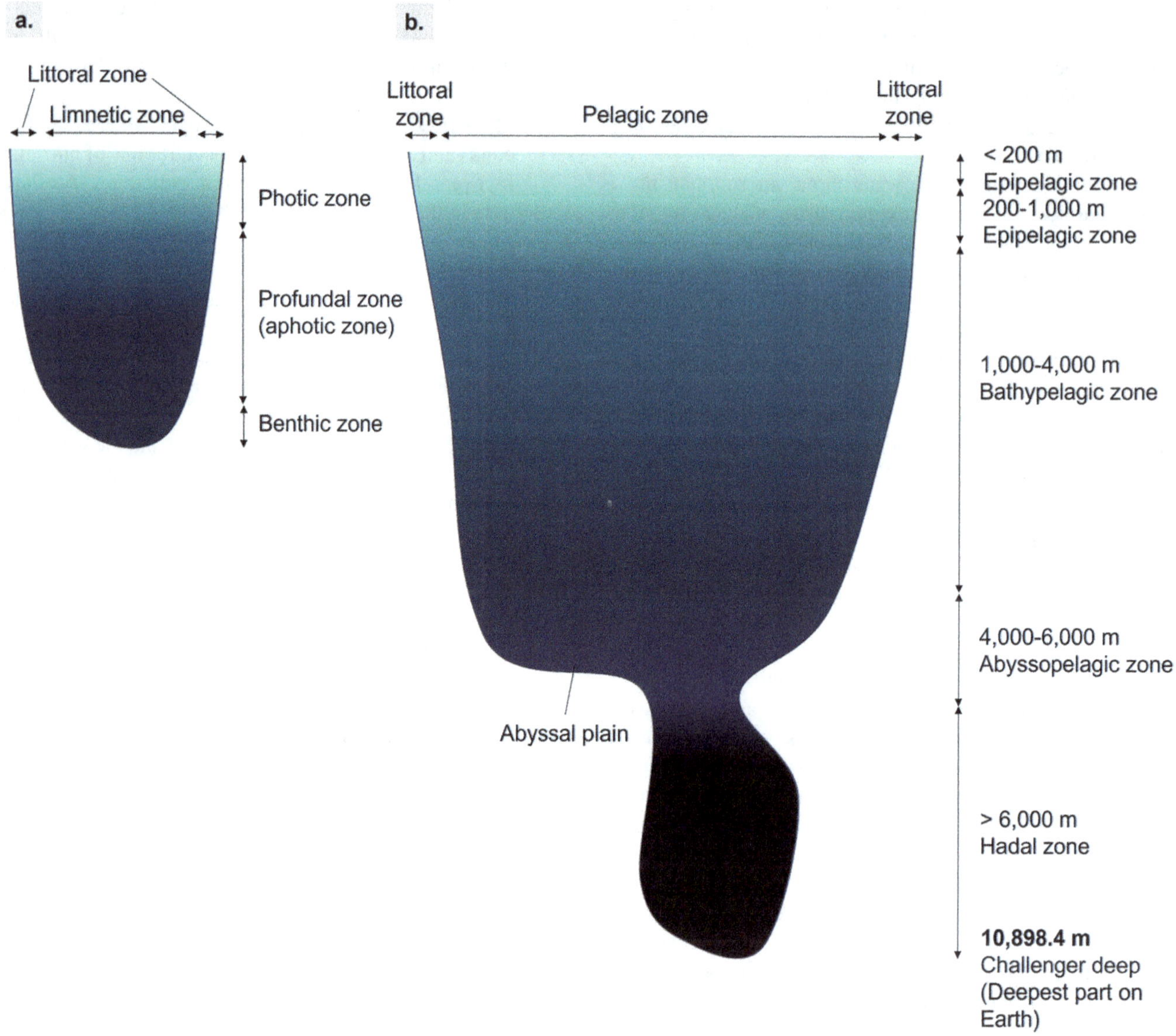

FIGURE 20.7 Horizontal and vertical zones in different aquatic ecosystems. a. Inland water. b. Seawater. (The littoral zone in the marine ecosystem is mainly termed the intertidal zone.)

biodiversity is also apparent in their wide stretch. The water current in upstream regions keeps the environment well-oxygenated. Also, it keeps fine solid particles like silt and clay suspended in the water column. However, the water current has a comparatively low impact on downstream regions. The high current of water causes physical damage to free-floating microorganisms in rivers and streams. The microbiota introduced from immediately adjacent habitats can also be swept away by the current. Most microorganisms in rivers and streams reside near submerged/exposed rocks and the land-water interface. The stable life forms of the lotic environment are periphytons and biofilms. Commonly growing bacteria are taxa belonging to Cyanobacteria and Bacteroides, as the **epilithons** (growing on the rocks). Freshwater microalgae and cyanobacteria thrive in the photic zone and contribute to primary production. Heterotrophic bacteria found in freshwater ecosystems belong to various groups, predominantly belonging to Proteobacteria, Actinobacteria, and Verrucomicrobia. Some heterotrophs on the water surface can even acquire energy from sunlight and are termed "**photoheterotrophs**". Examples of such planktonic forms are purple non-sulphur bacteria, green non-sulphur bacteria, and heliobacteria, which either utilise rhodopsin photosystems or aerobic anoxygenic reactions to acquire energy. This type of lifestyle is also quite common in marine ecosystems.

Fungi are quite common in freshwater but less abundant than bacteria. Both single-celled forms (yeasts) and multicellular filamentous forms can reside in the water bodies. They are saprophytic and parasitic, which break down organic matter and infect algae, plants, and even animals. Aquatic hyphomycetes (e.g., *Alatospora acuminata* and *Articulospora tetracladia*) are the most common and well-studied group of freshwater fungi. Most fungal inhabitants of freshwater ecosystems belong to Ascomycota. These fungi produce a high number of spores that are dispersed in the aquatic ecosystems through water currents. Chytrids are a group of primitive fungi which can also be reported in freshwater. Their taxa produce flagellated spores called zoospores that can swim through water to find new hosts or substrates.

Lake and pond ecosystems are examples of nonturbulent aquatic systems. They store the maximum volume of water in the inland aquatic ecosystems (i.e., about 87%). The biodiversity in these regions is highly influenced by nutritional or trophic status. The main inhabitants are plankton and benthos. These ecosystems team up with diverse bacterial communities, mainly belonging to Proteobacteria, Actinobacteria, Bacteroidetes, Cyanobacteria, and Acidobacteria.

Native microorganisms are specialists in decomposing organic debris present in the water column and sediments. In temperate environments, litter produced from the shaded plant parts of the overhead tree canopies can add organic matter to lakes. There are many lakes where the drainage input is quantitatively more important for the accumulation of organic matter. The biodegradation process also provides a good source of DOM in aquatic systems that influences microbial abundance and diversity. The hypersaline lakes host halophilic microorganisms. A green unicellular alga, *Dunaliella salina*, is a very common inhabitant in such environments and gives the lake a pink appearance due to the production of some pink colour pigments (Figure 20.8a). The hypersaline environments also have archaeal dominance (Haloarchaea), with a capacity to tolerate high salt concentrations. The most common representatives of those residing in high salt concentrations are *Halorubrum* and *Haloferax*. Many other inland water bodies also get colour due to the presence of distinct microorganisms. For example, Grand Prismatic Spring, USA, has variant colour patterns at both spatial and temporal scales due to different chlorophyll/carotenoid ratios and microbial mat of thermophilic bacteria and archaea (Figure 20.8b). Laguna Colorada, Bolivia, gets its red colour due to red algae and some microorganisms (Figure 20.8c). Lake Natron, Tanzania, also appears red, but that is due to the presence of a halophilic cyanobacterium, *Spirulina* (Figure 20.8d). Laguna La Brava, Chile, occasionally develops a purple colour mat due to purple nonsulphur bacteria (Figure 20.8e).

The wetlands are distinct ecosystems, lying at the interface of terrestrial and aquatic ecosystems. In the wetland, water tends to float on low-lying land. The covering of water on land can be either permanent or seasonal. The well-known patches of wetlands are swamps, marshes, bogs, and fens. The proximity of oxic-anoxic conditions in wetland ecosystems harbours both aerobic and anaerobic activities in microorganisms side by side. Such ecosystems host a variety of microorganisms, such as sulphate reducers, diazotrophs, methanotrophs, denitrifiers, phototrophs, and chemotrophs. These microorganisms have fast nutrient (re)cycling capacities and play a key role in global cycles of carbon, nitrogen, and sulphur. Despite covering less than 5% of the Earth's surface, they are the most dynamic and productive ecosystems on Earth. Wetland ecosystems are also prominent contributors to methane (CH_4) gas emissions (20-39% of the global emissions).

A large portion of the land ecosystem is covered with ice and snow. They are the storehouse of freshwater. The extreme environment of cold conditions supports a rich and stable microbial community. Snow microalgae are quite common on snow and ice surfaces and give snow various colours like red, green, and yellow. An example of colourfully striking organisms found in snow is *Chlamydomonas nivalis*. It produces red colour pigment and is responsible for a phenomenon called "watermelon snow". A different microalga, *Chloromonas* spp., shed a green colour on the snowfield (Figure 20.8f). Both examples of colourful microalgae come from the same family, Chlamydomonadaceae. The presence of excess microalgae on snowfields contributes to glacier melting. When microalgae cover the surface of the snow, they absorb light for certain metabolic events and raise the temperature of their surroundings, reducing the impact of the albedo effect (the surface's capacity to reflect sunlight). This results in the warming and melting of snow. The photosynthetic carbon produced by microalgae on snow also promotes the growth of heterotrophic microorganisms. Heterotrophic bacteria on snow reside as cold-adapted organisms (i.e., psychrotrophs and psychrophiles) and have slow metabolism. These microorganisms have developed various strategies to cope with extreme colds, including the production of antifreeze proteins, pigments to protect against UV radiation, and the ability to remain dormant during unfavourable conditions.

20.7.2 Microorganisms in Groundwater

The reservoir of groundwater is called an **aquifer**. Groundwater displays differences from other aquatic habitats in terms of boundary, relatively constant temperatures, poor availability of nutrients, and darkness. Despite their confined space and extreme situation, these regions represent a huge reservoir of microbiological species, some of which are endemic in their ecosystems. Examples of endemic microbial species of groundwater habitats are *Dehalococcoides* spp., *Geobacter* spp., and *Methanobacterium* spp. Microbial inhabitants in these subterranean regions have a major impact on groundwater chemistry, as they play roles in the mobility of metals, degradation of organic compounds, geological sequestration of greenhouse gases, and weathering of rocks.

Bacteria and archaea are the main residents of groundwater habitats due to the complexity of the environment. Many microbial residents in underground environments observe heterotrophic lifestyles, which heavily depend on organic carbon outsourced from the surface water. Phototrophs are rare due to their very poor ability to reach sunlight. The primary production in the deep subsurfaces relies on a chemotrophic lifestyle, utilising chemical energy. Most fungal residents of groundwater are yeast or yeast-like taxa. In addition, many groundwater fungi have representatives from Ascomycota and Basidiomycota. A considerable proportion of taxa belongs to unclassified aquatic fungi. Viruses multiply using bacterial cells and are present in high concentrations. In nutrient-poor

FIGURE 20.8 Distinct colour of inland water ecosystems due to the presence of diverse microorganisms. a. Lake Hillier, Australia, dominated by *Dunaliella salina*. b. Grand Prismatic Spring, dominated by microbial mats of thermophilic bacteria and archaea (variation in the colour is due to the chlorophyll/carotenoid ratio in different regions). c. Laguna Colorada, Bolivia, is dominated by red algae and microorganisms. d. Lake Natron, Tanzania, is dominated by the cyanobacterium *Spirulina*. e. Laguna La Brava, Chile, is dominated by purple non-sulphur bacteria. f. Snow algae bloom, Antarctica, dominated by green algae. (Image a obtained from Wikimedia/Kurioziteti123[4], c from Wikimedia/Atobit[5], d from iStock/derejeb[6], e adapted from Visscher et al. (2020)[7], f adapted from Gray et al. (2020).[8])

environments, microbial communities exist as microcolonies or biofilms attached to rock surfaces, sediments, and waste materials. The rock particles are composed of specific mineral constituents, which are sources of organics, H_2, and oxidants for microbial growth. Some inhabitants can be endoliths, living in the inner part of a rock and obtaining nutrition from the minerals.

The diversity is maximum immediately below ground and at the intersection of saturated and unsaturated zones. The water immediately below the ground has enough DO, which supports the growth of aerobes. Heterotrophic bacteria, fungi, and protozoa are common residents of this area and consume plenty of O_2 for respiration. Fungi utilise dead or decaying materials. Protozoans survive by feeding on bacterial cells. Degradation of organic matter by heterotrophs elevates CO_2 concentrations in aquatic ecosystems. This increases the CO_2 partial pressure in the unsaturated zone, which is comparatively higher than in the above ground zone. High CO_2 concentrations promote the formation of carbonic acid, which results in enhanced dissolution reactions and the leaching of minerals.

The deep subsurface of the aquifer contains little or no DO. The pores and spaces in subregions are filled with water. These regions have no air circulation and are completely saturated with water. The amount of organic matter in the deep zone is also small. Anaerobic bacteria and archaea are the dominant inhabitants of the saturated zone. These include anammox bacteria and methanogenic groups. Some heterotrophs found in such regions also display a facultative anaerobic lifestyle. In anaerobic respiration, NO_3^- is one of the best electron acceptors from the energy point of view. The threshold concentration of O_2 to perform anaerobic respiration is 0.5 mg/l. However, NO_3^- reduction can begin even at O_2 concentrations below 2.0 mg/l. Some anaerobes can utilise Mn(IV), Fe(III), SO_4^{2-}, and CO_2 when the concentrations of NO_3^- in the surroundings are depleted.

20.8 MICROORGANISMS IN MARINE ECOSYSTEMS

The key identity of the marine environment is a continuous body of salt water (i.e., seawater). The typical salt concentration in seawater is 3.1–3.8 g/l (i.e., an average of 3.5% of the weight of seawater). Chloride and sodium ions are the most abundant ions present in seawater. The compositions of salt also include some other elements, such as magnesium, calcium, potassium, and bromide ions. Similar to freshwater ecosystems, marine water bears different ecological zones. The littoral zones are prominent sea coasts, including sandy shores, rocky shores, estuaries, mudflats, continental shelves, and neritic zones (i.e., the shallow water over the continental shelf). The areas within the tidal range are called intertidal zones. The pelagic zones are the open areas in the marine ecosystem and contain the largest volume of water. Based on the penetration of light, it can be subdivided into Epipelagic (sunlight), Mesopelagic (twilight), Bathypelagic (midnight), and Abyssopelagic (abyssal zone). The abyssopelagic zone is equivalent to the benthic zone of the freshwater ecosystem. An extensive flat region of the deep ocean is an abyssal plain. The hadal zone is the deepest zone of the oceans, with a topography of long, narrow open containers emitting hot water. They are also known as oceanic trenches (Figure 20.7b).

The marine environment is the largest compartment of Earth and supports the most incredible biodiversity. Almost 90% of the total biomass of marine ecosystems is composed of microbial cells. Around 4×10^{29} microorganisms are dwelling in the seawater at any time. The estimated biomass is $\approx 3 \times 10^{17}$ g. In general, each ml of seawater contains millions of living microorganisms consisting of thousands of different species. Figure 20.9 illustrates some common taxa and microbial groups reported from seawater.

All the native microbial inhabitants of seawater are salt-loving (also known as halophiles) due to saline environments. Those microorganisms may also exhibit the properties of alkaliphiles (base-loving microbes), psychrophiles (cold-loving microbes), thermophiles (heat-loving microbes), and piezophiles (pressure-loving microbes), depending on their specific residences in the seawater. These terms are discussed in detail in Chapter 2. Seawater microorganisms account for the major activities in the seawater, serving as the engine of the marine ecosystems. The microbial density is highest at the top, which declines by more than 10-fold below 100 m. However, their density can be further high toward the outreach of the abyssal plain. About 70% of microbial life underwater thrives in the dark ocean. The subseafloor sediments and oceanic crust are distinct sites for microbial proliferation. The sea sediments have a very high amount of carbon, both in the reduced form (i.e., organic matter) and the oxidised forms (i.e., sedimentary carbonates). These regions also contain high amounts of nitrogenous substances. These substances make subseafloor regions an excellent reservoir of nutrients, making them very densely populated habitats, although the bioavailability of these nutrients can be a limiting factor. Many microorganisms have developed special mechanisms to survive in the deep sea's stressful conditions. The cell size of the deep-sea inhabitants is generally much smaller than they are found at the water's surface. Many bacteria have access to low bulk concentrations of nutrients and are slow-growing organisms because the total nutrient content does not necessarily translate to high bioavailable nutrient concentrations for individual microorganisms. These deep regions also do not contain sufficient DOM to support a heterotrophic lifestyle. Therefore, microorganisms in the deep regions have adapted to use unconventional electron donors and acceptors. The reduced iron and sulphur compounds are the main geofuels that act as the dark energy for microbial growth (detailed in Chapter 29).

The average depth of the seawater is about 3,800 m. Based on the penetration of the light, marine environments can be categorised into two vertical layers: photic and aphotic zones. Water below 200 m receives a limited amount of light, and below 1,000 m, it is completely dark. The in-between region is known as the disphotic zone (twilight zone). Particularly, the aphotic zones in the open sea have a low abundance of microorganisms due to the poor availability of nutrients. The oceanic crust is the solid body at or below the seafloor that harbours barophiles.

The hydrothermal vents are very unique sites for the marine microbiome. These vents emerge through fissures of the subduction zone and emit hot, chemically modified seawater. These extreme conditions support a thermophilic, chemosynthetically supported lifestyle. The hearty

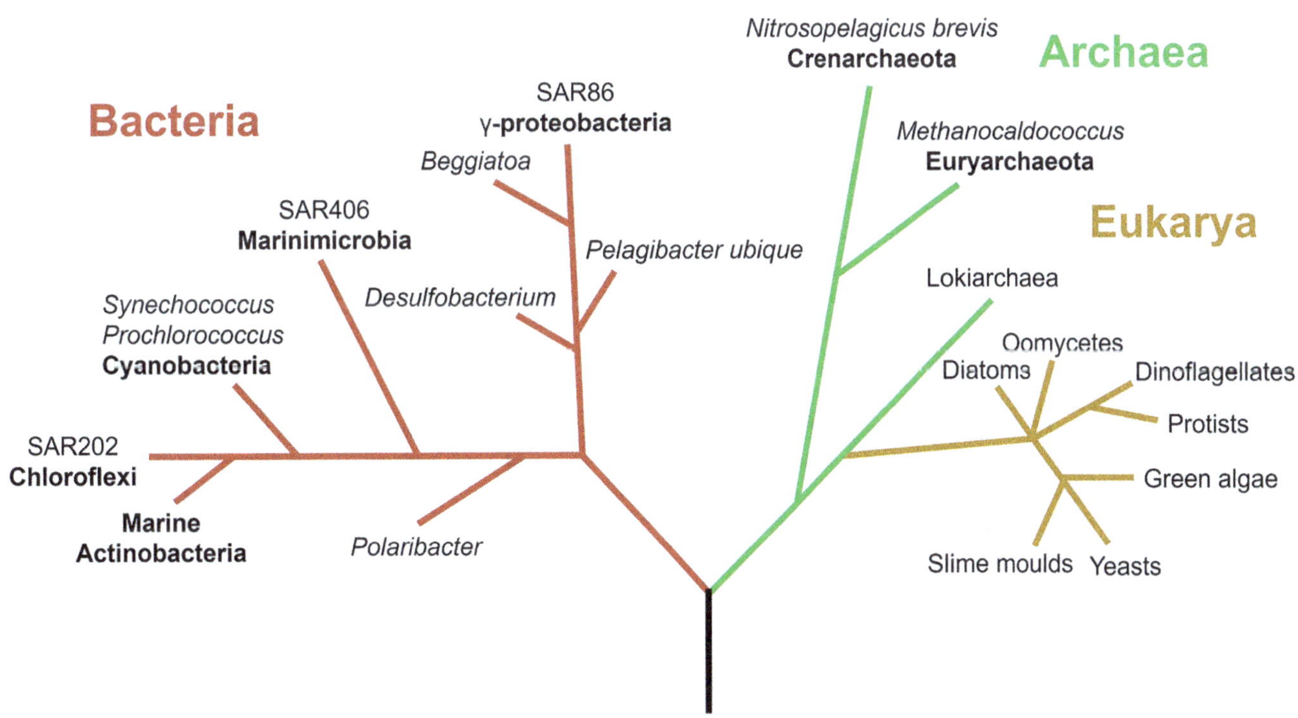

FIGURE 20.9 A tree of some important microbial forms found in the ocean.

survivors of the hydrothermal vents rely on symbiosis and chemolithotrophic lifestyles to meet their energy demands. Even the deepest point of the Earth, Challenger Deep, the Mariana Trench (~ 11,000 m below sea level), also catalogues microbial communities with some unusual metabolic capabilities. Such an environment was possibly the birthplace of life on Earth about 3.8 billion years ago and resembles the earliest microbial ecosystems. Since their origin, they have evolved in various physiological forms, with more than 90% of early microbial life having vanished. However, the basic cellular organisations of modern microbial cells do not seem to be very different from those of ancient life forms.

Photic zones contain abundant photoautotrophs that can fix carbon through photosynthesis. The organic matter produced in photic zones is the main source of nutrients for the heterotrophs. They can even support life in aphotic zones by their bulk deposition at the seafloor. *Prochlorococcus* and *Synechococcus* are two prominent inhabitants in the ocean euphotic zones and belong to the phylum Cyanobacteria. *Prochlorococcus* is the most abundant photoautotroph on Earth and is chiefly present in warm oligotrophic waters. A different photoautotroph, *Synechococcus,* is highly common in coastal turbid waters. A filamentous cyanobacterium, *Trichodesmium,* is capable of fixing nitrogen along with photosynthesis and exists in tropical and subtropical oceans. This phytoplankton alone is responsible for about half of the global ocean nitrogen supply. A particular species, *T. erythraeum,* turns the colour of the Red Sea red due to its seasonal bloom. Diatoms, coccolithophores, and dinoflagellates are predominant microalgae in seawater. Diatoms and dinoflagellates are particularly found attached to rocks, sand grains, and sediments. Many dinoflagellates growing on warm seawater surfaces exhibit the properties of bioluminescence. Many dinoflagellates exist as photosymbionts in a mutualistic relationship with corals. These symbiotic dinoflagellates are commonly known as **zooxanthellae**. They reside inside the translucent fleshy tissue of anemones, jellyfish, sponges, and flatworms, where zooxanthellae supply fixed carbon to the host organisms. In exchange, the coral reef provides protection, nutrients and a constant supply of CO_2. *Symbiodinium* is one commonly known taxon of zooxanthellae, which engages in symbiotic associations with marine invertebrates.

Compared to any other organisms, heterotrophic bacteria are the most widely distributed and abundant cellular organisms in the marine environment. ***Pelagibacter ubique* is the most dominant species in seawater**. This is also the most abundant organism on Earth (detailed in Box 20.1). A different widely distributed and abundant bacterium in seawater is *Roseobacter* (belonging to Alphaproteobacteria), which constitutes up to 30% of bacterial communities in the pelagic zone. The organisms belonging to the same clade are physiologically versatile and are major contributors to global carbon and sulphur cycles in the ocean.

A large number of bacterial taxa in seawater are light-producing **bioluminescent bacteria**. Bioluminescence is a natural process which relies on an enzyme, luciferase, that acts on substrate "luciferin" in the presence of O_2 to emit light (Figure 20.10). These bacteria in aquatic environments can be found free-living or symbiotically associated with different animals. In symbiotic association, the luminescence reaction of bacteria benefits host organisms for camouflage, prey and/or mate attraction, whereas the host organisms provide bacteria with protection and nutrients.

Members of the archaea can represent up to 40% of the marine microbial community. They are dominantly present in temperate marine coastal waters and the deep ocean. The planktonic marine archaea of the Marine Group I (MGI)

BOX 20.1. THE MOST ABUNDANT ORGANISMS ON THE EARTH

The most abundant and widely distributed organism on Earth comes from the unseen life in the aquatic ecosystem. Native to marine ecosystems, *Pelagibacter ubique* is the most numerous cellular form and is the resident of the nutrient-depleted open oceans. It belongs to the SAR11 clade (family Pelagibacteriaceae) in the phylum Alphaproteobacteria and grows globally. The genus name was coined from two Latin words, "pelagus" and "bacter", which means sea and bacterium, respectively— the literal meaning is a bacterium of the sea. The species name, ubiquie, is derived from a Latin epithet, which means "everywhere". The reason for its dominance over the vast oceanic body is its efficient metabolic adaptation. It can multiply even at low-nutrient conditions. Another reason for dominance is its photoheterotrophic lifestyle, which supports its energy supply from sunlight. It is not only the world's most abundant organism but also one of the smallest known cellular organisms (diameter ranging from 0.12-0.20 μm). They account for about 25% of all microbial cells present in marine ecosystems. The cell is rod or crescent-shaped and yields a volume of ~0.01 $μm^3$. This bacterium carries the smallest genome of any free-living bacterium (~1.3 Mbp). This bacterium has a streamlined genome, which includes the smallest number of predicted open reading frames, a complete absence of nonfunctional or redundant DNA (pseudogenes, introns, transposons, and plasmids), the shortest intergenic spacers, and few alternative metabolic pathways. The bacterium can synthesise all twenty different amino acids. Initially identified through non-culturing methods, the development of non-traditional media—which include ingredients like methionine, glycine, pyruvate, and artificial seawater—has made it feasible to grow them artificially. The doubling time for the cell is fairly slow (approximately 25-50 h) and observes seasonal abundance. A bacteriophage HTVC010P is the key predator of this bacterium. This phage is the commonest living agent in the entire biosphere.

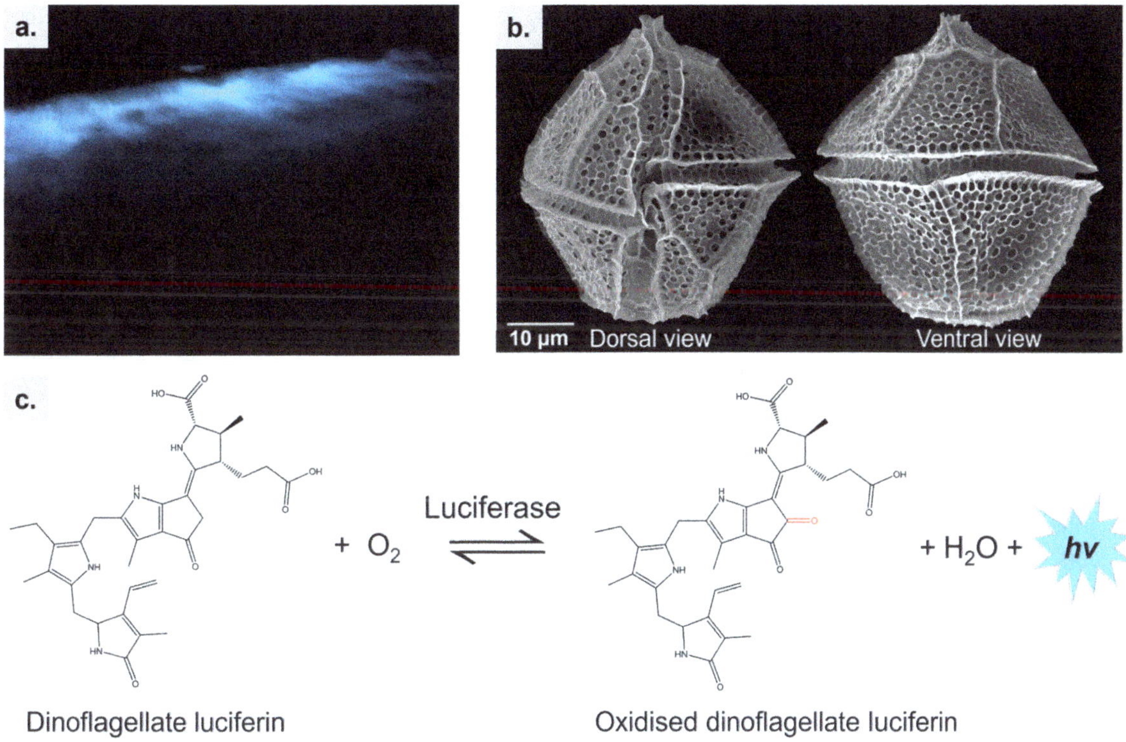

FIGURE 20.10 a. Bioluminescence in the ocean due to dinoflagellates. b. Ultrastructure of a bioluminescent dinoflagellate, *Lingulodinium polyedra*. c. Catalytic bioreaction of bioluminescence. (hv is the energy of a photon.) (Images a obtained from Wikimedia/Jed[9] and b from Tillmann et al. (2021)[10] with permission of © 2020 Elsevier.)

have cultured representatives and members from the phylum Thaumarchaeota. Many of them are ammonia-oxidising archaea and widely distributed pelagic chemoautotrophs that have control over a rate-limiting step of nitrification. As a result, they have long played a key role in the global nitrogen cycle. Representatives of marine group II (MGII) observe a photoheterotrophic lifestyle and prefer the photic zone. Discoveries suggest that a different group, Lokiarchaeota reported in a deep-sea hydrothermal vent share the closest evolutionary link between prokaryotes and eukaryotes. Some haloarchaea members (Phylum: Euryarchaeota) on the oceanic surface produce bacteriorhodopsin, which functions as a light-driven proton pump for ATP production. Bacteriorhodopsin gives haloarchaea a purple hue as it absorbs green colour and reflects red and blue when exposed to light.

Fungi are relatively less abundant in seawater, and those represented are called marine fungi. Based on ecological terms, marine fungi can be divided into two groups: obligate marine fungi (OMF) and marine-derived fungi (MDF). OMF are the permanent residents of saline environments, which multiply and sporulate exclusively in the marine and estuarine ecosystems. MDFs are also known as facultative marine fungi that can thrive in both marine and freshwater ecosystems. They are possibly native to freshwater ecosystems or terrestrial ecosystems, but can adapt to survive in marine ecosystems. Many fungal species in seawater are found attached to plant-decomposed materials or lie on rocks near shores. Some fungal spores possess appendages for attachment to the substratum. Some members are even responsible for parasitism of diatom species, causing shifts in diatom bloom dynamics. More than 1500 marine fungal taxa have been identified so far. Many marine fungi are filamentous and mainly belong to Ascomycota, Mitosporic fungi, and Basidiomycota. Recent studies reported that some members of chytrids were also residents of in deep-sea sediment, water column, and sea ice samples.

Communities of protozoans also exist in diverse sea environments, from the sea surface to deep vents. They constitute an important component of the marine environment and regulate the concentration of bacteria. They are single-celled eukaryotes, and those that reside in seawater can be flagellates, ciliates, amoebae, and foraminifera. Certain protozoans, such as foraminifera, contribute to sedimentary deposits on the seafloor. The members of this group are sensitive to environmental variations and act as bioindicators of seawater.

Viruses rule marine ecosystems through their diverse interaction with other organisms. They are noncellular biological agents and constitute the largest genetic reservoir in the ocean. They are even more abundant than bacteria or any other organisms. Their survival strictly requires a suitable host and they possess the ability to infect bacteria, fungi, seaweeds, and animals. They act as potential processors of bacterial growth. **The virus, HTVC010P, infecting *P. ubique*, is the most abundant biological agent on Earth**. A type of virus that attacks cyanobacteria is **cyanophages**, which are immensely present in seawater. Viruses in marine ecosystems can influence their target hosts through mechanisms like cell lysis, horizontal gene transfer, and modulating their metabolic pathways.

20.9 MICROORGANISMS IN ESTUARIES

The estuarine environment is a coastal water body that has the confluence of both marine and freshwater ecosystems. The influx of freshwater carries nutrient-rich sediments, and the seawater delivers saltwater at the margin of the sea. Such a distinct composition of water is termed brackish water. It has lower salt concentrations than seawater but higher than freshwater. The estuary may tend to accumulate high organic matter and provide shelter to numerous microorganisms. This transitional zone is a highly dynamic ecosystem because it constantly changes its physicochemical characteristics in response to tides, water currents, local climate, and freshwater input. These forces create many challenges for their residents, including microorganisms. River deltas, mangrove forests, river mouths, fjords, lagoons, tidal creeks, mud and sand flats, and seagrass beds are typical estuarine habitats. They are also notable examples of ecotones, which are transition zones between two different ecosystems. They are rich in biodiversity and so are applicable for microbial forms.

Phytoplankton are highly abundant in the photic zones. Heterotrophic activity in estuaries is also high due to the high magnitude of organic carbon and nutrient transport from terrestrial runoff and river inflow. The high heterotrophic activities lead to estuaries being in an oxygen-depleted condition throughout the year. The role of the detritus food chain in some estuarine environments (such as mangrove forests) is also apparent. In the detritus food chain, the dead remains of plants and animals lie at the base of the food web, which is processed by detritivores and decomposers.

Some of the largest sizes of bacteria on Earth reside in the sea sediments, particularly in the coastal regions. *Thiomargarita magnifica* is the largest known bacterium, which was recently discovered in the mangroves of Guadeloupe in the French Caribbean. This bacterium belongs to the phylum Proteobacteria and can attain as long as 2 cm in length, and can be seen with the naked eye. It has many peculiar characteristics, such as complex internal organisation, a dimorphic life cycle, and extremely large genomes, which are not common in other bacteria. A different species belonging to the same genus, *Thiomargarita namibiensis* (0.75 cm long), claimed to be the second largest bacterium, is also a resident of the estuarine environment, found on the coast of Namibia.

Archaea are diverse and prevalent in estuarine ecosystems. Particularly, ammonia-oxidising archaea (AOA) are abundant in ammonia-rich environments. Heterotrophic archaea are also highly abundant due to the presence of a high amount of organic matter in the regions. Fungi are comparatively more abundant on the coasts than in seawater because of their proximity to terrestrial ecosystems, sufficient nutrients, and substrata for attachment. Fungi in estuarine environments can be OMF and MDF. MDF are more abundant in the estuarine environments than in the open sea due to proximity to the land ecosystems and the inflow of freshwater. Members of Ascomycota, Basidiomycota, Glomeromycota, and Chytridiomycota are prominent in the estuaries. They are ecologically important for estuarine bioprocessing owing to their roles in decomposition, nutrient cycling, or food webs. Viruses are also highly abundant due to the abundance of other microbial groups, and they multiply by growing on suitable hosts. The composition of protozoans in estuaries is similar to that of other aquatic ecosystems.

20.10 ECOLOGICAL IMPLICATIONS

Microorganisms are vital for the functioning of all ecosystems. The aquatic environments also bear no exception for microbial roles in ecological implications. Their vast abundance, immense biomass, and complicated processes in aquatic environments have a significant impact on the structure and function of aquatic ecosystems. Both common and rare taxa of the community are important drivers of ecological processes. Common taxa are key players in processes, such as respiration, photosynthesis, and nutrient cycling. They also play a role in community stability in response to swift environmental variations. The assemblages of rare microbial taxa are pools of novel genes and functions. For example, a bacterium, *Methylophaga* (belonging to the phylum Gammaproteobacteria), has a very low abundance in marine ecosystems and is responsible for methanol assimilation. Also, some giant sulphur bacteria like *Beggiatoa*, *Thioploca*, and *Thiomargarita* are less numerically abundant but contribute significantly to sulphur cycling.

Some specific activities of aquatic microbiota that matter for ecosystem functioning are listed below.

- Photosynthetic activity in marine ecosystems accounts for one-half of global primary production, which immensely contributes to ecosystem services. Cyanobacteria and microalgae are the key contributors to this activity. Photoautotrophs of marine habitats are also responsible for producing approximately 50% of the Earth's O_2 supply.
- The chemosynthetic lifestyles contribute to the synthesis of a notable amount of organic matter in water. In the deep ocean, where there is no sunlight, the primary production is exclusively supported by chemosynthesis in microorganisms.
- The degradation of complex organic materials by decomposers in aquatic systems adds a substantial amount of DOM. DOM is the fuel for a heterotrophic lifestyle through microbial loops. A microbial loop is a trophic pathway of the water columns, where a group of microorganisms recycle the exudates of other organisms by feeding on them. These feeders subsequently become the resource and energy for the bigger organisms. This

cooperation of heterotrophic lifestyles supports the growth of diverse lifeforms and protects the water bodies from waste accumulation.

- Microorganisms can occupy the position at the base of aquatic food chains. Some lifeforms are responsible for primary production in the grazing food chains or act as detritivores for the decomposition of dead organic materials in the detritus food chains.
- The complex interaction of microbial communities with water parameters influences ecosystem health. They also play a dominant role in maintaining the water quality of lakes and ponds.
- The peculiar salty smell of seawater is due to a volatile compound, dimethyl sulphide (DMS), produced by a microbially-mediated reaction. DMS is a product of the degradation of a plant metabolite, dimethylsulphoniopropionate (DMSP), in the sea. The marine plants, including algae, produce as much as 6 x 10^7 gigatons of DMSP annually. About 90% of DMSP is consumed in assimilatory pathways, and the remaining 10% is broken down into DMS that emits a sea spray smell in the atmosphere.
- The immense growth of microalgae (e.g., *Chlorella*, diatoms, and dinoflagellates) and cyanobacteria in the surface water causes algal blooms. The degradation of these blooms that leads to a decrease in O_2 levels in the surroundings creates a situation of hypoxia, resulting in a dead zone for wildlife.
- Excessive growth of algae on the surface water discolours the water by the presence of their coloured pigments and prevents the passage of light to the underlying water column. Some species in algal blooms, such as *Gonyaulax polygramma, Gymnodinium*, *Microcystis aeruginosa*, etc., produce toxins that are harmful to aquatic wildlife.
- Microbial processes have direct impacts on the bioresources for fisheries, aquaculture, and coastal ecosystems. Microalgae are potential bioresources for biofuel production. Extremophiles produce enzymes and biomolecules that have tremendous applications in biotechnology, pharmaceuticals, and industries.
- Bloom events driven by increased nutrient influx and elevated temperature were widespread during the end-Permian epoch (EPE) (250 million years ago (mya)) in geological time scale, which has been linked to the largest mass extinction in Earth's history. Excessive blooms choked the aquatic ecosystems, making them uninhabitable for many aquatic creatures. This indicates that the unprecedented increase of nutrient input in the ocean under the stress of global warming can have a significant impact on accelerating the next mass extinction.
- Many chemical substances and elements in aquatic systems move through biogeochemical cycles. In this role, microorganisms are directly involved in various steps of the carbon, nitrogen, sulphur, oxygen, and phosphorus cycles for their mobility in various environmental compartments.
- Many microorganisms, dominantly chemolithotrophs, are responsible for the weathering and leaching of rocks and minerals in interfacial surface and below-ground water.
- An underwater ecosystem, coral reefs are strongly symbiotically associated with microbial communities in the sea for their nutrition and protection. The symbiotic organisms, zooxanthellae, offer a constant supply of organic matter for the fast growth of corals. The primary productivity of coral reefs in aquatic systems is comparable to that of rainforest ecosystems.
- Microorganisms can be potential indicators of ecosystem health and have a role in the assessment of the global impact of climate change. This is because MCCs and their function are susceptible and experience shifts to the effects of elevated water temperatures, anthropogenic disturbances, ecosystem stressors, and nutrient availability.
- Snow microalgae have contributed to the warming of the snow surface due to an increased rate of light energy absorption by algae, whereas algae-free snow reflects solar radiation to space. This process of snow warming has exacerbated glacier ice melting, causing sea-level rise.
- Some aquatic microorganisms are natural biocontrol agents of disease-causing organisms. A balanced microbial community restricts the spread of disease and maintains water quality for recreational or drinking reasons. For example, phages in lakes, ponds, and rivers can lyse bacterial pathogens and have the capacity to restore the water quality (as reported in the rivers Ganga and Yamuna in India).
- Aquatic microorganisms can be the potential candidates for the bioremediation of environmental contaminants such as petroleum hydrocarbons, polycyclic aromatic hydrocarbons (PAH), antibiotics, heavy metals, etc. In the past, they have been employed to treat oil spills from the water surface in seawater.

20.11 IMPACT OF ANTHROPOGENIC ACTIVITIES ON AQUATIC MICROBIOTA

According to the United Nations Children's Fund (UNICEF) and the World Health Organisation (WHO), around one-third of the world's population lacks access to safe drinking water. Increased anthropogenic activities in aquatic systems have posed a serious challenge to aquatic wildlife. Domestic and industrial wastes have contaminated the aquatic systems. The destruction of marine habitats is also very pronounced. Rising ocean temperatures have serious consequences for coral health and its native microbial communities. Excess amounts of CO_2 emissions in the environment have increased its concentration in aquatic environments, leading to ocean acidification. This process has threatened marine and inland water ecosystems, including the associated food chains. Accumulation of diverse inorganic and organic chemicals in aquatic ecosystems has harmed native microbial communities. The high-level environmental perturbations have induced a variation in the microbial compositions and their function. However, some microbial species can still be resilient (microbial ability to withstand disturbances) to a certain extent due to their metabolic flexibility.

Microbial response to anthropogenic disturbances mainly depends on the quality and quantity of contaminants. The increased aerobic respiration has reduced oxygen concentration and increased the load of CO_2 in aquatic ecosystems. Inorganic substances (like nitrogen and phosphorus) are nutrients for planktonic growth, which increases primary production in many aquatic ecosystems. This process has increased the incidence of eutrophication in many geographical regions. Many chemicals exert toxic effects on the natural microbiota and reduce their abundance. Microplastics in aquatic environments are highly pervasive and can be reported in floating water, seafloor, and riverbeds. The presence of microplastics in aquatic regions can also exert selective pressure on microbial communities. Many microbial species have been reported that can adhere to or consume plastics. The widespread use of antibiotics and their untreated discharge into water bodies raises concerns about the spread of pathogens and antibiotic resistance in water bodies. The contamination of antibiotics has increased the emergence of disease-causing bacteria in water, such as *Vibrio cholera* (diarrhoea), *Salmonella typhi* (typhoid), etc. Detailed information on water pollution and its treatment can be found elsewhere in the book (Unit 6).

SUMMARY

- Aquatic microbiology is the study of diverse microorganisms found in water bodies. They inhabit almost every compartment of the aquatic environment. They can float on the surface (neuston), float beneath the surface (plankton), be attached to fixed objects (periphyton), or reside at the bottom (benthos).
- Microorganisms can be reported from the surface layer to the deep oceans. The presence of sunlight on the surface water promotes the growth of phototrophs. The availability of nutrients leads to the overgrowth of algae and cyanobacteria, causing an algal bloom.
- Microorganisms are found in diverse aquatic zones, including seawater, lakes, rivers, estuaries, and hot springs. Microorganisms in seawater and inland water are distinct due to differences in the physicochemical properties of water.
- Inland water environments encompass surface water and groundwater. Surface water supports the growth of phototrophs and aerobes, while groundwater harbours chemoautotrophs and anaerobes. The microbial activities in wetlands are responsible for greenhouse gas emissions.
- The marine environment is extremely vast. About 70% of microbial inhabitants reside in completely dark regions. Deep hydrothermal vents represent unique microbial communities owing to their inhospitable conditions.
- Estuaries are coastal regions, having the confluence of both marine and freshwater ecosystems. This type of region serves as an ecotone and harbours distinct microbial communities due to the mixing of riverine and oceanic water.
- Aquatic microorganisms have many roles in ecosystem health, such as primary production, O_2 evolution and consumption, organic matter decomposition, nutrient cycling, bioremediation of chemicals, coral health, water quality improvement, and sea smell.

DIFFERENT TYPES OF PRACTICE QUESTIONS

Fill in the blanks

I. The frozen part of the hydrosphere is particularly referred to as the ………………..

II. ……………….. commonly found in the Alps and polar regions is responsible for the red colour of red snow.

III. The genus ……………….. is the most abundant oxygen-producing organism in the aquatic environment.

IV. When organisms grow at extremely high hydrostatic pressure, they are called ……………………

V. …………………………… turns the colour of the red sea red due to its seasonal bloom.

VI. Wetland ecosystems are prominent contributors to the global emission of ……………………

SOLUTION: I. cryosphere, II. *Chlamydomonas nivalis*, III. *Prochlorococcus*, IV. barophiles, V. *Trichodesmium erythraeum*, VI. methane (CH4) gas

Match the following

i. Plankton	a. Microorganisms stay at the bottom
ii. Neuston	b. Microorganisms in the surface microlayer at the air-water interface
iii. Benthos	c. Microorganisms growing on submerged surfaces
iv. Periphyton	d. Microorganisms found in a drifting state of water bodies
v. Virioplankton	e. Viruses of planktonic communities

SOLUTION: I-d, ii-b, iii-a, iv-c, v-e

Very short answer types of questions

I. Why do the microorganisms from freshwater habitats not survive in marine habitats?
II. Write the name of the most abundant cellular organism on Earth.
III. Differentiate between holoplankton and meroplankton.
IV. What is the reason for the pink colour in many salt lakes?
V. What are endolithic microorganisms?
VI. What is the peculiar salty smell of seawater?

Short answer types of questions

I. What are bioluminescent bacteria and their activity?
II. Write a short note on the ecological implications of aquatic microorganisms.
III. How does the Anthropocene influence the characteristics of water and its microbial inhabitants?
IV. What do you mean by microbial distribution pattern in the aquatic environment based on the phrase "Everything is everywhere, but the environment selects"?
V. What are the different ecological zones of the freshwater ecosystem and their native microbial forms?

Long answer type of questions

I. What is aquatic microbiology? Give a detailed account of different habitats for aquatic microorganisms.
II. Give a comparative account of the distribution of microorganisms in freshwater and marine water.
III. What are the different factors that influence aquatic microbiota under various sets of parameters?

NOTES

1 Jay Nadeau, Chris Lindensmith, Jody W. Deming, Vicente I. Fernandez, and Roman Stocker. Image courtesy of David Liittschwager. www.commons.wikimedia.org/wiki/File:Marine_microplankton.jpg
2 www.commons.wikimedia.org/wiki/File:Stromatolites_in_Sharkbay.jpg
3 www.oceanwiki.ethz.ch/lib/exe/detail.php?id=lecture2%3Abuoyancy&media=lecture2:thp.jpg
4 www.commons.wikimedia.org/wiki/File:LIQENI_HILLIER_-_ROZE.jpg
5 www.commons.m.wikimedia.org/wiki/File:Laguna_colorada_puntas_negras.jpg
6 www.istockphoto.com/photo/lake-natron-gm502130159-42198348
7 Visscher PT, Gallagher KL, Bouton A, et al. Modern arsenotrophic microbial mats provide an analogue for life in the anoxic Archean. *Communications Earth & Environment*. 2020;1(1):24.
8 Gray A, Krolikowski M, Fretwell P, et al. Remote sensing reveals Antarctic green snow algae as important terrestrial carbon sink. *Nature Communications*. 2020;11(1):2527.
9 www.commons.wikimedia.org/wiki/File:Dinoflagellate_bioluminescence.jpg
10 Tillmann U, Bantle A, Krock B, Elbrächter M, Gottschling M. Recommendations for epitypification of dinophytes exemplified by *Lingulodinium polyedra* and molecular phylogenetics of the Gonyaulacales based on curated rRNA sequence data. Harmful Algae. 2021;104:101956.

SUGGESTED READINGS

Bar-On YM, Phillips R, Milo R. The biomass distribution on Earth. *Proceedings of the National Academy of Sciences*. 2018;115(25):6506–11.

D'Hondt S, Pockalny R, Fulfer VM, Spivack AJ. Subseafloor life and its biogeochemical impacts. *Nature Communications*. 2019;10(1):3519.

Edwards KJ, Bach W, McCollom TM. Geomicrobiology in oceanography: mineral–microbe interactions at and below the seafloor. *Trends in Microbiology*. 2005;13:449–59.

Hirsch R, Ternes T, Haberer K, Kratz KL. Occurrence of antibiotics in the aquatic environment. *Science of the Total Environment*. 1999;225(1–2):109–18.

Hoshino T, Doi H, Uramoto GI, Wörmer L, Adhikari RR, Xiao N, Morono Y, D'Hondt S, Hinrichs KU, Inagaki F. Global diversity of microbial communities in marine sediment. *Proceedings of the National Academy of Sciences*. 2020;117(44):27587–97.

Huber JA, et al. Microbial population structure in the deep marine biosphere. *Science* 2007;318;97–100.

Kiene RP, Linn LJ, Bruton JA. New and important roles for DMSP in marine microbial communities. *Journal of Sea Research*. 2000;43(3–4):209–24.

Kilias ES, Junges L, Šupraha L, Leonard G, Metfies K, Richards TA. Chytrid fungi distribution and co-occurrence with diatoms correlate with sea ice melt in the Arctic Ocean. *Communications Biology*. 2020;3(1):183.

Langenheder S, Lindström ES. Factors influencing aquatic and terrestrial bacterial community assembly. *Environmental Microbiology Reports*. 2019;11(3):306–15.

Lindström ES, Langenheder S. Local and regional factors influencing bacterial community assembly. *Environmental Microbiology Reports*. 2012;4(1):1–9.

Mayol E, Jiménez MA, Herndl GJ, Duarte CM, Arrieta JM. Resolving the abundance and air-sea fluxes of airborne microorganisms in the North Atlantic Ocean. *Frontiers in Microbiology*. 2014,5:557.

Morris RM, Rappé MS, Connon SA, Vergin KL, Siebold WA, Carlson CA, Giovannoni SJ. SAR11 clade dominates ocean surface bacterioplankton communities. *Nature*. 2002;420:806–10.

Retter A, Nawaz A. The groundwater mycobiome: fungal diversity in terrestrial aquifers. In Mehner T, Tockner K (eds). *Encyclopedia of inland waters* (2nd edition, pp. 385–96). Elsevier; 2022.

Robinson C. Microbial respiration, the engine of ocean deoxygenation. *Frontiers in Marine Science*. 2019;5:533.

Ruiz GM, Rawlings TK, Dobbs FC, Drake LA, Mullady T, Huq A, Colwell RR. Global spread of microorganisms by ships. *Nature*. 2000;408(6808):49–50.

Santoro AE, Richter RA, Dupont CL. Planktonic marine archaea. *Annual Review of Marine Science*. 2019;11:131–58.

Shiklomanov I. World fresh water resources. In Gleick PH *(ed)*. *Water in crisis: A guide to the world's fresh water resources*. 1993. Oxford University Press.

Stocker R, Seymour JR. Ecology and physics of bacterial chemotaxis in the ocean. *Microbiology and Molecular Biology Reviews*. 2012;76(4):792–812.

Whitman WB, Coleman DC, Wiebe WJ. Prokaryotes: the unseen majority. *Proceedings of the National Academy of Sciences*. 1998;95(12):6578–83.

21 Methods for the Characterisation of Microorganisms

21.1 IDENTIFICATION AND CHARACTERISATION

The identification and characterisation of microorganisms are the primary objectives of a microbiologist, which is to understand their purview in the environment and for human welfare. A variety of techniques can identify microbial forms, from broad groupings like bacteria, archaea, fungi, and microalgae to narrow taxonomic lineages of genera, species, and strains. The identification of a microorganism is mainly built upon phenotypic, proteotypic, and genotypic variations. Traditional methods of characterising microorganisms include macroscopic and microscopic observations, as well as cultivation. A few macroscopic and microscopic microbial bodies in the form of fruiting bodies, biofilms, and other specialist structures are visible to the naked eye. They can be characterised to a certain extent through direct observation.

Culture-dependent methods are the most common practices to identify and characterise microorganisms. This method allows microorganisms to grow on a synthetic culture medium under aseptic artificial environments. In the culture-dependent methods, the visual appearance of the colonies grown on the culture plates is the first observation. Colour, form, size, elevation, opacity, and elevation are common parameters used to describe patches of bacteria and fungi. Some microscopic features of a microorganism can be distinguished under low-power and high-power magnifications of microscopes. In microscopic observation, staining allows better discrimination between various microbial forms, cell types, and different subcellular compartments within the cell. Biochemical tests are standard methods to describe the phenotypic characteristics of microbial isolates using chromogenic substances. Modern identification techniques of microorganisms rely on molecular tools involving nucleic acid and amino acid sequence analyses.

In recent years, culture-independent methods have emerged as a new strategy for studying microbial communities from any environment without isolating them in their physical forms. Because numerous microorganisms found in environmental samples cannot be grown using culture-based methods, culture-independent techniques have successfully identified many new microbial taxa from diverse ecological settings. This has become a driving force for several discoveries in microbial ecology that have offered in-depth information on microbial community compositions (MCCs) and their function. Most modern analytical methodologies involved in the study of microbiota are relatively cost-efficient, reliable, less time-consuming, and automated compared to the classical cumbersome methodologies. This chapter provides a general overview of various techniques for identifying and characterising microorganisms.

21.2 CULTURE-DEPENDENT METHODS

A culture-dependent method is a technique of microbial isolation and cultivation which allows the microorganisms to grow outside of their native environments. Microbially oriented industries are strictly dependent on culture-based approaches. Identifying a microorganism at the species level without using its cultivars is highly challenging. Cultivation practices require strict aseptic conditions and specific nutrient media under a fixed set of parameters. Aseptic conditions protect the environment from any contamination. Microbial cultivation requires nutrient media. Their growth strictly depends on their metabolic needs and origin. The proliferation of microbial cells on nutrient media requires a long incubation period (even several days). Microorganisms are purified from mixed colonies after repeated reculturing on a fresh medium. A thorough understanding of their isolation and cultivation in different growth media is detailed in the next chapter (Chapter 22). The purified colonies are further characterised using a wide range of techniques to understand their metabolic capabilities. Some common methods for characterising microorganisms include staining, biochemical testing, and molecular analysis. They are covered in the following subsections.

21.2.1 Staining

Staining facilitates the visualisation of structurally chemically different microorganisms and cellular structures using dyes. Dyes are colour compounds that react with specific components of cells and produce different shades under microscopic analysis. For example, a blue colour dye, cotton blue colour fungi. Saffranine is a red colour stain used to colour many microbial cells. Some dyes are instrument-sensitive rather than cell-specific. The staining procedures can also be different for different microscopic analyses. Acridine

DOI: 10.1201/9781032651095-26

Orange is a fluorescent dye that can selectively stain nucleic acids, making it useful for distinguishing between live and dead microbial cells under fluorescent microscopy. Some staining procedures require more than one type of stain, and the process is called differential staining. A Gram-staining procedure falls under the same category, which can be used to discriminate between two types of cells: Gram-positive and Gram-negative bacteria. Endospores and mycobacteria are also coloured through differential staining procedures.

Hans Christian Gram (1884) pioneered the Gram-staining technique based on the cell wall composition of bacteria. This technique uses two different dyes, namely, crystal violet and safranin. The Gram staining procedure starts with preparing a smear of microbial suspension on a clean glass slide, followed by air drying and gentle heating to fix the bacterial cells on the slide. The smear is flooded with crystal violet for 30-60 seconds (s) and then rinsed with distilled water. In the next step, the iodine solution is poured on the slide for 60 s and washed with distilled water. Iodine solution reacts with crystal violet to form a crystal violet-iodine complex called mordant. The mordant receives treatment with ethanol or acetone for 20-30 s, followed by washing. The last step involves colouring with safranin for 30-60 s and washing. The blot-dried slide is observed under a light microscope for bacterial cells of two distinct colours. Gram-positive bacteria appear blue/purple, while Gram-negative bacteria retain red/pink (Figure 21.1a and b).

The principle behind Gram-staining of bacteria stems from the differences in the cell wall compositions between two different types of cells. In the first step, when a smear is exposed to crystal violet, both types of bacterial cells take up the dye. This is followed by adding iodine solution, which fixes the dye inside the cells and prevents its easy removal. However, the next treatment with ethanol/acetone is particularly important in distinguishing between the two types of cells, as it specifically decolourises the Gram-negative cells. This occurs because Gram-negative cells' outer layer has a high lipid content and is dissolved in the ethanol/acetone. In contrast, the same treatment favours dehydration of the Gram-positive bacteria, promoting the closure of any pores and preventing the escape of crystal violet dye from cells. Gram-positive bacteria retain their blue/purple colour, but the other type loses it. In the last step, when the smear is exposed to the counterstain safranin, the decolourised Gram-negative cells take up this dye, colouring them red/pink.

Endospores are unique structural units and exclusive to some selected Gram-positive bacteria. They do not get a proper stain from Gram staining because they have a strong permeability barrier called a spore coat, which blocks the uptake of crystal violet/safranin dyes inside the cells. Therefore, they utilise a distinct staining procedure known as Schaeffer Fulton's method. The smear preparation is similar to Gram staining. This preparation is flooded with a dye, malachite green, for 5 min. A heat source in the form of steam is supplied, which ensures proper staining of both endospores and vegetative cells. The washing of the slide decolourises the vegetative cells, which can be stained using a counterstain, safranin. Under microscopic observation, endospores appear green, while vegetative cells give a red/pink colour (Figure 21.1.c).

Some bacterial cells, like *Mycobacterium,* do not retain colour after Gram staining due to specific cell layer compositions. The high content of lipids on the surface prevents the uptake of dyes, and the presence of mycolic acid on cells confers acid fastness (which resists decolourisation by acids during the staining process). Such bacteria are stained with a different method called Ziehl-Neelsen staining, named after two scientists, Franz Ziehl and Friedrich Neelsen, who made notable contributions to the development of the staining procedure. As per the protocol, the air-dried, heat-fixed smear is treated with a primary stain, carbol fuchsin, for 5 min. The dye is lipid-soluble and penetrates the cell wall. A steam treatment after primary staining ensures deep penetration of the stain, reaching the cytoplasm. After washing the stain, the smear is covered with the acid-alcohol for 5 min, followed by washing. Acid-alcohol decolourises the non-acid fast cells but does not affect acid-fast cells. The non-acid fast cells are coloured using a counterstain, methylene blue/malachite green, to see a blue/green colour appearance (Figure 21.1d). This procedure can be used to detect mycobacteria in histopathology specimens. The Ziehl-Neelsen staining is also compatible with some acid-fast fungi like *Histoplasma* and *Russula.*

There are many other types of stains for colouring microbial cells and cellular components. For example, cyanobacteria are stained with aniline blue or cotton blue. Microalgae are stained with iodine solution primarily due to the presence of starch (Figure 21.1e). However, many microalgae can be seen under the light microscope without staining due to the presence of many colourful pigments. Arbuscular mycorrhizal fungi (AMF) give a blue hue when stained with ink vinegar/trypan blue (Figure 21.1f). Table 21.1 provides an insight into the different types of dyes used to study diverse microbial forms under different microscopes.

21.2.2 Biochemical Testing

Biochemical testing is a classical and well-established method for the phenotypic identification and characterisation of microorganisms. The testing is based on the physiological potential of microorganisms, which determines the specific functional attributes (that are usually related to enzymatic activity). A wide range of biochemical assays is available to test cultured isolates as well as environmental and clinical samples. Tests are generally user-friendly and involve relatively easy procedures. Both qualitative and quantitative types of assays exist. The qualitative test is usually based on simple observations and does not require sophisticated instruments. In these tests, adding a few drops of liquid reagents conveys a visual change in the sample. On the other hand, quantitative assays allow the measurement of traits using analytical tools.

Some biochemical tests produce results in a short duration, while some can take a long incubation period for their growth in specific media. Some routine tests that are performed to identify Gram-positive bacteria include the catalase test, starch hydrolysis test, nitrate test, etc. Gram-negative bacteria

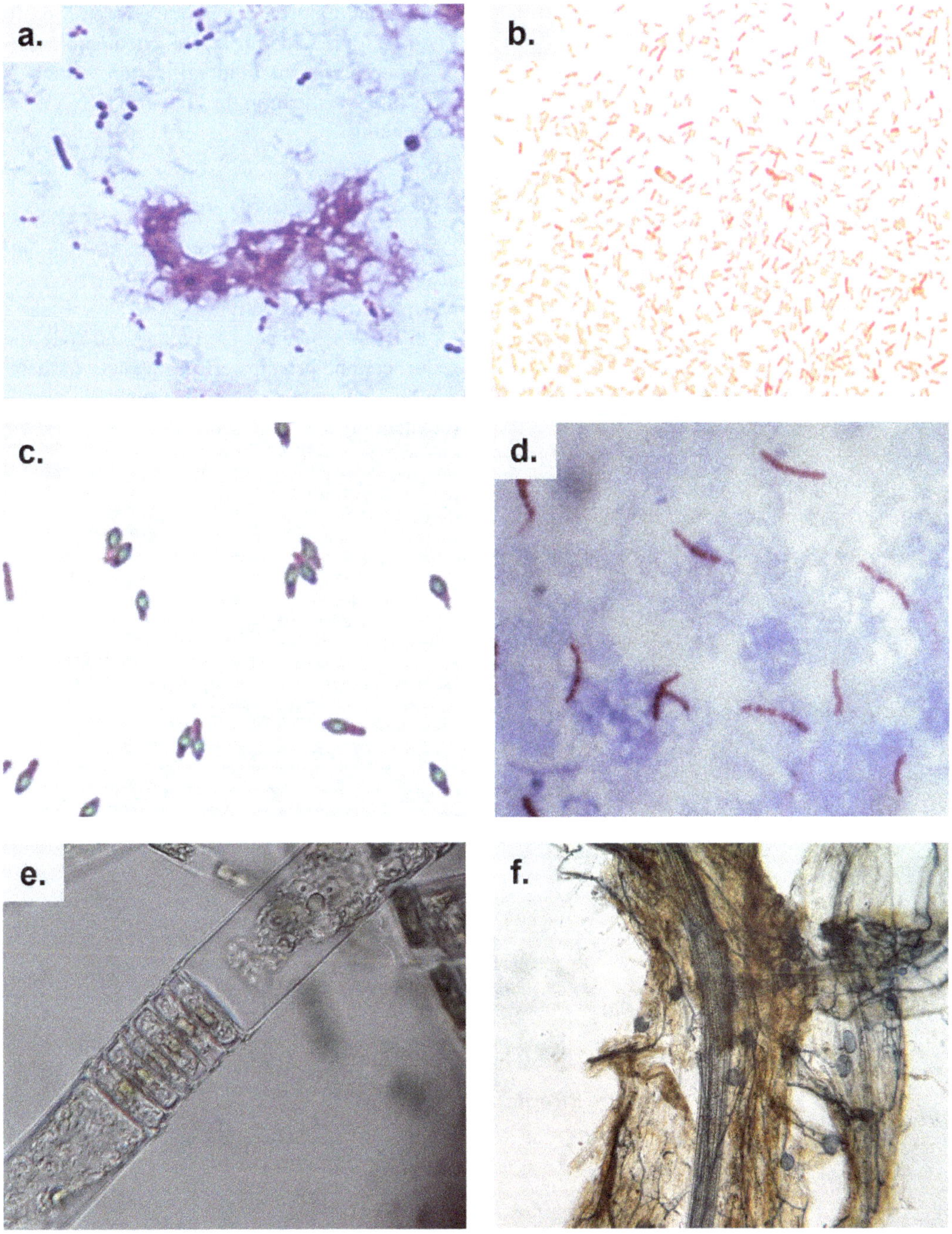

FIGURE 21.1 Microorganisms after staining. a. and b. Gram-positive and Gram-negative bacteria after Gram staining, respectively. c. Endospore stained with malachite green stain. d. Mycobacteria after Ziehl-Neelsen staining. e. Microalgae (*Oedogonium*) stained with iodine. f. Arbuscular mycorrhizal fungi (AMF) stained with ink vinegar. (Image credits: a and b from Nellie Laisram/Dayal Singh College and e from Savindra Kumar/Zakir Husain Delhi College (ZHDC).)

are generally identified using the oxidase test, McConkey agar test, Methyl Red/Voges-Proskauer (MR/VP) test, etc. Bergey's Manual of Determinative Bacteriology, published in several editions (1923–1994), provides different schemes for identifying unknown bacteria based on phenotypic characteristics. In addition, many investigators have prepared dichotomous keys and flowcharts to provide stepwise selections of biochemical tests to distinguish diverse bacteria. Gram staining, microscopic observation, and motility tests are some prior tests to biochemical methods that can eliminate different possibilities and allow the selection of a particular scheme. A different set of tests examines the carbon assimilation profile of bacteria and fungi that provides a lead in differentiating different species. A schematic outline of different dichotomous keys to select different tests for identifying bacteria is given in Figure 21.2.

Biochemical screening is also a popular method for diagnosing pathogenic microorganisms from clinical

TABLE 21.1
A list of some dyes for the visualisation of diverse microbial forms and their cellular structures under microscopes

Dyes	Microorganisms and their cellular structures
Crystal violet and safranin	Bacteria
Cotton blue	Fungi and cyanobacteria
Ink, vinegar/trypan blue	Arbuscular mycorrhizal fungi (AMF)
Iodine	Microalgae
Malachite green	Endospores
Carbol fuchsin	Mycobacteria
Silver nitrate	Bacterial flagella
Acridine orange	Nucleic acids

limitations. One of the major concerns is their misleading interpretation. Therefore, biochemical assays are usually used for preliminary identification, which is subjected to molecular verification through 16S rRNA or 18S rRNA gene sequencing.

21.2.3 Molecular Techniques

DNA sequences encode biological information; therefore, they also provide crucial information on biological systematics and physiological activities. The four bases (A, T, G, and C) in DNA sequences are unique and code for specific RNA sequences and proteins. The sequences of these bases in DNA can be determined using different sequencing techniques (detailed in subsection 21.2.3.1).

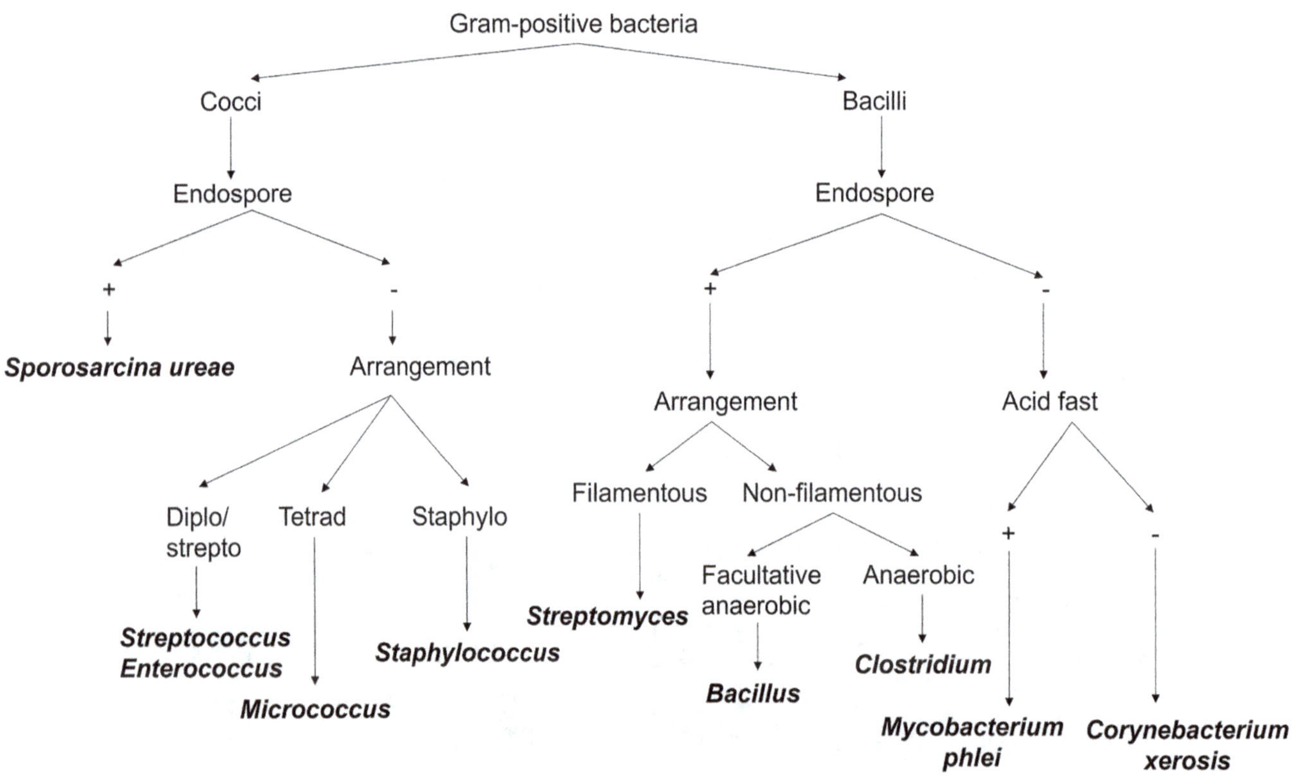

FIGURE 21.2 A dichotomous key for the identification of a few Gram-positive bacteria through various morphological and biochemical tests.

samples. As many as twenty different tests can be needed to detect clinically important bacterial taxa. Though individual tests are very time-efficient and economical, performing a set of biochemical tests following a scheme can be time-consuming and expensive. A new approach that provides a solution to laborious tasks is the application of phenotypic microarray, where a single experiment can detect multiple tests using multi-well plates by growing pure culture in each preconfigured well. This type of screening facilitates many parallel assays using limited resources and space. In addition, identification through biochemical tests has certain other

The advent of DNA-based techniques has drawn a common interest in the molecular identification of microorganisms. These methods are very sensitive, rapid, and reliable and provide accurate species identification. They not only help distinguish different species having the same phenotypic characteristics but also detect different strains of the same species. The ribosomal DNA sequences are the most commonly used markers for evolutionary comparisons. The 16S and 18S rRNA genes are the most commonly used pinched sequences for identification in prokaryotic and eukaryotic organisms, respectively. These ribosomal genes are functionally constant and universally distributed. They are also recognised as "evolutionary chronometers" and consist of both variable and conserved regions. A different gene, *RNA polymerase beta subunit* (*rpoB*), which encodes for a subunit of RNA polymerase, can be used for species-level discrimination in bacteria. *gyrB* exhibits even higher variability than 16S rDNA, allowing fine resolution of closely related species. In addition, some specific genes, like heat shock protein 60 (hsp60) and amoA, enable the identification of bacteria with specific traits. *Ribulose-1,5-bisphosphate carboxylase/oxygenase large subunit gene* (*rbcl*) encodes an enzyme essential for photosynthesis and can be used to identify cyanobacteria and microalgae. The internal transcribed spacer (ITS) sequence is also a suitable marker for identifying bacteria and fungi at species and subspecies levels. An ITS is a region of DNA found between the coding regions of small and large subunits of rRNA, which consists of the fastest evolving region and provides high taxonomic resolution for intraspecific variation in different microorganisms. However, the use of ITS as a stand-alone marker has some limitations, such as limited resolution and high variability.

Carl Woese and Fox (1977) were the main masterminds who used ribosomal DNA to trace evolutionary relationships through sequence comparisons. Sequence analysis of the 16S rRNA gene has led to the discovery of a new domain, archaea. The development of the polymerase chain reaction (PCR) technology in the 1980s was a technical breakthrough to capture the notion of sequence identification. This machine amplifies DNA using primers. PCR primers are short-stretch single-stranded DNA (generally 18 to 25 nucleotides long) and have complementary sequences to the target DNA for its binding. They recognise specific regions of DNA content and guide the initiation of DNA synthesis. The amplicon generated from the PCR is purified and sequenced using a DNA sequencer. The obtained sequence is identified using computational tools like the basic local alignment search tool (BLAST) by searching one or more databases. The parameters to assign an unknown sequence a particular identity (genus and species) are sequence homology followed by phylogenetic relationship. In addition, modern research studies have relied on quantitative real-time PCR (qPCR) to detect and identify targeted microorganisms and microbial groups from various environmental samples. This method is mainly used to quantify the abundance of specific microbial taxa and for their identification.

Molecular identification of a specific microbial isolate requires a pure culture. For the same, the microbial cells are subjected to DNA extraction using various physical and chemical treatments. The recovered DNA sequence is used as a template for the amplification of target-specific gene sequences. This calls for specific primers, and the process is run in a thermal cycler. The resultant amplicons are purified and sequenced. Further, the raw sequence data is processed and compared to sequences in a database for its valid identity.

21.2.4 Whole Genome Sequencing (WGS)

Whole genome sequencing (WGS) is a comprehensive tool to determine the complete DNA makeup of a cell or an organism. The high-resolution, base-by-base view of the entire DNA of the organisms is valuable for identifying variations within and between species. This method is also commonly associated with the detection of low-frequency mutations and genetic changes between different microbial strains. A classical method of WGS requires a cultivated microorganism, which is lysed to extract all its DNA, followed by its purification. The recovered DNA is fragmented, followed by library preparation. The DNA library is sequenced using a suitable sequencing technology.

21.2.4.1 Sequencing Technology

Sequencing technology has seen a lot of transitions since its inception in the 1970s, from manual sequencing (first-generation sequencing) to fully automated sequencing (second and third-generation sequencing) (Figure 21.3). The knowledge of long DNA sequences first emerged from Sanger sequencing, pioneered by Frederick Sanger (1977). The main principle of the Sanger sequencing is the random incorporation of chain-terminating fluorescently labelled dideoxynucleotides (ddNTPs) during PCR amplification, followed by electrophoresis of PCR-generated amplicons. However, these procedures are expensive and cumbersome for whole genome sequencing. This technology is restricted to the scrutiny of sequence lengths of less than 1 kb. Therefore, the early efforts on sequencing were focused on selected genes specifically for the identification or some gene characterisation (as mentioned in the previous section for ribosomal DNAs). The sequencing of larger DNA fragments has been performed using the shotgun method. The shotgun method involves random breaking of the genome into small DNA fragments. These fragments are sequenced, and the overlapping sequences are assembled into one long contiguous sequence.

The next-generation sequencing (NGS) (also known as high-throughput sequencing or second- and third-generation sequencing) technologies have broken many barriers of Sanger sequencing and allowed the sequencing of the entire DNA content of a cell (a genome) with high sensitivity, accuracy, and speed. NGS technologies are remarkably cheaper than Sanger sequencing. In the early 21st century, the advent of NGS technologies was a paradigm shift in sequencing methods, which has produced huge sets of genome sequences in microorganisms and higher organisms. The Human Genome Project (HGP) started in 1990 and concluded in 2003, spanning thirteen years. In this project, WGS was performed using Sanger sequencing under a budget of about 2.7 billion United States dollars (USD). On the contrary, contemporary NGS technologies take less than a day to sequence the human genome at an estimated cost of only around 1,000 USD (the analysis and interpretation costs are not included).

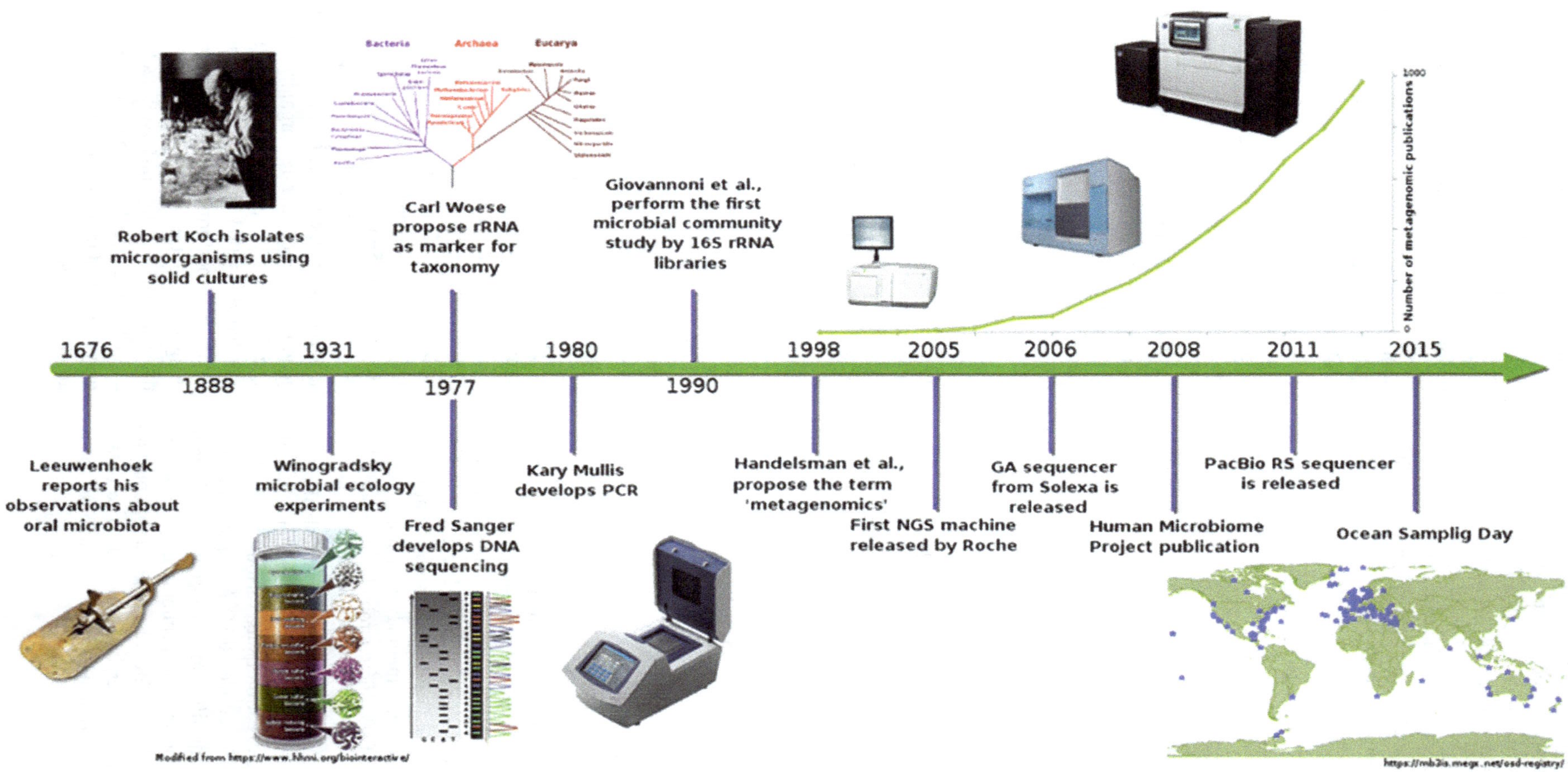

FIGURE 21.3 Timeline and milestone in microbial sequencing analysis. (Image adapted from Escobar-Zepeda et al. (2015).[1])

Second-generation sequencing came into existence with the discovery of pyrosequencing. This method is based on the detection of pyrophosphate through luminescence using two different enzymes, ATP sulphurylase and luciferase. Both Sanger sequencing and pyrosequencing are 'sequence-by-synthesis' techniques based on DNA polymerase, but the latter technique has precedence over others by producing output in real-time using natural nucleotides. In the coming years, Illumina sequencing gained more interest in sequencing projects and replaced pyrosequencing. Illumina sequencing also relies on the sequence-by-synthesis method, where fluorescently labelled nucleotides are incorporated into a DNA strand, and the sequence is determined by detecting the fluorescence signal. The commercial exploitation of Illumina sequencing expanded the WGS approach and drove a 'genomics revolution'. A different technique, Ion Torrent's sequencing, employs semiconductor chips to sequence DNA and detect hydrogen ions (H^+) during DNA polymerisation. However, this method has low throughput (i.e., less data per run) compared to Illumina sequencing.

In the modern world, third-generation sequencing is ready to replace second-generation sequencing. The key aspects of this forthcoming technology are the benefits of long-read and low capital investment. They offer more straightforward procedures to identify the sequences. The most anticipated approach is Oxford nanopore technology (ONT). It is also scalable using portable (the size of a mobile phone) and benchtop devices. The main principle of nanopore technology is the characteristic changes of electric current when a distinct nucleotide of a single-stranded DNA or RNA is driven through the nanopore across the bilayer membrane under an electric field. Another competitive technique in third-generation sequencing is PacBio sequencing based on SMRT (Single Molecule Real Time) technology. These two newly developed methodologies are independent of the requirement of DNA amplification and produce exceptionally long read lengths (about 20 kb) in a very short period (5-10 h). However, they are error-prone and may result in inaccurate taxonomic profiling and binning (about 5-15%).

21.3 MICROSCOPY

The phrase "seeing is believing", which emphasises the importance of direct visual evidence, is equally conceivable to microbial studies. The reason why the microbial life forms are less well known is that they are in the micro to nano size range— far from the eyesight limits of humans. In the 17th century, a Dutch scientist, Antonie van Leeuwenhoek, was the first person to observe microorganisms as "animalcules" using a simple single-lens microscope. This study changed the perspectives about the living world. The invention of advanced microscopes that could magnify objects several fold widened new dimensions to visualise several key features of microorganisms. Nowadays, the study of microbial structure is incomplete without integrating microscopes. Microscopes are used to produce magnified pictures of minute cells or particles. In recent years, microscopes have been integrated with new tools to elucidate various other aspects of microbial life. The staining of microbial cells is essential for microscopic analysis because most of the microorganisms are colourless. The staining of the cells also improves contrast and distinguishes different types of cells and cellular components. Fixing or smearing is an important prior step to staining. In some techniques, the observation of the microbial activity under a microscope does not require fixing and staining. However, the reasonable functional or taxonomic resolution observed under microscopes remains elusive due to some inherent limitations.

A microscope is composed of one or two lenses adhered to a body tube, an arm, and a base. These lenses are of the convex type, which acts as a magnifier. A convex lens converges the parallel rays of light to a focal point (F) through its passage. The distance between the optical centre and its focal point (F) is called the focal length (*f*). The magnifying power of a lens is inversely proportional to its focal length. The lens with a small aperture and a short focal length produces a sharp image of the object.

A simple dissecting microscope typically consists of only one lens with low levels of magnification. The layout of a dissecting microscope has a similar appearance to a simple compound microscope, but the properties are comparable to a simple hand lens. The dissecting microscope is rarely used in microbiology due to its poor magnification. Modern microscopes are composed of more than one lens and are known as compound microscopes. Some of the common variants of compound microscopes are the bright-field microscope (BFM), dark-field microscope (DFM), phase-contrast microscope (PCM), and fluorescence microscope. Electron microscopes are highly advanced and sophisticated instruments which achieve very high resolutions. These instruments rely on an electron beam to illuminate the object. They can be scanning electron microscope (SEM) and transmission electron microscope (TEM). A type of probe-associated microscope is a scanning probe microscope (SPM). A thorough description of different types of microscopes and their specifications is covered in the following subsections.

21.3.1 Bright-Field Microscope (BFM)

BFM is commonly known as a simple compound microscope. The early studies in microbiology heavily relied on BFM. This is still a common laboratory instrument to observe morphological characteristics. The instrument can be used for both fixed/unfixed and stained/unstained specimens. A stage is a platform to set the specimen for observation, which occupies the central position on the microscope. The body of the microscope consists of an arm and a stage. The optical setup of BFM includes an objective lens, one or more eyepieces, and a condenser (Figure 21.4a).

The objective lens is the most important part of the BFM, situated near the stage. It is the primary lens that is responsible for magnifying the image of a specimen. A BFM usually possesses 3-5 objective lenses of differing magnifying power (e.g., 4x, 10x, 40x, and 100x). They are fixed on a turret, a nose piece that can be rotated to position in front of the condenser

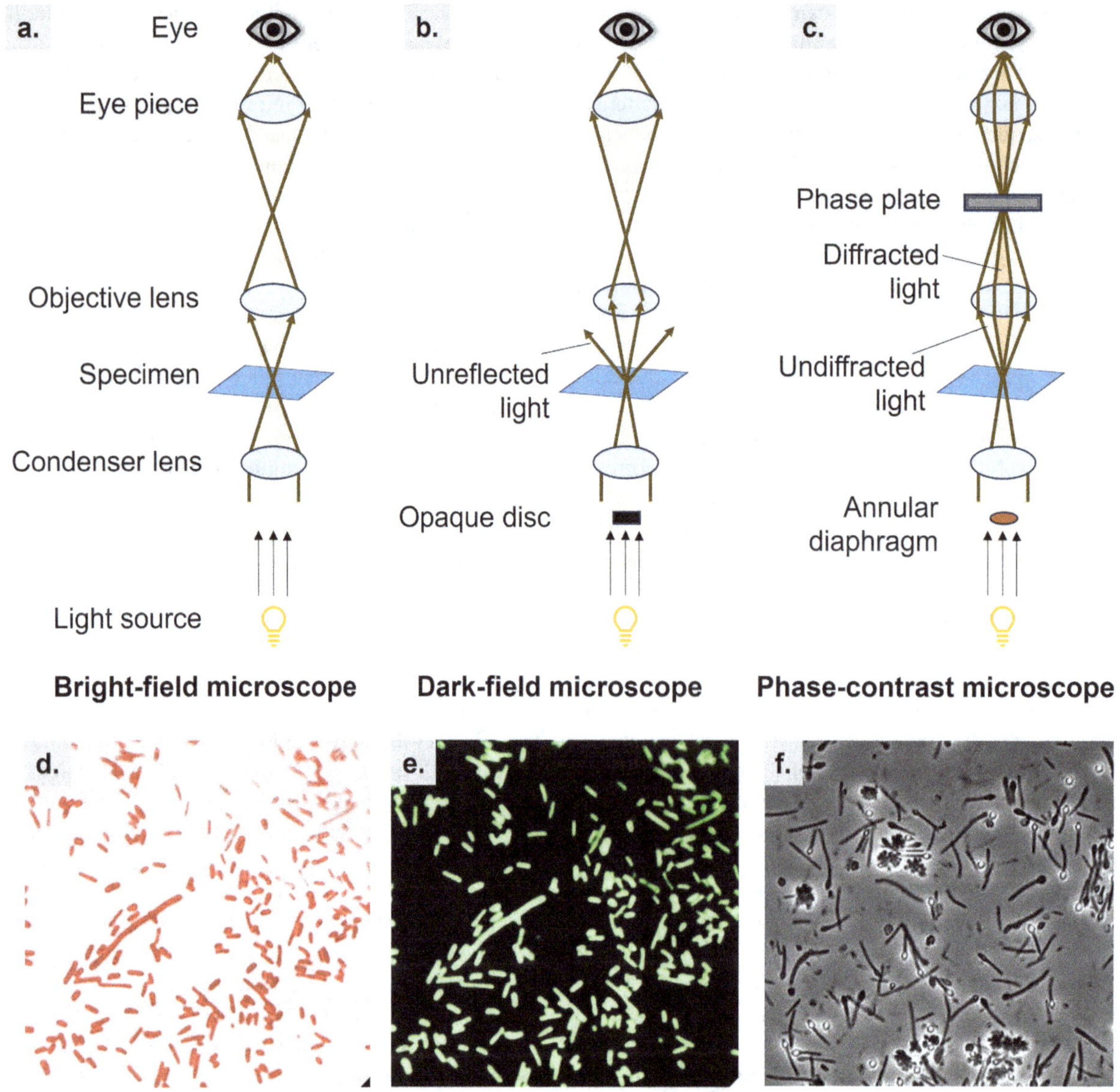

FIGURE 21.4 Three types of light microscopes. a. Bright-field microscope (BFM). b. Dark-field microscope (DFM). c. Phase-contrast microscope (PCM). d. Image of *Shigella dysenteriae* under BFM. e. Image of *Shigella dysenteriae* under DFM. f. Image of *Moorella thermoacetica* DSM 521^T under PCM. (Image f reproduced from Jiang et al. (2009)[2] with permission of Springer Nature.)

in order to study an object under different magnifications. In addition to magnifying the image, the objective lens creates a sharp image. The quality of an image depends on the resolution of the lens, which detects the minute differences between the articles of close proximity. The resolution of the objective lens depends on two physical factors: the wavelength of light and the numerical aperture of the lens. This relation can be numerically expressed using the Abbé equation:

$$d_{min} = 0.5\ \lambda/n.\sin\theta$$

where d_{min} is the minimum distance between two objects
λ is the wavelength of the light
$n.\sin\theta$ is the numerical aperture (NA)

The eyepiece is located at the top of the microscope and has a typical magnification power of 10x. The primary image produced by the objective lenses is received by the eyepiece to further enlarge the images, which is known as a "virtual image". The total magnification power of BFM is a multiplication of the magnification by the objective lens and the magnification by the eyepiece. For example, an objective lens of 40x, along with an eyepiece of 10x, will produce 400x magnification of an object. Therefore, the highest magnification a BFM can attain is 1000x. However, the best possible image under the highest magnification (100x) requires immersion oil between the objective lens and specimen. The refractive index of immersion oil is very close to that of glass, which allows minimum refraction of light at the specimen interface. This improves the lens's ability to capture fine detail from the specimen by increasing the numerical aperture and improving the resolving power of the objective lens. This whole setup can set a resolution limit of about 0.2 μm, which can visualise the shapes and general morphological features of a normal bacterial cell.

The side of the arm in the microscope is occupied with two focusing knobs, the fine and coarse adjustment knob. They can help move the stage up and down to focus on an image. A mirror or electric light is an illuminating source located at the base to capture the image on a bright background. A condenser is situated just below the stage to focus the light beam on the specimen. An aperture diaphragm is intimately associated with a condenser and controls the diameter of the light. The light that is reflected or transmitted through the object is perceived as an image. The difference in absorption and reflection of light over the subject body and its subcellular components produces bright and colourful images (Figure 21.4d). However, the transparent samples produce low-contrast images, which can be improved to a certain extent by the partial opening of the diaphragm. However, this arrangement causes the loss of resolution and sets diffraction artefacts.

21.3.2 Dark-Field Microscope (DFM)

DFM is a modified layout of BFM used to produce a contrast image of the specimen. This type of microscope is specifically equipped with an opaque disc between the source and condenser called a central aperture. The central aperture in the microscope blocks the direct light from being imaged while allowing the passage of light to the specimen through an oblique angle (Figure 21.4b). The scattered lights through the specimen are the only rays that can be received by the objective lens to image. The interaction of oblique rays with a specimen enhances contrast and produces a bright, illuminated image on a dark background (Figure 21.4e). Preparations like fixation and staining of the cells are not necessary for DFM. The technique is useful for easy and simple observation of living cells. The observation under DFM requires strict, strong illumination to achieve a high degree of contrast and a bright image. However, long exposure to intense light can damage living cells. In microbiology, living plankton in water samples is an example of a test specimen for study under DFM. This technique also allows the investigation of bacterial and protozoan mobility in their native states. Other microbiological samples, such as biological fluids, cell cultures, and clinical samples, can also be examined under DFM. In addition, DFM can be useful to illustrate the internal structures of a very thin section of large eukaryotic microorganisms with or without staining.

21.3.3 Phase-Contrast Microscope (PCM)

PCM is another method for studying living cells through high-contrast images. Frits Zernike received the Nobel Prize in Physics in 1953 for his invention of PCM. It is based on the principle of the phase shift in the light rays induced by variations in the refractive indices for different cell structures. This phase shift is a very minute change and not visible to the eye, but it can be translated into brightness variation for different cellular structures. This phenomenon can be achieved by scattering and diffraction of light rays from specimens. Light passing through a region of a low refractive index does not experience any phase change (deviation or retardation), whereas the same light passing through a high refractive index undergoes a deviation. The deviated and non-deviated rays are further segregated and approach the intermediate image plane. The final phase contrast image is a result of brightness differences observed in the eyepieces. The technique is suitable for studying the dynamics of growing cellular activities. This is also suitable for imaging the subcellular structures with similar transparency, which are often indistinguishable under BFM (Figure 21.4f).

A BFM can be converted into a PCM using two specialised optical accessories, *viz.*, a condenser annulus and a phase plate (Figure 21.4c). These two devices can translate the minute changes of phase shift into brightness differences. Other components of PCM are similar to BFM, including a light source, a condenser, an objective lens, and an eyepiece. The condenser annulus is a clear ring placed beneath the condenser. When a specimen is placed under PCM, the light from a source is primarily transferred to the condenser annulus, which transmits a hollow cone of light to the condenser and concentrates it on the specimen plane. Light emitted from limited angles can penetrate the specimen. Following that, the specimen emits some rays, which are collected on the objective lens. These rays are then retarded on a second object, the phase plate. The phase plate lies at the back focal plane of the objective lens and consists of a phase ring that causes a change in phase of light by $\lambda/4$. When the rays finally reach an eyepiece, they produce a highly contrasted image due to a complex interference pattern.

21.3.4 Fluorescence Microscope

A fluorescence microscope is a powerful imaging technique which mainly relies on fluorescence and phosphorescence rather than reflection and refraction. This technique is used to monitor cell physiology and observe different organelles and macromolecules using a variety of fluorochromes. The key components of a fluorescence microscope are a light source, an excitation filter, an objective lens, an emission filter, a dichroic mirror, an ocular, and a detector. When a fluorescent-labelled specimen is examined under a fluorescence microscope, a near-monochromatic light released from the light source passes through an excitation filter, which allows only the excitation wavelengths of light to pass through and blocks unwanted wavelengths. The dichroic mirror selectively forwards light of excitation wavelengths while reflecting others, and the objective lens steers these rays towards the specimen. When light of a specific wavelength falls on the specimen, it emits rays of longer wavelength, which are again gathered by the objective lens and then selectively directed toward a detector via a dichroic mirror. Finally, an emission filter specifically filters the emitted light that eventually reaches a detector, which is a digital imaging device that develops an image.

One of the most important steps in sample preparation is labelling the specimen with a fluorochrome. Fluorochromes (or fluorophores) are chemical stains (also called probes) which emit fluorescence of specific long wavelengths when excited using a laser of defined, relatively short wavelengths (usually blue or ultraviolet (UV)). Fluorochromes have a high degree of specificity in biological cells and possess a significant quantum yield (the ratio of photon absorption to emission). A vast array of fluorochromes with different spectral ranges is available to label different cellular, subcellular or molecular components. A green fluorescent protein (GFP) primarily isolated from a bioluminescent jellyfish, *Aequorea victoria,* is the most commonly used fluorochrome in biological studies. GFP is excited by violet/ultraviolet rays and emits green light. This reporter protein can be introduced into an organism via genetic modification. Some other important fluorochromes for fluorescence microscopy are Rhodamine, Fluorescein, Texas red, and Cy 3.18 (Table 21.2). They can be directly used to stain biological materials for observation under fluorescence microscopes.

The applications of fluorochromes are not limited to the fluorescence microscope. They exhibit modalities for various other techniques such as confocal laser scanning microscopy (CLSM), two-photon excitation microscopy (TPEM), fluorescence lifetime imaging microscopy (FLIM), förster or fluorescence resonance energy transfer (FRET) microscopy, fluorescence recovery after photobleaching (FRAP), etc.

Some biological cells and their cellular components have a natural ability to autofluorescence when irradiated. They can be observed under a fluorescence microscope without using any external stains. In contrast to light microscopes, this instrument does not provide spatial resolution but is used to image specific features of the cells. The photophysical phenomenon of fluorescence microscopy depends on the probe's intrinsic selectivity attached to biomolecules, which produces bright signals against a very dark (or black) background. The instrument can also be used to locate diverse labelling in the cells simultaneously using different fluorochromes. The instrument is so sensitive that a single fluorescent molecule can be detected in a dark background. The use of fluorescent protein (FP) tags makes it possible to visualise the dynamics of living cells.

21.3.4.1 Confocal Laser Scanning Microscope (CLSM)

CLSM is a technique of high-resolution, high-contrast optical three-dimensional (3D) imaging. This technique examines the dynamic molecular and physiological processes of living cells and locates proteins in specific cellular compartments on a microscope stage. Unlike many other microscopic techniques, the visualisation using CLSM is based on a non-invasive process that requires no cross-sectioning and requires minimal sample preparation. The credit for the development of the first CLSM goes to Marvin Minsky (1957). He devised this tool to eliminate the out-of-focus light in a wide-field fluorescence microscope.

The original design of CLSM is mainly equipped with a laser source, a beam expander, beam splitters (dichroic mirrors), an objective lens, pinholes, emission filters, and a photomultiplier (Figure 21.5a). During observation, a light source emits the laser beam that passes through a beam expander to increase the diameter of the input beam. The ray path is directed toward a primary dichroic mirror (Hauptfarbteiler (HFT)) and a secondary dichroic mirror (Nebenfarbteiler (NFT)). HFT separates excitation and emission light, whereas NFT causes a spectral division of fluorescence lights. The focused excitation light passes through the objective lens and reaches the focal

TABLE 21.2
A list of some common dyes/fluorochromes used in fluorescence microscopy, confocal laser scanning microscopy (CLSM), and flow cytometry, including their applications

Dyes/fluorochromes	Excitation wavelength (λ_{max}) (nm)	Emission wavelength (λ_{max}) (nm)	Colour	Labelling	Techniques used
GFP	~488	~509	Green	Gene expression monitoring	Fluorescence microscopy, CLSM, and flow cytometry
DAPI	~358	~461	Blue	DNA	Fluorescence microscopy and CLSM
Fluorescein	~495	~519	Green	Proteins, nucleic acids, and antibodies	Fluorescence microscopy, CLSM, and flow cytometry
Rhodamine	~530–560	~580–620	Red	Proteins and actin filaments	Fluorescence microscopy and CLSM
Texas Red	~596	~615	Red	Protein	Fluorescence microscopy and CLSM
Cy3	~550	~570	Yellow	Nucleic acids and antibodies	Fluorescence microscopy, CLSM, and flow cytometry
Cy5	~650	~670	Far-red	Proteins	Fluorescence microscopy, CLSM, and flow cytometry

plane to scan over the object. The fluorescent object emits a beam of different wavelengths that travels backwards through the objective lens and dichroic mirrors and moves toward the detection system. The non-confocal beams are blocked near the detection system at the entrance of the pinhole. The pinhole acts as a physical barrier, which has a very small aperture size that allows the fluorescence of only central spots and rejects out-of-focus light. Only the confocal beam passes through the pinhole and can be detected on the photomultiplier. An emission filter allows only selected wavelengths to reach the detector. The photomultiplier amplifies the signals of low intensity, where the amplification limit can be as high as one million times. The output of the optical signals is in the form of electric signals that are converted into digital numbers. Finally, an image is developed on the screen after a software-based computational analysis.

The construction of 3D images is based on high-resolution stacks of images obtained from the scanning of a sample point by point from different focal planes. The detector also detects signals in a pointwise fashion. The process of scanning is slow to achieve a better signal-to-noise ratio with high-depth selectivity. In addition to 3D imaging, the system can be used for 4D imaging by tracking the fluorescence at regular intervals and for 5D imaging by associating the multiple channels. The resolution of CLSM is comparable to that of other optical microscopes, but it is relatively low compared to electron microscopes. Figure 21.5b presents micrographs of CLSM of microbial colonisation of seed radicle using different stains, showing red-stained total bacteria and green-stained *Massilia* spp. However, the technique is highly suitable for live imaging. The intrinsic optical properties of the specimen can also be relevant for light emission under CLSM. Phycobiliproteins (PBPs) and chlorophyll-*a*, which are abundantly present in cyanobacteria, are two examples of intrinsic fluorescent markers for cell imaging under CLSM. However, most biological cells are incapable of having such markers. Fluorescent dyes are, therefore, necessary to visualise the selected biological objects. The staining of samples requires a very low concentration of dyes for their minimal impact on the living cells.

In microbiology, CLSM aids in the study of a high degree of the spatial organisation of microbial communities in diverse microenvironments. One such prominent example is an investigation of bacterial biofilm assembly on various surfaces, including an examination of the micro-physicochemical environment of the biofilm. The image analysis provides information about the accumulation and orientation of cells on the surface. CLSM appropriately sheds light on the phagocytosis of bacteria by protozoa. The technique is useful for tracking the motility of microorganisms and assessing the effects of environmental stresses on microbial cells. The plant microbiome is also a relevant research area within the context of CLSM.

21.3.5 Electron Microscopes

An electron microscope is a non-optical microscope which employs an electron beam and electromagnetic lens to produce a high-resolution image of a specimen. This could be possible with the advent of electromagnetic lenses in 1926 by Hans

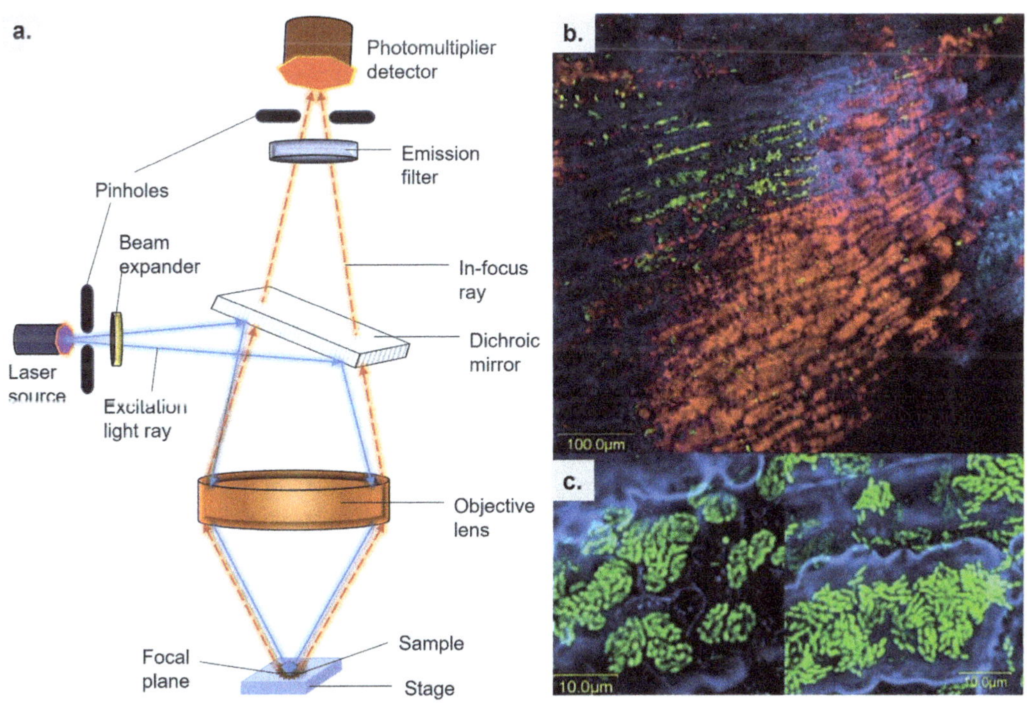

FIGURE 21.5 a. A schematic diagram of a Confocal Laser Scanning Microscope (CLSM). b-c. FISH-CLSM analysis of the microbial colonisation in the seed radicle, showing red-stained total bacteria, and green-stained cells of *Massilia* spp. (Image b-c adapted from Ofek et al. (2010).[3])

Busch. However, the first design of an electron microscope was constructed by Ernst Ruska and Max Knoll in 1931. This discovery earned its inventors a Nobel Prize in Physics in 1986. An electron microscope surpassed the limitations of optical microscopy, giving better resolution and magnification that can be used to study very minute objects. This soon became a powerful tool in microbiology to capture images of small microorganisms and their sub-structures. However, the main principle of electron microscopes does not differ much from that of an optical microscope.

An electromagnetic lens is typically made up of a solenoid of copper wire wound around a soft iron core, placed between two or more parallel iron poles. A current through a wire generates a magnetic field concentrated in the gap between the two poles. Magnetic lenses are convex types and aim to control and bend the beam of electrons, similar to a glass lens. The condenser is the first electromagnetic lens in the entire setup through which the electron beam passes. It focuses the accelerated beam on a specimen. An objective lens is the second lens in the setup and assists in magnifying the image by capturing the electrons that are scattered from the samples.

An electron gun in the machine is a source of the electron beam. An anode placed ahead of the electric gun is used to attract the maximum electrons towards the electromagnetic lenses. Electrons have much shorter wavelengths than visible light, which increases the resolving power of an electron microscope, giving minute details from the specimens. The use of electrons in place of light in microscopy improves the resolution 1000 times. The operation runs under a high vacuum to prevent electrical discharge. A pumping system is attached to the main instrument to provide a vacuum. Based on the penetration of the electron and other specifications, an electron microscope can be of two different types: SEM and TEM.

21.3.5.1 Scanning Electron Microscope (SEM)

SEM uses a fine electron beam to scan the surface of the specimen and produce a topographical image. The magnification in SEM can be adjusted up to 2×10^6 folds. SEM analysis of biological materials typically employs low accelerating voltages (in the range of 5-20 kV) because the high-voltage electrons can penetrate the cells and interfere with the actual morphological detail. The instrument is specifically equipped with a pair of scanning coils, which assist in focusing the beam. These coils allow beams to scan the specimen in a raster pattern (a rectangular left-to-right line from top to bottom). The interaction of the electron beam with the specimen causes a loss of energy due to scattering and absorption.

The image of the specimen is obtained by detecting two different types of reflected electrons: secondary electrons (SEs) and backscattered electrons (BSEs). These electrons are captured by two different types of detectors in SEM. An inelastic interaction between the beam and specimen generates SEs from the surface of the specimen, which are captured by a secondary electron detector (SED). An SED generates a topographic image of the specimen, where the resolution is independent of the components of the specimen. On the other hand, an elastic interaction between the beam and specimen produces the BSE. BSE originates from the interior of the sample and is captured by a BSE detector. The density of the specimen and applied voltage have significant roles in obtaining such signals (Figure 21.6a). A BSE detector produces bright spots for elements with high atomic numbers and dark spots for elements with low atomic numbers. The interaction between the electron beam and specimen also results in an emission of X-rays, which produces a characteristic yield for different elements. It is a phenomenon where an electron moves from an outer to an inner shell in response to an electron beam. A different device, energy-dispersive X-ray spectroscopy (EDX or EDS), can be installed to detect X-ray emission, which can be used for the elemental identification of the specimen. Figure 21.7 provides an SEM-EDX image of a bacterium for the effect of lead (Pb), giving valuable insights into the interaction of a heavy metal with bacterial cells.

The biological samples under SEM are hard to image in their native states due to poor electron density. Therefore, the sample is primarily treated with fixatives like glutaraldehyde and formaldehyde to stabilise the ultrastructure. A secondary fixative, osmium tetroxide (OsO_4), prevents the extraction of bilipid membranes by solvents. The sample is further treated with a series of solvents (ethanol or acetone) for dehydration, followed by drying. Thereafter, the specimen is placed on a metallic stub and coated with a thin layer of gold or palladium for observation. The specimen chamber is situated at the base of the electron column. The specimen stage allows the examination of specimens of sizes up to 200×80 mm^2.

21.3.5.2 Transmission Electron Microscope (TEM)

TEM involves the transmission of electrons through a sample and provides details on internal structure. The image dimension of TEM is 2D. The study under TEM requires complex sample preparation. After stabilising the cell structure using chemical fixatives and dehydration, the specimen is embedded in plastic. A specific instrument, the ultramicrotome, is used to cut the embedded specimen in electron-transparent areas using a glass or diamond knife. The sample must be ultra-thin, and the thickness of the section is below ~0.1 μm. The section is mounted on a metal grid (usually copper) and is stained. The biological cells do not produce sufficient contrast photographs under TEM because of low electron density. A type of staining called negative staining helps distinguish the edges and other structures. Uranyl acetate and uranyl formate are the staining reagents, which also serve as fixatives. The grid is now placed inside the TEM for observation. Similar to a light microscope, the specimen stage in TEM is situated between the condenser and objective lenses (Figure 21.6b).

Viewing the image requires a high accelerated voltage to penetrate the electron deep inside the specimen. The setup

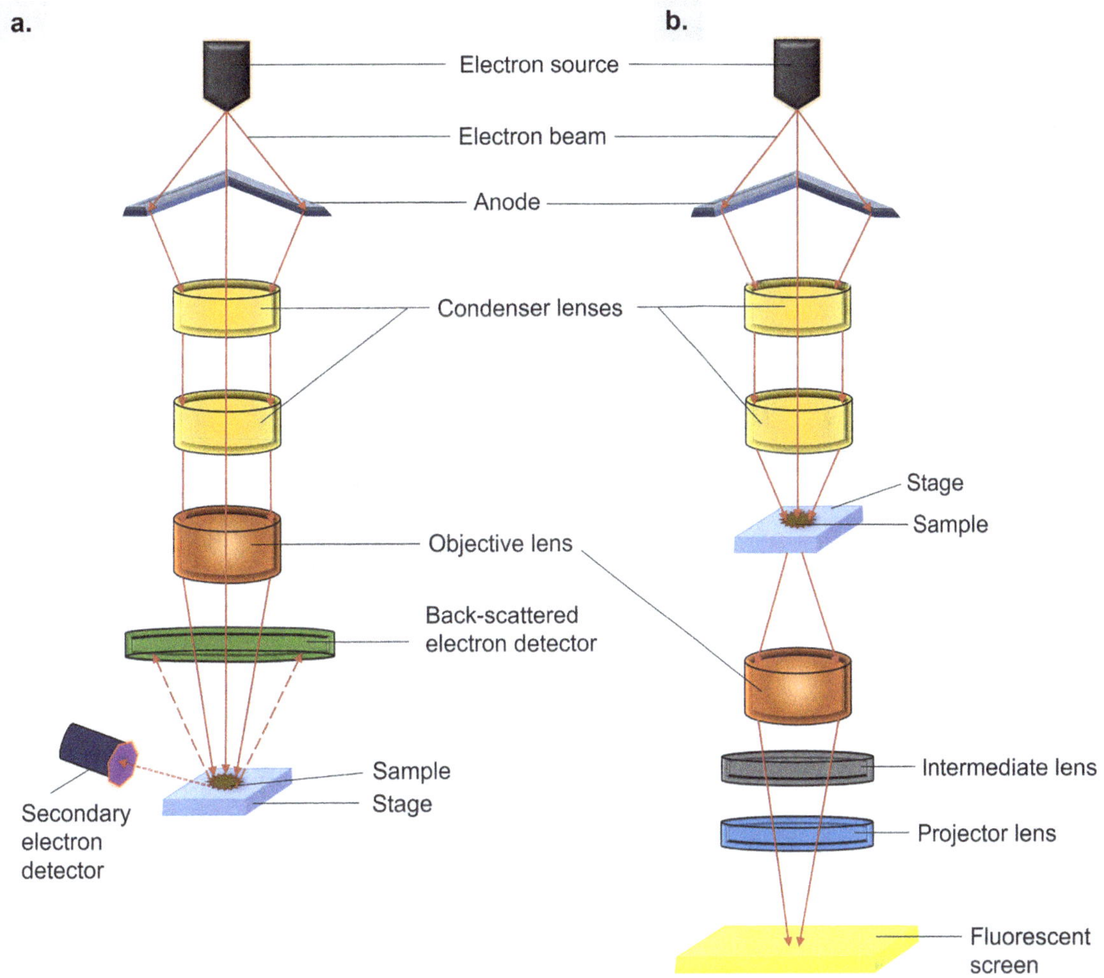

FIGURE 21.6 Schematic diagrams of electron microscopes. a. Scanning electron microscope (SEM). b. Transmission electron microscope (TEM).

achieves nanoscale resolution, producing magnification up to 5×10^7 times. In microbiology, this instrument is popularly used in the structural and chemical characterisation of bacteria, archaea, virus particles, and nanomaterials. Figure 21.7b provides a TEM image of a cross-section of a bacterium.

21.3.6 Scanning Probe Microscope (SPM)

SPM is a unique and powerful tool for studying the surface structure of the sample using a probe. The criterion for scanning the object using SPM is the "feels" of the surface of a sample rather than "looking" through the lens. The instrument is equipped with a physical probe, which is a short needle (around 2 μm), including a sharp tip (having widths of less than 100 Å). A probe is attached to a flexible cantilever that can bend or deflect to approach the surface profile of the sample. The instrument does not keep any lenses; hence, the diffraction limit does not apply to this microscope. During scanning, the tip of the probe lies in very close proximity to the surface of the sample through which its vertical motion is recorded. The tip of the probe also examines various forces, currents, capacitance, and interactions of the sample. The acquired digital data is translated into a 3D image with high spatial resolution in real-time using specific software on the computer. The final main output of the examination is a topographic map.

The two most popular established operating modes of SPM are scanning tunnelling microscopy (STM) and atomic force microscopy (AFM). These instruments attain a resolution of 0.1 Å and a vertical resolution of 0.05 Å, which is suitable for studying even individual atoms. STM was invented before AFM. In particular, AFM resolution is better than STM and has broad applications in nanosciences. The advent of SPM technology has made possible the study of microbial surfaces at high resolution. An investigation of bacterial biofilm topography is another possible application for SPM, providing detailed information on bacterial cell arrangements, extracellular polymeric substances (EPS), and biofilm architecture.

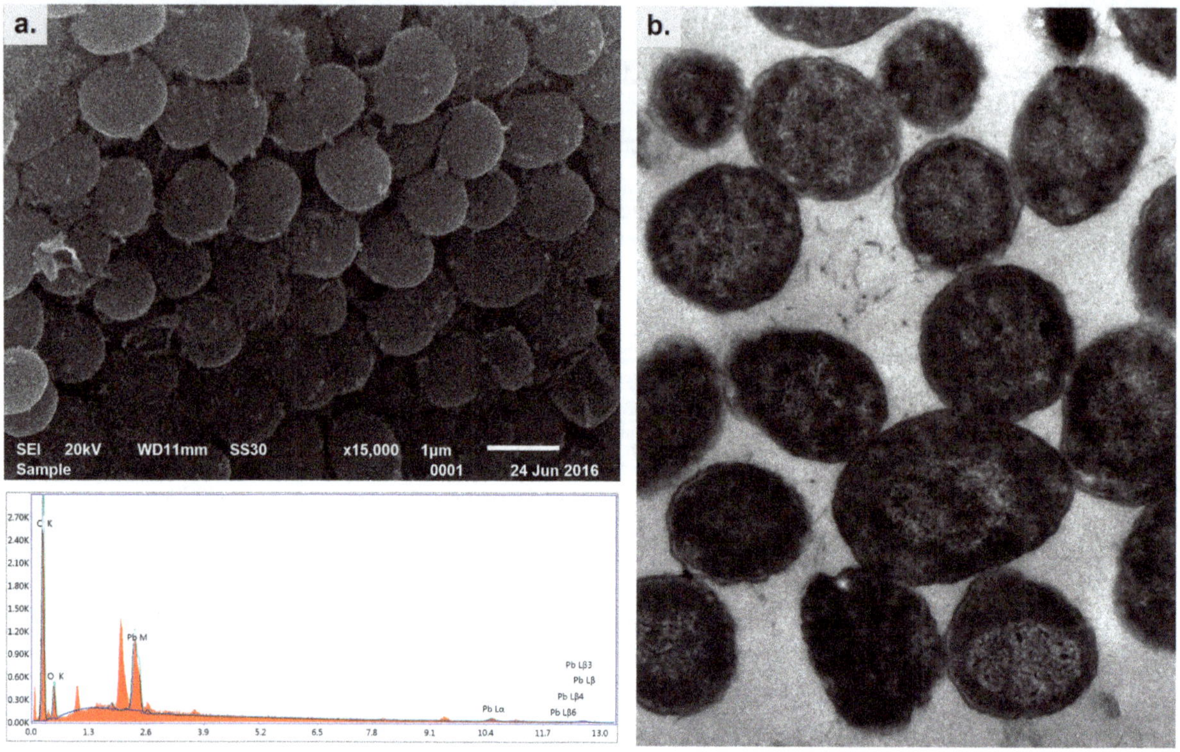

FIGURE 21.7 Electron micrographs of different bacteria. a. SEM-EDX image of *Staphylococcus* sp. after Pb treatment, providing valuable insights for morphological detail and Pb interaction with bacterial cells (showing Pb peaks in the graph). b. TEM image of cross-sectioned *Enterobacter* sp.

21.4 FLUORESCENCE *IN SITU* HYBRIDISATION (FISH)

In microbiology, fluorescence *in situ* hybridisation (FISH) is a nucleic acid-based detection method to identify and enumerate bacteria and archaea from an environmental sample. This is a tool for cultivation-independent identification, which provides visual images of microorganisms through a non-PCR-based technique. It is a rapid and sensitive method for an *in situ* study of microbial communities. The identification of microorganisms through FISH does not require prior knowledge of the environmental samples. The process relies on oligonucleotide probes (of 15-30 base pairs) labelled with a fluorochrome, which are complementary RNA sequences. These probes are available for different hierarchical sets ranging from domain to species-specific probes. They are attached to the oligonucleotide through covalent binding. The classical target sequence of the prokaryotic cells is the 16S rRNA sequence because it is considered the most useful marker for phylogenetic distinction and is relatively stable and abundantly present in an active cell. The dyes (or fluorochromes) used for the fluorescence can be fluorescein, Texas red, indocarbocyanine dyes, cys3, 4′,6-diamidino-2-phenylindole (DAPI), etc. When a probe is bound with specific genetic regions of an organism, it gives a fluorescent signal, which can be observed under a microscope and analysed. FISH can also be used for the detection and localisation of different functional genes and RNA sequences in higher organisms for genetic counselling and molecular diagnosis.

The process of FISH can typically be divided into five different sequential steps, including fixation, permeabilisation, hybridisation, washing, and visualisation. In the first step, the sample is treated with a fixing agent, followed by its transfer onto a glass slide. This is the foundation step, which prevents the degradation of a target sequence in the cells. Paraformaldehyde is a very commonly used fixing agent, which enables chemical cross-linking and induces partial cell lysis. In the next step, the active cells are mechanically disturbed, which allows probes to penetrate inside the bacterial cells. An ethanol treatment ensures sufficient permeability in the cells to deliver the probe. Gram-positive bacteria can be specifically treated with lysozyme to break the thick boundary of peptidoglycan. This is followed by denaturation and hybridisation reaction between genetic probes and target sequence under dark conditions at 35-55 °C, incubated for 15 min to several hours. The probe annealed with targeted sequences establishes a strong binding, and the unbound probes are removed through a washing step. In the final stage, the dyed cells are detected using a confocal laser scanning microscope (CLSM) (Figure 21.8).

This technique falls short for those cells which are slow-growing or metabolically inactive due to the low rRNA content. However, this limitation can be overcome to a certain extent by the use of several probes for one target organism, which can achieve better signals. In a different approach, the use of antibiotics, chloramphenicol, warrants the inhibition of

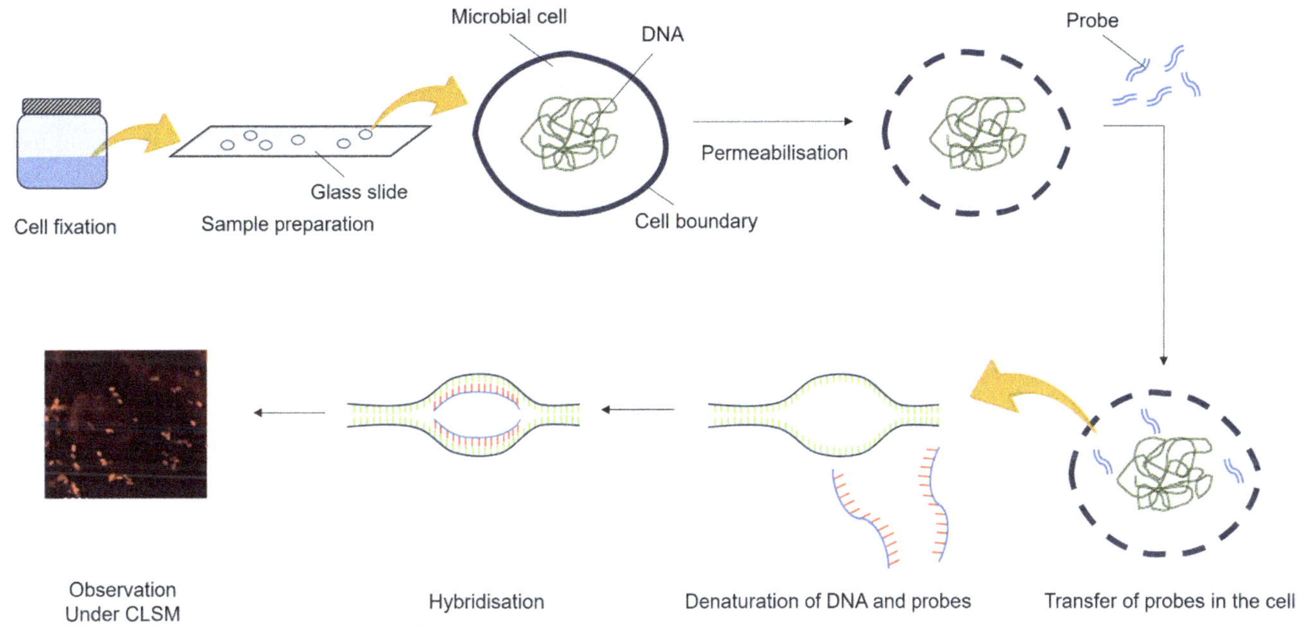

FIGURE 21.8 A workflow of fluorescence *in situ* hybridisation (FISH) for microbial community analysis and microbial observation under the confocal laser scanning microscope (CLSM).

protein synthesis, which encourages the accumulation of RNA content for better hybridisation. In addition, a non-labelled probe (so-called helper probe) can be used to increase the accessibility of nucleic acids through its binding adjacent to the probe target site. This ensures the unfolding of target sites for effective probe hybridisation.

21.4.1 Catalysed Reporter Deposition (CARD)-FISH

An advanced variant of FISH involves the integration of Catalysed Reporter Deposition (CARD), also known as tyramide signal amplification. This setup helps in amplifying the signal for the detection of environmental microorganisms. FISH associated with CARD achieves 10-fold stronger signals.

The process is initiated with the fixation and permeabilisation of the cells, similar to FISH. There are several methods for fixation and permeabilisation. The selection of the best methods depends on the types of samples, the specific target organisms, and the molecules being studied. In the next step, a native enzyme peroxidase of the cells is deactivated to avoid a false-positive signal. CARD reaction is carried out by the oxidation of hydrogen peroxide (H_2O_2), which is also a substrate for peroxidase. The oligonucleotide probes are tagged with horseradish peroxidase (HRP). The molecular weight of HRP is 40 kDa, which is much bigger than fluorochromes (500-1,000 Da). Therefore, transporting HRP across the membrane is challenging and requires a very effective permeabilisation technique. After its passage inside the cells, the probes hybridise with target sequences. The non-binding probes from the system are removed by washing. The next step is tyramide signal amplification using fluorescein-labelled tyramide. When fluorescein-labelled tyramide is introduced into the system, it is converted into a radical intermediate by HRP in the presence of H_2O_2. The radical intermediate reacts with different aromatic compounds, resulting in the deposition of fluorescein-labelled tyramide around HRP molecules. After the removal of the non-binding dye from washing, the slide is observed under a microscope for the accumulation of fluorescein-labelled tyramide attached to HRP. Sometimes, a second signal amplification can be performed to increase the sensitivity (more intense signals) for the detection of ultramicrobacteria (extremely small-sized bacteria or nanobacteria) in environmental samples. This additional step involves the deposition of HRP-labelled antibodies followed by a second amplification with fluorescein-labelled tyramides.

21.5 FLOW CYTOMETRY (FCM OR FC)

In microbiology, flow cytometry (FCM or FC) is a powerful tool for the rapid detection of microbial cells based on specific optical properties. The key advantage of this technique is the multi-parametric analysis at the single-cell level. FCM can identify quantitative (counting) as well as qualitative (several cell subgroups) traits of a cell suspension. Considering the automation and possible sample throughput, the instrument has a very high productivity and efficiency. The microbial cells in FCM can be measured in a size range between 0.2 to 150 μms. A typical analysis of a liquid sample can determine around 10,000 cells per second.

The instrument of FCM is mainly composed of three different components, including fluidics, optics, and electronics. Fluidics is a primary component that accurately injects the sample into the flow chamber, where cells pass to the centre of the laser intercept through a sheath fluid (physiological buffer), lined up in single files for detection. Optics measures the illuminance of the laser beam after interrogating the cells. It is composed of three different items: excitation optics, collection optics, and a detector. The excitation optics consists of a laser beam source, which emits a uniform light of a specific wavelength. The collection optics have collection lenses to collect the emitted light from the cells. An associated filter favours the passage of only the desired wavelength. The detector is made up of photomultiplier tubes (PMTs) and a photodiode to collect the signals. The electronic system is the brain of FCM, which converts the photocurrent collected from the detector into electric currents and amplifies the pulses into digital numbers. Cells are identified by labelling them with particular dyes (Table 21.2). When a stained cell passes through a detector, a laser beam excites the dye, and the detector measures the scattered light (intrinsic parameters) and fluorescence emissions. The computer converts these estimates into statistically relevant data sets, which are subject to data acquisition and analysis.

FCM is mainly used to assess and evaluate the microbial load in suspension. Soil samples can also be used for FCM analysis after preparing a suspension in a liquid solution. FCM significantly expands its applications in microbiology using two specialised settings, including fluorescence-activated cell sorting (FACS) and an imaging flow cytometer. FACS is a combination of FCM and a cell sorter, which facilitates the physical separation based on specific properties of the cells (such as cell densities, colour, and size). On the contrary, an imaging flow cytometer combines fluorescent microscopy with FCM for the morphological analysis of individual cells alongside the data of FCM. This procedure also allows observation of the physiological and metabolic status of bacterial cells.

FCM can be used to monitor microbial cells in both natural and artificial planktonic systems. It offers a broad spectrum of analysis, such as the estimation of the heterogeneity in the microbial communities and their assessment of abundance, biomass and activity in diverse aquatic systems. This is a powerful tool to discriminate among dormant, viable, and nonviable cells. FCM can be additionally employed in research to investigate microbial cell cycles and to evaluate *in vitro* antimicrobial activity (e.g., bacterial antibiotic susceptibility). The machine can also detect viruses, but is less sensitive toward ultra-small sizes (10-20 nm in diameter). During World War II, the United States (US) Army used this technology to detect germs and spores in aerosols.

21.6 MICROBIAL COMMUNITY ANALYSIS

In an environment, a microbial community is an assemblage of co-occurring and potentially interacting microorganisms such as viruses, bacteria, archaea, fungi, and microalgae. Microbial communities in their natural habitats are the primary driving force behind the ecological dynamics. Because of their very diverse makeup, they exhibit a wide range of effects on their surroundings. However, a wealth of genetic variety of natural microbiota is untapped and represents "**dark matter**". These microorganisms are highly mysterious, dwelling in conditions that are difficult to recreate in a laboratory context. Researchers have developed novel methods for examining and characterising previously elusive microorganisms. Continuous breakthroughs in sequencing technologies and analytics tools have shed light on the black matter of microbiology.

Microbial community analysis mainly relies on culture-independent approaches, where molecular tools and sequencing techniques recover and recognise the biological molecules (DNA, RNA, protein, and fatty acids) of micro-inhabitants in the natural environments. The community-level analysis provides detailed information about the complex relationship between ecosystem function and environmental parameters. The following are the most well-known methods for studying microbial communities.

21.6.1 Metagenomics

Metagenomics is an ideal tool for studying uncultivated microbial communities in environmental and clinical samples. This field was envisioned by Norman E. Pace in 1985 for microbial community analysis using rRNA gene sequences. However, the term "metagenomics" was coined years later by Handelsman and coworkers in 1998. This technique mainly involves the identification of DNA sequences directly recovered from environmental samples using a set of molecular techniques and bioinformatics tools. Direct access to the entire genetic content of microbial communities from a sample provides critical new insights into the structure and function of microbial composition. In contrast, traditional cultivation fails to grow more than 98% of the microorganisms present on Earth. Around 70-80% of the plant endophytes cannot be recognised in their physical forms due to the limitations of cultivation practices. Human microbiota is also very little understood in terms of their physical forms, they are as abundant as somatic cells and constitute far more genes than those of the human body. Many extreme environments, such as salt lakes, marine water, deep sea, hot water springs, contaminated sites, etc., also host many unique microorganisms that cannot be cultivated but have been recognised using metagenomics.

The procedure can be divided into two main steps: extraction of DNA and sequencing. The extraction of DNA from the environment is carried out using two different methods. In one method, the cells are recovered from the samples before DNA extraction. Another method involves cell lysis in the sample matrix followed by recovery of DNA. DNA obtained from the extraction procedures is crude. The co-extraction of humic substances and other hindering molecules interferes with cloning and sequencing. Hence, their removal is a prerequisite for handling genetic constituents. DNA is purified through different mechanical and chemical treatments. When the initial DNA concentration is very low and the study targets a specific region, the sequence can be amplified. This process can improve the resolution of downstream analysis.

Following that involves library preparation and sequencing, which relies on two different approaches. For traditional Sanger sequencing, long sequences can be cloned into a vector (such as a plasmid, cosmid, or bacterial artificial chromosome (BAC)), followed by fragmentation and sequencing. This approach has some inherent limitations due to the cumbersome efforts in the production and screening of cloned libraries. However, modern sequencing strategies like pyrosequencing and Illumina sequencing technologies (detailed in section 21.2.4) allow the sequencing of DNA without cloning but produce short sequences. The sequence libraries are prepared by end repair, A-tailing, and adapter ligation. The newer technology (e.g., ONT) has addressed many limitations of old sequencing techniques and has gained the confidence to unveil new knowledge of dark matter (which includes the vast array of unidentified uncultured microorganisms).

The analysis of sequence data involves assembly, binning, annotations, and metabolic reconstructions using computational resources (Figure 21.9a). Regardless of the sequencing methods, the analysis of sequence data is based on comparison with already known sequences from various sequence databases. The predictions about sequence identification are based on homology. A considerable amount of DNA sequences discovered do not match known sequences in existing databases due to novel sequences. Further, the output of results and datasets is stored in a database for future research directions.

Early research investigations involved only prokaryotic DNA as the tractable component of metagenomic analysis. However, the development of high-throughput sequence technologies has widened the scope of metagenomics to study large genome sequences of eukaryotic cells. The random sequencing of the DNA fragments provides details of the complete or near-complete genome sequences of the dominant taxa in the samples. Metagenomic tools have also been used to study fungal and microalgal communities in various environmental sites.

The advent of metagenomics has changed the profile of microbiology, making it realise that the unseen world of bacteria, archaea, and viruses is far outsized compared to their culture representatives. Also, it gave insights into the extent of our prior ignorance of organismal diversity and microbial community size in diverse environmental sites. Metagenomics has also uncovered functional diversities of the microbial world by analysing different genes. For example,

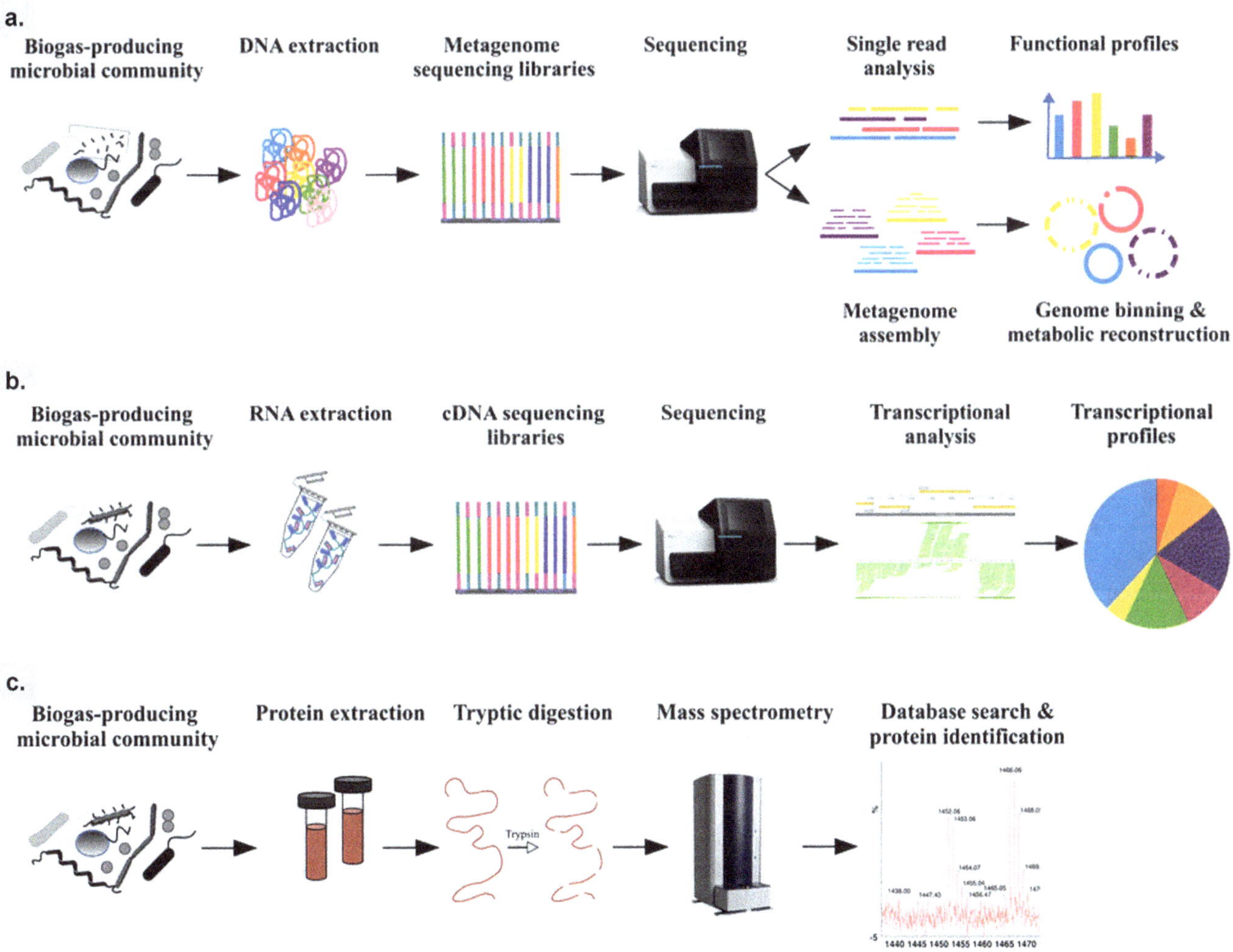

FIGURE 21.9 Workflows of different techniques using an example of a biogas-producing microbial community. a. Metagenomics. b. Metatranscriptomics. c. Metaproteomics. (Image adapted from Hassa et al. (2018).[4])

the discoveries of proteorhodopsin-based photoheterotrophy and ammonia-oxidising archaea using metagenomic tools are remarkable findings in the field of microbial ecology.

21.6.2 Denaturing Gradient Gel Electrophoresis (DGGE)

Denaturing gradient gel electrophoresis (DGGE) is a molecular fingerprinting technique to detect variable sequence composition of nearly identical lengths in environmental samples without sequencing. A single experiment can run multiple samples in parallel and provide comparative accounts. The method is robust, reproducible, and relatively inexpensive, providing a quick profile of the diversity and structure of microbial communities. This technique was developed to cumbersome the cloning practices in the phylogenetic identification of microbial consortia.

DGGE works in the combination of three different techniques, *viz.* PCR amplification, denaturation, and gel electrophoresis. The initial step of DGGE is similar to metagenomics, i.e., the extraction of DNA from the environmental samples followed by its purification. A set of primers can be used to amplify the target gene sequences using PCR. The most common marker in DGGE is the 16S rDNA. The PCR products are run on a polyacrylamide gel containing linear denaturant gradients to separate fragments. The most common chemical denaturants are formamide (0-40%) and urea (0-7 M), and their gradient in gel pace from low to high. When a gradient is a temperature, the technique is specifically called temperature gradient gel electrophoresis (TGGE). The linear gradient of DNA denaturants causes the melting of double-stranded DNA based on the difference in base pairing and develops bands at different heights on the gel as they move. These bands are visualised using stains (ethidium bromide, SYBR Green I, or silver stain), which represent the unique operational taxonomic units (OTUs) (Figure 21.10). Bands showing unique sequences can be excised from the gel, re-amplified, and sequenced for taxonomic identification.

DGGE analysis of PCR-amplified 16S rDNA fragments is commonly employed to compare or monitor the phylogenetic diversity between different samples. For example, DGGE analysis has been a pivotal tool in studying the diversity of highly slow-growing and complex groups, ammonia-oxidising bacteria (AOB) and ammonia-oxidising archaea (AOA) in polluted and unpolluted soils or waters.

21.6.3 terminal Restriction Fragment Length Polymorphism (t-RFLP)

t-RFLP is terminal restriction fragment length polymorphism, which is a variant of restriction fragment length polymorphism (RFLP). This is also a fingerprinting technology; however, the band is separated based on the digestion of fluorescently labelled PCR products using restriction endonucleases. The process begins with the extraction of DNA, followed by amplification of the target gene. The amplification of the DNA extract utilises either one or both primers with a different fluorescent dye. The fluorescently labelled amplicons are digested with restriction enzymes. Enzymes like *AluI*, *HhaI*, and *ThaI* are suitable restriction enzymes to recognise and cleave various regions in DNA. The DNA fragments are separated using capillary-based electrophoresis. When the gel is exposed to the laser, it detects only the terminal, end-labelled restriction DNA fragments. The numbers and sizes of different terminal restriction fragments correspond to the variations in the genetic profile.

21.6.4 Other Fingerprinting Methods

There are many other alternative fingerprinting methods to identify the genetic differences between diverse species in the environmental samples. Amplified ribosomal DNA restriction analysis (ARDRA) is also a variant of RFLP that relies on one to three tetra cutter restriction enzymes (e.g., *AluI* and *HaeIII*) for digesting PCR amplicons. However, ARDRA has low discriminating power to differentiate different taxa.

A different technique to study microbial communities is ribosomal RNA intergenic spacer analysis (RISA), which focuses on the 16S-23S intergenic spacer region in the rRNA operon as the target sequence. This relies on 16S and 23S rDNA primers for the amplification. The PCR amplicons are loaded on the gel and travel characteristic distances based on the size of the sequences. The intergenic region between 16S and 23S rRNA encodes sequences having high heterogeneity in terms of length and bases. The length heterogeneity is the parameter for distinguishing different taxa. A different analysis is automated ribosomal intergenic spacer analysis (ARISA), which is more sensitive than RISA. In ARISA, the amplification is carried out using fluorescence-tagged oligonucleotide primer, followed by the laser detection of fluorescent DNA fragments through an automated system.

21.6.5 Phospholipid Fatty Acid (PLFA) Analysis

Phospholipid fatty acid (PLFA) analysis is a prominent method to characterise microbial communities. Phospholipids are key components of the cellular membrane in microbial species. They are amphiphilic molecules comprising a polar head group and a lipophilic region of long fatty acid residues with 14-20 carbons. Many phospholipids have distinct fatty acid side chains, which are characteristic of different microbial species. These side chains can vary in terms of chain length, saturation, and branching. Since phospholipids can be quickly degraded after the death of cells, they can be potential markers of viable microbial forms. A list of different PLFA biomarkers in different microbial groups is mentioned in Table 21.3.

The methodology of PLFA analysis includes three major steps: extraction, fractionation, and analysis. The extraction method uses different organic solvents to collect the total or representative fraction of fatty acids from environmental samples. This preparation requires a mixture of chloroform, methanol, and buffer. The extraction procedure includes a brief sonication period to break the cell aggregates. The

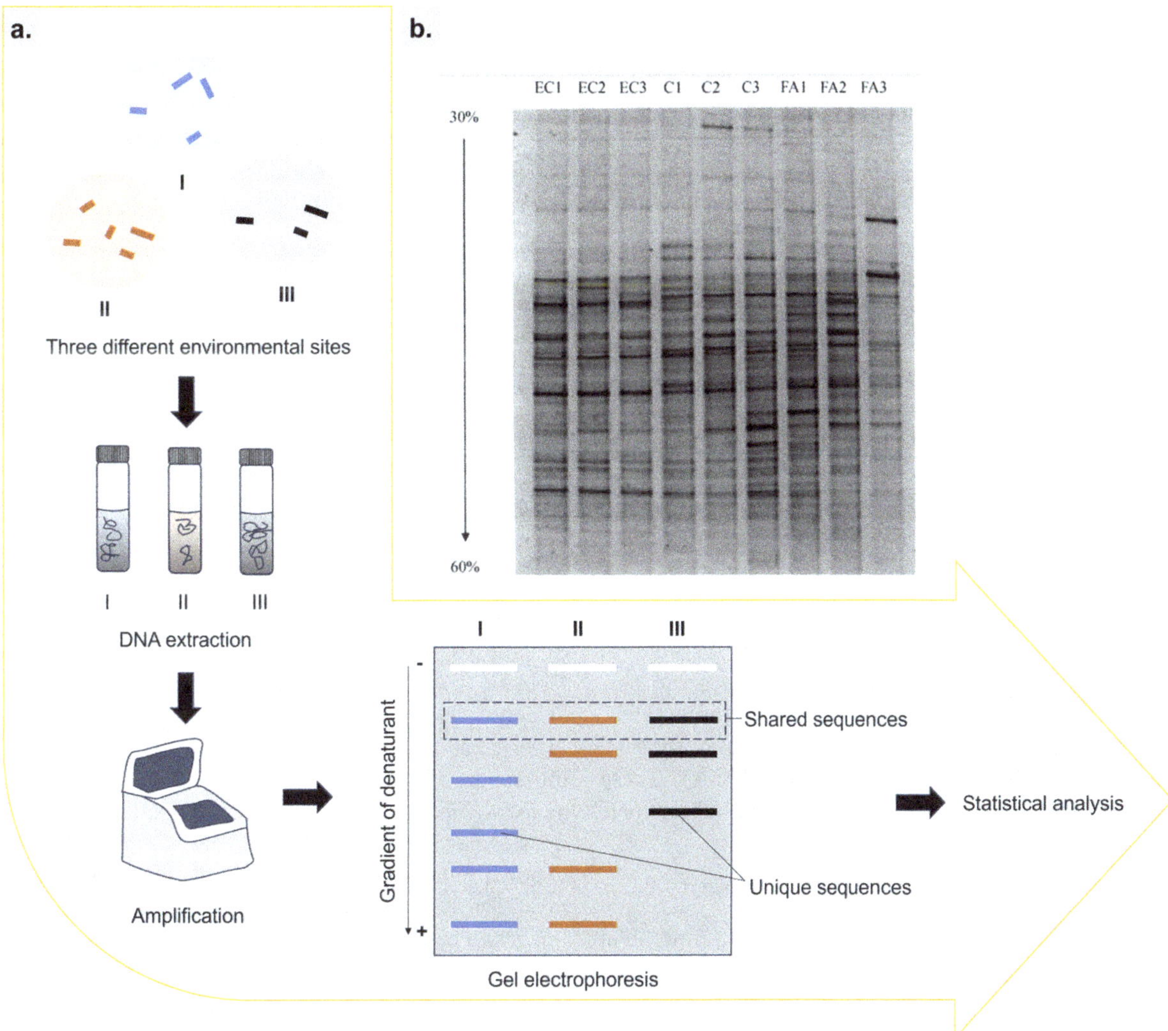

FIGURE 21.10 a. A scheme for the workflow of DGGE (denaturing gradient gel electrophoresis). b. A DGGE image of 16S rDNA of total soil bacterial communities in the control (C), pre-planting fertiliser application (FA), and ectomycorrhizal inoculation (EC) treatments. (Image b reproduced from Assigbetse et al. (2005)[5] with permission of © Springer Nature.)

next step is fractionation to separate the neutral, glyco- and phospholipids using chloroform, acetone and a methanol-chloroform-water solution in a silica gel. These fractions are esterified by alkaline methanolysis, which converts them into fatty acid methyl esters (FAMEs). The FAMEs are subjected to quantitative analysis using gas chromatography-mass spectrometry (GC-MS). The identification of FAMEs is carried out by comparing their mass spectra and retention times with commercially available standards.

PLFA analysis provides information on the quantitative description of microbial biomass from environmental samples. It can also be used as a biomarker to identify microbial taxa. However, it is valid only to a limited extent due to its low taxonomic resolution. The commonly detected microorganisms through PLFA analysis are sulphate-reducing bacteria and methane-oxidising bacteria (MOB) I and II. The composition of PLFA markers can also be employed as an indicator of stress signatures. In addition, the combination of PLFA analysis with other techniques offers knowledge of the functional component of the microbial community. However, the interpretations of PLFA data have some discrepancies and challenges, which hinder their applications for crucial insights.

21.7 ANALYSIS OF MICROBIAL ACTIVITIES

Microbial activities refer to biochemical processes and metabolic events of microbial communities in ecological settings. It has notable impacts on ecosystem services such as nutrient cycling, plant growth, pedogenesis, primary production, decomposition, etc. Metagenomic tools have been used to get valuable insight into the genetic potential of microbial communities and their functions. However, the functional capacity of a community does not always correlate with its composition because different microbial communities can sometimes perform similar functions (i.e., functional redundancy). Also, metagenomic tools fail to provide any information on dynamic changes in activities in response to variations in external factors. The information on the complex

TABLE 21.3
A list of phospholipid fatty acid (PLFA) biomarkers for various microbial groups

Microbial groups	PLFA biomarkers
Gram-positive bacteria	i13:0, a13:0, i14:0, i15:0, a15:0, i16:0, i17:0, a17:0, i18:0, and a18:0
Gram-negative bacteria	i15:0, i17:0, a15:0, i16:0, a17:0, 10Me17:0, 10Me18:0, and 18:1ω7c
Actinomycetes	10Me16:0, 10Me17:0, and 10Me18:0
Saprophytic and ectomycorrhizal fungi	18:2ω6,9c
Protozoa	20:2 ω6c, 20:3 ω6c, and 20:4 ω6c
Archaea	Phospholipid ether lipids (PLELs) instead of PLFA
Cyanobacteria	18:1ω7c
Arbuscular mycorrhizal fungi (AMF)	16:1ω5
Actinobacteria	10Me17:0 and 10Me18:0

i and a means iso and anteiso branching in fatty acids, respectively. c means cis configuration. Me indicates methyl group, whereas ω represents the position of the double bond relative to the Me end.

interactive effects of microbial communities requires new techniques. In recent years, various limitations of culture-independent techniques have been addressed by some new techniques such as metatranscriptomics, metaproteomics, stable isotope probing (SIP), microautoradiography (MAR), and enzyme-linked immunosorbent assay (ELISA). These techniques are discussed in detail in the following subsections.

21.7.1 Metatranscriptomics

Metatranscriptomics is a powerful approach in microbial ecology that sheds light on the active functional profile of microbial communities. This involves the study of the entire set of RNA transcripts (the transcriptome) from various microorganisms within a complex environmental sample at a specific time. This type of study bridges the gap between genetic potential (as seen in metagenomics) and actual functional activities, providing a more crucial understanding of how microbial communities interact in their natural environments. While the basic approaches employed in metagenomics and metatranscriptomics are similar, the fundamental difference lies in the information they capture. However, the processing of environmental RNA samples is highly challenging due to the short half-lives of RNA species, limited recovery of high-quality RNA, and difficulty distinguishing mRNA from other RNA species. Environmental metatranscriptomics has recently received a lot of attention using DNA microarrays. The main steps of metatranscriptomics are illustrated in Figure 21.9b.

21.7.1.1 DNA Microarray

A DNA microarray is a tool for hybridising complementary nucleic acid sequences on a solid surface for their detection. This technique provides a panoramic outlook of the microbial transcript in response to certain stimuli on a transcriptomic scale. The analysis of molecular sequences not only detects functional profiles but also measures the levels of expression. It utilises a chip with distinct spots for the collection of single-stranded DNA sequences called probes. Each spot of the chip consists of thousands of sequences of specific DNA (usually 10^{-12} moles). These probes are either complementary DNA (cDNA) obtained by reverse transcription or artificially designed and synthesised short oligonucleotides. A single experiment can measure the gene expression levels from thousands of genes (up to 200,000 spots/cm^2) in parallel analysis. The target sequences are labelled with fluorescent dyes. After proper incubation and washing, the spots are exposed to laser beams, which excite the dyes and emit some rays. The rays are collected on a scanner and generate the images. The image of different spots can be distinguished based on different intensities. The robust computational methods convert their intensities into numerical forms, which are directly proportional to the concentrations of target sequences.

This tool is used as one of the most straightforward methods to investigate bacterial metabolism under different stimuli. The analysis provides a comparison of gene expression between wild-type and mutant cells. Microarray technology also offers services for the identification of microorganisms and the comparison of transcripts from different environments. Using the probes of many target functional genes, the analysis can monitor the physiological status and functional activities of natural microbial communities. The popular molecular signatures of natural environments can be related to carbon, nitrogen, sulphur, and metal cycles. Furthermore, microarrays have many applications in clinical microbiology, such as pathogen identification and molecular diagnosis. However, this technique cannot detect low-abundance transcripts. The expression levels may also not be accurately measured.

In recent years, RNA sequencing (RNA-Seq) technologies have replaced the DNA microarray owing to their higher sensitivity, dynamic range, and ability to detect novel transcripts. RNA-Seq is based on reverse transcriptase activity, which allows transcripts to transform into cDNA, followed by amplification and library preparation. The prepared library is sequenced using an NGS platform.

21.7.2 Metaproteomics

Metaproteomics is used as a non-targeted method to identify different proteins associated with microbial activities. It can detect thousands of proteins of many dominant and subdominant microbial populations of natural sites in a short period. The procedure of metaproteomics includes four main steps: extraction, digestion, separation, and identification. The extraction of proteins from cells can be achieved through cell lysis. Following that, proteins are digested using suitable enzymes. The small fragments of proteins are separated using two different techniques, including gel-based and gel-free methods. A gel-based method involves one-dimensional and two-dimensional polyacrylamide gel electrophoresis (2D-PAGE), whereas the gel-free method relies on liquid chromatography (LC). After this, the proteins are identified using MALDI-TOF LC-MS/MS and bioinformatic tools (Figure 21.9c). MALDI-TOF LC-MS/MS combines three powerful analytical techniques: Matrix-Assisted Laser Desorption/Ionisation (MALDI), Time-of-Flight (TOF) mass spectrometry, and Liquid Chromatography-Mass Spectrometry (LC-MS/MS). The gel-based methods facilitate the separation of different proteins on the gel, and the spots of proteins can be identified using MS. In the gel-free method, the proteins are digested into small peptides using tryptic enzymes and then separated under LC, which further undergoes sequence analysis. The identification of peptide and protein sequences is based on the genomic information and gene annotation on which the protein-sequence database is constructed.

Metaproteomics offers deeper insights than metagenomics and metatranscriptomics as it reveals noble aspects of functional composition within microbial habitats. The analysis covers a wide array of ecophysiological traits in the local regions to planetary homeostasis at a grand scale. Though metaproteomics has proven its value in deciphering the metabolic details, it has mainly been employed in sites such as acid mine drainage (AMD) biofilms and enhanced biological phosphorus removal (EBPR) sludge, which are relatively less complex and have only a few taxa or a small number of dominant populations. For these environments, the well-established reference databases aid in protein identification. The complex systems, such as soil, sediments, freshwater, seawater, etc., comprise large biodiversity (species richness and abundance) and have a correspondingly large number of unique proteins at variable conditions (dynamic range), which do not fit for proteomic fingerprinting. Doing so creates many technical challenges for large-scale analysis of protein sequences. Nevertheless, metaproteomics has enormous potential to reflect specific microbial activities in a given ecosystem, which is free from the presumptions about the relevance of nucleic acid sequences. Proteomic analyses of the microbial community in response to environmental stimuli also give an unprecedented view of cellular physiology. Also, the accumulated protein sequence can fill the shortfall of protein databases for better predictions of the structure and function of different genes and proteins.

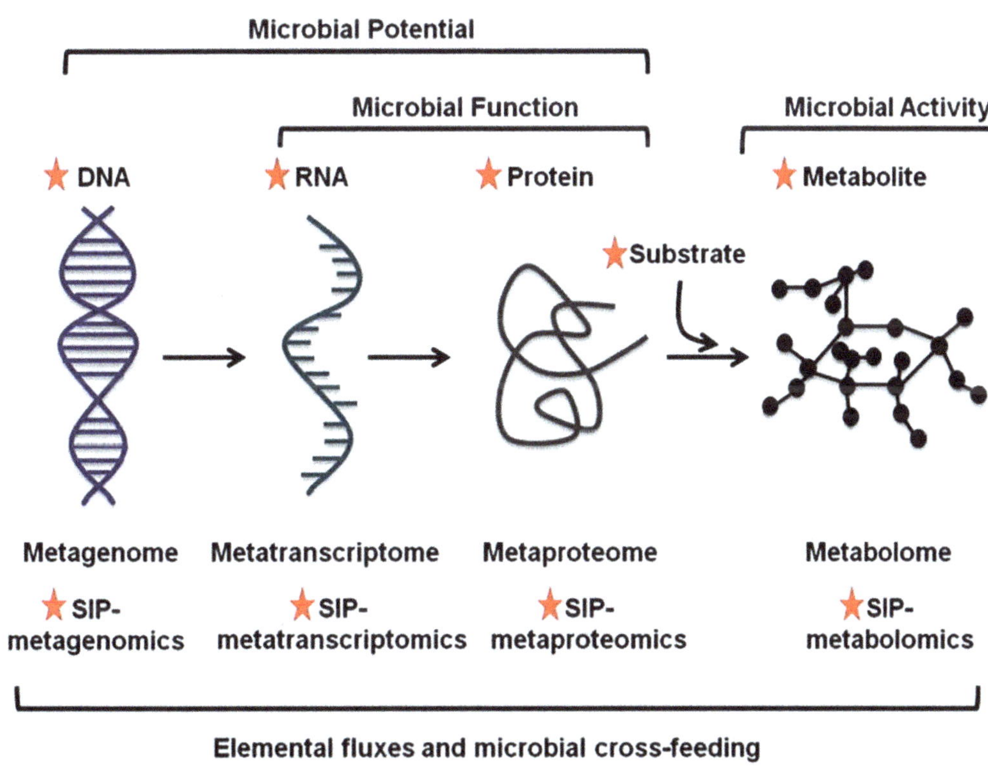

FIGURE 21.11 Overview of the multi-omics approach. Each level of information (DNA, RNA, proteins and metabolites) provides a different level of characterisation of the microbial community. SIP stands for stable isotope probing. (Image reproduced from Abram (2015).[6])

21.7.3 Stable Isotope Probing (SIP)

Stable isotope probing (SIP) is a technique that uses stable isotope-enriched substrates to assess the bioassimilation capacity of various chemicals in microbial communities. This technique provides valuable insight for estimating microbial activities and their interactions with nutrients. SIP involves the stable isotopes such as ^{2}H or D,^{13}C, ^{15}N, and ^{18}O-labelled substrates and examines biomolecules like DNA, RNA, phospholipids, and proteins (Figure 21.11). Of all types of isotopes, ^{13}C is the most widely utilised.

When the isotope-labelled substrate is introduced into the subject environment, indigenous active microorganisms in the community metabolise the labelled substrate and incorporate it into their biomolecules. Density-gradient centrifugation can be used to separate the biomass based on the difference in weight due to incorporated isotopes (Figure 21.12a). The fractions recovered from the centrifuge are subject to scrutiny through microarray, environmental genomics, or qPCR. The analysis of isotope-labelled biomolecules allows researchers to identify the specific microorganisms and their metabolic functions. The coupling of DNA with stable isotopes recognises the assimilation of specific growth substrates. The gene probing and sequence analysis of ^{13}C-DNA fractions can also be used for the taxonomic identification of the biomass. The recovery and analysis of RNA molecules can be employed to study the functionally active members of a microbial community. The labelled fatty acid in the biomass reflects taxa identification through distinct PLFAs.

The quantitative estimation of isotopes incorporated in biomass enables direct comparison of the rates of growth and substrate assimilation in microbial communities under specified conditions. SIPs can be commonly employed for biogeochemical analysis, quantification of the biodegradation potential of organic pollutants (such as ^{13}C-naphthalene), and identification of the microbial metabolic potential. The technique provides crucial information on methylotrophs using ^{13}C-methanol ($^{13}CH_3OH$) as a methylotrophic substrate. This technology can also be used in cross-kingdom plant-microbe symbiosis. Plants cultivated in the presence of ^{13}C-carbon dioxide ($^{13}CO_2$) can be used to identify labelled carbon uptake in microbial communities such as endophytes and rhizospheres. The Meselson-Stahl experiment of 1958, using a stable isotope 15N, is a landmark experiment in the history of molecular biology that provides strong evidence for the semiconservative model of DNA replication.

21.7.4 Microautoradiography (MAR)

MAR is a direct visualisation method of radionuclide-labelled compounds in cells on a microscopic slide. This focuses on the radiolabelling of targeted compounds by incorporating radioactive atoms. Compounds are frequently labelled with $_{3}H$ or $_{14}C$. The choice of radioactive atoms is determined by the research objectives that occur at the cellular and subcellular levels within biological systems. For example, $_{3}H$-thymidine can be used to track the synthesis of DNA. Similarly, $_{3}H$-uridine traces RNA synthesis. The use of radiolabeled amino acids (e.g., $_{3}H$- or $_{14}C$-leucine), fatty acids, glucose, sugars, and acetate provides valuable insights into the metabolic processes, pathways, and interactions in the cells.

The radiolabeled biomolecules emit rays inside the cells, which can be visualised by exposure to a thin film of photographic emulsion under a microscope. This technique has applications in measuring cell activities in ecological studies. Moreover, this method can also quantify the cells in the samples. Though this technique specifically responds to the metabolic state of microbial communities, it does not provide taxonomic resolution of the cells. This represents a major shortcoming of this technique, which can be effectively addressed by combining MAR with FISH.

MAR-FISH has paved the way for a simultaneous study of MCCs and their function to understand the ecophysiology of microorganisms in their natural environment with single-cell resolution. The procedure consists of five sequential steps, including sample incubation with radiolabeled substrates, fixation, sectioning/filtration, hybridisation, and visualisation (Figure 21.12b). The labelling of samples with radiolabeled substrates requires certain incubation periods to allow their assimilation inside the cells. The incubation time can range from a few minutes to several hours, depending on various factors. Microbial communities showing few activities and low abundance require incubation for a long duration. However, prolonged incubation may cause harm to the microbial communities due to the emission of harmful radiation. The labelled sample is fixed and then embedded in a stabilising agent such as agarose or resin to provide structural support, which is subjected to thin sectioning using a microtome or other devices. The ultra-thin sections are mounted on microscope slides and hybridised with fluorescently labelled oligonucleotides targeting specific rRNA probes. This is followed by autoradiographic visualisation, which involves the exposure of a radioactive labelled slide to a thin layer of photographic emulsion for the detection of radioactive decay. This experiment is strictly performed in darkness. The release of beta decay particles from the labelled probes into the emulsion reduces the silver ions, resulting in the formation of silver grains around the radioactive cells. The probe-stained cells can be observed using an epifluorescence microscope or CLSM, whereas silver grains adjacent to radiolabeled cells are detected under the transmission mode of microscopy. Figure 21.13 illustrates MAR-FISH analyses of ammonia-oxidising bacteria (AOB) and amoA-encoding members of the Thaumarchaeota (AEA) in a wastewater treatment plant under nitrifying conditions.

The MAR-FISH technique is widely employed to elucidate the ecophysiological features of diverse aquatic microbial habitats such as marine and freshwater ecosystems, hot-spring microbial mats, activated sewage sludge, etc. It provides crucial insights into the specialised resource utilisation in cosmopolitan members of aquatic microbial communities. In addition, this technique potentially addresses relevant

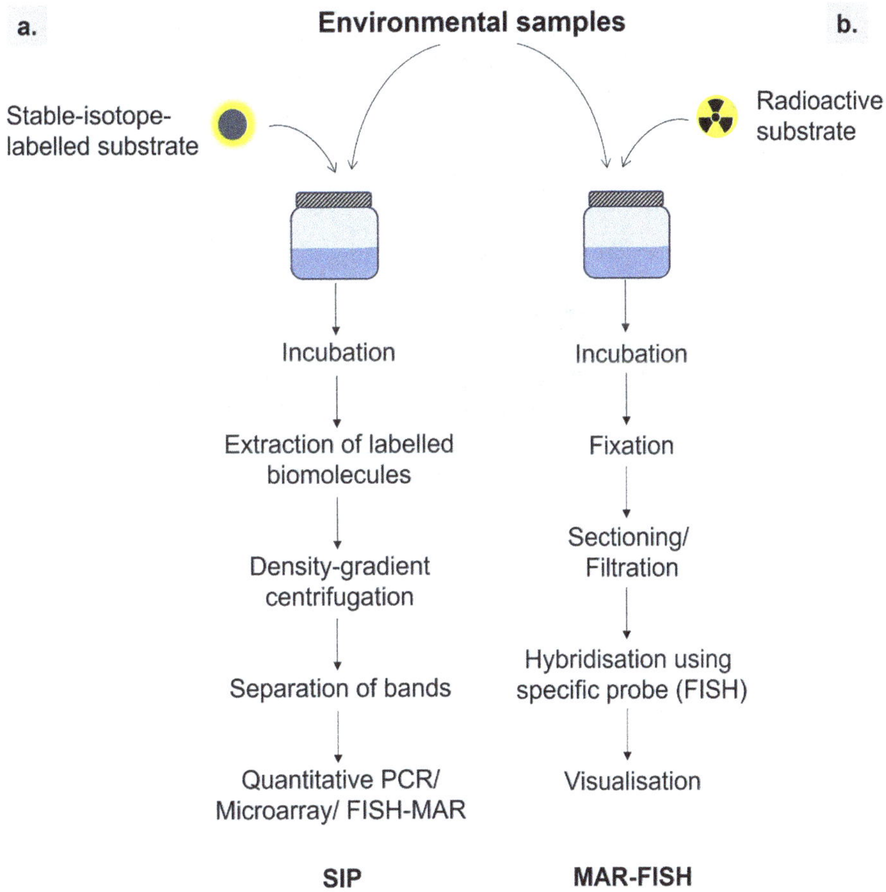

FIGURE 21.12 Methods for the study of microbial communities in environmental samples. a. Stable-Isotope Probing (SIP). b. Microautoradiography-fluorescent *in situ* hybridisation (MAR-FISH).

ecological questions, such as the production of microbial biomass and the contribution of diverse bacterioplankton in the biogeochemical cycles.

21.7.5 Enzyme-linked Immunosorbent Assay (ELISA)

ELISA is an analytical biochemical assay, which is a common tool in immunological studies. This mainly involves a plate-based assay and can be employed as a rapid screening method for the detection and quantification of specific antigens or antibodies related to bacterial or viral presence in environmental and clinical samples. One such example includes the examination of total *E. coli* in aquatic environments to identify the presence of faecal contamination and the possible presence of pathogens. This is a common assay in medical diagnostics and pathogen detection. ELISA can be used for the primary screening of several infectious diseases, such as acquired immunodeficiency syndrome (AIDS), coronavirus disease 19 (COVID-19), and tuberculosis. It can also be used for quality control, screening for contamination in food products, and monitoring of environmental pathogens.

ELISA requires a crude preparation of the sample and can be performed on 96-well or 384-well polystyrene plates. The method depends on the affinity between two molecules, which involves the multi-characters antigens and polyclonal antibodies attached with enzyme-conjugated antibodies. ELISA comes in several variants, such as direct, indirect, sandwich, and competitive ELISA. Indirect ELISA is a two-step ELISA that involves binding processes of antigen with primary antibody, followed by labelled secondary antibody. The indirect ELISA approach is the preferred method for the detection of microbial activities from environmental samples of all ELISA formats because of its easy operation, simple procedure, high sensitivity, relatively low cost, and quick analysis.

SUMMARY

- Culturing, biochemical tests, microscopic analysis, spectrography, and molecular-based approaches are common tools for the detection and characterisation of microorganisms in the environment. This is paramount for understanding microbial diversity and identifying their potential applications.
- Culture-dependent and culture-independent methods are two distinct but equally important approaches for the study of microorganisms in the environment. The scrutiny through the latter technique revealed that

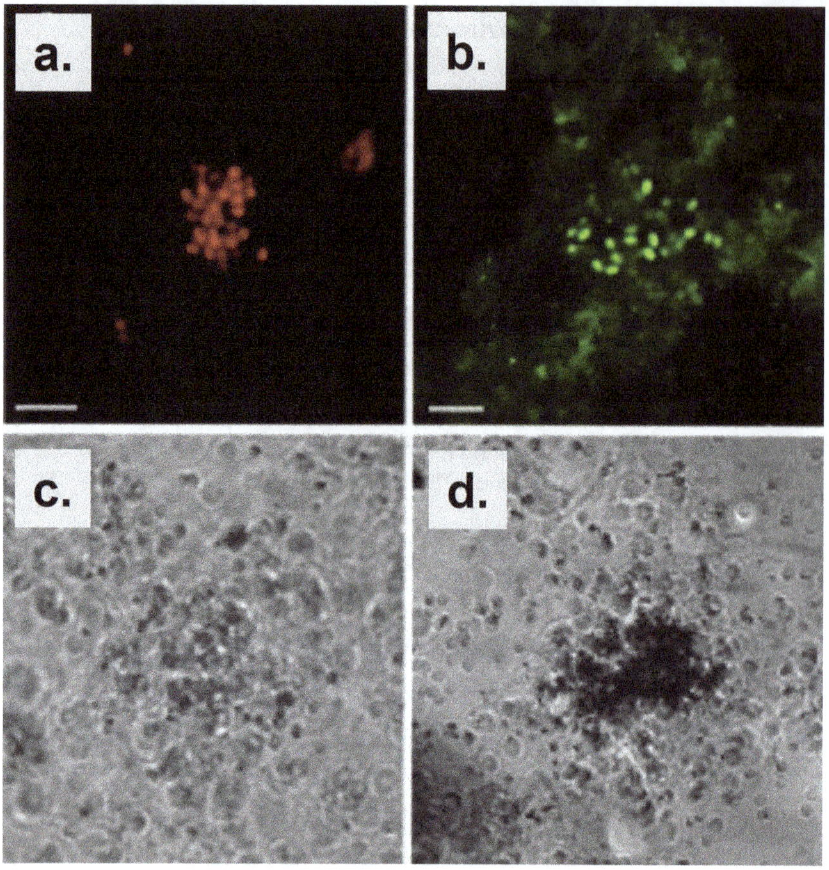

FIGURE 21.13 MAR-FISH analyses of ammonia-oxidising bacteria (AOB) and amoA-encoding members of the Thaumarchaeota (AEA) in a wastewater treatment plant under nitrifying conditions in the presence of $^{14}CO_2$. a. and c. represent FISH signals using different probes, whereas b. and d. are corresponding MAR signals. (Image adapted from Mußmann et al. (2011).[7])

more than 98% of microorganisms found in the natural environment cannot be cultivated on synthetic media.

- Culture-based techniques aim to grow microorganisms on artificial media. A traditional technique for their phenotypic characterisation is staining and biochemical testing. The molecular identification of microbial isolates is based on ribosomal DNA sequencing. WGS has become a routine tool for their characterisation.
- The viewing of micro to nano-sized range biological cells requires sophisticated microscopic techniques such as BFM, DFM, PCM, fluorescence microscope, CLSM, SEM, TEM, and SPM. Most microscopes consist of optical lenses, whereas SEM and TEM have electromagnetic lenses, and SPM possesses scanning probes.
- Culture-independent techniques like FISH and CARD-FISH are based on probe hybridisation and detection under a microscope. Metagenomics, DGGE, and t-RFLP are sequence-based analyses. PLFA analysis deals with distinct fatty acid side chains of microorganisms. FCM involves the rapid detection of microbial cells and cell sorting.
- Techniques like metatranscriptomics, DNA microarray, metaproteomics, SIP, MAR, MAR-FISH, and ELISA allow studies on the functional potential and ecophysiological aspects of microbial communities in natural environments.

DIFFERENT TYPES OF PRACTICE QUESTIONS

Fill in the blanks

I. .... works in a combination of three different techniques, *viz.* PCR amplification, denaturation, and gel electrophoresis.

II. In PLFA analysis, the quantitative estimation of fatty acid methyl esters (FAMEs) relies on the technique..........................

III. .... is a microscopic technique to study the structure of the sample using a physical probe.

IV. .... is an optical microscope to produce high-contrast 3D images.

V. In microautoradiography, $_3$H-thymidine can be used to track the synthesis of ....

SOLUTION: I. Denaturing gradient gel electrophoresis, II. gas chromatography mass spectrometry (GC- MS), III. Scanning probe microscope, IV. Confocal laser scanning microscope, V. DNA

Match the following

i. Crystal violet	a. Microalgae
ii. Iodine	b. Arbuscular mycorrhizal fungi
iii. Acridine orange	c. Endospore
iv. Trypan blue	d. DNA
v. Malachite green	e. Gram-positive bacteria

SOLUTION: i-e, ii- a, iii- d, iv- b, v- c

Very short answer type of questions

I. What is biochemical testing for microbial samples?
II. Write the names of two microscopes used for imaging living microbial cells.
III. Expand MAR-FISH.
IV. What is dark matter in the environmental microbiome?
V. What is the electromagnetic lens and its applications in microscopy?
VI. Write the names of two dyes and their ligands in flow cytometry.

Short answer type of questions

I. Why can mycobacteria and endospores not be stained using Gram staining? What are the suitable methods for their staining?
II. What is DNA sequencing? What are the different sequencing methods and their implications in microbial studies?
III. Write a note on the use of flow cytometry in microbiology.
IV. What is metagenomics? How has the advent of metagenomic tools altered the profile of microbiology?
V. Write the differences between the bright-field microscope, the dark-field microscope, and the phase contrast microscope.

Long answer type of questions

I. What are the culture-dependent and culture-independent methods in microbiology? What are their advantages and disadvantages? Shed light on two techniques of each category.
II. What are the different microscopic techniques employed in microbiology? Provide details on different types of electron microscopes.
III. What are the different techniques that unravel the genetic diversity and metabolic processes within microbial communities? How can we integrate these techniques to gain a comprehensive understanding of microbial ecology?

NOTES

1 Escobar-Zepeda A, Vera-Ponce de León A, Sanchez-Flores A. The road to metagenomics: from microbiology to DNA sequencing technologies and bioinformatics. *Frontiers in Genetics*. 2015;6:155161.
2 Jiang B, Henstra AM, Paulo PL, Balk M, van Doesburg W, Stams AJ. Atypical one-carbon metabolism of an acetogenic and hydrogenogenic *Moorella thermoacetica* strain. *Archives of Microbiology*. 2009;191:123–31.
3 Ofek M, Hadar Y, Minz D. Ecology of root colonizing *Massilia* (Oxalobacteraceae). *PloS ONE*. 2012;7(7):e40117.
4 Hassa J, Maus I, Off S, Pühler A, Scherer P, Klocke M, Schlüter A. Metagenome, metatranscriptome, and metaproteome approaches unraveled compositions and functional relationships of microbial communities residing in biogas plants. *Applied Microbiology and Biotechnology*. 2018;102:5045–63.
5 Assigbetse K, Gueye M, Thioulouse J, Duponnois R. Soil bacterial diversity responses to root colonization by an ectomycorrhizal fungus are not root-growth-dependent. Microbial Ecology. 2005;50:350–9.
6 Abram F. Systems-based approaches to unravel multi-species microbial community functioning. Computational and Structural Biotechnology Journal. 2015;13:24–32
7 Mußmann M, Brito I, Pitcher A, et al. Thaumarchaeotes abundant in refinery nitrifying sludges express amoA but are not obligate autotrophic ammonia oxidizers. Proceedings of the National Academy of Sciences. 2011;108(40):16771–6.

SUGGESTED READINGS

Alvarez-Barrientos A, Arroyo J, Cantón R, Nombela C, Sánchez-Pérez M. Applications of flow cytometry to clinical microbiology. *Clinical Microbiology Reviews*. 2000;13(2):167–95.

Bordowitz JR, Montgomery BL. Exploiting the autofluorescent properties of photosynthetic pigments for analysis of pigmentation and morphology in live *Fremyella diplosiphon* cells. *Sensors*. 2010;10(7):6969–79.

Bottari B, Ercolini D, Gatti M, Neviani E. Application of FISH technology for microbiological analysis: current state and prospects. *Applied Microbiology and Biotechnology*. 2006;73(3):485–94.

Caldwell DE, Korber DR, Lawrence JR. Confocal laser microscopy and digital image analysis in microbial ecology. In Schink B (ed.) *Advances in microbial ecology* 1992 (pp. 1–67). Springer.

Dumont MG, Murrell JC. Stable isotope probing—linking microbial identity to function. *Nature Reviews Microbiology*. 2005;3(6):499–504.

Eickhorst T, Tippkötter R. Improved detection of soil microorganisms using fluorescence *in situ* hybridization (FISH) and catalyzed reporter deposition (CARD-FISH). *Soil Biology and Biochemistry*. 2008;40(7):1883–91.

Elliott AD. Confocal microscopy: principles and modern practices. *Current Protocols in Cytometry*. 2020;92(1):e68.

Fisher MM, Triplett EW. Automated approach for ribosomal intergenic spacer analysis of microbial diversity and its application to freshwater bacterial communities. *Applied and Environmental Microbiology*. 1999;65(10):4630–6.

Franco-Duarte R, Černáková L, Kadam S, S. Kaushik K, Salehi B, Bevilacqua A, Corbo MR, Antolak H, Dybka-Stępień K, Leszczewicz M, Relison Tintino S. Advances in chemical and biological methods to identify microorganisms—from past to present. *Microorganisms*. 2019;7(5):130.

Frostegård A, Bååth E. The use of phospholipid fatty acid analysis to estimate bacterial and fungal biomass in soil. *Biology and Fertility of Soils*. 1996 Apr;22(1):59–65.

Heather JM, Chain B. The sequence of sequencers: the history of sequencing DNA. *Genomics*. 2016;107(1):1–8.

Hugenholtz P, Tyson GW. Metagenomics. *Nature*. 2008; 455(7212):481–3.

Hugerth LW, Andersson AF. Analysing microbial community composition through amplicon sequencing: from sampling to hypothesis testing. *Frontiers in Microbiology*. 2017;8:1561.

Kubota K. CARD-FISH for environmental microorganisms: technical advancement and future applications. *Microbes and Environments*. 2013;28(1):3–12.

Lee N, Nielsen PH, Andreasen KH, Juretschko S, Nielsen JL, Schleifer KH, Wagner M. Combination of fluorescent *in situ* hybridization and microautoradiography—a new tool for structure-function analyses in microbial ecology. *Applied and Environmental Microbiology*. 1999;65(3):1289–97.

Lichtman JW, Conchello JA. Fluorescence microscopy. *Nature Methods*. 2005;2(12):910–9.

McKinnon KM. Flow cytometry: an overview. *Current Protocols in Immunology*. 2018;120(1):5–1.

Michaelsen A, Pinzari F, Ripka K, Lubitz W, Piñar G. Application of molecular techniques for identification of fungal communities colonising paper material. *International Biodeterioration & Biodegradation*. 2006;58(3–4):133–41.

Mrozik A, Nowak A, Piotrowska-Seget Z. Microbial diversity in waters, sediments and microbial mats evaluated using fatty acid-based methods. *International Journal of Environmental Science and Technology*. 2014;11(5):1487–96.

Murphy DB, Oldfield R, Schwartz S, Davidson MW. Introduction to phase contrast microscopy. *Microscopy, The Source for Microscopy Education*. 2024. www.microscopyu.com/techniques/phase-contrast/introduction-to-phase-contrast-microscopy

Neuenschwander SM, Salcher MM, Pernthaler J. Fluorescence *in situ* hybridization and sequential catalyzed reporter deposition (2C-FISH) for the flow cytometric sorting of freshwater ultramicrobacteria. *Frontiers in Microbiology*. 2015;31:6–247.

Nielsen JL, Nielsen PH. Advances in microscopy: microautoradiography of single cells. *Methods in Enzymology*. 2005;397:237–56.

Radajewski S, Ineson P, Parekh NR, Murrell JC. Stable-isotope probing as a tool in microbial ecology. *Nature*. 2000 Feb;403(6770):646–9.

Rogers SW, Moorman TB, Ong SK. Fluorescent *in situ* hybridization and micro-autoradiography applied to ecophysiology in soil. *Soil Science Society of America Journal*. 2007;71(2):620–31.

Schwarz P, Lortholary O, Dromer F, Dannaoui E. Carbon assimilation profiles as a tool for identification of zygomycetes. *Journal of Clinical Microbiology*. 2007;45(5):1433–9.

Shapiro HM. *Practical flow cytometry*. Wiley; 2005.

Thomas T, Gilbert J, Meyer F. Metagenomics-a guide from sampling to data analysis. *Microbial Informatics and Experimentation*. 2012;2(1):1–2.

Wang N, He M, Shi HC. Novel indirect enzyme-linked immunosorbent assay (ELISA) method to detect Total *E. coli* in water environment. *Analytica Chimica Acta*. 2007;590(2):224–31.

Wang Y, Hammes F, De Roy K, Verstraete W, Boon N. Past, present and future applications of flow cytometry in aquatic microbiology. *Trends in Biotechnology*. 2010;28(8):416–24.

Willers C, Jansen van Rensburg PJ, Claassens S. Phospholipid fatty acid profiling of microbial communities–a review of interpretations and recent applications. *Journal of Applied Microbiology*. 2015;119(5):1207–18.

Zhou J, Thompson DK. Microarray technology and applications in environmental microbiology. *Advances in Agronomy*. 2004;82(10.1016):S0065–2113.

Zwirglmaier K. Fluorescence *in situ* hybridisation (FISH)–the next generation. *FEMS Microbiology Letters*. 2005;246(2):151–8.

22 Microbial Cultivation

Culture techniques are essential tools for the identification of microorganisms and for studying their roles in human welfare and ecological relevance. Microbial cultivation methods are indispensable for many commercial sectors, such as food and beverages, pharmaceutical, and biofuel industries. The cultivation of microorganisms is vital for the preparation of enzymes, secondary metabolites, and vaccines. The development of genetically modified microorganisms (GMOs) also requires pre-grown cultures. The cultured representatives give important clues for understanding their phylogenetic and functional diversities. Moreover, microbial cultivars provide sufficient validity of *in silico* predictions.

Early microbial studies depended solely on observational microbiology, based on their growth on natural substances. The initial cultivation practices can be traced back to the nineteenth century, when different food materials and wastes were used to compose nutrient media for cultivation. The knowledge of cultivation techniques derived through the integration of diverse fields of biological sciences, *viz.*, biotechnology, microbial genetics, and medical microbiology. In the 1860s, Louis Pasteur developed a synthetic liquid culture medium, which was used for bacterial reproducibility. The discovery and optimisation of culture media for the cultivation of microorganisms was a milestone in the history of microbiology. The advent of cultural techniques also challenged the theory of spontaneous generation, which stated that life originated from nonliving.

Robert Koch (1881) used gelatin as a gelling agent to prepare the first solid medium, but it suffered many limitations. A year later, Angelina Hesse (1882) proposed the application of agar as an alternative gelling agent. This discovery became a landmark for cultivation practices in microbiology. Julius Richard Petri (1886) invented a circular culture box called the Petri plate, which provides an easy view of the growth stages of microorganisms on solid media. In the later years, the isolation of microorganisms in pure form became paramount in industrial applications and research.

Though microbial cultivation is a complex and long process, it has always been an essential tool to study microorganisms in their physical forms. Isolation, cultivation, and purification of a microbial form from an environmental sample require at least 1-5 days. The isolation of slow-growing microorganisms is even more tedious. Bacteria and fungi are the most commonly studied microorganisms for cultivation. However, conventional culturing practices do not fit well with many archaea. Artificial cultivation practices for many microalgae and protozoa on synthetic media are uneconomical and undesirable for commercial trials. Many useful strains are stored as culture stocks in various culture collections across the globe. There are more than 200,000 culture stocks of bacteria and fungi worldwide. However, the worldwide culture stocks of archaea include only ~2,000 strains.

The stocks of microalgae and protozoans are also very limited. The methods of microbial cultivation are manifold, and the suitable choice of cultivation method mainly depends on the purposes and sources of inoculum. A general perspective on the cultivation of microorganisms through some common steps is illustrated in Figure 22.1 and discussed in the following sections.

22.1 SAMPLING AND STORAGE

Sampling is the first step of the cultivation, where a miniature scale of the environment is brought into the laboratory. Sampling can be performed from any physical environment. The choice of environmental sites for enumerating different microorganisms is enormous. Samples can be collected from soil, water, plant materials, compost, rhizosphere, phyllosphere, animal waste, industrial waste, agricultural residues, etc. Environmental sites have specific physicochemical conditions and harbour different microbial communities. When these samples are used for cultivation, culture media procure microorganisms with distinct traits. For example, agricultural soil, plant residues, grains, cereals, and potato wastes are the principal sources of starch hydrolysing bacteria and fungi. Metal-resistant microorganisms can be recovered from heavy metal-contaminated sites such as tanneries, electroplating industries, smelters, incinerators of municipal wastes, and e-waste. Psychrophiles or cryophiles prevail in cold conditions and can be isolated from permafrost, polar ice, glaciers, and the deep ocean. The samples for piezophiles (microorganisms that thrive at high pressures) come from the deep sea.

Sampling can be performed by two different methods: direct and indirect sampling (Table 22.1). Direct sampling involves the collection of visible materials directly from a natural site in a container. This sampling method does not require

DOI: 10.1201/9781032651095-27

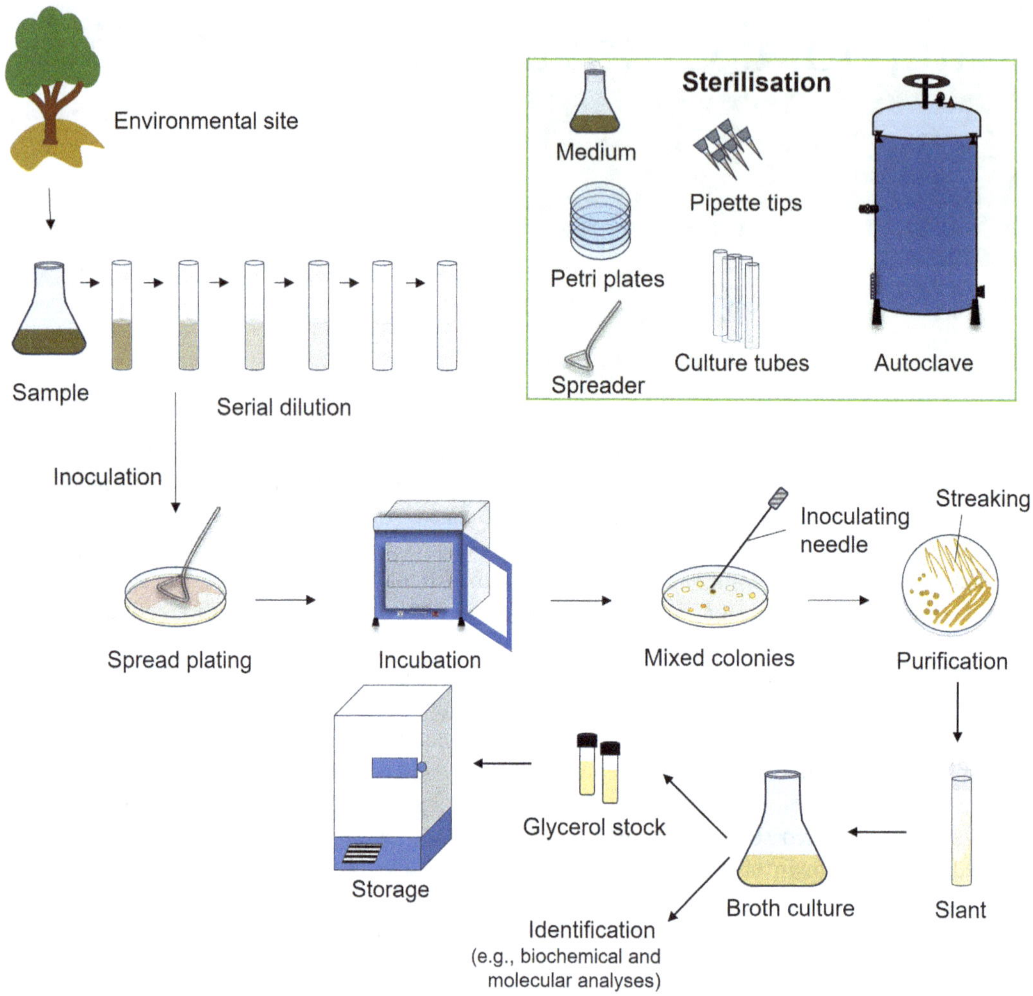

FIGURE 22.1 A schematic representation of the different steps of isolation of microorganisms from an environmental sample.

TABLE 22.1
Different types of sampling for the cultivation of microorganisms

Type of sampling	Method/device	Examples of samplers used
Direct sampling	Collection of physical materials from the environment	Vials, plastic containers, and polythene bags
	Passive air sampling	Petri plates with culture media
Indirect sampling	Collectors	Cotton swabs, tapes, wipes, and sponges
	Probes	Pressure-retaining sampler
	Sampling devices	Impactor and impinger
Direct exposure	Contact plate method	Convex agar medium
	Dip slip	Plastic paddles
	Petrifilm	Paper with a film of agar medium

much processing and can be directly used for dilution and culturing. Containers should essentially be aseptic, which can be a vial or polythene bag. These containers are usually sterilised by X-ray/UV ray treatment or autoclaving. During sample collection, the container should be exposed to air for a minimum time to prevent microbial contamination. After collection, samples are subjected to immediate transfer to the laboratory. Long transport of samples requires freezing conditions at temperatures below 10 °C (in an ice box or fridge) to preserve microbial activities. However, prolonged storage of samples poses a threat to the viability of microorganisms. Sensitive microorganisms are most vulnerable to physical stresses. Low-temperature transport is usually incompatible with thermophiles because contrasting temperature environments convey shock to the native microorganisms, resulting in a significant loss of their viability. That is why immediate culturing helps maintain optimal living conditions and ensures a high number of cultivars.

Indirect sampling methods use sterile physical supports to pick microbial cells from the surface environment, which are subject to further treatment for microbial extraction for culturing on synthetic media. This type of sampling is convenient for those environments that are physically inseparable (e.g., the study of surface microorganisms and biofilms). This method is also widely used for the detection of pathogenic microorganisms in humans. The physical supports for the sample collection in indirect sampling are collectors, probes, or other sampling devices. The collectors can be cotton swabs, tapes, wipes, and sponges. A swab is very commonly used to study the mouth microbiota. The collection of sediment and water from deep aquatic regions requires samplers, which are attached to different probes. In this preparation, a sampler ensures the safe passage of samples collected from deep regions through the water columns. The screening of piezophiles requires a pressure-retaining sampler to maintain high pressure for the sample. However, indirect sampling often allows limited recovery of the microbial cells due to various atrocities involved in extraction procedures. Therefore, indirect sampling is suitable for those environments where direct sampling is not possible.

Air is also a dynamic reservoir of microorganisms. The physical isolation of microorganisms from the air can be achieved through a different type of sampler, which typically relies on an external force to gather air particles. This type of sampling is called active air sampling. An active air sampler (AAS) is a pumping device that passes a continuous force of air (of a known volume) directly on Petri plates or a collection device. A different but most commonly used device for air sampling is the impactor. Impactor operates on the principles of impaction and inertia, where particles in an airstream stray from their natural course, causing airflow to create a steep bend to settle on solid or semisolid substratum. A different device for air sampling is the impinger, which impinges the air particles in a liquid medium. The solution can directly be used as an inoculum for liquid or solid medium to isolate microorganisms. This type of sampling requires a minimum duration of air exposure and is suitable for a site that has low microbial density. Another method for the isolation of airborne microorganisms is the settle plate method (exposure plate technique), where the agar plates are directly exposed to air by removing the lid of the plates for a specific duration. This allows air microorganisms to settle on the plates for cultivation. This method does not require any external force for the settlement of microorganisms, and it is also called passive sampling (detailed in Chapter 18).

Sample collection from extreme environments may require an enrichment setup, where a sampling device is exposed to the sampling site for hours to days to acclimate and augment microorganisms inside the container. For example, a past research investigation processed the sample of a hot spring in an enrichment setup in Hungate tubes for six days before it was transferred to the laboratory at ambient temperature for the culturing of a thermophilic archaeon, *Thermococcus*.

In a different method, the agar medium can be directly exposed to the study surface. This type of culturing method is free from any sample collection and can be of three different types: contact plate, dip-slip, and Petrifilm (Table 22..1). The contact plate method relies on a convex agar medium, which can be directly pressed onto a selected flat surface to gather microbial cells. Dip-slips are plastic paddles with two separate sides coated with agar media. Two faces of a paddle allow the study of the same sample on two different media or two different samples on the same media. Unlike contact plates, dip slips can be used for uneven surfaces. Petrifilm is a paper with a film of agar medium on it. This preparation is an easy-to-use, sustainable approach that offers great flexibility for attaching to the topography of a surface. Compared to conventional plating techniques, these methods directly procure a study surface, consume relatively less time, and offer selective recoveries of microbial cells. These techniques have applications specifically for the detection, isolation, and estimation of microbial load from surface environments.

22.2 CULTURE MEDIA

The culture media (or growth media) are central to the culturing of microorganisms. They are composed of a mixture of different substances that promote the rapid growth of microorganisms. However, a single medium cannot support the growth of all types of microorganisms. The growth media for bacteria, archaea, fungi, microalgae, and protozoa are distinct and tailored to meet their growth requirements. Different taxa from the same group can also have distinct requirements, and the compositions of their growth media are different. Culture media used for different purposes also contain different constituents (Figure 22.2). Even the same organisms grown for different purposes require different culture media. Media composition, including the types of ingredients and their concentrations, exerts a substantial influence on cultivation efficiencies. Most general-purpose culture media are formulated in such compositions that they tend to lean towards the growth of non-fastidious microorganisms. However, such media leave little or no space for the growth of slow-growing or unique microorganisms.

Of different microorganisms, the cultivation of bacteria is of prime interest. However, synthetic media have procured not more than 2% of the total bacterial diversity on Earth. The usage of artificial media presents several constraints for microbial cultivation. Synthetic chemicals often fail to meet the needs of natural environments. Therefore, many microorganisms cannot adjust to immediate variations in artificial environments. In some instances, the absence of spatial organisation of natural habitat in culture media can lead to abnormalities in microbial development. This is mainly due to the poor knowledge of specific growth conditions. For these reasons, numerous bacterial forms cannot be cultivated.

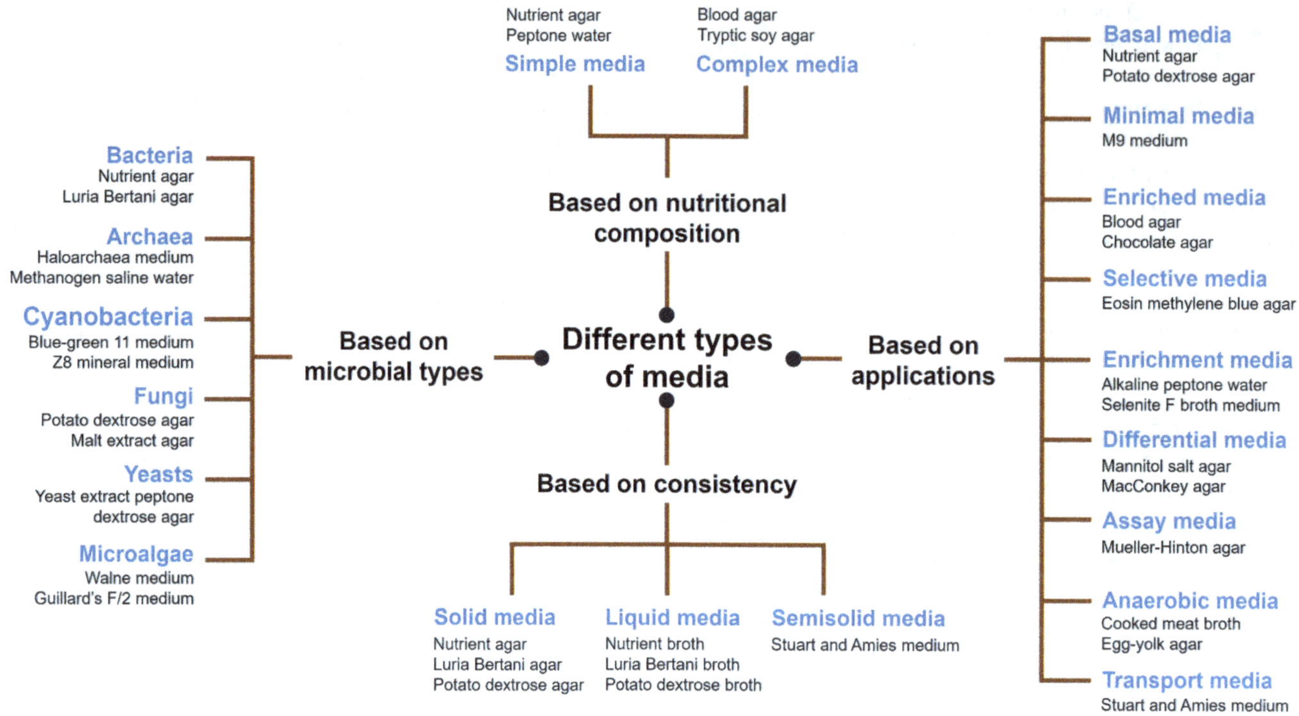

FIGURE 22.2 Different types of media based on microbial types, nutritional composition, consistency, and applications.

In recent years, the integration of modern technology with traditional cultivation methods has led to the expansion of new culturing methods for previously uncultured microorganisms. This growing interest in microbial cultivation incorporating new techniques constitutes a distinct field called **culturomics** (Box 22.1). One of the primary goals of culturomics is the streamlined preparation of media. A huge number of culture media are designed for the growth of many microorganisms previously considered uncultured.

In light of this piece of information on culture media, the Leibniz-Institute DSMZ-German Collection of Microorganisms and Cell Cultures has developed a comprehensive database called Media*Dive* to store information on different media used for the cultivation of microorganisms from all three domains of life (www.mediadive.dsmz.de/). This web-based resource contains information on more than 3,300 standardised cultivation media for the cultivation of over 46,000 microbial strains.

22.2.1 Formulations

A growth medium typically contains carbon, nitrogen, growth factors, trace elements, and buffers. The former two ingredients are major components, and other substances are added in small quantities as minor components. A medium can additionally contain substances like growth inhibitors, enzymes, and nutrients for growing specific microorganisms. The use of an anti-foaming agent prevents the formation of foam in the medium. Water is a universal solvent and essential for the preparation of any medium. A sufficient volume of water is required to dissolve all the components. Laboratory culturing practices generally use distilled water due to the requirements of strictly controlled parameters. On the contrary, most industrial fermentation setups utilise tap water, which is not only cost-effective, but the natural minerals present in tap water are beneficial for microbial growth. Table 22.2 provides information on different ingredients of culture media for the cultivation of bacteria and fungi.

All types of biological mass have a carbon backbone, and carbon is required in sufficient quantities for growth and development. There is a plethora of organic matter sources for media preparation that provide carbon for the cultivation of microorganisms. Simple and complex carbohydrates are the most commonly utilised carbon sources in growth media. Examples of simple carbohydrates for fermentation media are glucose, fructose, sucrose, lactose, glycerol, mannitol, sorbitol, trehalose, semi-purified polysaccharides, and sugar alcohol. Some by-products of industries such as blackstrap molasses and beet molasses (from the sugar industry), corn molasses, sulphite waste liquor (from the pulp and paper industry), and cannery waste (from bulk food processors) are commercial sources of simple carbohydrates. Examples of complex carbohydrates are starch, cellulose, and xylan. Their main sources are potatoes, corn, rice, and maize. They can also be derived from agricultural residues such as straws, wood waste, fruit pulp, corn cobs, etc. However, the carbon sources are not limited only to carbohydrates. Many hydrocarbons and vegetable oils can also be used as carbon sources in the fermentation medium. The degree of unsaturation in hydrocarbons has great nutritional value. The types and concentrations of carbon added to the media have a significant impact on the production of secondary metabolites.

TABLE 22.2
Some common ingredients in media composition for the cultivation of bacteria and fungi

Key sources	Ingredients
Carbon	Simple carbohydrates: Glucose, fructose, sucrose, lactose, glycerol, mannitol, sorbitol, and trehalose Complex carbohydrates: Starch, cellulose, and xylan Industrial wastes: Blackstrap molasses and beet molasses, corn molasses, sulphite waste liquor, and cannery waste Agricultural residues: Straws, wood waste, fruit pulp, and corn cob Many hydrocarbons and vegetable oils
Nitrogen	Inorganic substances: Ammonium chloride (NH_4Cl), ammonium sulphate $(NH_4)_2SO_4$, potassium nitrate (KNO_3), and ammonia (NH_3) Organic compounds: Casein, peptones, protein hydrolysate, yeast extract, beef extract, and urea Plant sources: Cereal grains and soybean meal Industrial wastes: corn-steep liquor, soybean meal, wheat bran, rice hull, corn cob, and cottonseed flour
Other nutrients	Phosphorus (in the form of KH_2PO_4 and K_2HPO_4), magnesium (Mg), calcium (Ca), potassium (K), iron (Fe), and manganese (Mn)
Vitamins	B1, B2, B7, B12, and pyridoxine (through yeast extract)
Precursors	Varies for different products
Antimicrobial compounds	Cycloheximide (for bacterial culture) and antibiotics like gentamicin, chloramphenicol, ciprofloxacin, and streptomycin (for fungal culture)
Antifoaming agent	Castor oil, cetostearyl alcohol, ether, and glycols
Buffer	K_2HPO_4 and KH_2PO_4, amino acids, peptides, proteins, and calcium carbonate ($CaCO_3$)

Microorganisms have numerous strategies to utilise organic matter for their metabolism. Glucose is a common ingredient in nutritional media. Glucose concentration is critical for the synthesis of different secondary metabolites. However, the same can also act as an inhibitory agent, as seen in the production of β-lactam antibiotic cephalosporin C by *Acremonium chrysogenum*. In bacteria, there are two different schemes for the consumption of carbon: diauxie and co-utilisation. Diauxie necessitates the use of two different carbons, one after the other, whereas co-utilisation permits the simultaneous use of both carbons without regard to preference. In contrast, autotrophic microorganisms, such as cyanobacteria, chemosynthetic bacteria, and microalgae, do not depend on organic matter for energy and carbon needs. Instead, they utilise carbon dioxide (CO_2) from the atmosphere and absorb sunlight to prepare their organic matter.

Nitrogen in media is essentially required for its assimilation in nucleic acids, proteins, and other structural and functional components. Several secondary metabolites produced by fungi and bacteria depend on nitrogen requirements. The nitrogen demand for the growth of microalgae is relatively higher than that of any other microorganisms. Certain bacteria require nitrogen in the form of nitrate (NO_3^-) and nitrite (NO_2^-) as terminal electron acceptors during anaerobic respiration. Nitrogen in the culture medium can be added in the form of inorganic or organic compounds. Ammonium chloride (NH_4Cl), ammonium sulphate $(NH_4)_2SO_4$, potassium nitrate (KNO_3), and ammonia (NH_3) can be among the inorganic nitrogen compounds present in the media. The organic sources of nitrogen in media are casein, peptones, yeast extract, beef extract, and urea. Protein hydrolysate is another common nitrogenous ingredient of microbial culture media, which is obtained by hydrolysing the complex protein using enzymes like trypsin and pancreatic juices. Plant raw materials, such as cereal grains and soybean meal, are notable substitutes for animal-derived nitrogen sources for the media. The residues of agroindustrial wastes such as corn-steep liquor, soybean meal, wheat bran, rice hull, corn cob, and cottonseed flour are widely used as cheap sources of nitrogen in the commercial preparation of media. Ammonia can serve as both a nitrogen source and a pH buffer for microbial cultivation. However, if it is present in slightly high concentrations, it hinders growth and impedes secondary metabolite production. In addition, NH_3 can be the main energy source for some special microorganisms, such as ammonia-oxidising bacteria. An amino acid, proline, is a valuable ingredient of culture media, which offers diverse benefits to the cultivars. It aids both carbon and nitrogen in the medium, acts as an osmoprotectant, and stimulates secondary metabolite production. The slow assimilation of proline promotes the productivity of antibiotics or bioactive compounds by delaying their growth process.

Other nutrients such as phosphorus (P), magnesium (Mg), calcium (Ca), iron (Fe), potassium (K), and manganese (Mn) are also fundamental to structural components and enzymatic action of microorganisms. Particularly, phosphorus is essential for the synthesis of nucleic acids, which is augmented in the form of mineral salts of phosphates (e.g., potassium dihydrogen phosphate (KH_2PO_4) and dipotassium hydrogen phosphate (K_2HPO_4)). Vitamins are growth factors and are required in low concentrations in the medium. They act as coenzymes for microbial activities. Yeast extract is a key source of many vitamins in culture media. Also, many microorganisms possess the ability to synthesise vitamins like thiamine (B1), riboflavin (B2), biotin (B7), vitamin B12, and pyridoxine to meet their requirement. Some bacteria desire nucleic acid bases such as purines and pyrimidines from external sources. Certain bioreactions necessitate precursors in the media to bring about any desired change and improve the yield of the products. For example, cobalt (Co) acts as a precursor for the production of vitamin B12 by bacteria like *Pseudomonas denitrificans* and *Propionibacterium shermanii.* A fungus, *Penicillium chrysogenum,* utilises phenylacetic acid as a precursor for the synthesis of penicillin G, while the presence of phenoxyacetic acid in the medium induces penicillin V production. Bacterial media may additionally consist of an antifungal compound (e.g., cycloheximide), and fungal media include antibiotics like gentamicin, chloramphenicol, or streptomycin to prevent the growth of undesirable microorganisms. In industrial setups, chemical additives like castor oil, cetostearyl alcohol, ether, and glycols are used as antifoaming substances to control foam formation during fermentation.

The pH in the media can be adjusted using an acid, hydrochloric acid (HCl) and a base, sodium hydroxide (NaOH). pH not only influences the growth of microorganisms but also has a strong impact on physical forms and the metabolic potential of the isolates. Most bacteria are cultured at pH 6.5-7.0, and fungi, at around pH 4.5. A buffer can be added to the media to maintain its pH during fermentation. It offers a stable environment against variation in the basicity and acidity of the medium while maintaining the same concentration of chemical constituents. This maintains the metabolic responses while protecting the living cells. Chemicals like K_2HPO_4 and KH_2PO_4 are the most commonly used buffers in laboratory preparations. The amino acids, peptides, and proteins in the media also often meet the demand for buffers. One common buffer found in commercial fermentation media is calcium carbonate ($CaCO_3$).

22.2.2 Media for Different Microbial Groups

Depending upon the microbial types, the composition of the media can be different. Bacteria in the laboratory are cultured using media like nutrient agar (NA), Luria Bertani agar (LBA), peptone water, and tryptic soy agar (TSA). They are most suitable for non-fastidious growth forms. Reasoner's 2A agar (R2A agar) is a low-nutrient medium recommended for culturing slow-growing bacteria from potable water samples. The isolation of specific microorganisms may require some semi-selective media. For example, the rhizobia from root nodules are grown on yeast extract mannitol agar (YMA). Marine Agar 2216 E contains a high concentration of salts and enumerates marine bacteria. Blood and its derivatives in the medium boost the growth of fastidious bacteria like *Streptococcus pneumoniae*, *Listeria monocytogenes*, and *Clostridium difficile.* Some notable examples of blood-containing media are brain heart infusion broth, sheep blood agar, and chocolate agar.

Fungi can also be cultured on a variety of media. Potato dextrose agar (PDA) and Czapek's agar (CZA) are the general-purpose media to support the growth of non-fastidious fungi. Yeast Extract Peptone Dextrose Agar (YPD) gives luxurious growth of yeast cells. The presence of peptone and yeast extract in YPD particularly stimulates the growth of yeast. Malt Extract Agar (MEA) is chiefly used to promote the sporulation of moulds. Sabouraud Dextrose Agar (SDA) is especially used to cultivate dermatophytes that infect humans' and animals' skin, hair, and nails. The key ingredient of SDA is the enzymatic digests of casein and animal tissues. Brain-heart infusion (BHI) agar containing defibrinated sheep blood recovers fastidious pathogenic fungi such as *Histoplasma capsulatum*, *Coccidioides immitis*, and *Blastomyces dermatitidis*. Some culture media used for fungal and bacterial cultures have common ingredients. Therefore, both fungi and bacteria witness the growth, especially when the environmental samples are inoculated onto such growth media. In this situation, the addition of antibiotics or antifungal substances in the medium suppresses the growth of unwanted microorganisms.

On the contrary, the media composition for the cultivation of archaea is very distinct. However, the general-purpose media is difficult to formulate for the archaea due to their diverse metabolic needs. The dietary needs of archaea have tremendous variations due to their unique metabolic requirements and diverse ecological settings. Furthermore, the cultivation of certain archaeal strains frequently necessitates the adjustment of culture parameters, such as pH, temperature, and nutrient content. The information on culturing techniques and the composition of culture media for archaea is very limited. Saline or enhanced marine media are ideal for the proper cultivation of halophilic archaea. Methanogens are cultivated using components such as acetate, hydrogen, and methanol in Hungate anaerobic culture tubes to maintain the anaerobic environment. Sulphur-reducing archaea like *Sulfolobus* can be cultured on sulphur-reducing media supplemented with elemental sulphur or thiosulfate.

The culture media for cyanobacteria and microalgae typically consist of inorganic salts. Some basic ingredients of those media are nitrogen, phosphorus, Ca, K, Mg, micronutrients, and vitamins. Nitrogen and phosphorus in the media are required in bulk quantities. They are photoautotrophs, and the appropriate light and CO_2 are essential for photosynthesis. Blue-green 11 (BG-11) is a versatile medium used for the mass cultivation of especially freshwater cyanobacteria (e.g., *Nostoc*, *Synechocystis*, *Synechococcus elongatus*, etc.). This can also be used for the cultivation of many microalgae, such as *Chlorella, Scenedesmus, Botryococcus,* and *Chlorococcum*, for lipids and biomass production. Z8 mineral medium is a different artificial medium used to grow freshwater cyanobacteria. MA medium typically harbours marine and brackish water cyanobacteria. Nitrogen-fixing cyanobacteria are commonly grown in an ASN-III medium. Walne medium and Guillard's F/2 medium are classic culture media for microalgae cultivation.

Most free-living protozoans, like *Paramecium* and other ciliates, feed on bacteria. An *in vitro* cultivation of protozoans mainly grows on special hay infusion media. Hay infusion consists of a handful of dried grass soaked in pond water. This medium yields an enrichment of bacteria followed by a heterotrophic succession of protozoans. However, the culture of protozoa on synthetic media is not meaningful because of their limited applications in biotechnology.

22.2.3 Nutritional Components

Based on the specific components, culture media can be simple or complex. Simple media consist of one or a few simple natural substances. The composition of simple media is usually known. Examples of those include NA and peptone water. Complex media contain a variety of substances with their undefined assortments and concentrations. Most of the complex substances are nutrient-enriched and are easy to prepare. Animal or plant tissues are important supplements to the media. They are blood agar, chocolate agar, and TSA.

A different category of media based on nutritional components can be synthetic or crude (nonsynthetic) media. Synthetic media are chemically defined media that include pure or partially pure substances with precise quantities. This kind of preparation is primarily used in research studies. The defined compositions of synthetic media allow the culturing of microorganisms for specific purposes. Examples of synthetic media are NA, PDA, CDA, mineral glucose medium, and Raulin medium. On the other hand, crude media are composed of natural substances. The chemical constituents in the medium are not precisely defined. They are mainly used for fermentation at industrial scales. Crude media are economical and profitable for product formation. They contain sufficient nutrients, growth factors, vitamins, and precursors for the growth of microorganisms. A byproduct of the sugar industry, molasses is a popular renewable raw material used in crude media for the production of yeast, ethanol, and citric acids.

22.2.4 Consistency of Media

The culture media having the same ingredients can exist in three different consistencies, including liquid, solid, and semisolid. This depends on the use of gelling agents. A liquid medium is an aqueous solution of different substances and is devoid of any gelling agent. They are also known as broth media. The entire microbial industrial enterprise depends on liquid media for mass cultivation of microbial cells, fermentation, and various biochemical tests. The free movement of cells in the solution allows them to utilise maximum nutrients from the media. The growth of aerobic microorganisms in a liquid medium requires proper agitation of the solution for sufficient mixing of different components. Examples of liquid media are nutrient broth, Luria Bertani broth (LBB), and potato dextrose broth (PDB).

When a liquid medium includes a solidifying agent, it develops a firm, transparent substratum called a solid medium. Solid media are especially used to culture bacteria and fungi in their physical forms. Examples of solid media include NA, PDA, CDA, and TSA. The solidifying agents are chemically inert substances, which means they do not react with any ingredients of the medium. Gelatin was used as an original solidifying agent in the early days. Years later, it was replaced by a much more stable substance, agar. Agar continues to be a common gelling agent in industries and laboratories. It is composed of long-chain polysaccharides obtained from red algae. Around 1.5-2.0% of agar is sufficient to convert any liquid media into translucent solid media (Table 22.3). Agar is a thermoreversible substance that melts at around 85 °C and solidifies below 42 °C. It is commonly used to prepare agar plates (in Petri plates), agar slants (in culture tubes), and agar deeps (in culture tubes). Other natural substances, such as xanthan gum, gellan gum, carrageenan, isabgol, and guar gum, are alternatives to gelling agents. Gellan gum is produced by a bacterium, *Sphingomonas elodea,* which can be used in place of agar for the cultivation of thermophiles and hyperthermophiles. Gellan gum has the property to withstand thickness even at 120 °C. The common use of solid media is the isolation of cultures in pure forms. This can also be utilised for the qualitative tests in cultivars. However, solid media are unprofitable for industrial applications because microorganisms on solid media grow only on the top surface, and a major underlying portion remains underutilised.

TABLE 22.3
The composition of M9 mineral medium and nutrient agar (NA)

M9 mineral medium	
Constituents	**Concentrations (g/l)**
$MgSO_4$	0.12
$CaCl_2$	0.011
Thiamine-HCl x 2 H_2O	0.373
Glucose	2
Nutrient agar*	
Constituents	**Concentrations (g/l)**
Peptone	5
Meat extract	3
Agar	15

Note: As per the record of MediaDive.
* 0.5% NaCl can be added to maintain the osmotic balance.

Semi-solid media are prepared by adding a low agar concentration (0.2-0.5%) to the liquid media. This type of media has limited applications. A typical use of a semi-solid medium is a bacterial motility test. Semisolid media can also be suitably used for the detection of a pathogenic bacterium, *Salmonella typhi*. Tubes containing semisolid media support the growth of microaerophilic bacteria. A semi-solid Stuart and Amies medium is recommended for the transport of clinical swab specimens.

22.2.5 Various Applications

Microbial media can be classified into different groups based on their various applications. The augmentation of a new element or the removal of an ingredient in the original recipe of a medium can modify its initial intended scope. Depending on specific constituents, different media can favour or inhibit the growth of microorganisms. While some media can be used for general purposes or industrial use, others discriminate among various taxa. One type of medium can offer rapid growth, and another can hamper the growth of microorganisms for the improved secretion of specific metabolites. These categories are covered under the following headings.

- Basal Media
  They are general-purpose media that are used for the identification, cultivation, and enumeration of bacteria and fungi. The ingredients mimic a general natural habitat that allows the growth of diverse microorganisms. Examples of basal media for bacteria are NA and LBA, and for fungi, are PDA, CDA, and Sabouraud dextrose agar (SDA).
- Minimal Media
  Minimal media are chemically defined composites that possess minimal nutrients for the possible growth of bacteria and yeasts. The main supplements of the media are low concentrations of inorganic salts and a simple low-energy carbon source such as succinate, glycerol, or glucose. Complex ingredients like proteins and complex carbohydrates are absent. This type of media is used for the cultivation of wild types of bacteria, selection of recombinants or exconjugants, and characterisation of auxotrophic mutants (inability to synthesise certain metabolites). One common example of a minimal medium is the M9 mineral medium (Table 22.3).
- Enriched Media
  Enriched media are highly nutritious media used for the cultivation of fastidious and nutritionally demanding microorganisms. Added nutrients of the enriched media generally include blood, serum, or growth factors. They are agar-based media, which are usually employed in the detection of pathogenic isolates. Examples of those include blood agar and chocolate agar. Blood agar typically utilises sheep's blood and is used in clinical microbiology. Chocolate agar also constitutes blood, but in the lysed form, which gives the medium a chocolaty appearance.
- Selective Media
  This type of media supports the growth of only selected microorganisms because of the presence of an inhibitory agent, which prevents the growth of other microorganisms. Inhibitory agents can be components like heavy metals, antibiotics, salts, dyes, phages, etc. In the same way, a medium with low pH will propagate acidophiles, inhibiting the growth of neutrophiles and alkaliphiles. The general-purpose bacterial media augmented with bile salts or some dyes suppress the Gram-positive bacteria. At the same time, the addition of antibiotics like nalidixic acid or colistin inhibits Gram-negative bacteria and allows for the growth of Gram-positive bacteria without competition. Bacitracin added to chocolate agar takes the form of a selective medium that isolates *Haemophilus*. The augmentation of heavy metal in the medium ensures the selective growth of heavy metal-resistant microorganisms.
- Enrichment Media
  Enrichment media are specifically liquid composites, which provide an environment favourable to the growth of specific microorganisms over others. This type of preparation mainly requires the supply of one or a few nutrients in high quantity to support the excess growth of specific microorganisms, which is contrary to the selective media that rely on an inhibitory agent to suppress unrequired microorganisms. Media having an augmented substance promotes the overgrowth of special microorganisms, causing discrimination against competitors. For example, the excess use of starch in media can harbour starch-utilising microorganisms. Synthetic substances like $(\text{d-Phe})_4$ and d-Ala-NH_2 promote the growth of *Bacillus cereus* DF4-B and *Ochrobactrum anthropi* C1-38, respectively. A different purpose of using enrichment media is to achieve the gradual adaptation of microbial species to fit in a modified environment.
- Differential Media
  Differential media are solid preparations that provide a visual distinction between different microbial taxa. The disparity is based on dissimilar biochemical reactions using the same substrates and reagents by different microbial strains. An example of a differential medium can be blood agar plates, where bacteria can develop three different types of impressions based on the haemolysis of blood. α-haemolysis yields partial haemolysis, producing a green colour in the medium. β-haemolysis is a hallmark of complete lysis, leaving a clear transparent zone. In γ-haemolysis, bacteria do not break down the blood cells, and the colour of the plates remains unchanged. Egg yolk agar discriminates between *Clostridium* spp. and other obligate anaerobes. The lipase-producing ability in bacteria is detected through a halo zone formation in tributyrin agar (using tributyrin oil).
  Some differential media exhibit the properties of those selective media as well. For example, mannitol salt agar (MSA) selectively promotes the growth of *Staphylococcus* and a few other halophilic Gram-positive bacteria. However, the high concentration of salt in the medium is detrimental to other bacteria. The same medium also allows the possibility of distinguishing *S. aureus* from other staphylococci. When *S. aureus* is grown in MSA, it ferments mannitol in the medium and produces acids, which turn the red colour of the medium into yellow due to the presence of an acid indicator (phenol red). Some species, on the other hand, do not ferment mannitol and retain the original shade of the medium. MacConkey agar is a selective medium that can be used to support the growth of only Gram-negative bacteria. It can also differentiate between lactose fermenting and lactose non-fermenting bacteria. Hektoen enteric agar selectively discriminates between

Salmonella and *Shigella* in faecal specimens. Another example of a differential-cum-selective medium is eosin methylene blue (EMB) agar, which supports the growth of Gram-negative bacteria and distinguishes different coliform bacteria on agar plates (detailed in Chapter 24).

- Assay Media
  This type of media can be a liquid or solid form and examines the specific traits of microorganisms. An assay medium can have a specific composition or a regular basal medium supplemented with a specific reagent. Mueller-Hinton agar medium is typically a specific assay medium used for the antibiotic susceptibility test through the disc diffusion method. In a different preparation, a vitamin B12 assay medium consists of all the necessary ingredients except vitamin B12 for the growth of a producer microorganism.
- Anaerobic Media
  Anaerobic media are used for the cultivation of strictly anaerobic bacteria and archaea. Most of them are freshly prepared. They consist of reducing agents such as Cysteine-HCl, thioglycolate, sodium-thioglycolate, and sodium dithionite to maintain a low redox potential. The oxygen-free environment in the medium is achieved using a flask sealed with gas-tight septa. The inoculation is performed by transferring the aliquot using a sterile syringe and needle to prevent air supply inside the vessel. The presence of redox indicators (e.g., resazurin) in the medium verifies the oxidation-reduction reaction inside the culture vessel. The common method of cultivation of anaerobic microorganisms is through enrichment methods, although some selective, nonselective, and differential media also exist. Examples of anaerobic broth media are cooked meat broth (CMB), peptone-yeast extract glucose (PYG) broth, and thioglycollate broth. Examples of some solid media for anaerobic microorganisms are egg-yolk agar (EYA) and anaerobic blood agar.
- Transport Media
  A transport medium is not a nutrient formulation compared to other media. This type of preparation supports the viability of existing cells during the transport of the samples away from their original environment. A transport medium (e.g., Stuart and Amies medium) is composed of a few salts and a buffer, of which salts help to regulate the osmotic balance, and the buffer avoids any change in the pH. The microbial cells are unable to proliferate in the medium due to the absence of any nutrients. At times, charcoal is added to the medium to ensure that nutrients and toxins present in the specimens are immobilised and do not affect the viability of microbial cells in the sample.

22.3 STERILISATION

In culturing practice, sterilisation is a process of eradicating any viable microbial cells or spores from the surface or object before inoculation. This is an essential step in microbial culturing to ensure a contamination-free environment. All equipment, glassware, media solutions, and chemicals must be sterilised before inoculation. However, the methods of sterilising different articles can be distinct. An autoclave is a quite common microbiology laboratory instrument for sterilising heat-insensitive solutions. This instrument is designed to operate at high temperatures and pressure to kill microbial cells and spores without affecting their volume and concentrations. It can also handle heat-resistant glassware and small equipment. In a different setting, glassware and powder can be sterilised in a hot air oven at very high temperatures. The heat-sensitive chemicals can be treated through filter sterilisation using filter syringes (a pore size of around 0.20 µm, suitable to remove all microbial cells). Heat-sensitive equipment can be sterilised using 70% ethanol and other chemicals. The glass spreader, forcep, blade, inoculating needle, and inoculating loop can be sterilised for instant use by soaking them in alcohol and heating them until they are red hot. A detailed overview of different sterilisation methods is given in Chapter 5. An important requirement in the culturing procedure is a laminar airflow chamber, which is mentioned in the following subheading.

22.3.1 Laminar Airflow Chamber

A Laminar airflow chamber is a partially enclosed cabinet that provides an aseptic environment for inoculation. The chamber is made up of stainless steel or plastic. It has a flat platform in the chamber to keep all the materials and carry out the experiments. The front side has a transparent sliding door called a **sash**. The sash can be adjusted for the opening to access the cabinet. The ceiling of the cabinet has two different types of lamps. One is a common illuminating system that allows workflow, while another is a germicidal UV lamp, which creates a germ-free environment. A UV lamp emits short wavelength UV-C rays, which interfere with the DNA base pairing of the microbial cells and cause structural and functional damage that eventually kills or inactivates them. A 10-15 min exposure to UV rays in the cabinet before the experiment is adequate to sterilise the surface and other objects inside the chamber. The critical workflow inside laminar airflow requires a bottle of ethanol, a source of open flame, and dry napkins to wipe the surface.

However, the most important part of the cabinet is a mechanical air blower attached to the rear side of the cabinet. The air blower is equipped with a mechanical air filter called a high-efficiency particulate air (HEPA) filter. The air blower passes outside air at a speed of about 0.3-0.5 m/s through the HEPA filter and discharges it inside the cabinet chamber. This unidirectional steady movement of filtered air sweeps out the particle contamination of the chamber and creates a germ-free environment. A HEPA filter is composed of a dense, haphazardly or well-organised layer of fine fibres, which form an intricate web of pathways to capture tiny particles and for air to travel through. The pore size in the filter is about 0.3 µm, which can remove airborne particles, including microbial cells, from the air with 99.99% efficiency.

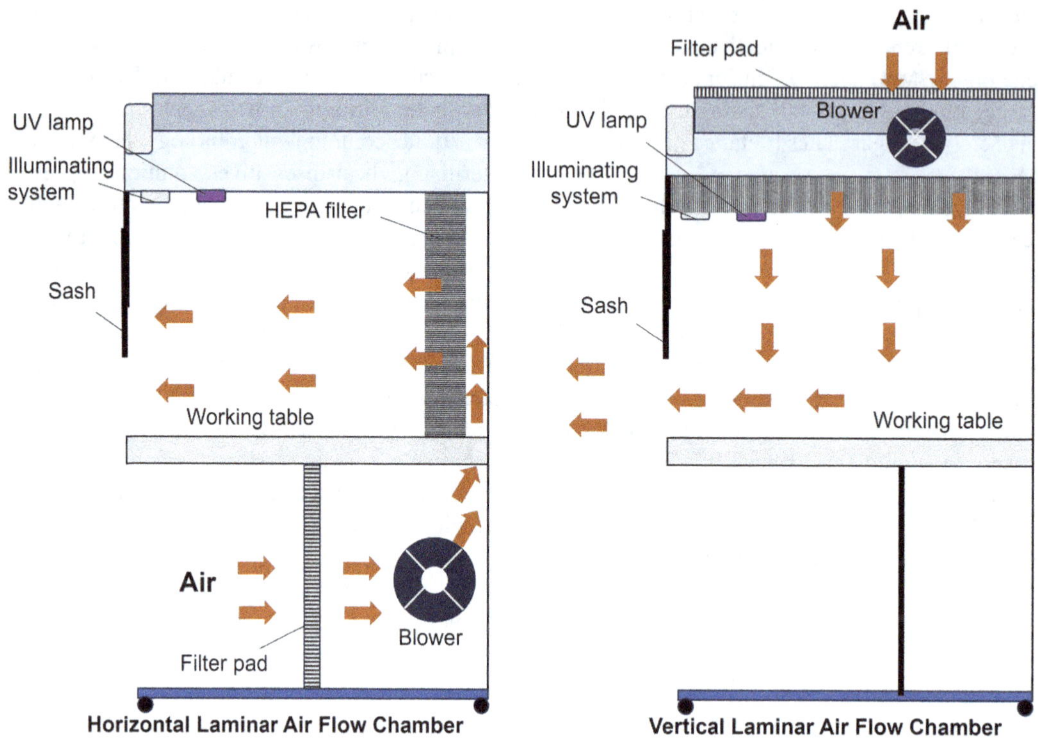

FIGURE 22.3 Diagrammatic side views of horizontal and vertical laminar airflow chambers and the movement of air. (HEPA: high-efficiency particulate air, UV: Ultraviolet.)

Based on the attachment sides of the air filter, the cabinet can be of two different types: horizontal and vertical cabinets. In the horizontal laminar flow chamber, an air filter is fitted at the back side of the sash, and air flows horizontally. In contrast, the vertical laminar flow chamber has an air filter fitted on the ceiling of the cabinet. The air flows in a vertical direction (Figure 22.3). These two different types of cabinets have different objectives. The vertical cabinet ensures the user's safety because downward movement and front access do not blow the air on the operator directly. This type of cabinet also has less cross-contamination. However, the continuous downward flow of air creates some turbulent effects, and many particles released inside the chamber may settle on the workbench.

22.4 SERIAL DILUTION

Serial dilution is a process of sample dilution to reduce the density of cells in both liquid and solid specimens. This technique was pioneered by Robert Koch in 1883. It is a very reliable method for the isolation of bacteria and fungi from an unknown specimen on a solid medium. The preparation employs a diluent in which the original inoculum is added through sequential resuspension. The diluent is usually saline water that is made using 0.85% salt (most often NaCl) in water. This use of salt in the diluent helps in maintaining the osmotic balance of microbial cells during dilution.

A tenfold dilution is widely used for microbial cultivation and is known as a log dilution, where the number of cells decreases by a factor of ten. When one part of the original stock is dissolved in nine parts of the diluent, it prepares a dilution of 10^{-1}. Further, one part of the solution from a 10^{-1} dilution transferred to nine parts of the diluent creates a dilution of 10^{-2}. Dilution at each step requires proper mixing for the even distribution of cells in the solution, which can be performed by vortexing or vigorous shaking. This step is repeated many times to reduce the concentration of microbial cells in the solution. This manifold dilution of the original stock prepares serial solutions of 10^{-1}, 10^{-2}, 10^{-3}, 10^{-4}, 10^{-5}, 10^{-6}, up to 10^{-n}. The formula for the dilution can be assembled as:

Dilution (D) = sample size/(sample size + volume of diluent)

$$D_n = D_1 \times D_2 \times \ldots\ldots \times D_n$$

Where D_n is the dilution of the nth dilution and D_1, D_2, and D_n are individual dilutions up to the nth series.

The dilution (D) is the reciprocal of the dilution factor (DF), which can be represented as:

$$D = 1/DF$$

This means if the dilution factor is 10, the dilution is 1/10.

Following that, a known volume from diluted samples is poured on a series of agar plates and then spread on the Petri plate using a spreader. These plates are incubated in the growth chamber for microbial cultivation. The volume size of the

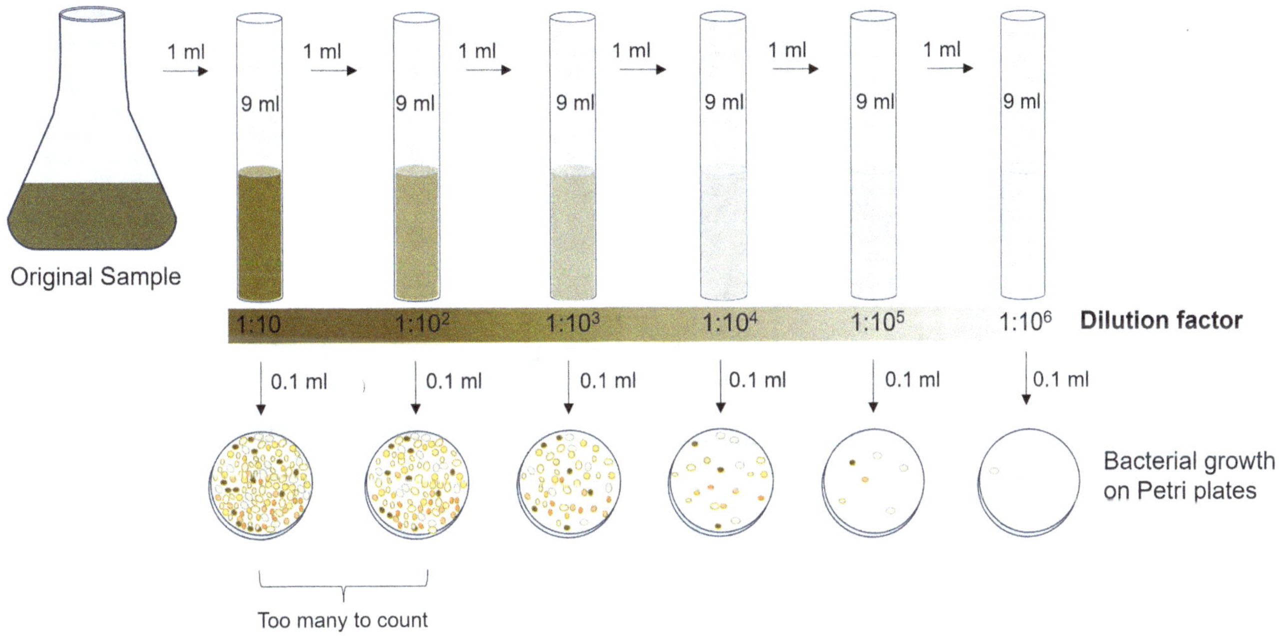

FIGURE 22.4 Serial dilution of a sample for the cultivation of bacteria, showing a high number of colonies to one colony on Petri plates, taking an aliquot from 1:10 to 1:10^6 dilution factors.

aliquot for spreading can be less than or exactly 1 ml. These plates produce growth after a certain period of incubation.

An unknown sample from a series of dilutions yields a series of culture plates having too many to count (TMTC) to an easily countable number of colonies (Figures 22.4 and 22.5). The colony count from the Petri plates of different dilutions provides an estimation of the total number of viable cells present in an unknown sample. A single microbial cell on an agar plate forms a single colony. This can be used to estimate the density of culturable microorganisms in the sample and is measured as colony-forming units (CFU)/ml and CFU/g for water and soil samples, respectively. The formula to determine the cell density (CD) in a sample can be framed as:

$$\text{CD (in CFU/ml)} = X_n/(D_n) \times v \text{ (in 1 ml of original stock solution)}$$

Where X_n denotes the number of colonies on a plate from the nth dilution,
D_n is the dilution from the nth dilution, and
v is the volume of an aliquot added to the plates.

For this exercise, plates that have a reasonable number of colonies to count are used for cell density calculation. The colony count of 30–300 is considered standard for CFU calculation. A Petri plate that represents fewer than thirty colonies is inclined to sampling error. However, when it is excessively loaded (i.e., more than 300 colonies per plate), it becomes difficult to reliably identify and count.

22.5 INOCULATION

A microbiological meaning of inoculation is an aseptic transfer of inoculum in a sterile culture medium for culturing. Inoculum can be any active entity that has a population of microorganisms or cells. The primary inoculum is derived from a natural environmental setting, while the secondary inoculum can be any pregrown culture that is used for subculturing. The microorganisms growing on a plate in mixed colonies can be purified using the streak plate method and their storage on a slant (Figure 22.6). Inoculum can also be transferred from solid culture to broths. The transfer of an aliquot from a liquid culture is usually performed through pipetting. However, microbial inoculation between two solid media requires an inoculating needle or loop. They are both hand-held portable devices that come in both reusable and disposable forms. A reusable inoculating needle is a straight metallic rod attached to a thin nichrome or platinum wire with a sharp, pointed edge at the terminal end. The pointed needle attempts to pick only a few cells from a single colony from a mixed culture plate to obtain a pure strain. However, an inoculating loop consists of a thin metal wire that is bent into a loop at the terminal end, which picks and transfers a relatively large amount of culture into a fresh medium (Figure 22.7). The disposable inoculating needles and loops are made up of plastic resins. Pre-sterilised toothpicks are an inexpensive substitute for single-use inoculation needles.

The plating technique is the technical term for the procedure of disseminating a cluster of microbial cells over the solid surface of the agar in the Petri plates in order to produce a macroscopic visible growth. The plating techniques come in many forms, including spread plating, streak plating, pour plating, and replica plating (Figure 22.8). They have different objectives in culturing practices and are mainly used for the ease of culturing. They are further discussed as follows.

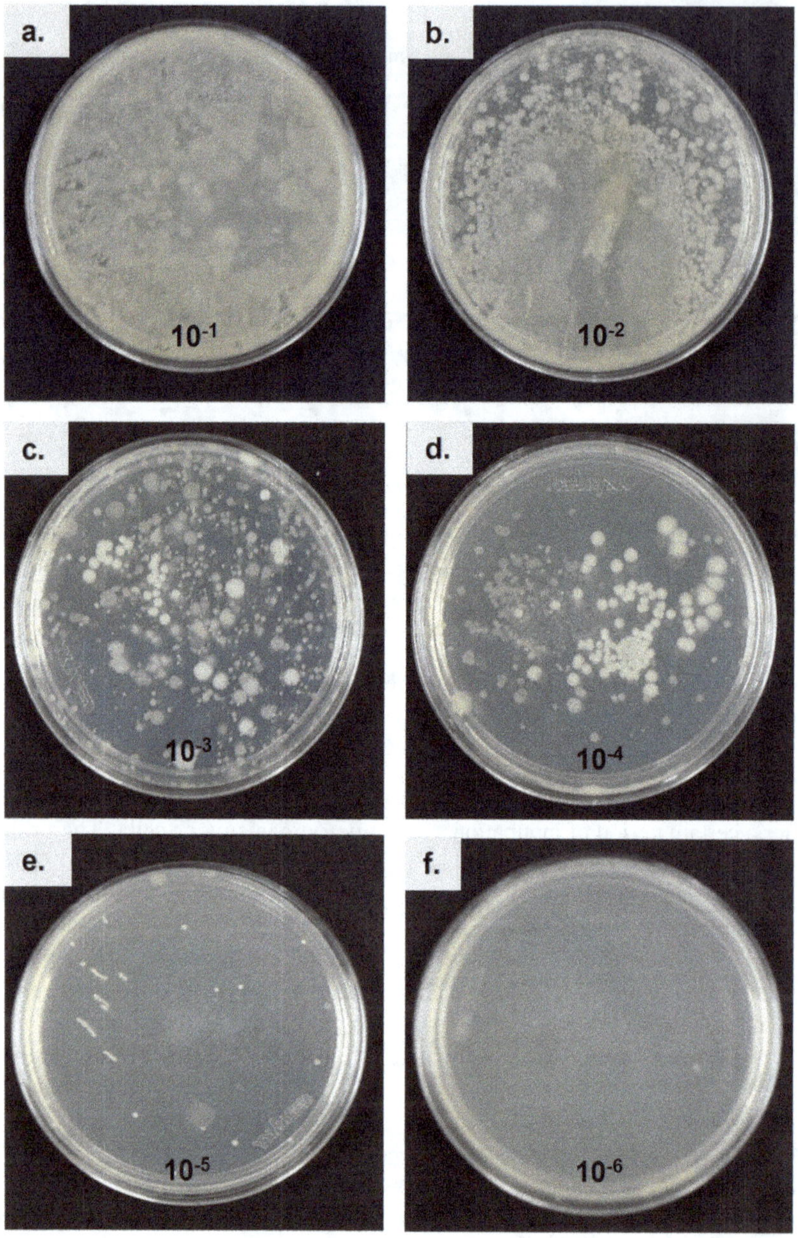

FIGURE 22.5 a-f. Petri plates showing mixed colonies of bacterial cultures isolated from an environmental site after serial dilution (from 10^{-1} to 10^{-6} dilution factors.)

- **Spread plating method** employs a known volume of aliquot (diluted inoculum) (usually 0.1 ml) that is poured onto the surface of a solid medium and spread using a sterile spreader. A spreader ensures a homogeneous distribution of microbial cells on the surface. The primary application of this type of plating is the screening of bacterial and fungal colonies from environmental samples (i.e., primary culturing). It can also be used to check the purity of an isolate and estimate the cell density in broth culture.
- **Streak plating method** is involved in secondary culturing, where the sample is streaked on an agar plate in a certain pattern across the surface of the plate using a loop or an inoculating needle. A needle or loop carrying a set of microbial clusters is moved from side to side on agar plates to generate a parallel stripe. This process is repeated with the initial inoculum moving over different quadrants on the same medium (Figure 22.6a). This method is used to obtain pure colonies from a mixed culture. The pure culture on the plates, after streaking, generates clones of the same colonies. This method and the one discussed before are not suitable for mass cultivation, as these approaches utilise only the surface part of the medium.
- **Pour plating** is a straightforward technique of culturing environmental bacteria on Petri plates, where the sterile molten cool media are mixed with inoculum and poured

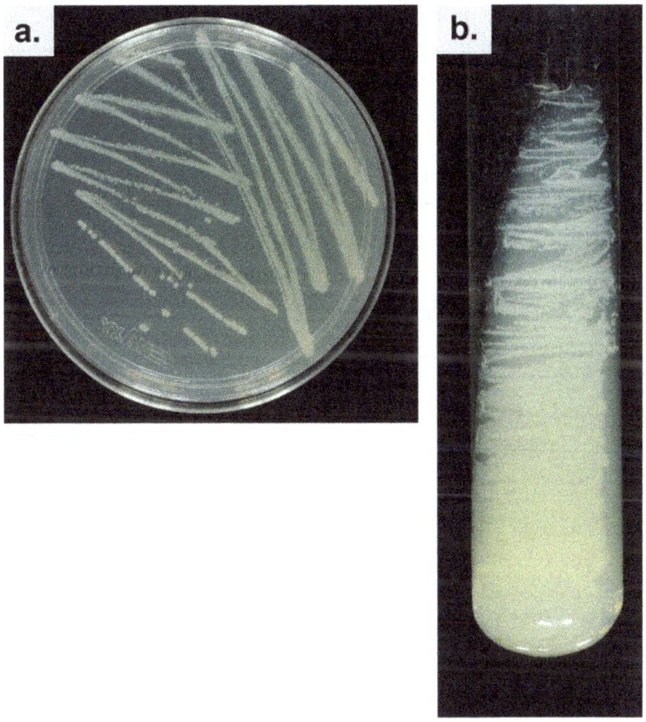

FIGURE 22.6 a. Image of pure bacterial colonies grown on a Petri plate through the streak plate method. b. Image showing an agar slant of a pure bacterial culture in a culture tube.

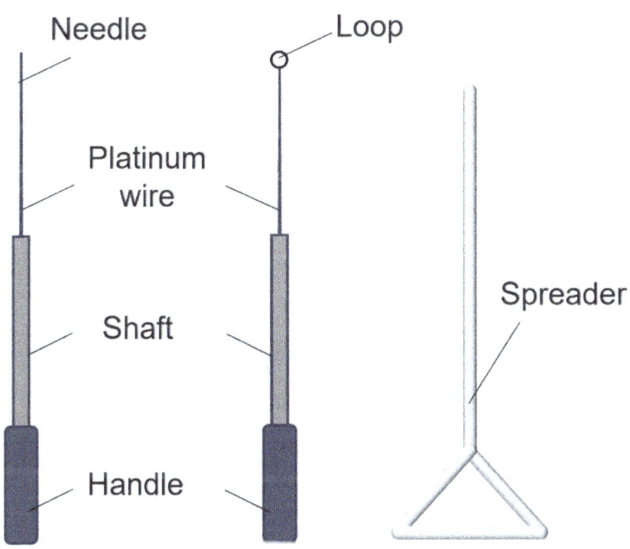

FIGURE 22.7 Diagrams of inoculating needle, inoculating loop, and spreader.

on the plates. As bacteria can grow on the surface as well as inside the medium, the volume of inoculum is larger than the spread plating method. This preparation utilises the full space of solidified media for their growth. Similar to the spread plate method, the pour plating method can be suitably used to calculate the CFU. However, this method does not fit for bulk production due to difficulties in the recovery of cultivars.

- **Replica plating** is a technique of imprinting the original spatial pattern of colonies from master Petri plates (i.e., primary plate) onto target plates (secondary plate). This requires a replica tool, which is generally a flat surface of wood or metal covered with a velvet pad or a cotton swab at the base. When a sterilised replica tool is pressed on a master plate, it duplicates the colony pattern. This velvet pad is pressed onto the surface of the target plate to receive the impression of each colony as they appear on the master plate. This technique allows the screening of a large number of colonies for a particular trait (such as antibiotic resistance and heavy metal resistance) in a single exercise. This type of plating can also be used in developing replicas of the original cultures and for the screening of auxotrophs.

22.6 INCUBATION

Incubation is a period of uniform conditions for the multiplication of microbial cells to form visible colonies or produce measurable growth for analysis. Different microorganisms require different incubation periods for their propagation. Primary inoculation from an environmental sample takes longer to produce colonies than secondary inoculation from an established culture. Bacteria that belong to rarely isolated groups take a long time to appear on Petri plates. An incubation period requires controlled environmental conditions (such as temperature, humidity, light, and aeration) to achieve optimal growth. These conditions are satisfied by an instrument called an **incubator**. It is an insulated, double-walled chamber that maintains a desirable physical environment for the rapid multiplication of microorganisms. An incubator must be regularly cleaned and sanitised to maintain a germ-free environment. The sophisticated incubator can be equipped with a dry heat or moist heat sterilisation unit to prevent contamination from microbial culturing. The modern design also comprises an in-chamber HEPA filter for contamination control during operation. Based on various requirements, incubators come in various shapes and specifications as follows.

22.6.1 Biochemical Oxygen Demand (BOD) Incubator

The BOD incubator is the most commonly used laboratory incubator for microbial cultivation. It was initially devised to measure the levels of organic matter and nitrogen in wastewater, but soon became a vital instrument for microbial cultivation. The chamber is a double-walled structure, where the outer wall is usually made up of thick mild steel and the inner wall is stainless steel. In between the outer and inner walls, the high-pressure in-place insulator prevents heat loss. The opening of the chamber consists of two doors: the outer one is a mechanical door, and the inner one is a glass door for observation. The interior of the chamber is occupied with light for illumination and perforated racks and trays to keep the materials. An axial blower fan provides a constant supply of air to maintain the circulation of heat inside the body. A thermometer is positioned on the interior wall to monitor and control the temperature of the chamber. An inbuilt condenser provides low temperatures. The system can run as low as 5 °C. A heater assists in heating the chamber. The temperature inside the incubator is maintained by the principle of a thermostat, where a device senses the temperature in its physical environment and intends to maintain it by turning on and off the heating/cooling system. A microprocessor-based circuit provides accurate, controlled and uniform temperature inside the chamber. All specific parameters of the incubator can be managed by a control panel present outside of the chamber, which consists of a power indicator, temperature regulator, heating switch, and main switch (Figure 22.9).

22.6.2 Bacteriological Incubator

The bacteriological incubator is a relatively simple instrument, mainly used for the growth of bacteria and yeasts above room temperature. Unlike a BOD incubator, this instrument lacks any cooling device. The main gadget is a U-shaped heater, which plays a role in increasing the temperature. The operation principle is based on the thermostat. Other features of the incubator are similar to a BOD incubator.

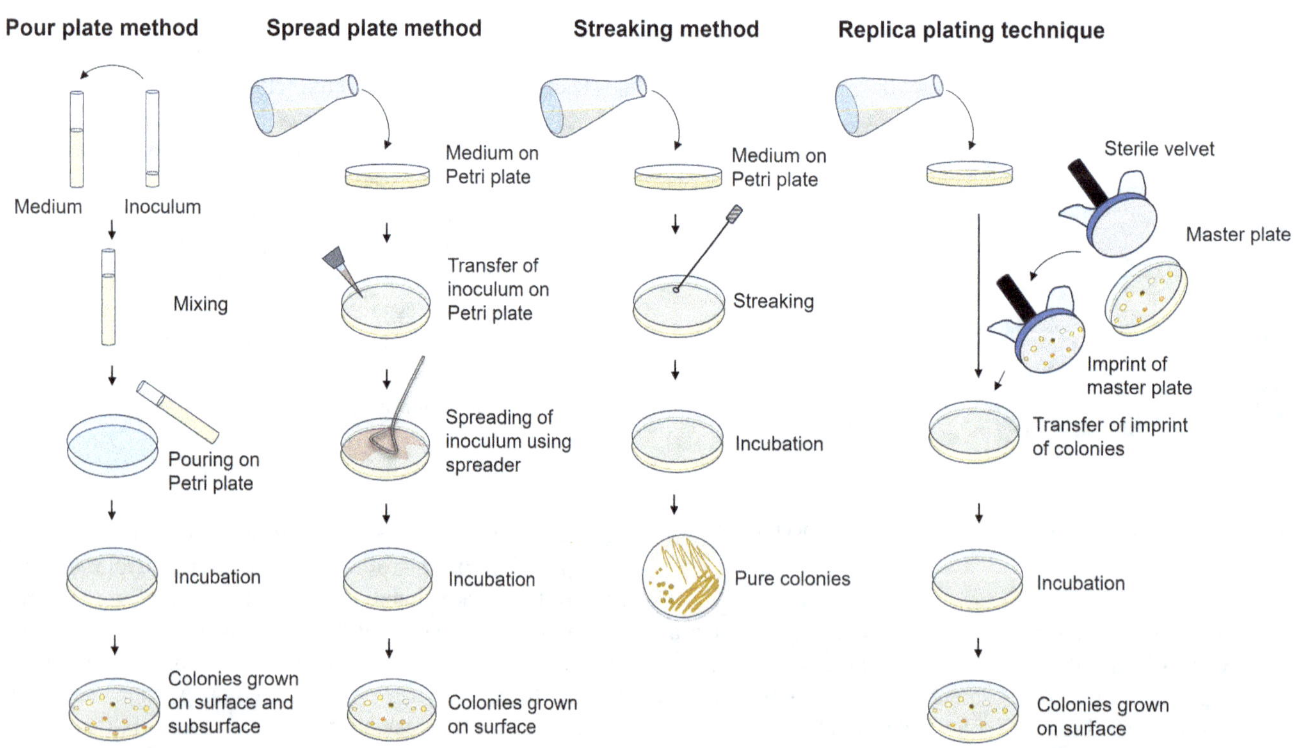

FIGURE 22.8 Different plating techniques for microbial cultivation.

22.6.3 Shaker Incubator

The Shaker incubator provides a facility for mixing the solution under a controlled environment that is essential for the growth of aerobic microorganisms in broth media. The instrument is equipped with a horizontally oscillating tray, which is driven by a motor. The shaking speed range of the tray can be 20-300 rotations per minute (rpm). The tray consists of clamps/clips to provide firm support for different-sized assorted vials/tubes/flasks during shaking. This setup can be implemented in a BOD incubator or exist as a separate variant of a BOD incubator. Shaking is primarily used to provide a steady oxygen supply in the medium.

22.6.4 CO_2 Incubator

This instrument is used for mammalian cell cultivation and includes the panel to control CO_2, along with other necessary parameters of an incubator. The 5-10% CO_2 concentrations inside the chamber mimic the natural condition of human blood and maintain a pH of around 7.0, which provides stable growth environments. This range of CO_2 concentration also helps maintain the required O_2 level in the cell. The chamber can even regulate the variable O_2 levels, simulating hypoxia (1-21%) to hyperoxia (5-90%) conditions for various research studies.

22.7 PURIFICATION

Purification is a step of obtaining a pure culture that is only one type of cell from a mixed culture of two or more colonies. The streak plate method is a popular and simple method to obtain a pure culture. In pure culture, the colonies are visual clones of each other. A highly specialised method of purification is micromanipulation. A device called a micromanipulator, equipped with a microscope and microinjector, picks the individual cells from mixed populations and inoculates them into a culture medium for their growth in pure forms. A different technique is dilution to extinction, where the sample is progressively diluted to the point that the diluted solution contains only a single microorganism.

22.8 PRESERVATION

The long-term survival of microbial cells necessitates highly effective storage facilities. The microbial forms can be preserved as whole bacterial cells/endospores and fungal

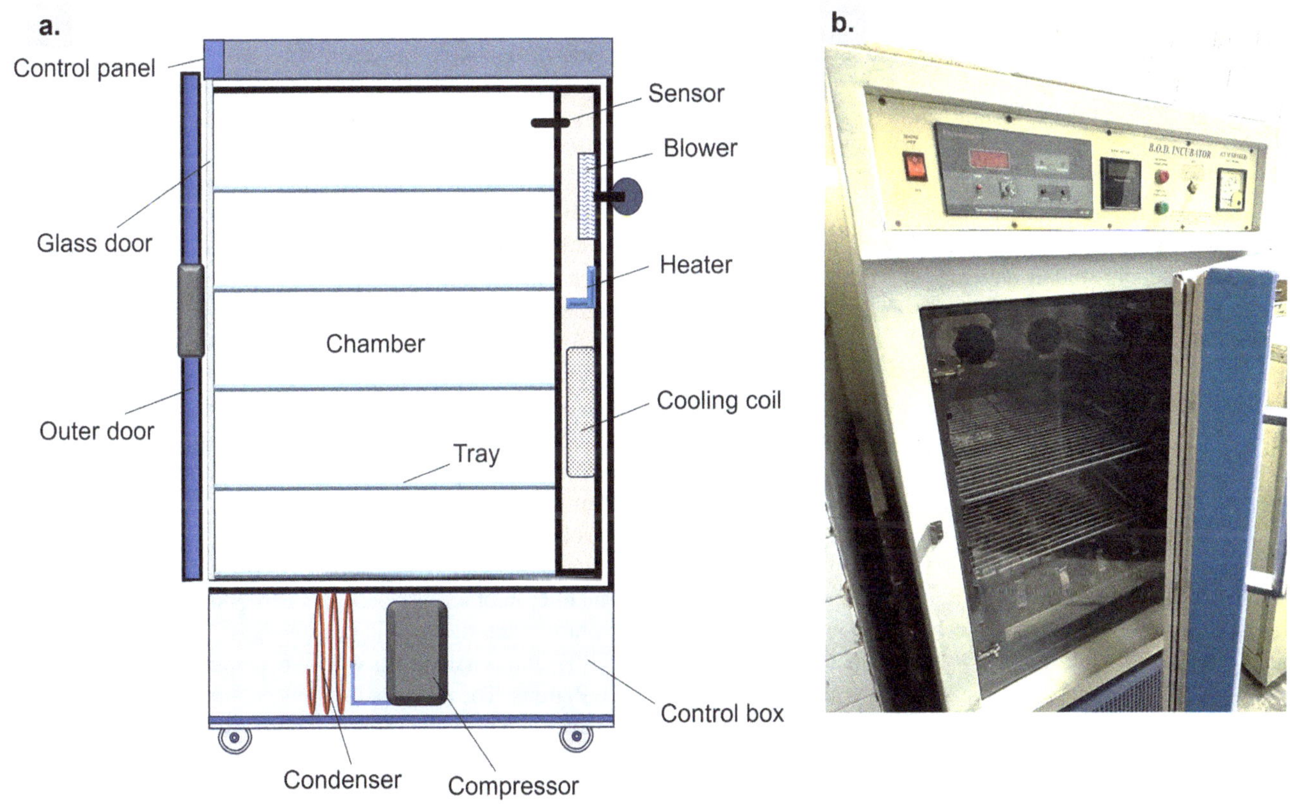

FIGURE 22.9 a. An illustration of the BOD (Biological oxygen demand) incubator from the side view. b. An original photo of the BOD from the front view.

cells/hyphae/spores. The non-fastidious bacterial isolates are stored in agar slants or stabs in a standard refrigerator at 4 °C for their use within a few weeks (Figure 22.6b). Long-term storage requires maintenance below -20 °C or -80 °C. The culture stocks stored at low temperatures have long-term viability. However, freezing of culture can cause some damage to microbial cells. Some cryoprotectants, such as glycerol or dimethyl sulphoxide (DMSO), can be used to prevent desiccation to a greater extent when stored below -20 °C. The use of glycerol stock for the same is widespread, which is prepared by suspending log-phase cells in the autoclaved glycerol solution (concentration of 5-20% (v/v)) in a ratio of 1:1. Glycerol-sensitive strains are stored in 50% of filter-sterilised DMSO. The density of approximately 10^7 cells/ml in the stock yields adequate recovery. They can also be kept in liquid nitrogen freezers at extremely low temperatures below -130 °C. A different long-term storage technique is freeze-drying or lyophilisation, which is equally competent for both non-fastidious and fastidious microorganisms. The recovery of microbial cells requires rapid thawing at ambient temperature and immediately transferring the aliquot to growth media. A detailed description of various preservation techniques is mentioned in Chapter 5.

22.9 CHARACTERISATION AND IDENTIFICATION

After culturing, bacteria and fungi are characterised and identified based on morphological, biochemical, microscopic, and molecular methods. Identification of prokaryotic cells mainly relies on 16S rRNA gene sequencing and comparison of amplicons to a reference database. The biochemical analysis provides information about their physiological traits and potential industrial relevance. Fungi are characterised using microscopic studies and various biochemical analyses. However, the authentic identification is based on their gene sequencing. The 18S rRNA gene and the internal transcribed spacer (ITS) region are the most commonly used molecular markers to unravel fungal diversity. Chapter 21 provided extensive coverage of the identification and characterisation of microorganisms.

22.10 ENDOPHYTES AND THEIR CULTIVATION

A wide range of bacteria and fungi reside inside and outside the plant body, both above- and below ground, and form a distinct branch of microbiology, "plant microbiome". Some microbial forms, such as archaebacteria, microalgae, and protozoa, can also sporadically colonise the plants. Numerous microorganisms colonise inside the plant tissues of roots, shoots, leaves, inflorescence, and fruit, which constitute "endophytes" (Figure 22.10). This term was coined by de Bary in 1866. Among all plant parts, the root is the favourite site for both microbial abundance and diversity. The diversity of root endophytes even surpasses that of the surrounding rhizosphere. A plant microbial endosphere may develop symbiotic relationships within the host body without causing any symptoms of the disease. Though microorganisms can colonise roots selectively, certain opportunistic infections can penetrate through the root wall and interact with endophytes. Microbial colonisation inside the plant tissues dates back more than 400 million years, and they have invaded every plant on Earth since then.

Endophytes assist plants in nutrition acquisition and help them withstand environmental stresses such as drought, salinity, and harsh temperatures. Their association can be critical for the growth, protection, development, and nourishment of the plant body. They can also be the source of many allelopathic substances and other secondary metabolites. Root microbiology even matters for soil health. Their capacity to produce valuable chemicals has enabled them to engage in human welfare through agriculture, pharmaceutical, and biofuel industries. Microorganisms that reside in the rhizosphere are extensively addressed in Chapter 19.

Plant-endophyte relationships are pervasive and exhibit a wide range of complexity. Endophytes invade the host in two ways: vertical transfer from a parent to its progeny through seeds and horizontal transfer from the surrounding environment. *Epichloë* and *Neotyphodium* fungal endophytes of grasses are examples of vertical transmission of plant endophytes. Some studies have documented *Bacillus* and *Pseudomonas* as seed-borne bacteria, enabling colonisation of the next generation of plants. In addition, the *Azolla-Anabaena* association is a wonderful example of vertical transfer of plant endophytes. Horizontal transfer, in the context of endophytes, generally refers to the transmission of endophytes from air, soil or other plants. A variety of plants can even select the endophytes that offer them advantages from their coexistence (e.g., legume-rhizobia symbiosis and *Frankia*-actinorhizal plants).

A large number of endophytes observe commensalism, which means that one species receives the profit while another species neither profits nor suffers harm. Endophytes can form either short- or long-term associations with plants as non-systematic or systematic endophytes. The specific developmental stage, environmental influences, and other biotic factors, such as pathogen attacks, occasionally attract non-systematic endophytes. However, the systematic endophytes attempt to find long-term advantageous cooperation through coevolution with their plant hosts.

Bacterial endophytes are extremely diverse and dominant in plant microbiomes. Various plant parts, but chiefly root systems, colonise bacteria as plant growth-promoting rhizobacteria (PGPR) and benefit plants with various plant growth-promoting (PGP) traits. The most useful traits are indole acetic acid (IAA) production, siderophore production, phosphate solubilisation, and 1-aminocyclopropane-1-carboxylate (ACC) deaminase activity. The typical representatives of bacterial endophytes for PGP activities are *Pseudomonas*, *Bacillus*, *Stenotrophomonas*, *Enterobacter*, *Paenibacillus*, and *Micrococcus*. Additionally, endophytic bacteria can indirectly accelerate plant growth by stimulating plant defence responses and suppressing plant diseases (though antagonistic activity) (e.g., *Chitinophaga* and

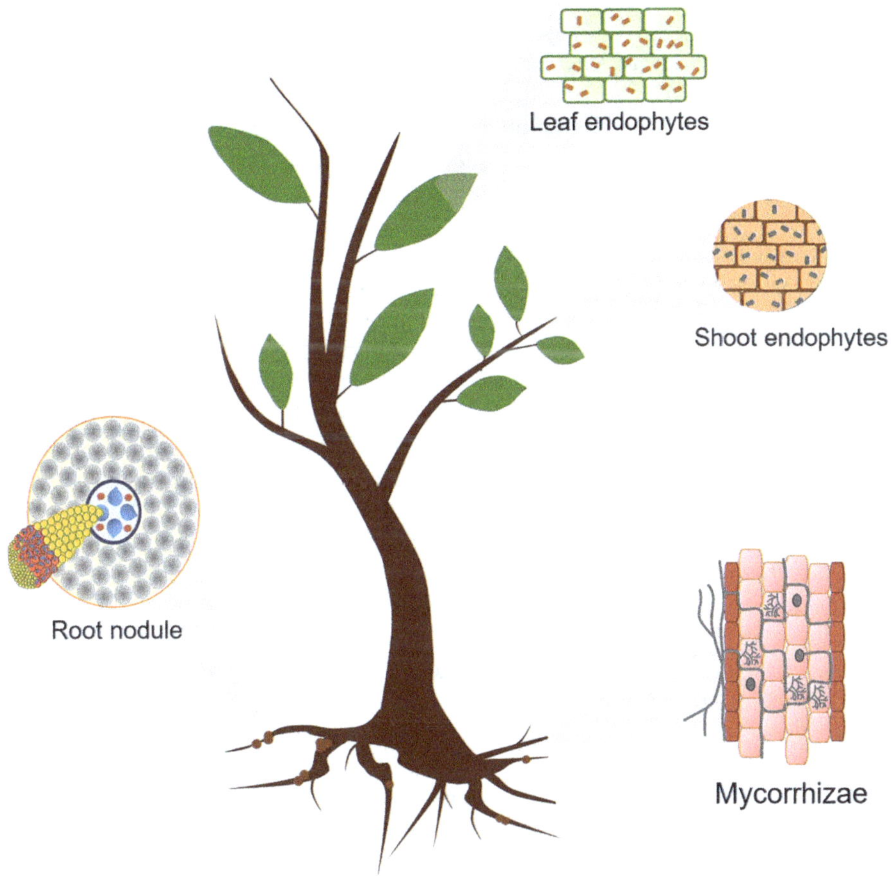

FIGURE 22.10 An illustration of a plant body showing endophytes in different plant parts.

Flavobacterium). *Bacillus thuringiensis* is also an indigenous resident of wild plants and is widely known for its insecticidal properties. Several host-associated bacteria are nitrogen-fixing species (diazotrophs). Many bacterial genera belonging to the phylum Proteobacteria are recognised as rhizobia. Some examples of endophytic rhizobia are *Rhizobium*, *Mesorhizobium*, and *Burkholderia*. They develop a specialised structure called a nodule in the legume roots (e.g., soybeans, clover, alfalfa, etc.). These bacteria fix nitrogen only when they exist as endosymbionts. A plant species, *Parasponia*, that is not a legume yet is susceptible to a rhizobium for nodule formation. A distinct filamentous Gram-positive bacterium, *Frankia* (belonging to Actinomycetes), draws the attention of numerous trees and shrubs for nodule formation and nitrogen fixation. These plants are termed actinorhizal plants (e.g., *Alnus*, *Betula*, *Casuarina*, etc.). Grass roots are localised by diazotrophic endophytes such as *Azospirillum*, *Azoarcus*, and *Gluconacetobacter diazotrophicus*. A gymnosperm, *Cycas,* invades some light-harvesting cyanobacteria like *Nostoc*, *Oscillatoria*, etc., in its specialised apogeotropic root called a coralloid root. *Nostoc* is a quite common inhabitant of living tissues in diverse plants such as the underground stems of *Gunnera manicata* and the thallus of *Anthoceros*. A fern, *Azolla,* colonises *Anabaena azollae* in its leaf pockets for nitrogen fixation.

Endophytic fungi (EF) belong to different taxa and offer multidimensional interactions. They are the potential sources of anticarcinogenic molecules and antibacterial substances. An endophytic fungus of *Taxus wallachiana* (Himalayan Yew), *Pestalotiopsis microspora* (phylum: Ascomycota) is a noble source of the anticancer drug Taxol. EF can be divided into two categories: the Clavicipitaceous group (C-group) and the non-Clavicipitaceous group (NC-group). The fungal endophytes of Ascomycota-members, *Neotyphodium* and *Epichloë,* are examples of Clavicipitaceous endophytes, which spread their hyphae throughout aerial portions, including inflorescences, in a variety of cool and warm season grasses and establish an obligate mutualistic relationship. Clavicipitaceous endophytes mainly colonise through vertical transfer. Fungi defend their plant hosts against herbivores and also help them survive in drought situations by secreting various bioactive compounds. However, NC-group endophytes have a wide diversity that colonises a wide range of hosts, including nonvascular, vascular, and woody plants. They are mainly transmitted to the hosts through horizontal propagation. Fungal taxa like *Alternaria*, *Penicillium*, *Aspergillus*, *Fusarium*, *Mucor*, and *Trichoderma* are a few that represent NC-group endophytes. A root colonising fungus, *Piriformospora indica,* mimics the abilities of typical arbuscular mycorrhiza (AM) for nutrient transport in plants.

BOX 22.1 CULTUROMICS

The vast repertoire of metagenomics data has generated a dark matter for cultivation techniques. Uncultured microorganisms are mainly bacteria and archaea that do not grow under laboratory culturing environments. Their proportion in the natural environment is much larger than that of the culturable isolates. Microscopic techniques and other molecular tools also provide an array of evidence about their existence (detailed in Chapter 21). The reason why many microorganisms do not grow in the laboratory is the lack of essential growth conditions in artificial conditions. Many advanced molecular strategies have shown the possibilities of coding the metabolic potential of uncultured microorganisms, which has significantly expanded the knowledge about novel situations for cultivation. Recent studies are exploring new ways to grow uncultured microorganisms in artificial environments. The approach is cited as the rebirth of culture techniques, "culturomics", which has given proof of the existence of many unseen biological entities.

Culturomics is a recent concept for optimising culture conditions for the growth of previously known uncultured bacteria. The main aim of culturomics is to find out the right conditions and tools for cultivation. Culturomics has led to the discovery of many new taxa from the human microbiota. The key advances made in this direction are through the use of modified growth media, modification in growth conditions, optical tweezers, and laser microdissection (for selective single-cell isolation). Coculturing is an established method for the growth of many unculturable bacteria, as many natural co-inhabitants are dependent on the activities of helper microorganisms for their growth. For example, cell extract from *Geobacillus toebii* serves as the growth-promoting factor for culturing of many thermophilic anaerobic bacteria in the family Clostridiaceae. A high-throughput culturing based on dilution-to-extinction has offered a competition-free growth of bacteria, where the dilution process in an environmental sample achieves only 1-5 cells per inoculum volume. Analysis of high-throughput sequencing data of RNA transcripts of uncultured microorganisms has provided information on particular requirements for their growth. In a study, knowledge of the requirement for mucin as an essential carbon source has led to the successful growth of a previously uncultured *Rikenella*-like bacterium on synthetic media. In a different approach, the encapsulation of cells as microdroplets is emerging as an advanced technique for cell culturing. The streamlined preparation of media for the cultivation and identification of rare and slow-growing bacteria from environmental samples is a profitable deal in industrial biotechnology.

22.10.1 Isolation and Cultivation of Endophytes

Any plant part can be used for the isolation and cultivation of endophytes. However, plant parts with superficial injury and disease are not suitable due to contamination from the surrounding environment. The cultivation of endophytes typically involves standard culture-based approaches with two additional steps, including surface sterilisation of plant materials and grinding plant tissues.

In surface sterilisation, plant parts are initially rinsed with water to remove all the dirt from their surface. This is preceded by the use of disinfectants to remove any microbial cells from the plant surface. One of the most common treatments is the sequential treatment of 70% alcohol for 1 min and sodium hypochlorite (NaOCl) (2.5%) for 4 min, followed by ethanol for 30 s. In a different treatment, mercuric chloride ($HgCl_2$) can be used in place of NaOCl. A different batch of disinfectant treatment includes treatment with commercial bleach (5.25% available chlorine) for 3 min, followed by 3% hydrogen peroxide (H_2O_2) for 3 min. A teepol solution (liquid detergent) is a suitable alternative for cleaning the surface of plant materials. The final treatment is three times washing with sterile distilled water for the complete removal of chemicals from the surface. The aliquot of sterile water from the last washing or the imprint of the sterilised tissues can be used for plating to ascertain the success of disinfection.

The next step is microbial extraction. For the same, the surface-sterilised plant material is transferred to a sterile mortar and ground to homogenise the tissue using sterile distilled water. Saline water (~0.9% NaCl aqueous solution) in place of distilled water is a suitable choice for extraction because this creates an isotonic environment and protects the indigenous microbial cells in a new environment. Also, the use of multiple salts in the solution (e.g., Ringer's solution) assists in maintaining the pH equilibrium. Efficient grinding ensures maximum extraction of endophytic bacteria from the host tissue. Further, the aliquots of tissue extract can be serially diluted and then plated on the agar plates.

The cultivation of specific endophytes calls for specific culture media and appropriate growth conditions. For example, the isolation of rhizobia from root nodules or soil requires yeast extract mannitol (YEM) agar or arabinose-gluconate (AG) medium. After inoculation on a suitable medium, they are incubated for a set period to allow for colony development. Following that, the colonies are purified by co-culturing, similar to the protocols mentioned in another section of the chapter (section 22.5).

SUMMARY

- Microbial cultivation is a method of propagation of microorganisms in an artificial environment. This involves many steps, such as sampling, culture media preparation, sterilisation, serial dilution, inoculation, incubation, purification, identification, and storage.

- Sampling is the first step in microbial cultivation, where a subset of the environmental sample is collected in a container. Prolonged storage of samples requires a temperature below 10 °C to preserve biological activities.
- Culture media consist of a mixture of different substances that promote the rapid growth of microorganisms. The compositions of media can be different for different microorganisms and their applications.
- An autoclave is widely used to sterilise small equipment and solutions through moist heat. Heat-sensitive fluid can be treated through filter sterilisation. Small objects like glass spreaders, wire needles and loops can be instantly sterilised using 70% ethanol and exposure to flame. A laminar airflow chamber is used to create an aseptic environment for the workflow.
- Serial dilution is a process of sample dilution to reduce the density of cells in the original sample. This procedure helps in growing a countable number of cells on Petri plates.
- Inoculation is an aseptic transfer of inoculum in a sterile culture medium for culturing. This is followed by incubation in an insulated chamber for a certain duration, which provides a uniform condition for microbial proliferation.
- The mixed colonies grown on the Petri plates are purified by repeated streaking on Petri plates. The purified colonies are identified, characterised, and then stored as culture stocks at low temperatures.
- A group of bacteria and fungi reside inside the plant bodies as endophytes. They can be selectively isolated by including two additional steps before serial dilution. These include surface sterilisation of plant materials and grinding of plant tissues for the extraction of the inside microorganisms.

DIFFERENT TYPES OF PRACTICE QUESTIONS

Fill in the blanks

I. Prolonged storage of microbial culture requires a temperature below ....
II. The main genetic marker for the identification of prokaryotic cells is ....
III. .... media supports the growth of only selected microorganisms because of the presence of an inhibitory agent.
IV. Chemicals like K_2HPO_4 and KH_2PO_4 are commonly used as .... for biological reactions.
V. .... are chemically defined media that include pure chemical substances with precise quantities.
VI. In laminar flow, the air blower is equipped with a mechanical air filter called ....

SOLUTION: I. -20°C, II. 16S rRNA gene, III. Selective, IV. buffer, V. Synthetic media, VI. high-efficiency particulate air (HEPA) filter

Match the following

i. Sampling	a. Refrigerator
ii. Sterilisation	b. BOD incubator
iii. Incubation	c. Autoclave
iv. Inoculation	d. Container
v. Storage	e. Needle

SOLUTION: i-d, ii-c, iii-b, iv-e, v-a

Very short answer type of questions

I. Expand CFU.
II. What is the use of saline solution in serial dilution?
III. Write the formula to estimate microbial density in a sample using the culturing technique.
IV. Write the name of an endophytic bacterium and fungus.
V. What is the main application of the replica technique?

Short answer type of questions

I. What are the different methods of sampling and their applications for the cultivation of microorganisms?
II. Elaborate on different plating techniques that have different objectives for microbial culturing.
III. Write a short note on the ingredients of culture media for culturing of bacteria and fungi.
IV. A serial dilution of 1 ml water sample enumerated 200 bacterial colonies onto a Petri plate from a 0.1 ml aliquot of 10^{-4} dilution. Calculate the CFU/ml.
V. In a series of 10-fold dilutions of 1 g of garden soil, a 0.5 ml aliquot from 10^{-1}, 10^{-2}, 10^{-3}, 10^{-4}, 10^{-5}, and 10^{-6} dilution was spread on plates, which enumerated too many to count, 654, 59, 5, 2, and 1 colonies on respective plates. Calculate the cell density of culturable bacteria in soil.
VI. Write the differences in cultivation procedures between endophytes and non-endophytes.

Long answer type of questions

I. The microbial growth requirements of nutrients can be general or strict depending on the type of microorganisms and their applications. Elaborate on this with suitable examples.

II. What are the applications of cultivation techniques in microbiology? Prepare a workflow layout for the isolation and characterisation of heat-resistant microorganisms from the hot water spring sediments.

SUGGESTED READING

American Type Culture Collection (ATCC). *The premier guide for culturing bacteria.* ATCC Bacteriology Culture Guide. 2022. www.atcc.org/resources/culture-guides/bacteriology-culture-guide

Bonnet M, Lagier JC, Raoult D, Khelaifia S. Bacterial culture through selective and non-selective conditions: the evolution of culture media in clinical microbiology. *New Microbes and New Infections*. 2020;34:100622.

Das N, Tripathi N, Basu S, Bose C, Maitra S, Khurana S. Progress in the development of gelling agents for improved culturability of microorganisms. *Frontiers in Microbiology*. 2015;6:698.

Gavrilov SN, Stracke C, Jensen K, Menzel P, Kallnik V, Slesarev A, Sokolova T, Zayulina K, Bräsen C, Bonch-Osmolovskaya EA, Peng X. Isolation and characterization of the first xylanolytic hyperthermophilic euryarchaeon *Thermococcus* sp. strain 2319x1 and its unusual multidomain glycosidase. *Frontiers in Microbiology*. 2016;7:552.

Hohnadel M, Maumy M, Chollet R. Development of a micromanipulation method for single cell isolation of prokaryotes and its application in food safety. *PLoS One*. 2018;13(5):e0198208.

Kim S, Park MS, Song J, Kang I, Cho JC. High-throughput cultivation based on dilution-to-extinction with catalase supplementation and a case study of cultivating acI bacteria from Lake Soyang. *Journal of Microbiology*. 2020;58(11):893–905.

Rawlinson S, Ciric L, Cloutman-Green E. How to carry out microbiological sampling of healthcare environment surfaces? A review of current evidence. *Journal of Hospital Infection*. 2019;103(4):363–74.

Stewart EJ. Growing unculturable bacteria. *Journal of Bacteriology*. 2012;194(16):4151–60.

Strobel HJ. Basic laboratory culture methods for anaerobic bacteria. *Biofuels*. 2009;247–61.

Unit VI

Microbial Indicators and Treatment of Water Pollution

23 Water Pollution

23.1 WATER AND WASTEWATER

Water is a vital resource for all living beings. The supply of accessible, adequate, and safe water is a crucial challenge of the 21st century. Water that has been contaminated or polluted has lost its quality, changed in its composition for physical, chemical, and biological characteristics, and is no longer fit for drinking and other necessities. This kind of water body is called wastewater. Being a universal solvent, water dissolves a variety of matter compared to any other liquid on Earth and, therefore, gets easily polluted. The easy accessibility of surface water sources like lakes, dams, canals, rivers, and streams makes them vulnerable to discharge sites for wastewater.

Wastewater poses a serious environmental problem to natural water bodies and depletes clean water resources. Sewage contamination of surface waters is a global threat to public health, including widespread human diseases. Many bacterial diseases, such as cholera, gastroenteritis, typhoid, bacillary dysentery, and acute diarrhoea, are transmitted from faecal-contaminated drinking water. Excessive nutrients like phosphate and nitrate in water cause eutrophication. Other notable repercussions of wastewater are degraded aquatic ecosystems, low biological diversity, greenhouse effects, and deforestation. Altered hydrological characteristics and low water quality also lead to habitat modification, increased flooding, sedimentation, and erosion.

23.2 POINT SOURCES *VS.* DIFFUSE SOURCES

Pollutants can enter aquatic environments from two broad categories of sources: point and diffuse (Table 23.1). A single pollutant can have point or diffuse sources, depending on its origin. The following subsections give detailed accounts of these two categories, including information about their sources and environmental impact.

23.2.1 Point Sources

When water contamination originates from one specific discernible place, it is called point source pollution. This comprises wastewater discharge from industrial facilities, storm sewers, oil refineries, wastewater treatment facilities, etc. This can even include contaminated water from illegal dumping, chemical and oil spills, and leaking septic systems (Figure 23.1). Point sources are identifiable as pipes, drains, ditches, channels, tunnels, conduits, and vessels, which are suitable conveyance for the wastewater. Despite originating from a single source, point source pollution can impact several kilometres of water bodies and oceans when they meet the natural ecosystems. Examples of point sources are mentioned as follows.

- Industrial effluent: Various chemical manufacturers, food processors, oil refineries, pharmaceutical companies, fermentation industries, power plants, pulp and paper mills, and automobile industries are the possible point sources of industrial effluent. Industrial facilities discharge immense amounts of chemicals into the environment during manufacturing, which eventually directly or indirectly enter the rivers, lakes, or oceans. According to the United Nations, industries dispose of about 300-400 megatons of contaminated waste in water bodies every year. Industrial effluent contains pollutants, such as heavy metals, biphenyls, detergents, dyes, fertilisers, hydrocarbons, greases, oils, pesticides, pharmaceuticals, phenols, and plasticisers.
- Domestic or municipal wastewater: It is the contaminated water released from housing societies, offices, and academic institutions. According to the United Nations World Water Development Report (WWDR) 2015, developing countries dump 90% of their untreated or insufficiently treated domestic and municipal wastewater into natural water bodies. Municipal wastewater can be categorised as greywater and blackwater. Greywater encompasses wastewater released from sinks, household baths, showers, dishwashers, and clothes washers. Chemicals in domestic wastewater are present in low concentrations and have low toxicity. The possibility of waterborne pathogens in such wastewater is also relatively low. Greywater, after some minor treatment, can be used for irrigation in gardens and fields. This type of approach features sustainability in wastewater management. On the contrary, blackwater is wastewater released from flush toilets and kitchens and requires proper attention at treatment facilities. Faecal matter and urine from the

DOI: 10.1201/9781032651095-29

TABLE 23.1
Point sources *vs.* diffuse sources

Characteristics	Point sources	Diffuse sources
Source	A single identifiable location	Large areas
Discharge options	Through pipes, drains, ditches, channels, tunnels, conduits, and vessels	Through runoff, surface spread, and combined sewer overflows
Dispersion	Simple (point/linear)	Complex
Initial impact	Localised and concentrated	Widespread and dispersed
Monitoring and regulation	Easy	Difficult
Response	Targeted and localised remediation	Comprehensive and integrated approaches
Natural causes	Volcanic eruption	Forest fire, snow meltdown, and landslides
Anthropogenic causes	CAFOs, industrial and domestic sludge, power plant, oil spill, and wastewater treatment plant	Agricultural fields, vehicular pollution, acid rain, stormwater, and mining

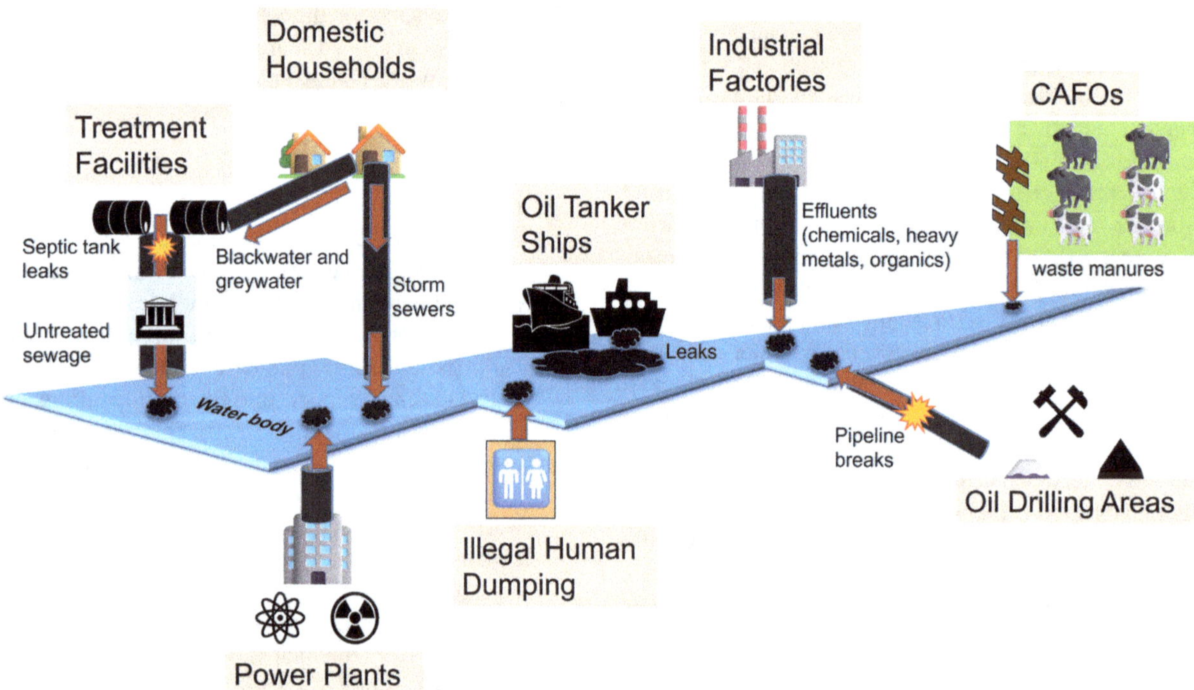

FIGURE 23.1 Various point sources of water pollution. (CAFOs: Concentrated animal feeding operations.)

toilets and organic wastes from the kitchens contain fastidious microbes that can have pathogenic concerns. Blackwater requires safe passage into a sewer and then to a treatment facility from the plumbing system before they are released into the natural environment.

- Oil spillage: An oil spill is the leakage of crude oil onto the surface of a large water body. This is a major environmental issue in the ocean and coastal regions. Oil spillage commonly occurs during drilling, transportation, or refuelling operations. Tonnes of crude oil pour into the ocean yearly due to oil spills. Many such incidents occur on a small scale, especially when ships are being refuelled and have local consequences. Big oil spills typically occur when pipelines break, large oil tanker ships sink, or drilling operations are hampered. Large oil spills can have long-term effects on aquatic ecosystems and economies. Oil spills can cause severe damage to sensitive beaches, mangroves, and wetlands. Exposure of toxic compounds present in the oil to aquatic wildlife has direct effects through skin absorption, inhalation of volatile compounds, and ingestion. This event even induces many negative effects on human health through foodborne illness and interferes with professional and recreational activities.
- Thermal pollution: Thermal pollution is a phenomenon of sudden variation in the water temperature due to anthropogenic or natural reasons. The prominent sources of thermal pollution in water bodies are power

plants (such as nuclear, coal-fired, and electric plants), industrial boilers, crude oil refineries, and steel melting factories. These industrial steps use water as a coolant that is released at high temperatures into rivers, streams, lakes, and ponds. Volcanic eruption is a natural cause of thermal pollution. It spews molten lava and magma into water pools, increasing their temperature manifold. A large amount of heat alters the water's physical, chemical, and biological characteristics. High water temperature also reduces the dissolved oxygen level in the aquatic systems and adversely affects their biodiversity. Extreme situations can even cause the death of wildlife.

- Radioactive wastes: Radioactive waste from nuclear power plants, radioactive material consumption, and radioisotopes used in medicine and research contaminate groundwater and rivers. Radioactivity from these contaminated waters could then enter the food chains, causing a large population at high risk for genetic variation. The waste products of nuclear technology and power generation are typically kept in suitable disposal facilities, such as deep burial, for an adequate duration to decay naturally. However, the environmental leakage of radionuclides is prone to the risk of radiation exposure.
- Concentrated Animal Feeding Operation (CAFO): These are livestock farms spread in confined areas that house animals to produce meat, milk products, and eggs. Concentrated animal feeding operation (CAFO) contributes to point source pollution as the animals raised in CAFOs prefer readymade food instead of grazing in fields, pastures, or range areas. The main concerns of CAFOs are their three types of wastes, *viz*., nutrients, chemicals, and pathogens. The use of growth hormones and antibiotics in CAFOs raises concerns for antimicrobial resistance in microorganisms. The spread of animal blood poses health risks. Surface waterways around CAFOs frequently contain ammonia, which depletes the oxygen level in the water or oxidises into nitrates, resulting in nutrient overload. High levels of phosphorus and nitrogen lead to water becoming unfit for fish and other aquatic species. The raw sewage produced from the farmland contaminates nearby water bodies when storage lagoons (pits in the ground where liquid manure is stored) overflow or spill into adjacent bodies of water as a result of intense rainfall or flooding. Surface discharge raises the concern of widespread contamination. CAFOs can also contaminate groundwater by runoff from land and leaching from improperly spread manure.
- Wastewater Treatment Plant (WWTP): A wastewater treatment plant (WWTP) is a site for wastewater treatment. However, this facility can also emerge as a point source of pollution due to overflow of wastewater in the surroundings and insufficiently treated effluents containing elevated levels of contaminants and water-borne pathogens. Other reasons for them to be a point source of pollution are plant malfunctions due to equipment failures, power outages, and operational mistakes. Also, during excessive precipitation or melting, sewer systems can overflow and discharge raw or partially treated effluent into natural aquatic systems.

23.2.2 Diffuse Sources

Diffuse (or nonpoint) sources are the indirect inputs of pollutants occurring along a wide area, which are not easily attributed to individual point source discharge. Diffuse source pollution is difficult to regulate since it comes from diverse sources and locations. Various phenomena, such as stormwater (excess water from heavy rain or snowstorms), surface run-off, precipitation, atmospheric deposition, snowmelt, drainage, seepage modification (water flow in permeable soil layers), and debris blown into waterways from land, are responsible for contaminant distribution over a wide area. Particularly, stormwater, surface run-off, and snowmelt pick up natural and human-made pollutants moving along/from the ground and carry them away to long distances. Religious and societal behaviours have been responsible for adding many contaminants to natural water bodies. Natural activities such as algal blooms, grazing, and silt from storms and floods also induce diffuse source pollution (Figure 23.2). Affected hydrology and water quality result in habitat modification and loss, aquatic biological diversity modification, intense flooding, sedimentation, and erosion. Some important diffuse sources are discussed as follows.

- Agricultural activities: Agricultural activities significantly impact and impair the water quality of rivers, streams, lakes, groundwater, and wetlands. Pathogenic microbes from livestock and poultry manure deteriorate drinking water quality. Agricultural runoff and irrigation return flow contribute to the discharge of fertilisers (nitrogen and phosphorus), livestock manure, and pesticides into water bodies. Excessive accumulation of organic matter and nutrients raises the concern of eutrophication and stimulates algal blooms in lakes and rivers. This condition develops hypoxia (low oxygen) that threatens aquatic wildlife.
- Urbanisation: The expansion of urban and suburban areas has vastly enhanced the volume and density of heavily polluted stormwater runoff. They carry a variety of pollutants, including oil, grease, toxic chemicals, plastics, sediment, pesticides, nutrients, pathogenic microbes, viruses, pharmaceuticals, road salts, heavy metals, etc. This nonpoint source of water pollution injures wildlife populations and native vegetation, fouls drinking water, and makes recreational areas vulnerable and unpleasant.
- Vehicular pollution: Vehicular pollution enters water bodies through runoff, atmospheric deposition, and direct discharge. Fine particulates from tyre wear and exhaust emissions can land on roadways and wash into rivers. These particles carry numerous contaminants,

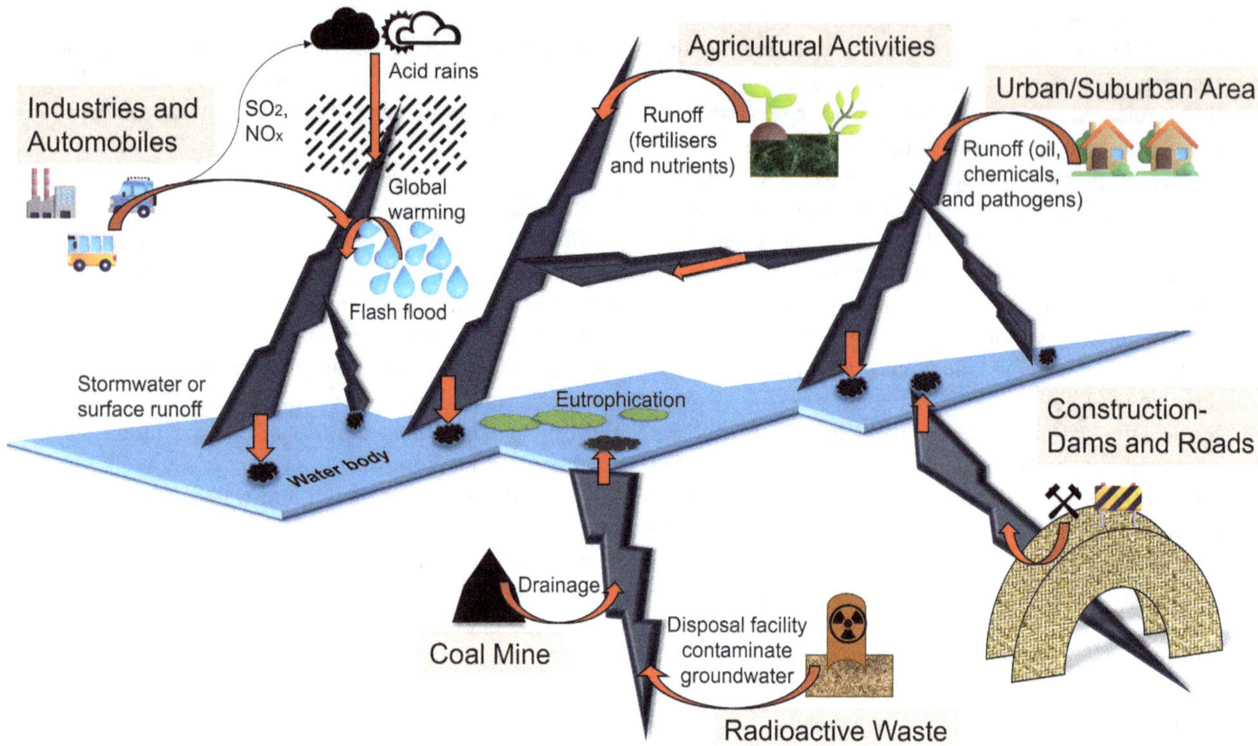

FIGURE 23.2 Various diffuse sources of water pollution.

including organic chemicals, engine oil leaks, and heavy metals.

- Acid rain: Acid rain is a process where gases of sulphur dioxide (SO_2) and nitrogen oxides (NO_x) react with water, oxygen, and other chemicals in the atmosphere and form sulfuric and nitric acids, which then mix with rain, snow, fog, or hail before falling to the ground. It has a pH of around 4-5.5. Burning fossil fuels in electric power generators, manufacturing industries, oil refineries, automobiles, and the processing of natural gas are the major emission sources of these gases in the atmosphere. These gases can be transported with wind and air currents from their emitted sources to distant places. Acidic particles and gases can also fall as dry deposition onto the surfaces of land and water bodies when the moisture content is low. Acid rain and acid runoff cause damage to aquatic bodies, soil, forest ecosystems, and natural and manmade infrastructure.
- Climate change: Climate change, driven by the accumulation of greenhouse gases in the atmosphere, is responsible for the rise in the Earth's temperature. This elevated temperature has altered the natural water cycle and affected water resources by enhanced evaporation, geographical modification of precipitation intensity, and variability of droughts and floods. The warming of the Earth has pushed off the snowline, reduced the capacity of natural reservoirs, increased the risk of flash floods in the moist season, and enhanced eutrophication in wetlands and marinas.
- Construction: The construction of roads, highways, and bridges contributes to polluted runoff in lakes, rivers, surface waters, and groundwater. Erosion from construction activities contributes to sediment and silt accumulation in runoff waters, deteriorating water quality.
- Mine drainage: Water bodies near coal mining activities and abandoned mines are nonpoint sources of water pollution. The most common problem from mine drainage originates in sulfuric acid- or pyrite-containing acid mine drainage. The chemical reaction and biological activities within the surface and shallow subsurface waters with sulphur-bearing mineral rocks enhance the leaching of heavy metals from the rocks into the acid-drainage water. The contaminated water infiltrates groundwater, surface water and soil, posing challenges to the environment and human health. Alkaline mine drainage from calcite/dolomite rocks and metal mine drainage also contribute to water pollution with similar repercussions.
- Hydromodification activities: Hydromodification activities cover channelisation, dams, and erosion of stream banks and shorelines. The siting, construction, and functioning of dams can adversely impact the local hydraulic regime, surface water quality, and habitat of the stream or river. Excessively high sediment loads from erosion can stifle submerged aquatic vegetation, cover shellfish beds, develop tidal flats, and increase levels of turbidity and nutrients in the water bodies.

- Marinas and boating: Recreational boating and marinas in coastal regions are potential nonpoint sources of water pollution. Various activities that occur at marinas, such as fueling, boat cleaning, parking lots, hull maintenance, and repair areas, contribute to water pollution. These pollutants are not usually traced back to a single boat but rather accumulate from many boats.

23.3 CONTROL MEASURES

Water pollution control strategies attempt to reduce contaminants entering water bodies while enhancing water quality. It requires a holistic approach, which addresses the complex and interrelated factors contributing to water contamination. The modern world seeks sustainable solutions, integrating multiple aspects of water management, pollution prevention, and ecosystem protection to control water pollution. Some key components of waste pollution control measures are as follows.

23.3.1 Monitoring

Water monitoring is a repeated analysis of water quality and the processing of its data. Detection systems allow discrete sections of water bodies to be monitored regularly. These systems can instantaneously identify biological and chemical contamination in the distribution system. Water monitoring data support actions on interception and remediation to combat the adverse impacts of pollution on the aquatic environment. Identifying the contaminants and their concentrations in the environment allows the implementation of specific treatment processes. Wastewater monitoring is also an essential control parameter for many stages of treatment in wastewater and sewage treatment plants.

23.3.2 Source Water Protection

Examples of potable water sources are rivers, streams, lakes, reservoirs, springs, and groundwater. Source water protection prevents exposure to contaminated water and reduces runoff pollution, risks of diseases, and treatment costs. This also protects water quality for wildlife, recreational use, and water supplies. Source water protection can be driven by three activities, including protection, improvement, and maintenance of water quality. These often include restoration of riparian zones (lands that occur along the edges of water bodies), stream bank stabilisation, land protection, best management practices (BMPs) for agricultural and forestry activities, and stormwater control. Regulatory bodies also attempt water protection by restricting certain polluting activities and establishing emergency response plans. In addition, education on water resources and conservation to local enterprises, industry, and residents positively impacts water protection.

23.3.3 Stormwater Control

Stormwater refers to the water runoff from rainwater and the snowmelt, which flows over land and does not soak into the ground. While passing through surface channels, stormwater gathers pollutants, which ultimately enter natural water bodies. The protection of natural water bodies from the harmful effects of stormwater runoff calls for various BMPs through environmental agencies and construction projects. The stormwater control facilities involve underground detention systems to remove nutrients, sediment, and debris from stormwater runoff and underground retention systems to reduce stormwater runoff volume and rate into surface waters.

23.3.4 Proper Use and Disposal of Harmful Materials

Chemicals like cleansing substances, mothballs, paints, batteries, motor oils, medicines, pesticides, and fertilisers contain hazardous or toxic substances. Avoiding their disposal down the drain, on the ground, and into storm sewers could prevent contamination of nearby surface waters, groundwater, and wastewater treatment systems. Careful instructions for their use, storage, and disposal on product labels can prevent many accidents.

23.3.5 Maintenance of Septic System

A septic system is an onsite wastewater treatment facility for municipal wastewater in suburban and rural areas. Malfunctioning of septic systems poses dangers of contamination to drinking water supplies and the environment. Faulty septic systems can accidentally release wastewater to local aquifers and waterways that contain pathogenic bacteria, viruses, pharmaceuticals, and chemicals. The household septic systems require frequent pumping, professional inspection, and maintenance. Efficient water use with high-efficiency toilets, showerheads, faucet aerators, and washing small loads of laundry can enhance the functioning and reduce the failure risk of a septic system.

23.3.6 Laws and Regulations

Several countries have addressed water pollution in their regions with national policies, standards, and practices. Efforts have aimed to identify pollution control, evaluate water quality, reduce and eliminate water pollution, enhance water security, and develop a green environment. The World Health Organisation (WHO) has set certain threshold limits for several variables, including salinity, pH, oxygen demands, and toxic chemicals (metals, organic chemicals, and microplastics) in water to curb their effects on water systems (detailed in Chapter 24). The United Nations Environment Programme (UNEP) is a global authority that sets the agenda to address environmental challenges and promotes nature-based solutions to water resources management, including water quality. The United States Environmental Protection Agency (US EPA) is a national agency that enforces Drinking Water Regulations. This agency issues and enforces licences/certificates of authorisation for wastewater discharges from owned sewerage systems to the aquatic environment. The

Clean Water Act (1972) is a landmark law enforcement in the US, which has regulated quality standards for surface waters and the discharge of pollutants into the waters and implemented pollution control programs.

Water law in the Eastern Europe, Caucasus, and Central Asia (EECCA) region also encompasses a diverse array of legal frameworks and policies to manage water resources, ensuring their sustainable use and protecting water quality. In Australia, the Water Act 2007 provides a framework for managing water resources. The Central Pollution Control Board (CPCB), India, passed the Water (Prevention and Regulation of Pollution) Act (1974) to prevent and regulate water pollution. The water regulations and acts framed by different agencies are subject to frequent revision according to the growth and requirements of the population and industries. There are also regulations on environmental topics like acid rain reduction and wetlands restoration.

23.3.7 Wastewater Treatment Processes

WWTPs employ various physicochemical and biological methods to restore water quality. The conventional methods for water treatment are aeration, screening, coagulation/flocculation, disinfection, filtration, sedimentation, and straining. Biological treatment focuses on activated sludge, trickling filters, septic tanks, etc. In a different study, an artificial wetland is a wetland designed for wastewater treatment, which employs natural biogeochemical processes to eliminate organic substances and nutrients. They are used in nearly fifty countries across the world. Special treatment methods cover activated carbon adsorption, chlorination/fluoridation, electrodialysis, ion trapping, osmosis, pre-disinfection, and pre-sedimentation. However, conventional methods have certain limitations, such as the production of potentially toxic secondary wastes during coagulation and flocculation and the requirement of longer duration for biological and ecological methods. Several new technologies, such as nanotechnology and microbial fuel cells (MFCs), are now emerging for wastewater treatment, which offer better remedies while reducing the limitations of conventional techniques. This can involve the use of nanomaterial-based methods for water pollution treatment. Differently, bacterial viruses (such as phages) native to the water system kill bacterial pathogens. Chapter 25 provides a full review of the various treatment options for wastewater.

23.4 FERMENTATION INDUSTRY EFFLUENT AND ITS MANAGEMENT

Although fermentation industries are recognised for their sustainable approaches, the large amount of liquid waste stream they release still poses many environmental concerns. The liquid waste stream of the fermentation industry is called fermentation industry effluent. These effluents have intense odour and colour. The requirements of specific acidic/basic chemicals and heat during fermentation reactions alter the pH and temperature of the water used. Fermentation broths are typically rich in nitrogen and phosphorus nutrients that can also be accumulated in the waste stream. When wastewater enters the natural water bodies without adequate treatment, it raises the concern of eutrophication. Fermentation effluents also contain certain toxic substances, which present a threat to aquatic wildlife. Particles from undissolved raw materials, by-products, and microbial cells and their fragments in the waste stream introduce solid residues in the water bodies.

In fermentation industries, the upstream and downstream processing also consume a high volume of water and release a high amount of effluent (Figure 23.3). In upstream processing, the cleaning of vessels and equipment requires a high amount of water, leading to the generation of wastewater. The downstream processing also releases an enormous amount of wastewater at several steps of recovery and purification. The main concerns are techniques like filtration and centrifugation. When microbial cells are separated from the culture medium

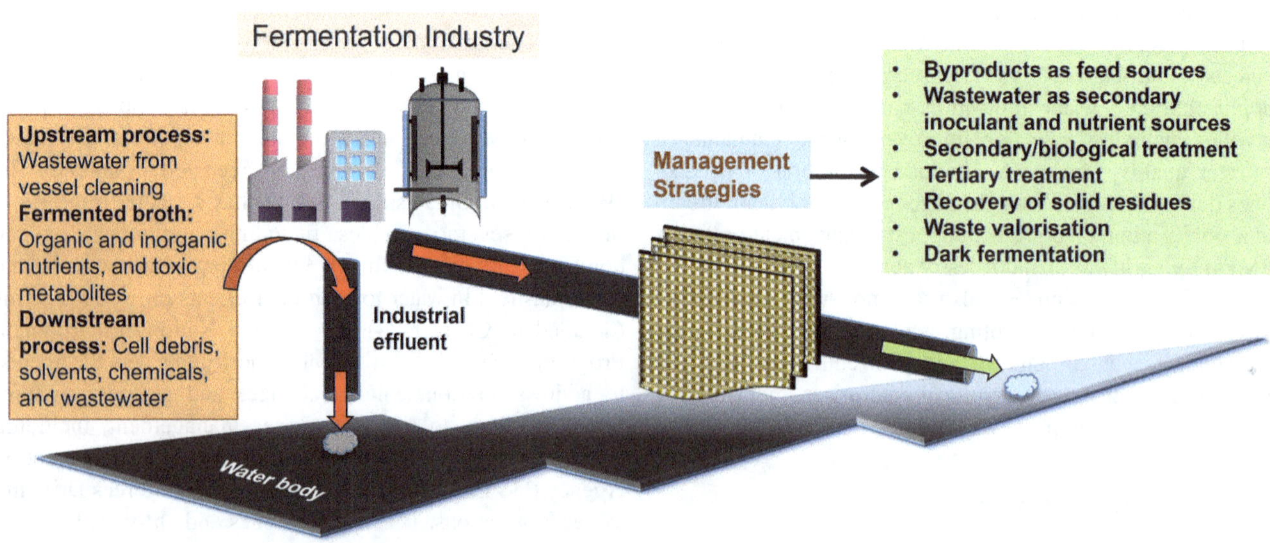

FIGURE 23.3 Different management strategies of fermentation industry effluents.

using these techniques, a large volume of liquid is released as effluent. Solvent extraction methods cause a huge loss of organic solvents during recovery, which enters the waste stream. The precipitation step consumes a high volume of chemicals (e.g., salts and solvents), releasing effluent with residual chemicals and precipitates. The distillation of alcoholic beverages, like whisky, can leave about 90% of the initial raw organic components in the waste stream. ABE (acetone-butanol-ethanol) fermentation also generates a high amount of wastewater from fermentation broth.

In the fermentation process, the product often represents a small portion of the initial raw materials. For example, ethanol can make up only 8-12% of the total fermentation broth in terms of volume. Similarly, the concentration of glutamic acid could be around 8-10% of the total fermentation volume. Antibiotics like penicillin in fermentation broth are present at concentrations of 1-5 g/l. The fermentation broth has even lower concentrations of useful enzymes (less than 1 g/l). The low-quantity production in fermentation industries is a sign that a high concentration of partially utilised substances exists in their effluents. A significant load of waste from the fermentation industry is derived from the microbial mass.

Microbial mass, along with unconsumed nutrients and metabolic by-products, increases the organic load in the effluent, which eventually elevates the biochemical oxygen demand (BOD) of water. The effluent produced during the distillation of alcohol (also called alcohol stillage) has a BOD in the range of 10,000-20,000 mg/l, while the same in the brewery is 1,400-1,800 mg/l. The effluent released from citric acid production generates BOD in the range of 8,000-12,000 mg/l. The sulphite paper mill liquor is highly challenging in terms of its organic load, resulting in very high BOD levels (i.e., 20,000-40,000 mg/l). The antibiotic waste and yeast production waste have BOD in the range of 5,000-30,000 and 3,000-14,000 mg/l, respectively.

The production of huge volumes of fermentation industry effluents calls for effective management strategies before they are released into the natural ecosystem (Figure 23.3). In one scheme, some byproducts from fermentation industries can be used as feed sources for cattle and other animals. Some amount of wastewater generated after fermentation can be used as secondary inoculants and a source of nutrients for the next round of fermentation. Wastewater can be treated in WWTPs. In the process of wastewater treatment, the preprimary and primary treatment steps ensure the recovery of solid residues in the effluent. Secondary treatment relies on biological treatment, which reduces the organic load of the effluent. Tertiary methods can be employed to remove specific contamination. The sludge produced from these industries is treated in composting and anaerobic digesters. In recent years, waste valorisation of industrial effluent has received huge popularity, which focuses on the transformation of waste streams into generating high-value products. This kind of approach promotes sustainability in waste treatment. Chapters 25 and 31 provide detailed information of diverse aspects of wastewater treatment and its management, respectively.

SUMMARY

- When chemicals, pathogens, wastes, animal manure, and temperature contaminate a water body, it results in water pollution. This polluted water body is referred to as wastewater. Pollutants can enter aquatic systems via two primary categories of sources: point and diffuse.
- Point source pollution occurs through a single, easily identifiable location. Some notable examples of point sources are industrial effluent, domestic wastewater, oil spillage, thermal pollution, and CAFOs.
- In diffuse source pollution, pollutants are released from a wide region rather than from a single source. This includes agriculture fields, stormwater, vehicular pollution, acid rain, etc.
- Water pollution control measures aim to reduce contaminants from entering water bodies and improve water quality. It covers wastewater monitoring, source water protection, careful waste disposal, septic system maintenance, enforcement of national policies, and wastewater treatment processes.
- Fermentation industries generate significant amounts of effluent, which requires effective management. Wastewater released from the fermentation industry can be treated through biological decomposition, owing to the presence of large amounts of biodegradable substances.

DIFFERENT TYPES OF PRACTICE QUESTIONS

Fill in the blanks

I. When water contamination originates from one specific discernible place, it is called.................. pollution.

II. .... refers to the water runoff from rainwater and the snow meltdown, which flows over land and does not soak on the ground.

III. .... is a repeated analysis of water quality and processing of their data.

IV. Excessive accumulation of organic matter and nutrients in the water bodies raises the concern of ....

V. The distillation of alcoholic beverages, like whisky, leaves about .... of the initial raw organic components in the waste stream.

SOLUTION: I. point source, II. Stormwater, III. Water monitoring, IV. eutrophication, V. 90%

Very short answer type of questions

I. Define wastewater.

II. Expand WWTP and CAFO.

III. Write the names of two bacterial diseases transmitted from faecal-contaminated drinking water.
IV. What are the two main gases responsible for acid rain?
V. What are the reasons for high BOD in the fermentation industry effluents?

Short answer type of questions

I. Describe point sources of pollution with some suitable examples.
II. How can diffuse sources of pollution impact the biodiversity of neighbouring terrestrial ecosystems?
III. Differentiate between greywater and blackwater.
IV. Write a short note on the fermentation industry effluent and its management schemes.

Long answer type of questions

I. What are the environmental impacts of point versus diffuse (nonpoint) source pollution on aquatic ecosystems? What are the challenges in monitoring and mitigating each type of pollution?
II. Give a detailed account of different control measures of water pollution.

SUGGESTED READINGS

Burkholder J, Libra B, Weyer P, Heathcote S, Kolpin D, Thorne PS, Wichman M. Impacts of waste from concentrated animal feeding operations on water quality. *Environmental Health Perspectives*. 2007;115(2):308–12.

Deletic A, Wang H. Water pollution control for sustainable development. *Engineering*. 2019;5(5):839–40.

EAP Task Force. 2008. Surface Water Quality Regulation in EECCA Countries: Directions for Reform 1–13 EAP Task Force, O. for economic cooperation and development (OECD). www.oecd.org/env/outreach/41832129.pdf

Goel PK. *Water pollution: causes, effects and control.* New Age International; 2006.

Hairom NH, Soon CF, Mohamed RM, Morsin M, Zainal N, Nayan N, Zulkifli CZ, Harun NH. A review of nanotechnological applications to detect and control surface water pollution. *Environmental Technology & Innovation*. 2021;24:102032.

Ji M, Liu Z, Sun K, Li Z, Fan X, Li Q. Bacteriophages in water pollution control: advantages and limitations. *Frontiers of Environmental Science & Engineering*. 2021;15:1–5.

Mason SA, Garneau D, Sutton R, Chu Y, Ehmann K, Barnes J, Fink P, Papazissimos D, Rogers DL. Microplastic pollution is widely detected in US municipal wastewater treatment plant effluent. *Environmental Pollution*. 2016;218:1045–54.

Petrie B, Barden R, Kasprzyk-Hordern B. A review on emerging contaminants in wastewaters and the environment: current knowledge, understudied areas and recommendations for future monitoring. *Water Research*. 2015;72:3–27.

Salaria N, Kumar A, Singh S, Rajput A, Shukla M, Tripathi SK. 2024. Biotreatability of industrial hazardous wastes. In Bhat RA, Dar GH, Hajam YA (eds). *Zero waste management technologies*. Springer.

Singh J, Yadav P, Pal AK, Mishra V. Water pollutants: origin and status. In Pooja D, Singh P, Kumar P, Patil S (eds.) *Sensors in water pollutants monitoring: role of material.* 2020 (pp. 5–20). Springer.

Speight JG. *Natural water remediation.* 2020. Elsevier.

Wu H, Wang R, Yan P, Wu S, Chen Z, Zhao Y, Cheng C, Hu Z, Zhuang L, Guo Z, Xie H. Constructed wetlands for pollution control. *Nature Reviews Earth & Environment*. 2023;4(4):218–34.

Zhang LL, Zhang R, Wang ZZ. Intergovernmental collaboration, instrument adaptation and embedded synergistic governance: based on 1984–2020 water pollution control policy research. *Environment, Development & Sustainability*. 2023;8:1–23.

24 Water Quality Parameters and Indicators

24.1 WATER QUALITY PARAMETERS

Everyone is entitled to sufficient, safe, accessible, and affordable water for personal and domestic uses. However, the global water crisis is a threat to the world's water supply, which is a result of various factors, including rising temperatures, inadequate sanitation and access to safe drinking water, and pollution. Over half of the world's population (around 4.5 billion people) lives without adequate sanitation worldwide[1]. The bad quality of a water body affects the whole surroundings, local biotas, and possible water uses. Particularly, the contamination of potable (drinking) water sources with complex sewage mixture and inadequate sanitation are linked to many lethal human infections such as cholera, typhoid, dysentery, hepatitis A, intestinal nematode infections, etc.

The health of an aquatic ecosystem is monitored based on certain parameters called environmental indicators. Environmental indicators can be physical, chemical, and biological. Colour, turbidity, temperature, electrical conductivity (EC), total suspended solids (TSS), and total dissolved solids (TDS) represent the physical characteristics of water. Dissolved oxygen (DO), biochemical oxygen demand (BOD), chemical oxygen demand (COD), heavy metals, ammoniacal nitrogen, and phosphorus reflect the chemical parameters of the water. Faecal coliforms, protozoan parasites, and aquatic fauna represent the biological conditions of water bodies. Various tests are conducted to identify the presence of contaminants and their levels in water bodies. Many parameters require physical and chemical analyses. A range of water testing kits is now available for on-site monitoring. Certain parameters require sophisticated tools and processing. Some water quality parameters can also be measured using satellite and aerial images. For example, remote sensing data on algal and phytoplankton concentrations, chlorophyll-*a* (chl-*a*) (measures phytoplankton population density), turbidity, suspended inorganic, coloured dissolved organic matter, etc., are useful parameters to estimate water quality standards. Figure 24.1 illustrates satellite data of algal blooms (using chl-*a*) on oceanic surfaces.

A water quality index (WQI) is a common means to estimate water quality by combining various characteristics into a single numerical value. Many countries have a regional water quality standard and determine WQI using multiple parameters. The most common parameters of WQI are temperature, turbidity, pH, EC, TDS, DO, BOD, COD, ammoniacal nitrogen, and phosphorus. These parameters reflect physical characteristics, chemical reactions and biological processes in water and indicate water conditions that significantly impact its potability. The calculation of WQI generates a score between 0 and 100. The high value of WQI indicates it is suitable for all uses, taking close to the pristine conditions. However, the values toward the lower sides indicate poor water quality.

Many leading international and national bodies have set the regulations and proposed guidelines related to water pollution. The United Nations has set seventeen Sustainable Development Goals (SDGs) addressing various global challenges, including the water crisis. One of its goals, SDG 6, emphasises the need for clean water and sanitation for all. This goal also aims to ensure the availability and sustainable management of water and sanitation for all. The World Health Organisation (WHO) has created guidelines for water quality that serve as the foundation for laws and standards in many nations across the globe (Table 24.1). In the United States (US), over ninety pollutants in drinking water are subject to regulation limitations imposed by the Environmental Protection Agency (EPA). In India, drinking water quality standards have been set by the Bureau of Indian Standards (BIS). The European Union (EU) policy (recast Drinking Water Directive) on drinking water quality ensures high-quality, safe human consumption and promotes health protection. The Standards for Drinking Water Quality (SDWQ) are the water quality standards set by the People's Republic of China (PRC) for drinking water and environmental pollutants. Many nations have set water quality standards and permissible limits for various contaminants in water bodies (Table 24.1). For instance, the WHO advises drinking water to have a concentration of 0.5 mg/l or less of ammoniacal nitrogen, but contaminated water can have levels above 3 mg/l. The US EPA has set regulations for total phosphorus concentrations of 0.10 mg/l to prevent eutrophication of flowing water.

This chapter provides an overview of different parameters that matter for water quality. This includes information on diverse physical, chemical, and biological indicators discussed in various sections below.

DOI: 10.1201/9781032651095-30

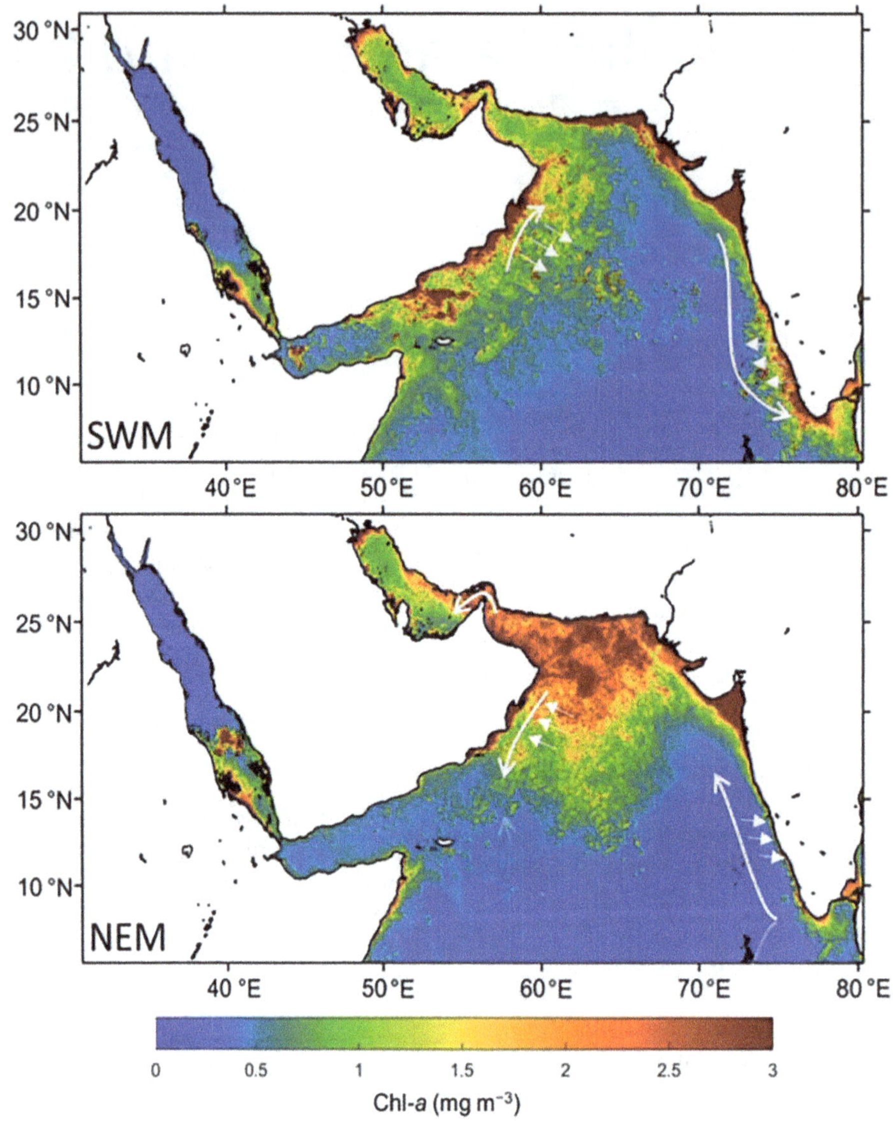

FIGURE 24.1 A remote sensing data of maximum surface chlorophyll-*a* (chl-*a*) concentrations in the North Indian Ocean during southwest and northeast monsoons (SWM and NEM) for the years 2002–2020. The red colour represents the highest and the blue colour depicts the lowest chl-*a* concentration. White arrows depict the directions of the horizontal movements of algae during the monsoons. (Image adapted from Al-Shehhi and Abdul Samad (2022).[3])

24.2 PHYSICAL INDICATORS

Physical indicators provide information about the physical state of an ecosystem. The physical attribute of water reflects the myriad of natural and human-induced activities. Some important identifiable and reliable physical pollution indicators of water are:

- Colour- It is often used as a quantitative assessment of the presence of potentially toxic materials in aquatic ecosystems. Chemicals such as metals, organics, and biological materials (e.g., algae, plankton, or humus) impart different shades to natural aquatic ecosystems. Coloured water reduces light transmission, which can affect the growth of aquatic biota.
- Taste and odour- The presence of various pollutants from domestic and industrial effluents, decomposition of natural materials, and different biological activities gives water a different taste and odour. Simple organic compounds produce sour or salty taste (e.g., lactic acid, acetic acid, and citric acid), while complex organic compounds develop sweet or bitter taste (e.g., phenols, polyphenols, and flavonoids).
- Temperature- Daily and seasonal fluctuations and thermal pollution greatly influence the temperature of water bodies. Elevated temperatures affect many properties of aquatic ecosystems, such as alkalinity, salinity, EC, thermal capacity, density, viscosity, solubility of dissolved gases, and rate of biochemical

TABLE 24.1
The recommended standard parameters and their permissible limit values are set by the World Health Organisation (WHO), the US Environmental Protection Agency (US EPA), the Bureau of Indian Standards (BIS), the European Union (EU), and the People's Republic of China (PRC) for drinking water

Parameters	WHO	US EPA	BIS	EU	PRC
Colour (Platinum-Cobalt or Pt-Co scale)	15	15	15	Acceptable to consumers and no abnormal change	<15
Taste	Tasteless	Unobjectionable	Agreeable	Acceptable to consumers and no abnormal change	-
Odour	Odourless or unobjectionable	Unobjectionable	Agreeable	Acceptable to consumers and no abnormal change	Lack of substances that produce a disgusting smell
Turbidity (Nephelometric turbidity unit)	1	4	1	Acceptable to consumers and no abnormal change	<3
TDS (mg/l)	500	500	500	-	1,000
EC (μS/cm)	750	<1,000	300	2,500	-
pH	7.0-8.5	6.5-8.5	6.5-8.5	6.5-9.5	6.5-8.5
Alkalinity total as $CaCO_3$ (mg/l)	10-500	10-400	200	0-546	450
Sodium (mg/l)	200	30-60	200	250	200
Chloride (mg/l)	250	250	250	250	250
Chromium as Cr^{3+} (mg/l)	0.05	0.1	0.05	0.05	0.05
Pb (μg/l)	0.01	15	0.01	0.01	0.05
Hg (mg/l)	0.001	0.002	0.001	0.001	0.00005
NO_3^- (mg/l)	50	10	45	50	10
SO_4^{2-} (mg/l)	200	250	200	250	250
Fluoride (mg/l)	1.5	4	1	1.5	1
Radioactive materials (becquerel/l or Bq/l): I. Alpha emitters II. Beta emitters	I. 0.5 II. 1.0	I. 0.55 (15 picocurie/l or pCi/l) II. 1.0	I. 0.1 II. 1.0	I. 0.1 II. 1.0 Tritium- 100 Bq/l	I. 0.1 II. 0.1
COD (mg/l)	10	10	10	-	<15
BOD (mg/l)	<5	<5	30	<3	<3
Total coliforms (counts/100 ml)	0	0	0	0	0

activities (such as metabolism, growth, reproduction, and photosynthesis).

- Transparency and turbidity- Transparency refers to the clarity of the water, while turbidity refers to the relative clarity of a liquid. Polluted water containing an excess amount of suspended solids, such as free-floating algae or fine pollution sediments, has high turbidity (with higher intensity of scattered light) and low transparency. Turbidity is considered an effective indicator of harmful events, such as the presence of pathogenic microorganisms in the water. The Jackson turbidimeter and photometer are common devices to measure turbidity in water.
- Conductivity- Measurement of electrical conductance gives an indication of the quantity of dissolved inorganic acids, bases, and salts in the water sample. Water bodies of pristine ecosystems have very low conductivity, whereas industrial effluents show very high conductivity. The outsourced nutrients in the freshwater aquatic system can increase the water conductivity. Elevated temperatures also increase the water conductivity. Many biological activities in freshwater ecosystems are inhibited when the EC exceeds 4 dS/m (decisiemens per metre).
- Salinity- The salinity represents the total amount of inorganic mineral ion compositions. It is typically expressed in parts per thousand (ppt) or ‰. Electrical conductivity indirectly estimates the total salinity and is frequently linked to extreme levels of ecosystem pollutants.
- Total Dissolved Solids (TDS)- It refers to all dissolved substances present in water. The diameter of TDS

particles is generally lower than 2 µm. TDS of water can be measured through gravimetric analysis. This is performed by filtering water samples in a pre-weighed porcelain dish using a filter. The filter is usually made of glass fibre or cellulose ester membranes and has pore sizes around 2.0 µm. After filtration, the filtered water in a porcelain dish is subjected to evaporation at 103-105 °C to a constant weight. The dish is placed in a desiccator to cool down at room temperature, followed by weighing. The cooling period in a desiccator prevents moisture absorption in the dish. An increase in the porcelain dish's weight represents the water sample's TDS. TDS can be calculated using the following formula:

$$TDS = \frac{(A - B)}{V}$$

where A is the weight of the dish with solid residues
B is the weight of the dish without solid residues
V is the volume of the sample (l)

Differently, a digital device TDS metre gives instant readings of TDS and is appropriate for on-site measurement. A high TDS level in tap water gives an unpleasant odour and bitter taste. TDS consists of a major amount of inorganic and some small-sized organic substances. Sodium, magnesium, calcium, potassium, carbonates, bicarbonates, nitrates, chlorides, and sulphates are common ingredients of dissolved content in water. Both natural and anthropogenic activities can be responsible for high TDS in water. Mineral springs have water with high quantities of dissolved particles because it has passed through salty rocks. These mineral waters have naturally elevated levels of dissolved solids, which confer health benefits. The TDS content of bottled mineral water normally ranges between 400 and 650 mg/l. However, TDS of more than 300–500 mg/l in regular drinking water is not recommended. A TDS above 1,000 mg/l is unfit for human consumption. TDS also has a minimum limit in drinking water that should not go below 50 mg/l because a low concentration of some minerals in drinking water could be detrimental to health.

- Total Suspended Solids (TSS)- TSS in an aquatic ecosystem can be defined as the amount of particulate matter or solids in water that can be trapped using a filter. TSS particles are larger than 2 µm in diameter and often consist of inorganic materials, sediment, silt, sand, plankton, algae, and decomposing organic materials. Raw wastewater typically has a TSS value ranging between 155–330 mg/l. However, the TSS of wastewater after primary and secondary treatment processes remain around 65 mg/l and less than 20 mg/l, respectively. When the concentrations of suspended particles in a liquid are very high, they tend to settle at the bottom. The presence of suspended solids reduces the assimilation capacity and transparency of the aquatic ecosystems, which impedes underwater photosynthesis, oxygenation, and movement of aquatic biotas.
- TSS of water can be measured through gravimetric analysis similar to TDS. However, solids larger than 2 µm are captured by passing a water sample through a glass fibre filter that has been previously weighed. After water filtration, the settled residues on the filter are dried at 103–105 °C to a constant weight. An increase in the weight of the filter after water evaporation represents the TSS of the water sample.

24.3 CHEMICAL INDICATORS

Water pollution by different chemicals has become a serious worldwide environmental concern. Chemical pollution indicators are mostly associated with nutrient cycling, water relations, and buffering capacity of the ecosystem. Dyes, surfactants, organochlorinated pesticides, pharmaceutical contaminants, polycyclic aromatic hydrocarbons (PAHs), polychlorinated biphenyls (PCBs), and brominated flame retardants are some commonly known organic pollutants of water. Many heavy metals, radionuclides, and other inorganic substances in high concentrations coexist in aquatic ecosystems. Quantitative estimation of different chemicals in the water requires sensitive analytical techniques like atomic absorption spectroscopy (AAS), inductively coupled plasma mass spectrometry (ICP-MS), optical emission spectrometry (OES), UV-vis spectrophotometer, electrochemical methods, gas chromatography-mass spectroscopy (GC-MS), immunoassay, liquid chromatography, etc.

Some important characteristics of potential chemical pollution indicators of water are:

- pH- The pH is the negative logarithm of the concentration (mol/l) of hydrogen ions (H^+) in an aqueous solution. The pH of an aquatic ecosystem indicates the strength of acid or alkaline conditions from the H^+ activity. Extreme pH values reduce growth, reproduction, and biodiversity and damage organs. It can also affect the solubility and mobility of nutrients and potentially toxic elements (PTEs). Fertilisers, mineralisation, nitrification, acid-generating rocks or soils, industrial effluents, confined animal feeding operations (CAFOs), and acid rain are distinct sources of contaminants that can raise acidity in water. Excessive respiratory activities in aquatic systems can also reduce the pH (<6 or 6.5) of water due to assimilation of high amount of carbon dioxide (CO_2). Low pH results into nutrient runoff, corrosion, and biodiversity loss. On the contrary, the pH of water can be elevated due to agricultural lime, cement and soap manufacturing, industrial landfills, asphalt production or disposal, and the dissolution of soil/bedrock. Alkaline water (pH >8.5) tastes more bitter and causes the skin to become dry, itchy, and irritated. They are corrosive to pipes and plumbing systems. The pH of the water is also measured using a typical pH metre, consisting of a probe connected to an electronic metre. Colour comparators and pH strips are two other important components to determine the pH by comparing the resulting colour with standard colour charts.

- Reactive carbon- The reactive carbon represents permanganate oxidisable carbon. It is significantly related to fresh organic matter, particulate organic carbon (POC), microbial biomass, and loose carbon from ecosystem minerals. Reactive carbon has a turnover with a time interval of two to five years, whereas recalcitrant carbon, like humus, has a residence time of hundreds to thousands of years. These species can easily bind to minerals and can be readily decomposed by microorganisms.
- Carbon dioxide (CO_2)- This gas in an aquatic ecosystem is an indirect indicator of organic pollution because it comes from the biodegradation of organic matter and respiratory activities. The gaseous exchange at the air-water interface also adds sufficient CO_2 to the water.
- Nutrients- High concentrations of nitrogen (like nitrate (NO_3^-) and ammonium (NH_4^+)) and phosphorus in an aquatic ecosystem increase phytoplankton growth (pond scum) that leads to eutrophication, which consumes oxygen (O_2) and lowers the water quality (detailed in Chapter 20). They mainly come from the discharge of agricultural drainage effluents, domestic sewage, or nitrogen-containing industrial effluents.
- Heavy metals- Among various metals, toxic heavy metals (THMs) in water bodies raise major concerns. THMs can be arsenic $(As)^2$, cadmium (Cd), lead (Pb), mercury (Hg), and chromium (Cr)(VI), which do not have any minimum threshold limit. Even their low concentrations pose significant challenges to the native lifeforms. Even many heavy metals (which are micronutrients) like copper (Cu), zinc (Zn), manganese (Mn), and iron (Fe) can also have a negative impact when their concentrations cross the threshold limits. Heavy metals are non-degradable and, therefore, remain in the environment for a very long time. They can be carcinogenic, mutagenic, or neurotoxic. Some of them (e.g., organomercuric compounds) have potential for biomagnification. Both natural reasons, such as rock abrasion, weathering and volcanic eruptions, and anthropogenic events like municipal and industrial wastewater, landfill leachates, and mining waste, contribute to heavy metal pollution in aquatic systems.
- Radionuclides- Radioactive materials (or radionuclides) are chemical substances with unstable atomic nuclei. Some examples of radionuclides that can be traced from surface waters and groundwater are uranium (U), radium (Ra), polonium (Po), and radon (Rn). When radionuclides decay, they emit harmful radiation (gamma rays, neutrons, or beta particles). Excessive radiation exposure raises health issues such as cancer.
- Total organic carbon (TOC)- It is a highly sensitive and non-specific measure of the entire organic matter present in the water sample. The two major sources of total organic carbon (TOC) in water are natural organic matter (NOM) and synthetic organic compounds. The NOM encompasses carbohydrates, proteins, amines, fats, humic acid, fulvic acid, urea, etc. Synthetic organic compounds can be chemicals like detergents, fertilisers, herbicides, organochlorine compounds, pesticides, dyes, etc. TOC measurements also cover volatile/purgeable organic carbon (VOC/POC) and non-purgeable organic carbon (NPOC). The TOC can be determined by adding the values of dissolved organic carbon (DOC) and non-dissolved organic carbon (NDOC). Differently, it can be measured by subtracting the values of total inorganic carbon (IC) (such as carbonate, carbonic acid, etc.) from total carbon (TC). A low TOC indicates the absence of potentially harmful chemicals in water.

  The TOC in water samples is measured using a TOC analyser, which operates on the principle of the IC-TC method. The detection of TOC involves three primary steps: acidification, oxidation, and detection/quantification. The acidification step allows the conversion of all inorganic carbon (such as bicarbonate and carbonate ions) in the sample into CO_2, which is then removed by inert gas sparging. This is followed by the oxidation step, which involves the transformation of organic fractions into CO_2 by various methods such as combustion, thermochemical oxidation, photooxidation, and persulphate oxidation. Carbon dioxide released from this step can be quantified using a non-dispersive infrared (ND IR) analysis or conductivity detectors. In a different method, organic fractions can be estimated based on absorbance characteristics using a UV-vis spectrophotometer and spectrofluorometer.

In addition, chemical parameters such as DO, BOD, and COD are crucial as water pollution indicators. The following subheadings give in-depth details about their importance and estimation.

24.3.1 Dissolved Oxygen (DO)

DO refers to the amount of O_2 present in water. DO in water bodies mainly comes from the photosynthetic activity of phytoplankton and macrophytes and the gaseous exchange with the atmosphere. Running water and moving streams dissolve more O_2 than the still water of a pond or lake. Clean aquatic ecosystems have a near-saturation of DO at their surface. The water of a healthy aquatic system has relatively high and stable levels of DO. High DO in the aquatic body is important to support aquatic wildlife. DO concentrations in healthy water should be greater than 6.5-8 mg/l. DO levels in water bodies fluctuate periodically, seasonally, and according to the natural ecology of the aquatic system. The DO is an important chemical pollution indicator during the wastewater treatment process control.

The discharge of a vast amount of wastewater into aquatic ecosystems reduces DO levels and deteriorates water quality. Pollution of water with organic waste (from sewage, manure, organic fibres from textile and paper processing, food wastes, etc.) and nutrients (from fertilisers and agricultural runoff) can have a severe impact on DO. Excess discharge of organic matter and nutrients promotes the biological degradation of organics, which enhances the growth of heterotrophic microorganisms and reduces O_2 concentration. Water bodies and chemical parameters (like TDS, organics, and heavy metals) can also react with each other to decrease O_2. When DO levels drop, the density of sensitive aquatic animals declines dramatically. Water with DO levels below 1 mg/l is considered hypoxic and usually devoid of life.

DO in water can be measured through titration, colourimetric method, and electrochemical/optical sensors. A standard titration-based iodometric method is most commonly used to determine DO concentrations in water samples. It was initially developed by a Hungarian chemist, L.W. Winkler (1888) and modified by Strickland and Parsons (1968). This method is based on the oxidising property of DO. The first step is the collection of a water sample in a BOD bottle without trapping any air bubbles and immediately covering it with a stopper (Figure 24.2). The water sample is treated with a manganese sulphate ($MnSO_4$) solution, followed by an alkali-iodide-azide reagent. In this process, $MnSO_4$ reacts with alkali (i.e., sodium hydroxide (NaOH)) and forms pale yellow precipitates of manganese hydroxide ($Mn(OH)_2$ or Mn^{2+}). The $Mn(OH)_2$ gets rapidly oxidised by an equivalent amount of DO into brown hydrated precipitates of manganese dioxide (MnO_2 or Mn^{4+}) (reactions 24.1 and 24.2). This process facilitates the fixation of DO present in the water sample. Fixation of samples is usually performed immediately after sample collection to prevent any change in DO levels from agitation or microbial respiration. The precipitate of MnO_2 settles to the bottom of the bottle. In the next step, concentrated sulphuric acid (H_2SO_4) is carefully added and mixed to dissolve the precipitate (Figure 24.2). At low pH, MnO_2 reacts with iodide ions (I^-) to form iodine (I_2) and divalent manganese (Mn^{2+}) (reaction 24.3). The amount of I_2 liberated in the reaction is equivalent to the total DO concentration of the water sample. Insoluble I_2 can form the triiodide ion (I_3^-) complex with iodide ions in the solution (reaction 24.4). A known low volume of the reaction mixture is then titrated with 0.025 N sodium thiosulphate ($Na_2S_2O_3$) solution, wherein the starch solution is used as an indicator that forms a blue-black colour with I_2. The $Na_2S_2O_3$ is a reducing agent, which reduces the oxidising agent I_2 to produce sodium tetrathionate ($Na_2S_4O_6$) and sodium iodide (NaI) (reaction 24.5). During this redox reaction, the blue-black colour disappears, and the solution becomes colourless (Figure 24.2). The volume of $Na_2S_2O_3$ titrant needed to complete the reaction is proportional to I_2 concentration and also the total DO of the water sample. This gives an estimate of the DO present in the water sample.

Fixation of DO:

$$MnSO_4 + 2NaOH \rightarrow Mn(OH)_2 + Na_2SO_4 \quad \text{reaction 24.1}$$

or, $$Mn^{2+} + OH^- \rightarrow Mn(OH)_2$$

$$Mn(OH)_2 + 0.5O_2 \rightarrow MnO_2 + H_2O \quad \text{reaction 24.2}$$

(Brown precipitate)

Liberation of iodine:

$$MnO_2 + 2I^- + 4H^+ \rightarrow Mn^{2+} + I_2 + 2H_2O \quad \text{reaction 24.3}$$

where H^+ is the hydrogen ion from H_2SO_4.

$$I_2 + I^- \rightarrow I_3^- \quad \text{reaction 24.4}$$

Titration:

$$2Na_2S_2O_3.5H_2O + I_2 \rightarrow Na_2S_4O_6 + 2NaI + 10H_2O \quad \text{reaction 24.5}$$

(blue-black (colourless)
colour in
presence of
starch indicator)

The presence of nitrite ions (NO_2^-) in water samples can interfere with titration. In acidic pH, NO_2^- oxidises I^- ions to I_2 and gets reduced to dioxohydrazine (N_2O_2) (reaction 24.6). The N_2O_2 is then further oxidised by DO into NO_2^- ion, which again starts the oxidation of I^- ions (reaction 24.7). The synthesis of I_2 allows the blue colour to reappear after titration and gives an erroneous result. However, the interference of NO_2^- can be overcome using sodium azide (NaN_3) in the reaction mixture, which is called azide modification. The NaN_3 forms hydrazoic acid or hydrogen azide (HN_3) in acidic conditions, which reacts with NO_2^- and releases nitrogen (N_2) and nitrous oxide (N_2O) gases (reactions 24.8 and 24.9).

Nitrite interference:

$$2NO_2^- + 2I^- + 4H^+ \rightarrow I_2 + N_2O_2 + 2H_2O \quad \text{reaction 24.6}$$

$$N_2O_2 + 0.5O_2 + H_2O \rightarrow 2NO_2^- + 2H^+ \quad \text{reaction 24.7}$$

Azide modification:

$$NaN_3 + H^+ \rightarrow HN_3 + Na^+ \quad \text{reaction 24.8}$$

$$HN_3 + NO_2^- + H^+ \rightarrow N_2 + N_2O + H_2O \quad \text{reaction 24.9}$$

After the titration of the sample, the DO can be estimated using the formula below:

$$\text{DO in mg/l} = \frac{\text{EW of } O_2 \times 1000 \times N \times v}{V}$$

Where N= Normality of $Na_2S_2O_3$
v = total volume (ml) of $Na_2S_2O_3$ consumed
V = Volume (ml) of sample used for titration
EW of O_2 = Equivalent weight of oxygen (8 g) (this factor may vary depending on the sample volume and reagent concentrations)
Based on the estimate, 1 ml of 0.025 N $Na_2S_2O_3$ is equivalent to 0.2 mg O_2.

In a different colourimetric method, DO can be estimated using indigo carmine and rhodazine D. The indigo carmine method produces a blue colour and detects DO in the range of

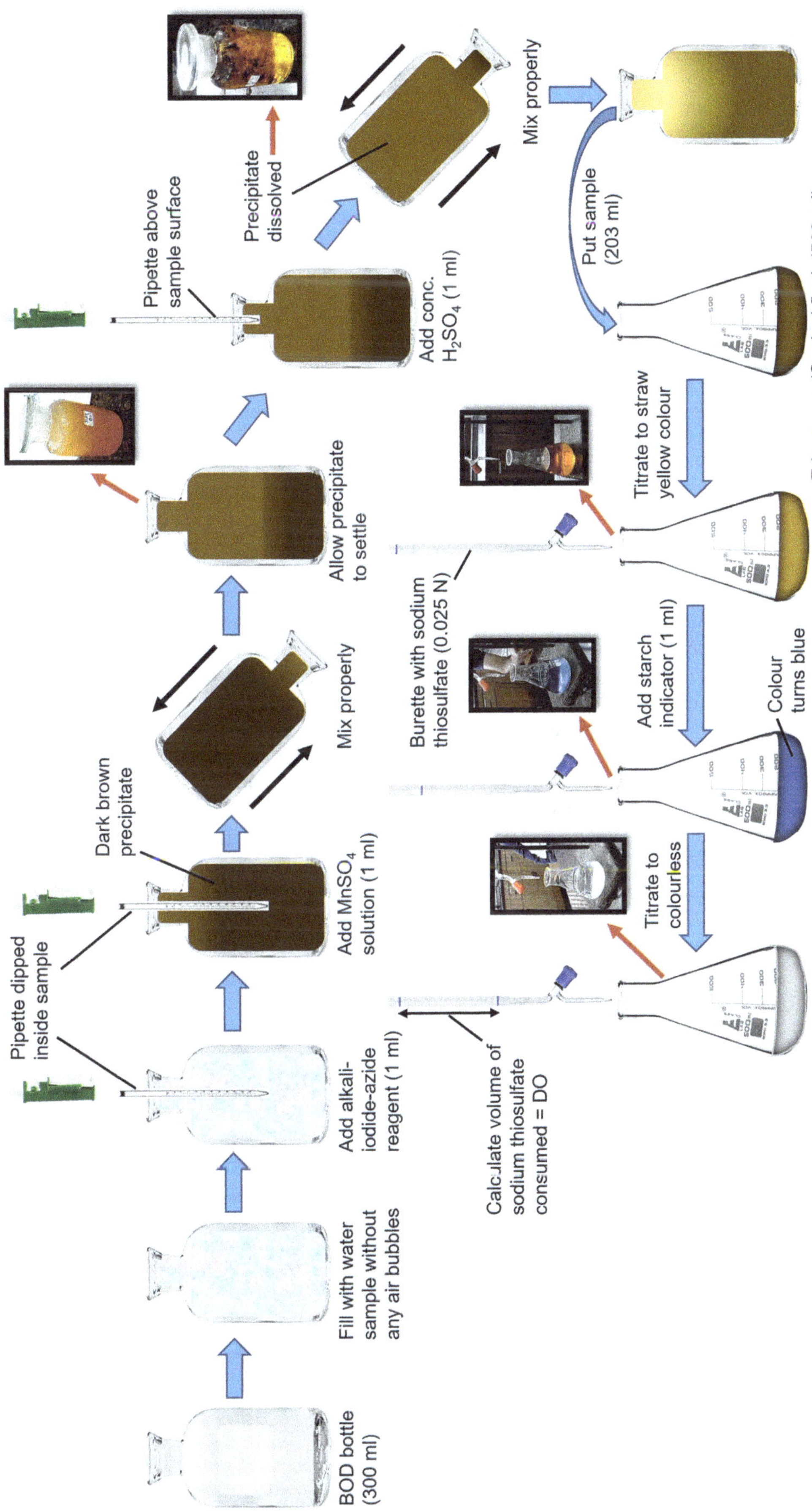

FIGURE 24.2 The schematic representation of the estimation of dissolved oxygen (DO) in water samples using the standard titration-based iodometric method. The red arrows indicate actual images at different stages of the procedure.

0.2 and 15 ppm (mg/l). When DO is in very low concentrations (in parts per billion (ppb)), it can be measured using the rhodazine D method, which develops a deep rose-coloured solution after the reaction. A digital DO metre provides a direct reading of DO concentration in mg/l, which employs electrochemical (polarographic and galvanic) or optical sensors. These sensors require power and only a few minutes to acquire a reading. This device is known for its accuracy, convenience, speed, portability, and low maintenance.

24.3.2 Biochemical Oxygen Demand (BOD)

BOD refers to the amount of molecular O_2 consumed by the biological oxidation of organic compounds in water during a specific time interval. The breakdown of organic matter in water samples by aerobic bacteria is the groundwork for oxygen demand. BOD represents the biodegradable organic load in water (of domestic sewage and industrial wastewater) and is a common index to assess water quality from the perspective of its O_2 concentration. A low BOD value indicates low concentrations of organic matter and low counts of heterotrophic microorganisms in the water sample. The testing of BOD is useful in the assessment of self-purification capacity and management of pollution control of water streams.

A 5-day incubation is conventional for the analysis of BOD, which is usually referred to as the BOD-5 (or BOD_5) method. This typically requires a temperature of 20 °C. Incubation in the dark prevents the synthesis of oxygen in the samples due to photosynthesis. During the incubation, the heterotrophic microorganisms present in the samples utilise DO to respire the biologically available organic carbon. The DO is measured on the first and last day, as mentioned in the previous section (section 24.3.1). The difference between initial and final DO concentrations gives the BOD value of the water sample. It can also be formulated as:

BOD (in mg/l) = $DO_{inital} - DO_{final}$, where

DO_{inital}: DO of water sample (in mg/l) when water sample was collected.

DO_{final}: DO of water sample (in mg/l) after 5-day incubation at 20 °C.

The high demand in contaminated water sample typically requires large dilution factors because the DO level in water does not exceed more than 9-10 mg/l. If the samples were diluted, the formula can be corrected for their dilution factors.

The oxidation of carbonaceous materials after 5 days is around 60-70%, and after 20 days, it is 95-99%. However, after 8-10 days of incubation, nitrifying bacteria start using oxygen (O_2) for nitrification, which constitutes nitrogenous BOD (NBOD). This is the reason why the 5-day incubation is optimal for the BOD test, which measures the initial carbonaceous BOD (CBOD) stage, ensuring a low population of nitrifying bacteria. However, if the sample has high ammonia (NH_3) concentrations, its oxidation can be prevented by adding allylthiourea to the water sample, which yields only carbonaceous oxidation.

The BOD values for different water bodies can be different. For example, pristine water has a BOD of around 1 mg/l, moderately polluted water, 2-8 mg/l, and treated municipal sewage, 20-30 mg/l. Municipal wastewater typically has BOD values of 200-300 mg/l, whereas food industry effluent has BOD values of 1,000-10,000 mg/l. The high value of BOD in food industry effluent is due to a relatively very high fraction of decomposable organic matter. The landfill leachate typically consists of high BOD ranging from 4,000-13,000 mg/l. Based on BOD values, the amount of aeration required at the biological treatment facilities can also be estimated.

24.3.3 Chemical Oxygen Demand (COD)

COD is an estimate of the amount of oxygen required for the oxidation of organic and inorganic compounds in water samples. Organic compounds are the primary contributors to COD, but some inorganics like NO_2^-, sulphide (S^{2-}), and ferrous salts also contribute to the overall COD. According to the basic principle of COD, an acidic environment facilitates the complete oxidation of almost all organic compounds into CO_2 in the presence of a potent oxidising agent (e.g., potassium dichromate ($K_2Cr_2O_7$)). However, substances like polycyclic aromatic hydrocarbons and dioxin-like compounds, being highly stable compounds and resistant to oxidation, cannot be measured accurately in COD. The COD represents a very important chemical pollution indicator, which reveals the water quality and pollution level.

The estimation of COD accounts for both biodegradable and non-biodegradable fractions of organic compounds. However, in a BOD test, only biologically reactive carbon is measured. This is the reason why the COD values are considerably higher than the BOD. The ratio of BOD to COD is a useful assessment measure to know whether waste can be treated biologically. When this ratio of a water sample equals or exceeds 0.8, biological treatment methods are appropriate. Some key differences between BOD and COD are given in Table 24.2.

The COD test is estimated using the dichromate reactor digestion method. This method utilises a strong oxidising agent, $K_2Cr_2O_7$, which oxidises the organic matter in the sample in the presence of H_2SO_4, mercuric sulphate ($HgSO_4$) and silver sulphate (Ag_2SO_4). The $HgSO_4$ reduces chloride interference in the water sample because it is commonly used as a disinfectant in water treatment (reaction 24.10). If the water sample consists of chloride ions and is not treated with mercuric salt, dichromate ($Cr_2O_7^{2-}$) ions will oxidise it into chlorine (Cl_2), resulting in higher COD than the actual value (reaction 24.11).

$$2Cl^- + 2HgSO_4 \rightarrow HgCl_2 + SO_4^{2-} \quad \text{reaction 24.10}$$

$$Cl^- + Cr_2O_7^{2-} + 14H^+ \rightarrow 3Cl_2 + 2Cr^{3+} + 7H_2O \quad \text{reaction 24.11}$$

Ag_2SO_4 acts as a catalyst and promotes the oxidation of certain straight-chain organic compounds, like diesel and motor oil. Ag_2SO_4 is added only after removing all chloride

TABLE 24.2
Differences between biochemical oxygen demand (BOD) and chemical oxygen demand (COD)

Characteristics	BOD	COD
Indicator of	Biodegradable organic pollutants	Both biodegradable and nonbiodegradable organic and inorganic
Determines the oxygen required for	Biological oxidation	Chemical oxidation
Dissolved oxygen consumed by	Microorganisms like bacteria	A strong oxidising agent, $K_2Cr_2O_7$, under acidic conditions
Duration to complete	5 days	3 h
Natural processes resemblance	Yes	No
Disposal of reagents used	Not needed	Required
Observed concentrations	Low. Only biologically reactive carbon is utilised	High. Total carbon (organic and inorganic) is utilised

ions. Otherwise, silver ions will be precipitated as silver chloride (AgCl) (reaction 24.12).

$$2Cl^- + Ag_2SO_4 \rightarrow 2AgCl + SO_4^{2-} \quad \text{reaction 24.12}$$

This is followed by the complete oxidation of organic matter using excess concentrations of acidified $K_2Cr_2O_7$ (reaction 24.13). However, this chemical oxidant is non-specific to organic or oxidisable inorganic compounds, which detects both sources of oxygen demand. This mixture is digested using a digester for a specified duration (usually two h at 150 °C). The sample consumes a certain amount of $K_2Cr_2O_7$, which is equivalent to the amount of O_2 required for the oxidation of organic materials. In this reaction, the dichromate ion initially produces nascent oxygen [O] while getting reduced into chromium(III) sulphate ($Cr_2(SO_4)_3$), which is a chromic ion (Cr^{3+}), producing a green hue to the solution. The nascent oxygen oxidises the organic load into CO_2 and water (reaction 24.14).

$$K_2Cr_2O_7 + 4H_2SO_4 \rightarrow K_2SO_4 + Cr_2(SO_4)_3 + 4H_2O + 3[O] \quad \text{reaction 24.13}$$

$$\text{Organic load} + [O] \rightarrow CO_2 + H_2O \quad \text{reaction 24.14}$$

The amount of unconsumed chromate ion (Cr^{6+}) is determined by titrating with a standard solution of the reducing agent ferrous ammonium sulphate (FAS) using ferroin [1,10-phenanthroline Fe(II) sulphate] ($[Fe(C1_2H_8N_2)_3]^{2+}$) as an indicator. Ferroin contains ferrous (Fe^{2+}) ions and is reddish brown. When ferroin is added to the wastewater sample, the unconsumed Cr^{6+} oxidises the Fe^{2+} of ferroin to ferric (Fe^{3+}) ions, which gives a bluish-green shade. During titration, the Fe^{2+} of the FAS then reduces the unconsumed Cr^{6+} to Cr^{3+} (reactions 24.15 and 24.16).

$$K_2Cr_2O_7 + 7H_2SO_4 + 6Fe(NH_4)_2(SO_4)_2 \rightarrow Cr_2(SO_4)_3 + 3Fe_2(SO_4)_3 + K_2SO_4 + 6(NH_4)_2SO_4 + 7H_2O \quad \text{reaction 24.15}$$

$$\text{or, } Cr_2O_7^{2-} + 14H^+ + 6Fe^{2+} \rightarrow 2Cr^{3+} + 6Fe^{3+} + 7H_2O \quad \text{reaction 24.16}$$

At the endpoint, after the full reduction of Cr^{6+} to Cr^{3+}, FAS reduces the ferroin indicator (Fe^{3+}) (blue-green colour) into the Fe^{2+} state, which gives a reddish-brown colour (reaction 24.17).

$$[Fe(C1_2H_8N_2)_3]^{3+} \text{ (bluish green)} + e^- \text{ (from FAS)} \rightarrow [Fe(C1_2H_8N_2)_3]^{2+} \text{ (reddish brown)} \quad \text{reaction 24.17}$$

In parallel, a reference blank solution is also maintained, which is treated as the test solution, but it contains distilled water rather than a wastewater sample. The distilled water is free of any waste; therefore, the entire $K_2Cr_2O_7$ added remains unoxidised. The total $K_2Cr_2O_7$ is available to react with the FAS in titration.

In the calculation, V1 represent the volume of FAS used for titrating unconsumed $K_2Cr_2O_7$ in the wastewater sample (ml). V2 is the volume of FAS used for titrating unconsumed $K_2Cr_2O_7$ in a blank sample (ml). The amount of $K_2Cr_2O_7$ consumed for oxidation of the total organic load in terms of FAS volume can be formulated as

$$\text{Total volume of FAS consumed} = (V2 - V1)$$

After the titration of the wastewater and blank sample, the COD can be estimated using the formula below.

$$\text{COD as } O_2 \text{ in mg/l} = \frac{(V2 - V1) \times N \times \text{EW of } O_2 \times 1000}{V}$$

Where V2 = Volume of FAS used for blank titration (ml)
V1 = Volume of FAS used for wastewater sample titration (ml)
N = Normality of FAS
V = Volume of water samples taken for titration analysis (in ml)
EW of O_2 = Equivalent weight of oxygen (8 g) [8 × 1000 mg]

If the original sample is diluted, the formula will include multiplication by the dilution factor.

A high COD value indicates a high concentration of organic pollutants, suggesting significant contamination. The COD levels (mg/l) in water vary with the types of different water sources, such as surface water (5-20), natural rivers (5-50), raw municipal sewage (500-1,200), industrial effluent (100-60,000), primary and secondary effluent (250-750), and treated effluent (25-250).

24.4 BIOLOGICAL INDICATORS

Biological indicators of water quality are the biological organisms and their responses that provide crucial information about the health and conditions of aquatic environments. These indicators include several types of organisms, including microbial forms (such as coliform bacteria, *Enterococci*, *Clostridium perfringens*, and pathogens), macroinvertebrates (such as insects, worms, and crustaceans), vertebrates (such as fish, amphibians, and reptiles), and macrophytes (such as algae and large plant bodies). The assessment of biological indicators can be based on their absence or presence. Some species are sensitive to pollutants, and their decline from their natural habitat is a strong indication of deteriorating water quality. Some biological indicators may exhibit high abundance in the contaminated environment. In addition, biological activities, growth rates, deformities, and fertility are different responses to biological indicators. However, the presence of an indicator organism doesn't always guarantee the presence of the target pathogens.

24.4.1 Indicator Microorganisms of Water Quality

Microorganisms in aquatic environments are ubiquitous and persist in modified or contaminated environmental conditions. The quality of water for human consumption and recreational use requires frequent analysis of pathogenic microorganisms owing to their pathogenic concerns. Faecal contamination leads to the most damaging effects in water. Some notable water-borne diseases are cholera (*Vibrio cholerae*), gastroenteritis (*Vibrio parahaemolyticus* and *Escherichia coli*), typhoid fever and serious salmonellosis (*Salmonella enterica* subsp. *enterica*), bacillary dysentery (*Shigella dysenteriae*), and acute diarrhoea. Some cyanobacteria present in water produce cyanotoxins, including microcystins, that are harmful to humans. The detection of pathogens in water samples raises pathogenic concerns. A routine evaluation for the presence of pathogens in water samples is ideal, but is time-consuming. Also, pathogens in water samples are often in low numbers, which necessitate specialised laboratories that add to the economic burden. Therefore, the potential presence of pathogens can be monitored through biological indicators.

Many microbial taxa residing in water can serve as indicators of the possible presence of pathogens and other contaminants in water. Looking for an indicator organism is a useful parameter to check water quality. These indicators can represent and quantify the concerns of faecal contamination and other environmental issues. There are certain key criteria for the selection of indicator microorganisms (Figure 24.3), which are mentioned below

- They should be easily isolated and quantified using simple microbiological methods.

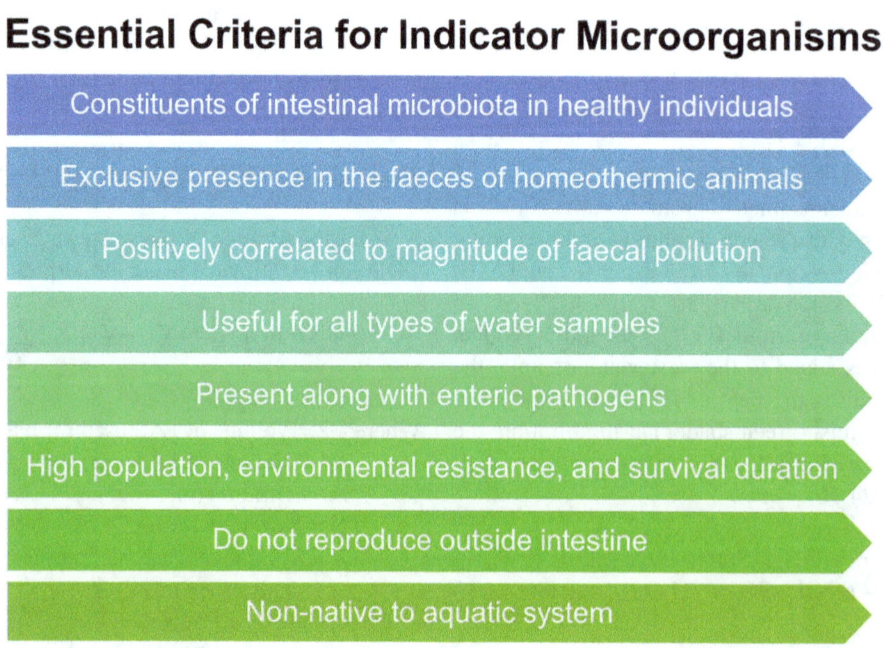

FIGURE 24.3 Essential criteria for indicator microorganisms.

- Indicator microorganisms should have a dense population, robustness, resistance to environmental factors, and longer survival duration than enteric pathogens.
- Their detection methods should be well-established, standardised, and relatively simple.
- Their tests should yield relatively fast results and incur low costs.

24.4.2 Groups of Microbial Indicators

Microbiological indicators of polluted water chiefly include bacteria (like *E. coli*, *Legionella*, *Salmonella*, *Shigella*, and *Vibrio*), protozoans (like *Cryptosporidium* and *Giardia*), and viruses (like enterovirus, hepatitis A virus, norovirus, poliovirus, and rotavirus). Bacterial indicators are often used as official standards of quality drinking water, recreational bathing water, and wastewater discharge. The microbial indicator groups of water quality can be further categorised according to their purposes. For example, faecal indicators in water samples indirectly indicate the possible presence of disease-causing pathogenic microorganisms, whereas the quality of freshwater is assessed by monitoring benthic algal growth and chl-*a*.

The faecal indicators are a group of organisms which indicate the presence of faecal contamination in water. Stormwater runoff, on-site septic systems, household and wild animal manure, and wastewater treatment facilities are some common sources of faecal contamination. Faecal streptococci and coliforms represent potential indicators of faecal contamination. Faecal streptococci typically reside in the digestive tract of humans and other warm-blooded animals. Faecal streptococci are potential indicator organisms because they are fewer common residents of water than other types of microorganisms, are highly resilient to environmental stress, and have a long environmental lifespan. Within the faecal streptococcus group, enterococci are a subgroup that is more particular to humans and thrives in saltwater. Faecal *Streptococcus bovis* and *Streptococcus equinus* are common inhabitants of warm-blooded animals, while *Enterococcus faecalis* and *Enterococcus faecium* typically reside in the human gut. The faecal coliform bacterial species *E. coli* is exclusive to human and other warm-blooded animal faeces. Section 24.8 gives a detailed description of coliform bacteria.

Raw sewage may contain some other widely distributed bacterial indicators. For instance, the Gram-positive, rod-shaped, anaerobic, sulphite-reducing and endospore-forming *Clostridium perfringens* has an exclusively faecal origin and is associated with human faeces. A Gram-negative, rod-shaped, non-sporulating *Pseudomonas aeruginosa* is an opportunistic pathogen, often present in sewage. A different Gram-negative, rod-shaped, facultatively anaerobic bacterium, *Aeromonas,* is pathogenic and infects warm-blooded and cold-blooded animals. It can be present in uncontaminated water and sewage-contaminated nutritious water. Anaerobic bacteria such as *Bifidobacterium* (*B. adolescentis* and *B. dentium*) and *Bacteroides* are other recommended human faecal pollution indicators. Elevated quantities of faecal bacteria in water also bring some observational changes, such as cloudiness, disagreeable odours, and increased O_2 consumption, which contribute to the potential health danger.

Phage viruses (coliphages or bacteriophages) are abundantly present in sewage and polluted water and are notable indicators of faecal and viral pollution. Human faecal viromes and faecal metagenomes frequently detect crAssphage (a double-stranded DNA bacteriophage that infects *Bacteroides intestinalis*). In addition, some human viruses like adenoviruses and polyomavirus can also be used to trace human faecal contamination in waters, considering their human specificity, environmental stability, and high frequency.

Some specific indicator organisms take part in more direct risk assessments and are termed index organisms. For example, *E. coli* is an index organism for the specific presence of *Salmonella*. The male-specific coliphage infects fertility (F)-factor-carrying *E. coli*, which can also be used as an index organism.

24.5 COLIFORMS

Coliforms are Gram-negative, rod-shaped, non-endospore forming, motile/non-motile, oxidase-negative, aerobic/facultatively anaerobic bacteria. They characteristically possess the enzyme β-galactosidase, which hydrolyses lactose into glucose and galactose intracellularly. These reducing sugars are further fermented into gases (CO_2 and H_2) and acids (lactic acid, acetic acid, propionic acid, butyric acid, etc.) that lower the pH of the medium.

Coliforms can be recognised as total coliforms, faecal coliforms, and non-faecal coliforms (Figure 24.4). Total coliforms include a vast collection of bacteria that are mostly innocuous and can be found in soil, plants, aquatic systems, human faeces, and animal faeces. On the other hand, faecal coliforms are a subgroup of total coliforms. Faecal coliforms are residents of the intestinal tract of warm-blooded animals that can enter the environment through faecal contamination. A different group, non-faecal coliforms in total coliforms, distinguish themselves from faecal coliforms, as

$$\text{Total coliforms} = \text{Faecal coliforms} + \text{Non faecal coliforms}$$

Coliforms typically include *Citrobacter*, *Enterobacter*, *Escherichia*, and *Klebsiella*. Several other genera, like *Hafnia*, *Serratia*, and *Yersinia*, can also be covered under the definition of coliforms. They all belong to the family Enterobacteriaceae. Although most coliforms themselves do not cause serious illness, their presence indicates the possible presence of pathogenic bacteria such as *Shigella*, *Salmonella*, and *Vibrio cholerae*, which have faecal origins. An absence of coliform bacteria indicates that the water is possibly bacteriologically safe for human consumption. The main advantage of using coliform as a water quality indicator is its simple and inexpensive testing methodology. They can

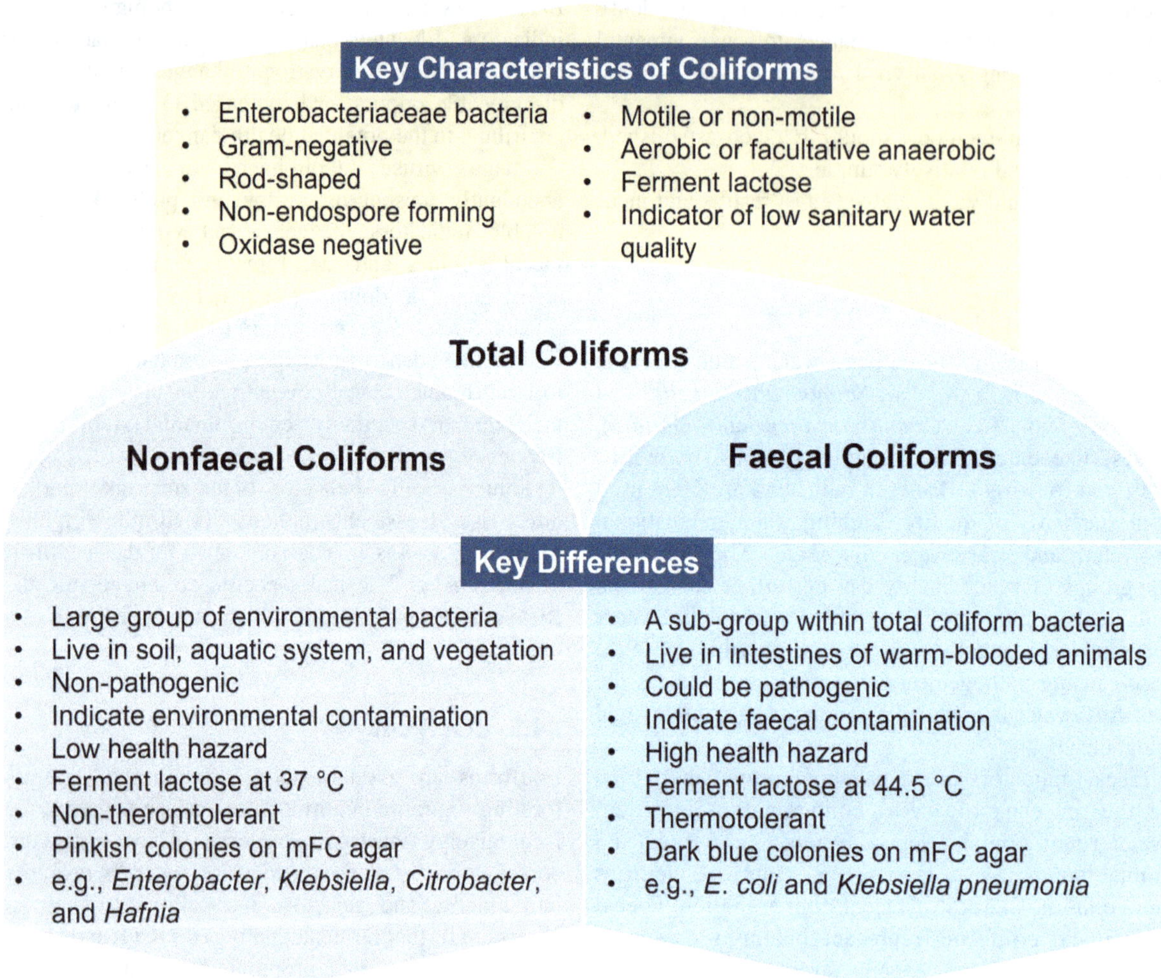

FIGURE 24.4 Key characteristics of coliforms and differences between total and faecal coliforms.

be relatively easily isolated compared to any gastrointestinal pathogens.

Total coliforms are routinely monitored to indicate the bacteriological quality of drinking water. However, the total coliform count is often inadequate to differentiate between faecal and non-faecal contamination. If drinking water possesses total coliforms, environmental factors rather than faecal contamination could likely be the reason for their presence in the water. It is possible that many bacteria in the total coliform do not have a faecal origin. The contamination of agricultural and stormwater runoff and the overflow of residential sewage or septic tanks can also be reasons for the possible presence of coliforms in the water stream. However, faecal pollution can typically occur from a discharge of human and animal faeces or untreated sewage.

Water samples that have tested positive for total coliforms are further examined for the presence of faecal coliform. When faecal coliform is detected in a drinking water sample, it is a sign of faecal contamination, which also increases the likelihood of pathogen presence. **In contrast to non-faecal coliforms, faecal coliforms are thermotolerant and ferment lactose at elevated temperatures of 44.5 °C to produce acid and different gases**. Even though faecal coliforms are strongly associated with faecal contamination, *Enterobacter cloacae* and *Klebsiella pneumoniae* can also be found in terrestrial and aquatic environments. However, *E. coli* is widely recognised as an indicator of faecal contamination due to its abundant occurrence in human and animal faeces. The intestines of humans and warm-blooded animals contain large amounts of *E. coli*. Most of their strains are harmless and maintain the physiology of the healthy host. Also, *E. coli* observe a stringent host specificity, which makes them ideal candidates as faecal source tracking markers. However, some strains of *E. coli* can be opportunistic pathogens and develop infections in immunocompromised hosts. Therefore, both faecal coliforms and *E. coli* are used as indicators of faecal pollution in water samples for routine practices.

The use of coliforms as indicators of water quality has certain limitations. They do not always guarantee the presence of pathogens in the water. The ability of coliforms to survive in non-intestinal environments for a considerable long duration is liable for their frequent natural presence in unpolluted tropical water. Multiplication of coliforms at an elevated temperature in natural surface water or as biofilms in the water distribution pipe systems may give a false indication of faecal contamination. An overgrowth of heterotrophic bacteria in the water might result in an underestimation of the true numbers of coliforms when isolated on selective media. Some pathogenic enteric viruses and protozoan parasites are better survivors than *E. coli* and feed on coliforms, reducing their density in water. Also, many coliforms are susceptible to drugs and may not grow in contaminated water.

National and international standards have predetermined and authorised acceptable limits for the density of microbiological indicators in water samples. Different regions of the world have different standards depending on their specific circumstances and risk assessments. For example, the WHO sets a limit of less than 20 CFU (colony-forming units)/ml of heterotrophic bacteria in drinkable water and prohibits the presence of coliform bacteria. The US EPA has set the standard that there should be no detectable coliform bacteria per 100 ml of drinking water, however, for every 100 ml of recreational water or wastewater discharges, there should be no more than 200 faecal coliforms. According to the European Union (EU) Drinking Water Directive, there should be no *E. coli* and faecal streptococci per 100 ml of tap water and 250 ml of bottled water.

24.6 DETECTION OF THE PRESENCE OF TOTAL COLIFORMS

The coliform group is extensively used as a water quality indicator. There are numerous techniques to monitor coliform contamination in drinking water. Most methods rely on biochemical tests based on microbial enzymatic activities and chromogenic reactions. This often requires the cultivation of microorganisms in a selective and differential medium. Detection of coliform bacteria further requires a series of tests to confirm their presence and differentiate between total coliforms and faecal coliforms (or *E. coli*). Of the different available tests, the most probable number (MPN) and the membrane filter techniques are well-approved methods for detecting coliforms in water samples. Some tests employ selective enzyme activity and novel chromogenic and fluorogenic substrates, which have high sensitivity for detection. Molecular methods like polymerase chain reaction (PCR), fluorescence *in situ* hybridisation (FISH), and immunological techniques are specific and rapid detection methods of coliforms without the need for cultivation steps. However, detection through molecular tools requires further scrutiny. Advanced methodologies have significantly enhanced the accuracy, efficiency, and ease of detecting coliform in water samples. Some well-known detection methods for coliforms are discussed in detail as follows.

24.6.1 Eosin-Methylene Blue (EMB) Agar

A water quality test frequently utilises the eosin-methylene blue (EMB) agar as a selective and differential medium to identify potential coliform contamination in water samples. When an aliquot of the water sample is spread, or pure colonies are streaked on EMB agar and incubated for 18–24 h, lactose fermenting bacteria in the medium develop coloured colonies. In contrast, lactose non-fermenting bacteria produce colourless colonies (Figure 24.5a). This colour formation allows differentiation between coliforms and non-coliforms. These growth forms can further be streaked on EMB agar to distinguish among coliforms.

EMB medium is mainly composed of lactose, sucrose, peptone (peptic digest of animal tissue), and dipotassium phosphate. The growth condition in the EMB agar medium requires a final pH of 7.2 ± 0.2. Lactose and sucrose in the medium are fermentable carbohydrates and energy sources. Peptone serves as a source of carbon, nitrogen, and other essential growth nutrients. Phosphate buffers the medium. EMB medium also specifically contains two dyes,

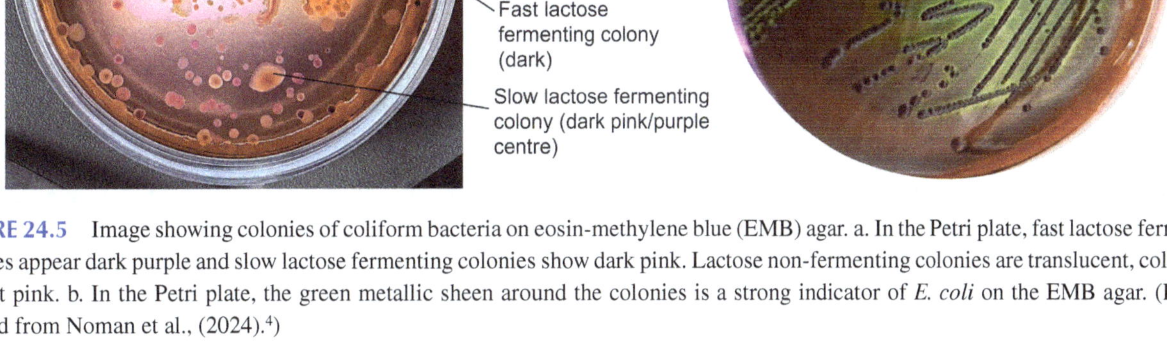

FIGURE 24.5 Image showing colonies of coliform bacteria on eosin-methylene blue (EMB) agar. a. In the Petri plate, fast lactose fermenting colonies appear dark purple and slow lactose fermenting colonies show dark pink. Lactose non-fermenting colonies are translucent, colourless, or light pink. b. In the Petri plate, the green metallic sheen around the colonies is a strong indicator of *E. coli* on the EMB agar. (Image b adapted from Noman et al., (2024).[4])

eosin-Y and methylene blue, that make the medium selective and differential. Eosin Y dye in the medium acts as a pH indicator that results in a dark purple to black colour formation at low pH. Methylene blue is a phenothiazine dye that has antimicrobial properties and particularly inhibits the growth of Gram-positive bacteria, such as *Staphylococcus aureus*. During the proliferation of coliforms on EMB agar medium, Gram-negative bacteria capable of fermenting lactose lower the pH of the medium, encouraging absorption of Eosin Y dye and forming a dark purple amide bond complex in the bacteria. Certain vigorous lactose-fermenting bacteria, acidifying the medium, produce flat, dark colonies with a green metallic sheen. Some slow lactose fermenters release a low amount of acid and produce large mucoid colonies that are often dark purple/pink in their centre or produce a brown-pink colouration. On the other hand, lactose non-fermenters increase the pH by deamination of proteins, and therefore, the dye is not absorbed. Their colonies will either be translucent, colourless or light lavender.

The EMB medium can differentiate between lactose fermenting (e.g., *E. coli*, *Enterobacter*, *Klebsiella*, and *Citrobacter*) and non-lactose fermenting (e.g., *Salmonella* and *Shigella*) enteric bacteria. Different coliforms produce different amounts of acids when fermenting lactose in EMB agar, which leads to colour changes due to the accumulation of dyes. When bacterial colonies on EMB agar have good growth, showing dark purple to black colonies with a green metallic sheen, they are *E. coli* (Figure 24.5b). Dark purple to black colouration, but without sheen on the medium, represents *Citrobacter*. On the other hand, luxuriant growth, dark-centred pink, mucoid colonies depict *Enterobacter aerogenes* (=*Aerobacter aerogenes*) and *Klebsiella pneumoniae*. Non-coliform bacteria like *Serratia*, *Pseudomonas aeruginosa*, *Shigella*, *Salmonella typhimurium*, and *Proteus mirabilis* remain colourless to light pink on the medium.

24.6.2 Most Probable Number (MPN) Test

The MPN test is a multi-step technique for estimating the presence and probable concentrations of total and faecal coliform populations in samples of soils, water, foods, and agricultural products. This test is also known as multiple tube fermentation (MTF). This method is based on statistical probability and involves multiple tubes with serial dilutions. A low number of faecal coliforms in a water sample indicates a low risk of faecal contamination, while a high number indicates a high risk of contamination.

In the MPN technique, three tests are conducted: presumptive, confirmatory, and completed (Figure 24.6). For the presumptive (a screening) test, the sample is diluted in an appropriate sterilised liquid culture medium (of 1X and 2X concentrations). A minimum of three dilutions having 3, 5, or 10 replicates per dilution are prepared in a set of test tubes (containing small Durham tubes to detect gas formation) as per the standard procedures. The suitable media for the test include lactose broth and lauryl sulphate broth (also called lauryl sulphate-tryptose-lactose broth). Lactose broth is a commonly used medium for identifying lactose-fermenting organisms, such as coliforms. Lactose broth typically contains lactose, beef extract, peptone, sodium chloride and phenol red (which indicates acid formation). However, lauryl sulphate broth mainly consists of sodium lauryl sulphate (bile salt), lactose, and an acidic dye. This medium is more selective for coliforms because the sodium lauryl sulphate inhibits the development of many other microorganisms. When an aliquot is inoculated in this medium and incubated at 35-37 °C for 24 to 48 h, a positive sample shows bacterial growth (which causes turbidity in the tube) and a recognisable biochemical reaction in the medium (through gas and acid production). However, an aliquot that has not received even a single viable cell does not produce any growth or biochemical reaction. Subsequently, the total number of positive and negative tubes for each dilution is noted and analysed using statistical MPN tables with a 95% confidence interval. According to the statistical probability of the coincidence of microorganisms in each aliquot, an index number is designated to represent the MPN of bacteria in the sample. A negative presumptive test indicates microbiologically safe water. If any tube shows acid and gas production, water is considered unsafe and is tested with a confirmed test.

Lactose fermentation produces gas and acids in several bacterial forms, not just coliforms. Therefore, a confirmatory test is performed to verify the existence of coliform. For this, a loopful of culture from positive fermentation tubes is inoculated into lactose broth or brilliant green lactose bile broth (BGLB), incubated at 35-37 °C for 24 h and observed for gas formation. It is also a selective medium for the coliforms. This test detects the confirmed presence of coliform through gas and acid production. A loopful of culture is also streaked on agar slants to examine the type of bacteria through Gram-staining. In addition, a loopful of culture can be inoculated into tryptone water and incubated at 35-37 °C for 24 h to differentiate between indole-positive (*E. coli*) and indole-negative (*Klebsiella*) coliforms. The positive test for indole production is indicated by the presence of a red or pink colour at the reagent layer after adding the Kovacs reagent (Figure 24.6).

To ensure positive results, a completed test is then performed by streaking inoculum from each positive tube (of lactose or BGLB broth) onto a differential and selective medium (EMB or Endo agar plates). The plates are incubated at 37 °C and 44.5 °C for 24 h. Their growth at two different temperatures allows for the detection of faecal and non-faecal coliforms, respectively (Figure 24.6).

The main advantages of the MPN test are the estimation of the total microbial count or population density of viable coliforms, ready identification of specific microbial groups with appropriate selective and differential media, and high uniformity among results obtained from different laboratories than those with a standard plate count. This is also suitable for microorganisms that cannot be grown on a solid medium or

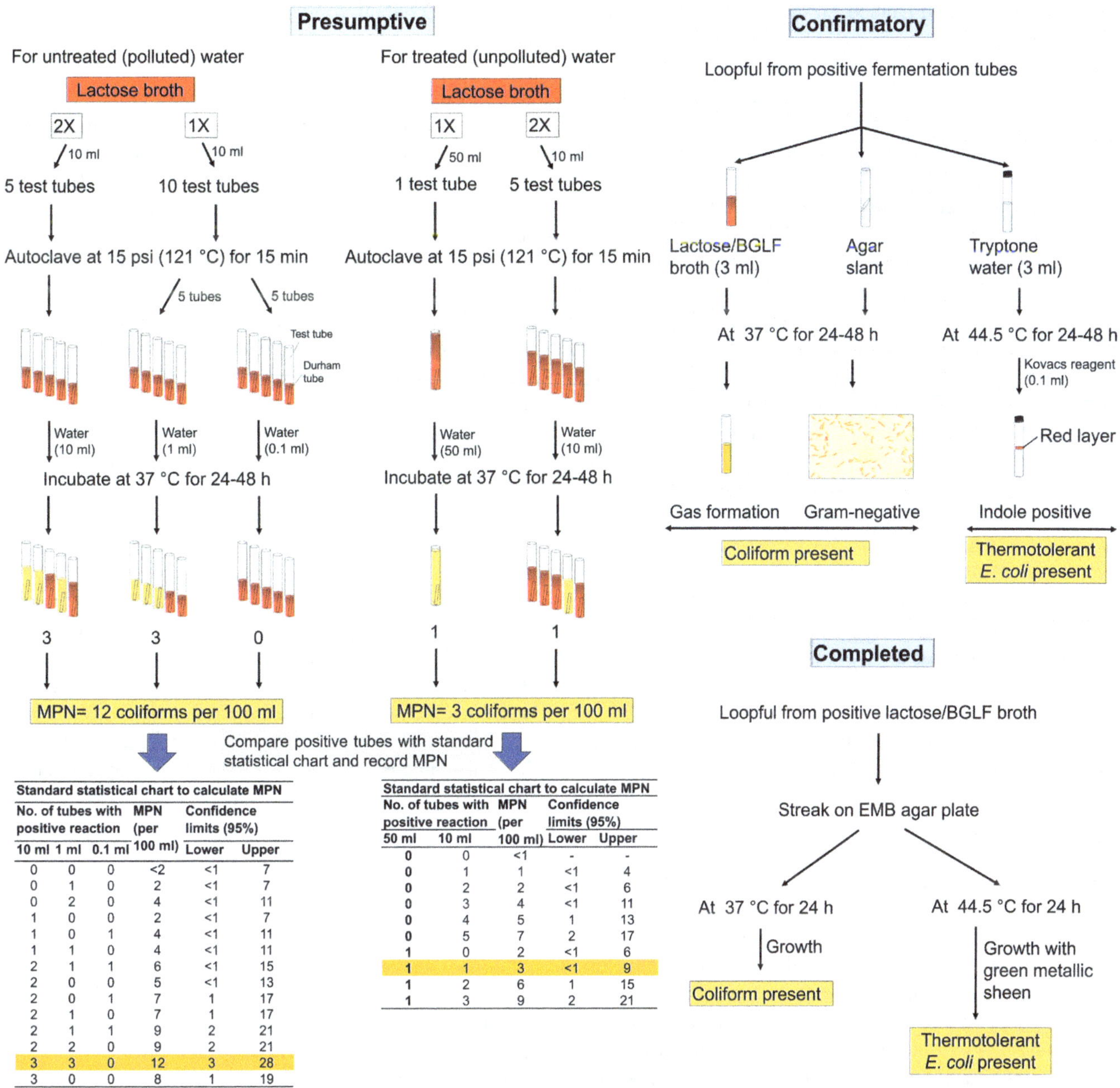

Standard statistical chart to calculate MPN

No. of tubes with positive reaction			MPN (per 100 ml)	Confidence limits (95%)	
10 ml	1 ml	0.1 ml		Lower	Upper
0	0	0	<2	<1	7
0	1	0	2	<1	7
0	2	0	4	<1	11
1	0	0	2	<1	7
1	0	1	4	<1	11
1	1	0	4	<1	11
2	1	1	6	<1	15
2	0	0	5	<1	13
2	0	1	7	1	17
2	1	0	7	1	17
2	1	1	9	2	21
2	2	0	9	2	21
3	3	0	12	3	28
3	0	0	8	1	19

Standard statistical chart to calculate MPN

No. of tubes with positive reaction		MPN (per 100 ml)	Confidence limits (95%)	
50 ml	10 ml		Lower	Upper
0	0	<1	-	-
0	1	1	<1	4
0	2	2	<1	6
0	3	4	<1	11
0	4	5	1	13
0	5	7	2	17
1	0	2	<1	6
1	1	3	<1	9
1	2	6	1	15
1	3	9	2	21

FIGURE 24.6 A schematic of the most probable number (MPN) test for estimating the total and faecal coliform populations in untreated (polluted) and treated (unpolluted) water samples. It consists of presumptive, confirmatory, and completed tests in sequence. A presumptive test involves the cultivation of bacteria (using water samples of different volumes) in different dilutions of lactose broth for acid and gas production, and the calculation of the MPN value using standard statistical tables. In the confirmatory test, bacteria from positive tubes (of the presumptive test) are evaluated through Gram staining and the production of gas and indole. The completed test involves the cultivation of bacteria from positive tubes (of the confirmatory test) on eosin methylene blue (EMB) or endo agar plates. (MPN values table inspired by Standard Methods for the Examination of Water and Wastewater (2023).[5])

are overgrown by other bacteria. However, the MPN technique also has some limitations. This technique does not allow for the morphological characteristics of microorganisms and requires numerous glassware and materials. It is also less sensitive towards estimating the exact number of microorganisms and can prevent microbial growth in the medium if any inhibitory agents are present.

24.6.3 The Presence-Absence Test

The presence-absence (P-A) test is a qualitative study that determines if an organism of interest is present in water samples. Detection of coliforms in drinking water with the P-A test involves an enzymatic assay with lauryl sulphate broth containing a particular substrate. For the same, the Colilert test is a US EPA-approved method for the simultaneous detection of total coliform and *E. coli* in food, water, and environmental samples.

The Colilert test is based on proprietary defined substrate technology, where the medium is designed to be highly selective for the target microorganisms, minimising false positives and false negatives. The main reagents of the Colilert test are two nutrient indicators, ortho-Nitrophenyl-β-galactoside (ONPG) and 4-methylumbelliferyl-β-D-glucuronide (MUG). Both substances are artificial carbon sources. ONPG is structurally similar to lactose and can enter the bacterial cell even in the absence of the transport protein permease. This is particularly useful for identifying bacteria with defective lactose permease but functional β-galactosidase. Inside the cells, β-galactosidase hydrolyses the β-galactoside bond in ONPG and releases the yellow-coloured compound o-nitrophenol (Figure 24.7a). On the other hand, MUG is metabolised using a different enzyme, β-D-glucuronidase, that is specifically present in *E. coli*. The hydrolysis of MUG by *E. coli* releases 4-methylumbelliferyl moiety, which fluoresces electric blue under long-wavelength ultraviolet light (Figure 24.7b). About 97% of *E. coli* give positive results for this test. Most non-coliforms lack these enzymes and do not grow on this medium, while those few that have these enzymes are selectively suppressed by the specifically formulated constituents of the media. This test is also well suited for confirmatory testing after a positive presumptive result in the MPN method.

24.6.4 Membrane Filter Test

The membrane filter test is a single-step technique for microbiological analysis of total coliform bacterial count in water sources. This technique involves membrane filtration of a liquid sample and incubation of the membrane filter on a selective and differential medium. In a typical experiment, a measured volume of water (usually 100 ml) is passed through a sterile membrane filter disc, having uniform porosity of a predetermined size, to trap bacterial forms in the water sample. A membrane filter composed of cellulose esters or similar polymeric materials, having a pore size of 0.45 micrometre (μm), is used to retain larger-sized biological and non-biological particles. After filtration, the membrane is placed on a sterile Petri dish containing a thin absorbent pad soaked with a specific medium. The preparation is incubated at 35-37 °C for 24 h. During incubation, the medium diffuses upwards into the filter and allows for the growth and differentiation of the microorganisms. The modified endo broth (M-Endo Broth MF) or agar medium, consisting of lactose as the fermentable carbohydrate, is a suitable medium for this experiment. In the medium, tryptose, casein enzymic hydrolysate, peptone, and yeast extract provide essential nutrients for the coliforms. Phosphates support buffering capacity. The presence of sodium sulphite, sodium deoxycholate, and basic fuchsin inhibits the growth of Gram-positive bacteria. The acids produced after lactose fermentation react with the dye basic fuchsin, which imparts pink or red colour to coliform colonies and the surrounding medium.

The membrane faecal coliform agar (m-FC agar) is a different selective medium for the cultivation and enumeration of faecal coliforms. The m-FC medium contains proteose peptone, tryptose, and yeast extract as complex sources of carbon, nitrogen, minerals, and vitamins. The fermentable carbohydrate in the m-FC medium is lactose. The methyl blue dye and bile salts in mFC agar prevent Gram-positive bacterial growth. A colour change of methyl blue and rosolic acid serves as the basis for the indicator system. On the m-FC medium, faecal coliform (e.g., *E. coli*) colonies appear blue, while non-coliforms (e.g., *Salmonella typhimurium* and *Shigella flexneri*) develop pinkish, grey, or cream-coloured colonies.

The advantages of membrane filter tests are accuracy, rapid availability, and reproducibility of results. This technique can process large volumes of samples using fewer resources, allowing the enumeration of coliforms. It can be suitably used for analysing water samples collected from water treatment plants, pharmaceutical and cosmetics industries, liquid foods with lower levels of particulate matter, and heat-labile substances (e.g., serum, certain carbohydrate solutions, and antibiotics). However, water samples that contain large quantities of suspended materials can easily clog filters. Diluting highly turbid water samples in a sterile diluent before membrane filtration can overcome the turbidity issue, but the dilution can be problematic if the original sample already has a low microbial load.

24.6.5 Genotypic and Phenotypic Profiling

Genotypic and phenotypic profiling are two modern microbial source tracking methods that take advantage of molecular techniques either independently or in conjunction with culture-based analysis. Genotypic methods differentiate the genetic patterns, whereas phenotypic methods differentiate the antibiotic resistance or carbon utilisation patterns of bacteria in the source sample.

These two methods can be bacterial library-dependent or library-independent. Library-dependent methods include ribotyping, repetitive intergenic DNA sequences (rep-PCR), pulsed-field gel electrophoresis (PFGE), multiple antibiotic resistance, and carbon source profiling. These methods require a comparison with a library or database of bacterial isolates of known origin. On the other hand, the library-independent approach eliminates the need for sample comparisons with a library of known bacterial isolates. A library-independent genotypic analysis might use bacterial host-specific markers and PCR testing. For example, source identification of the enteric anaerobic bacterial genera like *Bifidobacterium* relies on host-specific molecular markers using terminal-restriction fragment length polymorphism (T-RFLP). In a different approach, species-specific PCR primers of enterotoxin biomarkers can be used to amplify toxin genes of *E. coli*.

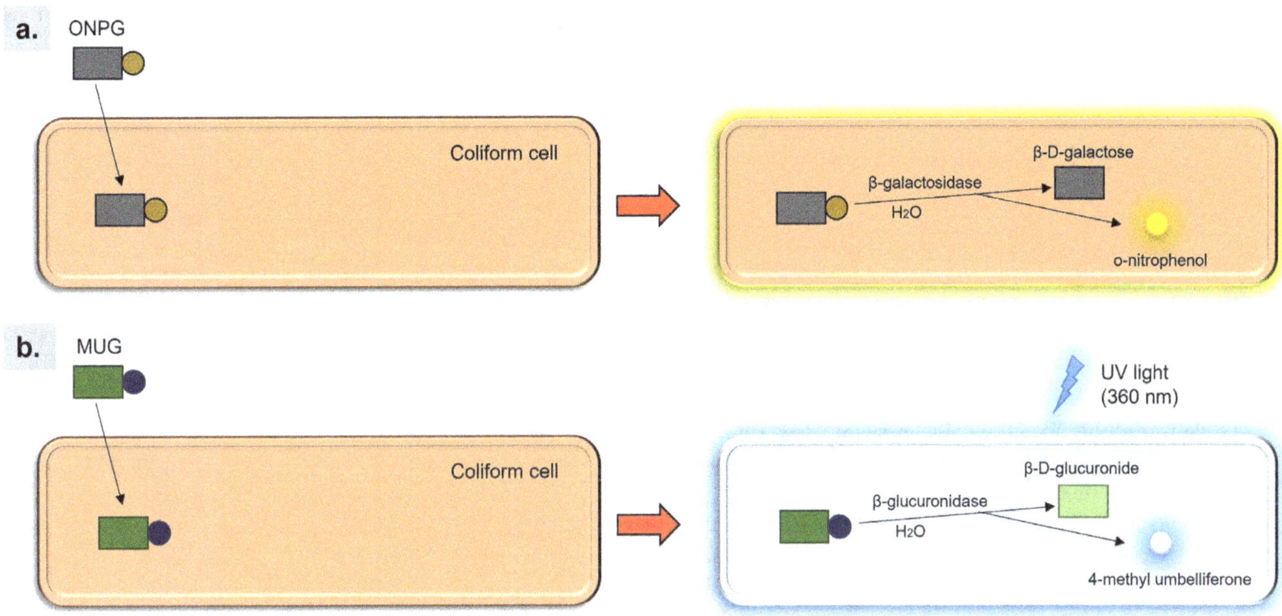

FIGURE 24.7 Detection of total coliform using the Colilert test. a. β-galactosidase cleaves the substrate o-nitrophenyl-β-D-galactopyranoside (ONPG) and releases a yellow product, o-nitrophenol. B. β-glucuronidase cleaves the substrate 4-methylumbelliferyl-β-D-glucuronide (MUG) and releases 4-methylumbelliferone that fluoresces under UV light.

The FISH technique relies on oligonucleotide probes to detect complementary sequences in cells, which specifically target Enterobacteriaceae 16S RNA molecules in water samples. Differently, the identification of total coliforms and *E. coli* is also possible using the DNA sequences coding for the *lacZ* (ß-galactosidase gene) and *uidA* (ß-D-glucuronidase gene). This is a viable alternative to conventional culture methods for detecting coliforms in drinking water, providing quick quantitative detection but still necessitating further research. However, molecular techniques based on PCR (qPCR) and metagenomics have shortcomings in their inability to discriminate between the DNA of living and dead cells.

24.6.6 Emerging Bioengineering Enumeration Methods

The labour-intensive and time-consuming nature of culture-based approaches (MPN test, EMB agar method, or membrane filtration technique) frequently provides results that are late in detecting unexpected bacterial outbreaks. It is plausible that the water under analysis has already been distributed and consumed by the time faecal pollution in drinking water is determined. Culture-based techniques only yield data on a fraction of the total bacterial populations because non-culturable bacterial cells are excluded. Many new methods have been developed and are under trial for coliform detection in water samples. For example, numerous antibodies against coliform bacteria have been developed in immunological surveillance. However, this test shows inadequate antibody specificity. In recent years, multidisciplinary systems combined with state-of-the-art methodologies from computer science, engineering, chemistry, and biology have emerged as strong alternatives or additions to existing approaches to detect the presence of coliform in water samples. Some ongoing developments in the detection systems of coliforms are-

- A time-lapse coherent imaging platform: This lens-free imaging device can identify premature faecal indicator bacteria on agar plates. An imaging system scans an agar plate every 30 minutes after inoculation. Captured high-resolution images are sent to software for time-lapse digital reconstruction, differentiating between static impurities (dust, air bubbles, and light speckles) and bacterial colonies.
- A real-time culture-independent imaging platform: It is an automated technology capable of enumerating total bacteria in a short duration. A platform is installed in drinking water pipeline systems, trapping 6 µl of water samples inside a flow cell. A 3D scanning system scans the sample at different focuses, analysing the captured images. The platform flushes the cell, preparing for subsequent measurements. This method eliminates the need for chemical supplies and hazardous waste, enabling real-time sending of wireless data to a server online. This method has some other advantages, such as the detection of sudden changes in the water quality, differentiation between biological and nonbiological particles with high accuracy, and high-frequency (10-min time resolution) monitoring.
- Microflow cytometry: A microfluidic-based portable microflow cytometry is a rapid, affordable, and efficient on-site *E. coli* enumeration without fluorescence

staining. The device features a four-layer panel with a 50-µm-wide microfluidic channel, allowing water samples and sheath fluid to pass unidirectionally. When a beam falls on a sample, microbial cells and particles differentially scatter the light. This can be recorded by an on-chip lens system.

SUMMARY

- Environmental indicators are effective sensors for capturing the important components of a complex ecosystem or environmental problems. The water quality indicators demonstrate the potential impact of irrigation, industrial effluents, domestic wastes, etc., on water bodies. This can be detected using physicochemical and biological methods.
- Physical indicators represent the physical state of water. Examples include colour, taste, odour, turbidity, temperature, transparency, TSS, and TDS. Chemical pollution indicators are linked to the ecosystem's buffering ability, water interactions, and nutrient cycling. These include various parameters such as pH, DO, BOD, COD, and chemicals.
- The term biological indicator is typically used for specific biological entities and their activities, mainly related to bacteria, viruses, and protozoa in a water sample. These indicators give information on the degree to which pollutants impact water quality on the biological parameters.
- Coliforms are rod-shaped, Gram-negative, non-endospore-forming, oxidase-negative, lactose hydrolysing bacteria that indicate faecal contamination in water. Faecal coliforms are frequently reported in human and animal faeces and are a reliable indicator of faecal contamination.
- Total coliforms are innocuous bacteria found in diverse environments. Faecal coliforms reside in warm-blooded animals' intestinal tracts and are released from faeces. *E. coli* represents a type species in the faecal coliform group. The presence of faecal coliforms in a water sample indicates insufficient sanitary conditions and contamination with faecal matter.
- Coliform contamination in drinking water is monitored using biochemical tests based on microbial enzymatic activities and chromogenic reactions. Techniques like multiple-tube fermentation and membrane filters with specialised media and incubation conditions are well-approved for detecting coliforms.

DIFFERENT TYPES OF PRACTICE QUESTIONS

Fill in the blanks

I. The health of the water ecosystem is monitored based on certain parameters called ....
II. Polluted water containing an excess amount of suspended solids has .... turbidity and .... transparency.
III. .... is the negative logarithm of the concentration (mol/l) of hydrogen ions (H^+) in an aqueous solution.
IV. .... is a differential and selective medium that can differentiate between lactose fermenting and non-lactose fermenting enteric bacteria.
V. The coliforms characteristically possess the enzyme ...., which hydrolyses lactose into glucose and galactose.
VI. The .... test is a multi-step technique used for estimating the presence and probable concentrations of coliform populations in environmental samples.
VII. Coliforms typically belong to the family ....

SOLUTION: I. environmental indicators, II. high, low, III. pH, IV. eosinmethylene blue (EMB) agar, V. β- galactosidase, VI. most probable number (MPN), VII. Enterobacteriaceae

Match the following

i. Heavy metals	a. IC-TC method
ii. TDS	b. Winkler method
iii. TOC	c. Gravimetric analysis
iv. BOD	d. Potassium dichromate method
v. COD	e. ICP-MS analysis

SOLUTION: i-e, ii-c, iii-a, iv-b, v-d

Very short answer type of questions

I. What is AQI?
II. Name any one physical and chemical indicator of water quality.
III. Expand TOC and TSS.
IV. A pond near an agricultural field was covered with a dense green film or scum on the surface. What could have happened?
V. The DO concentration of certain pond water was 0.7 mg/l. What might this indicate?
VI. The BOD values in two water samples, A and B, were reported as 1.2 mg/l and 260 mg/l, respectively. Which of these water samples would be more polluted? Give your answer with a suitable reason.
VII. Name any two bacteria typically included in the faecal coliform group.

VIII. How can we differentiate *Escherichia coli* from *Salmonella typhimurium* growing on an EMB agar medium?

Short answer type of questions

I. Write a short note on monitoring the water quality parameters.
II. Elaborate on the TDS estimation of water. How is it different from TSS?
III. What is DO? Write about its sources of water. Give a brief account of DO estimation.
IV. Write the differences between BOD and COD.
V. What are the essential criteria of the microorganisms that serve as biological indicators?
VI. Write the characteristics of coliforms. Differentiate between total and faecal coliforms.
VII. What are the reasons that genotyping could be a better alternative to the culture-dependent methods for coliform detection?

Long answer type of questions

I. Explain the significance of BOD as a water quality indicator. What is the chemical basis of detecting BOD? Why is an incubation of five days considered optimal for the BOD analysis?
II. Biological indicators of water quality provide information about the health of aquatic environments. Elaborate with suitable examples.
III. Give a detailed account of any three methods used for detecting coliforms in water samples.

NOTES

1 As per the report of WHO/UNICEF Joint Monitoring Programme (JMP) for Water Supply, Sanitation and Hygiene (2017).
2 The metalloid, As, is also considered a heavy metal.
3 Al-Shehhi MR, Abdul Samad Y. Identifying algal bloom 'hotspots' in marginal productive seas: a review and geospatial analysis. *Remote Sensing*. 2022;14(10):2457.
4 Al Noman Z, Anika TT, Safa UH, Alam S, Dey SS, Bhuiyan MN, Siddique MP. Antibiotic resistance risk assessment in Mymensingh, Bangladesh: Current scenario from human-animal-environmental interfaces viewpoint. *Heliyon*. 2024;10(15).
5 Standard Methods Committee of the American Public Health Association, American Water Works Association, and Water Environment Federation. 9221 Multiple-tube fermentation technique for members of the coliform group InStandard Methods For the Examination of Water and Wastewater. 2023. Lipps WC, Baxter TE, Braun-Howland E, editors. Washington DC: APHA Press.

SUGGESTED READINGS

Aziz KH, Mustafa FS, Omer KM, Hama S, Hamarawf RF, Rahman KO. Heavy metal pollution in the aquatic environment: efficient and low-cost removal approaches to eliminate their toxicity: a review. *RSC Advances*. 2023;13(26):17595–610.

Cabral JP. Water microbiology. Bacterial pathogens and water. *International Journal of Environmental Research and Public Health*. 2010;7(10):3657–703.

Chen S, Chen J, Wang X, Li J, Dong B, Xu Z. Footprints of total coliforms, faecal coliforms and *E. coli* in a wastewater treatment plant and the probabilistic assessment and reduction of *E. coli* infection risks. *Science of The Total Environment*. 2023;900:165845.

Cohen J, Shuval HI. Coliforms, fecal coliforms, and fecal streptococci as indicators of water pollution. *Water, Air, and Soil Pollution*. 1973;2:85–95.

Horvath RS, Ropp ME. Mechanism of action of eosin-methylene blue agar in the differentiation of *Escherichia coli* and *Enterobacter aerogenes*. *International Journal of Systematic and Evolutionary Microbiology*. 1974;24(2):221–4.

Kumar M, Puri A. A review of permissible limits of drinking water. *Indian Journal of Occupational and Environmental Medicine*. 2012;16(1):40–4.

Li E, Saleem F, Edge TA, Schellhorn HE. Biological indicators for fecal pollution detection and source tracking: A review. *Processes*. 2021;9(11):2058.

Lugo JL, Lugo ER, Puente MD. A systematic review of microorganisms as indicators of recreational water quality in natural and drinking water systems. *Journal of Water and Health*. 2021;19(1):20–8.

Maheux AF, Boudreau DK, Bisson MA, Dion-Dupont V, Bouchard S, Nkuranga M, Bergeron MG, Rodriguez MJ. Molecular method for detection of total coliforms in drinking water samples. *Applied and Environmental Microbiology*. 2014;80(14):4074–84.

Rompre A, Servais P, Baudart J, De-Roubin MR, Laurent P. Detection and enumeration of coliforms in drinking water: current methods and emerging approaches. *Journal of Microbiological Methods*. 2002;49(1):31–54.

Tsaridou C, Karabelas AJ. Drinking water standards and their implementation—A critical assessment. *Water*. 2021;13(20):2918.

Tyagi S, Sharma B, Singh P, Dobhal R. Water quality assessment in terms of water quality index. *American Journal of Water Resources*. 2013;1(3):34–8.

World Health Organization Progress on Sanitation and Drinking Water – 2015 Update and MDG Assessment. World Health Organization and UNICEF WHO Press, Geneva, Switzerland; 2015.

Zaghloul A, Saber M, El-Dewany C. Chemical indicators for pollution detection in terrestrial and aquatic ecosystems. *Bulletin of the National Research Centre*. 2019;43:1–7.

Zaghloul A, Saber M, Gadow S, Awad F. Biological indicators for pollution detection in terrestrial and aquatic ecosystems. *Bulletin of the National Research Centre*. 2020;44:1–1.

25 Wastewater Treatment Systems

25.1 WASTEWATER TREATMENT OBJECTIVES

Wastewater treatment is a method to restore the desirable quality of water contaminated by anthropogenic activities or natural processes. This is an effort to eradicate contaminants below the maximum allowable limit, maintaining micronutrients. Wastewater can be treated using various physical, chemical, biological, and integrated methods.

The treatment objectives often include stress management strategies before remedying the problem. The management schemes involve several arrangements such as waste valorisation, avoiding contamination of wastewater flow, and protecting public and environmental health. Various international and national agencies and governmental bodies have set stringent limits for the discharge of pollutants into the environment. Industries are obliged to comply with water quality standards for waste discharge into natural water bodies. This is particularly relevant for those natural systems that have poor assimilative or self-purification capacities. In addition, the preservation of natural resources, substantial financial savings, and the reduction of energy, waste, raw materials, and environmental emissions have also drawn significant attention to wastewater management.

Wastewater treatment is not a choice but an obligation to secure a sustainable environment. The content and characteristics of wastewater must be determined prior to planning and running a wastewater treatment facility. This could include the estimation of various physical, chemical, and biological parameters, which have been discussed in detail in Chapter 24. The primary aims of wastewater treatment are-

- Preventing the discharge of wastewater effluent into natural water bodies
- Separation of solids from liquid waste
- Reducing the size of solid waste gathered from wastewater
- Sludge waste management
- Supply of water bodies into the natural ecosystem after treatment

25.2 WASTEWATER TREATMENT PLANT (WWTP)

A wastewater treatment plant (WWTP) is a huge and complicated facility that receives effluents from many sources and reduces contaminants in wastewater to a level acceptable for environmental discharge (Figure 25.1). A WWTP is designed to be adaptable enough to handle seasonal variations, long-term changes in wastewater quality, and some additional requirements. A plant layout and operation requires a multidisciplinary approach, incorporating principles from engineering, chemistry, microbiology, geology, economics, law, and architecture. Based on the unit processes for the treatment of specific contaminants, WWTP can be classified as:

- Sewage treatment plant
- Industrial wastewater treatment plant
- Agricultural wastewater treatment plant
- Leachate treatment plant

Sewage treatment plant focuses on biological treatment, considering the bulk quantity of labile organic matter (a fraction of organic matter that can be easily broken down by native microflora). It can be a centralised or decentralised facility. Industrial wastewater consists of recalcitrant contaminants, and their removal requires special technical support in the industrial wastewater treatment plant. Agriculture wastewater has traces of fertilisers, pesticides, animal slurry, and crop residues, and they can be suitably treated in anaerobic lagoons and constructed wetlands. Such an industrial arrangement is called an agricultural wastewater treatment plant. A leachate treatment plant is a treatment facility designed for leachate, the contaminated liquid that drains from landfills. In recent years, the invention of integrated treatment techniques that blend the treatment of industrial wastes, agricultural wastes, and leachate has received wide attention for a successful and efficient waste management system.

25.3 WASTEWATER TREATMENT PROCESS

In wastewater treatment, the first challenge is the collection and conveyance of wastewater at the treatment facility. It is then stored in large tanks and basins to minimise environmental damage. This calls for a safety measure called spill retention, which focuses on the containment of hazardous spills within raised barriers such as berms, spill guards and spill pans and prevents them from spreading. A significant concern in spill retention is drainage leakage because if drain plugs and covers

DOI: 10.1201/9781032651095-31

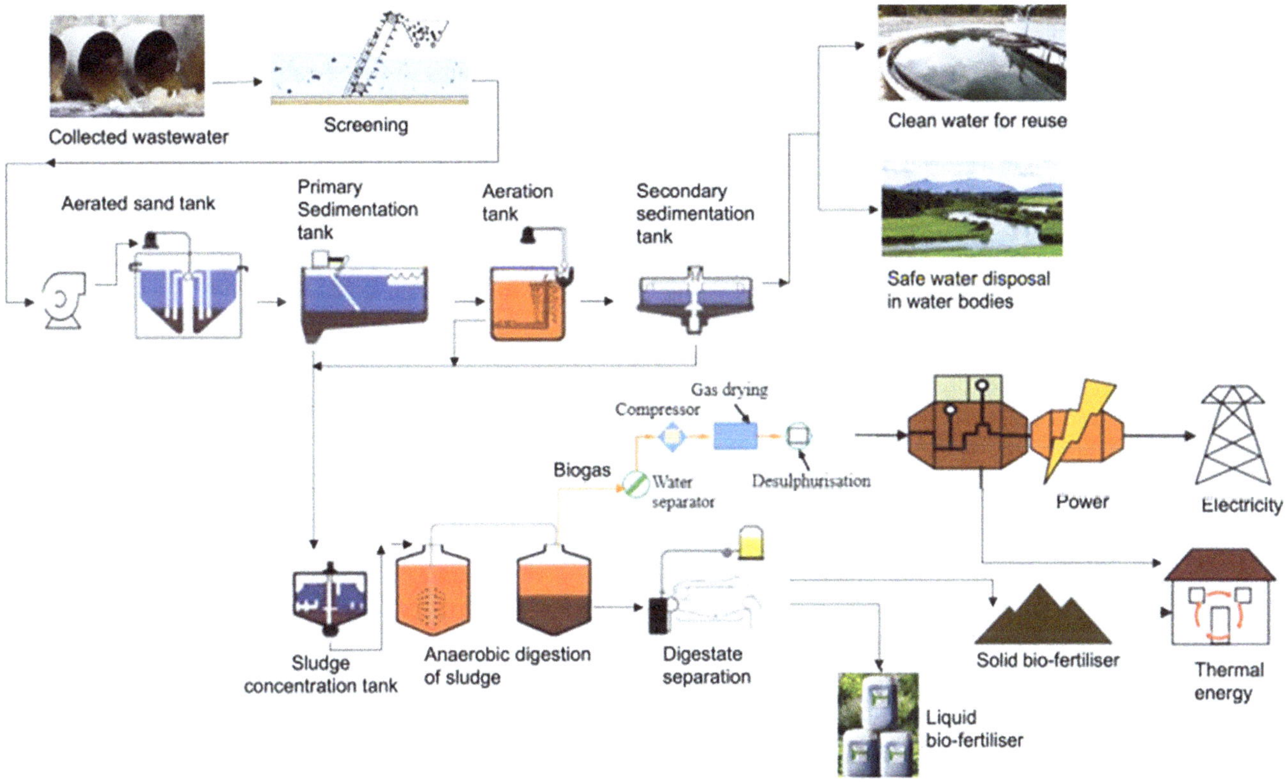

FIGURE 25.1 A flow chart of WWTP showing different treatment stages integrated with an anaerobic digestion (AD) system for biogas production and electricity generation. (Image reproduced from Ijoma et al. (2022)[1] with permission.)

are not functioning properly, wastewater effluent can leak out. It also helps prevent overloads of the treatment plant's capacity by containing the wastewater in the tank, allowing for controlled recovery and treatment.

Wastewater is initially cured in the pretreatment or primary treatment, followed by secondary treatment, which mainly relies on biological processes involving indigenous microbial forms. A part of the organic compounds that are not easily biologically oxidisable cause taste and odour problems in a water supply and produce toxicological effects in the aquatic environment and humans. Their treatment requires special procedures that constitute tertiary treatment. The choices of suitable techniques for tertiary treatment depend on the characteristics of the wastewater, the necessary effluent quality, and the availability of land. Table 25.1 sheds light on the main differences among primary, secondary, and tertiary treatment methods of wastewater. Wastewater treatment processes can be further extended into sludge treatment/stabilisation. The final step is the disposition of residuals and recirculation of water in the natural stream. The expenditure and efficiency of the selected process are significant concerns. Disposal of the solid debris or residuals is a major challenge post-wastewater treatment.

Many industries have facilities for source treatment before they release their wastewater. Source treatment mainly aims to remove specific or excess contaminants at the point sources to prevent their intermixing with waste streams or natural water bodies. The selection of source treatment technology varies across different industries and is based on the presence of targeted contaminants in wastewater. For example, when the alcohol manufacturing units release effluent (which is also called spent wash), it consists of high amount of organic matter and various other nutrients, which can be partially removed by aerobic/anaerobic treatment, chemical oxidation, granular activated carbon (GAC), polymeric resins, and wet air oxidation even at their source of origin. Tannery effluents are also treated at on-site treatment facilities. Their effluent contains high concentrations of chromium (Cr), organic matter, sulphides (S^{2-}), suspended solids, salts, dyes, ammonia (NH_3), fats, oils, grease, phenols, and solvents. They can be partially removed by pH adjustment, flocculation/coagulation, precipitation of Cr using lime, and the advanced oxidation process. The common methods and advantages of the wastewater treatment process, including pre-treatment, primary, secondary (biological), and tertiary (advanced) treatments, are given in Figure 25.2.

This chapter provides a detailed account of the principles, technical setups, and treatment options. The treatment of wastewater involves several sequential steps, including pretreatment, primary, secondary, and tertiary treatment, which are discussed in various sections below.

25.4 PRETREATMENT OF WASTEWATER

Pretreatment is a preliminary step for the removal of solids (e.g., sand, fibres, and trash), metals, synthetic polymers,

TABLE 25.1
A comparative account of primary, secondary, and tertiary treatment systems of wastewater

Parameters	Primary Treatment	Secondary Treatment	Tertiary Treatment
Stages	Initial treatment	Post-primary treatment	Final treatment
Types of methods	Physical, chemical, and mechanical	Mainly biological	Physical, chemical, and biological
Contaminants removed	Coarse matter, including large particles, floating debris, and solids (like gravel, grit, greasy substances, and sand)	Suspended and dissolved organic materials	Removal of residual organic matter, inorganic substances, and microorganisms
Processes	• Screening • Comminution • Grit removal • Primary sedimentation	• Biodegradation • Adsorption • Assimilation • Redox reaction • Bioprecipitation • Bioaccumulation • Nitrification • Denitrification • Methanogenesis	• Adsorption • Air stripping/steam stripping • Coagulation • Chemical oxidation • Chlorination • Desalination • Filtration • Ozone radiation • Photocatalysis • EBPR
Equipment used	• Coarse and fine screens • Comminutor and grinder • Aerated chamber Primary settling tank	• Activated sludge • Trickling filter • Rotating biological contactor • Septic tank • Oxidation pond • Membrane bioreactor	• Filtration system • Countercurrent extraction system • Sedimentation tank • Packed/tray tower • Ion permeable membrane • Semi-permeable membrane • Reverse osmosis
Maintenance costs	Low	Relatively high	Very high
Duration of treatment	Short	Long	Comparatively shorter
Complexity	Less	Fairly complicated	Highly complicated
Removal efficiency	35–40% of BOD, 30% of COD, 60–65% of settleable solids, 20% of nitrogen, 10% of phosphorus, 55% of faecal coliforms from municipal wastewater	85–95% of both BOD and suspended solids	About 99% of BOD, suspended solids, nitrogen, phosphorus, and pathogenic bacteria
Level of purity	Poor water quality and further treatment needed	A clarified effluent is enough to be discharged into natural water bodies	High-quality pure water for potability

oil, and grease from wastewater. The pretreatment methods can offer many advantages for further treatment. This step minimises the accumulation of solids, reduces blockage and abrasion of mechanical parts, and extends the life of the sanitation infrastructure. Pretreatment processes also make the wastewater compliant with the subsequent treatment processes. Some common methods for pretreatments are:

- Equalisation- It controls the fluctuation in the wastewater flow rate characteristics to provide relatively uniform conditions in the subsequent treatment. Treatment facilities are equipped with storage tanks or basins. The mixing of wastewater is customary to ensure adequate equalisation and prevent the accumulation of settleable solids, which is carried out using devices like baffling, turbine mixing, mechanical aeration, and diffused aeration. Equalisation minimises flow surges to the treatment systems, controls pH (to minimise the chemical requirements for neutralisation), prevents high concentrations of organic or toxic materials from entering the biological treatment system, provides continuous feed to biological systems, and controls the discharge of industrial waste to municipal sewers.
- Neutralisation- It routinely follows equalisation, in which acidic or alkaline wastewater is neutralised before being released for further treatment. Acidic wastewater can be neutralised with beds of lime (usually calcium oxide/calcium hydroxide) or limestone (calcium carbonate ($CaCO_3$)). The alkaline wastewater is treated with strong acids (sulphuric acid is economically preferred or hydrochloric acid). A different option for lowering water pH is through flue gases containing 14% carbon dioxide (CO_2). The optimum pH of water required across the treatment process is 6.5-8.5.
- Oil removal- The removal of excessive free-floating, emulsified, dissolved oil from wastewater is also

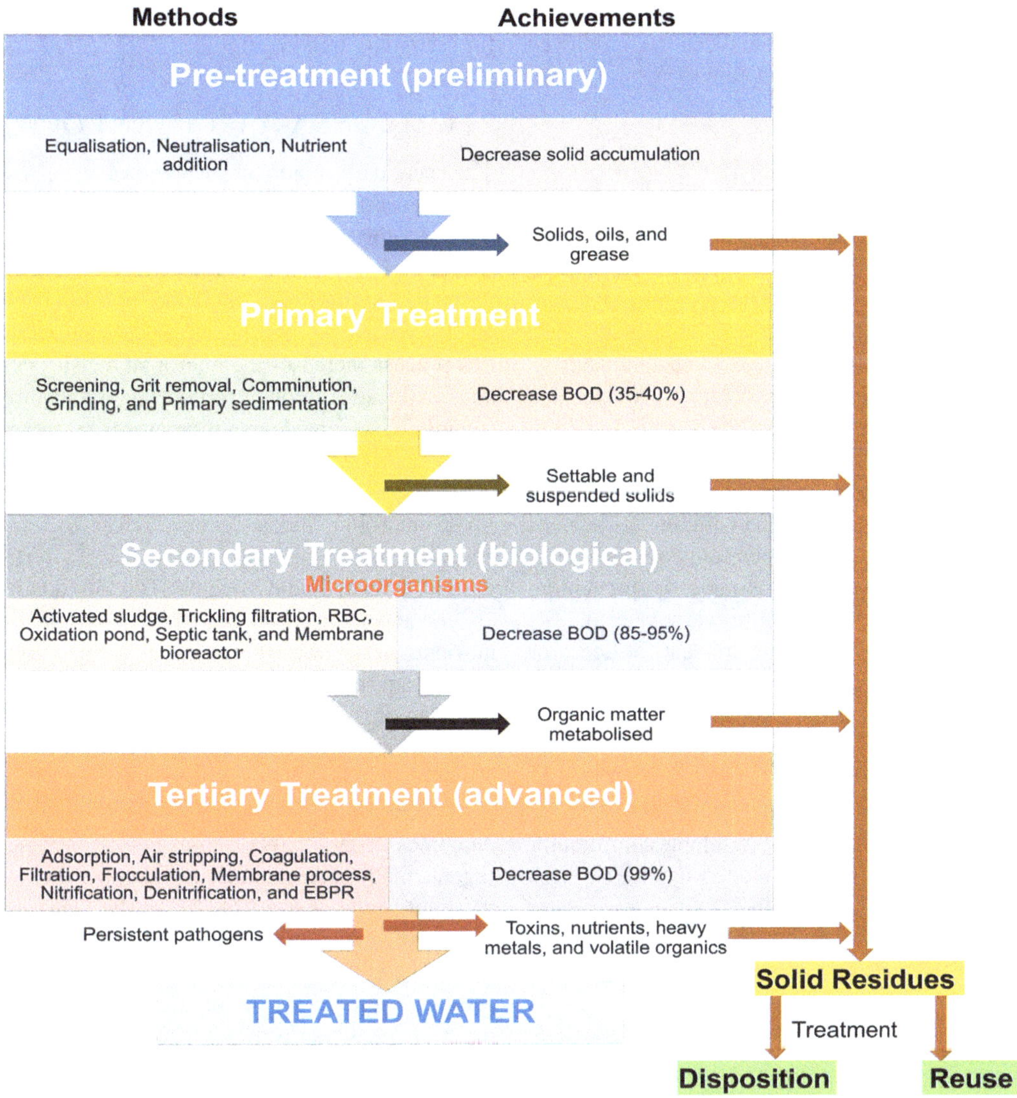

FIGURE 25.2 Main steps of wastewater treatment: primary, secondary, and tertiary treatment. Each step includes multiple treatment options and targets some specific achievements. Sludge produced from primary and secondary treatment undergoes further treatment for disposal or reuse.

an essential prerequisite for successful wastewater treatment. This step ensures minimum failure of downstream processing and effective removal. The American Petroleum Institute (API) separator is a useful device for removing free or floating oils based on the differences in the specific gravity of oil and water. In the removal process, lighter oils that remain at the top of the liquid are skimmed off from the surface in a skim tank, while the heavier oils and grease, which settle to the bottom, are removed by gravity separation. If wastewater contains any stable suspension of emulsified oil, it can be treated in an air flotation unit. The use of some chemicals (like detergents, alum or iron salts, acidification, or emulsion-breaking polymers), followed by injecting pressurised dissolved oxygen (DO) or nitrogen, can further restore the water quality. The dissolved oil recovered from this phase can be further treated separately.

25.5 PRIMARY TREATMENT OF WASTEWATER

After pretreatment, the wastewater reaches the primary treatment system. Primary treatment involves the physical separation of solids in wastewater by screening, grit removal, and sedimentation. The primary treatment predominantly separates the suspended solids via a sedimentation process in tanks. The primary treatment system can remove 60–65% of settleable solids, 50–70% of the total suspended solid (TDS) particles larger than 150 μm, about 35–40% of biochemical oxygen demand (BOD), about 30% of chemical oxygen demand (COD), about 20% of nitrogen, about 10% of phosphorus, and about 55% of faecal coliforms from municipal wastewater. Also, the removal of debris and pollutants ensures equipment safety and allows smooth flow for the secondary treatment phase. The primary treatment can be employed using different techniques, as discussed below.

25.5.1 Screening

Screening is the first step of primary treatment. In this process, large suspended solid matter and floating materials in wastewater, such as metal items, paper, plastics, rocks, plastic items, dead animals, branches of trees, and rags, are separated using vertical or inclined parallel bar screens. Modern WWTPs use two different types of screens: coarse and fine screens. Coarse screens have typical openings of around 6 mm that filter large debris, rags, and solids from wastewater, whereas fine screens have pore sizes ranging between 1.5 and 6 mm. Sometimes, a very fine screen, having a pore size of 0.2–1.5 mm, can also be placed in succession with the fine screens to remove suspended solids. These operations offer ease for downstream steps in wastewater treatment by preventing clogging and damage to equipment from wastewater debris. The recovered materials from screening are sent for decay or incineration.

25.5.2 Comminution

Debris that gets past the screens in wastewater is ground and shredded during the comminution process. A large comminutor or grinder that consists of a rotating slotted cylinder can be installed in the wastewater flow channel. When wastewater passes the comminutor, large solids that do not pass the slots are cut to smaller sizes by the blades (counter-rotating, intermeshing cutters) of the rotating cylinder.

25.5.3 Grit Removal

Grits are small, solid particles of sand, gravel, cinder, and mineral materials. These also include bone chips, coffee grounds, eggshells, and seeds in the wastewater with a specific gravity higher than that of the organic biodegradable solids. Grit removal typically requires an aerated chamber, which maintains a certain velocity of the liquid stream. This process facilitates the settling of gritty materials while maintaining the suspension of organic materials. The removal of grit prevents abrasion and damage to mechanical equipment because of its accumulation in aeration basins, anaerobic digesters, channels, and pipelines. The settled grit is then removed, washed and disposed of in the landfills or agricultural fields.

25.5.4 Primary Sedimentation

This is the last stage of primary treatment, where the wastewater rests in a large rectangular or circular tank, and gravity forces the solids to settle at the bottom. The solids (or sludge) are removed using a continuous belt device and then pumped into a sludge digester or other sludge treatment unit process. The settling of the solids can be accelerated by adding some chemicals. The effluent is now ready to proceed for secondary treatment.

25.6 SECONDARY TREATMENT OF WASTEWATER

The secondary treatment of wastewater exclusively relies on microbial activities. This works on the principle that various organic substances in wastewater serve as the food materials for the indigenous microorganisms. The microorganisms can perform activities like degradation, adsorption, assimilation and flocculation on the contaminants and multiply. Eventually, this process increases the microbial density in wastewater for an effective utilisation of carbonaceous and nitrogenous matter (mainly NH_3). The degradation process increases the diversity of the labile fraction of organic matter in the wastewater, which attracts new microorganisms from diverse environments. Some microbial species even promise to remove phosphate and drive off entrained gases (like CO_2, NH_3, and nitrogen gas (N_2)) from liquid streams. The degradation process can also be accelerated by augmenting some pregrown culture or nutrients in the wastewater. An efficient secondary treatment system can remove as much as 85–95% of BOD and the TDS particles smaller than 150 μm.

Biological treatment can be achieved by two different methods: aerobic or anaerobic methods. Each method relies on different types of microorganisms and gives different outcomes under distinct setups. During aerobic treatment, the organic matter is metabolised by aerobic microorganisms, and the final products are CO_2 and water. This process requires sufficient oxygen and nutrients. Activated sludge and trickling filtration are the two most commonly used designs for aerobic treatment. Anaerobic treatment or digestion relies on the biological breakdown of organic waste using anaerobic microorganisms under anaerobic conditions. This process releases gases like methane (CH_4) and CO_2. Examples of some common setups of anaerobic treatment are the up-flow anaerobic sludge blanket (UASB) reactor, expanded granular sludge bed (EGSB) reactor, down plug-flow anaerobic reactor (DPAR), fluidised-bed reactor (FBR), and septic tank. Anaerobic treatment is relatively inexpensive, requires less energy and nutrients, and generates less sludge. This process effectively stabilises biological sludge and high-strength industrial wastes (such as those from the food industry). Also, CH_4 gas produced during anaerobic digestion is a sustainable source of energy. However, CH_4 emission can be a great concern due to system failure, owing to its greenhouse effects.

Some common setups of secondary wastewater treatment are discussed in detail as follows.

25.6.1 Activated Sludge Process

Activated sludge is the flocculent culture of microorganisms developed in an aeration tank under controlled waste degradation conditions. It is the effluent of primary treated sewage/industrial wastewater that has a high density of microbial biomass, which can be used for secondary wastewater treatment. It is a mixed-culture biosystem consisting of various kinds of microorganisms, including bacteria (about 95%), fungi, protozoa, and even some invertebrates, occasionally.

The activated sludge system consists of three main components: an aeration tank, a settling tank, and return activated sludge (RAS) equipment (Figure 25.3). An aeration tank is a large, open vessel that works as a bioreactor. The space within the aeration tank facilitates suitable conditions for the degradation of organic wastes involving native microbes and activated sludge. Air supply is a critical need during the entire operation, and it is pumped using a diffuser. An agitator creates turbulent agitation in the tank, which improves the mixing of the wastewater with the activated sludge. After sufficient degradation in the aeration tank, the mixed liquor is discharged into settling tanks (a secondary sedimentation basin), where activated sludge settles to the bottom. The effluent is often treated with chlorine in order to eliminate pathogenic microorganisms and reduce odour before being released into the natural stream. A part of the settled activated sludge from settling tanks is sent to RAS equipment to re-seed the influent sewage of the aeration tank. RAS equipment maintains a desired flow rate in the aeration tank to achieve maximum treatment. Excess sludge is stored in storage tanks and can be treated by aerobic or anaerobic digestion before disposal.

The performance of the activated sludge process mainly depends on two factors: hydraulic residence time (HRT) and solid residence time (SRT). HRT is the average time that the wastewater spends in the system. The wastewater in the aeration tank, having a long HRT, maximises the intimate contact of wastewater with activated sludge, increasing the treatment efficiency. A typical activated sludge has an HRT of 4-8 h, which can be extended up to 36 h for maximum degradation. A different parameter, SRT, is the time that activated sludge solids (microorganisms) spend in the system. Long SRT has an impact on biological stability and efficiency. SRT is typically longer than HRT. The handling of low-strength wastewater and nitrification processes generally requires both long HRT and SRT.

Activated sludge unit comes in various configurations for different operational considerations. A set-up having a high-rate process maintains a high waste-to-activated sludge ratio that increases sludge production. A different configuration has a step-aeration process that involves the flow of both effluent and activated sludge at multiple points in the aeration tank along its length to ensure their homogenous distribution. A tapered aeration process meets varying demands of aeration through the injection of air along the length of the tank. A modified version of activated sludge employs granular sludge instead of flocculent sludge that settles quickly in the secondary clarifier. A hybrid system consists of fixed-film media (such as plastic carriers), which provides excess surface area for microbial growth in the aeration tank. The sequencing batch reactor (SBR) is another type of activated sludge process that operates in one tank and involves a batch mode of operation. It features a cyclic trend, having a sequence of feeding, aeration, reaction, sedimentation, and decanting of effluent in the same tank without the need for an external clarifier.

The biologically activated sludge process can be further modified with appropriate physicochemical conditions, allowing the biotransformation of nitrogen-containing organics into NH_3 (ammonification), NH_3 into nitrate (NO_3^-) (nitrification), and subsequently NO_3^- into N_2 (bacterial denitrification). Microorganisms (*Nitrosomonas* and *Nitrobacter*) involved in nitrification are autotrophic and

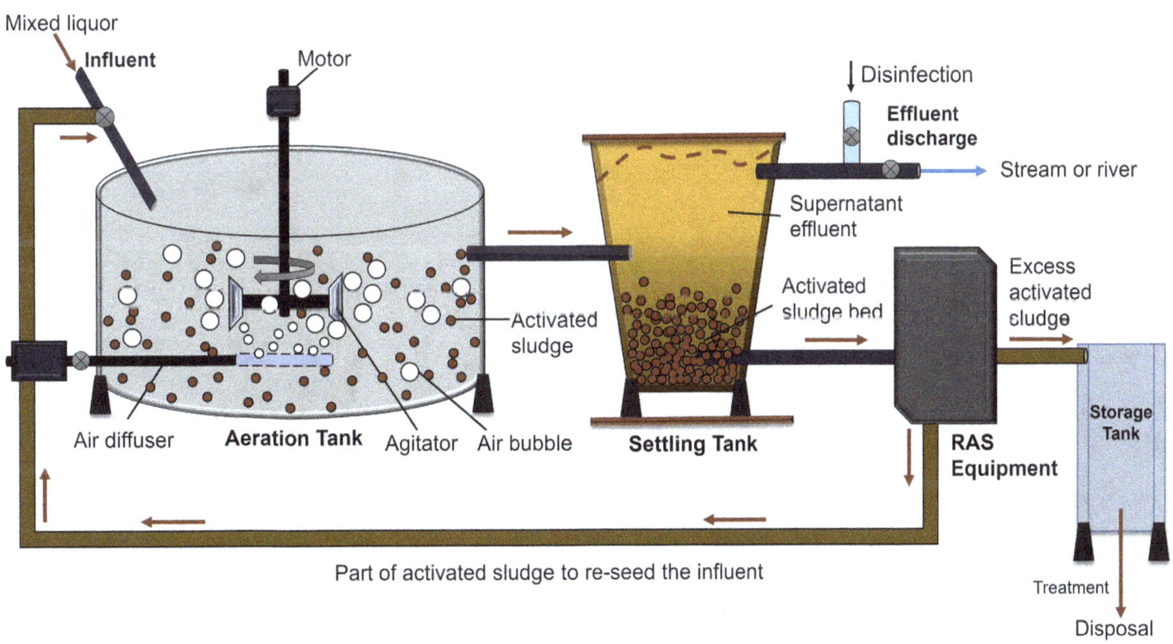

FIGURE 25.3 A schematic representation of an activated sludge system, showing an aeration tank, a settling tank, and RAS equipment for wastewater treatment (RAS: Return-activated sludge.)

slow-growing, and hence, the sludge in the tank is kept sufficiently long for their sufficient proliferation. Activated sludge can also be used to eliminate phosphorus contamination using chemicals or through the biological phosphorus removal process. Activated sludge treatment can reduce BOD and TSS by 85–95%, COD by 75–95%, and minerals by 50–90%.

25.6.2 Trickling Filter System

A trickling filter, biological trickling filter, or biofilter is a widely used conventional aerobic biological treatment system that removes or stabilises the organic materials from wastewater using microorganisms attached to a fixed filter bed. Like the activated sludge process, the trickling filtration process also requires a constant supply of food (organic wastes), sufficient oxygen, appropriate support, and nutrients. However, this process differs from the activated sludge process in terms of wastewater flow passing through the filter bed (Table 25.2). A trickling filter system typically consists of perforated rotary distributors, a filter bed, and an aeration system (Figure 25.4).

In the treatment process, wastewater from a primary clarifier (or sedimentation) tank is fed into the trickling filter unit via a dosing device (tipping bucket). Wastewater flows upwards from the dosing device into the arms of the perforated rotary distributor or static nozzles. This rotary distributor spreads the liquid wastewater over the surface of the filter bed. Filter beds are typically made up of cheap, inert, porous materials such as ceramic, coke, gravel, lava, peat moss, plastic media, or polyurethane foam. When the incoming wastewater trickles down the filter bed, the native microorganisms of the wastewater get attached to the filter bed and start multiplying, utilising the organic matter from the wastewater. The porous body of the bed provides a large surface area for the growth of microbial slime or the development of biofilm on it. These microorganisms require excess aeration to support their growth, which also ensures efficient organic

TABLE 25.2
Differences between activated sludge and biofilm processes for the secondary treatment of wastewater

Characteristics	Activated Sludge	Biofilm Process
Treatment system	Aerobic	Aerobic or anaerobic
Microbial growth system	Flocculent and suspended particles	Attached as a biofilm to a fixed filter bed
Wastewater flow	Into suspended sludge containing microorganisms	Microorganisms adhered to a stationary surface
Developed in	Aeration tanks	Systems like trickling filters and RBC
Aeration with	Diffuser	Blower (in trickling filter system) and stacked discs (in RBC)
Bacterial utilisation of particulate organic matter	High	Relatively low
HRT	Similar to SRT	Lower than SRT

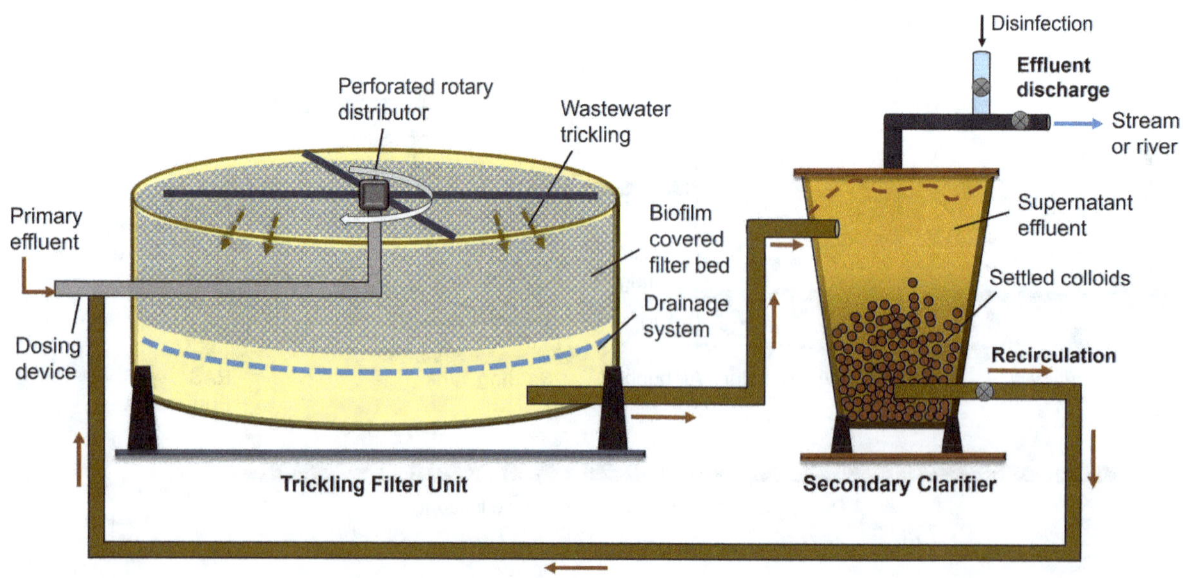

FIGURE 25.4 A schematic representation of a trickling filter system, showing the trickling filter unit and secondary clarifier for wastewater treatment.

matter decomposition. A blower (forced aeration) or natural air convection is used to supply air to the system. The treated wastewater goes down the filter bed into a drainage system. After treatment, the filter bed requires periodic cleaning and biomass removal to prevent clogging and channelling during operation. When the microbial slime in the filter bed becomes excessively thick and heavy, it detaches from the media. This sloughed-off material is also carried out of the bed by the wastewater flow and subsequently settles in the clarifier.

The treated wastewater enters a secondary clarifier (a sedimentation unit or settling tank). The secondary clarifier separates the wastewater into two components: sludge and supernatant. The sludge is the small colloids and suspended particles settled at the bottom. It can be treated through anaerobic digestion and is disposed of. The supernatant (effluent) from the clarifier is chlorinated, followed by dechlorination. The ultraviolet (UV) light treatment can also be used for disinfection. This process ensures the removal of pathogens and coliform bacteria in the water. Subsequently, the treated effluent flows along a step aerator where the DO levels are increased and finally released into a stream or river. Sometimes, the treated effluent, including the sloughed-off oxidised or decomposed organic matter from the secondary clarifier, is recirculated into the trickling filter unit via a return pump. This re-circulation elevates the DO level in the primary effluent and reduces the load of organic contaminants, which improves treatment efficiency.

Trickling filter systems have many advantages, such as high removal of BOD, including both carbonaceous and nitrogenous oxygen demand, low energy demand, production of the desired effluent quality with a short HRT, and good performance with semi-skilled labour. It can be used with other technologies, including rotating biological contactors (RBCs) and packed bed reactors (PBRs). Trickling filters are particularly useful for treating low-strength (COD< 1,000 mg/l) sewage or domestic wastewater. The low-rate filter configuration can achieve as high as 80–90% BOD_5 (5-day oxygen demand) removal.

25.6.3 Rotating Biological Contactor (RBC)

An RBC is a conventional aerobic biological wastewater treatment unit that relies on the biofilm process. The contactor is a fixed-bed reactor consisting of a series of circular stacked discs that are mounted on a horizontal shaft (Figure 25.5). The discs are made up of lightweight plastic sheets like polyethylene, polystyrene, or polyvinyl chloride (PVC). They have ridged patterns, which increase their specific surface area and provide sites for microbial attachment. The discs rotate at a speed of 1–2 rpm, partially submerged (40–80%) in the influent wastewater. When wastewater also flows along with the discs, the microorganisms are alternately exposed to the air and wastewater and develop the biofilm. Microorganisms in the biofilm consume organic matter from the wastewater, and the treated wastewater flows out of the contactor.

RBC can be a reliable tool for wastewater treatment, considering the high oxygen supply, low space requirement, prolonged contact of microbial biomass with pollutants, and low risk of channelling. This additionally supports enhanced process stability, resistance to varying organic loading, and low sludge production. Wastewater treatment using an RBC can achieve a BOD reduction of about 80-90% within two days. However, its operation requires a continuous energy supply and high maintenance costs. Anaerobic conditions due

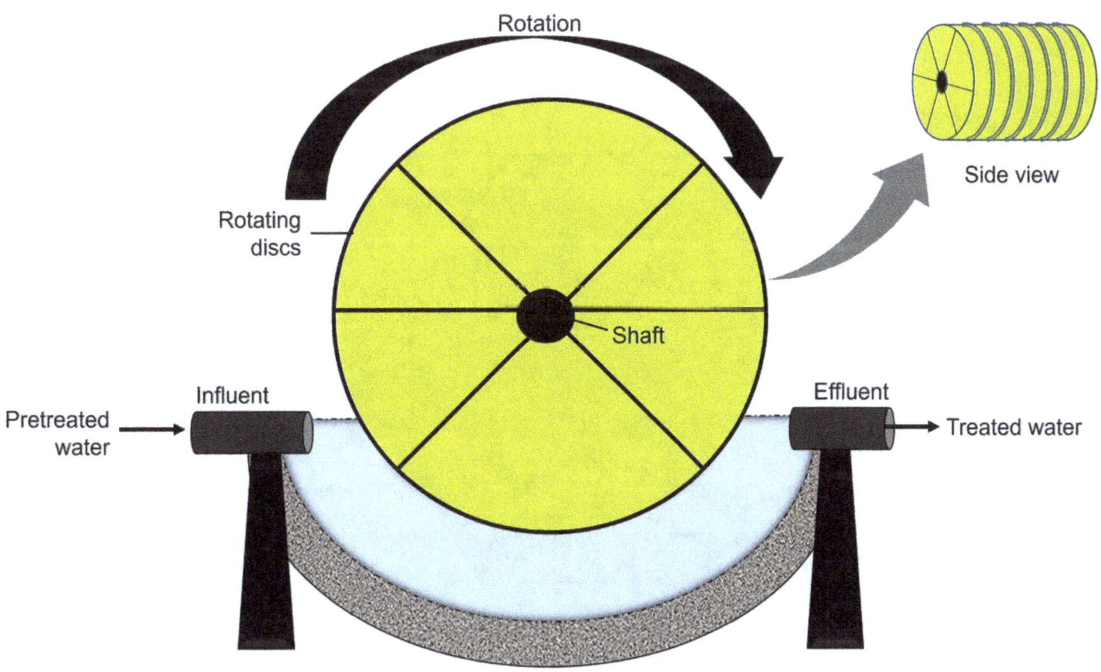

FIGURE 25.5 A schematic representation of a Rotating Biological Contactor (RBC) for biodegradation in aerobic situations.

to a large organic loading can result in an unpleasant odour and poor treatment efficiency. The operation can also be responsible for high aerosol concentrations in the surrounding area due to splashing and the release of droplets into the air.

25.6.4 Septic Tank

A septic tank is a small, underground, water-resistant tank built to receive, collect, and partially treat raw domestic sanitary wastewater near the source (Figure 25.6). This is one of the earliest wastewater treatment setups. It can be developed for small populations, such as campgrounds, household residences, institutions, and rural areas where soil is not suitable for sewer installation. The outer walls of the tank are water-sealed using bricks, mortar, or concrete to avoid underground infiltration. The wastewater inside the tank is held at low velocity under anaerobic conditions for a minimum period of 36 h. In this process, heavy solids settle at the bottom of the tank, while grease and lighter solids float to the top. The settled solids undergo microbial decomposition with the formation of gas. The entrained gases cause the solids to rise from the wastewater, which floats at the surface as a scum layer. When the gases escape from the scum layer, the solids re-settle again. Further, effluent from the septic tank is discharged into trenches or drain fields for natural subsoil infiltration or treatment into a WWTP.

This system has several limitations compared to more advanced treatment plants. A septic tank provides relatively weak purification compared to other setups. Wastewater treated in septic tanks removes about 50% of the organic matter and suspended solids in 2–4 days. The septage or slurry from the septic tank is often treated separately with lime or sent to wastewater treatment plants for volume reduction. The periodic removal of accumulated sludge and scum from the septic tank is necessary and can be further used. Also, there is a high risk of leakage in the tank, which might result in the spread of pathogenic microorganisms and contaminants to groundwater or nearby surface water.

25.6.5 Oxidation Pond

An oxidation pond, stabilisation pond, or lagoon is a large, shallow pond or basin designed to receive, hold, and treat wastewater for a specific duration. The oxidation pond can suitably work on partially treated wastewater from a septic tank. Wastewater treatment in oxidation ponds is among the oldest wastewater treatment processes, combining physical, biological, and chemical activities. The major forces involved during the treatment rely on natural schemes employing indigenous microbial communities, native chemicals, and sunlight. The algae in the oxidation pond use solar energy, CO_2, and inorganic compounds (pollutants) to perform photosynthesis. The photosynthetic product, oxygen, then supports the growth of aerobic bacteria, which break down the organic matter in the pond (Figure 25.7).

For some pond systems, mechanical aerators or aeration devices are installed to provide an excess oxygen supply in the water to treat high amounts of organic matter in the wastewater. Pond water may require additional support to remove nutrients and microbes before being returned to the environment. Any algae left in the pond effluent can be filtered out or eliminated through chemical treatment and settling. The bacterial sludge that deposits in the pond is usually removed by dredging. Wastewater in the oxidation pond lasts from 7 to

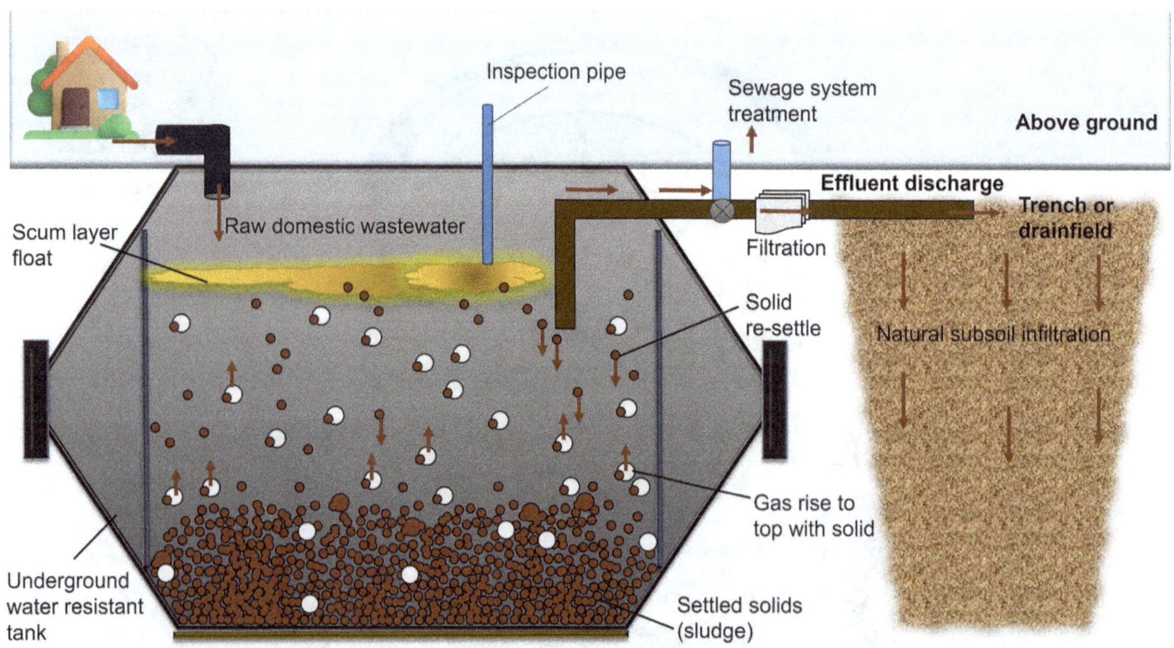

FIGURE 25.6 A schematic representation of an underground septic tank system showing the settling of sludge and effluent discharge into a trench or drain field.

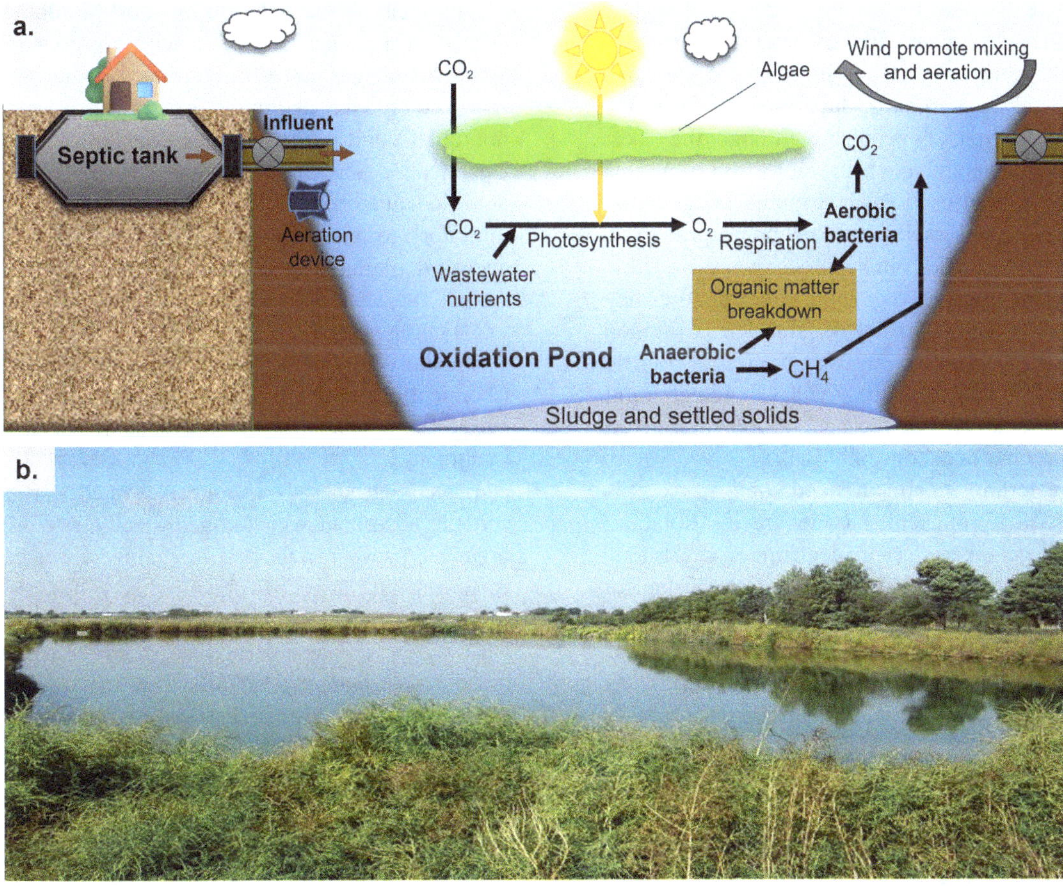

FIGURE 25.7 a. A schematic representation of an oxidation pond showing the biodegradation process. b. A setup of an oxidation pond in Wellington, TX, USA. (CO_2: carbon dioxide, O_2: oxygen gas, and CH_4: methane) (Image b reproduced from Butler et al. (2017).[2])

180 days. The retention time of wastewater in ponds depends on many factors, such as wastewater characteristics, pond design, climate conditions, and treatment objectives. After sufficient treatment, water can be discharged or used for domestic purposes.

Oxidation ponds can be of three types based on the oxygen conditions. These include facultative ponds, aerobic ponds, and anaerobic ponds. The most typical type of oxidation ponds is the facultative ones, having a depth of roughly 6–8 feet. This system features both aerobic and anaerobic environments. The upper layer supports aerobic activities and photosynthesis, and the lower regions favour anaerobic digestion. The wastewater treated in the oxidation ponds contains suspended solids levels and BOD values in the range of 30–150 mg/l and 20–60 mg/l, respectively. The aerobic ponds (or maturation ponds) have vertical layers that are oxygenated throughout. The anaerobic ponds are designed to have an anoxic situation for treatment using anaerobic microorganisms. An anaerobic pond unit is appropriate for the treatment of high-strength industrial or agricultural wastes (i.e., waste streams having high amounts of organic contaminants).

25.6.6 Membrane Bioreactor (MBR)

A membrane bioreactor (MBR) involves a combined effort of ultrafiltration/microfiltration membranes and activated sludge for treating municipal or industrial wastewater. This consists of a large bioreactor connected to a membrane filtration tank. A semipermeable membrane enables the selective passage of specific chemicals while keeping undesired particles on the source side. This process eliminates the need for secondary sedimentation. A part of the treated effluent can be pumped into the bioreactor using RAS. This type of setup features a longer SRT, high bacterial biomass concentration, and complete oxidation of organic matter. This process also ensures reduced biosolids generation. An MBR decreases nutrients like nitrogen and phosphorus to very low levels in the effluent. However, the high operation cost of this technique reduces its value for large-scale operations. Chapter 8 gives a detailed description of the design and principles of MBRs.

A related setup of the MBR is a moving bed biofilm reactor (MBBR). The MBBR contains carrier substances for microbial biofilm development. The carrier-attached microorganisms improve the contact with wastewater contaminants for their effective removal. The MBBR can be modified for aerated and unaerated systems, achieving enhanced nitrification or denitrification and priority pollutant removal.

25.7 TERTIARY TREATMENT OF WASTEWATER

Wastewater cleaned by primary and secondary treatment is only safe enough to release into the environment. The water that is sufficiently pure for drinking or other domestic uses requires tertiary treatment. Tertiary treatment employs sophisticated techniques to remove specific contaminants from the water. It is not a common facility in all WWTPs. Tertiary treatment is typically available through the "best available technology economically achievable" (BTEA). This includes procedures like adsorption, air stripping, coagulation, filtration, flocculation, ion exchange, membrane processes, nitrification, denitrification, and precipitation. These treatment options can achieve very high removal efficiency (about 99% of BOD). They remove toxins, nutrients (nitrogen and phosphorus), heavy metals, refractory biological organic polymers, volatile organic compounds (VOCs), and persistent pathogens like *Giardia duodenalis*, etc.

When wastewater has high concentrations of solid content, its removal before tertiary treatment stands obligatory. This can be accomplished by thickening (with gravity, flotation, and centrifugation), de-watering (with centrifuges, belt filters, and filter presses), and organic sludge stabilisation (biological treatment or with lime), followed by final disposition. However, these setups generally increase the overall budget for contaminant removal. The removal of specific contaminants in water calls for specific techniques, which are discussed as follows.

25.7.1 Suspended Solids Removal

- Physical straining- The suspended solids in wastewater of concentrations above 100 mg/l, carried from secondary clarification, are treated by physical straining methods like diatomaceous earth filtration, moving-bed filtration (a countercurrent extraction), sand filtration, or ultrafiltration (with pressure of about 50 lb/in^2). The filter used in straining requires replacement from time to time because it gets clogged (except for moving-bed filtration) due to the accumulation of solids that increase pressure and affect the filtration. The filter also needs frequent washing to produce high-quality effluent.
- Chemical treatment- It can be achieved through sequential steps of coagulation, flocculation, sedimentation, and filtration. In coagulation, chemicals like polymers and the oxides of aluminium (Al), iron (Fe), or calcium (Ca) are rapidly mixed with the wastewater for charge neutralisation of suspended solids. Floc particles are then aggregated, neutralised, and adsorbed, followed by settling in the sedimentation tank. The unsettled particles are filtered using porous media beds (such as sand).

25.7.2 Organic Matter Removal

- Stripping- Air stripping or steam stripping is used for the treatment of VOCs like benzene and toluene from wastewater. This is performed in packed or tray towers with countercurrent air-liquid flow by introducing steam. When the gas and liquid interact during stripping, the volatile compounds present in the liquid are transferred into the gaseous phase. This off-gas released from stripping can be treated with low-cost biofilters, which consist of a packed bed filled with a porous medium. When the off-gas passes through the biofilter, VOCs are adsorbed onto the medium's surface. This is followed by microbial stabilisation in the biofilter.
- Chemical oxidation- The contaminants which are non-biodegradable (refractory), odour-causing, and inhibitory to microbial growth are typically treated by chemical oxidation. Various oxidising agents such as ozone, titanium dioxide, hydrogen peroxide, ferrous ion (Fenton's reagent), permanganate ion, and chlorine compounds can be used for the oxidation of pollutants. Chemical treatment renders many priority pollutants harmless and makes them amenable to biological treatment.
- Activated carbon adsorption- Activated carbon is a suitable adsorbent which can recover many refractory organics, heavy metals, and inorganic substances from wastewater. It can remove organics like VOCs, pesticides, dyes, pharmaceuticals, fuels, solvents, and toxins from wastewater. Activated carbon assists in removing many priority pollutants and colours from the wastewater. It can also be suitably used to remove excess amounts of chlorine from potable water. Activated carbon has a high porosity and high surface area-to-unit weight ratio, which enhances its adsorption capacity for hydrophobic organics.

25.7.3 Inorganic Matter Removal

- Ion Exchange- Ion exchange is commonly employed for the removal of inorganic constituents like ions, heavy metals, and salts from water. When effluent is passed from an ion exchange medium, the specific ions present in water (with high molecular weight) are preferentially attached to the medium while exchanging some ions from the medium. This method also makes it possible to recover economically important elements from wastewater.
- Electrodialysis- It is a membrane-based process for demineralising wastewater and recovering heavy metals. The device consists of a chamber with two compartments, separated by anion and cation exchange membranes. When voltage is applied to a compartment containing contaminated water, the negatively charged anions present in the water are attracted to the positive electrode (anode), and the positively charged cations move to the negative electrode (cathode). This movement allows the ions to concentrate along anion or cation-permeable membranes, resulting in liquid dilution.
- Reverse Osmosis (RO)- It recovers more than 90% of inorganic ions, organic materials, and colloids from the water. In this process, impure water is placed in contact with a semi-permeable membrane under pressure higher than the osmotic pressure of the solution. This causes the water to permeate the membrane in the reverse direction to osmosis, while dissolved materials remain on their original side. The membrane tends to get foul after a

long run and requires renewal. This is a very common setup in commercial water purification systems for drinking purposes.

25.7.4 Nutrient Removal

- Nitrogen- Accumulation of nitrogenous compounds in water causes ammonia toxicity, eutrophication, high BOD, and methemoglobinemia in humans. A typical solution to remove nitrogen in water is the addition of lime to wastewater and then air stripping in a packed tower. Alternatively, specific natural media, like zeolite clinoptilolite (ZC), can be used to reduce NO_3^- concentration in water. A biological method for their removal is microbial nitrification, followed by denitrification in the modified activated sludge. A relatively new approach to nitrogen removal from wastewater is anaerobic ammonia oxidation (anammox), which allows the transformation of NH_3 into N_2 under anaerobic conditions.
- Phosphorus- Phosphorus in contaminated water bodies stimulates eutrophication and biological growth. Precipitation using chemicals like Al, Fe, or lime, followed by filtration, is a suitable treatment method. It can also be precipitated by lime at a high pH (>10.5). In biological treatment, phosphorus from wastewater can be specifically removed through the **enhanced biological phosphorus removal (EBPR)** process using **polyphosphate-accumulating bacteria (PAB)**. Examples of PAB are *Acinetobacter*, *Pseudomonas*, and *Rhodocyclus*. An effective treatment for phosphorus removal through biological methods requires both aerobic and anaerobic activities (Figure 25.8). An anaerobic situation stimulates the bacterial uptake of volatile fatty acids followed by their metabolism. This metabolic process encourages phosphorus accumulation as polyphosphate in aerobic conditions. In this process, PAB can store about 5–7% phosphorus of their total biomass. Some distinct situations favour phosphorus assimilation into adenosine triphosphate (ATP), nucleic acids, and other biomolecules.

25.8 SLUDGE TREATMENT

Waste-activated sludge (WAS) is the solid, semisolid, or slurry residual materials released as byproducts from primary and secondary wastewater treatment processes. Primary sludge is generated from processes like chemical precipitation and sedimentation, whereas secondary sludge is the activated biomass derived from biological treatments or septage from on-site septic tanks. The WAS may contain microorganisms, organic and inorganic particles, and extracellular polymeric substances (EPS). Primary and secondary sludges are generally mixed for further treatment and disposal. Sludge treatment is mainly employed to reduce its volume, remove water, eliminate pathogens, regulate decomposition, and stabilise organic materials. This process also immensely contributes to reducing expenses during transportation, pumping, and storage, minimising negative impacts on the environment and public health.

Sludge treatment combines the approaches of thickening, stabilisation, and dewatering processes. These processes run in a sequential order at the treatment plant. Thickening intends to minimise the volume of sludge slurry and concentrate the solids using a gravity thickener tank or dissolved air flotation. A gravity thickener tank is a cylindrical or conical body where

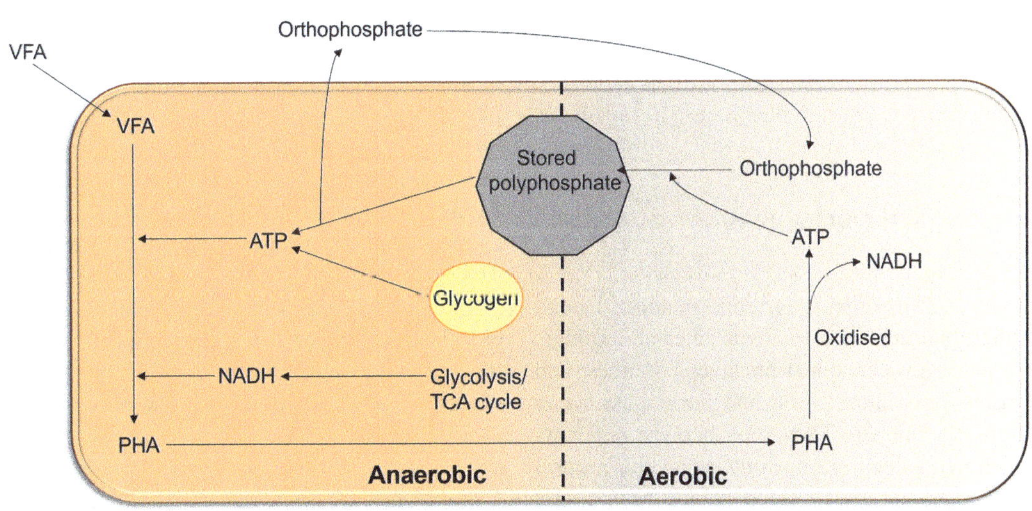

FIGURE 25.8 Processes involved in enhanced biological phosphorus removal (EBPR) in the cells of polyphosphate-accumulating bacteria (PAB). PAB absorbs VFA and stores them as PHA in anaerobic environments, which requires ATP and NADH. ATP is produced by polyphosphate hydrolysis (which releases orthophosphate) or glycogen breakdown, while NADH is produced through glycolysis or the TCA cycle. During the aerobic phase, an electron acceptor oxidises stored PHA, releasing energy for orthophosphate uptake and polyphosphate synthesis, leading to phosphate buildup. VFA- volatile fatty acids, PHA- polyhydroxyalkanoate, TCA- tricarboxylic acid, ATP- adenosine triphosphate, NADH- nicotinamide adenine dinucleotide hydrogen (reduced).

the suspended solid particles settle at the bottom and diluted liquid overflows at the top. In the dissolved air flotation, air bubbles are injected into the sludge, allowing the solid materials to rise to the surface. The layer of thickened sludge from dissolved air floatation is skimmed off. The clarified liquid effluent from the thickeners is then recirculated to the treatment plant influent.

Sludge stabilisation or digestion is a process of aerobic or anaerobic biological decomposition. During decomposition, the organic solids are broken into small, stable substances, leading to the reduction of the total solid mass, destruction of pathogens, and an easier drying of the sludge. Many treatment plants use a two-stage anaerobic system for bacterial decomposition, where thickened sludge (with 5% dry solids) is primarily heated and mixed in a closed tank for several days. This is followed by bacterial fermentation (at about 48–57 °C temperature) that produces fatty acids (acidogenesis). The fermented matter then enters a second tank, where bacteria induce anaerobic digestion of organic residues and produce biogas, including CH_4 (methanogenesis). The traditional two-stage anaerobic digestion process can be pretreated with thermal hydrolysis to improve the waste degradation efficiency. In thermal hydrolysis, thickened sludge is mixed with steam and fed to a reactor at 165 °C and high pressure for some duration. Pressurised sludge is then released into a flash tank, where there is a sudden pressure drop, causing most of the solid matter to burst. This hydrolysed sludge is cooled, diluted with water, and sent for biological decomposition.

The digested sludge is finally dewatered before disposal. Methods of dewatering include spreading on open drying beds of sand and the use of mechanical systems like the belt filter press, rotary drum vacuum pressure filter, and solid-bowl centrifuges. After treatment, the treated sludge can be disposed of by incineration and landfill. One alternative for managing this waste is land application on agricultural fields. Among these options, the use of treated sludge as soil conditioners and fertilisers on agricultural land is eco-friendly and sustainable, implying a circular economy approach. In contrast, incineration is the least suitable because it contributes to air pollution.

25.9 MICROBIAL INTEGRITY IN WASTEWATER TREATMENT

Biological wastewater treatment systems mainly depend on the native microbial activities of wastewater and sludge. Microorganisms in wastewater come from diverse sources such as domestic wastes, agriculture, food and beverages, urban runoff, industrial effluents, etc. They soon become residents of wastewater, allowing them to acclimatise to the natural wastewater environments. These microorganisms possess the ability to digest organic matter and assimilate/transform nutrients in wastewater for their growth and metabolism. This process aids in the degradation of organic matter, nutrient removal, and the transformation of toxic compounds into harmless products.

Microbial communities observe complex interactions with chemicals in the medium, which is very difficult to replicate in a synthetic environment. The concentrated sludge of the medium serves as the inoculum for wastewater treatment. Some cultured microbial sources have also shown potential for wastewater treatment and can be augmented in wastewater to accelerate the bioprocess or remove specific contaminants.

The active heterotrophic microorganisms in activated sludge undergo multistep reactions to transform complex organic matter into CO_2 and biomass. They initially break them down into smaller products, which can be digested into simpler compounds and followed by complete mineralisation. The overall degradation process of organic wastes in wastewater relies on bacterial consortia. A bacterial consortium is a group of different symbiotic bacteria coinhabiting in the same space. They can act synergistically or sequentially on the contaminants for their effective degradation.

Bacteria, moulds, and yeasts growing in wastewater and sludge have very powerful traits for the degradation of complex compounds using various enzymes. Depolymerisation requires extracellular hydrolases, which typically include amylases, proteases, and lipases. Oxidase enzymes are essential for the nitrification process, whereas reductases are involved in the reduction of nitrate to N_2. Laccases oxidise phenolic compounds, including the degradation of lignin. Phosphatases catalyse the removal of phosphate groups from organic molecules, aiding in the removal of phosphorus from wastewater. Respiratory enzymes are crucial for the mineralisation of carbohydrates into CO_2, water, and energy. Anaerobic conditions promote the growth of anaerobic bacteria and methanogens, which involve different sets of reactions and produce different outcomes.

25.9.1 Microorganisms in Wastewater Treatment

Among various microbial groups, bacteria are the most abundant in wastewater and play a major role in wastewater treatment. In activated sludge, the taxa belonging to the phylum Proteobacteria are predominant. They are also prevalent in sewage and other wastewater treatment settings, like constructed wetlands. Taxa belonging to Acidobacteria, Actinobacteria, Bacteroidetes, Chloroflexi, and Firmicutes coexist with Proteobacteria in many wastewater compositions. The most notable taxon in wastewater is *Zooglea* (Phylum: Proteobacteria). This heterotrophic bacterium has excellent properties for EPS production, which contributes to floc development and biofilm formation in activated sludge. This not only protects the bacterial viability against various stresses but also aids in the removal of various contaminants. Many other bacterial taxa, like *Pseudomonas*, *Comamonas*, *Treponema*, *Propionibacterium*, *Syntrophus*, and *Desulfobulbus*, also have similar properties and exhibit similar behaviour in WWTP. Filamentous bacteria also build biofloes in activated sludge and particularly remove organic matter under low COD or substrate-limiting conditions. Anaerobic bacteria like *Bifidobacterium*, *Clostridium*,

Propionibacterium, and *Peptostreptococcus* persist in anaerobic digesters and contribute to reducing sludge volume.

The biological oxidation of NH_3 demands proteobacterial ammonia oxidisers and anammox bacteria. Ammonia oxidising bacteria include *Nitrosomonas*, *Nitrosospira* and *Nitrosococcus*, and their metabolic product is nitrite. The oxidation of nitrite (NO_2^-) into NO_3^- requires nitrifying bacteria like *Nitrobacter*, *Nitrococcus*, and *Nitrospira*. The Anammox process releases N_2 from ammonia oxidation under anaerobic conditions. The anammox bacteria belong to a group, Planctomycetes and include *Candidatus* Brocadia, *Candidatus* Kuenenia, *Candidatus* Scalindua, *Candidatus* Anammoxoglobus, and *Candidatus* Jettenia. The denitrification process, where NO_3^- and NO_2^- are converted into nitrous oxide (N_2O) and then N_2, is also a crucial process for nitrogen removal. These reactions are performed by bacteria from diverse phylogenetic lineages, including chemoorganotrophs, lithoautotrophs, and phototrophs. Some prominent denitrifying species reported in wastewater are *Alcaligenes*, *Bacillus*, *Hyphomicrobium*, *Methylobacterium*, *Pseudomonas*, and *Paracoccus*.

Phosphorus can be removed using PAB, which performs EBPR (detailed in section 25.7.4). These bacteria have the unique ability to store large amounts of phosphorus within their cells as polyphosphate. In a different situation, bacteria like *Geobacter*, *Shewanella*, and *Rhodopseudomonas* demonstrate an ability for electron transport to a working electrode, which assists in electricity production during wastewater treatment. These microorganisms can be useful as microbial fuel cells (MFCs) for wastewater treatment and electricity production simultaneously (Box 25.1).

Archaea are mainly present in the anaerobic digestion process. Methanogens (Phylum: Euryarchaeota) are typically dominant Archaea in sludge. They use a few substrates, namely hydrogen gas, CO_2, methyl-group-containing compounds, and acetate, to generate CH_4 as a value-added by-product. This process overall reduces the amount of excess sludge. Molecular analysis of methanogenic sludges has identified archaeon taxa from orders Methanomicrobiales, Methanosarcinales (e.g., *Methanosaeta* and *Methanosarcina*), Methanobacteriales, and some uncultured groups. In addition, halobacteria, denitrifying methane-oxidising archaea and ammonia-oxidising archaea (e.g., *Nitrosopumilus maritimus*) assist in nitrogen removal. Archaea in wastewater also involve a possible symbiotic relationship with bacteria, which overall contribute to the floc structure, increase biological activities (nitrification and denitrification), and enhance organic matter mineralisation.

Fungi are relatively less common in wastewater treatment. However, when present, a core of common fungi belongs to some moulds and yeasts, including *Penicillium*, *Aspergillus*, *Candida*, and *Geotrichum*. They secrete diverse enzymes and can act on recalcitrant dyes, pesticides, organic matter, etc. A different microbial group of protozoa can directly take up particles in the micrometre range and perform their intracellular degradation. Protozoans excrete simple organic matter from their ingestion, which attracts bacterial colonisation for their further degradation.

25.9.2 Phage Treatment

In recent years, phage treatment has emerged as a novel bioweapon to control bacterial pathogens in water using

BOX 25.1 MICROBIAL FUEL CELL (MFC)

MFC is a renewable energy solution that treats organic and inorganic wastes while simultaneously producing electricity. This is a bioelectrochemical device consisting of two electrodes: an anode and a cathode. These electrodes are wired through an external electrical circuit and are separated by a proton exchange membrane (PEM). The anode chamber favours the anaerobic oxidation of organic wastes using microorganisms and generates free electrons. These electrons flow through the external circuit and reach the cathode chamber, where they reduce oxygen (or any other electron acceptors) in an abiotic condition, resulting in power generation. The reduction reaction at the cathode requires protons that come from the oxidation reaction in the anode, passing through the PEM. In some distinct situations, the MFC process can be employed for remediating copper (Cu), Fe, and mercury (Hg), where metal reduction replaces oxygen reduction in the cathode chamber. Many reduced heavy metals have low toxicity and low mobility.

Microorganisms involved in anodic reactions are called electrogens. Some common examples of electrogens are *Geobacter*, *Shewanella*, *Desulfuromonas*, and *Pseudomonas*. Electrogens show metabolic flexibility and have tremendous potential to degrade a wide variety of organic substances. These microorganisms have the ability to transfer electrons from their cells to the solid surface of the anode. The transfer of electrons from the oxidation reaction occurs either by adhering bacterial cells (through pili) on the anode or with exogenous mediators. Some bacteria can even develop biofilm on the anode surface, allowing for close contact between microbial cells and the electrode, which improves electron transfer efficiency.

MFC research is still in its early phases, and much more research is required before it can be used at industrial scales. The main hurdle in its commercialisation is limited power output, which can barely power an LED. While purifying wastewater, it can remove up to 50% of the COD and achieve power densities ranging from 420 to 460 mW/m^2. In recent years, a variety of configurations and modes of operation have emerged to maximise the efficiency of the MFC, demonstrating their potential for large-scale operations.

different bacteriophages (viruses that infect and kill bacteria). This method leverages the natural ability of some viruses to target specific bacteria, offering a potential alternative to traditional water disinfection methods. The key advantages of using phages are their wide distribution in the biosphere, high specificity towards their host, and rapid lysis of bacterial cells. Some phages even display the potential to disrupt the biofilm. When phages infect and lyse bacterial cells, they multiply exponentially, which produces a self-amplifying treatment effect. This implies that phages may produce a high number of offspring from a very small initial dosage. Their roles in water treatment also include tracking pollution sources, monitoring pathogenic bacteria and bloom-forming cyanobacteria, and controlling sludge. Examples of some phage indicators are *Escherichia coli* phages (somatic and male-specific (F+) coliphages), *Bacteroides* phages, and *Enterococcus* phages. In distinct situations, coliphages from the Podoviridae, Myoviridae, and Siphoviridae families are responsible for lowering the most probable number (MPN) of coliforms (e.g., *Citrobacter freundii*, *Klebsiella*, *Escherichia coli*, etc.) and pathogenic Enterobacteriaceae members (*Shigella*, *Vibrio cholerae*, and *Salmonella*) in sewage.

25.10 CARBON NEUTRALITY IN WASTEWATER TREATMENT

A WWTP is an essential unit of urban sanitation, but it can contribute 1–2% of the total greenhouse gas emissions in the environment. The major concerns come from gases like CO_2, CH_4, and N_2O during their operations. Increasing economic growth and urbanisation necessitate the implementation of more WWTPs in metropolitan areas. This has raised concerns about increased greenhouse gas emissions in the environment.

This calls for an approach called carbon neutrality in wastewater treatment, which facilitates the balancing of the amount of greenhouse gases (GHGs) emitted from the treatment by an equivalent amount of GHGs removed or offset in other processes, achieving a net zero carbon footprint. In this process, carbon neutrality can be partially met by the energy independence gained during treatment operations.

Carbon neutrality in wastewater treatment is an urgent demand to abate climate change amid the water crisis and resource scarcity. This involves a complex process and often requires reevaluation of already established setups and technological innovations. The microbial-assisted treatment processes have gained tremendous interest in an integrated biotechnological-based solution for CO_2 conversion in alternative energy drivers. The dual functionality of energy production and carbon sequestration of this wastewater treatment method has great potential to contribute to carbon neutrality. For example, the WAS is a vital byproduct of sewage treatment, having dual properties of contaminants and resources. The treatment of WAS needs substantial energy, which can be effectively compensated for through the recovery of biomass energy from sludge. In anaerobic digestion, organic matter is transformed into biogas, which can be used as a resource for heating and generating electricity. Methods such as sludge anaerobic co-digestion with exogenous substrates, anaerobic ammonia oxidation processes, and the use of UASB, EGSB, and MFC have shown some potential to reduce carbon emissions. The MFC technology further uses CO_2 released from the degradation process for sequestration into some organic compound or biomass.

SUMMARY

- Wastewater treatment aims to restore the quality of contaminated water using physicochemical and biological methods in a complex facility of WWTP. The treatment process can be divided into primary treatment, secondary treatment, and tertiary treatment.
- Primary treatment is the first stage of wastewater treatment that focuses on solid removal. This can be achieved through screening, grinding, grit removal, and sedimentation. Some WWTPs install pretreatment facilities before the main treatment stages for the ease of downstream processing.
- Secondary treatment mainly involves the biochemical oxidation of organic contaminants in wastewater. This facility mainly relies on activated sludge, a trickling filter, RBC, a septic tank, and a membrane bioreactor.
- Tertiary treatment is a complex process of purifying water. This is performed after secondary treatment to meet the drinking water standard. This calls for sophisticated techniques like ion exchangers, activated carbon adsorption, air stripping, reverse osmosis, chemical treatment, etc.
- Sludge produced by wastewater treatment processes can be treated by thickening, stabilisation, and dewatering processes. Stabilisation is a biological process which involves the anaerobic digestion of organic matter by microorganisms and produces biogas (e.g., CH_4).
- Biological treatment relies on native microbial communities (mainly bacteria) to digest organic matter, assimilate nutrients, and convert toxic compounds into harmless products. In a distinct approach, phages can be used to target and eliminate pathogenic bacteria in wastewater.
- The integrated biotechnological-based solutions have tremendous potential for CO_2 conversion into energy drivers for carbon neutrality.

DIFFERENT TYPES OF PRACTICE QUESTIONS

Fill in the blanks

I. The …………………….. process controls any fluctuation in the wastewater flow rate to provide uniform conditions in the subsequent treatment.

II. The secondary treatment system for wastewater can remove the BOD to about ………………………%.

III. Anaerobic digestion involves biodegradation of organic matter using microorganisms and releases gases like ……………. and ………………..

IV. A.......................... is an aerobic treatment system that consists of a large tank to remove the organic materials from wastewater using microorganisms attached to a fixed bed.
V. .... are large, shallow basins that are designed to receive, hold, and treat wastewater using algae and bacteria.

SOLUTION: I. Equalisation/ pretreatment, II. 85–95, III. methane, carbon dioxide, IV. trickling filter, biological trickling filter, or biofilter, V. oxidation pond/ stabilisation pond/ lagoon

Match the following

i. Pretreatment of wastewater	a. Anaerobic digestion
ii. Primary treatment of wastewater	b. Air stripping
iii. Secondary treatment of wastewater	c. Comminution
iv. Tertiary treatment of wastewater	d. Activated sludge process
v. Sludge treatment	e. Neutralisation

SOLUTION: i-e, ii-c, iii-d, iv-b, v-a

Match the following

i. Nitrifying bacterium	a. *Methanosarcina*
ii. Wastewater heterotrophic bacterium	b. *Geotrichum*
iii. Wastewater archaeon	c. *Zooglea*
iv. Wastewater fungus	d. Myoviridae
v. Phages	e. *Nitrospira*

SOLUTION: i-e, ii-c, iii-a, iv-b, v-d

Very short answer type of questions

I. Expand on WWTP, PAB, and MBR.
II. Name any two methods used for the pretreatment of wastewater.
III. What is a rotating biological contactor?
IV. Which bacterial phylum is most prevalent within activated sludge systems?
V. How can dissolved oil be removed during wastewater treatment?
VI. Why has sustainability become an urgent demand for wastewater treatment?

Short answer type of questions

I. What are the prime objectives of wastewater treatment?
II. Give a schematic representation of the wastewater treatment process.
III. Shed light on how primary treatment of wastewater involves the physical separation of solids in wastewater.
IV. Write the differences between the activated sludge and biofilm processes.
V. What are the roles of microorganisms in sludge treatment?
VI. What is the phage treatment in wastewater treatment?
VII. What is the significance of membrane bioreactors in wastewater treatment?

Long answer type of questions

I. Give a comparative account of primary, secondary, and tertiary treatment methods of wastewater.
II. Discuss any three methods of the secondary treatment of wastewater with suitable diagrams.
III. Why does the water used for drinking or other domestic purposes require tertiary treatment? What are the different methods to remove various contaminants in tertiary treatment?
IV. What are the different types of microorganisms and their respective roles in the wastewater treatment process?

NOTES

1 Ijoma GN, Mutungwazi A, Mannie T, et al. Addressing the water-energy nexus: A focus on the barriers and potentials of harnessing wastewater treatment processes for biogas production in Sub Saharan Africa. Heliyon. 2022.
2 Butler E, Hung YT, Suleiman, et al. Oxidation pond for municipal wastewater treatment. Applied Water Science. 2017;7:31–51.

SUGGESTED READINGS

Akbar M, Yaqoob A, Ahmad I, Ahmad A. Wastewater treatment: an overview. In Asiri A, Kamal T, Ahmad A, Tabassum S, Ahmad I (eds.) *Sodium alginate-based nanomaterials for wastewater treatment*. 2023 (Vol. 1, pp. 19–34). Elsevier.

Crini G, Lichtfouse E. Advantages and disadvantages of techniques used for wastewater treatment. *Environmental Chemistry Letters*. 2019;17:145–55.

Darra R, Hammad MB, Alshamsi F, Alhammadi S, Al-Ali W, Aidan A, Tawalbeh M, Halalsheh N, Al-Othman A. Wastewater treatment processes and microbial community. In Kumar V, Bilal M, Shahi SK, Garg VK (eds.) *Metagenomics to bioremediation*. 2023 (pp. 329–355). Academic Press.

Englande Jr AJ, Krenkel P, Shamas J. Wastewater treatment & water reclamation. In *Reference module in earth systems and environmental sciences*. 2015. Elsevier.

Ferrera I, Sanchez O. Insights into microbial diversity in wastewater treatment systems: How far have we come? *Biotechnology Advances*. 2016;34(5):790–802.

Gude VG. Wastewater treatment in microbial fuel cells–an overview. *Journal of Cleaner Production*. 2016;122:287–307.

Gujer W, Henze M, Mino T, Van Loosdrecht M. Activated sludge model No. 3. *Water Science and Technology*. 1999;39(1):183–93.

Hassard F, Biddle J, Cartmell E, Jefferson B, Tyrrel S, Stephenson T. Rotating biological contactors for wastewater treatment–a review. *Process Safety and Environmental Protection*. 2015;94:285–306.

Jain K, Patel AS, Pardhi VP, Flora SJ. Nanotechnology in wastewater management: a new paradigm towards wastewater treatment. *Molecules*. 2021;26(6):1797.

Jalili P, Ala A, Nazari P, Jalili B, Ganji DD. A comprehensive review of microbial fuel cells considering materials, methods, structures, and microorganisms. *Heliyon*. 2024;10:e25439.

Karia GL, Christian RA, Jariwala ND. *Wastewater treatment: concepts and design approach*. PHI Learning Pvt. Ltd.; 2023;1.

Li L, Wang X, Miao J, Abulimiti A, Jing X, Ren N. Carbon neutrality of wastewater treatment-A systematic concept beyond the plant boundary. *Environmental Science and Ecotechnology*. 2022;11:100180.

Nascimento AL, Souza AJ, Andrade PA, Andreote FD, Coscione AR, Oliveira FC, Regitano JB. Sewage sludge microbial structures and relations to their sources, treatments, and chemical attributes. *Frontiers in Microbiology*. 2018;9:1462.

Nguyen VT, Shieh WK. Secondary treatment. In Liu DHF, Liptak BG (eds.) *Wastewater treatment*. 2020 (pp. 179–230). CRC Press.

Qasim SR. *Wastewater treatment plants: planning, design, and operation*. Routledge; 2017.

Samer M. Biological and chemical wastewater treatment processes. *Wastewater Treatment Engineering*. 2015;150(10.5772):61250.

Seviour R, Nielsen PH, editors. *Microbial ecology of activated sludge*. IWA Publishing; 2010.

Shabir K, Kumar R, Maryam M, Firdous F. 19 Water waste treatment and reuse for sustainability. *Sustainability of Natural Resources: Planning, Development, and Management*. 2024;28:331.

Shah MP, Rodriguez-Couto S, editors. *Microbial wastewater treatment*. Elsevier; 2019.

Tejaswini V, Raju JT, Bhat SA. Water quality management models and systems. In Dar MUD, Shah AI, Bhat SA, Ali SR (eds.) *Advanced technologies for water quality treatment and management* 2023 (pp. 79–119). Apple Academic Press.

Zagklis DP, Bampos G. Tertiary wastewater treatment technologies: a review of technical, economic, and life cycle aspects. *Processes*. 2022;10(11):2304.

Unit VII

Microorganisms in Biogeochemical Cycles and Bioremediation

26 Biogeochemical Cycles and Microorganisms

Earth is home to billions of life forms. From the estimated mass of Earth, 6×10^{24} kg, only 1% of the mass is capable of biogeochemical reactions, but is essential for life and ecosystem balance. The biochemical capabilities of life and geochemical reactions of the environment are closely tied to the biogeochemistry of the Earth. Nature has many unforeseen metabolic capabilities that control chemical reactions. Likewise, geochemical reactions play an important role in the structure and function of ecosystems. **Biogeochemistry** is an interdisciplinary branch of science that deals with biology, geology, and chemistry to understand complex environmental processes. The foundation of "biogeochemistry" was laid down by **V Vernadsky in 1926**. This stream mainly focuses on the interactions between numerous chemicals and diverse life forms in various physical environments. This field also pays great attention to the study of the flow of chemical elements across different parts of the Earth, referred to as the "**biogeochemical cycle**".

The biogeochemical cycle is a two-way flux of chemicals involving various biochemical reactions in the closed system of Earth. It provides a basis for life support, the functioning of ecosystems, and the transformation of chemicals and energy. The four essential ingredients of the biogeochemical cycle are liquid water, sunlight, nutrients, and chemical disequilibrium. They are all equally important for a habitable environment on Earth. Liquid water is abundantly present on the surface as well as below ground. Sunlight is an external factor that plants and some bacteria require for photosynthesis (i.e., primary production). Nutrients are diverse chemical forms necessary for the growth and reproduction of living beings. Chemical disequilibrium refers to an unbalanced reaction where the concentrations of reactants and products in a chemical reaction are not constant. It is a powerful force of metabolic pathways and biosignatures (detailed in Box 26.1).

A high number of other forces can also have an impact on the biogeochemical cycle. A similar cycle that involves the mobility and interaction of an element but with no biological interplay constitutes a geochemical cycle. The geochemical cycle is mainly reported in Earth's crust, mantle, and core. Even a few elements in the environment do not have known biological influence and are regulated primarily by geochemical cycles rather than biogeochemical cycles. Examples of those biologically inactive elements are noble gases like helium (He), neon (Ne), and argon (Ar).

BOX 26.1. CHEMICAL DISEQUILIBRIUM

Chemical disequilibrium is a metric of disequilibrium of chemical substances (e.g., CH_4 and O_2), which can be used to detect life on other planets. The high degree of chemical disequilibrium in a sphere is a sign of life on a planet. On Earth, oxygenic photosynthesis has created a much larger chemical disequilibrium in the present atmosphere-ocean system than at any other time in the past. In the prebiotic Earth, the coexistence of water vapour and volcanic H_2, CO_2, and carbon monoxide (CO) also maintained high chemical disequilibrium. However, this chemical disequilibrium served as an anti-biosignature because the chemotrophic biosphere utilised them for their metabolism, leaving a small disequilibrium producing CH_4 and other gases. In early Earth, the rise of a chemotrophic lifestyle decreased the chemical disequilibrium due to highly redox-sensitive reactions with low activation energy barriers until oxygenic photosynthesis began to expand. The Earth's current disequilibrium with the chemical mixture of O_2, N_2, and liquid water is an ideal chemical disequilibrium for biosignature.

The quantification of chemical disequilibrium is an important method for the detection of life in exoplanets through remote spectroscopy. The biosignature gases like O_2, alongside CH_4 or others that are difficult to explain solely through abiotic processes, serve as a powerful tool in the search for extraterrestrial life. Other planets of the solar system include some disequilibrium due to the inflow of sunlight and mixing of atmospheric content through turbulence and transport. However, the atmospheric chemical disequilibrium is not strong enough to notice a biosignature. Therefore, the presence of biosignature gases is a promising biomarker for detecting life on other planets.

DOI: 10.1201/9781032651095-33

26.1 BIOGEOCHEMICAL CYCLES ON EARTH

From a very small-scale stretch to the entire globe, Earth is occupied with a complex mixture of ninety-two known natural elements. They are present in different compartments of the environment to varying extents. A total of twelve distinct elements are majorly involved in the structural organisation of Earth. These include oxygen (O), silicon (Si), aluminium (Al), iron (Fe), calcium (Ca), magnesium (Mg), sodium (Na), potassium (K), titanium (Ti), hydrogen (H), manganese (Mn), and phosphorus (P), which account for 99.23% of the Earth's crust and are known as abundant elements. Oxygen and Si occupy more than 70% of the continental crust. However, the movement of different elements in both directions between different reservoirs is not the same. Many substances rest in their reservoirs for thousands of years. On the contrary, certain elements are highly redox-sensitive and mobile. Many chemicals are critical to the existence of life and are present in all ecosystems.

The global metabolic map for Earth is majorly dependent on six elements. Elements like C, N, H, O, P, and S are considered the backbones of every living form and are involved in the formation of biomacromolecules. Nature has chosen many elements from the environment for biological integrity that are abundant. Many other elements, such as manganese (Mn), magnesium (Mg), nickel (Ni), copper (Cu), and zinc (Zn), also participate in biological activities. Nearly thirty elements essential for life are coupled and regulated by biogeochemistry. Several elements like arsenic (As), cadmium (Cd), chromium (Cr), mercury (Hg), and lead (Pb) are highly toxic and can negatively impact biomolecules and organisms. They also have active biogeochemical cycles involving resistant microorganisms as a result of their biotransformation and bioaccumulation.

The biogeochemical cycle is a theoretical concept that simplifies complex ecological processes of chemicals. Otherwise, many elements and compounds in the environment are interconnected and interdependent. The biogeochemical cycle of every chemical is a unique, self-perpetuating, and self-regenerating process, depending on its chemistry, distribution, and metabolic activities. The studies on biogeochemical cycles typically focus on individual elements such as H, O, N, and P and their movement in different ecosystems. However, some important compounds like water, methane (CH_4), and carbon dioxide (CO_2) that are composed of more than one element can be mobile in various ecosystems and form unique biogeochemical cycles. Often, many biogeochemical cycles are linked to each other in ecosystems. For instance, the nitrogen cycle is closely associated with the carbon, hydrogen, and oxygen cycles. Water is the universal solvent and is integral to all biogeochemical cycles. Also, every biogeochemical cycle involving water is associated with the hydrogen and oxygen cycles. However, indirect interactions between chemicals can significantly complicate our understanding of biogeochemical cycles. The coupled biogeochemical cycle is a relatively new concept and involves the interaction of more than one chemical. The biogeochemical cycles of rock, kerogen, and coal represent the continuous cycling of various minerals and elements that operate over much longer timescales.

The biogeochemical cycle is mainly a natural process, but can have a lot of human influence. Anthropogenic activities have released numerous substances like organic compounds, heavy metals, radionuclides, etc., which have induced great variation in their natural flux. The human-induced biogeochemical cycle is a relatively new concept for such modifications and has severely impacted the natural cycles. Climate change has changed the key facets of biogeochemical cycles. Industrial nitrogen fixation and deforestation have drastically modified the pattern of nutrient cycles, which has upset ecosystem functions. Changes in biogeochemical cycles have disrupted the natural balance in ecosystems and affected biodiversity. For example, the interruption in the sulphur cycle due to industrial emission of sulphur dioxide causes acid rain. Overuse of fertilisers containing phosphorus in agriculture has resulted in nutrient runoff in aquatic bodies, which triggered an overgrowth of algae, leading to eutrophication. In a key space, the severe wildfire in Australia in 2019–20 triggered an algal bloom in the Southern Ocean due to the deposition of iron-containing aerosol into iron-limited water.

26.2 BIOGEOCHEMICAL CYCLE ON GEOLOGICAL SCALE TIME

The evolution of the solar system, including Earth, dates back to ~4.5 billion years ago (bya). Elements like H and He are the lightest and originated much before the solar system, i.e., 16 billion years ago in the 'big bang'. The formation of other elements was the result of fusion, spallation, energetic particle interactions, neutron capture, and proton capture of light elements. The present composition of the Universe still includes approximately 73% H and 25% He, whereas the remaining 2% comprises all other elements. Most of the heavy elements reside on planets near the sun, making them heavy and dense, whereas low-weight elements are persistent in planets at farther distances.

Earth is a distinctive planet in the solar system compared to other planetary bodies. Since its initial formation, it has evolved considerably from a rocky planet to the only known habitable place. It represents a small portion of the universe that is composed of a variety of chemicals. Specifically, the origin of life at nearly 4.3 bya (based on geochemical evidence) was the major driver for excessive changes and chemical diversity on Earth. The liquid form of water on Earth's surface provided a unique environment for chemical reactions and mobilisation. Water was present on Earth 4.4 bya and was necessary for life. Also, the distance of Earth from the sun is very precise and provides a sufficient amount of radiant energy for primary production and to maintain optimum temperature for metabolic activities.

The earliest life on Earth evolved exclusively in prokaryotes, deriving a set of a few metabolic activities. This support system was the foundation of biogeochemical cycles on a planetary

FIGURE 26.1 a and b. Hydrothermal fields on the Mid-Atlantic Ridge mimic the primary sites for the origin of life on Earth. (MARUM– Center for Marine Environmental Sciences, University of Bremen.)

scale, which was invariably tied to various environmental changes. Life persisted after its first appearance and carried an imprint of biology in the form of chemical characteristics. Over the billions of years of the Earth's history, more than 99% of life forms have become extinct. However, several ancient biological pathways have remained conserved since the genesis. Microorganisms that thrived in the ancient past can be found with similar characteristics in the hydrothermal vents even today (Figure 26.1). They secure distinct life processes and play a pivotal role in biogeochemical cycles.

Modern organisms have evolved from microscopic single-celled organisms to continually adapting and diversifying to take more complex forms over a vast period of time. The tremendous diversity of life on Earth, in the present, is the consequence of a long and complicated evolutionary history. In geological scale time, several extraordinary challenges and extreme situations caused major extinctions of preexisting biological entities. However, the core biological machines have remained intact and regulated a set of nested biologically driven reactions even after at least five mass extinctions since 500 million years ago (mya).

Hydrothermal vents were the primary sites for the origin of life on Earth under oppressive situations of hot, sunless, anoxic, saline, and high-pressure environments. Primitive microorganisms have relied on sulphur compounds and methanogenesis for an autotrophic lifestyle. Methanogens (i.e., methane-producing archaea) represent one of the early life forms and survive by converting inorganic and organic compounds. Glycolysis is the most ancient metabolic pathway, which prevails in nearly all types of organisms. Glycolysis is a set of biochemical reactions of glucose oxidation that release energy. The intermediate compounds served as building blocks for adding chemical diversity in the early habitats. Photosynthesis was an essential process for the biological oxidation of Earth. Anoxygenic photosynthesis was the earliest form of photosynthesis. This type of metabolic pathway still exists on Earth, using compounds like hydrogen sulphide (H_2S), other sulphur compounds, hydrogen gas (H_2), and Fe. In the ancient Earth, these compounds were the key electron sources for photosynthesis, but did not produce oxygen gas (O_2). The first oxygenic photosynthetic organisms evolved around 3.4 bya. However, the oxygenation of the atmosphere, which is considered the permanent transition and turning point in the Earth's history, was a little delayed and happened roughly 2.4 bya. This time interval is called the great oxygenation event (GOE) (detailed in Box 26.2). This could be possible due to several taxa belonging to the phylum Cyanobacteria, which were able to utilise water molecules, CO_2, and sunlight to produce simple sugars and release O_2. The buildup of O_2 in the Earth's atmosphere was a consequence of photosynthesis, but it was a paradigm shift in evolutionary history. The main component of what made modern Earth much more habitable and diverse than its previous times is O_2. However, the credit for the functioning of most of the biological activities goes to biological nitrogen fixation, which started even before the GOE (as early as 3.2 bya) and is considered a primordial process because all life forms depend on fixed sources of nitrogen for the production of enzymes and genetic materials. The co-evolution of life successively generated sophisticated life forms (i.e., eukaryotic life forms). Further, ecological processes like natural selection, genetic drift, and mutation played critical roles in their establishment and diversification in various ecosystems.

26.3 KEY CHARACTERISTICS OF BIOGEOCHEMICAL CYCLES

The best way to study the biogeochemical cycle is through a box model where the distinct boxes represent the storage sites of the materials and arrows indicate their flow in the environment (Figure 26.2a). A mathematical framework and reductive approach in box modelling are central to the understanding and drawing of biogeochemical cycles. The coupled biogeochemical cycles can be illustrated by a matrix using columns and rows (Figure 26.2b). A typical quantitative biogeochemical cycle is described in terms of reservoir,

BOX 26.2. OXYGENATION IN EARTH'S ATMOSPHERE

The ancient Earth prevailed in a reducing environment, which was highly unsupportive of life events. The major oxidation process began with oxygenic photosynthesis, releasing free O_2 on Earth's surface. It took almost a billion years for the permanent rise of atmospheric O_2 after its first emergence. This long-term persistence of O_2 in the atmosphere is termed a 'Great Oxidation Event' (GOE). In this evolutionary past, many contemporary activities also influenced the rise of O_2 in the atmosphere. For example, Ni is essential for the growth of methane-producing microorganisms, but when its concentration was reduced, the atmospheric CH_4 concentration also declined. Methane production is inversely proportional to the oxygenation of the Earth's atmosphere, as CH_4 can react with free O_2 molecules.

The role of photosynthetic pigments, particularly chlorophyll, was also instrumental in the process of oxygenation. These pigments were characteristic of the newly evolved organisms, cyanobacteria. In photosynthesis, chlorophyll pigments capture sunlight and facilitate the photolysis of water, which releases electrons for carbon fixation and evolved oxygen as the waste product. However, the photosynthetic microorganisms in early Earth were capable of using electron donors such as H_2S, ferrous ions, and H_2 rather than water, which did not evolve O_2. When oxygenic photosynthesis was underway, anoxygenic photosynthesis posed a high threat to it, which could be a reason for the delay in oxygenation of the Earth's atmosphere. However, electron donors for anoxygenic photosynthesis were exhausted to maintain prolonged primary production. This also encouraged the emergence of oxygenic photosynthesis. Water was ubiquitous and offered a competitive advantage over other electron donors, exhibiting profound biogeochemical shifts for oxygenic photosynthetic niches.

Excess O_2 in the environment was then a poison, which eradicated anaerobic life and caused an extinction disaster. The rise of O_2 concentration supported the explosion of new life on Earth and served as the driver for biological diversification and abundance. The GOE set the stage for the evolution of heterotrophic organisms. The emergence and ecological expansion of complex organisms such as algae, fungi, plants, and animals was a momentous transition to the use of O_2 in cellular respiration. Their ability to utilise O_2 as a terminal electron acceptor in cellular respiration revolutionised the efficiency of energy production. In the electron transport chain, O_2 acted as a strong terminal electron acceptor due to its strong electronegative nature. This resulted in a significantly higher energy yield per molecule than anaerobic respiration. This process also recycled the O_2 into a reduced form (i.e., CO_2) for its next round of oxidation into oxygenation. This process has continued since then (i.e., GOE) and maintained the life balance as we know it.

exchange pool, fluxes, budget, residence time, source, sink, and feedback control. Many of these parameters can be expressed in quantitative terms. These technical terms are common in the biogeochemical cycles and are discussed in the following subsections.

26.3.1 Reservoir and Exchange Pool

The reservoir is an environmental site for the mass storage of a chemical, and is a relatively less active zone. On the contrary, an exchange pool represents a site that allows rapid exchange of a chemical with other environments (Figure 26.3a). The reservoirs and the exchange pools have contrasting natures but are essential for the immobilisation and mobilisation of chemicals in diverse environments. The same environment can be a reservoir or an exchange pool, depending on the context of the substance. For example, the atmosphere is a reservoir of nitrogen (N_2), storing vast amounts of this gas in inert form, and the same environment serves as the exchange pool of water, allowing exchange through condensation, evaporation, and precipitation. Similarly, the hydrosphere and the pedosphere can also be reservoirs or exchange pools for different substances.

The sub-crust and upper mantle zones in the geosphere and deep ocean are ultimate reservoirs and store various minerals, heavy metals, gases, and fossil fuels. These regions are highly inactive and take centuries to millennia to release their matter into the environment. The oceans, lakes, rivers, groundwater, ice caps, and glaciers are the biggest reservoirs of water. The atmosphere and hydrosphere are exchange pools of carbon. In the carbon cycle, CO_2 in water and air maintains a constant exchange with biotic activities such as photosynthesis and respiration. The biosphere chiefly operates as the exchange pool because the nutrients and organic matter are the temporary residents in the biological systems.

26.3.2 Fluxes

Flux is the rate of flow of chemicals from one compartment to another. It can occur on a regional to global scale. Flux can be classified as influx (or inflow) (moving towards the compartment) and efflux (going away from the compartment). If the difference between the two is zero, the ecosystem exists in an equilibrium or steady state. The flux of chemicals in the environment serves as the foundation for the dynamics of matter and energy in ecosystems and maintains thermodynamic disequilibrium. Flux is heavily influenced by various physicochemical and biological characteristics of environmental compartments. The quantitative label of flux is expressed in terms of mass, energy or volume per unit time. In the box model, fluxes are represented by arrows, which signify the direction of the movement and often the rate of a.

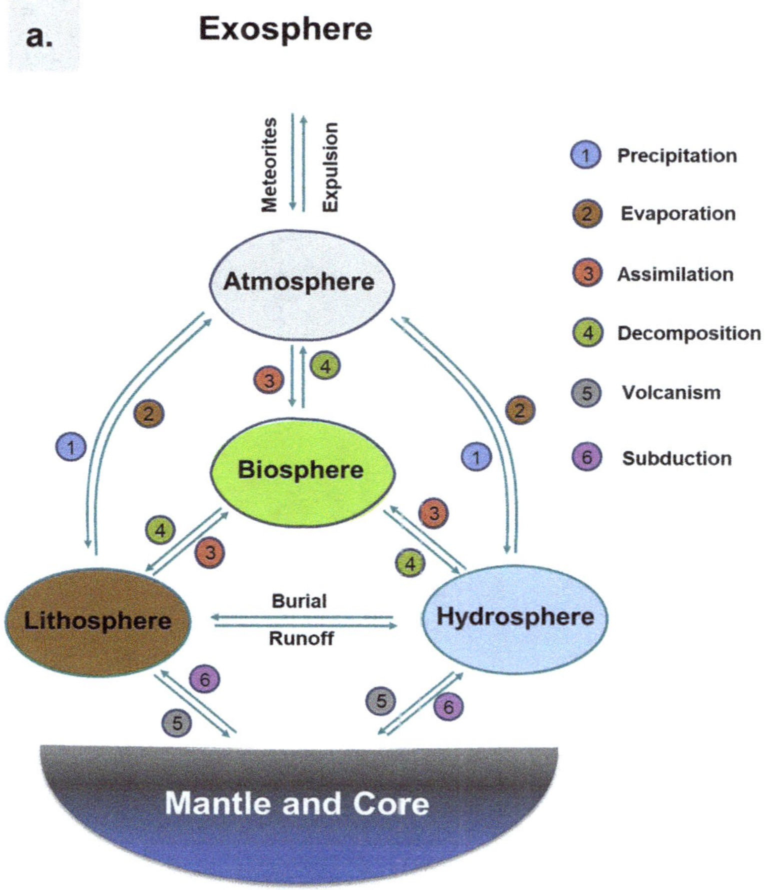

b.

Cycle	Atmosphere	Hydrosphere	Lithosphere	Biosphere
Carbon	CO_2 and CH_4	Dissolved CO_2, bicarbonate	Fossil fuels, biomass, and limestone	Organic matter and living organisms
Nitrogen	N_2, NO, NO_2 and NH_3	Dissolved nitrogen compounds	Nitrate deposits and ammonium ions	Proteins and nucleic acids
Phosphorus	None	Dissolved phosphate	Phosphate rocks	DNA and ATP
Sulphur	SO_2, H_2S and DMS	SO_4^{2-}, SO_3^{2-}, and S^{2-}	Pyrite and gypsum	Proteins and amino acids

FIGURE 26.2 a. A general box model of a biogeochemical cycle of a chemical, where the arrow depicts the movement of the chemical from one sphere to another due to some biogeochemical activities. b. A matrix of a coupled biogeochemical cycle of carbon, nitrogen, phosphorus, and sulphur, showing rows and columns.

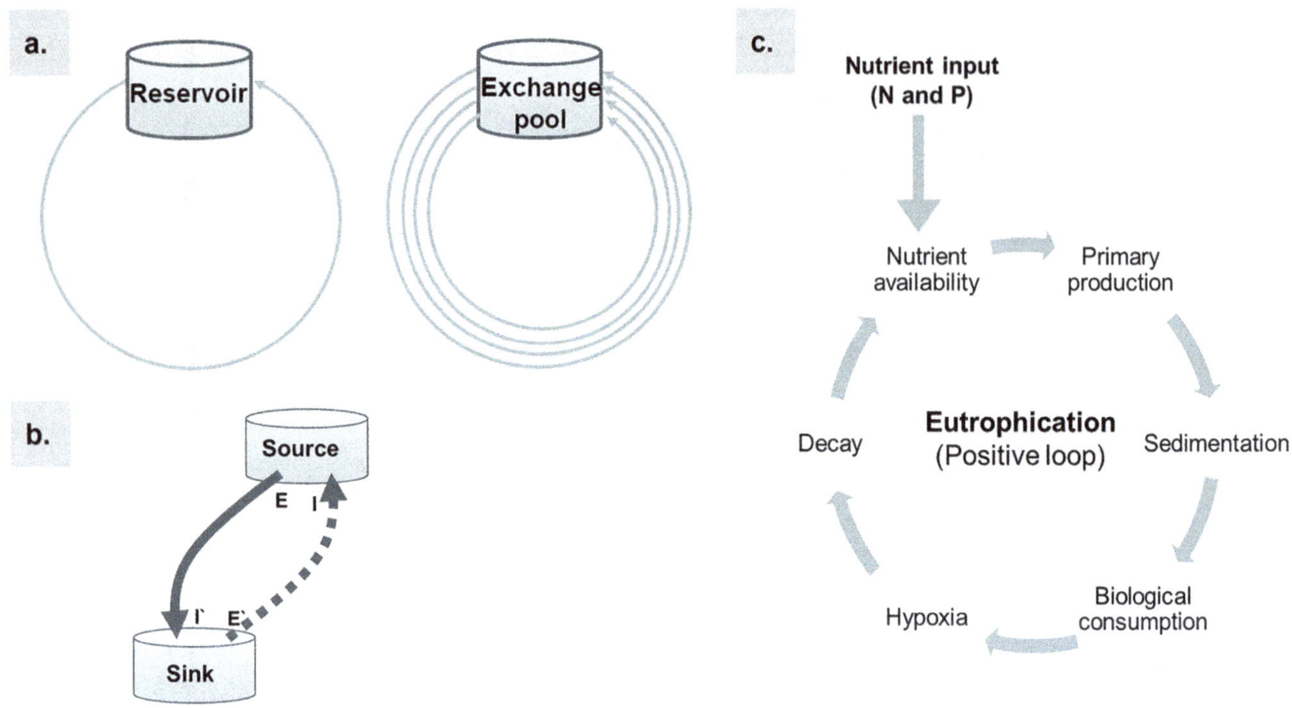

FIGURE 26.3 a. Flow of a chemical in a reservoir (depicting rare circulation) and an exchange pool (showing frequent circulation). b. A sink represents high influx (I`) and low efflux (E`) of a chemical, whereas a source represents high efflux (E) and low influx (I) of a chemical. The continuous arrow denotes a strong transition of matter, whereas the dotted arrow represents a low level of influence. c. An example of feedback control through eutrophication, showing a positive loop.

transfer. A heavy arrow depicts a strong transition of the materials, whereas a dotted arrow represents a low magnitude of movement (Figure 26.3b). Sometimes, the rates of transfer between compartments in the box model can be represented using numerical fluxes. The numerical fluxes of the model can efficiently be used to simulate the dynamics of biogeochemical cycles.

26.3.3 Sink and Source

When the influx of a chemical exceeds its efflux, the environmental compartment is referred to as a sink. It represents the deep burial site of a chemical on the Earth. On the contrary, if an environmental compartment has more efflux of a chemical than its influx, that site is the source (Figure 26.3b). The circulation of chemicals and energy greatly depends on the characteristics of the sinks and sources. The deep ocean that absorbs more carbon than it releases is an ideal example of a carbon sink. Forests also represent carbon sinks, owing to the immense sequestration of carbon through photosynthesis. Coal and fossil fuels are the rapid sources of carbon in the environment. Their combustion releases carbon as CO_2 in the atmosphere.

26.3.4 Residence Times

Residence time is a measure of the time spent by a chemical in an environmental compartment. It is numerically denoted in units of different temporal scales (from seconds to years), depending on the extent of their residence in a system. The properties of the chemical and the characteristics of the reservoir both influence the residence period of a chemical. Some reservoirs sequester chemicals very strongly, and chemicals spend time in isolation for a long duration. They can have a residence time of hundreds to thousands of years. However, some materials rest in an environment for a brief duration and have a short residence time (some days to a few years). Even some chemicals residing in the atmosphere and the hydrosphere travel in fractions of minutes to hours, exhibiting a very short residence time. For example, in the water cycle, a substantial amount of water is bound to rocks and old glaciers, where it has typically long residence times. Conversely, water in the human body rests for a few hours at most. The residence time of a chemical in a pool can be calculated as

$$\textit{Residence time in the pool} = \frac{\textit{Amount in the pool}}{\textit{Influx or efflux at equilibrium}}$$

If the reservoir has multiple influxes and effluxes, the sum of influx or efflux is applicable for the calculation of residence time. This formula provides an easy estimation of residence time, influx/efflux, or amount of chemicals in the pool if any two parameters are known. For example, in the carbon cycle, the atmosphere stores about 750 Gt[1] carbon in the form of CO_2, and it effluxes approximately 190 Gt carbon/year. Based on these observations, the average residence time of carbon in the atmosphere will be 4 years. In a different example, the oceans occupy 38,100 Gt of carbon, and there is an influx of 0.6 Gt carbon/yr via rivers (ignoring other exchange and internal cycling). The residence time of carbon in oceans is ~358 years.

26.3.5 Turnover

Turnover is the rate of movement of matter or energy at which the contents of a reservoir/pool are replaced. It is also known by other names, such as turnover rate. The turnover rate is the mathematical inverse of residence time, and the value is represented by per unit of time. Turnover provides information on the time required to fill or empty the reservoir. It also aids in estimating the pool size, fluxes, and timescale required to achieve a steady state.

26.3.6 Budget

All biogeochemical cycles maintain a budget. It is a balance sheet of a chemical for its input, output, and retention in different environmental compartments. This is generally expressed as the amount per annum. The budget of a chemical provides a simple understanding of the quantitative measure of its disposition in the sphere. It is also used to calculate the temporal evolution of the cycles. However, the high complexity of the global system makes it almost impossible to get an accurate measure of the chemical budget. For example, a carbon budget for a forest ecosystem involves an assessment of the forest's carbon flux and residence time. The sequestration of CO_2 by photoautotrophs is an input to the plant compartments, whereas CO_2 released from biomass through respiration and decomposition is an input to the atmosphere. The storage accounts for carbon uptake in trees, soil, and organic matter. Several factors, like nutrient and water availability, biodiversity, pests and diseases, and many other parameters, also contribute to the overall carbon budget. Therefore, the accurate quantification of the carbon budget in the carbon cycle is extremely challenging due to the complexity and variability of natural processes. The notable variations due to the impact of anthropogenic activities such as environmental pollution, recreation and land use, and agricultural setups also influence the annual budgets of chemicals. Figure 26.4 provides a schematic illustration of the overall carbon budget due to anthropogenic activities. For example, the synthesis of artificial nitrogen fertilisers and industrial CO_2 emissions have increased the annual budgets of nitrogen and carbon, respectively.

26.3.7 Feedback Control

Feedback control is a process of regulating an activity by its output as input for further progress. Positive feedback accelerates the amplification of the original process, enabling

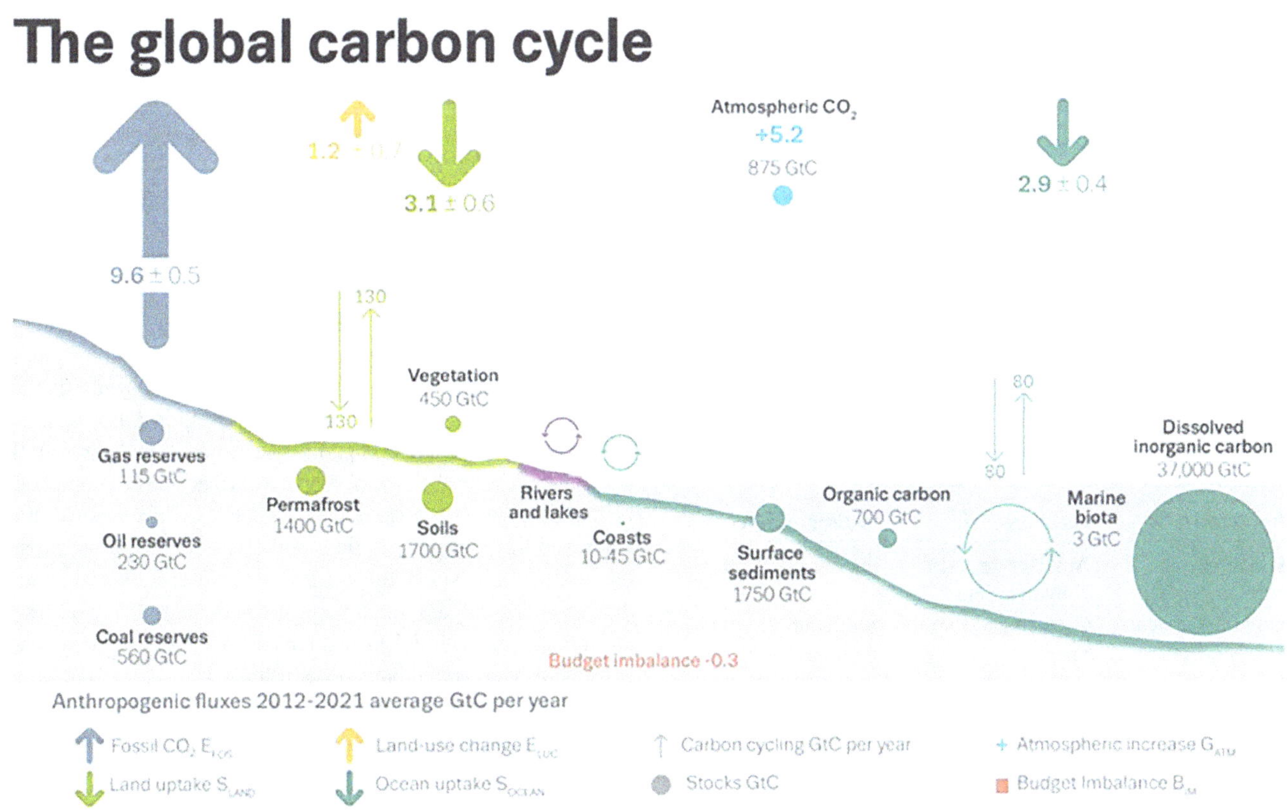

FIGURE 26.4 Schematic representation of the overall budget, efflux, and influx of carbon in the carbon cycle due to anthropogenic activities, averaged globally for the decade 2012–2021. (Friedlingstein et al. (2023).[2])

the destabilisation of the system. However, negative feedback is regulated in the opposite direction of the response to the initial input. The negative feedback tries to stabilise the system and is self-regulating. A classic example of a feedback loop is eutrophication (Figure 26.3c). The addition of nutrients in the lake causes an increase in the growth of phytoplankton and macrophytes, which depicts positive feedback. This response drastically reduces the O_2 concentration (by aerobic microbial digestion of decayed plant materials), causing a dead zone for aquatic animals. However, when the ratio of nitrogen to phosphorus in the lake is exhausted, the growth of phytoplankton and macrophytes is inhibited, showing characteristic negative feedback. This process increases the O_2 concentration of the lake and provides suitable conditions for the growth of fish and macroinvertebrates. In a different process, biogenic calcite-phosphorus precipitation in lakes represents negative feedback that inhibits plant growth by reducing the phosphorus availability in water.

26.3.8 Gaseous and Sedimentary Cycles

The Earth's system has two different categories of biogeochemical cycles: gaseous and sedimentary cycles. In the gaseous cycle, chemicals mainly move in the gaseous or solution forms and circulate above ground. Examples of those include the nitrogen cycle, oxygen cycle, and carbon cycle. On the other hand, in the sedimentary cycle, a chemical spends maximum time in minerals or rocks, such as the sulphur cycle and phosphorus cycle. The chemicals exist in the solid state. The key differences between the gaseous and sedimentary cycles are mentioned in Table 26.1.

26.4 ENVIRONMENTAL COMPARTMENTS

Based on the physical boundaries, Earth can be categorised into four distinct "spheres" or "environmental compartments". They are the geosphere, the atmosphere, the hydrosphere, and the biosphere (Figure 26.5). Every sphere of Earth is a complex reservoir of diverse chemicals and their compositions. The abundance and proportion of chemicals and the magnitudes of their impact significantly vary in each sphere. A simple example of this situation is different concentrations of three essential elements for life: carbon (C), nitrogen (N), and phosphorus (P) in different compartments of the environment. The relative orders of their abundance in the biosphere and the atmosphere are C>N>P and N>C>P, respectively, whereas their orders in the lithosphere are reversed from that of the atmosphere. These variations are characteristic of different spheres. However, their quantities and proportions can be inconsistent in local scales, even at near distances.

FIGURE 26.5 The four basic components of Earth, including the atmosphere, the hydrosphere, the geosphere, and the biosphere. The sun is the ultimate source of energy for Earth, regulating various processes on the planet.

TABLE 26.1
Differences between gaseous and sedimentary cycles

Characteristics	Gaseous cycle	Sedimentary cycles
Main reservoir	Atmosphere and hydrosphere	Earth crust
Phase of chemical	Gaseous and solution	Sediment
State of chemical	Free	Earthbound
Fluxes	High	Low
Residence time in the reservoir	Short period	Prolonged period
Impact on ecosystem function	Considerably high	Relatively low
Examples	N, O, C, and water	P, S, Ca, and Fe

Environmental compartments are interdependent and interconnected at local, regional, and global scales. A stand-still tree is a wonderful system for understanding the interconnected compartments on a local scale. Roughly half of the plant body rests beneath the soil (the root system) (i.e., the lithosphere), and the other half (i.e., stem, leaves, and inflorescence) is aboveground and raised in the air (i.e., the atmosphere). The soil provides a substratum, minerals, and water for plants, whereas the aboveground part is essential for gaseous exchange and photosynthesis. The movement of birds sets an example of the regional association of different environmental compartments. A bird spends time in the atmosphere and rests in the nest on a tree (that is also a component of the biosphere). The same bird visits the hydrosphere to drink water and the geosphere to find food. These observations expand our understanding that life on Earth is closely associated with diverse biological and nonbiological components. They represent an intricate and interlinked relationship, which is a fundamental concept in ecology and biogeochemistry.

All environmental compartments are distinct spaces on Earth, having special features, which are discussed as follows.

26.4.1 Atmosphere

The atmosphere is a region at the interface of Earth's surface and outer space, storing an envelope of gases and tiny suspended particles. Nitrogen (N_2) and O_2 are the most common gases in the air and cover 78% and 21% of the total atmosphere, respectively. The remaining 1% of the atmosphere is composed of Ar, CO_2, and other gases. The bulk gases are homogeneously mixed. Water vapour is also a notable constituent, but its concentration varies depending on the weather conditions. Some small particles, such as soot, dust, ash, smoke, microbial cells, spores, pollen, etc., also reside in the atmosphere and are collectively referred to as "aerosols". Carbon dioxide in the atmosphere is essential for photosynthesis. The presence of O_2 in the atmosphere makes the air breathable. Air is also an important transport medium for many low-buoyant, chemical, and biological agents at spatial scales. Chapter 18 provides details on the atmosphere and airborne microorganisms.

The extent of the Earth's atmosphere is as far as 10,000 km above the surface. The atmosphere is a dynamic subsystem of the Earth system, which is mainly driven by differential heating and cooling. Gravitational pull is the primary force that regulates all gases in the atmosphere. Several factors, like air pressure, humidity, cloud cover, wind direction, and speed, also have a significant impact on how particulate matter is distributed and interacts in the atmosphere.

Based on the temperature profile, the vertical structure of the atmosphere can be divided into five different layers. They sit as the layers of the troposphere, stratosphere, mesosphere, thermosphere, and exosphere, one above the other. Each layer has a different chemical makeup. The high altitude creates a drop in air pressure, which causes a decrease in air density. This means less matter per volume at every layer above the atmosphere. The troposphere is the bottommost layer that occupies more than three-quarters of the total atmosphere and provides the most hospitable situation in the air. However, the maximum mass of the Earth's troposphere is stored below 100 km, the Kármán line. This layer also consists of a high amount of water vapour that supports life in the air. The layer above this is the stratosphere, which features an ozone shield that offers protection from the Sun's ultraviolet (UV) rays by preventing them from reaching Earth's surface. The next level is the mesosphere— the coldest layer, where most of the meteors and asteroids burn up. The thermosphere is the second most distant layer, having the warmest temperatures in the atmosphere, which offers protection from high energy radiations (i.e., X-rays and UV rays). The exosphere is the outermost layer of the atmosphere and provides the first line of defence from meteors, asteroids, and harmful rays. In the last place, the exosphere disintegrates into outer space.

26.4.2 Hydrosphere

Water is a single essential component of the hydrosphere that can reside in three different states: vapour, liquid, and ice. It is the most abundant molecule on Earth. Some 1.385 billion km^3 of water is spread across the globe. Almost 71% of the Earth's surface area is covered with water. Earth appears blue from space because of the reflection of light from water on Earth. Reservoirs of water are highly diverse and manifold. These include oceans, estuaries, rivers, groundwater, ponds, streams, lakes, and glaciers. Inland water differs from seawater mainly in terms of its boundary. Most of the regions in inland water are freshwater. The maximum volume of liquid water is present in the ocean, which accounts for over 96% of all the Earth's water. Inland water accounts for nearly 3% of the total volume.

The meeting points of freshwater and saline water create estuaries. They are the unique sites for hydrological processes and biochemical reactions. A low volume of water resides as water vapour in the atmosphere and is bound with soil particles in the lithosphere. The solid form of water is confined to the discrete zone of Earth, called the cryosphere. About 13% of the Earth's surface area is in the cryosphere. Snow, permafrost, ice, and glaciers are solid forms of water bodies, which are biologically unavailable forms of water. The massive bodies of ice and snow are permanently located in polar regions and high altitudes. Also, many parts of the world witness temporary snowfall, depending on the climate. Life in the cryosphere is less common, but whatever is present is chiefly composed of microorganisms. The inhabitants have a very slow growth rate due to harsh environmental conditions, water stress, and limited nutrient availability. Some species have an ability to produce antifreeze proteins to prevent cell damage during extremely cold temperatures.

The liquid state of water is essential for the life support system. Almost 70% of biological cell volume is composed of water. Water is essentially required for many biological processes, such as oxygenic photosynthesis, respiration, transpiration, osmosis, pH regulation, and body temperature regulation. The water bodies are unique habitats for aquatic life. A significant portion of Earth's biodiversity resides in water. The coral reef is a wonderful example of a sea environment. It is often referred to as "tropical rainforests of the sea" due to its incredible biodiversity and ecological importance. Life in seawater and freshwater is highly distinct. Microorganisms in surface water and the deep sea have different mechanisms for primary production. The study on microbial life in water can be found in Chapter 20.

26.4.3 Geosphere

The geosphere is the solid part of the Earth, including rocks, surface landforms, sediments, and soils. Inorganic natural minerals are the building blocks of the geosphere. More than 90% of the solid body of Earth is made up of elements such as O, Fe, and Si. However, the majority of the mass on Earth is buried deep beneath the surface and cut off from other spheres of the globe. Based on the geochemical properties, the geosphere can be divided into three zones: core, mantle, and crust. The interior part is a superheated core— the subdivision is the inner core (solid) and outer core (molten metal). The main ingredient of the core is Fe, which is responsible for the magnetic field of Earth. Mantel lies above the core— the hot rock, mainly composed of silicates. The crust is a thin outermost layer and consists of continental and oceanic crust. The lithosphere is the outermost layer of the geosphere, distinguished by mechanical properties. This comprises the full width of the thin crust and the upper mantle zone.

The Earth's land surface is covered with soil, known as the pedosphere. It exists at the interface of the biosphere, the hydrosphere, and the atmosphere. Approximately 50% of soil consists of solid matter, of which 45% consists of inorganic materials and 5% organic matter. The other 50% is water and air pores in almost equal proportions. The pedosphere hosts incredibly diverse organisms and biological activities (detailed in Chapter 19).

26.4.4 Biosphere

The biosphere is the region on Earth that encompasses biotic components. It is composed of diverse creatures, including microorganisms, plants, animals, and humans. The biosphere reclines at the interface of the lithosphere, the hydrosphere, and the atmosphere. All life depends on the hydrosphere for water, the lithosphere for minerals, and the atmosphere for gases. The biosphere cannot exist in solitude. Living organisms are distinguished from non-living substances by their metabolism, growth, reproduction, and evolution. Living matter is an extraordinary substance and is present on only Earth's surface— surviving under a relatively narrow set of conditions. The mass of non-decomposed organic matter of the biotic system is also a pool of the biosphere.

A subdiscipline of biosphere, biogeography is central to understanding the geographical distribution and abundance of living organisms on Earth. One of the most essential units of biogeography is the biome. Biome is an extended approach to ecosystems that deals with the interrelationship of all biotic and non-biotic components in a unit of space. The most evident biomes are grassland, forest, desert, tundra, wetlands, ocean, and coral reefs. The characteristic communities of each biome reflect particular assemblages of microorganisms, vegetation and wildlife within geographical constraints. The building blocks of the entire biosphere are inorganic nutrients. All the organic pools of biotic components are derived from processes like carbon fixation, nitrogen fixation, and nutrient assimilation. The nutrient cycle allows the recirculation of various elements from complex remains of dead matter and the excrement of living forms with the help of decomposers.

Biological systems exhibit a hierarchical organisation in which living things are structured and arranged at various levels of complexity. This organisation includes several components in order of cell, tissue, organ, organ system, organism, population, community, ecosystem, biome, and biosphere. Every level is related to the ones above and below it and impacts each other. A cell is a basic unit of life and fundamental to self-sustaining processes. A group of cells forms tissue, and the tissues gather to develop an organ and the organ system. An organism maintains life through its organ systems and responds to stimuli. Organisms can be unicellular to multicellular and represent a huge diversity. Multiple individuals belonging to same species constitutes a population, and the population of more than one species creates a community. An ecosystem is the interaction of its community with its surroundings. The two or more ecosystems spreading over a wide geographical area form a biome. The quantitative and qualitative accounts of each taxon of all biomes make up the biosphere.

26.4.5 Anthroposphere

The anthroposphere includes all the impressions of human activities in environmental spheres. Human activities have had an overriding influence on the Earth's system and have altered the biogeochemical cycles of many chemicals. The controlled use of fire was the first technological breakthrough for early humans. Since then, numerous technological establishments have offered diverse benefits along with some negative consequences for natural ecosystems. As a result, many ecosystem services are hampered.

The geosphere has experienced surface erosion and biodiversity loss as a result of farming, deforestation, mining, and excavation activities. The excessive emission of greenhouse gases in the atmosphere has had a severe impact on climate and associated activities. In the global water cycle, water has received drastic variations in its fluxes, residence time, and overall budget. Heavy rainfall, prolonged drought, and floods are immediate concerns due to climate change. Though natural attenuation and restoration practices are trying to revive the imbalance of environmental conditions, the magnitude of environmental contaminants is far greater than their remedial efficiencies.

The excess release of chemicals into the environment has caused a major shift in their budget. The classic examples are the disparities in the budgets of nitrogen and carbon post-1950–60s. Particularly, after World War II, the revolution of industrial nitrogen fixation (Haber-Bosch process) has led to an unprecedented rise of synthetic derivatives of nitrogen compounds in the environment, which extended their influence far beyond their threshold limits. This process has had a severe impact on the nitrogen cycle, leading to its global-scale alteration. Urban and rural runoff of nitrogen-containing fertilisers and detergent-containing water increased the incidences of eutrophication in lakes and oceans. The intense combustion of fossil fuels altered the global CO_2 budget and drastically modified the carbon cycle, leading to global warming. In addition, many other chemicals have seen immense differences in their global pools, fluxes, and overall budgets due to anthropogenic influences.

26.5 DYNAMIC PROCESSES OF BIOGEOCHEMICAL CYCLES

Biogeochemical cycles involve several dynamic processes for the active movement of a chemical from one environment to another. Various physical, chemical, and biological factors are the critical parameters that influence the movement and distribution of chemicals. Both natural and anthropogenic activities take part in the transformation reactions. This often includes ecosystem interactions, feedback mechanisms, and climate and weather patterns. The interaction of these factors in the environment plays a critical role in maintaining the stability and functionality of Earth's ecosystems. Both long-term and short-term processes have roles in balancing the biogeochemical cycles. Some important processes that matter for the dynamics of biogeochemical cycles are discussed as follows.

26.5.1 Impact of Solar Radiation

The energy of sunlight is one of the main drivers of climate regulation, the hydrological cycle, and life. A slight fluctuation in solar irradiance can cause major shifts in the Earth's climate. The solar energy reaching the globe is circulated by various mechanisms, such as absorption, conduction, convection, radiation, and ocean currents. Solar radiation has a very important role in energy circulation, which is influenced by latent heat flux. This involves the process of energy gain during evaporation from the Earth's surface and energy loss during condensation of water in the atmosphere. It plays a vital role in Earth's climate system. For example, water in lakes, oceans, and other aquatic bodies evaporates due to heat generated by solar radiation. The water vapour eventually condenses and forms clouds to precipitate on Earth. Solar irradiance also causes the melting and freezing of water and can be responsible for its runoff. This continuous flow of water across different spheres from one form to another, creates a hydrological cycle or water cycle. The greenhouse gas effect is a consequence of solar energy, where greenhouse gases like CO_2 and CH_4 trap the energy of sunlight on Earth's surface, leading to global climate change. In a different role, sunlight is responsible for photochemical reactions. When the sunlight falls on a substance, it can transfer the radiant energy into chemical energy. This process is fundamental to photosynthesis and aids in biological productivity. Solar energy is also important for many other biological processes, such as transpiration and the regulation of circadian rhythms.

26.5.2 Geochemical Processes

Geochemical processes perform reactions between different elements for mineral formation and their degradation. The interaction between different chemicals produces new products by introducing new chemical bonds and breaking the existing bonds. Weathering, erosion, volcanic eruption, and subduction of tectonic plates have transferred huge amounts of material from Earth's crust to the environment. These physical movements play a major role in the distribution of chemicals across different spheres. In addition, some atmospheric processes like radiation, convection, precipitation, and cloud formation have immense contributions to the mobilisation of chemicals.

26.5.3 Biological Processes

Biological processes are the most diverse catalytic repertoires of nature. All types of organisms, including microorganisms to macroorganisms, contain hundreds to thousands of distinct bioreactions. These actions dictate which chemical reactions will fall in a distinct cell. Some bioreactions offer benefits for an integral part of metabolism (using intracellular enzymes), and some bioreactions can perform reactions even outside cells (involving extracellular enzymes). They have evolved since the beginning of life and are responsible for chemical diversity in nature. Certain bioreactions play a key role in driving biogeochemical cycles. For example, inert nitrogen gas is transformed into a reactive substance in the nitrogen cycle. The transformation of CO_2 into carbohydrates is an integral component of the carbon cycle for carbon fixation. Several pathways in respiration (such as glycolysis, citric acid pathway, and gluconeogenesis) are universal examples of diverse biochemical reactions. Microorganisms have the enormous genetic potential to perform numerous reactions for environmental benefits.

The enzymatic and nonenzymatic reactions are the fundamental driving forces behind all biological activities. Enzymes are widely distributed and provide compelling evidence for biogeochemical activities. They are a class of nano-biological devices that exhibit specificity for biological reactions. They are complex, multimeric protein compounds often associated with distinct prosthetic groups. Most enzymes execute only a single task, but some can perform more than one type of reaction. Enzymes are necessary for the synthesis of all primary and secondary metabolites. In theory, all enzymatic reactions can also occur non-enzymatically, but such reactions will run at much slower rates and often under harsh conditions.

Nonenzymatic reactions also win out in some situations. They can be driven by electrochemical gradients, pressure, or concentration gradients. Most non-enzymatic reactions often display broad substrate and product spectra but are free from thermosensitivity, energy-consuming cellular processes, and ageing. For some biological reactions, the choice of non-enzymatically or enzymatically driven reaction also exists. For example, the breakdown of hydrogen peroxide (H_2O_2), which yields water and O_2, can be a chemical intermolecular redox reaction or an enzyme-mediated process (through catalase). On the other hand, certain biological reactions are exclusively non-enzymatic. Biosorption is a suitable example of a non-enzymatic process, which involves physicochemical interactions between chemicals and biological materials.

26.5.4 Ecosystem-Level Approaches

Ecosystem-level approaches in a biogeochemical cycle gain insight into the flow of energy and matter and the trophic-level organisation. Energy flows through an ecosystem from one trophic level to the next and dissipates through heat and respiration into the environment. The complex network of feeding habits and the movement of nutrients at different trophic levels provides an environment with a balanced ecosystem. These aspects are discussed in the following subheadings.

26.5.4.1 Flow of Energy and Matter

Energy and matter are the two basic components of the system. All biogeochemical reactions are governed by the laws of conservation of matter and of thermodynamics. Based on these principles, there is a constant flow of matter and energy in the system. All the matter in Earth's subsystems is confined and maintains a continuous loop as it can be either transformed into other forms or transferred to different environmental compartments. The biogeochemical cycles maintain a constant flow of matter in the form of various physical and chemical natures in living organisms and the environment. Earth is considered a near approximation to a closed system because the movement of matter is restricted within the boundaries of Earth. Though meteorites and comet materials that fall upon Earth are extraterrestrial inputs, their contribution is minuscule.

The sunlight is the ultimate source of energy on Earth. The total amount of the sun's radiation that falls on the Earth's surface in one second is 1370 W/m^2 (W is the watt, and 1 watt is equal to 1 joule per second). Most of the energy returns to outer space as long-wavelength radiation, but some of it stays on Earth. Therefore, Earth is not an ideal example of a closed system in terms of energy. The plant body stores a major proportion of radiant energy in the form of chemical energy, through photosynthesis. This chemical energy is used in the metabolic activities of organisms and is transferred into higher trophic levels (according to the First Law of Thermodynamics). Some form of chemical energy is lost in the form of heat (according to the second law of thermodynamics). 10% of chemical energy is passed to the next trophic level for all succeeding levels of consumers. Energy follows unidirectional flow irrespective of the circular path for matter in the system. The other sources of energy on Earth are geothermal energy, tidal energy, hydroelectric energy, and wind energy.

26.5.4.2 Trophic Level of Organisation

A fundamental aspect of trophic-level organisation is the movement of chemicals from abiotic reservoirs to lower trophic organisms (e.g., microorganisms and plants), which in turn move to the higher trophic-level organisms (e.g., animals) through herbivory and carnivory. Typically, plants, algae, and cyanobacteria (i.e., primary producers) observe autotrophic lifestyles and fix inorganic carbon (i.e., CO_2) in organic matter and lie at the base of the trophic level. These organisms obtain inorganic nutrients from soil and water and store them in their body. The primary consumers of plants are herbivores (primary consumers), which enable the transfer of a part of organic matter and nutrients from primary producers into their bodies (i.e., second-level organisation). Carnivores obtain energy and nutrients by consuming other animals (i.e., third-level organisation). The dependency of higher trophic levels on the preceding trophic levels results in the transfer of energy and nutrients. The decomposition process returns organic components and nutrients from deceased individuals to the surrounding environment. Microorganisms have a very prominent role in decomposition. The death and decay of the living bodies attract microbial colonisation, followed by the recycling of the nutrients.

An organism may absorb certain substances more often than it excretes them, which prolongs the chemical residence in the body, and this process signifies bioaccumulation. However, when the concentration of a substance increases at higher trophic levels in a food chain, it leads to biomagnification. It is a meaningful event for the biological regulation of chemicals, where higher trophic levels accumulate higher concentrations of the substance compared to organisms that lie at lower trophic levels in the food chain. For example, the contamination of organomercury compounds in the water bodies tends to biomagnify in marine wildlife.

The interactions between diverse organisms at different levels constitute food chains. The complex feeding habits and interactions in the ecosystem form the food web. Though the food web is more intertwined with biogeochemical activities, the box models of biogeochemical cycles illustrate simple bridging steps of the organismal walk-in for chemical transfer. A different related concept is an ecological pyramid, which can be used to visualise the relative abundance, nutrients, and energy in the food chains through different trophic levels (Figure 26.6).

26.6 MICROBIAL SPHERE

Every area of Earth where microorganisms are present makes up the microbial sphere. The microbial component of any region is quite versatile. They are the biggest and most diverse reservoirs in the biosphere. Microbial signatures can be reported from as high as 40 km in the stratosphere, and the deepest known place on Earth for a microbial residence is 11 km below sea level in the Mariana Trench. The microbial sphere includes organisms from all three domains of life (including Bacteria, Archaea, and Eukarya). The first two groups are known for their simple organisation and lack certain cellular structures. Eukaryotic microorganisms are relatively complex and encompass fungi, protozoa, and microalgae (Figure 26.7).

The total microbial communities thriving in a specific habitat are called "microbiota". For example, soil microbiota and water microbiota refer to the vast and diverse communities of microorganisms that inhabit the soil and water. In the human body, different body parts host distinct microorganisms and constitute gut microbiota, skin microbiota, oral microbiota, vaginal microbiota, etc.

"Microbiome" is a related term broadly used to identify the types and relative abundance of microorganisms in an ecosystem. The scrutiny of microorganisms in microbiome analysis mainly relies on molecular sequences. The main output of the investigation is an understanding of their functions, relationships, and contributions to the overall ecosystem. The study on microbial communities from various objects, such as the human body, plant system, coral reef, and soil, constitutes the human microbiome, plant microbiome,

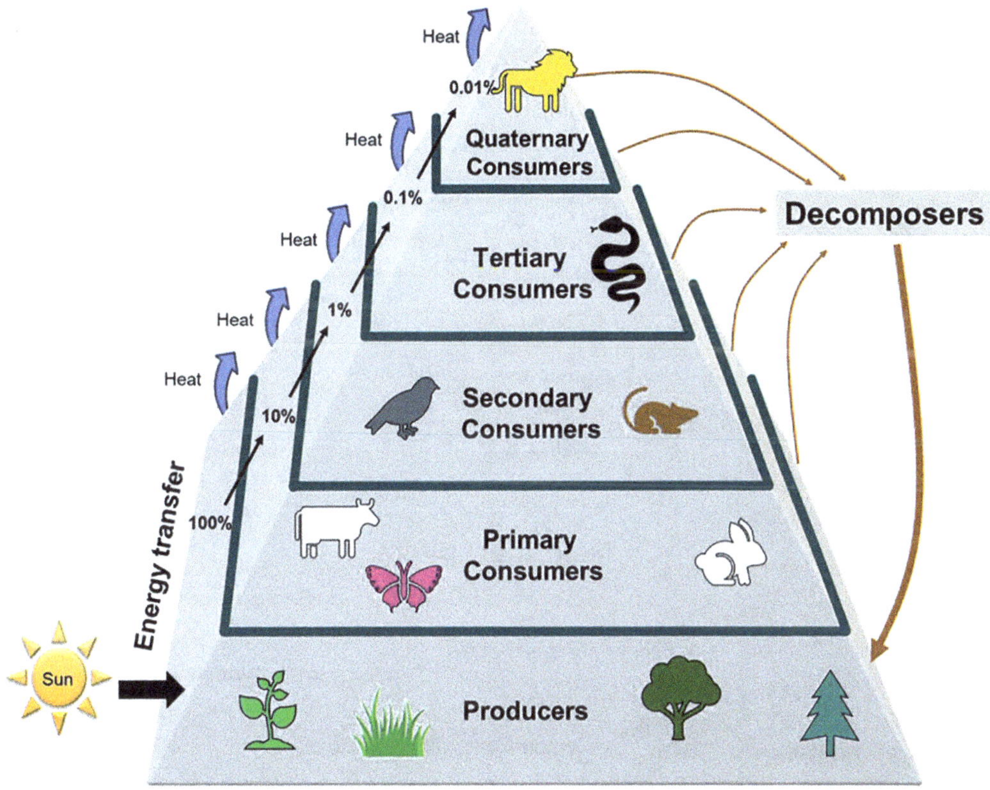

FIGURE 26.6 An ecological pyramid, showing trophic level organisation and energy transfer at each trophic level. The amount of energy that moves up to the next trophic level each time is approximately 10% of the previous level. Dead and decayed organisms from each trophic level are decomposed by decomposers and recycled in the soil.

coral reef microbiome, and soil microbiome, respectively. The study of microbial communities from different plant body parts generates root microbiome, seed microbiome, rhizospheric microbiome, phyllosphere microbiome, etc. Inbuilt environments such as houses, offices, and other manmade buildings have their microbiomes. Wastewater and treatment plants are home to distinct microbial forms that constitute unique microbiomes. Extreme environments, such as hot springs, deep-sea hydrothermal vents, and polar areas, also feature different microbiomes that have adaptations to survive harsh conditions. Microbiome research is a fast-expanding area, with scientists constantly discovering new insights into how these microbial populations and their functions affect people and the environment. In the recent past, the Human Microbiome Project (HMP) was a large-scale research initiative to sequence the microbial communities that inhabit the human body.

Microbial composition in any ecosystem has a significant impact on biogeochemical cycles, as every cubic centimetre on Earth is ruled by microorganisms. Of the six major building blocks, including H, C, N, O, S, and P, that are fundamental for biological systems, the biogeochemical cycles of the first five elements are mainly driven by microbial activities. They possess a relatively stable set of core metabolic pathways, which have enormous potential to produce and degrade a diverse array of chemicals. They allow various biochemical reactions and biotransformation of compounds. The rate at which microorganisms use a nutrient reservoir affects the turnover period of chemicals. In many ecosystems, native microbial species perform respiration, decomposition, assimilation, and reduction-oxidation as key metabolic processes. As discussed before, most of the biochemical reactions are enzymatic dependent, and this equally applies to microorganisms. The non-enzymatic reactions take place between minerals and their metabolic products. Even some geomicrobial activities can distinguish stable isotopes. A slight shift in microbial community composition can cause subtle variation in its surroundings. Environmental factors can also induce variation in geomicrobial activities. In addition, nutrient gradients in an environment can also influence species assemblages. The microbial activities in an ecosystem play a key role in the dynamics of chemical equilibrium.

Complex symbiotic associations (e.g., symbiotic nitrogen fixation, mycorrhizae, and lichens) add efforts to channel the chemicals for their partners. Lichens serve as pioneer communities in bare land and help in weathering rock. Microbial exudates like citric acid, fumaric acid, etc., promote weathering of rock and support pedogenesis.

Among all microbial groups, the bacterial forms are known for their immense genetic diversity. They have enormous potential for environmental implications and acquire immense genetic diversity. Since life originated until today,

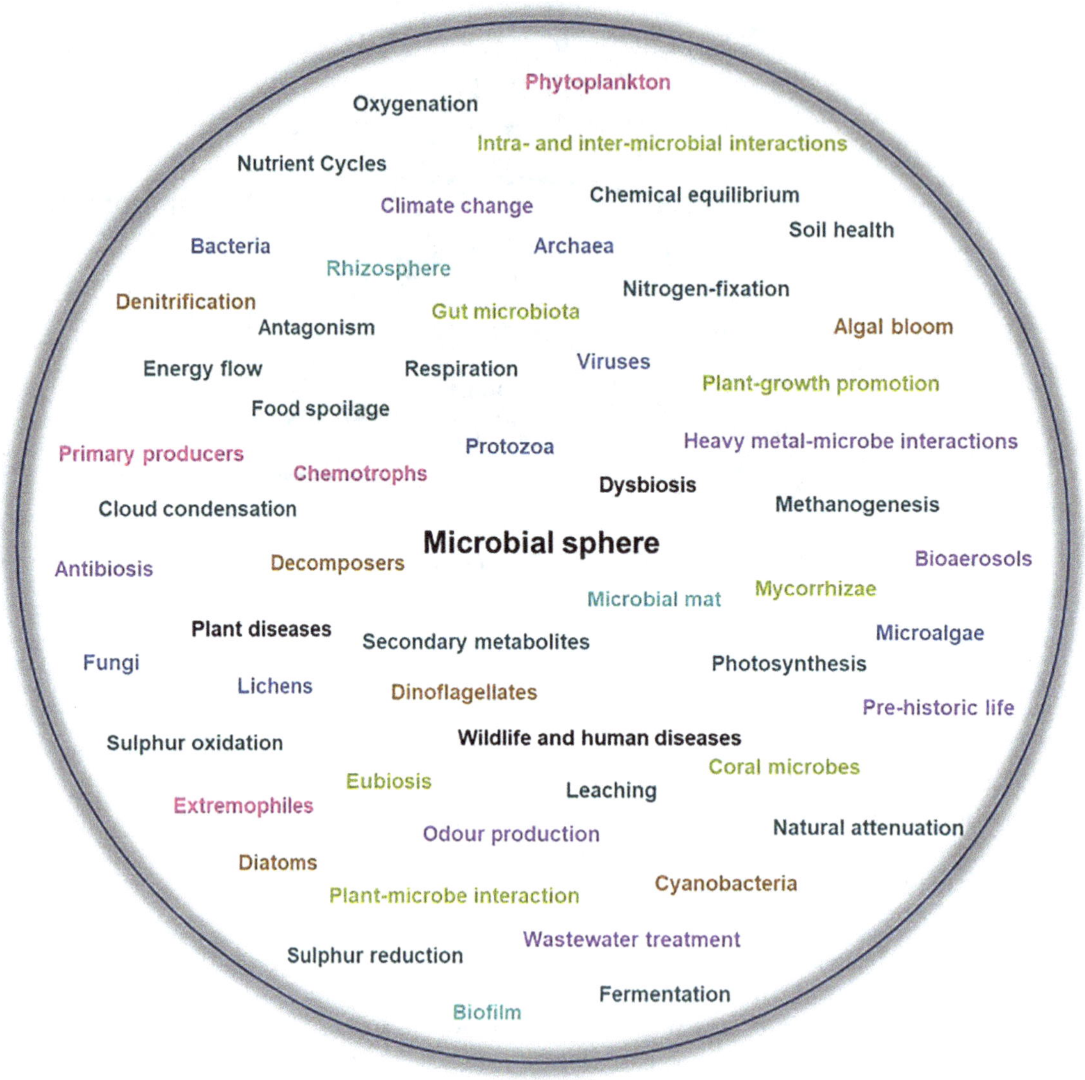

FIGURE 26.7 An illustration of the microbial sphere, including diverse microbial forms and their activities.

bacteria have evolved along with biogeochemical pathways for environmental regulation. Vertical gene transfer is the traditional mode of inheritance and has passed traits from parent to offspring. Horizontal gene transfer has played a significant role in heavy metal and antibiotic resistance in bacteria. Also, they have evolved to utilise available materials—that could be a metabolic product of any other bioreaction. They can be photosynthetic, chemosynthetic, or heterotrophic, contributing to the cycling of many nutrients. The anabolic pathways of cyanobacteria are credited to the photosynthesis and oxygenation of Earth. Biological nitrogen fixation is mainly restricted to certain bacteria and archaea, and is essential for the synthesis of protein, nucleic acid, and other molecules like chlorophyll and ATP. Various steps of the nitrogen cycle, such as nitrification, ammonification, and denitrification, are heavily dependent on microbial metabolism. Anoxygenic photosynthetic bacteria execute sulphur oxidation. Methanogens strictly belong to archaea and prevail in anoxic environments for the degradation of organic matter. Fermentation is a different anaerobic metabolic pathway— required for energy supply and production of secondary metabolites. Phosphate-solubilising microorganisms improve phosphorus bioavailability and alleviate soil phosphorus deficiency. A chemotrophic lifestyle is another unique characteristic of prokaryotes found in hostile environments. They can derive energy by oxidising inorganic nutrients such as Fe and Mn.

The extreme conditions of the environment inhabit special microorganisms and harbour unique survival potential under challenging conditions. The inhospitable situations could be extremely low and high temperatures and pH, the presence of non-threshold toxins (e.g., toxic heavy metals like As and Hg), chemicals in high concentrations (e.g., CO_2, Cu, and salt), and drought. The bulk of resistant bacteria may interact with extreme circumstances. For example, heavy metal-resistant bacteria in the contaminated regions facilitate the mitigation and removal of heavy metals. This process alters the mobility, bioavailability, and toxicity of elements. Examples include the transformation of heavy metals Cr(VI) into Cr(III), As(V) into As(III) and vice versa, Hg(II) into Hg(0). Even the redox activity can also influence the mobility and toxicity of radionuclides like uranium (U) and thorium (Th).

SUMMARY

- The "biogeochemical cycle" is the study of the movement of chemical elements through various compartments of the environment, such as the atmosphere, hydrosphere, geosphere, and biosphere.
- The biogeochemical cycle is a theoretical notion that simplifies complicated ecological processes and investigates the Earth's global metabolic map. Depending on the chemistry, distribution, and metabolic integrity, each biogeochemical cycle is a distinct, self-sustaining, and self-regenerating process.
- Biogeochemical cycles across geological time scales have changed the chemical and physical environments of Earth. Several distinct biological processes, such as biological nitrogen fixation (BNF) and the Great Oxidation Event (GOE), have dynamic effects on the biosphere.
- The easiest way to study how a biogeochemical cycle works is a box model, in which different boxes represent material storage locations, and arrows represent their fluxes. A standard quantitative biogeochemical cycle is explained in terms of source, sink, feedback control, budget, residence time, fluxes, reservoir, and exchange pool.
- The influence of solar radiation, geochemical reactions, and biological activities is the primary factors that regulate biogeochemical cycles.
- Microbial compositions have a huge influence on biogeochemical cycles as microorganisms control many geochemical activities. The microbial regulation of biogeochemical cycles of hydrogen, carbon, nitrogen, oxygen, and sulphur is apparent in many ecosystems.

DIFFERENT TYPES OF PRACTICE QUESTIONS

Fill in the blanks

I. The first oxygen on Earth could be produced by organisms belonging to phylum.....................

II. .... is the rate of movement of matter or energy in or away from a nutrient reservoir.

III. .... were the primary sites for the origin of life on Earth under oppressive situations of hot, sunless, anoxic, saline, and high-pressure environments.

IV. The Earth's land surface is covered with soil, lying at the interface of the atmosphere and lithosphere, known as the ....

V. .... is a single essential component and the most abundant molecule on Earth.

SOLUTION: I. Cyanobacteria, II. Flux, III. Hydrothermal vents, IV. pedosphere, V. Water

Match the following

i. Residence time	a. Input rate
ii. Influx	b. Storage house
iii. Reservoir	c. Control system
iv. Feedback loop	d. Duration of stay
v. Source	e. Run-off site

SOLUTION: i-d, ii-a, iii-b, iv-c, v-e

Very short answer type of questions

I. What is the main difference between geochemical and biogeochemical cycles?

II. What are the four essential ingredients of the biogeochemical cycles?

III. Calculate the residence time of carbon in oceans if it occupies 38,100 Gt C and there is an influx of 0.6 Gt C/yr via rivers (ignoring other exchange and internal cycling).

IV. Write the names of four basic environmental compartments of Earth.

V. What is a budget in the biogeochemical cycle?

VI. Write an example of a feedback control mechanism in the environment.

VII. What are the methods for illustration of biogeochemical cycles?

Short answer type of questions

I. How have different biological activities evolved in the geological time scale?

II. How do ecosystem-level approaches contribute to the biogeochemical cycles?

III. How are the terms microbial spheres, microbiota, and microbiome related to each other?

IV. What are the roles of enzymes in biogeochemical cycles? Give your statement with suitable examples.

V. Anthroposphere has left a strong impression on the biogeochemical cycles and overall environment. Justify.

VI. Write the differences between gaseous and sedimentary cycles.

Long answer type of questions

I. What is biogeochemistry? Discuss the key characteristics of the biogeochemical cycle.

NOTES

1 Gt stands for gigaton and 1 Gt equals 1 billion metric tonnes.
2 Friedlingstein P, O'sullivan M, Jones MW, et al. Global carbon budget 2023. *Earth System Science Data*. 2023;15(12):5301–69.

SUGGESTED READINGS

Bar-On YM, Phillips R, Milo R. The biomass distribution on Earth. *Proceedings of the National Academy of Sciences*. 2018;115(25):6506–11.

Brusseau ML. Ecosystems and ecosystem services. In Brusseau ML, Pepper I, Gerba C (eds). *Environmental and pollution science* 2019 (pp. 89–102). Academic Press.

Falkowski PG, Fenchel T, Delong EF. The microbial engines that drive Earth's biogeochemical cycles. *Science*. 2008;320(5879):1034–9.

Fernández-Remolar DC. Biogeochemical cycles. In Gargaud M. et al. (eds). *Encyclopedia of Astrobiology* 2011. Springer. doi.org/10.1007/978-3-642-11274-4_173

Finzi AC, Austin AT, Cleland EE, Frey SD, Houlton BZ, Wallenstein MD. Responses and feedbacks of coupled biogeochemical cycles to climate change: examples from terrestrial ecosystems. *Frontiers in Ecology and the Environment*. 2011;9(1):61–7.

Jabłońska J, Tawfik DS. The evolution of oxygen-utilizing enzymes suggests early biosphere oxygenation. *Nature Ecology & Evolution*. 2021;5(4):442–8.

Jacobson M, Charlson RJ, Rodhe H, Orians GH. *Earth System Science: from biogeochemical cycles to global changes*. Academic Press; 2000.

Keller MA, Piedrafita G, Ralser M. The widespread role of non-enzymatic reactions in cellular metabolism. *Current Opinion in Biotechnology*. 2015;34:153–61.

Lyons TW, Reinhard CT, Planavsky NJ. The rise of oxygen in Earth's early ocean and atmosphere. *Nature*. 2014;506(7488):307–15.

Madsen EL. Microorganisms and their roles in fundamental biogeochemical cycles. *Current Opinion in Biotechnology*. 2011;22(3):456–64.

Pérez-Guzmán L, Bogner KR, Lower BH. Earth's Ferrous Wheel. *Nature Education Knowledge*. 2010;3(10):32.

Reinhard CT, Planavsky NJ, Olson SL, Lyons TW, Erwin DH. Earth's oxygen cycle and the evolution of animal life. *Proceedings of the National Academy of Sciences*. 2016;113(32):8933–8.

Schlesinger WH, Cole JJ, Finzi AC, Holland EA. Introduction to coupled biogeochemical cycles. *Frontiers in Ecology and the Environment*. 2011;9(1):5–8.

Speight JG. *Natural water remediation: chemistry and technology*. Butterworth-Heinemann; 2019.

Tang W, Llort J, Weis J, Perron MM, Basart S, Li Z, Sathyendranath S, Jackson T, Sanz Rodriguez E, Proemse BC, Bowie AR. Widespread phytoplankton blooms triggered by 2019–2020 Australian wildfires. *Nature*. 2021;597(7876):370–5.

Wogan NF, Catling DC. When is chemical disequilibrium in Earth-like planetary atmospheres a biosignature versus an anti-biosignature? Disequilibria from dead to living worlds. *The Astrophysical Journal*. 2020;892(2):127.

27 Carbon Cycle and Microorganisms

27.1 CARBON IN THE ENVIRONMENT

Carbon belongs to the p-block element in the periodic table. The atomic number and atomic mass of carbon are 6 and 12.0107 u, respectively. It is the fourth most abundant element in the Universe and the fifteenth most abundant element in the Earth's crust. The major reservoirs of carbon on the Earth are rock, soil, ocean, and the biosphere. Carbon is a tetravalent element and can establish four bonds with other atoms. When a carbon atom creates covalent bonds with carbon and other atoms like oxygen, hydrogen, and nitrogen, it makes a large class of organic compounds (with few exceptions, such as cyanide, carbonate, and carbides). Many organic compounds form complex structures and contain diverse properties. The long chains of organic compounds form organic polymers, which are the backbone of life. Many bonds in organic polymers have high bond energy (i.e., the energy required to create bonds) (Table 27.1), contributing to their stability. Also, organic polymers have high energy density, which means they release a high amount of energy when they break. This property of the carbonic chain makes them an excellent source of fuel in the biosphere. Organic carbon has a wide distribution on Earth and different attributes, which can be expressed in some colour-based descriptions (detailed in Box 27.1).

Carbon mainly exists as carbon dioxide (CO_2) in the atmosphere, carbonic anions in water, and organic matter in soil and biomass. Many of Earth's minerals are entirely made up of carbon, such as diamond and graphite. They are called carbon allotropes. Calcareous rock contains the highest amounts of carbon ($\approx 10^8$ Gt C)[1] in the form of calcium carbonate ($CaCO_3$) and other minerals. The ocean represents the most abundant exchange pool of organic carbon with a global stock of ≈38,000 Gt C. The ocean stores 50 times more carbon than the atmosphere and 20 times more carbon than soil and vegetation on land. Soil constitutes ≈2,500 Gt C. The sum of all biomass on the Earth comprises ≈550 Gt C. Fossil fuels are hydrocarbon substances formed from dead and decayed biomass that lived millions of years ago. They also account for a high amount of organic carbon (≈4,000 Gt C). The atmosphere has only ≈840 Gt C but serves as a main exchange pool of carbon in the carbon cycle (Table 27.2).

The amount of carbon on the Earth is constant, but changes its states and reservoirs. More than 99% of Earth's carbon in the carbon cycle is stored in its crust. These reserves are mainly the deposits of living beings, including the decaying component of the marine inhabitants and the $CaCO_3$ from their skeletons and shells. The accumulation of $CaCO_3$ in the ocean over a long time is responsible for the formation of a sedimentary rock called **limestone**. A significant amount of decomposed organic matter is buried, compressed, and heated under many layers of sediment, which forms a waxy, insoluble hydrocarbon called **kerogen**. It serves as the precursor to natural gas and oil in the sediments and other hydrocarbons found in sedimentary rocks.

Carbon is the most crucial element for the biosphere. Approximately 50% of all dry biomass on the Earth has carbon constituents. Biological systems do not have any choice when it comes to carbon-based feeding. All types of organisms use carbon for respiration. Plants store the maximum amount of carbon (about 82% of all organic matter), while bacteria contain roughly 13% carbon. The remaining 5% of carbon is stored in other lifeforms (Figure 27.1).

27.2 CARBON SINKS AND SOURCES

On Earth, carbon is reserved as carbon sink and source based on its flow and residence time in their reservoirs. A carbon sink is a natural pool which stores more carbon than it releases. The carbon source is reversed to a carbon sink, which releases more carbon than it stores. Forests and oceans are the best examples of carbon sinks. Forests continually sequester carbon in the form of visible vegetation through photosynthesis and as humus in soil through decomposition. Of different vegetation types, tropical rainforests and boreal forests have the maximum storage of carbon. Plant bodies of the dense vegetation are the primary sink of carbon in tropical rainforests. However, the soil in tropical forests does not store a significant quantity of carbon compared to the soil in boreal and temperate forests. This is because the high decomposition rate in tropical forests leads to a dynamic equilibrium in carbon cycling. Tropical savannas and temperate grasslands have also an abundant stock of carbon. In the terrestrial ecosystem, plants occupy about 19% of the carbon, while the soil contains as much as 81% of the carbon.

The atmosphere is the most dynamic source of carbon on the Earth. It maintains a continuous exchange of carbon among oceans, soil, and living organisms. The concentration

DOI: 10.1201/9781032651095-34

TABLE 27.1
Bond energies of selected bonds of carbon with other atoms in organic matter

Bond	Bond energy (kJ/mol)
C-C	347
C=C	614
C-H	414
C-O	360
C=O	799
C-N	305
C-S	259

of CO_2 in the environment is continually rising because anthropogenic activities have turned the carbon sinks into carbon sources. Deforestation and land use changes have released a high amount of fixed carbon. Burning vegetation has increased the emission of CO_2 into the atmosphere. The combustion of fossil fuels and livestock operations are the most dominant sources of CO_2 in the environment. Volcanic eruption is an instantaneous natural carbon source, releasing huge volumes of CO_2 and other gases. Respiration in living organisms contributes to a steady and continuous supply of CO_2 in the environment. Decomposition and methanogenesis also account for the release of CO_2 in the environment.

BOX 27.1 COLOUR-BASED SPECTRUM OF CARBON IN THE ENVIRONMENT

The understanding of the carbon cycle can involve a full spectrum of colour-based (green, blue, teal, black, brown, and red) explanations of carbon based on their roles, characteristics, and sites in the environment (Box Figure 27.1.1). Green carbon represents carbon stored in plants and soil in terrestrial ecosystems, Blue carbon is carbon sequestered in marine ecosystems such as mangrove forests, seagrass meadows, tidal marshes, seaweed, and sediments. The teal carbon covers inland freshwater and wetlands. Black and brown types of carbon are light-absorbing atmospheric carbonaceous components built by the incomplete combustion of organic matter. The main difference between them lies in their optical properties. The red carbon represents secondary pigments of microalgae (belonging to the family Chlamydomonaceae) found in the cryosphere. Red snow (or colourful and black snow) melts more quickly because it absorbs more heat and reflects less sunlight (i.e., greenhouse effect). Snow and ice produce other colours like yellow, pink, grey, orange, and purple due to diverse microbial activities and are covered in red carbon.

Green, blue, and teal carbon are associated with carbon sequestration and climate change mitigation, while black, brown, and red carbon play key roles in heat balance and the melting of glacier surfaces. Of all types of colour-based carbon, green and blue types of carbon are dominant in the environment.

BOX FIGURE 27.1.1 Full-colour spectrum of carbon in the environment. (Image of red snow adapted from Jeff Having, University of Minnesota, taken in July 2019.)

TABLE 27.2
Global carbon stocks of Earth

Stores	Carbon stocks (values in approximation) (Unit is Gt C)
Crust and mantle	100,000,000
Ocean	38,000
Fossil fuel	4,000
Soil	2,500
Permafrost	1,600
Atmosphere	840
Biomass	550

Note: Gt stands for gigaton.

CO_2 is a fundamental molecule for life that is exchanged and cycled across different ecosystems. The carbon cycle involves multi-level processes, evident from the sphere to the cellular levels (Figure 27.2). In the lithosphere, sedimentary rocks and fossil fuels contain the maximum amount of carbon. A high amount of carbon resides in the ocean, which descends to the deep oceans, where it rests for a very long time—possibly thousands of years. At the cellular level, biological organisms involve carbon-fixing pathways, respiratory activities, transformation, and decomposition (Figure 27.2).

27.3 CARBON CYCLE

The carbon cycle is fundamental to the biogeochemical activities on Earth. It was pioneered by the joint efforts of

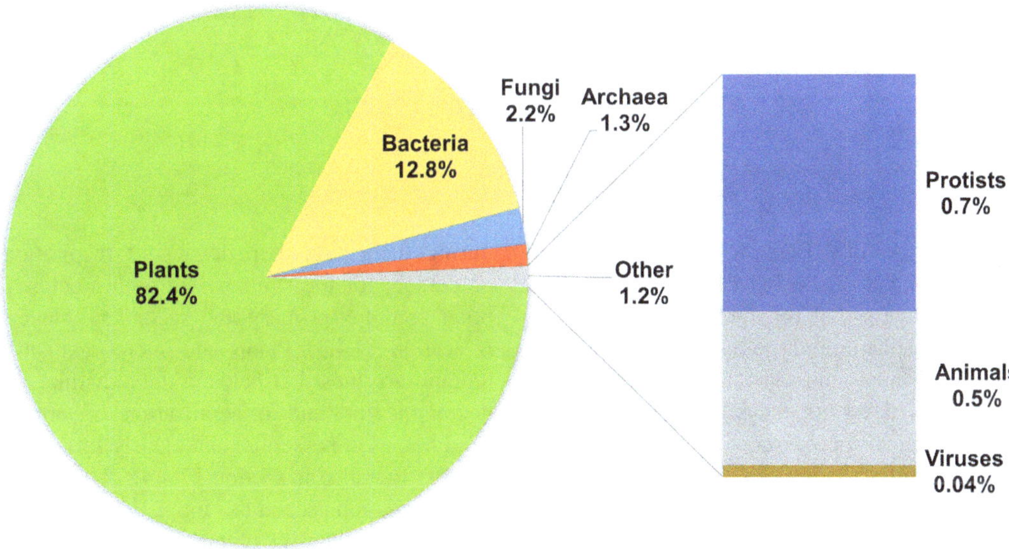

FIGURE 27.1 Proportion of carbon biomass in different types of organisms.

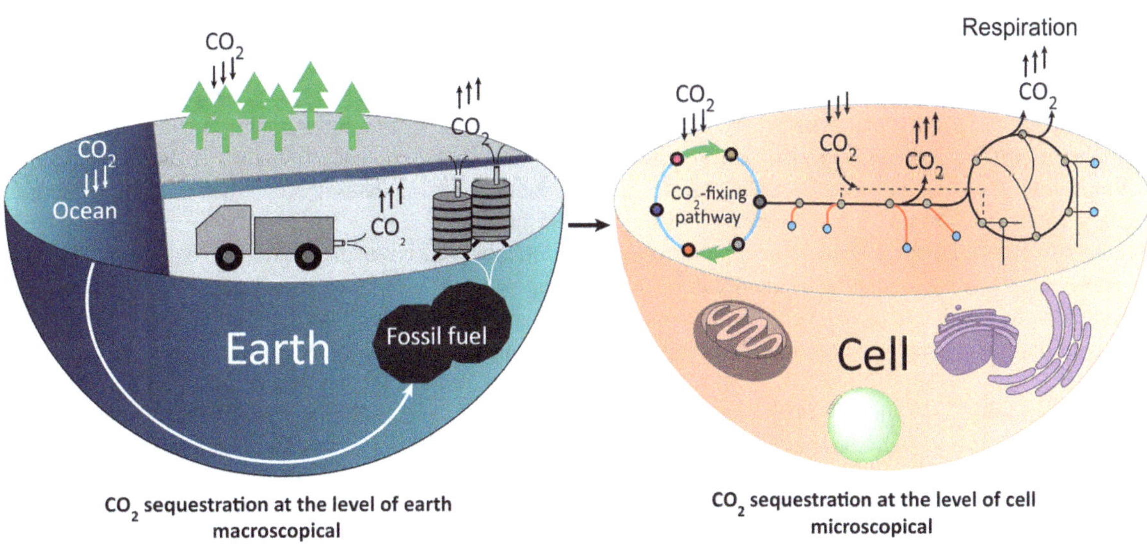

FIGURE 27.2 Carbon dioxide (CO_2) sequestration from Earth to the cell. (Image adapted from Hu et al. (2019)[2] with permission from Elsevier.)

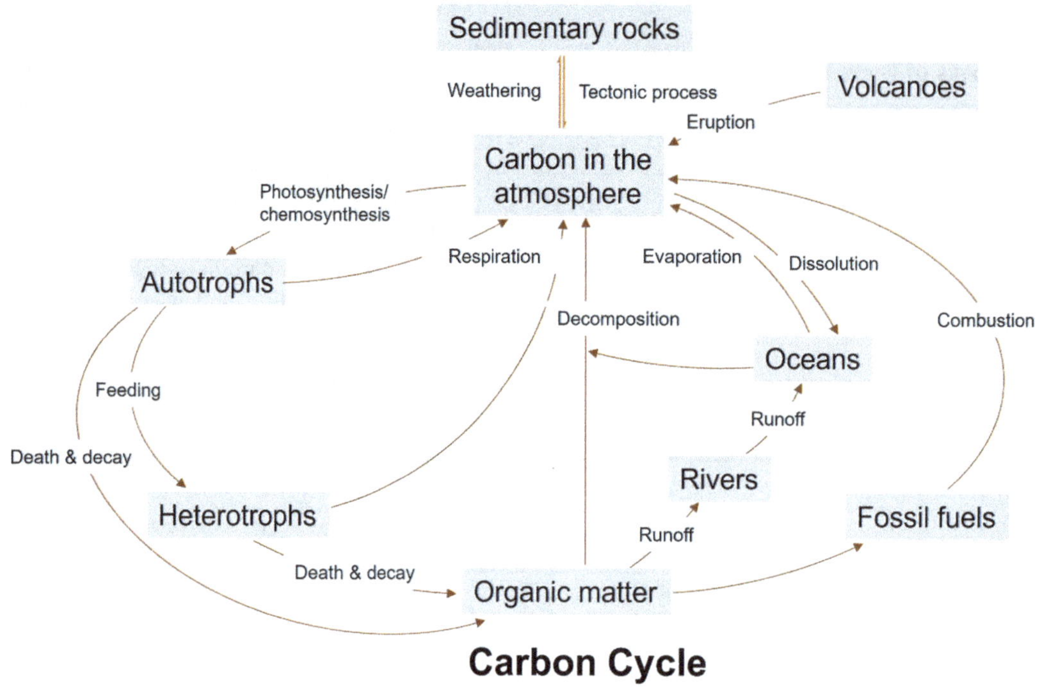

FIGURE 27.3 An illustration of the carbon cycle. (C represents carbon.)

two French chemists, Antoine-Laurent de Lavoisier and Armand Séguin, by the end of the 18th century. The carbon cycle is a dynamic process that regulates the atmosphere, the biosphere, the lithosphere, and the hydrosphere through various biological events and geochemical processes (Figure 27.3). Biological events play active roles in the carbon flow. Geochemical processes are generally slow but contribute significantly to the long-term carbon cycle. There are seven key activities in the carbon cycle, including photosynthesis, respiration, decomposition, combustion, dissolution, sedimentation, and extraction. These are briefly discussed as follows.

Photosynthesis is a cornerstone for the assimilatory flux of carbon. The photosynthetic conversion of CO_2 into glucose is the primary step of sequestering carbon into an organic form. In aquatic habitats, the abundant nutrients in the water bodies encourage the growth of phytoplankton and macrophytes. This process speeds up photosynthesis in the surface water and pulls more atmospheric carbon, enabling high primary production (detailed in section 27.4.1.1). The simple carbohydrates and other reaction intermediates are the elementary substrates for the synthesis of other biomolecules. Grazing and feeding allow the transfer of a certain amount of fixed carbon from primary production to a higher trophic level of organisms. The primary production of autotrophs is the main building block of the biosphere.

Cellular respiration is the opposite process of photosynthesis. Respiratory reactions consume the fixed carbon and release CO_2 into the atmosphere (detailed in section 27.4.2), which balances the fixed carbon achieved through photosynthesis. The various intermediates produced during the catabolic pathways of respiration serve as the precursors for the synthesis of amino acids, fatty acids and other complex primary and secondary metabolites (detailed in section 27.4.3). Plants and microorganisms possess many unique enzymes that help produce distinct metabolites. The decomposition and mineralisation of organic wastes and dead bodies are essential and closely linked to the respiratory process (detailed in section 27.4.4).

Combustion accounts for CO_2 emissions through the uncontrolled burning of biomass. This is purely an abiotic process and requires oxic conditions. This mechanism is analogous to the respiration process in biology. The primary cause of combustion is human activities, such as the use of fossil fuels for heating and vehicle transportation. Wildfire is a natural event for the combustion of biomass. The excessive release of CO_2 has severe implications for the environment and is the main driver of climate change, discussed in Box 27.2. Global Carbon Project is an initiative to monitor the release of CO_2 and its causes (detailed in Box 27.3). This project produces a large dataset, which helps formulate climate action programmes.

Carbon dioxide dissolution is a process of transporting atmospheric CO_2 into the water. A general gas solubility, Henry's Law, applies to this process, which facilitates equilibrium between gas dissolved in a liquid and its partial pressure in the atmosphere. Factors like temperature, pressure, turbulence, and biological activities are important regulators of CO_2 dissolution in water. Low temperature tends to promote CO_2 dissolution in water, while high temperatures decrease the solubility of gases. The high dissolution of CO_2 in the ocean results in the formation of carbonic acid. This process lowers the pH of the ocean, leading to ocean acidification. Freshwater absorbs more CO_2 than saline water because the ionic strength of saline conditions interferes with CO_2's ability to dissolve in the water.

Another key event in the carbon cycle is sedimentation, which involves the accumulation of organic and inorganic carbon from the water column to the bottom. The settled particles tend to form sedimentary rocks over a period. However, the magnitude of sedimentation is extremely slow. The subduction process enables the transfer of a notable amount of carbon from its sedimentary rocks into the mantle zone for its prolonged stay. Years later, carbon can be released from deep burial sites through geological processes like volcanic eruption, tectonic activities, erosion, and mantle degassing.

Based on the residence time of carbon in their reservoirs and its flux, the carbon cycle can be categorised as the fast and slow carbon cycles, which are discussed as follows.

27.3.1 The Slow Carbon Cycle

The slow carbon cycle mainly involves geological processes. The penetration of dissolved CO_2 in the deep ocean takes centuries because the mixing of gas and water due to molecular diffusion is slow. The formation of carbon-containing rock is also a long-term operation. This involves the cementation and compression of calcified shells and carbon-containing sediments. Carbon in the rock is transformed into coal, oil, and gases, which remain bound for millions of years. A large amount of carbon disappears in the mantle zone by the process

BOX 27.2 IMPACT OF CARBON DIOXIDE (CO_2) ON CLIMATE CHANGE

A Swedish chemist, **Svante Arrhenius,** in 1896 was the first to shed light on the role of CO_2 in climate change (i.e., global warming). Carbon dioxide, along with CH_4 and nitrous oxide (N_2O), account for 80% of the total radiative force for climate change. The current global rise of CO_2 in the atmosphere has established a new approximate balance in the carbon cycle. Eventually, the residence times of different carbon species in their reservoirs have also changed. The change is also inevitable for the Earth's carbon balance, and for the Earth's crust. The excess concentration of CO_2 has altered the carbon exchange process in the atmosphere, the surface ocean, and the vegetation. High dissolution of CO_2 in seawater has increased the acidification of marine ecosystems. Other greenhouse gases have a heat retention capacity greater than CO_2 in warming the Earth's surface. However, CO_2 poses a high accountability for climate change due to its excess emissions and prolonged stay in the atmosphere.

Global anthropogenic activities have a tremendous influence on CO_2 emissions. The carbon cycle was roughly balanced before human interruption. The beginning of the industrial era has increased the CO_2 in the atmosphere from approximately 277 ppm (in 1750) to 422 ppm (in 2024)— about a 50% expansion. The last decade has witnessed annual global CO_2 emissions of around 40 Gt C. The burning of fossil fuels is the most dominant source, emitting approximately 30 billion tonnes of CO_2 in the air every year (about 35% of all CO_2 emissions). Changes in land use patterns have dramatically induced shifts in carbon fluxes (contributing about 12% of total CO_2 emissions), with deforestation being the major force. Cement plants have released massive amounts of carbon into the atmosphere. Limestone is the principal component of cement manufacturing and has contributed about 8% of total CO_2 emissions. Although terrestrial biomass and the ocean have shown potential for the absorption of a high concentration of CO_2 (about 55%), a significant amount of CO_2 is still disseminated in the atmosphere. Elevated concentrations of CO_2 in the atmosphere have already caused a huge loss of biodiversity, and the increasing trend poses a grave threat to the biosphere. Owing to greater demand for industrialisation and urbanisation, the concentration of CO_2 continues to expand in the coming years, which can lead to serious consequences, including rising global temperatures, extreme weather events, and sea-level rise.

BOX 27.3 GLOBAL CARBON PROJECT

The Global Carbon Project (GCB) is an initiative of Future Earth in association with the World Climate Research Programme (WCRR) and has been active since 2001. It conducts an annual research evaluation of the global carbon cycle. GCP publishes their results on the carbon budget every year. "Global Carbon Atlas" is an online platform under GCP that disseminates information related to the carbon budget (www.globalcarbonatlas.org/). This webpage has options to explore, visualise, and interpret the most up-to-date data on carbon fluxes. This program has involved the collaboration of more than fifty research institutes across the world. This comprises a network of scientists, researchers, and innovators who have gathered data and performed analyses for comprehensive and accurate information on the global carbon cycle. The assessment mainly aims to quantify the imbalance between emission and storage through the carbon budget and trend. The project covers the studies on the inflow and outflow of CO_2 into the atmosphere and its balance in different reservoirs (Box Figure 27.3.1). The project also underpins the climate crisis due to the global rise of CO_2. Additionally, this project is concerned about the two other important greenhouse gases, including CH_4 and N_2O, for their contributions to climate change and focuses on efforts to reduce their emissions.

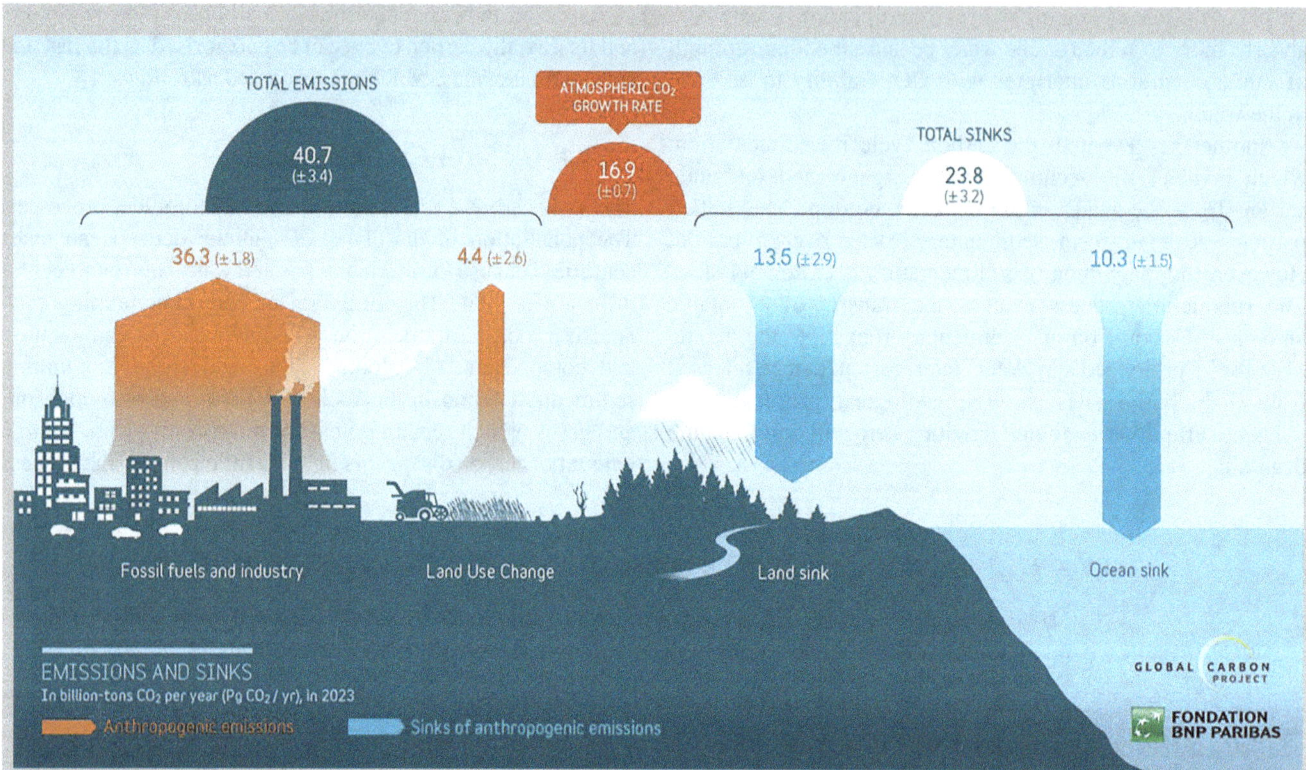

BOX FIGURE 27.3.1 Global carbon budget 2023. (Image adapted from Friedlingstein et al. (2023)[3] Global Carbon Budget 2023. Earth System Science Data.)

of subduction. The dissolved carbon and other ions are distributed through rivers and ultimately flow into the oceans. In the marine environment, dissolved CO_2 and calcium ions (Ca^{2+}) combine as $CaCO_3$ and build the shells of marine organisms like shellfish and some phytoplankton. When these organisms die, their shells rest in the sea sediments for a long period. The resurfacing of these buried and preserved carbon in the environment also takes a long time. In terrestrial ecosystems, the discharge of rock-bound carbon is associated with weathering, volcanism, and tectonic activities. During rainfall, water droplets combine with CO_2 and precipitate carbonic acid, which aids in the chemical breakdown of rock. Biological weathering employs native microorganisms and releases acids for the dissolution of rock— this is also a slow process. The transport of CO_2 back to the atmosphere from terrestrial ecosystems is also a long-term process. However, anthropogenic activities like mining and the extraction of fossil fuels have significantly disrupted the slow carbon cycle (detailed in Box 27.2).

27.3.2 The Fast Carbon Cycle

The fast carbon cycle is 10 to 1,000 times faster than the slow carbon cycle. Once the CO_2 is emitted into the atmosphere, it spreads rapidly in the surroundings and is distributed in the biosphere, the hydrosphere, and the lithosphere. The upper ocean and vegetation offer a constant exchange of CO_2 from the atmosphere. Biological activities take less time to redistribute the carbon in different reservoirs than geological activities. The influx of CO_2 through photosynthesis and the efflux of CO_2 through respiration and decomposition constitute a prompt supply of C, establishing a circular pathway (Figure 27.4). The dissolution of CO_2 in marine water is a fast process. The fast carbon cycle also accommodates anthropogenic activities

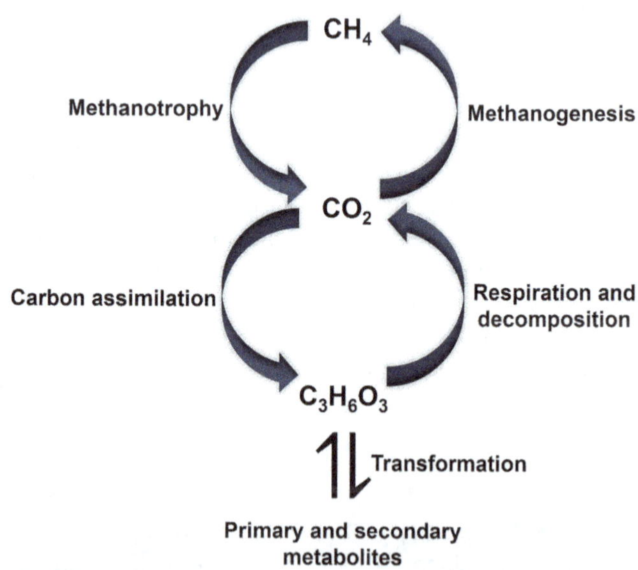

FIGURE 27.4 The passage of carbon assimilation, decomposition, and transformation through different biological activities in the carbon cycle. ($C_3H_6O_3$: carbohydrate, CO_2: carbon dioxide, and CH_4: methane.)

TABLE 27.3
Differences between the fast carbon cycle and the slow carbon cycle

Parameters	Fast carbon cycle	Slow carbon cycle
Flux of carbon	Fast	Slow
Amount of carbon flux (Gt C per year)	10-100	0.01-0.1
Rebalance time	Minutes to days	Hundred to thousand years
Sequestration	Photosynthesis and Chemosynthesis	Sedimentation and subduction
Extraction/emission	Respiration, decomposition, and combustion	Weathering, volcanic eruption, tectonic events
Interaction with the marine ecosystem	Dissolution of CO_2 at the ocean surface	Penetration of dissolved CO_2 in the deep ocean
Main sphere involved	Biosphere	Geosphere and hydrosphere (mainly oceans)
Main reservoir	Forest	Sedimentary rocks

TABLE 27.4
Nutritional classification of microorganisms based on their sources of energy, electrons, and carbon, including suitable examples of microorganisms

Energy sources	Electron sources	Carbon sources	Nutritional types	Examples
Light	Inorganic substances	CO_2	Photolithoautotroph	Microalgae, purple sulphur bacteria, green sulphur bacteria, and cyanobacteria
	Inorganic substances	Organic	Photolithoheterotroph	Purple nonsulphur bacteria and green nonsulphur bacteria
Light	Organic substances	CO_2	Photoorganoautotroph	Some Haloarchaea
	Organic substances	Organic	Photoorganoheterotroph	Some green nonsulphur bacteria
Chemical reaction	Inorganic substances	CO_2	Chemolithoautotroph	Methanogens, nitrifying bacteria, sulphur-oxidising bacteria, and hydrogen bacteria
	Inorganic substances	Organic	Chemolithoheterotroph	*Oceanithermus profundus*
Chemical	Organic substances	CO_2	Chemoorganoautotroph	Anaerobic methanotrophic archaea (ANME)
reaction	Organic substances	Organic	Chemoorganoheterotroph	Fungi, heterotrophic bacteria and protozoa

such as deforestation, urbanisation, and hydraulic activities for intense carbon transport. A direct comparison between the slow and fast carbon cycle can be studied in Table 27.3.

27.4 KEY MICROBIAL ACTIVITIES

The carbon cycle greatly benefits microorganisms in the processing of carbon. They serve as an engine to the carbon cycle and try to establish a balance of carbon in the environment. Microorganisms present in soil play a great role in the ecosystem's carbon budget. The quality and quantity of organic matter present in an ecosystem can have a strong impact on the microbial community compositions and their function. Microbial-mediated processes can also influence the residence period and flux of carbon in their reservoirs. In an ecosystem, the biomass of microbial cells can account for a high quantity of organic matter. For example, more than 90% of organic matter in the ocean is composed of microbial biomass, which represents the assimilative abilities of microorganisms (anabolism). The feed on fixed carbon is representative of the catabolic pathway. Catabolic pathways of microbes also have tremendous potential for degradation and transformation.

Microbial activities have several key moments in the carbon cycle that have wide to narrow impacts on the global system. Microorganisms can compose, decompose, and even transform organic materials. These activities are possible in microorganisms because they observe different nutritional modes (Table 27.4). Some important microbial pathways relevant to the carbon cycle are illustrated in Figure 27.5. Microbial-mediated processes associated with the carbon cycle are discussed in detail as follows.

27.4.1 Carbon Assimilation

In the global carbon cycle, carbon assimilation (or carbon fixation/carbon sequestration) is a process to reduce CO_2 into cellular carbon. Electrons and potential energy are two other basic requirements of carbon fixation. Carbon-fixing organisms are called autotrophs. When an autotroph is exposed to an energy source (e.g., light) and the environment consists of suitable electron donors, the biological cell allows the conversion of inorganic carbon into organic carbon. In microbes, photosynthesis and chemosynthesis are two key pathways of carbon assimilation. Photosynthesis is prevalent in

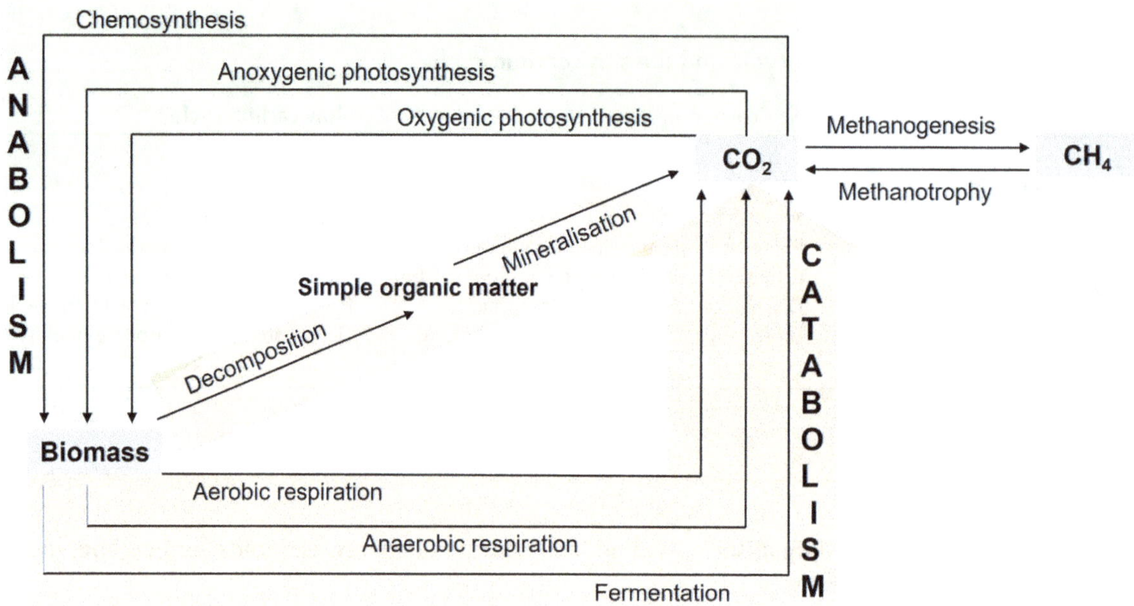

FIGURE 27.5 Different catabolic and anabolic reactions for carbon metabolism in the microbial system.

an environment that has sufficient sunlight. Chemosynthesis is the mode of autotrophic nutrition in dark regions. Figure 27.6a illustrates photosynthetic and chemosynthetic zones in an aquatic region. In the global platform, photosynthesis accounts for more than 90% of carbon assimilation, whereas chemosynthesis contributes about 10%. These processes are further discussed in the following subsections.

27.4.1.1 Photosynthesis

Photosynthesis is an essential endothermic natural process for carbon fixation. Photosynthetic microorganisms can belong to all three domains of life, including bacteria, archaea, and eukarya. They are responsible for high carbon sequestration in diverse habitats. All phototrophs in the aquatic body reside in the photic zone (also known as the epipelagic zone), which is not deeper than 200 m from the water surface. The marine environment is the chief carbon-sequestering environment, mainly performed by phytoplankton. Photoautotrophic microorganisms also reside on land surfaces. They grow in soil, on walls, and on other wet surfaces. Some phototrophic taxa establish symbiotic relationships with other organisms (e.g., fungi and plant roots and leaves) for mutual benefits. Based on the sources of electron donors, photosynthesis can be oxygenic and anoxygenic.

a. Oxygenic photosynthesis uses water molecules as electron sources and releases oxygen (O_2) as a byproduct. Oxygenic photosynthesis is reported in cyanobacteria, microalgae, and plants. They are dominantly present in both aquatic and terrestrial habitats. However, cyanobacteria and microalgae are more efficient for CO_2 fixation than plants. The marine cyanobacteria, *Prochlorococcus* and *Synechococcus,* mainly exist in nutrient-poor tropical and subtropical oceans. *Prochlorococcus* is among the smallest but the most abundant photosynthetic organisms in the marine environment. They account for more than 10^{24} individuals in the sea. Cyanobacteria like *Nostoc*, *Anabaena*, *Oscillatoria*, and *Gloeobacter* have ubiquitous distribution in land and freshwater ecosystems. Microalgae like *Chlamydomonas*, *Chlorella*, *Dunaliella*, dinoflagellates, and diatoms also prevail in diverse aquatic sites, including both wet and dry seasons.

Both microorganisms and plants share a common pathway of oxygenic photosynthesis. It consists of two stages: light-dependent and light-independent reactions. The light-dependent reaction facilitates the hydrolysis of water molecules in the presence of sunlight and releases O_2, electrons, and hydrogen ions. These ions are used in the production of high-energy molecules, ATP and NADPH + H^+. Further, light-reaction pulls CO_2 from the atmosphere and converts it into glucose and other sugars in the Calvin cycle, using ATP and NADPH + H^+ from the light-dependent reaction. Figure 27.6b illustrates light-dependent and light-independent reactions.

Photosynthesis requires chlorophyll pigments to capture light energy. These pigments give photosynthetic organisms a green hue. Microalgae and plants store the chlorophyll pigments in a unique double membrane-bound organelle, the chloroplast. RUBISCO (ribulose-1,5-bisphosphate carboxylase/oxygenase) is an essential enzyme required for the transformation of inorganic carbonic ions into organic form. They are found in the stroma of the chloroplast.

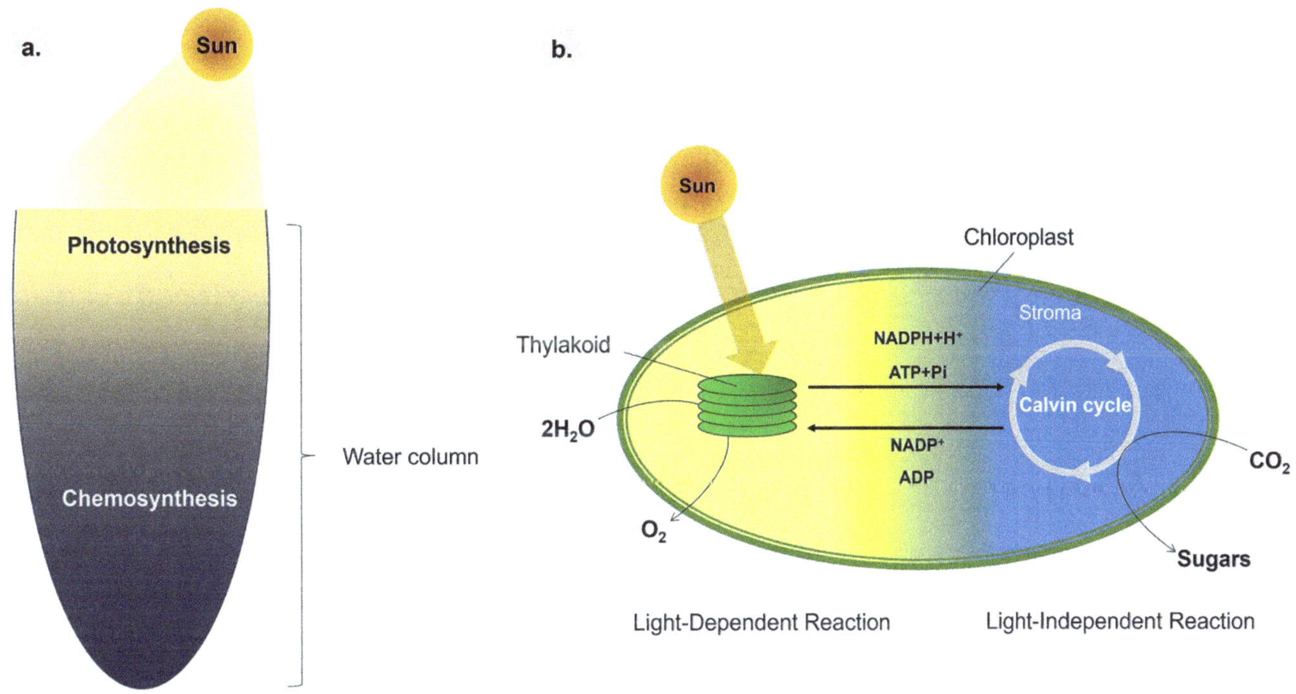

FIGURE 27.6 a. Ecological zones for photosynthesis and chemosynthesis in the water column. b. A schematic diagram of the carbon fixation in a chloroplast showing light-dependent and light-independent reactions.

However, in cyanobacteria, chlorophyll pigments are present in flattened, disc-like structures, called thylakoids, whereas RUBISCO resides in polyhedral compartments, called carboxysomes (Figure 27.7). RUBISCO is the most abundant protein on Earth, which reflects its critical role in carbon assimilation. This enzyme has an affinity for both CO_2 and O_2. The high concentrations of CO_2 and low temperature promote carboxylation (blinding with CO_2).

The photosynthetic product, glucose, is the building block of life. It serves as a precursor for the synthesis of complex carbohydrates. Starch and cellulose are among the most common polymers and serve as reserve materials in photoautotrophs. Starch consists of multiple glucose units bound through α-1,4 and α-1,6 glycosidic bonds. Cellulose is the most abundant organic compound in the entire biosphere, composed through β-1,4 glycosidic linkages and has a role in structural integrity.

b. Anoxygenic photosynthesis uses reduced sulphur species like hydrogen sulphide (H_2S) and elemental sulphur (S^0) in place of water to generate electrons. Examples of anoxygenic photosynthesis organisms are phylogenetically diverse bacteria, including green sulphur bacteria (GSB) (e.g., *Ignavibacterium* and *Chlorobium*), green nonsulphur bacteria (Chloroflexi, like *Chloroflexus* and *Chloronema*), purple bacteria (e.g., *Rhodospirillum* and *Rhodobacter*), Acidobacteria, and Heliobacteria. They possess a distinct light-harvesting pigment, bacteriochlorophylls, which shares similarities with oxygenic photosynthetic chlorophyll pigments (Chl *a* and *b*). In the light reaction, H_2S breaks down into S_0 and sulphate (SO_4^{2-}) and releases electrons. Electrons and hydrogen ions assist in the production of energy carriers for their use in the carbon assimilation process. The carbon assimilation process in anoxygenic photosynthetic microorganisms exhibits great variations. Green sulphur bacteria perform the reductive tri-carbonic acid (rTCA) cycle, green non-sulphur bacteria employ 3-hydroxypropionate (3HP) cycle and purple bacteria rely on the Calvin cycle. These pathways are discussed separately elsewhere in the chapter (section 27.5). Anoxygenic photosynthesis is considered one of the most ancient pathways for carbon assimilation on Earth. This type of photosynthesis was prevalent on the Earth's surface before the great oxygenation event (GOE). Sulphur-dependent photosynthesis exists even today, but only in some extreme environments. Hot springs and stagnant water are notable sites for the anoxygenic phototrophs.

27.4.1.2 Chemosynthesis

Some bacteria and archaea rely on chemosynthesis for carbon sequestration as primary production. Chemosynthetic pathways were among the earliest metabolic events in the origin of life and still exist in some extreme parts. This type of reaction prevails in the regions where the sunlight does not penetrate. The chemoautotrophic lifestyle is the only mode of

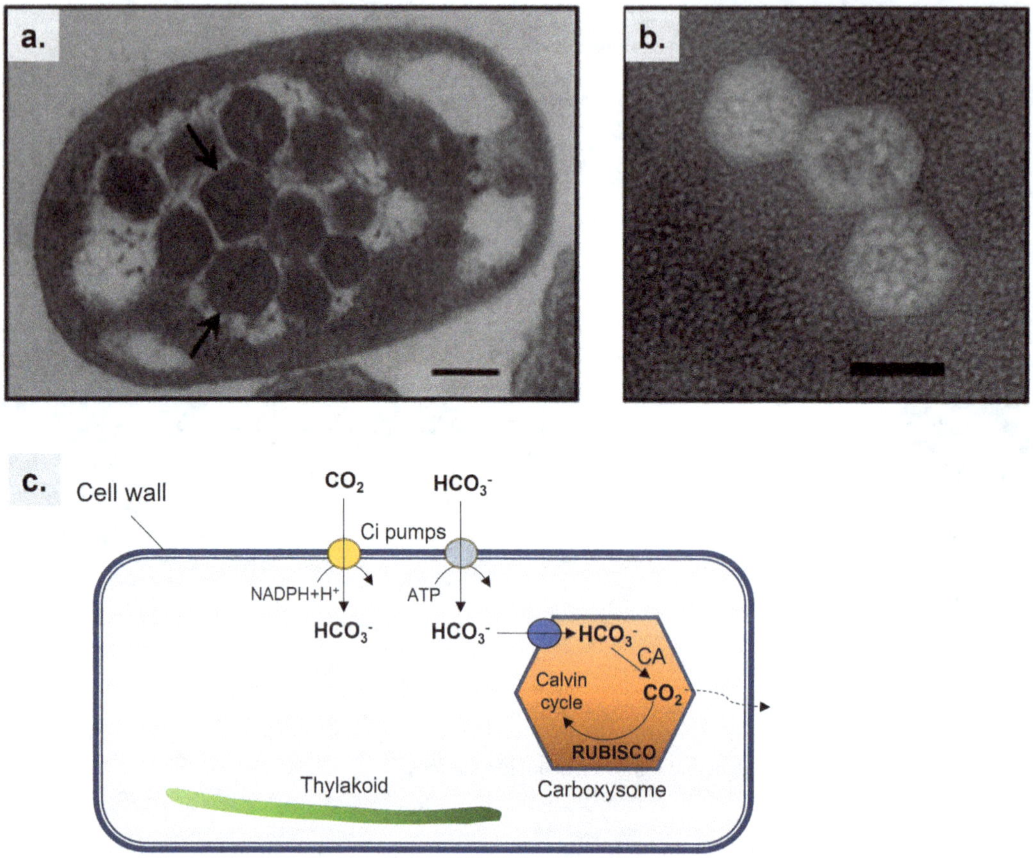

FIGURE. 27.7 Structure and function of carboxysome. a. Transmission electron micrograph (TEM) of a thin section of a wild-type *Halothiobacillus neapolitanus* cell showing multiple carboxysomes (arrows). b. Negatively stained purified carboxysomes. c. A simple illustration of a cyanobacterial cell showing a single carboxysome and its function. (Images a and b adapted from Menon et al. (2008)[4])

primary production in deep hydrothermal vents and below the seabed. The chemotrophic pathways use reduced inorganic (e.g., H_2, H_2S, S(0), NH_4^+, NO_2^-, and Fe^{2+}) and organic compounds (e.g., methane (CH_4) and methanol) to derive energy (Table 27.5). However, the fundamental requirement of CO_2 for carbon fixation is similar to photosynthesis. Examples of chemotrophic bacteria are hydrogen-oxidising bacteria, sulphur-oxidising bacteria, nitrifying bacteria, iron-oxidising bacteria, methane-oxidising bacteria, and carbon monoxide-oxidising bacteria. Archaea such as methanogens and sulphur-reducing archaea also exhibit chemotrophic lifestyles. The carbon fixation pathways in chemotrophs vary across different microbial groups. Bacteria like *Thiobacillus* and *Hydrogenovibrio* employ the Calvin cycle, whereas some chemotrophic archaea rely on a 3HP bicycle for carbon fixation. The reductive acetyl-CoA pathway is unique for a sulphate-reducing bacterium, *Desulfobacterium autotrophicum* and a methanogenic archaeaon like *Methanobacterium* for the same (detailed in section 27.5).

27.4.2 Respiration

Respiration plays a pivotal role in the carbon cycle. It involves a series of chemical reactions to convert organic compounds into CO_2. In biological systems, this process is required to transform the chemical energy into potential energy (i.e., ATP) to meet the energy requirement for metabolic events. The substrates for respiration could be carbohydrates, proteins, and fats, but glucose is the favourite material (Figure 27.8). Respiration pathways consist of numerous steps which generate the chemical diversity of organic compounds in the environment. Microorganisms have tremendous abilities to perform cellular respiration. Depending on the terminal electron acceptors, microbial respiration can be of two types: aerobic and anaerobic.

In aerobic respiration, the breakdown of organic compounds requires O_2. This is an essential process required for cellular activities in all types of higher organisms, such as plants and animals. Aerobic respiration is consistent with many microorganisms, including bacteria and fungi. Aerobic respiration can be divided into four main stages: glycolysis (the Embden-Meyerhof-Parnas (EMP) pathway), oxidative decarboxylation, tricarboxylic acid cycle (the Krebs cycle), and oxidative phosphorylation. Glycolysis is the first stage, where a six-carbon-containing sugar compound breaks down and produces two molecules of pyruvate (a three-carbon compound). This pathway does not incur any loss of carbon from organic forms. Some prokaryotic microorganisms

TABLE 27.5
Chemotrophic reactions in different types of microorganisms with suitable examples

Reactions	Types of microorganisms	Examples of microorganisms
$H_2 + O_2 \rightarrow H_2O$	Hydrogen-oxidising bacteria	*Hydrogenobacter thermophilus* and *Cupriavidus necator*
$H_2S + O_2 \rightarrow SO_4^{2-}$	Sulphur-oxidising bacteria	*Acidithiobacillus ferrooxidans*
$Fe^{2+} + O_2 \rightarrow Fe^{3+}$	Iron-oxidising bacteria	*Acidithiobacillus ferrooxidans*
$NH_4^+ + O_2 \rightarrow NO_2^- + H_2O$	Nitrifying bacteria	*Nitrosomonas*
$NO_2^- + O_2 \rightarrow NO_3^-$	Nitrifying bacteria	*Nitrobacter*
$CO_2 + H_2 \rightarrow CH_4$	Methanogens	*Methanosarcina* and *Methanobacterium*
$CH_4 + SO_4^{2-} \rightarrow H_2S + CO_2$	Syntrophic association of methanotrophs and sulphate-reducing bacteria	-
$H_2S + NO^-_3 \rightarrow SO_4^{2-} + N_2$	Sulphur-oxidising bacteria	*Thiobacillus denitrificans*
$H_2 + Fe^{3+} \rightarrow Fe^{2+}$	Iron-reducing bacteria	*Acidothiobacillus* spp.
$H_2 + SO_4^{2-} \rightarrow H_2S$	Sulphate-reducing bacteria	*Desulfovibrio* and *Desulfobacter*
$NH_4^+ + NO_2^- \rightarrow N_2 + H_2O$	Anammox	*Brocadia* and *Anammoxoglobus*
$CO + H_2O \rightarrow CO_2$	Carbon monoxide-oxidising bacteria	*Pseudomonas carboxydovorans* and *Clostridium thermoaceticum*

possess a different glucose-catabolic pathway, called the Entner-Doudoroff pathway, in place of glycolysis. The next step is oxidative decarboxylation, which initiates the removal of one carbon atom from pyruvate and releases CO_2. The main product of this reaction is a two-carbon compound, acetyl-CoA (AcC). This product enters the Krebs cycle and loses both carbon atoms in two distinct steps through decarboxylation. The complete oxidation of one glucose molecule requires two complete rounds of the oxidative decarboxylation and Krebs cycle (Figure 27.8). Many reactions of these pathways generate energy carriers and electrons that translate into energy (ATP) during oxidative phosphorylation.

A different pathway, the pentose phosphate pathway (PPP), runs parallel to glycolysis and consumes glucose to produce ribose 5-phosphate and NADPH. This pathway provides precursors for the biosynthesis of amino acids and nucleotides and also helps maintain carbon homeostasis. In different situations, some bacteria, fungi, protozoans, and plants employ the glyoxylate cycle to grow on fatty acids or acetate, using them as the sole carbon source for their survival. This pathway shares similarities with the Krebs cycle. The glyoxylate pathway involves anaplerotic reactions that bypass the two steps of the decarboxylation of the Krebs cycle. This pathway prevents the loss of carbon as CO_2 and has roles in carbon conservation and biosynthesis.

Anaerobic respiration involves non-oxygen-based pathways. However, it is less energy efficient than aerobic respiration. It is mainly reported in bacteria and archaea. Anaerobic respiration is the only mode of energy production for organisms in an anaerobic environment. This process shares some similarities with aerobic respiration, including the glycolytic pathway. Aerobic and anaerobic pathways mainly differ in terms of the final electron acceptors. A different type of anaerobic process called fermentation can be present in some bacteria and fungi. Some microorganisms observe strict anaerobiosis, while others live as microaerophiles or facultative anaerobes.

Some strict anaerobic bacteria and archaea follow the Krebs cycle and use low-oxidising agents such as nitrate (NO_3^-), arsenate (As(V)), ferric ion (Fe(III)), SO_4^{2-}, and S^0 as terminal electron acceptors. This type of respiration represents an ancient mode of energy production and still exists in some extreme environments. Examples of bacteria that use inorganic species as terminal electron acceptors for anaerobic respiration are *Geobacter*, *Desulfovibrio*, *Nitrobacter*, *Shewanella*, and *Acidithiobacillus*. Archaea, such as sulphate-reducing and nitrate-reducing archaea, also exhibit similar pathways and utilise sulphate and nitrate as terminal electron acceptors, respectively. This process is coupled with organic carbon decomposition (heterotrophic mode of nutrition) and CO_2 emission. Some microorganisms can even use carbonic substances as terminal electron acceptors. These may include organic compounds like fumarate, dimethyl sulphoxide (DMSO), and organic halogens, as well as inorganic carbon like CO_2 and carbon monoxide (CO).

In fermentation, pyruvate can find different routes to terminate its electrons. This pathway utilises endogenous electron acceptors to complete the oxidation of pyruvate. A few specific fungi (e.g., *Saccharomyces*) and bacteria (e.g., lactic acid bacteria (LAB), including *Lactobacillus* and *Streptococcus*) observe this metabolic route for energy production. These taxa observe a facultative anaerobic lifestyle. This means they prefer to live in an oxic environment, but when they experience an anoxic situation. It allows them to switch from an aerobic to an anaerobic lifestyle. Fermentation procedures are untapped sources of diverse secondary metabolites such as ethanol, lactic acid, acetic acid, formaldehyde, etc. Chapter 4 provides a comprehensive understanding of the fermentation process and its industrial applications.

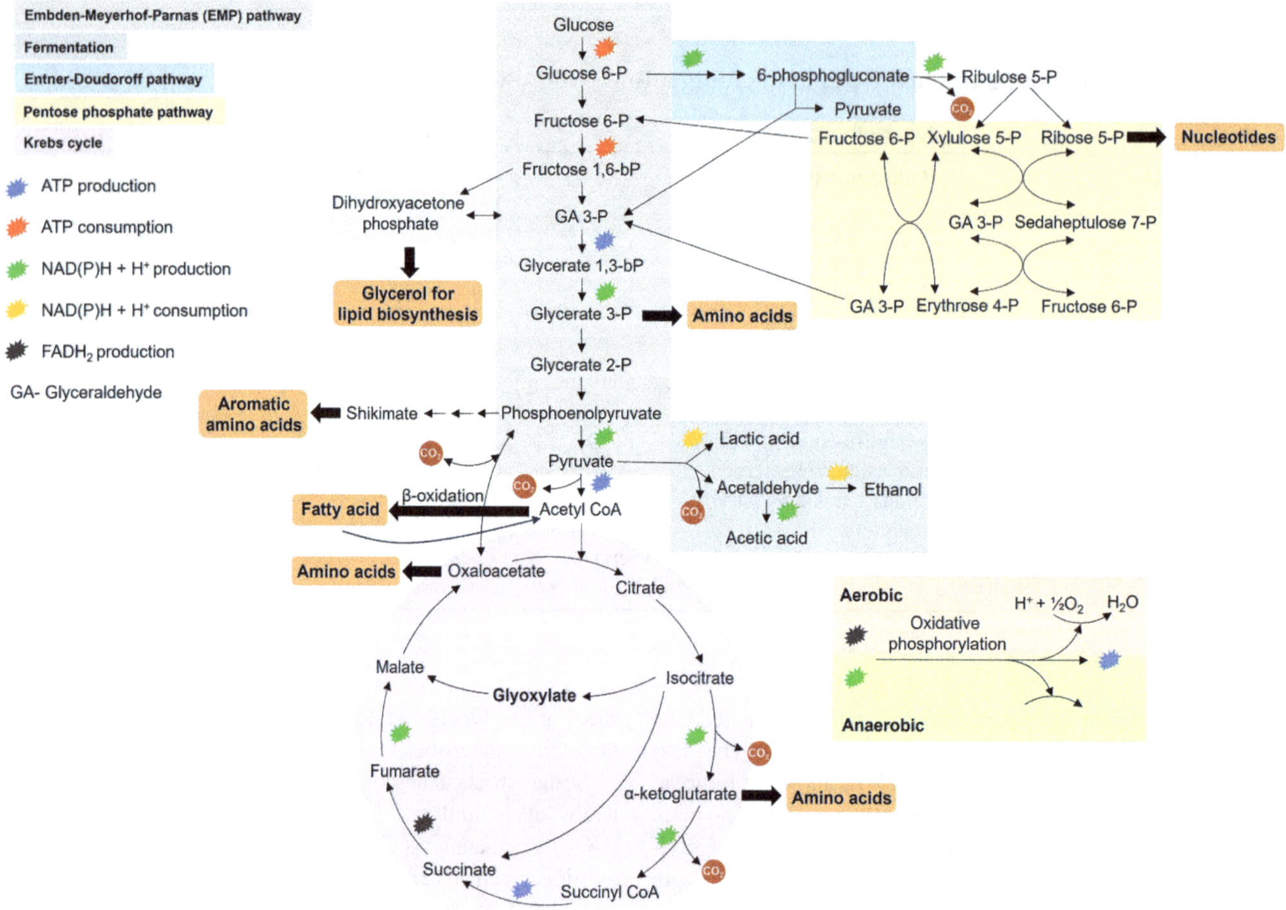

FIGURE 27.8 Central metabolic pathways of bacteria showing production of different organic compounds.

27.4.3 Biotransformation

Biotransformation is a process of chemical modification of a substance into new compounds. Microorganisms can convert a wide range of organic substances, including pollutants, natural products, and synthetic compounds, into diverse products. Biotransformation reactions are the main drivers that produce a wide array of organic compounds in biological systems called metabolites. They can be intermediate or end products of a metabolic event and can be categorised as primary and secondary metabolites. Primary metabolites are essentially required during the complete life cycle of an organism, whereas the production of secondary metabolites occurs under specific conditions. Enzymes are the main catalysts for the biotransformation reactions and play central roles in the metabolism of carbohydrates, fats, and proteins, as well as the biosynthesis of secondary metabolites (Figure 27.8). The enzymes found in respiratory pathways and carbon assimilatory pathways have major roles in transformation reactions. Some chemical substrates of bioreactions could themselves be products of previous microbial transformation.

Biotransformation can involve three different types of reactions: degradation, synthesis, and isomerisation, which are discussed below-

- Degradation is a process of breakdown of organic materials. Hydrolase enzymes such as amylases, proteases, lipases, cellulases, xylanases, tannases, cutinases, glycosidases, and nucleosidases are chief contributors to degradation reactions. Carbohydrases are particularly useful for the breakdown of complex carbohydrates into oligosaccharides or hexose/pentose units by hydrolysing their glycosidic bonds. The oxygenase enzymes desulphurise, dehalogenate, and hydroxylate organic pollutants like aromatic and aliphatic compounds. Laccases oxidise a wide range of natural and synthetic substrates, including phenols, polyphenols, and polycyclic aromatic hydrocarbons (PAHs). A recently discovered microbial enzyme, PETase (microbial source: *Ideonella sakaiensis*), can even degrade plastic.
- In the synthesis process, microorganisms build new compounds from simple substances. For example, the

carbon fixation pathway synthesises glucose using CO_2. The polymerisation of carbohydrates, amino acids, and nucleotides is also an example of a synthesis type of biotransformation. The ligase enzymes are responsible for the synthesis of new bonds, as is possible through DNA ligases. The production of antibiotics can also be included in the synthesis category.

- In isomerisation, organic compounds rearrange their elements and bonds and create new structures with new properties. For example, glucose isomerase transforms glucose and xylose into fructose and xylulose, respectively.

Natural transformation by native microbial communities is a slow, steady, and nonspecific process. Many native microorganisms can be cultured on synthetic media and used for the production of diverse chemicals at industrial scales. This approach has built the empire of industrial biotechnology for organic transformation by using microbial potential. Industrial biotechnology has developed various innovative solutions using microorganisms and their enzymes to reduce carbon footprint and promote sustainability. The transformation reactions help ecosystems to balance their carbon content. They also clean up contaminated sites and assist in wastewater treatment through biodegradation of pollutants.

27.4.4 Decomposition

Decomposition is the process of degradation of dead and decaying organic matter into simple constituents. Decomposition can be of two types: abiotic and biotic. The abiotic methods can be chemical and physical processes. The biotic methods involve the metabolic breakdown of organic matter for microbial heterotrophic nutrition. Large quantities of plant litter, animal carcasses, and their faeces are the main sources of organic matter in the environment. A high amount of synthetic organic substances present as environmental contaminants can also be degraded through the decomposition process.

The decomposition process requires cooperation among various types of organisms, such as invertebrates, protozoa, fungi, bacteria and even some animals. Invertebrates (mainly arthropods) are the primary ecosystem engineers, which act on large, complex organic matter and degrade it into simple compounds. They help with the early breakdown via oral feed and are also termed **detritivores**. Some animals act as scavengers, which digest complex material before detritivores show an action upon them. This process transforms the flesh or decaying plant materials into digestible organic matter. Decomposers feed on the remains of scavengers and detritivores by secreting many hydrolysing enzymes. They can be diverse heterotrophic bacteria, fungi, and protists, and are also termed saprophytes or saprobes. These organisms are the key drivers of the detritus food chain and brown food webs. In forest ecosystems, the fungal saprobes are essential for the decomposition process. Some commonly known fungal saprobes are *Rhizopus*, *Mucor*, *Penicillium*, *Aspergillus*, and *Alternaria*. Mushrooms also play a critical role in the degradation of organic matter. Bacterial decomposers take a huge responsibility for the decomposition of the bulk quantities of organic matter. They are highly diverse and chiefly include *Bacillus*, *Pseudomonas*, *Streptomyces*, and *Cellulomonas*. They have repositories of a wide array of enzymes, which break down various simple and complex organic compounds.

Microbial decomposition of organic matter in nature is a self-regulatory process. Proteins, lipids, carbohydrates, and nucleic acids are the major constituents of organic matter for their biodegradation. In the decomposition process, microorganisms involve numerous pathways for the biodegradation and mineralisation of organic compounds. Oxygen and moisture content are the two crucial requirements for the microbial decomposition of organic matter. The β-oxidation pathway in decomposers leads to the degradation of fatty acids and lipids. This process shortens the long chain of saturated fatty acids by the oxidation of the carbonyl group and yields acetyl-CoA, which can be further metabolised through the Krebs cycle. The complex structure of cellulose, hemicellulose, proteins, and lignin in plant cell walls can be degraded by the diverse microbial enzymes.

Decomposition is completed in five different stages: fragmentation, leaching, catabolism, humification, and mineralisation. Fragmentation is a process of breaking down organic compounds into small pieces. Leaching facilitates the release of water-soluble content. Catabolism involves the enzymatic breakdown of organic substances. Humification is a transformation reaction to produce humus, which shows resistance to further microbial breakdown. Lately, mineralisation favours the complete degradation of organic matter and allows the release of carbon as CO_2 and other elements, such as nitrogen, sulphur, and hydrogen, in their inorganic forms.

Each stage of decomposition harbours a distinct microbial composition. The types of microbial enzymes also vary according to the degree of decomposition. The microbial composition of decomposers can even vary from one ecosystem to another. For instance, the microorganisms present in ocean habitats and their rate of decomposition differ from those found in terrestrial environments. Compared to decomposers in forest settings, those in desert ecosystems are more acclimatised to surviving in low humidity levels. Decomposition is crucial not only for breaking down organic residues and recycling nutrients but also for the flow of energy and soil health. A composting facility is an infrastructure to speed up the decomposition of plant debris to produce compost for agricultural benefits.

A different biological process is anaerobic decomposition, which occurs when O_2 is insufficient or lacking. This is methanogenesis (detailed in section 27.4.5).

27.4.5 Methanogenesis

Methanogenesis is a biological process of CH_4 production in anoxic situations. This process is a means of energy conservation in certain microorganisms called methanogens. They belong to a phylum, Euryarchaeota, from the domain

Archaea. The evolutionary history of methanogens dates back to 2.5 mya before the great oxidation event (GOE). This means this process has played a dominant role in the carbon cycle in ancient times. In the modern world, methanogens can be reported from desert soil, landfill sites, wetlands, rice paddies, deep ocean, and animal guts and tissues. Methanogenesis controls CO_2 emissions in the environment.

Methanogens account for approximately 70% of CH_4 emissions. CH_4 production in the biological system is possible by three different pathways, including acetotrophic methanogenesis, hydrogenotrophic methanogenesis, and methylotrophic methanogenesis (Figure 27.9). In the first pathway, the byproduct of the fermentation reaction, acetic acid, acts as a substrate to produce CH_4 and CO_2. This type of reaction can be present in *Methanosarcina* and *Methanosaeta*. Hydrogenotrophic methanogenesis involves the reduction of CO_2 by a fermentation product, H_2, and produces CH_4. This process is reported in *Methanococcus* and *Methanobacterium*. Methylotrophic methanogenesis utilises methylated compounds such as methylated amines, methylated sulphides, or methanol for CH_4 production. This process prevails in an environment that has high ambient sulphate concentrations. Examples of microbes that have this type of reaction are *Methanosarcina* spp., *Methanomicrobium* spp., and *Methanobacterium* spp.

Dead and decaying matter are the main sources of organic matter, which various microorganisms can degrade to release various substrates for methanogenesis. The complete decay of organic matter by methanogens requires an obligatory mutual relation called **a syntrophic association**. In this process, the cooperation of more than one microorganism falls prey to the organic matter one after the other to complete a metabolic function. The metabolic end-product of a primary metaboliser serves as a substrate for secondary metabolisers (that are methanogens). Various microbial enzymatic, respiratory, and fermentation reactions are the primary syntrophic processes for the degradation of organic matter into its monomers that can be used as methanogenic substrates. Methanogenesis is the terminal step in the anaerobic breakdown, where methanogenic archaea use carbon as the final electron acceptor to produce CH_4.

Methane is a greenhouse gas that has a larger warming potential than CO_2 (28 times more potent), but it can be used as a potential bioresource when produced under controlled conditions. In recent years, methanogenesis has received wide recognition for manufacturing renewable fuel. This process is commercialised for biogas production using organic wastes. Additionally, methanogens are involved in treating domestic and industrial wastes in aerobic digestion. Methane can also be biologically transformed into economically important biomolecules, such as methanol, fatty acids, and biopolymers.

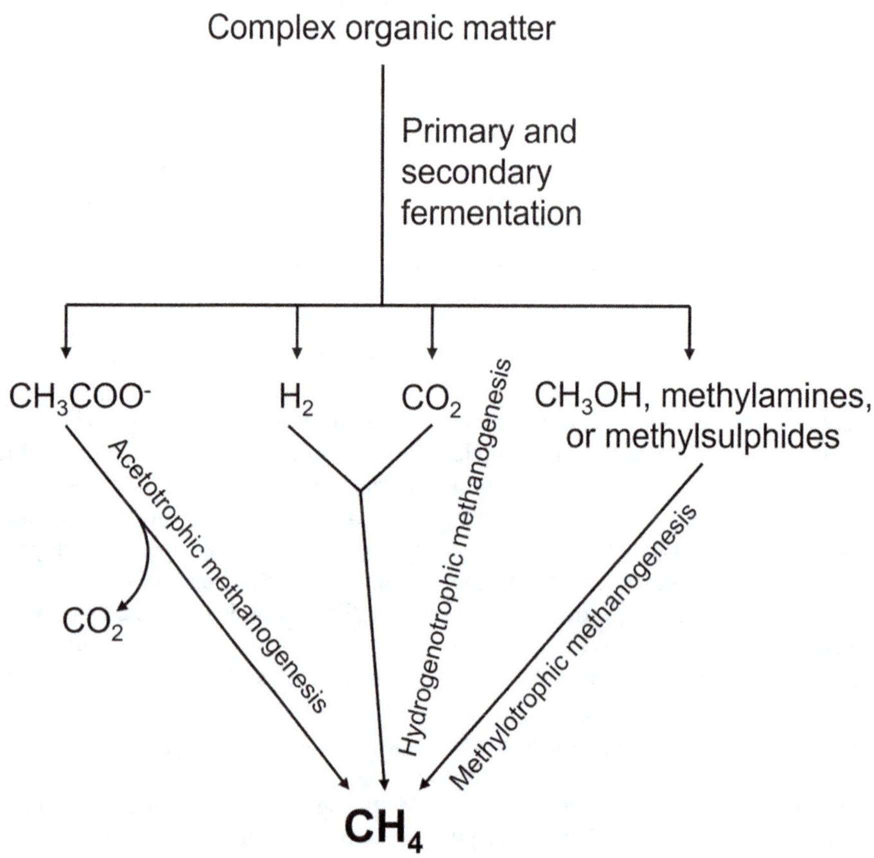

FIGURE 27.9 Different pathways of methanogenesis and the production of methane gas.

27.4.6 Methanotrophy and Methylotrophy

Methanotrophy is a biological process that performs the oxidation of CH_4. Methanotrophic reactions counteract CH_4 emissions and release CO_2 in the carbon cycle. Microorganisms involved in these reactions are termed methanotrophs or methanophiles. They have representatives from both the bacterial and archaeal domains. Methanotrophic reactions are possible in both aerobic and anaerobic situations. In aerobic conditions, the reaction of CH_4 oxidation is coupled with O_2 reduction. Examples of aerobic methane oxidisers are bacteria like *Methylomonas*, *Methylococcus*, and *Methylosinus*. Methane oxidation in anaerobic situations is coupled with sulphate reduction or nitrate/nitrite reduction. The organisms that facilitate such types of reactions are mainly the residents of marine sediments and the seafloor. They can be anaerobic methanotrophic archaea (ANME) such as *Methanobacterium* and *Methanoperedens*. They observe syntrophic interactions with sulphate-reducing bacteria (SRB). The protocooperation between two types of organisms promotes sulphate reduction by receiving electrons from the anaerobic oxidation of CH_4. In some environments, iron or manganese oxides serve as electron acceptors for CH_4 transformation into CO_2 (Figure 27.10).

Methylotrophy is a broad, specialised metabolic capability of microorganisms that use reduced carbon substrates with no carbon-carbon bonds (i.e., single carbon-containing compounds like methanol, formate, CH_4 and multicarbon compounds like dimethyl ether) for carbon and energy. Methanotrophs can be considered a subclass of methylotrophs that expand their cellular growth using CH_4 as their source of carbon and energy.

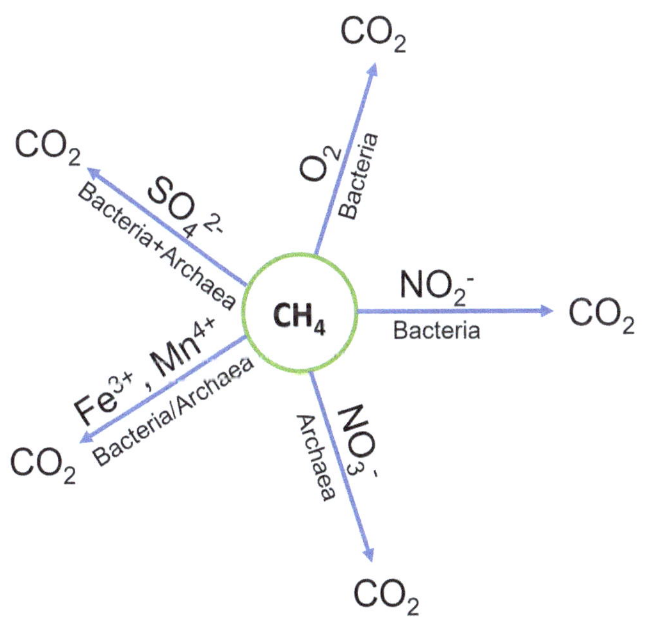

FIGURE 27.10 Methanotrophic reactions by microorganisms under different conditions.

27.5 CARBON FIXATION PATHWAYS

Carbon fixation is a multistep reaction in which atmospheric CO_2 is converted into organic compounds, passing through a series of intermediate compounds. These organics are assimilated into biomass. In microorganisms, as many as seven different carbon fixation pathways have been elucidated (Figure 27.11). These pathways share some common steps and also have differences. The differences in these pathways are primarily due to the specific enzymes involved. These pathways can be reported in different microorganisms and have different energy inputs (Table 27.6). They are briefly discussed as follows.

a. Calvin-Benson-Bassham (CBB) cycle or Calvin cycle
CBB is the most common type of carbon fixation pathway. The most important enzyme is RUBISCO. The overall reaction of CBB consumes six molecules of CO_2 to produce one molecule of glucose using 18 molecules of ATP and 12 molecules of NADPH + H^+. This pathway exists in cyanobacteria, purple bacteria, algae, and plants.

b. Reductive tri-carbonic acid (rTCA) cycle
The rTCA cycle is a variant of the tricarboxylic acid (TCA) cycle, which yields rTCA. Contrary to the TCA cycle, which has a catabolic route, the rTCA cycle is an anabolic pathway and involves the synthesis of organic matter using CO_2. The rTCA cycle consists of four specific enzymes, including pyruvate synthase (PyrS), PEP carboxylase (PEPC), α-ketoglutarate synthase (KGS), and isocitrate dehydrogenase (ICDH). This cycle operates in anaerobic conditions because PyrS and KGS are susceptible to O_2. This process requires 2 NAD(P)H + H^+ and 1 ATP per carbon. Certain archaea, green sulphur bacteria, and microaerophilic bacteria feature carbon fixation through the rTCA cycle.

c. Reductive acetyl-CoA pathway or Wood-Ljungdahl pathway (WLP)
It is the most powerful non-photosynthetic carbon fixation pathway. This pathway enables the reduction of CO_2 into CO and formic acid (HCOOH), which are converted into the formyl group and reduced to a methyl group ($-CH_3$). This intermediate combines with CO and coenzyme A to produce acetyl CoA. This pathway spends only 0.5 ATP and 1.67 NAD(P)H + H^+ per carbon unit. The synthesis of one molecule of pyruvate requires only 1 ATP. Some stages of the WLP are oxygen-sensitive. As a result, the cycle operates exclusively under anaerobic conditions. This type of pathway is found in acetogens (e.g., clostridia), which are anaerobic bacteria. This pathway has tremendous potential in industrial waste valorisation through gas fermentation using syngas and industrial waste gas. Some sulphate-reducing bacteria and methanogens also rely on WLP for carbon sequestration.

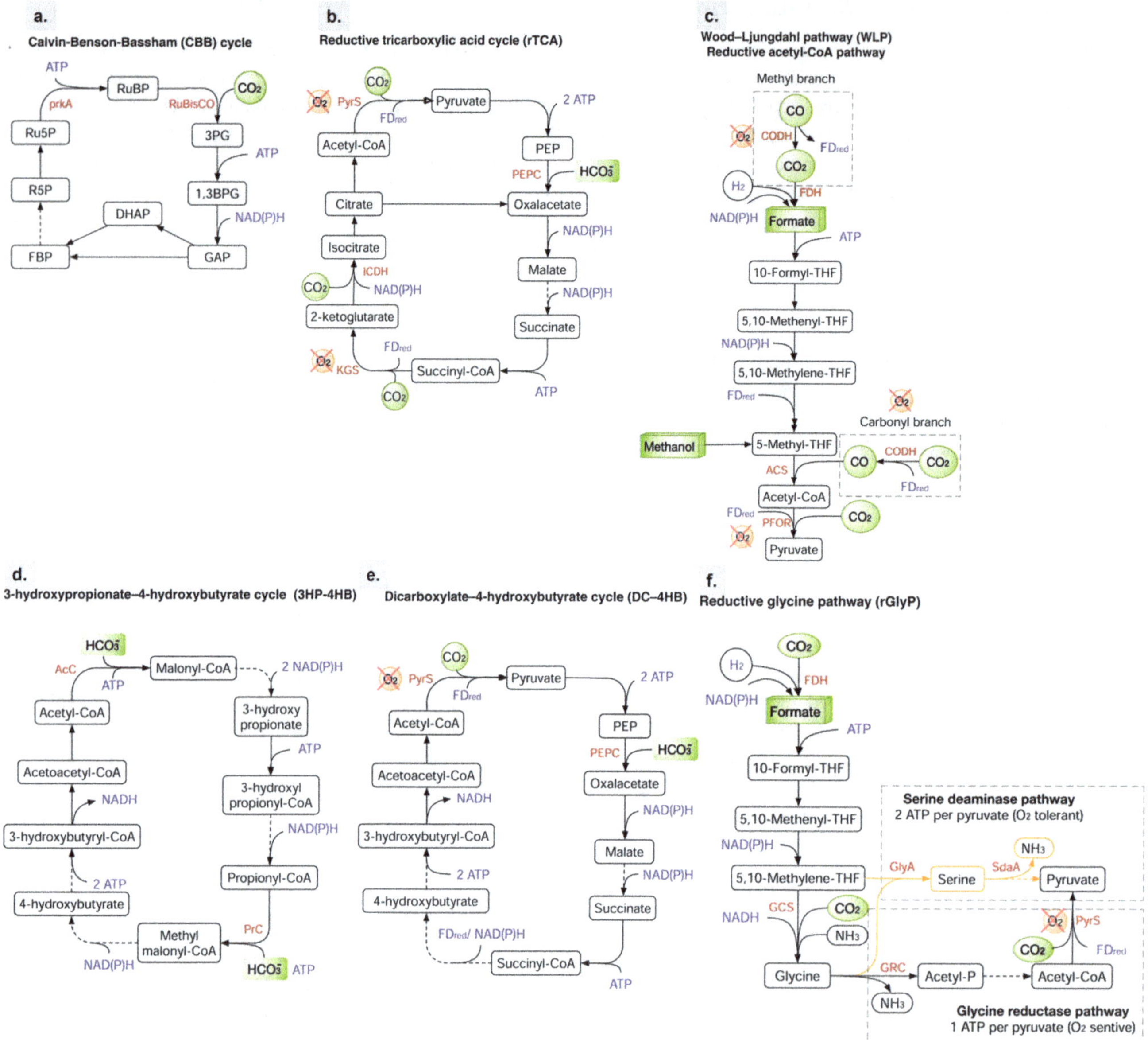

FIGURE 27.11 Different carbon fixation pathways in microorganisms. a. The simplified Calvin-Benson-Bassham (CBB) cycle. b. Reductive tricarboxylic acid cycle. c. Wood-Ljungdahl pathway. d. 3-hydroxypropionate/4-hydroxybutyrate cycle. e. Dicarboxylate/4-hydroxybutyrate cycle, and f. Reductive glycine pathway. RuBisCO, ribulose-1,5-bisphosphate carboxylase/oxygenase; prkA, phosphoribulokinase; RuBP, ribulose-1,5-bisphosphate; Ru5P, ribulose-5-phosphate; 3PG, 3-phosphoglycerate; 1,3BPG, 1,3-diphosphoglycerate; GAP, glyceraldehyde-3-phosphate; DHAP, dihydroxyacetone phosphate; FBP, fructose 1,6-bisphosphatase; R5P, ribose 5-phosphate; ACS, acetyl-CoA synthase; CODH, carbon monoxide dehydrogenase; FDH, ormatedehydrogenase; PFOR, pyruvate:ferredoxin oxidoreductase. Metabolites/cofactors: THF, tetrahydrofolate; PyrS, pyruvate synthase; PEPC, PEP carboxylase; KGS, α-ketoglutarate synthase; ICDH, isocitrate dehydrogenase; PEP, phosphoenolpyruvate. AcC, acetyl-CoA carboxylase; PrC, propionyl-CoA carboxylase; PyrS, pyruvate synthase; PEPC, PEP carboxylase; FDH, formate dehydrogenase; GCS, glycine cleavage/synthase system; GlyA, serine hydroxymethyltransferase; Sda, serine deaminase; GRC, Glycine reductase complex. Cofactors (ATP, NAD(P)H) are in blue and genes/enzymes are in red. (Adapted from Zhang et al. (2022).[5])

d. 3-hydroxypropionate-4-hydroxybutyrate cycle (3HP-4HB cycle)

The 3-hydroxypropionate-4-hydroxybutyrate (3HP-4HB) cycle is an energetically intensive scheme of carbon fixation. Biotin-dependent acetyl-CoA (AcC) and propionyl-CoA carboxylases (PrC) are the two main enzymes in this pathway. They are oxygen-tolerant and fix two molecules of inorganic bicarbonate into a single molecule of acetyl-CoA. The entire process consumes 5 ATP (or 9 ATP in some microorganisms)

TABLE 27.6
Carbon fixation pathways in microorganisms and the required energy input

CO_2 fixation pathways	Microbial groups	ATP/carbon	NAD(P)H + H^+/carbon
Calvin-Benson-Bassham (CBB) cycle	Cyanobacteria, purple bacteria, and microalgae	3	2
Reductive tri-carbonic acid (rTCA) cycle	Green sulphur bacteria	1	2
Wood-Ljungdahl pathway (WLP)	Acetogens	0.5	1.67
3-hydroxypropionate-4-hydroxybutyrate cycle (3HP-4HB cycle)	Autotrophic members of the Sulfolobales	2.5	2
Dicarboxylate-4-hydroxybutyrate cycle (DC-4HB cycle)	A hyperthermophilic archaeon, *Ignicoccus hospitalis*	1.5	2
3-hydroxypropionate bicycle (3HP bicycle)	Green non-sulphur bacteria	1.67	1.67
Reductive glycine pathway (rGlyP)	*Desulfovibrio desulfurican*s	0.5	1.67

and 4 NAD(P)H + H^+ for the synthesis of one acetyl-CoA. This pathway is reported in thermoacidophilic archaea belonging to the autotrophic members of the Sulfolobales.

e. Dicarboxylate-4-hydroxybutyrate cycle (DC-4HB cycle)
The dicarboxylate-4-hydroxybutyrate cycle (DC-4HB cycle) is a distinct autotrophic CO_2 fixation process, and it has less energy requirement compared to the CBB cycle. In this pathway, pyruvate synthase and PEP carboxylase are two recognisable CO_2-fixing enzymes. The cycle utilises reduced ferredoxin as the electron donor and CO_2 as the active inorganic carbon species. Ferredoxin-dependent carboxylases consist of oxygen-sensitive iron-sulphur clusters, and they require strict anaerobic environments. The DC-4HB cycle was possibly a well-integrated process in an iron-sulphur planet, which emerged in an early Earth environment. In the modern world, this cycle is used by a anaerobic, hyperthermophilic archaeon *Ignicoccus hospitalis*.

f. 3-hydroxypropionate bicycle (3HP bicycle)
The 3HP bicycle and the 3HP-4HB cycle share half of their metabolic pathway. They follow the same track from malonyl-CoA to succinyl-CoA, including the two essential CO_2 fixation enzymes, PrC and AcC. The 3HP bicycle is an energetically costly process and consumes 7 ATP for every pyruvate. 3HP bicycle can be reported in a nonsulphur photosynthetic bacterium, *Chloroflexus* and some chemotrophic archaea.

g. Reductive glycine pathway (rGlyP)
The reductive glycine pathway (rGlyP) is nearly identical to the WLP and is another energy-efficient carbon fixation pathway. The presence of oxygen-tolerant enzymes allows rGlyP to operate in aerobic conditions. The sulphate-reducing bacterium *Desulfovibrio desulfurican*s is a natural possessor of the rGlyP.

27.6 ASSOCIATION WITH OTHER ELEMENTAL CYCLES

Organic matter includes carbon-based compounds, typically containing hydrogen and oxygen along with carbon. They often contain nitrogen, sulphur, and phosphorus species. Owing to their various metabolic activities, nitrogen and sulphur cycles are closely associated with the carbon cycle. Carbon and nitrogen cycles are deeply interlinked through processes like assimilation, ammonification, and mineralisation. The sulphur cycle and carbon cycle in the marine ecosystem run side by side for dissimilatory sulphate reduction (DSR). Approximately 50% of organic remineralisation in marine sediment exhibits syntropic association with DSR. The mineralisation and synthesis of organophosphorus compounds show the interconnectedness of carbon and phosphorus in the environment, showing the apparent roles of microbes in their metabolism. The carbon cycle is also closely linked to the oxygen and water cycles. Methanogenesis and methanotrophy are strongly coupled with hydrogen and oxygen. In aerobic respiration, O_2 is a chief oxidant for the metabolism of organic substances. Oxygen also travels along with carbon in CO_2 and in organic matter from assimilation. The water cycle is the key regulator of the carbon cycle that allows for the movement of carbon in different ecosystems.

SUMMARY

- Carbon can form complex structures and long chains and constitutes organic matter. Carbon is highly mobile and changes its forms through the carbon cycle. Based on the time frames for its residence and transport in the reservoirs, the carbon cycle can be of two types: the slow carbon cycle and the fast carbon cycle.
- Microbial groups have an immense contribution to the carbon cycle. Their activities are important regulators of the balance of the carbon cycle and the dynamics of ecosystems.

- In the carbon cycle, carbon sequestration in microorganisms lays the foundation for primary production. This is possible through photosynthesis and chemosynthesis. This helps create building blocks, energy, and nutrients for living beings.
- Cellular respiration is a process of energy production that actively contributes to the carbon cycle. Aerobic respiration uses O_2, while anaerobic respiration recruits other substances to transform organic matter into CO_2. In a different process, fermentation converts organics into simple substances.
- Microbes are transformers and employ different enzymes to transform organic matter into various organic compounds, which is a major driver of chemical diversity. This process also contributes to biodiversity and environmental well-being.
- Decomposition has a central role in recycling organic matter and liberating carbon and other nutrients from dead and decaying bodies. Decomposers like bacteria and fungi break down complex organic compounds into simple substances, followed by the emission of CO_2.
- Methanogenesis and methanotrophy are two notable microbial pathways of the carbon cycle that counterbalance each other. Methanogenesis is a biogenic delivery of CH_4 using CO_2 or other simple organic compounds in anaerobic situations. On the contrary, methanotrophic reactions offset CH_4 emissions and produce CO_2.

DIFFERENT TYPES OF PRACTICE QUESTIONS

Fill in the blanks

I. In the biological system, respiration is the opposite process of …………………
II. Oxygenic photosynthesis is reported in a prokaryotic microbial group …………………..
III. All methanogens belong to only one domain of life, ……………..
IV. In cyanobacteria, the carbon fixation site is the …………………
V. ………………… is a distinct autotrophic CO_2 fixation process reported in the hyperthermophilic archaeon *Ignicoccus hospitalis*.

SOLUTION: I. photosynthesis, II. cyanobacteria, III. Archaea, IV. carboxysomes, V. Dicarboxylate- 4-hydroxybutyrate cycle (DC-4HB cycle)

Match the following

i. Methanogens	a. Rely on light for energy
ii. Methanotrophs	b. Degradation of nitrogen-containing organics
iii. Chemotrophs	c. Use inorganic substances for energy
iv. Photoautotrophs	d. Consume methane
v. Ammonification	e. Release methane

Very short answer type of questions

I. Write the names of any two biological activities responsible for the fast carbon cycle.
II. Define chemosynthesis with suitable examples.
III. Mention the names of any three different carbon fixation pathways.
IV. What is the role of the Wood-Ljungdahl pathway in industrial applications?
V. How is the carbon cycle associated with the sulphur cycle through biological processes?

SOLUTION: i-e, ii- d, iii-c, iv- a, v-b

Short answer type of questions

I. What are the carbon sinks and sources in the environment?
II. What is respiration, and how does microbial respiration play a role in the carbon cycle?
III. Respiration pathways are instrumental in driving chemodiversity in biological systems. Comment on it.
IV. How is methanogenesis closely associated with decomposition?
V. Write the differences between methanogens and methanotrophs with suitable examples.

Long answer type of questions

1. Why is carbon called the backbone of life? Discuss the roles of microorganisms in the carbon cycle.
2. How have human activities influenced the carbon cycle? Discuss the roles of microorganisms in mitigating the effects of carbon imbalance.

NOTES

1 Gt C is gigatonnes carbon, 1Gt C = 10^{12} kg carbon.

2 Hu G, Li Y, Ye C, et al. Engineering microorganisms for enhanced CO_2 sequestration. *Trends in Biotechnology*. 2019;37(5):532–47.
3 Friedlingstein P, O'sullivan M, Jones MW, et al. Global carbon budget 2023. *Earth System Science Data*. 2023;15(12):5301–69.
4 Menon BB, Dou Z, Heinhorst S, Shively JM, Cannon GC. *Halothiobacillus neapolitanus* carboxysomes sequester heterologous and chimeric RubisCO species. PLoS One. 2008;3(10):e3570.
5 Zhang C, Ottenheim C, Weingarten M, Ji L. Microbial utilization of next-generation Feedstocks for the biomanufacturing of value-added chemicals and food ingredients. Frontiers in Bioengineering and Biotechnology. 2022;10:874612.

SUGGESTED READINGS

Bar-On YM, Phillips R, Milo R. The biomass distribution on Earth. *Proceedings of the National Academy of Sciences*. 2018;115(25):6506–11.

Cox PM, Betts RA, Jones CD, Spall SA, Totterdell IJ. Acceleration of global warming due to carbon-cycle feedbacks in a coupled climate model. *Nature*. 2000;408(6809):184–7.

Friedlingstein P, Jones MW, O'Sullivan M, Andrew RM, Bakker DC, Hauck J, Le Quéré C, Peters GP, Peters W, Pongratz J, Sitch S. Global Carbon Budget 2021. *Earth System Science Data Discussions*. 2021:1–91. www.essd.copernicus.org/preprints/essd-2021-386/essd-2021-386.pdf

Guerrero-Cruz S, Vaksmaa A, Horn MA, Niemann H, Pijuan M, Ho A. Methanotrophs: discoveries, environmental relevance, and a perspective on current and future applications. *Frontiers in Microbiology*. 2021;12:1057.

IPCC Report. 2013. Chapter 6: Carbon and Other Biogeochemical Cycles. www.ipcc.ch/report/ar5/wg1; www.climatechange2013.org/images/report/WG1AR5_ALL_FINAL.pdf

Jurtshuk P Jr. Bacterial metabolism. In Baron S (ed). *Medical microbiology*. 1996 (4th ed.). University of Texas Medical Branch at Galveston. Chapter 4. www.ncbi.nlm.nih.gov/books/NBK7919/

Lal R. Carbon sequestration. *Philosophical Transactions of the Royal Society B: Biological Sciences*. 2008;363(1492):815–30.

Nakagawa S, Takai K. Deep-sea vent chemoautotrophs: diversity, biochemistry and ecological significance. *FEMS Microbiology Ecology*. 2008;65(1):1–4.

National Oceanic and Atmospheric Administration. Carbon cycle. 2025. www.noaa.gov/education/resource-collections/climate/carbon-cycle

Zhang C, Ottenheim C, Weingarten M, Ji L. Microbial utilization of next-generation feedstocks for the biomanufacturing of value-added chemicals and food ingredients. *Frontiers in Bioengineering and Biotechnology*. 2022;10:874612.

28 Nitrogen Cycle and Microorganisms

28.1 NITROGEN—AN ESSENTIAL ELEMENT

Nitrogen (N) is a p-block element with atomic number 7 and an atomic mass of 14.0067 u. It exists in a wide range of oxidation states, ranging from -3 to +5. It is the fifth most abundant element in the solar system. All of Earth's compartments contain sufficient quantities of nitrogen. The atmosphere is the most abundant reservoir of nitrogen, containing a total mass of about 3.8×10^{21} g. The lithosphere contains only marginally less nitrogen than the atmosphere. However, the bulk of nitrogen found in the sedimentary and metasedimentary rocks in the lithosphere is highly inactive. A notable amount of nitrogen (2.3×10^{18} g) is present in the hydrosphere, representing a significant but smaller portion than the atmospheric nitrogen. Nitrogen in the environment binds with other elements through various chemical and biological reactions, creating non-living and living materials. Nitrogen-based biological activities are crucial for life. However, such conversions on the primordial Earth were exclusively dependent on abiotic reactions.

In the biosphere, nitrogen is the fourth most abundant element after hydrogen, carbon, and oxygen. The organic forms of nitrogen occupy a central role because they are essential for the functional and structural integrity of organisms. The most common nitrogen-containing macromolecules in biological systems are proteins/peptides, nucleic acids, amino sugars, and energy currency (i.e., adenosine triphosphate (ATP)). The protein frameworks are an enzyme inventory crucial to all biological reactions. They are equally important as structural components and regulators. Nucleic acids are fundamental to life as they code information for biological functions and are passed on heredity. Chitin and peptidoglycans are examples of amino sugars found as the outer covering of many microorganisms. Chitin is composed of repeating β(1,4)-N-acetylglucosamine units and is the second most abundant organic compound on Earth. It is chiefly present in the cell walls of many fungi and the exoskeleton of arthropods. Peptidoglycan is a polymer of N-acetylglucosamine (GlcNAc) and N-acetylmuramic acid (MurNAc) found in the bacterial cell wall, which maintains cellular structure and protects it against various stresses. Some inorganic nitrogen species, such as nitrate (NO_3^-), ammonium (NH_4^+), and nitrite (NO_2^-), act as signalling molecules and growth-regulating factors in plants and microorganisms. In addition, nitrous oxide (N_2O) and ammonia (NH_3) are fundamental for studying the biosignature of life on planetary surfaces, where "biosignature" refers to indicators of past or present life.

28.2 NITROGEN CYCLE

The nitrogen cycle is principal to the global biogeochemical cycle. It exemplifies a cyclic process, where different nitrogen species pass through a series of events, involving various transformation steps. (Figure 28.1). The three characteristics of nitrogen, including the multiple oxidation states, the versatile redox chemistry, and the redox capabilities, ensure huge chemical diversity and drive its mobility. Both abiotic and biotic reactions actively participate in the nitrogen cycle. Some activities are responsible for their transport, while others contribute to their stability in the environment over the long term. The nitrogen cycle facilitates the mobility of nitrogenous compounds across different spheres of the globe. This process ensures the continuous supply of different nitrogen species on Earth.

Earth's atmosphere is composed of approximately 78% nitrogen, which represents the most dynamic reservoir. Nitrogen in the air resides as inert dinitrogen gas (N_2). It is very stable due to a powerful triple bond and, therefore, is inaccessible to biological functions in plants and animals. The breakdown of the triple bond of nitrogen gas is an energy-demanding process.

Earth's geological processes and the global nitrogen cycle demonstrate a complex interplay and are highly interlinked. Volcanism and weathering release nitrogen compounds from the Earth's interior into the atmosphere. The organic nitrogen buried in deep oceans can be sequestered in the sediments. At the air-land interface, NH_3 and NO_3^- react with soil particles, making them temporarily unavailable for plants. Diagenesis and metamorphism are two important geological processes found in sedimentary rocks. In diagenesis, nitrogen-containing sediments are compacted and chemically altered over time to turn into rock. In metamorphism, the hydrothermal process transfers nitrogen from sedimentary rocks to metamorphic rocks. Nitrogen in the subduction zones is transported downward in the mantle zone and upward in the Earth's crust.

In biology, particularly, microbially derived processes are the important drivers for the transformation and mobility of

DOI: 10.1201/9781032651095-35

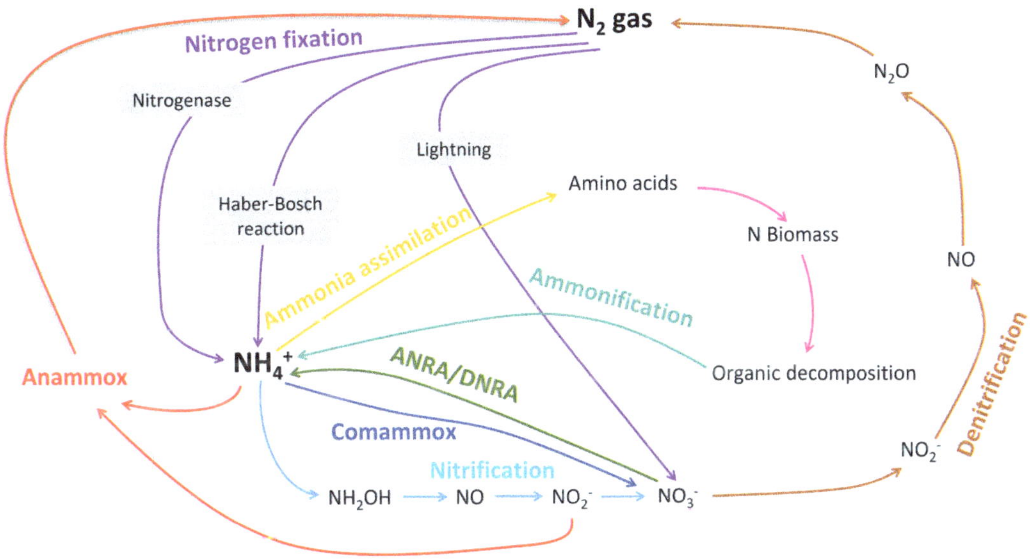

FIGURE 28.1 An illustration of the nitrogen cycle. Arrows depict the direction of nitrogen flow between different species. Colours are codes for different processes such as nitrogen fixation, nitrification, denitrification, anammox, comammox, assimilatory nitrate reduction to ammonia/dissimilatory nitrate reduction to ammonia (ANRA/DNRA), ammonia assimilation, and ammonification.

nitrogen species. As many as fourteen discrete redox reactions are responsible for the existence of different inorganic nitrogen species (e.g., N_2, NO_3^-, NO_2^-, NO, N_2O, NH_2OH, N_2H_4, and NH_4^+) in the biological systems. These reactions are catalysed by distinct nitrogen-converting enzymes (Figure 28.2). Many other elemental cycles (such as the carbon and hydrogen cycles) are also closely tied to the nitrogen cycle through redox and assimilation reactions. The biogeochemical processes of the nitrogen cycle can be divided into seven distinct nitrogen-transforming steps, including nitrogen fixation, nitrification, ammonia assimilation, nitrate reduction, depolymerisation and ammonification, denitrification, and anammox (Table 28.1). They are discussed in detail as follows.

28.2.1 Nitrogen Fixation

Nitrogen fixation is the process that involves the dissociation of the triple bond of nitrogen gas and its reaction with either hydrogen (H_2) or oxygen (O_2) to form NH_3 or different oxides of nitrogen (NO_x) (e.g., NO_2 or NO), respectively. Nitrogen fixation occurs through three different processes, including physical activity (by lightning), chemical synthesis, and biological fixation. Among all these processes, biological fixation and chemical synthesis involve a similar reduction reaction, whereas lightning facilitates oxidation reactions. The lightning in the atmosphere is a rare event for fixing nitrogen. A lightning bolt is a storehouse of extremely high temperatures and pressures, allowing the transformation of atmospheric nitrogen gas into NO_x in the air. It has a minuscule contribution compared to other processes. Industrial and biological methods of nitrogen fixation are two independent but crucial processes for the agricultural system. Industrial nitrogen fixation mainly relies on the Haber-Bosch process for large-scale chemical synthesis (detailed in Box 28.1). In the biological world, the process of nitrogen fixation is catalysed by an enzyme complex, and the process is called **biological nitrogen fixation** (BNF). Section 28.3 provides an overview of the biochemical process and the microorganisms involved in BNF.

The annual budget of nitrogen from all sources of nitrogen fixation is approximately 413 teragrams (Tg) N/yr. Lightning is responsible for only ~5 Tg N/yr. This process has helped maintain the oxidising capacity of the atmosphere[1]. BNF and lighting together account for a yearly budget of 203 Tg N. They were the only sources of readily available nitrogen in the environment prior to the Industrial Revolution. Nitrogen in the oceanic and terrestrial environments is mainly fixed through BNF. However, the annual budget of fixed nitrogen in the ocean (140 Tg N/yr) is much more than that of the terrestrial ecosystem (58 Tg N/yr). Approximately the same amount of nitrogen in the terrestrial ecosystem (60 Tg N/yr) is produced using biofertilisers for agriculture. The enormous use of chemical fertilisers and ammonification has increased the global nitrogen budget on Earth. Chemical synthesis contributes more than half (210 Tg N/yr) of the overall production (Table 28.2).

28.2.2 Nitrification

Nitrification is the microbially-mediated oxidation of NH_3 into NO_2^- and then NO_2^- into NO_3^- under oxic conditions. Ammonia for these sequential reactions comes from processes like nitrogen fixation and ammonification. Nitrification reactions are carried out by specific organisms called nitrifying microorganisms. They observe chemoautotrophic metabolism, and the nitrification process is coupled to ATP synthesis. These microorganisms are ubiquitous in soil, freshwater, and the ocean. The nitrification process is significantly influenced

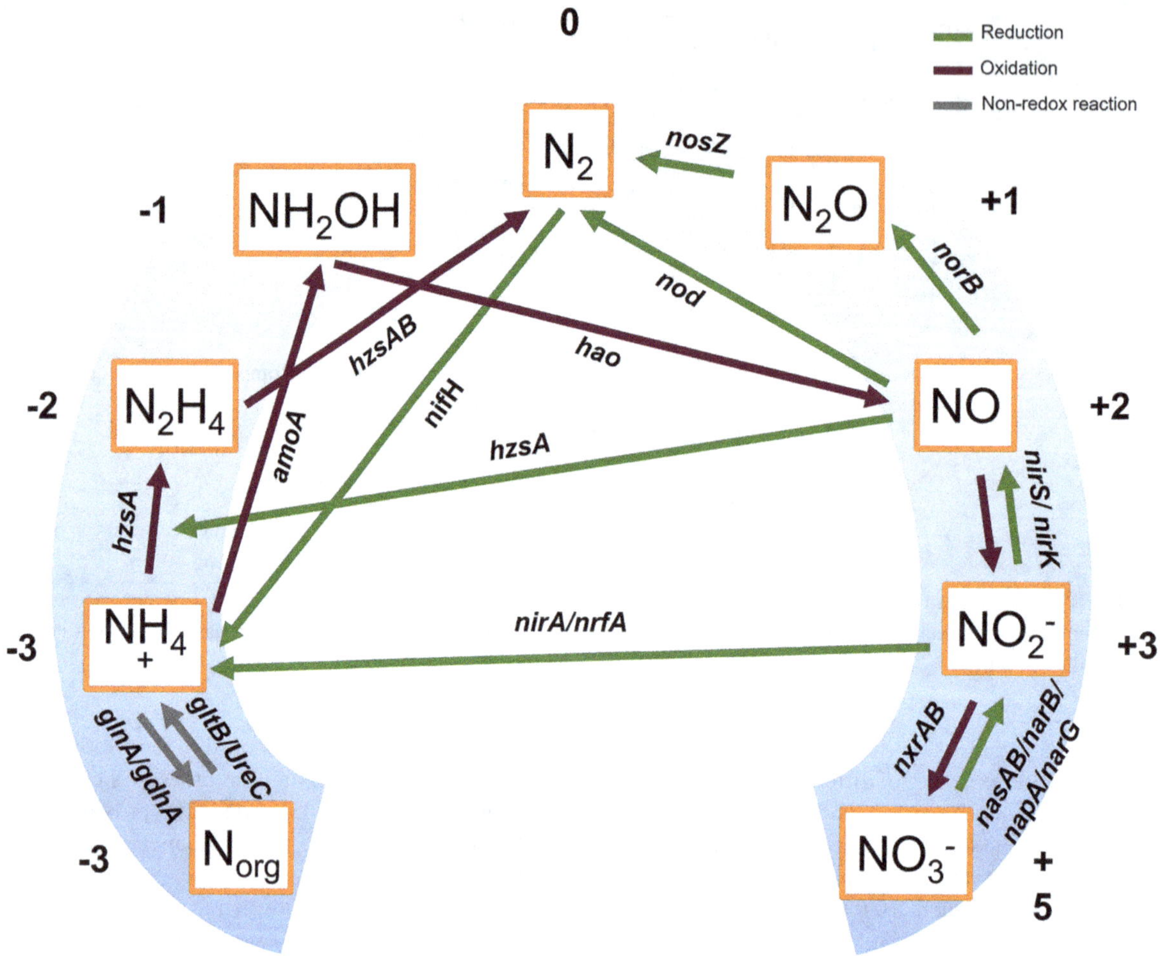

FIGURE 28.2 Redox and non-redox reactions of various nitrogen species in the biological system. Nitrogen species involved in the biological reactions are N_2 (dinitrogen), NH_2OH (hydroxylamine), N_2H_4 (hydrazine), NH_4^+ (ammonium), N_2O (nitrous oxide), NO (nitric oxide), NO_2^- (nitrite), and NO_3^- (nitrate). The numbers mentioned near different nitrogen species correspond to their oxidation numbers. Gene determinants labels for different reactions produce different enzymes: *nosZ* (nitrous oxide reductase), *norB* (nitric oxide reductase), *nirS/nirK* (nitrite reductase), *nxrAB* (nitrite reductase), *nasAB/narB/napA/narG* (nitrate reductase), *hzsA/hzsAB* (hydrazine synthase), *nirA/nrfA* (nitrite reductase), *hao* (hydroxylamine oxidoreductase), *amoA* (ammonia monooxygenase), *nifH* (nitrogenase), *glnA/gdhA* (glutamine synthetase/ glutamate dehydrogenase), and *gltB/UreC* (glutamate synthase).

by various environmental factors such as nitrogen and oxygen (O_2) concentrations, moisture, pH, temporal dynamics, and microbial community compositions (MCCs).

The first step of NH_3 oxidation produces an intermediate compound, hydroxylamine (NH_2OH), followed by a stable product, NO_2^-. These reactions are catalysed by enzymes ammonia monooxygenase (AMO) and hydroxylamine dehydrogenase (HAO), respectively. This type of metabolism is reported in ammonia-oxidising bacteria (AOB) and ammonia-oxidising archaea (AOA). AOB belong to members of the phylum Proteobacteria, such as *Nitrosomonas*, *Nitrosospira*, and *Nitrosococcus*. Examples of AOA are *Nitrosopumilus*, *Nitrososphaera*, and *Nitrosoarchaeum*. Both AOB and AOA are slow-growing organisms because a high proportion of energy produced from NH_3 oxidation is diverted into carbon fixation. This is the main reason why NH_3 oxidation is a rate-limiting step[2] of nitrification. Ammonia oxidation is also responsible for the emission of N_2O due to some side reactions.

In the second step of nitrification, NO_2^- is oxidised into NO_3^- by nitrite oxidising bacteria (NOB). An enzyme, nitrite oxidoreductase, is the main player in this oxidation reaction. NOB belong to a wide range of phyla, including Nitrospirae, Nitrospinae, Proteobacteria, and Chloroflexi. Examples of those include *Nitrobacter*, *Nitrospira*, and *Nitrococcus*. The final product of the nitrification process, NO_3^-, is a relatively stable form, and plants can readily absorb this ion as a nutrient.

In a different process, both steps of nitrification can be completed by a single bacterium, and this process is known as **comammox** (Complete Ammonia Oxidation). A bacterium, *Nitrospira*, belonging to the phylum Nitrospirae, is the only representative of this capability, encoding genes for NH_3 oxidation and NO_2^- oxidation. This bacterium has a

TABLE 28.1
A list of different biological steps of the nitrogen cycle showing processes, enzymes and responsible microorganisms

Steps	Processes	Enzymes	Organisms
Nitrogen fixation	Reduction of dinitrogen gas into ammonia	Nitrogenase	Diazotrophs
Nitrification	Oxidation of ammonia to nitrite and nitrite to nitrate	Ammonia monooxygenase, hydroxylamine oxidoreductase, and nitrite oxidoreductase	"nitroso" and "nitro" bacteria, including Comammox
Ammonia assimilation	Integration of ammonia into organic compounds	Glutamine synthetase (GS) and glutamate synthase (GOGAT)	Bacteria, archaea. fungi, and plants
Assimilatory/Dissimilatory nitrate reduction to ammonia (ANRA/ DNRA)	ANRA: Reduction of nitrate into nitrite and nitrite into ammonia using dehydrogenase or ferredoxin as an electron donor under aerobic and reduced nitrogen-limiting conditions	Cytoplasmic siroheme-containing nitrite reductase (NAR) and cytoplasmic assimilatory nitrite reductase (NIR)	Bacteria, fungi and plants
	DNRA: Reduction of nitrate into nitrite and then nitrite into ammonia under oxygen-limiting conditions	Periplasmic nitrate reductase (NAP) and cytochrome c nitrite reductase (ccNIR)	heterotrophic prokaryotes, diatoms, and fungi
Mineralisation (Ammonification)	Conversion of organic nitrogen (R–NH_2) into ammonia	Various enzymes of decomposition	Ammonifying bacteria and fungi
	Urea hydrolysis	Urease	All types of life forms
Denitrification	Reduction of nitrate into dinitrogen gas through a series of reductive steps under oxygen-limiting conditions	Nitrate reductase, nitrite reductase, nitric oxide reductase, and nitrous oxide reductase	Bacteria, archaea, and fungi
Anaerobic ammonium oxidation (Anammox)	Anoxic conversion of ammonia and nitrite into dinitrogen gas and water	Nitrite reductase, hydrazine synthase, and hydrazine dehydrogenase	Five bacterial genera belonging to the order Planctomycetales

BOX 28.1 CHEMICAL SYNTHESIS OF AMMONIA AND ITS IMPACT

The discovery of artificial nitrogen fixation was a paradigm shift for its industrial-level production at the beginning of the 20th century. A German patriot, Fritz Haber, was the original inventor of this process and filed a patent on 13 October 1908, identifying the process as a high-pressure chemical engineering process. The process was not noticed as a breakthrough until Carl Bosch recognised its production on an industrial scale. This led to the building of city-sized factories to produce NH_3. Both Haber and Bosch received Nobel Prizes in Chemistry for their contributions to ammonia synthesis in 1918 and 1931, respectively. The process got the name Haber-Bosch process. This artificial synthesis involved two gases, H_2 and N_2 and produced NH_3 in the presence of a catalyst, Fe, and a promoter, Mo. The reaction required very high pressure and temperature.

The main motivation for industrial-scale production was to fill the growing demand for food when global starvation was about to become a reality. A different reason for its large-scale synthesis was to transform it into various reactive nitrogen species for their use as raw materials in explosives during World War I. Though the chemical synthesis of nitrogen has fed billions of people, its intense applications in ammunition have led to the deaths of millions of people in forthcoming wars. Due to these reasons, nitrogenous compounds in the environment have accumulated at a rapid pace in the last two centuries.

In recent years, the total production of nitrogenous compounds has been around 210 Tg N/yr, of which more than 70-80% is employed in the production of N-based fertilisers to meet the needs of approximately half of the global food supply. The consumption rate of nitrogen fertilisers is even projected to trend upward in the coming years to fulfil the demand of the increased population. Nitrogen fertilisers in agriculture tend to be absorbed into plants with low efficiency (40–50%), resulting in an accumulation of enormous amounts of reactive nitrogen into the environment. The excess release of nitrogen through chemical fertilisers has raised many environmental concerns. The intensified nitrogen-related activities have disrupted the global nitrogen cycle. The major concern is their contamination of soil and water bodies. Nitrogen-containing residues move downstream in rivers and run into the ocean. Atmospheric deposition contributes some fixed nitrogen to the land and oceans. Greenhouse gas emissions, soil acidification, land degradation, depletion of non-renewable resources, and eutrophication are some notable problems associated with nitrogen contamination.

TABLE 28.2
Different sources of nitrogen fixation and their contribution at a global scale

Sources	Amount (in Tg N/yr)
Terrestrial BNF*	58
Marine BNF	140
Lightning fixation of nitrogen	5
Total global natural annual sources of N_r **	203
Haber Bosch process	210

* BNF is biological nitrogen fixation
** Nr is a reactive nitrogen

slow growth rate but has a high affinity to NH_3 and a high energy yield.

28.2.3 Ammonia Assimilation

Ammonia assimilation is the biological conversion of NH_3 into organic nitrogen. Various life forms, such as bacteria, archaea, fungi, and plants, can participate in ammonia assimilation. This biosynthetic pathway employs two sequential reactions involving the enzymes glutamine synthetase (GS) and glutamate synthase (GOGAT). The main substrate for the reaction is 2-oxoglutarate and requires the reducing power (NAD(P)H + H^+ or ferredoxin (Fd)) and energy (ATP). Ammonia assimilation produces two important amino acids: glutamine and glutamate. They serve as important precursors for the synthesis of other amino acids (e.g., alanine, aspartate, and asparagine) and other organic nitrogen (e.g., amino sugars). Ammonia assimilation in microorganisms is dominant in oxygen-limited environments such as wetlands and saturated and submerged areas (i.e., paddy fields). Ammonia is a highly toxic species, and therefore, it is rapidly assimilated. The incorporation of inorganic nitrogen species in organic residues through ammonia assimilation helps mitigate the toxic effects of nitrogen. This process also improves soil health, reduces nitrogen pollution, and enhances overall ecosystem resilience. Heterotrophic lifeforms such as wildlife and humans are highly dependent on plants and microorganisms for these nitrogen sources, owing to their ability to assimilate ammonia.

28.2.4 Nitrate Reduction

Nitrate reduction can be categorised into two different processes: assimilatory nitrate reduction to ammonia (ANRA) and dissimilatory nitrate reduction to ammonia (DNRA). ANRA can be reported in a wide range of organisms, including bacteria, fungi, algae, and plants, where NO_3^- can be assimilated into the biomass of the organisms. In bacteria, NO_3^- is converted into NO_2^- using an enzyme, cytoplasmic siroheme-containing nitrate reductase (NAR), which is further transformed into NH_3 using an enzyme, ferredoxin-dependent nitrite reductase (NIR). These enzymes are sensitive to high concentrations of O_2. The process of ANRA is followed by ammonia assimilation, which allows the incorporation of nitrogen into the structural proteins and enzymes. When the environment has high NH_3 concentrations, most heterotrophic microorganisms prefer ammonia assimilation rather than ANRA because the conversion of NO_3^- into NH_3 is an energy-intensive process. On the other hand, plants facilitate ANRA because they are photosynthetic bodies, and the energy shortfall does not apply to them.

DNRA is a type of nitrate respiration that allows organisms to generate energy. This process represents a chemoheterotrophic lifestyle and is primarily found in the bacterial domain. Some well-known DNRA-containing bacteria are *Beggiatoa*, *Thioploca*, and *Shewanella*. This process is coupled with the oxidation of organic matter that releases electrons, and nitrate acts as the terminal electron acceptor. DNRA involves a two-step reaction. An enzyme, periplasmic nitrate reductase (NAP), mediates nitrate reduction into NO_2^-, and cytochrome c nitrite reductase (ccNIR) catalyses NO_2^- reduction into NH_3. The full reduction of NO_3^- into NH_3 utilises a total of eight electrons. The overall process generates a proton motive force across the membrane and yields energy for microbial metabolism. Some bacterial representatives can participate in both denitrification and DNRA, depending on the prevailing conditions in the environment. They can switch between the two distinct nitrate reduction pathways. Generally, denitrification is dominant over DNRA, but the latter takes primacy when the environment has a low availability of NO_3^- or a high availability of labile carbon. Though DNRA occupies less importance in terms of quantitative contribution, it can still be crucial to maintain nitrogen in an active form.

28.2.5 Depolymerisation and Ammonification

Depolymerisation and ammonification are two different processes of decomposition, where complex dissolved organic nitrogenous compounds (having an amine/amide group) are converted into simple compounds. Depolymerisation is considered a rate-limiting step of the nitrogen cycle, whereas the ammonification process depends on the efficiency of depolymerisation. These reactions are mainly driven by decomposers involving various enzymes in heterotrophic bacteria and fungi. Some complex biomolecules require physical disintegration before decomposition. This process facilitates the fragmentation of large particles, making them accessible to biological reactions. Depolymerisation mainly involves the digestion of excretes or dead and decaying materials, allowing the breakdown of macromolecules like proteins, nucleic acids, chitin, and peptidoglycan. Enzymes like proteases, nucleases, and chitinases actively participate in depolymerisation reactions that degrade proteins, nucleic acids, and chitin, respectively. These reactions produce compounds like amino acids, purines, pyrimidines, and amino sugars.

These low molecular weight substances are subsequently deaminated using some microbial routes and yield NH_3.

This process is termed **ammonification**. Some bacteria have taken this responsibility, and they are called ammonifying bacteria. They can be reported in diverse habitats, including soil, sediments, and aquatic ecosystems. Some prominent examples of ammonifying bacteria are *Bacillus vulgaris*, *Bacillus ramosus*, *Clostridium*, *Pseudomonas*, and *Streptomyces*. Both aerobic and anaerobic situations can facilitate ammonification. The microbial enzymes responsible for ammonification are diverse and depend on specific organic nitrogen compounds. Aminopeptidases and carboxypeptidases remove the N-terminal and C-terminal amino groups (-NH_2) from a peptide chain, respectively. In addition, enzymes like deaminases and amidases act on peptides and cleave amide and amino groups, respectively. A different example of an ammonification reaction is the hydrolysis of urea into NH_3 and carbon dioxide (CO_2) in the presence of the enzyme urease. Ammonification is a crucial step in nitrogen mineralisation and generates an NH_3 pool. Ammonia regenerated by the ammonification process can be consumed in ammonia assimilation and nitrification.

28.2.6 Denitrification

Denitrification is a microbial-mediated multistep reduction process where reactive nitrogen species like NO_3^- and NO_2^- are converted into an inert nitrogen gas. This process represents a mode of microbial respiration, which occurs in conditions when oxygen is depleted. It involves a sequential conversion, reducing NO3- to NO_2^- followed by the synthesis of nitric oxide (NO), N_2O, and ultimately N_2. The enzyme inventory for the respective reductive steps is nitrate reductase (NAR), nitrite reductase (NIR), nitric oxide reductase (NOR), and nitrous oxide reductase (NOS). All-inclusive enzymes are localised in the cellular membranes and regulated by the oxygen gradient. A different standard condition required for the reactions is the availability of organic carbon for the oxidation reaction.

Denitrification is a primary mechanism for nitrogen loss that is essential to maintaining the budget of fixed nitrogen in the biogeochemical cycle. However, excess nitrogen loss causes fertility impoverishment in the fields. An intermediate denitrification step emits a greenhouse gas, N_2O, which is also an ozone layer depletion agent.

Denitrifying microorganisms are called denitrifiers. They prevail in a variety of environments, such as soil, wetland, groundwater, freshwater, ocean, and sea sediments. They mainly belong to bacteria and archaea. Most denitrifying bacteria are facultative anaerobes and utilise NO_3^- as a terminal electron acceptor under microoxic or anoxic conditions. This process is coupled with energy conservation (i.e., ATP production) for bacterial metabolism. Typical examples of denitrifying bacteria are *Pseudomonas*, *Bacillus*, *Micrococcus denitrificans*, and *Alcaligenes*. A denitrifying bacterium, *Thiobacillus denitrificans*, observes a strict chemolithoautotrophic lifestyle, which can be employed in engineered water treatment systems for NO_3^- removal. A nitrogen-fixing soil bacterium, *Bradyrhizobium japonicum*, has also demonstrated denitrification traits when grown in the presence of NO_3^- under anoxic situations. Some strains of *Paracoccus denitrificans* and *Thiosphaera pantotropha* can perform denitrification in even oxic conditions. Examples of denitrifying archaea are *Haloferax*, *Haloarcula*, and *Natrinema*, which have distinct canonical genes for denitrification. Recent evidence claimed that even some fungal species, such as *Fusarium oxysporum* and *Cylindrocarpon tonkinense*, also have the potential for denitrification.

28.2.7 Anammox

Anammox stands for Anaerobic Ammonium Oxidation, which involves the direct conversion of NH_3 and NO_2^- into N_2 and water. The current estimates suggest that anammox could be responsible for up to 67% loss of fixed nitrogen from oceans. This operation runs under anaerobic situations and utilises NO_2^- as an oxidant for the reduction of NH_3. The entire process involves three different enzymes, including NIR, hydrazine synthase, and hydrazine dehydrogenase. This type of metabolism is reported in a single order of bacteria, Planctomycetales, with only five bacterial representatives: *Brocadia*, *Kuenenia*, *Anammoxoglobus*, *Jettenia*, and *Scalindua*. These microorganisms have unique morphological and physiological capabilities. The anammox reaction occurs in a unique membrane-bound compartment called an **anammoxosome** (Figure 28.3). The membrane of an anammoxosome is characterised by the presence of ladderane moieties, which consist of two or more fused cyclobutane rings. The cell walls of these bacteria are devoid of peptidoglycan. These bacteria produce a toxic intermediate, hydrazine (N_2H_4), during the catabolic reaction. Given the bioprocess engineering practices, anammox can be employed as a cost-efficient, sustainable treatment solution for nitrogen removal from wastewater. One significant advantage of nitrogen removal through anammox is that it does not release N_2O.

28.3 BIOLOGICAL NITROGEN FIXATION (BNF)

More than half of the global supply of fixed nitrogen in the natural environment comes from BNF. In the evolutionary sequence, this process appeared before the Great Oxygenation Event (GOE). Nitrogen fixation back on primordial Earth was a strict requirement to meet the need for active nitrogen forms. The origin of BNF was essential for the evolution, establishment, and growth of diverse life forms. BNF emerged with the advent of nitrogen-fixing microorganisms. These organisms further evolved and diversified into a wide range of forms. They are further discussed in the next section (section 28.4).

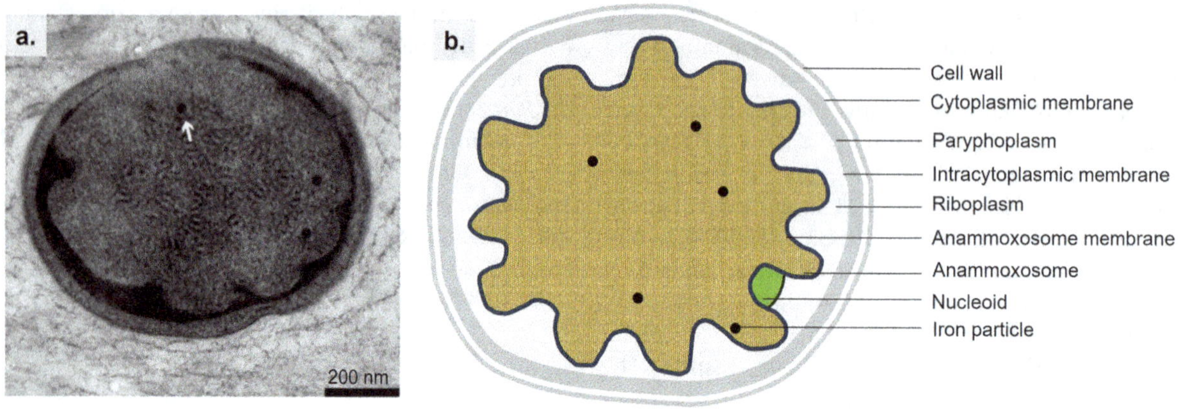

FIGURE 28.3 Structure of anammoxosome. a. Transmission electron micrograph of an anammox bacterium showing an anammoxosome. The arrow represents an iron particle inside an anammoxosome. b. A simple diagram. (Image adapted from van Niftrik et al. (2008)[3] with permission from Elsevier.)

28.4 DIAZOTROPHS

Only certain bacteria and archaea have the ability for nitrogen fixation and are particularly termed **'diazotrophs'**. The discovery of diazotrophs is attributed to S. Winogradsky for the report of *Clostridium pasteurianum* in 1893 and to M.W. Beijerinck for *Azotobacter chroococcum* in 1901. Nitrogen-fixing microorganisms are widely distributed on land and aquatic ecosystems (Figure 28.4). Recent research reports suggest that nitrogen-fixing organisms in natural communities play a keystone role in ecosystem function. Nitrogen-fixing properties in microorganisms have profound implications for ecosystem functioning, agriculture and the global nitrogen cycle. Figure 28.5 provides information on different types of diazotrophs based on microbial groups and functional attributes.

In bacteria, both Gram-positive and Gram-negative types can have the properties of nitrogen fixation. They can be obligate aerobes, microaerophiles, and facultative/obligate anaerobes. Nitrogen-fixing bacteria belong to a wide range of groups, such as Cyanobacteria, Proteobacteria, Actinobacteria, Firmicutes, Chlorobiales (green sulphur bacteria), Chromatiales, and Spirochetes. In contrast, nitrogen-fixing archaea are relatively less recognised for nitrogen fixation. Diazotrophic archaea are restricted to a few methanogens (Euryarchaeota). They show evolutionary relatedness with nitrogen-fixing bacteria and rely on the same mechanism for nitrogen fixation. They can be found in subtropical wetlands and rice fields. They also reside in soil and sea sediments. Some well-known representatives of nitrogen-fixing methanogens are *Methanococcus maripaludis*, *Methanococcus thermolithotrophicus*, and *Methanosarcina barkeri*.

Based on their association with other organisms, diazotrophs can be divided into two groups: non-symbiotic diazotrophs and symbiotic diazotrophs. Nonsymbiotic diazotrophs do not require hosts for nitrogen fixation. In symbiotic association, diazotrophs establish obligate or facultative relationships with other organisms. These two categories are further discussed as follows.

28.4.1 Non-symbiotic Diazotrophs

Non-symbiotic diazotrophs are free-living organisms and can fix nitrogen independently. These microorganisms can be heterotrophs and autotrophs.

- Heterotrophic diazotrophs are potential contributors to global bioavailable nitrogen. Examples of heterotrophic diazotrophs are *Azotobacter*, *Azospirillum*, *Klebsiella*, *Burkholderia*, *Azoarcus*, *Beijerinckia*, *Clostridium*, and *Geobacter*. Particularly, *Azotobacter* species such as *A. chroococcum* and *A. vinelandii* have demonstrated high nitrogen-fixing efficiency, contributing to plant nutrition and biological soil fertility. Soil N input rates from *Azotobacter* can range from 0.3 to 15 kg/ha/yr, reaching up to 60 kg/ha/yr. The application of these diazotrophs in crop-farming systems with different plants (e.g., rice, wheat, maize, sugarcane, and cotton) improves crop yields, reducing the impact of environmental degradation.
- Autotrophic diazotrophs are photosynthetic and chemotrophic organisms. Photosynthetic diazotrophs belong to cyanobacteria, green sulphur bacteria (e.g., *Chlorobium*), and purple sulphur bacteria (e.g., *Chromatium*). In these organisms, the photosynthetic products provide support to the energy-intensive nitrogen fixation process. These organisms have the ability to resist desiccation and survive many harsh conditions. They colonise environments like freshwater marine ecosystems, wet rocks, phylloplanes, and top rhizospheres. Many nitrogen-fixing cyanobacteria are part of the phytoplankton communities and play a key role in primary production for aquatic ecosystems. They also contribute as biofertilisers in agriculture (detailed in section 28.6). When applied to the agriculture field, they improve plant health and crop yield. Based on the presence of heterocysts, diazotrophic cyanobacteria can be non-heterocystous or heterocystous.

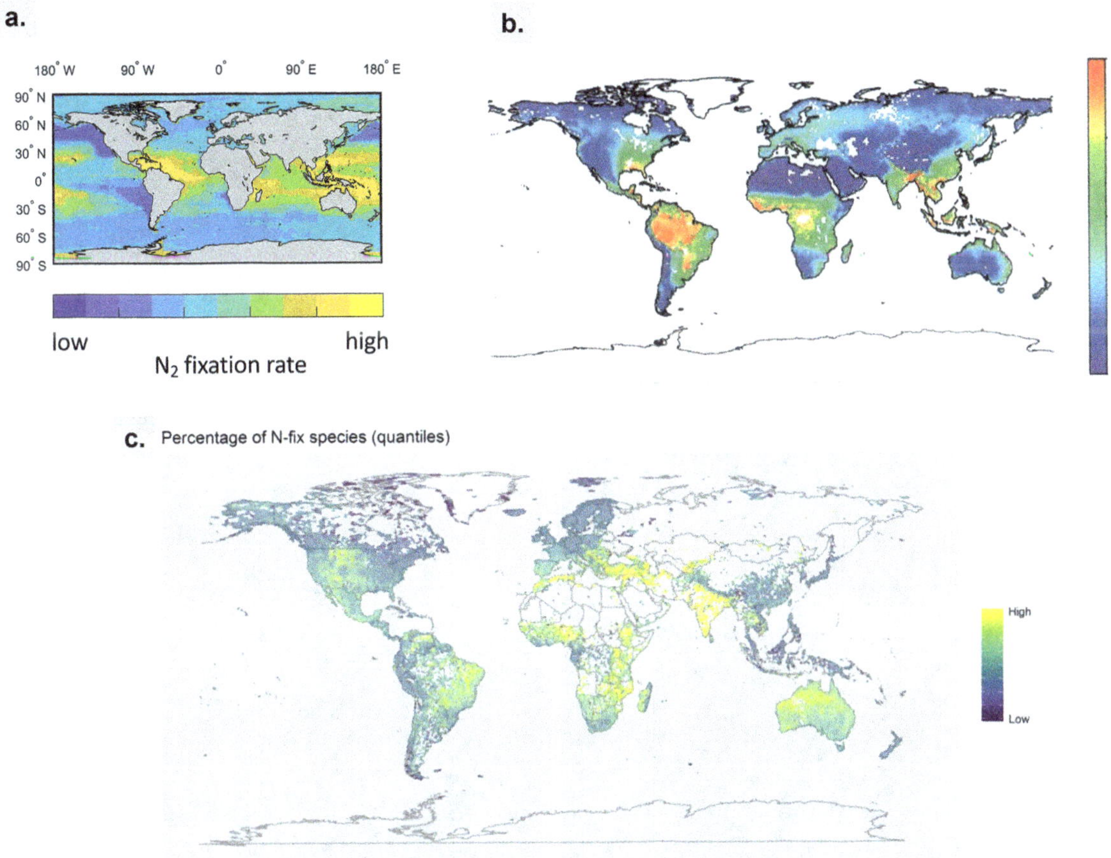

FIGURE 28.4 Global pattern of biological nitrogen fixation (BNF). a. In marine ecosystems. b. In terrestrial ecosystems. c. Species richness of nitrogen-fixing vascular plants relative to total species richness[4]. (Images a, b, and c were adapted from Tang and Cassar (2019)[5] with permission of © 2019 American Geophysical Union, Cleveland et al. (1999)[6] with permission of © 1999 American Geophysical Union, and Tedersoo et al. (2017)[7] with permission of © 2018 International Association for Vegetation Science, respectively.)

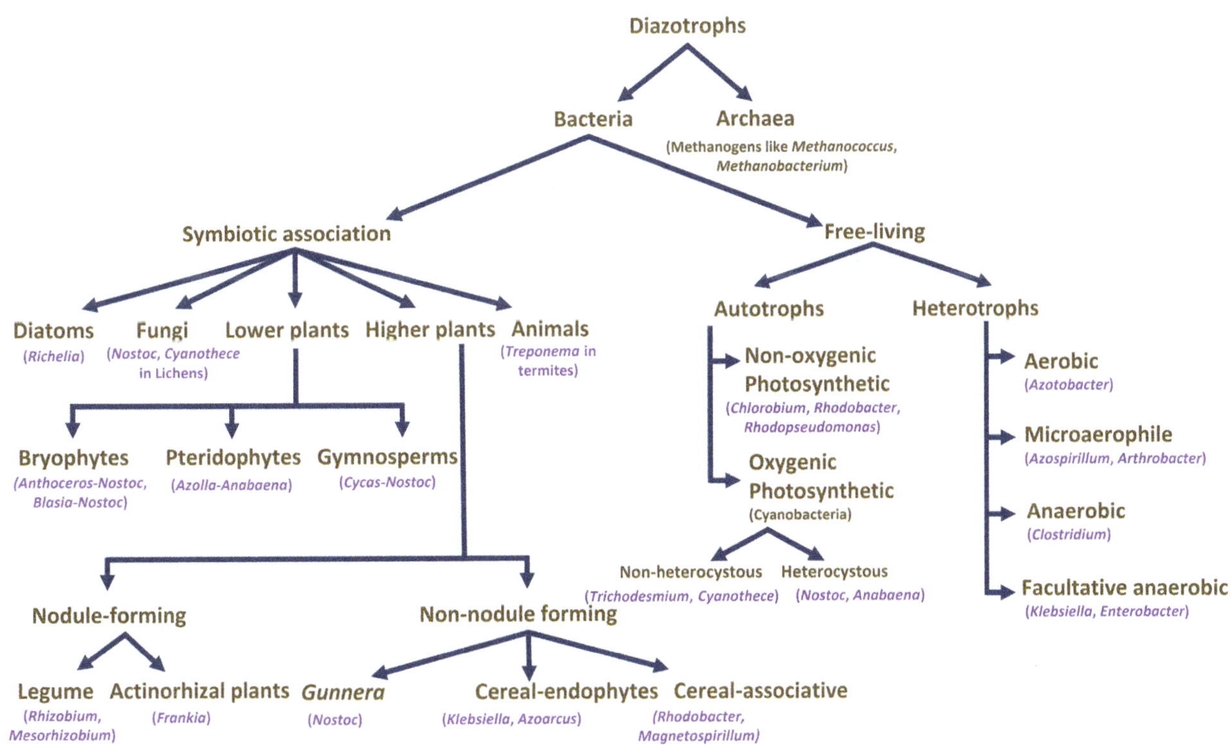

FIGURE 28.5 Different types of diazotrophs based on microbial types and functional characteristics with suitable examples.

FIGURE 28.6 Nitrogen-fixing cyanobacteria and their natural habitats. a. Micrograph of *Trichodesmium* under a dissecting microscope. b. Bloom of *Trichodesmium* in the coast region of the Canary Islands. c. Micrograph of *Anabaena* filament (at 40x). d. Dark-green *Anabaena* "spires" growing on top of senescing macro-algae *Cladophora glomerata* in the Eel River in California, USA. (Image a adapted from Momper et al. (2015)[8] with permission of © 2014, International Society for Microbial Ecology. Images b and d adapted from Benavides and Arístegui (2020)[9] and Bouma-Gregson et al. (2018)[10], respectively.)

- Some well-known non-heterocystous nitrogen-fixing cyanobacteria are *Oscillatoria*, *Trichodesmium*, *Gloeothece*, *Aphanothece*, *Gloeocapsa*, and *Plectonema*. A filamentous cyanobacterium, *Trichodesmium*, that fixes nitrogen in aerobic conditions alone makes a significant contribution to nitrogen fixation in tropical or subtropical regions with an estimated supply of up to half of the oceanic N_2 fixation (Figure 28.6a, b).
- Examples of heterocystous cyanobacteria are *Nostoc*, *Anabaena*, and *Aulosira*. In heterocystous forms, nitrogen fixation takes place inside the thick-walled structure of heterocysts. They can fix nitrogen both in free-living states and in associations with plants. An example of symbiotic association is *Anabaena* inhabiting with a green alga, *Cladophora glomerata*. They form dense green blooms in nutrient-rich aquatic systems (Figure 28.6c, d). Figure 28.7 provides a transmission electron micrograph (TEM) of a heterocyst.

28.4.2 Symbiotic Diazotrophs

Symbiotic diazotrophs establish partnerships with other organisms for nitrogen fixation. This phenomenon was first reported by two German agronomists, H Hellriegel and H Willfarth, in 1886–1888. In this symbiotic relationship, both partners have a mutual agreement to receive benefits through their interaction. All types of hosts require fixed nitrogen from diazotrophs. However, symbionts (i.e., microorganisms living in close association with other organisms) can have different requirements from hosts, depending on the organisms involved and environmental conditions. In a symbiotic relationship, the infection of a diazotroph in the host body is highly specific and strictly regulated. Some well-known nitrogen symbioses are rhizobia-legume symbiosis, *Frankia*-actinorhizal symbiosis, cyanobacteria-lower plant association, diazotrophic endophytes of grasses, and lichens (Table 28.3). In addition, diazotrophs can reside inside the bodies of a few macroalgae, protists, fungi, and animals. They are further discussed as follows.

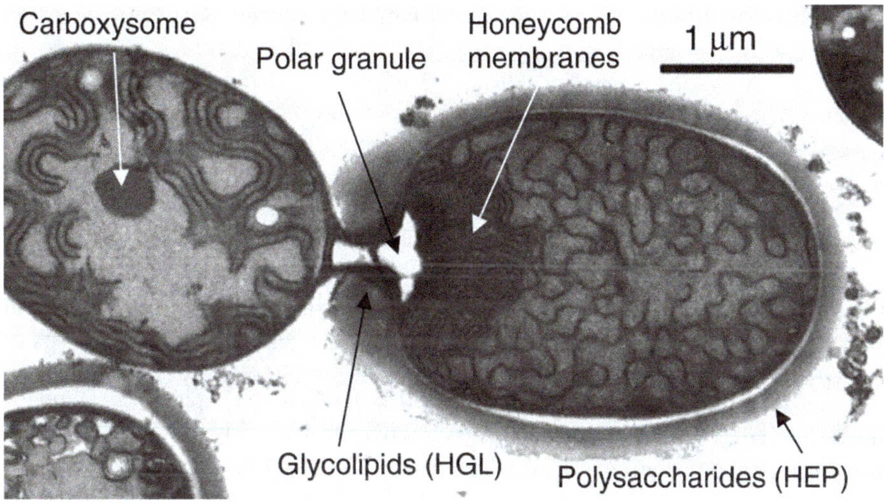

FIGURE 28.7 An ultrastructure of a heterocyst in *Nostoc* sp. Developed from photosynthesising vegetative cells, these thick-walled heterocysts give the oxygen-labile nitrogenase a microoxic environment for nitrogen fixation. (Image reproduced from María and Maldener (2019)[11] with permission of © 2019 John Wiley and Sons.)

TABLE 28.3
A list of various nitrogen-fixing organisms belonging to different groups from the domains Bacteria and Archaea

Domain	Phylum	Class/Order	Selected species	Hosts (if any)
Archaea	Euryarchaeota	Methanosarcinales	*Methanosarcina*	-
		Methanobacteriales	*Methanothermobacter* and *Methanobacterium*	-
Bacteria	Cyanobacteria	Oscillatoriales	*Trichodesmium*	-
			Cyanothece	Lichens
		Nostocales	*Nostoc*	Lichens, diatoms, bryophytes (*Blasia*, *Anthoceros*), pteridophyte (*Azolla*), gymnosperms (*Cycas*, *Macrozamia*), and angiosperm (*Gunnera*)
			Anabaena	Pteridophyte (*Azolla*)
			Richelia	Diatoms
			Calothrix	Lichens and gymnosperm (*Ceratozamia*)
			Scytonema	Lichens
	Actinobacteria	Frankiales	*Frankia,*	Actinorhizal plants
			Streptomyces and *Arthrobacter*	-
	Proteobacteria	α-Proteobacteria	*Azorhizobium* and *Gluconacetobacter*	Grasses
			Beijerinckia, *Magnetospirillum*, *Rhodopseudomonas,* and *Rhodobacter*	-
			Rhizobium, *Sinorhizobium*, *Mesorhizobium*, and *Bradyrhizobium*	Legumes
		β-Proteobacteria	*Burkholderia*	Legumes and grasses
			Azoarcus, Azospira, and *Herbaspirillum*	Grasses
		γ-Proteobacteria	*Azotobacter*	-
			Klebsiella, Enterobacter, and *Ervinia*	Grasses
		δ-Proteobacteria	*Desulfovibrio*	Human gut
			Geobacter	-
	Firmicutes	Bacillales	*Bacillus*, *Paenibacillus*, and *Clostridium*	-
	Bacteroidetes	Chlorobiales	*Chlorobium*	-
	Spirochaetes	Spirochaetales	*Treponema*	Termites
	Chloroflexi	Dehalococcoidia	*Dehalococcoides*	-

Note: Information on biological hosts is also included if they exist.

28.4.2.1 Rhizobia-Legume Association

Rhizobia-legume symbiosis is the most studied nitrogen-fixing biosystem. Almost 90% of leguminous plants can establish a symbiosis with a few bacterial lineages of diazotrophs. Plants like alfalfa, peas, clovers, soybeans, mungbeans, etc., are common examples of rhizobia-legume symbiosis. Plant roots are the main sites of infection and form a specialised structure called a **nodule**. A similar root nodule can also be observed in a non-legume plant, *Parasponia* (Family: Cannabaceae). A leguminous plant, *Sesbania,* can develop nodules in its stem and hosts a bacterium, *Azorhizobium caulinodans*.

The diazotrophic inhabitants of legume roots are aerobic, Gram-negative bacteria, broadly referred to as **rhizobia** (*sing.* rhizobium). Approximately seventeen genera belonging to α and β proteobacteria can establish nodules with legumes. Some important rhizobial genera are *Rhizobium, Azorhizobium, Mesorhizobium, Sinorhizobium,* and *Bradyrhizobium.* Rhizobia usually inhabit as free-living bacteria in soils, but they can establish a symbiotic relationship when they find suitable hosts. Nitrogen fixation in rhizobia takes place only when they reside inside roots. The rhizobia-legume association is also host-specific. For example, *Cajanus cajan* root nodules host *Rhizobium* sp. (Figure 28.8a, b). In this type of mutualism, plants get a supply of fixed nitrogen from rhizobia. In return, rhizobia have access to the plant's photosynthetic output— utilising up to 7–12% fixed carbon. This nitrogen-fixing biosystem in legumes can produce an average of 227 kg N/ha/yr, which can reach up to 300 kg N/ha/yr. In contrast, free-living diazotrophs fix only up to 50-60 kg N/ha/yr.

The nodule is considered a true organ of the plant root system and exhibits cellular specialisation. The process of nodulation in legumes involves a series of biochemical signals. The infection begins with the secretion of secondary plant metabolites like flavonoids and betains to attract host-compatible rhizobia. These metabolites act as signalling molecules that attract rhizobia to plant roots. This process triggers the synthesis of lipo-chitooligosaccharides (LCOs) or nod factors (or nodulation factors) coded by different *nod* genes in specific rhizobia. The release of LCOs or nod factors stimulates root-hair deformation and curling, activating the calcium-dependent signal transduction pathway. These chemical signals stimulate the formation of an infection thread— allowing the passage of rhizobia inside the epidermal cell and their transfer into cortical cells. The plant hormone cytokinins play an important role in the root system architecture to accommodate bacterial cells. Inside the nodules, bacteria are enclosed in a host-derived membrane "**symbiosome**" and are morphologically distinct and differentiated into so-called **bacteroids** (Figure 28.9).

28.4.2.2 *Frankia*-Actinorhizal Association

A different reciprocal advantageous nitrogen fixation partnership is the *Frankia*-actinorhizal association (Figure 28.8c). This can be reported in approximately two hundred different plants distributed in twenty-five genera and eight plant families in orders: Fagales, Rosales, and Cucurbitales. Examples of those plants are *Myrica, Alnus, Casuarina,* and *Cercocarpus.* A Gram-positive filamentous bacterium, *Frankia* (belonging to actinobacteria), is the only representative of diazotrophs that establishes an association with actinorhizal plants. This type of association is widely distributed in boreal forests.

Frankia is a non-agricultural taxon that fixes nitrogen in both free-living and symbiotic conditions. This bacterium is an aerobic, spore-forming, facultative saprophyte and grows in hyphal form. The terminal swellings of the hyphal filaments form a specialised spherical chamber for nitrogen fixation called vesicles (Figure 28.8d). The vesicle is surrounded by a laminated lipid envelope that maintains low partial O_2 pressure. Some actinorhizal plants like *Casuarina glauca* and *Myrica gale* can also produce heme protein (similar to leghaemoglobin in legumes) to create microaerophilic conditions around nitrogenase.

Similar to rhizobia-legume relationships, this partnership also develops root nodules below ground. The magnitudes of nitrogen-fixing capabilities in both types of symbiotic associations are also somewhat similar. However, the mode of bacterial infection in the host root in the *Frankia*-actinorhizal association is relatively complex, and the crucial information about the same is still not well determined. The signalling molecules in the *Frankia*-actinorhizal association are quite variable. In many situations, actinorhizal host plants attract *Frankia* using chemical signals of flavonoids and isoflavonoids. The canonical nod factors are also not often present in *Frankia.* Bacterial infection in the host root can have two different pathways: intracellular and intercellular modes. In the former mode, *Frankia* secretes root hair deforming factors (RHDFs), which induce root hair deformation for nodule formation. In contrast, the latter mode involves the infiltration of *Frankia* filaments through intercellular gaps and bypasses the root hair deformation process. Table 28.4 provides a comparative account of rhizobia and *Frankia* for their symbiotic association.

The major significance of the *Frankia*-actinorhizal association is ecological succession and land reclamation. Many actinorhizal plants are pioneer species that colonise barren and infertile lands. *The Frankia*-actinorhizal association also contributes to erosion control, water filtration, soil stabilisation, plant tolerance to various stresses, and the upkeep of wildlife habitats. They also offer various ecosystem services such as crop protection, timber production, fuelwood generation, and the manufacturing of other forest products.

28.4.2.3 Other Diazotrophic Symbioses

Many diazotrophs reside as epiphytes and endophytes of roots, stems, and leaves of many monocots and are a potential source of nitrogen in nitrogen-deficient soil. For example, an endophytic nitrogen-fixing bacterium, *Gluconacetobacter diazotrophicus*, can partner with sugarcane. A different nitrogen-fixing

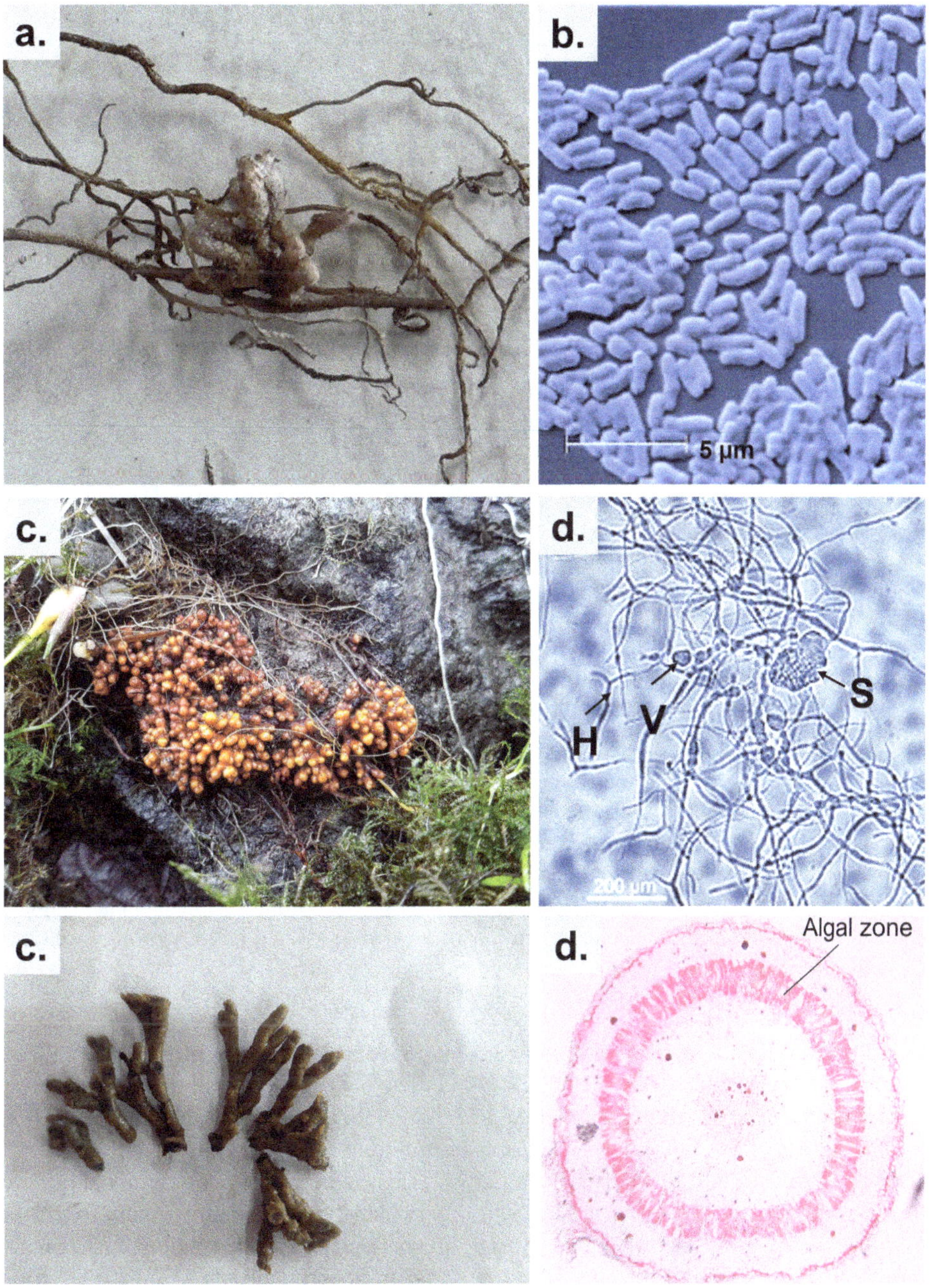

FIGURE 28.8 Root-diazotroph associations. a. *Cajanus cajan* root nodules inhabited *Rhizobium* sp. b. SEM image of *Rhizobium leguminosarum*. c. Alder root nodules inhabited *Frankia*. d. Structures of *Frankia* under light microscopy. e. Coralloid roots of *Cycas*. f. T.S. of coralloid root showing an algal zone in the middle cortex. (H: hyphae, V: vesicle, and S: sporangia) (Images b and d adapted from Reeve et al. (2013)[12] and Diagne et al. (2013)[13], respectively.)

bacterium, *Herbaspirillum* (Phylum Proteobacteria), colonises the cortical cells and vascular tissues of roots of Poaceae members such as maize, rice, sorghum, and sugarcane. Even many free-living nitrogen fixers can also reside as endophytes. Examples include *Azoarcus*, *Azospira*, *Azorhizobium*, *Geobacter*, *Klebsiella*, and *Enterobacter*. Several nitrogen fixers can inhabit bioenergy crops like *Miscanthus*, *Spartina*, and *Pennisetum*, contributing to their biomass production. A key characteristic of these endophytic bacteria is that they have a low dependence on plant metabolism for their growth. Some nitrogen-fixing cyanobacteria, such as *Nostoc*, *Ocillatoria*, *Scytonema*, and *Stigonema*, reside as epiphytes in the phyllosphere of rainforest plants. Cyanobacteria colonise different plant organs except for underground parts because they are photosynthetic symbionts and cannot grow in the absence of sunlight.

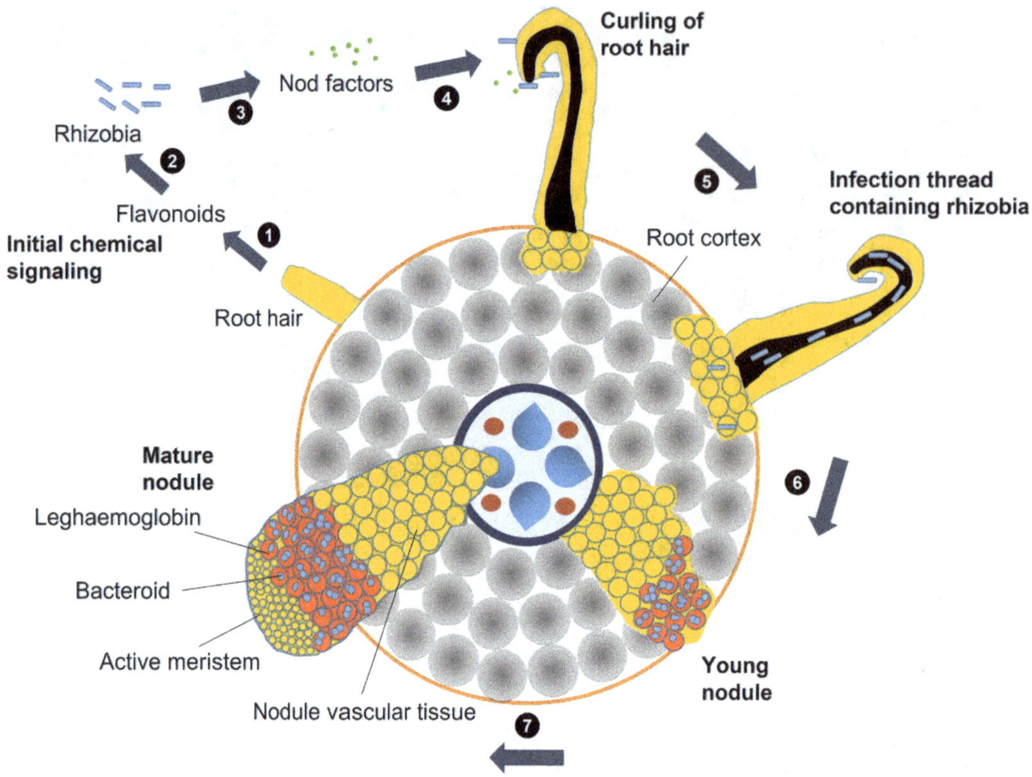

FIGURE 28.9 A general mechanism of nodule formation in legumes. Steps 1 and 2 show the release of flavonoids from root hair to attract rhizobia. Step 3 represents the synthesis of nod factors by rhizobia. Step 4 demonstrates the interaction of rhizobial cells with root hair for its curling. This step also demonstrates the formation of infection thread (a tube-like structure) in root hair. Step 5 visualises the transfer of rhizobial cells from the root hair to the root cortex. Step 6 depicts the stages of cell differentiation and nodule formation. The red colour structure in the young nodule represents leghaemoglobin. In step 7, a mature nodule has developed, showing a well-developed network with the vascular system for nutrient exchange.

Bryophytes, pteridophytes, and gymnosperms can also establish nitrogen-fixing symbiosis (Table 28.3). The diazotrophic partners in these plants are mainly the members of Nostocales (Cyanobacteria). In bryophytes, *Nostoc* thrives as epiphytic/endophytic in thalli of liverworts (e.g., *Cavicularia* and *Blasia*) and hornworts (e.g., *Anthoceros*). In pteridophytes, *Azolla* is the only example that hosts a nitrogen-fixing organism, *Anabaena azollae*. It is a strict obligate symbiosis and resides in the cavities of the dorsal lobes of *Azolla* leaves. This symbiotic association has enormous potential as a biofertiliser in the paddy fields and can fix nitrogen as high as 150 kg N/ha/crop. In gymnosperms, cycads like *Cycas*, *Dioon*, and *Macrozamia* establish a symbiotic relationship with *Nostoc* and other cyanobacteria. They reside in a specialised apogeotropic structure called coralloid roots, which typically grow upward in the soil. They are small, coral-shaped structures housing symbiotic nitrogen-fixing cyanobacteria. These symbionts develop a green differentiated zone in the cortical layer of the coralloid roots and fix nitrogen, which helps these cycads to grow in nitrogen-deficient soil (Figure 28.6e, f). An angiosperm, *Gunnera,* is an example of a higher plant that partners with *Nostoc*. It resides in special glands at the base of the petioles. The primary advantage to cyanobacteria in this partnership is shelter because they are photosynthetic and do not depend on plants for food. However, it is possible that plants provide some nutrients to the photosynthetic symbionts.

Lichen is a symbiotic association between two different organisms: fungi and algae. Almost 10% of all known lichens possess nitrogen-fixing photobionts, which can extract nitrogen from rock and atmospheric nitrogen for nitrogen fixation. For example, *Peltigera* and *Lobaria* host a diazotrophic taxon, *Nostoc,* and the lichen *Erioderma* hosts *Scytonema.*

In marine ecosystems, diazotrophs can establish symbiotic associations with some diatoms. For example, a cyanobacterium, *Richelia*, can partner with as many as thirteen different diatoms in tropical and subtropical surface waters. A different partnership exists between the unicellular cyanobacterium, *Crocosphaera watsonii* and diatoms. In the cold ocean, unicellular cyanobacterial group A (UCYN-A) (Candidatus *Atelocyanobacterium thalassa*) and UCYN-A2 are two different lineages of uncultured nitrogen-fixing cyanobacteria that establish symbiotic relationships with open ocean picoplanktonic prymnesiophyte algae and coastal nanoplanktonic coccolithophore *Braarudosphaera bigelowii*, respectively. Some studies claimed that they may contribute just as much, if not more, to nitrogen fixation than a free-living

TABLE 28.4
Similarities and differences between rhizobia and *Frankia* in their symbiotic association

Similarities between rhizobia and *Frankia*

- Nitrogen fixation process using nitrogenase
- Infect roots of higher plants
- Magnitude of nitrogen-fixing capabilities
- Symbiotic association with related plants to a monophyletic group of rosids
- Formation of nodules with host plants
- Heme-containing proteins

Differences between rhizobia and *Frankia*

Rhizobia	*Frankia*
Gram-negative and non-spore-forming	Gram-positive and spore-forming
They are a group of bacteria belonging to the Proteobacteria	It is a single genus with different species from the Actinobacteria
Infect plant species of only one family, Fabaceae (Legume plants)	Infect plant species of eight families belonging to three related orders: Fagales, Rosales and Cucurbitales (Actinorhizal plants)
Distribution mainly in tropical regions	Distribution mainly in temperate regions
Infection causes morphological changes in bacteria and the cellular structure known as bacteroids.	No such phenomenon is known.
Canonical *nodABC* genes are required for nodule formation	Canonical *nodABC* genes are uncommon for nodule formation
Nitrogen fixation only occurs in symbiotic conditions	Nitrogen fixation in both free-living and symbiotic situations
No special chamber for nitrogenase	Location of nitrogenase in a subcellular location called a vesicle.
Shielding the nitrogenase from oxygen by leghaemoglobin	Shielding the nitrogenase from oxygen by related heme-containing proteins
Extensive agricultural application and ecological role	No agriculture application, but a great ecological role

Trichodesmium does. An immediate research investigation reports that *B. bigelowii* possesses a specialised organelle for nitrogen fixation called **nitroplast**. This is the first report on a eukaryotic organism with nitrogen-fixing ability.

Many coral compartments in the coral reefs, such as the mucus, tissue, and skeleton, host diazotrophic bacteria and archaea. Most of the diazotrophs in corals are heterotrophic, but some cyanobacteria can also be present. The diazotrophic communities in coral are species-specific and even share the potential of vertical transfer from the parent colonies of coral to their larvae. However, corals prefer nitrogen from other sources rather than obtaining it from diazotrophic partners because nitrogen fixation is a high-energy intensive process.

The termite-diazotroph association is also significant as they thrive on a nitrogen-poor diet. The termite gut hosts bacteria like *Treponema*, *Spirochaeta*, and *Clostridium* to compensate for nitrogen requirements. Reports of diazotrophic colonisation in humans and ruminant animals also exist; however, the nitrogen status of higher organisms is not affected by bacteria because heterotrophic feeding habits consist of sufficient fixed nitrogen. A sulphate-reducing taxon, *Desulfovibrio diazotrophicus,* is a notable example of a diazotrophic bacterium that lives within mammals.

28.5 NITROGENASE

All nitrogen-fixing organisms employ an enzyme complex called nitrogenase (EC 1.18.6.1), which allows the conversion of N_2 into NH_3. It is mainly composed of two metalloproteins—a dinitrogenase reductase (Fe protein) and a dinitrogenase (Mo-Fe protein) (Figure 28.10). Dinitrogenase reductase contains an γ_2 homodimer and one Fe_4S_4 cluster, which transfers electrons from ferredoxin to dinitrogenase using the chemical energy of ATP. Some enzymes use flavodoxin as their electron donor instead of ferredoxin. Dinitrogenase is the main catalytic centre for the reduction reaction, which is composed of an $\alpha_2\beta_2$ heterotetramer and a variable number of Fe-S clusters (also called P cofactors and FeMo cofactors). Few bacterial (e.g., *Azotobacter vinelandii*) and archeal species (e.g., *Methanosarcina* spp.) contain a vanadium-containing nitrogenase that is a homologous alternative to Mo-Fe nitrogenase; however, this form is relatively less efficient for nitrogen fixation.

The main genetic determinants of nitrogenase are three *nif* genes, including *nifH*, *nifD*, and *nifK*. The first gene encodes Fe protein, while the latter two code for molybdenum-iron (Mo-Fe) protein. However, the synthesis and regulation of nitrogenase and its cofactors can involve as many as twenty related *nif* genes. The expression of *nif* genes is tightly regulated, and the organisation of *nif* genes can vary in nitrogen-fixing organisms.

Nitrogen fixation is an energy-intensive reaction due to the large **activation energy barrier**. Nitrogenase enzyme overcomes this barrier and enables the energetically demanding process of nitrogen fixation under biological conditions. The reaction consumes eight electrons and 16 ATP per molecule of N_2 to produce two molecules of NH_3 (reaction 28.1). The transfer of each electron consumes two ATP molecules. The

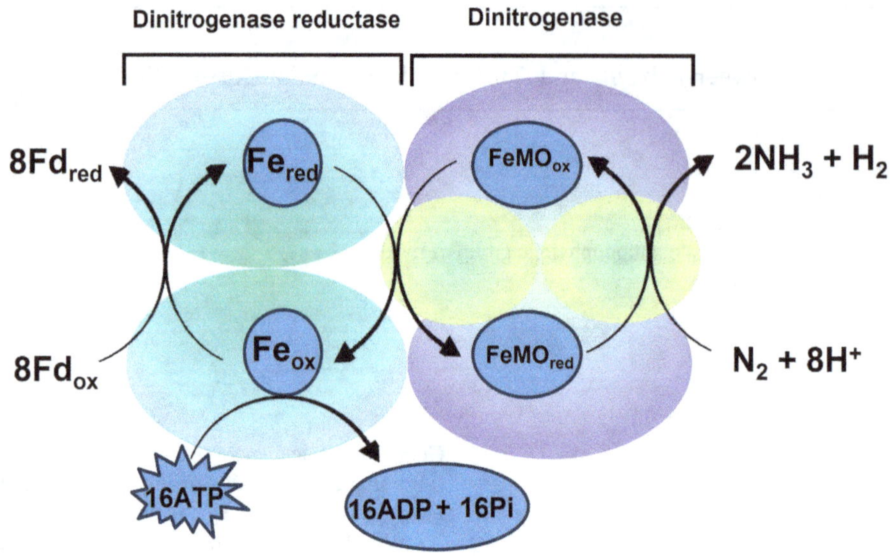

FIGURE 28.10 A schematic illustration of biological nitrogen fixation (BNF) using nitrogenase enzyme complex. It has two main components, including dinitrogenase reductase and dinitrogenase. Nitrogenase reductase is a homodimer and is responsible for the transport of electrons from Fd_{red} (ferredoxin) to dinitrogenase using 16 ATP. Dinitrogenase is a heterotetramer which facilitates the conversion of nitrogen gas (N_2) into ammonia (NH_3) and hydrogen gas (H_2).

reaction also releases H_2 as a byproduct, which is recycled by a different enzyme, hydrogenase. Hydrogenase is also a metalloenzyme with a complex nickel-iron (Ni-Fe) cofactor, and it mainly plays a role in improving the efficiency of nitrogen fixation.

$$N_2 + 8H^+ + 8e^- + 16\,ATP \rightarrow 2NH_3 + H_2 + 16ADP + 16P_i$$
reaction 28.1

Nitrogenase has a very low turnover (less than 10 s^{-1}) due to complex reactions. The enzyme complex is susceptible to an oxic environment and eventually becomes inactive when exposed to O_2 because it contains Fe. Nitrogen-fixing microorganisms, especially those that are obligate aerobes, acquire strict protection mechanisms to achieve anoxic conditions around the enzyme complex in an oxygen-rich environment. These can include habitat modification (anaerobiosis), physical protection (vesicle and heterocyst), symbiotic relationship (legume and actinorhizal plants), temporal separation (rhythmic activity), and protective mechanisms (high respiratory activity and intracellular oxygen management). These processes are distinct in different microorganisms and are discussed briefly as follows.

- Heterocyst- Some cyanobacteria like *Nostoc*, *Anabaena*, and *Gloeocystis perform nitrogen fixation in specialised cells called heterocysts*. These cells have very fast O_2 consumption (i.e., high respiratory rates), which creates an oxygen-limited environment for nitrogen fixation. In addition, the heterocysts do not consist of photosystem II (an oxygen-evolving complex), which prevents excess accumulation of O_2 inside the heterocyst.
- Rhythmic activity- Some non-filamentous, non-heterocystous cyanobacteria observe diurnal activity for nitrogen fixation (e.g., *Cyanothece*). In this process, cells perform nitrogen fixation at night, when there is low O_2 tension, and photosynthesis occurs during the daytime.
- Intracellular oxygen management- In some cyanobacteria like *Trichodesmium*, the incompatible reactions of photosynthesis and nitrogen fixation occur in the same cell at the same time by scavenging or detoxifying O_2. The Mehler reaction is one such method, where O_2 produced from photosystem II can be photo-reduced, which creates low intracellular O_2 tensions for nitrogen fixation. In addition, superoxide dismutase and other antioxidant enzymes neutralise reactive oxygen species (ROS), such as superoxide radicals, which could pose a threat to nitrogenase.
- Symbiotic association- A heme-containing protein, leghaemoglobin, present in legume roots, protects bacterial cells (i.e., rhizobia) from the deactivation of the nitrogenase enzyme. This scheme also protects nitrogenase in *Frankia*-actinorhizal plants. The cytoplasm of the infected cells is the main locale of its production and imparts the red/pink colour of the intracellular structure. Leghaemoglobin is a phytoglobin consisting of a heme bound to a ferrous ion (Fe^{2+}), which controls O_2 regulation in the microenvironment. The reduced O_2 tension inside the nodule prevents the deactivation of nitrogenase, while its controlled diffusion supports bacterial respiration for their growth and other activities.
- Vesicles- They are the terminal specialised, nitrogenase-containing cells in the hyphae of *Frankia*, which

are surrounded by a thick laminated lipid envelope. This scaffold maintains partial O_2 pressure in the microenvironment, only enough for respiratory activity and physically shields the nitrogenase enzyme from O_2.

- A high respiratory rate in a bacterium like *Azotobacter vinelandii* protects nitrogenase in an oxygenated atmosphere.
- Anaerobiosis- Methanogens like *Methanococcus* and *Methanothermococcus,* and bacteria like *Klebsiella pneumonia* and *Rhodobacter capsulatus* are strict anaerobes and can suitably fix nitrogen in anoxic environments. They do not require special protection for nitrogen fixation.

28.6 NITROGEN-FIXING BIOFERTILISERS

In the last two decades, synthetic nitrogen fertilisers have significantly increased agricultural productivity. However, excessive use of chemicals has compromised food quality, food safety, and environmental well-being. Growing concerns about these consequences have increased the demand for organically grown plant products. The use of nitrogen-fixing biofertilisers is in high demand for the same. Government interventions in sustainable development and eco-friendly approaches have also added fuel to it. Farmers have also started relying on organic manure and biofertilisers as an alternative to conventional farming practices. While nitrogen-fixing biofertilisers rely on natural processes, their application in agriculture involves a significant human intervention. They are grown in bioreactors and delivered to the agricultural system. However, the supply of NH_3 and its derivative products through biological processes still has much fewer negative effects than the chemically stimulated process.

The biofertilisers have been categorised based on crop types (cereals and grains, pulses and oilseeds, fruits and vegetables, and others), applications (seed and soil treatment), and geographical regions. Market analysis suggests that the industry of nitrogen-fixing biofertilisers is expected to touch USD 2.80 billion by 2028, with an annual growth rate of 10.9% from 2021 to 2028, from a value of USD 0.69 billion in 2020, as per the report of Data Bridge Market Research 2021. The USA, Canada, and Mexico regions are the biggest markets for nitrogen-fixing biofertilisers. Diazotrophs like *Rhizobium*, *Azotobacter*, *Azospirillum,* and different cyanobacteria are powerful inoculants for agricultural production. *Rhizobium* has emerged as the second most-used biofertiliser after mycorrhizae. The market for a biofertiliser based on *Azotobacter* is also very high and is projected to grow at a compound annual growth rate. *Azospirillum* has become a major player in agricultural productivity. The inoculation of cyanobacteria-based fertiliser (based on the *Azolla-Anabaena* partnership) is a traditional practice for paddy fields in Southeast Asia.

Nitrogen-fixing organisms offer great benefits for agriculture, but their use without assessment and monitoring can pose potential risks to the soil ecosystem and plant health. The introduction of foreign diazotrophs in a given site can cause an imbalance in natural communities and the environment. Excessive use of biofertilisers can disrupt the natural balance of nutrients, reduce soil biodiversity, and damage plants.

SUMMARY

- Nitrogen is abundantly present in the atmosphere as an inert N_2 and is an essential requirement for all life forms. Nitrogen gas can be converted into reactive forms by non-biotic reactions (Haber Bosch process and lightning) and a biotic reaction (biological nitrogen fixation (BNF)).
- The nitrogen cycle maintains the balance of nitrogen in ecosystems. It has seven key biologically relevant nitrogen-transforming steps, including nitrogen fixation, nitrification, comammox, ammonia assimilation, nitrate assimilation, denitrification, and anammox.
- The microbial system has an enormous ability to oxidise and reduce nitrogen compounds through a multitude of enzymes. Bacteria and archaea are the key players in the transformation, assimilation, and decomposition of nitrogen species.
- BNF is a process of converting N_2 into NH_3 using nitrogen-fixing bacteria and archaea. These organisms are collectively termed diazotrophs. They can be heterotrophic or photosynthetic and aerobic/anaerobic. Some diazotrophic species establish symbiotic relationships with other organisms for nitrogen supply.
- The most widely recognised partnerships are the rhizobia-legume symbiosis and the *Frankia*-actinorhizal symbiosis. A nodule is a specialised organ in the plant roots which hosts nitrogen-fixing bacteria. Other well-known symbiotic relations include the *Azolla-Anabaena* association, lichens, and microbial associations with lower plants.
- Diazotrophs catalyse the reaction using a specific metalloenzyme complex called nitrogenase. The genetic determinants of nitrogenase are the *nif* genes. Nitrogenase is an oxygen-sensitive enzyme which acquires various mechanisms for its protection in different diazotrophs.
- Growing demand for organic food has sparked the use of biofertilisers for nitrogen fixation. Some diazotrophs, such as *Azotobacter*, *Rhizobium*, *Azospirillum*, and different cyanobacteria, are commercially grown inoculants for nitrogen supply in agricultural production.

DIFFERENT TYPES OF PRACTICE QUESTIONS

Fill in the blanks

I. .... and .... are two different processes of decomposition of organic nitrogenous compounds.

II. The anammox reaction occurs in a unique membrane-bound compartment called an ....

III. Microorganisms that are capable of fixing atmospheric nitrogen gas into ammonia are called ....
IV. Inside the nodules, diazotrophs are differentiated into morphologically distinct forms called ....
V. .... is a Gram-positive filamentous bacterium that fixes nitrogen in symbiotic association with plants.
VI. .... is a microbial process for the reduction of nitrate to nitrogen gas.

SOLUTION: I. Depolymerisation, ammonification, II. anammoxosome, III. diazotrophs, IV. bacteroids, V. *Frankia*, VI. Denitrification

Match the following

i. Nitrite reductase	a. Nitrogen fixation
ii. Urease	b. Ammonification
iii. Nitrite oxidoreductase	c. Ammonia assimilation
iv. Glutamine synthetase	d. Nitrification
v. Dinitrogenase reductase	e. Denitrification

SOLUTION: i-e, ii-b, iii-d, iv-c, v-a

Very short answer type of questions

I. Expand commonox.
II. What is an alternative process of nitrification in biological systems?
III. Write the name of a nitrogen-fixing archaeon.
IV. Why is the nitrogen fixation process highly sensitive to oxygen?
V. Write the names of any two nitrogen-fixing biofertilisers.

Short answer type of questions

I. What are the different mechanisms for the protection of nitrogenase from oxygen?
II. Write a short note on diazotrophic cyanobacteria.
III. How are ammonification and depolymerisation steps in the nitrogen cycle closely associated?
IV. What are the different methods of nitrate reduction in microorganisms, and how do they contribute to the nitrogen cycle?
V. What are nitrogen-fixing biofertilisers and their applications in agricultural systems?

Long answer type of questions

I. Discuss the roles of microbial activities in the nitrogen cycle.
II. What is biological nitrogen fixation? Provide a detailed account of symbiotic and nonsymbiotic nitrogen fixation.

NOTES

1 Oxidising capacity of the atmosphere is an ability of the atmosphere to support oxidation reactions, particularly those that convert nitrogen oxides into other reactive forms.
2 The rate-limiting step refers to the slowest step in a reaction sequence, which determines the overall speed of the nitrification process.
3 van Niftrik L, Geerts WJ, van Donselaar EG, et al. Combined structural and chemical analysis of the anammoxosome: a membrane-bounded intracytoplasmic compartment in anammox bacteria. *Journal of Structural Biology*. 2008;161(3):401–10.
4 Based on all GBIF vascular plant records with coordinates, GBIF.org, 10th November 2017, GBIF Occurrence Download doi.org/10.15468/dl.4nqoev
5 Tang W, Cassar N. Data-driven modeling of the distribution of diazotrophs in the global ocean. *Geophysical Research Letters*. 2019;46(21):12258–69.
6 Cleveland CC, Townsend AR, Schimel DS, et al. Global patterns of terrestrial biological nitrogen (N_2) fixation in natural ecosystems. *Global Biogeochemical Cycles*. 1999;13(2):623–45.
7 Tedersoo L, Laanisto L, Rahimlou S, et al. Global database of plants with root-symbiotic nitrogen fixation: Nod DB. *Journal of Vegetation Science*. 2018;29(3):560–8.
8 Momper LM, Reese BK, Carvalho G, et al. A novel cohabitation between two diazotrophic cyanobacteria in the oligotrophic ocean. *The ISME Journal*. 2015;9(4):882–93.
9 Benavides M, Arístegui J. Untreated sewage outfalls do not promote *Trichodesmium* blooms in the coasts of the Canary Islands. *Scientific Reports*. 2020;10(1):18386.
10 Bouma-Gregson K, Kudela RM, Power ME. Widespread anatoxin-a detection in benthic cyanobacterial mats throughout a river network. *PLoS One*. 2018;13(5):e0197669.
11 Muro-Pastor AM, Maldener I. Cyanobacterial heterocysts. eLS. 2019:1–0.
12 Reeve W, Drew E, Ballard R, et al. Genome sequence of the clover-nodulating *Rhizobium leguminosarum* bv. *trifolii* strain SRDI565. *Standards in Genomic Sciences*. 2013;9(2):220–31.
13 Diagne N, Arumugam K, Ngom M, et al. Use of *Frankia* and actinorhizal plants for degraded lands reclamation. *BioMed Research International*. 2013;2013.

SUGGESTED READINGS

Aasfar A, Bargaz A, Yaakoubi K, Hilali A, Bennis I, Zeroual Y, Meftah Kadmiri I. Nitrogen fixing *Azotobacter* species as potential soil biological enhancers for crop nutrition and yield stability. *Frontiers in Microbiology*. 2021;12:354.
Barton LL, Northup DE. *Microbial Ecology*. Wiley-Blackwell; 2011.

Bergman B, Gallon JR, Rai AN, Stal LJ. N2 fixation by non-heterocystous cyanobacteria. *FEMS Microbiology Reviews*. 1997;19(3):139–85.

Bowen JL, Spivak AC, Bernhard AE, Fulweiler RW, Giblin AE. Salt marsh nitrogen cycling: where land meets the sea. *Trends in Microbiology*. 2024;32(6):565–76.

Boyd E, Peters JW. New insights into the evolutionary history of biological nitrogen fixation. *Frontiers in Microbiology*. 2013;4:201.

Chun SC. The Diazotroph as an endophyte and how a Diazotroph interacts with its plant host. In *Microbes: The Foundation Stone of the Biosphere* 2021 (pp. 391–423).

Erisman JW, Sutton MA, Galloway J, Klimont Z, Winiwarter W. How a century of ammonia synthesis changed the world. *Nature Geoscience*. 2008;1(10):636–9.

Fisher K, Newton WE. *Nitrogen fixation—a general overview. Nitrogen fixation at the millennium*. Elsevier; 2002:1–34.

Fowler D, Coyle M, Skiba U, Sutton MA, Cape JN, Reis S, Sheppard LJ, Jenkins A, Grizzetti B, Galloway JN, Vitousek P. The global nitrogen cycle in the twenty-first century. *Philosophical Transactions of the Royal Society B: Biological Sciences*. 2013;368(1621):20130164.

Gruber N, Galloway JN. An Earth-system perspective of the global nitrogen cycle. *Nature*. 2008;451(7176):293–6.

Gu B, Chang J, Min Y, Ge Y, Zhu Q, Galloway JN, Peng C. The role of industrial nitrogen in the global nitrogen biogeochemical cycle. *Scientific Reports*. 2013;3(1):1–7.

Hillel D, Hatfield JL, editors. *Encyclopaedia of soils in the environment*. Elsevier; 2005.

Kneip C, Lockhart P, Voß C, Maier UG. Nitrogen fixation in eukaryotes–new models for symbiosis. *BMC Evolutionary Biology*. 2007;7(1):1–2.

Kuypers MM, Marchant HK, Kartal B. The microbial nitrogen-cycling network. *Nature Reviews Microbiology*. 2018;16(5):263.

Lehtovirta-Morley LE. Ammonia oxidation: ecology, physiology, biochemistry and why they must all come together. *FEMS Microbiology Letters*. 2018;365(9):fny058.

Rajaneesh KM, Naik RK, Roy R, D'Costa PM. Cyanobacteria in tropical and subtropical marine environments: bloom formation and ecological role. In Singh PK, Kumar A, Singh VK, Shrivastava AK (eds). *Advances in cyanobacterial biology*. 2020 (pp. 35–46). Academic Press.

Shu D, Banerjee S, Mao X, Zhang J, Cui W, Zhang W, Zhang B, Chen S, Jiao S, Wei G. Conversion of monocropping to intercropping promotes rhizosphere microbiome functionality and soil nitrogen cycling. *Science of The Total Environment*. 2024;949:174953.

Stacey G. The *Rhizobium*-legume nitrogen-fixing symbiosis. In Bothe H, Ferguson SJ, Newton WE (eds). *Biology of the nitrogen cycle*. 2007 (pp. 147–163). Elsevier.

Stankiewicz BA, van Bergen PF. Nitrogen and N-containing macromolecules in the bio-and geosphere: an introduction. In: Stankiewicz BM, van Bergen PF (eds). Nitrogen-containing macromolecules in the bio- and geosphere. 1998 (pp. 1–18).

Stein LY, Klotz MG. The nitrogen cycle. *Current Biology*. 2016;26(3):R94–8.

Stüeken EE, Buick R, Guy BM, Koehler MC. Isotopic evidence for biological nitrogen fixation by molybdenum-nitrogenase from 3.2 Gyr. *Nature*. 2015;520(7549):666–9.

Tiedje JM, Sørensen J, Chang YY. Assimilatory and dissimilatory nitrate reduction: perspectives and methodology for simultaneous measurement of several nitrogen cycle processes. *Ecological Bulletins*. 1981;33:331–42.

Zhang X, Ward BB, Sigman DM. Global nitrogen cycle: critical enzymes, organisms, and processes for nitrogen budgets and dynamics. *Chemical Reviews*. 2020;120(12):5308–51.

29 Sulphur Cycle and Microorganisms

29.1 SULPHUR IN THE ENVIRONMENT

Sulphur is a p-block non-metal element with atomic number 16. This is the 9th and 14th most abundant element on Earth's surface and in Earth's crust, respectively. Sulphur can be reported from various compartments of the Earth. However, it exists in different forms and species in diverse compartments. Inorganic species are dominantly located in terrestrial and aquatic ecosystems, but much of their proportion is unavailable to the biosphere. Sulphur consists of eight oxidation states, ranging from the oxidation state of -2 in sulphide (S^{2-}) to +6 in sulphate (SO_4^{2-}). A variety of intermediate species, elemental sulphur (S^0), thiosulphate ($S_2O_3^{2-}$), dithionate, and sulphite (SO_3^{2-}), lie in a suite of various biotic and abiotic pathways (Table 29.1). The intermediate species can act as both electron donors and acceptors. Sulphate is commonly present in oxic environments, while S^{2-} is usually in anoxic environments.

The lithosphere is the largest sulphur reservoir, occupying approximately 95% of the Earth's sulphur. Salt domes are a good source of S^0. Minerals of sulphur with heavy metals and metalloids reside in the deep subsurface. More than 1,000 different sulphur minerals exist on the Earth. Pyrite is the most abundant of all. The hydrosphere is the second largest reservoir of sulphur (nearly 5%), where the ocean is the major pool. Sulphate is the second most abundant species in marine ecosystems after chloride and in freshwater ecosystems after carbonate. The sediments of seabeds also have sulphur zones. Sulphur-containing hydrothermal vents and sulphur springs are sulphide-containing water bodies. They are natural ecosystems with high amounts of dissolved sulphur compounds, giving them a unique rotten egg odour due to the presence of hydrogen sulphide (H_2S) gas. The atmosphere also contains a low concentration of sulphur in gaseous forms or aerosols (occupying less than 0.001% of the total sulphur). Sulphur in the atmosphere floats as H_2S, sulphur dioxide (SO_2), volatile organics, and particulate matter. Volatile organic sulphide compounds (VOSC) can be dimethyl sulphide (DMS), methanethiol (MT) and dimethyl disulphide (DMDS).

Sulphur is a vital element required for the growth of all types of organisms. Sulphur species serve as building blocks of structural components and provide room for redox centres. The synthesis of organic sulphur compounds is critical for the biosphere because every organism needs them in protein and secondary metabolites. They can be found in the carbon backbone of living organisms. Cysteine and methionine are two essential amino acids that contain sulphur. Methionine is the first amino acid incorporated in protein synthesis. Cysteine plays a central role in protein structure and function. Cystine is an amino acid dimer produced by the oxidation of two cysteine molecules, contributing to protein folding by cross-linking. Cysteine extends its role as a precursor to antioxidant molecules, glutathione, and iron-sulphur clusters. The disulphide bonds of cystine in thioredoxin and glutaredoxin have cytoprotective roles against oxidative stress. Sulphur-containing compounds such as biotin, coenzyme A, coenzyme M, lipoic acid, and thiamine serve as cofactors for numerous enzymes. Some natural antibiotics (e.g., penicillin and cephalosporin) and antimicrobial compounds also contain sulphur. In mammals, the protein hormone insulin is a sulphur-mediated structure that has a role in maintaining glucose homeostasis in the body. In addition, S^{2-} acts as a signalling molecule in neurotransmission, anti-inflammatory responses, etc.

Plants acquire sulphur from the soil in the form of SO_4^{2-}, which is assimilated into organic compounds through a series of biochemical reactions. Sulphur has an active role in chlorophyll synthesis and nitrogenase activity. Organosulphur compounds in plants give a distinct flavour and aroma. A class of sulphur-containing compounds, glucosinolates are characteristic of the plant family Brassicaceae (such as mustard and radish), which prevent herbivory and pathogen attacks in plants. Marine plants and phytoplankton produce a special organosulphur compound, dimethylsulphoniopropionate (DMSP), which helps regulate their internal osmotic environment in saline environments. This compound is highly abundant on the ocean surface and is the precursor for the climate-active gas DMS.

29.2 SULPHUR CYCLE

The sulphur cycle mainly runs in the sedimentary phase. It can be observed in different biogeochemical regions. Sulphur enters the terrestrial ecosystem from the atmosphere through dry deposition and wet precipitation. Sulphur dioxide in the atmosphere reacts with water (H_2O) and oxygen gas (O_2) to produce sulphuric acid and sulphurous acid, which result in acid rain. Inland runoff transports sulphur into the aquatic bodies. Hydrothermal reactions and chemical oxidation

DOI: 10.1201/9781032651095-36

produce S^0. Hydrogen sulphide tends to oxidise into SO_2 through photochemical reactions. Sulphate ions take part in cloud condensation. Reactive iron (Fe) species react with S^{2-} to produce different minerals such as ferrihydrite, goethite, hematite, etc. A reaction between iron (Fe^{2+}) and S^{2-} under specific geological conditions produces pyrite that is easily oxidised when exposed to O_2. Sedimentary rocks accumulate sulphur through the burial of organic matter and diagenetic processes. The geothermal activity in the high-sulphur deposits dissolves the sulphur compounds such as H_2S and S^0. When they come into contact with subsurface water bodies, these compounds escape to the atmosphere through sulphur springs. The burning of sulphur-containing substances results in the formation of SO_2. This is a major problem associated with coal-based power plants and the petroleum industries. Industrial emissions of SO_2 and H_2S and the burning of fossil fuels are the major causes of vertical transport to the atmosphere.

TABLE 29.1
A list of some sulphur species, their formulas, and oxidation states

Species	Formulas	Oxidation states
Sulphate	SO_4^{2-}	+6
Dithionate	S_2O^{2-}	+5
Sulphite	SO_3^{2-}	+4
Thiosulphate	$S_2O_3^{2-}$	+3
Tetrathionate	$S_4O_6^{2-}$	+2.5
Elemental sulphur	S^0	0
Organic sulphur	R-S-S-R*	-1
Pyrite	FeS_2	-1
Organic sulphur	R-SH	-2
Sulphide	S^{2-}	-2

* R is an organic group.

The death and decay of biological organisms release inorganic and organic sulphur compounds into the environment. Certain microorganisms feed on these compounds and mineralise sulphur species in the environment. The decomposition of biological matter releases some sulphur into the soil. The decomposed organic compounds serve as both a source and sink of sulphur in the soil systems. Microorganisms and plants can assimilate the inorganic sulphur and transform it into organic forms. However, plants can not readily assimilate organic sulphur from the soil. Organic sulphur compounds must first mineralise for plants to mobilise them. Humans and animals fulfil the requirements of sulphur mainly by ingesting plant products. Figure 29.1 illustrates a sulphur cycle showing the passage of sulphur compounds through terrestrial, aquatic, aerial, and biological systems.

Microbially catalysed reactions are one of the main drivers of sulphur biogeochemical processes. Many microbial pathways observe stepwise coupled redox reactions and play a major role in the global sulphur cycle. The microbial biomass consists of as much as 1% sulphur in its dry weight. Even certain microbial forms depend exclusively on sulphur species for their growth and metabolism. This chapter provides in-depth information on sulphur metabolism in various microorganisms.

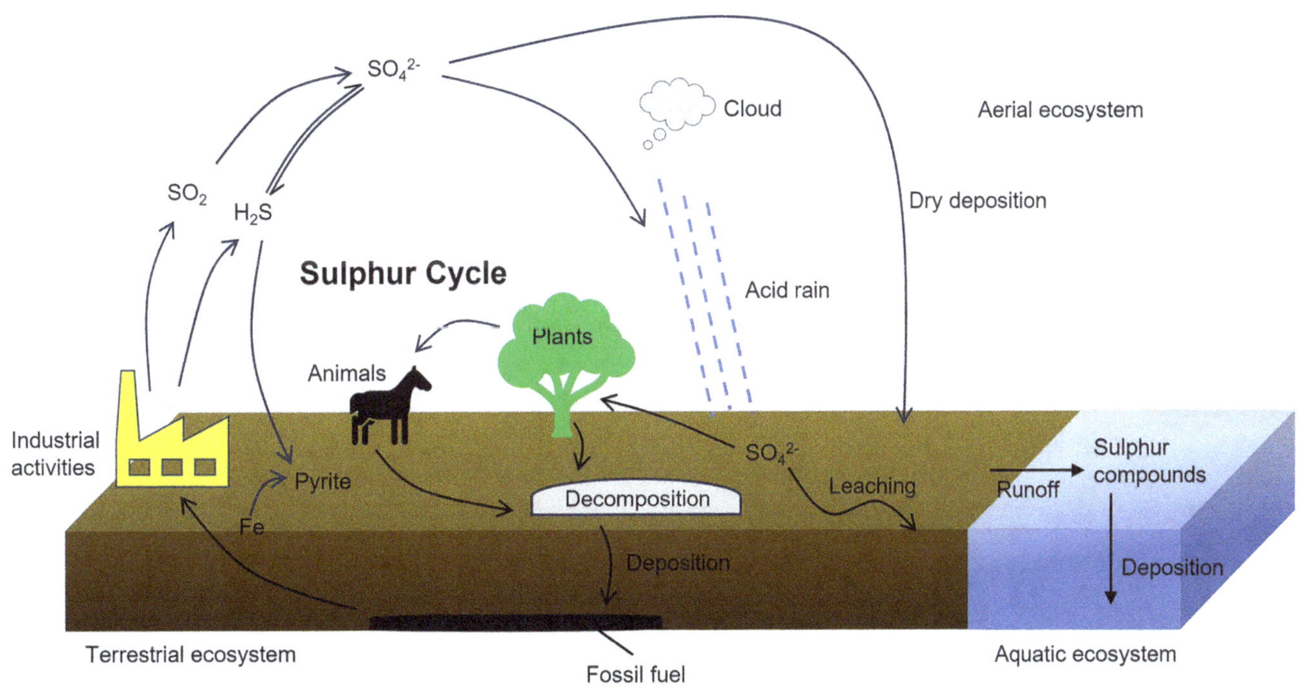

FIGURE 29.1 An illustration of the sulphur cycle, showing the transport of different sulphur species in different compartments of the environment.

29.3 SULPHUR METABOLISING MICROORGANISMS AND THEIR ACTIVITIES

The microbial sulphur metabolism is vital for the dynamics of the sulphur cycle (Figure 29.2). This mainly involves the reduction of sulphur species using sulphate-reducing microorganisms (SRMs) and the oxidation of sulphur species using sulphur-oxidising microorganisms (SOM). In distinct situations, S^{2-} serves as an energy source for chemotrophic reactions. The oxidation-reduction reaction between SO_4^{2-} and S^{2-} produces various intermediate species such as S^0, polysulphide, SO_3^{2-}, and $S_2O_3^{2-}$. Among all forms, S^0 is the least reactive because it has low solubility in H_2O.

The marine ecosystems observe an enigmatic biochemical process for sulphur metabolism in the midwater depths. Under an anoxic environment, a closely defined space facilitates the concurrent reduction and oxidation of different sulphur species. This process is termed the "cryptic sulphur cycle". This sort of sulphur metabolism is characterised by the quick conversion of different sulphur species (Figure 29.3). It is closely associated with the nitrogen metabolising process, anammox, and, therefore, linked with the nitrogen cycle. These processes are also often coupled with other biogeochemical cycles such as the carbon cycle, metal cycles, etc. A detailed description of the reduction and oxidation of sulphur compounds and other microbial-relevant processes is discussed in other sections of this chapter.

Many bacteria identified during the emergence of the discipline, microbiology in the 17th century were sulphur-metabolising taxa. A sulphur bacterium, *Beggiatoa,* is one of the first few newly discovered bacterial species. The discovery of sulphate-reducing bacteria (SRB) dates back to 1895, when Beijerinck demonstrated the role of sulphur in the microbial metabolic process. The discovery of a chemolithotrophic lifestyle was first observed in sulphur bacteria.

Sulphur bacteria are among the largest bacteria ever discovered. Because of their relatively big size, they could be easily observed at low magnification. They can attain a diameter of 100-750 μm, which is about 10-100 times bigger than the average size of any bacteria (5-10 μm). Examples of those include *Thiomargarita namibiensis*, *Beggiatoa* spp., *Thioploca*, and *Thiovulum majus* (Campylobacterota). A sulphur oxidising bacterium, *Thiomargarita magnifica* (Gammaproteobacteria), is a new name in the list of large bacteria, which broke all the records and boundaries of bacterial size. It measures up to 2 cm in length, which is a thousand times bigger than the size of a normal bacterium and is presently the biggest known bacterium on Earth. This bacterium was discovered in the mangrove sediments of Guadeloupe in the southern Caribbean Sea in 2022. This eyelash-sized bacterium has surpassed the theoretical limits of bacterial size, and the term "microorganism" has become a misnomer for this creature. *T. magnifica* is an example of a bacterium that has developed to a greater level of complexity. The interior of the large cell of *T. magnifica* consists of a large vacuole filled with fluid. The cytoplasm is a thin region around the periphery of the cell. Nucleic acid and ribosomes are present in membrane-bound structures called **pepins**.

Sulphur bacteria exhibit distinct routes of metabolism in the narrow set of chemical gradients of sulphur. Many species observe a strict chemolithotrophic lifestyle, thrive in a sulphur-rich environment, and obtain energy using inorganic sulphur compounds. The large sizes represent an adaptation to store enough nutrients and space for the biological reaction under oxygen-limited conditions.

The biogeochemical sulphur cycle in natural systems demonstrates an intricate relationship between chemistry

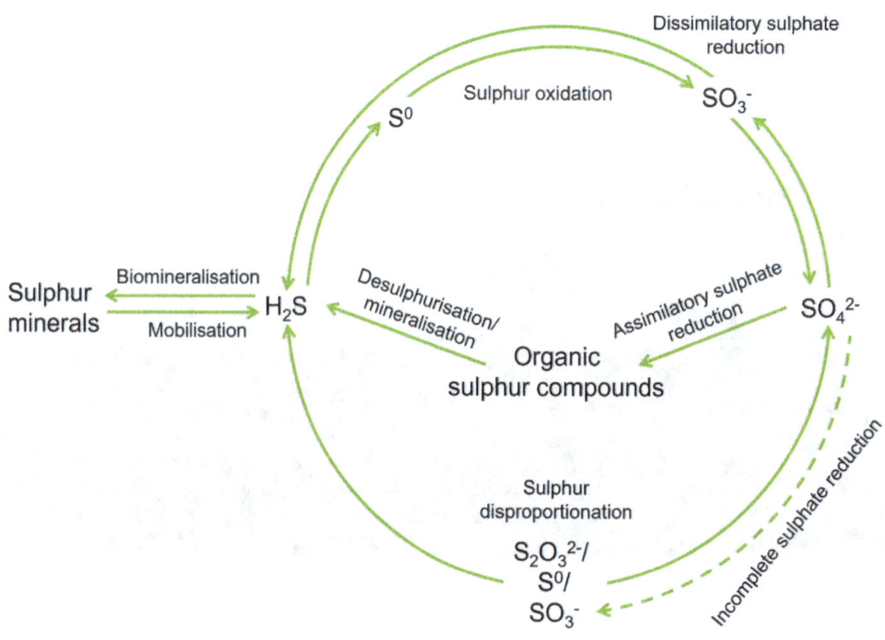

FIGURE 29.2 Microbially-mediated pathways in the sulphur cycle.

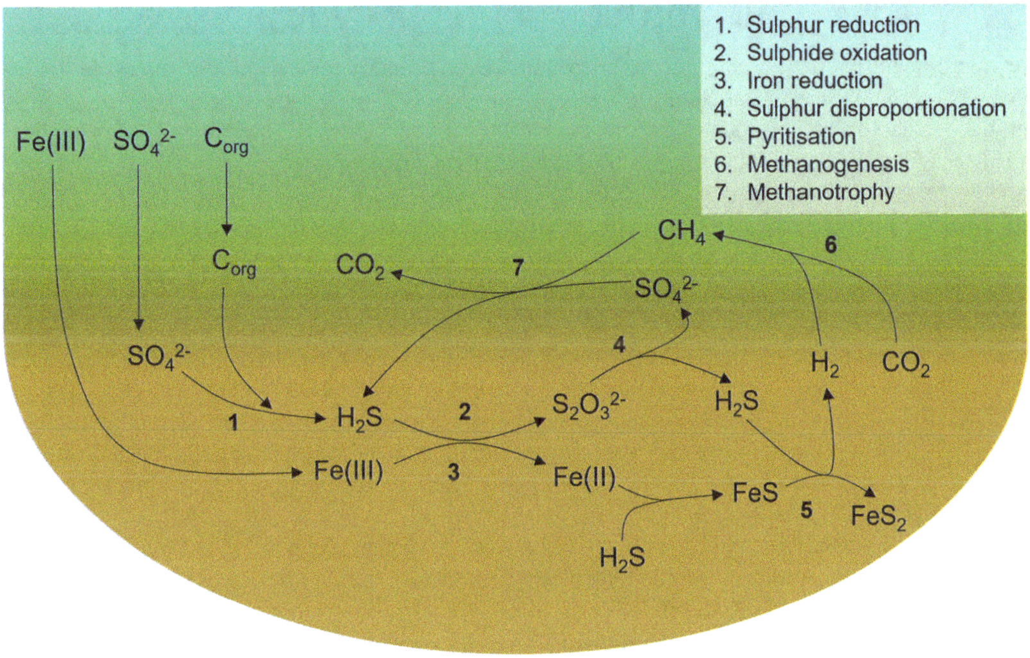

FIGURE 29.3 A "cryptic sulphur cycle" in marine sediments, where sulphate (SO_4^{2-}) is reduced to sulphide (S^{2-}) and then reoxidised involving various microbial reactions.

and biology. Since the origin of life, sulphur metabolism has played a key role in the growth and survival of organisms (Box 29.1). The modern world also witnesses sulphur-metabolising microorganisms in different biogeoactive zones. Oceans are areas of vast habitats for phylogenetically diverse sulphur-metabolising microorganisms. However, increasing the depth reduces the abundance of microorganisms, including sulphur-metabolising microorganisms. However, the diversity of SRMs is primarily dominated at the seafloor. The ocean surface is typically oxic (oxygen-rich) and has fewer reduced sulphur compounds and their metabolising microorganisms. The chemolithoautotrophic lifestyle harbours energy and electrons from sulphur oxidation for carbon fixation. In sulphur-respiring bacteria, sulphur is coupled to chemiosmotic ATP synthesis through an electron transport system (ETS). Sulphur-metabolising microorganisms also contribute to heavy metal leaching in acid mine drainage (AMD) sites.

Based on their functions, sulphur-metabolising microorganisms can involve dissimilatory and assimilatory processes. Dissimilatory pathways play a role in energy conservation coupled to microbial metabolism. These include sulphate reduction, sulphide oxidation, and sulphur disproportionation. However, the assimilatory pathway involves the assimilation of organic sulphur compounds in biological organisms for growth and metabolism.

In addition, mineralisation and desulphurisation facilitate the decomposition of sulphur compounds and the release of sulphur from materials. Some microorganisms also play roles in the synthesis of sulphur-containing minerals. Table 29.2 provides a list of various bioreactions of the sulphur cycle, including information on their environmental sites and examples of microorganisms. Sulphur metabolism in microorganisms features diverse mechanisms, discussed in the following sections.

29.3.1 Dissimilatory Reduction of Sulphur Compounds

Dissimilatory reduction of sulphur compounds is a process of energy production in certain microorganisms. This type of process represents a mode of anaerobic respiration in environments where O_2 is limited and oxidised sulphur species are abundant. Sulphur compounds serve as terminal electron acceptors in the respiratory pathway. Different sulphur compounds, including SO_4^{2-}, S^0, SO_3^{2-}, and $S_2O_3^{2-}$, can participate in this reduction pathway. Of different compounds, SO_4^{2-} is the most commonly available substrate, and the final product is S^{2-}. This process is specifically termed dissimilatory sulphate reduction (DSR). DSR is the main driver of the sulphur cycle in marine sediments. This process can also be reported from freshwater, hypersaline environments, polar regions, and hydrothermal sediments. Sulphate reduction by microorganisms is responsible for approximately 97% of total S^{2-} production on Earth.

This metabolism is characteristic of the many microbial groups collectively called SRMs. These include sulphate-reducing bacteria (SRB) and sulphate-reducing archaea. SRB is ubiquitous in marine sediments. Examples of SRB include *Desulfovibrio*, *Desulfobacter*, *Desulfobacterium*, *Desulfococcus*, *Desulfotomaculum*, and *Thermodesulfovibrio*. In archaea, genera like *Archaeoglobus*, *Pyrobaculum*, *Thermocladium*, and *Caldivirga* are notable candidates for the dissimilatory reduction of different sulphur compounds. Many research studies claim that SRMs are keystone organisms in

BOX 29.1 THE EVOLUTION OF LIFE USING SULPHUR SPECIES

The sulphur microorganisms have a magnificent history and were possibly responsible for the evolution of life around 3.8 billion years ago (bya). Sulphur metabolism is an ancient process. During the Archean Eon, which is over 4.0 bya, volcanic activities released a huge amount of H_2S on the Earth's surface. Due to its abundance and metabolic significance, early life forms heavily relied on sulphur species for their growth and metabolism.

Before the evolution of O_2, sulphur species were essential nutrients for metabolic activities such as photosynthesis and respiration. The autotrophic lifestyle was dependent on sulphur oxidation for electron donors (equivalent to water hydrolysis of oxygenic photosynthesis) and energy requirements (as a substitute for sunlight for photolysis in dark regions). The reduced sulphur species were also the key potential energy source for chemotrophs in aphotic environments. The chemolithoautotrophic lifeforms that survived on sulphur in early Earth are among the oldest microfossils recorded so far. Organismal evolution in sulphur microorganisms led to innovations of many metabolic pathways using sulphur species. In this series, sulphide oxidation induced the synthesis of a variety of inorganic and organic products. The reaction of S^{2-} with Fe and other d-block elements produced many catalysts for chemical reactions. An example of a sulphur catalyst is zinc sulphide (ZnS), which leads to the synthesis of organic bonds such as carbon-hydrogen bonds. A different catalyst, iron sulphide (FeS_2), has enhanced the phosphate solubility in the soil. In addition, bioactive sulphur species also assisted the carbon and nitrogen cycles.

These reactions elevated the concentrations of oxidised sulphur species, such as S^0, $S_2O_3^{2-}$, SO_3^{2-}, and SO_4^{2-}, in the environment. These metabolic products of photosynthesis and chemosynthesis served as a terminal electron acceptor in anaerobic respiration (equivalent to the role of O_2 in aerobic respiration). Sulphur reduction associated with sulphur disproportionation was a signature of the predominant pathway for early microbial metabolisms in the anoxic seabed.

The great oxidation event (GOE) around 2.3 bya marginalised the sulphur metabolism. Around 600 million years ago (mya), when O_2 concentration in the atmosphere increased several fold, sulphur metabolism was greatly shrunken and escaped to isolated niches. The remnants of the early metabolic capabilities of sulphur still exist at different corners of the globe. Phylogenetic studies have shown that many modern sulphur-metabolising microorganisms belong to the same lineages of early metabolism.

TABLE 29.2
A list of different bioreactions in the sulphur cycle, including information on microorganisms, nutritional modes, and environmental sites

Types of bioreactions	Microbial types	Examples of microorganisms*	Nutritional modes	Environmental sites
Dissimilatory sulphate reduction	Bacteria and archaea	*Desulfovibrio, Desulfobacter, Desulfobacterium, Desulfococcus, Desulfotomaculum, Thermodesulfovibrio* *Archaeoglobus, Pyrobaculum, Thermocladium, Caldivirga*	Lithotrophs or organotrophs	Anaerobic situations
Assimilatory sulphate reduction	All types of organisms	-	All modes	Aerobic and anaerobic situations
Sulphur oxidation	Bacteria and archaea (order Sulfolobales)	Green sulphur bacteria, purple sulphur bacteria, Heliobacteria, *Thiobacillus, Beggiatoa* *Paracoccus, Sulfolobus*	Photolithotrophs, chemotrophs, mixotrophs, and heterotrophs	Aerobic and anaerobic situations
Mineralisation/ desulphurisation	Bacteria, archaea, and fungi	*Thiobacillus, Rhodococcus,* *Sulfolobus* Ascomycetes and Basidiomycetes members	Chemolithotrophs and heterotrophs	Aerobic and anaerobic situations
Sulphur disproportionation	Bacteria and archaea	*Desulfobacter, Desulfococcus, Desulfotomaculum, Desulfotomaculum, Desulfovibrio* *Acidianus*	Chemolithotrophs	Anoxic aquatic habitats
Biomineralisation	Magnetotactic bacteria	*Magnetospirillum, Magnetococcus, Magnetospira*	Chemotrophs	The oxic-anoxic transition zone of aquatic habitats

* The black, blue, and brown coloured words in the third column represent the examples of bacteria, archaea, and fungi, respectively.

energy-limited conditions, contributing to energy flow in anaerobic ecosystems.

DSR involves a complex pathway involving many enzymes. However, three enzymes, sulphate adenylyltransferase, adenosine 5' phosphosulphate (APS) reductase, and dissimilatory sulphite reductase (DsrAB), are crucial for the reduction process and are found in SRMs. Sulphate adenylyltransferase contributes to the activation of SO_4^{2-}, catalysing the conversion of SO_4^{2-} and ATP into APS and pyrophosphate. The activation process reduces the standard redox potential of SO_4^{2-} and allows the transfer of two electrons to form an intermediate product, SO_3^{2-}. This step involves a different enzyme, APS reductase. This is followed by the action of DsrAB, which transfers six electrons to SO_3^{2-} and produces S^{2-} (Figure 29.4). Under substrate-limiting conditions, SRB can facilitate incomplete reduction of SO_4^{2-} and produce $S_2O_3^{2-}$.

SRMs can be lithotrophs or organotrophs based on the sources of electrons. In the marine ecosystem, DSR by organotrophs is closely associated with the carbon cycle. The conversion of one molecule of SO_4^{2-} into S^{2-} consumes eight electrons, deriving from the oxidation of organic compounds. Approximately 50% of organic mineralisation in marine sediment is associated with a syntropic relationship with DSR.

SRB are potential bioremediators for treating acid mine drainage. Many of them exhibit resistance to multiple heavy metals and can be used for their treatment in contaminated soil. Also, they can be employed as effective tools for the bioleaching of heavy metals.

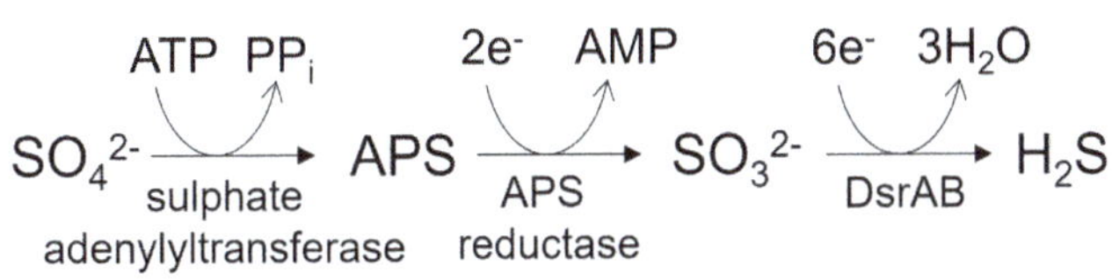

FIGURE 29.4 Dissimilatory sulphate reduction (DSR) pathway. (APS: adenosine 5' phosphosulphate, DsrAB: dissimilatory sulphite reductase) (SO_4^{2-}: sulphate, SO_3^{2-}: sulphite: SO_3^{2-}, H_2S: hydrogen sulphide)

29.3.2 Assimilatory Sulphate Reduction (ASR)

Assimilatory sulphate reduction (ASR) is a general process of sulphur assimilation that facilitates the conversion of inorganic sulphur species into organic forms. The organic sulphur compounds are incorporated into the biomass of organisms. This process can be reported in all types of microorganisms and plants. ASR and DSR share a common pathway for the reduction reaction. However, the final product, S^{2-}, in the ASR pathway, undergoes further modification to produce organic sulphur compounds. Microorganisms have a tremendous ability for ASR to produce biologically relevant substances.

Sulphur compounds can be assimilated into both inorganic and organic frameworks (Figure 29.5). The inorganic frameworks include iron-sulphur clusters, which are cofactors in several metalloproteins such as ferredoxins, nitrogenase, NADH dehydrogenase, and hydrogenase. The organic sulphur compounds are building blocks of sulphur-containing amino acids (e.g., cysteine and methionine), vitamins (e.g., biotin and thiamine), oligopeptides (e.g., glutathione (GSH)), antibiotics (e.g., penicillin and cephalosporins), etc. The synthesis of cysteine involves a step of sulphur incorporation, allowing the

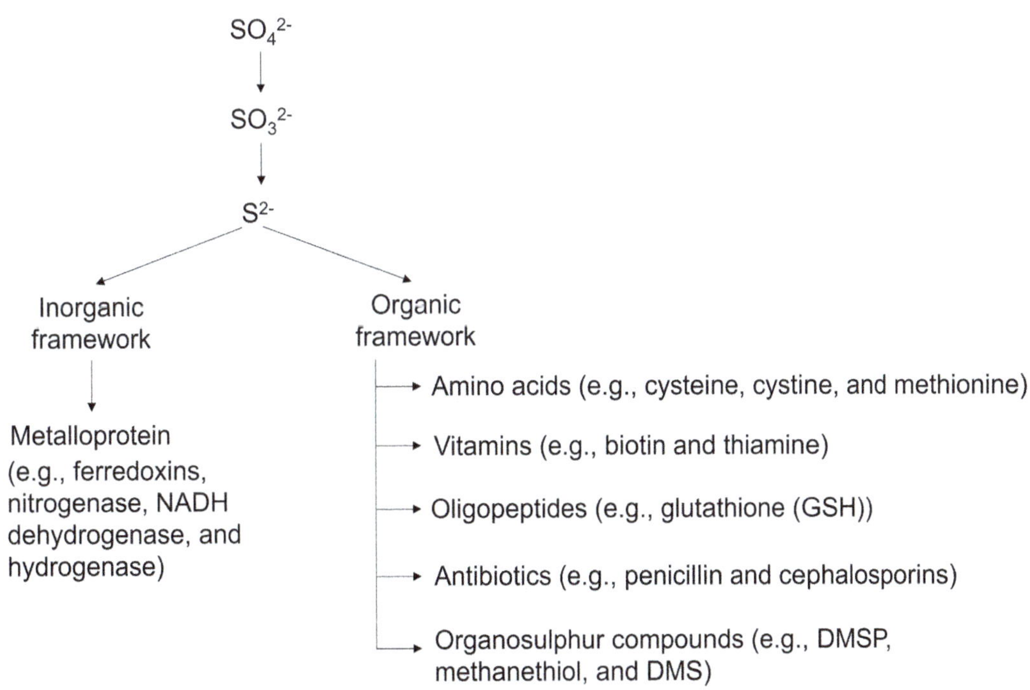

FIGURE 29.5 Assimilation of sulphur in a biological system (SO_4^{2-}: sulphate and SO_3^{2-}: sulphite.)

reaction between sulphide and O-acetyl serine. This amino acid acts as a central compound that produces many other organic sulphur compounds, including homocysteine and methionine. The synthesis of organic sulphur compounds, DMSP, in marine ecosystems is also a biologically mediated process. Methionine is the precursor for the synthesis of DMSP. Further, DMSP can be degraded into DMS using microorganisms, which is ecologically relevant for aquatic food webs and atmospheric chemistry (detailed in section 29.5).

29.3.3 Oxidation of Sulphur Compounds

In oxidative environments, SOMs facilitate oxidation reactions for the reduced sulphur species and transform them into different sulphur oxyanions. A microbial oxidation pathway typically involves the conversion of S^{2-} into S^0, which can be further converted into SO_3^{2-} and ultimately SO_4^{2-} (the most oxidised form of sulphur). Both aerobic and anaerobic situations can harbour SOM. Aerobic situations employ O_2 for the oxidation reactions, while anaerobic situations depend on other electron donors such as NO_3^- and nitrite (NO_2^-). SOM are diverse groups of bacteria and archaea showing photolithotrophic and chemotrophic lifestyles for sulphur metabolism. In archaea, they particularly belong to the order Sulfolobales (e.g., *Paracoccus* and *Sulfolobus*).

The process of sulphur oxidation can vary across different microorganisms. Microbial oxidation of sulphur occurs through three different pathways, *viz.* the sox pathway, the tetrathionate (SI4) pathway, and the branched pathway for thiosulphate oxidation (Figure 29.6). The sox pathway utilises three different substrates, including S^{2-}, $S_2O_3^{2-}$, and S^0. It involves sox enzymes (soxXA and soxYZ) and produces active S^0. A different set of enzymes (soxCD and soxB) transforms S^0 into SO_4^{2-}. These reactions generate free electrons, which can be channeled into cytochrome c (cyt c) for energy production. Such activity is prominent in bacteria like *Acidithiobacillus*, *Beggiatoa*, and *Paracoccus denitrificans*. The tetrathionate (SI4) pathway involves a distinct step and produces an intermediate tetrathionate ($S_4O_6^{2-}$) from $S_2O_3^{2-}$, which is further oxidised into SO_3^{2-}. Following

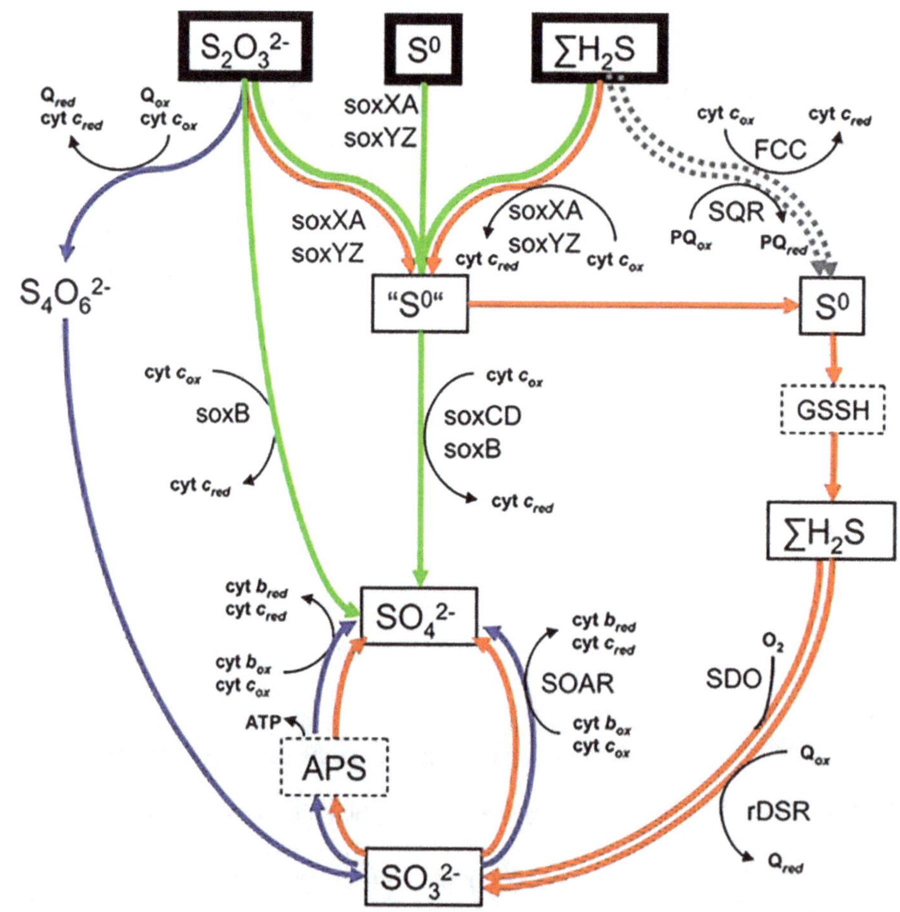

FIGURE 29.6 Microbial oxidation of sulphur through different pathways. Green arrows represent the sox pathway, blue arrows the tetrathionate (SI4) pathway and red arrows correspond to the branched pathway for thiosulfate oxidation. Possible entry sites of H_2S that are not part of these traditional pathways are shown with a grey dotted arrow. (FCC: flavocytochrome c oxidoreductase, SQR: sulphide: quinine: oxidoreductase, GSSH: S-sulphanyl glutathione, APS: adenosine-5′-phosphosulphate, SOAR: sulphite: cyt c: oxidoreductase, rDSR: reverse dissimilatory sulphate reductase, SDO: sulphur dioxygenase, soxXA, soxYZ, and soxB are subunits of the thiosulphate-oxidising multi-enzyme (TOMES) complex) (Image reproduced from Klatt and Polerecky (2015).[3])

that, SO_3^{2-} is transformed into SO_4^{2-} using different enzymes, and this reaction produces energy. This type of pathway is reported in *Thiobacillus thioparus* and *Acidithiobacillus* spp. The branched pathway for thiosulphate oxidation starts with the oxidation of $S_2O_3^{2-}$ into S^0, followed by its transformation into S-sulphanyl glutathione (GSSH). The GSSH is converted into S^{2-} that can be oxidised into SO_3^{2-} and SO_4^{2-} using enzymes like reverse dissimilatory sulphate reductase (rDSR) and sulphur dioxygenase (SDO). Differently, in the bacterium *Allochromatium vinosum*, S^{2-} is converted into SO_3^{2-} using flavocytochrome c sulphide dehydrogenase (FCC) and then SO_3^{2-} into S^0 using sulphide dehydrogenases (SQR). The electrons released from this reaction are shuttled into cyt c in ETS for energy production. The green sulphur bacterium, *Chlorobium limicola*, oxidises S^{2-} into S^0 using SQR and donates its electrons to PQ for photosynthesis.

In a photolithotrophic lifestyle, sulphur oxidation is a source of electrons for anoxygenic photosynthesis. Sulphur microorganisms use light-driven energy to stimulate the photosynthetic reaction centre, where sulphur metabolism satisfies the need for electrons. This process is equivalent to the photolysis of H_2O molecules that release O_2 in oxygenic photosynthesis. The taxonomically diverse sulphur bacteria, such as green sulphur bacteria, purple sulphur bacteria, and Heliobacteria, are obvious examples of obligatory sulphur photoautotrophy. They particularly belong to the phyla Chlorobi and Gammaproteobacteria. They possess distinct photosynthetic pigments, bacteriochlorophyll, which are analogous to chlorophyll. Bacteriochlorophylls capture light energy in the specific wavelengths of light (in the far-red region of the light spectrum) and transfer it to the reaction centre for the release of electrons from electron donors for carbon fixation. These sulphur-oxidising bacteria, along with cyanobacteria, can form multilayered sheets as the microbial mat (Figure 29.7). They can also be found growing on a moist or submerged surface and require the presence of sunlight.

Sulphur oxidation can also be a mode of producing energy in chemolithotrophs. This is a critical process for primary production in aphotic sulphur-containing aquatic zones. This process is analogous to the energy requirement from sunlight for light reaction in photosynthesis (Figure 29.8a–d). Even some chemolithotrophs can meet the requirements of both energy and electrons from reduced sulphur compounds (Figure 29.8e). This type of sulphur-dependent metabolism (i.e., chemolithoautotrophic lifestyle) was probably the first self-regulating biological activity on the primordial Earth. They were also possibly the progenitor of all life species. In modern times, *Acidithiobacillus thiooxidans, Thiobacillus denitrificans*, and *Beggiatoa* are the most well-known chemolithotrophs that oxidise sulphur.

Sulphur oxidation can also be reported from some mixotrophs (that utilise both autotrophic and heterotrophic lifestyles) and heterotrophs, but sulphur oxidation through these microorganisms lacks energy efficiency. Heterotrophic bacteria such as *Arthrobacter*, *Pseudomonas*, *Bacillus*, and *Micrococcus* can contribute to the oxidation of S^0 into $S_2O_3^{2-}$/SO_4^{2-} and oxidation of $S_2O_3^{2-}$ into SO_4^{2-}. The biooxidation of S^{2-} into SO_4^{2-} is dominant in soil and prevails in conditions that have high pH and warm temperatures. This type of reaction is particularly relevant in agricultural land for the sulphur uptake by the crop plants.

Sulphide-rich aquatic and coastal sediments possess a unique electrically conductive system, facilitating sulphur oxidation via **cable bacteria** (Figure 29.9). The cable bacteria are a long chain of hundreds to thousands of cells, spanning several centimetres. This type of structure fills the gap between the deep sediment layer and the sediment-water interface. When sediment regions do not contain enough O_2 and other terminal electron acceptors, they employ cable bacteria to couple the sulphide oxidation, reducing O_2 or NO_3^- from the surface layer.

29.3.4 Sulphur Disproportionation

Sulphur disproportionation is a microbially-mediated reaction that couples the oxidation and reduction of inorganic sulphur

FIGURE 29.7 a. Diagram of a microbial mat showing multiple layers of biofilms embedded in a matrix. Size ranges from a few millimetres to several centimetres. The top zone represents cyanobacteria and diatoms, followed by zones of cyanobacteria, purple sulphur bacteria, and sulphate-reducing bacteria. The top zone is rich in oxygen (O_2), while the bottom zone has high sulphite (SO_3^{2-}) concentrations, recognised as the sulphidic zone. b. and c. Images of cross-sections of microbial mats showing distinct zones. (Images b and c adapted from Prieto-Barajas et al. (2018)[4] with permission from © 2017 Pontificia Universidad Católica de Valparaíso. Production and hosting by Elsevier B.V.)

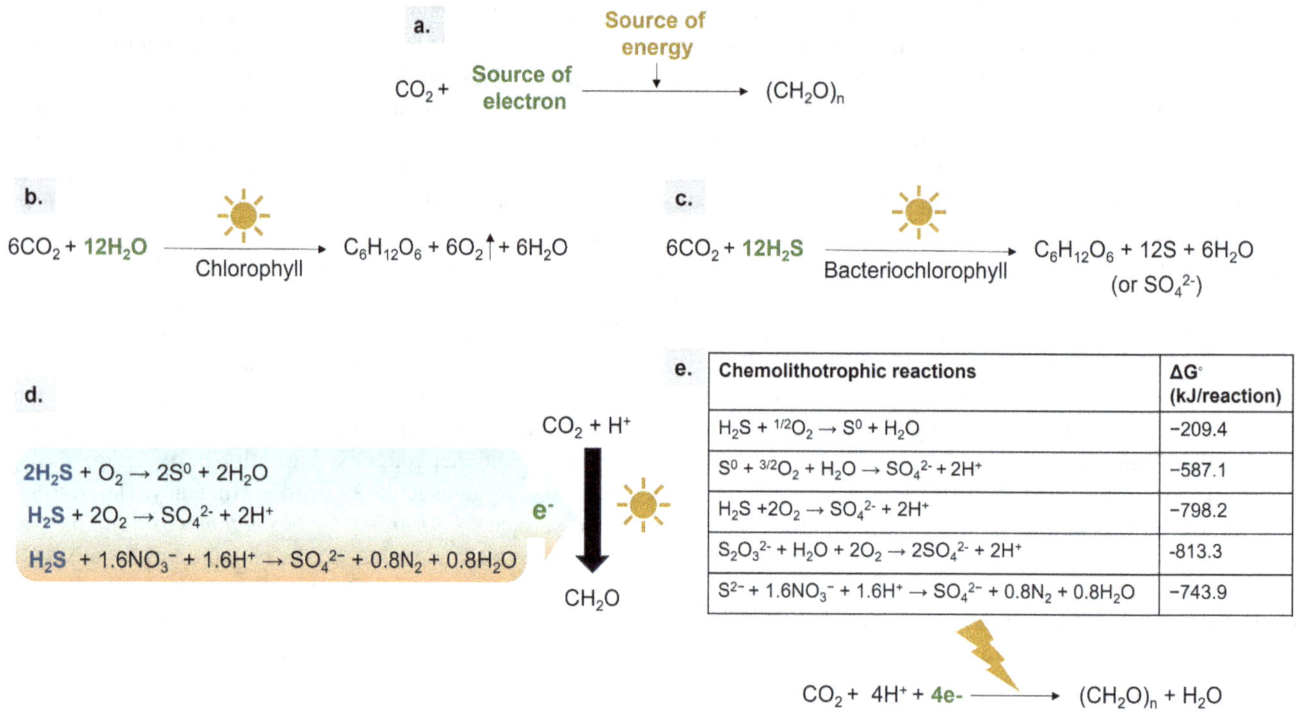

FIGURE 29.8 a. An illustration of the reaction of primary production in different environments. b. Reaction of oxygenic photosynthesis. c. Reaction of anoxygenic photosynthesis using hydrogen sulphide (H_2S). d. Reactions of sulphide oxidation under aerobic (green) and anaerobic situations (brown), providing electrons for carbon fixation. e. Chemolithotrophic reactions of different sulphur species, providing energy for carbon fixation.

into two different products. Thiosulphate, SO_3^{2-}, or S^0 can be different substrates for the reaction, and the end products are essentially SO_4^{2-}/SO_3^{2-} and S^{2-}. This process is an example of a chemolithotrophic reaction that supports primary production in the dark zones. This process also encourages the rate of pyrite formation.

The process of sulphur disproportionation is also termed "inorganic fermentation" or "dismutation" because it shares similarities with the fermentation of organic compounds. Both processes involve the redistribution of electrons in single elements through redox reactions (sulphur in the case of disproportionation and carbon in the case of fermentation). A different common characteristic is that they both produce energy without integrating with the electron transport system and proton gradient.

Only a subset of the metabolic pathways and enzymatic processes behind sulphur disproportionation is understood. Sulphur disproportionation can involve different enzymes for different substrates and produce different products. When the substrate is $S_2O_3^{2-}$, it is cleaved into S^{2-} and SO_3^{2-} using thiosulphate reductase. In a different reaction, SO_3^{2-} is broken into SO_4^{2-} using sulphite oxidoreductase. The sulphur intermediate of this reaction involves a different enzyme, sulphite reductase and produces S^{2-}. The pathway of S^0 disproportionation is currently unresolved. Figure 29.10 provides a list of different reactions of sulphur disproportionation.

Sulphur disproportionation is one of the most ancient biological processes. The evidence can be traced back to approximately 3.5 bya. The sulphide-mineral imprints in ancient and young sediments are the isotopic signatures of sulphur disproportionation. In the modern world, this reaction prevails in anoxic aquatic habitats like marine sediments, deep-sea hydrothermal vents, groundwater, and freshwater basins, but does not entirely preclude the oxic situations. In marine sediments and groundwater, sulphur bacteria conserve energy using sulphur disproportionation. Sulphur disproportionation has been reported from as many as forty-three SRB. Examples of those include *Desulfobacter*, *Desulfococcus*, *Desulfotomaculum*, and *Desulfovibrio*. This process can be reported in other types of bacteria belonging to Deltaproteobacteria and Proteobacteria. A few taxa showing this reaction can belong to Thermodesulfobacteria, Epsilonproteobacteria, Nitrospirae, and Firmicutes. New studies have reported that groundwater cable bacteria also use sulphur disproportionation to store energy. Some archaea, like *Acidianus*, have displayed disproportionation reactions even in aerobic situations.

29.3.5 Decomposition and Mineralisation

In the biogeochemical cycle of sulphur, decomposition and mineralisation are the processes of removing sulphur content from the bulk quantities of organic and inorganic matter

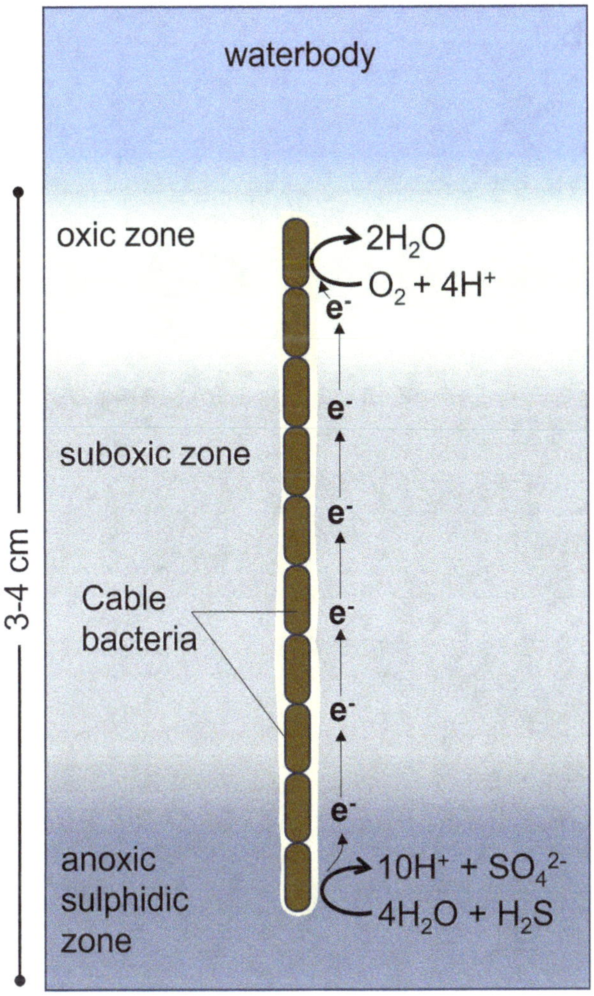

FIGURE 29.9 Cable bacteria showing conductivity of electrons in sediments.

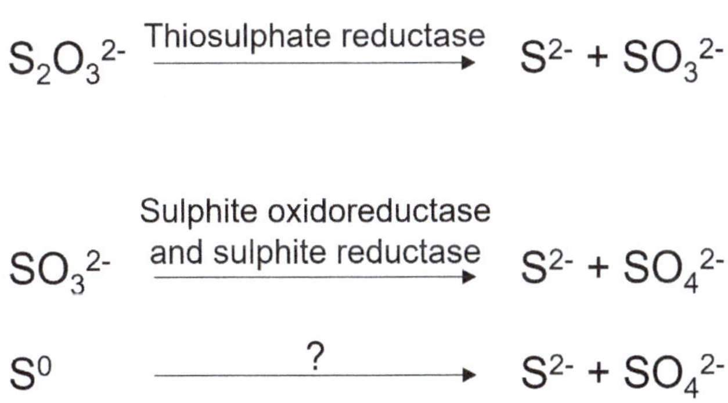

FIGURE 29.10 Different reactions of sulphur disproportionation.

using decomposers. Both terrestrial and aquatic ecosystems accumulate a large amount of organic sulphur compounds and are the dominant sites for such reactions. The list of decomposers that act on sulphur-containing compounds includes various bacteria and fungi. Among many suitable microorganisms, *Thiobacillus* spp. do an excellent job in biodegradation and mineralisation of sulphur-containing compounds. Certain thermophilic bacteria display dissimilative organic-sulphur oxidation. Enzymes are the main players in the removal of sulphur species from organic compounds. The proteolytic enzymes depolymerise sulphur-containing amino acids in proteins, followed by the release of sulphur-containing amino acids. The enzymes like cysteine desulphurase and 3-mercaptopyruvate sulphurtransferase have the properties of catabolism of cysteine. Sulphatases catalyse the hydrolysis of sulphate ester bonds from organic sulphur compounds, releasing SO_4^{2-}. Dimethylsulphide monooxygenase (DmoA) and methanesulphonate monooxygenase (MsmA) are responsible for the oxidation of DMS and methanesulphonate, respectively. An organosulphur compound, dibenzothiophene (DBT), which is dominantly present in marine water, is a target for microbial desulphurisation using dibenzothiophene (DBT) degrading enzymes (detailed in section 29.5).

29.3.6 Formation of Sulphur Minerals

The synthesis of metal sulphides at low-temperature conditions is likely a biologically-induced process. The site for the mineral formation can be extracellular or intracellular. In an extracellular process, microbial sulphate-reducing activity produces S^{2-} in the environment, which interacts with metals like Fe to form metal sulphides. In a different scheme, metals can bind to cell walls or exopolysaccharides (EPS) and react with S^{2-}. This type of mineral formation is termed biomineralisation (Figure 29.2). The synthesis of a sulphur-containing mineral, pyrite, involves a similar mechanism in anoxic deep seabeds.

Some bacterial forms called **magnetotactic bacteria** possess a specialised bilayered membrane-enclosed structure called a **magnetosome** for sulphur mineral formation (Figure 29.11). *Magnetospirillum*, *Magnetococcus*, and *Magnetospira* are some notable examples of magnetotactic bacteria. These bacteria predominantly exist near the oxic-anoxic transition zone of marine and freshwater ecosystems. The intracellular synthesis of iron sulphur minerals produces crystals of greigite (Fe_3S_4) and mackinawite ($FeS_{(1-x)}$). The interaction between Fe and biogenic S^{2-} within the cells involves an exergonic reaction. The energy produced during the reaction can be used for various metabolic events. In a different setting, the reduced sulphur products from sulphur disproportionation and other redox reactions involving polysulphide formation can also play a role in the synthesis of sulphur-containing minerals.

29.4 DIMETHYL SULPHIDE (DMS) IN THE MARINE SULPHUR CYCLE

The DMS production is critical to the marine sulphur cycle. It is a volatile organic compound that resides on the ocean surface and can escape into the atmosphere. DMS is found in nearly all seawater, giving a characteristic seawater smell. Seawater releases about 28–45 Tg of DMS into the atmosphere each year. A report by NASA (2023)[1] claimed that the entire DMS present on Earth is a product of only the biological process in the marine ecosystem. It is a product of the biodegradation

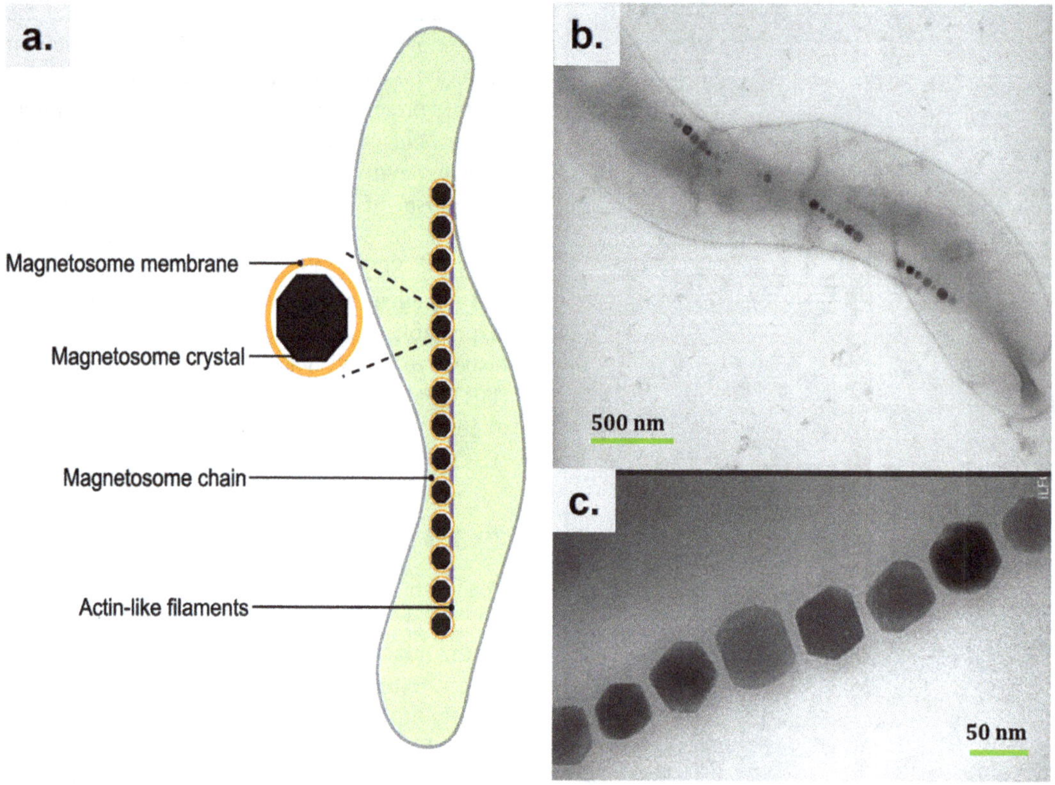

FIGURE 29.11 Illustration of magnetotactic bacteria. a. Diagram of membrane-bounded magnetosome chain inside the cell. b. Transmission electron micrograph (TEM) image of magnetosomes in *Magnetospirillum magneticum* AMB-1 cells. The magnetite crystals are aligned within the cytoplasm of the cells on a cytoskeleton made of actin-like proteins. c. An enlarged view of magnetite crystals. (Image a adapted from Lin et al. (2020)[5], and b and c from Ginet et al. (2011).[6])

of DMSP, which is chiefly produced by phytoplankton, plant halophytes, and marine algae. A sulphur-containing amino acid, methionine, is the precursor of DMSP. The massive quantities of DMSP in marine environments stimulate the synthesis of DMS. When DMSP is discharged into the water column, phytoplankton (chiefly coccolithophores) and chemotrophic bacteria act on this organosulphur compound and split it into DMS using enzymes called DMSP lyases. A giant coccolithophore, *Gephyrocapsa huxleyi* is a major contributor to the massive amount of accumulation of DMS in seawater. The degradation process also releases a different organosulphur compound, methanethiol (CH_3SH), which is assimilated in microbial biomass. DMS can also be assimilated, but it has a slow rate of digestion, resulting in its huge accumulation (about 10 times higher than methanethiol) in seawater.

DMS is the most prevalent biological source of sulphur in the atmosphere. An estimate suggests that atmospheric DMS contributes to ~43% of global SO_4^{2-} aerosols. A group of scientists has developed a CLAW hypothesis[2], which states that DMS-producing microorganisms induce a negative feedback loop during DMS production. An intense amount of energy from the sunlight promotes the growth of DMSP- and DMS-producing organisms, leading to large-scale DMS production followed by its emission into the atmosphere. The atmospheric sulphur chemistry mediates some photochemical reactions and converts organic sulphur compounds into inorganic sulphur compounds. The oxidation reaction turns DMS into SO_2 and SO_4^{2-}. These ions act as cloud condensation nuclei, which increases the liquid water content of clouds and cloud areas. Following that, it has an impact on Earth's radiative balance, resulting in high reflection of sunlight. This ultimately damps the growth of photosynthetic organisms and inhibits the DMSP and DMS production (Figure 29.12).

29.5 DESULPHURISATION

Desulphurisation is an industrial process that removes excess sulphur from fossil fuels such as coal, petroleum, and other waste products. The organic sulphur content in crude petroleum oil can be around 0.1–5.0 wt%, and its removal during oil purification is obligatory. This improves oil efficiency and reduces SO_2 emissions during consumption. Desulphurisation can be achieved using chemical or biological methods. In recent years, researchers have identified that the natural abilities of SRMs have tremendous potential for the same. These microorganisms remove sulphur as S^{2-}/SO_3^{2-}. *Rhodococcus* is a notable candidate for sulphur removal from both inorganic and organic compounds, owing to its ability to withstand extreme conditions. This bacterium consists of a 4S pathway involving a four-step enzymatic reaction. The four enzymes of the pathway are DszC, DszA, DszB, and DszD, typically found in the *dsz* operon.

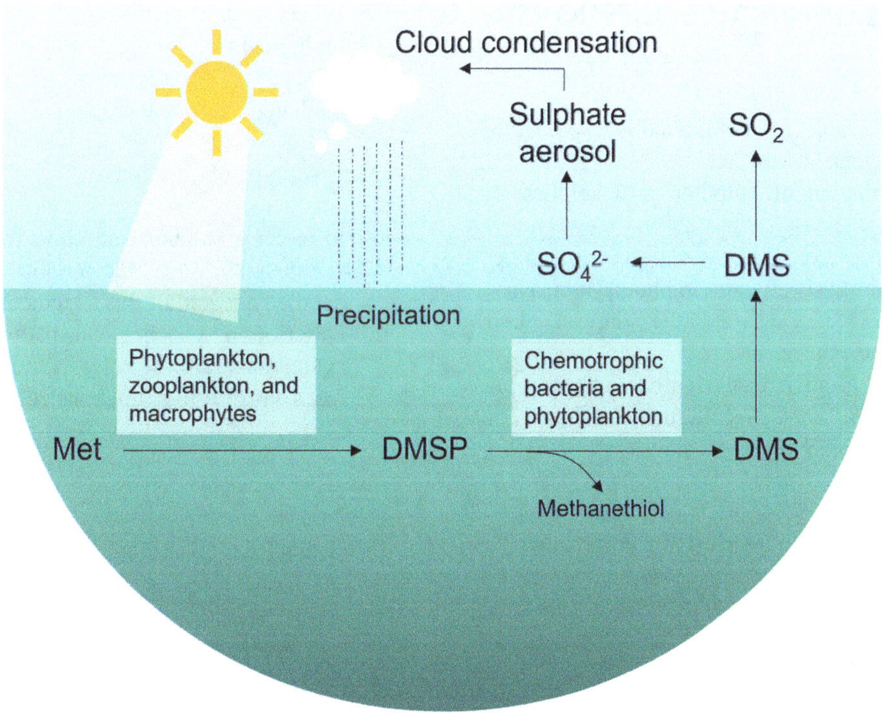

FIGURE 29.12 DMS feedback loop, showing synthesis and degradation of DMS in the marine ecosystem. (Met: methionine, DMSP: dimethylsulphoniopropionate, DMS: dimethyl sulphide.)

Some species of *Gordonia* (a Gram-positive bacterium) have also shown excellent capacity for desulphurisation. This bacterium is renowned for its hydrocarbon-degrading capabilities and can also be recruited to improve the quality of fossil fuels. Particularly, *Gordonia amicalis* is a potential contributor to desulphurisation. Genetic engineering approaches in *Gordonia* have aligned the 4S pathway and targeted specific sulphur compounds. Advances in bioprocess engineering have made biological desulphurisation a highly efficient, cost-effective, and sustainable technology for the refining of fossil fuels.

SUMMARY

- The sulphur cycle mainly runs as the sedimentary cycle. Sulphur in the environment exists from the most reducing species of sulphide (S^{2-}) to the most oxidising species of sulphate (SO_4^{2-}). The intermediate sulphur species include sulphite (SO_3^{2-}), elemental sulphur (S^0), thiosulphate ($S_2O_3^{2-}$), and dithionate.
- Microbial sulphur metabolism is one of the most ancient biological signatures on Earth. They play a prominent role in the regulation of the sulphur cycle. In particular, sulphur-reducing microorganisms (SRMs) and sulphur-oxidising microorganisms (SOMs) facilitate the mobility of different sulphur species in the environment.
- SRMs can display either dissimilatory or assimilatory sulphur reduction. When oxygen (O_2) is scarce and sulphur compounds are plentiful, microorganisms may acquire a dissimilatory reduction of sulphur to conserve energy. In a different setting, the assimilatory reaction promotes the assimilation of sulphur species in the biomass.
- Sulphur oxidation by SOM is equivalent to the photolysis of water (H_2O) molecules in anoxygenic photosynthesis. In a chemolithotrophic process, the oxidation of sulphur species provides energy for primary production in the aphotic zone.
- Sulphur disproportionation combines the oxidation and reduction of inorganic sulphur species ($S_2O_3^{2-}$, SO_3^{2-}, or S^0) to produce two distinct products (SO_4^{2-}/SO_3^{2-} and S^{2-}). Many heterotrophic microorganisms contribute to the degradation of bulk quantities of sulphur-containing organics and inorganics.
- Sulphur mineralisation is a process of the synthesis of sulphur-rich mineral deposits, such as pyrite. A unique bilayered membrane-enclosed structure known as a magnetosome found in some bacterial species (magnetotactic bacteria) exerts some efforts in the formation and storage of FeS minerals.
- DMS is a volatile sulphur compound generated predominantly by marine phytoplankton (mainly coccolithophores). This compound is also responsible for the characteristic odour of seawater.
- Biological desulphurisation is a process in industrial biotechnology to remove excess sulphur from fossil fuels like coal and oil. One notable mechanism for removal is their bioreduction into hydrogen sulphide (H_2S) using SRMs.

DIFFERENT TYPES OF PRACTICE QUESTIONS

Fill in the blanks

I. .... compound is responsible for the specific odour of seawater.
II. The transformation of sulphide into sulphate is a process of ....
III. In the sulphur cycle, ... is an analogous process of photolysis of water in photosynthesis.
IV. .... is the largest known bacterium on Earth.
V. .... is an industrial process that removes excess sulphur from fossil fuels such as coal, petroleum, and other waste products.

SOLUTION: I. Dimethyl sulphide (DMS), II. oxidation, III. sulphur oxidation, IV. *Thiomargarita magnifica*, V. Desulphurisation

Match the following

i. Green sulphur bacteria	a. Biomineralisation
ii. Magnetotactic bacteria	b. Photosynthesis
iii. Cable bacteria	c. Sulphur oxidation in sediments
iv. Sulphate dissimilatory reduction	d. Cloud formation
v. DMS production	e. Respiration

SOLUTION: i-b, ii-a, iii-c, iv-e, v-d

Very short answer type of questions

I. What is inorganic fermentation?
II. What are the sulphur bacteria?
III. Write the names of any two bacteria that observe a chemotrophic lifestyle using sulphur compounds.
IV. What are cable bacteria and their roles in the sulphur cycle?
V. What is a cryptic sulphur cycle?
VI. What is the reason that the first few microorganisms identified during the inception of microbiology belonged to sulphur bacteria?

Short answer type of questions

I. What is the role of sulphur in anaerobic respiration? What are the situations that promote this type of reaction?
II. What are magnetotactic bacteria? What are their roles in the sulphur cycle?
III. What is the negative feedback loop during DMS production?
IV. What is desulphurisation? What are the different biological activities responsible for desulphurisation?
V. What is the evolutionary significance of sulphur-mediated biological reactions?

Long answer type of questions

I. In biology, sulphur can play a role in photosynthesis, chemosynthesis, and anaerobic respiration. Provide a detailed account of this statement.
II. Describe biogeochemical pathways in the sulphur cycle. What are the roles of the associated biological reactions in ecosystem balance?

NOTES

1 NASA Webb Telescope Team. Webb Discovers Methane, Carbon Dioxide in Atmosphere of K2-18 b. 2023
2 This name was derived from the first letters of surnames of the four investigators, including Robert Jay Charlson, James Lovelock, Meinrat Andreae and Stephen G. Warren.
3 Klatt JM, Polerecky L. Assessment of the stoichiometry and efficiency of CO_2 fixation coupled to reduced sulfur oxidation. *Frontiers in Microbiology*. 2015;6:484.
4 Prieto-Barajas CM, Valencia-Cantero E, Santoyo G. Microbial mat ecosystems: Structure types, functional diversity, and biotechnological application. *Electronic Journal of Biotechnology*. 2018;31:48–56.
5 Lin W, Zhang W, Paterson GA. Expanding magnetic organelle biogenesis in the domain Bacteria. *Microbiome*. 2020;8:1–3.
6 Ginet N, Pardoux R, Adryanczyk G. Single-step production of a recyclable nanobiocatalyst for organophosphate pesticides biodegradation using functionalized bacterial magnetosomes. *PloS One*. 2011;6(6):e21442.

SUGGESTED READINGS

Anantharaman K, Hausmann B, Jungbluth SP, et al. Expanded diversity of microbial groups that shape the dissimilatory sulphur cycle. *The ISME Journal*. 2018;12(7):1715–28.

Bak F, Cypionka H. A novel type of energy metabolism involving fermentation of inorganic sulphur compounds. *Nature*. 1987;326(6116):891–2.

Caballero B, Trugo LC, Finglas PM. *Encyclopedia of food sciences and nutrition*. Academic; 2003.

Ghosh W, Dam B. Biochemistry and molecular biology of lithotrophic sulphur oxidation by taxonomically and ecologically diverse bacteria and archaea. *FEMS Microbiology Reviews*. 2009;33(6):999–1043.

Hinckley EL, Crawford JT, Fakhraei H, Driscoll CT. A shift in sulphur-cycle manipulation from atmospheric emissions to agricultural additions. *Nature Geoscience*. 2020;13(9):597–604.

Hou J, Deng HK, Liu ZX, Xu P, Wang LJ. Sulfur metabolism in *Rhodococcus* species and their application in desulfurization of fossil fuels. *Journal of Applied Microbiology*. 2023;134(3):lxad048.

Kertesz MA. Riding the sulphur cycle–metabolism of sulphonates and sulphate esters in Gram-negative bacteria. *FEMS Microbiology Reviews*. 2000;24(2):135–75.

Kjeldsen KU, Schreiber L, Thorup CA, et al. On the evolution and physiology of cable bacteria. *Proceedings of the National Academy of Sciences*. 2019;116(38):19116–25.

Koch T, Dahl C. A novel bacterial sulphur oxidation pathway provides a new link between the cycles of organic and inorganic sulphur compounds. *The ISME Journal*. 2018 Oct;12(10):2479–91.

Li Y, Gong X, Zhao Z, Shen Q, Zhang L. Distribution and release of volatile organic sulfur compounds in Yangcheng Lake. *Water*. 2022;14(8):1199.

LokaBharathi PA. Sulphur cycle. In Jørgensen SE and Fath BD (eds.) *Encyclopedia of ecology*. 2008 (Vol. 2). Elsevier.

Müller H, Marozava S, Probst AJ, Meckenstock RU. Groundwater cable bacteria conserve energy by sulphur disproportionation. *The ISME Journal*. 2020;14(2):623–4.

NASA Webb Telescope Team. Webb Discovers Methane, Carbon Dioxide in Atmosphere of K2-18 b 2023. www.nasa.gov/universe/exoplanets/webb-discovers-methane-carbon-dioxide-in-atmosphere-of-k2-18-b/

Rickard D. *Sulphidic sediments and sedimentary rocks*. Newnes; 2012.

30 Bioremediation of Contaminated Soil

30.1 ENVIRONMENTAL CONTAMINATION

Environmental contamination is a critical global challenge that has an impact on ecosystems, human health, and biodiversity. It arises from the accumulation of foreign agents or natural substances in the environment. Environmental contaminants can be chemical (e.g., heavy metals), biological (e.g., pathogens), and radiological (e.g., radioactive isotopes) that can enter the environment accidentally or deliberately. Chemical substances, in particular, pose significant challenges to the natural ecosystems. Human activities, such as industrial and domestic processes, agriculture, mining, chemical spills, fossil fuel combustion, and nuclear operations, are the primary sources of environmental chemical contaminants. Natural processes, including forest fires, volcanic eruptions, and sea spray, also release substantial quantities of contaminants.

Two different categories of chemical contaminants are organic compounds and inorganic substances. Organic contaminants include volatile organic compounds (e.g., toluene), organic contaminants (e.g., polychlorinated biphenyls (PCBs)), and complex compounds (e.g., polycyclic aromatic hydrocarbons (PAHs) and their derivatives). Some organic contaminants are persistent chemicals and are called persistent organic pollutants (POPs) and are very difficult to break down. High concentrations of antibiotics in the environment have emerged as a global issue due to their excessive use in health care and agriculture. Many organic contaminants tend to get biomagnified in the food web, causing harm to plants, animals, and humans. Inorganic contaminants include salt, minerals, heavy metals, metalloids, and radionuclides. Inorganic ions like nitrate, sulphate, chloride, etc., contaminate water, air, and soil from industrial wastes and effluents and adversely affect ecological systems. Heavy metals are elements which have a density above 5 g/cm^3. Some heavy metals like mercury (Hg), chromium (Cr), cadmium (Cd), copper (Cu), zinc (Zn), manganese (Mn), cobalt (Co), and lead (Pb) are notable contaminants of the environment. Arsenic (As) is a metalloid, and its exposure to humans is a global concern. Radioactive materials such as uranium (U), thorium (Th), etc., emit harmful radiation that is hazardous and causes ecological disturbance. Inorganic contaminants are non-degradable and exist in the environment for an extended period. Many chemicals, such as Cd, Hg, and benzene, are even non-threshold toxins, which have deleterious effects even at very low concentrations.

Global organisations, such as the World Health Organisation (WHO), the United Nations Environment Programme (UNEP), and the United States Environmental Protection Agency (US EPA), have implemented stringent regulations to control the emission of contaminants and mitigate their effects. Many other geographical regions, including the European Union, Australia, Canada, and New Zealand, also have agencies for pollution control, which have strict legislation on contamination. However, excessive consumption, high energy use, waste generation, and unsustainable practices have exacerbated environmental degradation due to the heavy load of chemical pollution. Many middle-income and low-income countries have failed to manage their pollution crisis. Developed countries have also contributed immensely due to industrialisation and high consumption.

30.2 CONTAMINATION IN SOIL

Soil, a non-renewable resource, is a vital component of terrestrial ecosystems, providing a foundation for biodiversity, agriculture, and climate regulation. It takes approximately 1,000 years for 2-3 cm of soil to form. Soil has a solid core, and about half of the matrix is porous and occupied with water and gases. The rise of different contaminants in soil has made it one of the world's most disrupted spheres. The porous matrix of soil facilitates the movement of contaminants, allowing chemical wastes to penetrate and disperse. Soil particles can also absorb or adsorb a portion of these wastes. Contaminants can migrate vertically through soil layers such as topsoil, subsoil, and sediments. Soil contamination is much more common in urban areas than in rural areas. However, all farmed soils that have high uptake of chemical fertilisers accumulate high concentrations of Cd and other metals and contain pesticides as well.

Both soil and chemical properties of contaminants can have impact on soil particle-contaminant interaction. The soil parameters that can play roles in soil pollution include its permeability, nutrient content, water content, pH, clay content, and humus content. The parent materials, climate conditions, and biodiversity, including vegetation, can also have dominant effects on the distribution of contaminants. Chemical properties of contaminants, such as solubility, mobility, reactivity, speciation, and degradability, control their distribution in the land ecosystem.

DOI: 10.1201/9781032651095-37

Soil contamination can be driven by activities such as chemical spillage, industrial discharge, and improper waste disposal. Particularly, the improper disposal of chemical wastes into the soil is a serious concern for land ecosystems. As per the Food and Agriculture Organisation (FAO) and the United Nations Educational, Scientific, and Cultural Organisation (UNESCO), approximately 33% of the world's land is already degraded due to various chemical contaminants, and current trends suggest that over 90% of land may face degradation by 2050. Land degradation has impacted farming and habitats. Excessive soil contamination has hampered plant growth, with prolonged exposure resulting in plant mortality. They have put a heavy burden on animals and humans and induced sublethal to lethal effects. While contamination disrupts microbial community structure and function, some resilient species exhibit adaptations that enable them to neutralise contaminants, highlighting their potential for bioremediation efforts. Overall, the degradation of land ecosystems due to the contamination of diverse chemicals has threatened the entire biological system. In this context, the treatment of contaminated soil using effective, cost-efficient, and eco-friendly methods is an urgent requirement.

30.3 REMOVAL OF ENVIRONMENTAL CONTAMINANTS

A variety of processes can be used for the treatment of contaminants, and such a process is called environmental remediation. It is mainly a human intervention to offset the environmental impairment caused by anthropogenic activities. When the treatment site is the soil matrix, the process is called soil remediation. A similar approach for water bodies is known as water remediation. This chapter deals with soil remediation. A detailed overview of wastewater treatment can be found elsewhere (Chapter 25). Many techniques, such as chemical precipitation and adsorption, can be used for both soil and water and are covered in this chapter.

The remediation approaches belong to physical, chemical, and biological processes. It can also be an integrated technique involving two or more processes. Physical methods employ soil washing, vitrification, electrokinesis, thermal desorption, and excavation. Chemical treatments depend on various chemical reagents for their precipitation, leaching, adsorption, and extraction. Biological methods make use of several biological materials, employing various metabolic-dependent and -independent processes. Biological processes are increasingly favoured due to their cost effectiveness, sustainability, and minimal environmental impact. Most physical and chemical treatment methods face significant scrutiny due to environmental safety concerns and high cost. "Bioremediation" is a more common term for biologically mediated contaminant treatment. The following sections provide a thorough understanding of diverse routes for the bioremediation of contaminated soil.

30.4 BIOREMEDIATION

Bioremediation can be defined as the treatment process of contaminants using biological materials, especially microorganisms. The key players in bioremediation are called "bioremediators" and chiefly include bacteria and fungi. Other microorganisms, such as microalgae, archaea, and protozoa, can have mild effects on contaminants. When the remediating microorganisms are exclusively fungal species, the process is termed "mycoremediation". Likewise, many algae have shown the possibility to remediate wastes and wastewater, and the process is called "phycoremediation". A different method of biotreatment relies on earthworms under the term "vermiremediation". This process exploits the burrowing and feeding habits of earthworms to degrade organic matter. When they excrete the nutrient-rich castes in the soil, the structure and fertility of the soil are enhanced. The billions of bacteria found in the earthworms' digestive tract also co-digest the hydrocarbons. Effective soil remediation is crucial not only for ecosystem restoration but also for ensuring food security and mitigating climate change impacts. Bioremediation is a hot topic in research and academia, but it is still in its nascent stage for practical implications and has several merits and demerits (Table 30.1).

The principal events of bioremediation are biodegradation, mobilisation, and immobilisation (Figure 30.1). Biodegradation is the process of breaking down chemicals using enzymes, which is mainly applicable to certain organic compounds. In mobilisation, a mobilising agent facilitates the transport of contaminants away from their existing environments (e.g. bioleaching). On the contrary, immobilisation is a process where the contaminants reside in the same space, but they are either sequestered, absorbed, or adsorbed in the soil (e.g. bioprecipitation and biosorption). For persistent contaminants like heavy metals and radionuclides, biological processes such as bioprecipitation, biosorption, and bioleaching provide effective containment solutions. These processes are equally effective for the treatment of POPs. Table 30.2 gives a comparative account of mobilisation, immobilisation, and biodegradation for the biotreatment of various contaminants.

TABLE 30.1
Merits and demerits of bioremediation

Merits	Demerits
• Eco-friendly approach • Sustainability • Cost-effective process • Natural technology • Restoration of contaminated sites • Easy manipulation of bioprocess	• Sensitive to high levels of toxicity • Relatively slow process • Full-scale field operation is difficult and complex.

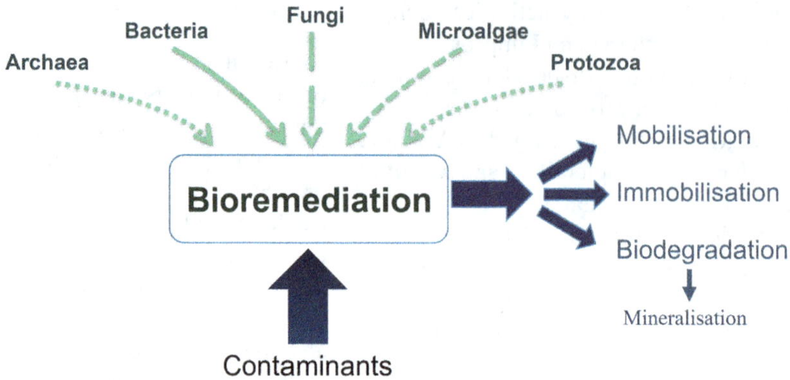

FIGURE 30.1 Microorganisms involved in bioremediation of environmental contaminants. The complete and dotted arrows correspond to the relative contributions of microorganisms.

TABLE 30.2
A comparative study on different approaches to bioremediation

Parameters	Mobilisation	Immobilisation	Degradation
Treatment of organic contaminants	Many hydrocarbons	All types of hydrocarbons	Many hydrocarbons
Treatment of heavy metals and radionuclides	Possible	Possible	Not possible
Concentrations of contaminants after treatment	Reduced	Not reduced	Reduced
Mechanisms	Increasing mobility and bioavailability	Decreasing mobility and bioavailability	Breakdown of contaminants
Types of biotreatment	Displacement-based removal	Risk-based approaches	Concentration-based removal
Examples of bioactivities	Bioleaching and biosurfactant production	Biostabilisation and biosorption	Mineralisation

30.5 MICROBIAL ACTIVITIES FOR BIOREMEDIATION

Microbial activities are the cornerstone of bioremediation, utilising the natural capabilities of microorganisms to treat diverse contaminants in the environment. These activities include biodegradation, redox reactions, biosorption, bioaccumulation, biomineralisation, bioprecipitation, biovolatilisation, biosurfactant production, and bioleaching (Figure 30.2). Table 30.3 enlists different bioactivities for the removal of contaminants using different microorganisms/enzymes. Also, they are briefly discussed as follows.

- Biodegradation involves the microbial decomposition of organic matter into simpler compounds, making it one of the most widely studied and applied processes in bioremediation (detailed in Section 30.6).
- Microbial redox reactions alter the oxidation states of contaminants, affecting their mobility, toxicity, and bioavailability. Many heavy metals, such as Cr(VI), As(III), As(V), Hg(II), and Cu(II), can be suitably detoxified using oxidation-reduction reactions. For example, the microbial enzyme chromium reductase reduces the highly toxic and mobile Cr(VI) into the immobile and less harmful Cr(III). In this process, the reactant is highly mobile and toxic, whereas the product is immobile and harmless at biological pH. Some microbial forms transform U(VI) into U(IV), reducing its mobility and solubility. A bacterium, *Deinococcus radiodurans,* is beneficial for the removal of U due to its extreme resistance to radiation. This bacterium also displays its usefulness for the treatment of other heavy metals from environments that have high radiation levels. In a different mode, certain bacteria (e.g., *Geobacter metallireducens*, *Shewanella oneidensis*, and many Deltaproteobacteria) employ heavy metals like Mn(IV), Co(III), Cr(VI), As(V), and Fe(III) in the anaerobic respiratory pathway as terminal electron acceptors and reduce them into less toxic forms. The redox reaction can also be useful for the degradation of organic compounds and to neutralise inorganic substances like nitrates and sulphates.
- Biosorption is a non-metabolic process that relies on the natural adsorptive properties of biological materials to remove contaminants like heavy metals, dyes, and pharmaceuticals from the solution. Different

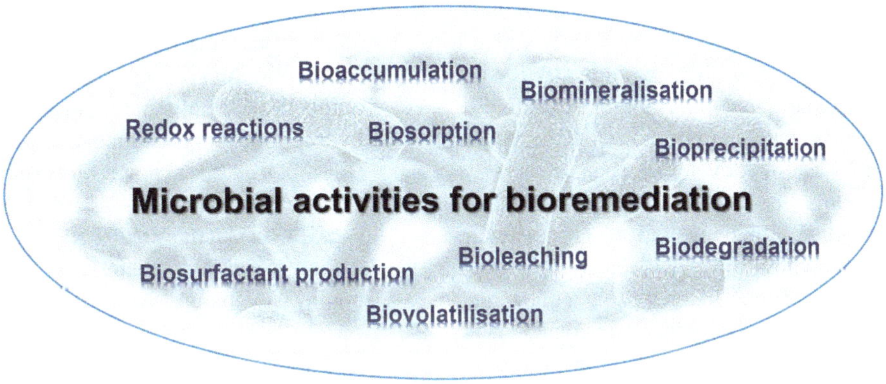

FIGURE 30.2 Various microbial activities for bioremediation.

TABLE 30.3
A list of different bioactivities for the removal of contaminants using different microorganisms/enzymes

Activities	Contaminants	Microorganisms	Specifications
Biodegradation (aerobic)	PET plastics	*Ideonella sakaiensis*	PETase enzyme
	Endosulfan and acaricide	*Trichoderma harzianum*	Oxidative and hydrolytic pathways
Biodegradation (anaerobic)	Toluene	A denitrifying bacterium, *Thauera* sp.	Benzylsuccinate synthase
Redox reaction (oxidation-reduction)	Cr(VI) into Cr(III)	Various bacteria and some fungi	Chromate reductases
	As(III) into As(V) and *vice versa*	Various bacteria and some fungi	Arsenite reductase, arsenate reductase and arsenite oxidase
Biosorption	Organic and inorganic contaminants	Both living and dead microbial cells	Passive process
Bioaccumulation	Some organic dyes and As, Pb, Cr(VI), and Cd	Mainly by contaminants-resistant bacteria and fungi	Uptake systems of essential nutrients
Biomineralisation	Cr(VI) and Pb	*Penicillium chrysogenum*	Microbially induced calcite precipitation (MICP)
Bioprecipitation	As, Cd, Cu, Fe, and Zn	*Desulfovibrio vulgaris*, *Desulfomicrobium baculatum*	Precipitation as metal sulphides
Biovolatilisation	As	A variety of microorganisms	Conversion into organic As
	Hg(II)	Bacteria and archaea	Reduction into Hg(0)
Biosurfactants	Petroleum hydrocarbons (PHCs) and heavy metals	Rhamnolipid from *Pseudomonas*	Washing and flushing for the desorption of contaminants from soil
Bioleaching	Co, Cu, U, and Zn	*Acidithiobacillus ferrooxidans* and *Acidithiobacillus thiooxidans*	Dissolution of heavy metals

functional groups found on the microbial cell walls and exopolysaccharides enable efficient and reversible chemical binding with contaminants. This process is independent of any metabolic activities and can involve living as well as dead biomasses. The biological matrices used in biosorption are called biosorbents, and the target chemicals to be removed are termed adsorbates. Cell wall materials of bacteria, fungi, and algae have a variety of functional groups that recover chemicals effectively. In this process, adsorbents can be recycled for multiple uses, making them a sustainable and cost-effective solution for wastewater treatment.

- Bioaccumulation involves the active uptake and storage of contaminants within living cells, though its application is limited by the toxic effects of accumulated contaminants on cell metabolism. Also, the cell wall and membrane act as barriers for many biological organisms and inhibit the transport of chemicals inside the cells. However, bioaccumulation is inevitable when contaminants mimic essential nutrients and pass through specific channels. Toxin-resistant bacteria have properties to sequester harmful chemicals, reducing their bioavailability. The intracellular architecture of toxin-resistant bacteria effectively stores foreign chemicals along with native chemical species, thereby reducing their toxicity.
- Biomineralisation is a process that transforms chemical contaminants into stable mineral forms,

such as oxides, carbonates, phosphates, sulphides, and oxalate. In this process, the chemical exudates of microorganisms react with contaminants to get deposited as nanoscale crystals. A notable example is microbially induced calcite precipitation (MICP), where urease-producing bacteria develop carbonate crystals to immobilise metals like Cd, Pb, and Cr(VI). MICP is also referred to as "biocementation". A different process of biomineralisation involves phosphatase activity in microorganisms. In anaerobic conditions, microbial phosphatase activity induces the synthesis of U(VI) phosphate minerals, which is a strong force for radioactive waste containment action.

- Bioprecipitation induces the conversion of soluble contaminants into insoluble forms, such as metal sulphides. Sulphate-reducing bacteria are key players in this process, offering a sustainable treatment option for acid mine drainage. These microorganisms produce sulphides in anaerobic situations, which react with metals to precipitate as metal sulphides (this reaction can also be included as an example of biomineralisation). In different situations, enzymatic redox reactions can convert soluble metal ions into precipitates. In many cases, the reduced form is less soluble, making it prone to precipitation. For example, Cr(VI) bioreduction into Cr(III) tends to precipitate at a pH below 5.
- Biovolatilisation converts certain metal(loid)s, like As and Hg, into volatile derivatives *via* microbial enzymatic activity. The element As can be transformed into its volatile forms, trimethylarsine, dimethylarsine, and arsine, in the presence of microbial enzymes, arsenic methyltransferases. In a different reaction, a bacterial enzyme, mercury reductase, catalyses the reduction of toxic Hg(II) into the inert, volatile Hg(0). Despite its promise, this method concerns over air pollution caused by the release of volatile compounds.
- Bioleaching mobilises metals through microbial production of organic acids or sulphuric acids. This is a popular method used for the extraction of metals from mines. Soil bacteria such as *Acidithiobacillus ferrooxidans* and *Acidithiobacillus thiooxidans* are primary candidates for bioleaching. These bacteria at contaminated sites stimulate the synthesis of sulphuric acid that enables the dissolution of heavy metals. The organic acids of fungal origin can also be used for bioleaching. The citric acid produced by *Penicillium* spp., *Aspergillus* spp., and *Yarrowia lipolytica*; the fumaric acid by *Mucor* spp., *Rhizopus* spp.; the malic acid by *Schizophyllum commune*, *Penicillium* spp., and *Aspergillus* spp. are some common examples of fungal exudates, which can leach metal ions.
- Biosurfactants are surface-active amphiphilic compounds that are used as washing and flushing agents for the recovery of contaminants from soil. They belong to structurally diverse groups of compounds such as glycolipids, phospholipids, fatty acids, lipopeptides, and lipoproteins. Aerobic bacteria and fungi are the most common sources of biosurfactants. Biosurfactants reduce interfacial tension and desorb heavy metals and hydrophobic contaminants like petroleum hydrocarbons (PHCs). The best-known biosurfactant for biotreatment processes is rhamnolipid, isolated from *Pseudomonas* spp. Biosurfactants are excellent alternatives to synthetic surfactants due to their nontoxic, biodegradable, and noninvasive nature.

30.6 PRINCIPLE OF BIODEGRADATION

Biodegradation is a natural process in which microorganisms break down organic compounds into simpler, less harmful substances. It is the most commonly employed method for hydrocarbon removal, particularly in environmental cleaning programme. While natural organic compounds are readily degraded by indigenous microorganisms, synthetic and complex hydrocarbons require specialised microbial enzymes due to their structural complexity and resistance to enzymatic activity. The biodegradation of recalcitrant hydrocarbons is even more complicated. The number of double and triple bonds, polymerisation, functional groups, aromaticity, chlorination, and nitration affect the degradability of hydrocarbons.

The transformation of complex materials into simple compounds through biological methods depends on the presence of specific enzymes in microorganisms. For example, the depolymerisation of plant-based polymers such as cellulose and lignin requires enzymes like cellulases and lignin peroxidases, respectively. Laccase is responsible for oxidising phenolic compounds, PAHs, dyes, and pesticides. Nitrilases catalyse the hydrolysis of toxic nitrile compounds such as herbicides, polymers, and plastics. Organophosphorus hydrolase (source: *Pseudomonas diminuta*) degrades organophosphate compounds found in pesticides and warfare. An enzyme, esterase, enforces the breakdown of ester bonds in organic compounds. The disruption of a carbon-halogen bond in halogenated substances requires a bacterial enzyme, dehalogenase. Many POPs rest in the soil for a long period (such as organochlorine pesticides (OCPs), PAHs, PCBs, dioxins, and dibenzofurans) because they resist microbial degradation due to their stability and low bioavailability. Microorganisms can involve co-metabolism for the degradation of some POPs, where the metabolism of the primary substrate (such as glucose) induces the degradation of the contaminants. However, co-metabolism often leads to partial degradation and produces toxic substances.

Advances in microbial research have uncovered new pathways and enzymes to degrade POPs. For example, an enzyme, PETase, has revolutionised the biodegradation of polyethylene terephthalate (PET) (a type of plastic commonly found in packaging and water bottles). This enzyme was isolated from *Ideonella sakaiensis* and has shown promising features for the recycling of plastic waste.

Similarly, enzymes from *Trichoderma harzianum* are effective against organochlorine compounds like Endosulfan. An insecticide, dichlorodiphenyltrichloroethane (DDT), can be partially degraded through reductive dichlorination using bacteria like *Bacillus*, *Aerobacter*, *Klebsiella*, etc. Enzymes like biphenyl dioxygenase (BPH), phosphotriesterases, and oxygenases reported in some specific microorganisms participate in bioremediation for the degradation of various POPs. Microorganisms with bioremediation potential can be genetically modified for effective treatment. Genetically modified microorganisms (GMOs) enhance bioremediation efficiency by offering greater stability, tolerance to contaminants, and multi-functionality. For instance, a genetically engineered *Pseudomonas putida* strain, equipped with multiple plasmids, can degrade two-thirds of hydrocarbons in oil spills.

Microorganisms can participate in the degradation of organic contaminants through aerobic or anaerobic pathways. Figure 30.3 illustrates the key steps involved in aerobic and anaerobic biodegradation pathways. Aerobic pathways are more energetically favourable and effective for most organic compounds. Biodegradation of organic compounds can be a multistep reaction and requires multiple enzymes and cofactors. Simple sugars can directly participate in the respiratory pathway, enabling their complete oxidation into carbon dioxide (CO_2). Some carbon units of sugar compounds can be assimilated into biomass and contribute to the carbon backbone of microorganisms. Complete degradation of glucose through aerobic respiration generates approximately 30–32 ATP and releases six molecules of CO_2, passing through glycolysis, the citric acid cycle, and oxidative phosphorylation. These processes involve at least twenty-three different enzymes. Complex and synthetic organic compounds require additional enzymes and specific metabolism for aerobic degradation pathways in microbes. The substantial microbial populations and oxygen levels promote the biodegradation of PHCs. The biodegradation of 1 kg of petroleum requires approximately 3-3.5 kg of oxygen (O_2).

Anaerobic degradation is driven by obligate and facultative anaerobes. This is an effective and sustainable method for sludge treatment and biogas production through a technique called anaerobic digestion (detailed in Chapter 31). However, it is a relatively slow process and requires a controlled operation. Some common steps of anaerobic degradation are hydrolysis, acidogenesis, acetogenesis, and methanogenesis. The process releases methane (CH_4), CO_2, and hydrogen (H_2) along with intermediate products such as organic acids and ethanol.

30.7 FACTORS INFLUENCING BIOREMEDIATION

In bioremediation, the interplay between chemicals and microorganisms is of central importance to their successful treatment. They hinge on the soil matrix and establish a complex interaction. In general, microbial activities play dominant roles in shaping physical and chemical changes of a chemical in the environment. However, soil contamination of any chemical exerts toxic effects on native microbiota and their activities. High concentrations of contaminants at the sites can be detrimental to the native species. In this situation, only microorganisms resistant to chemicals can withstand contaminated environments. Only a few representatives of resistant microorganisms contribute to the treatment of

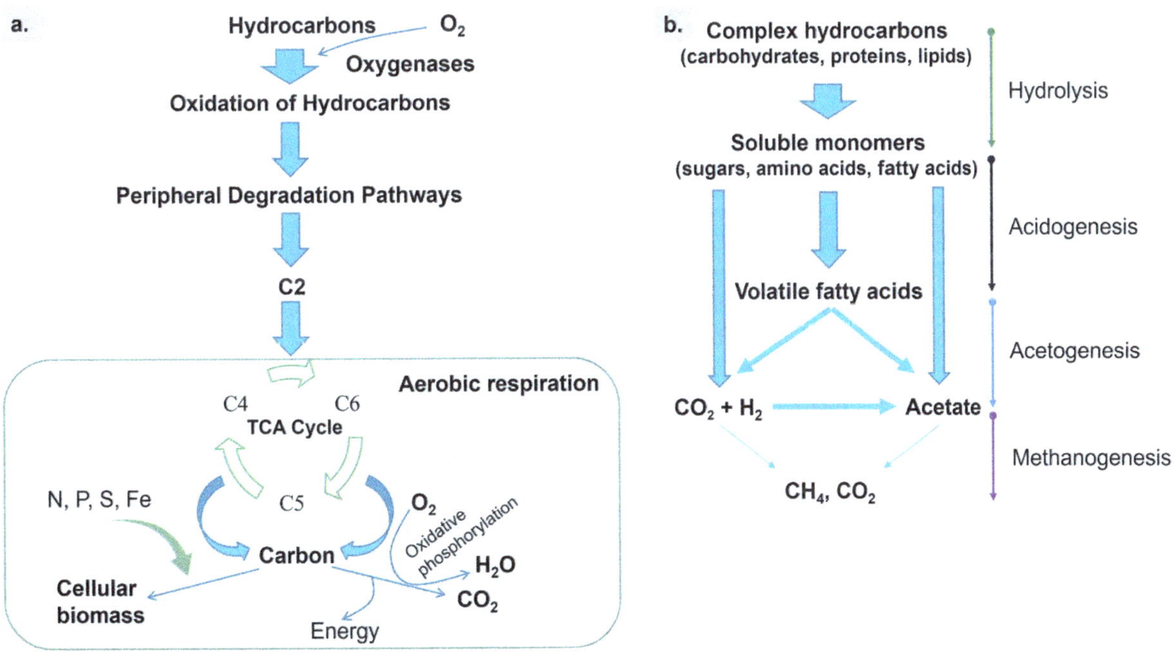

FIGURE 30.3 Biodegradation pathways in microorganisms. a. Aerobic biodegradation. b. Anaerobic biodegradation. C2, C4, C5, and C6 are different carbon units in organic compounds. TCA cycle stands for the tricarboxylic acid cycle.

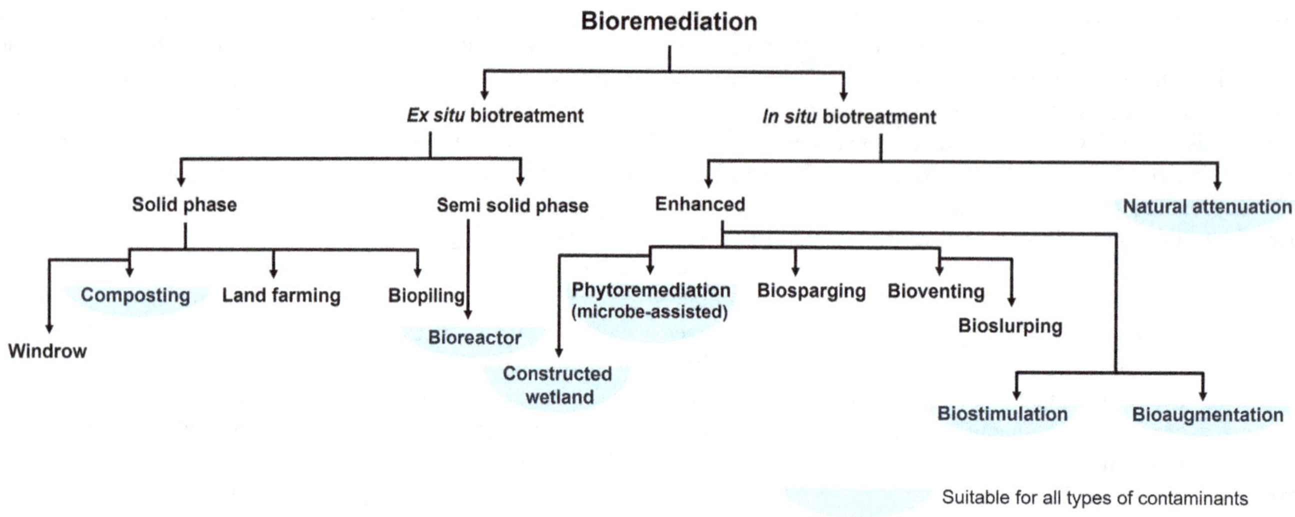

FIGURE 30.4 Different setups of bioremediation.

contaminants at the sites. The detoxification of toxic chemicals is much more difficult than the removal of regular contaminants.

In many microorganisms, plasmid-encoded genes produce relevant enzymes to metabolise or mobilise the contaminants. Examples include xylene monooxygenase and naphthalene dioxygenase for the degradation of aromatic hydrocarbons. The lateral gene transfer allows these determinants to disseminate between various microbial species, which provides an adaptation to susceptible microorganisms in a harsh environment. However, many relevant traits are unique and can be found in some selected microbial species. The molecular variation in microorganisms through genetic engineering and mutation has also improved their roles for enhanced treatment practices. For example, genetic engineering in *Dehalococcoides mccartyi* improved the ability to degrade chlorinated solvents, such as trichloroethylene (TCE). Differently, some microbial strains can be engineered with completely new traits for the removal of specific contaminants. For example, *E. coli* is equipped with genes that can degrade plastic.

In many biotreatment options, microbial consortia (groups of microorganisms in the same culture) show more robustness toward contaminants than an axenic culture (a culture of only one microbial form). Single microbial strains are often unable to bear the metabolic burden of a chemical species due to metabolic constraints. In a consortium, the populations of different microorganisms coexist in the same environment and support each other's growth and proliferation. Microbial consortia show a better ability to be domesticated at affected sites. Also, they exhibit better tolerance to contaminants than axenic cultures.

Various environmental factors significantly influence the microbial action on contaminants. Changes in physical factors such as temperature, and moisture content have an impact on microbial response to chemicals. Chemical factors such as pH, salinity, organic matter, nitrogen content, electron acceptors and donors, and aeration can change the properties of contaminants, their association with the soil matrix, and the metabolic processes of microorganisms. For example, the presence and absence of O_2 in the environment promote aerobic and anaerobic microorganisms, respectively. The role of growth supplements for microorganisms is highly prominent for microbial proliferation and certain biochemical reactions. Site characteristics, such as depth, geographical area, and degree of contamination, also affect microbial outreach and performance. The treatment of saturated and unsaturated zones in soil involves different strategies. Many non-technical factors, such as long-term effectiveness, overall cost, stakeholder perception, environmental risk, environmental policies, etc., are also notable measures for the overall effectiveness of a remediation process.

30.8 DIFFERENT SETUPS FOR BIOREMEDIATION

Based on the sites of operation, two broad categories of bioremediation are *ex situ* and *in situ* biotreatment (Figure 30.4). The choice between the two depends on several factors, including the nature of contaminants, site characteristics, financial challenges, and time frames. The details about each type of operation and their various setups are as follows.

30.8.1 *Ex Situ* Biotreatment

Ex situ biotreatment refers to the off-site removal of contaminants. This type of operation is administered through the excavation and transport of contaminated materials away from affected regions. Most *ex situ* treatments are designed to provide high aeration in the contaminated soil, which stimulates aerobic biodegradation. Some procedures use foreign microorganisms and chemicals to accelerate the degradation processes. The same reaction system also allows the removal of other contaminants, such as heavy metals, involving diverse biological activities. However, the transfer of polluted soil away from its locations is costly, time-consuming, and dangerous. Also, it has climate change implications due to high energy consumption and greenhouse gas emissions associated with transportation and treatment processes.

Ex situ options become necessary for the treatment of the soil matrix that has severe contamination and/or requires rapid intervention. The safe and effective treatment is a critical requirement. For the same reason, the containment of waste materials is a strict requirement that can be partially achieved using big containers and through their storage confined to a restricted area. The main advantages of *ex situ* technologies are that they are amenable to a variety of contaminants, accelerate the removal process, maintain the uniformity of treatment, and allow easy monitoring and control. Also, the extensive preliminary assessment of the contaminated sites for *ex situ* treatment is not necessary. There are various setups for *ex situ* biotreatment, which are discussed as follows.

30.8.1.1 Composting

Composting is an oxygen-required process for the biological decomposition of agricultural and municipal wastes by mixing them with bulking agents and animal manure. These agents enhance microbial access to wastes and stimulate their growth by consuming organic matter in the wastes. Materials such as wood chips, sawdust, shredded plant matter, straw, and paper are frequently used as bulking agents. They increase porosity, ventilation, and moisture retention in the compost. Animal manures of horses, cows, and chickens are good sources of organic matter and healthy microorganisms for composting. The moisture content of compost must be sufficient to meet the water requirement for microbial metabolic activity. Regular mixing of contents ensures proper aeration in the microenvironment of the soil. The shortfall of aeration would otherwise encourage methanogens, leading to anaerobic digestion of solid wastes and causing the release of CH_4. Prolonged biodegradation can produce excess heat, which increases the temperature of the system to as high as 50-60 °C. This may additionally contribute to the co-metabolism of the hydrocarbons for effective biodegradation. Composting can also be effectively used for the removal of recalcitrant contaminants such as PAHs, petroleum, pesticides, and chlorophenols. The role of composting in the treatment of heavy metal contamination is also evident. Composting of heavy metal-contaminated waste mainly imparts stabilisation of heavy metals through complexation, adsorption, and (co) precipitation. Composts that are rich in decomposed matter can be employed as substitutes for chemical fertilisers in agricultural fields. Composting is a powerful tool for waste management and waste valorisation, as discussed in the next chapter (Chapter 31).

30.8.1.2 Landfarming

Landfarming is a management strategy for the remediation of petroleum products above ground. The process involves on-site excavation of buried soil through drilling and its spreading in a thin layer onto the surface (generally less than 0.3 m) (Figure 30.5). This setup enhances the aeration in contaminated soil and stimulates the growth

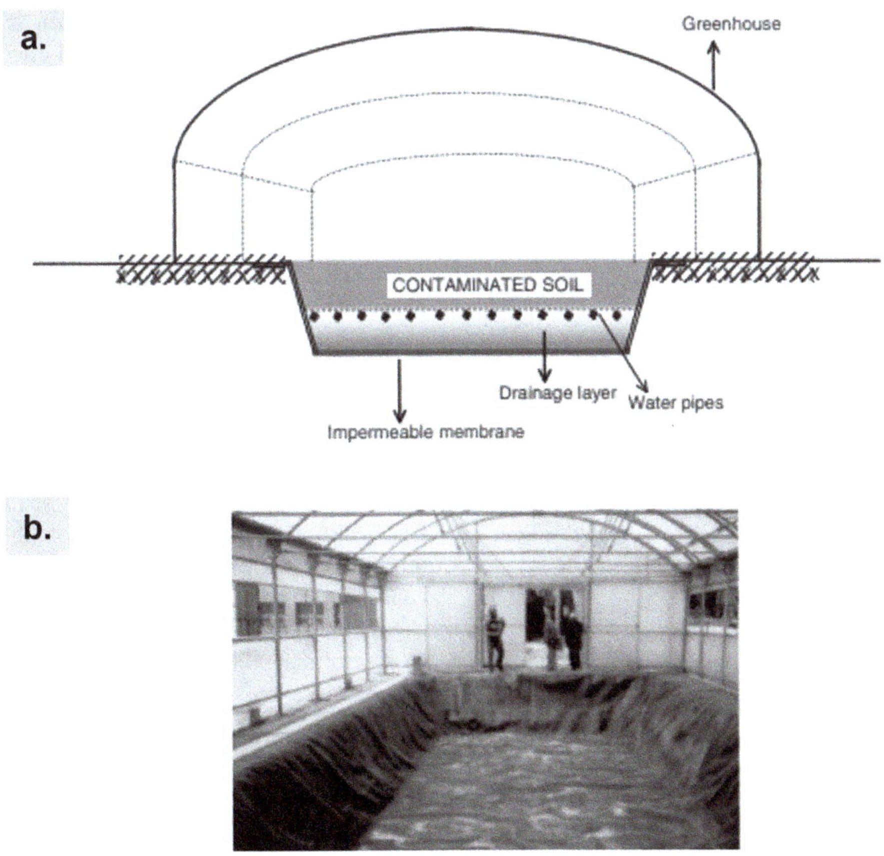

FIGURE 30.5 a. Landfarming treatment facility. a. Diagram. b. A pilot-scale setup. (Images adapted from Picado et al. (2001)[1] with permission of Taylor & Francis.)

of indigenous aerobic bacteria for degradation. The tilling process (turning of soil) also increases the oxygen supply in the soil. Heterotrophic aerobic bacteria are primarily responsible for the breakdown of petroleum products. The augmentation of minerals, nutrients, and moisture can bring additional benefits through the stimulation of microbial activity. This strategy is suitable where the depth of the contaminated site exceeds 1.0 m. As the management is on-site, this type of treatment requires less labour and is less expensive than other *ex situ* treatment options. The major shortcoming of this technique is the requirement of a large area for the spreading of soil. Contamination can spread widely if contaminants leach from the surface. The idea that volatile substances evaporate rather than biodegrade is a major concern. Also, this technique can be ineffective for heavily contaminated sites (e.g., when the PHCs concentration is greater than 50,000 mg/l).

30.8.1.3 Biopiling

Biopiling is an assembly of excavated soil (containing petroleum products) into piles for their biodegradation. The enforced aeration, addition of nutrients, and moisture into piles stimulate microbial activity for the breakdown of organic contaminants. The ideal conditions for the biopiling of contaminated soil are pH 6–8, moisture content of 40–85%, soil temperature of 10–45 °C, and a C: N: P ratio of 100: 1: 1 to 100: 1: 0.5. Blowers or vacuum pumps carry out the injection of aeration into the pile. This force-vented biopile technology is mainly employed for the treatment of non-halogenated volatile organic compounds (VOCs) and fuel hydrocarbons. Halogenated VOCs, semivolatile organic compounds (SVOCs), and pesticides can also be treated to some extent. However, the vapour release from the volatile constituents into the atmosphere is a major shortcoming of this technology.

30.8.1.4 Turned Windrows

Turned windrows involve periodic turning of piles. In large-scale operations, air often cannot penetrate deep into the enclosed materials. Periodic turning of the large volume of piles ensures uniform distribution of aeration, which enhances the biodegradation of hydrocarbons. Mixtures and the turning process require large tracts of land and sturdy equipment. This activity speeds up the degradation process by exposing hidden materials to aeration.

30.8.1.5 Bioreactors

Bioreactors are large tanks or vessels for the activities of whole microbial cells or cell-free enzymes in the treatment of contaminated soil. Bioreactor-based bioremediation uses excavated materials and transfers them into an enclosed chamber. The augmentation of inocula and nutrients promotes biological reactions for the degradation or detoxification of contaminants. A bioreactor is a highly sophisticated and engineered system for microbial breakdown of soil contaminants. It maintains the strictly controlled reactions and produces the desired conditions. The biotreatment of soil in a bioreactor can occur in batch, semi-continuous, or continuous mode. Of them, batch mode is the most popular. Unit 2 provides a detailed account of the specifications of bioreactors (specifically called fermenters in microbiology) and their applications. Of any other biotreatment options, bioreactors have the potential to remove high amounts of contaminants in a short period.

The full-scale equipment for the biotreatment of contaminated soil is a "**slurry bioreactor**". This bioreactor uses a slurry of contaminated soil (10–30% w/v) suspended in the aqueous medium or wastewater. The coarse particles can be crushed into small particles as a pretreatment before slurry preparation. The next step is the augmentation of contaminant-specific microorganisms to the aqueous suspension of soil. A mechanical device, an agitator in the bioreactor, facilitates the proper mixing of different components in a slurry. In addition, the mixing of microorganisms with slurry grants microbes easy access to contaminants and enhances mass transfer rates, resulting in their fast removal.

A different type of bioreactor consists of a large media bed known as a **biofilter**. The filter medium is a porous structure, which houses diverse microorganisms. When a contaminated substance passes through the filter bed, the contaminants are trapped in the bed, and the microorganisms within the biofilter degrade or detoxify them. This process is called biofiltration. A biofilter unit in wastewater treatment can be used for a broad range of contaminants and filter the water (discussed in Chapter 25). A similar device can be prepared to filter and purify air. The air filter is useful for the biotreatment of air contaminants such as VOCs and odorous compounds.

When comparing these two types of bioreactors, slurry bioreactors can be used for diverse contaminants and have high treatment efficiency, while biofilters are mainly installed for the removal of specific contaminants in wastewater and air purification. The process in slurry bioreactors is highly scalable to treat large amounts of contaminants. On the other hand, biofilters benefit from lower initial capital and operating costs, relying on natural microbial communities and low power input.

While bioreactor technology is a very good way to meet specific treatment needs, it is not practical to use it on a wide scale to clean up contaminated soil, as the excavation and transportation of soil matrices are highly cost-intensive. The accumulation of biomass or sediments in bioreactors can occasionally cause clogging or plugging problems that impair material flow and treatment effectiveness. Maintaining consistent conditions throughout the entire system is also challenging. Moreover strictly controlled operation in a bioreactor may not be suitable for the natural traits of soil.

30.8.2 *In Situ* Biotreatment

The *in situ* technique is an on-site clean-up strategy by stimulating microbial activities at contaminated sites. This approach can be used for the biodegradation of organic contaminants and in the immobilisation of both organic and inorganic contaminants. Cost efficiency and noninvasive procedures are the most important traits of on-site treatments. *In situ* biotreatment options have been used to address some maritime disasters. Examples include the treatment of oil spills at the *Deepwater Horizon*, Gulf of Mexico (2010) and the *Exxon Valdez*, Prince William Sound, Alaska (1989) using indigenous oil-eating bacteria. These biotreatments not only degraded the oil but also offered regular monitoring of the affected regions.

The effectiveness and sustainability of microbial applications at contaminated sites are the major concerns for long-term effects. Input of nutrients or exogenous microorganisms at contaminated sites could lead to soil degradation and pose risks to human health and the environment, owing to the disruption of indigenous microbial communities, the emergence of antibiotic-resistant bacteria, and the production of toxic byproducts. Also, in some cases, the on-site installation of equipment for the biotreatment can be a formidable task. Ongoing research and development are improving *in situ* bioremediation techniques, making them more efficient, effective, and adaptable to a wider range of environmental challenges. *In situ* biotreatment encompasses a lot of variations in techniques, which are suitable to meet the requirements of specific sites, targeting contaminant types. These methods are discussed in detail as follows.

30.8.2.1 Natural Attenuation

Natural attenuation is a self-operating on-site natural process to neutralise the effects of contaminants. In this scheme, indigenous microorganisms of the site are responsible for the reclamation of the land. This type of treatment does not involve any human intervention. The effectiveness of this self-regulatory process highly depends on the types of contaminants and the specific conditions of the site. It is the most cost-effective option for biotreatment, but takes a very long period (several years). PHCs and non-halogenated solvents are readily biodegradable organic contaminants in natural soil environments. A monitoring study reported that the natural attenuation took about four years to remove about 60% of PHCs. Regular monitoring of the site is required for risk assessments and endpoint forecasting. However, natural attenuation is usually not effective at very high or very low concentrations of contaminants. Moreover, this process does not effectively act on POPs.

30.8.2.2 Biostimulation

Biostimulation is a process of on-site supply of nutrients for the stimulation of indigenous microorganisms. The key principle of this process is the rejuvenation of natural microflora. The selection of appropriate nutrients supports the desired microbial activity that can ensure the success of biological treatment at target sites. For example, the supply of various limiting nutrients, like carbon, nitrogen, metal, etc., in contaminated soil accelerates microbial proliferation and/or enhances biochemical activities to eliminate the contaminants. Particularly, elemental sulphur acts as an energy substrate for microorganisms, which acidifies the environment and achieves the leaching of heavy metals. Additives like acetate, Tween 80, glucose, and molasses serve as growth stimulators for microorganisms at the contaminated sites and boost their degradative capacity. The composts, cow dung, and plant wastes are good sources of organic nutrients and act as stimulants in polluted soil to remediate PHCs. Sufficient oxygen levels in the soil also encourage microorganisms to accelerate the bioremoval process. However, the delivery and permeability of nutrients to the tight and impermeable subsurface areas are the main challenges to this technique. Also, microbial exposure to nutrients can induce the growth of some undesirable microorganisms, which can compete with useful microorganisms and raise other concerns.

30.8.2.3 Bioaugmentation

Bioaugmentation is the addition (or augmentation) of pregrown microbial cultures in contaminated soil. The most successful way to apply bioaugmentation is through the use of highly efficient microbial strains. The pre-grown microbial cultures possess special abilities to target contaminants, which can be pure strains or mixed cultures (microbial consortia). The use of single strains is usually ineffective because they possess a narrow range of metabolic capabilities, while microbial consortia involve the cooperation of many microbial species in degrading a contaminant. Based on the origin of microorganisms, the process can be recognised as bioaugmentation using autochthonous (i.e., indigenous) microorganisms and bioaugmentation using allochthonous (i.e., foreign) microorganisms.

An augmentation of indigenous microbial culture has the advantage of prior adaptation to survive in the same environment. For example, the augmentation of the green nonsulphur bacterium, *Dehalococcoides*, confers dechlorination and mineralisation of chlorine-containing solvents at contaminated sites. However, the functional deficiency in the native strains can be a major drawback to effective treatment. On the other hand, the augmentation of foreign microorganisms in the soil grants novel properties for the cleanup of contaminants. However, foreign microorganisms can have difficulty cooperating with native microorganisms and other factors of the soil. The limited dispersal of an inoculant in the soil matrix hampers the treatment in subsurface environments through bioaugmentation. Also, the inoculation of foreign microorganisms can interfere with the natural balance of the ecosystem, leading to unpredictable situations.

An integrated approach to biostimulation and bioaugmentation is also evident in the effective treatment of soil. Both chemical and biological inoculants can induce greater effects in the contaminated system than their counterparts. The nutrient applications enhance the biochemical activities in preexisting microbial communities, whereas the augmentation of new strains creates new possibilities for effective treatment. The combined biostimulation-bioaugmentation strategy has shown accelerated degradation of PHCs in soil.

30.8.2.4 Bioventing

Bioventing is the process of oxygen supply in the unsaturated zone of soil for the biodegradation of organic compounds. It is one of those methods of bioremoval that has the potential for large-scale applications. This method combines the approaches of physical, chemical, and biological treatments. In the treatment, the biological process relies on a controlled airflow rate only enough to sustain microbial activity in the subsurface regions (Figure 30.6a). The low airflow rate ensures the biodegradation of VOCs while minimising volatilisation. The addition of hydrogen peroxide at the site can be used as an alternative source of aeration in the unsaturated zone, as the breakdown of hydrogen peroxide produces water and O_2. Some nutrients for microbial growth may also provide benefits to the bioventing process. Bioventing has been primarily used to clean petroleum-contaminated sites, but can also be implemented for non-chlorinated solvents, pesticides, and other organic chemicals.

30.8.2.5 Biosparging

Biosparging also shares the same goal as bioventing, attempting to promote the biological activity of the indigenous aerobic microorganisms. In this process, air is injected into a saturated zone for the removal of contaminants (Figure 30.6b). Biosparging promotes the biodegradation of hydrocarbons. A similar process is air sparging, which promotes the volatilisation of VOCs through physical processes.

30.8.2.6 Bioslurping

Bioslurping is an integrated technology of bioventing and vacuum-enhanced free-product recovery. This combined treatment offers dual benefits through the breakdown of hydrocarbons followed by the recovery of their free product. Bioventing stimulates biodegradation, while vacuum-enhanced recovery employs a pump for the extraction of free products after degradation. The recovery pump applies negative pressure to develop a limited vacuum, which uplifts non-aqueous phase liquids from the capillary fringes. This process is suitable for the treatment of both unsaturated and saturated zones (Figure 30.6c).

Bioventing, biosparging, and bioslurping have one common approach, which is the supply of oxygen below

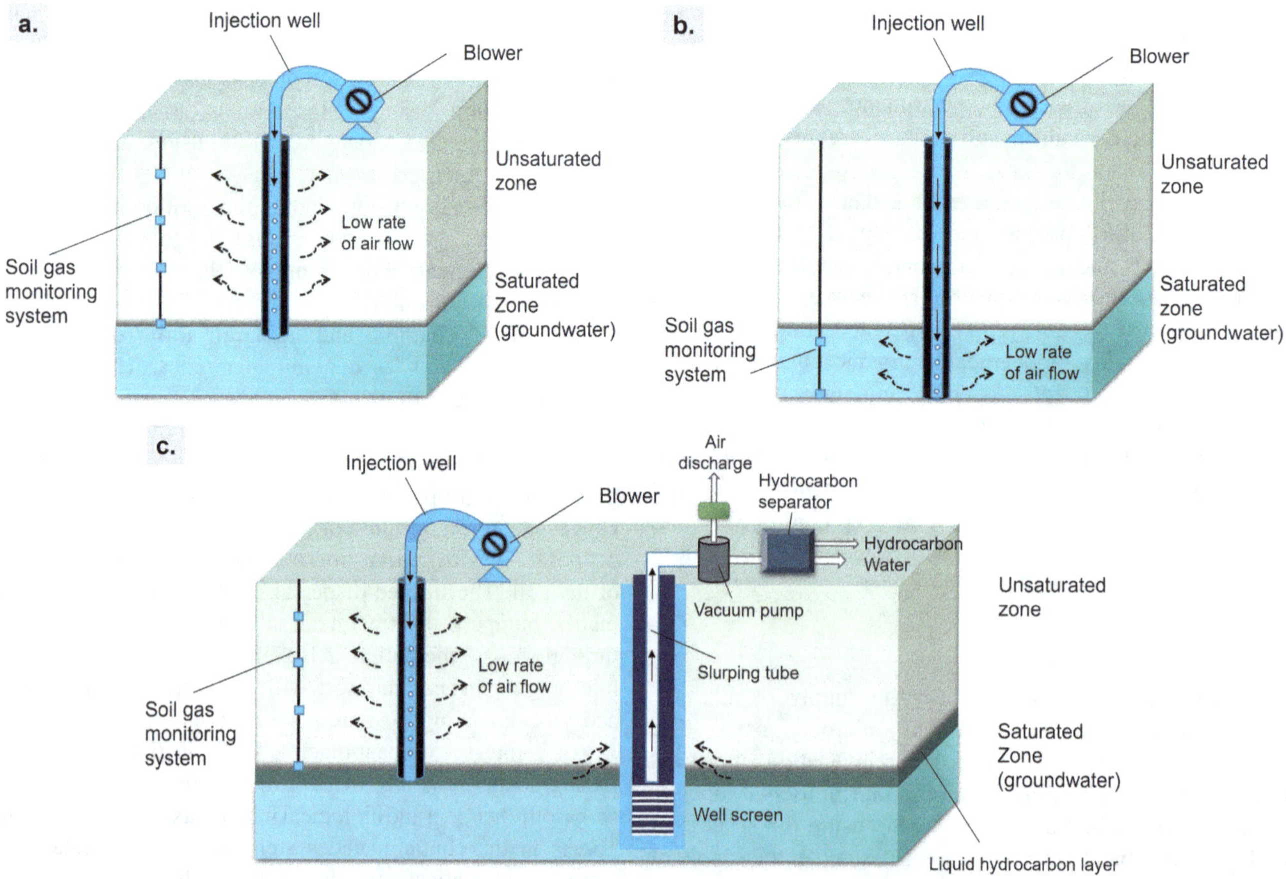

FIGURE 30.6 Different biotreatments of organic contaminants below ground. a. Bioventing. b. Biosparging. c. Bioslurping.

the ground surface for enhanced biodegradation. These are traditional methods of treating subterranean regions, and their development date back to the early to mid-1980s. The rate of biodegradation in a subsurface environment depends upon the heaviness of petroleum products. The low-weight derivatives (e.g., gasoline) tend to break down at a very fast rate, whereas mid-weight (e.g., diesel fuel) and heavier (e.g., lubricating oils) derivatives take a longer time. Because of their ease of installation, low cost, and minimal disruption, these techniques have an appeal for *in situ* treatment of contaminants. However, they are unsuitable for contaminated sites which have low permeability, low moisture content, insufficient delineation, and low ambient temperatures. The vacuum applied during bioslurping can lead to soil drying, and excessive removal of fluid from below ground can reduce the groundwater level.

30.8.2.7 Phytoremediation and Microbial Assistance

A living plant body can participate in soil treatment through its root system. The plant-based approach to the cleanup of contaminants is known as phytoremediation. This method uses various plant species to eliminate, break down, or detoxify environmental contaminants. The hyperaccumulator plants can accumulate different toxicants inside the plant body, and the process is known as **phytoextraction/ phytoaccumulation**. The plant exudates from different plant species can assist in the degradation of organic compounds through **phytodegradation**. A different method focuses on a plant's ability to reduce the mobility of contaminants by phytostabilisation. **Phytovolatilisation** involves the conversion of a chemical into its volatile form using plants. **Rhizofiltration** is the absorption/adsorption of chemicals in roots from solutions. These processes are eco-friendly solutions for the rejuvenation of degraded land, but are slow.

The protocooperation of soil microorganisms with plants can accelerate the process of phytoremediation, and the process is distinguished as "**microbe-assisted phytoremediation**". Microorganisms can help some plant species to live in soil with high amounts of contaminants. Both root endophytes and soil microorganisms have roles in plant survival potential at contaminated sites and in phytoremediation. Plant growth-promoting rhizobacteria (PGPR) like *Bacillus*, *Rhizobium*, *Paenibacillus*, *Pseudomonas*, and *Rhodococcus* possess diverse capabilities to assist phytoremediation. Plant growth-promoting (PGP) traits of PGPR, such as the production of phytohormones, siderophores, and chelating substances, promote plant growth and assist in phytoremediation. This kind of synergism is feasible when bacteria and fungi colonise the plant roots. Plants also release exudates and enzymes that activate microbial and biochemical activities to promote the bioremediation process in the soil.

All different microbial activities of bioremediation, including detoxification, transformation, extraction, sequestration, and assimilation, apply to microbe-assisted phytoremediation. Microbial biosurfactants assist in phytoextraction by making metals more soluble and bioavailable. Exopolysaccharides have significant metal-binding properties, which reduce their mobility and bioavailability to encourage phytostabilisation. The biostimulation and bioaugmentation in the rhizosphere also expand the role of microbial contributions to enhanced plant growth and the removal of contaminants.

30.8.2.8 Constructed Wetland

A constructed wetland is an artificial arrangement of soil, aquatic plants (i.e., reeds, duckweed), microorganisms, and a filter bed (usually of sand, soil, and/or gravel) for the treatment of wastewater in natural environments. It is a low-cost passive treatment that acts as a filtration unit or purification system for inland water in controlled environments. The flow regime of wetlands and the retention time of water in the wetland play a key role in the sedimentation of contaminants. Bacteria and plants act upon contaminants present in the wastewater during its retention, which separate or stabilise them from the wastewater. The outflow of the water after treatment in a constructed wetland can be used for domestic purposes (Figure 30.7). However, constructed wetland suffers from some limitations, such as space requirements, maintenance needs, and climate-related constraints.

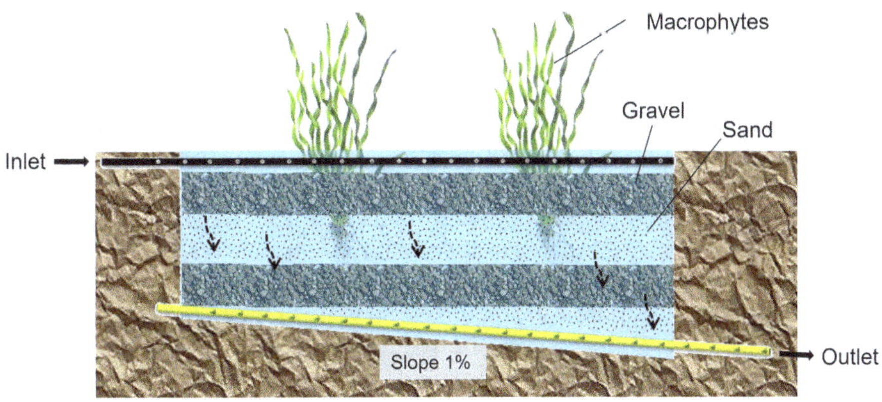

FIGURE 30.7 An illustration of a constructed wetland.

30.9 INTEGRATED APPROACHES OF TREATMENT

The combination of different biotechniques and physicochemical processes has emerged as a viable option for cleaning up contaminated sites. This kind of approach is in demand for the breakdown of recalcitrant contaminants, their removal from remote locations, and the simultaneous treatment of multiple contaminants. The combined use of different approaches enhances the efficiency and effectiveness of each technique involved and boosts the overall removal process. Moreover, the integrated approaches allow for more customised and thorough cleaning solutions, considering the variety and complexity of contaminants in the environment.

A specific treatment requires some chemicals, such as electron donors, redox mediators, and other additives, along with microorganisms, to neutralise the contaminants. Such a process can be called a chemical-assisted microbially mediated treatment. Fenton-mediated biodegradation is a novel method for the removal of POPs, where the hydroxyl radical generated from the Fenton chemistry assists in the biodegradation of chemicals. The knowledge about the suitable nutrients for biotreatment inspires the removal rate and efficiency. The microbial fuel cell (MFC) is an example of an integrated technology that combines biological activity with an electrochemical treatment system. This protocooperation not only provides benefits for treatment but also produces electricity. This biotreatment option is currently in the developmental stage for large-scale operation.

Nanobioremediation combines the approaches of nanotechnology and bioremediation to clean contaminated sites through sustainable solutions. In this process, nanoparticles carrying enzymes or microorganisms can be employed as a targeted treatment to clean the less permeable sections of soil, increasing the treatment's efficiency (Box 30.1). In a different process, a membrane bioreactor (MBR) constitutes a bioreactor and a membrane filtration unit for an *ex situ* treatment. Microbial cells in the bioreactor digest organic matter in the wastewater and break it down into minute particles. The membrane filters allow the selective transport of fluid for its purification. Both municipal and industrial wastewater treatment plants frequently utilise MBR technology. Electrokinetically enhanced bioremediation (or bioelectrokinetic pathway) leverages the electric field for the mobility of contaminants, making them more accessible to bioremediation.

SUMMARY

- Chemicals like organic compounds, heavy metals, radionuclides, and other inorganic compounds are notable environmental contaminants and can be present in the soil. They pose a serious threat to all life forms and exert deleterious effects on various ecosystems.
- Bioremediation is an eco-friendly and sustainable solution to reduce environmental contaminants, mainly using microorganisms. The main events of

BOX 30.1 NANOTECHNOLOGY IN BIOREMEDIATION

The application of nanotechnology in the treatment of environmental contaminants is a modern setup of remediation, which involves the use of nanomaterials or nanoparticles for economical and accelerated decontamination. Nanomaterials are small particles (typically <100 nm) with high surface area and possess distinctive chemical characteristics. Nanoparticles can be designed to have certain qualities that make them highly effective and efficient in eliminating heavy metals, organic contaminants, and even some types of pathogens. In bioremediation, nanoparticles can act as catalysts or adsorbents. Elemental or zero-valent metal nanoparticles of iron, nickel, and palladium are suitable choices of nanoparticles. The choice of suitable nanoparticles in remediation depends on the types of contaminants, the environmental circumstances, and the desired outcomes.

The nanoscale zero-valent iron (nZVI) or nano iron has the most common applications owing to its low toxicity and low production cost, which can practically be injected into the subsurface environments of the contaminated sites. Several studies have reported using nZVI particles to remediate sites contaminated with aromatic polycyclic hydrocarbons, herbicides, metals, and pesticides. The use of nanoparticles integrated with biological activities is a viable, practical, and sustainable method to remediate a contaminant.

Based on distinctive characteristics associated with nanoparticles in bioremediation, biological treatment can be categorised into two different processes: nanobioremediation and bionanoremediation. Nanobioremediation typically involves an initial treatment of high concentrations of contaminants using nanoparticle physicochemical technology, followed by bioremediation mediated by biological technology. In a different approach, microbial cells, including bacterial, fungal, and algal cultures, can be used for the green synthesis of nanoparticles. When these particles are involved in the treatment of contaminants, the technology is referred to as "bionanoremediation".

The use of nanoparticles for remediation also comes with some drawbacks. Nanoparticles tend to lose reactivity with ageing. They show clustering, which affects their transport processes into the targeted contaminated zones. However, the coating of the surface of nanoparticles with substances like sodium carboxymethyl cellulose, guar gum, and lactate can overcome this limitation to some extent. The coating allows an electronic repulsion between nanoparticles, allowing their improved mobility, stability, and reactivity.

bioremediation are biodegradation, mobilisation, and immobilisation.

- Microorganisms possess a variety of abilities such as biodegradation, redox reaction (oxidation-reduction), biosorption, bioaccumulation, biomineralisation, bioprecipitation, biovolatilisation, biosurfactants, and bioleaching that can neutralise the toxic effects of chemicals.
- Biodegradation is an enzyme-mediated process for the breakdown of organic matter involving aerobic or anaerobic reactions. The complex organic matter requires unique enzymes for its degradation.
- The types and concentrations of contaminants, the microorganisms and bioactivities involved, site characteristics, and many non-technical factors have a direct impact on the efficacy of biotreatment.
- Two categories of bioremediation are *ex situ* and *in situ* biotreatments. *Ex situ* options cover composting, landfarming, biopile, turned windrows, and bioreactors. *In situ* options are natural attenuation, biostimulation, bioaugmentation, bioventing, biosparging, bioslurping, microbial-phytoremediation, and constructed wetlands.
- Integrated approaches such as chemical-assisted bioremediation, Fenton-mediated biodegradation, MBR, bioelectrokinetic pathway, and nanobioremediation have promising outcomes for the treatment of many contaminants.

DIFFERENT TYPES OF PRACTICE QUESTIONS

Fill in the blanks

I. The microorganisms which perform bioremediation are called……………………

II. ……………….. and ……………………… are uptake processes of biological organisms for bioremediation.

III. Bioventing involves the supply of aeration into ………………. zone of soil for *in situ* bioremediation.

IV. ………………….. is the process by which hydrocarbons are broken down into simple carbonic compounds by microorganisms.

V. Bioslurping is an integrated technology of ……………… and ……………………..

VI. …………………… is a bioremediation process to convert organic contaminants and heavy metals into their volatile derivatives.

SOLUTION: I. bioremediators, II. Bioaccumulation, biosorption, III. unsaturated, IV. Biodegradation, V. bioventing, vacuumenhanced free- product recovery, VI. Biovolatilisation

Match the following

i. Natural attenuation	a. Mining
ii. Landfarming	b. Tank/vessel
iii. Bioslurping	c. Self-operating process
iv. Bioreactor	d. Non-aqueous phase liquid
v. Bioleaching	e. Tilling

SOLUTION: i-c, ii-e, iii-d, iv-b, v-a

Very short answer type of questions

I. Define bioremediation.

II. Write the names of two different soil organic contaminants.

III. What are the three main events of bioremediation?

IV. Expand the term MICP.

V. Name any two bioprocesses that can be used in the treatment of heavy metals.

VI. What is a turned windrow, and where is it used?

Short answer type of questions

I. Write a brief note on the merits and demerits of bioremediation.

II. What are the different factors that influence bioremediation?

III. Give a comparative account of biostimulation and bioaugmentation.

IV. How are mobilisation, immobilisation, and biodegradation responsible for the bioremediation of soil?

V. What is the use of a bioreactor in bioremediation?

VI. Write a short note on microbial assistance in phytoremediation.

VII. What are the main differences between *in situ* and *ex situ* biotreatment?

Long answer types of questions

I. What are the different types of microorganisms involved in bioremediation? Discuss various types of microbial activities responsible for bioremediation.

II. Give a detailed, illustrative description of *ex situ* bioremediation for soil treatment.

III. What is *in situ* biotreatment? Discuss the different options for *in situ* treatment of soil.

NOTE

1 Picado A, Nogueira A, Baeta-Hall L, et al. Landfarming in a PAH-contaminated soil. *Journal of Environmental Science and Health, Part A*. 2001;36(9):1579–88.

SUGGESTED READINGS

Adams GO, Fufeyin PT, Okoro SE, Ehinomen I. Bioremediation, biostimulation and bioaugmentation: a review. *International Journal of Environmental Bioremediation & Biodegradation.* 2015;3(1):28–39.

Azubuike CC, Chikere CB, Okpokwasili GC. Bioremediation techniques–classification based on site of application: principles, advantages, limitations and prospects. *World Journal of Microbiology and Biotechnology*. 2016;32(11):1–8.

Cecchin I, Reddy KR, Thomé A, Tessaro EF, Schnaid F. Nanobioremediation: Integration of nanoparticles and bioremediation for sustainable remediation of chlorinated organic contaminants in soils. *International Biodeterioration & Biodegradation*. 2017;119:419–28.

da Silva S, Gonçalves I, Gomes de Almeida FC, et al. Soil bioremediation: overview of technologies and trends. *Energies*. 2020;13(18):4664.

Das N, Chandran P. Microbial degradation of petroleum hydrocarbon contaminants: an overview. *Biotechnology Research International*. 2011;2011. Article ID: 941810.

Gadd GM. Geomycology: biogeochemical transformations of rocks, minerals, metals and radionuclides by fungi, bioweathering and bioremediation. *Mycological Research.* 2007;111(1):3–49.

Gadd GM. Metals, minerals and microbes: geomicrobiology and bioremediation. *Microbiology*. 2010;156(3):609–43.

Gaur N, Narasimhulu K, PydiSetty Y. Recent advances in the bio-remediation of persistent organic pollutants and its effect on environment. *Journal of Cleaner Production.* 2018;198:1602–31.

Hinchee RE, Leeson A. *Soil bioventing: principles and practice.* CRC Press; 1996.

Huang H, Ye L. Biological technologies for cHRPs and risk control. In Ren, H, Zhang, X (eds.) *High-risk pollutants in wastewater.* 2020 (pp. 209–236). Elsevier.

Kumar S. *Modern treatment strategies for marine pollution.* Elsevier; 2020.

Kumar V, Shahi SK, Singh S. Bioremediation: an eco-sustainable approach for restoration of contaminated sites. In *Microbial bioprospecting for sustainable development.* 2018 (pp. 115–136). Springer.

Megharaj M, Naidu R. Soil and brownfield bioremediation. *Microbial Biotechnology*. 2017;10(5):1244–9.

Megharaj M, Ramakrishnan B, Venkateswarlu K, Sethunathan N, Naidu R. Bioremediation approaches for organic pollutants: a critical perspective. *Environment International.* 2011;37(8):1362–75.

Mousavi SM, Hashemi SA, Iman Moezzi SM, Ravan N, Gholami A, Lai CW, Chiang WH, Omidifar N, Yousefi K, Behbudi G. Recent advances in enzymes for the bioremediation of pollutants. *Biochemistry Research International.* 2021;2021:5599204.

National Research Council. *In situ bioremediation: when does it work?* National Academies Press; 1993.

Picado A, Nogueira A, Baeta-Hall L, Mendonça E, de Fátima Rodrigues M, do Céu Sàágua M, Martins A, Anselmo AM. Landfarming in a PAH-contaminated soil. *Journal of Environmental Science and Health, Part A*. 2001;36(9):1579–88.

Rahman Z, Singh VP. Bioremediation of toxic heavy metals (THMs) contaminated sites: concepts, applications and challenges. *Environmental Science and Pollution Research.* 2020;27(22):27563–81.

Speight JG. *Reaction mechanisms in environmental engineering: analysis and prediction.* Butterworth-Heinemann; 2018.

Tomei MC, Daugulis AJ. *Ex situ* bioremediation of contaminated soils: an overview of conventional and innovative technologies. *Critical Reviews in Environmental Science and Technology.* 2013;43(20):2107–39.

Wang M, Zhang DQ, Dong JW, Tan SK. Constructed wetlands for wastewater treatment in cold climate—A review. *Journal of Environmental Sciences*. 2017;57:293–311.

Ward CH. *Subsurface restoration*. CRC Press; 1997.

Yan A, Wang Y, Tan SN, Yusof ML, Ghosh S, Chen Z. Phytoremediation: a promising approach for revegetation of heavy metal-polluted land. *Frontiers in Plant Science*. 2020;11:359.

Federal Remediation Technologies Roundtable. Technology Screening Matrix: Bioventing. 2020. https//frtr.gov/matrix/Bioventing/

31 Industrial Waste Management and Environmental Monitoring

31.1 WASTES AND THEIR VARIOUS TYPES

Wastes are undesirable substances in the environment. They enter the natural ecosystem from small-scale household activities to large-scale industrial production. Excessive population growth, increased consumption, and intense industrial activities have generated massive waste in the environment. Wastes exist in all states of matter, including solids, liquids, or gases. Waste can belong to inorganic and organic streams. Inorganic streams encompass sand, salt, metals, and radionuclides. Organic streams can be of two different types, *viz.*, organic contaminants and organic matter. These two exhibit differences based on their origins and their nature. Organic contaminants are synthetic carbon materials such as organic solvents, dyes, phenolic compounds, surfactants, pesticides, pharmaceuticals, petroleum-based wastage, plastics, and other industrial substances (such as polychlorinated biphenyls (PCBs) and polycyclic aromatic hydrocarbons (PAHs)). However, the term organic matter is generally used for biobased substances, including biomass residues, excretes, and other biological wastes. For example, when the body of an organism decays, it releases a large amount of organic matter into the environment. Green plants and autotrophic microorganisms are enormous sources of organic matter, mainly derived through primary production.

Based on their degradation properties, wastes can be persistent and nonpersistent. Persistent chemicals rest in the environment because they are resistant to degradation and persist for a longer duration. Nonpersistent chemicals are highly degradable and disappear quickly from the environment. Biodegradable organics tend to convert into small, nontoxic substances using biological treatment. Organic contaminants can be biodegradable or nonbiodegradable, depending on their susceptibility to biological decomposition, whereas most organic matter is biodegradable.

Based on their environmental effects, wastes can be categorised into two potential variants: hazardous and non-hazardous. Hazardous wastes are ignitable, corrosive, radioactive, and toxic materials that have an impact on both human health and the environment. Non-hazardous wastes include plastics, paper, metals, glass, and organic wastes, and they do not pose a direct threat to humans and the environment (Figure 31.1). These substances do not react violently, but their stay in the environment can have long-term repercussions.

Based on their source of origin, wastes can be classified as municipal waste, agricultural waste, medical waste, e-waste, construction and demolition waste, and industrial waste. Municipal wastes include trash and garbage released from households, gardens, yards, and small-scale businesses. An estimate by the World Bank in 2018 states that global municipal solid waste production is expected to nearly double by 2050 (growing to 3.4 billion tonnes). Agricultural activities like farming and livestock rearing generate an immense amount of agricultural waste. Crop residues, animal manure, and agricultural chemicals have added to the burden of solid wastes on land and water ecosystems. Medical wastes include pathological wastes (like human and animal body parts, tissues, fluids, and blood), sharp items (like blades, scissors, forceps, and needles), pharmaceutical wastes (like drugs and creams), and chemical liquid waste (like disinfectants). Their management requires intensive attention for human health care. E-waste is discarded electronic items that have accumulated in the environment at a very rapid pace in recent years. The construction and demolition of buildings and other infrastructure release debris of bricks, metals, glass, plastic, paints, etc., which are considered construction and demolition waste. They accumulate in significant amounts, raising concerns about their proper disposal. Industrial wastes are materials produced by industries. These are discussed in detail in the next section, including information on their management.

31.2 INDUSTRIAL WASTES AND THEIR MANAGEMENT

Industrial wastes are unwanted substances generated by factories, mills, mining activities, food manufacturing, farming, health care facilities, and consumer goods. They pose a severe contamination risk to our environment when discharged into aquatic ecosystems and dumped onto terrestrial ecosystems. Socio-economic development in a society is a key contributor to the amount of waste produced worldwide. Industrial waste exists in two physical forms: solid debris and wastewater (aqueous discard). Industrial waste occupies a relatively high amount of solid residues and wastewater.

Industrial waste management is a scheme for various actions required to manage wastes released by industries

DOI: 10.1201/9781032651095-38

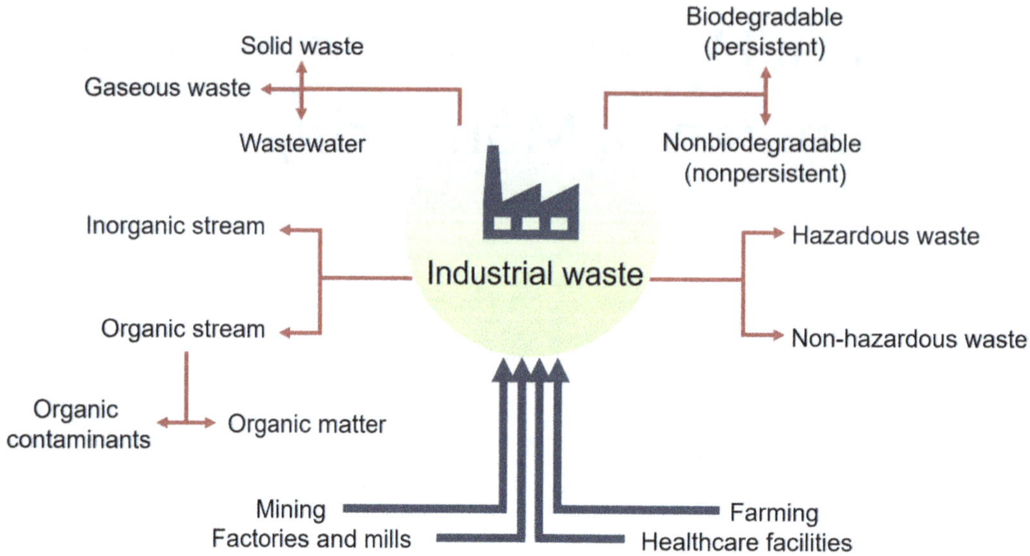

FIGURE 31.1 Various sources of industrial waste and their different categories.

that protect human health and the environment. The common waste management practices for industrial wastes are segregation, land application, landfills, and recycling. However, it is an extremely challenging task due to its excessive amount and heterogeneous nature. One strategy to put the waste management scheme into enforcement is the "7R" policy. It advocates for refuse, reduce, reuse, repair, repurpose, recycle, and recovery. This is an extension of the 3R rule, which concentrates on reduce, reuse, and recycle. In recent years, industrial wastes have been repurposed as raw materials for manufacturing valuable goods through bioprocess engineering, leveraging cost-competitive advantages. When the waste stream consists of a high amount of organic substances, management strategies often rely on the biodegradation process to implement the waste management scheme

A different viewpoint of waste management is the waste hierarchy (Figure 31.2). The European Commission (EC), the United States Environmental Protection Agency (US EPA), the Waste Management and Resource Recovery Association of Australia (WMRR), and other leading organisations around the World have acknowledged this waste management framework. Waste hierarchy is an integrated approach which sets targets to achieve the maximum practical benefits from waste and minimise waste generation. It establishes an order of preference for action to source reduction, prevention, reusing, recycling, reprocessing, energy recovery, treatment, and waste disposal, according to what fits best for the environment. The first two options are highly preferable and managed by avoidance. Waste disposal is the least suitable action on the list and is administered through incineration, pyrolysis, gasification, and landfilling. All other intermediate actions are methods of resource recovery. Many innovative industrial settings have recognised the roles of microorganisms in the waste management framework. Even during waste disposal, the material is spontaneously ceded to the microbial communities of the landfill.

31.3 CIRCULAR ECONOMY APPROACH (CEA)

The circular economy approach (CEA) is a waste management scheme which maintains the circularity of materials in the system through the prolonged use of products, restoration and regeneration of the associated substances, and elimination of waste. This approach aims to create value from finite resources and mitigate their impact on biodiversity and the environment. The Ellen MacArthur Foundation (EMF) has played a key role in supporting the circular economy approach involving people in business, academia, policymakers, and institutions across the globe. The economic system of this closed loop is the antithesis of the traditional linear economy, which is short-sighted and relies on a throw-away concept (take-make-waste). The conventional linear economic approach generates waste, contributes to the pollution crisis, and raises other environmental issues such as biodiversity loss, depletion of nonrenewable resources, emission of greenhouse gases, etc. (Figure 31.3).

CEA typically involves the collection of raw materials from the Earth to make products through a smart product design and keep them in service at their highest value for a prolonged period. Even when a product's life is over, the materials affixed to the product are recycled and used for other purposes. This whole approach facilitates long-term growth and creates new opportunities. CEA can be employed in various sectors such as food industries, textiles, energy production, agriculture practices, building construction, etc. The circular economy is crucial even for many stages of the product lifecycle, including design, manufacturing, distribution, and disposal across various industries. The adoption of CEA in the food sector alone has the potential to reduce about 49% of GHGs.

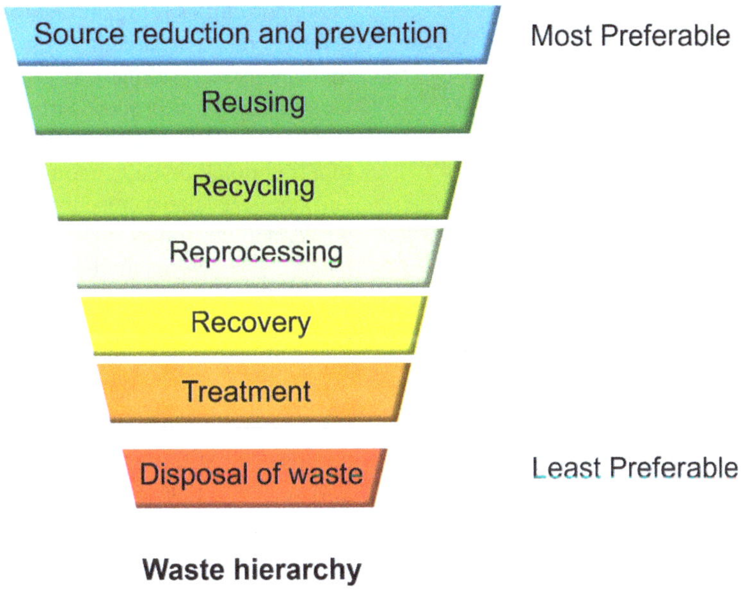

FIGURE 31.2 A model of the waste hierarchy.

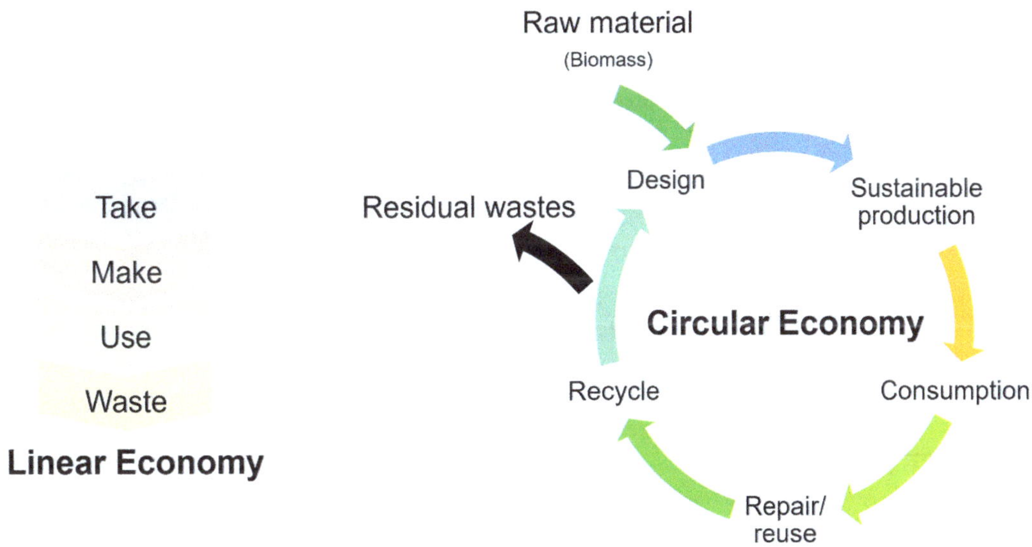

FIGURE 31.3 Linear economy *vs*. circular economy.

31.3.1 CEA Integration with Microbial Technology

The vast reservoirs of microorganisms and their diverse metabolic activities have offered promising opportunities for the circular economy. Microbial solutions require much less energy and envision true CEA. A plethora of microbial enzymes contribute to industrial manufacturing and waste treatment, reducing harmful gas emissions and improving resource efficiency. Bacteria and fungi show tremendous potential to transform various chemicals, thereby lessening their adverse effects on the environment. Microalgal and cyanobacterial cultivation offers several benefits owing to their rapid growth rate and multifaceted properties. These small photosynthetic units participate in carbon sequestration, wastewater treatment, environmental management, food and feed production, pharmaceuticals, fertilisers, and cosmetics (Figure 31.4).

The idea of circularity to regenerate the natural systems is incomplete without the redevelopment of contaminated sites. The use of microorganisms as bioremediators in the management of contaminants offers sustainable solutions to address environmental issues while mitigating the problem of bulk amounts of waste in the environment. Biodegradation and biotransformation are two invaluable tools for environmental clean-up, where the transformed and degraded materials can be used as bioenergy, biomaterials, or biopharmaceuticals, drawing into circular loops.

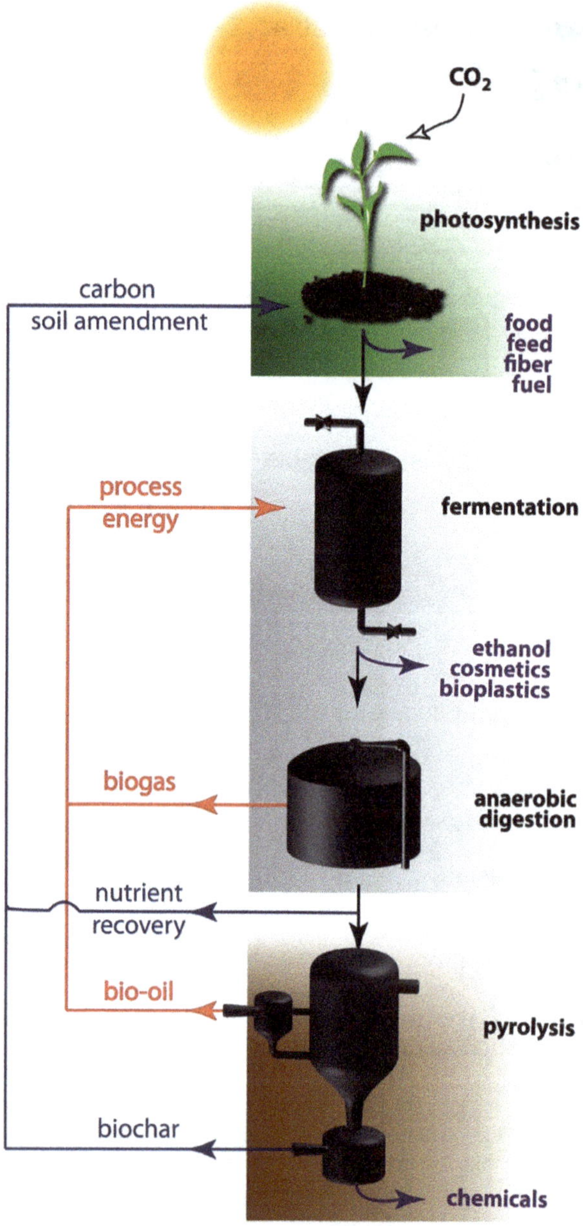

FIGURE 31.4 Recycling of energy and nutrients in a bio-based economy through photosynthesis, fermentation technology, and thermochemical conversion. (Image reproduced from Vanholme et al. (2013).[1])

Lignocellulosic wastes and organic wastes are the two most commonly used feedstocks for the production of resources using microbial technology. Lignocellulosic wastes are the residues left over from agricultural industries, and the organic wastes are released from domestic waste. Both can be used by fungal and bacterial cultures to generate bioenergy. The lignocellulosic biomass has great potential in bioethanol production without competing with land use and food production. The anaerobic digestion of organic wastes using microbial metabolic pathways is a promising way of energy recovery. The biogas produced during anaerobic digestion is a form of renewable energy source. Several industries extend the opportunities for manufacturing diverse bioproducts using a variety of waste streams. This type of framework has paved the way to achieve many targets of Sustainable Development Goals (SDGs), as proposed by the United Nations in 2015.

For example, ongoing textile manufacturing is a set example of a traditional linear economy, which involves nonrenewable and non-recyclable resources, produces vast amounts of waste, and consumes high volumes of water, chemicals, and energy. The new methods of making cloth have seen interest in biofabrication and synthetic biology, which involves less dependency on chemicals, water, and energy. Biofabrication utilises biological building blocks and fermentation processes to develop biopolymers and engineered fibres. Microbial silk and bacterial cellulose are sustainable alternatives to traditional textile materials. In addition, microbial enzymes have replaced many chemicals at several processing steps (such as pretreatment, desizing, scouring, bleaching, polishing, and pigmentation) of traditional textile manufacturing. The high reliance on microbial enzymes over the use of chemicals in other industries, such as detergent industries, tanneries, and paper and pulp industries, has a similar purview (discussed in Chapter 1). These schemes have minimised the environmental impact and offered novel pathways to circular textiles.

31.4 MICROORGANISMS IN WASTE MANAGEMENT

Microorganisms act on both inorganic and organic streams. Microbial treatment of waste offers marked environmental and economic welfare. Their involvement has diverse outcomes for waste management. The microbial workforce on inorganic streams demonstrates capabilities in biomineralisation, metal recovery, nutrient removal, bioconcrete production, and wastewater treatment. This involves the treatment of metals, non-metals, and radionuclides. Microbial tolerance to inorganic contaminants is often a critical requirement for their action because many chemicals pose negative effects on the cellular organisation. A notable scheme of their treatment is their transformation into less toxic forms, enabling them to detoxify or be immobilised. These methods are a primary goal of bioremediation and are discussed in the previous chapter (detailed in Chapter 30).

Microbial integrity-based organic matter breakdown is also an essential waste management strategy. Bacteria and fungi are the primary decomposers of organic wastes. This process helps microorganisms to gain energy and carbon to build their biomass. They perform a diverse array of metabolisms to degrade them, which also assists in maintaining the balance in the carbon, nitrogen, and sulphur cycles. The principal activities of microbial decomposition include the depolymerisation of long-chain polymers, leaching, humification, catabolism, and mineralisation. The feasibility of organic degradation by microorganisms contributes to a sustainable economy. Conversion of organic wastes into beneficial products is a win-win situation for the zero-waste movement. The most convincing application of microbial technology for waste management is the bioconversion of wastes into bioenergy (detailed in section 31.6).

Both biotic and abiotic factors can influence the decomposition of organic substances. In the biotic factor, microbes and their activities have a major influence on the breakdown of carbonic compounds. The microbial succession (the change in the species composition and abundance over time) sets the platform for the biodegradation of degraded products by previously existing communities. During this process, the microbial community keeps changing at different stages and constructs complex microbial ecosystems. This process can be continued until the system's carbon deposits are exhausted. On the other hand, the feedstock, nutritional content, and environmental conditions (e.g., pH, temperature, and moisture) are some notable abiotic factors that affect microbial utilisation of contaminants.

31.5 MICROBIAL ROUTES FOR WASTE VALORISATION

Waste valorisation is a promising approach to transforming waste streams into economically beneficial materials. The excess amount of waste stream, their underutilised nature, and weighty disposal practices incite waste valorisation. This is a highly efficient strategy, where the residual wastes of one industry could be used as raw materials or feedstocks for another. Industrial wastes are more productive resources than domestic wastes for waste valorisation because the industrial manufacturing process produces invariable, consistent, and predictable waste. This approach not only minimises the adverse effects of waste on the environment but also promotes environmental and economic sustainability. Moreover, waste valorisation addresses the issue of natural resource scarcity by repurposing waste. This process is appraised for its low treatment costs.

The sustainability in waste valorisation finds routes through microbial conversion. The conversion of the waste can be a multistep bioreactions, which require a group of microorganisms (microbial consortia). They can co-act or perform one after another, involving different sets of enzymes. The division of labour by two or more interacting microbial populations in a consortium facilitates an accelerated degradation of complex materials. The valuable outputs of these processes can be bioenergy, fertilisers, and animal feed. The biosynthesis of pharmaceuticals, biosurfactants, biodegradable polymers, and new bioenergy are other growing areas in waste valorisation through microbial fermentation.

In ethanol production, carbohydrate-rich food wastes, as low-cost supplements of carbon and other nutrients, are useful feedstocks for waste valorisation. This often includes waste materials such as molasses and bagasse from the sugar mills and whey/acid whey from the dairy industry. Based on the consumption of different feedstocks, biofuel production has been classified into first, second, third, and fourth generations. The first generation of biofuel represents sustainable challenges owing to the competition for resources for food and land. Second-generation biofuel relies on microbial activities. Third and fourth-generation biofuels can be produced using algae and genetically modified organisms. Particularly, the production of second-generation fuel aligns with the concept of waste valorisation using agricultural wastes. This approach has received wide recognition and is currently shifting to industrial-scale production.

Plastic wastes previously considered non-biodegradable are now biodegradable and can be converted into value-added products through bio-upcycling plastic waste. The identification of new microbial pathways and enzymes has found routes for the depolymerisation of a variety of plastics (polyethylene (PE), polyethylene terephthalate (PET), polyvinyl chloride (PVC), polyurethane (PUR), etc.). Particularly, a new microbial enzyme, PETase, has shown the potential to break down a special type of plastic (found in most consumer packaging, called polyethylene terephthalate (PET)) into smaller parts in hours or days through depolymerisation. The depolymerised materials can be repolymerised into new substances, and the entire process of depolymerisation-repolymerisation is crucial for the circular economy. This practice has great biotechnological potential for the biosynthesis of valuable chemicals such as polyhydroxyalkanoate (PHA), succinic acid, and biosurfactants (Figure 31.5). The synthesis of biodegradable polymers has significant potential for manufacturing new plastic materials (denominated as bioplastic). The options of microbial integrity in plastic degradation and synthesis of bioplastics have paved the way to recycle plastic and reduced the reliance on non-renewable feedstocks for their synthesis.

The industrial production of antibiotic penicillin depends on industrial waste, corn steep liquor, and relies on a microbial process (involving *Penicillium chrysogenum*). Corn steep liquor is a byproduct of corn wet-milling and has many nutrients, which makes it a valuable resource in industrial microbiology. In a different setup, food and beverage waste are suitable raw materials for sugar syrup production (e.g., high fructose corn syrup (HFCS-42)). The α-amylase production through fermentation relies on agro-industrial waste substrates such as sugarcane molasses and bagasse, wheat bran, coconut oil cake, corn cob, and rice husk. Vitamin B12 production typically recruits industrial wastes such as beet molasses and corn-steep liquor. High-quality citric acid under solid-state fermentation (SSF) is produced using agro-industrial residues such as banana peel, cotton waste, apple pomace, and orange peel.

A different microbial technique that has grabbed the attention of waste valorisation is microbial electrosynthesis (MES). The setup consists of an electrochemical cell, including a cathode and an anode. The microorganisms reside at the cathodic end and allow electricity-driven bioconversion of carbon dioxide (CO_2) to yield organic products. MES is a robust technology that recycles CO_2 into a range of commercially available carbon-neutral chemicals, such as isobutanol, 3-methyl-1-butanol, hexanol, caproic acid, and poly-3-hydroxybutyrate (PHB). Bacteria *Rhodopseudomonas palustris*, *Cupriavidus necator*, *Sporomusa ovata*, and enriched methanogenic communities can be suitably

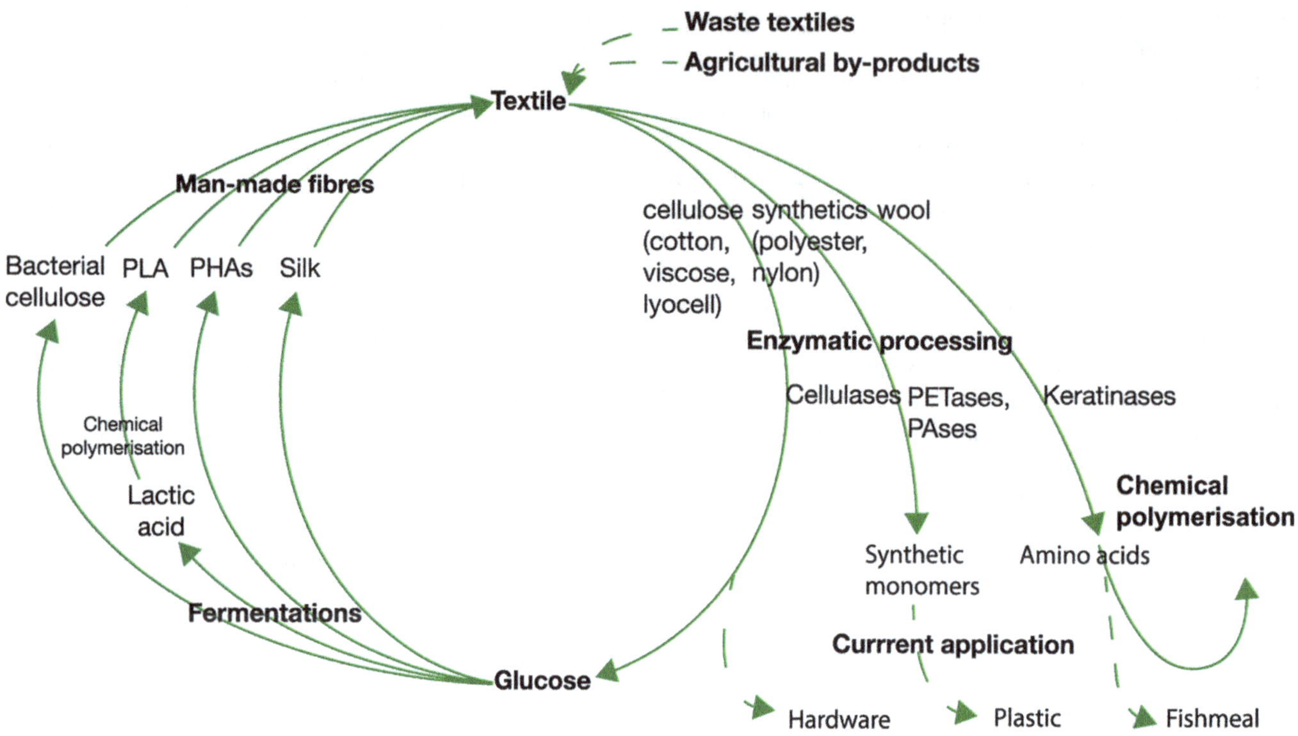

FIGURE 31.5 An example of closed-loop bio-based recycling in a circular economy of the textile industry. (PLA: Polylactic acid, PHAs: Polyhydroxyalkanoates, PET in PETases: Polyethylene terephthalate, and PAses: Phosphatases) (Image reproduced from Ribul et al. (2021).[2])

employed in the MES. This research initiative on bio-electro CO_2 recycling is expanding into the technological viability and feasibility of manufacturing useful biochemicals from scratch. BioRECO$_2$VER, SCALIBUR, BIOCON-CO_2, BAC-TO-FUEL, and Celbicon are some well-known projects that are working to make it possible.

The conversion of waste streams into useful products is often performed in **biorefineries**. The International Energy Agency (IEA) Bioenergy Task 42 states that "biorefinery is a facility for the sustainable processing of biomass into a spectrum of marketable products (food, feed, materials, and chemicals) and energy (fuels, power, and heat)". The utilisation of the full potential of biorefineries calls for a multidisciplinary approach, which could address the complex challenges of industrial facilities to transform them into a bio-based economy. Biorefineries involve intricate systems, and the fermentation process is only one part of biorefineries. This can additionally include facilities for feedstock preparation, gasification, pyrolysis, purification, etc. The ethanol production in biorefineries has seen tremendous upscaling for large-scale production. Anaerobic digestion and composting of waste streams are also among the most common practices under the broad framework of biorefineries. Despite significant progress, biorefineries encounter many challenges during the processing of wastes. These include high processing costs, the heterogeneous nature of feedstock, the complex process of biodegradation, low biomass yield, and the presence of inhibitory agents.

The practice of waste valorisation is expanding beyond using various traditional industrial wastes. The remnants of the many fermentation industries can also be further utilised for the biosynthesis of biofuel, bioplastics, and other materials. For example, a large amount of wastewater generated from bioethanol production can be used to produce bio-based chemicals. This type of process can be used in the synthesis of fatty acid methyl esters from yeast, the production of oleaginous biomass using algae, and the removal of odour, colour, and contaminants.

31.6 BIOCONVERSION OF WASTE INTO BIOENERGY

An energy crisis is one of the major global issues of the present world. The shortfall of primary energy resources (e.g., depletion of fossil fuels) and their adverse environmental impact place great responsibility on research communities to find alternative sources. Renewable sources such as solar, wind, geothermal, hydropower, and biomass energy are promising choices that have offered environmentally and economically friendly solutions. In biomass energy, several discarded materials, such as food waste, agricultural waste, municipal solid waste, animal manure, and human sewage, are renewable sources and can be used for bioenergy production. These waste materials as feedstock are potentially cost-competitive, considering their abundance, renewability, low cost, and high degradability. The main forms of bioenergy

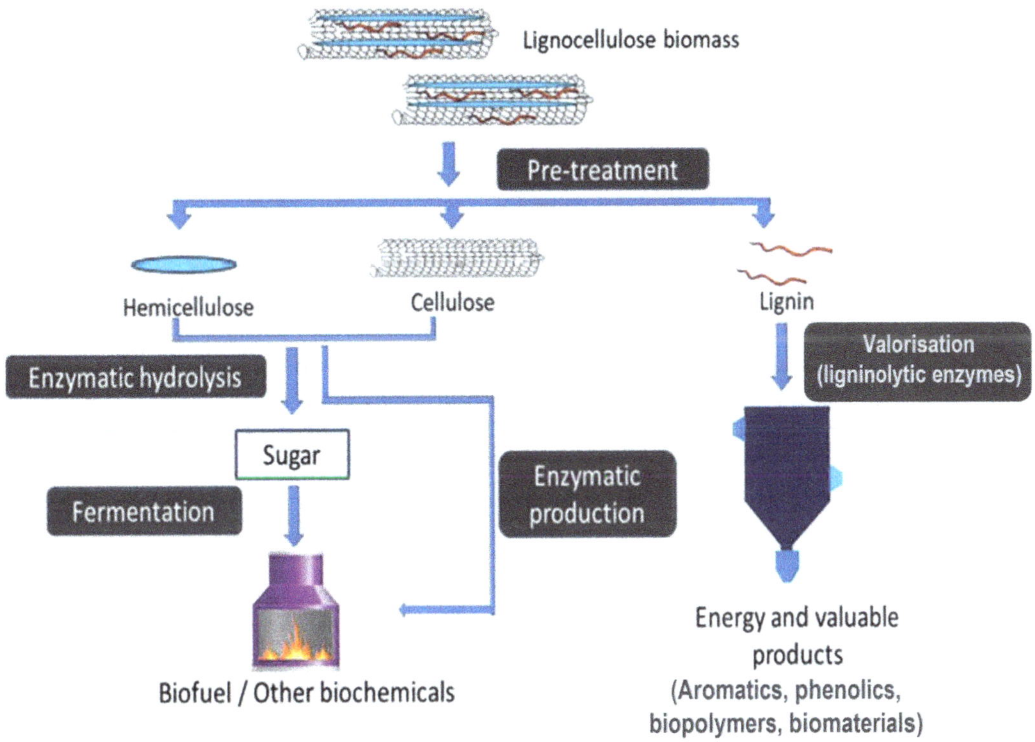

FIGURE 31.6 Flow chart of bioconversion of lignocellulosic biomass for the production of biofuel and other valuable products. (Image adapted from Sankaran et al. (2021).[3])

produced from these waste biomasses are biogas, bioethanol, biodiesel, biohydrogen, syngas, bio-oil, biochar, and bioelectricity (Figure 31.6).

Biomass is a repository of chemical energy that can be burned to produce heat or converted into fuel through specific biochemical processes. The burning process involves the thermochemical conversion of waste (under high pressure (1–240 bar) and temperature (150–1600 °C)). Standard methods of burning are gasification, liquefaction, pyrolysis, or combustion. Although thermochemical conversion is an efficient and sustainable method for fuel production, high investment costs and low energy efficiency hinder its commercial upscaling.

The waste-to-bioenergy technological options using biochemical processes are eco-friendly, economical, and sustainable. These processes mainly utilise bacteria and fungi, implying their microbial pathways and enzymes. Modern methods of bioenergy conversion from waste include anaerobic digestion, composting, fermentation, biohydrogen production, and microbial fuel cells (MFCs).

A prominent source of bioenergy is bioethanol, produced through the alcoholic fermentation of fermentable sugars (detailed in Chapter 13). Complex polysaccharides found in waste materials are suitable raw materials for bioethanol production. They require some pretreatment before full utilisation in their bioconversion into ethanol. Some complex microbial communities of water and soil involve dark fermentation and produce hydrogen (H_2) gas for sustainable bioenergy generation. This biomass conversion is mainly catalysed by an enzyme, hydrogenase, in the absence of light and oxygen (O_2). In a different process, MFC is a device that involves electrochemically active bacteria (EAB) for the breakdown of waste materials and generates bioelectricity. This technology shows the possibility of simultaneous wastewater treatment and bioenergy production (detailed in Chapter 25).

Anaerobic digestion has gained popularity in waste management by converting waste into biogas. Methane (CH_4) gas is the most valuable product in this type of bioconversion. Composting is a process of digestion of wastes by aerobic microorganisms, and the product, compost, can be employed as an organic fertiliser, humus, and soil conditioner for agriculture. The overproduced and low-quality compost can be utilised as a potential energy source after thermochemical conversion. They are further discussed in the following subsections.

31.6.1 Anaerobic Digestion

Anaerobic digestion is a process of the breakdown of organic matter by microorganisms in anaerobic situations. Biodegradable materials such as livestock manure, industrial waste, municipal waste, food waste, and other organics are suitable feedstock for the reaction. The biological reactions of anaerobic digestion are carried out in a large tank called a bioreactor (or digester). During digestion, a digester loaded with organic wastes is left to rot in a completely air-tight condition. The system is strictly controlled and kept away from the outer environment because the presence of O_2 in the reactor can disrupt the entire process. The native microbial communities proliferate under the controlled parameters of the digester and consume the organic matter present in the wastes. The two valuable outputs of anaerobic digestion are biogas and digestate. The biogas is composed of CH_4, CO_2, hydrogen sulphide (H_2S), water vapour, etc. Methane is the primary component in this mixture, constituting about 50-75%, followed by CO_2 (25-50%). The leftover material from digestion is digestate, which can be either liquid or solid or a combination of both forms (Figure 31.7).

Digesters are available in a wide range of designs depending on different parameters. Based on flow, the digesters can be tailored to operate under batch (e.g., Anaerobic Sequencing Batch Reactors (ASBR) and Plug Flow Reactors (PFR)) or continuous operation (e.g., up-flow anaerobic sludge blanket (UASB) reactor and continuous stirred tank reactor (CSTR)). The biological treatment of waste can be improved by processing wastes through heat treatment, maceration, or sonication before digestion.

Wastes with high solids content (20-50%) are treated using a dry digester, where the feedstock is placed in a container, and a stimulant is sprayed on top of it for microbial activation. On the other hand, low-solids wastewater treatment calls for a wet digester, which is designed to introduce feedstock with normal pumps and consists of a homogeniser to mix the content perfectly. A wet digester requires more space than a dry digester due to the high-volume liquid-to-feedstock ratio.

The digestive process involves a complex series of bioactivities. This involves sequential reactions of hydrolysis, acidogenesis, acetogenesis, and methanogenesis. The first bioreaction in digestion is hydrolysis using hydrolytic bacteria, which turn the organic streams into sugars, amino acids, and fatty acids. The resultant substances are handled by acidogenic (fermentative) bacteria, which yield organic acids, CO_2, H_2, and ammonia (NH_3). In the next step, acetogenic bacteria transform organic acids into acetic acid and release additional CO_2, H_2, and NH_3. The final reaction is the conversion of CO_2 into CH_4 using methanogens (Figure 31.8). The byproduct, H_2S, in the biogas during digestion has unintended adverse effects, which can be removed by desulphurisation. It can be achieved through biological methods involving sulphide oxidation using chemolithoautotrophic metabolism and anoxygenic photosynthetic activity.

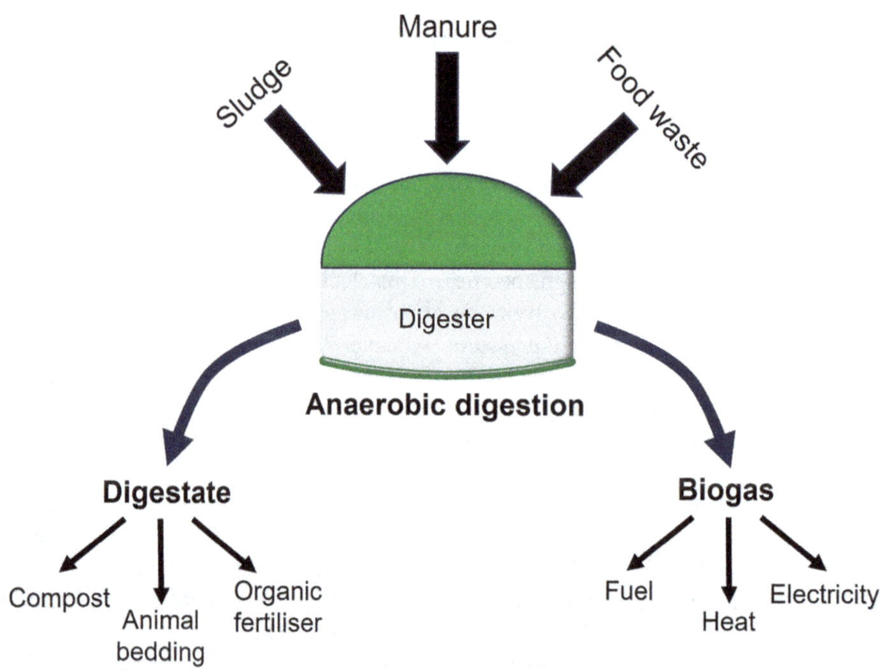

FIGURE 31.7 An illustration of anaerobic digestion and the key benefits.

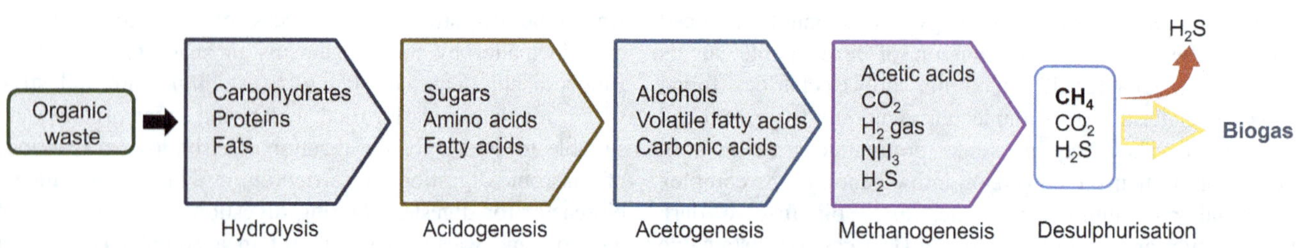

FIGURE 31.8 A step-wise decomposition process in anaerobic digestion.

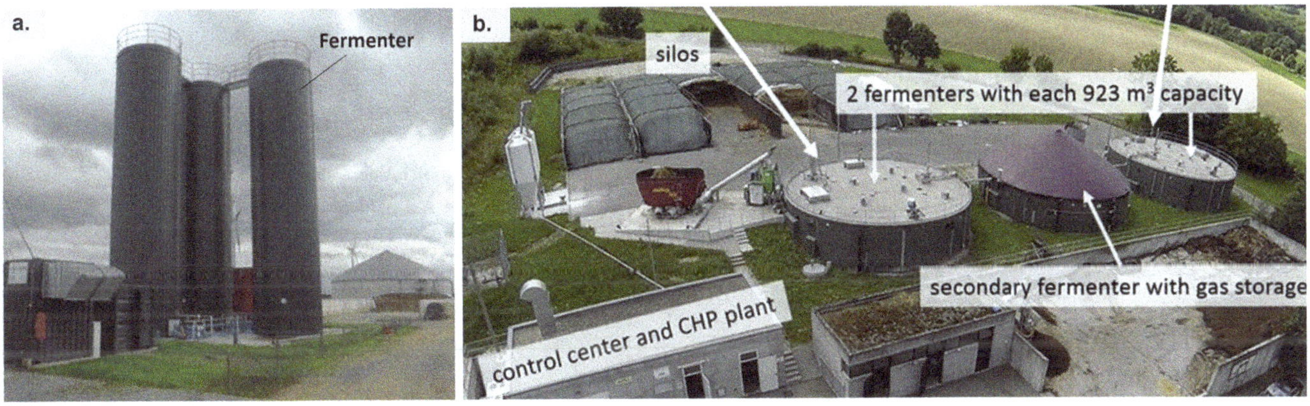

FIGURE 31.9 a. A picture of a biogas plant. b. Aerial view of a biogas plant at Eningen, Germany. (CHP: combined heat and power). (Images adapted from Maus et al. (2016)[4] and image b adapted from Bierer et al. (2019)[5] with permission of © 2019 WILEY-VCH Verlag GmbH & Co. KGaA, Weinheim.)

FIGURE 31.10 A composting unit. (Image captured by Rajan and Anchal Singh/Zakir Husain Delhi College.)

Bioenergy and digestate produced from an anaerobic digester contribute to a circular and sustainable bioeconomy. With appropriate treatment, biogas can be utilised in direct combustion systems and electricity production. It can also be used to power motor vehicles and in cooling systems. Digestate can be employed as a soil conditioner, compost, fertiliser, animal bedding, and other bio-based products. Some prominent industrial setups that are heavily moved by anaerobic digestion for waste management include BIOGEST PowerRing, EnviWaste, Fitec Self-Cleaning Digester, and Memthane®. Figure 31.9 shows two different setups of biogas plants for anaerobic digestion.

31.6.2 Composting

Composting is a sustainable way to convert organic matter into value-added products in an oxic environment. Healthy compost teams up with nutritionally diverse microorganisms, including bacteria, archaea, fungi, and protozoa. Composting passes through three sequential stages, predominated by microbial communities of mesophiles and thermophiles at different stages. The first stage is the mesophilic phase (10–42 °C), which is distinguished by the rapid increase in temperature for the decomposition process. This lasts for a couple of days after the initiation. The prolonged high temperature encourages the growth of thermophiles and shifts into the second stage, the thermophilic phase (45-70 °C). After extensive treatment by thermophilic microbial communities, a compost enters the stage of cooling, called the maturation phase (50-23 °C). This phase stabilises the organic matter and heat and lasts a few weeks to months. This stage hosts a second round of the mesophilic stage with distinct microbial communities. A complete composting cycle takes around one month to two years, depending on the feedstock, size and turning of the pile. The final form of compost is composed of rich brown colour, soil-like particles (Figure 31.10). This process can be accelerated by involving earthworms during composting. The process is called **vermicomposting**.

TABLE 31.1
Anaerobic digestion *vs*. composting.

	Anaerobic digestion	Composting
Biodegradation	Strict anaerobic conditions	Strict aerobic condition
Gases produced	Mainly CH_4 and CO_2	Mainly CO_2
Physical output	Digestate	Compost
Microorganism involved	Bacteria and archaea	Bacteria, archaea, fungi, and protozoa
Key benefits	Bioenergy and fertiliser	Stable organic end product to restore soil fertility
Operational modes	Industrial scale	Local farmland and industrial scale
Process monitoring	Highly controlled	Less controlled
Odour	Less odour nuisance	High odour nuisance
Residual output	Nitrogen-heavy state fertilisers	Balanced carbon and nitrogen
Risk	Pasteurisation ensures the removal of all infectious agents	Possible spread of disease from compost
Resource recovery	Much efficient	Less efficient
Investment	High level of investment	Low capital investment
Energy input	Energy surplus	Energy demand

During the initial stage, mesophilic bacteria like *Pseudomonas*, *Bacillus*, *Actinobacteria*, and *Micrococcus* are the dominant residents in the organic materials. The saprophytic fungi, such as Aspergillus and Penicillium, also cooperate in the degradation process. The second stage attracts thermophilic bacteria like *Bacillus*, *Geobacillus*, and thermophilic actinomycetes to degrade complex matter. This situation also promotes the proliferation of thermophilic fungi like *Thermoascus* and *Humicola*. Further, the long mesophilic stage again sets the stage for mesophilic microorganisms to digest the remaining matter. This stage supports the growth of nitrifying bacteria such as *Nitrosomonas* and *Nitrobacter* for the transformation of NH_3 into nitrate (NO_3^-). The saprophytic fungi degrade the organic matter to produce humus.

Both anaerobic digestion and composting may treat similar wastes. Even the digestate produced from anaerobic digestion requires composting before it can be applied to agricultural land. Composting is mainly performed to recover a stable organic final product from organic wastes, which can be used to improve soil fertility. About 50% of the carbon in compost supports plant growth, while the remaining 50% is released as CO_2 through mineralisation. The low capital investment, low-level training, and simple process monitoring promote its setup at local farmlands. However, composting is less sustainable than anaerobic digestion and demands for energy supply in large-scale operations. Table 31.1 provides a comparative account between anaerobic digestion and composting.

31.7 BIOPROCESS ENGINEERING AND SYNTHETIC BIOLOGY

In commercial treatment facilities, where microbial processes are employed, there is a need for improvement in waste bioconversion. This is a stringent requirement in biorefineries, which calls for bioprocess engineering. It involves an interdisciplinary approach, including biology and engineering. It applies some engineering principles for biological innovations and the optimisation of biological treatment. The foundation of bioprocessing is based upon the transformational power of fermentation strategies coupled with upstream and downstream processing.

Synthetic biology is a technique for designing custom-made organisms. This often comprises redesigning an organism, introducing new features in an organism or engineering new biological systems for their novel, efficient, and predictable causes. The term "**bioengineering**" is also used synonymously with synthetic biology. Synthetic biology frequently uses iterative design, where researchers continually enhance and optimise the built biological systems by testing, analysing, and changing designs in response to the results. DNA is an interchangeable piece at the molecular level that allows scientists to introduce new metabolic pathways to improve the functional organisation of natural systems. The ability to engineer biological systems through mutations and genetic recombination techniques has evolved as an art to generate variability and diversity. These frameworks hold promise to suppress or prevent the expression of particular genes to uncover improvements in their activities. The relatively new gene silencing techniques, such as RNA interference (RNAi) technology and clustered regularly interspaced short palindromic repeats (CRISPR)-based methods, have added more fuel to modify gene expression in industrially useful microorganisms.

One of the most well-known applications of the synthetic biology approach is the modification of microbial capacity to increase the yield of desired products. The insertion of genes transforms a single microorganism into numerous strains. A genetically modified organism can digest many xenobiotics simultaneously. An engineered form also ensures their survival in harsh, contaminated environments. These approaches to synthetic biology have been a paradigm shift in achieving the maximum benefits of biotechnology.

Process evaluation is the backbone of bioprocess engineering, which involves systematic assessment and analysis of the various aspects of bioprocesses. Strict compliance with operations allows highly reproducible processes and product consistency. Bioreactor design is another parameter for bioprocess engineering that offers support in creating, controlling, and programming cellular activities. The combination of various engineering promises to improve yield, quality, and efficiency.

31.8 MICROBIOLOGICAL ENVIRONMENTAL MONITORING

Environmental monitoring is a systematic observation to assess the quality of natural ecosystems at temporal and spatial scales. Environmental monitoring plays a vital role in giving essential information on ecological disturbances caused by humans and the effects produced by natural stressors. The main objectives of the monitoring programmes are to generate a database of environmental conditions, examine the trend, ensure compliance with laws, and support policymakers in formulating a regulation for mitigating the risk of adverse effects.

Environmental monitoring characterises all types of sites, including soil, atmosphere, and water. The main monitoring repertoires are physical, chemical, and biological parameters. Physical parameters comprise the elements of temperature, turbidity, conductivity, etc. Chemical parameters involve the qualitative and quantitative analyses of various natural and artificial traits like pH, salinity, heavy metals, organic substances, etc. Likewise, many biological activities or biological entities are critical signatures for environmental situations and are typically referred to as **bioindicators**. Biological testing can adequately keep track of a robust understanding of environmental conditions through interactions between organisms and stressors. Biological indicators include different organisms belonging to microorganisms, plants, and animals. The community composition shift, species distribution, physiology, and appearance of indicators are suitable parameters to depict the change in response to environmental stimulus. The rise or fall of the population of indicator species provides information on the variation in environmental traits.

The elements of microbial activities as bioindicators are particularly evident and have received major attention to gauge the patterns and trends of human-based pressures. They play fundamental roles, owing to their ability to quickly respond to environmental variations, biomass turnover, and ubiquitous existence. Their presence and functional attributes are the major players in the assessment of the environment and health. Microbial regulation involving metabolic pathways and enzymatic activity rates can be used as a multi-metric index to assess ecological integrity. Natural microorganisms are conservative biomarkers for the description and prediction of a biological or chemical change. Microbial communities can be used to study functional redundancy, which involves the study of the relationship between biodiversity and ecosystem functioning. This type of investigation provides a clue about ecological resilience and stability. A highly functional, redundant community is generally highly robust, which keeps the functional attributes intact and maintains overall stability in an ecosystem even when some native species are lost.

The indicator microorganisms provide crucial insights into functional attributes, early warning signs, and stress response. For example, the ratio of archaeal and bacterial ammonia-oxidisers is a suitable indicator of wetland health conditions. The soil microbiological attributes, such as basal soil respiration and microbial biomass carbon soil, are notable parameters for monitoring soil quality. Coliform bacteria act as indicators of faecal contamination in water and provide the status of the potential presence of pathogenic bacteria (detailed in Chapter 24). The presence of phages is an indicator of water health, owing to their roles in the identification of bacterial pathogens and their treatment. A large number of microorganisms contribute to mineral exploration and are reported in selected sites.

Indicator microorganisms also detect harmful environmental changes and mitigate pollution problems. For example, natural attenuation is a remediation process which is based on a monitoring approach of natural processes for attenuating environmental pollutants at contaminated sites (detailed in Chapter 30). Differently, lichens are good bioindicators of air quality because they are sensitive to air pollutants (e.g., sulphur dioxide (SO_2), NO_3^-, and NH_3). In lichens, the photosynthetic pigment chlorophyll of algal partners is vulnerable to excess deposition of harmful chemicals. This process interrupts their association with fungal partners and impacts their overall distribution in the affected regions.

The involvement of microbial indicators has many innovative multidisciplinary solutions over traditional techniques. Scientists have designed an artificial system called biosensors, combining biological activities with mechanical and electrical engineering. This integrated system meets the regulatory pressure and offers sustainable and eco-friendly solutions to recognise a target analyte. The bioengineering of microorganisms and their molecules is a driving force for the emergence of new tools for detection assays.

31.9 BIOSENSORS

A biosensor is a compact analytical device containing a biological recognition element (BRE) integrated with a physical transducer. It detects biological or chemical reactions. A BRE can be whole microbial cells, organelles, enzymes, cofactors, antibodies, receptors, or nucleic acids. When a BRE responds to a target compound, the transducer converts a biorecognition event into measurable signals (Figure 31.11). A stable and strong connection between the transducer and BRE in the biosensor is an essential requirement to enhance its analytical performance and storage stability.

A biosensor is often fabricated using microorganisms. Microalgae, bacteria, and yeasts are excellent choices of microbial cells for biosensors because they can be grown in large quantities using low-cost cultivation procedures. Both

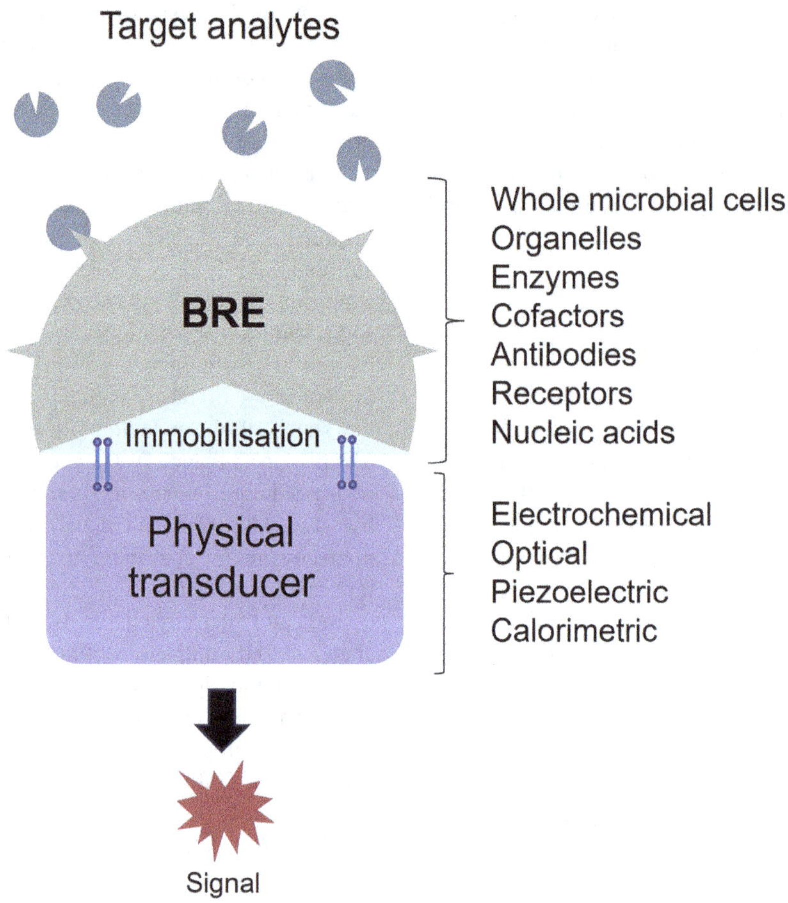

FIGURE 31.11 An illustration of a biosensor. (BRE stands for biological recognition element.)

whole microbial cells and enzymes offer different advantages and disadvantages in biosensor technology. Whole microbial cells exhibit properties like robustness and tolerance. However, they have limited selectivity and low specificity owing to their diverse metabolic pathways. In this direction, the tools of genetic manipulation and metabolic engineering in microbial cells can enhance their selectivity and sensitivity by blocking or inducing different metabolic pathways. In contrast, enzyme-based biosensors have broad scopes and are highly prominent in many industrial and environmental applications. The enzyme can be suitably immobilised onto a suitable support material, which ensures high enzyme stability, integrity, and sensitivity. However, the applications of enzymes under *in vivo* conditions frequently hamper their activity. Also, the processes of enzyme purification and immobilisation raise concerns about the overall costs of setups.

The transducer systems in biosensors can be electrochemical, optical, piezoelectric, or calorimetric systems. These systems operate on different principles and are briefly discussed as follows.

- Electrochemical techniques for the transducer system are based on electrochemistry and typically involve amperometry, conductometry, potentiometry, or voltammetry. This approach can also utilise an integrated microbial process (e.g., MFC).
- Optical methods rely on optical signals, which can be accomplished in label-free or label-based sensing. The label-free method is based on the direct interaction of the analyte with the transducer (e.g., surface plasmon resonance (SPR)-based biosensors). On the other hand, the label-based method involves the labelling of an analyte, and the optical signalling is measured in the form of fluorescence, luminescence, and colours. Some microbial cells can be engineered to produce optical signals for biosensors (detailed in Box 31.1).
- Piezoelectric biosensors work on the principle of electric current production in response to mechanical stress applied to a material.
- Calorimetric biosensors detect the variation in temperature due to specific reactions taking place in the solution, which is proportional to the analyte concentration.
- A relatively new type of biosensor is a lab-on-a-chip (LOC). It is a miniature-sized apparatus which involves the microfabrication of a microarray and microfluidic techniques onto an integrated electronic chip. This framework provides an automated, high-throughput platform for on-site monitoring of environmental parameters.

BOX 31.1 OPTICAL SIGNALS IN BIOSENSORS USING MICROORGANISMS

Optical signals in biosensors detect variations in light properties to measure biological or chemical substances. Some biological compounds exhibit properties that can be used to develop optical signals for the presence and concentration of specific substances. Microorganisms can be genetically engineered to produce such compounds to generate optical signals. The optical signal generation in microbial biosensors mainly involves heterologous transformation, where genes encoding signal-generating molecules are introduced into the microorganism. Some notable biological optical signals are fluorescence, bioluminescence, colourimetric change, and refractive index change.

The fluorescence signal is mainly generated from green fluorescent protein (GFP), which is native to the jellyfish *Aequorea victoria*. The bacteria can be engineered to express GFP to detect environmental contaminants. Some other fluorescent proteins, such as red fluorescent protein (RFP) and blue fluorescent protein (BFP), have been discovered to produce new colours in the biosensors. Aequorin, firefly luciferase, and bacterial luciferase (Lux) are notable examples of bioluminescence signals. Aequorin is reported in the jellyfish *Aequorea victoria* and firefly luciferase in fireflies. Aequorin-based biosensors are mainly used to monitor calcium dynamics in cells, while firefly-based biosensors facilitate protein-protein interactions. The bacterial luciferase is naturally harboured in bioluminescent bacteria such as *Vibrio*, *Photobacterium*, and *Photorhabdus*. The full gene cassette of bioluminescent reaction in bacteria consists of five genes (*luxCDABE*), of which only the former two genes are responsible for the synthesis of luciferase protein (a heterodimer). Bacterial luciferase-based biosensors generate signals for various contaminants in water and soil samples. A different sensing procedure to detect a particular analyte is a colourimetric technique, which involves identification through the conversion of a chromogenic substance into a colourful substance by the naked eye or simple portable optical detectors. In addition, a technique, surface plasmon resonance (SPR), that detects changes in the refractive index can be used to observe changes in the refractive index for the effects of the analyte in the biosensors. This has applications in drug discovery, environmental monitoring, and diagnostics.

The microbial biosensor is usually a low-cost, portable, easy-to-use, sensitive, and precise field monitoring probe for manifold applications in the food industry, agriculture, medicine, and environmental monitoring. This technique can even be useful to monitor multiple analytes simultaneously and has roles in high-throughput screening for scientific discoveries. In the food sector, microbial sensors can be used to ensure food safety by monitoring their nutritional content, pathogens, additives, and shelf lives. The target analytes in food analyses can be ethanol, glucose, fructose, amino acids, long-chain fatty acids, vitamins, steroids and cholesterol. The miniaturised sensing platforms in the healthcare sector offer services in biopharmaceutical manufacturing, clinical analysis, disease screening, and diagnosis.

Biosensors also monitor various pollutants, such as pesticides, toxins, and pathogens, in soil, water, and air. The technique can be suitably used to determine the BOD in water samples. The conventional approach of BOD determination takes five days and involves an off-site monitoring method. In contrast, a biosensor-based estimation allows on-site detection and assesses BOD within five to forty-five minutes. An example of a commercially available BOD monitoring device is MB-BOD from Biosensors SL, Spain. In recent years, a new biosensing device has been designed to identify the presence of arsenic (As). Scientists have engineered bacteria (such as *E. coli*) with different levels of As sensitivity that can even detect concentrations as low as 10 ppb (which is the threshold limit according to the World Health Organisation (WHO)). This ultrasensitive biosensor can be connected to a smartphone, which offers exciting aptness for the new generation of monitoring devices.

SUMMARY

- Waste materials released from various industries pose significant challenges to health and the environment. They belong to the inorganic and organic streams. Industrial waste management employs different schemes to manage waste, such as the waste hierarchy and the "7R" strategy.
- A circular economy is a system of production and consumption of products where resources are used efficiently and waste is minimised by creating a closed-loop system. This benefits both the economy and the environment. The vast reservoirs of microorganisms and their metabolic activities contribute to the circular economy through bioconversions.
- Waste valorisation is a method that turns waste materials into valuable products, aligning with the circular economy. The microbial workforce can be used to transform waste streams into economically beneficial materials.
- Industrial waste materials can be converted into biogas, bioethanol, biodiesel, biohydrogen, syngas, bio-oil, charcoal, and bioelectricity. They represent renewable forms of bioenergy. Anaerobic digestion and composting are the two primary microbial-based methods to decompose organic matter.
- Bioprocess engineering, such as metabolic engineering, bioreactor design, and process optimisation, grafts new abilities for the waste treatment process. Synthetic biology frequently entails bioengineering in organisms to improve their performance.

- Many microorganisms are critical signatures of environmental conditions and can be used for environmental monitoring. A biosensor is an analytical device that consists of a BRE and a physical transducer. The signals in biosensors can be measured through electrochemical, optical, piezoelectric, or calorimetric transducer systems.

DIFFERENT TYPES OF PRACTICE QUESTIONS

Fill in the blanks

I. Industrial wastes exist in two physical forms of .... and ....
II. .... maintains the circularity of materials in the system through the prolonged use of products and restoration and regeneration of materials.
III. The two valuable outputs of anaerobic digestion are .... and ....
IV. .... is a remediation process based on a monitoring approach of natural processes for attenuating environmental pollutants at contaminated sites.
V. A microbial biosensor is a compact analytical device containing a .... integrated with a physical transducer.

SOLUTION: I. solid debris, wastewater, II. circular economy approach, III. biogas, digestate, IV. natural attenuation, V. biological recognition element (BRE)

Very short answer type of questions

I. Give two examples of each inorganic waste and organic waste.
II. Name any two industries that have the possibility of replacing their nonrenewable and nonrecyclable approaches using microbial technology.
III. What are the different forms of bioenergy?
IV. How does bioenergy differ from nonrenewable energy?
V. Why are lichens treated as bioindicators of air pollution?
VI. What is the role of a biological recognition element in a biosensor?

Short answer type of questions

I. What is the "7R" policy? How is it different from the 3R rule and waste hierarchy?
II. Write a short note on the circular economy approach and its integration with microbial technology.
III. What are the advantages of using microbial processes in industrial waste management?
IV. Write the differences between composting and anaerobic digestion.
V. How do bioprocess engineering and synthetic biology graft new abilities for waste treatment?
VI. What are biosensors? What are the advantages of using microbial biosensors?

Long answer type of questions

I. What is industrial waste and its sources? What are the emerging trends in microbial applications for industrial waste management?
II. What are the key challenges in environmental monitoring? Write a note on the roles of microorganisms in monitoring environmental samples with suitable examples.

NOTES

1 Vanholme B, Desmet T, Ronsse F, et al. Towards a carbon-negative sustainable bio-based economy. *Frontiers in Plant Science*. 2013;4:174.
2 Ribul M, Lanot A, Pisapia CT, et al. Mechanical, chemical, biological: Moving towards closed-loop bio-based recycling in a circular economy of sustainable textiles. *Journal of Cleaner Production*. 2021;326:129325.
3 Sankaran R, Markandan K, Khoo KS, et al. The expansion of lignocellulose biomass conversion into bioenergy via nanobiotechnology. *Frontiers in Nanotechnology*. 2021;3:96.
4 Maus I, Koeck DE, Cibis KG, et al. Unraveling the microbiome of a thermophilic biogas plant by metagenome and metatranscriptome analysis complemented by characterization of bacterial and archaeal isolates. *Biotechnology for Biofuels*. 2016;9:1–28.
5 Bierer B, Kress P, Nägele HJ, et al. Investigating flexible feeding effects on the biogas quality in full-scale anaerobic digestion by high resolution, photoacoustic-based NDIR sensing. *Engineering in Life Sciences*. 2019;19:700–10.

SUGGESTED READING

Ayilara MS, Olanrewaju OS, Babalola OO, Odeyemi O. Waste management through composting: challenges and potentials. *Sustainability*. 2020;12(11):4456.

Bian B, Bajracharya S, Xu J, Pant D, Saikaly PE. Microbial electrosynthesis from CO_2: Challenges, opportunities and perspectives in the context of circular bioeconomy. *Bioresource Technology*. 2020;302:122863.

D'souza SF. Microbial biosensors. *Biosensors and Bioelectronics*. 2001;16(6):337–53.

Ewing TA, Nouse N, van Lint M, van Haveren J, Hugenholtz J, van Es DS. Fermentation for the production of biobased chemicals in a circular economy: a perspective for the period 2022–2050. *Green Chemistry*. 2022;24(17):6373–405.

Gil A. Challenges on waste-to-energy for the valorization of industrial wastes: Electricity, heat and cold, bioliquids and biofuels. *Environmental Nanotechnology, Monitoring & Management*. 2022;17:100615.

Holm-Nielsen JB, Al Seadi T, Oleskowicz-Popiel P. The future of anaerobic digestion and biogas utilization. *Bioresource Technology*. 2009;100(22):5478–84. www.ellenmacarthurfoundation.org/

Khan S, Anjum R, Raza ST, Bazai NA, Ihtisham M. Technologies for municipal solid waste management: current status, challenges, and future perspectives. *Chemosphere*. 2022;288:132403.

Koul B, Yakoob M, Shah MP. Agricultural waste management strategies for environmental sustainability. *Environmental Research*. 2022;206:112285.

Kumar R, Kumar P. Future microbial applications for bioenergy production: a perspective. *Frontiers in Microbiology*. 2017 Mar 21;8:450.

Lad BC, Coleman SM, Alper HS. Microbial valorization of underutilized and nonconventional waste streams. *Journal of Industrial Microbiology and Biotechnology*. 2022;49(2):kuab056.

Logar RM, Vodovnik M. The applications of microbes in environmental monitoring. In Méndez-Vilas A (ed). *Communicating current research and educational topics and trends in applied microbiology*. 2007(pp. 577–85). Formatex.

Wang B. Arsenic: a smartphone-friendly biosensor to tackle an insidious global threat. *The Microbiologist*. 2022. www.the-microbiologist.com/features/arsenic-a-smartphone-friendly-biosensor-to-tackle-an-insidious-global-threat/45.article

Wong JW, Tyagi RD, Pandey A. *Current developments in biotechnology and bioengineering: solid waste management*. Elsevier; 2016.

Xu X, Ying Y. Microbial biosensors for environmental monitoring and food analysis. *Food Reviews International*. 2011;27(3):300–29.

Index

Note: Figures are indicated by *italics*. Tables are indicated by **bold**.

For Product Safety Concerns and Information please contact our EU
representative GPSR@taylorandfrancis.com
Taylor & Francis Verlag GmbH, Kaufingerstraße 24, 80331 München, Germany

www.ingramcontent.com/pod-product-compliance
Lightning Source LLC
LaVergne TN
LVHW080115011026
849160LV00013B/1105